Vitamins and Minerals

Vitamin E RDA (mg/day)[e]	Vitamin K AI (µg/day)	Calcium AI (mg/day)	Phosphorus RDA (mg/day)	Magnesium RDA (mg/day)	Iron RDA (mg/day)	Zinc RDA (mg/day)	Iodine RDA (µg/day)	Selenium RDA (µg/day)	Copper RDA (µg/day)	Manganese AI (mg/day)	Fluoride AI (mg/day)	Chromium AI (µg/day)	Molybdenum RDA (µg/day)
4	2.0	210	100	30	0.27	2	110	15	200	0.003	0.01	0.2	2
5	2.5	270	275	75	11	3	130	20	220	0.6	0.5	5.5	3
6	30	500	460	80	7	3	90	20	340	1.2	0.7	11	17
7	55	800	500	130	10	5	90	30	440	1.5	1.0	15	22
11	60	1300	1250	240	8	8	120	40	700	1.9	2	25	34
15	75	1300	1250	410	11	11	150	55	890	2.2	3	35	43
15	120	1000	700	400	8	11	150	55	900	2.3	4	35	45
15	120	1000	700	420	8	11	150	55	900	2.3	4	35	45
15	120	1200	700	420	8	11	150	55	900	2.3	4	30	45
15	120	1200	700	420	8	11	150	55	900	2.3	4	30	45
11	60	1300	1250	240	8	8	120	40	700	1.6	2	21	34
15	75	1300	1250	360	15	9	150	55	890	1.6	3	24	43
15	90	1000	700	310	18	8	150	55	900	1.8	3	25	45
15	90	1000	700	320	18	8	150	55	900	1.8	3	25	45
15	90	1200	700	320	8	8	150	55	900	1.8	3	20	45
15	90	1200	700	320	8	8	150	55	900	1.8	3	20	45
15	75	1300	1250	400	27	13	220	60	1000	2.0	3	29	50
15	90	1000	700	350	27	11	220	60	1000	2.0	3	30	50
15	90	1000	700	360	27	11	220	60	1000	2.0	3	30	50
19	75	1300	1250	360	10	14	290	70	1300	2.6	3	44	50
19	90	1000	700	310	9	12	290	70	1300	2.6	3	45	50
19	90	1000	700	320	9	12	290	70	1300	2.6	3	45	50

[e] Vitamin E recommendations are expressed as α-tocopherol.

SOURCE: Adapted with permission from the *Dietary Reference Intakes* series, National Academy Press. Copyright 1997, 1998, 2000, 2001, by the National Academy of Sciences. Courtesy of the National Academy Press, Washington, D.C.

Minerals

Zinc (mg/day)	Iodine (µg/day)	Selenium (µg/day)	Copper (µg/day)	Manganese (mg/day)	Fluoride (mg/day)	Molybdenum (µg/day)	Boron (mg/day)	Nickel (mg/day)	Vanadium (mg/day)
4	—	45	—	—	0.7	—	—	—	—
5	—	60	—	—	0.9	—	—	—	—
7	200	90	1000	2	1.3	300	3	0.2	—
12	300	150	3000	3	2.2	600	6	0.3	—
23	600	280	5000	6	10	1100	11	0.6	—
34	900	400	8000	9	10	1700	17	1.0	—
40	1100	400	10,000	11	10	2000	20	1.0	1.8
40	1100	400	10,000	11	10	2000	20	1.0	1.8
34	900	400	8000	9	10	1700	17	1.0	—
40	1100	400	10,000	11	10	2000	20	1.0	—
34	900	400	8000	9	10	1700	17	1.0	—
40	1100	400	10,000	11	10	2000	20	1.0	—

NOTE: An Upper Limit was not established for vitamins and minerals not listed and for those age groups listed with a dash (—) because of a lack of data, not because these nutrients are safe to consume at any level of intake. All nutrients can have adverse effects when intakes are excessive.

SOURCE: Adapted with permission from the *Dietary Reference Intakes* series, National Academy Press. Copyright 1997, 1998, 2000, 2001, by the National Academy of Sciences. Courtesy of the National Academy Press, Washington, D.C.

SIXTH EDITION

Nutrition and Diet Therapy

PRINCIPLES AND PRACTICE

Corinne Balog Cataldo
Linda Kelly DeBruyne
Eleanor Noss Whitney

THOMSON

WADSWORTH

Australia • Canada • Mexico • Singapore • Spain • United Kingdom • United States

Dedication

To life and its unexpected twists and turns.
Corkie

To my husband, Tom DeBruyne, the adventure and the magic continue. With love.
Linda

To all of my grandchildren and stepgrandchildren: Max, Zoey, Emily, Rebecca, Sarah, Will, Toot-Toot, and Jacob.
Ellie

Nutrition Publisher: Peter Marshall
Development Editor: Elizabeth Howe
Assistant Editor: John Boyd
Editorial Assistant: Madinah Chang
Technology Project Manager: Star MacKenzie
Marketing Manager: Jennifer Somerville
Marketing Assistant: Mona Weltmer
Project Managers, Editorial Production: Sandra Craig, Jerilyn Emori
Print/Media Buyer: Barbara Britton
Permissions Editor: Joohee Lee
Production Service: Dusty Friedman, The Book Company

Text and Cover Designer: Carolyn Deacy
Photo Researcher: Myrna Engler
Copy Editor: Patricia Lewis
Illustrators: Hespenheide Design, Impact Publications, McMahon Medical Art
Photographer and Food Stylist: Polara Studios, Inc., and Carol Ladd
Cover Image: Jennie Oppenheimer
Cover Printer: Phoenix Color Corp.
Compositor: Parkwood Composition Service, Inc.
Printer: Courier Companies, Inc., Kendallville

Printed in the United States of America
3 4 5 6 7 06 05 04 03

For more information about our products, contact us at:
Thomson Learning Academic Resource Center
1-800-423-0563
For permission to use material from this text, contact us by:
Phone: 1-800-730-2214
Fax: 1-800-730-2215
Web: http://www.thomsonrights.com

Library of Congress Control Number: 2002111506

ISBN 0-534-57691-5

Wadsworth/Thomson Learning
10 Davis Drive
Belmont, CA 94002-3098
USA

Asia
Thomson Learning
5 Shenton Way #01-01
UIC Building
Singapore 068808

Australia
Nelson Thomson Learning
102 Dodds Street
South Melbourne, Victoria 3205
Australia

Canada
Nelson Thomson Learning
1120 Birchmount Road
Toronto, Ontario M1K 5G4
Canada

Europe/Middle East/Africa
Thomson Learning
High Holborn House
50/51 Bedford Row
London WC1R 4LR
United Kingdom

Contents in Brief

PART THREE

Nutrition in Health Care

PART FOUR

Diet Therapy

Contents

PART THREE

Nutrition in Health Care

How to Features Appear on the Following Pages

Case Studies Appear in Chapters of Parts Two, Three, and Four

Preface

Nutrition is a young and rapidly expanding science with many unanswered questions and new "facts" surfacing every day. This sixth edition of *Nutrition and Diet Therapy* reflects the many changes that have occurred in the field of nutrition since the last edition. The goal of the text is to provide basic nutrition information and advice on how to apply that information to everyday situations. A second mission is to help students evaluate and interpret new nutrition research.

The book is structured in a deliberate way. It first introduces the basics of nutrition and shows how nutrition supports health. It then describes how nutrient needs change throughout the life cycle. The second half of the book begins by looking at some ways that poor nutrition may lead to disease and then describes the potential impact of illnesses, medications, and complementary therapies on nutrient needs and nutritional health. The remaining chapters of the book focus on medical nutrition therapy and its role in a variety of medical conditions.

Each chapter includes a wealth of nutrition tools to facilitate teaching and learning. Definitions of key terms appear at the bottom of each page, and notes in the margins clarify nutrition information, remind readers of previously defined terms, and provide cross-references. "How to" skill boxes help readers work through calculations or give practical suggestions for applying nutrition information. The "Self Check" feature at the end of each chapter helps readers review and test their understanding of the chapter material. "Critical Thinking" questions encourage readers to synthesize information from the chapter or from previous chapters. The "Nutrition on the Net" feature at the end of each chapter provides website addresses and content descriptions for key nutrition-related organizations. New "In Summary" statements have been added at the end of each major chapter section to help students assimilate the material and assess reading comprehension.

Special features help apply the chapter concepts and provide a rich learning experience for students. The applied and personal "Self-Study" exercises in Chapters 1–10 ask students to focus on the evaluation of a single dietary concept. Case studies in the later chapters challenge readers to apply chapter information to clinical situations. "Clinical Applications" provide practice with mathematical calculations and help students understand the impact of nutrition-related issues on health care professionals and their clients. Later chapters also include "Nutrition Assessment Checklists," which summarize assessment parameters relevant to different stages of the life cycle or groups of disorders. New to this edition are "Diet-Drug Interaction" boxes, which point out interactions relevant to the medications described in each chapter.

A hallmark of the text is the "Nutrition in Practice" section located at the end of every chapter. These sections provide coverage of current research topics, advanced subjects, or specialty areas such as nutrition and dental health. This edition includes new Nutrition in Practice sections examining fat replacers, community nutrition, phytochemicals, the immune system, genetics and its implications for nutrition care, probiotics, hypoglycemia, the metabolic syndrome, and multiple organ system failure.

A robust set of appendixes supports the book. The appendixes include a wealth of information on the contents of foods and enteral formulas, U.S. nutrient intake recommendations and the exchange system, additional information about nutrition assessment, aids to calculations, and nutrition resources.

We hope that as you discover the many fascinating aspects of nutrition, you will enthusiastically apply the concepts in both your professional and your personal life. For nutrition updates and other resources, we invite you to visit our website: http://www.wadsworth.com/nutrition.

Acknowledgments

Among the most difficult words to write are those that express the depth of our gratitude to the many dedicated people whose efforts have made this book possible. A special note of appreciation to Sharon Rolfes for her numerous contributions to the chapters and Nutrition in Practice sections as well as to the Dietary Reference Intakes on the inside front cover and the Body Mass Index table on the inside back cover. Many thanks to Fran Webb for sharing her knowledge, ideas, and resources about the latest nutrition developments. Special thanks to Jayme Ebaugh for her expert word processing. Thanks to Janet Isaacs and Sharon Rolfes for their work on the Self-Study activities. Thanks also to Joohee Lee for her help in securing permissions, Pat Lewis for copyediting our manuscript, and Myrna Engler for finding the photographs that grace the pages of our text. We also wish to acknowledge the efforts of Elizabeth Hand, Bob Geltz, and their staff at ESHA for creating an extremely accurate and practical food composition appendix. We are indebted to our editorial team, Peter Marshall and Elizabeth Howe, and our production team—Sandra Craig, Jerilyn Emori, and Dusty Friedman—for seeing this project through from start to finish. We would also like to acknowledge Jennifer Somerville for her marketing efforts. To the many others involved in designing, indexing, typesetting, dummying, and marketing, we offer our thanks. We are especially grateful to our associates, family, and friends for their continued encouragement and support and to our reviewers who consistently offer excellent suggestions for improving the text.

Reviewers

Sara Long Anderson
Southern Illinois University

Paul Araujo
Baltimore City College

Christina Brecht
East Stroudsburg University

Frieda F. Brown
Georgia Southern University

Anita Dalis
CS Mott Community College

Maggi Dorsett
Butte Community College

Janet W. Gloeckner
James Madison University

Georgette Howell
SUNY Rockland Community College

Eleanor B. Huang
Orange Coast College

Sandra Jentzen
Mott Community College

Kathy Jones
Guilford Technical Community College

Judy Kaufman
Monroe Community College

Mary Pat Maciolek
Middlesex County College

Nelda W. Malm
Seminole Community College

Myrtle McCulloch
Georgetown University

Marjorie McNairn
Butte Community College

Marcia Nahikian-Nelms
Southeast Missouri State University

Jillann Neely
Orondoga Community College

Linda P. Pickner
Greenville Technical College

Linda Pope
Southwest Tennessee Community College

Tonia Reinhard
Wayne State University

Lori Roth-Yousey
University of Minnesota

Judith Sadler
Western Michigan University

Janet Sass
Northern Virginia Community College

Nancy Shaw
Southeastern Louisiana University

Wendy Stevens
Luther College

Janet Tanner
Midlands Technical College

Norman Temple
Athabasca University

About the Authors

Corinne Balog Cataldo, M.M.Sc., R.D., C.N.S.D., received her B.S. in community health nutrition from Georgia State University in 1976 and her M.M.Sc. in clinical dietetics from Emory University in 1979. She has worked in private practice in Atlanta, as a clinical dietitian and metabolic support nutritionist at Georgia Baptist Medical Center in Atlanta, as a faculty member and dietetic internship coordinator at Emory University, and as a nutritionist for the Infant Formula Council. She has made numerous presentations, and in addition to this book, she has written a manual on tube feedings and the books *Understanding Normal and Clinical Nutrition, Nutrition for Health and Health Care,* and *Understanding Clinical Nutrition.* She maintains professional memberships in the American Dietetic Association, the American Society for Parenteral and Enteral Nutrition, the American Diabetes Association, and Dietitians in Nutrition Support.

Linda Kelly DeBruyne, M.S., R.D., received her B.S. in 1980 and her M.S. in 1982 in nutrition and food science at Florida State University. She is a founding member of Nutrition and Health Associates, an information resource center in Tallahassee, Florida, where her specialty areas are life cycle nutrition and fitness. Her other publications include the textbooks *Nutrition for Health and Health Care, Life Span Nutrition: Conception through Life, Health: Making Life Choices,* and *The Fitness Triad* and a multimedia CD-ROM called *Nutrition Interactive.* As a consultant for a group of Tallahassee pediatricians, she teaches infant nutrition classes to parents. She maintains a professional membership in the American Dietetic Association.

Eleanor Noss Whitney, Ph.D., received her B.A. in biology from Radcliffe College in 1960 and her Ph.D. in biology from Washington University in St. Louis in 1970. Formerly on the faculty at Florida State University, and a dietitian registered with the American Dietetic Association she now devotes full time to research, writing, and consulting. Her earlier publications include articles in *Science, Genetics,* and other journals. Her textbooks include *Understanding Nutrition, Nutrition Concepts and Controversies, Life Span Nutrition: Conception through Life, Understanding Normal and Clinical Nutrition, Nutrition for Health and Health Care,* and *Essential Life Choices* for college students and *Making Life Choices* for high school students. Her most intense interests currently include energy conservation, solar energy uses, alternatively fueled vehicles, and ecosystem restoration.

CHAPTER 1 | PERSPECTIVES ON
HEALTH AND NUTRITION

*E*very day, several times a day, you make choices that will either improve your **health** or harm it. Each choice may influence your health only a little, but when these choices are repeated over years and decades, their effects become significant.

The choices people make each day affect not only their physical health, but also their **wellness**—all the characteristics that make a person strong, confident, and able to function well with family, friends, and others. People who consistently make poor lifestyle choices, on a daily basis, increase their risks of developing diseases. Figure 1–1 shows how a person's health can fall anywhere along a continuum, from maximum wellness on the one end to total failure to function (death) on the other.

As health care professionals, when you take responsibility for your own health by making daily choices and practicing behaviors that enhance your well-being, you prepare yourself physically, mentally, and emotionally to meet the demands of your profession. As health care professionals, however, you have a responsibility to your clients as well as to yourselves. You have unique opportunities to make your clients aware of the benefits of positive health choices and behaviors, to show them how to change their behaviors and make daily choices to enhance their own health, and to serve as role models for those behaviors.

This text focuses on how nutrition choices affect health and disease. The early chapters introduce the basics of nutrition to support good health. The later chapters emphasize diet therapy and its role in supporting health and treating diseases and symptoms.

health: a range of states with physical, mental, emotional, spiritual, and social components. At a minimum, *health* means freedom from physical disease, mental disturbances, emotional distress, spiritual discontent, social maladjustment, and other negative states. At a maximum, *health* means "wellness."

wellness: maximum well-being; the top range of health states; the goal of the person who strives toward realizing his or her full potential physically, mentally, emotionally, spiritually, and socially.

Figure 1–1

The Health Line

No matter how well you maintain your health today, you may still be able to improve tomorrow. Likewise, a person who is well today can slip by failing to maintain health-promoting habits.

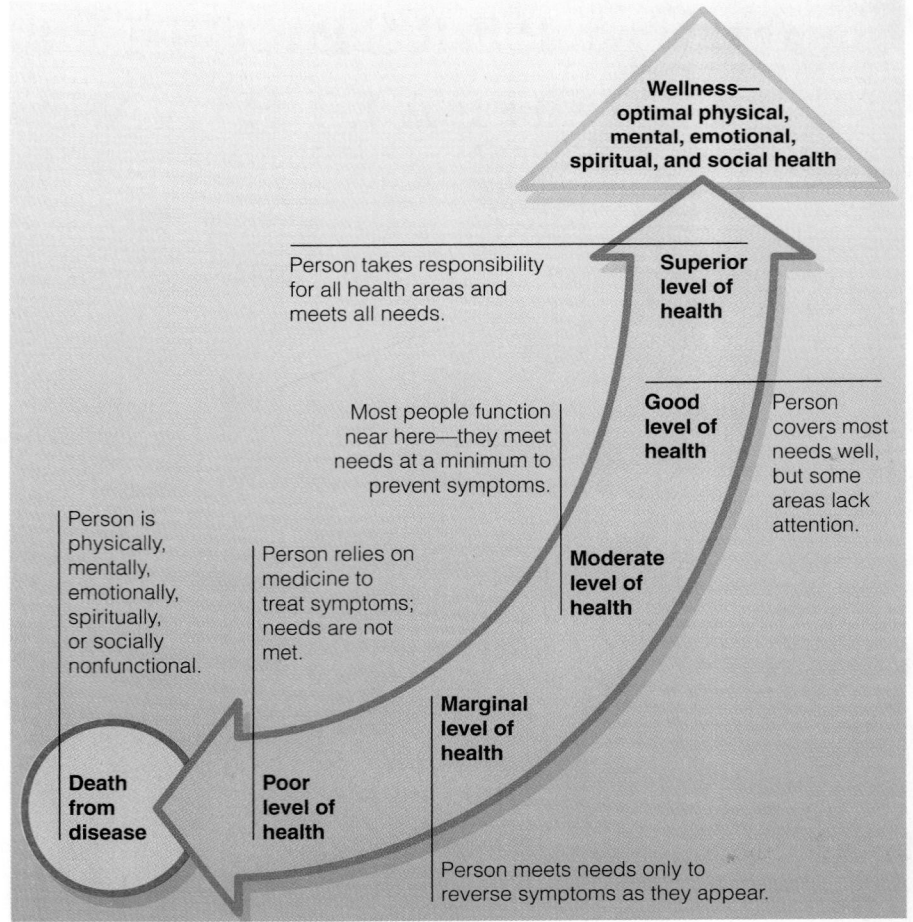

Food Choices

Sound **nutrition** throughout life does not ensure good health and long life, but it can certainly help to tip the balance in their favor. Nevertheless, most people choose foods for reasons other than their nourishing value. Even people who claim to choose foods primarily for the sake of health or nutrition will admit that other factors also influence their food choices. Because food choices become an integral part of people's lifestyles, they sometimes find it difficult to change their eating habits. Health care professionals who help clients make diet changes must understand the dynamics of food choices, because people will alter their eating habits only if their preferences are honored.

Preference Why do people like certain foods? One reason, of course, is their preference for certain tastes.[1] Some tastes are widely liked, such as the sweetness of sugar and the zest of salt. Research suggests that genetics may influence people's taste preferences, a finding that may eventually have implications for clinical nutrition.[2] For example, sensitivity to bitter taste is an inheritable trait. People born with great sensitivity to bitter tastes tend to avoid foods with bitter flavors such as broccoli, cabbage, brussels sprouts, spinach, and grapefruit juice.[3] These foods, as well as many other fruits and vegetables, contain compounds called **phytochemicals** that may reduce the risk of cancer. Thus the role that genetics may play in food selection is gaining importance in cancer research.

Habit Sometimes habit dictates people's food choices. People eat a sandwich for lunch or drink orange juice at breakfast simply because they have always done so.

Associations People also like foods with happy associations—foods eaten in the midst of warm family gatherings on traditional holidays or given to them as children by someone who loved them. By the same token, people can attach intense and unalterable dislikes to foods that they ate when they were sick or that were forced on them when they weren't hungry.

Ethnic Heritage and Tradition Every country, and every region of a country, has its own typical foods and ways of combining them into meals. The foodways of North America reflect the many different cultural and ethnic backgrounds of its inhabitants. Many foods with foreign origins are familiar items on North American menus: tacos, egg rolls, lasagna, and gyros, to name a few. Still others, such as spaghetti and croissants, are almost staples in the "American diet." North American regional cuisines like Cajun and TexMex blend the traditions of several cultures. Table 1–1 (pp. 4–5) presents profiles of selected **ethnic diets,** together with comments on their nutritional merits and drawbacks.

Values People's values, environmental ethics, religious beliefs, and political views also influence their food choices. By choosing to eat some foods or avoid others, people make statements about themselves that reflect their values. For example, people may select only foods that come in containers that can be reused or recycled. A political activist may boycott fruit or vegetables picked by migrant workers who have been exploited. Some people choose only brands of canned tuna fish that state "dolphin safe" on the label, meaning that dolphins were not killed when the tuna were netted. Religion also influences many people's food choices. Jewish law sets forth an extensive set of dietary rules. Many Christians forgo meat during Lent, the period prior to Easter. Other faiths prohibit some dietary practices and promote others. Diet planners can foster sound nutrition practices only if they respect and honor each person's values.

Social Pressure Social pressure is another powerful influence on people's food choices. Such pressure operates in all circles and across all cultural lines. It is often considered rude to refuse food or drink being shared by a group or offered

Nutrition is only one of the many factors that influence people's food choices.

Ethnic meals and family gatherings nourish the spirit as well as the body.

nutrition: the science of foods and the nutrients and other substances they contain, and of their ingestion, digestion, absorption, transport, metabolism, interaction, storage, and excretion. A broader definition includes the study of the environment and of human behavior as it relates to these processes.

phytochemicals: nonnutrient compounds in plant-derived foods that have biological activity in the body.

ethnic diets: foodways and cuisines typical of national origins, races, cultural heritages, or geographic locations.

Perspectives on Health and Nutrition

Table 1–1 Characteristics of Selected Ethnic Diets

Staple Foods	Strengths of the Diet	Weaknesses of the Diet
Hispanic Americans from Cuba, Haiti, Puerto Rico		
Include: ■ Steamed white rice; wheat breads. ■ Starchy vegetables (beans, cassavas, yuccas); plantains; green peppers; tomatoes; garlic. ■ Dried, salted fish; chicken; pork. ■ Lard; olive oil; sugar; jams and jellies; sweet pastries; sugared fruit juices; coffee. **Exclude:** ■ Green, leafy vegetables. ■ Milk as a beverage for adults. ■ Fish other than dried and salted.	Provides adequate protein, many other nutrients, and fiber.	May provide too much fat, especially animal fat; may lack calcium.
Hispanic Americans from Mexico, Central America		
Include: ■ Steamed rice, corn products such as tortillas. ■ Many varieties of beans; chili peppers; tomatoes; mangoes; prickly pear fruit; potatoes. ■ Meat and sausages; fish; poultry; eggs. ■ Lard; chocolate and coffee drinks; cakes; pastries. **Exclude:** ■ Green, leafy vegetables; yellow vegetables. ■ Milk as a beverage for adults.	Most nutrients can be obtained.	Is high in kcalories and fat, especially saturated fat, and high in sugar.
Black Americans from West Indies, Central or South America and Recent African Immigrants		
Include: ■ Millet, corn, wheat, rice, or barley. ■ Starchy roots such as cassavas, yams; plantains; bananas; coconuts; peanuts; fresh fruits; hot peppers; tomatoes; onions; okra. ■ Palm oil; fruit wine; tea; coffee; honey; molasses. **Exclude:** ■ Milk and milk products (meat and fish limited use).	Is low in fat and salt; is high in fiber.	Is low in calcium, iron, and vitamin B_{12}; is potentially low in protein, depending on availability of foods.
Southern Black Americans from West Africa (Many Generations in United States)		
Include: ■ Rice; hominy grits; biscuits; cornmeal and cornbread. ■ Legumes; potatoes; onions; tomatoes; hot peppers; green, leafy vegetables; okra; sweet potatoes; squashes; corn; cabbage; melons; peaches. ■ Smoked pork; meats and poultry; fish; thick stews. ■ Pecans; butter, shortening, and lard; sugar; bread puddings, pies, and sweets. **Exclude:** ■ Milk and milk products. ■ Yeast breads.	Provides ample nutrients of meat.	Provides excess protein; is high in kcalories; provides excess fat, especially saturated fat; is high in salt; is low in calcium.
Chinese Americans from China (Diets Sometimes Vary with Region)		
Include: ■ Rice and rice gruel; wheat noodles; soybean noodles. ■ Corn; vegetables from the cabbage family; squashes; cucumbers; eggplant; leafy vegetables; various shoots (bamboo, mung, and soy); sweet potatoes; radishes; onions; peas and pods; mushrooms; roots; local vegetables; pickled vegetables; sea vegetables; plums; peaches; tangerines; kumquats; other citrus fruits; litchis; longans; mangoes; papayas; pomegranates. ■ Soybean products (tofu and soy milk); meat; fish with bones; poultry; seafood.	Is low in fat; is high in fiber and many nutrients.	Depending on availability of protein-rich foods, protein and iron may be low; high in salt.

Table 1–1   Characteristics of Selected Ethnic Diets—*continued*

Staple Foods	Strengths of the Diet	Weaknesses of the Diet
■ Soup or tea as beverage; soy sauce; sugar.		
Exclude:		
■ Milk and most milk products.		

Japanese Americans from Japan

Staple Foods	Strengths of the Diet	Weaknesses of the Diet
Include:	Provides abundant nutrients with little fat.	Is high in salt.
■ Rice.		
■ Vegetables (including pickled and sea); fruits; salads.		
■ Soy (miso, tofu, bean paste); fish with bones; seafood.		
■ Sugars as seasoning; ginseng; soy sauce.		
Exclude:		
■ Milk and milk products.		

Korean Americans from South Korea

Staple Foods	Strengths of the Diet	Weaknesses of the Diet
Include:	Provides adequate protein.	Is high in fat; monotonous in winter (kimchi is served at each meal, to the exclusion of other vegetables); without the traditional small fish with bones, calcium can be lacking.
■ Rice; noodles.		
■ Leafy vegetables; kimchi (hot pickled cabbage); sea vegetables; hot peppers; seasonal fruits; mushrooms.		
■ Small fish with bones; grilled beef; chicken; squid, octopus, and lobster; mussels; eggs.		
■ Lard and vegetable fat for frying; sesame oil; nuts and seeds; ginger; sugar as seasoning.		
Exclude:		
■ Milk and milk products.		

Vietnamese Americans from Vietnam

Staple Foods	Strengths of the Diet	Weaknesses of the Diet
Include:	Provides adequate protein and vitamins.	Can be low in iron or calcium.
■ Rice, rice noodles; french bread and croissants.		
■ Hot peppers; curries of asparagus and potatoes; salads; tropical fruits and vegetables; lemons and limes.		
■ Small portions of poultry; eggs; fish pâtés; nuoc nam (a strong, fermented fish sauce).		
■ Sweets, candies, sweetened drinks; coffee; tea; butter.		
Exclude:		
■ Milk and milk products.		

Native Americans

Staple Foods	Strengths of the Diet	Weaknesses of the Diet
Include:	Varies with region; may provide adequate protein and fiber.	Varies with region; may be low in calcium.
■ *Southeast:* corn; cornmeal; coontie (flour from a palmlike plant); fried breads; pumpkins; squashes; papayas; alligator, snake, wild hog, duck, fish, and shellfish.		
■ *Northeast:* blueberries; cranberries; beans; corn; pumpkins; fish; lobster; wild game; maple syrup.		
■ *Midwest:* bison; beans; corn; melons; squashes; tomatoes.		
■ *Southwest:* corn (many varieties); beans; squash; pumpkins; chili peppers; melons; pinenuts; cactus.		
■ *Northwest:* salmon; caviar; other fish; otter; seal; elk; whale; bear; other game; wild fruits, nuts, and greens.		
Exclude:		
■ Milk and milk products.		

Italian Americans from Italy

Staple Foods	Strengths of the Diet	Weaknesses of the Diet
Include:	Provides adequate protein and most nutrients.	Varies with region; can be high in fat.
■ *Northern Italy:* egg-based, ribbon-shaped pastas; cheese; cream; butter; meat; eggs.		
■ *Southern Italy:* wheat pastas; artichokes, eggplants, peppers, and tomatoes; beans; olive oil.		

by a host. Sometimes you become accepted as a member of a social gathering only when you "break bread" with the others.

Emotional State People may eat in response to emotional stimuli—for example, to relieve boredom or depression or to calm anxiety. A lonely person may choose to eat rather than to call a friend and risk rejection. A person who has returned home from an exciting evening out may unwind with a late-night snack. Eating in response to emotions can easily lead to overeating and obesity, but may be appropriate at times. For example, sharing food at times of bereavement serves both the givers' need to provide comfort and the receiver's need to be cared for and to interact with others.

In contrast, some people do not eat at all, or eat very little, in response to emotions. A depressed person may simply have no appetite for food. A person overcome with grief can easily lose all interest in food. As long as a person's appetite or interest in food returns quickly, little harm is done.

Availability, Convenience, and Economy The influence of these factors on people's food selections is clear. You cannot eat foods if they are not available, if you cannot get to the grocery store, if you do not have the time or skill to prepare them, or if you cannot afford them. Convenience plays a major role in many people's food selections today, but along with convenience, consumers want great taste and quality—foods that are fast, delicious, and nutritious.[4] The demand for foods that are ready to eat or can be easily prepared in a microwave oven demonstrates this influence. Consumers today spend more than half their food budget on meals that require little, if any, further preparation.[5] They frequently eat out, bring ready-to-eat meals home, or have food delivered.

Age Age influences people's food choices. Infants, for example, depend on others to choose foods for them. Older children also rely on others, but become more active in selecting foods that taste sweet and are familiar to them and rejecting those whose taste or texture they dislike. In contrast, the links between taste preferences and food choices in adults are less direct than in children.[6] Adults often choose foods based on health concerns such as body weight. Indeed, adults may avoid sweet or familiar foods because of such concerns.

Occupation Some people have jobs that keep them away from home for days at a time, or require them to conduct business in restaurants or at conventions, or involve hectic schedules that allow little or no time for meals at home. For these people, the kinds of restaurants available to them, and the cost of eating out so often, may limit food choices.

Body Image Sometimes people select foods that they associate with ideals of body image. The fashion and movie industries, not the medical community, have defined what people believe to be the ideal body—sometimes an excessively thin body for women or an excessively muscular body for men. Both men and women seek "beautiful bodies," and in doing so, they select or avoid foods that they believe will improve or impair their physical appearance. Such intentions are rational when based on sound nutrition and fitness knowledge, but when based on faddism or carried to extremes, they undermine good health.

Medical Conditions Sometimes medical conditions and their treatments (including medications) limit the foods a person can select. For example, a person with heart disease might need to adopt a low-fat diet. The chemotherapy needed to treat cancer can interfere with a person's appetite or limit food choices by causing vomiting. Allergy to certain foods can also limit choices. The second half of this text discusses how diet can be modified to accommodate different medical conditions.

Health and Nutrition Consumers today cite nutrition as a primary concern in making food choices, yet the foods they choose do not always reflect this concern. Food manufacturers and restaurant chefs have responded to scientific findings linking health with nutrition by offering an abundant selection of health-promoting foods and beverages. In some cases, the health-promoting foods are as simple and familiar as oatmeal or tomatoes. In other cases, the foods have been processed or prepared in a way that provides health benefits, perhaps by lowering the fat content. In still other cases, manufacturers have developed **functional foods**—products that contain physiologically active ingredients that may provide health benefits. Examples of functional foods include a margarine made with plant sterolesters that lowers blood cholesterol and orange juice fortified with calcium to help build strong bones. More and more functional foods are being developed—a trend that seems likely to continue as consumer demand for such products grows.[7]

Consumers may be led to believe that functional foods are a new category of foods. The American Dietetic Association, however, classifies all foods as functional at some physiological level.[8] All foods contain thousands of nonnutrient compounds that are biologically active in the body, so virtually all of them have some special value in supporting health.

In Summary A person selects foods for many different reasons. Whatever those reasons may be, food choices influence health—both positively and negatively. Individual food selections neither make nor break a diet's healthfulness, but the balance of foods selected over time can make an important difference to health. For this reason, people are wise to think "nutrition" when making their food choices.

The Nutrients

You are a collection of molecules that move. All these moving parts are arranged in patterns of extraordinary complexity and order—cells, tissues, and organs. Although the arrangement remains constant, the parts are continually changing, using **nutrients** and energy derived from nutrients.

Almost any food you eat is composed of dozens or even hundreds of different kinds of materials. Spinach, for example, is composed mostly of water (95 percent), and most of its solid materials are the compounds carbohydrates, fats (properly called lipids), and proteins. If you could remove these materials, you would find a tiny residue of minerals, vitamins, and other compounds. Water, carbohydrates, fats, proteins, vitamins, and minerals are the six classes of nutrients commonly found in spinach and other foods. Some of the other materials in foods, such as the pigments and other phytochemicals, are not nutrients, but may still be important to health. The body can make some nutrients for itself, at least in limited quantities, but it cannot make them all, and it makes some in insufficient quantities to meet its needs. Therefore, the body must obtain many nutrients from foods. The nutrients that foods must supply are called **essential nutrients.**

Carbohydrates, Fats, and Proteins Four of the six classes of nutrients (carbohydrates, fats, proteins, and vitamins) contain carbon, which is found in all living things. They are therefore **organic** (meaning, literally, "alive"). During metabolism, three of these four (carbohydrates, fats, and proteins) provide energy the body can use. These **energy-yielding nutrients** continually replenish the energy you spend daily. Carbohydrates and fats meet most of the body's energy needs; proteins make a significant contribution only when other fuels are unavailable.

The six classes of nutrients are water, carbohydrates, fats, proteins, vitamins, and minerals.

Metabolism, the set of processes by which nutrients are rearranged into body structures or broken down to yield energy, is described in Chapter 6.

functional foods: foods that contain physiologically active compounds that may provide health benefits beyond their nutrient contributions.

nutrients: substances obtained from food and used in the body to provide energy and structural materials and to serve as regulating agents to promote growth, maintenance, and repair. Nutrients may also reduce the risks of some diseases.

essential nutrients: nutrients a person must obtain from food because the body cannot make them for itself in sufficient quantities to meet physiological needs.

organic: carbon containing. The four organic nutrients are carbohydrate, fat, protein, and vitamins.

energy-yielding nutrients: the fuel nutrients, those that yield energy the body can use. The three energy-yielding nutrients are carbohydrate, protein, and fat.

Perspectives on Health and Nutrition

ow to

Calculate the Energy a Food Provides

*T*he following example shows how to calculate the energy available from 1 slice of bread that has 1 teaspoon of butter on it. From food tables such as Appendix A in this book, you can determine that buttered bread contains 15 grams carbohydrate, 2 grams protein, and 5 grams fat:

$$15 \text{ g carbohydrate} \times 4 \text{ kcal/g} = 60 \text{ kcal.}$$
$$2 \text{ g protein} \times 4 \text{ kcal/g} = 8 \text{ kcal.}$$
$$5 \text{ g fat} \times 9 \text{ kcal/g} = 45 \text{ kcal.}$$
$$\text{Total} = 113 \text{ kcal.}$$

From this information, you can calculate the percentage of kcalories each of the energy nutrients contributes to the total. To determine the percentage of kcalories from fat, for example, divide the 45 fat kcalories by the total 113 kcalories:

$$45 \text{ fat kcal} \div 113 \text{ total kcal} = 0.398 \text{ (round to 0.40).}$$
Then multiply by 100 to get the percentage:
$$0.40 \times 100 = 40\%.$$

Health recommendations that urge people to limit fat intake to 30 percent of kcalories refer to the day's total energy intake, not to individual foods. Still, if the proportion of fat in each food choice throughout a day exceeds 30 percent of kcalories, then the day's total surely will, too. Knowing that this snack provides 40 percent of its kcalories from fat alerts a person to the need to make lower-fat selections at other times that day.

Vitamins, Minerals, and Water Vitamins are organic but do not provide energy to the body. They facilitate the release of energy from the three energy-yielding nutrients. In contrast, minerals and water are **inorganic** nutrients. Minerals yield no energy in the human body, but like vitamins, they help to regulate the release of energy, among their many other roles. As for water, it is the medium in which all of the body's processes take place.

kCalories: A Measure of Energy The amount of energy carbohydrates, fats, and proteins release can be measured in calories (or more properly, kilocalories, or **kcalories**). kCalories are not constituents of foods; they are a measure of the energy foods provide. It is as incorrect to refer to the kcalories in a food as it is to refer to the inches in a person. It is correct to refer to the energy a food provides and to the height of a person. The energy a food provides depends on how much carbohydrate, fat, and protein the food contains.

Carbohydrate yields 4 kcalories of energy from each gram, and so does protein. Fat yields 9 kcalories per gram. If you know how many grams of each nutrient a food contains, you can derive the number of kcalories potentially available from the food. Simply multiply the carbohydrate grams times 4, the protein grams times 4, and the fat grams times 9, and add the results together (the accompanying "How to" box describes how to calculate the energy a food provides).

Energy Nutrients in Foods Practically all foods contain mixtures of the energy-yielding nutrients, although foods are sometimes classified by their predominant nutrient. To speak of meat as "a protein" or of bread as "a carbohydrate," however, is inaccurate. Each is rich in a particular nutrient, but a protein-rich food such as beef contains a lot of fat along with protein, and a carbohydrate-rich food such as cornbread also contains fat (corn oil) and protein. Only a few foods are exceptions to this rule, the common ones being sugar (which is pure carbohydrate) and oil (which is pure fat).

Energy Storage in the Body The body first uses the energy-yielding nutrients to build new compounds and fuel metabolic and physical activities. Excesses are then rearranged into storage compounds, primarily body fat, and put away for later use. Thus, if you take in more energy than you expend, whether from car-

Food energy can also be measured in kilojoules (kJ). The kilojoule is the international unit of energy. One kcalorie equals 4.2 kJ.

inorganic: not containing carbon or pertaining to living things.

kcalories: units by which energy is measured. One kcalorie (kcal) is the amount of heat necessary to raise the temperature of 1 kilogram (kg) of water 1°C. Most people speak of these units simply as calories, but on paper the word *calorie* is prefaced by a *k* for kilocalorie. We use kcalories and kcal throughout this book.

bohydrate, fat, or protein, the result is usually a gain of body fat. Too much meat (a protein-rich food) is just as fattening as too many potatoes (a carbohydrate-rich food).

Alcohol, Not a Nutrient   When taken in excess of energy need, alcohol, too, is converted to body fat and stored. The body derives energy from alcohol at the rate of 7 kcalories per gram. Alcohol is not a nutrient, however, because it cannot support the body's growth, maintenance, or repair. When alcohol contributes too much energy to a person's diet, the harm it does extends far beyond the problems of excess body fat. Nutrition in Practice 23 discusses alcohol's effects on nutrition.

In Summary   Foods provide nutrients—substances that support the growth, maintenance, and repair of the body's tissues. The six classes of nutrients are water, carbohydrates, fats, proteins, vitamins, and minerals. Foods rich in the energy-yielding nutrients(carbohydrates, fats, and proteins) provide the major materials for building the body's tissues and yield energy the body can use or store. Energy is measured in kcalories. Vitamins, minerals, and water facilitate a variety of activities in the body.

Nutrition Standards and Guidelines

How well you nourish yourself depends on the many different foods you select at numerous meals over days, months, and years. Several nutrition standards and dietary guidelines have been developed to help you whenever you are selecting foods—whether shopping at the grocery store, choosing from a restaurant menu, or preparing a home-cooked meal.

Dietary Reference Intakes

Defining the amounts of energy, nutrients, and other dietary components that best support health is a huge task. For more than 60 years, nutrition experts produced a set of energy and nutrient standards known as the Recommended Dietary Allowances (RDA) for people in the United States. The Canadian equivalent of the RDA was the RNI (Recommended Nutrient Intakes). Over the years, these standards were revised periodically as new evidence became available, but each revision maintained the original goal of protecting against nutrient deficiencies. Given the abundance of research now linking diet and health, that goal has been broadened to include supporting optimal activities within the body and preventing chronic diseases as well. Previous editions also focused narrowly on nutrients known to be essential. With recent research revealing the health benefits of other dietary components such as phytochemicals, their recommended intakes need to be addressed as well.

To that end, a major revision of dietary recommendations was implemented. The revised recommendations are called **Dietary Reference Intakes (DRI)**. The first in a series of reports was published in 1997 and presents both the framework for developing the DRI and the revised recommendations for the five nutrients that play key roles in bone health—calcium, phosphorus, magnesium, vitamin D, and fluoride.[9] A second report came out in 1998 and features revised recommendations for the eight B vitamins and a related compound, choline.[10] A third report, released in 2000, presents recommendations for the antioxidant nutrients—vitamin C, vitamin E, selenium, and beta-carotene.[11] The most recent report, published in 2001, presents recommendations for

Dietary Reference Intakes (DRI): a set of values for the dietary nutrient intakes of healthy people in the United States and Canada. These values are used for planning and assessing diets.

Figure 1-2

Nutrient Recommendations and the Energy Recommendation Compared

Nutrient recommendations are set high enough to cover nearly everyone's requirements (the boxes represent people). The energy recommendation is set at the average, or mean, so that half the population's requirements fall below and half above it.

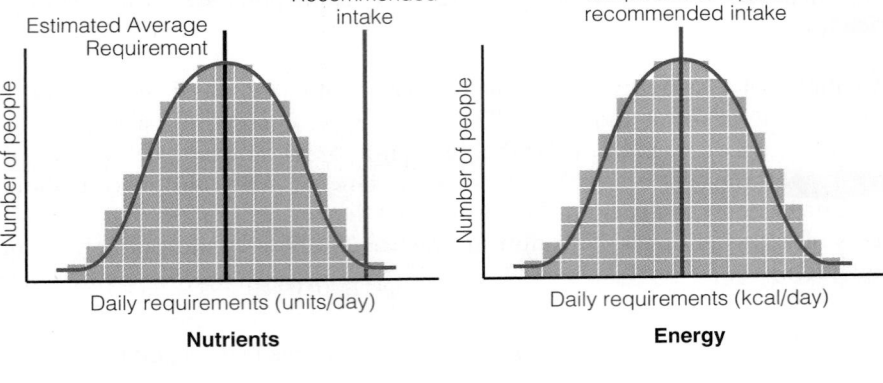

Look for upcoming DRI reports on energy and energy-yielding nutrients, fluids and their associated minerals, and other food components.

Recommended Dietary Allowances (RDA): a set of values reflecting the average daily amounts of nutrients considered adequate to meet the known nutrient needs of practically all healthy people; a goal for dietary intake by individuals.

Adequate Intake (AI): a value that is used as a guide for nutrient intake when scientific evidence is insufficient for determination of an RDA.

requirement: the lowest continuing intake of a nutrient that will maintain a specified criterion of adequacy.

deficient: in regard to nutrient intake, the amount below which almost all healthy people can be expected, over time, to experience deficiency symptoms.

Estimated Average Requirement (EAR): the amount of a nutrient that will maintain a specific biochemical or physiological function in half the people of a given age and gender group.

Tolerable Upper Intake Level (UL): the maximum amount of a nutrient that appears safe for most healthy people and beyond which there is an increased risk of adverse health effects.

vitamins A and K and for a dozen minerals.[12] Recommendations for all these nutrients appear on the inside front cover.

Setting Nutrient Recommendations: RDA and AI   One advantage of the DRI is that they apply to the diets of individuals. The committee offers two sets of values to be used as nutrient intake goals by individuals: a revised set of **Recommended Dietary Allowances (RDA)** and a set called **Adequate Intakes (AI).**

Based on solid experimental evidence and other reliable observations, the RDA are the foundation of the DRI. The AI values are also based on scientific findings as much as possible, but estimating their values requires some educated guesswork. The committee establishes an AI value whenever scientific evidence is insufficient to generate an RDA.[13]

In the last decade, abundant new research has linked nutrients in the diet with the promotion of health and the prevention of chronic diseases. An advantage of the DRI is that they take into account disease prevention where appropriate, as well as nutrient adequacy. For example, the AI for calcium is based on intakes thought to reduce the likelihood of osteoporosis-related fractures later in life.

To ensure that the vitamin and mineral recommendations meet the needs of as many people as possible, the recommendations are set near the top end of the range of the population's estimated average requirements (see Figure 1–2). Small amounts above the daily **requirement** do no harm, whereas amounts below the requirement lead to health problems. When people's intakes are consistently **deficient,** their nutrient stores decline, and over time this decline leads to deficiency symptoms and poor health. In contrast to the vitamin and mineral recommendations, the energy recommendation is not generous, but rather is set at the population's estimated average requirement (see Figure 1–2). Although not enough energy may cause undernutrition, too much energy is as bad for health as too little because excess energy leads to obesity.

Facilitating Nutrition Research and Policy: EAR   In addition to the RDA and AI, the DRI committee has established another set of values: **Estimated Average Requirements (EAR).** These values establish average requirements for given life stage and gender groups that researchers and nutrition policymakers need in their work. Nutrition scientists may use the EAR as standards in research. Public health officials may use them to assess nutrient intakes of populations and make recommendations. The EAR values form the scientific basis upon which the RDA are set.

Establishing Safety Guidelines: UL   The DRI committee also establishes upper limits of intake for nutrients posing a hazard when consumed in excess. These values, the **Tolerable Upper Intake Levels (UL),** are indispensable to consumers

Table 1–2   The DRI and Their Uses

Reference Value	Description
Recommended Dietary Allowances (RDA)	Nutrient intake goals for individuals. RDA values are derived from the Estimated Average Requirements (see below).
Adequate Intakes (AI)	Nutrient intake goals for individuals. AI values are set whenever scientific data are insufficient to allow establishment of an RDA value.
Tolerable Upper Intake Levels (UL)	Suggested upper limits of intakes of potentially toxic nutrients. Intakes above the UL are likely to cause illness from toxicity.
Estimated Average Requirements (EAR)	Population-wide average nutrient requirements for nutrition research and policy making. The basis upon which RDA values are set.

SOURCE: Adapted from Food and Nutrition Board, Institute of Medicine, National Academy of Sciences, Dietary Reference Intake (DRIs) for calcium, phosphorus, magnesium, vitamin D and fluoride, *Nutrition Today* 32 (1997): 182–190.

who take supplements. Consumers need to know how much of a nutrient is too much. The UL are also of value to public health officials who set allowances for nutrients that are added to foods and water. The UL values are listed on the inside front cover.

Using Nutrient Recommendations   Each of the four DRI categories serves a unique purpose. For example, the EAR are most appropriately used to develop and evaluate nutrition programs for *groups* such as schoolchildren or military personnel. The RDA (or AI if an RDA is not available) can be used to set goals for *individuals.* The UL help to keep nutrient intakes below the amounts that increase the risk of toxicity. With these understandings, professionals can use the DRI for a variety of purposes. Table 1–2 sums up the names and purposes of the nutrient intake standards just introduced.

In addition to understanding the unique purposes of the DRI, it is important to keep their uses in perspective. Consider the following:

- The values are recommendations for optimal and safe intakes, not minimum requirements; except for energy, they include a generous margin of safety. Figure 1–3 on the next page presents an accurate view of how a person's nutrient needs fall within a range, with marginal and danger zones both below and above the range.

- The values reflect daily intakes to be achieved on average, over time. They assume that intakes will vary from day to day, and they are set high enough to ensure that body nutrient stores will meet nutrient needs during periods of inadequate intakes lasting a day or two for some nutrients and up to a month or two for others.

- The values are chosen in reference to specific indicators of nutrient adequacy, such as blood nutrient concentrations, normal growth, and reduction of certain chronic diseases or other disorders.

- The recommendations are designed to meet the needs of most *healthy* people. Medical problems alter nutrient needs as later chapters describe.

- The recommendations are specific for people of both genders as well as various ages and stages of life: infants, children, adolescents, men, women, pregnant women, and lactating women.

Figure 1–3

Naive versus Accurate View of Nutrient Recommendations

The RDA for a given nutrient represents a point that lies within a range of appropriate and reasonable intakes between toxicity and deficiency. The AI also falls within this range, but its determination is not as exact as an RDA's. Both of these recommendations are high enough to provide reserves in times of short-term dietary inadequacies, but not so high as to approach toxicity. Nutrient intakes above or below this range may be equally harmful.

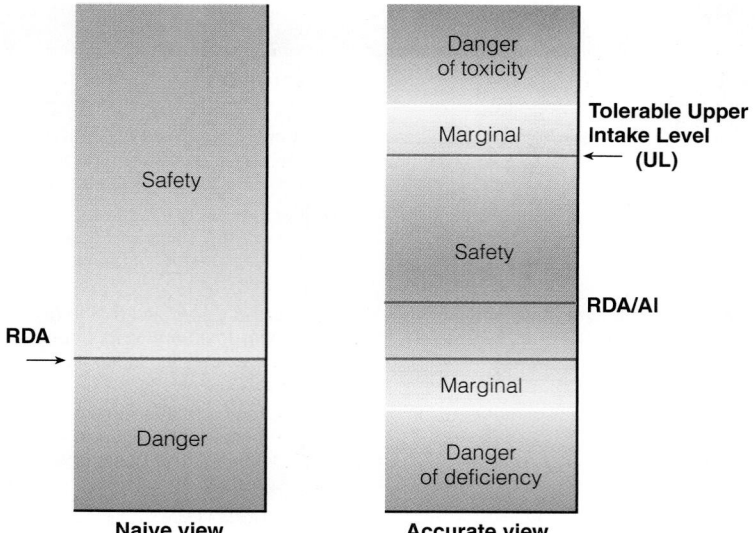

Table 1–3 Leading Causes of Death in the United States, 1999

The five diseases shaded red are nutrition related. Those shaded green are alcohol related.

1. Heart disease
2. Cancers
3. Strokes
4. Chronic lower respiratory diseases
5. Motor vehicle and other accidents
6. Diabetes mellitus
7. Pneumonia and influenza
8. Alzheimer's disease
9. Kidney disease
10. Infections of the blood
11. Suicide
12. Liver disease and cirrhosis
13. Hypertension (high blood pressure)
14. Homicide
15. Aneurysm (internal hemorrhage)

SOURCE: National Center for Health Statistics, 2001.

overnutrition: overconsumption of food energy or nutrients sufficient to cause disease or increased susceptibility to disease; a form of malnutrition.

undernutrition: underconsumption of food energy or nutrients severe enough to cause disease or increased susceptibility to disease; a form of malnutrition.

chronic diseases: degenerative diseases characterized by deterioration of the body organs; also called chronic, **noncommunicable diseases (NCD).** Examples include heart disease, cancer, and diabetes.

Chapter One

Dietary Guidelines

Nutrient recommendations such as the DRI were developed to ensure adequate nutrient intakes, but they do little to protect people from excess intakes of fat, cholesterol, sugar, salt, and alcohol. Government authorities are now as much concerned about **overnutrition** as they once were about **undernutrition.** Research confirms that dietary excesses, especially of energy, fat, and alcohol, contribute to many **chronic diseases,** including heart disease, cancer, stroke, diabetes, and liver disease.[14] Only two common lifestyle habits are more influential on health than a person's choice of diet: smoking and other tobacco use, and excessive drinking of alcohol. Table 1–3 lists the leading causes of death in the United States and shows that the top three are nutrition related (and related to tobacco use), while others, such as accidents, suicide, and liver disease, are alcohol related. And, although diet is a powerful influence on these diseases, they cannot be prevented by a healthy diet alone; genetics, physical activity, and lifestyle also play a role. Within the range set by genetic inheritance, however, disease development is strongly influenced by the foods a person chooses to eat. The *Dietary Guidelines for Americans* originated from the awareness that overnutrition contributes to disease.

Figure 1–4 presents the 2000 *Dietary Guidelines for Americans.*[15] In general, the *Dietary Guidelines* answer the question, What should an individual eat to stay healthy? The first two guidelines encourage people to aim for fitness by combining sensible eating with regular physical activity to achieve and maintain a healthy weight. The next four guidelines urge people to build a healthy base by using the Food Guide Pyramid in meal planning; choosing a variety of grains, vegetables, and fruits daily; and keeping foods safe. The last four guidelines encourage people to choose sensibly in their use of fats, sugars, salt, and alcoholic beverages for those who partake. Together, these ten guidelines point the way toward better health. Table 1–4 presents *Canada's Guidelines for Healthy Eating.*

Diet-Planning Principles

How can health care professionals help people select foods to create a diet that supplies all the needed nutrients in the appropriate amounts for good health? The principle is simple enough: encourage clients to eat a variety of foods that supply all the nutrients the body needs. In practice, how do people do this? It helps to keep in mind six basic diet-planning principles, which are listed in the margin on p. 13 in alphabetical order for ease in remembering them.

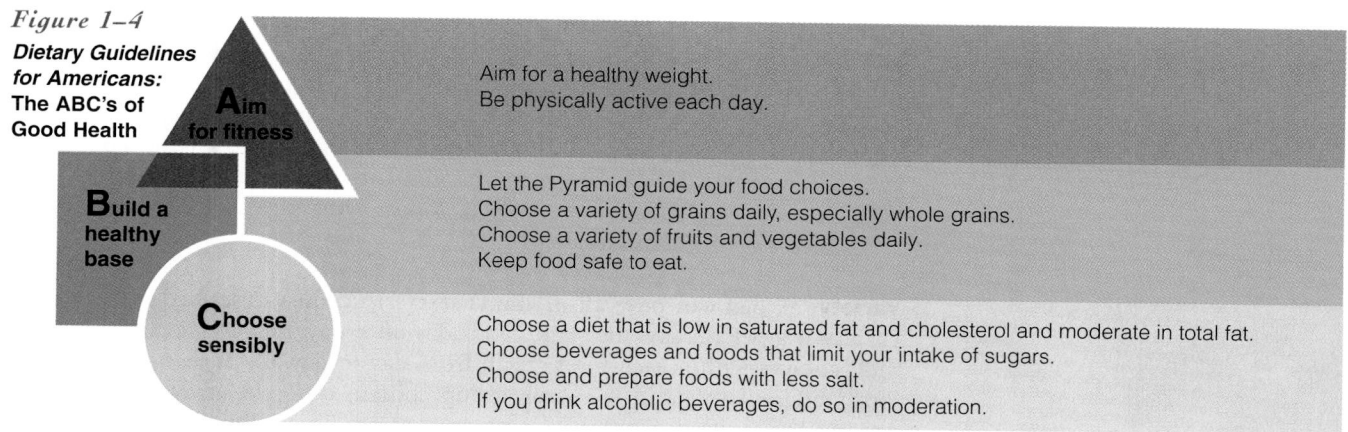

Figure 1–4

Dietary Guidelines for Americans: The ABC's of Good Health

Aim for fitness

Aim for a healthy weight.
Be physically active each day.

Build a healthy base

Let the Pyramid guide your food choices.
Choose a variety of grains daily, especially whole grains.
Choose a variety of fruits and vegetables daily.
Keep food safe to eat.

Choose sensibly

Choose a diet that is low in saturated fat and cholesterol and moderate in total fat.
Choose beverages and foods that limit your intake of sugars.
Choose and prepare foods with less salt.
If you drink alcoholic beverages, do so in moderation.

NOTE: These guidelines are intended for adults and healthy children ages 2 and older.
SOURCE: U.S. Department of Agriculture and U.S. Department of Health and Human Services, *Nutrition and Your Health: Dietary Guidelines for Americans*, Home and Garden Bulletin no. 232 (Washington, D.C.: 2000).

Adequacy The earlier discussion on the DRI already addressed the ideal of **dietary adequacy.** An adequate diet provides enough energy and enough of every nutrient to meet the needs of healthy people.

Balance As for **dietary balance,** the essential minerals calcium and iron illustrate its importance. Meats, fish, and poultry are rich in iron but poor sources of calcium. Similarly, milk and milk products are rich in calcium but poor sources of iron. In fact, milk (except breast milk) and milk products are so low in iron that overuse of these foods can actually lead to iron-deficiency anemia by displacing iron-rich foods from the diet. Yet milk is the single most nutritious food for infants and can be an important calcium source for people of all ages.

Advise clients to use some meat and meat alternates for iron and to use some milk and milk products for calcium. They need to save some space for other foods, too, since a diet consisting only of milk and meat would not be adequate. To obtain the other needed nutrients, people have to eat vegetables, fruits, grains, and other foods. In short, balance in the diet helps to ensure adequacy.

kCalorie (Energy) Control While it takes thought and skill to design an adequate, balanced diet, incorporating **kcalorie control** presents an added challenge—to eat enough without eating too much. Energy balance and weight control are discussed in Chapter 7, but the key to controlling energy intake is to select foods of high nutrient density.

Nutrient Density To eat well without overeating, select foods that deliver the most nutrients per kcalorie, a concept known as **nutrient density.** For example, among foods containing calcium, a 1½-ounce portion of cheddar cheese and 1 cup of fat-free milk both provide about the same amount of calcium; but the cheese contributes twice as much food energy (kcalories) as the fat-free milk (see Appendix A). The fat-free milk, then, is twice as calcium dense as the cheddar cheese: it offers the same amount of calcium for half the kcalories. Both foods are excellent choices for adequacy's sake alone, but to achieve adequacy while controlling kcalories, the fat-free milk is the better choice.

Moderation Foods rich in fat and sugar provide enjoyment and energy but relatively few nutrients. In addition, they promote weight gain when eaten in excess. A person practicing **moderation** would eat such foods only on occasion and would regularly select foods low in fat and sugar, a practice that automatically improves nutrient density. Returning to the example of cheddar cheese and fat-free milk, the fat-free milk not only offers the same amount of calcium for less

Table 1–4 *Canada's Guidelines for Healthy Eating*

- Enjoy a variety of foods.
- Emphasize cereals, breads, other grain products, vegetables, and fruits.
- Choose lower-fat dairy products, leaner meats, and foods prepared with little or no fat.
- Achieve and maintain a healthy body weight by enjoying regular physical activity and healthy eating.
- Limit salt, alcohol, and caffeine.

SOURCE: These guidelines derive from *Action Towards Healthy Eating: The Report of the Communications/Implementation Committee and Nutrition Recommendations . . . A Call for Action: Summary Report of the Scientific Review Committee and the Communications/ Implementation Committee,* which are available from Branch Publications Unit, Health Services and Promotion Branch, Department of Health and Welfare, 5th Floor, Jeanne Mance Building, Ottawa, Ontario K1A 1B4.

Diet-planning principles:
- Adequacy.
- Balance.
- kCalorie control.
- Density (nutrient)
- Moderation.
- Variety.

dietary adequacy: the characteristic of a diet that provides all the essential nutrients, fiber, and energy necessary to maintain health and body weight.

dietary balance: providing foods of a number of types in balance with one another such that foods rich in one nutrient do not crowd out foods that are rich in another nutrient.

Perspectives on Health and Nutrition

© Polara Studios, Inc.

This cola and bunch of grapes illustrate nutrient density. Each provides about 150 kcalories, but the grapes offer a trace of protein, some vitamins, minerals, and fiber along with the energy; the cola beverage offers only "empty" kcalories. Grapes, or any fruit for that matter, are more nutrient dense than cola beverages.

energy, but it contains far less fat than the cheese. When helping clients to plan a diet that is low in fat and sugar, remember that the most nutrient-dense foods are the best choices for the sake of both kcalorie control and moderation.

The reason may not be obvious, but choosing nutrient-dense, low-fat, low-sugar foods can also help control salt intake. Although added salt does not influence the nutrient density of foods, manufacturers usually add all three—salt, sugar, and fat—to enhance the taste of foods. Thus pursuing nutrient density helps keep salt intake moderate.

Variety A diet can have all of the characteristics just described but still lack **variety** if a person eats the same foods day after day. Encourage clients to vary their choices within each food group from day to day, for at least two reasons. First, different foods in the same group contain different arrays of nutrients. Among the fruits, for example, strawberries are especially rich in vitamin C, while mangoes are rich in vitamin A. Thus variety helps ensure adequacy. Second, no food is guaranteed entirely free of substances that in excess could cause harm. Choosing strawberries today, mangoes tomorrow, and watermelon the day after ensures that the total diet will have diluted concentrations of any contaminants that may be present in the foods.

These diet-planning principles offer a framework of excellence to strive for in planning diets; they are dietary ideals. To plan a diet that achieves these ideals, the planner needs to know which foods offer which nutrients and how many daily servings of these foods are recommended.

Food Group Plans

The human body needs about 40 different vitamins and minerals. Each nutrient has its own unique pattern of distribution in foods. Although working all the nutrients into the meals a person eats might seem quite a challenge, people all over the world meet their nutrition needs from an amazing variety of diets.

Food Groups For many people, **food group plans** such as the **Daily Food Guide** serve as the basis for planning adequate, balanced diets. Imagine shopping for groceries in a store with no signs and no organization. You might find canned peaches next to a box of taco shells and find canned pears several aisles over next to a box of cereal. To pick up even a few groceries in such a store would be quite a task. You might even forget which groceries you wanted and, instead, simply grab whatever you could find. To ease the task of shopping, grocers group similar foods together and provide signs so that you can find them. In much the same way, food group plans ease the task of planning nutritious diets by sorting foods according to their key nutrient contributions. For example, cheese and yogurt fall within the milk group because each is notable for calcium, riboflavin, and protein, as well as small amounts of other nutrients. Legumes are included in the meat, fish, and poultry group because these foods are notable for their protein, iron, and zinc. Legumes are unique, however, because they are also grouped with vegetables by virtue of their fiber, carbohydrate, and vitamins.

Food group plans exclude some foods from the main clusters because they make no notable nutrient contributions to people's intakes. These foods, called "miscellaneous" foods, are included in a diet plan only as extras in small quantities once basic nutrient needs have been met by nutritious foods. Examples of miscellaneous foods are soft drinks, jams, salad dressings, and butter.

The Daily Food Guide Figure 1–5 on pp. 16–17 presents the Daily Food Guide, a food group plan that assigns foods to five major food groups. The figure lists the number of servings recommended, the most notable nutrients of each food

kcalorie (energy) control: management of food energy intake.

nutrient density: a measure of the nutrients a food provides relative to the energy it provides. The more nutrients and the fewer kcalories, the higher the nutrient density.

moderation: providing enough, but not too much of a substance.

variety (dietary): eating a wide selection of foods within and among the major food groups (the opposite of monotony).

food group plans: diet-planning tools that sort foods of similar origin and nutrient content into groups and then specify that people should eat certain numbers of servings from each group.

Daily Food Guide: the USDA's food group plan for ensuring dietary adequacy that assigns foods to five major food groups.

group, the serving sizes, and the foods within each group categorized by nutrient density. It also includes the Food Guide Pyramid, which presents the Daily Food Guide in pictorial form. These guides apply to older children and adults only; young children have their own pyramid, which is presented in Chapter 13. The pyramid shape, with grains at the base, is designed to convey the message that grains should be abundant and form the foundation of a healthy diet. Fruits and vegetables share the next level of the pyramid, indicating that they, too, should have a prominent place in the diet. Meats and milks appear in a smaller band near the top. A few servings of each can provide valuable nutrients, such as protein, vitamins, and minerals, without too much fat and cholesterol. Fats, oils, and sweets occupy the tiny apex, indicating that they should be used sparingly and only after basic nutrient needs have been met by the foundation foods.

Variety helps to ensure an adequate and balanced diet.

Nutrient Density   The Daily Food Guide offers a strong foundation for a healthy diet, but it fails to specify food energy intakes. Large fat and energy differences exist within a single food group—for example, between fat-free milk and ice cream, baked chicken and bacon, green beans and french fries, apples and avocados, or bread and croissants. Yet, according to the Daily Food Guide, any of these choices would be acceptable. Figure 1–5 provides a nutrient density key for foods *within* each group. By using this color-coded key, people can control their food energy intakes while using the Daily Food Guide. The foods identified by green icons are the highest in nutrient density; they contribute the most nutrients for the fewest kcalories. Those with yellow icons are moderate in nutrient density; and those with red icons are lowest in nutrient density. People who have low energy allowances are advised to choose the most nutrient-dense foods within each group—those with green icons. People with high energy needs may choose some of the less nutrient-dense, higher-kcalorie foods.

Flexibility in Food Choices   The beauty of the Daily Food Guide is that it is easy to learn and simple to use. It may appear rigid, but it actually offers great flexibility once its intent is understood. For example, cheese can be substituted for milk because both supply the key nutrients for the milk group (protein, calcium, and riboflavin) in about the same amounts. To limit kcalories, choose fat-free milk; to add kcalories, choose cheese. Legumes are alternative choices for meats, so vegetarians can adapt the pattern by using legumes in place of meat selections. Nutrition in Practice 4 shows how to plan healthy vegetarian diets.

Recommended Servings   Each food group offers valuable nutrients, and people should select foods from each group daily. The recommended numbers of daily servings are:

- 6 to 11 servings of breads and cereals.
- 3 to 5 servings of vegetables.
- 2 to 4 servings of fruits.
- 2 to 3 servings of meat and meat alternates.
- 2 servings of milk and milk products. (Teenagers and young adults, women who are pregnant or breastfeeding, and women past menopause are advised to have 3 servings, and teenagers who are pregnant or breastfeeding should have 4.)

Table 1–5 on p. 18 shows the approximate food energy provided by choosing the lower, middle, or upper number of servings from each food group. Find yourself at the top of the table to see which level of energy intake may be most appropriate.

Perspectives on Health and Nutrition

Breads, Cereals, and Other Grain Products: 6 to 11 servings per day.

These foods are notable for their contributions of complex carbohydrates, riboflavin, thiamin, niacin, iron, protein, magnesium, and fiber.
Serving = 1 slice bread; ½ c cooked cereal, rice, or pasta; 1 oz ready-to-eat cereal; ½ bun, bagel, or English muffin; 1 small roll, biscuit, or muffin; 3 to 4 small or 2 large crackers.

◆ Whole grains (wheat, oats, barley, millet, rye, bulgur, couscous, polenta), enriched breads, rolls, tortillas, cereals, bagels, rice, pastas (macaroni, spaghetti), air-popped corn.
◇ Pancakes, muffins, cornbread, crackers, cookies, biscuits, presweetened cereals, granola, taco shells, waffles, french toast.
◆ Croissants, fried rice, doughnuts, pastries, cakes, pies.

Vegetables: 3 to 5 servings per day (use dark green, leafy vegetables and legumes several times a week).

These foods are notable for their contributions of vitamin A, vitamin C, folate, potassium, magnesium, and fiber, and for their lack of fat and cholesterol.
Serving = ½ c cooked or raw vegetables; 1 c leafy raw vegetables; ½ c cooked legumes; ¾ c vegetable juice.

◆ Bamboo shoots, bean sprouts, bok choy, broccoli, brussels sprouts, cabbage, carrots, cauliflower, corn, cucumbers, eggplant, green beans, green peas, leafy greens (spinach, mustard, and collard greens), legumes, lettuce, mushrooms, okra, onions, peppers, potatoes, pumpkin, scallions, seaweed, snow peas, soybeans, tomatoes, water chestnuts, winter squash.
◇ Candied sweet potatoes.
◆ French fries, tempura vegetables, scalloped potatoes, potato salad.

Fruits: 2 to 4 servings per day.

These foods are notable for their contributions of vitamin A, vitamin C, potassium, and fiber, and for their lack of sodium, fat, and cholesterol.
Serving = typical portion (such as 1 medium apple, banana, or orange, ½ grapefruit, 1 melon wedge); ¾ c juice; ½ c berries; ½ c diced, cooked, or canned fruit; ½ c dried fruit.

◆ Apples, apricots, bananas, blueberries, cantaloupe, grapefruit, guava, oranges, orange juice, papaya, peaches, pears, plums, strawberries, watermelon; unsweetened juices.
◇ Canned or frozen fruit (in syrup); sweetened juices.
◆ Dried fruit, coconut, avocados, olives.

Meat, Poultry, Fish, and Alternates: 2 to 3 servings per day.

Meat, poultry, and fish are notable for their contributions of protein, phosphorus, vitamin B$_6$, vitamin B$_{12}$, zinc, iron, niacin, and thiamin; legumes are notable for their protein, fiber, thiamin, folate, vitamin E, potassium, magnesium, iron, and zinc, and for their lack of fat and cholesterol.
Servings = 2 to 3 oz lean, cooked meat, poultry, or fish (total 5 to 7 oz per day); count 1 egg, ½ c cooked legumes, 4 oz tofu, or ⅓ c nuts, seeds, or peanut butter as 1 oz meat (or about ⅓ serving).

◆ Poultry (light meat, no skin), fish, shellfish, legumes, egg whites.
◇ Lean meat (fat-trimmed beef, lamb, pork); poultry (dark meat, no skin); ham; refried beans; whole eggs, tofu, tempeh.
◆ Hot dogs, luncheon meats, ground beef, peanut butter, nuts, sausage, bacon, fried fish or poultry, duck.

Key:
◆ Foods generally highest in nutrient density (good first choice).
◇ Foods moderate in nutrient density (reasonable second choice).
◆ Foods lowest in nutrient density (limit selections).

Figure 1–5
The Daily Food Guide

© Polara Studios, Inc. (all)

Milk, Cheese, and Yogurt: 2 servings per day.

3 servings per day for teenagers and young adults, pregnant/lactating women, women past menopause; 4 servings per day for pregnant/lactating teenagers. These foods are notable for their contributions of calcium, riboflavin, protein, vitamin B_{12}, and, when fortified, vitamin D and vitamin A.
Serving = 1 c milk or yogurt; 2 oz process cheese food; ½ oz cheese.

- ◆ Fat-free and 1% low-fat milk (and fat-free products such as buttermilk, cottage cheese, cheese, yogurt); fortified soy milk.
- ◆ 2% reduced-fat milk (and low-fat products such as yogurt, cheese, cottage cheese); chocolate milk; ice milk.
- ◆ Whole milk (and whole-milk products such as cheese, yogurt); custard; milkshakes; ice cream.

NOTE: These serving recommendations were established before the 1997 DRI, which raised the recommended intake for calcium; meeting the calcium recommendation may require an additional serving from the milk, cheese, and yogurt group.

Fats, Sweets, and Alcoholic Beverages: Use sparingly.

These foods contribute sugar, fat, alcohol, and food energy (kcalories). They should be used sparingly because they provide food energy while contributing few nutrients. Miscellaneous foods not high in kcalories, such as spices, herbs, coffee, tea, and diet soft drinks, can be used freely.

- ◆ Foods high in fat include margarine, salad dressing, oils, lard, mayonnaise, sour cream, cream cheese, butter, gravy, sauces, potato chips, chocolate bars.
- ◆ Foods high in sugar include cakes, pies, cookies, doughnuts, sweet rolls, candy, soft drinks, fruit drinks, jelly, syrup, gelatin, desserts, sugar, and honey.
- ◆ Alcoholic beverages include wine, beer, and liquor.

© Polara Studios, Inc. (both)

KEY

⬤ Fat (naturally occurring and added)
▽ Sugars (added)
These symbols show fats, oils, and added sugars in foods.

Food Guide Pyramid
A Guide to Daily Food Choices
The breadth of the base shows that grains (breads, cereals, rice, and pasta) deserve most emphasis in the diet. The tip is smallest: use fats, oils, and sweets sparingly.

Fats, Oils & Sweets
Use sparingly

Milk, Yogurt & Cheese Group
2–3 servings

Meat, Poultry, Fish, Dry Beans, Eggs & Nuts Group
2–3 servings

Vegetable Group
3–5 servings

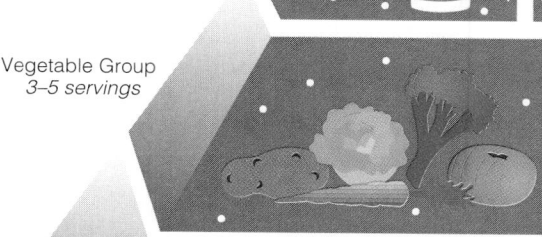

Fruit Group
2–4 servings

Bread, Cereal, Rice & Pasta Group
6–11 servings

Table 1–5 Recommended Servings for Different Energy Intakes

	Sedentary Women, Some Older Adults	Children, Teenage Girls, Active Women, Sedentary Men	Teenage Boys, Active Men
kCalories[a]	About 1600	About 2000	About 2800
Breads, cereals, rice, and pasta group (especially whole grains)	6	9	11
Vegetable group	3	4	5
Fruit group	2	3	4
Milk, yogurt, and cheese group (preferably fat-free or low fat)[b]	2–3	2–3	2–3
Meat, poultry, fish, dry beans, eggs, and nuts group (preferably lean or low fat)	2 (5 oz total)	2 (6 oz total)	3 (7 oz total)

SOURCE: Adapted from U.S. Department of Agriculture, Center for Nutrition Policy and Promotion, *The Food Guide Pyramid,* Home and Garden Bulletin no. 252 (Washington, D.C.: 1996).

[a]Assumes mostly low-fat and low-kcalorie food choices.
[b]Women who are pregnant or lactating, teenagers, and young adults to age 24 need 3 servings. In fact, given the 1997 DRI, which raised the calcium recommendation, all individuals may need an additional milk serving to meet their calcium need.

Serving Sizes To use the Food Guide Pyramid effectively, a person needs to know what constitutes a serving. The serving sizes in the Food Guide Pyramid are specific and precise, but they differ for each food group and for various foods within a group. For example, a serving of milk is 1 cup, whereas a serving of fruit juice is ¾ cup. A serving of legumes (dry beans, peas, and lentils) differs depending on whether it is counted as a vegetable serving or a meat alternate serving. As a vegetable, ½ cup of cooked beans or peas counts as a serving. As a meat substitute, ½ cup of cooked legumes counts as 1 ounce of meat, but a *serving* of meat is 2 to 3 ounces: 1 to 1½ cups of cooked legumes therefore counts as a serving of meat.

Serving sizes for the Food Guide Pyramid are, for the most part, smaller than what most people put on their plate, pour into a glass, or read on a food label. For example, a serving of bread is one slice, but when people eat a sandwich, they usually eat two slices, or two servings. As mentioned, a serving of milk is 1 cup, meaning 8 ounces. Many drinking glasses hold 12 ounces or more. A serving of rice, pasta, or cereal on a food label is 1 cup. In the Food Guide Pyramid, it is ½ cup.

Restaurants, knowing that their customers want their money's worth, often deliver portions of food much larger than those specified in the Food Guide Pyramid. The trend in the United States has been toward consuming larger food portions.[16] At the same time, people are gaining more and more body weight, suggesting the need to control serving sizes. Figure 1–5 lists the serving sizes for standard foods within each food group.

Nutrition Surveys

Researchers use nutrition surveys to determine which foods people are eating, to assess people's nutritional health, and to measure people's knowledge, attitudes, and behaviors about nutrition and how these relate to health.[17] The resulting wealth of information can be used for a variety of purposes. For example, Congress uses this information to establish public policy on nutrition education, assess food assistance programs, and regulate the food supply. Scientists use the information to establish research priorities. One of the first nutrition surveys, taken before World War II, suggested that up to a third of the U.S. population might be eating poorly. Programs to correct **malnutrition** have been evolving ever since.

Food Consumption Surveys **Food consumption surveys** determine the kinds and amounts of foods people eat. Researchers calculate the energy and nutrients in the foods and compare the amounts consumed with a standard such as the DRI. An example of this type of survey is the Continuing Survey of Food Intakes by Individuals (CSFII).

Nutrition Status Surveys **Nutrition status surveys** examine the people themselves, using nutrition assessment methods. The data provide information on several nutrition-related conditions, such as growth retardation, heart disease, and nutrient deficiencies. The National Health and Nutrition Examination Survey (NHANES) is an example of a nutrition status survey. Both the CSFII and the NHANES oversample high-risk groups (low-income families, infants and children, and the elderly) in order to glean an accurate estimate of their health and nutrition status.

Nutrition assessment methods are described in Chapter 16.

Coordinating Nutrition Survey Data Until 1990, findings from the nation's many nutrition surveys were almost impossible to compare and synthesize into a single cohesive report. Then the National Nutrition Monitoring and Related Research Act was enacted to coordinate the many nutrition-related activities that had been under way within 22 different federal agencies. The law mandated that the U.S. Department of Agriculture (USDA) and the Department of Health and Human Services (DHHS) establish and implement a Ten-Year Comprehensive Plan for nutrition monitoring and related research. All major reports that examine the contribution of diet and nutrition status to the health of the people of the United States depend on information collected and coordinated by this national program. For example, Congress uses this information to establish public policy on nutrition education, food assistance programs, and the regulation of the food supply. Scientists use the information to establish research priorities. These data also provide the basis for developing and monitoring national health goals.

malnutrition: any condition caused by deficient or excess energy or nutrient intake or by an imbalance of nutrients.

food consumption surveys: surveys that measure the amounts and kinds of food people consume (using diet histories), estimate the nutrient intakes, and compare them with a standard such as the DRI.

nutrition status surveys: surveys that evaluate people's nutrition status using nutrition assessment methods.

***Healthy People* Reports** *Healthy People,* a program that sets goals for improving the nation's health was initiated over 20 years ago. At the close of the twentieth century, the nation's progress toward meeting its *Healthy People 2000* goals was mixed.[18] For almost 60 percent of the objectives, the population either met the target or was moving in the right direction. Successes included reductions in the incidences of food- and water-borne infections and infant mortality. On the downside, the population was moving in the opposite direction of several key objectives, most notably for reducing weight and increasing physical activity. Table 1–6 on the next page lists the nutrition-related objectives for 2010.

In Summary The Dietary Reference Intakes (DRI) are a set of four nutrient intake values that can be used to plan and evaluate diets for healthy people. The *Dietary Guidelines* offer practical advice on how to eat for good health. A well-planned diet delivers adequate nutrients, a balanced array of nutrients, and an appropriate amount of energy. It is based on nutrient-dense foods, moderate in substances that can harm health, and varied in its selections. Food group plans serve as the basis for planning adequate, balanced, and varied diets. Nutrition surveys measure people's food consumption and evaluate the nutrition status of populations. Information gathered from nutrition surveys serves as the basis for many major diet and nutrition reports, including *Healthy People 2010.*

Table 1–6 Healthy People 2010 Nutrition-Related Objectives

- Increase *nutrition education* among consumers and in educational settings at all levels.
- Increase the proportion of adults who are at a *healthy weight*.
- Reduce the proportion of adults who are *obese*.
- Reduce the proportion of children and adolescents who are *overweight* or *obese*.
- Reduce *growth retardation* among low-income children under age 5 years.
- Increase the proportion of mothers who *breastfeed* their infants.
- Increase the proportion of persons aged 2 years and older who consume at least two daily servings of *fruit*.
- Increase the proportion of persons aged 2 years and older who consume at least three daily servings of *vegetables,* with at least one-third being dark green or orange vegetables.
- Increase the proportion of persons aged 2 years and older who consume at least six daily servings of *grain products,* with at least three being whole grains.
- Increase the proportion of persons aged 2 years and older who consume less than 10 percent of kcalories from *saturated fat.*
- Increase the proportion of persons aged 2 years and older who consume no more than 30 percent of kcalories from *total fat.*
- Increase the proportion of persons aged 2 years and older who consume 2400 milligrams or less of *sodium.*
- Increase the proportion of adults with *high blood pressure* who are taking action (for example, losing weight, increasing physical activity, or reducing sodium intake) to help control their blood pressure.
- Increase the proportion of persons aged 2 years and older who meet dietary recommendations for *calcium.*
- Reduce *iron deficiency* among young children, females of childbearing age, and pregnant females.
- Reduce *anemia* among low-income pregnant females in their third trimester.
- Increase the proportion of pregnancies begun with an optimum *folic acid* level.
- Increase the proportion of children and adolescents aged 6 to 19 years whose intake of *meals and snacks at school* contributes to good overall dietary quality.
- Increase the proportion of worksites that offer *nutrition or weight management classes or counseling.*
- Reduce the proportion of adults with *osteoporosis.*
- Increase the proportion of physician office visits made by patients with a diagnosis of cardiovascular disease, diabetes, or hyperlipidemia that include *counseling or education related to diet and nutrition.*
- Reduce deaths from anaphylaxis caused by *food allergies.*
- Reduce *food-borne illness.*
- Increase *food security* among U.S. households and in so doing reduce hunger.

SOURCE: Details about these and many other objectives are available from the U.S. Department of Health and Human Services, *Healthy People 2010,* 2nd ed. (Washington, D.C.: Government Printing Office, 2000), online at **www.health.gov/healthypeople**.

Food Labels

Today consumers know more about the links between diet and disease than they did in the past, and they are demanding still more information on disease prevention. Many people rely on food labels to tell them which substances to select and which ones to limit for health reasons. Figure 1–6 features the requirements for label information. Most food labels must conform to all these requirements. Exceptions include plain coffee, tea, spices, and other foods contributing few nutrients; foods produced by small businesses; packages with fewer than 12 square inches of surface area; and those prepared and sold in the same establishment as long as the foods do not make nutrient or health claims.[19]*

*For example, restaurants need not provide complete nutrition information unless they are making "heart healthy" claims for menu items.

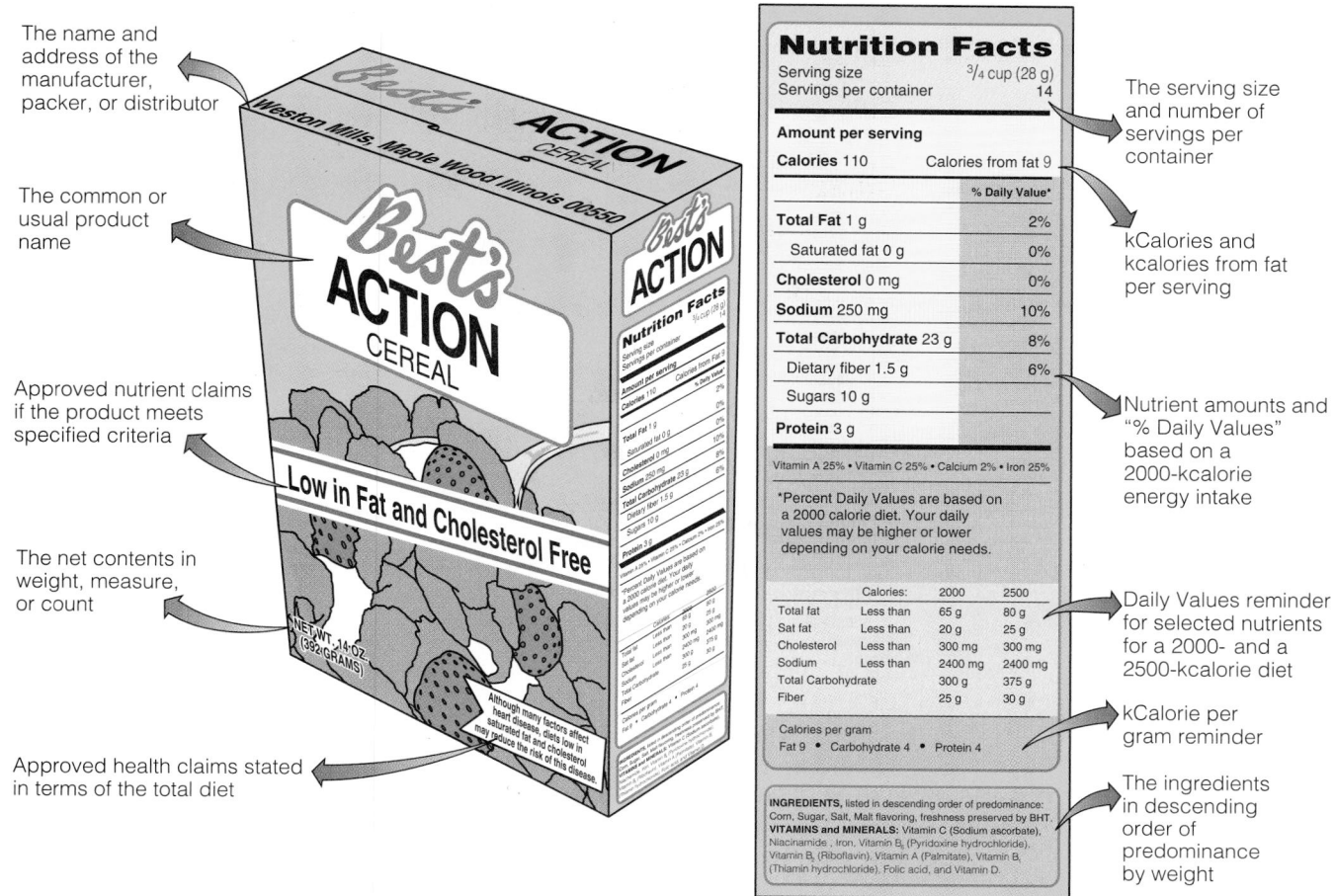

The name and address of the manufacturer, packer, or distributor

The common or usual product name

Approved nutrient claims if the product meets specified criteria

The net contents in weight, measure, or count

Approved health claims stated in terms of the total diet

The serving size and number of servings per container

kCalories and kcalories from fat per serving

Nutrient amounts and "% Daily Values" based on a 2000-kcalorie energy intake

Daily Values reminder for selected nutrients for a 2000- and a 2500-kcalorie diet

kCalorie per gram reminder

The ingredients in descending order of predominance by weight

Figure 1–6
Example of a Food Label

The Daily Values on Labels   The **Daily Values** (inside back cover) are a set of nutrient standards created by the Food and Drug Administration (FDA) for use on food labels. The Daily Values do two things: they set adequacy standards for nutrients that are desirable in the diet such as protein, vitamins, minerals, and fiber, and they also set moderation standards for other nutrients that must be limited, such as fat, cholesterol, and sodium, according to the *Dietary Guidelines*.

Nutrient Claims   The FDA defines the **nutrient claims** a label may use to describe the contents of a product (see Table 1–7 on p. 23). Definitions include the conditions under which each term can be used. For example, in addition to having less than 2 milligrams of cholesterol, a "cholesterol-free" product may not contain more than 2 grams of saturated fat per serving.

Some descriptions imply that a food contains, or does not contain, a nutrient. Implied claims are prohibited unless they meet specified criteria. For example, a claim that a product "contains no oil" *implies* that the food contains no fat. If the product is truly fat-free, then it may make the no-oil claim, but if it contains another source of fat, such as butter, it may not.

Health Claims   The FDA has established strict guidelines pertaining to claims about health on food labels. Table 1–8 on p. 24 presents the **health claims** that have been approved to date and the criteria they must meet. A review of the table reveals that the fats play a key role in whether a food can make a health claim. Whole milk, even though it is high in calcium, may not make a claim about osteoporosis because it contains too much saturated fat to qualify. Low-fat and fat-free milk, however, do qualify to bear the calcium and osteoporosis claim.

Daily Values: reference values developed by the FDA specifically for use on food labels.

nutrient claims: statements that characterize the quantity of a nutrient in a food.

health claims: statements that characterize the relationship between a nutrient or other substance in food and a disease or health-related condition.

Perspectives on Health and Nutrition

21

Grocers often post nutrition-information placards or pamphlets near nonpackaged items such as raw fruits, vegetables, and seafood.

Those who design food labels must proceed carefully. For example, use of the word *healthy* in a name such as "Healthy Start" or the use of a heart-shaped logo may imply that a food is health promoting. Foods bearing such words or logos must not exceed limits set for fat, saturated fat, cholesterol, and sodium contents.

Consumers may wonder whether they are reducing their health risks if they eat only "healthy" foods. They may be, and that is all that a label is allowed to say: that a substance "may" or "might" reduce disease risks. This wording is tentative because scientists are still accumulating evidence concerning the roles of diet in disease. A claim must also state that the development of a disease rests on many factors. A permissible health claim reads like this: "Development of heart disease depends on many factors. A healthful diet low in saturated fat and cholesterol may lower blood cholesterol levels and may reduce the risk of heart disease." Health claims on labels are so carefully controlled that they can be an asset to consumers who would rather not worry about grams, percentages, and other mathematical stumbling blocks as they shop for foods.

Structure-Function Claims Consumers need to be aware that a different kind of claim, known as a "structure-function" claim, may also appear on food or dietary supplement labels. **Structure-function claims** are statements about a food substance's effect on a structure or function of the body—for example, "antioxidants support heart health." Structure-function claims are not required to have FDA approval, but the claims may not refer to the reduction of disease risk.[20] Structure-function claims must also be accompanied by the FDA disclaimer statement shown in the margin. The claims must be carefully worded to avoid any mention of a specific disease. Typical claims are "slows aging," "improves memory," and "builds strong bones." To make a more specific claim such as "prevents osteoporosis," the manufacturer would have to submit to the rigorous requirements for health claims or meet the even stricter safety and efficacy standards applied to drugs. These rules ensure that consumers can have confidence that when a claim names a specific disease, it means that there is substantial scientific agreement that the food, in the context of a healthy diet, may help protect against that disease.

This statement has not been evaluated by the Food and Drug Administration. This product is not intended to diagnose, treat, mitigate, cure, or prevent any disease.

structure-function claims: statements that describe how a product may affect a structure or function of the body; for example, "calcium builds strong bones." Structure-function claims do not require FDA authorization.

Table 1-7   Terms Used on Food Labels

General Terms

Free, without, no, zero: Contains no amount or a trivial amount. *kCalorie-free* means containing fewer than 5 kcal per serving; *sugar-free* or *fat-free* means containing less than 0.5 g per serving.

Fresh: Raw, unprocessed, or minimally processed (blanched or irradiated) with no added preservatives.

Good source: Provides 10 to 19% of the Daily Value of a given nutrient per serving.

Healthy: A food that is low in fat, saturated fat, cholesterol, and sodium and that contains at least 10% of the Daily Value for vitamin A, vitamin C, iron, calcium, protein, or fiber.

High: Provides 20% or more of the Daily Value per serving.

Less, fewer, reduced: Provides 25% less of a nutrient or kcalories than a reference food. This may occur naturally or as a result of altering the food. For example, pretzels, which are usually low in fat, can claim to provide less fat than potato chips, a comparable food.

Light: This descriptor has three meanings on labels:

- A serving provides one-third fewer kcalories or half the fat of the regular product.
- A serving of a low-kcalorie, low-fat food provides half the sodium normally present.
- The product is light in color and texture; the label must make this intent clear, as in "light brown sugar."

More: Contains at least 10% more of the Daily Value for a given nutrient than a comparable food. The nutrient may be added or may occur naturally.

Carbohydrates: Fiber and Sugar Terms

High fiber: Provides 5 g or more fiber per serving; a high-fiber claim made on a food that contains more than 3 g fat per serving and per 100 g of food must also declare total fat.

Sugar-free: Provides less than 0.5 g sugar per serving.

Energy Terms

kCalorie-free: Contains fewer than 5 kcal per serving.

Low kcalorie: Contains 40 kcal or less per serving.

Reduced kcalorie: Contains at least 25% fewer kcalories per serving than the comparison product.

Fat Terms (Meat and Poultry Products)

Extra lean: A product contains:

- Less than 5 g fat.
- Less than 2 g saturated fat.
- Less than 95 mg cholesterol per serving.

Lean:[a] A product contains:

- Less than 10 g fat.
- Less than 4.5 g saturated fat.
- Less than 95 mg cholesterol per serving.

Fat and Cholesterol Terms (All Products)

Cholesterol-free: Contains less than 2 mg cholesterol per serving and 2 g or less saturated fat per serving.

Fat-free: Contains less than 0.5 g fat per serving.

Low cholesterol: Contains 20 mg or less cholesterol per serving and 2 g or less saturated fat per serving.

Low fat: Contains 3 g or less fat per serving.

Low saturated fat: Contains 1 g or less saturated fat per serving.

Percent fat free: May be used only if the product meets the definition of low fat or fat-free. Requires disclosure of grams fat per 100 g food.

Reduced or less cholesterol: Contains 25% or less cholesterol than the comparison food and 2 g or less saturated fat per serving.

Reduced or less fat: Contains 25% or less fat than the comparison food.

Reduced or less saturated fat: Contains 25% or less saturated fat than the comparison food and reduced by more than 1 g per serving.

Saturated fat-free: Contains less than 0.5 g saturated fat and less than 0.5 g *trans*-fatty acids.

Sodium Terms

Low sodium: Contains 140 mg or less sodium per serving.

Sodium-free: Contains less than 5 mg sodium per serving.

Very low sodium: Contains 35 mg or less sodium per serving.

---

[a]The word *lean* as part of the brand name (as in "Lean Supreme") indicates that the product contains fewer than 10 grams of fat per serving. *Lean* ground beef can contain up to 22.5 percent fat by weight.

SOURCES: The new food label, *FDA Backgrounder,* December 10, 1992; Nutrition labeling of meat and poultry products, *FSIS Backgrounder,* January 1993.

Table 1–8 Food Label Health Claims and Their Criteria

Health Claim	Criteria
Calcium and reduced risk of osteoporosis	■ High in calcium (≥20% DV) ■ No more phosphorus than calcium
Sodium and reduced risk of hypertension	■ Low in sodium (≤140 mg/serving)
Dietary saturated fat and cholesterol and reduced risk of coronary heart disease	■ Low in saturated fat (≤1 g/serving) ■ Low in cholesterol (≤20 mg/serving) ■ Low in fat (≤3 g/serving)
Dietary fat and reduced risk of cancer	■ Low in fat (≤3 g/serving)
Fiber-containing grain products, fruits, and vegetables and reduced risk of cancer	■ Low in fat (≤3 g/serving) ■ Good source of dietary fiber (≥10% DV)
Fruits, vegetables, and grain products that contain fiber, particularly soluble fiber, and reduced risk of coronary heart disease	■ Low in saturated fat (≤1 g/serving) ■ Low in cholesterol (≤20 mg/serving) ■ Low in fat (≤3 g/serving) ■ Soluble fiber (≥0.6 g/serving)
Fruits and vegetables and reduced risk of cancer	■ Low in fat (≤3 g/serving) ■ Good source of vitamin A, vitamin C, or dietary fiber (≥10% DV)
Folate and reduced risk of neural tube defects	■ Good source of folate (≥10% DV) ■ Limited in vitamin A and vitamin D (≤100% DV)
Sugar alcohols and reduced risk of tooth decay	■ Sugar-free ■ Cannot lower dental plaque pH below 5.7 by bacterial fermentation
Soluble fiber from whole oats and from psyllium seed husk and reduced risk of heart disease	■ Soluble fiber from oats (≥0.75 g/serving) or from psyllium seed husk (≥1.7 g/serving) ■ Low in saturated fat (≤1 g/serving) ■ Low in cholesterol (≤20 mg/serving) ■ Low in fat (≤3 g/serving)
Soy protein and reduced risk of heart disease	■ Soy protein (6.25 g/serving) ■ Low in saturated fat (≤1 g/serving) ■ Low in cholesterol (≤20 mg/serving) ■ Low in fat (≤3 g/serving)
Whole grains and reduced risk of heart disease and certain cancers	■ Whole-grain (≥51%) ■ Low in saturated fat (≤1 g/serving) ■ Low in cholesterol (≤20 mg/serving) ■ Low in fat (≤3 g/serving)
Plant sterol and plant stanol esters and heart disease	■ Plant sterol esters (0.65 g/serving) or plant stanol esters (1.7 g/serving) ■ Low in saturated fat (≤1 g/serving) ■ Low in cholesterol (≤20 mg/serving) ■ Low in fat (≤3 g/serving)
Potassium and reduced risk of hypertension and stroke	■ Good source of potassium (≥10% DV) ■ Low in sodium (≤140 mg/serving) ■ Low in saturated fat (≤1 g/serving) ■ Low in cholesterol (≤20 mg/serving) ■ Low in fat (≤3 g/serving)

NOTE: With the exception of sugar alcohols and dental caries, all other health claims must also meet two additional criteria. First, a food making a health claim must be a naturally good source (containing at least 10 percent of the Daily Value) of at least one of the following nutrients: vitamin A, vitamin C, iron, calcium, protein, or fiber. Second, foods are disqualified from making health claims if a standard serving contains more than 20 percent of the Daily Value for total fat, saturated fat, cholesterol, or sodium.

In Summary Food labels provide consumers with information they need to select foods that will help them meet their nutrition and health goals.

Self Study

MAKING FOOD CHOICES

We decide what to eat, when to eat, and even whether to eat for a variety of reasons. Examine the factors that influence your food choices by keeping a food diary for 24 hours. Record the times and places of meals and snacks, the types and amounts of foods eaten, and a description of your thoughts and feelings when eating. Now examine your food record and consider your choices.

- Which, if any, of your food choices were influenced by emotions (happiness, boredom, or disappointment, for example)?
- Was social pressure a factor in any food decisions?
- Which, if any, of your food choices were influenced by marketing strategies or food advertisements?
- How large a role do availability, convenience, and economy play in your food choices?
- Do your age, ethnicity, or health concerns influence your food choices?

- How many times did you eat because you were truly hungry? How often did you think of health and nutrition when making food choices? Were those food choices different from others made during the day?

Compare the choices you made in your 24-hour food diary to the food guide pyramid.

- Do you eat at least the minimum number of servings from each of the five food groups daily?
- Do you try to vary your choices within each food group from day to day?
- What dietary changes could you make to improve your chances of enjoying good health?

Food groups	Suggested servings	Servings consumed
Bread, cereal, rice, and pasta	6 to 11 servings	
Vegetable	3 to 5 servings	
Fruit	2 to 4 servings	
Milk, yogurt, and cheese	2 to 3 servings	
Meat, poultry, fish, dry beans, eggs, and nuts	2 to 3 servings	
Fats, oils, and sweets	Use sparingly	

Self Check

1. When people eat the foods typical of their families or geographic region, their choices are influenced by:
   a. occupation.
   b. nutrition.
   c. emotional state.
   d. ethnic heritage or tradition.

2. The energy-yielding nutrients are:
   a. fats, minerals, and water.
   b. minerals, proteins, and vitamins.
   c. carbohydrates, fats, and vitamins.
   d. carbohydrates, fats, and proteins.

3. The inorganic nutrients are:
   a. proteins and fats.
   b. vitamins and minerals.
   c. minerals and water.
   d. vitamins and proteins.

4. Alcohol is not a nutrient because:
   a. the body derives no energy from it.
   b. it is organic.
   c. it is converted to body fat.
   d. it does not contribute to the body's growth or repair.

5. DRI stands for:
   a. Daily Recommended Intakes.
   b. Dietary Requirements for Individuals.
   c. Dietary Reference Intakes.
   d. Daily Recommendations for Individuals.

6. Which of the following is consistent with the *Dietary Guidelines for Americans*?
   a. Choose a diet moderate in saturated fat and cholesterol.
   b. Be physically active each day.
   c. Choose a diet with plenty of milk products and meats.
   d. Eat an abundance of foods to ensure nutrient adequacy.

7. A slice of apple pie supplies 350 kcalories with 3 grams of fiber; an apple provides 80 kcalories and the same 3 grams of fiber. This is an example of:
   a. kcalorie control.
   b. nutrient density.
   c. variety.
   d. essential nutrients.

8. The diet-planning principle that provides enough, but not too much, of a constituent is:
   a. adequacy.
   b. balance.
   c. moderation.
   d. variety.

9. Foods within a given food group are similar in their contents of:
   a. energy.
   b. proteins and fibers.
   c. vitamins and minerals.
   d. carbohydrates and fats.

10. According to the Food Guide Pyramid, which foods should form the foundation of a healthy diet?
    a. vegetables
    b. breads, cereals, rice, and pasta
    c. fruits
    d. milk, yogurt, and cheese

Critical Thinking

1. Calculate the energy provided by a food from its energy-nutrient contents. A cup of fried rice contains 5 grams protein, 30 grams carbohydrate, and 11 grams fat.
   a. How many kcalories does the rice provide from these energy nutrients?

   —————————— = —— kcal protein.
   —————————— = —— kcal carbohydrate.
   —————————— = —— kcal fat.
   Total = —— kcal.

   b. What percentage of the energy in the fried rice comes from each of the energy-yielding nutrients?

   —————————— = —— % kcal from protein.
   —————————— = —— % kcal from carbohydrate.
   —————————— = —— % kcal from fat.
   Total = —— %
   Note: The total should add up to 100%; 99% or 101% due to rounding is also acceptable.

Answers to these questions appear in Appendix G.

Clinical Applications

1. Review the list of foods and beverages consumed in your 24 hour food diary (see Self Study, p. 25). Look at each item on your list and consider why you chose the particular food or beverage you did. In going down your list, you may be surprised to discover exactly why you chose certain foods.

2. As a health care professional, you can uncover clues about a client's food choices by paying close attention. You may be surprised to discover why a client chooses certain foods, but you can then use this knowledge to serve the best interests of the client. For example, an elderly, undernourished widower may eat the same sandwich for lunch every day. In talking with the client, you discover that is what he and his wife fixed together each day. Consider ways you might be able to help the client learn to eat other foods and vary his choices.

Nutrition on the Net

For further study of the topics of this chapter, access these websites.

Find updates and quick links to these and other nutrition-related sites at our website:
www.wadsworth.com/nutrition

Search for nutrition at the U.S. Government health information site:
www.healthfinder.gov

Review the Dietary Reference Intakes:
www.nap.edu/readingroom or **www2.nas.edu/fnb**

Review nutrient recommendations from the Food and Agriculture Organization and the World Health Organization:
www.fao.org and **www.who.org**

Learn more about the *Dietary Guidelines for Americans:*
health.gov/dietaryguidelines

View Canadian information on nutrition guidelines and food labels at:
www.hc-sc.gc.ca

Visit the Food Guide Pyramid section (including its ethnic/cultural pyramids) of the U.S. Department of Agriculture:
www.nal.usda.gov/fnic

See food pyramids for various ethnic groups at Oldways Preservation and Exchange Trust:
www.oldwayspt.org

View *Healthy People 2010:*
web.health.gov/healthypeople

Review the Canadian *National Plan of Action for Nutrition:*
www.hc-sc.ca/datahpsb/npu

Get information from the Food Surveys Research Group:
www.barc.usda.gov/bhnrc/foodsurvey

Learn more about food labeling from the Food and Drug Administration:
www.cfsan.fda.gov or **vm.cfsan.fda.gov/label.html**

Learn about NHANES:
www.cdc.gov/nchs/nhanes.htm

Notes

1. K. Glanz and coauthors, Why Americans eat what they do: Taste, nutrition, cost, convenience, and weight control concerns as influences on food consumption, *Journal of the American Dietetic Association* 98 (1998): 1118–1126.

2. L. L. Birch, Development of food preferences, *Annual Review of Nutrition* 19 (1999): 41–62.

3. A. Drewnowski, Taste preferences and food intake, *Annual Review of Nutrition* 17 (1997): 237–253.

4. A. E. Sloan, America's appetite '96: The top ten trends to watch and work on, *Food Technology* 50 (1996): 55–71.

5. F. Katz, "How nutritious?" meets "How convenient?" *Food Technology* 53 (1999): 44–50.

6. M. Nestle and coauthors, Behavioral and social influence on food choice, *Nutrition Reviews* 56 (1998): S50–S74.

7. Position of The American Dietetic Association: Functional foods, *Journal of the American Dietetic Association* 99 (1999): 1278–1285.

8. Position of The American Dietetic Association: The role of nutrition in health promotion and disease prevention programs, *Journal of the American Dietetic Association* 98 (1998): 205–208.

9. Standing Committee on the Scientific Evaluation of Dietary Reference Intakes, Food and Nutrition Board, Institute of Medicine, *Dietary Reference Intakes for Calcium, Phosphorus, Magnesium, Vitamin D, and Fluoride* (Washington, D.C.: National Academy Press, 1997).

10. Standing Committee on the Scientific Evaluation of Dietary Reference Intakes, Food and Nutrition Board, Institute of Medicine, *Dietary Reference Intakes for Thiamin, Riboflavin, Niacin, Vitamin B_6, Folate, Vitamin B_{12}, Pantothenic Acid, Biotin, and Choline* (Washington, D.C.: National Academy Press, 1998).

11. Standing Committee on the Scientific Evaluation of Dietary Reference Intakes, Food and Nutrition Board, Institute of Medicine, *Dietary Reference Intakes for Vitamin C, Vitamin E, Selenium, and Carotenoids* (Washington, D.C.: National Academy Press, 2000).

12. Standing Committee on the Scientific Evaluation of Dietary Reference Intakes, Food and Nutrition Board, Institute of Medicine, *Dietary Reference Intakes for Vitamin A, Vitamin K, Arsenic, Boron, Chromium, Copper, Iodine, Iron, Manganese, Molybdenum, Nickel, Silicon, Vanadium, and Zinc* (Washington, D.C.: National Academy Press, 2001).

13. Standing Committee on the Scientific Evaluation of Dietary Reference Intakes, *Applications in Dietary Assessment* (Washington, D.C.: National Academy Press, 2000).

14. F. B. Hu and coauthors, Diet, lifestyle, and the risk of type 2 diabetes mellitus in women, *New England Journal of Medicine* 345 (2001): 790–797; A. K. Kant and coauthors, A prospective study of diet quality and mortality in women, *Journal of the American Medical Association* 283 (2000): 2109–2115; M. Y. Hwang, Benefits and dangers of alcohol, *Journal of the American Medical Association* 282 (1999): 104.

15. U.S. Department of Agriculture and U.S. Department of Health and Human Services, *Nutrition and Your Health: Dietary Guidelines for Americans,* Home and Garden Bulletin no. 232 (Washington, D.C.: 2000).

16. L. R. Young and M. Nestle, The contribution of expanding portion sizes to the U.S. obesity epidemic, *American Journal of Public Health* 92 (2002): 246–249.

17. R. R. Briefel, Nutrition monitoring in the United States, in *Present Knowledge in Nutrition,* 8th ed., ed. B. A. Bowman and R. M. Russell (Washington, D.C.: International Life Sciences Institute Press, 2001), pp. 617–635.

18. D. S. Hatcher, Healthy People at 2000, *Public Health Reports* 114 (1999): 563–564.

19. P. Kurtzweil, Today's special: Nutrition information, *FDA Consumer,* May/June 1997, pp. 21–25.

20. Food and Drug Administration, FDA finalizes rules for claims on dietary supplements, *FDA Talk Paper,* **http://vm.cfsan.fda.gov/~lrd**, accessed May 12, 2001.

Nutrition in Practice

▪ FINDING THE TRUTH ABOUT NUTRITION ▪

Nutrition and health receive so much attention on television, in the popular press, and on the Internet that it is easy to be overwhelmed with conflicting, confusing information. Determining whether nutrition information is accurate can be a challenging task. It is also an important task because nutrition affects a person both professionally and personally.

A person watches a nutrition report on television and then reads a conflicting report the next day in the newspaper. Why do nutrition news reports and claims for nutrition products seem to contradict each other so often?

The problem of conflicting messages arises for several reasons:

- Popular media, often faced with tight deadlines and limited time or space to report new information, rush to present the latest "breakthrough" in a headline or a 60-second spot. They can hardly help omitting important facts about the study or studies that the "breakthrough" is based on.

- Despite tremendous advances in the last few decades, scientists still have much to learn about the human body and nutrition. Scientists themselves often disagree on their first tentative interpretations of new research findings, yet these are the very findings that the public hears most about.

- The popular media often broadcast preliminary findings in hopes of grabbing attention and boosting readership or television ratings.

- Commercial promoters turn preliminary findings into advertisements for products or supplements long before the findings have been validated—or disproved. The scientific process requires many experiments or trials to confirm a new finding. Seldom do promoters wait as long as they should to make their claims.

- Promoters are aware that consumers like to try new products or treatments even though they probably will not withstand the tests of time and scientific scrutiny.

Table NP1–1   FANSA's Red Flags of Junk Science

1. Recommendations that promise a quick fix.
2. Dire warnings of danger from a single product or regimen.
3. Claims that sound too good to be true.
4. Simplistic conclusions drawn from a complex study.
5. Recommendations based on a single study.
6. Dramatic statements that are refuted by reputable scientific organizations.
7. Lists of "good" and "bad" foods.
8. Recommendations made to help sell a product.
9. Recommendations based on studies published without peer review.
10. Recommendations from studies that ignore differences among individuals or groups

SOURCE: Reprinted with permission from B. Hansen. President's address, 1996: A virtual organization for nutrition in the 21st century, *American Journal of Clinical Nutrition* 64 (1996): 796–799. © American Society for Clinical Nutrition.

So how can a person tell what claims to believe?

The Food and Nutrition Science Alliance (FANSA), whose partners include the American Dietetic Association (ADA), the American Society for Nutritional Sciences (ASNS), and the Institute of Food Technologists (IFT), attempts to help consumers distinguish valid from misleading nutrition information. FANSA has created a list of ten red flags for detecting "junk science" (see Table NP1–1).[1]

Because nutrition misinformation harms the health and economic status of consumers, the ADA works with health care professionals and educators to present sound nutrition information to the public and to actively confront nutrition misinformation.[2] Table NP1–2 offers a list of credible sources of nutrition information.

Nutrition in Practice

Table NP1–2 Credible Sources of Nutrition Information
Professional health organizations, government health agencies, volunteer health agencies, and consumer groups provide consumers with reliable health and nutrition information. Credible sources of nutrition information include: ■ Professional health organizations, especially the American Dietetic Association's National Center for Nutrition and Dietetics (NCND) **www.eatright.org/ncnd.html**; also the Society for Nutrition Education **www.sne.org** and the American Medical Association **www.ama~assn.org**. ■ Government health agencies such as the Federal Trade Commission (FTC) **www.ftc.gov**, the U.S. Department of Health and Human Services (DHHS) **www.os.dhhs.gov**, the Food and Drug Administration (FDA) **www.fda.gov**, and the U.S. Department of Agriculture (USDA) **www.usda.gov**. ■ Volunteer health agencies such as the American Cancer Society **www.cancer.org**, the American Diabetes Association **www.diabetes.org**, and the American Heart Association **www.americaheart**. ■ Reputable consumer groups such as the Better Business Bureau **www.bbb.org**, the Consumers Union **www.consumersunion.org**, the American Council on Science and Health **www.acsh.org**, and the National Council on Science and Health Fraud **www.ncahf.org**. Appendix D provides addresses and websites for these and other organizations.

SOURCE: Data from Position of The American Dietetic Association: Food and nutrition misinformation, *Journal of the American Dietetic Association* 102 (2002): 260–266.

What about nutrition and health information found on the Internet? How does a person know whether the websites are reliable?

With hundreds of millions of websites on the Internet, searching for nutrition and health information can be daunting. The Internet offers no guarantee of the accuracy of the information found there, and much of it is pure fiction. Websites must be evaluated for their accuracy, just like every other source.

To help users find reliable nutrition information on the Internet, Tufts University maintains an online rating and review guide called the Nutrition Navigator (see Table NP1–3 on p. 30). The ratings reflect the opinions of a panel of nutrition experts who have scored selected websites on the basis of their accuracy, depth, and ease of use. In addition to a rating, the Nutrition Navigator provides a review of the website's content and usefulness. Table NP1-3 provides other clues to identifying reliable nutrition information sites and lists some credible sites.

The Federal Trade Commission (FTC), the Food and Drug Administration (FDA), and other law enforcement agencies have launched "Operation Cure.All" (see Table NP1–3 on the next page) to take action against fraudulent marketing of supplements and health products on the Internet.[3] The latest actions target unscrupulous companies that use the Internet to promote products to the most vulnerable consumers—

those with diseases such as AIDS, Alzheimer's, and cancer. Of greatest concern are those products that not only make false promises, but also are potentially dangerous. For example, herbal products touted as safe treatments for serious illnesses such as AIDS may interact with medications and impair the effectiveness of the medicines. The FTC advises consumers to be suspicious of:

■ Claims that a product is "natural" or "nontoxic." "Natural" or "nontoxic" does not always mean safe.

■ Claims that a product is a "scientific breakthrough," "miraculous cure," "secret ingredient," or "ancient remedy."

■ Claims that a product cures a wide range of illnesses.

■ Claims that use impressive-sounding medical terms.

■ Claims of a "money-back" guarantee.

Everyone seems to be giving advice on nutrition. How can a person tell whom to listen to?

Registered dietitians (R.D.'s) and nutrition professionals with advanced degrees (M.S., Ph.D.) are experts (see the glossary on p. 30). These professionals are probably in the best position to answer a person's nutrition questions. On the other hand, **"nutritionists"** may be experts or quacks, depending on the state where they practice. Some states require people who use this title to meet strict standards. In other states, a "nutritionist" may be any individual who claims a career connection with the nutrition field.

Other purveyors of nutrition information may also lack credentials. A health food store owner may be in the nutrition business simply because it is a lucrative market. The owner may have a background in business or sales and no education in nutrition at all. Such a person is not qualified to provide nutrition information to customers. For accurate nutrition information, seek out a trained professional with a college education in nutrition—an expert in the field of **dietetics**.

Nutrition in Practice

Table NP1–3 Evaluating the Reliability of Websites

To determine whether a website offers reliable nutrition information, ask the following questions:

- **Who?** Who is responsible for the site? Look for the author's name and credentials. Is the person qualified to speak on nutrition?

- **When?** When was the site last updated? Because nutrition is an ever-changing science, sites need to be dated and updated frequently.

- **Where?** Where is the information coming from? The three letters following the dot in a Web address identify the site's affiliation. Addresses ending in "gov" (government), "edu" (educational institute), and "org" (organization) generally provide reliable information; "com" (commercial) sites represent businesses and, depending on their qualifications and integrity, may or may not offer dependable information.

- **Why?** Why is the site giving you this information? Is the site providing a public service or selling a product? Many commercial sites provide accurate information, but some do not. When money is the prime motivation, be aware that the information may be biased.

Some credible websites include:

National Council Against Health Fraud
www.ncahf.org

Stephen Barrett's Quackwatch
www.quackwatch.com

Centers for Disease Control and Prevention's Current Health Related Hoaxes and Rumors
www.cdc.gov/hoax_rumors.htm

Tufts University
www.navigator.tufts.edu

Federal Trade Commission's Operation Cure.All
www.ftc.gov/opa/2001/06/cureall.htm

What about other health care professionals?

All members of the health care team share responsibility for helping each client to achieve optimal health, but the registered dietitian is usually the primary nutrition expert. Each of the other team members has a related specialty. Some physicians are specialists in clinical nutrition and are also experts in the field. Other physicians, nurses, and **dietetic technicians (D.T.R.'s)** often assist dietitians in providing nutrition information and may help to administer direct nutrition care. Nurses play central roles in client care management and client relationships. Visiting nurses and home health care nurses may become intimately involved in clients' nutrition care at home, teaching them both theory and cooking techniques. Physical therapists can provide individualized exercise programs related to nutrition—for example, to help control obesity. Social workers may provide practical and emotional support.

What roles might these other health care professionals play in nutrition care?

Some of the responsibilities of the health care professional might be:

- Helping people understand why nutrition is important to them.
- Answering questions about food and diet.
- Explaining to clients how modified diets work.
- Collecting information about clients that may influence their nutritional health.
- Identifying clients at risk for poor nutrition status (see Chapter 16) and recommending or taking appropriate action.

Glossary of Nutrition Experts

dietetic technicians registered (D.T.R.'s): professionals who have earned an associate degree or higher; have completed a dietetic technician program approved by the American Dietetic Association (ADA); have passed a national registration exam; and assist in planning, implementing, and evaluating nutritional care.

dietetics: the practical application of nutrition, including the assessment of nutrition status, recommendation of appropriate diets, nutrition education, and the planning and serving of meals.

nutritionists: persons who specialize in the study of nutrition. Some nutritionists are registered dietitians, but others are self-described experts whose training may be minimal or nonexistent. Some states make the term meaningful by allowing it to apply only to people who have master's (M.S.) or doctoral (Ph.D.) degrees from institutions accredited to offer such degrees in nutrition or related fields.

registered dietitians (R.D.'s): dietitians who have graduated from a university or college after completing a program of dietetics that has been accredited by the American Dietetic Association (or Dietitians of Canada). The dietitian must serve in an approved internship or coordinated program to practice the necessary skills, pass the association registration examination, and maintain competency through continuing education. Many states require licensing for practicing dietitians. Licensed dietitians (L.D.'s) have met all *state* requirements to offer nutrition advice.

Nutrition in Practice

- Recognizing when clients need extra help with nutrition problems (in such cases, the problems should be referred to a dietitian or physician).

Health care professionals may routinely perform these nutrition-related tasks:

- Obtaining diet histories.
- Taking weight and height measurements.
- Feeding clients who cannot feed themselves.
- Recording what clients eat or drink.
- Observing clients' responses and reactions to foods.
- Helping clients mark menus.
- Monitoring weight changes.
- Monitoring food and drug interactions.
- Encouraging clients to eat.
- Assisting clients at home in planning their diets and managing their kitchen chores.
- Alerting the physician or dietitian when nutrition problems are identified.

- Charting actions taken and communicating on these matters with other professionals as needed.

Thus, although the dietitian assumes the primary role as the nutrition expert on a health care team, other health care professionals play important roles in administering nutrition care.

Notes

1. B. Hansen, President's address, 1996: A virtual organization for nutrition in the 21st century, *American Journal of Clinical Nutrition* 64 (1996): 796–799.

2. Position of The American Dietetic Association: Food and nutrition misinformation, *Journal of the American Dietetic Association* 102 (2002): 260–266.

3. Federal Trade Commission, "Operation Cure.All" wages new battle in ongoing war against Internet health fraud, **www.ftc. gov/opa/2001/06/cureall.htm**, site visited on October 26, 2001.

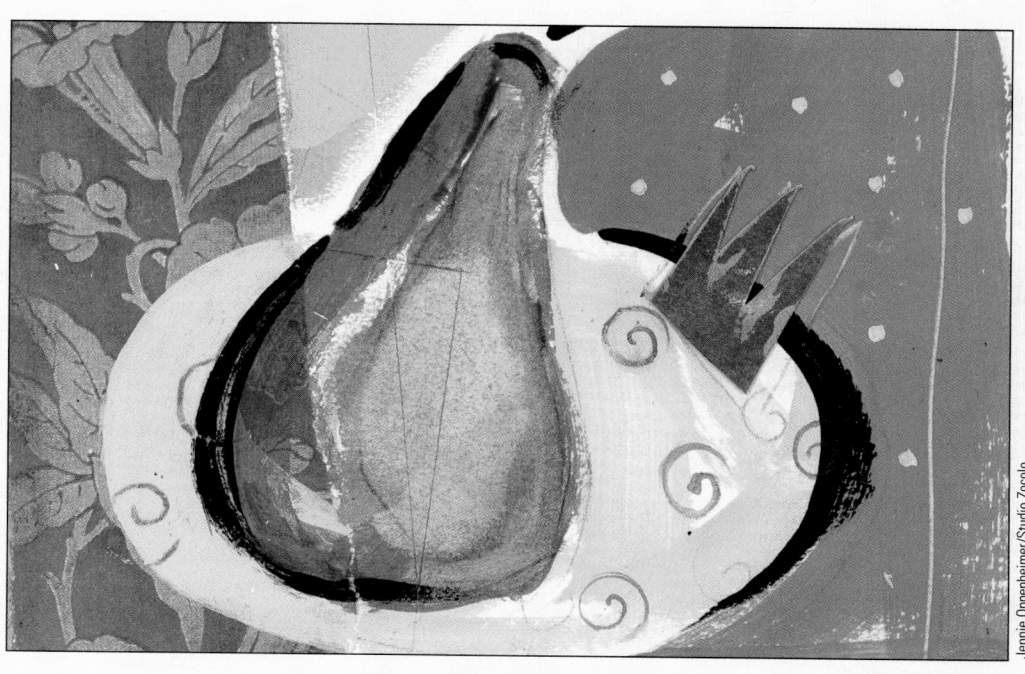

Jennie Oppenheimer/Studio Zocolo

CHAPTER 2 | CARBOHYDRATES

Grains, vegetables, legumes, fruits, and milk offer ample carbohydrate.

carbohydrates: energy nutrients composed of monosaccharides.

carbo = carbon

hydrate = water

simple carbohydrates: the monosaccharides (glucose, fructose, and galactose) and the disaccharides (sucrose, lactose, and maltose); also called **sugars.**

monosaccharides: single sugar units.

mono = one

saccharide = sugar

disaccharides: pairs of sugar units bonded together.

di = two

complex carbohydrates: long chains of sugars arranged as starch or fiber; also called **polysaccharides.**

poly = many

glucose: a monosaccharide, the sugar common to all disaccharides and polysaccharides; also called *blood sugar* or *dextrose.*

homeostasis: the maintenance of constant internal conditions (such as chemistry, temperature, and blood pressure) by the body's control system.

homeo = the same

stasis = staying

insulin: a hormone secreted by the pancreas in response to high blood glucose. It promotes cellular glucose uptake for use or storage.

glucagon (GLOO-ka-gon): a hormone that is secreted by special cells in the pancreas in response to low blood glucose concentration and elicits release of glucose from storage.

fructose: a monosaccharide; sometimes known as fruit sugar. It is abundant in fruits, honey, and saps.

fruct = fruit.

*M*ost people would like to feel good all the time. Part of the secret of feeling well is replenishing the body's energy supply with food. That means choosing foods that contain the energy nutrients—carbohydrate and fat, primarily. But which to choose?

Carbohydrate is the preferred energy source for most of the body's functions. As long as carbohydrate is available, the human brain depends exclusively on it as an energy source. Athletes eat a "high-carb" diet to store as much muscle fuel as possible, and dietary recommendations urge people to eat carbohydrate-rich foods for better health. Many people, however, mistakenly think of carbohydrate-rich foods as "fattening" and avoid them. In truth, people who wish to lose fat and maintain lean tissue can best do so by controlling kcalories, choosing high-carbohydrate, high-fiber foods, limiting fat-rich foods, and being physically active. All unrefined plant foods—grains, vegetables, legumes, and fruits—provide ample carbohydrate and fiber with little or no fat. Milk is the only animal-derived food that contains significant amounts of carbohydrate.

Carbohydrate shares its fuel-providing responsibility with fat. Fat, however, has disadvantages: it normally is not used as fuel by the brain and central nervous system, and diets high in fat, especially saturated fat, are associated with chronic diseases. The other energy sources available to the body—protein and alcohol—offer no advantage as fuels. Protein is best left to serve its own diverse functions, as discussed in Chapter 4. Alcohol, of course, has well-known undesirable side effects when used in excess. Alcohol abuse and its relationships with nutrition are the subject of Nutrition in Practice 23.

The Chemist's View of Carbohydrates

Chemists divide the **carbohydrates** into two categories: simple and complex. The **simple carbohydrates** (sugars) are:

- **Monosaccharides** (single sugars).
- **Disaccharides** (double sugars).

The **complex carbohydrates** (polysaccharides) are:

- Starch.
- Glycogen.
- Most fibers.[*]

All of these carbohydrates are composed of the simple sugar **glucose** and other compounds that are much like glucose in composition and structure. Figure 2–1 shows the chemical structure of glucose.

Monosaccharides (Single Sugars)

Three monosaccharides are important in nutrition: glucose, fructose, and galactose. Almost all the body's cells use glucose as their chief energy source. The body can obtain this glucose from all carbohydrates. In the body, blood glucose **homeostasis** is regulated primarily by two hormones: **insulin,** which moves glucose from the blood into the cells, and **glucagon,** which brings glucose out of storage when blood glucose falls (as occurs between meals).

Plants also make **fructose,** which is the sweetest of the sugars. It is abundant in fruits, honey, and saps. Glucose and fructose are the most common monosac-

[*]Lignin is the exception among fibers. Lignin is a polyphenol, not a polysaccharide.

charides in nature. The third single sugar, **galactose,** appears in nature only as part of lactose, a disaccharide also known as milk sugar. Only during digestion is galactose freed as a single sugar.

Disaccharides (Double Sugars)

In disaccharides, pairs of single sugars are linked together. Three disaccharides are important in nutrition: maltose, sucrose, and lactose. All three have glucose as one of their single sugars. As Table 2–1 shows, the other monosaccharide is either another glucose (in maltose), or fructose (in sucrose), or galactose (in lactose). The shapes of the sugars in Table 2–1 reflect their chemical structures as drawn on paper.

Sucrose: Table Sugar Sucrose (table, or white, sugar) is the most familiar of the three disaccharides and is what people mean when they speak of "sugar." This sugar is usually obtained by refining the juice from sugar beets or sugarcane to provide the brown, white, and powdered sugars available in the supermarket, but it occurs naturally in many fruits and vegetables.

When a person eats a food containing sucrose, enzymes in the digestive tract split the sucrose into its glucose and fructose components. Because the body can convert fructose to glucose, one molecule of sucrose can ultimately yield two molecules of glucose.

Honey versus Sugar People often ask: What is the difference between honey and white sugar? Is honey more nutritious? Honey, like white sugar, contains glucose and fructose. The difference is that in white sugar, the glucose and fructose are bonded together in pairs, whereas in honey some of them are paired and some are free single sugars. When you eat either white sugar or honey, though, your body breaks all of the sugars apart into single sugars. It ultimately makes no difference, then, whether you eat single sugars linked together, as in white sugar, or the same sugars unlinked, as in honey; they will end up as single sugars in your body. True, honey contains trace amounts of a few vitamins and minerals, but to say that honey is nutritious is misleading.

Fruits versus Sugar Some sugar sources are more nutritious than others, though. Consider a fruit such as an orange. The orange provides the same sugars and about the same energy as a tablespoon of sugar or honey, but the packaging makes a big difference in nutrient density. The sugars of the orange are diluted in a large volume of fluid that contains valuable vitamins and minerals, and the flesh and skin of the orange are supported by fibers that also offer health benefits. A tablespoon of honey offers no such bonuses.

Cola Beverages and Sweets A cola beverage, containing many teaspoons of sugar, offers no advantages either. Table 2–2 on the next page shows sample nutrients supplied by some sugar sources; note the "0s" and "traces" by honey, sugar,

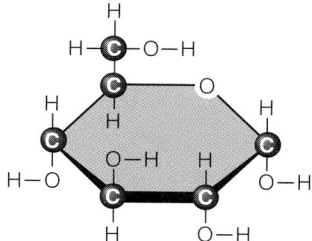

Figure 2–1
Chemical Structure of Glucose
On paper, the stucture of glucose has to be drawn flat, but in nature the five carbons and oxygen are roughly in a plane, with the H, OH, and CH$_2$OH extending out above and below it.

Honey, like sugar, contains glucose and fructose.

Digestion and absorption are the topics of Chapter 5.

A 12 oz can of cola contains the equivalent of 10 tsp sugar.

galactose: a monosaccharide; part of the disaccharide lactose.

sucrose: a disaccharide composed of glucose and fructose; commonly known as table sugar, beet sugar, or cane sugar.
sucro = sugar

Table 2–1	The Major Sugars	
Monosaccharides	**Disaccharides**	
Glucose	Sucrose (glucose + fructose)	
Fructose	Lactose (glucose + galactose)	
Galactose	Maltose (glucose + glucose)	
(found only as part of lactose)		

Carbohydrates

Table 2–2   Sample Nutrients in Sugars and Other Foods

The indicated portion of any of these foods provides approximately 100 kcalories. Notice that for a similar number of kcalories and grams of carbohydrate, milk, legumes, fruits, grains, and vegetables offer more of the other nutrients than do the sugars.

	Size of 100 kcal Portion	Carbohydrate (g)	Protein (g)	Calcium (mg)	Iron (mg)	Vitamin A (µg)	Vitamin C (mg)
Foods							
Milk, 1% low-fat	1 c	12	8	300	0.1	144	2
Kidney beans	½ c	20	7	30	1.6	0	2
Apricots	6	24	2	30	1.1	554	22
Bread, whole wheat	1½ slices	20	4	30	1.9	0	0
Broccoli, cooked	2 c	20	12	188	2.2	696	148
Sugars							
Sugar, white	2 tbs	24	0	trace	trace	0	0
Molasses	2 tbs	28	0	84	2	0	0
Cola beverage	1 c	26	0	6	trace	0	0
Honey	1½ tbs	26	trace	2	0.2	0	trace

An orange provides the same sugars as honey or sugar, but the packaging makes a big nutritional difference.

The short chains of glucose units that result from the breakdown of starch are known as **dextrins.** *The word sometimes appears on food labels because dextrins can be used as thickening agents in foods.*

lactose: a disaccharide composed of glucose and galactose; commonly known as milk sugar.
lact = milk

maltose: a disaccharide composed of two glucose units; sometimes known as malt sugar.

starch: a plant polysaccharide composed of glucose and digestible by human beings.

and the cola beverage and the substantial numbers by the others. Sucrose is often the principal ingredient of carbonated beverages, candy, cakes, frostings, cookies, and other concentrated sweets, so they are not nutritious carbohydrate choices.

Lactose: Milk Sugar   **Lactose** is the principal carbohydrate of milk. Most human babies are born with the digestive enzymes necessary to split lactose into its two monosaccharide parts, glucose and galactose, so as to absorb it. Breast milk thus provides a simple, easily digested carbohydrate that meets an infant's energy needs; most formulas do, too, because they are made from milk.

Maltose   The third disaccharide, **maltose,** is a plant sugar that consists of two glucose units. Maltose appears at only one stage in the life of a plant—when the plant is breaking down its stored starch for energy and starting to sprout.

Starch and Glycogen (Energy-Yielding Polysaccharides)

Unlike the sugars, which contain the three monosaccharides—glucose, fructose, and galactose—in different combinations, the polysaccharides, starch and glycogen, are composed entirely of glucose. They differ from each other only in the nature of the bonds that link the glucose units together.

Starch   **Starch** is a long, straight or branched chain of hundreds of glucose units linked together. These giant molecules are packed side by side in a rice grain or potato root—as many as a million per cubic inch of food. When a person eats the plant, the body splits the starch into glucose units and uses the glucose for energy.

Starchy Foods   All starchy foods are plant foods. Grains are the richest food source of starch. In most societies, people depend on a staple grain for much of their food energy: rice in Asia; wheat in Canada, the United States, and Europe; corn in much of Central and South America; and millet, rye, barley, and oats elsewhere. A second important source of starch is the legume (bean and pea) family. Legumes include peanuts and "dry" beans such as butter beans, kidney beans, "baked" beans, black-eyed peas (cowpeas), chickpeas (garbanzo beans), and soybeans. Root vegetables (tubers) such as potatoes and yams are a third major source of starch, and in many non-Western societies, they are the primary starch sources.

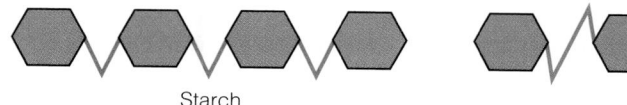

Starch   Cellulose

Figure 2–2

Starch and Cellulose Molecules Compared (Small Segments)
The bonds that link the glucose units together in cellulose are different from the bonds in starch (and glycogen). Human enzymes cannot digest cellulose.

Grains, legumes, and tubers not only are rich in starch, but also contain abundant dietary fiber, protein, and other nutrients. When nutrition experts advise people to seek out carbohydrate-rich foods to meet most of their energy needs, these are the foods they are recommending.

Glycogen Glycogen molecules, which are also made of chains of glucose, are more highly branched than starch molecules. Glycogen stores energy for human beings and animals, just as starch stores energy for plants. Because glycogen does not occur in plants and is found in meats only to a limited extent, it is not important as a nutrient. Nevertheless, glycogen plays an important role in the body as a readily available source of glucose, especially during physical activity.

Glycogen use during physical activity is discussed in Chapter 10.

The Fibers

The **fibers** of plants are constituents of plant cell walls. Most fibers are polysaccharides—chains of sugars—just as starch is, but in fibers the sugar units are held together by bonds that human digestive enzymes cannot break. Figure 2–2 shows the difference between starch and the plant fiber cellulose. In addition to cellulose, fibers include the polysaccharides hemicellulose, pectins, gums, and mucilages, as well as the nonpolysaccharide lignins.

Fibers in Foods Cellulose is the main constituent of plant cell walls, so it is found in all vegetables, fruits, and legumes. Hemicellulose is the main constituent of cereal fibers. Pectins are abundant in vegetables and fruits, especially citrus fruits and apples. The food industry uses pectins to thicken jelly and keep salad dressing from separating. Gums and mucilages have similar structures and are used as additives or stabilizers by the food industry. Lignins are the tough, woody parts of plants; few foods people eat contain much lignin.

Fibers in the Body Although cellulose and other fibers are not broken down by human enzymes, some fibers can be digested by bacteria in the human digestive tract. Bacterial digestion of fibers can generate some absorbable products that can yield energy when metabolized. Food fibers, therefore, can contribute some energy, depending on the extent to which they break down in the body.

Soluble versus Insoluble Fibers Fibers are classified according to several characteristics, including their chemical structure, their digestibility by bacterial enzymes, and their solubility in water. Some fibers are **insoluble,** meaning that they do not dissolve readily in water; other fibers are **soluble**—they do dissolve in water. These distinctions influence the health effects of fibers, which are discussed in a later section.

glycogen (GLY-co-gen): a polysaccharide composed of glucose, made and stored by liver and muscle tissues of human beings and animals as a storage form of glucose. Glycogen is not a significant food source of carbohydrate and is not counted as one of the complex carbohydrates in foods.

fibers: a general term denoting in plant foods the polysaccharides cellulose, hemicellulose, pectins, gums, and mucilages, as well as the nonpolysaccharide lignins, that are not attacked by human digestive enzymes.

insoluble fibers: the tough, fibrous structures of fruits, vegetables, and grains; indigestible food components that do not dissolve in water.

soluble fibers: indigestible food components that readily dissolve in water and often impart gummy or gel-like characteristics to foods. An example is pectin from fruit, which is used to thicken jellies.

In Summary Carbohydrate is the body's preferred energy source. Six simple sugars are important in nutrition: the three monosaccharides (glucose, fructose, and galactose) and the three disaccharides (sucrose, lactose, and maltose). The three disaccharides are pairs of monosaccharides, with each containing glucose paired with one of the three monosaccharides. The complex carbohydrates are the polysaccharides (chains of monosaccharides): glycogen, starches, and fibers. Both glycogen and starch are storage forms of glucose—glycogen in the body, starch in plants—and both yield energy for human use. The fibers also contain glucose (and other monosaccharides), but their bonds cannot be broken by human digestive enzymes, so they yield little, if any, energy.

Health Effects of Sugars and Alternative Sweeteners

Starch-rich and fiber-rich foods such as vegetables, grains, legumes, and fruits should predominate in people's diets; concentrated sweets should account for only 10 percent or less of total kcalories. For most people, this means that total carbohydrate intake should increase while sugar intake should decline. The sugars in vegetables, fruits, grains, and milk are acceptable because they are accompanied by many nutrients. In contrast, concentrated sweets contribute many kcalories, but relatively few nutrients, and so should be limited. People who want to limit their use of sugar may choose from two sets of alternative sweeteners: sugar alcohols and artificial sweeteners.

Sugars

In the United States, it is estimated that each man, woman, and child consumes over 100 pounds of sugar per year, or a little less than 2 pounds per week.[1] This number may be somewhat higher than actual intakes because it does not account for waste, such as the syrup drained from sweet pickles or jam that molds and is thrown away. The steady upward trend in the consumption of sugar (see Figure 2–3) is largely the result of a dramatic increase in consumption of commercially prepared foods and beverages. Food manufacturers are adding sugars to foods during processing. In contrast, people are adding less sugar in the kitchen. Recommendations that people reduce their consumption of concentrated sugars to 10 percent or less of total kcalories stem from widely published reports of research conducted during the 1970s. These reports implicated sugars (mainly sucrose) as a possible contributing factor in several diseases. Since then, many accusations have been made against the sugar sucrose, but the Food and Drug Administration (FDA) and the National Academy of Sciences have concluded that in the amounts people currently consume, sugar poses no major health risk. Still, some controversy continues. A brief description of the accusations pertaining to sugar's effects on health may help to clarify the issues. Most commonly, sugar is accused of causing obesity, diabetes, hyperactivity, and aggressive behavior in children, and dental caries.

Sugar and Obesity On the first accusation—that sugar causes obesity—the evidence shows a supportive role, but not a direct cause-and-effect relationship. The incidence of obesity rises as a population's sugar consumption increases, but because sweet treats are often high-kcalorie and high-fat, it is unclear whether the obesity develops because of the high sugar intake, the accompanying fat, the excess energy—or quite likely, all of these. In addition, physical activity usually declines as well. Population studies simply cannot separate the effects of sugar from those of too great an energy intake or too little energy expenditure. Furthermore, obesity may also occur where sugar intakes are low.

Concentrated sweets and soft drinks do make it easy to exceed energy needs quickly, however, and an excessive intake of food energy can cause obesity. In the United States, obesity has reached epidemic proportions, while intakes of both food energy and sugar-sweetened beverages, especially soft drinks, have skyrocketed.[2] Thus, to the extent that sugar contributes to an excessive energy intake, it can play a role in the development of obesity.

Starch- and fiber-rich foods are the foods to emphasize.

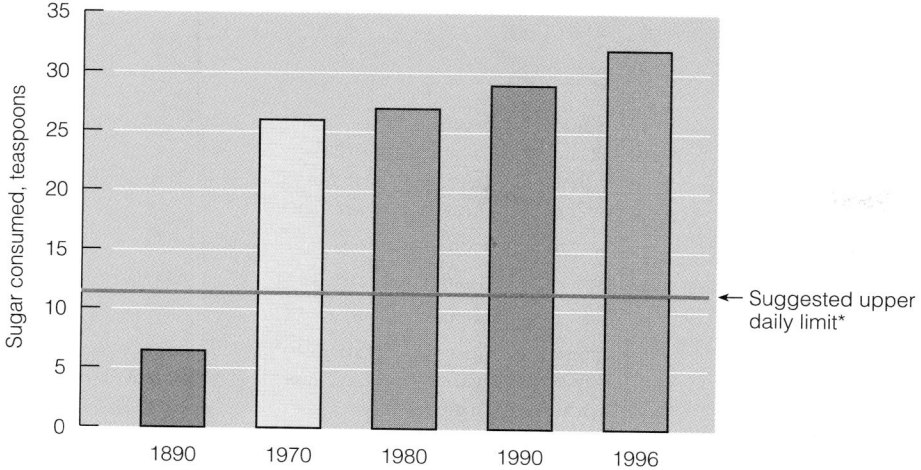

← Suggested upper daily limit*

Figure 2–3

Daily Sugars Consumption in the United States, 1890–1996
*Recommended upper limit for a 2200-kcalorie diet.

SOURCE: U.S. Department of Agriculture, Economic Research Service, *A Dietary Assessment of the U.S. Food Supply: Comparing Per Capita Food Consumption with the Food Guide Pyramid Serving Recommendations* (Washington, D.C.: Government Printing Office 1998), AER no. 772, p. 25.

Sugar and Diabetes   On sugar's relation to diabetes, the evidence is conflicting and interesting. In many parts of the world, as sugar consumption has increased, a profound increase in the incidence of one type of diabetes (type 2) has occurred. Yet, in other populations, no relationship has been found between diabetes and sugar consumption. Body fatness is one known risk factor: the majority of people with **type 2 diabetes** are overweight or obese, and evidence shows that weight reduction helps to prevent diabetes or relieve its symptoms. Genetics, age, and physical inactivity are also considered risk factors for type 2 diabetes.[3] Thus sugar may be a causative factor only when it contributes to obesity.

Sugar and Behavior   The accusation that sugar causes hyperactive or aggressive behavior in children stands unproven. Scientific research has failed to demonstrate any consistent effect of sugar on behavior in either normal or hyperactive children.[4] If sugar is related to behavior problems in children, it may be because the sugary foods replace nutrient-dense foods in children's diets, making nutrient deficiencies likely. Many different nutrient deficiencies adversely affect behavior. A lack of nutrients in children's diets, not sugar itself, can in some cases contribute to undesirable behavior.

Chapter 13 offers further discussion of children's nutrition.

Sugar and Dental Caries   As to whether sugar contributes to **dental caries,** the evidence says yes. Any carbohydrate-containing food, including bread, bananas, or milk, as well as sugar, can support bacterial growth in the mouth. These bacteria produce the acid that eats away tooth enamel. Of major importance is the length of time the food stays in the mouth. This, in turn, depends on the composition of the food, how sticky the food is, how often a person eats the food, and especially whether the teeth are brushed afterward.[5] Total sugar intake still plays a major role in caries incidence; populations with diets providing more than 10 percent of kcalories from sugar have an unacceptably high incidence of dental caries.[6] Nutrition in Practice 2 discusses nutrition and dental health.

Recommended Sugar Intakes   Moderate sugar intakes (5 to 10 percent of total kcalories)—enough for pleasure, but not enough to displace more nutritious foods—are not harmful. Sugar is a delicious, concentrated source of food energy, but it contains no protein, vitamins, or minerals. Eaten in place of nutrient- and fiber-rich foods, it makes malnutrition likely.

type 2 diabetes: the more common type of diabetes in which the body cells resist insulin.

dental caries: the gradual decay and disintegration of a tooth.

Recognizing Sugars   People often fail to recognize sugar in all its forms and so fail to realize how much they consume. To help your clients estimate how much

Carbohydrates

brown sugar: refined white sugar crystals to which manufacturers have added molasses syrup with natural flavor and color; 91 to 96 percent pure sucrose.

concentrated fruit juice sweetener: a concentrated sugar syrup made from dehydrated, deflavored fruit juice, commonly grape juice; used to sweeten products that can then claim to be "all fruit."

confectioners' sugar: finely powdered sucrose; 99.9 percent pure.

corn sweeteners: corn syrup and sugars derived from corn.

corn syrup: a syrup containing mostly glucose, produced by the action of enzymes on cornstarch. See also *high-fructose corn syrup (HFCS)*.

dextrose: an older name for glucose and the name used for glucose in intravenous solutions.

fructose, galactose, glucose: already defined (pp. 34 and 35).

granulated sugar: common table sugar, crystalline sucrose; 99.9 percent pure.

high-fructose corn syrup (HFCS): the predominant sweetener used in processed foods today. HFCS is mostly fructose; glucose makes up the balance.

honey: sugar (mostly sucrose) formed from nectar gathered by bees. An enzyme splits the sucrose into glucose and fructose. Composition and flavor vary, but honey always contains a mixture of sucrose, fructose, and glucose.

invert sugar: a mixture of glucose and fructose formed by splitting sucrose in a chemical process; sold only in liquid form, sweeter than sucrose. Invert sugar is used as an additive to help preserve food freshness and prevent shrinkage.

lactose: already defined (p. 36).

levulose: an older name for fructose.

maltose: already defined (p. 36).

maple sugar: a sugar (mostly sucrose) purified from concentrated sap of the sugar maple tree. Maple sugar is expensive compared with other sweeteners.

molasses: a thick brown syrup, left over from sugarcane juice during sugar refining. Blackstrap molasses contains iron, which comes from the machinery used to process it. This iron is not as well absorbed as the iron in meats and other foods.

raw sugar: the first crop of crystals harvested during sugar processing. Raw sugar cannot be sold in the United States because it contains too much filth (dirt, insect fragments, and the like). Sugar sold domestically as raw sugar has actually gone through about half of the refining steps.

sucrose: already defined (p. 35).

turbinado (ter-bih-NOD-oh) **sugar:** raw (brown) sugar from which the filth has been washed; legal to sell in the United States.

white sugar: pure sucrose, produced by dissolving, concentrating, and recrystallizing raw sugar.

---

sugar they consume, tell them to treat all of the following concentrated sweets as equivalent to 1 teaspoon of white sugar:

- 1 teaspoon brown sugar, candy, jam, jelly, any corn sweetener, syrup, honey, molasses, or maple sugar.
- 1 tablespoon catsup.
- 1½ ounces carbonated soft drink.

These portions of sugar all provide about the same number of kcalories. Some are closer to 10 kcalories (for example, 14 kcalories for molasses), while some are over 20 (22 kcalories for honey), so an average figure of 20 kcalories is an acceptable approximation. The accompanying glossary presents the multitude of names that denote sugar on food labels. The next section discusses sugar substitutes.

In Summary Sugars pose no major health threat except for an increased risk of dental caries. Excessive intakes may displace needed nutrients and fiber and may contribute to obesity. A person deciding to limit daily sugar intake should recognize that not all sugars need to be restricted, just concentrated sweets, which are relatively empty of other nutrients and high in kcalories. Sugars that occur naturally in fruits, vegetables, and milk are acceptable.

---

sugar alcohols: sugarlike compounds; like sugars, they are sweet to taste but yield 2 to 3 kcal per gram, slightly less than sucrose. Examples are maltitol, mannitol, sorbitol, isomalt, lactitol, and xylitol.

nutritive sweeteners: sweeteners that yield energy, including both the sugars and the sugar alcohols.

Alternative Sweeteners: Sugar Alcohols

The **sugar alcohols** are carbohydrates, but they yield slightly less energy (2 to 3 kcalories per gram) than sucrose (4 kcalories per gram) because they are not absorbed completely.[7] The sugar alcohols are sometimes called **nutritive sweeteners.**

Artificial Sweeteners

acesulfame (AY-sul-fame) **potassium:** a 0-kcalorie artificial sweetener that tastes 200 times as sweet as sucrose; also known as acesulfame-K because K is the chemical symbol for potassium. Acceptable Daily Intake (ADI) = 15 milligrams/kilogram body weight.

alitame (AL-ih-tame): a low-kcalorie artificial sweetener that is composed of two amino acids (alanine and aspartic acid) and tastes 2000 times as sweet as sucrose; FDA approval pending.

aspartame (ah-SPAR-tame or ASS-par-tame): a low-kcalorie artificial sweetener that is composed of two amino acids (phenylalanine and aspartic acid) and tastes 200 times as sweet as sucrose. ADI = 50 milligrams/kilogram body weight.

cyclamate (SIGH-kla-mate): a 0-kcalorie artificial sweetener that tastes 30 times as sweet as sucrose; FDA approval pending in the United States; available in Canada in grocery stores but only as a tabletop sweetener, not as an additive.

saccharin (SAK-ah-ren): a 0-kcalorie artificial sweetener that tastes 500 times as sweet as sucrose; approved in the United States, but available in Canada only in pharmacies and only as a sweetener, not as an additive. ADI = 5 milligrams/kilogram body weight.

sucralose (SUE-kra-lose): a 0-kcalorie artificial sweetener that tastes 600 times as sweet as sucrose. ADI = 5 milligrams/kilogram body weight.

The sugar alcohols occur naturally in fruits; they are also used by manufacturers as a low-energy bulk ingredient in many products. Unlike sucrose, sugar alcohols are fermented to gases in the large intestine by intestinal bacteria. Consequently, side effects such as gas, abdominal discomfort, and diarrhea make the sugar alcohols less attractive than the artificial sweeteners.

The advantage of using sugar alcohols is that they do not contribute to dental caries. Bacteria in the mouth metabolize sugar alcohols much more slowly than sucrose, thereby inhibiting the production of acids that promote caries formation. The sugar alcohols are therefore valuable in chewing gums, breath mints, and other products that people keep in their mouths a while. The FDA allows food labels to carry a health claim (see p. 24 in Chapter 1) about the relationship between sugar alcohols and the nonpromotion of dental caries as long as the FDA criteria for sugar-free status and other criteria are met.[8]

These chewing gums contain sugar alcohols, which are better than sugar for the teeth, but are not kcalorie-free.

Alternative Sweeteners: Artificial Sweeteners

The **artificial sweeteners** are not carbohydrates. They yield virtually no energy in the amounts typically used and are sometimes called nonnutritive sweeteners. Like the sugar alcohols, artificial sweeteners make foods taste sweet without promoting tooth decay. The accompanying glossary offers details on six artificial sweeteners of interest.

Saccharin Saccharin has been used for more than 100 years in the United States and is currently used by more than 50 million people, mainly in soft drinks and as the tabletop sweetener Sweet'n Low or Sweet Thing. Questions about the safety of saccharin arose in 1977 when experiments suggested that it caused bladder tumors in second-generation rats fed high doses. The FDA proposed banning saccharin as a result. Public outcry in favor of saccharin was so loud that Congress declared a moratorium on the ban, a moratorium that has been repeatedly extended. In 1991, the FDA withdrew its proposal to ban saccharin, but labels still had to carry a consumer warning about saccharin as a cancer hazard. Saccharin remained on the government's roster of "anticipated carcinogens" until the year 2000. At that time, government officials reviewed the research and reversed their opinion, removing saccharin from the carcinogen list and freeing it from the labeling requirement.[9] Opponents to these changes, however, maintain that saccharin causes cancer in mice and rats and so should be avoided.

Does saccharin cause cancer? The largest population study to date, involving 9000 men and women, showed overall that saccharin use did not raise the risk of cancer. Among certain small groups of the population, however, such as those

artificial sweeteners: noncarbohydrate, nonkcaloric synthetic sweetening agents; sometimes called **nonnutritive sweeteners.**

Carbohydrates

who both smoked heavily and used saccharin, the risk of bladder cancer was slightly greater. Other studies involving more than 5000 people with bladder cancer showed no association between bladder cancer and saccharin use.[10] Common sense dictates that consuming large amounts of saccharin is probably not safe, but at moderate intake levels, saccharin is currently assumed to be safe for most people. The FDA has set **Acceptable Daily Intake (ADI)** levels for the artificial sweeteners used in the United States.

Aspartame **Aspartame** is the active ingredient in NutraSweet, which is used in many commercially prepared foods, and in Equal, a tabletop sweetener. Aspartame was approved by the FDA in 1981 and currently dominates the world market for artificial sweeteners. Aspartame is one of the most studied of all food additives: extensive animal and human studies document its safety. Long-term consumption of aspartame is safe and is not associated with any adverse health effects.[11] Aspartame is approved for use in more than 90 countries.

The FDA's approval of aspartame is based on the assumption that no one will consume more than the ADI of 50 milligrams per kilogram of body weight in a day. This daily intake is indeed a lot: for a 132-pound person, it adds up to 80 packets of Equal or 15 soft drinks sweetened only with aspartame. Most people consume between 2 and 10 milligrams per kilogram of body weight per day. Still, a child who drinks a quart of Kool-Aid sweetened with aspartame on a hot day, and who also has pudding, chewing gum, cereal, and other products sweetened with aspartame, takes in more than the ADI. Although this presents no proven hazard, it seems wise to offer children other foods so as not to exceed the limit. Infants or toddlers under two years old should not be fed artificially sweetened foods and drinks.

Aspartame and PKU Although aspartame is considered safe for most people, individuals with the metabolic disorder phenylketonuria (PKU) are an exception. The labels of products that contain aspartame must include information for individuals with PKU. Aspartame contains the amino acid phenylalanine, and people with PKU cannot dispose of it efficiently. Adults with PKU can use some aspartame, but children with PKU need to get all their phenylalanine from nutrient-rich foods such as milk and meat (see Nutrition in Practice 20).

Acesulfame Potassium The FDA approved **acesulfame potassium** (acesulfame-K) in 1988 after reviewing more than 90 safety studies, conducted over 15 years. Marketed under the trade names Sunette and Sweet One, acesulfame-K is about as sweet as aspartame. It is used in chewing gum, beverages, instant coffee and tea, gelatins, and puddings, as well as for table use. Unlike aspartame, acesulfame-K holds up well during cooking.

Sucralose Recently approved for use as a sweetener in the United States, **sucralose** is the only artificial sweetener made from sucrose. Many years of testing have deemed sucralose safe to use and, specifically, not a cause of cancer. Sucralose is not recognized by the body as sugar and therefore passes through unchanged. Sucralose is heat stable and so is useful for cooking and baking; it is used in commercially prepared products and as the tabletop sweetener Splenda. Its sugarlike taste and versatility are earning sucralose some popularity among consumers.

Other Artificial Sweeteners FDA approval for two other sweeteners—**cyclamate** and **alitame**—is still pending. To date, no safety issues have been raised for alitame. Cyclamate, on the other hand, has been battling safety issues for 50 years. Approved by the FDA in 1949, cyclamate was banned in 1970 because of evidence indicating that it caused bladder cancer in rats. In 1985, the National Academy of Sciences concluded that evidence to date indicated that cyclamate did not cause cancer in human beings, but warranted further studies. In Canada, cyclamate is

Acceptable Daily Intake (ADI): the amount of a sweetener that individuals can safely consume each day over the course of a lifetime without adverse effects. It includes a 100-fold safety factor.

restricted to use as a tabletop sweetener on the advice of a physician. In the United States, the FDA is currently reviewing a petition to reapprove the use of cyclamate.

Artificial Sweeteners and Weight Control Many people eat and drink products sweetened with artificial sweeteners to help them control weight. Does this work? Ironically, a few studies have reported that after consuming such products, people experience heightened feelings of hunger. Despite these reports, most studies find that artificial sweeteners do not heighten feelings of hunger, enhance food intake, or cause weight gain in people. When people reduce their energy intakes by replacing sugar in their diets with artificial sweeteners, and then compensate for the reduced energy at later meals, energy intake may stay the same or increase. Using artificial sweeteners will not automatically lower energy intake; to successfully control energy intake, a person needs to make informed diet and activity decisions throughout the day.

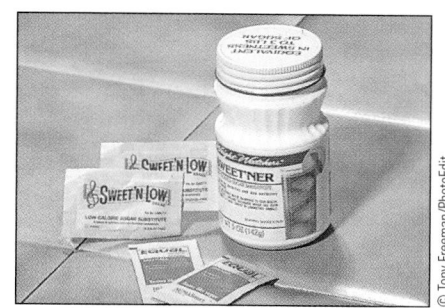

Artificial sweeteners offer the sweet taste of sugar without the kcalories.

In Summary Two types of alternative sweeteners are sugar alcohols and artificial sweeteners. Sugar alcohols are carbohydrates, but they yield slightly less energy than sucrose. Sugar alcohols do not contribute to dental caries. The artificial sweeteners are not carbohydrates and yield no energy. Like the sugar alcohols, artificial sweeteners do not promote tooth decay.

Health Effects of Complex Carbohydrates

Despite dietary recommendations that people should eat generous servings of complex carbohydrate–rich foods for their health, many people still believe that carbohydrate is the "fattening" component of foods. Gram for gram, carbohydrates contribute fewer kcalories to the body than dietary fat, so a diet of high-carbohydrate foods is likely to be lower in kcalories than a diet of high-fat foods.

The health benefits a person can expect from a diet high in complex carbohydrate-rich foods and low in concentrated sugar are many. Such a diet is almost invariably low in fat and saturated fat, low in food energy, and high in fiber, vitamins, and minerals. All these factors working together can help reduce the risks of obesity, cancer, cardiovascular disease, diabetes, dental caries, gastrointestinal disorders, and malnutrition.

It is difficult to sort out which complex carbohydrates contribute to which health benefits. Starch and fibers almost always occur together in foods (except refined foods), so it is hard to distinguish their effects. Some health effects appear to be especially closely associated with fibers, however, and these are discussed next.

Fibers

Fiber-rich foods benefit health in many ways (see Table 2–3 on the next page). Fibers in foods are thought to play a beneficial role in the prevention or management of:

- *Weight control.* Fiber-rich foods contribute little energy and take longer to eat than low-fiber foods, thereby enhancing satiety. High-fiber foods also promote a feeling of fullness as they absorb water. In addition, soluble fibers in a meal slow the movement of food through the upper digestive tract, so a person feels fuller longer.[12] In a study of more than 200 men, researchers assessed body fat and diet composition (energy, carbohydrate, fat, protein, and fiber). Dietary fiber had the strongest correlation with body fat: the men with the highest percentage of body fat consumed significantly less fiber than leaner men.[13] A diet high in fiber-rich foods can promote weight loss if those foods displace concentrated fats and sweets.

Table 2–3 Water Solubilities, Sources, and Health Effects of Fiber

Fiber Type	Major Food Sources	Possible Health Effects
Soluble Gums, mucilages, pectins, psyllium,[a] some hemicellulose	Barley, fruits, legumes, oats, oat bran, rye, seeds, vegetables	These fibers lower blood cholesterol; slow glucose absorption; slow transit of food through upper digestive tract; hold moisture in stools, softening them; are partly fermentable into fragments the body can use.
Insoluble Cellulose, lignin, some hemicellulose	Brown rice, fruits, legumes, seeds, vegetables, wheat bran, whole grains	These fibers soften stools, regulate bowel movements; speed transit of material through small intestine; increase fecal weight and speed fecal passage through colon; reduce risks of diverticulosis, hemorrhoids, and appendicitis; may reduce risk of colon cancer.

[a]Psyllium, a fiber laxative and a cereal additive, has both soluble and insoluble properties.

- *Constipation, hemorrhoids, and diarrhea.* Fibers that both enlarge and soften stools (such as insoluble wheat bran) ease elimination for the rectal muscles, thereby alleviating or preventing constipation and hemorrhoids. Other fibers help to solidify watery stools.

- *Appendicitis.* Some fibers (again, such as wheat bran) help keep the contents of the intestinal tract moving easily. This action helps prevent compaction of the intestinal contents, which could obstruct the appendix and permit bacteria to invade and infect it.

- *Diverticulosis.* Fibers stimulate the muscles of the digestive tract so that they retain their health and tone. This prevents the muscles from becoming weak and the lining of the digestive tract from bulging out in places, as occurs in diverticulosis. Insoluble fiber, particularly cellulose, seems to be most beneficial in lowering the risk of diverticulosis.[14] Chapter 19 describes diverticulosis.

- *Heart disease.* Diets high in fiber, especially cereal fiber, may reduce the risk of heart disease.[15] Soluble fibers recommended for lowering blood cholesterol may do so by delaying absorption in the digestive tract and by binding with cholesterol compounds and carrying them out of the body with the feces.[16] High-fiber foods may also lower blood cholesterol indirectly by displacing fatty, cholesterol-raising foods from the diet. Even when dietary fat intake is low, research shows that high intakes of soluble fiber (such as that in apples and oat bran) exert separate and significant blood cholesterol–lowering effects.[17]

- *Diabetes.* Some fibers delay the passage of nutrients from the stomach to the small intestine. This delay slows glucose absorption, thus eliciting a moderate insulin response and a moderate rise in blood glucose. This slow, sustained rise in blood glucose is desirable. The term **glycemic effect** describes the effect of food on blood glucose; a high glycemic effect reflects fast glucose absorption and a surge in blood glucose, which is undesirable. An extensive long-term study of more than 65,000 women showed that women whose diets had the highest glycemic effect and the lowest cereal fiber content were most vulnerable to type 2 diabetes independent of other dietary constituents or known risk factors.[18]

Some research has shown that high-fiber diets protect against colon cancer.[19] Two recent studies, however, found no correlation between fiber-rich diets and low rates of colon cancer.[20] Perhaps future research will clarify this issue, but in

glycemic effect: a measure of the extent to which a food raises the blood glucose concentration and elicits an insulin response, as compared with pure glucose.

the meantime, there are plenty of reasons to eat a high-fiber diet, including the ones listed above.[21] Fiber-rich whole grains, fruits, vegetables, and legumes are concentrated sources of antioxidants and other phytochemicals, vitamins, and minerals that may be important in cancer prevention.[22]

Fibers and Health Claims   The FDA authorizes three health claims on food labels concerning fiber. One is for "fiber-containing grain products, fruits, and vegetables and reduced risk of cancer." Another is for "fruits, vegetables, and grain products that contain fiber, and reduced risk of coronary heart disease," and a third is for "soluble fiber from whole oats and from psyllium seed husk and reduced risk of coronary heart disease." Chapter 1 describes the criteria foods must meet to bear these health claims.

Even with all these advantages, carbohydrate in the form of raw fiber is not a wonder cure. In some cases it can be detrimental. When too much fiber is consumed, some essential vitamins and minerals may bind to it and be excreted with it without becoming available for the body to use. Also, consuming purified fiber such as cellulose may not confer the same health benefits as consuming cellulose from a food source such as whole grains. Athletes may want to avoid bulky, fiber-rich foods just prior to competition.

Fibers in Foods   Meals selected according to the Food Guide Pyramid deliver ample fiber. Most people in the United States do not eat this way, though, and their average fiber intakes are about half of the recommended 20 to 35 grams daily.[23] Research shows that only when people meet the grain recommendation and one or both of the vegetable and fruit recommendations (based on the Food Guide Pyramid), do they achieve fiber intakes higher than 20 grams per day.[24] As health care professionals, you can advise clients to obtain enough fiber by eating ample servings of whole foods, using Figure 2–4 on the next page as a guide.

Miscellaneous whole foods also add some fiber: you can get about 1 gram from ½ ounce of nuts, a tablespoon of peanut butter, a large pickle, or 5 olives. A day's meals based on the Daily Food Guide, such as those shown in Figure 2–5 (on p. 48), not only meet carbohydrate recommendations, but provide abundant fiber, too.

Carbohydrates: Food Sources and Recommendations

Grains, vegetables, fruits, and milk are noted for the valuable energy-yielding carbohydrates they contain: starches and dilute sugars. Each class of foods makes its own typical carbohydrate contribution. The Daily Food Guide and the Food Guide Pyramid in Chapter 1 can help you and your clients obtain carbohydrate-rich foods.

Breads, Cereal, Rice, and Pasta   A serving of most foods in this group—a slice of whole-wheat bread, half an English muffin or bagel, a 6-inch tortilla, or ½ cup of rice, pasta, or cooked cereal—provides abundant carbohydrate as starch.[*] Some foods in this group, especially baked goods such as biscuits, croissants, muffins, and snack crackers, also contain sugar, fat, or both.

Vegetables   Some vegetables are major contributors of starch in the diet. Just a small white or sweet potato or ½ cup of cooked dry beans, corn, peas, plantain, or winter squash provides 15 grams of carbohydrate, as much as in a slice of bread, though as a mixture of sugars and starch. A ½-cup portion of carrots, okra, onions, tomatoes, cooked greens, or most other nonstarchy vegetables or a cup of salad greens provides about 5 grams as a mixture of starch and sugars. Each of these foods also contributes a little protein and no fat.

---

[*]Gram values in this section are adapted from the 1995 Exchange System.

Bread, Cereal, Rice, and Pasta Group

Whole-grain products provide 1 to 2 g of fiber or more per serving:

- 1 slice whole-wheat or rye bread (2 g).
- 1 slice pumpernickel bread (2 g).
- ½ c ready-to-eat 100% bran cereal (10 g).
- ½ c cooked barley, bulgur, grits, oatmeal (2 to 3 g).
- ½ c cooked pasta (1 g).
- ½ c white rice (<1 g).

Vegetable Group

Most vegetables contain 2 to 3 g of fiber per serving:

- 1 c raw bean sprouts.
- ½ c cooked broccoli, brussels sprouts, cabbage, carrots, cauliflower, collards, corn, eggplant, green beans, green peas, kale, mushrooms, okra, parsnips, potatoes, pumpkin, spinach, sweet potatoes, swiss chard, winter squash.
- ½ c chopped raw carrots, peppers.

Fruit Group

Fresh, frozen, and dried fruits have about 2 g of fiber per serving:

- 1 medium apple, banana, kiwi, nectarine, orange, pear.
- ½ c applesauce, blackberries, blueberries, raspberries, strawberries.
- Fruit juices contain very little fiber.

Meat and Meat Alternates Group

Many legumes provide about 8 g of fiber per serving:

- ½ c cooked baked beans, black beans, black-eyed peas, kidney beans, navy beans, pinto beans.

Some legumes provide about 5 g of fiber per serving:

- ½ c cooked garbanzo beans, great northern beans, lentils, lima beans, split peas.

Most nuts and seeds provide 1 to 3 g of fiber per serving:

- 1 oz almonds, cashews, hazelnuts, peanuts, pecans, pumpkin seeds, sunflower seeds.

NOTE: Appendix A provides fiber grams for over 2000 foods.

Figure 2–4
Fiber in Selected Foods

Fruits The size of a typical serving of fruit varies depending on the form of the fruit: ¾ cup of juice; a small banana, apple, or orange; ½ cup of most canned or fresh fruit; or ¼ cup of dried fruit. A typical fruit serving contains an average of about 15 grams of carbohydrate, mostly as sugars, including the fruit sugar fructose. Fruits vary greatly in their water and fiber contents, and therefore their sugar concentrations vary also. With the exception of avocado, which is high in fat, the fruits contain insignificant amounts of fat and protein.

Milk, Cheese, and Yogurt One cup of milk or yogurt or the equivalent (1 cup of buttermilk, ⅓ cup of dry milk powder, or ½ cup of evaporated milk) provides a generous 12 grams of carbohydrate. Among cheeses, cottage cheese provides about 6 grams of carbohydrate per cup, while most other types contain little, if any, carbohydrate. All milk products vary in fat content, an important consideration in choosing among them; Chapter 3 provides the details.

Cream and butter, although dairy products, are not equivalent to milk because they contain little or no carbohydrate and insignificant amounts of the other nutrients important in milk. They are appropriately placed with the fats at the top of the Food Guide Pyramid.

Meat, Poultry, Fish, Dry Beans, Eggs, and Nuts With two exceptions, foods of this group provide almost no carbohydrate to the diet. The exceptions are nuts, which provide a little starch and fiber along with their abundant fat, and dry beans, which are excellent sources of both starch and fiber. Just a ½-cup serving of beans provides 15 grams of carbohydrate, an amount equal to the richest sources in the Food Guide Pyramid.

Recommendations Carbohydrate has no RDA, but there is a recommendation that 55 to 60 percent of total kcalories should come from carbohydrate. The FDA used this guideline in establishing a Daily Value for carbohydrate of 300 grams per day, or 60 percent of kcalories based on 2000 kcalories per day. For people who eat about 1700 kcalories a day, the recommended intake translates to some 900 to 1000 kcalories from carbohydrate per day or, at 4 kcalories per gram, some 225 to 250 grams. Most of this carbohydrate should be starch, not sugars; thus many servings of starchy foods are needed to meet this recommendation.

Figure 2–5 on the next page offers an example of meals that provide about 1700 kcalories and 225 grams of carbohydrate from nutrient-dense, fiber-rich foods. The carbohydrate content of a diet can be determined by using a nutrient composition table such as that found in Appendix A, the exchange list system described in Chapter 24, or a computer diet analysis program.

Carbohydrates on Food Labels Food labels list the amount, in grams, of total carbohydrate—including starch, fibers, and sugars—per serving. Fiber grams are also listed separately, as are the grams of sugars. (With this information, consumers can calculate starch grams by subtracting the grams of fibers and sugars from the total carbohydrate.) Sugars on the Nutrition Facts panel of a food label reflect both added sugars and those that occur naturally in foods. Total carbohydrate and dietary fiber are also expressed as "% Daily Values" for a person consuming 2000 kcalories; there is no Daily Value for sugars.

In Summary Clearly, a diet rich in complex carbohydrates—starches and fibers—supports efforts to control body weight and prevent heart disease, diabetes, gastrointestinal disorders, and possibly some types of cancer. For these reasons, recommendations urge people to eat plenty of whole grains, vegetables, legumes, and fruits—enough to provide 55 to 60 percent of the daily energy intake from carbohydrate.

Energy Nutrients in Perspective

An uninterrupted flow of energy is so vital to life that other functions are sometimes sacrificed to maintain it. For example, when a child is fed too little food, the food the child does consume will be used for energy to keep the heart and

Before heading off to classes, a student eats breakfast:

2 shredded wheat biscuits.

1 c 1% low-fat milk.

½ banana (sliced).

Then goes home for a quick lunch:

1 turkey sandwich on whole-wheat bread with mayonnaise and mustard.

1 c vegetable juice (canned).

While studying that afternoon, the student eats a snack:

4 whole-wheat crackers.

1 oz cheddar cheese.

1 apple.

That night, the student makes dinner:

A salad made with:

  1 c raw spinach leaves, shredded carrots, and sliced mushrooms.

  ⅓ c garbanzo beans.

  5 lg olives.

  1 tbs ranch salad dressing.

A dinner of:

  1 c spaghetti with meat sauce.

  ½ c green beans.

  2 tsp butter.

And, for dessert:

  1¼ c strawberries (fresh).

Later that evening, the student enjoys a bedtime snack:

3 graham crackers.

1 c 1% low-fat milk.

> **Total kcal: 1725**
> 53% kcal from carbohydrate
> 29% kcal from fat
> 18% kcal from protein

Figure 2–5

A Day's Meals That Offer 1700 kCalories and Meet the Carbohydrate Recommendation

lungs going, while growth comes to a standstill. To go totally without an energy supply, even for a few minutes, is to die. The urgency of the need for energy has ensured that all creatures have developed ways of accumulating built-in reserves to protect themselves from being deprived of energy. One major provision against this sort of emergency is glycogen, the storage form of glucose. (The other is body fat, about which the next chapter says more.)

Glycogen Used for Energy When a person does not eat carbohydrate, the person's body rapidly devours its glycogen stores. Stored glycogen can return glucose to the blood whenever the supply runs short, but the liver cells can store only limited amounts of glycogen. Once this supply is depleted, the body must turn to the other energy nutrients—fat and protein—to meet its energy needs.

Fat Used for Energy Unlike the liver cells, which can store only a limited amount of glycogen, the body's fat cells can store virtually unlimited quantities of fat, so supplies almost never run out. Fat normally is used to meet about half of the body's energy needs, and most tissues can use it as is. The brain and nerves, however, need their energy as glucose, and the body cannot convert fat to glucose. After a long period of glucose deprivation, brain and nerve cells develop the ability to derive about half of their energy from a special form of fat known as **ketones,** but they still require glucose as well. This means that people wanting to lose weight need to eat a certain minimum amount of carbohydrate to meet their energy needs, even when they are limiting their food intakes.

Protein Used for Energy During a fast, when the available glycogen is gone and no carbohydrate is coming in from food, brain and nerve cells demand the glucose they need from the only available source—protein. The body begins to dismantle its own muscles and other lean tissues to generate glucose. Only adequate dietary carbohydrate can prevent this use of protein for energy, and its action in doing so is known as the **protein-sparing effect** of carbohydrate.

Ultimately, after half of the body's protein is used, death occurs. Death from loss of lean body tissues will occur even in an obese person who fasts too long. It should be clear, then, that although carbohydrate is an ideal energy source, fat and sometimes protein are also extremely important in meeting energy demands.

Chapter 7 offers guidelines for weight loss.

ketones (KEY-tones): acidic, fat-related compounds formed from the incomplete breakdown of fat when carbohydrate is not available; technically known as *ketone bodies.*

protein-sparing effect: the effect of carbohydrate in providing energy that allows protein to be used for other purposes.

Self Check

1. Carbohydrates are found in virtually all foods **except:**
   a. milks.
   b. meats.
   c. breads.
   d. vegetables.

2. Complex carbohydrates include:
   a. galactose, starch, and glycogen.
   b. starch, glycogen, and fiber.
   c. lactose, maltose, and glycogen.
   d. sucrose, fructose, and glucose.

3. The chief energy source of the body is:
   a. sucrose.
   b. starch.
   c. glucose.
   d. fructose.

4. The primary form of stored glucose in animals is:
   a. glycogen.
   b. cellulose.
   c. starch.
   d. lactose.

5. The polysaccharide that helps form the cell walls of plants is:
   a. cellulose.
   b. starch.
   c. glycogen.
   d. lactose.

6. Which of the following items may denote sugar on food labels?
   a. corn syrup
   b. aspartame
   c. xylitol
   d. cellulose

7. The two types of alternative sweeteners are:
   a. saccharin and cyclamate.
   b. sugar alcohols and artificial sweeteners.
   c. sorbitol and xylitol.
   d. sucrose and fructose.

8. A diet high in complex carbohydrates is:
   a. most likely low in fat.
   b. most likely low in fiber.
   c. most likely poor in vitamins and minerals.
   d. most likely disease promoting.

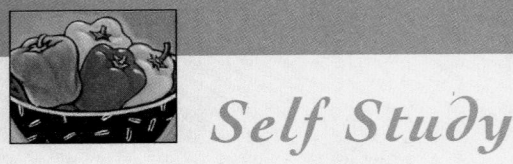

CARBOHYDRATES

Most of the energy we receive from foods comes from carbohydrates. Healthy choices provide complex carbohydrates or naturally occurring simple carbohydrates, rich in water-soluble vitamins and dietary fiber. A diet that is consistently low in dietary fiber and high in added sugar can lead to health problems. Consider the following examples of related foods and identify which are most similar to your food choices.

- Do you select whole-grain products and fresh fruits and vegetables regularly?
- Do you choose foods that increase your intake of fiber and limit your intake of sugars?

High in Fiber/Low in Added Sugar	Intermediate	Low in Fiber/High in Added Sugar
Apple with peel	Applesauce, sweetened	Fruit drink, 10% apple juice
Brown rice	Cream of rice cereal	Rice crispy treat
Pumpernickel bread	Bagel, plain	Danish pastry
Baked sweet potato	Candied sweet potato casserole	Sweet potato pie
Corn on the cob	Creamed corn	Frosted corn flakes
Oatmeal	Granola	Granola breakfast bar

9. A fiber-rich diet may help to prevent or control:
   a. diabetes.
   b. heart disease.
   c. constipation.
   d. all of the above.

10. The recommended fiber intake is:
    a. 10 to 15 grams per day.
    b. 15 to 25 grams per day.
    c. 20 to 35 grams per day.
    d. 40 to 55 grams per day.

Critical Thinking

Health recommendations suggest that 55 to 60 percent of the daily energy intake should come from carbohydrates, but stating recommendations in terms of percentage of energy intake is meaningful only if energy intake is known. The following exercises illustrate this concept.

1. Consider a student who has a high carbohydrate intake (70 percent of energy intake) and a moderate energy intake (2000 kcalories a day). How many grams of carbohydrates does the student consume each day?
   a. 160 grams
   b. 300 grams
   c. 350 grams
   d. 200 grams

2. Now consider an athlete who eats twice as much carbohydrate (in grams) as the student and has a much higher energy intake (6000 kcalories a day). What percentage does carbohydrate contribute to this person's daily intake?
   a. 70 percent
   b. 47 percent
   c. 57 percent
   d. 60 percent

Answers to these questions can be found in Appendix G.

Clinical Applications

1. Considering the health benefits of carbohydrate-rich foods, especially those that provide starch and fiber, what suggestions would you offer to a client who reports the following:

   - Eats only 3 servings of refined, sugary breads or cereals each day.
   - Eats one serving of vegetables (usually french fries) each day.
   - Drinks fruit juice once a day, but never eats fruit.
   - Eats cheese at least twice a day, but does not drink milk.
   - Eats large servings of meat at least twice a day.
   - Eats hard candy two or three times a day.

Nutrition on the Net

Notes

1. U.S. Department of Agriculture, Economic Research Service, *A Dietary Assessment of the U.S. Food Supply: Comparing Per Capita Food Consumption with the Food Guide Pyramid Serving Recommendations* (Washington, D.C.: Government Printing Office, 1998), AER no. 772.

2. A. M. Coulston and R. K. Johnson, Sugar and sugars: Myths and realities, *Journal of the American Dietetic Association* 102 (2002): 351–353; J. F. Guthrie and J. F. Morton, Food sources of added sweeteners in the diets of Americans, *Journal of the American Dietetic Association* 100 (2000): 43–48.

3. J. Mann, Carbohydrates, in *Present Knowledge in Nutrition,* 8th ed., ed. B. A. Bowman and R. M. Russell (Washington, D.C.: Int'l Life Sciences Institute Press, 2001), pp. 59–71; A. R. Folsom and coauthors, Increase in fasting insulin and glucose over seven years with increasing weight and inactivity of young adults: The CARDIA study, *American Journal of Epidemiology* 144 (1996): 235–246.

4. M. L. Wolraich, D. B. Wilson, and J. W. White, The effect of sugar on behavior or cognition in children: A meta-analysis, *Journal of the American Medical Association* 274 (1995): 1617–1621.

5. S. Gibson and S. Williams, Dental caries in pre-school children: Associations with social class, toothbrushing habit and consumption of sugars and sugar-containing foods. Further analysis of data from the National Diet and Nutrition Survey of children aged 1.5–4.5 years, *Caries Research* 32 (1999): 101–113.

6. K. G. Konig and J. M. Navia, Nutritional role of sugars in oral health, *American Journal of Clinical Nutrition* 62 (1995): 275S–283S.

7. Position of The American Dietetic Association: Use of nutritive and nonnutritive sweeteners, *Journal of the American Dietetic Association* 98 (1998): 580–587; S. S. Natah and coauthors, Metabolic response to lactitol and xylitol in healthy men, *American Journal of Clinical Nutrition* 65 (1997): 947–950.

8. Food and Drug Administration, Food labeling: Health claims—Sugar alcohols and dental caries, *Federal Register* 61 (August 23, 1996): 43433–43447.

9. Fact Sheet: The Report on Carcinogens, 9th ed., National Institutes of Health News Release, available at **www.nih.gov/news/pr/may2000/niehs-15.htm.**

10. Position of The American Dietetic Association, 1998.

11. Position of The American Dietetic Association, 1998.

12. A. Sparti and coauthors, Effect of diet high or low in unavailable and slowly digestible carbohydrates on the pattern of 24-h substrate oxidation and feelings of hunger in humans, *American Journal of Clinical Nutrition* 72 (2000): 1461–1468.

13. L. H. Nelson and L. A. Tucker, Diet composition related to body fat in a multivariate study of 203 men, *Journal of the American Dietetic Association* 96 (1996): 771–777.

14. E. Cunningham and W. Marcason, What role does fiber play in diverticular disease? *Journal of the American Dietetic Association* 102 (2002): 225; M. A. S. Van Duyn and E. Pivonka, Overview of the health benefits of fruit and vegetable consumption for the dietetics professional: Selected literature, *Journal of the American Dietetic Assocaition* 100 (2000): 1511–1521.

15. D. J. A. Jenkins and coauthors, Soluble fiber intake at a dose approved by the US Food and Drug Administration for a claim of health benefits: Serum lipid risk factors for cardiovascular disease assessed in a randomized controlled crossover trial, *American Journal of Clinical Nutrition* 75 (2002): 834–839. E. B. Rimm and coauthors, Vegetable, fruit, and cereal fiber intake and risk of coronary heart disease among men, *Journal of the American Medical Association* 275 (1996): 447–451.

16. L. Van Horn and N. Ernst, A summary of the science supporting the new National Cholesterol Education program dietary recommendations: What dietitians should know, *Journal of the American Dietetic Association* 101 (2001): 1148–1154; L. Brown and coauthors, Cholesterol-lowering effects of dietary fiber: A meta-analysis, *American Journal of Clinical Nutrition* 69 (1999): 30–42.

17. Brown and coauthors, 1999.

18. J. Salmeron and coauthors, Dietary fiber, glycemic load, and risk of noninsulin-dependent diabetes mellitus in women, *Journal of the American Medical Association* 277 (1997): 472–477.

19. B. S. Reddy, Role of dietary fiber in colon cancer: An overview, *American Journal of Medicine* 106 (1999): S16–S19; D. Kritchevsky, Protective role of wheat bran fiber: Preclinical data, *American Journal of Medicine* 106 (1999): S28–S31.

20. A. Schatzkin and coauthors, Lack of effect of a low-fat, high-fiber diet on the recurrence of colorectal adenomas, *New England Journal of Medicine* 342 (2000): 1149–1155; D. S. Alberts and coauthors, Lack of effect of a high-fiber cereal supplement on the recurrence of colorectal adenomas, *New England Journal of Medicine* 342 (2000): 1156–1162.

21. J. A. Story and D. A. Savaino, Dietary fiber and colorectal cancer: What is appropriate advice? *Nutrition Reviews* 59 (2001): 84–86.

22. J. L. Slavin, Mechanisms for the impact of whole grain foods on cancer risk, *Journal of the American College of Nutrition* 19 (2000): S300–S307.

23. Position of The American Dietetic Association: Health implications of dietary fiber, *Journal of the American Dietetic Association* 102 (2002): 993–1000.

24. S. M. Krebs-Smith and coauthors, Characterizing food intake patterns of American adults, *American Journal of Clinical Nutrition* 65 (1997): 1264S–1268S.

Nutrition in Practice

■ NUTRITION AND DENTAL HEALTH ■

Chapter 2 emphasized the health benefits of eating complex carbohydrate–rich foods. Complex carbohydrates may support overall health, but they do not necessarily promote dental health. The carbohydrates people eat and the times they eat them play a major role in the development of **dental caries**—a pervasive health problem throughout the world.

What is dental caries?

Dental caries is an infectious oral disease that develops in the tooth **enamel** (see the accompanying glossary and Figure NP2–1). Caries develops when bacteria that reside in the **dental plaque** consume and metabolize carbohydrates, producing acids that attack the tooth enamel. Thus at least two main ingredients are required to make dental caries: bacteria and carbohydrates. In addition, factors such as heredity, nutrition status during early tooth development, dental hygiene practices, and fluoride intake influence a person's susceptibility to caries. Poor nutrition during pregnancy, infancy, or early childhood can impair the development of healthy teeth, making caries likely.[1] Table NP2–1 shows the effects of specific nutrient deficiencies on tooth development.

How do carbohydrate-rich foods promote caries development?

The bacteria that promote dental caries thrive on food particles that contain carbohydrate. Both sugar and starch can support bacterial growth. Equally important is the length of time the food stays in the mouth, and this depends on how soon the teeth are brushed after eating and how sticky the food is. The damage a food does relates to both its carbohydrate content and its stickiness. For example, raisins and granola, which adhere to the teeth, cause more caries than a food that is easily rinsed off such as a sugary beverage.

Sugar can be eaten without inviting tooth decay if it is removed from tooth surfaces promptly. Bacterial action is maximal in the first 20 minutes after the first contact. If immediate brushing is not possible, water or other beverages swished in the mouth after a meal can effectively rinse the teeth. Once-a-day flossing may also effectively control formation of caries, regardless of the carbohydrate content of the diet. Some people may never get caries because they have inherited resistance to them.

Doesn't saliva rinse the mouth and protect the teeth?

Yes, and some foods stimulate more saliva flow than others. Saliva protects against caries formation in several ways. It not only rinses the mouth, but also dilutes the caries-causing acid produced by bacteria, exerts antibacterial action, and provides protective minerals such as calcium and phosphorus that promote the repair (remineralization) of tooth enamel.[2] Foods that elicit saliva flow may therefore defend against caries formation, but not all of them are protective; foods that also contain sugar may promote acid formation. Apples are an example: they stimulate saliva flow, but they also release sugar, so they have both caries-preventing and caries-promoting effects. Clearly, many different factors influence caries development, making it difficult to predict exactly which foods are **cariogenic**.

Do any foods prevent caries?

Yes. Some foods stimulate saliva flow and do not contribute to acid formation in the mouth: cheese is an example. Such foods are good choices to eat at the end of a meal. Cheese is a powerful saliva stimulant and

Glossary of Dental Caries Terms

cariogenic (KARE-ee-oh-JEN-ik): conducive to dental decay.

dental caries (KARE-eez): the gradual decay and disintegration of a tooth.

dental plaque (PLACK): a gummy mass of bacteria that grows on teeth and can lead to dental caries and gum disease.

enamel: the hard, white, dense substance that forms a covering for the crown of a tooth.

Nutrition in Practice

Table NP2–1 Nutrient Deficiencies Affecting Tooth Development

Nutrient Deficiency	Effect on Tooth Development
Protein	Small, irregularly shaped teeth; delayed eruption; high caries susceptibility
Vitamin C	Disturbance of dentin formation
Vitamin A	Disturbance of enamel formation, delayed eruption
Vitamin D	Poor mineralization, pitting, striations
Calcium	Poor mineralization
Phosphorus	Poor mineralization
Magnesium	Enamel underdeveloped
Iron	High caries susceptibility
Zinc	High caries susceptibility
Fluoride	High caries susceptibility

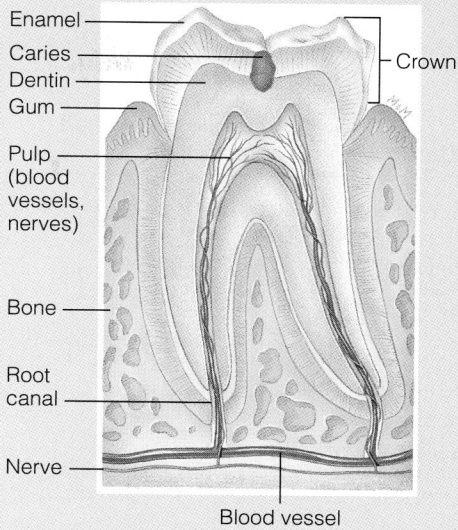

Figure NP2–1

A Tooth

The inner layer of dentin is bonelike material. The outer layer is enamel, which is harder than bone. Caries begins when acid dissolves the enamel that covers the tooth. If it is not repaired, the decay may penetrate the dentin and spread into the pulp of the tooth, causing inflammation and an abscess.

does not promote acid formation, so it reduces the cariogenicity of a meal. Furthermore, the high calcium and phosphorus contents of cheese support dental health.

High-fiber foods are, in general, anticariogenic, especially if their carbohydrate content is low. For example, raw vegetables do not stick to the teeth, and because they require vigorous chewing, they stimulate saliva flow. Cocoa products (including chocolate), coffee, tea, and beer all contain tannin, an acid that prevents caries formation. Table NP2–2 on the next page lists some dietary recommendations for controlling dental caries.

Besides foods, what other factors protect against dental caries development?

Research shows that when fluoride is added to the water supply, the children in the community have fewer dental caries than children who drink nonfluoridated water. Water fluoridation is the most effective, least expensive way to provide dental care to everyone.[3] Fluoride increases the mineralization of teeth, helps prevent tooth decay, and promotes tooth enamel remineralization throughout life.[4] The following recommendations will maximize protection against dental caries:

- Use sugars sparingly; watch for hidden sugars in foods; and use low-sugar or sugar-free products whenever possible.

- Restrict sweets to mealtimes.
- After eating a meal or a between-meal snack, brush and floss or, at least, rinse with water.
- Limit the time that teeth are exposed to sticky foods.
- In any case, brush at least twice daily, and floss at least once daily.
- Visit a dentist regularly; repair damaged teeth.
- Drink fluoridated water; provide infants and children with fluoride supplements when such water is not available.
- Eat a balanced diet composed of a variety of foods that will maintain adequate nutrition status.
- Eat foods that are rich in calcium and phosphorus.
- Eat a variety of firm, fibrous foods that will stimulate saliva flow.

In summary, learning and practicing sound dental hygiene habits, as well as developing eating habits that are consistent with both dental health and nutritional health, will serve a person throughout life.

Nutrition in Practice

Table NP2–2 Dietary Recommendations for Controlling Dental Caries

Food Group	Low Cariogenicity (Use When Teeth Cannot Be Brushed Immediately)	High Cariogenicity (Do Not Use unless Followed by Prompt and Thorough Dental Hygiene)
Dairy	Milk, cheese, plain yogurt	Ice cream, ice milk, milkshakes, fruited yogurts, eggnog
Meats/meat alternates	Meat, fish, poultry, eggs, legumes	Peanut butter with added sugar, luncheon meats with added sugar, meats with sugared glazes
Fruits[a]	Fresh, packed in water	Dried (raisins, figs, dates), packed in syrup or juice, jams, jellies, preserves, fruit juices and drinks
Vegetables	Most vegetables	Candied sweet potatoes, glazed carrots
Breads/cereals[b]	Popcorn, toast, hard rolls, pretzels, pizza, bagels	Cookies, sweet rolls, pies, cakes, dry sugared cereals as between-meal snacks, doughnuts, potato chips, granola bars, oatmeal, oat cereals, oatmeal baked goods[c]
Other	Sugarless gum, coffee or tea without sugar, nuts	Sugared soft drinks, candy, fudge, caramels, honey, sugars, syrups

[a]Tiny particles of bananas can get lodged between teeth and decompose, increasing risk of caries.
[b]Tiny particles of breads, crackers, and chips can also become lodged in teeth, promoting caries formation.
[c]The soluble fiber in oats makes this grain particularly sticky and therefore cariogenic.

Notes

1. D. Fitzsimons and coauthors, Nutritional and oral health guidelines for pregnant women, infants, and children, *Journal of the American Dietetic Association* 98 (1998): 182–189.

2. D. P. DePaola and C. F. Schachtele, Diet and oral health, in *Biochemical and Physiological Aspects of Human Nutrition,* M. H. Stipanuk (Philadelphia: W. B. Sanders, 2000), pp. 866–881; International Food Information Council Foundation, Nutrition and oral health: Making the connection, *IFIC Review* May 1998, pp. 1–7.

3. Position of The American Dietetic Association: The impact of fluoride on health, *Journal of the American Dietetic Association* 101 (2001): 126–132.

4. Position of The American Dietetic Association, 2001.

Jennie Oppenheimer/Studio Zocolo

CHAPTER 3 | LIPIDS

*M*ost people know that too much fat in the diet imposes health risks, but they may be surprised to learn that too little does, too. People in the United States, however, are more likely to eat too much fat than too little.

Fat is a member of the class of compounds called **lipids.** The lipids in foods and in the human body include triglycerides (**fats** and **oils**), phospholipids, and sterols.

Energy from Fat   Lipids perform many tasks in the body, but most importantly, they provide energy. A constant flow of energy is so vital to life that, in a pinch, any other function is sacrificed to maintain it. Chapter 2 described one safeguard against such an emergency—the stores of glycogen in the liver that provide glucose to the blood whenever the supply runs short. The body's stores of glycogen are limited, however. In contrast, the body's capacity to store fat for energy is virtually unlimited due to the fat-storing cells of the **adipose tissue.** Unlike most body cells, which can store only limited amounts of fat, the adipose, or fat, cells seem able to expand almost indefinitely. The more fat they store, the larger they grow. Fat cells are more than just storage depots, however; fat cells secrete hormones and produce enzymes that influence the body's intake of food and its use of energy nutrients.[1] Figure 3–1 shows a fat cell.

The fat stored in fat cells supplies 60 percent of the body's ongoing energy needs during rest. During physical activity or prolonged periods of food deprivation, fat stores may make an even greater energy contribution.

Roles of Body Fat   In addition to supplying energy, fat serves other roles in the body. Natural oils in the skin provide a radiant complexion; in the scalp, they help nourish the hair and make it glossy. The layer of fat beneath the skin insulates the body from extremes of temperature. A pad of hard fat beneath each kidney protects it from being jarred and damaged, even during a motorcycle ride on a bumpy road. The soft fat in a woman's breasts protects her mammary glands from heat and cold and cushions them against shock. The fat embedded in muscle tissue

Chapter 10 discusses fat use during physical activity.

Chapter 6 discusses fat use during fasting.

lipids: a family of compounds that includes triglycerides (fats and oils), phospholipids, and sterols.

fats: lipids that are solid at room temperature (70°F or 25°C).

oils: lipids that are liquid at room temperature (70°F or 25°C).

adipose tissue: the body's fat, which consists of masses of fat-storing cells called adipose cells.

Figure 3–1
A Fat Cell
Within the fat, or adipose, cell, lipid is stored in a droplet. This droplet can greatly enlarge, and the fat cell membrane will grow to accommodate its swollen contents.

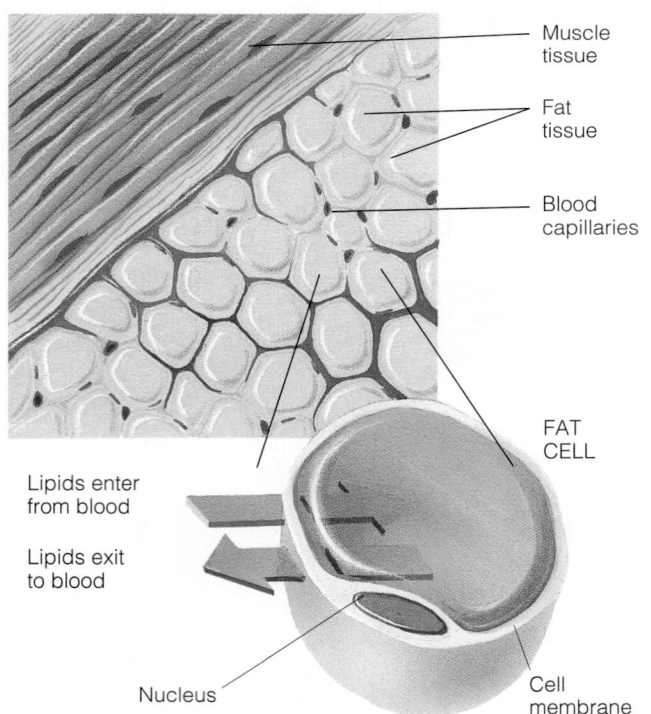

Muscle tissue

Fat tissue

Blood capillaries

FAT CELL

Lipids enter from blood

Lipids exit to blood

Nucleus

Cell membrane

shares with muscle glycogen the task of providing energy when the muscles are active. The phospholipids and the sterol cholesterol are cell membrane constituents that help maintain the structure and health of all cells. Table 3–1 summarizes the major functions of fats in the body.

In Summary Lipids in the body not only serve as energy reserves but also protect the body from temperature extremes, cushion the vital organs, and provide the major material of cell membranes.

The Chemist's View of Lipids

The diverse and vital functions that lipids play in the body reveal why eating too little fat can be harmful. As mentioned earlier, though, too much fat in the diet seems to be the bigger problem for most people. To understand both the beneficial and harmful effects that fats exert on the body, a closer look at the structure and function of members of the lipid family is in order.

Triglycerides

When people talk about fat—for example, "I'm too fat" or "That meat is fatty"—they are usually referring to triglycerides. Among lipids, **triglycerides** predominate—both in the diet and in the body. The name *triglyceride* almost explains itself: three **fatty acids** *(tri)* attached to a **glycerol** "backbone." Figure 3–2 shows how three fatty acids combine with glycerol to make a triglyceride.

Fatty Acids

When energy from any energy-yielding nutrient is to be stored as fat, the nutrient is first broken into small fragments. Then the fragments are linked together into chains known as fatty acids. The fatty acids are then packaged, three at a time, with glycerol to make triglycerides.

Chain Length and Saturation Fatty acids may differ from one another in two ways—in chain length and in degree of saturation. The chain length refers to the number of carbons in a fatty acid. Saturation also refers to its chemical structure—specifically, to the number of hydrogens the carbons in the fatty acid are holding. If every available bond is filled to capacity with hydrogen, the chain is

triglycerides (try-GLISS-er-ides): one of the three main classes of lipids; the chief form of fat in foods and the major storage form of fat in the body; composed of glycerol with three fatty acids attached.
  tri = three
  glyceride = a compound of glycerol

fatty acids: organic compounds composed of a chain of carbon atoms with hydrogens attached and an acid group at one end.

glycerol (GLISS-er-ol): an organic compound, three carbons long, that can form the backbone of triglycerides and phospholipids.

Figure 3–2
Triglyceride Formation
Glycerol, a small, water-soluble compound, plus three fatty acids, equals a triglyceride.

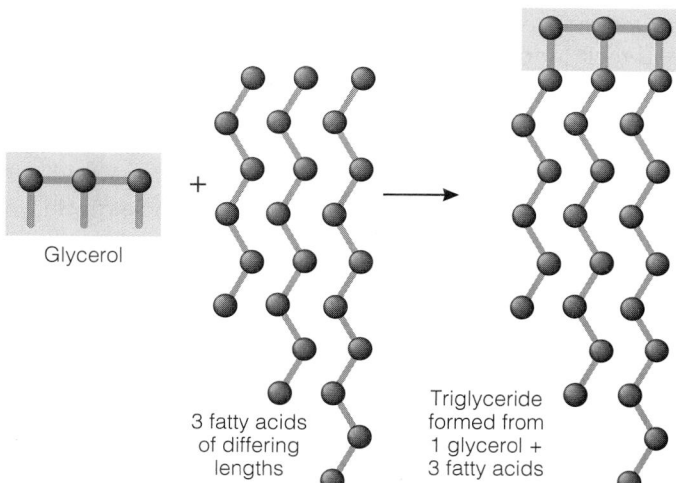

Glycerol

3 fatty acids of differing lengths

Triglyceride formed from 1 glycerol + 3 fatty acids

Figure 3–3

Three Fatty Acids
The more carbon atoms in a fatty acid, the longer its chain. The more hydrogen atoms attached to those carbons, the more saturated the fatty acid.

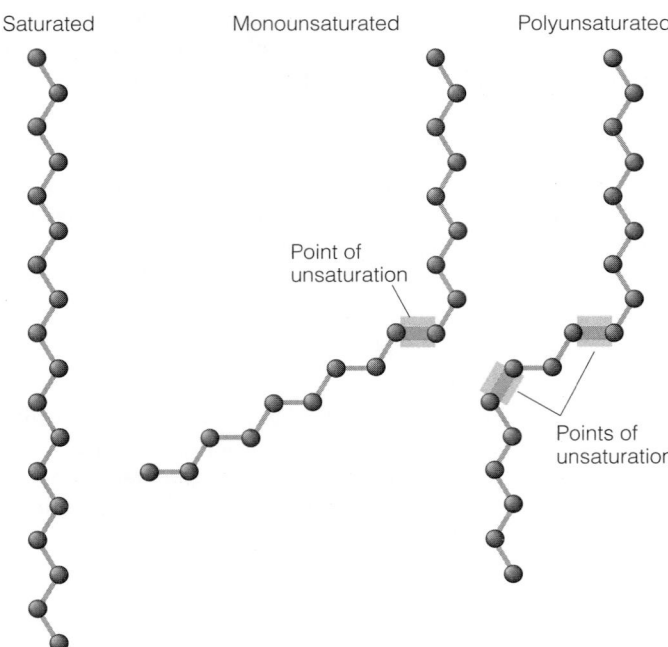

Saturated   Monounsaturated   Polyunsaturated

Point of unsaturation

Points of unsaturation

saturated fatty acid: a fatty acid carrying the maximum possible number of hydrogen atoms (having no points of unsaturation). A saturated fat is a triglyceride that contains three saturated fatty acids.

unsaturated fatty acid: a fatty acid with one or more points of unsaturation where hydrogens are missing (includes monounsaturated and polyunsaturated fatty acids).

monounsaturated fatty acid: a fatty acid that has one point of unsaturation; for example, the oleic acid found in olive oil.

polyunsaturated fatty acids (PUFA): fatty acids with two or more points of unsaturation. For example, linoleic acid has two such points, and linolenic acid has three. Thus polyunsaturated *fat* is composed of triglycerides containing a high percentage of PUFA.

rancid: the term used to describe fats when they have deteriorated, usually by oxidation. Rancid fats often have an "off" odor.

oxidation (OKS-ee-day-shun): the process of a substance combining with oxygen.

antioxidants: compounds that protect others from oxidation by being oxidized themselves.

called a **saturated fatty acid.** A saturated fatty acid is fully loaded with hydrogens and has only single bonds between its carbons. The first zigzag structure in Figure 3–3 represents a saturated fatty acid.

Unsaturated Fatty Acids   In some fatty acids, especially those of plants and fish, hydrogens are missing in the fatty acid chains. The points where the hydrogens are missing are called points of unsaturation, and a chain containing such points is called an **unsaturated fatty acid.** An unsaturated fatty acid has at least one double bond between its carbons. If there is one point of unsaturation, the chain is **monounsaturated.** The second structure in Figure 3–3 is an example.

If there are two or more points of unsaturation, then the fatty acid is polyunsaturated (see the third structure in Figure 3–3). **Polyunsaturated fatty acids** are often abbreviated on food labels as PUFA.

Hard and Soft Fat   A triglyceride can contain any combination of fatty acids—long chain or short chain and saturated, monounsaturated, or polyunsaturated. The degree of saturation of the fatty acids in a fat influences the health of the body (discussed in a later section) and the characteristics of foods. Fats that contain the shorter-chain or the more unsaturated fatty acids are softer at room temperature and melt more readily. Thus, looking at three fats—lard (which comes from pork), chicken fat, and safflower oil—lard is the most saturated and the hardest; chicken fat is less saturated and somewhat soft; and safflower oil, which is the most unsaturated, is a liquid at room temperature.

Stability   Saturation also influences stability. Fats can become **rancid** when exposed to oxygen. Polyunsaturated fatty acids spoil most readily because their double bonds are unstable. The **oxidation** of unsaturated fats produces a variety of compounds that smell and taste rancid; saturated fats are more resistant to oxidation and thus less likely to become rancid. Other types of spoilage can occur due to microbial growth.

Manufacturers can protect fat-containing products against rancidity in three ways—none of them perfect. First, products may be sealed air-tight and refrigerated—an expensive and inconvenient storage system. Second, manufacturers may add **antioxidants** to compete for the oxygen and thus protect the oil (examples are the additives BHA and BHT and vitamins C and E). Third, manufactur-

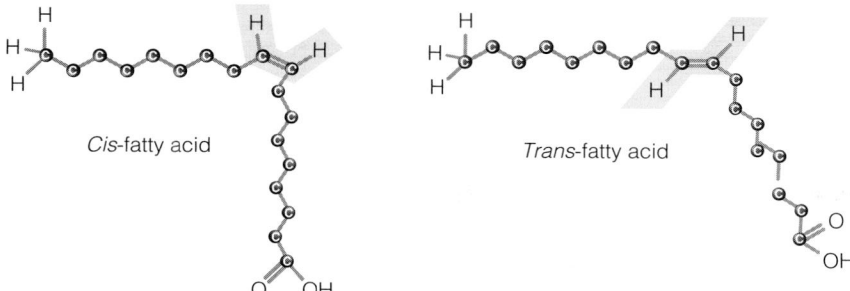

Figure 3–4

Cis- and Trans- Fatty Acids Compared

Cis-fatty acid

Trans-fatty acid

ers may saturate some or all of the points of unsaturation by adding hydrogen molecules—a process known as hydrogenation.

Hydrogenation Hydrogenation offers two advantages: it protects against oxidation (which prolongs shelf life) and alters the texture of foods. When partially hydrogenated, vegetable oils become spreadable margarine. Hydrogenated fats make pie crusts flaky and puddings creamy. A disadvantage is that hydrogenation makes polyunsaturated fats more saturated. Consequently, any health advantages of using polyunsaturated fats instead of saturated fats are lost in hydrogenation.

***Trans*-Fatty Acids** Another disadvantage of hydrogenation is that some of the molecules that remain unsaturated after processing change shape from *cis* to *trans*. In nature, most unsaturated fatty acids are *cis*-fatty acids—meaning that the hydrogens next to the double bonds are on the same side of the carbon chain. Only a few (notably those found in milk and butter) are ***trans*-fatty acids**—meaning that the hydrogens next to the double bonds are on opposite sides of the carbon chain (see Figure 3–4). These arrangements result in different configurations for the fatty acids, and this difference affects function: in the body, *trans*-fatty acids behave more like saturated fats than like unsaturated fats. The relationship between *trans*-fatty acids and heart disease has been the subject of much recent research, as a later section describes.

Essential Fatty Acids The human body can synthesize all the fatty acids it needs from carbohydrate, fat, or protein except for two—**linoleic acid** and **linolenic acid.** Both linoleic acid and linolenic acid are polyunsaturated fatty acids. Because they cannot be made from other substances in the body, they must be obtained from food and are therefore called **essential fatty acids.** Linoleic acid and linolenic acid are found in small amounts in plant oils, and the body readily stores them, making deficiencies unlikely. From both of these essential fatty acids, the body makes important substances that help regulate a wide range of body functions: blood pressure, clot formation, blood lipid concentration, the immune response, the inflammatory response to injury, and many others.[2] These two essential nutrients also serve as structural components of cell membranes.

Linoleic Acid: An Omega-6 Fatty Acid Linoleic acid is an **omega-6 fatty acid** found in the seeds of plants and in the oils produced from the seeds. Any diet that contains vegetable oils, seeds, nuts, and whole-grain foods provides enough linoleic acid to meet the body's needs. Researchers have long known and appreciated the importance of the omega-6 fatty acid family.

Linolenic Acid and Other Omega-3 Fatty Acids Linolenic acid belongs to a family of polyunsaturated fatty acids known as **omega-3 fatty acids,** a family that also includes **EPA** and **DHA.** EPA and DHA are found primarily in fish oils. As mentioned, the human body cannot make linolenic acid, but given dietary linolenic acid, it can make EPA and DHA, although the process is slow.

Food additives, including BHA and BHT, are discussed in Chapter 11.

hydrogenation (high-dro-gen-AY-shun): a chemical process by which hydrogens are added to monounsaturated or polyunsaturated fats to reduce the number of double bonds, making the fats more saturated (solid) and more resistant to oxidation (protecting against rancidity). Hydrogenation produces *trans*-fatty acids.

***trans*-fatty acids:** fatty acids with an unusual configuration around the double bond.

linoleic acid, linolenic acid: polyunsaturated fatty acids, essential for human beings.

essential fatty acids: fatty acids that the body requires but cannot make in amounts sufficient to meet its physiological needs.

omega-6 fatty acids: a polyunsaturated fatty acid with its endmost double bond six carbons back from the end of its carbon chain; long recognized as important in nutrition. Linoleic acid is an example.

omega-3 fatty acids: polyunsaturated fatty acids in which the endmost double bond is three carbons back from the end of the carbon chain; relatively newly recognized as important in nutrition. Linolenic acid is an example.

EPA, DHA: omega-3 fatty acids made from linolenic acid. The full name for EPA is *eicosapentaenoic* (EYE-cosa-PENTA-ee-NO-ick) *acid.* The full name for DHA is *docosahexaenoic* (DOE-cosa-HEXA-ee-NO-ick) *acid.*

Table 3–2   Sources of Omega-6 and Omega-3 Fatty Acids

Omega-6	
Linoleic acid	Leafy vegetables, seeds, nuts, grains, vegetable oils (corn, cottonseed, safflower, sesame, soybean, sunflower), poultry fat

Omega-3	
Linolenic acid	Oils (canola, flaxseed, soybean, walnut, wheat germ; liquid or soft margarine made from canola or soybean oil) Nuts and seeds (butternuts, walnuts, soybean kernels) Vegetables (soybeans)
EPA and DHA	Human milk Coldwater fish[a] (mackerel, salmon, bluefish, mullet, sablefish, menhaden, anchovy, herring, lake trout, sardines, tuna) (or can be made from linolenic acid)

[a]All of these fish except tuna provide at least 1 gram of omega-3 fatty acids in 100 grams of fish (3.5 ounces); the fish oil content of each species varies with the season and site of harvest. Tuna provides fewer omega-3 fatty acids, but because it is commonly consumed, its contribution can be significant.

Chemists use the term omega, the last letter of the Greek alphabet, to refer to the position of the last double bond in a fatty acid.

The importance of omega-3 fatty acids was first recognized during the 1980s when research began to unveil impressive roles for EPA and DHA in metabolism and disease prevention. DHA is one of the most abundant structural lipids in the brain, and both EPA and DHA are needed for normal brain development.[3] EPA and DHA are also especially active in the rods and cones of the retina of the eye.[4] Today researchers know that these omega-3 fatty acids are essential for normal growth and development and that they may play an important role in the prevention and treatment of heart disease, hypertension, arthritis, and cancer.[5]

Recommendations and Intake   Official recommendations for omega-6 and omega-3 fatty acids do not yet exist, but they soon will. The U.S. diet is high in omega-6 fatty acids (due to the increased production and use of vegetable cooking oils such as corn and cottonseed oils) and low in omega-3 fatty acids.[6] Experts recommend a more balanced intake. The best way for people to increase their intakes of omega-3 fatty acids is to follow the advice of the American Heart Association: eat at least two servings of fish each week (see Table 3–2).[7] Even one fish meal per week has been associated with a reduced risk of heart attack.[8]

Omega-3 Supplements   Omega-3 fatty acid supplements are being aggressively marketed as a cure-all for many different diseases without regard for consumer safety. Although some claims are based on research, confirmation that fish oil can prevent or treat heart disease, cancer, or other diseases in individuals or populations is lacking. Experts agree that adding *fish* to the diet two or three times per week may help to prevent heart disease, but the idea that fish oil *supplements* are beneficial and safe in any amount is erroneous.

In the first place, the supplements themselves may carry hazards. They may contain toxic amounts of fat-soluble vitamins, pesticide residues, or heavy metals such as mercury that may be concentrated in the pills. Unfortunately, this caution must also be applied to some fish living in mercury-polluted waters (see Chapter 11). Even when the supplements are not toxic, they may have harmful effects on the body. One potential problem is that the supplement form may not function the same way as the form available directly from foods. Furthermore, omega-3 and omega-6 fatty acids compete for the same slots in the body. Consequently, taking supplements of one can easily induce a deficiency of the other. Still another drawback is that fish oil can cause or aggravate illness by altering blood lipids or blood clotting. Finally, the quantities of omega-3 fatty acids in supplements vary widely from what the labels say they contain.

Phospholipids

Up to now, this discussion has focused on one of the three classes of lipids, the triglycerides (fats and oils), and their component parts, the fatty acids (see Table 3–3). The other two classes of lipids, the phospholipids and sterols, make up only 5 percent of the lipids in the diet, but they are nevertheless worthy of attention. Among the **phospholipids,** the lecithins are of particular interest.

Structure of Phospholipids Like the triglycerides, the **lecithins** and some other phospholipids have a backbone of glycerol; they differ from the triglycerides in having only two fatty acids attached to the glycerol. In place of the third fatty acid, they have a phosphate group and a molecule of **choline** or a similar compound. The fatty acids make phospholipids soluble in fat; the phosphate group enables them to dissolve in water. Such versatility benefits the food industry, which uses phospholipids as **emulsifiers** to mix fats with water in such products as mayonnaise and candy bars.

Roles of Phospholipids Lecithins and other phospholipids are important constituents of cell membranes. They also act as emulsifiers in the body, helping to keep other fats in solution in the watery blood and body fluids.

Phospholipids: Not Essential Lecithins periodically receive attention in the popular press. People may hear that lecithins are a major constituent of cell membranes (true), that the functioning of all cells depends on the integrity of the cell membranes (true), and that consumers must therefore take lecithin supplements (false). The body digests lecithins before it absorbs them, so the lecithins people eat do not reach the body tissues intact. Instead, the lecithins used for building cell membranes are made from scratch by the liver. In other words, the lecithins are not essential nutrients.

Sterols

Sterols are large, complex molecules consisting of interconnected rings of carbon. Cholesterol is the most familiar sterol, but others, such as vitamin D and the sex hormones (for example, testosterone), are important, too.

Sterols in Foods Both plant and animal-derived foods contain sterols, but only animal-derived foods contain cholesterol: meats, eggs, fish, poultry, and dairy products. Organ meats, such as liver and kidneys, and eggs are richest in cholesterol; cheeses and meats have less. Shellfish contain many sterols, but much less cholesterol than was thought in the past.

Cholesterol Synthesis Like the lecithins, cholesterol can be made by the body, so it is not an essential nutrient. Your liver is manufacturing it now, as you read, at the rate of perhaps 50,000,000,000,000,000 molecules per second. The raw materials that the liver uses to make cholesterol can all be taken from glucose or saturated fatty acids. In other words, cholesterol can be made from either carbohydrate or fat. Most of the body's cholesterol ends up in the cells, where it performs vital structural and metabolic functions.

Cholesterol's Two Routes in the Body After being made, cholesterol either leaves the liver or is transformed there into related compounds such as vitamin D. Cholesterol leaves the liver by two routes:

1. It may be made into **bile,** stored in the gallbladder, and delivered to the intestine.

2. It may travel, via the bloodstream, to all the body's cells.

Table 3–3	The Lipid Family

Triglycerides (fats and oils)
- Glycerol (1 per triglyceride)
- Fatty acids (3 per triglyceride)
  - Saturated
  - Monounsaturated
  - Polyunsaturated
    - Omega-6
    - Omega-3

Phospholipids (such as the lecithins)

Sterols (such as cholesterol)

phospholipids: one of the three main classes of lipids. These compounds are similar to triglycerides, but have choline (or another compound) and a phosphorus-containing acid in place of one of the fatty acids.

lecithins: one type of phospholipid.

choline: a nonessential nutrient that can be made in the body from an amino acid.

emulsifiers: substances that mix with both fat and water and that permanently disperse the fat in the water, forming an emulsion.

sterols: one of the three main classes of lipids. Sterols include cholesterol, vitamin D, and the sex hormones (such as testosterone).

bile: a compound made by the liver from cholesterol and stored in the gallbladder. Bile prepares fat for digestion.

Cholesterol Recycled   The bile that is made from cholesterol in the liver is released into the intestine to aid in the digestion and absorption of fat (see Chapter 5). After bile does its job, some is reabsorbed into the body and recycled; the rest is excreted in the feces.

Cholesterol Excreted   While bile is in the intestine, some of it may be trapped by soluble fibers or by some medications, which carry it out of the body in feces. The excretion of bile reduces the total amount of cholesterol remaining in the body.

Both the intestine and the liver make lipoproteins. Chapter 5 tells the story of lipid transport.

Cholesterol Transport   Some cholesterol, packaged with other lipids and protein, leaves the liver via the arteries and is transported to the body tissues by the blood. These packages of lipids and proteins are called lipoproteins. As the lipoproteins travel through the body, tissues can extract lipids from them. Cholesterol's harmful effects in the body occur when it forms deposits in the artery walls. These deposits lead to **atherosclerosis,** a disease that can cause heart attacks and strokes.

In Summary   Table 3–3 on the previous page summarizes the members of the lipid family. The predominant lipids both in foods and in the body are triglycerides: glycerol backbones with three fatty acids attached. Fatty acids vary in the length of their carbon chains and their degree of unsaturation. Those that are fully loaded with hydrogens are saturated; those that are missing hydrogens and therefore have double bonds are unsaturated (monounsaturated or polyunsaturated). Most triglycerides contain more than one type of fatty acid. Fatty acid saturation affects fats' physical characteristics and storage properties. Hydrogenation, which makes polyunsaturated fats more saturated, gives rise to *trans*-fatty acids, altered fatty acids that may have health effects similar to those of saturated fatty acids. Linoleic acid and linolenic acid are essential nutrients. In addition to serving as structural parts of cell membranes, they make powerful substances that help regulate blood pressure, blood clot formation, and the immune response. Phospholipids, including the lecithins, have a unique chemical structure that allows them to be soluble in both water and fat. In the body, phospholipids are part of cell membranes; the food industry uses phospholipids as emulsifiers. Sterols include cholesterol, bile, vitamin D, and the sex hormones. Only animal-derived foods contain cholesterol.

Fats and Health

Of all the dietary factors related to chronic diseases prevalent in developed countries, fat is by far the most significant. Excessive intakes of dietary fat contribute to obesity, diabetes, cancer, **cardiovascular disease,** and probably other diseases and disorders as well. Heart disease is the number one killer of adults in the United States. In the interest of good health and disease prevention, one change that most people should make in their diets is to limit their intakes of total fat. It is especially important to limit saturated fat because the cholesterol that accumulates in arteries is manufactured largely from fragments derived from saturated fat.

Fats and Fatty Acids

Most people realize that elevated blood cholesterol is an important risk factor for heart disease. The higher the blood cholesterol, the greater the risk of heart disease. Most people may not realize, though, that cholesterol in *food* is not the main

atherosclerosis (ath-er-oh-scler-OH-sis): a type of artery disease characterized by accumulations of lipid-containing material on the inner walls of the arteries (see Chapter 25).

cardiovascular disease (CVD): a general term for all diseases of the heart and blood vessels (see Chapter 25).

influential factor in raising *blood* cholesterol. It is total fat, especially saturated fat, that raises blood cholesterol. An extensive review of well-controlled studies on the effects of dietary fats on blood cholesterol reached the following conclusions:

- Saturated fatty acids elevate blood cholesterol and are the main dietary determinants of blood cholesterol levels.[9]*

Less influential, but still significant, were the following factors:

- Polyunsaturated fatty acids lower blood cholesterol.
- Monounsaturated fatty acids lower blood cholesterol.

Saturated Fat and Blood Cholesterol Because the dietary factor thought to be most influential in raising blood cholesterol is saturated fat, reducing only the total fat in the diet may not lower cholesterol.[10] It is better to reduce total fat *and* replace cholesterol-raising saturated fat with polyunsaturated or monounsaturated fat. A meta-analysis of more than 30 studies lends further support to the beneficial effects of reducing dietary saturated fatty acids.[11]

Omega-3 Fatty Acids and Blood Cholesterol Of the polyunsaturated fatty acids, the omega-3 fatty acids appear to reduce blood cholesterol the most.[12] This effect was first noticed when researchers learned that the Inuit peoples of Alaska and Greenland, despite high-energy, high-fat, high-cholesterol diets, enjoyed relative freedom from heart diseases—especially atherosclerosis. Analysis of the foods common in Inuit diets, which derive primarily from marine animals, revealed that they were rich in omega-3 fatty acids, particularly EPA and DHA.

***Trans*-Fatty Acids and Blood Cholesterol** *Trans*-fatty acids, like saturated fatty acids, elevate blood cholesterol and thus raise the risk of heart disease and heart attack.[13] When news of *trans*-fatty acids' effects on heart health was first emerging, some people hastily switched from using margarine back to butter, believing oversimplified reports that margarine provided no heart health advantage over butter. It is true that most margarines and virtually all shortenings are made from mostly hydrogenated fats and therefore contain substantial *trans*-fatty acids—up to 40 percent. Some margarines, however, especially the soft or liquid varieties, are made from unhydrogenated oils. These have long proved to be less likely to elevate serum cholesterol than the saturated fats of butter. When oils (but not hydrogenated oils) are the first ingredient listed on a margarine label, that margarine is probably low in both *trans*-fatty acids and saturated fat.

In addition to soft and liquid margarine choices, some margarines contain few or no *trans*-fatty acids. Some of these also contain a functional food ingredient, **sterol esters,** that reduces blood cholesterol when consumed in addition to a low-fat diet.† Sterol esters are not recognized by the intestine and therefore are not absorbed, and they also block the absorption of cholesterol.[14] Simply adding the margarine to a high-fat diet is unlikely to bring benefits, however. Sterol esters work only when people cut their fat intakes as well. Drawbacks include the price (three or four times higher than regular margarine), a high fat content (the full-fat kind equals the fat in regular margarine), and an unproven record of safety for use by consumers of all ages.

Foods other than margarine also contribute *trans*-fatty acids. Fast foods, chips, baked goods, and other commercially prepared foods are high in fats containing up to 50 percent *trans*-fatty acids. Fast-food chains fry foods in hydrogenated vegetable oil that contains abundant *trans*-fatty acids. Overall, consumers are now eating more fats containing *trans*-fatty acids than ever before, amounting to about 3 percent of daily calories, and they are eating these *trans*-fatty acids in the form of processed foods (see the margin list).[15]

*It should be noted that not all saturated fatty acids have the same cholesterol-raising effect. For example, stearic acid, an 18-carbon fatty acid, does not raise blood cholesterol.
†Two brand names of margarines with sterol esters currently on the market are *Benecol* and *Take Control.*

Enjoy fish and low-fat foods for good health.

The words *hydrogenated vegetable oil* or *shortening* in an ingredients list indicate trans-fatty acids in the product.

Some common food sources of trans-fatty acids:
- *Most hardened margarines and shortenings.*
- *Salad dressing, mayonnaise.*
- *Biscuits, rolls, cakes, crackers.*
- *Corn snacks and chips.*
- *Other fried snacks and chips.*
- *Cookies, doughnuts.*
- *French fries, fried chicken or fish.*
- *Fried fast foods, even those fried in commercial "vegetable oils."*

sterol esters compounds belonging to the sterol family of lipids, derived from plants, that have been shown experimentally to reduce blood cholesterol when consumed in place of other fats in a low-fat diet.

These foods are major contributors of *trans*-fatty acids.

Current food labels can be misleading in this regard, but changes are in the works. Now, grams of *trans*-fatty acids are counted with the polyunsaturated fats from which they arose, and not with the saturated fats whose health effects they mimic. Soon, a separate statement of *trans*-fatty acids on food labels may help consumers make informed choices. Reducing total fat and replacing both saturated and *trans* fats with monounsaturated and polyunsaturated fats may be the wisest strategies for preventing heart disease.[16]

Recommendations

The American Heart Association (AHA) recommends that *total* fat intake should not exceed 30 percent of the day's total energy intake. Currently, saturated fat contributes about 13 percent of total energy intake. The AHA recommends reducing saturated fat to less than 10 percent of total energy intake.[17] The major sources of saturated fat are meat and dairy products such as whole milk, cream, and butter, so limiting these foods is advised. Foods high in saturated fat are often high in cholesterol. Thus, although dietary cholesterol exerts a smaller effect on blood cholesterol than saturated fat, by limiting saturated fat, cholesterol intake will be reduced as well. The AHA recommends limiting total cholesterol intake to less than 300 milligrams a day.[18] Recommendations to limit fat are not intended for infants or children under two years old.

Limiting dietary fat intake to 30 percent or less of total energy requires careful planning at every meal. According to research on U.S. food choices, people typically eat about 34 percent of their energy intakes as fat.[19] Even though this is less than in the recent past, it is more than people were eating a century ago when foods were less highly processed, and it is more than people need. A later "How to" box offers specific suggestions for putting these guidelines into practice.

In Summary Health authorities point to high fat intakes as a major flaw in the North American diet: excess fat contributes to heart disease, obesity, and other health problems. High blood cholesterol, specifically, poses a risk of heart disease, and high intakes of saturated fat contribute most to high blood cholesterol. Omega-3 fatty acids appear to be protective. High intakes of *trans*-fatty acids also appear to raise blood cholesterol. Cholesterol in foods presents less of a risk. The AHA recommends limiting total fat intake to 30 percent or less of total energy intake and saturated fat to less than 10 percent.

Fats in Foods

Fats are important in foods as well as in the body. Many of the compounds that give foods their flavors and aromas are found in fats and oils. The delicious aromas associated with bacon, ham, and other meats, as well as with onions being sautéed, come from fats. Fats also influence the texture of many foods, enhancing smoothness, creaminess, moistness, or crispness.[20] In addition, four vitamins—A, D, E, and K—are soluble in fat. When the fat is removed from a food, many fat-soluble compounds, including these vitamins, are also removed. Table 3–4 summarizes the roles of fats in foods.

Fats are also an important part of most people's ethnic or national cuisines. Each culture has its own favorite food sources of fats and oils. In Canada, canola oil (also known as rapeseed oil) is widely used. In the Mediterranean area, Greeks, Italians, and Spaniards rely heavily on olive oil. Both canola oil and olive oil are rich sources of monounsaturated fatty acids. Asians use the polyunsaturated oil of soybeans. Jewish people traditionally employ chicken fat. Everywhere in North America, butter and margarine are widely used.

These cultural associations along with the very real pleasures fats contribute to meals help explain why fat consumption is so high. Nevertheless, for most people, reducing dietary fat is important for health. The first step to reducing dietary fat is finding out where the fat is.

Finding the Fats in Foods

Of the groups in the Food Guide Pyramid, the fats at the tip and the meats (and nuts) one step below always contain fat, but two other food groups—the milk, cheese, and yogurt group and the breads and cereal group—sometimes contain fat as well. Vegetables and fruits, if unprocessed, are virtually fat-free, with two notable exceptions: avocados and olives, which are rich in monounsaturated fat. Grains, too, in their natural state contain little or no fat.

Added Fats in Foods A dollop of dessert topping, a spread of butter on bread, oil or shortening in a recipe, dressing on a salad—all of these are examples of *added* fats. Indeed, all sorts of fats can be added to foods during commercial or home preparation or at the table. The following amounts of these fats contain about 5 grams of pure fat, providing 45 kcalories and negligible protein and carbohydrate:

- 1 teaspoon of oil or shortening.
- 1½ teaspoons of mayonnaise, butter, or margarine.
- 1 tablespoon of regular salad dressing, cream cheese, or heavy cream.
- 1½ tablespoons of sour cream.

These foods provide the majority of added fats to the diet. They are the fats of fried foods or baked goods, sauces and mixed dishes, and dips and spreads.

Hidden Fats in Foods Other foods that are significant contributors of fat include convenience foods, lunch meats, and other prepared meats. The fat in these foods is sometimes referred to as invisible fat because it does not have the obvious appearance of fat. An ounce of lean meat or low-fat cheese supplies about half its kcalories from fat (28 kcalories from protein and 27 kcalories from fat). An ounce of high-fat meat (such as bologna) or most cheeses supplies 72 percent of its energy from fat (28 kcalories from protein and 72 kcalories from fat). Two tablespoons of peanut butter supply 72 percent of their energy as fat (32 kcalories from protein, 24 kcalories from carbohydrate, and 140 kcalories from fat)! Thus foods that are usually thought of as protein-rich foods may actually contain more fat energy than protein energy. Note that the values for meat given here are for 1-ounce portions. An average serving of hamburger is usually 3 or 4 ounces. An average dinner steak may be 8 ounces or larger.

Fats in the Milk, Yogurt, and Cheese Group The fat in milk is about 63 percent saturated fat; the cholesterol content is 33 milligrams per cup for whole milk or 4 milligrams for fat-free milk. Thus choosing fat-free in place of whole milk reduces your intake of cholesterol as well as of saturated fat.

Note that cream and butter do not appear in the milk group. Milk and yogurt are rich in calcium and protein, but cream and butter are not. Cream and butter are fats, as are whipped cream, sour cream, and cream cheese. That is why the food group that includes milk is carefully labeled the "milk, yogurt, and cheese group," not the "dairy group."

Fats in the Meat, Poultry, Fish, Dry Beans and Nuts, and Eggs Group The fats in meats and eggs are about half saturated. The fats in poultry, fish, and nuts are more unsaturated than saturated—a healthier balance. Eating fish instead of meat two or three times a week supports heart health. Fish is not only leaner than most other animal-protein sources, but it is a source of omega-3 fatty acids as

Table 3–4   The Functions of Fats in Foods
Fats in foods:
■ Contribute flavor and aroma.
■ Influence the texture, adding creaminess, smoothness, moistness, or crispness.
■ Help make foods tender.
■ Carry fat-soluble vitamins.
■ Provide essential fatty acids.

Remember that fat is a more concentrated energy source than the other energy nutrients: 1 g carbohydrate or protein = 4 kcal, but 1 g fat = 9 kcal.

Remember that an ounce of meat is not an ounce of protein. An ounce (30 g) of lean meat contains 7 g protein and 3 g fat. The other 20 g are largely water with associated vitamins and minerals.

1 c whole milk:
- *8 g fat.*
- *5 g saturated fat.*
- *33 mg cholesterol.*

1 c reduced-fat milk:
- *5 g fat.*
- *3 g saturated fat.*
- *18 mg cholesterol.*

1 c low-fat milk:
- *3 g fat.*
- *1.5 g saturated fat.*
- *10 mg cholesterol.*

1 c fat-free milk:
- *<1 g fat.*
- *<1 g saturated fat.*
- *4 mg cholesterol.*

Pork chop with a half-inch of fat (275 kcal and 19 g fat).

Pork chop with fat trimmed off (165 kcal and 8 g fat).

Potato with 1 tbs butter and 1 tbs sour cream (350 kcal and 14 g fat).

Plain potato (220 kcal and <1 g fat).

Whole milk, 1 c (150 kcal and 8 g fat).

Fat-free milk, 1 c (90 kcal and <1 g fat).

Figure 3–5
Food Fat and kCalories

At room temperature, unsaturated fats (such as those found in oil), are usually liquid, whereas saturated fats (such as those found in butter) are solid.

well. As noted earlier, diets rich in fish oils can lower blood cholesterol, just as diets low in fat and saturated fat can.

Fats in the Bread, Cereal, Rice, and Pasta Group   Grains and cereals in their natural state are very low in fat, but fat may be added during processing or cooking. The fat in these foods can be particularly hard to detect, so people must remember which foods stand out as being high in fat. Notable are granola, croissants, biscuits, cornbread, dinner rolls, quick breads, snack and party crackers, muffins, pancakes, and waffles. Packaged breakfast bars often resemble candy bars in their fat and sugar contents.

Cutting Fat Intake and Choosing Unsaturated Fats

Knowing which foods contain the most fat is the first step toward meeting the recommendation to reduce dietary fat in general and saturated fat in particular. As a general rule, a person who eats meat and wishes to reduce both saturated fat and cholesterol intake can eat fewer high-fat meats and dairy foods, fewer eggs, and more poultry (without the skin), fish, and fat-free dairy products. A vegetarian who eats dairy products and eggs can shift to fat-free milk and low-fat cheeses and limit butter and egg intake. Vegetarians who omit animal-derived foods generally eat less saturated fat and consume no cholesterol because plant foods do not contain cholesterol. The "How To" box on p. 67 offers strategies for lowering fat, food group by food group.

Fats and kCalories   Removing fat from food also removes energy as Figure 3–5 shows. A small pork chop with the fat trimmed to within a half-inch of the lean provides 275 kcalories; with the fat trimmed off completely, it supplies 165 kcalories. A baked potato with butter and sour cream (1 tablespoon each) has 350 kcalories; a plain baked potato has 220 kcalories. The single most effective step you can take to reduce the energy value of a food is to eat it with less fat.

Choosing Unsaturated Fats   When a person does eat fats, those to choose are the unsaturated ones. Remember, the softer a fat is, the more unsaturated it is. Generally speaking, vegetable and fish oils are rich in polyunsaturates, olive oil and canola oil are rich in monounsaturates, and the harder fats—animal fats—are more saturated (see Figure 3–6 on page 68).

Don't Overdo Fat Restriction   Some people actually manage to eat too *little* fat—to their detriment. Among them are young women and men with eating disorders, described in Nutrition in Practice 7. As a practical guideline, it is wise to include the equivalent of at least a teaspoon of fat in every meal.

Cautions   If your clients wish to make choices consistent with current recommendations, they need to learn how to read food labels, limit fat in general, and seek out the polyunsaturated and monounsaturated fats in preference to the saturated ones. They must be wary, however: vegetable fat or vegetable oil doesn't always mean unsaturated fat. Both coconut oil and palm oil, for example, which are often used in nondairy creamers, are saturated fats, and both raise blood cholesterol.

Chapters 2 and 3 have looked briefly at the two major energy fuels in the body—carbohydrate and fat. When used for energy, each has desirable characteristics. The glucose derived from carbohydrate is needed by the brain and nerve tissues and is easily used for energy in other cells. Fat is a particularly useful fuel because the body stores it efficiently and in generous amounts. Chapter 4 looks at protein, a nutrient that can be used as fuel, but whose primary role is to provide machinery for getting things done.

How to

Lower Fat Intake—by Food Group

*F*ats can sneak into the diet in every food group. Inspect each group to discover where the fats are, and control them as follows.

MEAT, FISH, AND POULTRY

- Choose fish, poultry, or lean cuts of pork or beef; look for cuts named *round* or *loin* (eye of round, top round, round tip, tenderloin, sirloin, and top loin).
- Trim the fat from pork and beef; remove the skin from poultry.
- Grill, roast, broil, bake, stir-fry, stew, or braise meats. (Don't fry.) When possible, place meat on a rack while cooking so that fat can drain.
- Use lean ground turkey instead of hamburger in recipes.
- Brown ground meats without added fat; then drain off fat.
- Refrigerate meat pan drippings and broth; when the broth solidifies, remove the fat and use the defatted broth in recipes.
- Select tuna packed in water; rinse oil-packed tuna with hot water to remove much of the fat.
- Fill kabob skewers with lots of vegetables and slivers of meat; create main dishes and casseroles by combining a little meat, fish, or poultry with a lot of pasta, rice, or vegetables.
- Make meatless spaghetti sauces and casseroles.
- Eat a meatless meal or two daily (use the milk and other food groups with care as suggested next).

MILK AND CHEESES

- Drink fat-free, low-fat, and reduced-fat milk instead of whole milk.
- Use fat-free, low-fat, and reduced-fat cheeses (such as part-skim ricotta and low-fat mozzarella) instead of regular cheeses.
- Use fat-free yogurt or sour cream instead of regular sour cream.
- Use evaporated fat-free milk instead of cream.
- Enjoy fat-free frozen yogurt, sherbet, or ice milk instead of ice cream.

FRUITS AND VEGETABLES

- Use butter-flavored granules instead of butter or margarine on vegetables.

- Use fat-free yogurt or fat-free salad dressing instead of sour cream, cheese, mayonnaise, or other sauces on vegetables and in casseroles.
- Select fat-free or low-fat mayonnaise or salad dressings, or use herbs, lemon juice, and spices instead of regular salad dressing.
- Add a little water to thick, bottled salad dressings to dilute the amount of fat each serving provides.
- Eat at least two vegetables (in addition to a salad) with dinner.
- Snack on raw vegetables or fruits instead of high-fat foods like potato chips.
- Enjoy fruit for dessert.

BREADS AND CEREALS

- Use fruit butters or jellies instead of butter or margarine on bread.
- Select breads, cereals, and crackers that are low in fat (for example, bagels instead of croissants).

OTHER FOODS AND COOKING TIPS

- Use a nonstick pan or coat the pan lightly with cooking oil.
- Use egg substitutes in recipes instead of whole eggs or use 2 egg whites in place of each whole egg.
- Use half the margarine, butter, or oil called for in a recipe. (The minimum amount of fat for muffins, quick breads, and biscuits is 1 to 2 tablespoons per cup of flour; for cakes and cookies, 2 tablespoons per cup.)
- Select whipped types of butter, margarine, or cream cheese for use at the table; they contain half the kcalories of the regular types.
- Use wine, lemon juice, or broth instead of butter or margarine when cooking.
- Stir-fry in a small amount of oil; add moisture and flavor with broth, tomato juice, or wine.
- Use variety to enhance enjoyment of the meal: vary colors, textures, and temperatures—hot cooked versus cool raw foods—and use garnishes to complement the food.

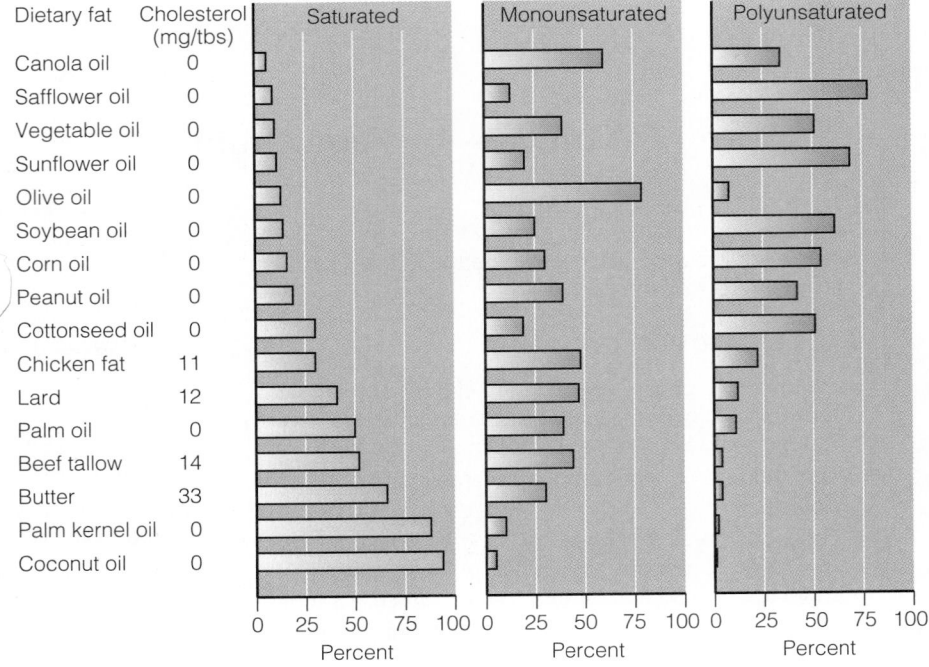

Figure 3–6

Comparison of Dietary Fats

Most fats are mixtures of saturated, monounsaturated, and polyunsaturated fatty acids.

Dietary fat	Cholesterol (mg/tbs)
Canola oil	0
Safflower oil	0
Vegetable oil	0
Sunflower oil	0
Olive oil	0
Soybean oil	0
Corn oil	0
Peanut oil	0
Cottonseed oil	0
Chicken fat	11
Lard	12
Palm oil	0
Beef tallow	14
Butter	33
Palm kernel oil	0
Coconut oil	0

Self Study

LIPIDS

Fats give foods their flavor, texture, and palatability. Unfortunately, these same characteristics entice people to eat too much from time to time. Do you know how to select low-fat foods that will help you meet dietary fat recommendations? Consider the following examples of foods and think about how often you select the item that is lower in fat. In each of these pairs, which food are you more likely to select?

- Peanuts or pretzels?
- Hot dog or turkey sandwich?
- Whole milk or low-fat milk?
- Fried chicken or baked chicken?

- Tuna packed in oil or tuna packed in water?
- Spaghetti with alfredo sauce or with marinara sauce?
- Croissants or bagels?
- Sausage pizza or mushroom pizza?

The second item in each pair is lower in fat, and making such fat-free or low-fat food choices regularly can help you meet dietary fat recommendations. In addition, eating plenty of whole-grain products, fresh vegetables, legumes, and fruits daily will help to keep your fat intake under control.

Self Check

1. Three classes of lipids in the body are:
   a. triglycerides, fatty acids, and cholesterol.
   b. triglycerides, phospholipids, and sterols.
   c. fatty acids, phospholipids, and cholesterol.
   d. glycerol, fatty acids, and triglycerides.

2. Fat in the body cannot:
   a. provide energy.
   b. insulate the body against extreme temperatures.
   c. form a part of cell membranes.
   d. make glucose.

3. A fatty acid that has the maximum possible number of hydrogen atoms is known as a(n):
   a. saturated fatty acid.
   b. monounsaturated fatty acid.
   c. PUFA.
   d. essential fatty acid.

4. Essential fatty acids:
   a. are used to make substances that regulate blood pressure, among other functions.
   b. can be made from carbohydrates.
   c. include lecithin and cholesterol.
   d. cannot be found in commonly-eaten foods.

5. To include omega-3 fatty acids in the diet, the American Heart Association recommends eating:
   a. cholesterol-free margarine.
   b. fish oil supplements.
   c. hydrogenated margarine.
   d. at least two fish meals per week.

6. Lecithins and other phospholipids in the body function as:
   a. emulsifiers.
   b. enzymes.
   c. temperature regulators.
   d. shock absorbers.

7. Excess dietary fat contributes to:
   a. heart disease, obesity, diabetes, and cancer.
   b. liver disease, sickle-cell anemia, fatty liver, and ulcerative colitis.
   c. iron-deficiency anemia, Wilson's disease, and fatty liver.
   d. food allergies, hyperglycemia, hepatitis, and ulcerative colitis.

8. Some examples of foods with hidden fats are:
   a. cheese, lettuce, and fruit juices.
   b. peanut butter, cheese, and lunch meats.

   c. fish, rice, and potatoes.
   d. baked potatoes, vegetables, and fruits.

9. Generally speaking, vegetable and fish oils are rich in:
   a. polyunsaturated fat.
   b. saturated fat.
   c. cholesterol.
   d. *trans*-fatty acids.

10. Two ways to lower fat intake are to:
    a. eat no red meat or bread.
    b. fry meat for a shorter time so that it absorbs less fat and substitute sour cream for butter.
    c. look for cuts of meat named round or loin and remove the skin from poultry.
    d. use nuts instead of meat and drink no milk.

Critical Thinking

1. A person consuming 2200 kcalories a day who wants to meet the American Heart Association recommendations should limit daily fat intake to:
   a. 25 grams or less.
   b. 73 grams or less.
   c. 98 grams or less.
   d. 123 grams or less.

2. The difference between *cis-* and *trans*-fatty acids is:
   a. the number of double bonds.
   b. the length of their carbon chains.
   c. the location of the first double bond.
   d. the configuration around the double bond.

Answers to these questions can be found in Appendix G.

Clinical Applications

The connection between the overconsumption of fats and chronic diseases (obesity, diabetes, cancer, and cardiovascular disease) emphasizes how important it is to be alert to a client's fat intake.

1. What advice would you offer a client who reports the following:

   ▪ Eats 2 or more 6-ounce servings of meat each day.

   ▪ Drinks whole milk and eats regular cheddar cheese each day.

   ▪ Eats 4 to 5 servings of breads and cereals each day, including a bagel with cream cheese for breakfast, a bologna sandwich

on white bread for lunch, and biscuits or cornbread with butter to accompany dinner.

   ▪ Eats one serving of fruit each day and eats vegetables only on occasion.

2. Make a list of low-fat and fat-free foods, beverages, and seasonings that your client can substitute in place of higher-fat items.

Nutrition on the Net

For further study of the topics of this chapter, access these websites.

Find updates and quick links to these and other nutrition-related sites at our website: **www.wadsworth.com/nutrition**

Search for cholesterol and dietary fat at the U.S. Government health information site: **www.healthfinder.gov**

Review the American Dietetic Association's *ABC's of Fats, Oils, and Cholesterol:* **www.eatright.org/nfs2.html**

Search for fat at the International Food Information Council site: **ificinfo.health.org**

Notes

1. R. L. Bradley, K. A. Cleveland, and B. Cheatham, The adipocyte as a secretory organ: Mechanisms of vesicle transport and secretory pathways, *Recent Progress in Hormone Research* 56 (2001): 329–358; E. Faloia and coauthors, Adipose tissue as an endocrine organ? A review of some recent data, *Eating and Weight Disorders* 5 (2000): 116–123.

2. D. Hwang, Fatty acids and immune responses—A new perspective in searching for clues to mechanism, *Annual Review of Nutrition* 20 (2000): 431–456; W. E. Connor, α-Linolenic acid in health and disease, *American Journal of Clinical Nutrition* 69 (1999): 827–828; E. J. Schaefer, Effects of dietary fatty acids on lipoproteins and cardiovascular disease risk: Summary, *American Journal of Clinical Nutrition* 65 (1997): 1655S–1656S.

3. D. L. O'Connor and coauthors, Growth and development in preterm infants fed long-chain polyunsaturated fatty acids: A prospective, randomized controlled trial, *Pediatrics* 108 (2001): 359–371; R. Uauy and coauthors, Role of essential fatty acids in the function of the developing nervous system, *Lipids* 31 (1996): S167–S176.

4. J. P. SanGiovanni and coauthors, Meta-analysis of dietary essential fatty acids and long-chain polyunsaturated fatty acids as they relate to visual resolution acuity in healthy preterm infants, *Pediatrics* 105 (2000): 1292–1298; Uauy and coauthors, 1996.

5. H. Iso and coauthors, Intake of fish and omega-3 fatty acids and risk of stroke in women, *Journal of the American Medical Association* 285 (2001): 304–312; D. S. Siscovick and coauthors, Dietary intake of long-chain n-3 polyunsaturated fatty acids and the risk of primary cardiac arrest, *American Journal of Clinical Nutrition* 71 (2000): 208S–212S; W. E. Connor, Importance of n-3 fatty acids in health and disease, *American Journal of Clinical Nutrition* 71 (2000): 171S–175S; N. F. Sheard, Fish consumption and risk of sudden cardiac death, *Nutrition Reviews* 56 (1998): 177–179; A. P. Simopoulos, Epidemiological aspects of omega-3 fatty acids in disease states, in *Handbook of Lipids in Human Nutrition,* ed. G. A. Spiller (Boca Raton, Fla.: CRC Press, 1996), pp. 75–89.

6. A. P. Simopoulos, Omega-3 fatty acids, part I: Metabolic effects of omega-3 fatty acids and essentiality, in *Handbook of Lipids in Human Nutrition,* ed. G. A. Spiller (Boca Raton, Fla.: CRC Press, 1996), pp. 51–73.

7. Fish oil, American Heart Association Recommendation, 2000, available at **www.americanheart.org**;

8. Siscovick and coauthors, 2000.

9. P. M. Kris-Etherton and S. Yu, Individual fatty acid effects on plasma lipids and lipoproteins: Human studies, *American Journal of Clinical Nutrition* 65 (1997): 1628S–1644S; R. McPherson and G. A. Spiller, Effects of dietary fatty acids and cholesterol on cardiovascular disease risk factors in man, in *Handbook of Lipids in Human Nutrition,* ed. G. A. Spiller (Boca Raton, Fla.: CRC Press, 1996), pp. 41–49.

10. M. Law, Dietary fat and adult diseases and the implications for childhood nutrition: An epidemiologic approach, *American Journal of Clinical Nutrition* 72 (2000): 1291S–1296S; S. M. Grundy, The optimal ratio fat-to-carbohydrate in the diet, *Annual Review of Nutrition* 19 (1999): 325–341; F. B. Hu and coauthors, Dietary fat intakes and the risk of coronary heart disease in women, *New England Journal of Medicine* 337 (1997): 1491–1499.

11. S. Yu-Poth and coauthors, Effects of the National Cholesterol Education Program's Step I and Step II dietary intervention programs on cardiovascular disease risk factors: A meta-analysis, *American Journal of Clinical Nutrition* 69 (1999): 632–646.

12. M. T. Montoya and coauthors, Fatty acid saturation of the diet and plasma lipid concentrations, lipoprotein particle concentrations, and cholesterol efflux capacity, *American Journal of Clinical Nutrition* 75 (2002): 484–491.

13. R. Krauss and coauthors, AHA Dietary Guidelines, Revision 2000: A statement for healthcare professionals from the Nutrition Committee of the American Heart Association, *Circulation,* published online on October 5, 2000, **http://circ.ahajournals.org/cgi/content/full/4304635102**; G. J. Nelson, Dietary fat, *trans* fatty acids, and risk of coronary heart disease, *Nutrition Reviews* 56 (1998): 250–252; Hu and coauthors, 1997; A. Ascherio and W. C. Willett, Health effects of *trans* fatty acids, *American Journal of Clinical Nutrition* 66 (1997): 1006S–1010S.

14. M. A. Hallikainen and M. I. Uusitupa, Effects of 2 low-fat stanol ester–containing margarines on serum cholesterol concentrations as part of a low-fat diet in hypercholesterolemic subjects, *American Journal of Clinical Nutrition* 69 (1999): 403–410.

15. S. L. Elias and S. M. Innis, Bakery foods are the major dietary source of *trans*-fatty acids among pregnant women with diets providing 30 percent energy from fats, *Journal of the American Dietetic Association* 102 (2002): 46–51; D. B. Allison and coauthors, Estimated intakes of *trans* fatty and other fatty acids in the US population, *Journal of the American Dietetic Association* 99 (1999): 166–174; A. P. Simopoulos, Trans fatty acids, in *Handbook of Lipids in Human Nutrition,* ed. G. A. Spiller (Boca Raton, Fla.: CRC Press, 1996), pp. 91–99.

16. A. H. Lichtenstein and coauthors, Effects of different forms of dietary hydrogenated fats on serum lipoprotein cholesterol levels, *New England Journal of Medicine* 340 (1999): 1933–1940; Hu and coauthors, 1997; M. B. Katan, High-oil compared with low-fat, high-carbohydrate diets in the prevention of ischemic heart disease, *American Journal of Clinical Nutrition* 66 (1997): 974S–979S.

17. R. P. Lauber and N. F. Sheard, The American Heart Association Dietary Guidelines for 2000: A summary report, *Nutrition Reviews* 59 (2001): 298–306.

18. Lauber and Sheard, 2001.

19. Federation of American Societies for Experimental Biology, Executive summary from the third report on nutrition monitoring in the United States, *Journal of Nutrition* 126 (1996): 1907S–1936S.

20. A. Drewnowski, Why do we like fat? *Journal of the American Dietetic Association* 97 (1997): S58–S62.

Nutrition in Practice

▪ FAT REPLACERS ▪

Today consumers can choose from thousands of reduced-fat products on grocery shelves. Fat-free bakery products, cheeses, frozen desserts, and many other products are available that taste rich but offer less than half a gram of fat in a serving. Some of these products use traditional ingredients such as sugar, starch, fat-free milk, soluble fiber, or egg whites in place of fat whereas others are using specially designed **fat replacers**.[1]

What kinds of fat replacers are used in food today?

Food chemists have been working for decades on ways to reduce the fat in foods. Because the functions of fat in food are both diverse and desirable, it is not an easy ingredient to replace. In general, fat replacers fall into three categories: carbohydrate based (Oatrim and Z-Trim), protein based (Simplesse), and fat based (Salatrim and olestra).[2] The glossary and Table NP3–1 on the next page describe the various fat replacers.

Please tell me about the carbohydrate-based fat replacers.

Researchers at the U.S. Department of Agriculture (USDA) developed Oatrim, which is derived from oat fiber. Oatrim can be used as a fat substitute to reduce the energy content of foods such as frozen desserts, salad dressings, soups, and high-fiber baked goods by half. Oatrim replaces the fat in food by way of a chemical process that combines the starch and fiber in the oat flour to produce a shortening-like gel or a powder that gives the taste and texture of fat to foods.[3] Oatrim is stable when heated, but it cannot be used for frying. Unlike other fat substitutes, Oatrim lowers cholesterol not only by replacing saturated fat, but also by providing fiber.

Z-Trim is made from the seed hulls of oats, peas, soybeans, or rice or the bran from corn or wheat. Z-Trim provides no kcalories and imparts no flavor to foods. It can be used in baked goods, cheese, and meat products.[4]

What can you tell me about the protein-based fat replacer Simplesse? I think I've eaten ice cream made with it.

The Food and Drug Administration (FDA) declared Simplesse safe for use in ice cream and frozen desserts in 1990. Simplesse is made from protein—either egg white or milk—which is processed into mistlike particles similar in consistency to fat. Because the components of Simplesse have long been used in foods, safety studies were not required. This fat substitute mimics the rich taste and texture of fat but cuts the kcalorie content by up to 80 percent. Because proteins coagulate at high temperatures, Simplesse cannot be used for frying or cooking, but it can be used in ice creams, yogurts, salad dressings, mayonnaise, and butter.

How do fat-based replacers differ from fat?

Fat-based replacers can be categorized into modified and synthetic. Modified fat-based replacers are triglycerides that have been altered to contain specific mixtures of fatty acids or specific arrangements of fatty acids. Salatrim is an example of a modified fat-based replacer. **Olestra** is the only synthetic fat-based replacer approved so far. Olestra is synthetic because its chemical configuration does not occur in nature.

Olestra, which was formerly known as sucrose polyester, is a synthetic combination of sucrose and

Glossary

fat replacers: ingredients that replace some or all of the functions of fat and may or may not provide energy. In this text, the term *fat replacer* is used interchangeably with **fat substitute,** which technically applies only to an ingredient that replaces all of the functions of fat and provides no energy.

olestra: a synthetic fat made from sucrose and fatty acids that provides 0 kcalories per gram; formerly known as **sucrose polyester.**

Nutrition in Practice

Table NP3–1 A Sampling of Fat Replacers

For comparison, remember that fat has 9 kcalories per gram.

Fat Replacers	Energy (kcal/g)	Properties	Uses in Foods
Carbohydrate-Based Fat Replacers			
Fruits purees and pastes of apples, bananas, cherries, plums, or prunes; add bulk and tenderness to baked goods.	1–4	Replace bulk of fat; add moisture and tenderness.	Baked goods, candy, dairy products.
Gels derived from cellulose or starch to mimic the texture of fats in regular margarine and other products.	0–4[a]	Replace bulk; lend thickness.	Fat-free margarines, salad dressing, frozen desserts.
Gums extracted from beans, sea vegetables, or other sources.	0–4	Add bulk; thicken salad dressings.	Salad dressing, processed meats, desserts.
Maltodextrins made from corn; powdered and flavored to resemble the taste of butter.	1–4	Add "buttery" flavor.	Butter-flavored "sprinkles" for melting on hot foods.
Oatrim derived from oat fiber; has the added advantage of providing satiety.	4	Creamy, replaces bulk of fat; can be used in baking but not frying.	Dips, dressings, baked goods.
Z-Trim a modified form of insoluble fiber; is powdered and feels like fat in the mouth.	0	Creamy, replaces bulk of fat; can be used in baking but not frying.	Cheese, ground beef, chocolates, baked goods.
Fat-Based Fat Replacers			
Olestra[b] a noncaloric artificial fat made from sucrose and fatty acids; formerly called *sucrose polyester*.	0	Same properties as fats; heat stable in frying, cooking, and baking.	Potato chips, tortilla chips, crackers.
Salatrim[c] derived from fat and contains short- and long-chain fatty acids.	5	Same properties as fats; can be used in baking but not frying.	Chocolate coatings, dairy products, spreads.
Protein-Based Fat Replacers			
Microparticulated protein[d] processed from the proteins of milk or egg white into mistlike particles that roll over the tongue, making it feel and taste like fat.	4	Creamy; heat stable in some cooking and baking but not frying.	Ice cream, dairy products, mayonnaise, salad dressing, baked goods, spreads.

[a]Energy made available by action of colonic bacteria.
[b]Trade name: Olean.
[c]Trade name: Benefat.
[d]Trade names: Simplesse and K-Blazer.

fatty acids that looks, feels, and tastes like food fat. Unlike sucrose or fatty acids, though, olestra is indigestible; the body has no way to digest it. Olestra can therefore be substituted for fats without adding kcalories or raising a person's blood lipids.

From some points of view, olestra is the most successful of the fat replacers. Its properties are identical to those of fats and oils when used in frying, cooking, and baking. It can be heated to frying temperatures without breaking down; it performs all of the functions of fat in cakes, pie crusts, and other baked goods; and

most remarkably—aside from a slight aftertaste—it tastes like fat. Before jumping for joy, however, the attributes of olestra must be weighed against evidence concerning its safety.

Didn't the FDA require evidence of olestra's safety before approving it?

More than two decades of research have revealed that olestra is safe in most regards, but when consumed in large quantities, it causes digestive distress, nutrient losses, and losses of phytochemicals.

Nutrition in Practice

Pros of Olestra

- Zero kcalories
- Zero fat and saturated fat
- Zero cholesterol
- Withstands frying
- Withstands baking
- Tastes like fat

Cons of Olestra

- Vitamin losses
- Phytochemical losses
- Possible digestive upset
- Possible anal leakage
- Slight aftertaste
- Expensive

Figure NP3–1
Olestra's Pros and Cons

The presence of olestra in the large intestine causes diarrhea, gas, cramping, and an urgent need for defecation. Oily olestra can creep through the feces and leak uncontrollably from the anus. No one yet knows who is most likely to encounter these effects, but some of these symptoms almost always occur when olestra is consumed in large quantities. The FDA decided that these digestive problems were unpleasant, but did not constitute a safety problem. When researchers recently provided olestra-containing snacks or ordinary snacks to 3000 volunteers, they found no significant increase in digestive distress with olestra.[5]

Olestra in the digestive tract dissolves fat-soluble substances in foods. Consequently, the absorption of the fat-soluble vitamins (vitamins A, D, E, and K) is greatly reduced when olestra is present in a meal. Thus, when olestra-containing chips are eaten with other foods, the olestra traps the vitamins those foods contain, robbing the eater of those vitamins. To compensate for this effect, olestra is fortified with vitamins A, D, E, and K. The FDA ruled that fortification removes the threat of harm from malnutrition that olestra could otherwise cause. The FDA requires olestra-containing foods to bear this warning: **"This Product Contains Olestra.** Olestra may cause abdominal cramping and loose stools. Olestra inhibits the absorption of some vitamins and other nutrients. Vitamins A, D, E, and K have been added." Olestra's pros and cons are summed up in Figure NP3–1.

Do Fat Substitutes Work?

People hope, of course, that fat replacers will help them fight both obesity and heart disease by lowering fat intakes. People who choose foods in which fat content is reduced, but not necessarily foods made with fat replacers, generally consume fewer kcalories, less fat, and more nutrients than nonusers of such foods.[6] Whether eating fat replacers will assist in weight control, however, is an open question. As for blood lipids, people who were "heavy users" of olestra, that is, they consumed more than 2 grams each day for a year, were found to have significantly lowered their blood cholesterol compared with their values before consuming olestra, but their weights did not change significantly.[7]

Some experts are concerned that people may feel at liberty to eat *more* high-fat foods by rationalizing that they obtain fat "credit" when they eat foods containing fat substitutes. Indeed, this seems to be the case with artificial sweeteners. Although consumption of sugar substitutes has risen since their introduction, sugar consumption has risen as well. Another concern is that people may become so carried away with eating foods containing fat substitutes that they will neglect to eat more nutrient-dense foods such as fresh fruits and vegetables. It seems that with fat substitutes, as with most things in life, moderation is the key to appropriate use.

Notes

1. Position of The American Dietetic Association: Fat replacers, *Journal of the American Dietetic Association* 98 (1998): 463–468.

2. N. I. Hahn, Replacing fat with food technology, *Journal of the American Dietetic Association* 97 (1997): 15–16.

3. G. F. Inglett, New grain products and their beneficial components, *Nutrition Today* 36 (2001): 66–68.

4. Hahn, 1997.

5. R. S. Sandler and coauthors, Gastrointestinal symptoms in 3181 volunteers ingesting snack foods containing olestra or triglycerides: A 6-week randomized, placebo-controlled trial, *Annals of Internal Medicine* 130 (1999): 253–261.

6. E. Kennedy and S. Bowman, Assessment of the effect of fat-modified foods on diet quality in adults, 19 to 50 years using data from the Continuing Survey of Food Intake by Individuals, *Journal of the American Dietetic Association* 101 (2001): 455–460.

7. R. E. Patterson and coauthors, Changes in diet, weight, and serum lipid levels associated with olestra consumption, *Archives of Internal Medicine* 160 (2000): 2600–2604.

CHAPTER 4 | # PROTEINS AND AMINO ACIDS

*P*eople think of proteins as bodybuilding nutrients, the material of strong muscles, and rightly so. No new living tissue can be built without them. Some proteins form structures such as muscle, bone, skin, and other tissues. Other proteins do the cells' work. The energy to fuel that work comes from carbohydrates and fats.

The Chemist's View of Proteins

Proteins are chemical compounds that contain the same atoms as carbohydrates and lipids—carbon (C), hydrogen (H), and oxygen (O)—but proteins are different in that they also contain nitrogen (N) atoms. These nitrogen atoms give the name *amino* (nitrogen containing) to the amino acids that form the links in the chains we call proteins.

The Structure of Proteins

About 20 different **amino acids** may appear in proteins.* All amino acids share a common chemical "backbone," and it is these backbones that are linked together to form proteins. Each amino acid also carries a side group, which varies from one amino acid to another (see Figure 4–1). The side group makes the amino acids differ in size, shape, and electrical charge. The side groups on amino acids are what make proteins so varied in comparison with either carbohydrates or lipids.

Protein Chains   The 20 amino acids can be linked end-to-end in a virtually infinite variety of sequences to form proteins. When two amino acids bond together, the resulting structure is known as a **dipeptide.** Three amino acids bonded together form a **tripeptide.** As additional amino acids join the chain, the structure becomes a **polypeptide.** Most proteins are polypeptides 100 to 300 amino acids long.

Protein Shapes   Polypeptide chains twist into complex shapes. Each amino acid has special characteristics that attract it to, or repel it from, the surrounding fluids and other amino acids. Because of these interactions, polypeptide chains fold and intertwine into intricate coils (see Figure 4–2). The amino acid sequence of a protein determines the specific way the chain will fold.

Protein Functions   The dramatically different shapes of proteins enable them to perform different tasks in the body. Some, such as hemoglobin in the blood (see Figure 4–3 on p. 78), are globular in shape; some are hollow balls that can carry and store materials within them; and some, such as those that form tendons, are more than ten times as long as they are wide, forming stiff, sturdy, rodlike structures.

*Besides the 20 common amino acids, which can all be components of proteins, others occur individually (for example, ornithine).

---

proteins: compounds composed of carbon, hydrogen, oxygen, and nitrogen atoms arranged into strands of amino acids. Some amino acids also contain sulfur atoms.

amino (a-MEEN-oh) **acids:** building blocks of protein. Each has an amino group and an acid group attached to a central carbon, which also carries a distinctive side chain.

  amino = containing nitrogen

dipeptide: two amino acids bonded together.

  di = two
  peptide = amino acid

tripeptide: three amino acids bonded together.

  tri = three

polypeptide: ten or more amino acids bonded together. An intermediate strand of between four and ten amino acids is an *oligopeptide.*

  poly = many
  oligo = few

Figure 4–1

Amino Acid Structure and Examples of Amino Acids

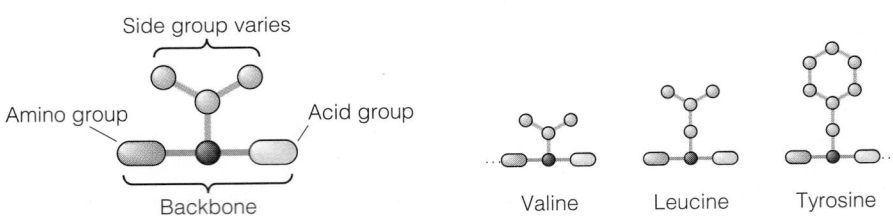

All amino acids have a "backbone" made of an amino group (which contains nitrogen) and an acid group. The side gorup varies from one amino acid to the next.

Note that the side group is a unique structure that differentiates one amino acid from another.

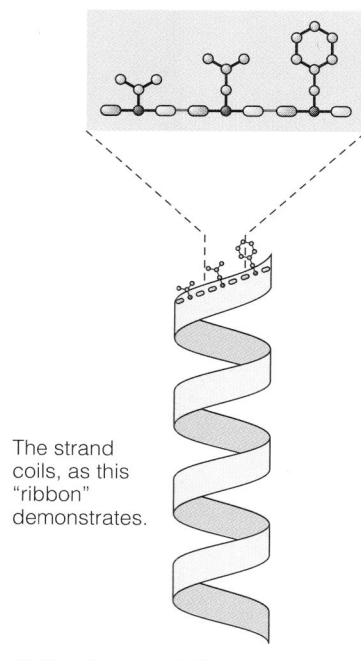
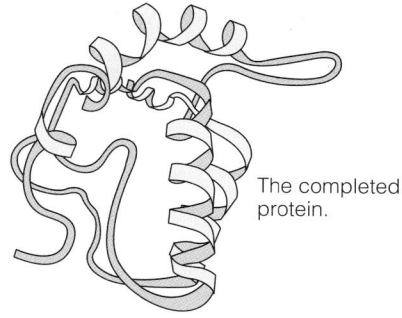

A portion of a strand of amino acids.

Figure 4–2

The Coiling and Folding of a Protein Molecule

The completed protein.

The strand coils, as this "ribbon" demonstrates.

Folding the coil. Once coiled and folded, the protein may be functional as is, or it may need to join with other proteins or add a vitamin or mineral to become active.

Coiling the strand. The strand of amino acids takes on a spring-like shape as their side groups variously attract and repel each other.

Essential Amino Acids

Proteins in foods do not provide body proteins directly, but rather supply the amino acids from which the body makes its own proteins. The body can make over half of the amino acids for itself; the proteins in foods do not need to supply these. But there are other amino acids that the body cannot make at all, and some that it cannot make fast enough to meet its needs. The proteins in foods must supply these nine amino acids to the body; they are therefore called **essential amino acids.**

Sometimes a nonessential amino acid can become essential. During illness or conditions of trauma, or in other special circumstances such as premature birth, the need for an amino acid that is normally nonessential may become greater than the body's ability to produce it. In such circumstances, that amino acid becomes a **conditionally essential amino acid.** Research suggests that glutamine, normally a nonessential amino acid, may be a conditionally essential amino acid for critically ill people.[1]

In Summary Chemically speaking, proteins are more complex than carbohydrates or lipids; proteins are made of some 20 different amino acids, 9 of which the body cannot make (they are essential). Each amino acid contains an amino group, an acid group, a hydrogen atom, and a unique side group. The distinctive sequence of amino acids in each protein determines its shape and function.

Proteins in the Body

What distinguishes you chemically from any other human being are minute differences in your particular body proteins (enzymes, antibodies, and others). These differences are determined by your proteins' amino acid sequences, which are written into the genes you inherited from your parents and ancestors. The genes direct the making of all the body's proteins.

essential amino acids: amino acids that the body cannot synthesize in amounts sufficient to meet physiological need; also called *indispensable amino acids*. Nine amino acids are known to be essential for human adults:

- *histidine (HISS-tuh-deen).*
- *isoleucine (eye-so-LOO-seen).*
- *leucine (LOO-seen).*
- *lysine (LYE-seen).*
- *methionine (meh-THIGH-oh-neen).*
- *phenylalanine (fen-il-AL-uh-neen).*
- *threonine (THREE-oh-neen).*
- *tryptophan (TRIP-toe-fane, TRIP-toe-fan).*
- *valine (VAY-leen).*

conditionally essential amino acid: an amino acid that is normally nonessential but must be supplied by the diet in special circumstances when the need for it becomes greater than the body's ability to produce it.

Figure 4-3
A Portion of a Hemoglobin Molecule
This model represents a portion of a
hemoglobin molecule magnified millions
of times.

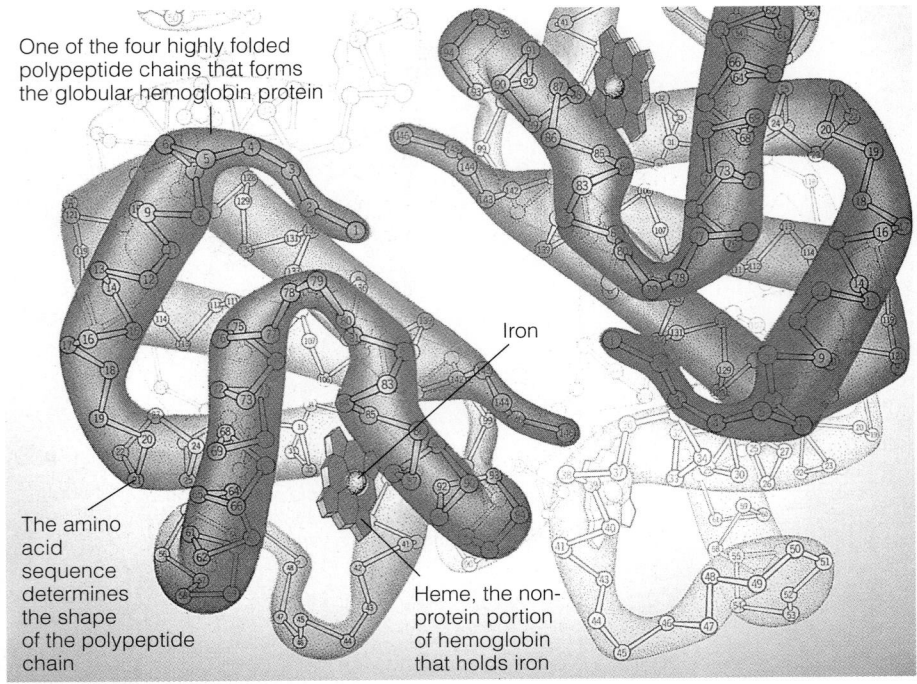

One of the four highly folded
polypeptide chains that forms
the globular hemoglobin protein

Iron

The amino
acid
sequence
determines
the shape
of the polypeptide
chain

Heme, the non-
protein portion
of hemoglobin
that holds iron

The human body contains an estimated 10,000 to 50,000 different kinds of proteins. The roles of more than 1000 of these proteins are now known. Only a few of the many roles proteins play are described here, but these should serve to illustrate proteins' versatility, uniqueness, and importance.

Enzymes **Enzymes** are catalysts that are essential to all life processes. Enzymes in the cells of plants or animals put together the pairs of sugars that make disaccharides and the long strands of sugars that make starch, cellulose, and glycogen. Enzymes also dismantle these compounds to free their constituent parts and release energy. Enzymes also assemble and disassemble lipids, assemble all other compounds that the body makes, and disassemble all compounds that the body can use for building tissue and other metabolic work. As Figure 4–4 shows, enzymes themselves are not altered by the reactions they facilitate. All enzymes are proteins, and when amino acids have to be put together to make proteins, it is enzymes that put them together, too. In other words, these proteins can even make other proteins.

The protein story moves in a circle. To follow the circle in nutrition, start with a person eating food proteins. The food proteins are broken down by the body's proteins (digestive enzymes) into amino acids. The amino acids enter the cells of the body, where proteins (enzymes) put the amino acids together in long chains whose sequences are specified by the genes. The chains fold and twist back on themselves to form proteins, and some of these proteins become enzymes themselves. Some of these enzymes break apart compounds; others put compounds together. Day by day, in billions of reactions, these processes repeat themselves, and life goes on. Only living systems can achieve such self-renewal. A toaster cannot produce another toaster; a car cannot fix a broken-down car. Only living creatures and the parts they are composed of—the cells—can duplicate and repair themselves.

Fluid and Electrolyte Balance Proteins help maintain the body's **fluid and electrolyte balance.** As Figure 4–5 shows, the body's fluids are contained in three major body compartments: (1) the spaces inside the blood vessels; (2) the spaces within the cells; and (3) the spaces between the cells (the interstitial spaces outside the blood vessels). Fluids flow back and forth between these compartments, and proteins in the fluids, together with minerals, help to maintain the needed distribution of these fluids.

Minerals are helper nutrients. The attraction of protein and mineral particles to water is due to osmotic pressure (see Chapter 9).

enzymes: protein catalysts. A catalyst is a compound that facilitates chemical reactions without itself being changed in the process.

fluid and electrolyte balance: maintenance of the necessary amounts and types of fluid and minerals in each compartment of the body fluids.

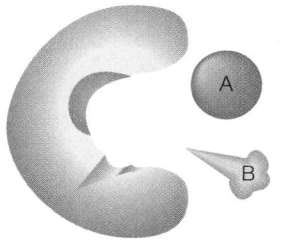 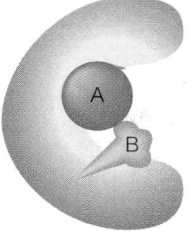 

Enzyme plus
two compunds
A and B

Enzyme
complex with
A and B

Enzyme plus
new compound
AB

Figure 4–4
Enzyme Action
Enzymes are catalysts: they speed up reactions that would happen anyway, but much more slowly. This enzyme works by positioning two compounds, A and B, so that the reaction between them will be especially likely to take place.

Compounds A and B are attracted to the enzyme's active site and park there for a moment in the exact position that makes the reaction between them most likely to occur. They react by bonding together and leave the enzyme as the new compound, AB.

A single enzyme can facilitate several hundred such synthetic reactions in a second. Other enzymes break compounds apart into two or more products or rearrange the atoms in one compound to make another one.

Proteins are able to help determine the distribution of fluids in living systems for two reasons: first, proteins cannot pass freely across the membranes that separate the body compartments, and second, they are attracted to water. A cell that "wants" a certain amount of water in its interior space cannot move the water around directly, but it can manufacture proteins, and these proteins will hold water. Thus the cell can use proteins to help regulate the distribution of water indirectly. Similarly, the body makes proteins for the blood and the interstitial (intercellular) spaces. These proteins help maintain the fluid volume in those spaces. When too much fluid collects in the interstitial spaces, **edema** results.

Not only is the quantity of the body fluids vital to life, but so is their composition. Special transport proteins in the membranes of cells continuously transfer substances into and out of cells to maintain balance. For example, sodium is concentrated outside the cells, and potassium is concentrated inside. The balance of these two electrolytes is critical to nerve transmission and muscle contraction. Any disturbance in this balance triggers a major medical emergency. Such imbalances can cause irregular heartbeats, kidney failure, muscular weakness, and even death.

Acid-Base Balance Proteins also help maintain the balance between **acids** and **bases** within the body's fluids. Normal body processes continually produce acids and bases, which must be carried by the blood to the kidneys and lungs for excretion. The blood must do this without upsetting its own **acid-base balance.** Blood **pH** is one of the most tightly controlled conditions in the body. If the blood becomes too acidic, vital proteins may undergo **denaturation,** losing their shape and ability to function. A similar situation arises when the balance tips too far toward base. These imbalances are known as **acidosis** and **alkalosis,** respectively, and both can be fatal. Figure 4–6 on the next page shows the normal and abnormal pH ranges of body fluids, as well as the pHs of some common substances.

Proteins such as albumin help to prevent acid-base imbalances. In a sense, the proteins protect one another by gathering up extra acid (hydrogen) ions when there are too many in the surrounding medium and by releasing them when there are too few. By accepting and releasing hydrogen ions, proteins act as **buffers,** maintaining the acid-base balance of the blood and body fluids.

Antibodies Other proteins in the blood—the **antibodies**—act against viruses, bacteria, and other disease agents. The antibodies work so efficiently that if a million bacterial cells are injected into the skin of a healthy person, fewer than ten are likely to survive for five hours, which explains why most diseases never have a chance to get started. Without sufficient protein, the body cannot maintain its resistance to disease.

Hormones The blood also carries messenger molecules known as **hormones,** and *some* hormones are proteins. (Recall that some hormones are sterols, members of the lipid family.) Among the proteins that act as hormones are the thyroid

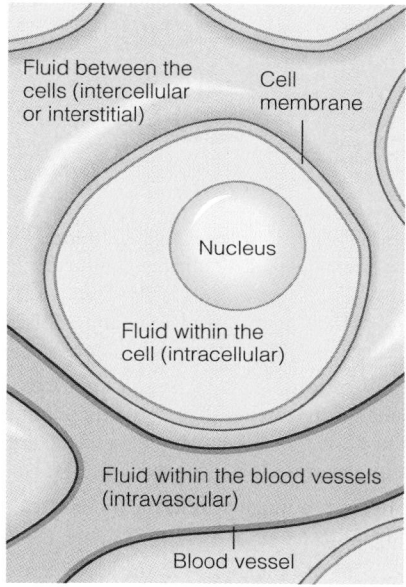

Fluid between the cells (intercellular or interstitial)

Cell membrane

Nucleus

Fluid within the cell (intracellular)

Fluid within the blood vessels (intravascular)

Blood vessel

Figure 4–5
One Cell and Its Associated Fluids

edema (eh-DEEM-uh): the swelling of body tissue caused by leakage of fluid from the blood vessels and accumulation of the fluid in the interstitial spaces.

acids: compounds that release hydrogen ions in a solution.

bases: compounds that accept hydrogen ions in a solution.

acid-base balance: the balance maintained between acid and base concentrations in the blood and body fluids.

Proteins and Amino Acids

Figure 4–6

The pH Scale

A substance's acidity or alkalinity is measured in pH units. Each step down the scale indicates a tenfold increase in the concentration of hydrogen ions. Notice how small the range of normal blood pH is.

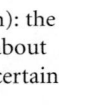

pH's of common substances:

Basic

14	Concentrated lye
13	
12	
11	Household ammonia
10	
9	Baking soda
8	Pancreatic juice / Blood
7	Water / Milk
6	Urine
5	Coffee
4	Orange juice
3	Vinegar
2	Lemon juice / Gastric juice
1	
0	Battery acid

pH neutral

Acidic

Normal and abnormal pH ranges of blood

8.00 — Death
— Alkalosis
7.45 —
— Normal
7.35 —
— Acidosis
6.8 — Death

hormones and insulin. Hormones have many profound effects, which will become evident in subsequent chapters.

Transport Proteins   Some proteins move about in the body fluids, transporting nutrients and other molecules from one organ to another. The protein hemoglobin, which carries oxygen from the lungs to the body's cells, is a prime example. The lipoproteins transport lipids around the body. In addition, special proteins also carry vitamins and minerals.

Growth, Maintenance, and Repair   The body uses amino acids to build the proteins of all its new tissues. The new tissues may be in an embryo, in a growing child, or in new hair and nails. Proteins also help replace worn-out cells in everyone's body all the time. For example, the millions of cells that line the intestinal tract live for three days; they are constantly being shed and must be replaced. The cells of the skin die and rub off, and new ones grow from underneath. The body uses amino acids to repair damaged tissues, too. The protein collagen serves as the mending material of torn tissue, forming scars to hold the separated parts together.

Both inside and outside the body, then, cells constantly make and break down their proteins. When proteins break down, their component amino acids are liberated to join the general circulation. Some of these amino acids may be promptly recycled into other proteins; others may be stripped of their nitrogen and used for energy. By reusing amino acids to build proteins, however, the body conserves and recycles a valuable commodity.

People need to eat protein-rich foods every day to replace the protein they continuously lose. If the body is growing, it needs more protein than is necessary just for maintenance. Children end each day with more blood cells, more muscle

pH: the concentration of hydrogen ions. The lower the pH, the stronger the acid. Thus pH 2 is a strong acid; pH 6 is a weak acid; pH 7 is neutral; and a pH above 7 is alkaline.

denaturation (dee-nay-cher-AY-shun): the change in a protein's shape brought about by heat, acid, or other agents. Past a certain point, denaturation is irreversible.

acidosis: too much acid in the blood and body fluids.

alkalosis: too much base in the blood and body fluids.

buffers: compounds that can reversibly combine with hydrogen ions to help keep a solution's acidity or alkalinity constant.

antibodies: large proteins of the blood and body fluids, produced in response to invasion of the body by unfamiliar molecules (mostly proteins) called *antigens.* Antibodies inactivate the invaders and so protect the body.

anti = against

hormones: chemical messengers. Hormones are secreted by a variety of glands in the body in response to altered conditions. Each travels to one or more target tissues or organs and elicits specific responses to restore normal conditions.

cells, and more skin cells than they had at the beginning of the day. So protein is needed both for routine maintenance (replacement) and for growth (addition).

Nitrogen Balance   If the body maintains the same amount of protein in its tissues from day to day, it is in **nitrogen balance.** If the body adds protein, it is in positive nitrogen balance; if it loses protein, it is in negative nitrogen balance.

Normally, healthy adults are in nitrogen balance; that is, their nitrogen intakes equal their nitrogen outputs. In other words, protein intake from food balances with nitrogen excretion in the urine, feces, and sweat. Growing children and pregnant women are in positive nitrogen balance because they are adding new blood, bone, and muscle cells to their bodies. People who are fasting or starving, such as those with anorexia nervosa, and people who suffer from traumas, such as burns (see Chapter 27), are in negative nitrogen balance because their bodies are forced to use protein for energy.

Energy   Even though amino acids are needed to do the work that only they can perform—build vital proteins—they will be sacrificed to provide energy and glucose if need be. Keeping energy and glucose available is one of the body's highest priorities: without energy, cells die; without glucose, the brain and nervous system falter. When glucose or fatty acids are limited, cells are forced to use amino acids for energy and glucose. The body does not make a specialized storage form of protein as it does for carbohydrate and fat. Glucose is stored as glycogen in the liver and muscles, fat as triglycerides in the adipose tissue, but body protein is available only as the working and structural components of the tissues. When the need arises, the body dismantles its tissue proteins and uses them for energy. Thus, over time, energy deprivation (starvation) always incurs wasting of lean body tissue as well as fat loss.

In Summary   The list of protein functions discussed here and summarized in Table 4–1 is by no means exhaustive. Nevertheless, it does give some sense of the immense variety of proteins and their importance in the body.

Protein and Health

In the short time that scientists have been studying nutrition, no nutrient has been more intensely scrutinized than protein. As you know by now, it is indispensable to life. And it should come as no surprise that protein deficiency can have devastating effects on people's health. But as with the other nutrients, protein in excess can also be harmful; the end of this section discusses the consequences of protein excess.

Protein-Energy Malnutrition

When people are deprived of food and suffer an energy deficit, they degrade their own body protein for energy and indirectly suffer a protein deficiency, as well as an energy deficiency. Because protein and energy deprivation go hand in hand, public health officials have adopted an abbreviation for the overlapping pair: **protein-energy malnutrition (PEM).** PEM often strikes early in childhood, but it endangers many adults as well. PEM is the most widespread form of malnutrition in the world today. Most of the 33,000 children who die each day are malnourished and suffer from infectious diseases.[2] PEM is prevalent in Africa, Central America, South America, the Middle East, and East Asia, but developed countries including the United States are not immune to it. PEM is common among some population groups in the United States: impoverished people living

Growing children end each day with more bone, blood, muscle, and skin cells than they had at the beginning of the day.

Nitrogen equilibrium (zero nitrogen balance): N in = N out.
Positive nitrogen balance: N in > N out.
Negative nitrogen balance: N in < N out.

nitrogen balance: the amount of nitrogen consumed (N in) as compared with the amount of nitrogen excreted (N out) in a given period of time. The laboratory scientist can estimate the protein in a sample of food, body tissue, or excreta by measuring the nitrogen in it.

protein-energy malnutrition (PEM): a deficiency of protein and food energy; the world's most widespread malnutrition problem, including both marasmus and kwashiorkor.

mal = bad, poor

Proteins and Amino Acids

Table 4-1	**Summary of Functions of Proteins**		
■ *Enzymes.* Proteins facilitate chemical reactions.		■ *Transportation.* Proteins transport substances such as lipids, minerals, and oxygen around the body.	
■ *Fluid and electrolyte balance.* Proteins help to maintain the distribution and composition of various body fluids.		■ *Growth and maintenance.* Proteins form integral parts of most body structures such as skin, tendons, ligaments, membranes, muscles, organs, and bones. As such, they support the growth and repair of body tissues.	
■ *Acid-base balance.* Proteins help maintain the acid-base balance of body fluids by acting as buffers.			
■ *Antibodies.* Proteins act against disease agents to fight diseases.		■ *Energy.* Proteins provide some fuel for the body's energy needs.	
■ *Hormones.* Proteins regulate body processes. Some, but not all, hormones are made of protein.			

PEM can be a consequence of many different conditions. PEM has been recognized in people with many chronic diseases such as cancer and AIDS and in those who have severe stresses such as burns or extensive infections (see Chapter 27). The consequences of PEM as a world malnutrition problem are considered here; the problems associated with PEM and illness are described throughout later chapters.

on U.S. Indian reservations, in inner cities, and in rural areas; many elderly people; homeless children; and those suffering from the eating disorder anorexia nervosa.

Of all population groups, children are most seriously affected by malnutrition. Children who are thin for their heights may have recently developed PEM whereas children who are short for their ages may have experienced PEM for extended periods of time. Stunted growth due to PEM is easy to overlook because a small child may look quite normal, but it may be the most common sign of malnutrition in the developing countries. Experts estimate that 226 million children are stunted, shorter than they should be for their age.[3]

Marasmus and Kwashiorkor   PEM takes two different forms, with some cases exhibiting a combination of the two. In one form, the person is shriveled and lean all over—this disease is called **marasmus.** In the second, a swollen belly and skin rash are present, and the disease is named **kwashiorkor.**[4*] In the combination, some features of each type are present. Marasmus reflects a chronic inadequate food intake and therefore inadequate energy, vitamins, and minerals, as well as too little protein. Kwashiorkor may result from severe acute malnutrition, with too little protein to support body functions.

Marasmus   Marasmus commonly occurs in children from 6 to 18 months of age in all the overpopulated urban slums of the world. Children in impoverished nations subsist on a weak cereal drink that supplies scant energy and protein of low quality: such food can barely sustain life, much less support growth. Consequently, marasmic children look like little old people—just skin and bones.

Without adequate nutrition, muscles, including the heart muscle, waste and weaken. Because the brain normally grows to almost its full adult size within the first two years of life, marasmus impairs brain development and learning ability. Reduced synthesis of key hormones leads to a metabolism so slow that body temperature drops below normal. There is little or no fat under the skin to insulate against cold. Hospital workers find that the primary need of marasmic children is to be wrapped up and kept warm.

The starving child faces this threat to life by engaging in as little activity as possible—not even crying for food. The body gathers all its forces to meet the crisis, so it cuts down on any expenditure of protein not needed for the heart, lungs, and brain to function. Growth ceases; the child is no larger at age four than at age two. Digestive enzymes are in short supply, the digestive tract lining deteriorates, and absorption fails. The child cannot assimilate what little food is eaten.

Blood proteins, including hemoglobin, are no longer synthesized. Antibodies to fight off invading bacteria are degraded to provide amino acids for other uses, rendering the child vulnerable to infection. Then **dysentery,** an infection of the digestive tract, causes diarrhea, further depleting the body of nutrients. In the marasmic child, once infection has set in, kwashiorkor often follows. The infec-

marasmus (ma-RAZZ-mus): a disease related to PEM. Marasmus results from severe deprivation, or impaired absorption, of protein, energy, vitamins, and minerals.

kwashiorkor (kwash-ee-OR-core or kwash-ee-or-CORE): a disease related to PEM.

dysentery (DIS-en-terry): an infection of the gastrointestinal tract caused by an amoeba or bacterium that gives rise to severe diarrhea.

dys = bad
entery = intestine

*A term gaining acceptance for use in place of kwashiorkor is *hypalbuminemic-type PEM.*

Table 4–2 Features of Marasmus and Kwashiorkor in Children

Separating PEM into two classifications oversimplifies the condition, but at the extremes, marasmus and kwashiorkor exhibit marked differences. Marasmus-kwashiorkor mix presents symptoms common to both marasmus and kwashiorkor. In all cases, children are likely to develop diarrhea, infections, and multiple nutrient deficiencies.

Marasmus	Kwashiorkor
Infancy (less than 2 yr)	Older infants and young children (1 to 3 yr)
Severe deprivation or impaired absorption of protein, energy, vitamins, and minerals	Inadequate protein intake or, more commonly, infections
Develops slowly; chronic PEM	Rapid onset; acute PEM
Severe weight loss	Some weight loss
Severe muscle wasting with fat loss	Some muscle wasting, with retention of some body fat
Growth: <60% weight-for-age	Growth: 60 to 80% weight-for-age
No detectable edema	Edema
No fatty liver	Enlarged, fatty liver
Anxiety, apathy	Apathy, misery, irritability, sadness
Appetite may be normal or impaired	Loss of appetite
Hair is sparse, thin, and dry; easily pulled out	Hair is dry and brittle; easily pulled out; changes color; becomes straight
Skin is dry, thin, and wrinkled	Skin develops lesions

tion that occurs with malnutrition is responsible for two-thirds of the deaths of young children in developing countries.[5]

If caught in time, a child's starvation may be reversed by careful nutrition therapy. The fluid balances are most critical. Diarrhea will have depleted the body's potassium and disturbed other electrolyte balances. Careful correction of fluid and electrolyte imbalances usually raises the blood pressure and strengthens the heart. After the first 24 to 48 hours, protein and energy may be given in small quantities, with intakes gradually increased as tolerated. Years after PEM is corrected, however, a child may still experience deficits in thinking and achievement in school compared with well-nourished peers.[6]

Kwashiorkor Kwashiorkor was originally a Ghanaian word meaning an "evil spirit that infects the first child when the second child is born." If you consider how kwashiorkor often develops, you can easily see how the Ghanaians arrived at this name for the disease. When a mother who has been nursing her first child bears a second child, she weans the first child and puts the second one on the breast. The first child, suddenly switched from nutrient-dense, protein-rich breast milk to a starchy, protein-poor gruel, soon begins to sicken and die. Kwashiorkor typically sets in at about the age of two. Though rare in the United States, kwashiorkor has recently been diagnosed in more than a dozen children fed ill-conceived vegetarian or "anti-allergy" diets or given a protein-poor "health-food" rice drink instead of cow's milk.[7]

Some symptoms of kwashiorkor resemble those of marasmus (see Table 4–2), but without severe wasting of body fat. Proteins and hormones that previously maintained fluid balance diminish, and fluid leaks into the interstitial spaces. The child's limbs and face become swollen with edema, a distinguishing feature of kwashiorkor; the belly bulges with a **fatty liver** caused by lack of the protein carriers that transport fat out of the liver. The child's hair loses its color; the skin becomes patchy and scaly, sometimes with ulcers and sores that fail to heal.

Experts assure us that we possess the knowledge, technology, and resources to end hunger. Programs that have involved the local people in the process of identifying the problem and devising its solution have met with some success. But

When two variables interact so that each increases the other, **synergism** (SIN-er-jiz-um) is said to be occurring. Malnutrition and infection are a deadly combination because they work in this way.

syn = with, together
ergism = work

fatty liver: an accumulation of fat in the liver. In PEM, fat accumulates in the liver because no protein is available to form the lipoproteins that normally escort fat molecules in the blood (see Chapter 23).

Proteins and Amino Acids

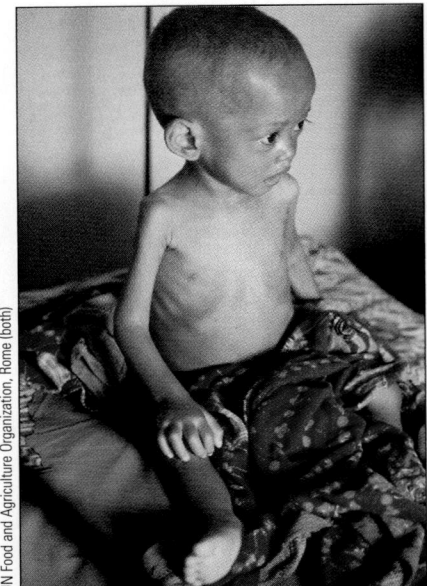

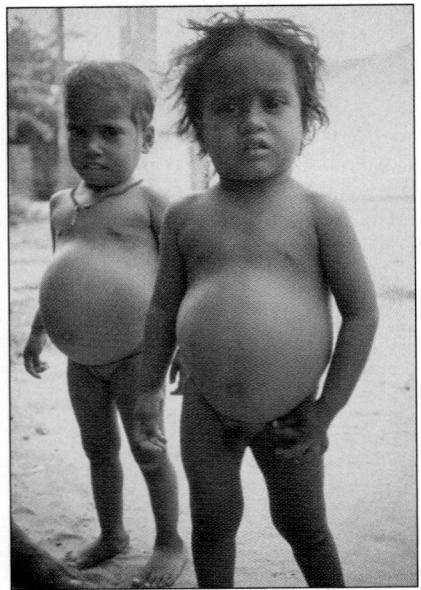

In the photo on top, the extreme loss of muscle and fat characteristic of marasmus is apparent in the child's matchstick arms and legs. In contrast, the edema and enlarged liver characteristic of kwashiorkor are apparent in the swollen bellies of the children in the photo on bottom.

Muscle work builds muscle; protein supplements do not, and athletes do not need them. Protein supplements and other ergogenic aids are discussed in Nutrition in Practice 10.

until those who have the food, technology, and resources make fighting hunger a priority, the war on hunger will not be won.

Protein Excess

While many of the world's people struggle to obtain enough food and enough protein to survive, in the developed nations protein is so abundant that problems of protein excess are seen. Overconsumption of protein offers no benefits and may pose health risks. For example, protein-rich foods are often high-fat foods that contribute to obesity with its accompanying health risks. The higher a person's intake of protein-rich foods such as meat and milk, the more likely that fruits, vegetables, and grains will be crowded out, making the diet inadequate in other nutrients.

Excretion of the end products of protein metabolism depends, in part, on an adequate fluid intake and healthy kidneys. A high protein intake increases the work of the kidneys.[8] In fact, one of the most effective ways to slow the progression of kidney disease is to restrict dietary protein.[9]

Some research suggests that diets high in protein promote calcium excretion, depleting the bones of their chief mineral. When high protein intakes are accompanied by low calcium intakes, as is typical of many women's diets in the United States, calcium losses incurred by excess protein may compromise bone health. One study of the relationship between protein intake and bone fractures in women reported a higher risk of arm fractures in women with high animal-protein intakes.[10] There are evidently no benefits to be gained by consuming a diet that derives more than 15 percent of its energy from protein.

In Summary Protein deficiencies arise from both energy-poor and protein-poor diets and lead to the devastating diseases of marasmus and kwashiorkor. Together, these diseases are known as PEM (protein-energy malnutrition), a major form of malnutrition causing death in children worldwide. Excesses of protein offer no advantage; in fact, overconsumption of protein-rich foods may incur health problems as well.

Amino Acid Supplements

In view of the high protein intakes of people in the United States and other developed nations, it is surprising that many people feel compelled to take protein and amino acid supplements. Why do people take protein or amino acid supplements? Athletes take them to build muscle. Dieters take them to spare their bodies' protein while losing weight. Popular reports that the amino acid tryptophan could relieve pain and cure depression and insomnia led to widespread public use. More than 1500 people who elected to take tryptophan developed an illness called EMS (short for *eosinophilia-myalgia syndrome*). EMS is characterized by severe muscle and joint pain, limb swelling, an elevated white blood cell count, extremely high fever, and, in at least 38 cases, death.[11] Contaminants in the supplement were determined to be the cause of the disease, and the Food and Drug Administration (FDA) issued a recall of all products containing tryptophan except for specific medical formulas.

A few years ago the FDA asked a panel of scientists from a well-known research group to review the safety of amino acid supplements. The scientists concluded that any use of amino acids as dietary supplements is inappropriate and that some (serine and proline) present a high risk of toxicity. The panel also singled out some groups of people whose growth or altered metabolism makes them especially likely to suffer harm from amino acid supplements:

- All women of childbearing age.
- Pregnant or lactating women.
- Infants, children, and adolescents.

- Elderly people.
- People with inborn errors of metabolism that affect their bodies' handling of amino acids.
- Smokers.
- People on low-protein diets.
- People with chronic or acute mental or physical illnesses who take amino acids without medical supervision.

Also, because weight lifters and bodybuilders may take frequent, massive doses of amino acids, they may suffer harm from the supplements while believing false promises of benefits. A recent review of the literature has concluded that not enough research exists to support recommending long-term consumption of amino acid supplements by healthy people.[12] Anyone considering taking amino acid supplements should check with a physician first.

Protein Recommendations

The committee that established the RDA states that a generous daily protein allowance for a healthy adult is 0.8 gram per kilogram (2.2 pounds) of healthy body weight. The protein RDA is adjusted to cover additional needs for building new tissue and so is higher for infants, children, and pregnant and lactating women.

In setting the RDA, the committee assumes that the protein eaten will be of high quality, that it will be consumed together with adequate energy from carbohydrate and fat, and that other nutrients in the diet will be adequate. The committee also assumes that the RDA will be applied only to healthy individuals with no unusual alteration of protein metabolism. Most people in this country receive much more protein than they need.

In Summary Normal, healthy people do not need amino acid or protein supplements. Optimally, the diet will be adequate in energy from carbohydrate and fat and will deliver 0.8 gram of protein per kilogram of healthy body weight each day.

1989 RDA for protein (adult) = 0.8 g/kg.

To figure your protein need:
1. Find your body weight in pounds (or kilograms).
2. Convert pounds to kilograms, if necessary (pounds divided by 2.2).
3. Multiply kilograms by 0.8 to find total grams of protein recommended.

For example:
1. Weight = 110 lb.
2. 110 lb ÷ 2.2 lb/kg = 50 kg.
3. 50 kg × 0.8 g/kg = 40 g.

Protein in Foods

To make body protein, a cell must have all the needed amino acids available simultaneously. Therefore, the first important requirement for protein in the diet of a healthy adult is that it should supply at least the 9 essential amino acids and enough nitrogen and energy for the synthesis of the other 11. A protein that fits this description is called a complete protein.

Complete Protein A **complete protein** contains all the essential amino acids in amounts adequate for human use; it may or may not contain all the others. Generally, proteins derived from animal foods (meats, fish, poultry, eggs, and milk) are complete, although gelatin is an exception. Proteins derived from plant foods (legumes, grains, and vegetables) vary more. Some plant proteins are notoriously **incomplete**—for example, corn protein. Others are complete—for example, soy protein. As discussed in Nutrition in Practice 4, the educated vegetarian can design a diet that is adequate in protein by choosing a variety of legumes, whole grains, nuts, and vegetables. Table 4–3 on the next page lists the protein contents of foods based on the food groups of the Daily Food Guide in Chapter 1. Fruits are not included in Table 4–3 because they contribute only small amounts of protein.

Protein Digestibility and Quality Ideally, a protein should be not only complete, but highly digestible as well so that sufficient numbers of amino acids

complete protein: a protein containing all the amino acids essential in human nutrition in amounts adequate for human use.

incomplete protein: a protein lacking or low in one or more of the essential amino acids.

Vegetarians obtain their protein from whole grains, legumes, nuts, vegetables, and, in some cases, eggs and milk products.

Reminder: Carbohydrate and fat allow amino acids to be used to build body proteins. This is known as the *protein-sparing effect* of carbohydrate and fat.

high-quality protein: an easily digestible, complete protein.

limiting amino acid: the essential amino acid that is present in dietary protein in the shortest supply relative to the amount needed for protein synthesis in the body.

reach the body's cells to permit them to make the proteins they need. Such a protein is called a **high-quality protein.** One of the finest proteins available by these standards is egg protein. Eggs are a highly valued protein source in developing nations where protein-rich foods are scarce.

Limiting Amino Acids Ideally, dietary protein supplies each amino acid in the amount needed for protein synthesis in the body. If one amino acid is supplied in an amount smaller than is needed, then the total amount of protein that can be synthesized will be limited. The body makes only complete proteins; if one amino acid is missing, the others cannot form a "partial" protein. An essential amino acid supplied in less than the amount needed to support protein synthesis is called a **limiting amino acid.** (By analogy, suppose that a sign maker plans to make 100 identical signs saying "LEFT TURN ONLY." The sign maker needs 200 Ls, 200 Ns, 200 Ts, and 100 of each of the other letters. If only 20 Ls are available, only 10 signs can be made, even if all the other letters are available in unlimited quantities. Suppose further that the sign maker has no place to keep leftover letters—just as the body has no storage place for extra amino acids. If the sign maker doesn't get some additional Ls right away, he will have to throw away all the other letters).

Mutual Supplementation and Complementary Proteins If the body does not receive all the essential amino acids it needs, the supply of essential amino acids will dwindle until body organs are compromised. Obtaining enough essential

Table 4–3   Protein-Containing Foods
Milk, Cheese, and Yogurt
Each of the following provides about 8 grams of protein:[a] ■ 1 c milk, buttermilk, or yogurt (choose reduced-fat, low-fat, or fat-free). ■ 1 oz regular cheese (for example, cheddar or swiss).[b] ■ ¼ c cottage cheese (choose reduced-fat, low-fat, or fat-free).
Meat, Poultry, Fish, and Alternates
Each of the following provides about 7 grams of protein: ■ 1 oz meat, poultry, or fish (choose lean meats to limit fat intake). ■ ½ c legumes (navy beans, pinto beans, black beans, lentils, soybeans, and other dried beans and peas). ■ 1 egg.[b] ■ ½ c tofu (soybean curd).[b] ■ 2 tbs peanut butter.[b] ■ 1 to 2 oz nuts or seeds.[b]
Breads, Cereals, and Other Grain Products
Each of the following provides about 3 grams of protein: ■ 1 slice of bread. ■ ½ c cooked rice, pasta, cereals, or other grain foods.
Vegetables
Each of the following provides about 2 grams of protein: ■ ½ c cooked vegetables. ■ 1 c raw vegetables.

[a]For reference, an adult might need 40 to 100 grams of protein in a day.
[b]These are medium-fat or high-fat choices.

amino acids presents no problem to people who regularly eat complete proteins, such as those of meat, fish, poultry, cheese, eggs, milk, and many soybean products. The proteins of these foods contain ample amounts of all the essential amino acids. An equally sound choice is to eat two incomplete protein foods from plants so that each supplies the amino acids missing in the other. In this strategy, called **mutual supplementation,** the two protein-rich foods are combined to yield **complementary proteins** (see Figure 4–7)—proteins containing all the essential amino acids in amounts sufficient to support health. This concept is illustrated in Figure NP4–1 on p. 92. The two proteins need not even be eaten together, so long as the day's meals supply them both, and the diet provides enough energy and total protein from a variety of sources.

Protein Sparing Dietary protein—no matter how high the quality—will not be used efficiently and will not support growth when energy from carbohydrate and fat is lacking. The body assigns top priority to meeting its energy need and, if necessary, will break down protein to meet this need. After stripping off and excreting the nitrogen from the amino acids, the body will use the remaining carbon skeletons in much the same way it uses those from glucose or fat. A major reason why people must have ample carbohydrate and fat in the diet is to prevent this wasting of protein.

In Summary A diet inadequate in any of the essential amino acids limits protein synthesis. The best guarantee of amino acid adequacy is to eat foods containing complete proteins or mixtures of foods containing incomplete but complementary proteins so that each can supply the amino acids missing in the other. Vegetarians can meet their protein needs by eating a variety of whole grains, legumes, seeds, nuts, and vegetables.

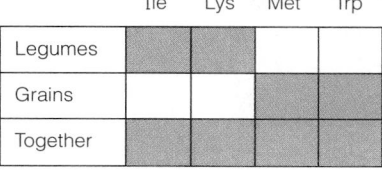

Figure 4–7

An Example of Mutual Supplementation
In general, legumes provide plenty of isoleucine (Ile) and lysine (Lys), but fall short in methionine (Met) and tryptophan (Trp). Grains have the opposite strengths and weaknesses, making them a perfect match for legumes.

mutual supplementation: the strategy of combining two incomplete protein sources so that the amino acids in one food make up for those lacking in the other food. Such protein combinations are sometimes called *complementary proteins.*

complementary proteins: two or more proteins whose amino acid assortments complement each other in such a way that the essential amino acids missing from one are supplied by the other.

Self Study

PROTEINS AND AMINO ACIDS

Most people in the United States and Canada consume more protein than they need. This is not surprising considering the abundance of food eaten and the central role meats hold in the North American diet. Using our food diary from the Self Study in Chapter 1, estimate your protein intake for the day. Multiply the number of servings you consumed by the estimated protein per serving to guesstimate your total protein intake.

The protein RDA for young adults (19 to 24 years old) is 46 grams for women and 58 grams for men. Health experts advise people to maintain moderate protein intakes—between the RDA and twice the RDA.

- Do you consume enough, but not too much, protein daily?
- How often do you select plant-based protein foods?

Food Groups	Servings Consumed	Estimated Protein	Totals
Bread, cereal, rice, and pasta	_____	3 g/serving	_____
Vegetable	_____	2 g/serving	_____
Fruit	_____	0 g/serving	_____
Milk, yogurt, and cheese	_____	8 g/serving	_____
Meat, poultry, fish, dry beans, eggs, and nuts	_____	7 g/oz	_____
Fats, oils, and sweets	_____	0 g/serving	_____
Total estimated protein intake			_____

Self Check

1. Proteins are chemically different from carbohydrates and fats because they also contain:
   a. iron.
   b. sodium.
   c. nitrogen.
   d. phosphorus.

2. The basic building blocks for protein are:
   a. side groups.
   b. amino acids.
   c. glucose units.
   d. saturated bonds.

3. Enzymes are proteins that, among other things:
   a. defend the body against disease.
   b. regulate fluid and electrolyte balance.
   c. facilitate chemical reactions by changing themselves.
   d. help assemble disaccharides into starch, cellulose, or glycogen.

4. Functions of proteins in the body include:
   a. supplying omega-3 fatty acids for growth, lowering serum cholesterol, and helping with weight contol.
   b. supplying fiber to aid digestion, digesting cellulose, and providing the main fuel source for muscles.
   c. protecting organs against shock, helping the body use carbohydrate efficiently, and providing triglycerides.

   d. supporting growth and maintenance, supplying hormones to regulate body processes, and maintaining fluid and electrolyte balance.

5. The swelling of body tissue caused by the leakage of fluid from the blood vessels into the interstitial spaces is called:
   a. edema.
   b. anemia.
   c. acidosis.
   d. sickle-cell anemia.

6. Major proteins in the blood that protect against bacteria and other disease agents are called:
   a. acids.
   b. buffers.
   c. antigens.
   d. antibodies.

7. Marasmus can be distinguished from kwashiorkor because in marasmus:
   a. only adults are victims.
   b. the cause is usually an infection.
   c. severe wasting of body fat and muscle occurs.
   d. the limbs and face swell with edema, and the belly bulges with a fatty liver.

8. The RDA for protein for a healthy adult is _____ gram(s) per kilogram of appropriate or average body weight for height.

a. 0.5
b. 0.8
c. 1.1
d. 1.4

9. Generally speaking, from which of the following foods are complete proteins derived?
   a. milk, gelatin, and soy
   b. rice, potatoes, and eggs
   c. meats, fish, and poultry
   d. vegetables, grains, and fruits

10. An incomplete protein lacks one or more:
    a. hydrogen bonds.
    b. essential fatty acids.
    c. saturated fatty acids.
    d. essential amino acids.

Critical Thinking

1. The protein RDA for a healthy adult who weighs 180 pounds is:
   a. 50 milligrams per day.
   b. 65 grams per day.
   c. 180 grams per day
   d. 2000 milligrams per day.

2. Which of these foods has the least protein per serving?
   a. rice
   b. broccoli
   c. pinto beans
   d. orange juice

Answers to these questions can be found in Appendix G.

Clinical Applications

1. Considering the health effects of too little dietary protein, what suggestions would you have for a teenage girl who reports the following information about her food intake:

   ■ She never eats any meat or other animal-derived foods because she is a vegan. On a typical day, she eats toast and juice for breakfast; chips, a soft drink, and a piece of fruit for lunch; and a small serving of plain pasta with tomato sauce or steamed vegetables for dinner, along with a glass of water or tea.

   ■ She takes amino acid supplements because a friend told her that the only way to get amino acids if she doesn't eat meat is to take them as supplements.

2. Considering the health effects of excess dietary protein, what advice would you have for a college athlete who tells you he wants to bulk up his muscles and reports the following information about his food intake:

   ■ He eats large servings of meat (usually red meat) at least twice a day. He drinks whole milk two or three times a day and eats eggs and bacon for breakfast almost every day.

   ■ He avoids breads, cereals, and pasta in order to save room for protein-rich foods such as meat, milk, and eggs.

   ■ He eats a piece of fruit once in a while, but seldom eats vegetables because they are too time-consuming to prepare.

Nutrition on the Net

For further study of the topics of this chapter, access these websites.

Find updates and quick links to these and other nutrition-related sites at our website:
www.wadsworth.com/nutrition

To learn more about protein in foods, visit the American Dietetic Association (ADA) site:
www.eatright.org

Search the World Health Organization (WHO) site for information on protein-energy malnutrition:
www.who.org

For more on vegetarian diets, search the Food and Drug Administration (FDA) site for foods:
www.fda.gov

Search among thousands of current scientific and medical abstracts for any topic related to protein at:
www.ncbi.nlm.nih.gov/PubMed/

Learn more about protein-energy malnutrition and world hunger from the World Health Organization Nutrition Programme:
www.who.ch/nut/prot

Search for amino acid supplements at the National Council for Reliable Health Information site:
www.ncrhi.org

Notes

1. J. C. Teran, K. D. Mullen, and A. J. McCullough, Glutamine—A conditionally essential amino acid in cirrhosis? *American Journal of Clinical Nutrition* 62 (1995): 897–900.

2. D. G. Schroeder and R. Martorell, Enhancing child survival by preventing malnutrition, *American Journal of Clinical Nutrition* 65 (1997): 1080–1081.

3. The State of the World's Children 1998: A UNICEF Report, Malnutrition: Causes, consequences, and solutions, *Nutrition Reviews* 56 (1998): 115–123.

4. B. Woodward, Protein, calories, and immune defenses, *Nutrition Reviews* 56 (1998): 584–592.

5. P. W. Yoon and coauthors, The effect of malnutrition on the risk of diarrhea and respiratory mortality in children <2 y of age in Cebu, Philippines, *American Journal of Clinical Nutrition* 65 (1997): 1070–1077.

6. S. M. Grantham-McGregor, S. P. Walker, and S. Chang, Nutritional deficiencies and later behavioral development, *Proceedings of the Nutrition Society* 59 (2000): 47–54.

7. T. Liu and coauthors, Kwashiorkor in the United States: Fad diets, perceived and true milk allergy, and nutritional ignorance, *Archives of Dermatology* 137 (2001): 630–636; N. F. Carvalho and coauthors, Severe nutritional deficiencies in toddlers resulting from health food milk alternatives (electronic article), Pediatrics 107 (2001): e46.

8. A. R. Skov and coauthors, Changes in renal function during weight loss induced by high vs. low-protein low-fat diets in overweight subjects, *International Journal of Obesity and Related Metabolic Disorders* 23 (1999): 1170–1177; E. Brändle, H. G. Sieberth, and R. E. Hautmann, Effect of chronic dietary protein intake on the renal function in healthy subjects, *European Journal of Clinical Nutrition* 50 (1996): 734–740.

9. M. T. Pedrini and coauthors, The effect of dietary protein restriction on the progression of diabetic and nondiabetic renal diseases: A meta-analysis, *Annals of Internal Medicine* 124 (1996): 627–632.

10. D. Feskanich and coauthors, Protein consumption and bone fractures in women, *American Journal of Epidemiology* 143 (1996): 472–479.

11. Food and Drug Administration, Impurities confirmed in dietary supplement 5-hydroxy-L-tryptophan, FDA Talk Paper, August 1998, available from Food and Drug Administration, U.S. Department of Health and Human Services, Public Health Service, 5600 Fishers Lane, Rockville, MD 20857.

12. P. Garlick, Assessment of the safety of glutamine and other amino acids, *Journal of Nutrition* 131 (2001): S2556–S2561.

Nutrition in Practice

■ VEGETARIAN DIETS ■

Eating patterns all along the continuum of dietary choices—from one end, where people eat no foods of animal origin, to the other end, where they eat generous quantities of meat every day—can support or compromise nutritional health. The quality of the diet depends not on whether it consists of all plant foods or centers on meat, but on whether the eater's food choices are based on sound nutrition principles: adequacy of nutrient intakes, balance and variety of foods chosen, appropriate energy intake, and moderation in intakes of substances such as fat, sodium, alcohol, and caffeine that are harmful in excess. As mentioned in Chapter 3, however, because vegetarian diets exclude at least some animal-derived foods, they are usually lower in saturated fat and cholesterol than many meat-based diets.

People choose to exclude meat and other animal-derived foods from their diets for various reasons—philosophies, health attitudes, or convenience. Some believe that vegetarianism is better for the environment; some, that it is healthier; and some, that it is less costly than the meat-eating alternative. Some just like it better. Whatever the reasons, vegetarians and health professionals who work with them should be aware of the nutrition and health implications of vegetarian diets.

Because vegetarian diets vary in both the types and amounts of animal-derived foods they include, these differences must be considered when evaluating the health status of vegetarians. The glossary on this page defines the various kinds of vegetarian diets.

Are vegetarian diets nutritionally sound?

The American Dietetic Association takes the position that well-planned vegetarian diets offer nutrition and health benefits to adults in general.[1] Research suggests that meat-eating adults who switch to vegetarian diets reduce their risks of heart disease, hypertension, diabetes, some types of cancer, and obesity.[2]

What should be my main concerns when planning a nutritionally sound vegetarian diet?

A vegetarian diet planner faces the same task as other diet planners—obtaining a variety of foods that provide all the needed nutrients within an energy allowance that maintains a healthy body weight. The challenge is to do so using at least one less food group. Since all vegetarians omit meat, and some omit other animal-derived foods, protein, the nutrient that meat is famous for, merits some discussion here.

Isn't protein a problem in vegetarian diets?

No, protein is not the problem it was once thought to be in vegetarian diets. People who include animal-derived foods such as milk and eggs in their diets need not worry at all about protein deficiency. Even for those who eat only plant-derived foods, protein intakes are usually satisfactory as long as energy intakes are adequate and protein sources are varied.[3] A mixture of proteins from whole grains, legumes, seeds, nuts, and vegetables can provide adequate amounts of high-quality protein.

Glossary of Vegetarian Terms

lacto-ovo vegetarians: people who include milk or milk products and eggs, but omit meat, fish, shellfish, and poultry from their diets.

lacto-vegetarians: people who include milk or milk products, but exclude meat, poultry, fish, shellfish, and eggs from their diets.

semivegetarians: people who include some, but not all, groups of animal-derived foods in their diets; they usually exclude meat and may occasionally include poultry, fish, and shellfish; also called *partial vegetarians*.

vegans: people who exclude all animal-derived foods (including meat, poultry, fish, shellfish, eggs, cheese, and milk) from their diets; also called *strict vegetarians* or *total vegetarians*.

Nutrition in Practice

The idea persists that **vegans** must carefully combine their plant-protein foods in order to obtain the protein they need, but this is not necessary. Plant foods can provide more than enough high-quality protein and can sustain people in good health, as long as the diet supplies sufficient energy and does not include too many empty-kcalorie foods. As Chapter 4 explained, however, a meal delivers higher-quality protein when it combines two or more different individual plant-protein sources. Figure NP4–1 shows combinations of plant proteins that provide higher-quality protein than the individual foods alone could supply.

What sorts of food energy intakes do vegetarian diets provide?

Researchers find that vegetarians as a group are closer to a healthy body weight than nonvegetarians. Because obesity impairs health in a number of ways, vegetarians therefore have a health advantage. Vegetarian diets tend to be high in complex carbohydrates and low in fat, characteristics that are consistent with current dietary recommendations aimed at reducing the incidence of obesity and other degenerative diseases in this country.

Not all vegetarians fit the average pattern, though. Obesity does threaten vegetarians who include milk, eggs, and cheese in their diets. They can easily consume both a high-fat diet and excess food energy and so must be careful to select fat-free and low-fat dairy foods and to avoid relying too heavily on these foods in general.

In contrast, people who exclude all animal-derived foods (vegans) may have trouble obtaining *enough* food energy. This is especially true for children and pregnant and lactating women.[4] Vegan diets can fail to provide food energy sufficient to support the growth of a child within a bulk of food small enough for the child to eat.[5] Plant foods that are best suited to meeting energy needs in a small volume are cereals, legumes, and nuts; these foods should be emphasized in a vegan child's diet. Table NP4-1 offers a suggested daily food guide for vegetarians.

Tell me about vitamins and minerals. Does a person eating a vegetarian diet need to take vitamin supplements?

That depends on the kind of vegetarian diet. The diet of **lacto-ovo vegetarians** can be complete in all vitamins, but for vegans, several vitamins may be a problem. One such vitamin is B_{12}. Because vitamin B_{12} occurs only in animal-derived foods, supplements are necessary to prevent deficiency. Women who have adhered to all-plant diets for many years are especially likely to have low vitamin B_{12} stores. Pregnant vegan women, whose needs for vitamin B_{12} are especially high, find it virtually impossible to maintain adequate vitamin B_{12} status without taking supplements or including a reliable food source of the nutrient.

A vitamin B_{12} deficiency can take a long time to develop in adults because up to four years' worth of the vitamin can be stored in the body. But when the deficiency sets in, it does severe damage to the nervous system (see Chapter 8). In infants, deficiencies set in more rapidly and so threaten their nervous systems early. All vegan mothers must be sure to take the appropriate supplements or to use vitamin B_{12}–fortified products such as soy milk or breakfast cereals.

What other vitamins do vegans need?

Another vitamin of concern is vitamin D. The milk drinker

Grains	Legumes	Seeds and Nuts	Vegetables
Barley	Dried beans	Almonds	Broccoli
Bulgur	(pinto, kidney, navy, etc.)	Cashews	Cabbage
Oats	Dried lentils	Nut butters	Peppers
Pasta	Dried peas	Sesame seeds	Spinach
Rice	Peanuts	Sunflower seeds	Squash
Whole-grain breads	Soy products	Walnuts	

Figure NP4–1

Nonmeat Mixtures That Provide High-Quality Protein
Vegetarians who eat no foods from animal sources select foods from two or more of these columns to create high-quality protein combinations:

Black beans and rice, a favorite Hispanic combination.

Tofu and stir-fried vegetables with rice, an Asian dish.

Nutrition in Practice

is protected, provided the milk is fortified with vitamin D, but there is no practical source of vitamin D in plant foods. Regular exposure to the sun will prevent a deficiency, but vegans who are homebound or live in a northern climate or smoggy city probably should take vitamin D supplements. Excesses of vitamin D are toxic, and one should not exceed the recommended daily amount of 5 micrograms.

Riboflavin, another vitamin often obtained from milk, is not a problem for the vegan who eats dark greens frequently in ample servings. The vegan who doesn't consume a lot of greens, however, may not meet riboflavin needs. Nutritional yeast is a rich source of riboflavin for the vegetarian.

So, on a vegan diet, vitamin B_{12}, vitamin D, and riboflavin can be problems if a person is not careful. What about minerals?

For *all* vegetarians, not just the vegan, two minerals may be of concern—iron and zinc. Legumes are an important source of iron in the vegetarian diet. The iron in legumes, however, is not as absorbable as that in meat. In fact, people absorb three times as much iron from a meal that includes meat as from one that does not. For this reason, the iron recommendation for adult vegetarian men, premenopausal women, and adolescent girls is almost double the recommendation for meat eaters of the same gender and age.[6] Because vitamin C in fruits and vegetables can triple iron absorption from other foods eaten at the same meal, vegetarian meals should be rich in foods offering vitamin C.

Zinc may also be a problem nutrient for vegetarians. It is widespread in plant foods, but its availability may be hindered by the fibers and other binders found in fruits and vegetables. The zinc needs of vegetarians and the effects of mineral binders are subjects of intensive study at the present time. While research continues, vegetarians are advised to eat varied diets that include whole-grain breads well leavened with yeast, which improves the availability of their minerals.

Table NP4–1 Daily Food Guide for Vegetarian Meal Planning

Food Group	Suggested Daily Servings	Serving Sizes
Breads, cereals, rice, pasta, and other grain products	6 to 11	1 slice bread
		½ bun, bagel, or English muffin
		½ c cooked cereal, rice, or pasta
		1 oz ready-to-eat cereal
Vegetables	3 to 5[a]	½ c cooked or chopped raw vegetables
		1 c raw leafy vegetables
Fruits	2 to 4	1 medium-sized piece fresh fruit
		¾ c fruit juice
		½ c canned, cooked, or chopped raw fruit
		¼ c dried fruit
Legumes, nuts, seeds, eggs, and other meat substitutes	2 to 3	½ c cooked legumes
		¼ c tofu or tempeh
		1 c soy milk
		2 tbs peanut butter, nuts, or seeds (these tend to be high in fat, so use sparingly)
		1 egg or 2 egg whites
Milk, yogurt, cheese, and other milk products	2 to 3[b]	1 c milk
		1 c yogurt
		1½ oz cheese

[a]Include 1 cup of dark green vegetables daily to help meet iron requirements.
[b]People who do not use milk or milk products: use soy milk fortified with calcium, vitamin D, and vitamin B_{12}. Other nonmilk calcium-rich food sources are provided in Chapter 9.
SOURCE: Adapted from Position of The American Dietetic Association Vegetarian diets, *Journal of the American Dietetic Association* 97 (1997): 1320.

What about calcium for the vegan?

Good thinking. Yes, calcium is of concern. The milk-drinking vegetarian is protected from deficiency, but the vegan must find other sources of calcium. Some good calcium sources are regular and ample servings of stone-ground meal, self-rising flour and meal, legumes, calcium-fortified soy milk, calcium-fortified orange juice, some nuts such as almonds, and certain seeds such as sesame seeds. The choices should be varied because binders in some of these foods may hinder calcium absorption. The vegetarian is urged to use calcium-fortified soy milk in ample quantities regularly. This is especially important for children. Infant formula based on soy is fortified with calcium and can easily be used in cooking foods, even for adults.

Are there any other health advantages to the vegetarian diet?

Yes. Vegetarian protein foods are often higher in fiber, richer in certain vitamins and minerals, and lower in fat, especially saturated fat, than meats. Vegetarians can enjoy a nutritious diet very low in fat provided that they limit other high-fat foods such as butter, cream cheese, sour cream, and nuts. If vegetarians follow the guidelines presented here and plan carefully, they can support their health as well as, or perhaps better than, nonvegetarians.

Abundant evidence supports the idea that vegetarians may actually be healthier than meat eaters. Informed vegetarians are not only more likely to be at the desired weights for their heights, but to have lower blood cholesterol levels, lower rates of certain kinds of cancer, better digestive function, and more. Even among people who are health conscious, generally vegetarians experience fewer deaths from cardiovascular disease than meat eaters do. Since many vegetarians also abstain from smoking and the consumption of alcohol, dietary practices alone probably do not account for all the aspects of improved health. Clearly, however, they contribute significantly to it.

Notes

1. Position of The American Dietetic Association: Vegetarian diets, *Journal of the American Dietetic Association* 97 (1997): 1317–1321.

2. E. L. Ashton, F. S. Dalais, and M. J. Ball, Effect of meat replacement by tofu on CHD risk factors including copper induced LDL oxidation, *Journal of the American College of Nutrition* 19 (2000): 761–767; T. J. Key and coauthors, Mortality in vegetarians and nonvegetarians: Detailed findings from a collaborative analysis of 5 prospective studies, *American Journal of Clinical Nutrition* 70 (1999): 516S–524S; P. Walter, Effects of vegetarian diets on aging and longevity, *Nutrition Reviews* 55 (1997): S61–S68.

3. Position of The American Dietetic Association, 1997.

4. M. Hebbelinck, P. Clarys, and A. De Malsche, Growth, development, and physical fitness of Flemish vegetarian children, adolescents, and young adults, *American Journal of Clinical Nutrition* 70 (1999): 579S–585S.

5. Hebbelinck, Clarys, and De Malsche, 1999.

6. Standing Committee on the Scientific Evaluation of Dietary Reference Intakes, Food and Nutrition Board, Institute of Medicine, *Dietary Reference Intakes for Vitamin A, Vitamin K, Arsenic, Boron, Chromium, Copper, Iodine, Iron, Manganese, Molybdenum, Nickel, Silicon, Vanadium, and Zinc* (Washington, D.C.: National Academy Press, 2001), p. 9-45.

Jennie Oppenheimer/Studio Zocolo

CHAPTER 5

DIGESTION AND ABSORPTION

*T*he body's ability to transform the foods a person eats into the nutrients that fuel the body's work is quite remarkable. Yet most people probably give little, if any, thought to all the body does with food once it is eaten. This chapter offers the reader the opportunity to learn how the body digests, absorbs, and transports the nutrients and how it excretes the unwanted substances in foods. The next chapter shows how the body uses the nutrients once they have been absorbed and are traveling in the blood and lymph.

One of the beauties of the digestive tract is that it is selective. Materials that are nutritive for the body are broken down into particles that can be absorbed into the bloodstream. Most of the nonnutritive materials are left undigested and pass out the other end of the digestive tract.

Anatomy of the Digestive Tract

The **gastrointestinal (GI) tract** is a flexible muscular tube measuring about 15 feet in length from the mouth to the anus.[1] Figure 5–1 on pages 98 and 99 traces the path followed by food from one end to the other. The accompanying glossary defines GI anatomy terms. In a sense, the human body surrounds the GI tract. Only when a nutrient or other substance passes through the cells of the digestive tract wall does it actually enter the body.

The Digestive Organs

The process of **digestion** begins in the mouth. As you chew, your teeth crush and soften foods, while saliva mixes with the food mass and moistens it for comfortable swallowing. Saliva also helps dissolve the food so that you can taste it; only particles in solution can react with taste buds.

Mouth to the Esophagus Once a mouthful of food has been swallowed, it is called a **bolus.** Each bolus first slides across your **epiglottis,** bypassing the entrance to your lungs. During each swallow, the epiglottis closes off your air passages so that you do not choke.

Esophagus to the Stomach Next, the bolus slides down the **esophagus,** which conducts it through the diaphragm to the stomach. The **lower esophageal sphincter,** a band of muscle surrounding the esophagus where it enters the stomach, closes behind the bolus so that it cannot slip back. The stomach retains the bolus for a while, adds juices to it (gastric juices are discussed on p. 103), and grinds it into a semiliquid mass called **chyme.** Then, bit by bit, the stomach releases the chyme through another **sphincter,** the **pyloric sphincter,** which opens into the **small intestine** and then closes behind the chyme.

The Small Intestine At the top of the small intestine, the chyme passes by an opening from the common bile duct, which secretes fluids into the small intestine from two organs outside the GI tract—the **gallbladder** and the **pancreas.** The chyme travels on down the small intestine through its three segments—the **duodenum,** the **jejunum,** and the **ileum**—a total of about 10 feet of tubing coiled within the abdomen.[2]

The Colon (Large Intestine) Having traveled the length of the small intestine, the chyme passes through another sphincter, the **ileocecal valve,** into the beginning of the **colon (large intestine)** in the lower right-hand side of the abdomen. In the colon, the chyme travels up the right-hand side of the abdomen, across the

gastrointestinal (GI) tract: the digestive tract. The principal organs are the stomach and intestines.
  gastro = stomach

digestion: the process by which complex food particles are broken down to smaller absorbable particles.

bolus (BOH-lus): the portion of food swallowed at one time.

sphincter (SFINK-ter): a circular muscle surrounding, and able to close, a body opening.
  sphincter = band (binder)

chyme (KIME): the semiliquid mass of partly digested food expelled by the stomach into the duodenum (the top portion of the small intestine).

These terms are listed in order from the beginning of the digestive tract to the end.

epiglottis (epp-ee-GLOT-tiss):   a cartilage structure in the throat that prevents fluid or food from entering the trachea when a person swallows.
> *epi* = upon (over)
> *glottis* = back of tongue

esophagus (e-SOFF-uh-gus):   the food pipe; the conduit from the mouth to the stomach.

lower esophageal sphincter (SFINK-ter):   the sphincter muscle at the junction between the esophagus and the stomach (also called *cardiac sphincter*).

pyloric (pie-LORE-ic) **sphincter**:   the sphincter muscle separating the stomach from the small intestine (also called *pylorus* or *pyloric valve*).
> *pylorus* = gatekeeper

gallbladder: the organ that stores and concentrates bile. When it receives the signal that fat is present in the duodenum, the gallbladder contracts and squirts bile through the bile duct into the duodenum.

pancreas: a gland that secretes enzymes and digestive juices into the duodenum. (This is its exocrine function; it also has the endocrine function of secreting insulin and other hormones into the blood.)

small intestine: a 10-foot length of small-diameter (1-inch) intestine that is the major site of digestion of food and absorption of nutrients.

duodenum (doo-oh-DEEN-um or doo-ODD-ah-num):   the top portion of the small intestine (about "12 fingers' breadth" long, in ancient terminology).

jejunum (je-JOON-um):   the first two-fifths of the small intestine beyond the duodenum.

ileum (ILL-ee-um):   the last segment of the small intestine.

ileocecal (ill-ee-oh-SEEK-ul) **valve**:   the sphincter muscle separating the small and large intestines.

colon or **large intestine**:   the last portion of the intestine, which absorbs water. Its main segments are the ascending colon, the transverse colon, the descending colon, and the sigmoid colon.
> *sigmoid* = shaped like the letter **S**

appendix: a narrow blind sac extending from the beginning of the colon. The appendix stores lymphocytes.

rectum: the muscular terminal part of the GI tract extending from the sigmoid colon to the anus. The rectum stores waste prior to elimination.

anus (AY-nus):   the terminal sphincter muscle of the GI tract.

---

front to the left-hand side, down to the lower left-hand side, and finally below the other folds of the intestines to the back side of the body above the **rectum.**

The Rectum   During chyme's passage to the rectum, the colon withdraws water from it, leaving semisolid waste. The strong muscles of the rectum hold back this waste until it is time to defecate. Then the rectal muscles relax, and the last sphincter in the system, the **anus,** opens to allow the wastes to pass. Thus food follows the path shown in the margin.

The Involuntary Muscles and the Glands

You are usually unaware of all the activity that goes on between the time you swallow and the time you defecate. As is the case with so much else that happens in the body, the muscles and **glands** of the digestive tract meet internal needs without your having to exert any conscious effort to get the work done.

People consciously chew and swallow, but even in the mouth there are some processes over which you have no control. The salivary glands secrete just enough saliva to moisten each mouthful of food so that it can pass easily down your esophagus.

Gastrointestinal Motility   Once you have swallowed, materials are moved through the rest of the GI tract by involuntary muscular contractions. This motion, known as **gastrointestinal motility,** consists of two types of movement, peristalsis and segmentation (see Figure 5–2 on p. 100). Peristalsis propels, or pushes; segmentation mixes, with more gradual pushing.

The path of food through the digestive tract:

- Mouth.
- Esophagus.
- Lower esophageal sphincter (or cardiac sphincter).
- Stomach.
- Pyloric sphincter.
- Duodenum (common bile duct enters here), jejunum, ileum.
- Ileocecal valve.
- Colon.
- Rectum.
- Anus.

glands: single cells or groups of cells that secrete materials for special uses in the body. Glands may be *exocrine glands,* secreting their materials "out" (into the digestive tract or onto the surface of the skin), or *endocrine glands,* secreting their materials "in" (into the blood).
> *exo* = outside
> *endo* = inside
> *krine* = to separate

Digestion and Absorption

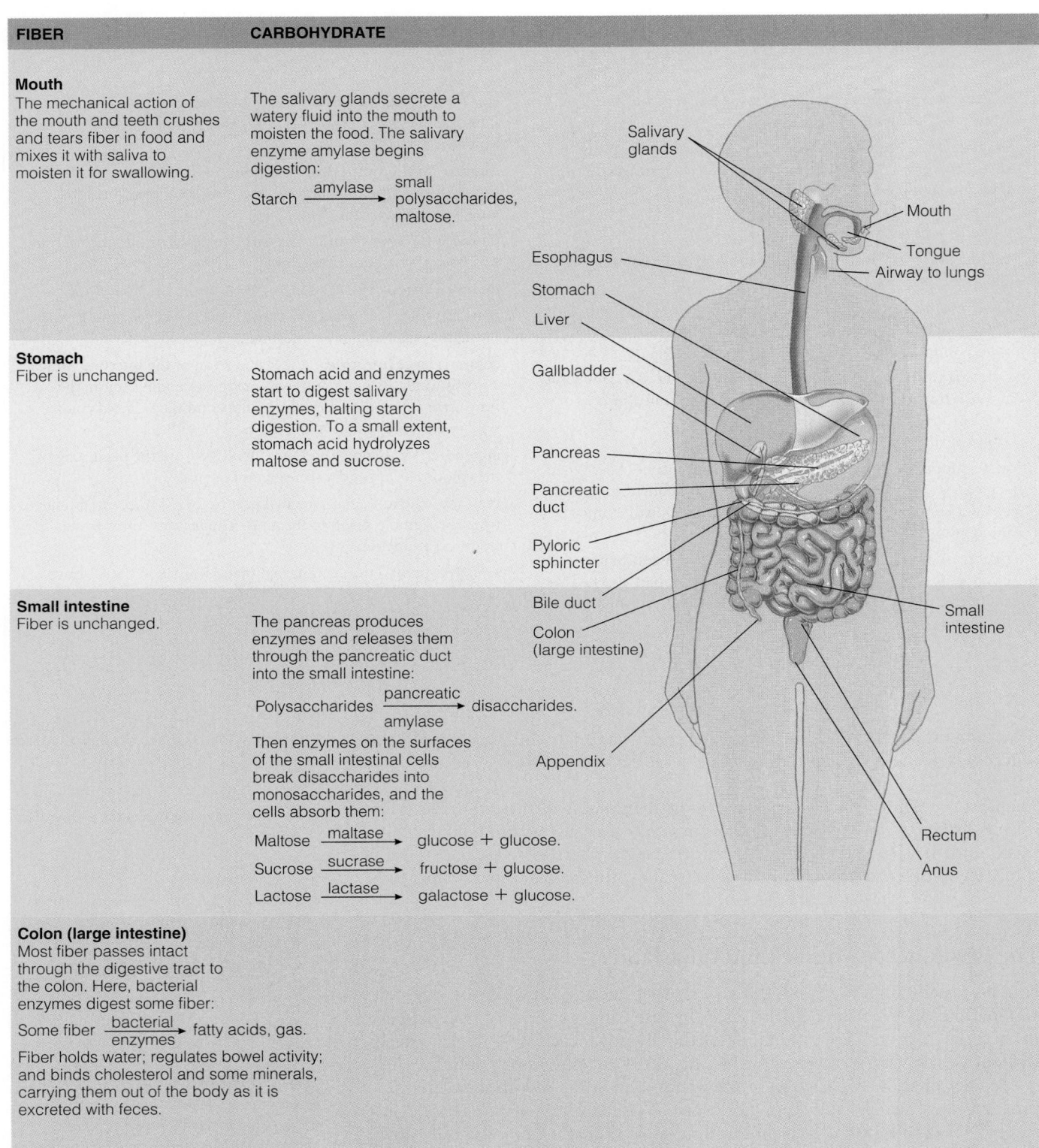

Mouth
The mechanical action of the mouth and teeth crushes and tears fiber in food and mixes it with saliva to moisten it for swallowing.

The salivary glands secrete a watery fluid into the mouth to moisten the food. The salivary enzyme amylase begins digestion:

$$\text{Starch} \xrightarrow{\text{amylase}} \text{small polysaccharides, maltose.}$$

Stomach
Fiber is unchanged.

Stomach acid and enzymes start to digest salivary enzymes, halting starch digestion. To a small extent, stomach acid hydrolyzes maltose and sucrose.

Small intestine
Fiber is unchanged.

The pancreas produces enzymes and releases them through the pancreatic duct into the small intestine:

$$\text{Polysaccharides} \xrightarrow{\text{pancreatic amylase}} \text{disaccharides.}$$

Then enzymes on the surfaces of the small intestinal cells break disaccharides into monosaccharides, and the cells absorb them:

$$\text{Maltose} \xrightarrow{\text{maltase}} \text{glucose} + \text{glucose.}$$

$$\text{Sucrose} \xrightarrow{\text{sucrase}} \text{fructose} + \text{glucose.}$$

$$\text{Lactose} \xrightarrow{\text{lactase}} \text{galactose} + \text{glucose.}$$

Colon (large intestine)
Most fiber passes intact through the digestive tract to the colon. Here, bacterial enzymes digest some fiber:

$$\text{Some fiber} \xrightarrow{\text{bacterial enzymes}} \text{fatty acids, gas.}$$

Fiber holds water; regulates bowel activity; and binds cholesterol and some minerals, carrying them out of the body as it is excreted with feces.

Salivary glands
Mouth
Tongue
Airway to lungs
Esophagus
Stomach
Liver
Gallbladder
Pancreas
Pancreatic duct
Pyloric sphincter
Bile duct
Colon (large intestine)
Small intestine
Appendix
Rectum
Anus

Figure 5–1
The Gastrointestinal Tract

FAT	PROTEIN	VITAMINS	MINERALS AND WATER
Mouth Glands in the base of the tongue secrete a fat-digesting enzyme known as lingual lipase. Some hard fats begin to melt as they reach body temperature.	Chewing and crushing moisten protein-rich foods and mix them with saliva to be swallowed.	No action.	The salivary glands add water to disperse and carry food.
Stomach The acid-stable lingual lipase splits one bond of triglycerides to produce diglycerides and fatty acids. The degree of hydrolysis is slight for most fats but may be appreciable for milk fats. The stomach's churning action mixes fat with water and acid. A gastric lipase accesses and hydrolyzes a very small amount of fat.	Stomach acid uncoils protein strands and activates stomach enzymes: Protein $\xrightarrow[\text{HCl}]{\text{pepsin}}$ smaller polypeptides.	Intrinsic factor (see Chapter 8) attaches to vitamin B_{12}.	Stomach acid (HCl) acts on iron to reduce it, making it more absorbable (see Chapter 9). The stomach secretes enough watery fluid to turn a moist, chewed mass of solid food into liquid chyme.
Small intestine Bile flows in from the liver and gallbladder (via the common bile duct): Fat $\xrightarrow{\text{bile}}$ emulsified fat. Pancreatic lipase flows in from the pancreas (via the pancreatic duct): Emulsified fat $\xrightarrow[\text{lipase}]{\text{pancreatic}}$ monoglycerides, glycerol, fatty acids (absorbed).	Pancreatic and small intestinal enzymes split polypeptides further: Polypeptides $\xrightarrow[\substack{\text{and intestinal}\\\text{proteases}}]{\text{pancreatic}}$ dipeptides, tripeptides, and amino acids. Then enzymes on the surface of the small intestinal cells hydrolyze these peptides, and the cells absorb them: Peptides $\xrightarrow[\substack{\text{dipeptidases}\\\text{and tripeptidases}}]{\text{intestinal}}$ amino acids (absorbed)	Bile emulsifies fat-soluble vitamins and aids in their absorption with other fats. Water-soluble vitamins are absorbed.	The small intestine, pancreas, and liver add enough fluid so that approximately 2 gallons are secreted into the intestine in a day. Many minerals are absorbed. Vitamin D aids in the absorption of calcium.
Colon Some fat and cholesterol, trapped in fiber, exit in feces.		Bacteria produce vitamin K, which is absorbed.	More minerals and most of the water are absorbed.

Figure 5–1

The Gastrointestinal Tract (continued)

Digestion and Absorption

Figure 5–2
Peristalsis and Segmentation

Esophagus
Liver
Stomach
Small intestine
Large intestine

Longitudinal muscles are outside.

Circular muscles are on the inside.

The small intestine has two muscle layers that work together in peristalsis and segmentation.

PERISTALSIS

SEGMENTATON

Chyme

Chyme

The inner circular muscle layer contracts, tightening the tube and pushing the food forward in the intestine.

Simultaneous circular muscle contractions occur in food-containing sections of the intestine, creating segments within the intestine.

When the outer longitudinal muscles contract, the circular muscles relax and the intestinal tube is loose.

Muscles circling the middle of each segment contract, and the first set of muscles relaxes. The chyme is broken up and mixed with digestive juices.

As the circular and longitudinal muscles tighten and relax, the food moves ahead of the constriction.

These alternating contractions, occurring 12 –16 times per minute, continue to mix the chyme and bring it into contact with the intestinal lining for absorption of nutrients.

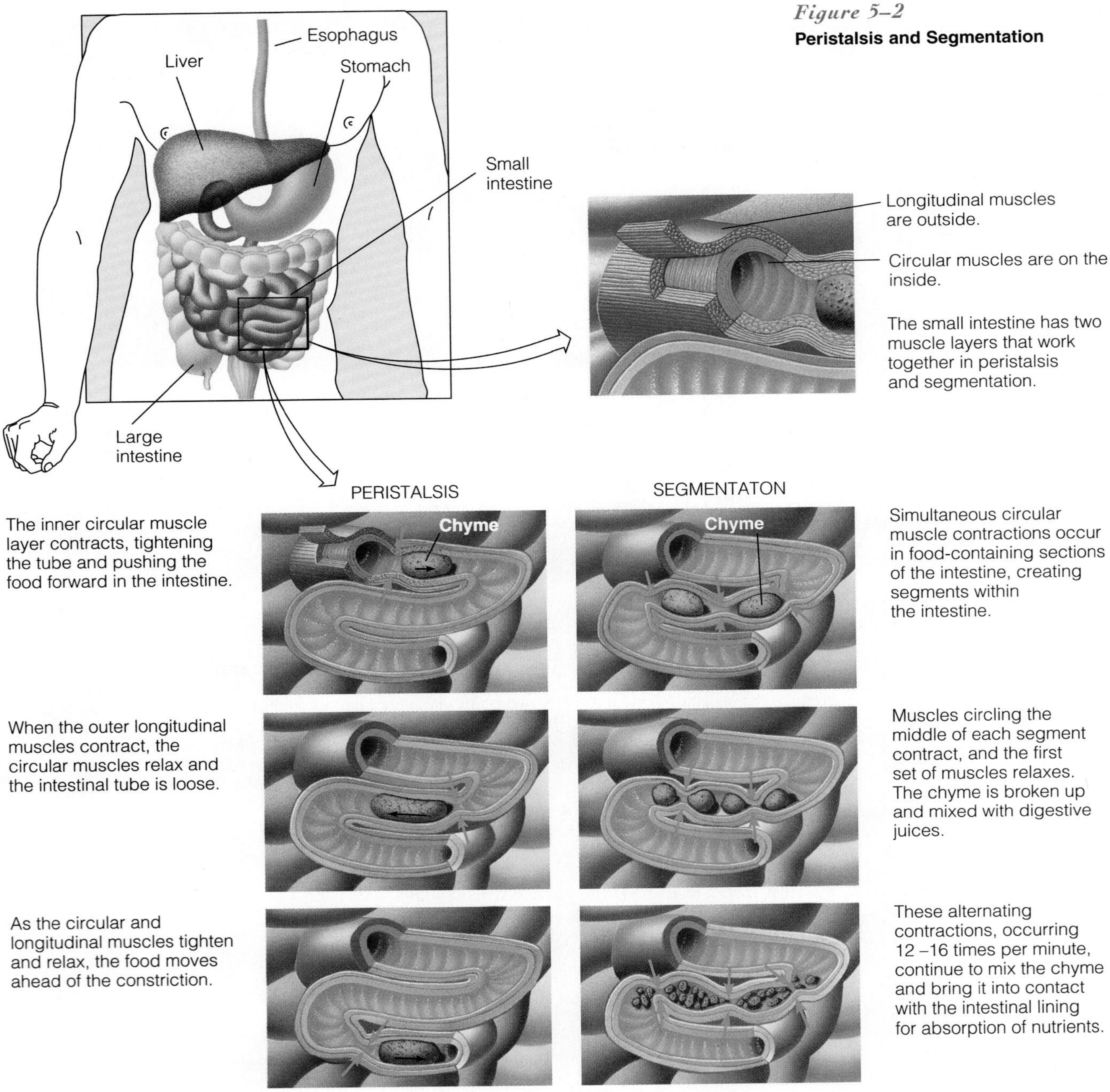

gastrointestinal motility: spontaneous motion in the digestive tract accomplished by involuntary muscular contractions.

peristalsis (peri-STALL-sis): successive waves of involuntary muscular contractions passing along the walls of the GI tract that push the contents along.

   peri = around
   stellein = wrap

Chapter Five
100

Peristalsis Peristalsis begins when the bolus enters the esophagus. The entire GI tract is ringed with circular muscles that can squeeze it tightly. Surrounding these rings of muscle are longitudinal muscles. When the rings tighten and the long muscles relax, the tube is constricted. When the rings relax and the long muscles tighten, the tube bulges. These actions follow each other continuously and push the intestinal contents along. If you have ever watched a bolus of food pass along the body of a snake, you have a good picture of how these muscles work. The waves of contraction ripple through the GI tract at varying rates and intensities depending on the part of the GI tract and on whether food is present. Peristalsis, aided by the sphincter muscles that surround the GI tract at key places, keeps things moving along.

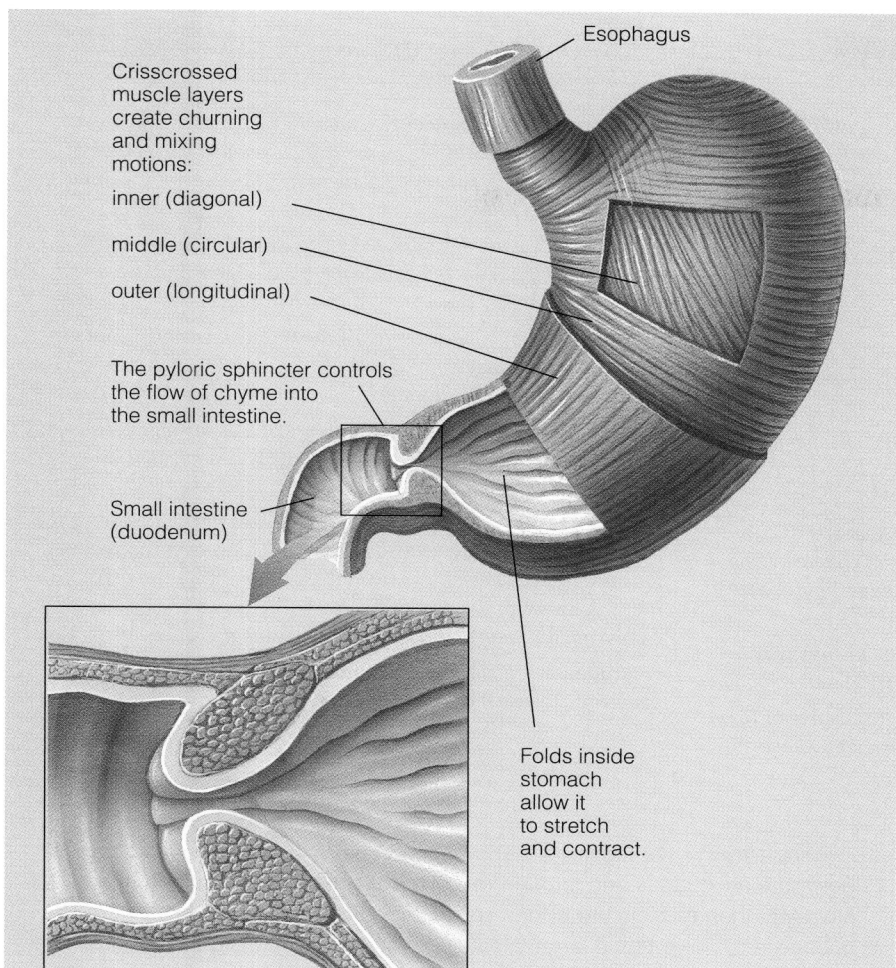

Crisscrossed
muscle layers
create churning
and mixing
motions:

inner (diagonal)

middle (circular)

outer (longitudinal)

The pyloric sphincter controls
the flow of chyme into
the small intestine.

Small intestine
(duodenum)

Esophagus

Folds inside
stomach
allow it
to stretch
and contract.

Figure 5–3
Stomach Muscles
The stomach has three layers of muscles.

Segmentation The intestines not only push but also periodically squeeze their contents as if a string tied around the intestines were being pulled tight. This motion, called **segmentation,** forces the contents back a few inches, mixing them and promoting close contact with the digestive juices and the absorbing cells of the intestinal walls before letting the contents slowly move along again.

Liquefying Process Besides forcing the intestinal contents along, the muscles of the GI tract help to liquefy them to chyme so that the digestive enzymes will have access to all their nutrients. The mouth initiates this liquefying process by chewing, adding saliva, and stirring with the tongue to reduce the food to a coarse mash suitable for swallowing. The stomach then further mixes and kneads the food.

Stomach Action The stomach has the thickest walls and strongest muscles of all the GI tract organs. In addition to the circular and longitudinal muscles, the stomach has a third layer of diagonal muscles that also alternately contract and relax (see Figure 5–3). These three sets of muscles work to force the chyme downward, but the pyloric sphincter usually remains tightly closed so that the stomach's contents are thoroughly mixed and squeezed before being released. Meanwhile, the gastric glands are adding juices. When the chyme is thoroughly liquefied, the pyloric sphincter opens briefly, about three times a minute, to allow small portions through. At this point, the intestinal contents no longer resemble food in the least.

segmentation: a periodic squeezing or partitioning of the intestine by its circular muscles that both mixes and slowly pushes the contents along.

Digestive Glands and Their Secretions

These terms are listed in order from the beginning of the digestive tract to the end.

salivary glands: exocrine glands that secrete saliva into the mouth.

saliva: the secretion of the salivary glands. The principal enzyme is salivary amylase.

amylase (AM-uh-lace): an enzyme that splits amylose (a form of starch). Amylase is a carbohydrase. The ending *-ase* indicates an enzyme; the root tells what it digests. Other examples: protease, lipase.

gastric glands: exocrine glands in the stomach wall that secrete gastric juice into the stomach.

> *gastro* = stomach

gastric juice: the digestive secretion of the gastric glands containing a mixture of water, hydrochloric acid, and enzymes. The principal enzymes are pepsin (acts on proteins) and lipase (acts on emulsified fats).

hydrochloric acid (HCl): an acid composed of hydrogen and chloride atoms; normally produced by the gastric glands.

mucus (MYOO-cuss): a mucopolysaccharide (a relative of carbohydrate) secreted by cells of the stomach wall that protects the cells from exposure to digestive juices (and other destructive agents). The cellular lining of the stomach wall with its coat of mucus is known as the mucous membrane. (The noun is *mucus;* the adjective is *mucous.*)

pepsin: a protein-digesting enzyme (gastric protease) in the stomach. It circulates as a precursor, pepsinogen, and is converted to pepsin by the action of stomach acid.

intestinal juice: the secretion of the intestinal glands; contains enzymes for the digestion of carbohydrate and protein and a minor enzyme for fat digestion.

bile: an emulsifier that prepares fats and oils for digestion; made by the liver, stored in the gallbladder, and released into the small intestine when needed.

pancreatic (pank-ree-AT-ic) **juice:** the exocrine secretion of the pancreas, containing enzymes for the digestion of carbohydrate, fat, and protein. Juice flows from the pancreas into the small intestine through the pancreatic duct. The pancreas also has an endocrine function, the secretion of insulin and other hormones.

bicarbonate: an alkaline secretion of the pancreas; part of the pancreatic juice. (Bicarbonate also occurs widely in all cell fluids.)

In Summary As Figure 5–1 shows, food enters the mouth and travels down the esophagus and through the lower esophageal sphincter to the stomach, then through the pyloric sphincter to the small intestine, on through the ileocecal valve to the large intestine, past the appendix to the rectum, ending at the anus. The wavelike contractions of peristalsis and the periodic squeezing of segmentation keep things moving at a reasonable pace.

The Process of Digestion

One person eats nothing but vegetables, fruits, and nuts; another, nothing but meat, milk, and potatoes. How is it that both people wind up with essentially the same body composition? It all comes down to the body rendering food—whatever it is to start with—into the basic units that make up carbohydrate, fat, and protein. The body absorbs these units and builds its tissues from them.

To digest food, five different body organs secrete digestive juices: the salivary glands, the stomach, the small intestine, the liver (via the gallbladder), and the pancreas. Each of the juices has a turn to mix with the food and promote its breakdown to small units that can be absorbed into the body. The glossary above defines some of the digestive glands and their juices.

Digestion in the Mouth

Digestion of carbohydrate begins in the mouth, where the **salivary glands** secrete **saliva,** which contains water, salts, and enzymes (including salivary **amylase**) that break the bonds in the chains of starch. Saliva also protects the tooth surfaces and linings of the mouth, esophagus, and stomach from attack by molecules that might harm them. The enzymes in the mouth do not, for the most part, affect the fats, proteins, vitamins, minerals, and fiber that are present in the foods people eat (review Figure 5–1).

Digestion in the Stomach

Gastric juice, secreted by the **gastric glands,** is composed of water, enzymes, and **hydrochloric acid.** The acid is so strong that it burns the throat if it happens to reflux into the upper esophagus and mouth. The strong acidity of the stomach prevents bacterial growth and kills most bacteria that enter the body with food. You might expect that the stomach's acid would attack the stomach itself, but the cells of the stomach wall secrete **mucus,** a thick, slimy, white polysaccharide that coats and protects the stomach's lining.

Digestive Activities   The major digestive event in the stomach is the initial breakdown of proteins. Other than being crushed and mixed with saliva in the mouth, nothing happens to protein until it comes in contact with the gastric juices in the stomach. There, the acid helps to uncoil (denature) the protein's tangled strands so that the stomach enzymes can attack the bonds. Both the enzyme **pepsin** and the stomach acid itself act as catalysts in the process. Minor events are the digestion of some fat by a gastric lipase, the digestion of sucrose (to a very small extent) by the stomach acid, and the attachment of a protein carrier to vitamin B_{12}.

The stomach enzymes work most efficiently in the stomach's strong acid, but salivary amylase, which is swallowed with food, does not work in acid this strong. Consequently, the digestion of starch gradually ceases as the acid penetrates the bolus. In fact, salivary amylase becomes just another protein to be digested. The amino acids in amylase end up being absorbed and recycled into other body proteins.

Antacids: Their Use and Misuse   Note that the strong acidity of the stomach is a desirable condition, television commercials for antacids notwithstanding. In a person who eats too quickly or eats too much, the stomach is likely to react with such violence as to cause regurgitation. When this happens, the stomach acid creates a bad taste in the mouth, which the person may interpret as "acid indigestion." Responding to television commercials, the overeater may take antacids to neutralize the stomach acid. But then the stomach will have to secrete more acid to enable the digestive enzymes to do their work. The consumer's stomach ends up with the same amount of acid but has had to work against the antacid to produce it.

Antacids are not designed to relieve the digestive discomfort of the hasty and abusive eater. Their proper use is to correct an abnormal condition, such as that of the person with ulcers, whose stomach or duodenal lining has been attacked by acid. To avoid falling into the same trap as our misguided consumer, remember to eat slowly, chew thoroughly, and perhaps consume less at a sitting.

Digestion in the Small and Large Intestines

By the time food leaves the stomach, digestion of all three energy-yielding nutrients has begun, but the process gains momentum in the small intestine. There, the pancreas and the liver contribute additional digestive juices through the duct leading into the duodenum, and the small intestine adds **intestinal juice.** These juices contain digestive enzymes, bicarbonate, and bile.

Digestive Enzymes   **Pancreatic juice** contributes enzymes that digest fats, proteins, and carbohydrates. Glands in the intestinal wall also secrete digestive enzymes. (Review the glossary of digestive glands and their secretions on p. 102 for details.)

Bicarbonate   The pancreatic juice also contains sodium **bicarbonate,** which neutralizes the acidic chyme as it enters the small intestine. From this point on, the contents of the digestive tract are neutral or slightly alkaline. The enzymes of both the intestine and the pancreas work best in this environment.

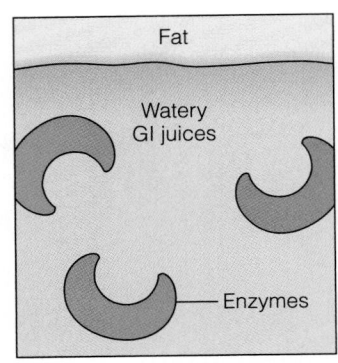 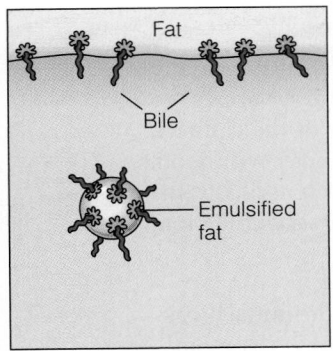 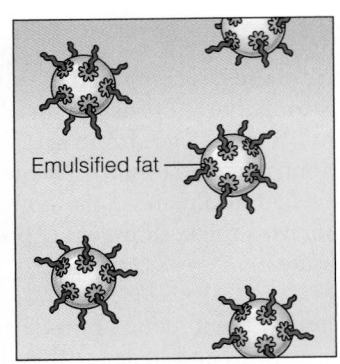 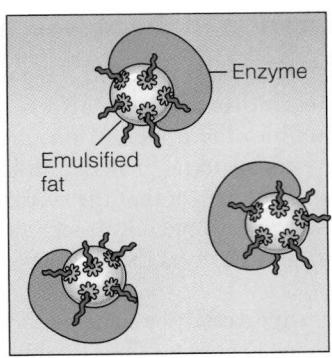

In the stomach, the fat and watery GI juices tend to separate. The enzymes are in the water and can't get at the fat.

When fat enters the small intestine, the gallbladder secretes bile. Bile has an affinity for both fat and water, so it can bring the fat into the water.

Bile's emulsifying action converts large fat globules into small droplets that repel each other.

After emulsification, the enzymes have easy access to the fat droplets.

Figure 5–4

Emulsification of Fat by Bile
Like bile, detergents are emulsifiers and work the same way, which is why they are effective at removing grease spots from clothes. Molecule by molecule, the grease is dissolved out of the spot and suspended in the water, where it can be rinsed away.

Reminder: An emulsifier is a substance that mixes with both fat and water and permanently disperses the fat in the water, forming an emulsion.

Mayonnaise, made from vinegar and oil, would separate as other vinegar-and-oil salad dressings do if food chemists did not blend the vinegar and oil with a third ingredient—an emulsifier. The emulsifier mixes well with the fatty oil and the watery vinegar. In the case of mayonnaise, the emulsifier is lecithin from egg yolks.

Bile Bile is secreted by the liver continuously and is concentrated and stored in the gallbladder. The gallbladder squirts bile into the duodenum whenever fat arrives there. Bile is not an enzyme but an emulsifier that brings fats into suspension in water (see Figure 5–4). After the fats are emulsified, enzymes can work on them, and they can be absorbed. Thanks to all these secretions, all three energy-yielding nutrients are digested in the small intestine.

The Rate of Digestion The rate of digestion of the energy nutrients depends upon the contents of the meal. If the meal is high in simple sugars, digestion proceeds fairly rapidly. On the other hand, if the meal is rich in fat, digestion is slower.

Protective Factors The intestine contains bacteria that produce a variety of vitamins, including biotin and vitamin K (although bacteria alone cannot meet the need for these vitamins). The GI bacteria also protect people from infections. Provided that the normal **intestinal flora** are thriving, infectious bacteria have a hard time getting established and launching an attack on the system. In addition, the small intestine and the entire GI tract manufacture and maintain a strong arsenal of defenses against foreign invaders. Several different types of defending cells are present there and confer specific immunity against intestinal diseases.

The Final Stage The story of how food is broken down into nutrients that can be absorbed is now nearly complete. The three energy-yielding nutrients—carbohydrate, fat, and protein—are disassembled to basic building blocks before they are absorbed. Most of the other nutrients—vitamins, minerals, and water—are absorbed as they are. Undigested residues, such as some fibers, are not absorbed but continue through the digestive tract, providing a semisolid mass that helps stimulate the muscles of the GI tract so that they will remain strong and perform peristalsis efficiently. Fiber also retains water, keeping the stools soft, and carries bile acids, sterols, and fat with it out of the body. Drinking plenty of water in conjunction with eating foods high in fiber supplies fluid for the fiber to take up. This is the basis for the recommendation to drink water and eat fiber-rich foods to relieve constipation.

The process of absorbing the nutrients into the body is discussed in the next section. For the moment, let us assume that the digested nutrients simply disappear from the GI tract as they are ready. Virtually all nutrients are gone by the time the contents of the GI tract reach the end of the small intestine. Little remains but water, a few undissolved salts and body secretions, and undigested materials such as fiber. These enter the large intestine (colon).

intestinal flora: the bacterial inhabitants of the GI tract.
flora = plant growth

In the colon, intestinal bacteria degrade some of the fiber to simpler compounds. The colon itself retrieves from its contents the materials that the body is designed to recycle—water and dissolved salts. The waste that is finally excreted has little or nothing of value left in it. The body has extracted all that it can use from the food.

In Summary To digest food, the salivary glands, stomach, pancreas, liver (via the gallbladder), and small intestine deliver fluids and digestive enzymes.

The Absorptive System

Within three or four hours after you have eaten a meal, your body must find a way to absorb some two hundred thousand million amino acid molecules one by one and a comparable number of monosaccharide, monoglyceride, glycerol, fatty acid, vitamin, and mineral molecules as well. The absorptive system is ingeniously designed to accomplish this task.

The Small Intestine

Most absorption takes place in the small intestine. The small intestine is a tube about 10 feet long and about an inch across, yet it provides a surface comparable in area to a tennis court. When nutrient molecules make contact with this surface, they are absorbed and carried off to the liver and other parts of the body.

Villi and Microvilli How does the intestine manage to provide such a large absorptive surface area? Its inner surface looks smooth, but viewed through a microscope, it turns out to be wrinkled into hundreds of folds. Each fold is covered with thousands of fingerlike projections called **villi.** The villi are as numerous as the hairs on velvet fabric. A single villus, magnified still more, turns out to be composed of several hundred cells, each covered with microscopic hairs called **microvilli** (see Figure 5–5 on p. 106).

The villi are in constant motion. A thin sheet of muscle lines each villus so that it can wave, squirm, and wiggle like the tentacles of a sea anemone. Any nutrient molecule small enough to be absorbed is trapped among the microvilli and drawn into a cell beneath them. Some partially digested nutrients are caught in the microvilli, digested further by enzymes there, and then absorbed into the cells.

Specialization in the Intestinal Tract As you can see, the intestinal tract is beautifully designed to perform its functions. A further refinement of the system is that the cells of successive portions of the tract are specialized to absorb different nutrients. The nutrients that are ready for absorption early are absorbed near the top of the tract; those that take longer to be digested are absorbed further down. The rate at which the nutrients travel through the GI tract is finely adjusted to maximize their availability to the appropriate absorptive segment of the tract when they are ready. The lowly "gut" turns out to be one of the most elegantly designed organ systems in the body.

The Myth of "Food Combining" The idea that people should not eat certain food combinations (for example, fruit and meat) at the same meal, because the digestive system cannot handle more than one task at a time, is a myth. The art of "food combining" (which actually emphasizes "food separating") is based on this idea, and it represents faulty logic and a gross underestimation of the body's capabilities. In fact, the contrary is often true; foods eaten together can enhance each other's use by the body. For example, vitamin C in a pineapple or other

villi (VILL-ee or VILL-eye): fingerlike projections from the folds of the small intestine. The singular form is **villus.**
  villus = shaggy hair

microvilli (MY-cro-VILL-ee or MY-cro-VILL-eye): tiny, hairlike projections on each cell of every villus that can trap nutrient particles and transport them into the cells. The singular form is **microvillus.**

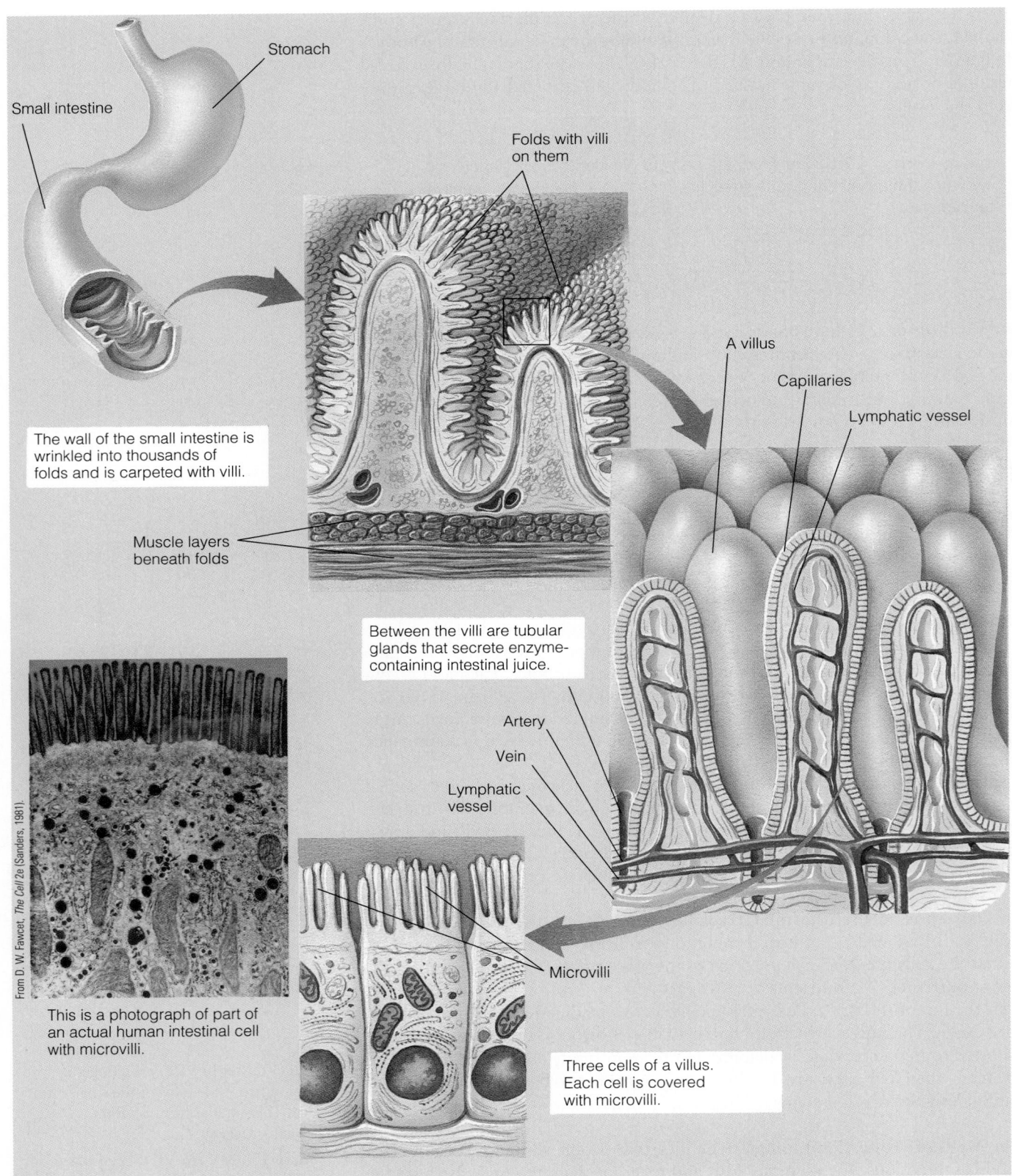

Stomach

Small intestine

Folds with villi on them

The wall of the small intestine is wrinkled into thousands of folds and is carpeted with villi.

Muscle layers beneath folds

A villus

Capillaries

Lymphatic vessel

Between the villi are tubular glands that secrete enzyme-containing intestinal juice.

Artery

Vein

Lymphatic vessel

From D. W. Fawcett, *The Cell* 2e (Sanders, 1981).

This is a photograph of part of an actual human intestinal cell with microvilli.

Microvilli

Three cells of a villus. Each cell is covered with microvilli.

Figure 5–5
The Small Intestinal Villi

citrus fruit can enhance the absorption of iron from a meal of chicken and rice or other iron-containing foods. Many other instances of mutually beneficial interactions are presented in later chapters.

Release of Absorbed Nutrients

Once a molecule has entered a cell in a villus, the next step is to transmit it to a destination elsewhere in the body by way of the body's two transport systems—the bloodstream and the **lymphatic system.** As Figure 5–5 shows, both systems supply vessels to each villus. Through these vessels, the nutrients leave the cell and enter either the **lymph** or the blood. In either case, the nutrients end up in the blood, at least for a while. The water-soluble nutrients (and the smaller products of fat digestion) are released directly into the bloodstream by way of the capillaries, but the larger fats and the fat-soluble vitamins find direct access into the capillaries impossible because these nutrients are insoluble in water (and blood is mostly water). They require some packaging before they are released.

The intestinal cells assemble the monoglycerides and long-chain fatty acids into larger triglyceride molecules. These triglycerides, fat-soluble vitamins (when present), and other large lipids (cholesterol and the phospholipids) are then packaged for transport. They cluster together with special proteins to form **chylomicrons,** one kind of **lipoproteins** (lipoproteins are described beginning on p. 108). Finally, the cells release the chylomicrons into the lymphatic system. They can then glide through the lymph spaces until they arrive at a point of entry into the bloodstream near the heart.

In Summary The many folds and villi of the small intestine dramatically increase its surface area, facilitating nutrient absorption. Nutrients pass through the cells of the villi and enter either the blood (if they are water soluble or small fat fragments) or the lymph (if they are fat soluble).

Transport of Nutrients

Once a nutrient has entered the bloodstream or the lymphatic system, it may be transported to any part of the body and thus becomes available to any of the cells, from the tips of the toes to the roots of the hair. The circulatory systems are arranged to deliver nutrients wherever they are needed.

The Vascular System

The vascular or blood circulatory system is a closed system of vessels through which blood flows continuously in a figure eight, with the heart serving as a pump at the crossover point. On each loop of the figure eight, blood travels a simple route: heart to arteries to capillaries to veins to heart.

The routing of the blood through the digestive system is different, however. The blood is carried to the digestive system (as it is to all organs) by way of an **artery,** which (as in all organs) branches into **capillaries** to reach every cell. Blood leaving the digestive system, however, goes by way of a **vein,** not back to the heart, but to the liver. This vein again branches into capillaries so that every cell of the liver has access to the newly absorbed nutrients that the blood is carrying. Blood leaving the liver then returns to the heart by way of another vein. The route is heart to arteries to capillaries (in intestines) to vein to capillaries (in liver) to vein to heart.

An anatomist studying this system knows there must be a reason for this special arrangement. The liver is located in the circulation system at the point where it will have the first chance at the materials absorbed from the GI tract. In fact, the liver is the body's major metabolic organ (see Figure 5–6 on page 108) and

The blood arriving at the intestines flows through the **mesentery** (MEZ-en-terry), a strong, flexible membrane that surrounds and supports the abdominal organs.
  mes = middle

The vein that collects blood from the mesentery and conducts it to capillaries in the liver is the **portal vein.**
  portal = gateway

The vein that collects blood from the liver capillaries and returns it to the heart is the **hepatic vein.**
  hepat = liver

The artery that delivers oxygen-rich blood from the heart to the liver is the **hepatic artery.**

lymphatic system: a loosely organized system of vessels and ducts that conveys the products of digestion toward the heart.

lymph (LIMF): the body fluid found in lymphatic vessels. Lymph consists of all the constituents of blood except red blood cells.

chylomicrons (kye-lo-MY-crons): the lipoproteins that transport lipids from the intestinal cells into the body. The cells of the body remove the lipids they need from the chylomicrons, leaving chylomicron remnants to be picked up by the liver cells.

lipoproteins: clusters of lipids associated with proteins that serve as transport vehicles for lipids in the lymph and blood.

artery: a vessel that carries blood away from the heart.

capillaries: small vessels that branch from an artery. Capillaries connect arteries to veins. Oxygen, nutrients, and waste materials are exchanged across capillary walls.

vein: a vessel that carries blood back to the heart.

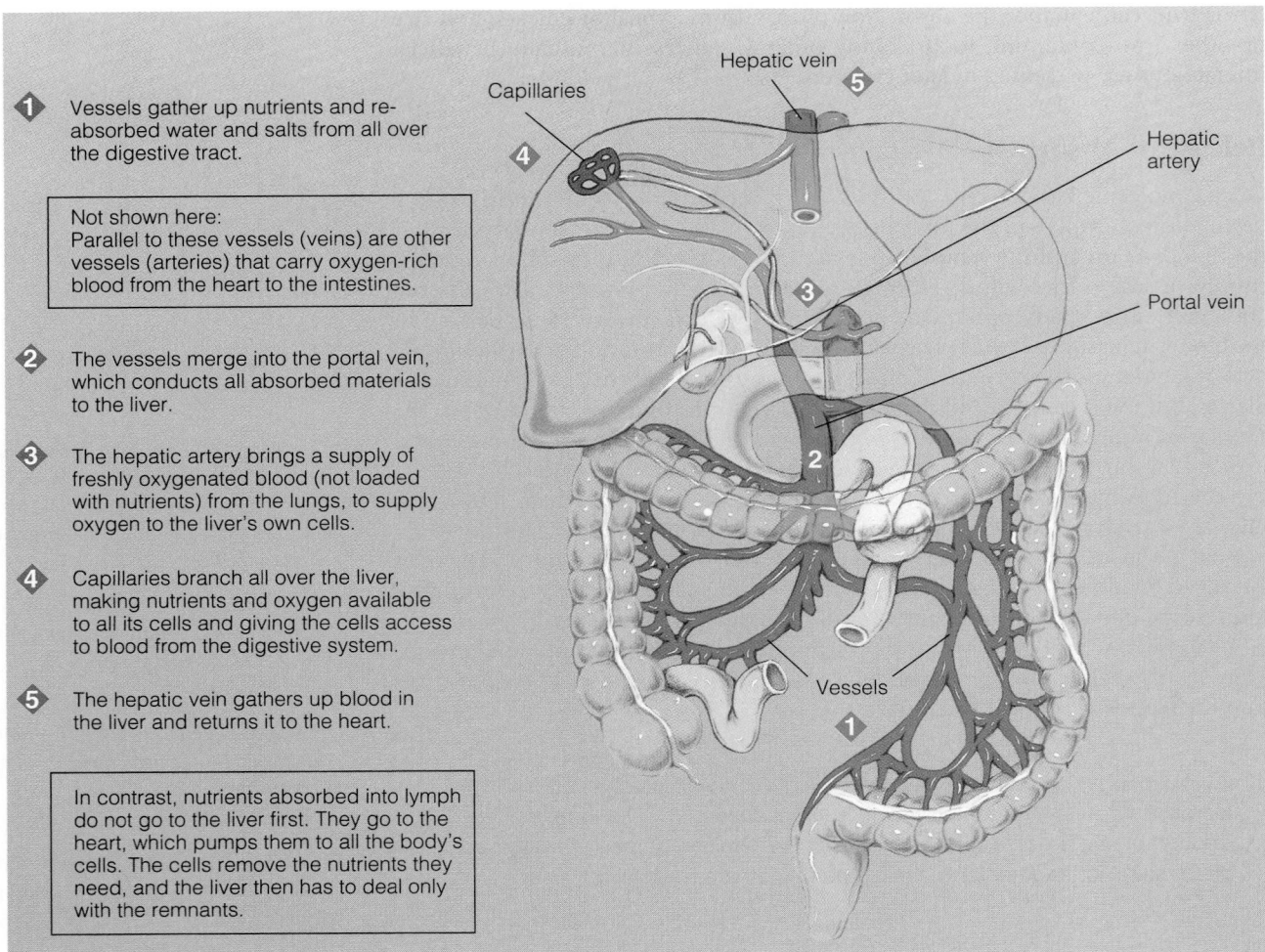

① Vessels gather up nutrients and re-absorbed water and salts from all over the digestive tract.

Not shown here:
Parallel to these vessels (veins) are other vessels (arteries) that carry oxygen-rich blood from the heart to the intestines.

② The vessels merge into the portal vein, which conducts all absorbed materials to the liver.

③ The hepatic artery brings a supply of freshly oxygenated blood (not loaded with nutrients) from the lungs, to supply oxygen to the liver's own cells.

④ Capillaries branch all over the liver, making nutrients and oxygen available to all its cells and giving the cells access to blood from the digestive system.

⑤ The hepatic vein gathers up blood in the liver and returns it to the heart.

In contrast, nutrients absorbed into lymph do not go to the liver first. They go to the heart, which pumps them to all the body's cells. The cells remove the nutrients they need, and the liver then has to deal only with the remnants.

Figure 5–6
The Liver and Its Circulatory System

The duct that conveys lymph toward the heart is the **thoracic** (thor-ASS-ic) **duct.** The **subclavian vein** connects this duct with the right upper chamber of the heart, providing a passageway by which lymph can be returned to the vascular system.

must prepare the absorbed nutrients for use by the rest of the body. Furthermore, the liver stands as a gatekeeper to waylay intruders that might otherwise harm the heart or brain. Chapter 23 offers more information about this noble organ.

The Lymphatic System

The lymphatic system is a one-way route for fluids to travel from tissue spaces into the blood. The lymphatic system has no pump; instead, lymph is squeezed from one portion of the body to another like water in a sponge, as muscles contract and create pressure here and there. Ultimately, the lymph collects in a large duct behind the heart. This duct terminates in a vein that conducts the lymph into the heart. Thus some materials from the GI tract enter the lymphatic system before entering the bloodstream.

Transport of Lipids: Lipoproteins

Within the circulatory system, lipids always travel from place to place bundled with protein, that is, as lipoproteins. When physicians measure a person's blood lipid profile, they are interested not only in the types of fat present (such as triglycerides and cholesterol) but also in the types of lipoproteins that carry them.

VLDL, LDL, and HDL As mentioned earlier, chylomicrons transport newly absorbed (*diet-derived*) lipids from the intestinal cells to the rest of the body. As chylomicrons circulate through the body, cells remove their lipid contents, so the chylomicrons get smaller and smaller. The liver picks up the chylomicron remnants and assembles new lipoproteins, which are known as **very-low-density lipoproteins (VLDL)**. As the body's cells remove triglycerides from the VLDL, the proportion of their contents shifts. As this occurs, VLDL become cholesterol-rich **low-density lipoproteins (LDL)**. Lipids returning to the liver for metabolism or excretion from other parts of the body are packaged in lipoproteins known as **high-density lipoproteins (HDL)**.

The more lipid in the lipoprotein molecule, the lower the density; the more protein, the higher the density. Both LDL and HDL carry lipids around in the blood, but LDL are larger, lighter, and more lipid filled; HDL are smaller, denser, and packaged with more protein. LDL deliver cholesterol and triglycerides from the liver to the tissues; HDL scavenge excess cholesterol and phospholipids from the tissues and return them to the liver for metabolism or disposal. Figure 5–7 (p. 110) shows the relative sizes and composition of the lipoproteins.

Health Implications of LDL and HDL The distinction between LDL and HDL has implications for the health of the heart and blood vessels. Elevated LDL concentrations in the blood are associated with a high risk of heart disease, and elevated HDL concentrations are associated with a low risk.[3] These associations explain why some people refer to LDL as "bad" cholesterol and HDL as "good" cholesterol. Keep in mind, though, that there is only *one* kind of cholesterol; the differences between LDL and HDL reflect *proportions* of lipids and proteins within them—not the type of cholesterol. Factors that improve the LDL-to-HDL ratio include:

- Weight control (see Chapter 7).
- Polyunsaturated or monounsaturated, instead of saturated, fatty acids in the diet (see Chapter 3).
- Soluble fibers (see Chapter 2).
- Physical activity (see Chapter 10).

Lipoproteins and heart disease are discussed in Chapter 25.

In Summary Nutrients leaving the digestive system via the blood are routed directly to the liver before being transported to the body's cells. Those leaving via the lymphatic system eventually enter the vascular system, but bypass the liver at first. Within the circulatory system, lipids travel bundled with proteins as lipoproteins. Different types of lipoproteins include chylomicrons, very-low-density lipoproteins (VLDL), low-density lipoproteins (LDL), and high-density lipoproteins (HDL). Elevated blood concentrations of LDL are associated with a high risk of heart disease. Elevated HDL are associated with a low risk of heart disease.

The System at Its Best

The GI tract is the first organ in the body to deal with the nutrients that will ultimately maintain the health and nutrition status of the whole body. The intricate architecture of the GI tract makes it sensitive and responsive to conditions in its environment. One condition indispensable to its performance is its own good health. Such lifestyle factors as sleep, physical activity, state of mind, and nutrition affect GI tract health. Adequate sleep allows for repair and maintenance of tissue. Physical activity promotes healthy muscle tone and may protect against cancer of the colon.[4] Mental state profoundly affects digestion and absorption through the activity of nerves and hormones that help regulate these processes. A relaxed, peaceful attitude during a meal enhances digestion and absorption.

very-low-density lipoproteins (VLDL): the type of lipoproteins made primarily by liver cells to transport lipids to various tissues in the body; composed primarily of triglycerides.

low-density lipoproteins (LDL): the type of lipoproteins derived from VLDL as cells remove triglycerides from them. LDL carry cholesterol and triglycerides from the liver to the cells of the body and are composed primarily of cholesterol.

high-density lipoproteins (HDL): the type of lipoproteins that transport cholesterol back to the liver from peripheral cells; composed primarily of protein.

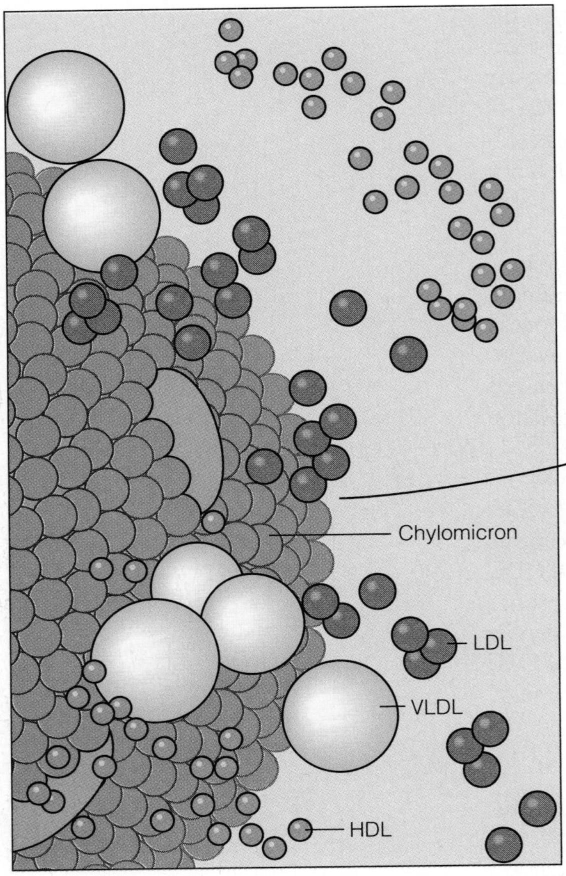

Phospholipid — **Protein**

Cholesterol

Triglyceride

Chylomicron

LDL

VLDL

HDL

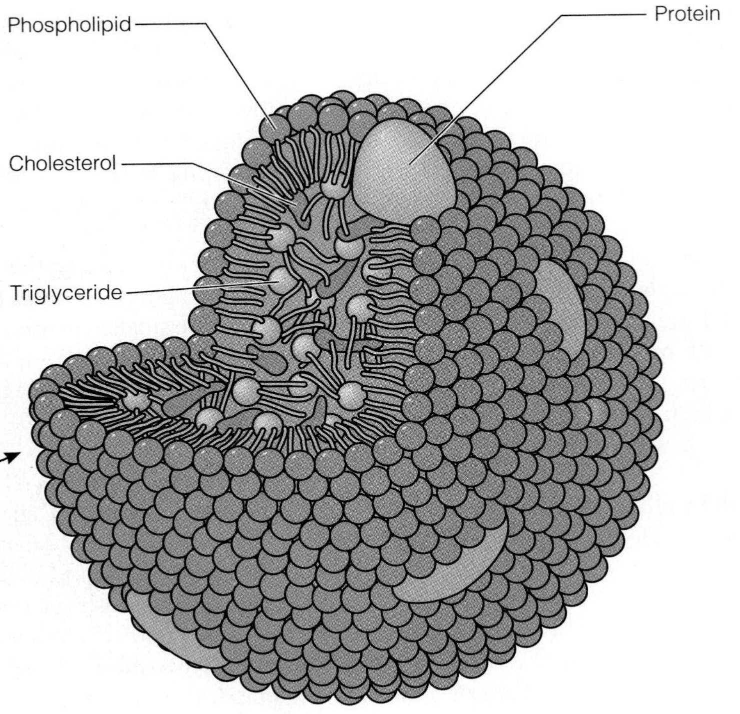

This solar system of lipoproteins shows their relative sizes. Notice how large the fat-filled chylomicron is compared with the others and how the others get progressively smaller as their proportion of fat declines and protein increases.

A typical lipoprotein contains an interior of triglycerides and cholesterol surrounded by phospholipids. The phospholipids' fatty acid "tails" point toward the interior, where the lipids are. Proteins near the outer ends of the phospholipids cover the structure. This arrangement of hydrophobic molecules on the inside and hydrophilic molecules on the outside allows lipids to travel through the watery fluids of the blood.

Chylomicrons contain so little protein and so much triglyceride that they are the lowest in density.

Very-low-density lipoproteins (VLDL) are half triglycerides, accounting for their low density.

Low-density lipoproteins (LDL) are half cholesterol, accounting for their implication in heart disease.

High-density lipoproteins (HDL) are half protein, accounting for their high density.

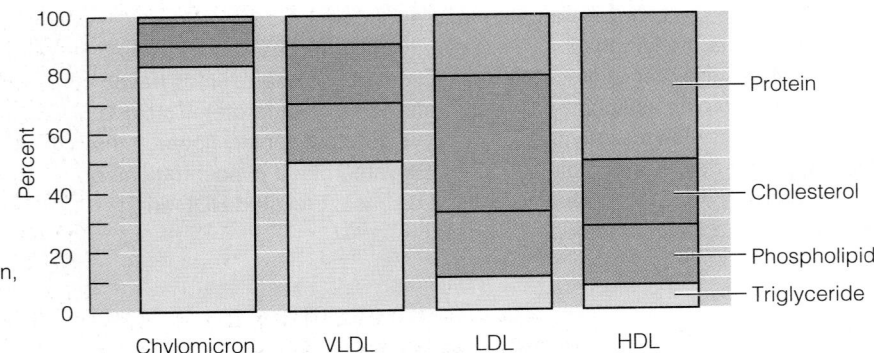

Protein

Cholesterol

Phospholipid

Triglyceride

Figure 5–7
The Lipoproteins

Self Study

DIGESTION AND ABSORPTION

Digestion transforms the foods we eat into nutrients, and absorption moves nutrients from the GI tract into the blood. Optimal digestion and absorption depend on the good health of the digestive tract, which is affected by such lifestyle factors as

sleep, physical activity, state of mind, and the meals you eat. Determine whether the following foods and food habits promote or impede healthy digestion and absorption.

Foods and Food Habits	Promote	Impede
Take small bites of food	_____	_____
Chew thoroughly before swallowing.	_____	_____
Exercise immediately after eating to prevent weight gain.	_____	_____
Eat a low-fiber diet.	_____	_____
Drink plenty of fluids.	_____	_____
Eat a few large meals instead of several smaller ones.	_____	_____
Eat quickly and then lie down to rest.	_____	_____
Create a meal using citrus fruits and meat.	_____	_____
Tackle family problems at the dinner table.	_____	_____

- Do you experience GI distress regularly?
- What changes can you make in your eating habits to promote GI health?

Self Check

1. Once food is swallowed, it travels through the digestive tract in this order:
   a. esophagus, stomach, colon, liver.
   b. esophagus, stomach, small intestine, colon.
   c. small intestine, stomach, esophagus, colon.
   d. small intestine, large intestine, stomach, colon.

2. Once chyme travels the length of the small intestine, it passes through the ileocecal valve at the beginning of the:
   a. colon.
   b. stomach.
   c. esophagus.
   d. jejunum.

3. The periodic squeezing or partitioning of the intestine by its circular muscles that both mixes and slowly pushes the contents along is known as:
   a. secretion.
   b. absorption.
   c. peristalsis.
   d. segmentation.

4. An enzyme in saliva begins the digestion of:
   a. starch.
   b. vitamins.
   c. protein.
   d. minerals.

5. Bile is:
   a. an enzyme that splits starch.
   b. an alkaline secretion of the pancreas.
   c. an emulsifier made by the liver that prepares fats and oils for digestion.
   d. a stomach secretion containing water, hydrochloric acid, and the enzymes pepsin and lipase.

6. Which nutrient passes through the large intestine mostly unabsorbed?
   a. fiber
   b. vitamins
   c. minerals
   d. starch

7. The two major nutrient transport systems in the body are:
   a. LDL and HDL.
   b. digestion and absorption.
   c. lipoproteins and chylomicrons.
   d. vascular and lymphatic systems.

8. Within the circulatory system, lipids always travel from place to place bundled with proteins as:
   a. microvilli.
   b. chylomicrons.
   c. lipoproteins.
   d. phospholipids.

9. Elevated LDL concentrations in the blood are associated with:
   a. a high-protein diet.
   b. a low risk of diabetes.
   c. too much physical activity.
   d. a high risk of heart disease.

10. Three factors that improve the LDL-to-HDL ratio include:
   a. polyunsaturated fat, rest, and dietary HDL.
   b. antioxidants, insoluble fibers, and dietary HDL.
   c. saturated fat, antioxidants, and insoluble fibers.
   d. weight control, soluble fibers, and physical activity.

Answers to these questions can be found in Appendix G.

Clinical Applications

1. What suggestions might you have for a client who eats antacids before and after every meal in the belief that they will prevent or relieve heartburn or acid indigestion?

2. People who experience malabsorption frequently have the most problem digesting fat. Considering the differences in fat, carbohydrate, and protein digestion and absorption, can you offer an explanation?

3. How might you explain the importance of dietary fiber to a client who frequently experiences constipation?

Nutrition on the Net

For further study of the topics of this chapter, access these websites.

Find updates and quick links to these and other nutrition-related sites at our website:
www.wadsworth.com/nutrition

Visit the Center for Digestive Health and Nutrition:
www.gihealth.com

Visit the Digest This! section of the American College of Gastroenterology:
www.acg.gi.org

Notes

1. W. F. Ganong, *Review of Medical Physiology* (Norwalk, Conn.: Appleton & Lang, 1993), pp. 438–465.

2. The small intestine in living adults is almost two and a half times shorter than at death, when muscles are relaxed and elongated. Ganong, 1993.

3. J. M. Dietschy, Theoretical considerations of what regulates low-density lipoprotein and high-density-lipoprotein cholesterol, *American Journal of Clinical Nutrition* 65 (1997): 1581S–1589S.

4. The American Cancer Society 1996 Dietary Guidelines Advisory Committee, *Guidelines on Diet, Nutrition, and Cancer Prevention: Reducing the Risk of Cancer with Healthy Food Choices and Physical Activity*, a booklet.

Nutrition in Practice

■ COMMON DIGESTIVE PROBLEMS ■

The facts of anatomy and physiology presented in Chapter 5 permit easy understanding of some common problems that occasionally arise in the digestive tract. Food may slip into the air passages instead of the esophagus, causing choking. Bowel movements may be loose and watery, as in diarrhea, or painful and hard, as in constipation. Some people complain about belching, while others are bothered by intestinal gas. This Nutrition in Practice describes some of the symptoms of these common digestive problems and offers strategies for preventing them (the accompanying glossary defines these terms).

What happens when a person chokes?

A person chokes when a piece of food slips into the **trachea** and cuts off breathing (see Figure NP5–1 on the next page). Food can lodge so securely that it cuts off all air. No sound can be made because the **larynx** is in the trachea and makes sounds only when air is pushed across it. For this reason, it is imperative that everyone learn to recognize the international signal for choking (shown in Figure NP5–2 on p. 115).

The choking scenario might read like this. A person is dining in a restaurant with friends. A chunk of food, usually meat, becomes lodged in his trachea so firmly that he cannot make a sound. Often he chooses to suffer alone rather than "make a scene in public." If he tries to communicate distress to his friends, he must depend on pantomime. The friends are bewildered by his antics and become terribly worried when he "faints" after a few minutes without air. They call for an ambulance, but by the time it arrives, he is dead from suffocation.

How can someone help a person who is choking?

To help a person who is choking, first ask this critical question: "Can you make any sound at all?" If so, relax. You have time to decide what you can do to help. Whatever you do, don't hit him on the back—the particle may become lodged more firmly in his air passage. If the person cannot make a sound, shout for help and perform the **Heimlich maneuver** (described in Figure NP5–2). You would do well to take a lifesaving course and practice

Glossary

belching: the expulsion of gas from the stomach through the mouth.

colonic irrigation: the popular, but potentially harmful practice of "washing" the large intestine with a powerful enema machine.

constipation: the condition of having infrequent or difficult bowel movements.

defecate (DEF-uh-cate): to move the bowels and eliminate waste.
  defaecare = to remove dregs

diaphragm (DYE-a-fram): the dome-shaped muscle forming a partition between the chest cavity and the abdominal cavity; the muscle responsible for breathing.

diarrhea: the frequent passage of watery bowel movements.

enemas: solutions inserted into the rectum and colon to stimulate a bowel movement and empty the lower large intestine.

heartburn: a burning sensation in the chest area caused by backflow of stomach acid into the esophagus.

Heimlich (HIME-lick) **maneuver (abdominal thrust maneuver):** a technique for dislodging an object from the trachea of a choking person (see Figure NP5–2); named for the physician who developed it.

hemorrhoids (HEM-oh-royds): painful swelling of the veins surrounding the rectum.

hiccups (HICK-ups): repeated cough-like sounds and jerks that are produced when an involuntary spasm of the diaphragm muscle sucks air down the windpipe; also spelled *hiccoughs*.

larynx: the voice box (see Figure NP5–1).

trachea (TRAKE-ee-uh): the windpipe; the passageway from the mouth and nose to the lungs.

vomiting: expulsion of the contents of the stomach up through the esophagus to the mouth.

Nutrition in Practice

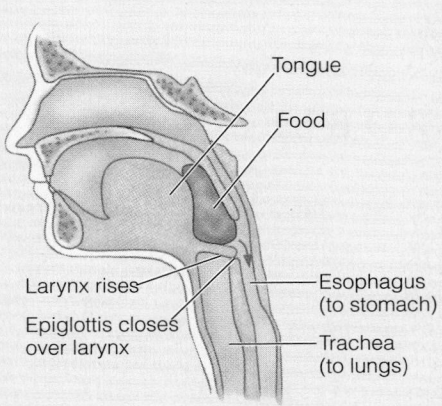

Tongue

Food

Larynx rises

Epiglottis closes
over larynx

Esophagus
(to stomach)

Trachea
(to lungs)

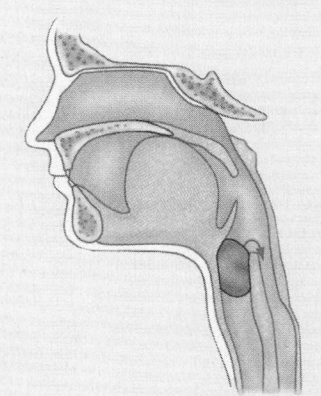

Swallowing. The epiglottis closes
over the larynx, blocking entrance
to the lungs via the trachea. The red
arrow shows that food is heading
down the esophagus normally.

Choking. A choking person cannot
speak or gasp because food lodged
in the trachea blocks the passage of
air. The red arrow points to where
the food should have gone to
prevent choking.

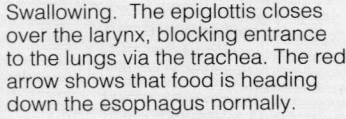

Figure NP5–1
Normal Swallowing and Choking

these techniques, for you will have no time for hesitation
once you are called on to perform this death-defying act.

Do some foods cause choking more than others?

Almost any food can cause choking, although some
are cited more often than others: tough meats, hot
dogs, nuts, grapes, carrots, marshmallows, hard candies, popcorn, and peanut butter eaten off of a spoon.
These foods are particularly difficult for young children to chew and swallow safely. Each year, more than
300 children in the United States choke to death.
Always remain alert to the dangers of choking whenever young children are eating. Children should be
seated, not running around, when eating or drinking.
To prevent choking, cut food into small pieces, chew
thoroughly before swallowing, don't talk or laugh
with food in your mouth, and don't eat when breathing hard.

Please explain what causes vomiting and when it is serious.

Another common digestive mishap is **vomiting.** Vomiting can be a symptom of many different diseases or
may arise in situations that upset the body's equilibrium, such as air or sea travel. For whatever reason, the
lower esophageal sphincter relaxes, the **diaphragm** and
the abdominal muscles contract, and the contents of

the stomach are propelled up
through the esophagus to the mouth
and expelled.

If vomiting continues long
enough or is severe enough, the
reverse peristalsis will extend beyond
the stomach and carry the contents
of the duodenum, with its green bile,
into the stomach and then up the
esophagus. Although certainly
unpleasant and wearying for the
nauseated person, vomiting such as
this is no cause for alarm. Vomiting
is one of the body's adaptive mechanisms to rid itself of something irritating. The best advice is to rest and
drink small amounts of fluids as tolerated until the nausea subsides.

Vomiting can be serious, however, when large quantities of fluid
are lost from the GI tract, causing
dehydration. As Chapter 18 describes, prompt emergency care may be needed in those cases.

What are some cases in which vomiting requires medical attention?

In an infant, vomiting is likely to become serious early
in its course, and a physician should be contacted soon
after onset. Infants have more fluid between their body
cells than adults do, so more fluid can move readily
into the digestive tract and be lost from the body. Consequently, the body water of infants becomes depleted
and their body salt balance upset faster than in adults.

Self-induced vomiting, such as occurs in bulimia
nervosa, also has serious consequences. In addition to
fluid and salt imbalances, repeated vomiting can cause
irritation and infection of the pharynx, esophagus, and
salivary glands; erosion of the teeth; and dental caries.
The esophagus may rupture or tear, as may the stomach.
Sometimes the eyes become red from pressure during
vomiting. Bulimic behavior reflects underlying problems
that require intervention. (Bulimia nervosa is discussed
in the Nutrition in Practice following Chapter 7.)

Projectile vomiting is also serious. The contents of
the stomach are expelled with such force that they leave
the mouth in a wide arc like a bullet leaving a gun. This
type of vomiting requires immediate medical attention.

What about diarrhea? What should be done about it, and when is medical help needed?

Diarrhea is characterized by frequent, loose, watery
stools. Such stools indicate that the intestinal contents

Figure NP5–2
First Aid for Choking

• The strategy most likely to succeed is abdominal thrusts, sometimes called the Heimlich maneuver.

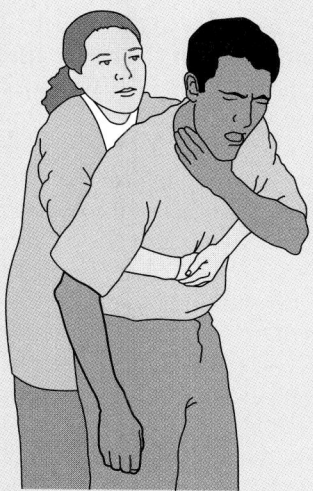

This universal signal for choking alerts others to the need for assistance. Stand behind the person, and wrap your arms around him. Place the thumb side of one fist snugly against his body, slightly above the navel and below the rib cage. Grasp your fist with your other hand and give him a sudden strong hug inward and upward. Repeat thrusts as necessary.

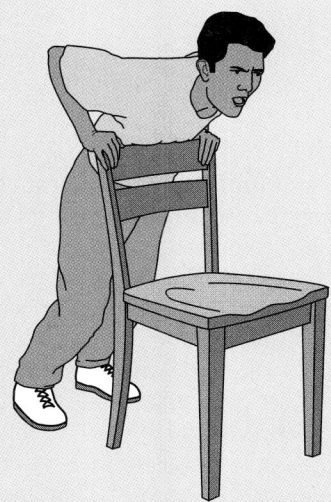

To self-administer first aid, place the thumb side of one fist slightly above the navel and below the rib cage, grasp the fist with your other hand, and then press inward and upward with a quick motion. If this is unsuccessful, quickly press your upper abdomen over any firm surface such as the back of a chair, a countertop, or a railing.

• If all else fails, open the mouth by grasping both the tongue and lower jaw and lifting. Then, and only if you can see the object, use your finger to sweep it out and begin rescue breathing.

have moved too quickly through the intestines for fluid absorption to take place, or that water has been drawn from the cells lining the intestinal tract and added to the food residue. Like vomiting, diarrhea can lead to considerable fluid and salt losses, but the composition of the fluids is different. Stomach fluids lost in vomiting are highly acidic, whereas intestinal fluids lost in diarrhea are nearly neutral. When fluid losses require medical attention, correct replacement is crucial.

Diarrhea is a symptom of a variety of medical conditions and treatments. It may occur abruptly in a healthy person as a result of infections (such as food poisoning) or as a side effect of medications. When used in large quantities, food ingredients such as the sugar alternative sorbitol and the fat alternative olestra may also cause diarrhea in some people. If a food is responsible, then that food must be omitted from the diet, at least temporarily. If medication is responsible, a different medicine, when possible, or a different form (injectable versus oral, for example) may alleviate the problem.

As you can see, treatment for diarrhea depends on its cause and its severity.[1] Mild diarrhea may remit without treatment; simply rest and drink fluids to replace losses. If diarrhea persists, though, especially in an infant, call a physician. Severe diarrhea can lead to dehydration and electrolyte imbalances.

What is constipation, and when does it require medical attention?

Like diarrhea, **constipation** describes a symptom, not a disease. Each person's GI tract has its own cycle of waste elimination, which depends on its owner's health, the type of food eaten, when it was eaten, and when the person takes time to **defecate.** For some people, bowel movements occur daily; for others, several days may pass between movements. Only when people pass stools that are difficult or painful to expel or when they experience a reduced frequency of bowel movements from their typical pattern are they constipated. Abdominal discomfort, headaches, backaches, and the passing of gas sometimes accompany constipation.

Often a person's lifestyle may cause constipation. Being too busy to respond to the defecation signal is a common complaint. If a person receives the signal to defecate and ignores it, the signal may not return for several hours. In the meantime, water continues to be withdrawn from the fecal matter, so when the person does defecate, the bowel movement is dry and hard. In such a case, a person's daily regimen may need to be revised to allow time to have a bowel movement when the body sends its signal. One possibility is to go to bed earlier in order to rise earlier, allowing ample time for a leisurely breakfast and a movement.

Another cause of constipation is lack of physical activity. Physical activity improves muscle tone, not just of the outer body, but also of the digestive tract.

Nutrition in Practice

Although constipation usually reflects lifestyle habits, in some cases it may be a side effect of medication or may reflect a medical problem such as tumors that are obstructing the passage of waste. If discomfort is associated with passing fecal matter, seek medical advice to rule out disease. Once this has been done, dietary or other measures for correction can be considered.

Does fiber, water, prune juice, or honey help relieve ordinary constipation?

Yes to all four. Fibers in cereal products help to prevent constipation by increasing fecal mass. In the GI tract, fiber attracts water, creating soft, bulky stools that stimulate bowel contractions to push the contents along. These contractions strengthen the intestinal muscles. The improved muscle tone, together with the water content of the stools, eases elimination, reducing the pressure in the rectal veins and helping to prevent **hemorrhoids.**

Drinking plenty of water in conjunction with eating high-fiber foods also helps with constipation. The increased bulk physically stimulates the upper GI tract, promoting peristalsis throughout.

Prunes are high in fiber and also contain a laxative substance.* If a morning defecation is desired, a person can drink prune juice at bedtime; if the evening is preferred, the person can drink prune juice with breakfast.

Honey can also have a laxative effect due to its incomplete absorption.[2] Although this characteristic may cause problems for people with GI tract disorders such as irritable bowel syndrome (see Chapter 19), eating honey may be an easy and effective treatment for those who are constipated. Honey should never be fed to infants, however, because of the risk of botulism (as explained in Chapter 11).

Another simple tactic that can help with some constipation is to add some fat to the diet. Fat summons bile into the duodenum. Bile's high salt content draws water from the intestinal wall, which stimulates peristalsis and softens the fecal matter.

Is it ever necessary to take laxatives, enemas, or mineral oil for constipation?

The changes in lifestyle or diet suggested above should correct chronic constipation without the use of laxatives, **enemas,** or mineral oil, although television commercials often try to persuade people otherwise. One of the fallacies often perpetrated by advertisements is that one person's successful use of a product is a good recommendation for others to use that product.

As a matter of fact, even diet changes that relieve constipation for one person may increase the constipation of another. For instance, increasing fiber intake stimulates peristalsis and helps the person with a sluggish colon. Some people, though, have a spastic type of constipation, in which peristalsis promotes strong contractions that close off a segment of the colon and prevent passage; for these people, increasing fiber intake would be exactly the wrong thing to do.

A person who seems to need products such as laxatives should seek a physician's opinion. Advice from friends or alternative medicine practitioners may cause more harm than good. One potentially harmful but currently popular practice that is being promoted by some alternative medicine practitioners is **colonic irrigation**— the internal washing of the large intestine with a powerful enema machine. Not only is such an extreme cleansing unnecessary, but the force of the machine can rupture the intestine. Less extreme practices can cause problems, too. Frequent use of laxatives and enemas can lead to dependency; upset the body's fluid, salt, and mineral balances; and, in the case of mineral oil, interfere with the absorption of fat-soluble vitamins. (Mineral oil dissolves the vitamins, but is not itself absorbed; instead, it leaves the body, carrying the vitamins with it.)

You promised to discuss belching and gas, too.

Many people complain of problems that they attribute to excessive gas. For some, **belching** is the complaint. Others blame intestinal gas for abdominal discomforts and embarrassment. Most people believe that the problems occur after they eat certain foods. This may be the case with intestinal gas, but belching results from swallowing air. The best advice for belching seems to be to eat slowly, chew thoroughly, and relax while eating.

Everyone swallows a little bit of air with each mouthful of food, but people who eat too fast may swallow too much air and then have to belch. Ill-fitting dentures, carbonated beverages, and chewing gum can also contribute to the swallowing of air with resultant belching. Occasionally, belching can be a sign of a more serious disorder, such as gallbladder disease or a peptic ulcer.

People who eat or drink too fast may also trigger **hiccups,** the repeated spasms that produce cough-like sounds and jerky movements. Normally, hiccups soon subside and are of no medical significance, but they can be bothersome. The most effective cure is to hold the breath for as long as possible, which helps to relieve the spasms of the diaphragm.

*This substance is dihydroxyphenyl isatin.

Nutrition in Practice

Beans, broccoli, cabbage, and onions produce gas in many people. People troubled by gas need to determine which foods bother them and then eat those foods in moderation.

Though expelling gas can be a humiliating experience, it is quite normal. (People experiencing painful bloating from malabsorption diseases, however, require medical treatment.) Healthy people expel several hundred milliliters of gas several times a day. Almost all (99 percent) of the gases expelled—nitrogen, oxygen, hydrogen, methane, and carbon dioxide—are odorless. The remaining "volatile" gases are the infamous ones.

Foods that produce gas usually must be determined individually. Table 19–2, later in the book, lists some of the likely candidates. The most common offenders are foods rich in the carbohydrates—sugars, starches, and fibers. When partially digested carbohydrates reach the large intestine, bacteria digest them, giving off gas as a by-product. People can test foods suspected of forming gas by omitting them individually for a trial period and seeing if there is any improvement.

What is heartburn?

Almost everyone has experienced heartburn at one time or another, usually soon after eating a meal. **Heartburn** is the painful sensation a person feels behind the breastbone when the lower esophageal sphincter fails to prevent the stomach contents from refluxing into the esophagus. This may happen if a person eats or drinks too much (or both). Tight clothing and even changes of position (lying down, bending over) can cause it, too, as can some medications and

smoking. A defect of the sphincter muscle itself is a possible, but less common, cause.

If the heartburn is not caused by an anatomical defect, treatment is fairly simple. To avoid such misery in the future, the person needs to learn to eat less at a sitting, chew food more thoroughly, and eat it more slowly.

What about more serious digestive problems?

Some illnesses seriously affect the digestion and absorption of nutrients. Individuals with these disorders risk severe malnutrition, and the diet must be altered to ensure health. The details of these disorders and their diet therapies are described in Chapters 18 through 28.

Table NP5–1 summarizes strategies to prevent or alleviate common GI problems. Many of these problems reflect hurried lifestyles. For this reason, many of their remedies require that people slow down and take the time to eat leisurely; chew food thoroughly to prevent choking and heartburn; rest until vomiting and diarrhea subside; and heed the urge to defecate. In addition, learn how to handle life's day-to-day problems

Table NP5–1 Strategies to Prevent or Alleviate Common GI Problems	
GI Problem	**Strategies**
Choking	▪ Take small bites of food. ▪ Chew thoroughly before swallowing. ▪ Don't talk or laugh with food in your mouth. ▪ Don't eat when breathing hard.
Diarrhea	▪ Rest. ▪ Drink fluids to replace losses. ▪ Call for medical help if diarrhea persists.
Constipation	▪ Eat a high-fiber diet. ▪ Drink plenty of fluids. ▪ Exercise regularly. ▪ Respond promptly to the urge to defecate.
Belching	▪ Eat slowly. ▪ Chew thoroughly. ▪ Relax while eating.
Intestinal gas	▪ Eat bothersome foods in moderation.
Heartburn	▪ Eat small meals. ▪ Drink liquids between meals. ▪ Sit up while eating. ▪ Wait 1 hour after eating before lying down. ▪ Wait 2 hours after eating before exercising. ▪ Refrain from wearing tight-fitting clothing. ▪ Avoid foods, beverages, and medications that aggravate your heartburn. ▪ Refrain from smoking cigarettes. ▪ Lose weight if overweight.

Nutrition in Practice

and challenges without overreacting and becoming upset; learn how to relax, to get enough sleep, and to enjoy life. Remember, "what's eating you" may cause more GI distress than what you eat.

Nutrition on the Net

For further study of the topics of this Nutrition in Practice, access these websites.

Find updates and quick links to these and other nutrition-related sites at our website:
www.wadsworth.com/nutrition

Search for choking, vomiting, diarrhea, constipation, heartburn, indigestion, and ulcers at the U.S. Government health information site:
www.healthfinder.gov

Visit the Center for Digestive Health and Nutrition:
www.gihealth.com

Visit the Digestive Diseases section of the National Institute of Diabetes & Digestive & Kidney Diseases:
www.niddk.nih.gov/health/health.htm

Visit the Digest This! section of the American College of Gastroenterology:
www.acg.gi.org

Notes

1. M. Donowitz, F. T. Kokke, and R. Saidi, Evaluation of patients with chronic diarrhea, *New England Journal of Medicine* 332 (1995): 725–729.

2. S. D. Ladas, D. N. Haritos, and S. A. Raptis, Honey may have a laxative effect on normal subjects because of incomplete fructose absorption, *American Journal of Clinical Nutrition* 62 (1995): 1212–1215.

Jennie Oppenheimer/Studio Zocolo

METABOLISM AND ENERGY BALANCE

*E*very organ, every tissue, and every cell of the body engages in **metabolism,** the chemical reactions involved in releasing energy, breaking down compounds, making new compounds, and transporting compounds from place to place. Viewed from this perspective, the body is a giant factory that works with astounding efficiency to produce a myriad of products and dispose of a myriad of wastes. All these processes are regulated by hormonal signals that coordinate supply and demand in much the same way as a superb communication system coordinates a smoothly functioning economy.

In disease, metabolic processes always become disturbed, and some diseases are caused by metabolic abnormalities. This chapter provides the normal background against which the disruptions caused by diseases can best be understood.

The Organs and Their Metabolic Roles

When the body is functioning normally, every organ plays a metabolic role that serves the others. Metabolic reactions also consume or release energy and therefore affect body weight, with consequences for health.

The Principal Organs

Of particular concern to metabolism are the digestive organs, the liver, the pancreas, the circulatory system, and the kidneys. Together, they perform much of the work of breaking down compounds, making new ones, transporting nutrients and oxygen throughout the body, and removing the waste products generated by metabolic processes.

The Digestive Organs   The digestive system has just received attention in Chapter 5. Notable among the digestive system's activities are the physical processes that allow foods to be transported throughout the GI tract; the production of digestive juices and enzymes; the absorption of nutrients; the making of transport proteins to carry lipids and vitamins to other sites in the body; and in the lower digestive tract, the reabsorption of salts and fluids. The digestive system also possesses the body's most rapidly multiplying cells: when healthy, they replace themselves every few days. Disorders affecting the GI tract lead to failures to ingest, digest, absorb, and metabolize nutrients, as described in Chapters 18 and 19.

The Liver   Nutrients absorbed into the bloodstream are conducted to the liver, as described in Chapter 5. The liver is one of the body's most active metabolic factories. It receives nutrients and metabolizes, packages, stores, or ships them out for use by other organs. It metabolizes and stores most vitamins and many minerals. It manufactures bile, which the body uses in emulsifying fat for digestion and absorption. It metabolizes and detoxifies drugs, prepares waste products for excretion, and participates in iron recycling and blood cell manufacture. It also makes many proteins necessary for health, including immune factors, transport proteins, and clotting factors. When liver disorders disrupt metabolism, they profoundly affect both nutrition and health status, as described in Chapter 23.

The Pancreas   The pancreas contributes digestive juices to the GI tract, but also has another metabolic function: it produces the hormones insulin and glucagon, which regulate the body's use of glucose. It is insulin that prompts cells to take glucose up and use it as fuel; insulin also prompts liver cells in particular to store

metabolism: the sum total of all the chemical reactions that go on in living cells.
> *meta* = among
> *bole* = change

120

glucose as glycogen. The liver cells can later release glucose back into the blood as needed. Glucose is an indispensable fuel for brain and nerve cells and for red blood cells as well. Its availability is therefore crucial to normal nervous system activity and normal blood chemistry. Abnormalities associated with the digestive functions of the pancreas are described in Chapter 23, and those associated with its hormonal functions are described in Chapter 24.

The Heart and Blood Vessels The heart and blood vessels conduct blood with its cargo of nutrients and oxygen to all other body cells and carry wastes from them. Diseases of the heart and arteries therefore affect the health of the whole body. Metabolic reactions that affect the heart and blood vessels include, most importantly, the making and transport of lipoproteins, the carriers of cholesterol and other lipids from the liver to the tissues and back again. When lipoproteins deposit cholesterol in the artery walls, coronary heart disease and high blood pressure can result, increasing the risk of disability or death from heart attacks and strokes. Chapter 25 is devoted to these conditions.

The Kidneys The kidneys are also active metabolic organs. Unceasingly, for 24 hours of every day, they filter waste products from the blood to be excreted in the urine and reabsorb needed nutrients, thereby maintaining the blood's delicate chemical balances. The kidneys' cells also produce compounds that help to regulate blood pressure and convert a precursor compound to active vitamin D, thereby helping to maintain the bones. Disorders of the kidneys nearly always involve the heart and the skeleton; kidney disorders are the subject of Chapter 26.

Energy for Metabolic Work

The metabolic work that the body's cells do, like all work, requires energy, and food supplies that energy. Food in turn gets its energy from the sun, either directly (in the case of photosynthesizing plants, which make carbohydrate) or indirectly (in the case of animals that eat plants). When chemical reactions in cells release stored energy from energy-yielding nutrients, that energy becomes available to do the cells' work.

Heat Energy and Body Temperature The cells of each organ conduct metabolic activities specific to that organ, but in addition, all cells must maintain themselves, and many must reproduce. To do this, they must have all the essential nutrients available to them: the energy nutrients, the vitamins, and the minerals, as well as water. As cells do their metabolic work, the chemical reactions involved release heat, and this heat keeps the body warm. By regulating the rates at which these metabolic reactions release heat energy, the body maintains its constant normal temperature of about 98.6°F.

Accelerated Metabolism During severe stress to the body, metabolism speeds up. Fever sometimes develops. An accelerated metabolism signifies that fuels are being burned at a rate more rapid than normal; this may lead to wasting of body organs and loss of weight including loss of vital lean tissue. Chapter 27 describes the metabolic consequences of severe stress and, in particular, of severe infections, major surgery, and burns. Chapter 28 describes the metabolic consequences of the wasting diseases—cancer and AIDS.

In Summary As this brief discussion has shown, metabolism occurs all through the body, all the time, and supports normal health. The remainder of this chapter delves into one aspect of metabolism—the metabolism of the energy nutrients and the resulting consequences of underweight and overweight.

The Body's Energy Metabolism

The body manages its energy supply with amazing precision. Consider, for example, that a consistent 1 percent error in energy intake can cause a person to become more than 200 pounds overweight in a lifetime. Yet most people maintain their weight within about a 10- to 20-pound range throughout their lives. How do they do this? How does the body manage excess energy? And how does it manage to do without food for prolonged periods—as when someone is starving or fasting? The answers to these questions lie in an understanding of metabolism.

Energy metabolism is defined as the sum total of all the chemical reactions that manage energy nutrients in the body. Earlier chapters introduced the energy-yielding nutrients—carbohydrate, fat, and protein—and showed how they are broken down into basic units that are absorbed into the blood. Picking up from there, what becomes of these nutrients? The question is important because it provides insight into proper and improper ways to aid the body in maintaining or losing weight.

Energy metabolism centers on four basic units:

- From carbohydrates: glucose.
- From lipids: glycerol and fatty acids.
- From proteins: amino acids.

Building up Body Compounds

When not needed by the cells for energy, the basic units of energy-yielding nutrients can be used to build body compounds. The building up of body compounds is known as **anabolism;** this book represents anabolic reactions, wherever possible, by "up" arrows in chemical diagrams (such as those shown in Figure 6–1). Glucose units can be strung together to make glycogen chains. Glycerol and fatty acids can be assembled into triglycerides. Amino acids can be linked together to make proteins. These anabolic reactions, in which simple compounds are put together to form larger, more complex structures, involve doing work and so require energy.

Breaking down Nutrients for Energy

The breaking down of body compounds is known as **catabolism;** catabolic reactions usually release energy and are represented, wherever possible, by "down" arrows in chemical diagrams (see Figure 6–1). Glycogen can be broken down to glucose, triglycerides to fatty acids and glycerol, and protein to amino acids. When the body needs energy, it breaks any or all of the four basic units—glucose, fatty acids, glycerol, and amino acids—into even smaller units.

Glucose Breakdown Glucose breakdown occurs via a pathway known as **glycolysis.** In glycolysis, glucose is broken down to **pyruvate.**[*] Most often, pyruvate is then converted to a smaller compound, **acetyl CoA.** In a series of metabolic reactions called the tricarboxylic acid (TCA) cycle, acetyl CoA splits, and its energy is donated to storage compounds, used to do the body's work, or used to produce heat.

The following sequence is central to an understanding of metabolism:

$$\text{Glucose} \leftrightarrow \text{pyruvate} \rightarrow \text{acetyl CoA} \rightarrow \text{energy.}$$

Notice the two-way arrow between glucose and pyruvate and the one-way arrows after pyruvate. They show that pyruvate can be reconverted to glucose but that acetyl CoA cannot. Any compound that can be converted to pyruvate can be used to make glucose. Any compound that is converted to acetyl CoA cannot be used to make glucose.

[*]The term *pyruvate* means a salt of pyruvic acid. Throughout this book, the ending *–ate* is used interchangeably with *–ic acid;* for our purposes, they mean the same thing.

---

energy metabolism: all the reactions by which the body obtains and spends the energy from food or body stores.

anabolism (an-ABB-o-lism): reactions in which small molecules are put together to build larger ones. Anabolic reactions consume energy.

  ana = up

catabolism (ca-TAB-o-lism): reactions in which large molecules are broken down to smaller ones. Catabolic reactions usually release energy.

  kata = down

glycolysis (gligh-COLL-uh-sis): the metabolic breakdown of glucose to pyruvate.

  glyco = glucose
  lysis = breakdown

pyruvate (PIE-roo-vate): pyruvic acid, a 3-carbon compound derived from glucose, glycerol, and certain amino acids in metabolism.

acetyl CoA (ASS-uh-teel or uh-SEET-ul co-AY): a compound made up of acetic acid (formed from the breakdown of pyruvate) with a molecule of CoA attached to it. **CoA**(co-AY) is a nickname for a small molecule (coenzyme A) that participates in metabolism.

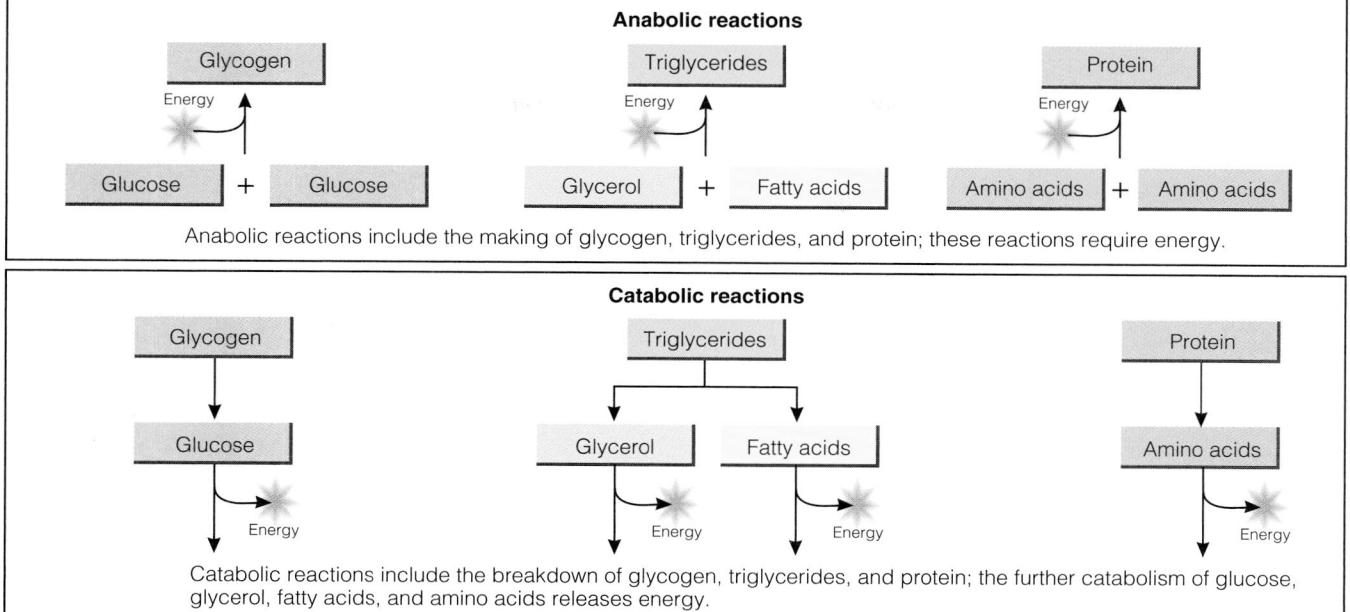

Anabolic reactions

Anabolic reactions include the making of glycogen, triglycerides, and protein; these reactions require energy.

Catabolic reactions

Catabolic reactions include the breakdown of glycogen, triglycerides, and protein; the further catabolism of glucose, glycerol, fatty acids, and amino acids releases energy.

Figure 6–1

Anabolic and Catabolic Reactions Compared

Fat Breakdown   Triglycerides (the primary form of fat in the body) cannot be converted to glucose, for the most part, because they consist mostly of fatty acids. Fatty acids are broken down into 2-carbon fragments that combine with CoA to form acetyl CoA. Because fatty acids are broken down to acetyl CoA, they cannot be used to make glucose. The glycerol portion of a triglyceride is interconvertible with pyruvate and can yield glucose, but glycerol represents only about 5 percent of the weight of a triglyceride molecule.* Thus fat is an inefficient source of glucose. About 95 percent of it cannot be converted to glucose at all; therefore fat, for the most part, cannot provide energy for the organs (brain and nervous system) that require glucose as fuel. This leaves the task of fueling the nervous system's activities mainly to carbohydrate and protein.

Amino Acid Breakdown   Ideally, amino acids are used to maintain supplies of needed body proteins and will not be used for energy. If amino acids are needed for energy, or if they are consumed in excess, they first undergo **deamination**, a reaction in which they are stripped of their nitrogen. The nitrogen can be used to make other compounds, including the nonessential amino acids, or it can be excreted. With nitrogen removed, about half of the amino acids can be converted to pyruvate and can therefore provide glucose. The other amino acids are converted to acetyl CoA directly or enter the TCA cycle at another point. Thus protein, unlike fat, is a fairly efficient source of glucose when carbohydrate is not available.

In Summary   Figure 6–2 on the next page depicts the major metabolic pathways involving carbohydrates, fats, and amino acids. Note the central pathway from glucose to pyruvate to acetyl CoA to energy. Also note that all carbohydrates, some amino acids, and the glycerol from fat can be converted to pyruvate and then to glucose. Finally, note that the vast majority of fragments from fat can be used only for energy and not to make glucose. With these understandings, you can follow the events that lead to weight gain and weight loss.

The reactions by which the complete oxidation of acetyl CoA is accomplished are those of the TCA cycle, or Krebs cycle, and oxidative phosphorylation. The net result is that acetyl CoA splits, and some of its energy is made available for the body's use.

The principal nitrogen-excretion product of metabolism is **urea** (you-REE-uh).

The making of glucose from protein or fat is **gluconeogenesis** (gloo-co-nee-o-JEN-uh-sis). About 5% of fat (the glycerol portion of triglycerides) and about 50% of protein (the glucogenic amino acids) can be converted to glucose.
gluco, glyco = glucose
neo = new
genesis = making

deamination: removal of the amino (NH_2) group from a compound such as an amino acid.

*Figure 3–2 in Chapter 3 showed glycerol (3 carbons) plus 3 fatty acids (most often 16 to 18 carbons) equals a triglyceride. Thus the small glycerol molecule represents only 3 of the 50 or so carbons in the triglyceride.

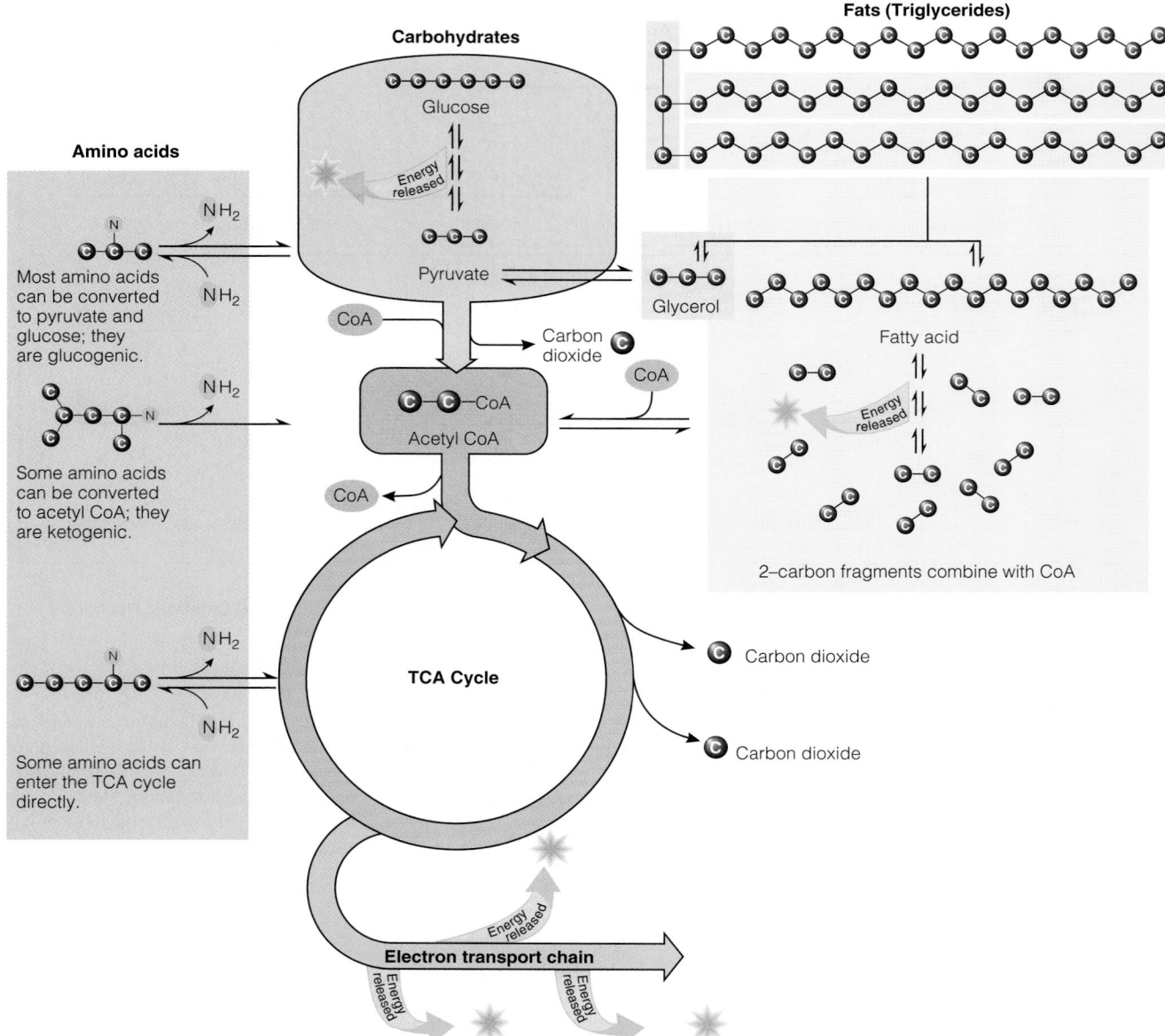

Amino acids

Most amino acids can be converted to pyruvate and glucose; they are glucogenic.

Some amino acids can be converted to acetyl CoA; they are ketogenic.

Some amino acids can enter the TCA cycle directly.

Carbohydrates

Glucose

Energy released

Pyruvate

CoA

Carbon dioxide

CoA

Acetyl CoA

CoA

TCA Cycle

Carbon dioxide

Carbon dioxide

Electron transport chain

Energy released

Fats (Triglycerides)

Glycerol

Fatty acid

Energy released

2–carbon fragments combine with CoA

Figure 6–2
The Central Pathways of Energy Metabolism

The Body's Energy Budget

The average person takes in close to a million kcalories a year and expends more than 99 percent of them, maintaining a stable weight for years on end. In other words, the body's energy budget is balanced. Some people, however, eat too much and get fat; others eat too little and get thin. This section examines metabolism from the perspective of the energy budget, looking first at the two forms of unbalanced budget, feasting and fasting, and then at a balanced budget.

The Economics of Feasting

Everyone knows that when people consume more energy than they expend, much of the excess is stored as body fat. Fat can be made from an excess of any energy-yielding nutrient. In addition, excess energy from alcohol is also stored as fat.[1] Fat cells enlarge as they fill with fat, and the body's fat-storing capacity seems to be able to expand indefinitely, as Figure 6–3 shows.

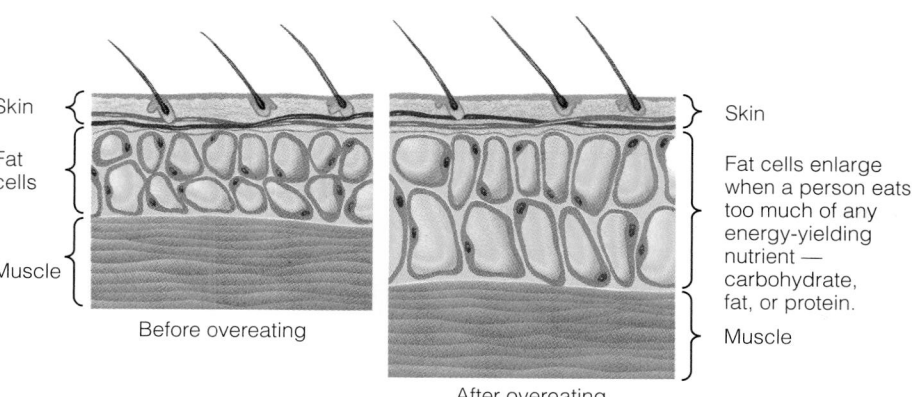

Figure 6–3
Fat Cell Enlargement

Skin

Fat cells

Muscle

Before overeating

Skin

Fat cells enlarge when a person eats too much of any energy-yielding nutrient — carbohydrate, fat, or protein.

Muscle

After overeating

Excess Carbohydrate Surplus carbohydrate (glucose) is first stored as glycogen in the liver and muscles, but the glycogen-storing cells have limited capacity. Once glycogen stores are filled, any additional carbohydrate is routed to fat. Thus excess carbohydrate can contribute to obesity.

Excess Fat Surplus dietary fat contributes easily to the body's fat stores. During digestion and metabolism, fat may break down into fragments, such as acetyl CoA, but if the flow of these fragments is rapid enough to meet the body's need for energy, any excess fragments that are available will be stored as triglycerides in the fat cells.

Excess Protein Surplus protein may encounter the same fate. If not needed to build body protein (as in response to physical activity) or to meet energy needs, amino acids will lose their nitrogens and will be converted through the intermediates, pyruvate and acetyl CoA, to triglycerides. These, too, swell the fat cells and add to body weight. Figure 6–4 on the next page shows the metabolic events of feasting.

In Summary Excess energy from carbohydrate, fat, protein, and alcohol will be stored in the body as fat.

The Economics of Fasting

The body spends energy all the time. Even when a person is asleep and totally relaxed, the cells of many organs are hard at work. In fact, this cellular work, which maintains all life processes, represents about two-thirds of the total energy a person spends in a day. (The other one-third is the work that a person's muscles do voluntarily during waking hours.)

Energy Deficit The body's top priority is to meet the energy needs for this ongoing cellular activity. Its normal way of doing so is by periodic refueling, that is, by eating several times a day. When food is not available, the body uses fuel reserves from its own tissues. If people choose not to eat, we say they are *fasting;* if they have no choice (as in a famine), we say they are *starving.* In the body, no metabolic difference exists between fasting and starving. In either case, the body is forced to switch to a wasting metabolism, drawing on its stores of carbohydrate and fat and, within a day or so, on its vital protein tissues as well.

Glycogen Used First As a fast or period of starvation begins, glucose from the liver's stored glycogen and fatty acids from the body's stored fat both flow into cells to fuel their work. Several hours later, most of the glucose is used up—liver glycogen is exhausted. Low blood glucose concentrations serve as a signal to promote further fat breakdown.

In *fasting*, a person voluntarily stops eating food, whereas in *starvation* the failure to eat is involuntary. The body, however, makes no distinction between the two—metabolically, fasting and starvation are identical.

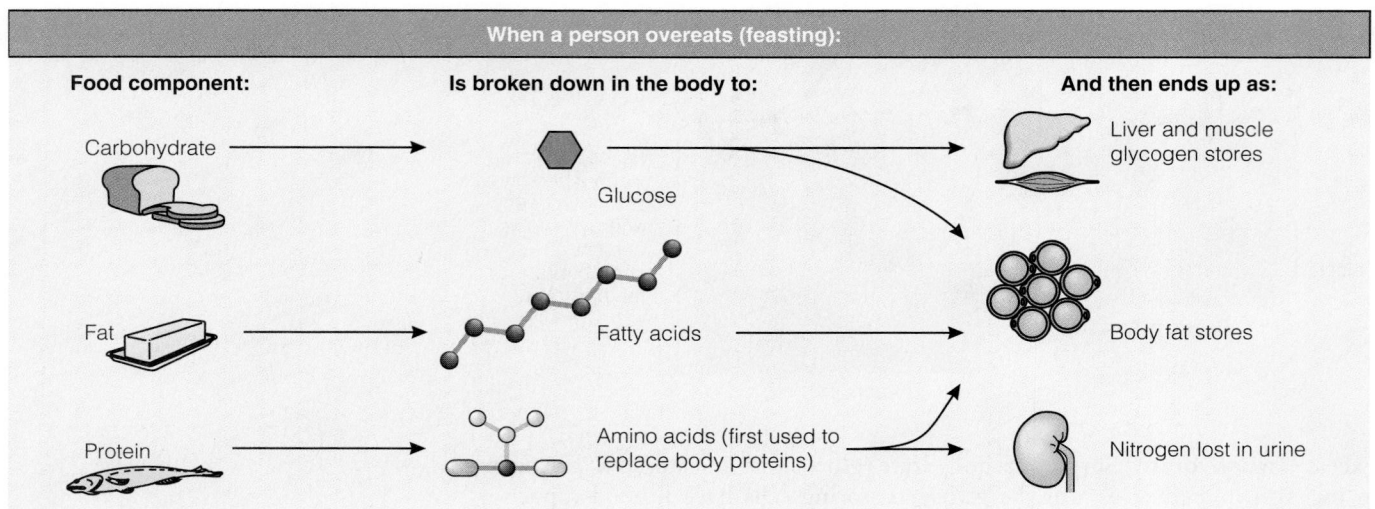

Food component: **Is broken down in the body to:**   **And then ends up as:**

Carbohydrate

Glucose

Liver and muscle glycogen stores

Fat

Fatty acids

Body fat stores

Protein

Amino acids (first used to replace body proteins)

Nitrogen lost in urine

Figure 6–4

Feasting

When people overeat, they store energy.

Reminder: The liver releases glucose, and the fat cells release fat to fuel the body's cells, but the brain can use only glucose.

Reminder: *Ketone bodies* are acidic, fat-related compounds formed from the incomplete breakdown of fat when carbohydrate is not available. Small amounts of ketone bodies are normally produced during energy metabolism, but when their blood concentration rises, they spill into the urine. The combination of high blood ketone bodies (*ketonemia*) and ketone bodies in the urine (*ketonuria*) is called *ketosis.*

In fasting, muscle and lean tissues give up protein to supply amino acids for conversion to glucose. This glucose, with ketone bodies produced from fat, fuels the brain's activities.

Fasting = living on the body's fat and protein.

Glucose Needed for Brain At this point, a few hours into a fast, most of the cells are depending on fatty acids to continue providing fuel. But the nervous system (brain and nerves) and red blood cells cannot use fatty acids; they still need glucose. Even if other energy fuel is available, glucose has to be present to permit the brain's energy-metabolizing machinery to work. Normally, the nervous system consumes about two-thirds of the total glucose used each day—about 400 to 600 kcalories' worth.

Protein Breakdown and Ketosis Because fat stores cannot provide the glucose needed by the brain and nerves, body protein tissues (such as liver and muscle) always break down to some extent during fasting. In the first few days of a fast, body protein provides about 90 percent of the needed glucose, and glycerol provides about 10 percent. If body protein losses were to continue at this rate, death would ensue within about three weeks. As the fast continues, however, the body finds a way to use its fat to fuel the brain. It adapts by condensing together acetyl CoA fragments derived from fatty acids to produce ketone bodies, which can serve as fuel for some brain cells. Ketone body production rises until, after several weeks of fasting, it is meeting much of the nervous system's energy needs. Still, many areas of the brain rely exclusively on glucose, and body protein continues to be sacrificed to produce it. Figure 6–5 shows the metabolic events that occur during fasting.

Slowed Metabolism As fasting continues and the body is shifting to partial dependence on ketone bodies for energy, the body simultaneously reduces its energy output (metabolic rate) and conserves both fat and lean tissue. Because of the slowed metabolism, the loss of fat falls to a bare minimum. Thus, although *weight* loss during fasting may be quite dramatic, *fat* loss may actually be less than when at least some food is supplied.

Hazards of Fasting The body's adaptations to fasting are sufficient to maintain life for a long period. Mental alertness need not be diminished. Even physical energy may remain unimpaired for a surprisingly long time. Still, fasting is not without its hazards. Among the many changes that take place in the body are:

- Wasting of lean tissues.
- Impairment of disease resistance.
- Lowering of body temperature.
- Disturbances of the body's fluid and electrolyte balances.

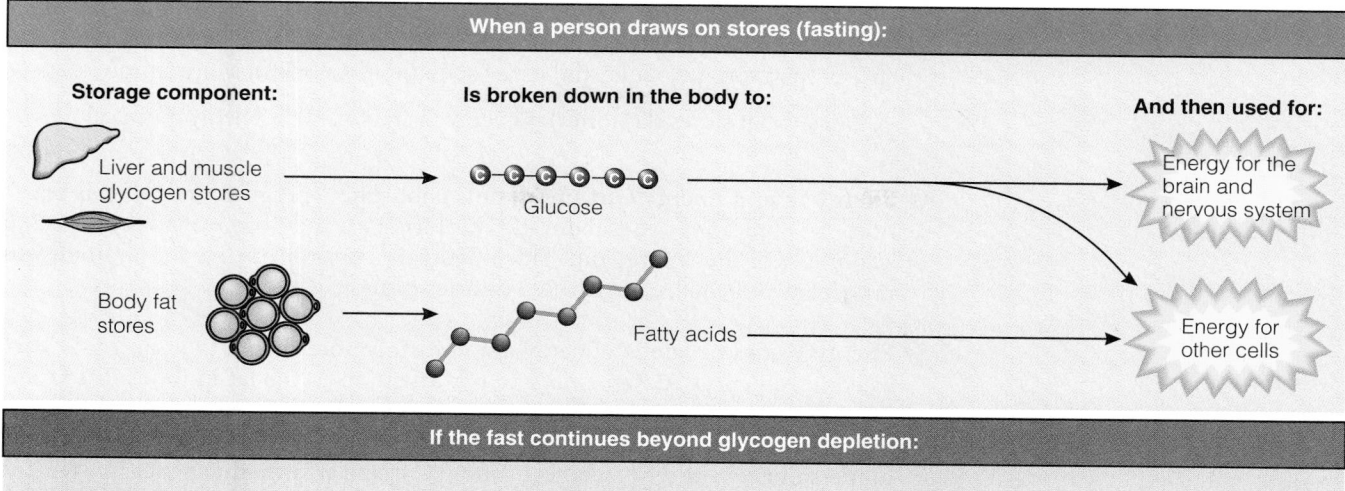

When a person draws on stores (fasting):

Storage component:

Liver and muscle glycogen stores

Body fat stores

Is broken down in the body to:

Glucose

Fatty acids

And then used for:

Energy for the brain and nervous system

Energy for other cells

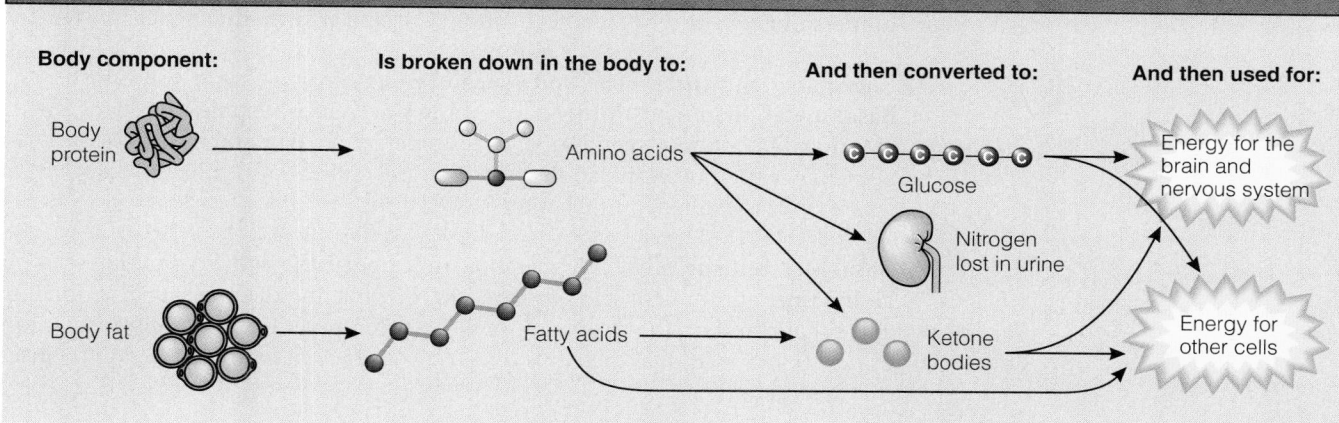

If the fast continues beyond glycogen depletion:

Body component:

Body protein

Body fat

Is broken down in the body to:

Amino acids

Fatty acids

And then converted to:

Glucose

Nitrogen lost in urine

Ketone bodies

And then used for:

Energy for the brain and nervous system

Energy for other cells

Figure 6–5

Fasting
When people are fasting, they draw on stored energy.

For the person who wants to lose weight, fasting is a dangerous way to go. The body's lean tissue continues to be degraded, sometimes amounting to as much as 50 percent of the weight lost. Over the long term, a diet only moderately restricted in energy can actually promote a greater rate of *weight* loss, a faster rate of *fat* loss, and the retention of more lean tissue than a severely restricted fast.

Alterations similar to those in fasting are seen in low-carbohydrate dieting. Renewed food intake, especially of carbohydrate, results in dramatic changes in the body's salt and water balance, accounting for most of the wide swings in body weight seen in people on fasts or low-carbohydrate diets.

In Summary When fasting, the body makes a number of adaptations: increasing the breakdown of fat to provide energy for most of the cells, using glycerol and amino acids to make glucose for the red blood cells and central nervous system, producing ketones to fuel the brain, and slowing metabolism. All of these measures conserve energy and minimize losses. In fact, metabolism slows to such an extent that the loss of fat eventually slows to less than would be achieved with a low-kcalorie diet.

Energy Balance

If a person maintains a healthy weight over time, the person has a balanced energy budget. Food energy intake has equaled energy expenditure, and so, deposits of fat made at one time have been compensated for by withdrawals made at another. In other words, the body uses fat as a savings account for energy. In the case of fat, though, unlike money, more is not better; there is an optimum.

A day's energy balance can be stated like this: Change in body fat stores (expressed in kcalories) equals the food energy taken in (kcalories) minus the energy spent on metabolic and other activities (kcalories). More simply:

Change in fat stores (kcalories) = energy in (kcalories) − energy out (kcalories).

Energy In and Energy Out You know about the "energy in" side of this equation. An apple gives you about 125 kcalories; a candy bar provides about 300 kcalories. On the "energy out" side, if you are physically active for an hour, you may spend 100, 300, or even 500 kcalories or more.

Energy Values of Foods Energy amounts for more than a thousand foods are listed in Appendix A. Remember to pay attention to the fat in foods. Fat kcalories add up quickly and may contribute more to body fat stores than do carbohydrate kcalories, and in the short term, fat contributes little to the feeling of fullness during a meal.[2] Chapter 3 offered strategies for reducing fat intake.

Energy Expenditures The body spends energy in two major ways: to fuel its **basal metabolism** and to fuel its **voluntary activities.** People can change their voluntary activities to spend more or less energy in a day, and over time they can also change their basal metabolism by building up the body's metabolically active lean tissue, as explained in Chapter 7.

Basal Metabolism The basal metabolism supports the body's work that goes on all the time without conscious awareness. The beating of the heart, the inhaling and exhaling of air, the maintenance of body temperature, and the transmission of nerve and hormonal messages to direct these activities are the basal processes that maintain life.

Basal Metabolic Rate The basal metabolic rate (BMR) is the rate at which the body spends energy for these maintenance activities. This rate varies from person to person and may vary for a single individual with a change in circumstance or physical condition. For example, an infant's metabolic rate relative to body weight is much faster than an adult's to support the infant's extraordinary growth rate. In general, BMR is fast in people with considerable lean body mass (growing children, physically active people, pregnant women, and males). One way to increase the BMR then is to maximize lean body tissue by participating regularly in endurance and strength-building activities.[3] BMR is also fast in people who are tall and so have a large surface area for their weight, in people with fever or under stress, in people taking certain medications, and in people with highly active thyroid glands. BMR is slowed down by loss of lean tissue and depression of thyroid hormone activity due to disease, inactivity, fasting, or malnutrition. Table 6–1 summarizes the factors that speed up and slow down the BMR.

Basal Metabolic Needs Basal metabolic needs are surprisingly large. A person whose total energy needs are 2000 kcalories a day spends 1200 to 1400 of them to support basal metabolism. The Clinical Applications feature at the end of this chapter shows how to estimate energy expenditure for basal metabolism.

Energy for Activities The number of kcalories spent on voluntary activities depends on three factors: muscle mass, body weight, and activity. The larger the muscle mass required for the activity and the heavier the weight of the body part being moved, the more kcalories are spent. The activity's duration, frequency, and intensity also influence energy costs: the longer, the more frequent, and the more intense the activity, the more kcalories spent per minute. Table 6–2 on page 131 shows the energy expended on various types of activities. The energy spent on activ-

basal metabolism: the energy needed to maintain life when a person is at complete rest after a 12-hour fast. Basal metabolism is normally the largest part of a person's daily energy expenditure.

voluntary activities: the component of a person's daily energy expenditure that involves conscious and deliberate muscular work—walking, lifting, climbing, and other physical activities. Voluntary activities normally require less energy in a day than basal metabolism does.

Table 6–1	Factors That Affect BMR
Factor	**Effect on BMR**
Age	In youth, the BMR is higher; lean body mass diminishes with age, slowing the BMR.[a]
Height	In tall, thin people, the BMR is higher.[b]
Growth	In children and pregnant women, the BMR is higher.
Body composition	The more lean tissue, the higher the BMR. The more fat tissue, the lower the BMR.[c]
Fever	Fever raises the BMR.[d]
Stresses	Some stresses and certain medications raise the BMR.
Environmental temperature	Both heat and cold raise the BMR.
Fasting/starvation	Fasting/starvation lowers the BMR.[e]
Malnutrition	Malnutrition lowers the BMR.
Hormones	The thyroid hormone thyroxine, for example, is a key BMR regulator; the more thyroxine produced, the higher the BMR.[f]

[a]The BMR begins to decrease in early adulthood (after growth and development cease) at a rate of about 2 percent/decade. A reduction in voluntary activity as well brings the total decline in energy expenditure to 5 percent/decade.

[b]If two people weigh the same, the taller, thinner person will have the faster metabolic rate, reflecting the greater skin surface through which heat is lost by radiation, in proportion to the body's volume.

[c]In general, males tend to have a higher BMR than females due to their greater lean body mass.

[d]Fever raises the BMR by 7 percent for each degree Fahrenheit.

[e]Prolonged starvation reduces the total amount of metabolically active lean tissue in the body, although the decline occurs sooner and to a greater extent than body losses alone can explain. More likely, the neural and hormonal changes that accompany fasting are responsible for changes in the BMR.

[f]The thyroid gland releases hormones that travel to the cells and influence cellular metabolism. Thyroid hormone activity can speed up or slow down the rate of metabolism by as much as 50 percent.

ities, added to the energy required for basal metabolism, equals the total energy spent in a day (see the Clinical Applications feature at the end of this chapter).

Energy to Manage Food One component of energy expenditure is not taken into account in the calculations in the Clinical Applications: the energy required for the body to process food. When food is taken into the body, many cells that have been dormant become active. The muscles that move the food through the intestinal tract speed up their rhythmic contractions, and the cells that manufacture and secrete digestive juices begin their tasks. All these and other cells need extra energy as they come alive to participate in the digestion, absorption, and metabolism of food. This stimulation of cellular activity produces heat and is known as the **thermic effect of food.** The thermic effect of food is generally thought to represent about 10 percent of the total food energy taken in. For purposes of rough estimates, though, the thermic effect of food can be ignored; the 10 percent it might contribute to total energy output is smaller than the probable errors involved in estimating energy input from food or output for activities.

thermic effect of food: an estimation of the energy required to process food (digest, absorb, transport, metabolize, and store ingested nutrients).

In Summary A person takes in energy from food and, on average, spends most of it on basal metabolic activities, some of it on physical activities, and a little on the thermic effect of food. Because energy requirements vary from person to person, such factors as age, gender, and weight must be considered when calculating energy spent on basal metabolism, and the intensity and duration of the activity must be taken into account when calculating expenditures on physical activities.

Physical activity spends energy and benefits health in many ways.

© 2000 PhotoDisc, Inc.

Self Study

METABOLISM AND ENERGY BALANCE

Metabolism explains how the cells in the body use nutrients to meet their needs. Cells may start with small, simple compounds and use them as building blocks to form larger, more complex structures (anabolism). These anabolic reactions involve doing work and so require energy. Alternatively, cells may break down large compounds into smaller ones (catabolism). Catabolic reactions usually release energy. Determine whether the following reactions are anabolic or catabolic.

Reaction	Anabolic	Catabolic
A cracker becomes glucose.	_____	_____
Glucose becomes glycogen.	_____	_____
You consume more energy than your body expends.	_____	_____
Fasting.	_____	_____
A piece of ham becomes amino acids.	_____	_____
Amino acids become your muscles.	_____	_____
A cookie becomes fatty acids.	_____	_____
Fatty acids become body fat.	_____	_____
Fatty acids provide energy.	_____	_____

Self Check

1. The principal organs of metabolism are:
   a. the stomach, heart, lungs, liver, and brain.
   b. the lungs, bloodstream, pancreas, liver, and heart.
   c. the brain, digestive system, heart, stomach, and lungs.
   d. the digestive organs, liver, pancreas, circulatory system, and kidneys.

2. Anabolism is defined as the:
   a. metabolic breakdown of glucose to pyruvate.
   b. removal of nitrogen from a compound such as an amino acid.
   c. reactions in which large molecules are broken down to smaller ones and energy is released.
   d. reactions in which small molecules are put together to build larger ones and energy is consumed.

3. During glycolysis and the tricarboxylic acid (TCA) cycle, glucose is first broken down into ___ , then ___ , and finally ___ for use in the body.
   a. glycerol, pyruvate, energy
   b. pyruvate, acetyl CoA, energy
   c. acetyl CoA, pyruvate, energy
   d. amino acids, acetyl CoA, energy

4. Fats are catabolized to ___ and ___ for use in the body.
   a. glucose, energy
   b. glycerol, fatty acids
   c. amino acids, energy
   d. glycogen, fatty acids

5. Two functions of protein in the body are:
   a. catabolism to glycogen and synthesis of triglycerides for storage.
   b. catabolism to glycerol and fatty acids for storage and synthesis of new proteins.
   c. maintenance of the body's supply of amino acids and conversion of protein to glucose for energy.
   d. anabolism of glucose to glycogen for storage and conversion of protein to fat for energy.

6. As carbohydrate and fat stores are depleted during fasting or starvation, the body then uses ____ as its fuel source.
   a. alcohol
   b. protein
   c. glucose
   d. triglycerides

7. When carbohydrate is not available to provide energy for the brain, as in starvation, the body produces ketone bodies from:
   a. glucose.
   b. glycerol.
   c. acetyl CoA.
   d. amino acids.

8. Three hazards of fasting are:
   a. water weight loss, decrease in mental alertness, and wasting of lean tissue.
   b. water weight gain, impairment of disease resistance, and lowering of body temperature.
   c. water weight gain, decrease in mental alertness, and impairment of disease resistance.
   d. wasting of lean tissue, impairment of disease resistance, and disturbances of the body's salt and water balance.

9. Two activities that contribute to the basal metabolic rate are:
   a. walking and running.
   b. maintenance of heartbeat and running.

   c. maintenance of body temperature and walking.
   d. maintenance of heartbeat and body temperature.

10. Three factors that affect the body's basal metabolic rate are:
    a. height, weight, and energy intake.
    b. age, body composition, and height.
    c. fever, body composition, and altitude.
    d. weight, fever, and environmental temperature.

Answers to these questions can be found in Appendix G.

Clinical Applications

This Clinical Applications feature shows one way of calculating energy needs. Another way often used in clinical practice is the BEE (basal energy expenditure) method, explained in Chapter 27.

1. *Basal metabolism.* Convert your own or your client's body weight from pounds to kilograms (if necessary). Then multiply by the factor 1.0 kcalorie per kilogram of body weight per hour for men or 0.9 for women.* Then multiply by the 24 hours in a day. For example, suppose your client is a 160-pound man:

   ■ Change pounds to kilograms: 160 lb ÷ 2.2 lb/kg = 72.7 kg.

   ■ Multiply weight in kilograms by the BMR factor: 72.7 kg × 1 kcal/kg/hr = 72.7 kcal/hr.

   ■ Multiply kcalories used in one hour by hours in a day: 72.7 kcal/hr × 24 hr/day = 1744.8 kcal/day.

   Energy for the BMR equals 1745 (rounded) kcalories per day.

*Men's metabolic energy needs are assumed to be higher than women's because their hormones induce them to develop more lean tissue than do most women, and lean tissue burns more energy per hour.

2. *Voluntary muscular activity.* To estimate the energy for activities, determine from Table 6-2 the level of intensity that typifies average daily activity. Then multiply the BMR kcalories by the corresponding activity factor. For example, if the 160-pound man engages in mostly light activity, his activity factor would be 1.6. Multiply this factor by his BMR kcalories:

   ■ 1.6 × 1745 kcal/day = 2792 kcal/day.

3. *Total energy needs.* The result, 2792 kcalories/day, expresses his total daily energy needs.

4. Alternatively, total energy expenditure can be estimated in one step based on body weight as shown in the last column of Table 6-2. As an example, for a 160-pound man engaged in mostly light activity:

   ■ 38 kcal/kg/day × 72.7 kg = 2763 kcal/day.

   ■ The difference between 2792 and 2763 is insignificant and acceptable. Either way, the man's energy needs are about 2800 kcalories per day.

Table 6–2	Estimating Daily Energy Expenditure at Various Levels of Physical Activity		
Level of Intensity	**Type of Activity**	**Activity Factor (x BMR)**	**Energy Expenditures (kcal/kg/day)**
Very light	Seated and standing activities, painting trades, driving, laboratory work, typing, sewing, ironing, cooking, playing cards, playing a musical instrument	1.3 Men 1.3 Women	31 30
Light	Walking on a level surface at 2.5 to 3 mph, garage work, electrical trades, carpentry, restaurant trades, housecleaning, child care, golf, sailing, table tennis	1.6 Men 1.5 Women	38 35
Moderate	Walking 3.5 to 4 mph, weeding and hoeing, carrying a load, cycling, skiing, tennis, dancing	1.7 Men 1.6 Women	41 37
Heavy	Walking with a load uphill, tree felling, heavy manual digging, basketball, climbing, football, soccer	2.1 Men 1.9 Women	50 44
Exceptional	Athletes training in professional or world-class events	2.4 Men 2.2 Women	58 51

SOURCE: Reprinted with permission from *Recommended Dietary Allowances: 10th edition.* Copyright 1989 by the National Academy of Sciences. Courtesy of the National Academy Press, Washington, D.C.

Notes

1. P. M. Suter, E. Häsler, and W. Vetter, Effects of alcohol on energy metabolism and body weight regulation: Is alcohol a risk factor for obesity? *Nutrition Reviews* 55 (1997): 157–171.

2. M. Yao and S. B. Roberts, Dietary energy density and weight regulation, *Nutrition Reviews* 59 (2001): 247–258; G. Zorrilla, Hunger and satiety: Deceptively simple words for the complex mechanisms that tell us when to eat and when to stop, *Journal of the American Dietetic Association* 98 (1998): 1111; A. Golay and E. Bobbioni, The role of dietary fat in obesity, *International Journal of Obesity and Related Metabolic Disorders* 21 (1997): S2–S11.

3. R. Maughan and K. P. Aulin, Energy costs of physical activity, *World Review of Nutrition and Dietetics* 82 (1997): 18–32.

Nutrition in Practice

■ HUNGER AND COMMUNITY NUTRITION ■

One person in every five worldwide experiences persistent hunger—not the healthy appetite triggered by anticipation of a hearty meal, but the painful sensation caused by a lack of food. Tens of thousands die of starvation each day: one every two seconds.

In the United States, about one in every nine households has one or more members who experience pain from hunger caused by lack of food. In these households, well over two and a half million children are hungry at least some of the time.[1] Given the enormous wealth and economic growth in this country, do these numbers surprise you? The limited or uncertain availability of nutritionally adequate and safe foods is known as **food insecurity** and is a major problem in our nation today.[2] The "How to" box on the next page describes how national surveys identify food insecurity in the United States, and Figure NP6–1 presents the most recent findings; the glossary defines related terms. Surveys like these provide crude, but necessary, data to estimate the degree of hunger in this country.

Why is hunger a problem in developed countries such as the United States where food is abundant?

Hunger has many causes, but in developed countries, the primary cause is **food poverty.** People are hungry not because there is no food nearby to purchase, but because they lack money. An estimated one out of eight people in the United States lives in poverty. Even those above the poverty line may not have food security. Physical and mental illnesses and disabilities, sudden job losses, and high living expenses threaten their financial stability. Further contributing to food poverty are other problems such as abuse of alcohol and other drugs; lack of awareness of available food assistance programs; and the reluctance of people, particularly the elderly, to accept what they perceive as "welfare" or "charity." Lack of resources remains the major cause of food poverty, and solving this problem would do a lot to relieve hunger.

In the United States, food poverty and hunger reach into many segments of society, affecting not only the chronic poor (migrant workers, the unskilled and unemployed, the homeless, and some elderly) but also the so-called working poor. Some are displaced farm families. Some are former blue-collar and white-collar workers forced out of their trades and professions into minimum-wage jobs. These people outnumber the chronic poor, and they are not on welfare—they have jobs, but the pay is too low to meet their needs. Families with incomes below a certain level are simply unable to buy sufficient amounts of nourishing foods, even if they are skilled in food shopping. For many of the children in these families, school lunch is their only meal of the day. Otherwise they go hungry, waiting for an adult to find money for food.

What U.S. food programs are directed at relieving hunger in the United States?

The American Dietetic Association (ADA) calls for aggressive action to bring an end to domestic hunger

Glossary

food bank: a facility that provides food to the hungry.

food insecurity: limited or uncertain access to foods of sufficient quality or quantity to sustain a healthy and active life.

food poverty: hunger occurring when enough food exists in an area but some of the people cannot obtain it because they lack money, are being deprived for political reasons, live in a country at war, or suffer from other problems such as lack of transportation.

food recovery: collecting food for distribution to low-income people who are hungry. Four common methods of food recovery are:

■ **field gleaning:** collecting crops from fields that either have already been harvested or are not profitable to harvest.

■ **perishable food salvage:** collecting perishable produce from wholesalers and markets.

■ **food rescue:** collecting prepared foods from commercial kitchens.

■ **nonperishable food collection:** collecting processed foods from wholesalers and markets.

How to

Identify Food Insecurity in a U.S. Household

Questions like these are asked on surveys to determine the extent of food insecurity in a household. The more questions answered "Yes," the more intense the hunger the household is experiencing.

- Do you often go hungry?
- Do you often have too little food to eat because you have no money, transportation, or kitchen appliances (stove, refrigerator)?
- Do you ever rely on nutritionally inferior foods to feed yourself or your children because you lack any of these resources?

- Do you ever eat less than you feel you should because you lack any of these resources?
- Do you ever skip meals or cut the size of meals because you lack any of these resources?
- Do you ever rely on neighbors, friends, relatives, or schools to feed any of your children because there is not enough food in the house?
- Do your children ever say they are hungry because there is not enough food in the house?
- Do you or any of your children ever go to bed hungry because there is not enough food in the house?

and to achieve food and nutrition security.[3] Many federal and local programs aim to prevent or relieve malnutrition and hunger in the United States.

An extensive network of federal assistance programs provides life-giving food daily to millions of U.S. citizens.[4] One out of every six Americans receives food assistance of some kind, at a total cost of almost $40 billion per year.[5] Even so, the programs are not fully successful in preventing hunger, even among those who receive their benefits.[6] Programs described in the life cycle chapters include the WIC program for low-income pregnant women, breastfeeding mothers,

and their young children (Chapter 12); the school lunch, and breakfast programs for children (Chapter 13); and the food assistance programs for older adults such as congregate meals and Meals on Wheels (Nutrition in Practice 14).

The centerpiece of food programs for low-income people in the United States is the Food Stamp Program, administered by the U.S. Department of Agriculture (USDA). The USDA issues food stamp coupons or debit cards through state agencies to households—people who buy and prepare food together. The amount a household receives depends on its size and income.

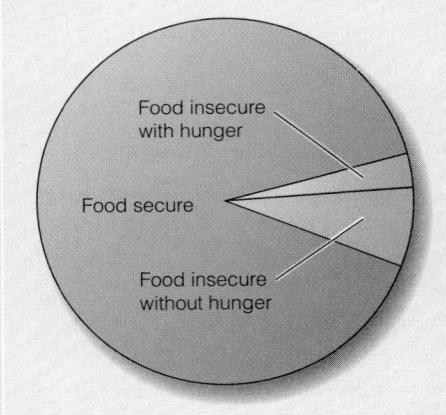

Figure NP6–1

Prevalence of Food Insecurity and Hunger in U.S. Households, 1999

These people and many others like them in the United States face food insecurity daily.

Nutrition in Practice

# *ow to*

Plan Healthy, Thrifty Meals

*C*hapter 1 introduced the Food Guide Pyramid and principles for planning a healthy diet. Meeting that goal on a limited budget adds to the challenge. To save money and spend wisely, plan and shop for healthy meals with the following tips in mind:

PLANNING

- Make a grocery list before going to the store to avoid expensive "impulse" items. Do not shop when hungry.
- Use leftovers.
- Center meals on rice, noodles, and other grains.
- Use small quantities of meat, poultry, fish, or eggs.
- Use legumes instead of meat, poultry, fish, or eggs several times a week.
- Use cooked cereals such as oatmeal instead of ready-to-eat breakfast cereals.
- Cook large quantities when time and money allow.
- Check for sales and clip coupons for products you need; plan meals to take advantage of sale items.

SHOPPING

- Buy day-old bread and other products from the bakery outlet.
- Select whole foods instead of convenience foods (potatoes instead of instant mashed potatoes, for example).

- Try store brands.
- Buy fresh produce that is in season; buy canned or frozen items at other times.
- Buy only the amount of fresh foods that you will eat before it spoils. Buy large bags of frozen items or dry goods; when cooking, take out the amount needed and store the remainder.
- Buy fat-free dry milk; mix and refrigerate quantities needed for a day or two. Buy fresh milk by the gallon or half-gallon.
- Buy less expensive cuts of meat. Chuck and bottom round roast are usually inexpensive; cover during cooking and cook long enough to make meat tender. Buy whole chickens instead of pieces.
- Compare the unit price (cost per ounce, for example) of similar foods so that you can select the least expensive brand or size.
- Buy nonfood items such as toilet paper and laundry detergent at discount stores instead of grocery stores.

For daily menus and recipes for healthy, thrifty meals, visit the USDA Center for Nutrition Policy and Promotion: **www.usda.gov/cnpp**

Recipients may use the coupons or cards like cash to purchase food and food-bearing plants and seeds, but not to buy tobacco, cleaning items, alcohol, or other nonfood items. The accompanying "How to" box offers shopping tips for those on a limited budget.

The Food Stamp Program is the largest of the federal food assistance programs, both in amount of money spent and in number of people participating. Over 17 million people receive food stamps at a cost of over $20 billion per year; more than half of the recipients are children.[7]

Although food assistance programs improve nutrient intakes significantly, hunger continues to plague the United States.[8] Of the estimated 2 million homeless people in the United States who are eligible for food assistance, only 15 percent of single adults and 50 percent of families receive food stamps. Health care pro-

fessionals working with clients who may be having financial problems can encourage them to talk with a social worker who can assess their eligibility for food assistance programs.

Are there other programs aimed at reducing hunger in the United States?

Efforts to resolve the problem of hunger in the United States do not depend solely on federal assistance programs. National **food recovery** programs have made a dramatic difference; the largest program, America's Second Harvest, coordinates the efforts of almost 200 food banks in providing more than 1 billion pounds of food to 45,000 local agencies that feed 26 million people a year. Table NP6–1 on the next page lists addresses, phone numbers, and websites for America's Second Harvest and other hunger-relief organizations.

Nutrition in Practice

Table NP6–1   Hunger-Relief Organizations

Action without Borders
350 Fifth Ave., Suite 6614
New York, NY 10118
(212) 843-3973
www.idealist.org

America's Second Harvest
35 E. Wacker Dr. #2000
Chicago, IL 60601
(800) 771-2303
www.secondharvest.org

Bread for the World
50 F St. NW, Suite 500
Washington, DC 20010
(800) 822-7323
(202) 639-9400;
fax (202) 639-9401
www.bread.org

Congressional Hunger
   Center
229½ Pennsylvania Ave.
Washington, DC 20003
(202) 547-7022
www.hungercenter.org

Foodchain
912 Baltimore, #300
Kansas City, MO 64105
(800) 845-3008
www.foodchain.org

OXFAM America
26 West St.
Boston, MA 02111-1206
(800) 77-OXFAM or
(800) 776-9326
www.oxfam.org

Pan American Health
Organization
525 23 St. NW
Washington, DC 20037
(202) 974-3000
www.paho.org

Society of St. Andrew
3383 Sweet Hollow Rd.
Big Island, VA 24526
(800) 333-4597
www.endhunger.org

United Nations Food and
Agriculture Organization (FAO)
1001 22nd St. NW, Suite 300
Washington, DC 20437
(202) 653-2400
www.fao.org

United Nations International
Children's Emergency Fund
(UNICEF)
3 United Nations Plaza
New York, NY 10017-4414
(212) 326-7035
www.unicef.org

World Food Program
Via Cesare Giulio Viola, 68
Parco dé Medici
Rome, Italy 00148
www.wfp.org

World Health Organization
(WHO)
525 23rd St. NW

Washington, DC 20037
(202) 861-3200
www.who.org

World Hunger Program
Brown University
Box 1831
Providence, RI 02912
(401) 863-2700
**www.brown.edu/Departments/
World_Hunger_Program/
hungerweb/WHP/
overview.html**

World Hunger Year
505 Eighth Ave., 21st Floor
New York, NY 10018-6582
(800) GleanIt
www.worldhungeryear.org

Each year, an estimated one-fifth of our food supply is wasted in fields, commercial kitchens, grocery stores, and restaurants—that's enough food to feed 49 million people. Food recovery programs collect and distribute good food that would otherwise go to waste.[9] Volunteers might pick corn left in an already harvested field, a grocer might deliver ripe bananas to a local **food bank,** and a caterer might take leftover chicken salad to a community shelter, for example. All of these efforts help to feed the hungry in the United States.

What about local efforts and community nutrition programs?

Food recovery programs depend on volunteers. Concerned citizens work through local agencies and churches to feed the hungry. Community-based food pantries provide groceries, and soup kitchens serve

Feeding the hungry in the United States.

Community-based efforts to feed citizens include food pantries that provide groceries.

Nutrition in Practice

prepared meals. Meals often deliver adequate nourishment, but most homeless people receive fewer than one and a half meals a day, so many are still inadequately nourished. Health care professionals can serve as valuable members of community groups seeking to provide food assistance.

Notes

1. Federal Interagency Forum on Child and Family Statistics, *America's Children: Key National Indicators of Well-Being*, 2001, available at **www.childstats.gov.**

2. Position of The American Dietetic Association: Domestic food and nutrition security, *Journal of the American Dietetic Association* 98 (1998): 337–342.

3. Position of The American Dietetic Association, 1998.

4. C. S. Kramer-LeBlanc and K. McMurry, Discussion paper on domestic food security, *Family Economics and Nutrition Review* 11 (1998): 49–78.

5. P. P. Basiotis, C. S. Kramer-LeBlanc, and E. T. Kennedy, Maintaining nutrition security and diet quality: The role of the Food Stamp Program and WIC, *Family Economics and Nutrition Review* 11 (1998): 4–16.

6. USDA Center for Nutrition Policy and Promotion, Could there be hunger in America? *Nutrition Insight*, September 1998.

7. USDA Food and Nutrition Service website, **www.usda.gov/fcs,** visited March 20, 2002.

8. D. Rose, J.-P. Habicht, and B. Devaney, Household participation in the food stamp and WIC programs increases the nutrient intakes of preschool children, *Journal of Nutrition* 128 (1998): 548–555.

9. U.S. Department of Agriculture, *A Citizen's Guide to Food Recovery*, December 1996.

Jennie Oppenheimer/Studio Zocolo

CHAPTER 7 OVERWEIGHT, UNDERWEIGHT, AND WEIGHT CONTROL

Are you pleased with your body weight? If you answered yes, you are a rare individual. Nearly all people in our society think they should weigh more or less (mostly less) than they do. Usually, their primary reason is appearance, but they often perceive, correctly, that their weight is also related to physical health. At the extremes, both overweight and underweight present health risks.

Overweight and underweight both result from unbalanced energy budgets. The simple picture is as follows. Overweight people have consumed more food energy than they have spent and have banked the surplus in their body fat. To reduce body fat, overweight people need to spend more energy than they take in from food. In contrast, underweight people have consumed too little food energy to support their activities and so have depleted their bodies' fat stores and possibly some of their lean tissues as well. To gain weight, they need to take in more food energy than they expend. As you will see, though, the details of the body's weight regulation are quite complex.

This chapter's missions are to examine the problems associated with excessive and deficient body fatness; to present strategies toward solving these problems; and to point out how appropriate body composition, once achieved, can be maintained. The chapter emphasizes overweight because it has been more intensively studied and is a more widespread health problem in the developed countries.

Body Weight and Body Composition

The body's weight reflects its composition—the proportions of its bone, muscle, fat, fluid, and other tissue. All of these body components can vary in quantity and quality: the bones can be dense or porous, the muscles can be well developed or underdeveloped, fat can be abundant or scarce, and so on. By far the most variable tissue, though, is body fat. More than any other component, fat responds to changes in food intake and physical activity, so it is fat that is usually the target of efforts at weight control.

Healthy Body Weight and the BMI

How much should a person weigh? How can a person know if her weight is appropriate for her height and age? How can a person know if his weight is jeopardizing his health? In the past, to determine if a person had a healthy body weight, the person's height and weight were compared to a table of suggested weights for heights associated with good health. Weight-for-height tables are no longer the recommended method of evaluation. Instead, health professionals and obesity experts use a single standard, derived mathematically from the height and weight measures—the **body mass index (BMI)**. The BMI has replaced weight-for-height tables in clinical settings. The margin provides the calculation to demonstrate how BMI values are derived (see the inside back cover for the BMI table). A person who takes measurements in pounds and inches can convert them to metric units or can use this modified equation:

$$BMI = \frac{weight \ (lb)}{height \ (in)^2} \times 705.$$

As the BMI table shows, healthy weight falls between a BMI of 18.5 and 24.9. Most people with a BMI within this range have few of the health risks typically associated with too-low or too-high body weight. Risks increase as BMI falls below 18.5 or rises above 24.9 (see Figure 7–1), reflecting the reality that both

$$BMI = \frac{weight \ (kg)}{height \ (m)^2} \ .$$

To convert pounds to kilograms, divide by 2.2.

To convert inches to meters, divide by 39.37.

The inside back cover shows weights for various heights using the BMI to define underweight, healthy weight, overweight, and obesity.

body mass index (BMI): an index of a person's weight in relation to height, determined by dividing the weight (in kilograms) by the square of the height (in meters).

underweight and overweight impair health status. Figure 7–2 presents visual images associated with various BMI values.

The BMI values are most accurate in assessing degrees of obesity and are less useful for evaluating nonobese people's body fatness. BMI values fail to reveal two valuable pieces of information in assessing disease risk. They don't reveal how much of the weight is fat, and they don't indicate where the fat is located. For this knowledge, measures of body composition are needed.

Body Composition

For many people, being overweight compared with the standard means that they are over*fat*. This is not the case, though, for athletes with dense bones and well-developed muscles; they may be over*weight* but carry little body fat. Conversely, inactive people may seem to have acceptable weights, but still may carry too much body fat. In addition, the distribution of fat on the body may be even more critical than overfatness alone.

Central Obesity Even more than total body fat, fat that collects deep within the central abdominal area of the body may be especially likely to lead to diabetes, stroke, hypertension, and coronary artery disease (see Figure 7–3 on the next page). The risk of death from all causes may be higher for those with **central obesity** than for those whose fat accumulates elsewhere in the body. Unlike the fat layers lying just beneath the skin of the abdomen and elsewhere, **intra-abdominal fat,** when mobilized, goes directly to the liver where it is made into cholesterol-carrying low-density lipoprotein (LDL). Fat from elsewhere may arrive in the liver eventually, but it takes a circuitous route that first allows other tissues the chance to pull it from the circulation and metabolize it.

Abdominal fat creates the "apple" profile of central obesity. Fat around the hips and thighs creates more of a "pear" profile. Abdominal fat is common in women past menopause and even more common in men. Even when total body fat is similar, men have more abdominal fat than either premenopausal or postmenopausal women. For those women with abdominal fat, the risks of

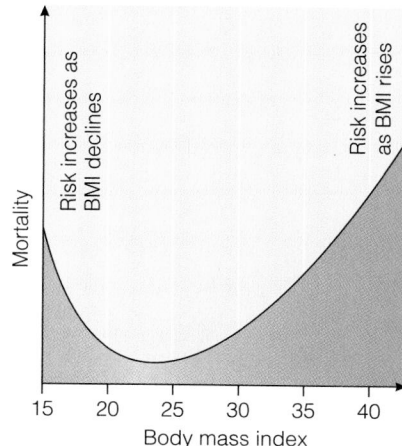

Figure 7–1

Body Mass Index and Mortality
This J-shaped curve describes the relationship between body mass index (BMI) and mortality and shows that both underweight and overweight present risks of a premature death.

central obesity: excess fat on the abdomen and around the trunk of the body.

intra-abdominal fat: fat stored within the abdominal cavity in association with the internal abdominal organs, as opposed to the fat stored directly under the abdominal skin (subcutaneous fat).

Figure 7–2

Silhouettes and BMI (Actual BMI Shown)

SOURCE: Reprinted from "The Body Test" (1988). © Dietitians of Canada.

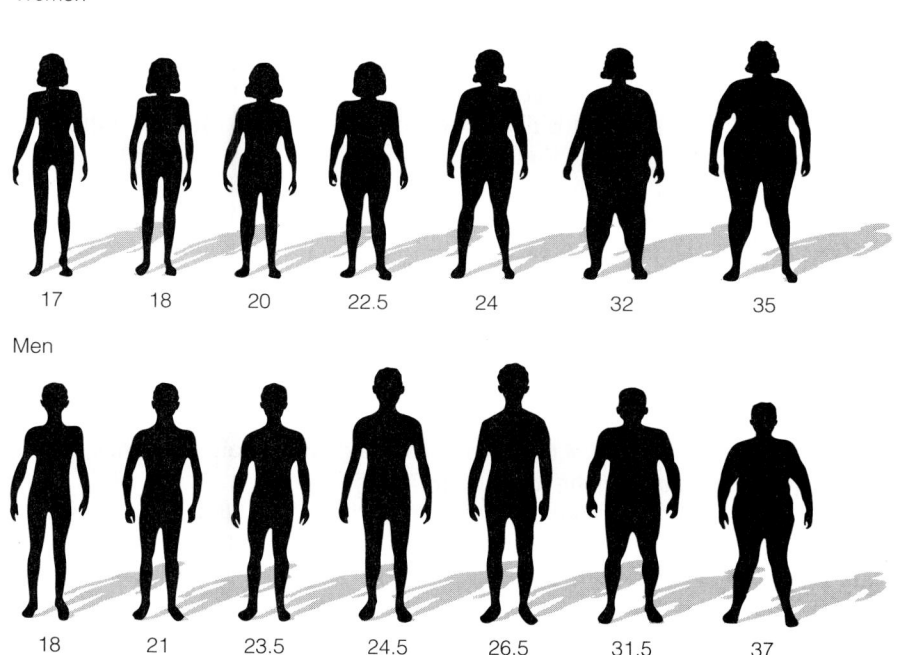

Women

| 17 | 18 | 20 | 22.5 | 24 | 32 | 35 |

Men

| 18 | 21 | 23.5 | 24.5 | 26.5 | 31.5 | 37 |

Overweight, Underweight, and Weight Control

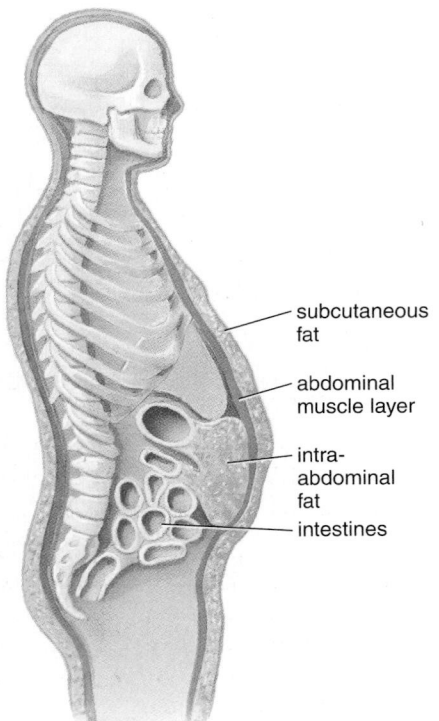

Figure 7–3

Intra-abdominal Fat and Subcutaneous Fat
The fat deep within the body's abdominal cavity may pose an especially high risk to health.

fatfold measure: a clinical estimate of total body fatness in which the thickness of a fold of skin on the back of the arm (over the triceps muscle), below the shoulder blade (subscapular), or in other places is measured with a caliper. (The older, less preferred, term is **skinfold test.**)

waist circumference: a measurement used to assess a person's abdominal fat.

overweight: body weight above some standard of acceptable weight that is usually defined in relation to height (such as BMI).

obesity: a chronic disease characterized by excessively high body fat in relation to lean body tissue.

cardiovascular disease and mortality are increased, just as they are for men.[1] Smokers, too, may carry more of their body fat centrally. A smoker may weigh less than the average nonsmoker, but the smoker's waist circumference may be greater, leading researchers to think that smoking may directly affect body fat distribution. Two other factors that may affect body fat distribution are intakes of alcohol and physical activity. Moderate-to-high alcohol consumption may favor central obesity. In contrast, regular physical activity seems to prevent abdominal fat accumulation.[2]

Fatfold Measures Fatfold measurements provide an accurate estimate of total body fat and a fair assessment of the fat's location. About half of the fat in the body lies directly beneath the skin, so the thickness of this subcutaneous fat is assumed to reflect total body fat. Measures taken from central-body sites (around the abdomen) better reflect changes in fatness than those taken from upper sites (arm and back). A skilled assessor can obtain an accurate **fatfold measure** and then compare the measurement with standards (see Appendix D).

Waist Circumference **Waist circumference** serves as an indicator of abdominal fatness (see Figure 7–4). Previously, a ratio of waist-to-hip measurements served this purpose, but waist circumference alone has been deemed a valid indicator for both men and women. In general, women with a waist circumference greater than 35 inches and men with a waist circumference greater than 40 inches have a high risk of central obesity–related health problems.[3]

How Much Is Too Much Body Fat?

The ideal amount of body fat depends partly on the person. A man with a BMI within the recommended range may have between 12 and 20 percent body fat; a woman, because of her greater quantity of indispensable fat, 20 to 30 percent. For many athletes, a lower-than-average percentage of body fat may be ideal—just enough fat to provide fuel, insulate and protect the body, assist in nerve impulse transmissions, and support normal hormone activity, but not so much as to burden the muscles. For athletes, then, ideal body fat might be 5 to 10 percent for men and 15 to 20 percent for women.

For an Alaskan fisherman, a higher-than-average percentage of body fat is probably beneficial because fat helps prevent heat loss in cold weather. A woman starting a pregnancy needs sufficient body fat to support conception and fetal growth. Below a certain threshold for body fat, individuals may become infertile, develop depression, experience abnormal hunger regulation, or become unable to keep warm. These thresholds differ for each function and for each individual; much remains to be learned about them.

Clearly, the most important criterion of appropriate fatness is health. Researchers find that health problems develop when body fat exceeds 22 percent in young men, 25 percent in men over age 40, 32 percent in young women, and 35 percent in women over age 40.

In Summary The BMI has replaced weight-for-height tables in evaluating whether a person has a healthy weight. Health risks increase with a BMI below 18.5 or above 24.9. Central obesity, in which excess fat is distributed around the trunk of the body, presents greater health risks than excess fat distributed on the lower body. The ideal amount of body fat varies from person to person, but researchers have found that body fat in excess of 22 percent for young men and 32 percent for young women (the levels rise slightly with age) poses health risks. Researchers use a number of techniques to assess body composition including waist circumference and fatfold measures.

Risks of Overweight and Obesity

Overweight is associated with disease risks. For example, it can precipitate hypertension and bring on strokes.[4] Often weight loss alone can normalize the blood pressure of an overfat person; some people with hypertension can tell you exactly at what weight their blood pressure begins to rise. Weight gain can also precipitate diabetes in genetically susceptible people.[5]

Despite our nation's preoccupation with body image and weight loss, the prevalence of overweight and obesity continues to rise dramatically (see Figure 7–5 on the next page).[6] During the previous decade, obesity increased in every state, in both genders, and across all ages, races, and education levels. An estimated 97 million U.S. adults (almost 55 percent) are **overweight** (BMI of 25 or greater), while about 20 percent are dangerously obese (BMI of 30 or greater).[7] The prevalence of overweight is especially high among women, the poor, and some ethnic groups. If this trend continues, some obesity experts predict that by the year 2230, every adult in the United States will be overweight. Such a dramatic statement sounds preposterous, but it speaks to the reality: obesity is a major public health problem without a solution.

Health Risks of Obesity   The health risks of **obesity** are so many that it has been declared a disease. In the United States, obesity is second only to tobacco use as the most significant cause of preventable death.[8] Besides diabetes and hypertension already mentioned, other risks threaten obese adults. Among them are high blood lipids, cardiovascular disease, sleep apnea (abnormal ceasing of breathing during sleep), osteoarthritis, abdominal hernias, some cancers, varicose veins, gout, gallbladder disease, respiratory problems (including Pickwickian syndrome, a breathing blockage linked with sudden death), liver malfunction, complications in pregnancy and surgery, flat feet, and even a high accident rate. Each year these obesity-related illnesses cost our nation billions of dollars.[9] The cost in terms of lives is also great. People with lifelong obesity are twice as likely to die prematurely as others.

People want to know exactly how much fat is too fat for health. Ideally, a person has enough fat to meet basic needs but not so much as to incur health risks. Some evidence indicates that being even moderately overweight aggravates the risk of heart disease, assuming that the excess weight is fat, not muscle.[10] Some obese people, however, seem to remain healthy and live long despite their body fatness. Physical fitness, despite body fatness, may protect against early death from disease.[11] It may be that genetics and other risk factors such as smoking help determine who among the overweight will be susceptible to diseases and who will stay healthy. Still, the majority of obese people do develop associated health problems.

National Guidelines for Identifying Those at Risk from Obesity   In 1998, obesity experts developed U.S. guidelines to identify and evaluate the risks to health from overweight and obesity as determined by three indicators.[12] The first indicator is a person's BMI. The BMI, which defines average relative weight for height in people older than 20 years, usually correlates with body fatness and degree of disease risks (see Table 7–1 on the next page).[13] As a general guideline, overweight for adults is defined as BMI of 25.0 through 29.9 and obesity as BMI equal to or greater than 30.

The second indicator is waist circumference, which reflects the degree of intra-abdominal fatness in proportion to body fatness. As Table 7–1 shows, women with a waist circumference greater than 35 inches, and men with a waist circumference greater than 40 inches, are at greater risk of type 2 diabetes, hypertension, and cardiovascular disease than women or men with waist circumferences equal to, or below, these measures. In other words, waist circumference is an independent predictor of disease risk.

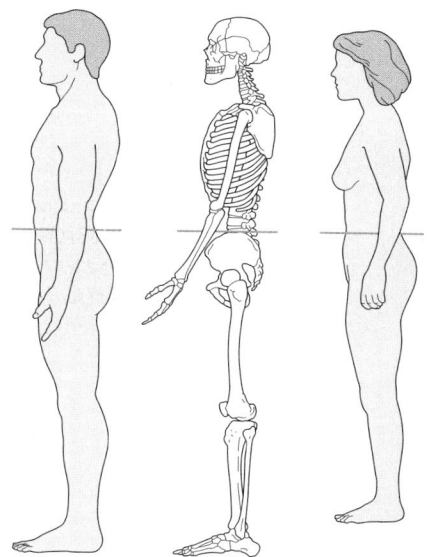

Figure 7–4

Measuring Waist Circumference
Using a nonstretching tape measure, measure the body around the point just above the iliac crest. Take the measure at the end of a normal expiration. A healthy waist circumference for men is no larger than 102 centimeters (40 inches); for women, no larger than 88 centimeters (35 inches).

SOURCE: National Heart, Lung, and Blood Institute Expert Panel, National Institutes of Health, *Clinical Guidelines on the Identification, Evaluation, and Treatment of Overweight and Obesity in Adults* (Washington, D.C.: Government Printing Office, 1998), p. 59.

At 6 feet 3 inches tall and 245 pounds, Mike O'Hearn has a BMI greater than 30 and would be considered overweight by most weight-for-height standards. Yet he is clearly not overfat. In fact, his body fat is only 8 percent.

Overweight, Underweight, and Weight Control

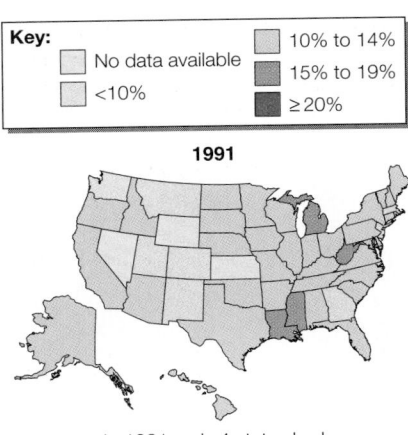

Key:
No data available	10% to 14%
<10%	15% to 19%
	≥20%

1991

In 1991, only 4 states had
obesity rates >15 percent.

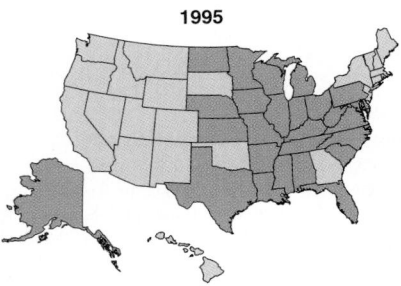

1995

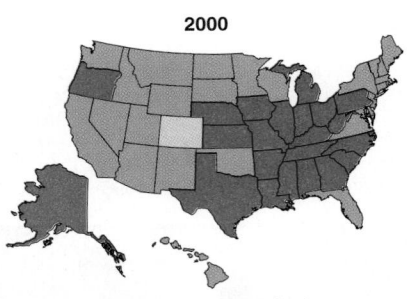

2000

By 2000, 22 states had
obesity rates >20 percent.

Figure 7–5

The Increasing Prevalence of Obesity among U.S. Adults

In 1991, only four states had an obesity rate (BMI ≥30) of greater than 15 percent. By 2000, all states except Colorado reported an obesity rate of at least 15 percent, many reported a rate of greater than 20 percent, and the trend is steadily upward.

SOURCE: U.S. Obesity Trends 1985 to 2000, Centers for Disease Control and Prevention, available at **www.cdc.gov/nccdphp/ dnpa/obesity/trend/maps/slide/001.htm**

Table 7–1 Disease Risks Based on BMI and Waist Circumference[a]

The degree of risk is heightened by the presence of specific disease or other risk factors, such as elevated blood LDL cholesterol or smoking (see the margin on page 145).

BMI	Waist ≤ 40 in. (Men) or ≤ 35 in. (Women)	Waist ≥ 40 in. (Men) or ≥ 35 in. (Women)
18.5 or less	Underweight	Low	—
18.5–24.9	Normal	Low	—
25.0–29.9	Overweight	Increased	High
30.0–34.9	Obese, class I	High	Very high
35.0–39.9	Obese, class II	Very high	Very high
40 or greater Extremely obese, class III	Extremely high	Extremely high

[a]Risk for type 2 diabetes, hypertension, and cardiovascular disease.

SOURCE: National Heart, Lung, and Blood Institute, National Institutes of Health, *The Practical Guide: Identification, Evaluation, and Treatment of Overweight and Obesity in Adults*, NIH publication no. 00-4084 (Washington, D.C.: Government Printing Office, 2000).

The third indicator is the person's disease risk profile. The categories of disease risk listed in Table 7–1 reflect disease risk *relative* to risk at normal weight. Relative risk is not the same as *absolute* risk, which is determined by the presence of certain obesity-related diseases or risk factors for disease.[14] People who have one or more of the diseases listed in the margin on p. 145, or three or more of the cardiovascular disease risk factors listed, have a very high absolute risk for disease complications and mortality that requires aggressive treatment to manage the disease or modify the risk factors.

Other Risks of Obesity   While some obese people seem to escape health problems, few in our society can avoid the social and economic handicaps. Our society places enormous value on thinness. Obese people are less sought after for romance, less often hired, and less often admitted to college. They pay higher insurance premiums, and they pay more for clothing.[15] This is especially true for women. In contrast, people with other chronic conditions such as asthma, diabetes, and epilepsy do not differ socially or economically from nonoverweight people. These social and economic disadvantages have psychological consequences, too: fat people often feel rejected and embarrassed, and this hurts self-esteem.

In Summary   The health risks of obesity are many and serious. The weight appropriate for an individual depends largely on factors specific to that individual, including BMI, waist circumference, family health history, and current health status. Obesity also incurs social, economic, and psychological risks.

Causes of Obesity

Henceforth, this chapter will use the term *obesity* to refer to excess body fat. Excess body fat accumulates when people take in more food energy than they spend. Why do they do this? Is it genetic? Metabolic? Psychological? Behavioral? All of these? Most likely, obesity has many interrelated causes: many experts in the field speak of several different *obesities*.

Genetics and Weight   A person's genetic makeup almost certainly influences the body's tendency to consume or store too much energy or to burn too little.[16] When both parents are obese, the chances that their children will be obese are quite high (up to 80 percent), whereas when neither parent is obese, the chances are relatively small (less than 10 percent). Adoption studies find a similarity in obesity between biological parents and their natural children, but not between adoptive parents and their adopted children.

To determine the relative contributions of genetic and environmental factors to body weight, one group of researchers studied identical and fraternal twins, some of whom were reared together and some apart. Like previous studies, this study found that identical twins were twice as likely to have similar weights as fraternal twins were—even when reared apart. These findings suggest an important role for genetics in determining a person's susceptibility to obesity.[17]

Genetics may influence the way energy is *stored*. When identical twins are given an extra 1000 kcalories a day for 100 days, some pairs gain less than 10 pounds while others gain up to 30 pounds. Within each pair, the amount of weight gain, percentage of body fat, and distribution of fat are similar.

Genetics may also influence how much energy the body *spends*. For example, the differences in basal metabolic rate (BMR) between individuals are greater than can be explained by age, gender, and body composition alone. Similarities within families suggest a genetic influence on BMR. A low metabolic rate is a major risk factor for weight gain.

Lipoprotein Lipase Some of the research investigating genetic influence on obesity focuses on the enzyme **lipoprotein lipase (LPL),** which promotes fat storage in fat cells and muscle cells. People with high LPL activity are especially efficient at storing fat. Obese people generally have much more LPL activity in their fat cells than lean people do.

Leptin Researchers have discovered a gene in humans called the obesity *(ob)* gene. The obesity gene codes for the protein **leptin.** Leptin is a hormone produced and secreted by the fat cells in proportion to the amount of fat stored.[18] A gain in body fatness stimulates the production of leptin, which, by way of the hypothalamus, suppresses the appetite, increases energy expenditure, and produces fat loss. Fat loss produces the opposite effect—suppression of leptin production, increased appetite, and decreased energy expenditure. As the accompanying photo shows, mice with a defective obesity gene do not produce leptin and can weigh up to three times as much as normal mice. When injected with leptin, the mice lose weight. (Because leptin is a protein, it would be destroyed during digestion if given orally; consequently, it must be given by injection.)

Researchers have identified a genetic deficiency of leptin in human beings as well.[19] An error in the gene that codes for leptin was discovered in two extremely obese children whose blood levels of leptin are barely detectable. Without leptin, the children have little appetite control; they are constantly hungry and eat considerably more than their siblings or peers. Given daily injections of leptin, one of these children has lost a substantial amount of weight, confirming leptin's role in regulating appetite and body weight.[20]

Most obese people do not have leptin deficiency, however. In fact, in obese people, the more body fat, the more leptin.[21] Researchers speculate that leptin rises in an effort to suppress appetite and inhibit fat storage when fat cells are ample. Obese people with elevated leptin concentrations may be resistant to its satiating effect.[22] The absence of or resistance to leptin in obesity parallels the scenario of insulin in diabetes: some people have an insulin deficiency (type 1), whereas many others have elevated insulin, but are resistant to its glucose-storing effect (type 2).

Research on leptin is ongoing, and scientists are exploring the possibility that leptin may one day help treat human obesity.[23] Even if leptin never proves useful as an antiobesity drug, its discovery has contributed much to our understanding of the complexities of the human body. For example, scientists no longer view adipose tissue as a metabolically sluggish storage depot for fat, but

The National Heart, Lung, and Blood Institute states that aggressive treatment is urgently needed for a clinically obese person (BMI ≥30) who also has any of the following:

- *Established cardiovascular disease (CVD).*
- *Established type 2 diabetes or impaired glucose tolerance.*
- *Sleep apnea, a disturbance of breathing in sleep, including temporary stopping of breathing.*

The same urgency for treatment exists for an obese person with any *three* of the following CVD risk factors:

- *Hypertension.*
- *Smoking.*
- *High LDL cholesterol.*
- *Low HDL cholesterol.*
- *Sedentary lifestyle.*
- *Age older than 45 years (men) or 55 years (women).*
- *Heart disease of an immediate family member before age 55 (male) or 65 (female).*

SOURCE: National Heart, Lung, and Blood Institute, National Institutes of Health, *The Practical Guide: Identification, Evaluation, and Treatment of Overweight and Obesity in Adults,* NIH publication no. 00-4084 (Washington, D.C.: Government Printing Office, 2000).

Genes instruct cells to make proteins, and each protein performs a unique function.

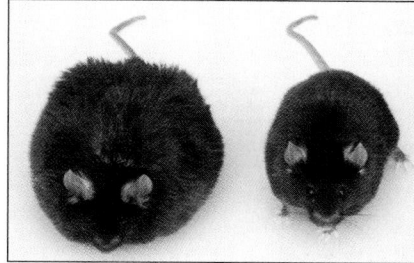

The mouse on the left is genetically obese—it lacks the gene for producing leptin. The mouse on the right is *also* genetically obese, but because it receives leptin, it eats less, expends more energy, and is less obese than it would be had it not received the leptin.

lipoprotein lipase (LPL): an enzyme mounted on the surface of fat cells (and other cells). It hydrolyzes triglycerides in the blood into fatty acids and glycerol for absorption into the cells. There they are metabolized or reassembled for storage.

leptin: a protein produced by fat cells under the direction of the obesity gene that increases satiety and energy expenditure.
leptos = thin

Overweight, Underweight, and Weight Control

rather as a hormonally active regulatory tissue with widespread effects on the body.[24] In addition to its appetite function, leptin may have roles in immunity, reproduction, bone formation, and sexual maturation.[25]

Fat Cell Development   Another cause of obesity may be the development of excess fat cells during childhood. The amount of fat on a person's body reflects both fat cell *number* and fat cell *size*. The number of fat cells increases most rapidly during the growing years of late childhood and early puberty. Fat cell number increases more rapidly in obese children than in lean children, and obese children entering their teen years may already have as many fat cells as do adults of normal weight.

Fat cells can also expand in size. Upon reaching their maximum size, the cells may divide. Thus obesity develops when a person's fat cells increase in number, in size, or quite often both. With fat loss, the size of the fat cells shrinks, but not their number. For this reason, people with extra fat cells may tend to regain lost weight rapidly. Prevention of obesity, then, is most critical during the growing years when fat cell number is increasing.

Set-Point Theory   One popular theory of why a person may store too much fat is the **set-point theory.** The set-point theory proposes that body weight, like body temperature, is physiologically regulated. Researchers have noted that most people who lose weight on reducing diets quickly regain all their lost weight. This suggests that somehow the body chooses a weight that it wants to be and defends that weight by regulating eating behaviors and hormonal actions. Research confirms that the body adjusts its metabolism whenever it gains or loses weight—in the direction that returns to the initial body weight: energy expenditure increases with weight gain and decreases with weight loss. These changes in energy expenditure are greater than those predicted based on body composition and help to explain why it is so difficult for an obese person to maintain weight losses. Researchers speculate that an individual's set point for body weight is adjustable, shifting over the life span in response to physiological changes and to genetic, dietary, and other factors.[26]

Environmental Stimuli   To some extent, obesity may be environmentally determined. People may overeat in a response to stimuli in their surroundings—primarily, the availability of many delectable foods. Most people in the United States find high-kcalorie foods readily available, relatively inexpensive, heavily advertised, and wonderfully delicious.[27] Food is available everywhere, all the time—thanks largely to fast food. Fast-food restaurants line our highways and crowd out mom and pop restaurants and businesses in our small towns and big cities. Convenience stores, service stations, malls, airports, and even our schools offer fast food as well. Most alarming are the extraordinarily large serving sizes and ready-to-go meals offered in supersize combinations.[28] People buy the large sizes and combinations, perceiving them to be a good value, but then they eat more than they need. Research shows that people eat more if they're served more.[29]

Fast food is often high in fat, and fat is perceived as especially palatable. People can easily overeat fat-rich foods because their delicious tastes stimulate eating, and each bite of food is kcalorie-dense.[30] Not only does fat deliver more than twice the kcalories, gram for gram, as protein and carbohydrate, but it also seems to be stored preferentially by the body, and with great efficiency in many people.[31] A steady diet of high-kcalorie, high-fat fast food, then, probably encourages obesity.[32]

Learned Behavior   Psychological stimuli also trigger inappropriate eating behaviors in some people. Appropriate eating behavior is a response to **hunger.** Hunger is a drive programmed into people by their heredity. **Appetite,** in con-

Reminder: The energy required to process food in the body is known as the *thermic effect of food.*

set-point theory: the theory that proposes that the body tends to maintain a certain weight by means of its own internal controls.

hunger: the physiological need to eat, experienced as a drive to obtain food; an unpleasant sensation that demands relief.

appetite: the psychological desire to eat; a learned motivation that is experienced as a pleasant sensation that accompanies the sight, smell, or thought of appealing foods.

trast, is learned and can lead people to ignore hunger or to overrespond to it. Hunger is physiological, whereas appetite is psychological, and the two do not always coincide.

Food behavior is also intimately connected to deep emotional needs such as the primitive fear of starvation. Yearnings, cravings, and addictions with profound psychological significance can express themselves in people's eating behavior. An emotionally insecure person might eat rather than call a friend and risk rejection. Another person might eat to relieve boredom or to ward off depression.

Lack of physical activity fosters obesity.

Physical Inactivity   The possible causes of obesity mentioned so far all relate to the input side of the energy equation. What about output? People may be obese not because they eat too much, but because they spend too little energy. More than one-third of the overweight population report no physical activity during their leisure time. Obese people observed closely are often seen to eat less than lean people, but they are sometimes so extraordinarily inactive that they still manage to accumulate an energy surplus. Reducing their food intake further would jeopardize health and incur nutrient deficiencies. Physical activity, then, is a necessary component of nutritional health. People must be physically active if they are to eat enough food to deliver all the nutrients needed without unhealthy weight gain.

One hundred years ago, 30 percent of the energy used in farm and factory work came from muscle power; today only 1 percent does. Modern technology has replaced physical activity at home, at work, and in transportation. Inactivity is an important contributor to obesity and poor health.[33] In turn, television watching may contribute most to physical inactivity.[34]

Watching television contributes to obesity in several ways. First, television viewing requires little energy beyond the resting metabolic rate. Second, it replaces time spent in more vigorous activities. Third, television influences family food purchases; viewers are more likely to engage in between-meal snacking and eat the high-kcalorie foods most heavily advertised on programs.

In Summary   Like all the other "causes" of obesity, inactivity alone fails to explain it fully. Genetics, fat cell development, set point, and overeating all offer possible, but still incomplete, explanations. Most likely, obesity has not one cause, but different causes and combinations of causes in different people. After all, no two people are alike either physically or psychologically. Some causes may be within a person's control, and some may be beyond it. In recent years, the view has been gaining ground that obesity is not simply a matter of undisciplined gluttony or laziness. Philosophies of weight management and treatment have been evolving to square with this view.

Obesity Treatment: Who Should Lose?

An estimated 30 to 40 percent of all U.S. women (and 20 to 25 percent of all U.S. men) are trying to lose weight at any given time, spending up to $40 billion each year to do so. Some of these people do not even need to lose weight. Others need to lose weight, but are not successful; few succeed, and even fewer succeed permanently. Only 6 percent of people who try to lose weight achieve long-term success.[35]

Many people assume that every overweight person can achieve slenderness and should pursue that goal. Consider, however, that most overweight people cannot become slender. People vary in their weight tendencies just as they vary

diet pills: pills that depress the appetite temporarily; often, physician-prescribed amphetamines (speed). It is generally agreed that these medications are of little value for weight loss and that their use can cause a dangerous dependency.

diuretic abuse: use of diuretics to promote water excretion by dieters who believe their weight excesses are due to water accumulation. Diuretics promote water loss, not fat loss, and their use can cause dehydration and mineral imbalances.

fad diets: diets based on exaggerated or false theories of weight loss. Such diets are usually inadequate in energy and nutrients. Most fad diets, including the currently popular Atkins Diet and Zone Diet, advocate essentially the same high-protein, low-carbohydrate diet. Such diets may offer short-term weight-loss success to some who try them, but they fail to produce long-lasting results for most people. Furthermore, high-protein, low-carbohydrate diets are often high in fat and low in fiber, vitamins, and some minerals. Long-term use of such diets may produce adverse side effects such as nausea, fatigue, constipation, and low blood pressure. Some fad diets are more dangerous to health than obesity itself.

herbal laxatives: laxatives containing senna, aloe, rhubarb root, castor oil, or buckthorn, which are commonly sold as "dieter's tea." Such products cause nausea, vomiting, diarrhea, fainting, and, in some users, possibly death.[a]

herbal products: substances extracted from plants to mimic medications that suppress appetite. Such substances may have dangerous side effects. St. John's wort, for example, is often prepared in combination with the herbal stimulant ephedrine, extracted from the Chinese plant ma huang. Ephedrine has been implicated in several cases of heart attacks and seizures (see Chapter 15).

low-carbohydrate diets: diets designed to bring about metabolic responses similar to those of fasting (see Chapter 6, pp. 125–127). Without sufficient carbohydrate, the body cannot use its fat in the normal way, and ketosis results. Many physiological hazards accompany low-carbohydrate diets: high blood cholesterol, mineral imbalances, hypoglycemia, and more.

[a]P. Kurtzweil, Dieter's brews make tea time a dangerous affair, *FDA Consumer,* July/August 1997, pp. 6–11.

in their potentials for height and degrees of health. The question of whether a person should lose weight depends on many factors: the extent of overweight, age, health, and genetics, to name a few. Weight-loss advice, then, does not apply equally to all overweight people. Some people may risk more in the process of losing weight than in remaining overweight. Others may reap significant health benefits with just modest weight loss.

The risks people incur in attempting to lose weight often depend on how they go about it. Weight-loss plans and obesity treatments abound—some are adequate, but many are ineffective and possibly dangerous. Inappropriate ways of treating obesity are listed in the accompanying glossary. The next section addresses aggressive approaches to obesity for those obese people who face high risks of medical problems and must lose weight rapidly. The section after that discusses reasonable approaches to overweight for those seeking safe, gradual weight loss.

In Summary Weight-loss advice does not apply equally to all overweight people. Some people may risk more in the process of losing weight than in remaining overweight. Weight-loss efforts are often misguided. When pursued via unwise weight-loss techniques, they can be physically and psychologically damaging.

Aggressive Treatments of Obesity

For some obese people, the medical problems caused by their obesity demand treatment approaches that may, themselves, incur some risks. The health benefits to be gained by weight loss, however, may make these risks worth taking.

Obesity Drugs

Several prescription medications for weight loss have been tried over the years. When used as part of a long-term comprehensive weight-loss program, medications can help obese people to lose approximately 10 percent of their weight and maintain that loss for at least a year.[36] Because weight regain commonly occurs with the discontinuation of drug therapy, treatment is long term. And the long-term use of medications poses risks. Medical experts do not yet know whether a person would benefit more from maintaining a 20-pound excess or from taking a medication for a decade to keep the 20 pounds off.

The challenge, then, is to develop an effective medication that can be used over time without adverse side effects or the potential for abuse. No such medication currently exists.[37] Two prescription medications, however, are currently in use; others, including leptin, are being studied.

Sibutramine Sibutramine is an appetite suppressant that works on the brain's neurotransmitters.[*] Sibutramine enhances satiety and elevates energy expenditure.[38] Sibutramine also raises blood pressure, however, so the Food and Drug Administration (FDA) cautions those with hypertension against using it. Anyone taking sibutramine should monitor blood pressure carefully. As more information becomes known about the molecular chemistry of appetite control, safer appetite suppressants may be developed.

Orlistat Orlistat, recently approved by the FDA, takes a different approach to weight loss.[†] Not an appetite suppressant, orlistat inhibits the production of fat-digesting enzymes in the pancreas and so reduces fat absorption by about 30 percent.[39] As a result of absorbing less fat, people often lose weight.[40] The problem with undigested fat is that it leaves the digestive tract intact, carrying with it fat-soluble vitamins and phytochemicals that would otherwise have been absorbed by the body. Possible side effects of orlistat resemble those of the artificial fat olestra—diarrhea and digestive distress.[41]

Surgery

Surgery as an approach to weight loss is justified in some specific cases of **clinically severe obesity.** Two **gastric partitioning** procedures, **gastroplasty** and **gastric bypass,** have gained wide acceptance. Both procedures limit food intake by effectively reducing the size of the stomach. They reduce the size of the outlet from the stomach into the intestine as well, so they delay the passage of food for digestion and absorption.

A new procedure, **gastric banding,** has been approved by the FDA and seems promising.[42] The new technique may be safer than the traditional surgeries because the surgeon makes only two tiny incisions in the abdomen and through them applies a restricting band or pouch around the stomach organ. So far, patients receiving this type of surgery seem to have less pain, faster recovery, and fewer complications than those having more extensive surgeries. Many lose weight because the band causes them to feel full after eating just a small amount of food.

The long-term safety and effectiveness of gastric surgery depend, in large part, on compliance with dietary instructions. Common immediate postsurgical complications include infections, nausea, vomiting, and dehydration; in the long term, vitamin and mineral deficiencies and psychiatric disorders are common. Lifelong medical supervision is necessary for those who choose the surgical route, but in suitable candidates the benefits of weight loss prove worth the risks.

> *In Summary* Obese people with high risks of medical problems may need aggressive treatment, including drugs or surgery. Others may benefit most from improving eating and exercise habits.

clinically severe obesity: a BMI of 40 or greater or a BMI of 35 or greater and one or more serious conditions such as hypertension. A less preferred term used to describe the same condition is *morbid obesity.*

gastric partitioning: a surgical procedure used to treat clinically severe obesity. The operation limits food intake by effectively reducing the size of the stomach and delays gastric emptying by restricting the outlet.

gastroplasty: surgery that partitions the stomach by stapling off a "pouch" or otherwise modifying the stomach, thereby reducing total food intake.

gastric bypass: surgery that reroutes food from the stomach to the lower part of the small intestine; creates a chronic, lifelong state of malabsorption by preventing normal digestion and absorption of nutrients.

gastric banding: a surgical means of producing weight loss by restricting stomach size with a constricting band or pouch; used in people whose severe obesity brings extreme health risks.

---

[*]Sibutramine's trade name is Meridia.
[†]Orlistat's trade name is XENICAL.

Overweight, Underweight, and Weight Control

149

A healthy body contains enough lean tissue to support health and the right amount of fat to meet body needs.

Weight-loss pointers:
- *Adopt reasonable expectations about health and weight goals and about how long it will take to achieve them.*

- *Be involved in planning.*
- *Keep in mind that you will want to maintain your lost weight. Practice needed behaviors as you go.*

- *Adopt a realistic plan.*

1 lb body fat = 3500 kcal.
To lose a pound a week, cut 500 kcal/day.

- *Make the diet adequate by emphasizing nutrient-dense foods.*

Reasonable Strategies for Weight Loss

The *Dietary Guidelines for Americans* (Figure 1–4 on p. 13) suggest that for good health, a person should "aim for a healthy weight." The focus is not so much on weight loss as on health gains. The *Guidelines* go on to say, "if you're overweight, try not to gain more weight and then lose weight to improve your health." Modest weight loss, even when a person is still overweight, can improve control of diabetes and reduce the risks of heart disease by lowering blood pressure and blood cholesterol, especially for those with abdominal fat.

Of course, the same eating and activity habits that improve health often lead to a healthier body weight and composition as well. Successful weight loss, then, is not defined by pounds lost, but by health gained. People less concerned with disease risks may prefer to set goals for personal fitness, such as being able to play with children or climb stairs without becoming short of breath.

Whether the goal is health or fitness, weight-loss expectations need to be reasonable. Unreachable targets ensure frustration and failure. Setting reasonable goals helps to achieve the desired result in managing weight. For example, obese people who must reduce their weight to lower their disease risks might set three broad goals:

1. Reduce body weight by about 10 percent over half a year's time.

2. Maintain a lower body weight over the long term.

3. At a minimum, prevent further weight gain.[43]

Such goals may be achieved or even exceeded, providing a sense of accomplishment instead of disappointment. The accompanying "How to" box offers a way of judging weight-loss diets and programs based on sound nutrition principles.

A Healthful Eating Plan

No particular eating plan is magical, and no particular food must be either included or avoided. You are the one who will have to live with the plan, so you had better be involved in its planning. Don't think of it as a diet you are going "on"—because then you may be tempted to go "off." The diet is successful only if you can maintain a healthy weight. Think of it as an eating plan that you will adopt for life. It must consist of foods that you like, that are available to you, and that are within your means.

A Realistic Energy Intake   Guidelines for a weight-loss diet are outlined in Table 7–2 on p. 152. For those with a BMI of 35 or greater, a deficit of 500 to 1000 kcalories per day will produce the desired loss. This amounts to an intake of 1000 to 1200 kcalories per day for most women and 1200 to 1600 kcalories per day for most men.[44] Figure 7–6 on p. 152 suggests daily food servings from which to build a balanced 1200-kcalorie diet. Diets providing energy intakes lower than 800 kcalories, called very-low-calorie diets (VLCD), are notoriously unsuccessful at achieving lasting weight loss and can be dangerous, and so are not recommended.[45] For those with a BMI ranging from 27 to 35, a deficit of 300 to 500 kcalories per day will result in a loss of ½ to 1 pound per week.[46] You will experience a healthier, more successful weight loss with a small energy deficit that provides an adequate intake than with a large energy deficit that creates feelings of starvation and deprivation, which can lead to an irresistible urge to binge.

Nutritional Adequacy   Nutritional adequacy should be a high priority. Take a look at the 1200-kcalorie food plan in Figure 7–6. Notice that this pattern offers the minimum number of servings suggested in the Daily Food Guide (intro-

How to

Rate Sound and Unsound Weight-Loss Schemes and Diets

Start by giving each diet or program 160 points. Subtract points as instructed, whenever a diet falls short of ideals.

> Scoring: 160 = fine.
> 140–150 = possibly safe with some disadvantage.
> 120–130 = needs improvement.
> 110 or below = dangerous to use.

Does the diet or program:

1. Provide a reasonable number of kcalories (not fewer than 1200 kcalories for an average-size person)? If not, give it a minus 10.

2. Provide enough, but not too much, protein (at least the recommended intake, but not more than twice that much)? If no, minus 10.

3. Provide enough fat for balance but not so much fat as to go against current recommendations (between 20 and 30 percent of kcalories from fat)? If no, minus 10.

4. Provide enough carbohydrate to spare protein and prevent ketosis (100 grams of carbohydrate for the average-size person)? Is it mostly complex carbohydrate (not more than 10 percent of the kcalories as concentrated sugar)? If no to either or both, minus 10.

5. Offer a balanced assortment of vitamins and minerals—that is, foods from all food groups? If it omits a food group (for example, meats), does it provide a suitable substitute? Count five food groups in all: milk/milk products, meat/fish/poultry/eggs/legumes, fruits, vegetables, and breads/cereals/grains. For *each* food group omitted and not adequately substituted for, subtract 10 points.

6. Offer variety, in the sense that different foods can be selected each day? If you'd class it as boring or monotonous, give it a minus 10.

7. Consist of ordinary foods that are available locally (for example, in the main grocery stores) at the prices people normally pay? Or does the dieter have to buy special, expensive, or unusual foods to adhere to the diet? If you would class it as "bizarre" or "requiring special foods," minus 10.

8. Promise dramatic, rapid weight loss (substantially more than 1 percent of total body weight per week)? If yes, minus 10.

9. Encourage permanent, realistic lifestyle changes, including regular physical activity and the behavioral changes needed for weight maintenance? If not, minus 10.

10. Misrepresent salespeople as "counselors" supposedly qualified to give guidance in nutrition and/or general health without a profit motive, or collect large sums of money at the start, or require that clients sign contracts for expensive long-term programs? If so, minus 10.

11. Fail to inform clients about the risks associated with weight loss in general or the specific program being promoted? If so, minus 10.

12. Promote unproven or spurious weight-loss aids such as starch blockers, diuretics, sauna belts, body wraps, passive exercise, ear stapling, acupuncture, electric muscle stimulating (EMS) devices, spirulina, amino acid supplements (e.g., arginine, ornithine), glucomannan, appetite suppressants, "unique" ingredients, and so forth? If so, minus 10.

duced in Chapter 1) and allows a teaspoon of fat at each of three meals. Such an intake would allow most people to lose weight at a satisfactory rate and still meet their nutrient needs with careful food selections. (Women might need an iron supplement.)

Small Portions Overweight people usually need to learn to eat less food at each meal—one piece of chicken for dinner instead of two, a teaspoon of butter on the vegetables instead of a tablespoon, and one cookie for dessert instead of six. Chew foods slowly and thoroughly. The goal is to eat enough food for energy, nutrients, and pleasure, but not more. This amount should leave a person feeling satisfied—not necessarily full. Keep in mind that even low-fat foods can deliver a lot of kcalories when a person eats large quantities.

- *Eat small portions of foods at each meal.*

Carbohydrates Center meals and snacks on complex carbohydrate foods. Fresh fruits, vegetables, legumes, and whole grains offer abundant vitamins and minerals. They also offer more fiber (which provides bulk and satiety) and less fat and food energy than refined foods. Limit, but don't eliminate, lean meats or other low-fat protein sources.

- *Make grains, legumes, vegetables, and fruits central to your diet plan.*

Delicious, low-fat, carbohydrate-rich foods such as fresh fruits, vegetables, whole grains, and legumes offer abundant vitamins, minerals, and fiber.

- *Limit concentrated sweets and alcoholic beverages.*

- *Eat frequent, small meals.*

- *Drink plenty of water (8 glasses or more a day).*

- *Learn, practice, and follow a healthful eating plan for the rest of your life.*

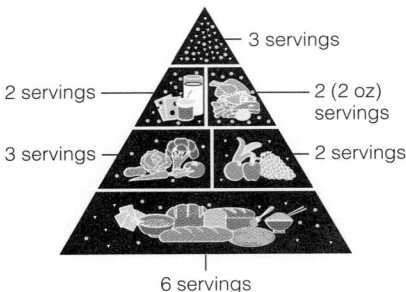

3 servings

2 servings

2 (2 oz) servings

3 servings

2 servings

6 servings

Figure 7–6

Suggested Daily Servings for an Adequate 1200-kCalorie Diet[a]
Choose the lowest-kcalorie, fat-free, or lowest-fat options from each group. Strictly limit serving sizes to those specified in Figure 1–5 of Chapter 1. To further reduce kcalories, reduce servings of fats and added sugars.

[a]Assumes no alcohol intake.

Chapter Seven

Table 7–2	Recommendations for a Weight-Loss Diet
Nutrient	**Recommended Intake**
kCalories	
For people with BMI ≥35	Approximately 500 to 1000 kcalories per day reduction from usual intake
For people with BMI between 27 and 35	Approximately 300 to 500 kcalories per day reduction from usual intake
Total fat	30% or less of total kcalories
Saturated fatty acids[a]	8 to 10% of total kcalories
Monounsaturated fatty acids	Up to 15% of total kcalories
Polyunsaturated fatty acids	Up to 10% of total kcalories
Cholesterol[a]	300 mg or less per day
Protein[b]	Approximately 15% of total kcalories
Carbohydrate[c]	55% or more of total kcalories
Sodium chloride	No more than 2400 mg of sodium or approximately 6 g of sodium chloride (salt) per day
Calcium	1000 to 1500 mg per day
Fiber[c]	20 to 30 g per day

[a]People with high blood cholesterol should aim for less than 7 percent kcalories from saturated fat and 200 milligrams cholesterol per day.
[b]Protein should be derived from plant sources and lean sources of animal protein.
[c]Carbohydrates and fiber should be derived from vegetables, fruits, and whole grains.
SOURCE: National Heart, Lung, and Blood Institute Expert Panel, National Institutes of Health, *Clinical Guidelines on the Identification, Evaluation, and Treatment of Overweight and Obesity in Adults* (Washington, D.C.: Government Printing Office, 1998), p. 74.

Sugar and Alcohol A person trying to achieve or maintain a healthy weight needs to pay attention not only to fat, but to sugar and alcohol, too. Using them for pleasure on occasion is compatible with health as long as most daily choices are of nutrient-dense foods.

Small Frequent Meals Three meals a day is standard in our society, but no law says you can't have four or five—be sure they are smaller, of course. People who eat frequent, small meals are reported to be successful at weight loss and maintenance.[47] Make sure that mild hunger, not appetite, is prompting you to eat. Eat regularly, and eat before you become extremely hungry.

Adequate Water Learn to satisfy thirst with water. Water fills the stomach between meals and dilutes the metabolic wastes generated from the breakdown of fat, easing their excretion. Water meets the fluid needs that were formerly met by eating extra food (remember that food provides water).

In summary, adopt an "eating plan for good health" rather than a "diet for weight loss." That way, you will be able to keep the lost weight off.

Physical Activity

Either dieting or physical activity alone can produce some weight loss. Clearly, however, the combination is most effective.[48] People who combine diet and physical activity are more likely to lose more fat, retain more muscle, and regain less weight than those who only diet.[49]

Energy Expenditure Physical activity makes many contributions to weight loss and maintenance. For one thing, it directly increases energy output by the muscles and cardiovascular system. A 150-pound person walking a brisk 4 miles per hour for 30 minutes spends an extra 185 kcalories on that activity. A football player may spend several thousand extra kcalories on a day of heavy training.

BMR Activity also contributes to energy output in an indirect way—by speeding up basal metabolism.[50] It does this both immediately and over the long term. On any given day, after intense and prolonged exercise, basal metabolism remains elevated for several hours. Over the long term, daily vigorous activity for many weeks gradually shifts body composition toward more lean tissue, which is more active metabolically than fat tissue. The ongoing metabolic rate rises accordingly, and this makes a contribution toward continued weight loss or maintenance.

The raised metabolic rate continues for as long as the person is physically active on a regular basis. The more energy expended in metabolic activities, the greater the energy requirement. This means that a person can eat more without gaining weight.

Appetite Control Physical activity also helps to control appetite. People think that exercising will make them want to eat, but this is not entirely true.[51] Yes, active people do have healthy appetites, but immediately after a good workout, most people do not feel like eating. They want to shower and may be thirsty, but they do not want to eat. The reason is that the body has responded to the stress of activity by mobilizing fuels from storage: glucose and fatty acids are abundant in the blood. At the same time, the body has suppressed its digestive functions. Hard physical work and eating are not compatible.

Psychological Benefits Physical activity helps especially to curb the inappropriate appetite that prompts a person to eat when bored, anxious, or depressed. Weight-management programs encourage people to go out and be active when they're tempted to eat but not really hungry.

Physical activity also helps to reduce stress. Since stress itself is a cue to inappropriate eating behavior for many people, activity can help here, too.

Activity offers still more psychological advantages. The fit person looks and feels healthy, and high self-esteem accompanies these benefits. High self-esteem tends to support a person's resolve to persist in a weight-control effort, rounding out a beneficial cycle.

Choosing Activities What kind of physical activity is best? People seeking to lose weight should choose activities that they enjoy and are willing to do regularly. Health care professionals frequently advise people who want to manage their body weight and lose fat to engage in activities of low-to-moderate intensity for a long duration, such as an hour-long fast-paced walk. The reasoning behind such advice is that people exercising at low-to-moderate intensity are likely to stick with their activity for longer times and are less likely to injure themselves. People who engage in regular, *vigorous* physical activities, however, have less body fat than those who engage in moderately intense activities. The conditioned body that is adapted to strenuous and prolonged aerobic activity uses more fat all day long, not just during activity. The bottom line on physical activity and weight and/or fat loss seems to be that total energy expenditure is the main factor, regardless of how a person does it.

In addition to activities such as walking or aerobic dance, there are hundreds of ways to incorporate energy-spending activities into daily routines: take the stairs instead of the elevator, walk to the neighbor's apartment instead of making a phone call, and rake the grass clippings instead of using a bagger. These activities burn only a few kcalories each, but over a year's time they become significant.

Spot Reducing People sometimes ask about "spot reducing." Unfortunately, no one part of the body gives up fat in preference to another. Fat cells all over the body release fat in response to demand, and the fat is then used by whatever muscles are

Regular physical activity helps people achieve and maintain healthy weights.

Benefits of physical activity in a weight-management program:
- *Favorable effects on disease risks.*
- *Short-term increase in energy expenditure (from exercise and from a slight rise in BMR).*
- *Long-term increase (slight) in BMR.*
- *Appetite control.*
- *Stress reduction and control of stress eating.*
- *Physical, and therefore psychological, well-being.*
- *High self-esteem.*

active. No exercise can remove the fat from any one particular area—and, incidentally, neither can a massage machine that claims to break up fat on trouble spots.

Physical activity can help with trouble spots in another way, though. Strengthening muscles in a trouble area can help to improve their tone; stretching to gain flexibility can help with posture problems. Thus cardiorespiratory endurance, strength, and flexibility workouts all have a place in fitness programs.

Behavior and Attitude

Behavior modification therapy provides ways to overcome barriers to making dietary changes and increasing physical activity. Behavior and attitude are important supporting factors in achieving and maintaining appropriate body weight and composition. Changing the behaviors of overeating and underexercising that lead to, and perpetuate, obesity requires time and effort.

Becoming Aware of Behaviors   A person who is aware of all the behaviors that create a problem has a head start on developing a solution. First, the person needs to establish a baseline (a record of present eating and physical activity behaviors) against which to measure future progress. It is best to keep a diary (see Figure 7–7) that includes the time and place of meals and snacks, the type and amount of foods eaten, the persons present when food is eaten, and a description of the individual's feelings when eating. The diary should also record physical activities: the kind, the intensity level, the duration, and the person's feelings about them. These entries will help the individual identify possible behaviors to change.

Making Small Changes   The "How to" box on p. 155 describes behavioral strategies to support weight management. A particularly attractive feature of these strategies is that they do not involve blaming oneself or putting oneself down—an important element in fostering self-esteem.

Maintaining Weight   Finally, be aware that it can be hard to maintain weight loss. On arriving at the goal weight after months of self-discipline and new habit formation, the victorious weight loser must not "celebrate" by resuming old eating habits. Membership in an ongoing weight-control organization and regular,

behavior modification: the changing of behavior by the manipulation of *antecedents* (cues or environmental factors that trigger behavior), the behavior itself, and *consequences* (the penalties or rewards attached to behavior).

Figure 7–7
Food and Activity Diary
A record of diet and physical activity habits reveals problem areas, the first step toward improving behaviors.

Time	Place	Activity or food eaten	People present	Mood
10:30	School vending machine	6 peanut butter crackers and 12 oz. cola	by myself	starved
12:15	Restaurant	Sub sandwich and 12 oz. cola	friends	relaxed & friendly
3:00	Gym	45 min weight training	workout partner	tired
4:00	Snack bar	Small frozen yogurt	by myself	OK

 **ow to**

Apply Behavior Modification to Manage Body Fatness

1. Eliminate inappropriate cues:
   - Don't buy problem foods.
   - Eat only in one room at the designated time.
   - Shop when not hungry.
   - Avoid vending machines, fast-food restaurants, and convenience stores.
   - Turn off television, video games, and computers.
2. Suppress the cues you cannot eliminate:
   - Serve individual plates; don't serve "family style."
   - Make small portions look large by spreading them over the plate.
   - Create obstacles to consuming problem foods—wrap them and freeze them, making them less quickly accessible.
   - Control deprivation; plan and eat regular meals.
   - Plan the time spent in sedentary activities, such as watching television or using a computer—don't use these activities just to fill time.
3. Strengthen cues to appropriate behaviors:
   - Share appropriate foods with others.
   - Store appropriate foods in convenient spots in the refrigerator.
   - Learn appropriate portion sizes.
   - Plan appropriate snacks.
   - Keep sports and play equipment by the door.
4. Repeat desired behaviors:
   - Slow down eating—put down utensils between bites.
   - Always use utensils.
   - Leave some food on your plate.
   - Move more—shake a leg, pace, stretch often.
   - Join groups of active people and participate.
5. Arrange negative consequences for negative behavior:
   - Ask that others respond neutrally to your deviations (make no comments—even negative attention is a reward).
   - If you slip, don't punish yourself.
6. Reward yourself personally and immediately for positive behaviors:
   - Buy tickets to sports events, movies, concerts, or other nonfood amusement.
   - Indulge in a new small purchase.
   - Get a massage; buy some flowers.
   - Take a hot bath; read a good book.
   - Treat yourself to a lesson in a new active pursuit such as horseback riding, handball, or tennis.
   - Praise yourself; visit friends.
   - Nap; relax.

continued physical activity can provide indispensable support for the formerly overweight person who wants to remain trim.

Personal Attitude For many people, overeating and being overweight may have become an integral part of their identity. Changing diet and activity behaviors without attention to a person's self-concept invites failure.

Many people overeat to cope with the stresses of life. To break out of that pattern, they must first identify the particular stressors that trigger their urges to overeat. Then, when faced with these situations, they must learn to practice problem-solving skills. When the problems that trigger the urge to overeat are dealt with in alternative ways, people may find that they eat less. The message is that sound emotional health supports the ability to take care of health in all ways—including nutrition, weight management, and fitness.

In Summary A person who adopts a lifelong "eating plan for good health" rather than a "diet for weight loss" will be more likely to keep the lost weight off. The margins provide several tips for successful weight management. Physical activity should be an integral part of a weight-management program. Physical activity can increase energy expenditure, improve body composition, help control appetite, reduce stress and stress eating, and enhance physical and psychological well-being. Behavior modification provides ways to overcome barriers to successful weight management.

Underweight

Underweight is far less prevalent than overweight, affecting no more than 10 percent of U.S. adults. The health risks associated with underweight are fewer than those that accompany overweight. Both underweight women and those who have lost a significant amount of weight, however, are more susceptible to osteoporosis. Underweight women may become infertile or may give birth to unhealthy infants. An underweight woman can improve her chances of bearing a healthy infant by gaining weight prior to conception, during pregnancy, or both.

Underweight becomes more hazardous when accompanied by undernutrition. An inadequate supply of nutrients and energy leaves the body underprepared to handle its many metabolic and physical tasks. A person without reserves has a particularly tough battle against medical stresses such as surgery or the wasting diseases of cancer and AIDS. Thus underweight people are urged to gain lean tissue and body fat (as an energy reserve) and to acquire protective amounts of all the nutrients that can be stored.

An extreme underweight condition known as anorexia nervosa is sometimes seen in young people who exercise unreasonable self-denial in order to control their weight. They go to such extremes that they become severely undernourished and underweight. The distinguishing feature of a person with anorexia nervosa, as opposed to other thin people, is that the starvation is intentional. Anorexia nervosa is a major eating disorder seen in our society today. Another is bulimia nervosa—compulsive overeating, usually with purging. Eating disorders are the subject of the Nutrition in Practice that follows this chapter.

In Summary Both the incidence of underweight and the health problems associated with it are less prevalent than overweight and its associated problems.

Strategies for Weight Gain

Weight gain, like weight loss, is an individual matter. People who are healthy at their present weights may stay there; those who are unhealthy might try to gain.

Some people are unalterably thin by reasons of genetics or early physical influences. Those who wish to gain weight for appearance's sake or to improve athletic performance should be aware that a healthful weight can be achieved only through physical activity, particularly strength training, combined with a high energy intake. Eating many high-kcalorie foods can bring about weight gain, but it will be mostly fat, and this can be as detrimental to health as being slightly underweight. In an athlete, such a weight gain can impair performance. Therefore, in weight gain, as in weight loss, physical activity is an essential component of a sound plan.

Physical Activity to Build Muscles The person who wants to gain weight should use **weight training** primarily. As activity is increased, energy intake must be increased to support that activity. Eating extra food will then support a gain of both muscle and fat. About 700 to 1000 kcalories a day above normal energy needs is enough to support both the activity and the building of muscle.

Energy-Dense Foods Energy-dense foods (the very ones eliminated from a successful weight-loss diet) hold the key to weight gain. Pick the highest-kcalorie items from each food group—that is, milkshakes instead of fat-free milk, peanut butter instead of lean meat, avocados instead of cucumbers, and blueberry muffins instead of whole-wheat bread. Because fat contains more than twice as many kcalories per teaspoon as sugar does, fat adds kcalories without adding much bulk.

Weight-gain pointers:
- *Be physically active and eat to build muscles.*

weight training: the use of free weights or weight machines to provide resistance for developing muscle strength and endurance; also called *resistance training*. A person's own body weight may also be used to provide resistance as when a person does push-ups, pull-ups, or abdominal crunches.

Be aware that health experts recommend a low-fat diet for the general U.S. population because the general population is overweight and at risk for heart disease. Consumption of high-fat foods is not healthy for most people, of course, but may be essential for an underweight individual who needs to gain weight. An underweight person who is physically active and eating a nutritionally adequate diet can afford a few extra kcalories from fat.

■ *Eat energy-dense foods regularly.*

Three Meals Daily People wanting to gain weight should eat at least three hearty meals a day. Many people who are underweight have simply been too busy (sometimes for months) to eat enough to gain or maintain weight. Therefore, they need to make meals a priority and plan them in advance. Taking time to prepare and eat each meal can help, as can learning to eat more food within the first 20 minutes of a meal. Another suggestion is to eat meaty appetizers or the main course first and leave the soup or salad until later.

■ *Eat at least three hearty meals a day.*

Large Portions It is also important for the underweight person to learn to eat more food at each meal: have two sandwiches for lunch instead of one, drink milk from a larger glass, and eat cereal from a larger bowl. Expect to feel full. Most underweight individuals are accustomed to small quantities of food. When they begin eating significantly more, they feel uncomfortable. This is normal and passes over time.

■ *Eat large portions of foods and expect to feel full.*

Extra Snacks Since a substantially higher energy intake is needed each day, in addition to eating more food at each meal, it is necessary to eat more frequently. Between-meal snacking offers a solution. For example, a student might make three sandwiches in the morning and eat them between classes in addition to the day's three regular meals.

■ *Eat snacks between meals.*

Juice and Milk Beverages provide an easy way to increase energy intake. Consider that 6 cups of cranberry juice add almost 1000 kcalories to the day's intake. kCalories can be added to milk by mixing in powdered milk or packets of instant breakfast.

■ *Drink plenty of juice and milk.*

For people who are underweight due to illness, concentrated liquid formulas are often recommended because a weak person can swallow them easily. A registered dietitian can recommend high-protein, high-kcalorie formulas to help the underweight person maintain or gain weight. Used in addition to regular meals, these formulas can help considerably.

In Summary To gain weight, a person must train physically and increase energy intake by selecting energy-dense foods, eating regular meals, taking larger portions, and consuming extra snacks and beverages.

Self Study

OVERWEIGHT, UNDERWEIGHT, AND WEIGHT CONTROL

Does your BMI fall between 18.5 and 24.9? If so, you may want to maintain your weight. If not, you may need to gain or lose weight to improve your fitness and health. Determine whether these food and activity choices are typical of your lifestyle.

■ On the average, do your lifestyle choices promote weight gain, weight loss, or weight maintenance?

Food and Activity Choices	Frequency per Week
Promote weight gain:	
Drink plenty of juice.	_____
Eat energy-dense foods.	_____
Eat large portions.	_____
Eat peanut butter crackers between meals.	_____
Eat three or more large meals a day.	_____
Promote weight loss:	
Drink plenty of water.	_____
Eat nutrient-dense foods.	_____
Eat slowly.	_____
Eat small portions.	_____
Limit snacks to healthful choices.	_____
Limit television watching.	_____
Participate in physical activity.	_____
Select low-fat foods.	_____
Share a restaurant meal or take home leftovers.	_____

Self Check

1. The BMI range that correlates with the fewest health risks is:
   a. 16.5 to 20.9.
   b. 18.5 to 24.9.
   c. 25.5 to 30.9.
   d. 30.5 to 34.9.

2. The profile of central obesity is sometimes referred to as a(an):
   a. beer.
   b. pear.
   c. apple.
   d. potato.

3. Which of the following health risks is **not** associated with being overweight?
   a. hypertension
   b. heart disease
   c. type 1 diabetes
   d. gallbladder disease

4. Two causes of obesity in humans are:
   a. set-point theory and BMI.
   b. genetics and physical inactivity.
   c. genetics and low-carbohydrate diets.
   d. mineral imbalances and fat cell imbalance.

5. The protein produced by the fat cells under the direction of the *ob* gene is called:
   a. leptin.
   b. orlistat.
   c. sibutramine.
   d. lipoprotein lipase.

6. The obesity theory that suggests the body chooses to be at a specific weight is the:
   a. fat cell theory.
   b. enzyme theory.
   c. set-point theory.
   d. external cue theory.

7. Which of the following is a reasonable treatment for obesity?
   a. Eat energy-dense foods regularly.
   b. Drink plenty of juice and milk.
   c. Use strength training to build muscles.
   d. Make legumes, grains, vegetables, and fruits central to your diet plan.

8. Physical activity does **not** help a person to:
   a. lose weight.
   b. retain muscle.
   c. maintain weight loss.
   d. lose fat in trouble spots.

9. Suggestions to change behaviors for successful weight control include:
   a. shop only when hungry.
   b. eat in front of the television for distraction.
   c. learn appropriate portion sizes.
   d. eat quickly.

10. Which strategy would **not** help an underweight person to gain weight?
    a. Exercise.
    b. Drink plenty of water.
    c. Eat snacks between meals.
    d. Eat large portions of foods.

Critical Thinking

1. Some obese people are at greater risk of disease than others. Which of the following sets of risk factors in an obese person necessitates aggressive treatment to reduce the risk?
   a. sedentary lifestyle, smoker, female 45 years of age
   b. hypertension, high LDL cholesterol, male 46 years of age
   c. sedentary lifestyle, low LDL cholesterol, female 60 years of age
   d. hypertension, smoker, low LDL cholesterol

Answers to these questions can be found in Appendix G.

Clinical Applications

1. Consider a female client with the following height and weight:

   ▪ Height: 65 inches (or 165 centimeters).
   ▪ Weight: 165 pounds (or 75 kilograms).
   ▪ Calculate your client's BMI using the equation on p. 140.
   ▪ Record your client's BMI: ____.
   ▪ Look up the disease risk for a person with your client's BMI value in Table 7–1 on p. 144.

   ▪ Record your client's risk of disease based on the BMI: _____.

2. What information might you want to learn about your client's health habits and family or personal medical history to help you in further determining the client's risk of disease?

Nutrition on the Net

For further study of the topics of this chapter, access these websites.

Find updates and quick links to these and other nutrition-related sites at our website:
www.wadsworth.com/nutrition

Search for obesity at:
www.ilsi.org/

Information on a variety of obesity topics is available at the American Obesity Association website:
www.obesity.org/

Search for obesity at:
www.eatright.org/

Peruse the offerings of the Division of Nutrition and Physical Activity, National Center for Chronic Disease Prevention and Health Promotion at their website:
www.cdc.gov/nccdphp/dnpa

Many materials to help teach others about obesity are available at the Weight Control Information Network:
www.niddk.nih.gov/health/nutrit/win.htm

Read guides on fitness and healthy weight and access many other materials at Shape Up America:
www.shapeup.org

Review a transcript of presentations and panel discussions of leading obesity experts and fad diet authors by searching for Symposium on the Great Nutrition Debate at the USDA's site:
www.usda.gov

Read the latest materials on obesity diagnosis and treatment here:
www.nhlbi.nih.gov/guidelines/obesity/practgde. htm

The Partnership for Healthy Weight Management offers practical advice on starting a weight-loss program at this site:
www.consumer.gov/weightloss/

Notes

1. M. J. Williams and coauthors, Regional fat distribution in women and risk of cardiovascular disease, *American Journal of Clinical Nutrition* 65 (1997): 855–860.

2. G. R. Hunter and coauthors, Fat distribution, physical activity, and cardiovascular risk factors, *Medicine and Science in Sports and Exercise* 29 (1997): 362–369.

3. National Heart, Lung, and Blood Institute Expert Panel, National Institutes of Health, *Clinical Guidelines on the Identification, Evaluation, and Treatment of Overweight and Obesity in Adults* (Washington, D.C.: Government Printing Office, 1998).

4. D. A. McCarron and M. E. Reusser, Body weight and blood pressure regulation, *American Journal of Clinical Nutrition* 63 (1996): 423S–425S.

5. American Diabetes Association Position Statement: Evidence-based nutrition principles and recommendations for the treatment and prevention of diabetes and related complications, *Journal of the American Dietetic Association* 102 (2002): 109–118; J. Tuomilehto and coauthors, Prevention of type 2 diabetes mellitus by changes in lifestyle among subjects with impaired glucose tolerance, *New England Journal of Medicine* 344 (2001): 1343–1350.

6. A. H. Mokdad and coauthors, The spread of the obesity epidemic in the United States, 1991–1998, *Journal of the American Medical Association* 282 (1999): 1519–1522.

7. A. H. Mokdad and coauthors, The continuing epidemics of obesity and diabetes in the United States, *Journal of the American Medical Association* 286 (2001): 1195–1200; National Heart, Lung, and Blood Institute Expert Panel, 1998.

8. National Heart, Lung, and Blood Institute Expert Panel, 1998; J. Albu and coauthors, Obesity solutions: Report of a meeting, *Nutrition Reviews* 55 (1997): 150–156.

9. L. K. Khan and B. A. Bowman, OBESITY: A major global public health problem, *Annual Review of Nutrition* 19 (1999): xiii–xvii; National Task Force on the Prevention and Treatment of Obesity, Long-term pharmocotherapy in the management of obesity, *Journal of the American Medical Association* 276 (1996): 1907–1915.

10. J. K. Alexander, Obesity and coronary heart disease, *American Journal of Medical Sciences* 321 (2001): 215–224; R. H. Eckel and R. M. Krauss, American Heart Association call to action: Obesity as a major risk factor for coronary heart disease, *Circulation* 97 (1998): 2099–2100.

11. P. T. Katzmarzyk and coauthors, Fitness, fatness, and estimated coronary heart disease risk: The HERITAGE Family Study, *Medicine and Science in Sports and Exercise* 33 (2001): 585–590; C. D. Lee, S. N. Blair, and A. S. Jackson, Cardiorespiratory fitness, body composition, and all-cause and cardiovascular disease mortality in men, *American Journal of Clinical Nutrition* 69 (1999): 373–380; S. W. Farrell and coauthors, Influences of cardiorespiratory fitness levels and other predictors of cardiovascular disease mortality in men, *Medicine and Science in Sports and Exercise* 30 (1998): 899–904.

12. National Heart, Lung, and Blood Institute Expert Panel, 1998.

13. J. Stevens and coauthors, Evaluation of WHO and NHANES II standards for overweight using mortality rates, *Journal of the American Dietetic Association* 100 (2000): 825–827; A. Must and coauthors, The disease burden associated with overweight and obesity, *Journal of the American Medical Association* 282 (1999): 1523–1529.

14. National Heart, Lung, and Blood Institute, National Institutes of Health, *The Practical Guide: Identification, Evaluation, and Treatment of Overweight and Obesity in Adults,* NIH publication no. 00-4084 (Washington, D.C.: Government Printing Office, 2000).

15. J. C. Seidell, Societal and personal costs of obesity, *Experimental and Clinical Endocrinology and Diabetes* 106 (1998): S7–S9.

16. L. Perusse and C. Bouchard, Genotype-environment interaction in human obesity, *Nutrition Reviews* 57 (1999): S31–S38.

17. P. Foreyt and W. S. C. Poston II, Diet, genetics, and obesity, *Food Technology* 51 (1997): 70–73; C. Bouchard, Human variation in body mass: Evidence for a role of the genes, *Nutrition Reviews* 55 (1997): S21–S30.

18. C. L. Baile and M. A. Della-Fera, Regulation of metabolism and body fat mass by leptin, *Annual Review of Nutrition* 20 (2000): 105–127.

19. C. T. Montague and coauthors, Congenital leptin deficiency is associated with severe early-onset obesity in humans, *Nature* 387 (1997): 903–908.

20. S. Farooqi and coauthors, Effects of recombinant leptin therapy in a child with congenital leptin deficiency, *New England Journal of Medicine* 341 (1999): 879–884.

21. H. Fors and coauthors, Serum leptin levels correlate with growth hormone secretion and body fat in children, *Journal of Clinical Endocrinology and Metabolism* 84 (1999): 3586–3590; G. Marchini and coauthors, Plasma leptin in infants: Relations to birth weight and weight loss, *Pediatrics* 101 (1998): 429–432; C. S. Fox and coauthors, Is a low leptin concentration, a low resting metabolic rate, or both the expression of the "thrifty genotype"? Results from the Mexican Pima Indians, *American Journal of Clinical Nutrition* 68 (1998): 1053–1057; J. M. Friedman, Leptin, leptin receptors, and the control of body weight, *Nutrition Reviews* 56 (1998): S38–S46.

22. Albu and coauthors, 1997.

23. M. S. Westerterp-Plantenga and coauthors, Effects of weekly administration of pegylated recombinant human OB protein on appetite profile and energy metabolism in obese men, *American Journal of Clinical Nutrition* 74 (2001): 426–434; S. B. Heymsfield and coauthors, Recombinant leptin for weight loss in obese and lean adults: A randomized, controlled, dose-escalation trial, *Journal of the American Medical Association* 282 (1999): 1568–1575.

24. R. L. Bradley, K. A. Cleveland, and B. Cheatham, The adipocyte as a secretory organ: Mechanism of vesicle transport and secretory pathways, *Recent Progress in Hormone Research* 56 (2001): 329–358; R. B. S. Harris, Leptin—Much more than a satiety signal, *Annual Review of Nutrition* 20 (2000): 45–75.

25. Harris, 2000.

26. R. E. Keesey and M. D. Hirvonen, Body weight set-points: Determination and adjustment, *Journal of Nutrition* 127 (1997): S1875–S1883.

27. K. Brownell, The pressure to eat—Why we're getting fatter, *Nutrition Action Healthletter,* July/August 1998, pp. 3–8.

28. L. R. Young and M. Nestle, The contribution of expanding portion sizes to the U.S. obesity epidemic, *American Journal of Public Health* 92 (2002): 246–249.

29. B. Liebman and D. Schardt, Diet and health: Ten megatrends, *Nutrition Action Healthletter,* January/February 2001, pp. 3–12.

30. S. M. Green and coauthors, Comparison of high-fat and high-carbohydrate foods in a meal or snack on short-term fat and energy intakes in obese women, *British Journal of Nutrition* 84 (2000): 521–530.

31. G. A. Bray and B. M. Popkin, Dietary fat intake does affect obesity! *American Journal of Clinical Nutrition* 68 (1998): 1157–1173.

32. J. K. Binkley, J. Eales, and M. Jekanowski, The relation between dietary change and rising U.S. obesity, *International Journal of Obesity and Related Metabolic Disorders* 24 (2000): 1032–1029.

33. U. G. Kyle and coauthors, Physical activity and fat-free and fat mass by bioelectrical impedance in 3853 adults, *Medicine and Science in Sports and Exercise* 33 (2001): 576–584; P. R. Steffen and coauthors, Effects of exercise and weight loss on blood pressure during daily life, *Medicine and Science in Sports and Exercise* 33

(2001): 1635–1640; P. T. Williams, Physical fitness and activity as separate heart disease risk factors: A meta-analysis, *Medicine and Science in Sports and Exercise* 33 (2001): 754–761; D. S. Michaud and coauthors, Physical activity, obesity, height, and risk of pancreatic cancer, *Journal of the American Medical Association* 286 (2001): 921–929.

34. R. E. Anderson and coauthors, Relationship of physical activity and television watching with body weight and level of fatness among children: Results from the Third National Health and Nutrition Survey, *Journal of the American Medical Association* 279 (1998): 938–942.

35. S. Sarlio-Lahteenkorva, A. Rissanen, and J. Kaprio, A descriptive study of weight loss maintenance: 6 and 15 year follow-up of initially overweight adults, *International Journal of Obesity and Related Metabolic Disorders* 24 (2000): 116–125.

36. National Task Force on the Prevention and Treatment of Obesity, 1996.

37. C. H. Halsted, Is blockade of pancreatic lipase the answer? *American Journal of Clinical Nutrition* 69 (1999): 1059–1060.

38. D. L. Hansen and coauthors, Thermogenic effects of sibutramine in humans, *American Journal of Clinical Nutrition* 68 (1998): 1180–1186.

39. L. J. Aronne, Modern medical management of obesity: The role of pharmaceutical intervention, *Journal of the American Dietetic Association* 98 (1998): S23–S26.

40. M. H. Davidson and coauthors, Weight control and risk factor reduction in obese subjects treated for 2 years with orlistat, *Journal of the American Medical Association* 281 (1999): 235–242.

41. National Heart, Lung, and Blood Institute Expert Panel, 1998.

42. B. A. Schwetz, New weight-reduction system, *Journal of the American Medical Association* 286 (2001): 527; FDA approves implanted stomach band to treat severe obesity, *FDA Talk Paper*, June 5, 2001, available at **www.fda.gov;** R. Weiner, D. Wagner, and H. Bockhorn, Laparoscopic gastric banding for morbid obesity, *Journal of Laparoendoscopic and Advanced Surgical Techniques* 9 (1999): 23–30.

43. National Heart, Lung, and Blood Institute Expert Panel, 1998.

44. National Heart, Lung, and Blood Institute, National Institutes of Health, 2000, p. 3.

45. National Heart, Lung, and Blood Institute Expert Panel, 1998, p. 75.

46. National Heart, Lung, and Blood Institute Expert Panel, 1998, p. 74.

47. W. J. McCarthy, Strategies for achieving long-term weight maintenance (letter), *Journal of the American Dietetic Association* 98 (1998): 1273.

48. American Diabetes Association Position Statement, 2002; National Heart, Lung, and Blood Institute, National Institutes of Health, 2000, p. 13; Position of The American Dietetic Association: Weight management, *Journal of the American Dietetic Association* 97 (1997): 71–74; S. N. Blair, Diet and activity: The synergistic merger, *Nutrition Today* 30 (1996): 108–112; J. H. Wilmore, Increasing physical activity: Alterations in body mass and composition, *American Journal of Clinical Nutrition* 63 (1996): 254–260.

49. American College of Sports Medicine Position Stand, Appropriate intervention strategies for weight loss and prevention of weight regain for adults, *Medicine and Science in Sports and Exercise* 33 (2001): 2145–2156.

50. H. M. Sjödin and coauthors, The influence of physical activity on BMR, *Medicine and Science in Sports and Exercise* 28 (1996): 85–91.

51. N. A. King, A. Tremblay, and J. E. Blundell, Effects of exercise on appetite control: Implications for energy balance, *Medicine and Science in Sports and Exercise* 29 (1997): 1076–1089.

Nutrition in Practice

▪ EATING DISORDERS ▪

An estimated 5 million people in the United States, primarily girls and young women, suffer from some form of an **eating disorder.**[1] About 5 percent of females and 1 percent of males have **anorexia nervosa, bulimia nervosa,** or **binge eating disorder.** Many more suffer from other related conditions that do not meet the strict criteria for anorexia nervosa, bulimia nervosa, or binge eating disorder, but still imperil a person's well-being. Characteristics of disordered eating such as restrained eating, binge eating, purging, fear of fatness, and distortion of body image are common, especially among young middle-class girls.[2] In most other societies, these behaviors and attitudes are much less prevalent.

Why do so many young people in our society suffer from eating disorders?

Excessive pressure to be thin is at least partly to blame. In one national survey of more than 6700 adolescents in grades 5 through 12, almost half of the girls and a fifth of the boys reported having dieted to lose weight.[3] By making thinness the ideal, society pushes people to view a healthy body of normal weight as too fat. Healthy people then take unhealthy actions to lose weight. Severe restriction of food intake may create intense hunger that leads to binges. Research confirms this theory, showing that unhealthy or dangerous diets often precede binge eating in adolescent girls.[4] Energy restriction followed by bingeing can set in motion a pattern of weight cycling, which may make weight loss and maintenance more difficult over time.

People who attempt extreme weight loss are dissatisfied with their bodies to begin with; they may also be depressed or suffer social anxiety. As weight loss becomes more and more difficult, psychological problems worsen, and the likelihood of developing full-blown eating disorders intensifies.

People with anorexia nervosa suffer from an extreme preoccupation with weight loss that seriously endangers their health and even their lives. People with bulimia engage in episodes of binge eating alternating with periods of severe dieting or self-starvation. Some bulimics also follow binge eating with self-induced vomiting, laxative abuse, or diuretic abuse to "undo the damage." The glossary on this page defines the relevant terms.

Are there other groups, besides girls and young women, who are vulnerable to anorexia nervosa and bulimia nervosa?

Yes. Athletes who participate in sports that emphasize leanness are at special risk for developing eating disor-

Glossary

amenorrhea (ay-MEN-oh-REE-ah):   the absence of or cessation of menstruation. **Primary amenorrhea** is menarche delayed beyond 16 years of age. **Secondary amenorrhea** is the absence of three to six consecutive menstrual cycles.

anorexia nervosa: an eating disorder characterized by a refusal to maintain a minimally normal body weight, self-starvation to the extreme, and a disturbed perception of body weight and shape; seen (usually) in adolescent girls and young women.
anorexia = without appetite
nervosa = of nervous origin

binge eating disorder: an eating disorder whose criteria are similar to those of bulimia nervosa, excluding purging or other compensatory behaviors.

bulimia (byoo-LEEM-ee-uh) **nervosa:**   recurring episodes of binge eating combined with a morbid fear of becoming fat, usually followed by self-induced vomiting or purging.

cathartic: a strong laxative.

cognitive therapy: psychological therapy aimed at changing undesirable behaviors by changing underlying thought processes contributing to these behaviors. In anorexia nervosa, a goal is to replace false beliefs about body weight, eating, and self-worth with health-promoting beliefs.

eating disorder: a disturbance in eating behavior that jeopardizes a person's physical and psychological health.

emetic (em-ETT-ic):   an agent that causes vomiting.

female athlete triad: a potentially fatal triad of medical problems: disordered eating, amenorrhea, and osteoporosis.

Nutrition in Practice

ders.[5] Athletes must often meet stringent weight requirements to compete in their sport. Many athletes report that they engage in behaviors that are typical of people with eating disorders. Female competitors often report being terrified of becoming fat, being obsessed with food, and using laxatives in attempting to control weight. Dancers, jockeys, wrestlers, distance runners, bodybuilders, divers, figure skaters, gymnasts, and others whose body weight and appearance are frequently judged in comparison with an "ideal" are especially prone to develop problems.[6] These athletes may engage in extreme weight-loss practices such as overtraining, prolonged fasting, vomiting, taking diet pills, and using steam baths and saunas to induce sweating.[7]

Men account for about 1 in 20 cases in the general population, but among male athletes and dancers, eating disorders are much more common. Male teenagers normally average about 15 percent of body weight as fat, but some high school athletes strive to carry only 5 percent or so of their body weight as fat.

Wrestlers, for example, are required to "make weight" to compete in the lowest weight class to face the smallest possible opponents. To that end, wrestlers starve themselves, don rubber suits, sweat in steam rooms, and take diuretics to shed water weight before weighing in for competition. These practices were responsible for the deaths of three college athletes in recent years and have caused untold misery and harm to many others.[8] Athletes engaging in these practices actually compromise their athletic abilities. The diminished anaerobic strength, reduced endurance, decreased oxygen capacity, and general weakness caused by food deprivation and dehydration can impair performance, an effect lasting days after food and water are replenished.

Even among athletes, however, women are most vulnerable to developing eating disorders. Many female athletes appear healthy, but in fact may easily develop the three interrelated components of the **female athlete triad:** disordered eating, **amenorrhea** (the absence of three or more consecutive menstrual cycles), and osteoporosis.[9]

How does the female athlete triad develop?

Many athletic women engage in self-destructive eating behaviors (disordered eating) because they and their coaches have adopted unsuitable weight standards. An athlete's body must be heavier for height than a nonathlete's body because the athlete's bones and muscles are denser. Weight standards that may be appropriate for others are inappropriate for athletes. Measures such as fatfold measures yield more useful information about body composition.

Many young female athletes severely restrict energy intakes to improve performance, enhance the aesthetic appeal of their performance, or meet the weight guidelines of their specific sports. They fail to realize that the loss of lean tissue that accompanies energy restriction actually impairs their physical performance. Risk factors for the female athlete triad include the following:

- Young age (adolescence).
- Pressure to excel at a chosen physical activity.
- Focus on achieving or maintaining an "ideal" body weight or body fat percentage.
- Participation in endurance sports or competitions that judge performance on aesthetic appeal such as gymnastics, figure skating, or dance.
- Dieting at an early age.
- Unsupervised dieting.

As for amenorrhea, its prevalence among premenopausal women in the United States is about 2 to 5 percent overall, but among female athletes it may be as high as 66 percent. Contrary to previous notions, amenorrhea is not a normal adaptation to strenuous physical training: it is a symptom of something going wrong. Amenorrhea is characterized by low blood estrogen, infertility, and often bone mineral losses.

In general, weight-bearing physical activity, dietary calcium, and the hormone estrogen protect against the bone loss of osteoporosis, but in women with disordered eating and amenorrhea, strenuous activity may impair bone health.[10] Vigorous training combined with low food energy intakes and other life stresses seems to trigger amenorrhea and promote bone loss. Low estrogen leads to diminished bone mass and increased bone fragility. Many amenorrheic athletes have decreased bone density, similar to that of 50- to 60-year-old women, when they should have dense, strong bones. Amenorrheic athletes should be encouraged to consume at least 1500 milligrams of calcium each day, to eat nutrient-dense foods, and to obtain enough food energy to cover the energy expended in physical activity. Hormone replacement therapy may be appropriate for some women.[11]

What can be done to prevent eating disorders in athletes and dancers?

To prevent eating disorders in athletes and dancers, both the performers and their coaches must be educated about links between inappropriate body weight ideals, improper weight-loss techniques, eating disorder development, adequate nutrition, and safe weight-control methods. Coaches and dance instructors

Nutrition in Practice

*H*ow to

Combat Eating Disorders

*T*he following guidelines may be useful in combating eating disorders:

- Never restrict food servings to below the numbers suggested for adequacy by the Food Guide Pyramid.
- Eat frequently. People often do not eat frequent meals because of time constraints, but eating can be incorporated into other activities, such as snacking while studying or commuting. The person who eats frequently never gets so hungry as to allow hunger to dictate food choices.
- If not at a healthy weight, establish a reasonable weight goal based on a healthy body composition.
- Allow a reasonable time to achieve the goal. A reasonable loss of excess fat can be achieved at the rate of about 1 percent of body weight per week.

- Establish a weight-maintenance support group with people who share interests.

Specific guidelines for athletes and dancers include:

- Replace weight-based goals with performance-based goals.
- Remember that eating disorders impair physical performance. Seek confidential help in obtaining treatment if needed.
- Restrict weight-loss activities to the off-season.
- Focus on proper nutrition as an important facet of your training—as important as proper technique.

should never encourage unhealthy weight loss to qualify for competition or to conform with distorted artistic ideals. Frequent weighings can push young people who are striving to lose weight into a cycle of starving to confront the scale, then bingeing uncontrollably afterward. The erosion of self-esteem that accompanies these events can interfere with the normal psychological development of the teen years and set the stage for serious problems later on.

The accompanying "How to" box provides some suggestions to help athletes and dancers protect themselves against developing eating disorders. The next sections describe eating disorders that anyone, athlete or nonathlete, may experience.

What are the characteristics of anorexia nervosa?

Most anorexia nervosa victims are females who come from middle- or upper-class families. Family patterns often include parents who oppose one another's authority and who vacillate between defending and condemning the anorexic child's behavior, confusing the child and disrupting normal parental control. The family values achievement and outward appearances more than an inner sense of self-worth and self-actualization.

The person with anorexia nervosa is often a perfectionist who works hard to please her parents. She may identify so strongly with her parents' ideals and goals for

her that she sometimes feels she has no identity of her own. She is respectful of authority but sometimes feels like a robot, and she may act that way, too: polite but controlled, rigid, and unspontaneous.[12] She earnestly desires to control her own destiny, but she feels controlled by others. When she does not eat, she gains control.

Although families of children with anorexia nervosa often have problems, blame is a useless concept, and parents may suffer deeply from being blamed for their child's illness. In truth, no one knows what causes anorexia nervosa, and it may turn out that the cause is a physical one, and not related to parenting. Rather than judging parents, a more useful tactic is to identify the family's strong points and resources and to prepare them for the job of helping their ill child benefit from treatment.

How does a person know when dieting is going too far?

When a person loses weight to well below the average for her height and is no longer slim, but too slim, and still doesn't stop, she has gone too far. Regardless of how thin she is, she looks in the mirror and sees herself as fat. Central to the diagnosis of anorexia nervosa is a distorted body image that overestimates body fatness. Table NP7–1 shows the criteria that professionals use to diagnose anorexia nervosa. Anorexia nervosa resembles an addiction. The characteristic behavior is obses-

Nutrition in Practice

Table NP7–1	Criteria for Diagnosis of Anorexia Nervosa

A person with anorexia nervosa demonstrates the following:

A. Refusal to maintain body weight at or above a minimal normal weight for age and height, e.g., weight loss leading to maintenance of body weight less than 85 percent of that expected; or failure to make expected weight gain during period of growth, leading to body weight less than 85 percent of that expected.

B. Intense fear of gaining weight or becoming fat, even though underweight.

C. Disturbance in the way in which one's body weight or shape is experienced; undue influence of body weight or shape on self-evaluation, or denial of the seriousness of the current low body weight.

D. In females past puberty, amenorrhea, i.e., the absence of at least three consecutive menstrual cycles. (A woman is considered to have amenorrhea if her periods occur only following hormone, e.g., estrogen, administration.)

Two types:

■ **Restricting type:** During the episode of anorexia nervosa, the person does not regularly engage in binge eating or purging behavior (i.e., self-induced vomiting or the misuse of laxatives, diuretics, or enemas).

■ **Binge eating/purging type:** During the episode of anorexia nervosa, the person regularly engages in binge eating or purging behavior (i.e., self-induced vomiting or the misuse of laxatives, diuretics, or enemas).

SOURCE: Reprinted with permission from the *Diagnostic and Statistical Manual of Mental Disorders*, 4th ed., Text Revision (Washington, D.C.: American Psychiatric Association, 2000). Copyright 2000 American Psychiatric Association.

sive and compulsive. Before drawing conclusions about someone who is extremely thin or who eats very little, remember that diagnosis of anorexia nervosa requires professional assessment.

What is the harm in being very thin?

Anorexia nervosa damages the body much as starvation does. In young people, growth ceases and normal development falters. They lose so much lean tissue that basal metabolic rate slows.[13] Additionally, the heart pumps inefficiently and irregularly, the heart muscle becomes weak and thin, the heart chambers diminish in size, and the blood pressure falls. Electrolytes that help to regulate heartbeat become unbalanced. Many deaths from heart failure occur in people with anorexia.

Starvation brings other physical consequences as well: impaired immune response, anemia, and a loss of digestive function that worsens malnutrition. Digestive functioning becomes sluggish, the stomach empties slowly, and the lining of the intestinal tract shrinks. The ailing digestive tract fails to sufficiently digest any food the victim may eat. The pancreas slows its production of digestive enzymes. The person may suffer from diarrhea, further worsening malnutrition.

What kind of treatment helps people with anorexia nervosa?

Treatment of anorexia nervosa requires a multidisciplinary approach that addresses two sets of issues and behaviors: those relating to food and weight, and those involving relationships with oneself and others. Teams

of physicians, nurses, psychiatrists, family therapists, and dietitians work together to treat people with anorexia nervosa. Appropriate diet is crucial for normalizing body weight and must be tailored individually to each client's needs.[14] Seldom are clients willing to eat for themselves, but if they are, chances are they can recover without other interventions.

Professionals classify clients based on the risks posed by the degree of malnutrition present. Clients

Women with anorexia nervosa see themselves as fat, even when they are dangerously underweight.

Nutrition in Practice

Table NP7–2   Criteria for Diagnosis of Bulimia Nervosa

A person with bulimia nervosa demonstrates the following:

A. Recurrent episodes of binge eating. An episode of binge eating is characterized by both of the following:

   1. eating, in a discrete period of time (e.g., within any two-hour period), an amount of food that is definitely larger than most people would eat during a similar period of time and under similar circumstances, and,

   2. a sense of lack of control over eating during the episode (e.g., a feeling that one cannot stop eating or control what or how much one is eating).

B. Recurrent inappropriate compensatory behavior in order to prevent weight gain, such as self-induced vomiting; misuse of laxatives, diuretics, enemas, or other medications; fasting; or excessive exercise.

C. Binge eating and inappropriate compensatory behaviors that both occur, on average, at least twice a week for three months.

D. Self-evaluation unduly influenced by body shape and weight.

E. The disturbance does not occur exclusively during episodes of anorexia nervosa.

Two types:

- **Purging type:** The person regularly engages in self-induced vomiting or the misuse of laxatives, diuretics, or enemas.
- **Nonpurging type:** The person uses other inappropriate compensatory behaviors, such as fasting or excessive exercise, but does not regularly engage in self-induced vomiting or the misuse of laxatives, diuretics, or enemas.

SOURCE: Reprinted with permission © 2000 American Psychiatric Association. *Diagnostic and Statistical Manual of Mental Disorders*, 4th ed., Text Revision (Washington, D.C.: American Psychiatric Association, 2000).

---

with low risks may benefit from family counseling, **cognitive therapy,** behavior modification, and nutrition guidance; those with greater risks may also need other forms of psychotherapy and supplemental formulas to provide extra nutrients and energy.

High-risk clients may require hospitalization and may need to be force-fed by tube at first to forestall death. This step causes psychological trauma. Medications are commonly prescribed, but to date, they play a limited role in treatment.

Denial runs high among those with anorexia nervosa. Few seek treatment on their own. Almost half of the women who are treated can maintain their body weight within 15 percent of healthy weight; at that weight, many of them begin menstruating again. The other half have poor or fair treatment outcomes, and two-thirds of those treated fight an ongoing mental battle with recurring morbid thoughts about food and body weight.[15] Many relapse into abnormal eating behaviors to some extent. About 5 percent die during treatment, 1 percent by suicide.[16]

How does bulimia nervosa differ from anorexia nervosa?

Bulimia nervosa is distinct from anorexia nervosa and is more prevalent. More men suffer from bulimia nervosa than from anorexia, but bulimia is still more common in women. The secretive nature of bulimic behaviors makes recognition of the problem difficult, but once it is recognized, diagnosis is based on the criteria listed in Table NP7–2.

The typical person with bulimia is well educated, in her early twenties, and close to ideal body weight. She is a high achiever, with a strong feeling of dependence on her parents. She experiences considerable social anxiety and has difficulty establishing personal relationships. She is sometimes depressed and often exhibits impulsive behavior.

Like the person with anorexia nervosa, the person with bulimia spends much time thinking about her body weight and food. Her preoccupation with food manifests itself in secretive binge-eating episodes followed by self-induced vomiting, fasting, or the use of laxatives or diuretics. Such behaviors typically begin in late adolescence after a long series of various unsuccessful weight-reduction diets. People with bulimia commonly follow a pattern of restrictive dieting interspersed with bulimic behaviors and experience weight fluctuations of more than 10 pounds up and down over short periods of time.

Unlike the person with anorexia nervosa, the person with bulimia is aware of the consequences of her behavior, feels that it is abnormal, and is deeply ashamed of it. She feels inadequate and unable to control her eating, so she tends to be passive and to look to men for confirmation of her sense of self-worth. When she is rejected, either in reality or in her imagination, her bulimia becomes worse. If her depression deepens, she may seek solace in drug or alcohol abuse or other addictive behaviors. Many studies show a link between bulimia nervosa and drug and alcohol dependency.[17]

Nutrition in Practice

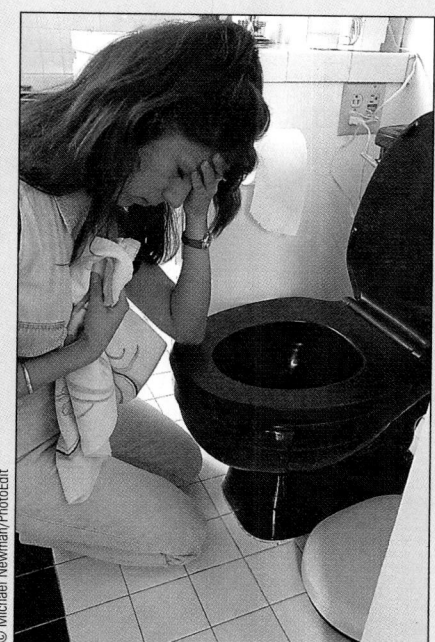

For many people with bulimia, guilt, depression, and self-condemnation follow a binge-eating episode.

What exactly is binge eating?

Binge eating is unlike normal eating, and the food is not consumed for its nutritional value. The binge eater has a compulsion to eat. A typical binge occurs periodically, is done in secret, usually at night, and lasts an hour or more. A binge frequently follows a period of rigid dieting, so the binge eating is accelerated by hunger. During a binge, the person with bulimia may consume from 1000 to many thousands of kcalories of food. The food typically contains little fiber or water, has a smooth texture, and is high in sugar and fat, so it is easy to consume vast amounts rapidly with little chewing.

What are the consequences of binge eating?

After a binge, the person may use a **cathartic**—a strong laxative that can injure the lower intestinal tract. Or the person may induce vomiting, using an **emetic**—a drug intended as first aid for poisoning.

On first glance, purging seems to offer a quick and easy solution to the problems of unwanted kcalories and body weight. Many people perceive such behavior as neutral or even positive, when, in fact, bingeing and purging have serious physical consequences. Fluid and electrolyte imbalances caused by vomiting or diarrhea can lead to metabolic alkalosis, a condition characterized by apathy, confusion, and muscle spasms. Vomiting causes irritation and infection of the pharynx, esophagus, and salivary glands; erosion of the teeth; and dental caries. The esophagus may rupture or tear, as may the stomach. Overuse of emetics depletes potassium concentrations and can lead to death by heart failure.

What is the treatment for bulimia?

As for people with anorexia nervosa, a team approach provides the most effective treatment for people with bulimia. Bulimia is easier to treat than anorexia nervosa in many respects because it seems to be more of a chosen behavior. People with bulimia know that their behavior is abnormal, and many are willing to try to cooperate.

The goal of the dietary plan to treat bulimia is to help clients gain control, establish regular eating patterns, and restore nutritional health. Energy intake should not be severely restricted. The person needs to learn to eat a quantity of nutritious food sufficient to nourish her body and to satisfy hunger (at least 1600 kcalories a day). The "How to" box on p. 168 offers some ways to begin correcting bulimia nervosa.

Anorexia nervosa and bulimia nervosa are distinct eating disorders, yet they sometimes overlap. Anorexia victims may purge, and victims of both conditions share an overconcern with body weight and the tendency to drastically undereat. The two disorders can also appear in the same person, or one can lead to the other.

At so tender an age as 12 years, beautifully growing, normal-weight female youngsters are already worried that they are too fat. Most are "on diets." Magazines, newspapers, and television all present the message that to be thin is to be beautiful and happy. Anorexia nervosa and bulimia are not a form of rebellion against these unreasonable expectations, but rather the exaggerated acceptance of them. Perhaps a person's best defense against these disorders is to learn to appreciate his or her own uniqueness.

Nutrition in Practice

How to

Establish Healthy Eating Patterns

*T*he following advice has proved useful for people fighting bulimia nervosa.

Planning principles:

- Plan meals and snacks; record plans in a food diary prior to eating.
- Plan meals and snacks that require eating at the table and using utensils.
- Refrain from finger foods.
- Refrain from "dieting" or skipping meals.

Nutrition principles:

- Eat a well-balanced diet and regularly timed meals consisting of a variety of foods.
- Include raw vegetables, salad, or raw fruit at meals to prolong eating times.
- Choose whole-grain, high-fiber breads, pasta, rice, and cereals to increase bulk.

- Consume adequate fluid, particularly water.

Other tips:

- Choose meals that provide protein and fat for satiety, and bulky, fiber-rich carbohydrates for immediate feelings of fullness.
- Choose portions that meet the definition of "a serving" according to the Daily Food Guide (pages 16–17).
- For convenience (and to reduce temptation) select foods that naturally divide into portions. Select one potato, rather than rice or pasta that can be overloaded onto the plate; purchase yogurt and cottage cheese in individual containers; look for small packages of precut steak or chicken; choose frozen dinners with metered portions.
- Include 30 minutes of physical activity every day—exercise may be an important tool in controlling bulimia.

Notes

1. Position of The American Dietetic Association: Nutrition intervention in the treatment of anorexia nervosa, bulimia nervosa, and eating disorders not otherwise specified (EDNOS), *Journal of the American Dietetic Association* 101 (2001): 810–819.

2. G. B. Schreiber and coauthors, Weight modification efforts reported by black and white preadolescent girls: National Heart, Lung, and Blood Institute Growth and Health Study, *Pediatrics* 98 (1996): 63–70.

3. D. Neumark-Sztainer and P. J. Hannan, Weight-related behaviors among adolescent girls and boys: Results from a national survey, *Archives of Pediatrics and Adolescent Medicine* 154 (2000): 569–577.

4. G. C. Patton and coauthors, Onset of adolescent eating disorders: Population-based cohort study over 3 years, *British Journal of Medicine* 318 (1999): 765–768; D. Neumark-Sztainer, R. Butler, and H. Palti, Dieting and binge eating: Which dieters are at risk? *Journal of the American Dietetic Association* 95 (1995): 586–588.

5. American Academy of Pediatrics, Committee on Sports Medicine and Fitness, Medical concerns in the female athlete, *Pediatrics* 106 (2000): 610–613.

6. American Academy of Pediatrics, 2000; A. Yates, Eating disorders in women athletes, *Eating Disorders Review*, July/August 1996, pp. 1–4.

7. Position stand from the Committee on Sports Medicine and Fitness on promotion of healthy weight-control practices in young athletes, *Pediatrics* 97 (1996): 752–753.

8. Hyperthermia and dehydration-related deaths associated with intentional rapid weight loss in three collegiate wrestlers—North Carolina, Wisconsin, and Michigan, November–December 1997, *Morbidity and Mortality Weekly Report* 47 (1998): 105–108.

9. American College of Sports Medicine, Position stand, The female athlete triad, *Medicine and Science in Sports and Exercise* 29 (1997): i–ix.

10. N. A. Armsey, Stress injury to bone in the female athlete, *Clinical Sports Medicine* 16 (1997): 197–224.

11. American Academy of Pediatrics, 2000.

12. T. Pryor and M. W. Weiderman, Personality features and expressed concerns of adolescents with eating disorders, *Adolescence* 33 (1998): 291–300.

13. A. Polito and coauthors, Basal metabolic rate in anorexia nervosa: Relation to body composition and leptin concentrations, *American Journal of Clinical Nutrition* 71 (2000): 1495–1502.

14. A. E. Becker and coauthors, Eating disorders, *New England Journal of Medicine* 340 (1999): 1092–1098.

15. B. Bower, Women with anorexia face ongoing problems, *Science News* 154 (1998): 39.

16. B. Herzog and coauthors, Mortality in eating disorders: A descriptive study, *International Journal of Eating Disorders* 28 (2000): 20–26.

17. B. Vastag, What's the connection? No easy answers for people with eating disorders and drug abuse, *Journal of the American Medical Association* 285 (2001): 1006–1007.

Jennie Oppenheimer/Studio Zocolo

CHAPTER 8 | THE VITAMINS

Table 8–1   Vitamin Names
Fat-Soluble Vitamins
Vitamin A
Vitamin D
Vitamin E
Vitamin K
Water-Soluble Vitamins
B vitamins
Thiamin
Riboflavin
Niacin
Pantothenic acid
Biotin
Vitamin B_6
Folate
Vitamin B_{12}
Vitamin C

Carbohydrate, fat, and protein are sometimes referred to as macronutrients because they are measured in foods in gram amounts. Vitamins and minerals are sometimes referred to as micronutrients because they are measured in foods in milligram and microgram amounts.

vitamins: essential, noncaloric, organic nutrients needed in tiny amounts in the diet.

bioavailability: the rate and extent to which a nutrient is absorbed and used.

precursors: compounds that can be converted into other compounds; with regard to vitamins, compounds that can be converted into active vitamins; also known as **provitamins.**

*E*arlier chapters focused primarily on the energy-yielding nutrients—carbohydrate, fat, and protein. This chapter and the next one discuss the nutrients everyone thinks of when nutrition is mentioned—the vitamins and minerals.

The Vitamins—An Overview

The **vitamins** occur in foods in much smaller quantities than do the energy-yielding nutrients, and they themselves contribute no energy to the body. Instead, they serve mostly as facilitators of body processes. They are a powerful group of substances, as their absence attests. Vitamin A deficiency can cause blindness; a lack of niacin can cause mental illness; and a lack of vitamin D can retard growth. The consequences of deficiencies are so dire and the effects of restoring the needed nutrients so dramatic that people spend billions of dollars each year on vitamin supplements to cure many different ailments. Vitamins certainly contribute to sound nutritional health, but supplements do not cure all ills. Actually, a vitamin can cure only the disease caused by a deficiency of that vitamin. The vitamins' roles in supporting optimal health extend far beyond preventing deficiency diseases, however. Emerging evidence points to relationships between low intakes of vitamins and chronic diseases such as cancer and heart disease.

A child once defined vitamins as "what, if you don't eat, you get sick." The description is both insightful and accurate. A more prosaic definition is that vitamins are potent, essential, noncaloric, organic nutrients needed from foods in trace amounts to perform specific functions that promote growth reproduction, and the maintenance of health and life. Two characteristics distinguish vitamins from energy nutrients:

1. Vitamins do not yield energy when broken down, but assist the enzymes that release energy from carbohydrate, fat, and protein.

2. Vitamins are needed in much smaller amounts than the energy nutrients.

As the individual vitamins were discovered, they were named or given letters, numbers, or both. This led to the confusion that still exists today. This chapter uses the names shown in Table 8–1; alternative names are given in Tables 8–2 and 8–3, which appear later in the chapter.

Bioavailability   The availability of vitamins from foods depends on two factors: the quantity provided by a food and the amount absorbed and used by the body (the vitamin's **bioavailability**). Researchers analyze foods to determine their vitamin contents and publish the results in tables of food composition such as Appendix A. Determining the bioavailability of a vitamin is more difficult because it depends on many factors, including:

- The efficiency of digestion.
- A person's previous nutrient intake and nutrition status.
- Other foods eaten at the same time.
- Method of food preparation (raw or cooked, for example).
- The source of the nutrient (naturally occurring, synthetic, or fortified).

Experts consider these factors when estimating recommended intakes.

Precursors   Some of the vitamins are available from foods in inactive forms known as **precursors,** or provitamins. Once inside the body, the precursor is converted to the active form of the vitamin. Thus, in measuring a person's vitamin intake, it is important to count both the amount of the actual vitamin and the potential amount available from its precursors.

Solubility Vitamins fall naturally into two classes—fat soluble and water soluble. The solubility of a vitamin confers on it many characteristics and determines how it is absorbed, transported, stored, and excreted. This discussion of vitamins begins with the fat-soluble vitamins.

In Summary Vitamins are essential, noncaloric nutrients that are needed in trace amounts in the diet to help facilitate body processes.

The Fat-Soluble Vitamins

The fat-soluble vitamins—A, D, E, and K—usually occur together in the fats and oils of foods, and the body absorbs them in the same way it absorbs lipids. Therefore, any condition that interferes with fat absorption can precipitate a deficiency of the fat-soluble vitamins. Once absorbed, fat-soluble vitamins are stored in the liver and fatty tissues until the body needs them. They are not readily excreted, and unlike most of the water-soluble vitamins, they can build up to toxic concentrations. Excesses of vitamins A, D, and K from supplements can reach toxic levels especially easily.

The capacity to store fat-soluble vitamins affords a person some flexibility in dietary intake. When blood concentrations begin to decline, the body can retrieve the vitamins from storage. Thus a person need not eat a day's allowance of each fat-soluble vitamin every day, but need only make sure that over time, average daily intakes approximate recommended intakes. In contrast, water-soluble vitamins must be consumed more regularly because the body does not store them to any great extent.

Vitamin A and Beta-Carotene

Vitamin A has the distinction of being the first fat-soluble vitamin to be recognized. Today, after almost a century of revelations, vitamin A and its plant-derived precursor, **beta-carotene,** are the focus of much research around the world. Much of this intensive research effort is based on accumulating evidence that both vitamin A and beta-carotene may protect against certain types of cancer.

Metabolic Roles of Vitamin A Vitamin A is a versatile vitamin, with roles in vision, protein synthesis and cell differentiation (thereby maintaining the health of body linings and skin), immunity, and reproduction and growth.[1] Three different forms of vitamin A are active in the body: retinol, retinal, and retinoic acid. Each form of vitamin A performs specific tasks. Retinol supports reproduction and is the major transport and storage form of the vitamin; the cells convert retinol to retinal or retinoic acid as needed. Retinal is active in vision, and retinoic acid acts as a hormone, regulating cell differentiation, growth, and embryonic development. A special transport protein, **retinol-binding protein (RBP),** picks up retinol from the liver where it is stored and carries it in the blood.

Vitamin A in Vision Vitamin A plays two indispensable roles in the eye. It helps maintain a healthy, crystal-clear outer window, the **cornea;** and it participates in the events of light detection at the **retina.** Figure 8–1 on the next page shows vitamin A's site of action inside the eye.

When vitamin A is lacking, the eye has difficulty adapting to changing light levels. At night, after the eye has adapted to darkness, a lag occurs before the eye can see again after a flash of bright light. This lag in the recovery of night vision is known as **night blindness.** Because night blindness is easy to test, it aids in the diagnosis of vitamin A deficiency. Night blindness is only a symptom, however, and may indicate a condition other than vitamin A deficiency.

vitamin A: a fat-soluble vitamin. Its three chemical forms are *retinol* (the alcohol form), *retinal* (the aldehyde form), and *retinoic acid* (the acid form).

beta-carotene: a vitamin A precursor made by plants and stored in human fat tissue; an orange pigment.

retinol-binding protein (RBP): the specific protein responsible for transporting retinol. Measurement of the blood concentration of RBP is a sensitive test of vitamin A status.

cornea (KOR-nee-uh): the hard, transparent membrane covering the outside of the eye.

retina (RET-in-uh): the layer of light-sensitive nerve cells lining the back of the inside of the eye; consists of rods and cones.

night blindness: the slow recovery of vision after exposure to flashes of bright light at night; an early symptom of vitamin A deficiency.

Figure 8–1

Vitamin A's Role in Vision
As light enters the eye, pigments within the cells of the retina absorb the light and generate nerve impulses that travel to the brain. Each pigment contains retinal, the active form of vitamin A.

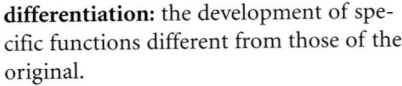

Pigment-containing retina cells (rods and cones).

Eye

Light energy

To the brain

Nerve impulses

differentiation: the development of specific functions different from those of the original.

epithelial (ep-i-THEE-lee-ul) **cells:** cells on the surface of the skin and mucous membranes.

epithelial tissue: tissue composing the layers of the body that serve as selective barriers between the body's interior and the environment (examples are the cornea, the skin, the respiratory lining, and the lining of the digestive tract).

mucous membrane: membrane composed of mucus-secreting cells that lines the surfaces of body tissues. (Reminder: *Mucus* is the smooth, slippery substance secreted by these cells.)

antioxidant (anti-OX-ih-dant): a compound that protects other compounds from oxygen by itself reacting with oxygen. *Oxidation* is a potentially damaging effect of normal cell chemistry involving oxygen.
anti = against
oxy = oxygen

free radicals: highly reactive chemical forms that can cause destructive changes in nearby compounds, sometimes setting up a chain reaction.

Vitamin A in Protein Synthesis and Cell Differentiation The role that vitamin A plays in vision is undeniably important, but only one-thousandth of the body's vitamin A is in the retina. Much more is in the skin and the linings of organs, where it works behind the scenes at the genetic level to promote protein synthesis and cell **differentiation.** The process of cell differentiation allows each type of cell to mature so that it is capable of performing a specific function.

All body surfaces, both inside and out, are covered by layers of cells known as **epithelial cells.** The **epithelial tissue** on the outside of the body is, of course, the skin. The epithelial tissues inside the body include the linings of the mouth, stomach, and intestines; the linings of the lungs and the passages leading to them; the lining of the bladder; the linings of the uterus and vagina; and the linings of the eyelids and sinus passageways. The epithelial tissues on the inside of the body must be kept smooth. To ensure that they are, the epithelial cells on their surfaces secrete a smooth, slippery substance (mucus) that coats and protects the tissues from invasive microorganisms and other harmful particles. The **mucous membrane** that lines the stomach also shields its cells from digestion by gastric juices. Vitamin A, by way of its role in cell differentiation, helps to maintain the integrity of the epithelial cells.

Vitamin A in Immunity Vitamin A's role in protecting the body's immune response was recognized more than 70 years ago when researchers noticed that vitamin A deficiency reduces resistance to infection.[2] Immune function depends on the growth, differentiation, and activation of the cells that defend the body against infectious agents.[3] The immune system is the topic of Nutrition in Practice 15.

Vitamin A in Reproduction and Growth Vitamin A also supports reproduction and growth. In men, vitamin A participates in sperm development, and in women, vitamin A promotes normal fetal growth and development.[4]

Metabolic Roles of Beta-Carotene For many years scientists believed beta-carotene to be of interest solely as a vitamin A precursor. Eventually, though, researchers began to recognize beta-carotene as an extremely effective **antioxidant** in the body. Antioxidants are compounds that protect other compounds (such as lipids in cell membranes) from attack by oxygen. Oxygen triggers the formation of compounds known as **free radicals** that can start chain reactions in cell membranes. If left uncontrolled, these chain reactions can damage cell structures and impair cell functions. Oxidative and free-radical damage to cells is suspected of instigating some early stages of cancer and heart disease. Research has identified links between oxidative damage and the development of many other diseases, including age-related blindness, arthritis, cataracts, diabetes, and kidney disease.[5]

Studies of populations suggest that people whose diets are low in beta-carotene–rich foods have higher incidences of certain types of cancer than those whose diets contain generous amounts of foods rich in beta-carotene. Based on findings that beta-carotene in foods may protect against cancer, researchers designed a study to determine the effects of beta-carotene *supplements* on the incidence of lung cancer among smokers. The researchers expected to see a beneficial effect, but instead found that smokers taking the beta-carotene supplements suffered a *greater* incidence of lung cancer than those taking placebos.[6] In the Physicians' Health Study, researchers tested the effects of beta-carotene supplements versus a placebo on cancer and heart disease risk in more than 20,000 male physicians for 12 years.[7] Beta-carotene neither decreased nor increased cancer or heart disease risk. The lead researcher on this study as well as on three other large-scale studies of antioxidant vitamins and disease risk (Dr. C. H. Hennekens) states that "the totality of evidence on beta-carotene supplements shows that well-nourished populations accrue no benefits on cancer or heart disease."[8] He emphasizes the need for more research on antioxidant vitamins and disease risk and reminds consumers that so far, there is no "magic pill" for reducing the risk of cancer and heart disease. Positive changes in diet and lifestyle, however, can reduce risk. Beta-carotene in foods is just one among many nutrients and compounds present in foods. As discussed in Nutrition in Practice 8, many other phytochemicals are also present in foods and may be responsible for some of the protective effects attributed to beta-carotene. Until more is known, eating beta-carotene–rich foods, not supplements, is in the best interest of health. Based on research so far, the DRI committee has not established a recommended intake value for beta-carotene.

Vitamin A Deficiency Up to a year's supply of vitamin A can be stored in the body, 90 percent of it in the liver. If a healthy adult were to stop eating vitamin A–rich foods, deficiency symptoms would not begin to appear until after stores were depleted, which would take one to two years. Then, however, the consequences would be profound and severe. Table 8–2, later in this chapter, lists some of them.

In vitamin A deficiency, cell differentiation and maturation are impaired. The epithelial cells flatten and begin to produce **keratin**—the hard, inflexible protein of hair and nails. In the eye, this process leads to drying and hardening of the cornea, which may progress to permanent blindness.

Blindness Vitamin A deficiency is the major cause of childhood blindness in the world, causing more than half a million children to lose their sight every year.[9] More than 200 million children worldwide endure less severe forms of vitamin A deficiency, making them vulnerable to infectious diseases.

Infections All body surfaces, both inside and out, maintain their integrity with the help of vitamin A. When vitamin A is lacking, cells of the skin harden and flatten, making it dry, rough, scaly, and hard. An accumulation of keratin makes a lump around each hair **follicle** (keratinization).

In the mouth, a vitamin A deficiency results in drying and hardening of the salivary glands, making them susceptible to infection. Secretions of mucus in the stomach and intestines are reduced, hindering normal digestion and absorption of nutrients. Infections of other mucous membranes also become likely.

Vitamin A's role in maintaining the body's defensive barriers may partially explain the relationship between vitamin A deficiency and susceptibility to infection.[10] In several studies, when children with measles complicated by diarrhea, infections such as pneumonia, or both were given vitamin A supplements, their overall survival rates were significantly greater than those of similar children who did not receive vitamin A.[11]

The evidence that vitamin A reduces the severity of measles and measles-related infections and diarrhea has prompted the World Health Organization (WHO) and UNICEF (the United Nations International Children's Emergency

Reminder: Phytochemicals are nonnutrient compounds found in plant-derived foods that have biological activity in the body.

The progressive blindness caused by vitamin A deficiency is called **xerophthalmia** (zer-off-THAL-mee-uh).
xero = dry
ophthalm = eye

An early sign of xerophthalmia is **xerosis** (drying of the cornea); the last and most severe stage is **keratomalacia** (kerr-uh-to-mal-AY-shuh), or total blindness.
malacia = softening, weakening

The accumulation of the hard material keratin around each hair follicle is **follicular hyperkeratosis.**

keratin (KERR-uh-tin): a water-insoluble protein; the normal protein of hair and nails. Keratin-producing cells may replace mucus-producing cells in vitamin A deficiency.

follicle (FOLL-i-cul): a group of cells in the skin from which a hair grows.

The Vitamins

Fund) to make control of vitamin A deficiency a major goal in their quest to improve child survival throughout the developing world. The American Academy of Pediatrics recommends vitamin A supplementation for certain groups of measles-infected infants and children in the United States.

Vitamin A Toxicity When the body stores excess vitamin A, toxicity is possible. Normally, toxicity symptoms are likely only when animal-derived foods or supplements are the source of the excess vitamin, for in these sources the vitamin is already in its active form, called **preformed vitamin A.** Plant foods contain the vitamin only as beta-carotene, its inactive, precursor form. The precursor does not convert to active vitamin A rapidly enough to cause toxicity.

Overdoses of vitamin A damage the same body systems that exhibit symptoms in vitamin A deficiency (see Table 8–2 later in the chapter). Children are most vulnerable to vitamin A toxicity because, being smaller, they need less than adults, and it is easy to give them too much in pill form. The availability of breakfast cereals, instant meals, fortified milk, and chewable candylike vitamins, each containing 100 percent or more of the recommended daily intake of vitamin A, makes it possible for a well-meaning parent to provide several times the daily allowance of the vitamin to a child within a few hours. Serious toxicity is seen in infants and young children when they are given more than ten times the recommended amount every day for weeks at a time.

Excessive vitamin A also poses a **teratogenic** risk.[12] In pregnant women, chronic use of vitamin A supplements providing three to four times the amount recommended for pregnancy can cause malformations of the fetus.[13]

Certain vitamin A relatives are available by prescription as acne treatments. When applied directly to the skin surface, these preparations help relieve the symptoms of acne. Taking massive doses of vitamin A internally will *not* cure acne, however, and may cause the miseries itemized in Table 8–2. Foods are always a better choice than supplements for needed nutrients. The best way to ensure a safe vitamin A intake is to eat generous servings of vitamin A–rich foods. Well-nourished, healthy people do not need vitamin A supplements.

Beta-Carotene Conversion and Toxicity When beta-carotene is converted to retinol in the body, losses occur. This is why nutrition scientists do not use micrograms to specify the quantity of beta-carotene in foods. Instead, they use a value known as **retinol activity equivalents (RAE),** which express the amount of retinol the body actually derives from a plant food after conversion. The body can make one unit of retinol from about 12 units of beta-carotene.

As mentioned earlier, beta-carotene from plant foods is not converted to the active form of vitamin A rapidly enough to be hazardous. It has, however, been known to turn people bright yellow if they eat too much. Beta-carotene builds up in the fat just beneath the skin and imparts a yellow cast.

Vitamin A in Foods Preformed vitamin A is found only in foods of animal origin. The richest sources of vitamin A are liver and fish oil, but milk, cheese, and fortified cereals are also good sources. Healthy people can eat vitamin A–rich foods in large amounts without risking toxicity with the possible exception of liver. Eating liver once every week or so is enough. Butter and eggs also provide some vitamin A to the diet.

Because vitamin A is fat soluble, it is lost when milk is skimmed. Fat-free milk is thus often fortified with vitamin A to compensate. Margarine is also usually fortified so as to provide the same amount of vitamin A as butter. Snapshot 8–1 shows a sampling of the richest food sources of both preformed vitamin A and beta-carotene.

Fast-food meals often lack vitamin A. When fast-food restaurants offer salads with cheese, carrots, and other vitamin A–rich foods, the nutritional quality of their meals greatly improves.

A unit used earlier to express vitamin A amounts in foods was the **IU (international unit).**

1 RAE = 1 μg retinol.
= 12 μg beta-carotene.

Yellowing of the skin caused by excess carotene in the blood is known as **carotenemia** (KAR-oh-teh-NEE-me-ah). Carotenemia can be distinguished from jaundice because the mucous membranes lining the eyelids do not turn yellow as they do in jaundice.

preformed vitamin A: vitamin A in its active form.

teratogenic (ter-AT-oh-jen-ik): causing abnormal fetal development and birth defects.
terato = monster
genic = to produce

retinol activity equivalents (RAE): a measure of vitamin A activity; the amount of retinol that the body will derive from a food containing preformed retinol or its precursor beta-carotene.

Snapshot 8–1

Vitamin A and Beta-Carotene
These foods provide 10 percent or more of the vitamin A Daily Value (DV = 900 µg/day).[a]

Carrots:[c] 106% DV per ½ c cooked

Mango: 28% DV per ½ c

Beef liver:[b] 1010% DV per 3 oz

Fortified milk:[b] 17% DV per cup

Spinach:[c] 41% DV per ½ c cooked

Sweet potato:[c] 108% DV per ½ c mashed

[a]The Daily Values are based on a 2000-kcalorie diet.
[b]Preformed vitamin A.
[c]Beta-carotene.

Beta-Carotene in Foods   Many foods from plants contain beta-carotene, the orange pigment responsible for the bold colors of many fruits and vegetables. Carrots, sweet potatoes, pumpkins, cantaloupe, and apricots are all rich sources, and their bright orange color enhances the eye appeal of the plate. Another colorful group, *dark* green vegetables, such as spinach, other greens, and broccoli, owe their color to both chlorophyll and beta-carotene. The orange and green pigments together impart a deep, murky green color to the vegetables. Other colorful vegetables, such as iceberg lettuce, beets, and sweet corn, can fool you into thinking they contain beta-carotene, but these foods derive their color from other pigments and are poor sources of beta-carotene. As for "white" plant foods such as grains and potatoes, they have none. Recommendations to eat *dark* green or *deep* orange vegetables and fruits at least every other day help people to meet their vitamin A needs.

Vitamin D

Vitamin D is different from all the other nutrients in that the body can synthesize it in significant quantities with the help of sunlight. Therefore, in a sense, vitamin D is not an essential nutrient. Given enough sun, people need no vitamin D from foods.

Vitamin D's Metabolic Conversions   The liver manufactures a vitamin D precursor, which migrates to the skin where it is converted to a second precursor with the help of the sun's ultraviolet rays. Next, the liver and then the kidneys alter the second precursor to produce the active vitamin. Vitamin D precursors from plants require the same two conversions by the liver and kidneys to become active. The biologic activity of the active vitamin is 500- to 1000-fold greater than that of its precursor. Diseases that affect either the liver or the kidneys may impair the transformations of precursor vitamin D to active vitamin D and therefore produce symptoms of vitamin D deficiency.

Vitamin D's Actions   Although known as a vitamin, vitamin D is actually a hormone—a compound manufactured by one organ of the body that has effects on another. The best-known vitamin D target organs are the intestine, the kidneys, and the bones, but scientists have discovered several other vitamin D target tissues, including the brain, the pancreas, the skin, the reproductive organs, and many cancer cells.[14] These discoveries suggest that numerous additional functions for vitamin D may surface, including regulation of the immune system.[15]

Sunlight promotes vitamin D synthesis in the skin. Exposure to the sun should be moderate, however; excessive exposure may cause skin cancer.

The precursor of vitamin D made in the liver is 7-dehydrocholesterol, which is made from cholesterol. This is one of the body's many "good" uses for cholesterol.

The final, active vitamin is 1-25 dihydroxycholecalciferol, or, more simply, dihydroxy vitamin D.

The Vitamins

Vitamin D's Roles in Bone Vitamin D is a member of a large, cooperative bone-making and maintenance team composed of nutrients and other compounds, including vitamins A, C, and K; the hormones parathormone and calcitonin; the protein collagen; and the minerals calcium, phosphorus, magnesium, and fluoride. Vitamin D's special role in bone growth is to make calcium and phosphorus available in the blood that bathes the bones. The bones grow denser and stronger as the minerals are deposited from the blood. Vitamin D acts in three ways to maintain blood concentrations of calcium and phosphorus: it stimulates their absorption from the GI tract; it mobilizes calcium and phosphorus from bones into the blood; and it stimulates their retention by the kidneys.

Vitamin D Deficiency The symptoms of vitamin D deficiency are those of calcium deficiency, as shown in Table 8–2. The bones fail to calcify normally and may grow so weak that they become bent when they have to support the body's weight. A child with the vitamin D–deficiency disease **rickets** who is old enough to walk characteristically develops bowed legs, often the most obvious sign of the disease. Rickets was a major pediatric health problem in the United States before vitamin D fortification of commercially prepared milk was introduced decades ago. Unfortunately, rickets seems to be making a comeback among African American breastfed infants who are exposed to little sunlight and receive no vitamin D supplements and among toddlers fed unfortified soy and rice beverages instead of milk.[16] Worldwide, rickets afflicts a large number of children because of inadequate food and lack of sunlight.

Adult rickets, or **osteomalacia,** occurs most often in women who have low calcium intakes and little exposure to sun and who go through repeated pregnancies and periods of lactation. The bones of the legs may soften to such an extent that a young woman who is tall and straight at age 20 may be condemned by repeated pregnancies to become bent, bowlegged, and stooped before she is 30.

Inadequate vitamin D is recognized as a risk factor in **osteoporosis** (reduced bone density). Without sufficient vitamin D, absorption of calcium is limited, and bone remodeling is impaired. This combination leads to a loss of bone mass.

Vitamin D Toxicity Whereas vitamin D deficiency depresses calcium absorption, blood calcium, and bone mineralization, an excess of vitamin D does the opposite, as shown in Table 8–2. It enhances calcium absorption, produces high blood calcium, and promotes return of bone calcium into the blood. The excess calcium then tends to precipitate in the soft tissues, forming stones, including kidney stones. Calcification may also harden the blood vessels and is especially dangerous in the major arteries of the heart and lungs, where it can cause death.

Vitamin D in excess is the most toxic of all the vitamins. The amounts of vitamin D in foods available in the United States and Canada are well within safe limits, but supplements containing the vitamin in concentrated form are not. Adults should use caution when taking vitamin D supplements and keep them out of the reach of children. The DRI committee has set a Tolerable Upper Intake Level for vitamin D at 50 micrograms per day (2000 IU on supplement labels).

Vitamin D from the Sun Most of the world's population relies on natural exposure to sunlight to maintain adequate vitamin D nutrition. The sun imposes no risk of vitamin D toxicity. Prolonged exposure to sunlight degrades the vitamin D precursor in the skin, preventing its conversion to the active vitamin. Even lifeguards on southern beaches are safe from vitamin D toxicity from the sun.

Prolonged exposure to sunlight has other undesirable consequences, however, such as premature wrinkling of the skin and the risk of skin cancer. These risks may be reduced by using sunscreens. Unfortunately, sunscreens with sun protection factors (SPF) of 8 and above also retard vitamin D synthesis.[17] A strategy to avoid this dilemma is to apply sunscreen after enough time has elapsed to provide

rickets: the vitamin D–deficiency disease in children.

osteomalacia (os-tee-oh-mal-AY-shuh): a bone disease characterized by softening of the bones; symptoms include bending of the spine and bowing of the legs. The disease occurs most often in adult women.
osteo = bone
mal = bad (soft)

osteoporosis: literally, porous bones; reduced density of the bones, also known as *adult bone loss.*

sufficient vitamin D. For most people, exposing hands, face, and arms on a clear summer day for 10 minutes, a few times a week, should be sufficient to maintain vitamin D nutrition. Dark-skinned people require longer exposure than light-skinned people, but by 3 hours, vitamin D synthesis in heavily pigmented skin arrives at the same plateau as in fair skin after 30 minutes.

The ultraviolet rays from tanning lamps and tanning booths may also stimulate vitamin D synthesis, but the hazards outweigh any possible benefits. The Food and Drug Administration (FDA) warns that if the lamps are not properly filtered, people using tanning booths risk burns, damage to the eyes and blood vessels, and skin cancer.

Heavy clouds, smoke, or smog may filter out the ultraviolet rays of the sun that promote vitamin D synthesis. Together with skin pigmentation, smog probably accounts for dark-skinned people in northern, smoggy cities developing rickets. For these people, and for those who are unable to go outdoors frequently, dietary vitamin D is especially important.

Vitamin D in Foods   Only a few animal foods, notably, eggs, liver, butter, some fish, and fortified milk, supply significant amounts of vitamin D. For those who use margarine in place of butter, fortified margarine is a significant source. Infant formulas are fortified with vitamin D in amounts adequate for daily intake. Breast milk is low in vitamin D, so vitamin D supplements are recommended for dark-skinned, breastfed infants who do not have adequate exposure to the sun.[18] These sources, plus any exposure to the sun, provide babies with more than enough of this vitamin.

The fortification of milk with vitamin D is the best guarantee that children will meet their vitamin D needs and underscores the importance of milk in children's diets. Unlike milk, cheese and yogurt are not fortified with vitamin D. Vegans, and especially their children, may have low vitamin D intakes because no fortified plant source except margarine exists. In the United States, breakfast cereals may be fortified with vitamin D, as their labels indicate.

Most adults, especially in sunny regions, need not make special efforts to obtain vitamin D in food. People who are not outdoors much or who live in northern or predominantly cloudy or smoggy areas, however, are advised to make sure their milk is fortified with vitamin D and to drink at least 2 cups a day.

Vitamin E

More than 80 years ago, researchers discovered a compound in vegetable oils necessary for reproduction in rats. The compound was named **tocopherol,** which means "offspring." Eventually, the compound was named vitamin E. When chemists isolated four tocopherol compounds, they designated them by the first four letters of the Greek alphabet: alpha, beta, gamma, and delta. Of these, alpha-tocopherol is the gold standard for vitamin E activity, and recommended intakes are based on it. Table 8–2 summarizes important information about vitamin E.

Vitamin E as an Antioxidant   Like beta-carotene, vitamin E is a fat-soluble antioxidant. It protects other substances from oxidation by being oxidized itself. If there is plenty of vitamin E in the membranes of cells exposed to an oxidant, chances are this vitamin will take the brunt of any oxidative attack, protecting the lipids and other vulnerable components of the membranes. Vitamin E is especially effective in preventing the oxidation of the polyunsaturated fatty acids (PUFA), but it protects all other lipids (for example, vitamin A) as well.

Vitamin E exerts an especially important antioxidant effect in the lungs, where the cells are exposed to high concentrations of oxygen. Vitamin E also protects the lungs from air pollutants that are strong oxidants.

Some evidence suggests that vitamin E may also offer protection against heart disease by protecting LDL (low-density lipoproteins) from oxidation.[19]

Vitamin D activity was previously expressed in international units (IU), but as of 1980, it is expressed in micrograms of cholecalciferol, the active form of vitamin D. To convert, use the following factor:
- 100 IU = 2.5 μg.
- 400 IU = 10 μg.

Reminder: *Oxidation* is a type of chemical reaction, so named because oxygen is one of the agents that often brings it about.

tocopherol (tuh-KOFF-er-ol): a general term for several chemically related compounds, one of which has vitamin E activity.

Vegetable oils, some nuts and seeds such as almonds and sunflower seeds, and wheat germ are vitamin E–rich.

The oxidation of LDL encourages the development of atherosclerosis. In at least some studies, vitamin E both from foods and from supplements seemed to be effective in defending against heart disease.[20] Other research, however, did not find that vitamin E reduces the risk of heart disease or heart attack. Researchers studying more than 9000 men and women at high risk for heart attack found that 10 micrograms of vitamin E daily for about four and a half years had no effect on heart attack or death from heart disease.[21] The researchers noted several possible reasons for the lack of effect of vitamin E in their study:

- The "high-risk" status of the population studied.
- The moderate duration (four and a half years) of vitamin E supplementation.
- The use of vitamin E alone, without other antioxidants.

Vitamin E supplements in this study had no adverse effects, an encouraging finding for future, longer-term studies that will address remaining questions about vitamin E supplements and disease risk. So far, the evidence is insufficient to recommend vitamin E supplements as a heart disease preventive.[22]

Vitamin E Myths Although research continues to reveal possible roles for vitamin E, it has also clearly discredited claims that vitamin E improves athletic skill, enhances sexual performance, or cures sexual dysfunction in males. Vitamin E also does not prevent or cure hereditary **muscular dystrophy,** nor does it slow or prevent processes of aging, such as graying of the hair, wrinkling of the skin, or reduced activity of body organs.

Vitamin E Deficiency When blood concentrations of vitamin E fall below a certain critical level, the red blood cells tend to break open and spill their contents, probably because the PUFA in their membranes oxidize. This classic vitamin E–deficiency symptom, known as **erythrocyte hemolysis,** is seen in premature infants born before the transfer of vitamin E from the mother to the fetus that takes place in the last weeks of pregnancy. Vitamin E treatment corrects erythrocyte hemolysis.

Two other conditions appear to respond to vitamin E therapy. One is a nonmalignant breast disease (**fibrocystic breast disease**), and the other (**intermittent claudication**) is an abnormality of blood flow that causes cramping in the legs.

In human beings, vitamin E deficiency is usually associated with diseases, notably those that cause malabsorption of fat. These include diseases of the liver, gallbladder, and pancreas, as well as various hereditary diseases involving digestion and use of nutrients.

On rare occasions, vitamin E deficiencies develop in people without diseases. Most likely, such deficiencies occur after years of eating diets extremely low in fat; using fat substitutes, such as diet margarines and salad dressings, as the only sources of fat; or consuming diets composed of highly processed or "convenience" foods. Extensive heating in the processing of foods destroys vitamin E.

Vitamin E Toxicity Vitamin E supplement use has increased in recent years as its antioxidant action against disease has been recognized. As a result, signs of toxicity are now known or suspected, although vitamin E toxicity is not nearly as common, and its effects are not as serious, as vitamin A or vitamin D toxicity. Extremely high doses of vitamin E interfere with the blood-clotting action of vitamin K and enhance the action of anticoagulant medications, leading to hemorrhage. For most individuals, however, ordinary supplemental doses (no more than 800 milligrams daily) taken over a period of months seem to have no adverse health effects.[23] The Tolerable Upper Intake Level for vitamin E (1000 milligrams) is more than 65 times greater than the recommended intake for adults (15 milligrams).

Caution: Other serious conditions can cause lumps in the breast and pain in the legs. Don't self-diagnose; see a physician.

muscular dystrophy (DIS-tro-fee): a hereditary disease in which the muscles gradually weaken; its most debilitating effects arise in the lungs. This disease should not be confused with *nutritional* muscular dystrophy, a vitamin E–deficiency disease of animals characterized by gradual paralysis of the muscles.

erythrocyte (er-REETH-ro-cite) **hemolysis** (he-MOLL-uh-sis): rupture of the red blood cells, caused by vitamin E deficiency.
erythro = red
cyte = cell
hemo = blood
lysis = breaking

fibrocystic breast disease: a harmless condition in which the breasts develop lumps, sometimes associated with caffeine consumption. In some, it responds to abstinence from caffeine; in others, it can be treated with vitamin E.
fibr = fibrous lumps
cystic = in sacs

intermittent claudication: severe calf pain caused by inadequate blood supply; it occurs when walking and subsides during rest.
intermittent = at intervals
claudicare = to limp

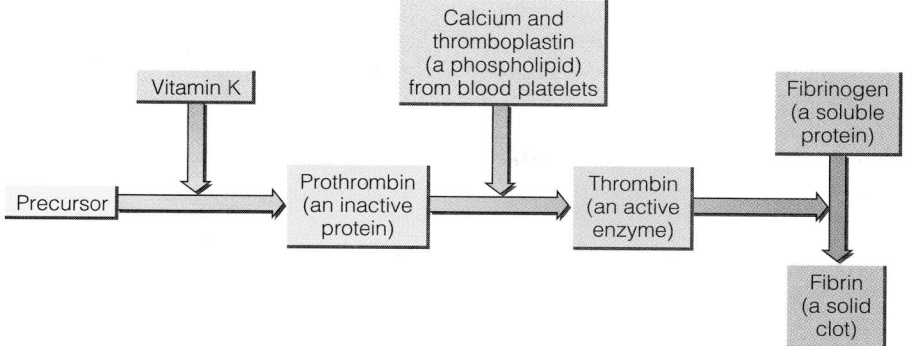

Figure 8–2
Blood-Clotting Process
When blood is exposed to air, foreign substances, or secretions from injured tissues, platelets (small, cell-like structures in the blood) release a phospholipid known as thromboplastin. Thromboplastin catalyzes the conversion of the inactive protein prothrombin to the active enzyme thrombin. Thrombin then catalyzes the conversion of the precursor protein fibrinogen to the active protein fibrin that forms the clot.

Vitamin E in Foods   Vitamin E is widespread in foods. About 20 percent of the vitamin E in the diet comes from vegetable oils and the products made from them, such as margarine, salad dressings, and shortenings. (Soybean oils and wheat germ oils have especially high concentrations of vitamin E.) Another 20 percent comes from fruits and vegetables although none of these is a good source by itself. Fortified cereals and other grain products contribute about 15 percent of vitamin E in the diet, and meats, poultry, fish, eggs, milk products, nuts, and seeds contribute smaller percentages. Because vitamin E is readily destroyed by heat processing and oxidation, fresh or lightly processed foods are the best sources of this vitamin.

Published values of the vitamin E in food reflect all of the different tocopherols and are expressed in "milligrams of tocopherol equivalents." These measures overestimate the amount of alpha-tocopherol. To estimate the alpha-tocopherol content of foods stated in tocopherol equivalents, multiply by 0.8.[24]

On vitamin bottles, vitamin E activity is often expressed in IU:

- 1 IU "natural" vitamin E = .67 mg alpha-tocopherol
- 1 IU synthetic vitamin E = .45 mg alpha-tocopherol

Vitamin K

Vitamin K has long been known for its role in blood clotting, where its presence can make the difference between life and death. The vitamin also participates in the synthesis of several bone proteins.[25] Research is under way to determine the specific roles of these vitamin K–dependent proteins in bone metabolism and the risk of osteoporosis.[26] Vitamin K may play a role in reducing the risk of hip fracture: in a large study of women, those who ate abundant green vegetables, good sources of vitamin K, suffered hip fractures less often than those with lower intakes.[27]

K stands for the Danish word *koagulation* (coagulation or "clotting").

Blood Clotting   At least 13 different proteins and the mineral calcium are involved in making blood clot. Vitamin K is essential for the activation of seven of these proteins, among them prothrombin, the precursor of the protein thrombin (see Figure 8–2 above).[28] When any of the blood-clotting factors is lacking, **hemorrhagic disease** results. If an artery or vein is cut or broken, bleeding goes unchecked. Of course, this is not to say that hemorrhaging is always caused by a vitamin K deficiency.

Intestinal Synthesis   Like vitamin D, vitamin K can be obtained from a nonfood source. Bacteria in the intestinal tract synthesize vitamin K that the body can absorb, but people cannot depend on this source alone for their vitamin K.

Reminder: The bacterial inhabitants of the digestive tract are known as the *intestinal flora.*
flora = plant inhabitants

Vitamin K Deficiency   Vitamin K deficiency is rare, but may occur in two circumstances. First, it may arise in conditions of fat malabsorption. Second, some medications interfere with vitamin K's synthesis and action in the body: antibiotics kill the vitamin K–producing bacteria in the intestine, and anticoagulant medications interfere with vitamin K metabolism and activity. When vitamin K deficiency does occur, it can be fatal.

hemorrhagic (hem-oh-RAJ-ik) **disease:** the vitamin K–deficiency disease in which blood fails to clot.

Notable food sources of vitamin K include milk, eggs, brussels sprouts, cabbage, and spinach.

Vitamin K for Newborns Newborn infants present a unique case of vitamin K nutrition. An infant is born with a **sterile** digestive tract, and some weeks pass before the vitamin K–producing bacteria become fully established in the infant's intestines. At the same time, plasma prothrombin concentrations are low (this helps prevent blood clotting during the stress of birth, which might otherwise be fatal). A single dose of vitamin K, usually in a water-soluble form, is recommended at birth to prevent hemorrhagic disease in the newborn.

Vitamin K Toxicity A high intake of vitamin K can reduce the effectiveness of anticoagulant medications used to prevent the blood from clotting. People taking these medications should eat consistent amounts of vitamin K–rich foods from day to day. Vitamin K–toxicity symptoms include red cell hemolysis, **jaundice,** and brain damage (see Table 8–2 on pp. 181–182). Reports of vitamin K toxicity among healthy adults are rare, however, and the DRI committee has set no Tolerable Upper Intake Level.

Vitamin K in Foods Many foods contain ample amounts of vitamin K, notably, green leafy vegetables, members of the cabbage family, and liver. Other vegetables, milk, meats, eggs, cereal, and fruits provide smaller, but still significant, amounts.

In Summary The fat-soluble vitamins are vitamins A, D, E, and K. Vitamin A is essential to vision, cell differentiation and integrity of epithelial tissues, immunity, and reproduction and growth. Vitamin A deficiency causes blindness, sickness, and death and is a major problem worldwide. Overdoses of vitamin A are possible and dangerous. Vitamin D raises calcium and phosphorus levels in the blood. A deficiency can cause rickets in children or osteomalacia in adults. Vitamin D is the most toxic of all the vitamins. People exposed to the sun make vitamin D in their skin; fortified milk is an important food source. Vitamin E acts as an antioxidant in cell membranes and is especially important in the lungs where cells are exposed to high concentrations of oxygen. Vitamin E may protect against heart disease, but evidence is not conclusive yet. Vitamin E deficiency is rare in healthy human beings. The vitamin is widely distributed in plant foods. Vitamin K is necessary for blood to clot. The bacterial inhabitants of the digestive tract produce vitamin K, but people need vitamin K from foods as well. Dark green, leafy vegetables are good sources of vitamin K. Table 8–2 offers a complete summary of the fat-soluble vitamins.

The Water-Soluble Vitamins

The B vitamins and vitamin C are the water-soluble vitamins. These vitamins, found in the watery compartments of foods, are distributed into water-filled compartments of the body. They are easily absorbed into the bloodstream and are just as easily excreted if their blood concentrations rise too high. Thus the water-soluble vitamins are less likely to reach toxic concentrations in the body than are the fat-soluble vitamins. Foods never deliver excessive amounts of the water-soluble vitamins, but the large doses concentrated in vitamin supplements can reach toxic levels.

The B Vitamins

Despite advertisements that claim otherwise, the B vitamins do not give people energy. Carbohydrate, fat, and protein—the *energy-yielding* nutrients—supply the fuel for energy. The B vitamins help to burn that fuel, but do not serve as fuel themselves.

sterile: free of microorganisms, such as bacteria.

jaundice: yellowing of the skin due to spillover of bile pigments from the liver into the general circulation.

Table 8–2 The Fat-Soluble Vitamins—A Summary

Vitamin A

Other Names	Deficiency Symptoms	Toxicity Symptoms

Retinol, retinal, retinoic acid; main precursor is beta-carotene

Chief Functions in the Body

Vision: health of cornea, epithelial cells, mucous membranes; skin health; bone and tooth growth; reproduction; hormone synthesis and regulation; immunity; cancer protection

2001 RDA

Men: 900 µg RAE/day
Women: 700 µg RAE/day

Upper Level

Adults: 3000 µg/day

Deficiency Disease Name

Hypovitaminosis A

Significant Sources

Retinol: fortified milk, cheese, cream, butter, fortified margarine, eggs, liver

Beta-carotene: spinach and other dark leafy greens; broccoli; deep orange fruits (apricots, cantaloupe) and vegetables (squash, carrots, sweet potatoes, pumpkin)

Deficiency and Toxicity Symptoms

	Deficiency Symptoms	Toxicity Symptoms
Blood/Circulatory System	Anemia (small-cell type)[a]	Red blood cell breakage, nosebleeds
Bones/Teeth	Cessation of bone growth, painful joints; impaired enamel formation, cracks in teeth, tendency to decay	Bone pain; growth retardation; increase of pressure inside skull mimicking brain tumor; headaches
Digestive System	Diarrhea, changes in lining	Abdominal cramps and pain, nausea, vomiting, diarrhea, weight loss
Immune System	Suppression of immune reactions; frequent respiratory, digestive, bladder, vaginal, and other infections	Overreactivity
Nervous/Muscular Systems		Blurred vision, pain in calves, fatigue, irritability, loss of appetite
Skin and Cornea	Night blindness, keratinization, corneal degeneration leading to blindness,[b] plugging of hair follicles with keratin, forming white lumps (hyperkeratosis)	Dry skin, rashes, loss of hair
Other	Kidney stones, impaired growth	Cessation of menstruation, liver and spleen enlargement

Vitamin D

Other Names	Deficiency Symptoms	Toxicity Symptoms

Calciferol, cholecalciferol, dihydroxy vitamin D; precursor is cholesterol

Chief Functions in the Body

Mineralization of bones (raises blood calcium and phosphorus via absorption from digestive tract, and by withdrawing calcium from bones and stimulating retention by kidneys)

1997 Adequate Intake (AI)

Adults: 5 µg/day (19–50 yr)
      10 µg/day (51–70 yr)
      15 µg/day (>70 yr)

Upper Level

Adults: 50 µg/day

Deficiency Disease Name

Rickets, osteomalacia

Significant Sources

Self-synthesis with sunlight; fortified milk, margarine, butter, and cereals; eggs, liver, small fish (sardines)

Deficiency and Toxicity Symptoms

	Deficiency Symptoms	Toxicity Symptoms
Blood/Circulatory System	Decreased blood calcium and/or phosphorus	Raised blood calcium and phosphorus
Bones/Teeth	Abnormal growth, misshapen bones (bowing of legs), soft bones, joint pain, malformed teeth	Increased calcium withdrawal
Nervous/Muscular Systems	Muscle spasms	Excessive thirst, headaches, irritability, loss of appetite, weakness, nausea
Other		Kidney stones, stones in arteries, death

(Continued on next page)

[a]Small-cell anemia is termed *microcytic anemia;* large-cell type is *macrocytic* or *megaloblastic anemia.*
[b]Corneal degeneration progresses from *keratinization* (hardening) to *xerosis* (drying) to *xerophthalmia* (thickening, opacity, and irreversible blindness).

Table 8–2   The Fat-Soluble Vitamins—A Summary—continued

Vitamin E

Other Names	Deficiency Symptoms	Toxicity Symptoms
Alpha-tocopherol, tocopherol	**Blood/Circulatory System**	
Chief Functions in the Body	Red blood cell damage, anemia	Augments the effects of anticlotting medication
Antioxidant (detoxification of strong oxidants), stabilization of cell membranes, regulation of oxidation reactions, protection of PUFA and vitamin A	**Digestive System**	General discomfort
	Nervous/Muscular Systems	
2000 RDA	Degeneration, weakness, difficulty walking, leg cramps	(No symptoms reported)
Adults: 15 mg/day		
Upper Level		
Adults: 1000 mg/day		
Deficiency Disease Name		
(No name)		
Significant Sources		
Polyunsaturated plant oils (margarine, salad dressings, shortenings), green and leafy vegetables, wheat germ, whole-grain products, nuts, seeds		

Vitamin K

Other Names	Deficiency Symptoms	Toxicity Symptoms
Phylloquinone, menaquinone, naphthoquinone	**Blood/Circulatory System**	
Chief Functions in the Body	Hemorrhaging	Interference with anticlotting medication; vitamin K analogues may cause jaundice, red cell hemolysis, and brain damage
Synthesis of blood-clotting proteins and a protein that binds calcium in the bones		
2001: Adequate Intake (AI)		
Men: 120 μg/day Women: 90 μg/day		
Deficiency Disease Name		
(No name)		
Significant Sources		
Bacterial synthesis in the digestive tract; liver, leafy green vegetables, cabbage-type vegetables, milk		

coenzyme (co-EN-zime): a small molecule that works with an enzyme to promote the enzyme's activity. Many coenzymes have B vitamins as part of their structure.

co = with

Coenzymes   The eight B vitamins were listed in Table 8–1. Each is part of an enzyme helper known as a **coenzyme.** Each B vitamin has other important functions in the body as well, but the roles these vitamins play as parts of coenzymes are the best understood. A coenzyme is a small molecule that combines with an enzyme to make it active. With the coenzyme in place, a substance is attracted to the enzyme, and the reaction proceeds instantaneously. Figure 8–3 illustrates coenzyme action.

A coenzyme already mentioned in Chapter 6 was coenzyme A, or CoA, made from the vitamin pantothenic acid. Various other coenzymes containing thiamin, riboflavin, niacin, or biotin also participate in the release of energy from glucose, amino acids, and fats. A coenzyme containing vitamin B_6 assists enzymes that metabolize amino acids. The making of new cells depends on a folate coenzyme, and the making of this coenzyme depends on vitamin B_{12}.

The eight B vitamins play many specific roles in helping the enzymes to perform thousands of different molecular conversions in the body. They must be present in every cell continuously for the cells to function as they should. As for vitamin C, its primary role, discussed later, is as an antioxidant.

B Vitamin Deficiencies In academic and clinical discussions of the vitamins, different sets of deficiency symptoms are ascribed to each individual vitamin. Such clear-cut symptoms, however, are found only in laboratory animals that have been fed contrived diets that lack just one nutrient. In reality, a deficiency of any single B vitamin seldom shows up in isolation because people do not eat nutrients one by one; they eat foods containing mixtures of many nutrients. If a major class of foods is missing from the diet, all of the nutrients delivered by those foods will be lacking to various extents.

In only two cases have dietary deficiencies associated with single B vitamins been observed on a large scale in human populations. Diseases have been named for these deficiency states. One of them, **beriberi,** was first observed in Southeast Asia when the custom of polishing rice became widespread. Rice contributed 80 percent of the energy intake of the people in these areas, and rice hulls were their principal source of thiamin. When the hulls were removed to make the rice whiter, beriberi spread like wildfire.

The niacin-deficiency disease, **pellagra,** became widespread in the southern United States in the early part of the twentieth century among people who subsisted on a low-protein diet with a staple grain of corn. This diet was unusual in that it supplied neither enough niacin nor enough of its amino acid precursor tryptophan to make the niacin intake adequate.

Even in the cases of beriberi and pellagra, the deficiencies were probably not pure. When foods were provided containing the one vitamin known to be needed, other vitamins that may have been in short supply came as part of the package.

Major deficiency diseases such as pellagra and beriberi no longer occur in the United States and Canada, but more subtle deficiencies of nutrients, including the B vitamins, are sometimes observed. When they do occur, it is usually in people whose food choices are poor because of poverty, ignorance, illness, or poor health habits such as alcohol abuse.

Interdependent Systems Table 8–3, at the end of this chapter, sums up a few of the better-established facts about B vitamin deficiencies. A look at the table will make another generalization possible. Different body systems depend to different extents on these vitamins. Processes in nerves and in their responding tissues, the muscles, depend heavily on glucose metabolism and hence on thiamin, so paralysis sets in when this vitamin is lacking; but thiamin is important in all cells, not just in nerves and muscles. Similarly, because the red blood cells and GI tract cells divide the most rapidly, two of the first symptoms of a deficiency of folate are a type of anemia and GI deterioration—but again, all systems depend on folate, not just these. The list of symptoms in Table 8–3 is far from complete.

B Vitamin Enrichment of Foods If the staple food of a region is made from **refined grain,** vitamin B deficiencies are especially likely. One way to protect people from deficiencies is to add nutrients to their staple food, a process known as **fortification** or **enrichment.** The enrichment of refined breads and cereals has drastically reduced the incidence of iron and B vitamin deficiencies.

The preceding discussion has shown both the great importance of the B vitamins in promoting normal, healthy functioning of all body systems and the severe

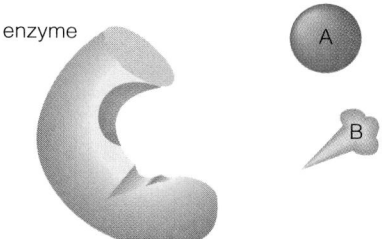

enzyme

Without the coenzyme, compounds A and B don't respond to the enzyme.

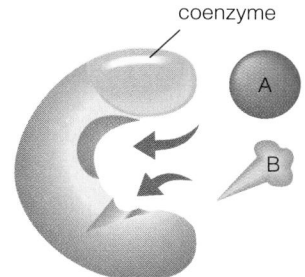

coenzyme

With the coenzyme in place, compounds A and B are attracted to the active site on the enzyme, and they react.

The reaction is completed with the formation of a new product. In this case the product is AB.

Figure 8–3
Coenzyme Action

Note: The terms *fortified* and *enriched* may be used interchangeably.

beriberi: the thiamin-deficiency disease; characterized by loss of sensation in the hands and feet, muscular weakness, advancing paralysis, and abnormal heart action.

pellagra (pell-AY-gra): the niacin-deficiency disease. Symptoms include the "4 Ds": diarrhea, dermatitis, dementia, and, ultimately, death.
pellis = skin
agra = seizure

The Vitamins

Nutritious foods such as pork, legumes, sunflower seeds, and enriched and whole-grain breads are valuable sources of thiamin.

consequences of deficiency. Now you may want to know how to be sure you and your clients are getting enough of these vital nutrients. The next sections present information on each B vitamin. While reading further, keep in mind that *foods* can provide all the needed nutrients and that supplements are a poor second choice. Some supplements are absurdly costly; but even if they are inexpensive, most people don't need them. Nutrition in Practice 9 discusses uses and choices of supplements in more detail.

Thiamin

All cells use thiamin, which plays a critical role in their energy metabolism. Thiamin also occupies a special site on nerve cell membranes. Consequently, processes in nerves and in their responding tissues, the muscles, depend heavily on thiamin.

Thiamin Need As long as people consume enough food to meet their energy needs—and obtain that energy from nutritious foods—thiamin needs will be met. People who derive a large proportion of their energy from empty-kcalorie items, like sugar or alcohol, however, risk thiamin deficiency, a condition that seems to be reappearing as the population of malnourished and homeless people rises. A person who is fasting or who has adopted a very-low-kcalorie diet needs as much thiamin as when eating enough to meet energy needs.

In developed countries today, abuse of alcohol often leads to a severe form of thiamin deficiency, Wernicke-Korsakoff syndrome.[29] Alcohol contributes energy, but carries almost no nutrients with it and often displaces food. In addition, alcohol impairs absorption of thiamin from the digestive tract and hastens its excretion in urine, tripling the risk of deficiency. Wernicke-Korsakoff syndrome is characterized by symptoms that are almost indistinguishable from alcohol abuse itself: mental confusion, disorientation, loss of memory, jerky eye movements, and staggering gait. Unlike alcohol toxicity, the syndrome responds quickly to an injection of thiamin, and some experts recommend a precautionary dose for any patients suspected of having the syndrome.[30]

Thiamin in Foods Thiamin occurs in small quantities in virtually all nutritious foods, but it is concentrated in only a few foods, of which pork is the most commonly eaten. A useful guideline for meeting thiamin needs is to keep empty-kcalorie foods to a minimum and to include ten or more different servings of nutritious foods each day, assuming that each serving will contribute, on the average, about 10 percent of needs. Foods chosen from the bread and cereal group should be either whole grain or enriched. Thiamin is not stored in the body to any great extent, so daily intake is important.

Riboflavin

Like thiamin, riboflavin facilitates energy production in the body. The needs of infants, children, and pregnant women rise rapidly during periods of active growth.

Riboflavin in Foods Unlike thiamin, riboflavin is not evenly distributed among the food groups. The major contributors of riboflavin to people's diets are milk, milk products, meats, and green vegetables (broccoli, turnip greens, asparagus, and spinach). The riboflavin richness of milk and milk products is a good reason to include these foods in every day's meals. No other commonly eaten food can make such a substantial contribution. People who omit milk and milk products from their diets can substitute generous servings of dark green, leafy vegetables. Among the meats, liver and heart are the richest sources, but all lean meats, as well as eggs, offer some riboflavin.

refined grain: a product from which the bran, germ, and husk have been removed, leaving only the endosperm.

fortification: the addition to a food of nutrients that were either not originally present or present in insignificant amounts. Fortification can be used to correct or prevent a widespread nutrient deficiency, to balance the total nutrient profile of a food, or to restore nutrients lost in processing.

enrichment: the addition to a food of nutrients to meet a specified standard. In the case of refined bread or cereal, five nutrients have been added: thiamin, riboflavin, niacin, and folate in amounts approximately equivalent to, or higher than, those originally present and iron in amounts to alleviate the prevalence of iron-deficiency anemia.

Effects of Light Riboflavin is light sensitive; the ultraviolet rays of the sun or of fluorescent lamps can destroy it. For this reason, milk is sold in cardboard or opaque plastic containers to protect the riboflavin in the milk from light. In contrast, riboflavin is heat stable, so ordinary cooking does not destroy it.

Niacin

Like thiamin and riboflavin, niacin participates in the energy metabolism of every body cell. Niacin is unique among the B vitamins in that the body can make it from protein. The amino acid tryptophan can be converted to niacin in the body: 60 milligrams of tryptophan yield 1 milligram of niacin. Recommended intakes are therefore stated in **niacin equivalents (NE),** reflecting the body's ability to convert tryptophan to niacin.

Niacin Used as a Medication Certain forms of niacin supplements in amounts ten times or more than the dietary recommendation cause "niacin flush," a dilation of the capillaries of the skin with perceptible tingling that, if intense, can be painful. The Tolerable Upper Intake Level (35 milligrams NE) is based on flushing as the critical adverse effect.[31] Physicians sometimes use diet and large doses of a form of niacin (nicotinic acid) to lower blood cholesterol in the treatment of atherosclerosis. When used this way, niacin leaves the realm of nutrition to become a pharmacological agent, a drug. As with any medication, self-dosing with niacin is ill advised; large doses may injure the liver, cause ulcers, and produce some symptoms of diabetes.[32]

Niacin in Foods Meat, poultry, and fish contribute about half the niacin equivalents most people consume; enriched breads and cereals contribute about a fourth. Among the vegetables, mushrooms, asparagus, and green leafy vegetables are the richest niacin sources. Niacin is less vulnerable to losses during food preparation and storage than other water-soluble vitamins. Being fairly heat-resistant, niacin can withstand reasonable cooking times, but like other water-soluble vitamins, it will leach into cooking water.

Pantothenic Acid and Biotin

Two other B vitamins—pantothenic acid and biotin—are also important in energy metabolism. Pantothenic acid was first recognized as a substance that stimulates growth. It is a component of a key enzyme that makes possible the release of energy from the energy nutrients. Pantothenic acid is involved in more than 100 different steps in the synthesis of lipids, neurotransmitters, steroid hormones, and hemoglobin. Biotin plays an important role in metabolism as a coenzyme that carries carbon dioxide. This role is critical in the TCA cycle.

Pantothenic Acid and Biotin in Foods Both pantothenic acid and biotin are more widespread in foods than the other vitamins discussed so far. There seems to be no danger that people who consume a variety of foods will suffer deficiencies. Claims that pantothenic acid and biotin are needed in pill form to prevent or cure disease conditions are at best unfounded and at worst intentionally misleading.

Biotin Deficiency Biotin deficiencies are rare, but have been reported in adults fed artificially by vein without biotin supplementation. Researchers can induce biotin deficiency in animals or human beings by feeding them raw egg whites, which contain a protein that binds biotin and prevents its absorption.* Long-term use of anticonvulsant medication may also lead to biotin deficiency, as may alcohol abuse.[33]

Vitamin B₆

Vitamin B_6 has been called the "sleeping giant" of vitamins. A surge of research interest in the last two decades has not only revealed new knowledge, but has also raised new questions. For example, unlike the other water-soluble vitamins,

*The protein **avidin** in egg whites binds biotin.

Milk and milk products supply much (about 50 percent) of the riboflavin in people's diets, but meats, eggs, green vegetables, and enriched and whole-grain beads and cereals are good sources, too.

A food containing 1 mg of niacin and 60 mg of tryptophan contains the niacin equivalent of 2 mg, or 2 mg NE.

When a normal dose of a nutrient clears up a deficiency condition, the effect is a **physiological** one. When a large dose of a nutrient overwhelms a body system and acts like a drug, the effect is a **pharmacological** one.

Niacin-rich foods include meat, fish, poultry, and peanut butter, as well as enriched breads and cereals and a few vegetables.

niacin equivalents (NE): the amount of niacin present in food, including the niacin that can theoretically be made from its precursor tryptophan present in the food.

Meat, chicken, and fish, as well as some fruits and vegetables, are good sources of vitamin B_6.

vitamin B_6 is stored extensively in muscle tissue. Most recently, research interest has centered on a possible role for vitamin B_6 in the treatment of disease.

Metabolic Roles of Vitamin B_6   Vitamin B_6 has long been known to play roles in protein and amino acid metabolism. In the cells, vitamin B_6 helps to convert one kind of amino acid, which the cells have in abundance, to another, which they need in larger amounts. It also aids in the conversion of the amino acid tryptophan to niacin and plays important roles in the synthesis of hemoglobin and neurotransmitters, the communication molecules of the brain. Vitamin B_6 also assists in releasing stored glucose from glycogen and thus contributes to the regulation of blood glucose. Research suggests new roles for vitamin B_6 in immune function and hormone response.[34] The association between vitamin B_6 and immune function is related to the critical role the vitamin plays in protein metabolism. Vitamin B_6 deficiency can significantly impair the immune response, perhaps by way of impaired antibody production.[35]

Vitamin B_6 status may be related to cardiovascular disease risk. Elevated blood levels of the amino acid homocysteine correlate with a high incidence of heart disease.[36] Evidence suggests that low blood concentrations of vitamin B_6 and folate are associated with elevated homocysteine concentrations.[37] Researchers conducted a study of more than 80,000 women to determine whether vitamin B_6 and folate intake might influence the risk of heart disease.[38] Women with the highest intakes of vitamin B_6 from food and supplements combined had lower risks of heart disease than women with low intakes. This was true of folate as well, and the risk was lower still in women with the highest intakes of both vitamins.

Vitamin B_6 Deficiency

Besides a weakening immune response, vitamin B_6 deficiency is expressed in general symptoms, such as weakness, irritability, and insomnia. Other symptoms include a greasy, flaky dermatitis; anemia; and, in advanced cases, convulsions.

Vitamin B_6 Toxicity

For years it was believed that vitamin B_6, like other water-soluble vitamins, could not reach toxic concentrations in the body. Toxic effects of vitamin B_6 became known when a physician reported them in women who had been taking more than 2 grams of vitamin B_6 daily for two months or more. Most of these women had been attempting to relieve premenstrual syndrome (PMS), the cluster of physical, emotional, and psychological symptoms that some women experience prior to menstruation. The first symptom of toxicity was numb feet; then the women lost sensation in their hands; then they became unable to walk. The women recovered after they discontinued the supplements.

The specific cause or causes of PMS remain undefined, although researchers agree that the hormonal changes of the menstrual cycle must be responsible. Despite a lack of conclusive evidence that vitamin B_6 is an effective treatment for PMS, the vitamin and many other unproven remedies remain popular among women suffering from PMS.

Vitamin B_6 and Protein Intake   Vitamin B_6's many roles in amino acid metabolism are reflected in dietary needs that are roughly proportional to protein intakes. The RDA for vitamin B_6 is more than adequate to handle average protein intakes of 100 grams per day for men and 60 grams per day for women.

Vitamin B_6 in Foods   The richest food sources of vitamin B_6 are protein-rich meat, fish, and poultry. Potatoes, a few other vegetables, and some fruits are good sources, too. Foods lose vitamin B_6 when heated.

Folate

The B vitamin folate is active in cell division. During periods of rapid growth and cell division, such as pregnancy and adolescence, folate needs increase, and deficiency is especially likely. When a deficiency occurs, the replacement of the rapidly dividing cells of the blood and the GI tract falters. Not surprisingly, then, two of the first symptoms of a folate deficiency are a type of anemia and GI tract deterioration (see Table 8–3).

Folate, Alcohol, and Drugs Of all the vitamins, folate appears to be the most vulnerable to interactions with alcohol and other drugs. As Nutrition in Practice 23 describes, alcohol-addicted people risk folate deficiency because alcohol impairs folate's absorption and increases its excretion. Furthermore, as people's alcohol intakes rise, their folate intakes decline. Many medications, including aspirin, oral contraceptives, and anticonvulsants, also impair folate status. Smoking exerts a negative effect on folate status as well.

Folate and Neural Tube Defects Research studies confirm the importance of folate in preventing **neural tube defects (NTD)**.[39] The brain and spinal cord develop from the neural tube, and defects in its orderly formation during the early weeks of pregnancy may result in various central nervous system disorders and death. Folate supplements taken before conception and continued throughout the first trimester of pregnancy can prevent NTD.[40] For this reason, the American Academy of Pediatrics and the Public Health Services recommend that all women of childbearing age who are capable of becoming pregnant take 0.4 milligram (400 micrograms) of folate daily.[41]

Folate status improves more with supplementation or fortification than with a dietary intake that meets recommendations.[42] Neural tube defects arise early in pregnancy before most women realize they are pregnant, and most women eat too few fruits and vegetables to supply even half the folate needed to prevent NTD. For these reasons, in the late 1990s, the FDA mandated that enriched grain products (flour, cornmeal, pasta, and rice) be fortified with an especially absorbable form of folate. Since then, women's intakes of enriched breads, cereals, and pasta prior to and during the first critical days of pregnancy have been increasing. During the same period, NTD births fell by almost 20 percent. Whether increases in folate intakes deserve the credit for the decline is not yet known.[43] Folate fortification also raises safety concerns, however. High doses of folate can complicate the diagnosis of vitamin B_{12} deficiency, as discussed later. The DRI committee set a Tolerable Upper Intake Level of 1000 micrograms per day from fortified foods or supplements.

Folate status, like vitamin B_6 status, may be related to cardiovascular disease risk.[44] As discussed earlier, elevated levels of homocysteine are associated with a greater risk of cardiovascular disease. One of folate's key roles in the body is to break down homocysteine. Without folate, homocysteine accumulates. Fortified foods and folate supplements raise blood folate and reduce homocysteine levels.[45]

Folate in Foods As Snapshot 8–2 on p. 188 shows, the best food sources of folate are liver, legumes, beets, and leafy green vegetables (the vitamin's name suggests the word *foliage*). Among the fruits, oranges, orange juice, and cantaloupe are the best sources. With fortification, grain products are good sources of folate, too. The bioavailability of added folate is good, and fortification is expected to increase women's average folate intakes by 100 micrograms per day.[46] Heat and oxidation during cooking and storage can destroy up to half of the folate in foods.

The difference in absorption between naturally occurring food folate and synthetic folate that enriches foods and is added to supplements necessitated a new unit of measurement for folate: the **dietary folate equivalents, or DFE.**[47] The dietary folate equivalents convert all forms of folate into units that are

The two main types of neural tube defects are **spina bifida** (literally "split spine") and **anencephaly** ("no brain").

Bread products, flour, corn grits, and pasta must be fortified with 140 µg per 100 g of food (about ½ c cooked food or 1 slice of bread).

neural tube defects (NTD): malformations of the brain, spinal cord, or both during embryonic development.

dietary folate equivalents (DFE): the amount of folate available to the body from naturally occurring sources, fortified foods, and supplements, accounting for differences in bioavailability from each source.

Snapshot 8–2
Folate
All of these foods provide 10 percent or more of the folate Daily Value (DV = 400 μg/day).[a]

Asparagus: 32% DV per ½ c cooked

Liver: 46% DV per 3 oz braised

Enriched cereal: 21% DV per ¾ c

Beets: 17% DV per ½ c cooked

Spinach: 33% DV per 1 c raw

Avocado 18% DV per ½ c

Pinto beans: 37% DV per ½ c cooked

[a]The Daily Values are based on a 2000-kcalorie diet.

equivalent to the folate in foods. Most food labels and tables of food composition express folate values in micrograms, however, so the accompanying "How to" box describes how to estimate dietary folate equivalents.

Vitamin B$_{12}$

Vitamin B$_{12}$ and folate share a special relationship: vitamin B$_{12}$ assists folate in cell division. Their roles intertwine, but each performs a specific task that the other cannot accomplish.

Vitamin B$_{12}$, Folate, and Cell Division   Vitamin B$_{12}$ (in coenzyme form) stands by to accept carbon groups from folate as folate removes them from other compounds. The passing of these carbon groups from folate to vitamin B$_{12}$ regenerates the active form of folate so that it can continue its dismantling tasks. In the absence of vitamin B$_{12}$, folate is trapped in its inactive, metabolically useless form, unable to do its job. When folate is either trapped due to a vitamin B$_{12}$ deficiency or unavailable due to a deficiency of folate itself, cells that are growing most rapidly, notably, the blood cells, are the first to be affected. Thus a deficiency of either nutrient—vitamin B$_{12}$ or folate—impairs maturation of the blood cells and produces anemia. The anemia is identifiable by microscopic examination of the blood, which reveals many large, immature red blood cells. Either vitamin B$_{12}$ or folate will clear up the anemia.

Vitamin B$_{12}$ and the Nervous System   Although either vitamin will clear up the anemia caused by vitamin B$_{12}$ deficiency, if folate is given when vitamin B$_{12}$ is needed, the result is disastrous, not to the blood but to the nervous system. The reason: vitamin B$_{12}$ also helps maintain nerve fibers. A vitamin B$_{12}$ deficiency can ultimately result in devastating neurological symptoms, undetectable by a blood test. A deceptive folate "cure" of the anemia in vitamin B$_{12}$ deficiency allows the nerve deterioration to progress, leading to paralysis and permanent nerve damage. This interaction between folate and vitamin B$_{12}$ raises safety concerns about the use of folate supplements and fortification of foods.

The way folate masks vitamin B$_{12}$ deficiency underlines a point already made several times: it takes a skilled diagnostician to make a correct diagnosis. A person who self-diagnoses on the basis of a single observed symptom takes a serious risk.

Vitamin B$_{12}$ Absorption   Vitamin B$_{12}$ requires an **intrinsic** factor—a compound made inside the body—for absorption from the intestinal tract into the bloodstream. This intrinsic factor is made in the stomach, where it attaches to the vitamin; the complex then passes to the small intestine and is gradually absorbed.

Large-cell anemia is known as **macrocytic** or **megaloblastic anemia.**
macro = large
cyte = cell
mega = large

intrinsic: inside the system. Anemia that reflects a vitamin B$_{12}$ deficiency caused by lack of intrinsic factor is known as **pernicious anemia.**

How to

Estimate Dietary Folate Equivalents

Folate is expressed in terms of DFE (dietary folate equivalents) because synthetic folate from supplements and fortified foods is absorbed at almost twice (1.7 times) the rate of naturally occurring folate from other foods. Use the following equation to calculate:

DFE = µg food folate + (1.7 × µg synthetic folate).

Consider, for example, a pregnant woman who takes a supplement and eats a bowl of fortified cornflakes, 2 slices of fortified bread, and a cup of fortified pasta. From the supplement and fortified foods, she obtains synthetic folate:

Supplement	100 µg folate
Fortified cornflakes	100 µg folate
Fortified bread	40 µg folate
Fortified pasta	60 µg folate
	300 µg folate

To calculate the DFE, multiply the amount of synthetic folate by 1.7:

300 µg × 1.7 = 510 µg DFE.

Now add the naturally occurring folate from the other foods in her diet—in this example, another 90 µg of folate.

510 µg DFE + 90 µg = 600 µg DFE.

Notice that if we had not converted synthetic folate from supplements and fortified foods to DFE, this woman's intake would appear to fall short of the 600 µg recommendation for pregnancy (300 µg + 90 µg = 390 µg). But as our example shows, her intake does meet the recommendation. At this time, supplement and fortified food labels list folate in µg only, not µg DFE, making such calculations necessary.

Loss of Intrinsic Factor In some cases, intrinsic factor production becomes inadequate or ceases altogether—for example, after surgical removal of the stomach. Some people inherit a defective gene for intrinsic factor. Because vitamin B_{12} deficiency in the body may be caused either by a lack of the vitamin in the diet or by the body's inability to absorb the vitamin, a change in diet alone may not correct the deficiency. When absorption failure is the problem, vitamin B_{12} must be supplied by injection.

Vitamin B_{12} in Foods A unique characteristic of vitamin B_{12} is that it is found almost exclusively in foods derived from animals. People who eat meat are guaranteed an adequate intake, and lacto-ovo vegetarians (who use milk, cheese, and eggs) are also protected from deficiency. It is a myth, however, that fermented soy products, such as miso (a soybean paste), or sea algae, such as spirulina, provide vitamin B_{12} in its active form. Extensive research shows that the amounts of vitamin B_{12} listed on the labels of these plant products are inaccurate and misleading because the vitamin B_{12} in these products occurs in an inactive, unavailable form. Vegans must take vitamin B_{12} supplements or find other sources of active vitamin B_{12}. Some loss of vitamin B_{12} occurs when foods are heated in microwave ovens.[48]

Vitamin B_{12} Deficiency in Vegans Vegans are at special risk for undetected vitamin B_{12} deficiency for two reasons: first, they receive none in their diets, and second, they consume large amounts of folate in the vegetables they eat. Because the body can store 1000 times the amount of vitamin B_{12} used each day, a deficiency may take years to develop in a new vegetarian. When a deficiency does develop, though, it may progress to a dangerous extreme because the deficiency of vitamin B_{12} may be masked by the high folate intake.

Non-B Vitamins

Other compounds are sometimes inappropriately called B vitamins because, like the true B vitamins, they serve as coenzymes in metabolism. Even if they were essential, however, supplements would be unnecessary because these compounds are abundant in foods.

Inositol, Choline, and Carnitine Among the non-B vitamins are a trio of coenzymes known as inositol, choline, and carnitine. Researchers are exploring the possibility that these substances may be essential. Thus far, only choline has been assigned an Adequate Intake (AI) value.

Other Non-B Vitamins Other substances have also been mistaken for essential nutrients. They include para-aminobenzoic acid (PABA), bioflavonoids (vitamin P or hesperidin), and ubiquinone. Other names you may hear are "vitamin B_{15}" (a hoax) and "vitamin B_{17}" (laetrile, a fake cancer-curing drug and not a vitamin by any stretch of the imagination). There is, however, one other water-soluble vitamin of great interest and importance—vitamin C.

Vitamin C

Two hundred years ago, any man who joined the crew of a seagoing ship knew he had only half a chance of returning alive—not because he might be slain by pirates or die in a storm, but because he might contract the dread disease **scurvy.** Then a physician with the British navy found that citrus fruits could cure the disease, and thereafter, all ships were required to carry lime juice for every sailor. (This is why British sailors are still called "limeys" today.) Nearly 200 years later, the antiscurvy factor in citrus fruits was isolated from lemon juice and named **ascorbic acid.** Today, hundreds of millions of vitamin C pills are produced in pharmaceutical laboratories.

Metabolic Roles of Vitamin C Vitamin C's action defies a simple, tidy description. It plays many important roles in the body, and its modes of action differ in different situations.

Vitamin C's Role in Collagen Formation The best-understood action of vitamin C is its role in helping to form **collagen,** the single most important protein of connective tissue. Collagen serves as the matrix on which bone is formed, the material of scars, and an important part of the "glue" that attaches one cell to another. This latter function is especially important in the artery walls, which must expand and contract with each beat of the heart, and in the walls of the capillaries, which are thin and fragile.

Vitamin C as an Antioxidant Vitamin C is also an important antioxidant. Recall that the antioxidants beta-carotene and vitamin E protect fat-soluble substances from oxidizing agents; vitamin C protects water-soluble substances the same way. Vitamin C's antioxidant action is twofold. First, by being oxidized itself, vitamin C regenerates already-oxidized substances such as iron and copper to their original, active form. Second, in the process, the vitamin removes the damaging oxidizing agent. In the intestines, it protects iron from oxidation and so enhances iron absorption. In the cells and body fluids, it helps to protect other molecules, including the fat-soluble compounds vitamin A, vitamin E, and the polyunsaturated fatty acids.

Vitamin C in Amino Acid Metabolism Vitamin C is also involved in the metabolism of several amino acids. Some of these amino acids end up being used to make hormones of great importance in body functioning, among them norepinephrine and thyroxine.

Role of Stress During stress, the adrenal glands release large quantities of vitamin C together with the stress hormones epinephrine and norepinephrine. What the vitamin has to do with the stress reaction is unclear, but it is known that stress increases vitamin C needs somewhat.

---

scurvy: the vitamin C–deficiency disease.

ascorbic acid: one of the two active forms of vitamin C. Many people refer to vitamin C by this name.
a = without
scorbic = having scurvy

collagen: the characteristic protein of connective tissue.
kolla = glue
gennan = produce

Vitamin C as a Possible Antihistamine Newspaper headlines touting vitamin C as a cure for colds and cancer have appeared frequently over the years. Some research suggests that vitamin C (2 grams per day for two weeks) may reduce the severity and duration of cold and allergy symptoms by reducing blood histamine concentrations. In other words, vitamin C acts as an antihistamine. If further research confirms vitamin C's antihistamine effect, its use may permit people to rely less heavily on antihistamine drugs when suffering from cold and allergy symptoms.

Vitamin C's Role in Cancer Prevention and Treatment The role of vitamin C in the prevention and treatment of cancer is still being studied. In a dozen or so different well-controlled studies, researchers identified individuals with and without cancer and assessed their dietary intakes of vitamin C. They found that people with high vitamin C intakes had lower risks of these cancers than did people with low intakes. The correlation may reflect not just an association with vitamin C, but the broader benefits of a diet rich in fruits and vegetables and low in fat. It does not support the taking of vitamin C supplements to prevent or treat cancer.

Vitamin C Deficiency When intake of vitamin C is inadequate, the body's vitamin C pool dwindles, and **latent** scurvy appears. The blood vessels show the first deficiency signs. The gums around the teeth begin to bleed easily, and capillaries under the skin break spontaneously, producing pinpoint hemorrhages. Then the symptoms of **overt** scurvy appear. Muscles, including the heart muscle, may degenerate. The skin becomes rough, brown, scaly, and dry. Wounds fail to heal because scar tissue will not form without collagen. Bone rebuilding falters; the ends of the long bones become softened, malformed, and painful; and fractures occur. The teeth may become loose in the jawbone and fall out. Anemia and infections are common. Sudden death is likely, perhaps because of massive bleeding into the joints and body cavities.

It takes only 10 or so milligrams of vitamin C a day to prevent overt scurvy, and not much more than that to cure it. Once diagnosed, scurvy is readily reversible with moderate doses, in the neighborhood of 100 milligrams per day. Such an intake is easily achieved by including vitamin C–rich foods in the diet.

Vitamin C Toxicity The easy availability of vitamin C in pill form and the publication of books recommending vitamin C to prevent everything from the common cold to life-threatening cancer have led thousands of people to take megadoses of vitamin C. Not surprisingly, instances of vitamin C causing harm have surfaced.

Some of the suspected toxic effects of vitamin C megadoses have not been confirmed, but others have been seen often enough to warrant concern. Nausea, abdominal cramps, and diarrhea are often reported. Several instances of interference with medical regimens are known. Large amounts of vitamin C excreted in the urine obscure the results of tests used to detect diabetes. People taking anticoagulants may unwittingly counteract the effect of these medications if they also take massive doses of vitamin C. Vitamin C megadoses can enhance iron absorption too much, resulting in iron overload (see Chapter 9).

People with sickle-cell anemia may be especially vulnerable to megadoses of vitamin C. Those who have a tendency toward **gout,** as well as those who have a genetic abnormality that alters the way they metabolize vitamin C, are more prone to forming stones if they take megadoses of vitamin C.

Recommended Intakes of Vitamin C The vitamin C RDA is 90 milligrams for men and 75 milligrams for women. These amounts are far higher than the 10 milligrams per day needed to prevent the symptoms of scurvy. In fact, they are close to the amount at which the body's pool of vitamin C is full to overflowing: about 100 milligrams per day.

Doses of 10 to 30 or more times the recommended intake of a nutrient are termed **megadoses.** In the case of vitamin C, current recommendations are 75 mg/day for women and 90 mg/day for men. The Tolerable Upper Intake Level for vitamin C is 2000 mg/day.

The anticoagulants with which vitamin C interferes are warfarin and dicumarol.

latent: the period in the course of a disease when the conditions are present but the symptoms have not begun to appear.
latens = lying hidden

overt: out in the open, full-blown.
ouvrire = to open

gout (GOWT): a metabolic disease in which crystals of uric acid precipitate in the joints.

The Vitamins

Special Needs for Vitamin C As is true of all nutrients, unusual circumstances may raise vitamin C needs. Among the stresses known to do so are infections; burns; surgery; extremely high or low temperatures; toxic doses of heavy metals, such as lead, mercury, and cadmium; and the chronic use of certain medications, including aspirin, barbiturates, and oral contraceptives. Smoking, too, has adverse effects on vitamin C status. Cigarette smoke contains oxidants, which deplete this potent antioxidant.[49] Accordingly, the vitamin C recommendation for smokers is set high, at 125 milligrams for men and 105 milligrams for women.

Safe Limits Few instances warrant the taking of more than 100 to 300 milligrams of vitamin C a day. The risks may not be great for adults who dose themselves with 1 to 2 grams a day, but those taking more than 2 grams, and especially those taking above 3 grams per day, should be aware of the distinct possibility of harm.[50]

Vitamin C in Foods The inclusion of intelligently selected fruits and vegetables in the daily diet guarantees a generous intake of vitamin C. Even those who wish to ingest amounts well above the RDA can easily meet their goals by eating certain foods (see Snapshot 8–3). Citrus fruits are rightly famous for their vitamin C contents. Certain other fruits and vegetables are also rich sources: cantaloupe, strawberries, broccoli, and brussels sprouts. No animal foods other than organ meats, such as liver and kidneys, contain vitamin C. The humble potato is an important source of vitamin C in Western countries, where potatoes are eaten so frequently that they make substantial vitamin C contributions overall. They provide about 20 percent of all the vitamin C in the average diet. Vitamin C in foods is easily oxidized, so store cut produce and juices in airtight containers.

Vitamin C and Iron Absorption Eating foods containing vitamin C at the same meal with foods containing iron can double or triple the absorption of iron from those foods. This strategy is highly recommended for women and children, whose energy intakes are not large enough to guarantee that they will get enough iron from the foods they eat. Table 8–3 (pp. 193–196) summarizes functions, deficiency and toxicity symptoms, and food sources of vitamin C and the other water-soluble vitamins.

Table 8–3   The Water-Soluble Vitamins—A Summary

Thiamin

Other Names	Deficiency Symptoms	Toxicity Symptoms
Vitamin B$_1$	**Blood/Circulatory System**	
Chief Functions in the Body	Edema, enlarged heart, abnormal heart rhythms, heart failure	(No symptoms reported)
Part of a coenzyme used in energy metabolism, supports normal appetite and nervous system function	**Nervous/Muscular Systems**	
	Degeneration, wasting, weakness, pain, low morale, difficulty walking, loss of reflexes, mental confusion, paralysis	(No symptoms reported)
1998 RDA		
Men: 1.2 mg/day Women: 1.1 mg/day		
Deficiency Disease Name		
Beriberi		
Significant Sources		
Occurs in all nutritious foods in moderate amounts; pork, ham, bacon, liver; whole grains, enriched, or fortified grain products; legumes, nuts		

Riboflavin

Other Names	Deficiency Symptoms	Toxicity Symptoms
Vitamin B$_2$	**Mouth, Gums, Tongue**	
Chief Functions in the Body	Cracks at corners of mouth,[a] magenta tongue	(No symptoms reported)
Part of a coenzyme used in energy metabolism, supports normal vision and skin health	**Nervous System and Eyes**	
	Hypersensitivity to light,[b] reddening of cornea	(No symptoms reported)
1998 RDA	**Other**	
Men: 1.3 mg/day Women: 1.1 mg/day	Greasy dermatitis	(No symptoms reported)
Deficiency Disease Name		
Ariboflavinosis		
Significant Sources		
Milk, yogurt, cottage cheese, liver, leafy green vegetables, whole-grain or enriched breads and cereals		

Niacin

Other Names	Deficiency Symptoms	Toxicity Symptoms
Nicotinic acid, nicotinamide, niacinamide, vitamin B$_3$; precursor is dietary tryptophan	**Digestive System**	
	Diarrhea	Diarrhea, heartburn, nausea, ulcer irritation, vomiting
Chief Functions in the Body	**Mouth, Gums, Tongue**	
Part of a coenzyme used in energy metabolism; supports health of skin, nervous system, and digestive system	Swollen, smooth, bright red tongue[c]	(No symptoms reported)

(Continued on next page)

[a]Cracks at the corners of the mouth are termed *cheilosis* (kee-LOH-sis).
[b]Hypersensitivity to light is *photophobia*.
[c]Smoothness of the tongue is caused by loss of its surface structures and is termed *glossitis* (gloss-EYE-tis).

Table 8–3   The Water-Soluble Vitamins—A Summary—continued

Niacin—continued

	Deficiency Symptoms	Toxicity Symptoms
1998 RDA		
Men: 16 mg NE/day Women: 14 mg NE/day	***Nervous System*** Depression, apathy, fatigue, headache, and loss of memory	Fainting, dizziness
Upper Level		***Skin***
Adults: 35 mg/day	Flaky skin rash on areas exposed to sun	Painful flush and rash ("niacin flush), sweating
Deficiency Disease Name		***Other***
Pellagra		Abnormal liver function, impaired glucose tolerance.
Significant Sources		
Milk, eggs, meat, poultry, fish, whole-grain and enriched breads and cereals, nuts, and all protein-containing food		

Pantothenic Acid

Other Names	Deficiency Symptoms	Toxicity Symptoms
(None)	***Digestive System***	
Chief Functions in the Body	Vomiting, intestinal distress	(No symptoms reported)
Part of a coenzyme used in energy metabolism	***Nervous System*** Insomnia, fatigue	(No symptoms reported)
1998 Adequate Intake (AI)		
Adults: 5 mg/day		
Deficiency Disease Name		
(No name)		
Significant Sources		
Widespread in foods		

Biotin

Other Names	Deficiency Symptoms	Toxicity Symptoms
(None)		***Blood/Circulatory System***
Chief Functions in the Body	Abnormal heart action	(No symptoms reported)
Part of a coenzyme used in energy metabolism, fat synthesis, amino acid metabolism, and glycogen synthesis	Loss of appetite, nausea	***Digestive System*** (No symptoms reported)
	Depression, lethargy, hallucinations	***Nervous/Muscular Systems*** (No symptoms reported)
1998 Adequate Intake (AI)		
Adults: 30 µg/day	Drying, rash, loss of hair	***Skin*** (No symptoms reported)
Deficiency Disease		
(No name)		
Significant Sources		
Widespread in foods		

Table 8–3 The Water-Soluble Vitamins—A Summary

Vitamin B$_6$

Other Names	Deficiency Symptoms	Toxicity Symptoms
Pyridoxine, pyridoxal, pyridoxamine	**Blood/Circulatory System**	
Chief Functions in the Body	Anemia (small-cell type)[d]	
Part of a coenzyme used in amino acid and fatty acid metabolism, helps convert tryptophan to niacin, helps make red blood cells	**Nervous/Muscular Systems**	
	Abnormal brain wave pattern, irritability muscle twitching, convulsions	Depression, fatigue, impaired memory, irritability, headaches, numbness, damage to nerves, difficulty walking, loss of reflexes, weakness, restlessness
1998 RDA		
Adults (19–50 yr): 1.3 mg/day		**Skin**
Upper Level	Irritation of sweat glands, rashes, greasy dermatitis	(No symptoms reported)
Adults: 100 mg/day		
Deficiency Disease Name		
(No name)		
Significant Sources		
Green and leafy vegetables, meats, fish, poultry, shellfish, legumes, fruits, whole grains		

Folate

Other Names	Deficiency Symptoms	Toxicity Symptoms
Folic acid, folacin, pteroylglutamic acid	**Blood/Circulatory System**	
Chief Functions in the Body	Anemia (large-cell type)[d]	(No symptoms reported)
Part of a coenzyme used in new cell synthesis	**Immune System**	
	Suppression, frequent infections	(No symptoms reported)
1998 RDA	**Mouth, Gums, Tongue**	
Adults: 400 µg/day	Smooth red tongue[c]	(No symptoms reported)
Upper Level	**Nervous System**	
Adults: 1000 µg/day	Depression, mental confusion, fainting	(No symptoms reported)
Deficiency Disease	**Other**	
(No name)		Masks vitamin B$_{12}$ deficiency
Significant Sources		
Fortified grains, leafy green vegetables, legumes, seeds, liver		

Vitamin B$_{12}$

Other Names	Deficiency Symptoms	Toxicity Symptoms
Cyanocobalamin	**Blood/Circulatory System**	
Chief Functions in the Body	Anemia (large-cell type)[d]	(No symptoms reported)
Part of a coenzyme used in new cell synthesis, helps maintain nerve cells	**Mouth, Gums, Tongue**	
	Smooth tongue[c]	(No symptoms reported)

(Continued on next page)

[d]Small-cell anemia is termed *microcytic* anemia; large-cell type is *macrocytic* or *megaloblastic anemia.*

Table 8–3   The Water-Soluble Vitamins—A Summary

Vitamin B₁₂

	Deficiency Symptoms	Toxicity Symptoms
1998 RDA	***Nervous/Muscular Systems***	(No symptoms reported)
Adults: 2.4 µg/day	Fatigue, degeneration progressing to paralysis	
Deficiency Disease		***Skin***
(No nameᵉ)	Hypersensitivity	(No symptoms reported)
Significant Sources		
Animal products (meat, fish, poultry, milk, cheese, eggs)		

Vitamin C

Other Names	Deficiency Symptoms	Toxicity Symptoms
Ascorbic acid	***Blood/Circulatory System***	(No symptoms reported)
Chief Functions in the Body	Anemia, atherosclerotic plaques, pinpoint hemorrhages	
Helps in collagen synthesis (strengthens blood vessel walls, forms scar tissue, provides matrix for bone growth), thyroxine synthesis, and amino acid metabolism; serves as an antioxidant; strengthens resistance to infection; helps in absorption of iron		***Digestive System***
		Nausea, abdominal cramps, diarrhea, excessive urination
	Immune System	
	Suppression, frequent infections	(No symptoms reported)
2000 RDA	***Mouth, Gums, Tongue***	
Men: 90 mg/day Women: 75 mg/day	Bleeding gums, loosened teeth	(No symptoms reported)
	Muscular/Nervous Systems	
Upper Level	Muscle degeneration and pain, hysteria, depression	Headache, fatigue, insomnia
Adults: 2000 mg/day	***Skeletal System***	
	Bone fragility, joint pain	(No symptoms reported)
Deficiency Disease Name		***Skin***
Scurvy	Rough skin, blotchy bruises	Rashes
Significant Sources		***Other***
Citrus fruits, cabbage-type vegetables, dark green vegetables, cantaloupe, strawberries, peppers, lettuce, tomatoes, potatoes, papayas, mangoes	Failure of wounds to heal	Interference with medical tests, aggravation of gout symptoms; deficiency symptoms may appear at first on withdrawal of high doses.

ᵉThe name *pernicious anemia* refers to the vitamin B₁₂ deficiency caused by lack of intrinsic factor, but not to that caused by inadequate dietary intake

In Summary The B vitamins and vitamin C are the water-soluble vitamins. Each B vitamin is part of an enzyme helper known as a coenzyme. As parts of coenzymes, the B vitamins assist in the release of energy from glucose, amino acids, and fats and help in many other body processes. Folate and vitamin B₁₂ are important in cell division. Vitamin C's primary role is as an antioxidant. Historically, famous B vitamin–deficiency diseases are beriberi (thiamin) and pellagra (niacin). The vitamin C–deficiency disease is known as scurvy.

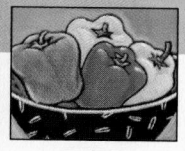

Self Study

VITAMINS

A diet that offers a variety of foods from each group, prepared with reasonable care, serves up ample vitamins. The cereal and bread group delivers thiamin, riboflavin, niacin, and folate. The fruit and vegetable groups excel in folate, vitamin C, vitamin A, and vitamin K. The meat group serves thiamin, niacin, vitamin B_6, and vitamin B_{12}. The milk group stands out for riboflavin, vitamin B_{12}, vitamin A, and vitamin D. Even the miscellaneous group with its vegetable oils provides vitamin E. Determine whether these food choices are typical of your diet.

Food Choices	Frequency per Week
Citrus fruit	_____
Dark green, leafy vegetables	_____
Deep yellow or orange fruits or vegetables	_____
Legumes	_____
Milk and milk products	_____
Vegetable oils	_____
Whole or enriched grain products	_____

- Do you eat dark green, leafy or deep yellow vegetables daily?
- Do you drink vitamin A– and D–fortified milk regularly?
- Do you use vegetable oils when you cook?
- Do you choose whole or enriched grains, citrus fruits, and legumes often?

Self Check

1. Which of the following vitamins are fat soluble?
   a. vitamins B, C, and E
   b. vitamins B, C, D, and E
   c. vitamins A, C, E, and K
   d. vitamins A, D, E, and K

2. Which of the following describes fat-soluble vitamins?
   a. They include thiamin, vitamin A, and vitamin K.
   b. They cannot be stored to any great extent and so must be consumed daily.
   c. Toxic levels can be reached by consuming citrus fruits and vegetables.
   d. They can be stored in the liver and fatty tissues and can build up toxic concentrations.

3. Night blindness and susceptibility to infection are the result of a deficiency of which vitamin?
   a. niacin
   b. vitamin C
   c. vitamin A
   d. vitamin K

4. Good sources of vitamin D include:
   a. eggs, fortified milk, and sunlight.
   b. citrus fruits, sweet potatoes, and spinach.
   c. leafy green vegetables, cabbage, and liver.
   d. breast milk, polyunsaturated plant oils, and citrus fruits.

5. Which of the following describes water-soluble vitamins?
   a. They include vitamins D and E.
   b. They are frequently toxic.
   c. They are stored extensively in tissues.
   d. They are easily absorbed and excreted.

6. A coenzyme is:
   a. a fat-soluble vitamin.
   b. an energy-yielding nutrient.
   c. a source of vitamin K.
   d. a molecule that combines with an enzyme to make it active.

7. Good food sources of folate include:
   a. citrus fruits, dairy products, and eggs.
   b. liver, legumes, and leafy green vegetables.
   c. dark green vegetables, corn, and cabbage.
   d. potatoes, broccoli, and whole-wheat bread.

8. Which vitamin is present only in foods of animal origin?
   a. riboflavin
   b. pantothenic acid

c. vitamin B_{12}

   d. the inactive form of vitamin A

9. Which of the following nutrients is an antioxidant that protects water-soluble substances from oxidizing agents?

   a. beta-carotene

   b. thiamin

   c. vitamin C

   d. vitamin D

10. Eating foods containing vitamin C at the same meal can increase the absorption of which mineral?

   a. iron

   b. calcium

   c. magnesium

   d. folate

Critical Thinking

1. A person who eats a protein-rich diet centered on meat is highly unlikely to be deficient in which of the following nutrients?

   a. folate

   b. vitamin C

   c. niacin

   d. vitamin E

Answers to these questions can be found in Appendix G.

Clinical Applications

1. How might a vitamin deficiency weaken a client's resistance to disease?

2. Pull together information from Chapter 1 about the different food groups and the significant sources of vitamins shown in the photos throughout this chapter. Consider which vitamins might be lacking in the diet of a client who reports the following:

   - Dislikes leafy, green vegetables.
   - Never uses milk, milk products, or cheese.
   - Follows a very-low-fat diet.
   - Eats a fruit or vegetable once a day.

What additional information would help you pinpoint problems with vitamin intake?

Nutrition on the Net

For further study of the topics of this chapter, access these websites. Be aware that many websites on the Internet are peddling vitamin supplements, not accurate information.

Find updates and quick links to these and other nutrition-related sites at our website:
www.wadsworth.com/nutrition

Search for "vitamins" at the American Dietetic Association:
www.eatright.org

Review the Dietary Reference Intakes for the water-soluble vitamins:
www.nap.edu/readingroom

Visit the World Health Organization to learn about "vitamin deficiencies" around the world:
www.who.int

Search for "vitamins" at the U.S. Government health information site:
www.healthfinder.gov

Learn more about neural tube defects from the Spina Bifida Association of America:
www.sbaa.org

Read about Dr. Joseph Goldberger and his groundbreaking discovery linking pellagra to diet by searching for his name at:
www.nih.gov or **www.pbs.org**

Learn how fruits and vegetables support a healthy diet rich in vitamins from the 5 A Day for Better Health program:
www.5aday.com

Notes

1. N. W. Solomons, Vitamin A and carotenoids, in *Present Knowledge in Nutrition*, 8th ed., ed. B. A. Bowman and R. M. Russell (Washington, D.C.: International Life Sciences Institute Press, 2001), pp. 127–145.

2. A. C. Ross, Vitamin A and retinoids, in *Modern Nutrition in Health and Disease*, 9th ed., ed. M.E. Shils and coeditors (Baltimore, Md.: Williams & Wilkins, 1999), pp. 305–327.

3Standing Committee on the Scientific Evaluation of Dietary Reference Intakes, Food and Nutrition Board, Institute of Health, *Dietary Reference Intakes for Vitamin A, Vitamin K, Arsenic, Boron, Chromium, Copper, Iodine, Iron, Manganese, Molybdenum, Nickel, Silicon, Vanadium, and Zinc* (Washington, D.C.: National Academy Press, 2001), pp. 4-1–4-61.

4. Ross, 1999.

5. J. K. Jarvis and K. Neville, Antioxidant vitamins; Current and future directions, *Nutrition Today* 35 (2000): 214–221; R. A. Jacob, Evidence that diet modification reduces in vivo oxidant damage, *Nutrition Reviews* 57 (1999): 225–258; B. Halliwell, Antioxidants and human disease: A general introduction, *Nutrition Reviews* 55 (1997): 544–549.

6. K. Smigel, Beta-carotene fails to prevent cancer in two major studies: CARET intervention stopped, *Journal of the National Cancer Institute* 88 (1996): 145; G. S. Omenn and coauthors, Effects of a combination of beta-carotene and vitamin A on lung cancer and cardiovascular disease, *New England Journal of Medicine* 334 (1996): 1150–1155.

7. C. H. Hennekens and coauthors, Lack of effect of long-term supplementation with beta carotene on the incidence of malignant neoplasms and cardiovascular disease, *New England Journal of Medicine* 334 (1996): 1145–1149.

8. Jarvis and Neville, 2000.

9. Standing Committee on the Scientific Evaluation of Dietary Reference Intakes, 2001.

10. N. S. Scrimshaw and J. P. SanGiovanni, Synergism of nutrition, infection, and immunity, *American Journal of Clinical Nutrition* 66 (1997): 464S–477S.

11. C. E. West, Vitamin A and measles, *Nutrition Reviews* 58 (2000): S46–S54.

12. Standing Committee on the Scientific Evaluation of Dietary Reference Intakes, 2001, pp. 4-30–4-36; K. J. Rothman and coauthors, Teratogenicity of high vitamin intake, *New England Journal of Medicine* 333 (1995): 1369–1373; D. R. Soprano and K. J. Soprano, Retinoids as teratogens, *Annual Review of Nutrition* 15 (1995): 111–132.

13. Rothman and coauthors, 1995.

14. A. W. Norman, Vitamin D, in *Present Knowledge in Nutrition,* 8th ed., ed. B. A. Bowman and R. M. Russell (Washington, D.C.: International Life Sciences Institute Press, 2001), pp. 146–155.

15. Norman, 2001; M. F. Holick, Vitamin D, in *Modern Nutrition in Health and Disease,* 9th ed., ed. M. E. Shils and coeditors (Baltimore, Md.: Williams & Wilkins, 1999), pp. 329–345.

16. N. F. Carvalho and coauthors, Severe nutritional deficiencies in toddlers resulting from health food milk alternatives, *Pediatrics* 107 (2001): E46; S. R. Kreiter and coauthors, Nutritional rickets in African American breast-fed infants, *Journal of Pediatrics* 137 (2000): 153–157.

17. Holick, 1999.

18. American Academy of Pediatrics, *Pediatric Nutrition Handbook,* 4th ed., ed. R. E. Kleinman (Elk Grove Village, Ill.: American Academy of Pediatrics, 1998), pp. 3–20.

19. Standing Committee on the Scientific Evaluation of Dietary Reference Intakes, Food and Nutrition Board, Institute of Health, *Dietary Reference Intakes for Vitamin C, Vitamin E, Selenium, and Carotenoids* (Washington, D.C.: National Academy Press, 2000), pp. 211–216; J. Regnström and coauthors, Inverse relation between the concentration of low-density lipoprotein vitamin E and severity of coronary artery disease, *American Journal of Clinical Nutrition* 63 (1996): 377–385; K. G. Losonczy, T. B. Harris, and R. J. Havlik, Vitamin E and vitamin C supplements use and risk of all-cause and coronary heart disease mortality in older persons: The Established Populations (or Epidemiologic Studies of the Elderly), *American Journal of Clinical Nutrition* 64 (1996): 190–196; L. H. Kushi and coauthors, Dietary antioxidant vitamins and death from coronary heart disease in postmenopausal women, *New England Journal of Medicine* 334 (1996): 1156–1162.

20. M. Meydani, Effect of functional food ingredients: Vitamin E modulation of cardiovascular diseases and immune status in the elderly, *American Journal of Clinical Nutrition* 71 (2000): 1665S–1668S; Kushi and coauthors, 1996.

21. The Heart Outcomes Prevention Evaluation Study Investigators, Vitamin E supplementation and cardiovascular events in high-risk patients, *New England Journal of Medicine* 342 (2000): 154–160.

22. Standing Committee on the Scientific Evaluation of Dietary Reference Intakes, 2000.

23. S. N. Meydani and coauthors, Assessment of the safety of supplementation with different amounts of vitamin E in healthy older adults, *American Journal of Clinical Nutrition* 68 (1998): 311–318.

24. Standing Committee on the Scientific Evaluation of Dietary Reference Intakes, 2000.

25. G. Ferland, The vitamin K–dependent proteins: An update, *Nutrition Reviews* 56 (1998): 223–230.

26. S. L. Booth and coauthors, Dietary vitamin K intakes are associated with hip fracture but not with bone mineral density in elderly men and women, *American Journal of Clinical Nutrition* 71 (2000): 1201–1208; L. J. Sokoll and coauthors, Changes in serum osteocalcin, plasma phylloquinone and urinary g-carboxyglutamic acid in response to altered intakes of dietary phylloquinone in human subjects, *American Journal of Clinical Nutrition* 65 (1997): 779–784.

27. D. Feskanich and coauthors, Vitamin K and hip fractures in women: A prospective study, *American Journal of Clinical Nutrition* 69 (1999): 77–79.

28. Ferland, 1998.

29. Standing Committee on the Scientific Evaluation of Dietary Reference Intakes, Food and Nutrition Board, Institute of Health, *Dietary Reference Intakes for Thiamin, Riboflavin, Niacin, Vitamin B_6, Folate, Vitamin B_{12}, Pantothenic Acid, Biotin, and Choline* (Washington, D.C.: National Academy Press, 1998), pp. 58–86; C. G. Harper and coauthors, Prevalence of Wernicke-Korsakoff syndrome in Australia: Has thiamin fortification made a difference? *Medical Journal of Australia* 168 (1998): 542–545.

30. J. B. Hack and R. S. Hoffman, Thiamine before glucose to prevent Wernicke encephalopathy: Examining the conventional wisdom (letter), *Journal of the American Medical Association* 279 (1998): 583–584.

31. Standing Committee on the Scientific Evaluation of Dietary Reference Intakes, 1998, pp.123–149.

32. D. Cervantes-Laurean, N. G. McElvaney, and J. Moss, Niacin, in *Modern Nutrition in Health and Disease,* 9th ed., ed. M. E. Shils and coeditors (Baltimore, Md.: Williams & Wilkins, 1999), pp. 401–411.

33. D. M. Mock, Biotin, in *Modern Nutrition in Health and Disease,* 9th ed., ed. M. E. Shils and coeditors (Baltimore, Md.: Williams & Wilkins, 1999), pp. 459–466.

34. J. E. Leklem, Vitamin B_6, in *Modern Nutrition in Health and Disease,* 9th ed., ed. M. E. Shils and coeditors (Baltimore, Md..: Williams & Wilkins, 1999), pp. 413–421.

35. Leklem, 1999.

36. S. E. Vollset, Plasma total homocysteine and cardiovascular and noncardiovascular mortality: The Hordaland Homocysteine Study, *American Journal of Clinical Nutrition* 74 (2001): 130–136; G. S. Omenn and coauthors, Preventing coronary heart disease: B vitamins and homocysteine, *Circulation* 97 (1998): 421–424; E. B. Rimm and coauthors, Folate and vitamin B_6 from diet and supplements in relation to risk of coronary heart disease among women, *Journal of the American Medical Association* 279 (1998): 359–364.

37. S. M. Saw and coauthors, Genetic, dietary, and other lifestyle determinants of plasma homocysteine concentrations in middle-aged and older Chinese men and women in Singapore, *American Journal of Clinical Nutrition* 73 (2001): 232–239; P. F. Jaques and coauthors, Determinants of plasma total homocysteine concentration in the Framingham offspring cohort, *American Journal of Clinical Nutrition* 73 (2001): 613–621; J. V. Woodside and coauthors, Effect of B-group vitamins and antioxidant vitamins on hyperhomocysteinemia: A double-blind, randomized, factorial-design, controlled trial, *American Journal of Clinical Nutrition* 67 (1998): 858–866.

38. Rimm and coauthors, 1998.

39. Committee on Genetics, American Academy of Pediatrics, Folic acid for the prevention of neural tube defects, *Pediatrics* 104 (1999): 325–327; C. E. Butterworth, Jr., and A. Bendich, Folic acid and the prevention of birth defects, *Annual Review of Nutrition* 16 (1996): 73–97.

40. Committee on Genetics, 1999.

41. Committee on Genetics, 1999; Centers for Disease Control and Prevention, Recommendations for use of folic acid to reduce number of spina bifida cases and other neural tube defects, *Journal of the American Medical Association* 269 (1993): 1233, 1236, 1238.

42. C. J. Cuskelly, H. McNulty, and J. M. Scott, Effect of increasing dietary folate on red-cell folate: Implications for prevention of neural tube defects, *Lancet* 347 (1996): 657–659.

43. M. A. Honein and coauthors, Impact of folic acid fortification of the US food supply on the occurrence of neural tube defects, *Journal of the American Medical Association* 285 (2001): 2981–2986.

44. Standing Committee on the Scientific Evaluation of Dietary Reference Intakes, 1998, pp. 260–264.

45. J. R. Backstrand, The history and future of food fortification in the United States: A public health perspective, *Nutrition Reviews* 60 (2002): 15–26. A; de Bree and coauthors, Association between B vitamin intake and plasma homocysteine concentration in the general Dutch population aged 20–65 y, *American Journal of Clinical Nutrition* 73 (2001): 1027–1033; J. A. Tice and coauthors, Cost-effectiveness of vitamin therapy to lower plasma homocysteine levels for the prevention of coronary heart disease—Effect of grain fortification and beyond, *Journal of the American Medical Association* 286 (2001): 936–943; D. L. McKay and coauthors, Multivitamin/mineral supplementation improves plasma B-vitamin status and homocysteine concentration in healthy older adults consuming folate-fortified diet, *Journal of Nutrition* 130 (2000): 3090–3096.

46. C. M. Pfeiffer and coauthors, Absorption of folate from fortified cereal-grain products and of supplemental folate consumed with or without food determined by using dual-label-stable isotope protocol, *American Journal of Clinical Nutrition* 66 (1997): 1388–1397.

47. C. W. Suitor and L. B. Bailey, Dietary folate equivalents: Interpretation and application, *Journal of the American Dietetic Association* 100 (2000): 88–94.

48. F. Watanabe and coauthors, Effects of microwave heating on the loss of vitamin B_{12} in foods, *Journal of Agricultural and Food Chemistry* 46 (1998): 206–210.

49. J. Lykkesfeldt and coauthors, Ascorbic acid and dehydroascorbic acid as biomarkers of oxidative stress caused by smoking, *American Journal of Clinical Nutrition* 66 (1997): 1388–1397.

50. Standing Committee on the Scientific Evaluation of Dietary Reference Intakes, 2000, p. 155.

Nutrition in Practice

■ PHYTOCHEMICALS ■

The wisdom of the familiar advice, "Eat your vegetables, they're good for you," stands on firmer scientific ground today than ever before as population studies around the world suggest that diets rich in vegetables and fruits protect against heart disease, cancer, and other chronic diseases.[1] We now know that the "goodness" of vegetables, fruits, and other whole foods such as legumes and grains comes not only from the nutrients they contain, but from the **nonnutrients** known as **phytochemicals** that they offer.

Vegetables, fruits, and other whole foods are the simplest examples of foods now known as **functional foods.** Functional foods provide health benefits beyond basic nutrition by altering one or more physiological processes. Modified foods, such as those that have been fortified with nutrients, phytochemicals, herbs, or other food components, also are functional foods.[2] Functional foods that fit this description include orange juice fortified with calcium, folate-enriched cereal, beverages with herbal additives, and even fat-modified milk (fat-free, low-fat, and reduced fat). This Nutrition in Practice focuses on the evidence concerning the effectiveness and safety of a few selected phytochemicals in the simplest of functional foods—vegetables, fruits, and other whole foods. Table NP8–1 on p. 202 begins by defining some terms. Table NP8–2 on p. 203 introduces the names, possible physiological effects, and food sources of phytochemicals.

Exactly what are phytochemicals, and what do they do?

Phytochemicals are nonnutrient compounds found in plants. In foods, phytochemicals impart tastes, aromas, colors, and other characteristics. They give hot peppers their burning sensation, garlic and onions their pungent flavor, chocolate its bitter tang, and tomatoes their dark red color. In the body, phytochemicals can have profound physiological effects, acting as antioxidants, mimicking hormones, and suppressing the development of diseases. Notably, cancer and heart disease are linked to processes involving oxygen compounds in the body, and antioxidants are thought to oppose these actions.

Why are phytochemicals receiving so much media attention these days, and what are some examples of those in the spotlight?

Diets rich in whole grains, legumes, vegetables, and fruits seem to be protective against heart disease and cancer, but identifying *the* specific foods or components of foods that are responsible is difficult.[3] Whenever bits of research news surface, however, new supplements appear—and terms like "antioxidants" and "phytochemicals" become buzzwords again. Meanwhile, scientists are conducting extensive research studies to discover phytochemical connections to disease prevention, but so far, solid evidence is generally lacking. Some of the likeliest candidates include **flavonoids** and **carotenoids** (including **lycopene**).

What are flavonoids, and which foods are they found in?

Flavonoids, a large group of phytochemicals known for their health-promoting qualities, are found in whole grains, vegetables, fruits, herbs, spices, teas, and red wine. A large body of population evidence spanning many countries reveals that deaths from cancer, heart disease, and heart attacks are less common wherever these foods are plentiful in the diet.[4] Flavonoids are powerful antioxidants that may help to protect LDL against oxidation and reduce blood platelet stickiness, making blood clots less likely.[5]

Flavonoids impart a bitter taste to foods, so manufacturers often refine away the natural flavonoids to please consumers who usually prefer milder flavors.[6] For example, the hearty taste of whole-wheat foods vanishes when whole wheat is refined into white flour by removing the tough brown parts that contain flavonoids. For white grape juice or white wine, manufacturers remove the red, flavonoid-rich grape skins to lighten the flavor and color of the product, while greatly reducing its beneficial flavonoid content. For example, one flavonoid of grapes and wine may have anticancer activity.[7]

What about carotenoids? I hear about them in news stories and read about them in magazines, on the Internet, and in supplement advertisements.

In addition to flavonoids, fruits and vegetables are rich in carotenoids—the red and yellow pigments of plants. Some carotenoids, such as beta-carotene, are vitamin A precursors. Studies suggest that a diet rich in carotenoids is associated with a lower risk of heart disease.[8] Among the carotenoids that may defend against heart disease is lycopene.[9] Researchers are investigating a tentative link between low lycopene in the blood and elevated incidence

Nutrition in Practice

Table NP8–1   Phytochemical and Functional Food Terms

- **carotenoids** (kah-ROT-eh-noyds): pigments commonly found in plants and animals, some of which have vitamin A activity. The carotenoid with the greatest vitamin A activity is beta-carotene.
- **flavonoids** (FLAY-von-oyds): yellow pigments in foods; phytochemicals that may exert physiological effects on the body.
  flavus = yellow
- **functional foods:** a general term for foods with beneficial physiological or psychological effects beyond providing essential nutrients. May also be referred to as *medical foods, foods for medical purposes,* or other terms with no legal or scientific definitions. Also defined in Chapter 1.
- **lycopene** (LYE-koh-peen): a pigment responsible for the red color of tomatoes and other red-hued vegetables; a phytochemical that may act as an antioxidant in the body.
- **nonnutrients:** compounds in foods that do not fit within the six classes of nutrients.
- **organosulfur compounds:** a large group of phytochemicals containing the mineral sulfur. Organosulfur phytochemicals are responsible for the pungent flavors and aromas of foods belonging to the onion, leek, chive, shallot, and garlic family; they are thought to stimulate cancer defenses in the body.
- **phytochemicals** (FIGH-toe-CHEM-ih-cals): biologically active compounds of plants believed to confer resistance to diseases on the eater; also defined in Chapter 1.
  phyto = plant
- **phytosterols** (FIGH-toe-STER-ols, figh-TOSS-ter-ols): phytochemicals structurally similar to mammalian steroid hormones, such as the female sex hormone estrogen. Phytosterols may or may not mimic hormone activity in the human body.

of heart disease, heart attack, and stroke.[10] Lycopene may also protect against certain types of cancer.

What is lycopene, and what foods contain it?

Lycopene is a red pigment with powerful antioxidant activity found in guava, papaya, pink grapefruit, tomatoes (especially cooked tomatoes and tomato products), and watermelon.[11] More than 80 percent of the lycopene consumed in the United States comes from tomato products such as tomato sauce, tomato juice, and catsup. Around the world, people who eat five or more tomato-containing meals per week are less likely to suffer from cancers of the esophagus, prostate, or stomach than those who avoid tomatoes.[12] Lycopene is a leading candidate for this protective effect. Researchers examined the relationship between lycopene intake specifically and the risk of lung cancer in a large group of men and women.[13] Men and women with high lycopene intakes had significantly lower risks of lung cancer than those with low intakes.

Lycopene inhibits the reproduction of cancer cells.[14] African American women with low lycopene intakes ran a three to four times greater risk of developing cancerous changes of the cervix than similar women with high lycopene intakes.[15] Studies also find an association between increased breast cancer risk and low intakes of lycopene and related compounds.[16]

Sounds as though I might want to go out and buy some lycopene supplements. What do you think?

Stick with the tried and true. Eat more fruits and vegetables. Cancer research favors eating foods high in lycopene, but research does not support the consumption of purified supplements of lycopene. Recall from Chapter 8 that diets high in the carotenoid beta-carotene often correlate with low rates of lung cancer. When given to smokers in studies, however, purified beta-carotene in supplement form *increases* lung cancer rates. The only safe option is to eat lycopene-rich foods such as tomatoes and tomato products and avoid concentrated supplements of lycopene until safety studies are completed.

What about other phytochemical supplements? Sometimes I just don't have the time to eat as many fruits and vegetables as I know I should.

Even if you don't always eat in the best interest of your health, taking supplements of purified phytochemicals is not the way to go. Phytochemicals can alter body functions, sometimes powerfully. Researchers are just beginning to understand how a handful of phytochemicals work, and what is current today may change tomorrow. The body is equipped to handle phytochemicals in diluted form, mixed with all of the other constituents of foods, but it is not adapted to phytochemicals in concentrated form. Fruits, vegetables, legumes, whole grains, nuts, and seeds contain a wide array of beneficial nutrients and thousands of phytochemicals. The best way to reap the benefits of phytochemicals is by eating foods, not supplements (see Figure NP8–1 on p. 204).

Nutrition in Practice

Table NP8–2 A Sampling of Phytochemicals: Possible Actions and Food Sources

Name or Class	Possible Effects	Food Sources
Capsaicin	Modulates blood clotting, possibly reducing the risk of fatal clots in heart and artery disease.	Hot peppers
Carotenoids (include beta-carotene, lutein, lycopene, and hundreds of related compounds)[a]	Act as antioxidants; possibly reduce risks of heart disease, age-related eye diseases,[b] cancer, and other diseases.	Deeply pigmented fruits and vegetables (apricots, broccoli, cantaloupe, carrots, pink grapefruit, pumpkin, spinach, sweet potatoes, tomatoes, and watermelon)
Curcumin	May inhibit enzymes that activate carcinogens.	Turmeric, a yellow-colored spice
Flavonoids (include flavones, flavonols, isoflavones, catechin, and others)[c]	Many flavonoids act as antioxidants, scavenge carcinogens, bind to nitrates in the stomach, preventing conversion to nitrosamines; inhibit cell proliferation; flavonoids of blueberries may improve memory.	Berries, black tea, celery, chocolate, citrus fruits, green tea, olives, onions, oregano, purple grapes, purple grape juice, soybeans and soy products, vegetables, whole wheat, wine
Indoles[d]	May trigger production of enzymes that block DNA damage from carcinogens; may inhibit estrogen action.	Broccoli and other cruciferous vegetables (brussels sprouts, cabbage, cauliflower), horseradish, mustard greens
Isothiocyanates (including sulforaphane)	Inhibit enzymes that activate carcinogens; trigger production of enzymes that detoxify carcinogens.	Broccoli and other cruciferous vegetables (brussels sprouts, cabbage, cauliflower), horseradish, mustard greens
Lignans[e]	Block estrogen activity in cells, possibly reducing the risk of cancer of the breast, colon, ovaries, and prostate.	Flaxseed, whole grains
Monoterpenes (include limonene)	May trigger enzyme production to detoxify carcinogens; inhibit cancer promotion and cell proliferation.	Citrus fruit peels and oils
Organosulfur compounds (including allicin)	May speed production of carcinogen-destroying enzymes; slow production of carcinogen-activating enzymes.	Chives, garlic, leeks, onions
Phenolic acids (including ellagic acid)[c]	May trigger enzyme production to make carcinogens water soluble, facilitating excretion.	Coffee beans, fruits (apples, blueberries, cherries, grapes, oranges, pears, prunes), oats, potatoes, soybeans, strawberries
Phytic acid	Binds to minerals, preventing free-radical formation, possibly reducing cancer risk.	Whole grains
Phytosterols (genistein and diadzein)	Estrogen inhibition may produce these actions: inhibit cell replication in GI tract; reduce risk of breast, colon, ovarian, prostate, and other estrogen-sensitive cancers; reduce cancer cell survival; may reduce risk of osteoporosis and heart disease.	Soybeans, soy flour, soy milk, tofu, textured vegetable protein, other legume products
Protease inhibitors	May suppress enzyme production in cancer cells, slowing tumor growth; inhibit hormone binding; inhibit malignant changes in cells.	Broccoli sprouts, potatoes, soybeans and other legumes, soy products
Resveratrol	Offsets artery-damaging effects of high-fat diets.	Red wine, peanuts
Saponins	May interfere with DNA replication, preventing cancer cells from multiplying; stimulate immune response.	Alfalfa sprouts, other sprouts, green vegetables, potatoes, tomatoes
Tannins[c]	May inhibit carcinogen activation and cancer promotion; act as antioxidants.	Black-eyed peas, grapes, lentils, red and white wine, tea

[a]Other carotenoids include alpha-carotene, beta-cryptoxanthin, and zeaxanthin.
[b]The age-related eye disease is macular degeneration.
[c]A subset of the larger group *phenolic phytochemicals.*
[d]Indoles include dithiothiones, isothiocyantes, and others.
[e]Lignans act as phytosterols, but their food sources are limited.

Nutrition in Practice

Broccoli sprouts contain an abundance of the isothiocyanate sulforaphane.

An apple a day—rich in flavonoids.

The phytosterols genistein and diadzein are found in soybeans.

Garlic provides abundant allicin, one of the organosulfur compounds.

The phytochemicals of grapes, red wine, and peanuts include resveratrol.

The ellagic acid of strawberries is a phenolic acid.

Citrus fruits provide limonene.

The flavonoids in black tea may protect against heart disease, whereas those in green tea may defend against cancer.

Tomatoes are famous for their abundant lycopene.

Flaxseed is the richest source of lignans.

Blueberries are a rich source of flavonoids.

Figure NP8–1

An Array of Phytochemicals in a Variety of Fruits and Vegetables
Consult Table NP8–2 for potential functions and related food sources.

Notes

1. L. Arab and S. Steck, Lycopene and cardiovascular disease, *American Journal of Clinical Nutrition* 71 (2000): 1691S–1695S; L. LeMarchand and coauthors, Intake of flavonoids and lung cancer, *Journal of the National Cancer Institute* 92 (2000): 154–160; P. Knekt and coauthors, Quercetin intake and the incidence of cerebrovascular disease, *European Journal of Clinical Nutrition* 54 (2000): 415–417.

2. J. A. Milner, Functional foods: The US perspective, *American Journal of Clinical Nutrition* 71 (2000): 1654S–1659S; Position of The American Dietetic Association: Functional foods, *Journal of the American Dietetic Association* 99 (1999): 1278–1285.

3. A. L. Normen and coauthors, Plant sterol intakes and colorectal cancer risk in the Netherlands Cohort Study on Diet and Cancer, *American Journal of Clinical Nutrition* 74 (2001):

141–148; C. M. Steinmaus, S. Nunez, and A. H. Smith, Diet and bladder cancer: A meta-analysis of six dietary variables, *American Journal of Epidemiology* 151 (2000): 693–702; M. R. Law and J. K. Morris, By how much does fruit and vegetable consumption reduce the risk of ischaemic heart diseases? *European Journal of Nutrition* 52 (1998): 549–556.

4. J. M. Geleijnse and coauthors, Inverse association of tea and flavonoid intakes with incident myocardial infarction: the Rotterdam Study, *American Journal of Clinical Nutrition* 75 (2002): 880–886; LeMarchand and coauthors, 2000; Knekt and coauthors, 2000; L. F. Macrae, Wheat bran fiber and development of adenomatous polyps: Evidence from randomized, controlled, clinical trials, *American Journal of Medicine* 106 (1999): S38–S42; S. V. Nigdikar and coauthors, Consumption of red wine polyphenols reduces the susceptibility of low-density lipoproteins to oxidation in vivo, *American Journal of Clinical Nutrition* 68 (1998): 258–265.

5. B. Fuhrman and M. Aviram, Flavonoids protect LDL from oxidation and attenuate atherosclerosis, *Current Opinion in Lipidology* 12 (2001): 41–48; Nigdikar and coauthors, 1998.

6. A. Drewnowski and C. Gomez-Caneros, Bitter taste, phytonutrients, and the consumer: A review, *American Journal of Clinical Nutrition* 72 (2000): 1424–1435.

7. Y. Scheider and coauthors, Anti-proliferative effect of resveratrol, a natural component of grapes and wine, on human colonic cancer cells, *Cancer Letter* 158 (2000): 85–91.

8. S. Liu and coauthors, Intake of vegetables rich in carotenoids and risk of coronary heart disease in men: The Physicians' Heart Study, *International Journal of Epidemiology* 30 (2001): 130–135; S. B. Kritchevsky, beta-Carotene, carotenoids and the prevention of coronary heart disease, *Journal of Nutrition* 129 (1999): 5–8.

9. Arab and Steck, 2000.

10. T. H. Rissanen and coauthors, Low serum lycopene concentration is associated with an excess incidence of acute coronary events and stroke: The Kupio Ischaemic Heart Disease Risk Factor Study, *British Journal of Nutrition* 85 (2001): 749–754.

11. G. R. Beecher, Nutrient contents of tomatoes and tomato products, *Proceedings of the Society for Experimental and Biological Medicine* 218 (1998): 98–100.

12. E. Giovannucci, Tomatoes, tomato-based products, lycopene, and cancer: Review of the epidemiologic literature, *Journal of the National Cancer Institute* 91 (1999): 317–331; S. K. Clinton, Lycopene: Chemistry, biology, and implications for human health and disease, *Nutrition Reviews* 56 (1998): 35–51.

13. D. S. Michaud and coauthors, Intake of specific carotenoids and the risk of lung cancer in 2 prospective US cohorts, *American Journal of Clinical Nutrition* 72 (2000): 990–997.

14. P. Prakash, R. M. Russell, and N. I. Krinsky, In vitro inhibition of proliferation of estrogen-dependent and estrogen-independent human breast cancer cells treated with carotenoids or retinoids, *Journal of Nutrition* 131 (2001): 1574–1580; H. Gerster, The potential role of lycopene for human health, *Journal of the American College of Nutrition* 16 (1997): 109–126.

15. P. A. Kantesky and coauthors, Dietary intake and blood levels of lycopene: Association with cervical dysplasia among non-Hispanic black women, *Nutrition and Cancer* 31 (1998): 31–40.

16. P. Toniolo and coauthors, Serum carotenoids and breast cancer, *American Journal of Epidemiology* 153 (2001): 1142–1147.

Jennie Oppenheimer/Studio Zocolo

WATER AND THE MINERALS

Water is the most indispensable nutrient of all.

*T*he body's water cannot be considered separately from the minerals dissolved in it. A person can drink pure water, but in the body, water mingles with minerals to become fluids in which all life processes take place. This chapter begins by discussing the body's fluids and their chief minerals. The focus then shifts to other functions of the minerals.

Water and Body Fluids

Water constitutes about 60 percent of an adult's body weight and a higher percentage of a child's. Every cell in the body is bathed in a fluid of the exact composition that is best for that cell. The body fluids bring to each cell the ingredients it requires and carry away the end products of the life-sustaining reactions that take place within the cell's boundaries. The water in the body fluids:

- Carries nutrients and waste products throughout the body.
- Participates in metabolic reactions.
- Serves as the solvent for minerals, vitamins, amino acids, glucose, and many other small molecules.
- Aids in maintaining the body's blood pressure and temperature.
- Acts as a lubricant and cushion around joints.
- Serves as a shock absorber inside the eyes, spinal cord, and amniotic sac surrounding a fetus in the womb.

To support these and other vital functions, the body actively regulates its **water balance.**

Water Balance

The cells themselves regulate the composition and amounts of fluids within and surrounding them. The entire system of cells and fluids remains in a delicate but firmly maintained state of dynamic equilibrium. Imbalances such as **dehydration** and **water intoxication** can occur, but the body quickly restores the balance to normal if it can. The body controls both water intake and water excretion.

Water Intake Regulation   The body can survive for only a few days without water. In healthy people, thirst and satiety govern water intake. Thirst is finely adjusted to ensure a water intake that meets the body's needs. When the blood becomes too concentrated (having lost water but not salt and other dissolved substances), the mouth becomes dry, and the brain center known as the **hypothalamus** initiates drinking behavior.

Thirst lags behind the lack of water. A water deficiency that develops slowly can switch on drinking behavior in time to prevent serious dehydration, but a deficiency that develops quickly may not. Also, thirst itself does not remedy a water deficiency; a person must pay attention to the thirst signal and take the time to get a drink. The long-distance runner, the gardener in hot weather, and the busy child at play can experience serious dehydration if they fail to drink promptly in response to their needs for water. With aging, thirst sensations may diminish. Dehydration can threaten elderly people who do not develop the habit of drinking water regularly.

Water Excretion Regulation   Water excretion involves the brain and the kidneys. The cells of the brain's hypothalamus, which monitor blood salts, stimulate the **pituitary gland** to release **antidiuretic hormone (ADH)** whenever the salts are too concentrated, or the blood volume or blood pressure is too low. ADH

water balance: the balance between water intake and water excretion that keeps the body's water content constant.

dehydration: the loss of water from the body that occurs when water output exceeds water input. The symptoms progress rapidly from thirst, to weakness, to exhaustion and delirium and end in death if not corrected.

water intoxication: the rare condition in which body water contents are too high. The symptoms may include confusion, convulsion, coma, and even death in extreme cases.

hypothalamus (high-poh-THALL-uh-mus): a part of the brain that helps regulate many body balances, including fluid balance.

pituitary (pit-TOO-ih-tary) **gland:** in the brain, the "king gland" that regulates the operation of many other glands.

antidiuretic hormone (ADH): a hormone released by the pituitary gland in response to high salt concentrations in the blood. The kidneys respond by reabsorbing water.

stimulates the kidneys to reabsorb water rather than excrete it. Thus the more water you need, the less you excrete.

If too much water is lost from the body, blood volume and blood pressure fall. Cells in the kidneys respond to the low blood pressure by releasing an enzyme. Through a complex series of events, involving the hormone **aldosterone,** this enzyme also causes the kidneys to retain more water. Again, the effect is that when more water is needed, less is excreted.

Minimum Water Needed These mechanisms can maintain water balance only if a person drinks enough water. The body must excrete a minimum of about 500 milliliters each day as urine—enough to carry away the waste products generated by a day's metabolic activities. Above this amount, excretion adjusts to balance intake, so the more a person drinks, the more dilute the urine becomes. In addition to urine, some water is lost from the lungs as vapor, some is excreted in feces, and some evaporates from the skin. A person's water losses from all of these routes total about 2½ liters (about 2½ quarts) a day on the average. Table 9–1 shows how fluid intake and output naturally balance out.

Water Recommendations and Sources Water needs vary greatly depending on the foods a person eats, the environmental temperature and humidity, the person's activity level, and other factors. Accordingly, a general water requirement is difficult to establish. Recommendations for adults are expressed in proportion to the amount of energy expended under normal environmental conditions. For the person who expends about 2000 kcalories a day, this works out to 2 to 3 liters, or about 8 to 12 cups. You can tell from the color of the urine whether a person needs more water. Pale yellow urine reflects appropriate dilution.

Plain water best meets people's fluid needs, but milk and fruit juice can also contribute to the day's recommended intake. Alcoholic beverages and those containing caffeine such as coffee, tea, and some sodas are not good water substitutes. Both alcohol and caffeine act as diuretics, causing the body to lose fluids. Many people who drink beer or coffee, then, may end up with a net fluid loss rather than a gain, to the detriment of the body's fluid balance.[1] Foods provide water, too. Most fruits and vegetables contain up to 95 percent water; many meats and cheeses contain at least 50 percent. The energy nutrients in foods also give up water during metabolism.

Fluid and Electrolyte Balance

When mineral **salts** dissolve in water, they separate (dissociate) into charged particles known as ions, which can conduct electricity. For this reason, a salt that partly dissociates in water is known as an **electrolyte.** The body fluids, which contain water and partly dissociated salts, are **electrolyte solutions.**

The body's electrolytes are vital to the life of the cells and therefore must be closely regulated to help maintain the appropriate distribution of body fluids. The major minerals form salts that dissolve in the body fluids; the cells direct where these salts go; and the movement of the salts determines where the fluids flow because water follows salt. Cells use this force to move fluids back and forth across their membranes. Thanks to the electrolytes, water can be held in compartments where it is needed.

Proteins in the cell membranes move ions in or out of the cells. These protein pumps tend to concentrate sodium and chloride outside cells and potassium and other ions inside. By maintaining specific amounts of sodium outside and potassium inside, cells can regulate the exact amounts of water inside and outside their boundaries. Physicians apply the same principle when treating kidney failure; they use an electrolyte solution to draw excess fluid out of the blood (see the discussion of *dialysis* in Nutrition in Practice 26).

Healthy kidneys regulate the body's sodium, as well as its water, with remarkable precision. The intestinal tract absorbs sodium readily, and it travels freely in the

Table 9–1	Water Balance
Water Source	**Amount (ml)**
Liquids	550 to 1500
Foods	700 to 1000
Metabolic water	200 to 300
	1450 to 2800
Water Output	
Kidneys	500 to 1400
Skin	450 to 900
Lungs	350
Feces	150
	1450 to 2800

The enzyme **renin** (REN-in), released by the kidneys in response to low blood pressure, aids the kidneys in retaining water through the **renin-angiotensin mechanism.**

500 ml = about ½ qt.

Exceptions: A compound in which the positive ions are hydrogen ions (H$^+$) is an acid (example: hydrochloric acid, or H$^+$Cl$^-$); a compound in which the negative ions are hydroxyl ions (OH$^-$) is a base (example: potassium hydroxide, or K$^+$OH$^-$).

The simple statement that water follows salt describes the force that chemists call *osmosis.*

aldosterone (al-DOS-ter-own): a hormone secreted by the adrenal glands that stimulates the reabsorption of sodium by the kidneys; also regulates chloride and potassium concentrations.

salts: compounds composed of charged particles (ions). An example of a salt is potassium chloride (K$^+$Cl$^-$).

electrolyte: a salt that dissolves in water and dissociates into charged particles called ions.

electrolyte solutions: solutions that can conduct electricity.

Table 9–2 The Major and Trace Minerals	
Major Minerals	**Trace Minerals**
■ Calcium	■ Arsenic
■ Chloride	■ Boron
■ Magnesium	■ Chromium
■ Phosphorus	■ Cobalt
■ Potassium	■ Copper
■ Sodium	■ Fluoride
■ Sulfur	■ Iodine
	■ Iron
	■ Manganese
	■ Molybdenum
	■ Nickel
	■ Selenium
	■ Silicon
	■ Zinc

Reminder: *Buffers* are compounds that help keep a solution's acidity or alkalinity constant; buffers are capable of neutralizing both acids and bases and thereby maintaining the original concentration of hydrogen ions (pH) in the solution.

extracellular fluid: fluid residing outside the cells; includes the fluid between the cells *(interstitial fluid),* plasma, and the water of structures such as the skin and bones. Extracellular fluid accounts for about one-third of the body's water.

blood, but the kidneys excrete unneeded amounts. The kidneys actually filter all of the sodium out of the blood; then, with great precision, they return to the bloodstream the exact amount the body needs to retain. Thus the body's total electrolytes remain constant, while the urinary electrolytes fluctuate according to what is eaten.

In some cases the body's mechanisms for maintaining fluid and electrolyte balances cannot compensate for a sudden loss of large amounts of fluid and electrolytes. Vomiting, diarrhea, heavy sweating, fever, burns, wounds, and the like may incur great fluid and electrolyte losses, precipitating an emergency that demands medical intervention.

Acid-Base Balance

The body uses ions not only to help maintain water balance, but also to regulate the acidity (pH) of its fluids. Electrolyte mixtures in the body fluids, as well as proteins, protect the body against changes in acidity by acting as buffers—substances that can accommodate excess acids or bases.

The body's buffer systems serve as a first line of defense against changes in the fluids' acid-base balance. The lungs, skin, GI tract, and kidneys provide other defenses. Of these organ systems, the kidneys play the primary role in maintaining acid-base balance. Disorders of the kidneys, therefore, impair the body's ability to regulate its acid-base balance, as well as its fluid and electrolyte balances.

In Summary Water makes up about 60 percent of the body's weight. It helps transport nutrients and waste products throughout the body, participates in metabolic reactions, acts as a solvent, assists in maintaining blood pressure and body temperature, acts as a lubricant and cushion around joints, and serves as a shock absorber. To maintain water balance, intake from liquids, foods, and metabolism must equal losses from kidneys, skin, lungs, and feces. Electrolytes help maintain the appropriate distribution of body fluids and help to maintain acid-base balance as well.

The Major Minerals

Table 9–2 lists the major and trace minerals in the body, and Figure 9–1 shows the amounts found in the body. As you can see, the most prevalent minerals are calcium and phosphorus, the chief minerals of bone. The distinction between the major and the trace minerals does not mean that one group is more important than the other. A deficiency of the few micrograms of iodine needed daily is just as serious as a deficiency of the several hundred milligrams of calcium. The major minerals are so named because they are present, and needed, in larger amounts in the body than the trace minerals.

Although all the major minerals influence the body's fluid balance, sodium, chloride, and potassium are most noted for that role. For this reason, these three minerals are discussed first. Each major mineral also plays other specific roles in the body. Sodium, potassium, calcium, and magnesium are critical to nerve transmission and muscle contractions. Phosphorus and magnesium are involved in energy metabolism. Calcium, phosphorus, and magnesium contribute to the structure of the bones. Sulfur helps determine the shape of proteins.

Sodium

Sodium is the principal electrolyte in the **extracellular fluid** (the fluid outside the cells) and the primary regulator of the extracellular fluid volume. When the blood concentration of sodium rises, as when a person eats salted foods, thirst prompts the person to drink water until the appropriate sodium-to-water ratio

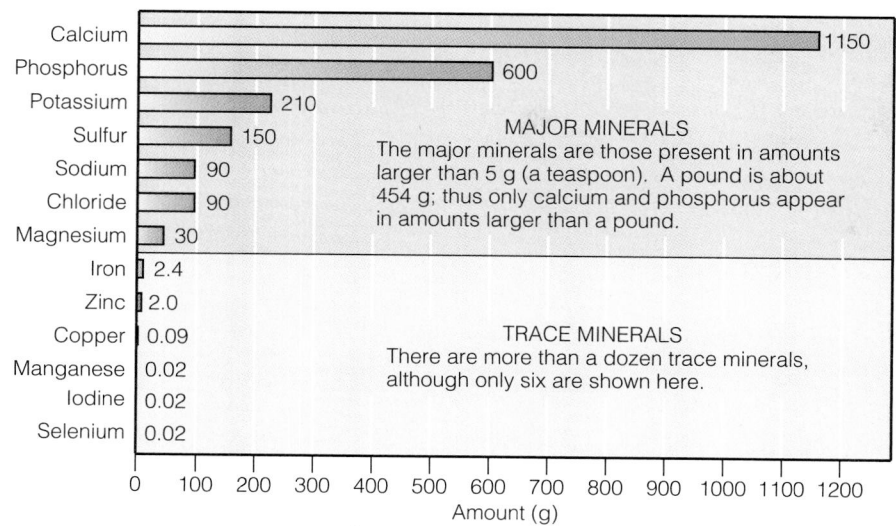

Figure 9–1

The Amounts of Minerals in a 60-kilogram (132-pound) Human Body

is restored. Sodium also helps maintain acid-base balance and is essential to muscle contraction and nerve transmission.

Sodium Recommendations and Food Sources Diets rarely lack sodium. For this reason, recommended intakes have not been set; instead an estimated *minimum* sodium requirement has been established. Health recommendations advise a *maximum* intake of *salt,* primarily to help prevent high blood pressure.[2]

Cultures vary in their use of salt. In the United States, men consume an average of 3300 milligrams of sodium (equivalent to about 8 grams of salt) a day. Asian people, whose staple sauces and flavorings are based on soy sauce and monosodium glutamate (MSG), consume the equivalent of about 30 to 40 grams of salt per day. In China, Japan, and Korea, high blood pressure is as prevalent, or more so, as in the United States.

Sodium intakes also vary widely. People who eat mostly processed foods have the highest sodium intakes, while those who eat mostly whole, unprocessed foods, such as fresh fruits and vegetables, have the lowest intakes. In fact, about three-fourths of the sodium in people's diets comes from salt added to foods by manufacturers. Figure 9–2 on p. 212 shows that processed foods contain not only more sodium but also less potassium than their less processed counterparts.

Sodium and Blood Pressure Sodium, or the salt that delivers it, contributes to high blood pressure in susceptible people.[3] Susceptibility may increase with potassium, calcium, and magnesium deficiencies. Chapter 25 offers suggestions for avoiding excessive salt intakes and describes the relationships of dietary factors to blood pressure.

Chloride

The chloride ion is the major negative ion of the extracellular fluids, where it occurs primarily in association with sodium. Chloride can move freely across cell membranes and so is also found inside the cells in association with potassium. Like sodium, chloride is critical to maintaining fluid, electrolyte, and acid-base balance in the body. In the stomach, the chloride ion is part of hydrochloric acid, which maintains the strong acidity of the gastric fluids.

Salt is a major food source of chloride, and as with sodium, processed foods are a major contributor of this mineral to people's diets. A chloride recommendation has not been established, but an estimated minimum requirement has been determined for adults.

Estimated *minimum* requirement for sodium: 500 mg/day.

Recommended *maximum* intake of salt: 6 g/day (2400 mg sodium).
- *5 g salt = about 2 g sodium.*
- *3 g salt = ½ tsp.*

Estimated minimum requirement for chloride: 750 mg/day.

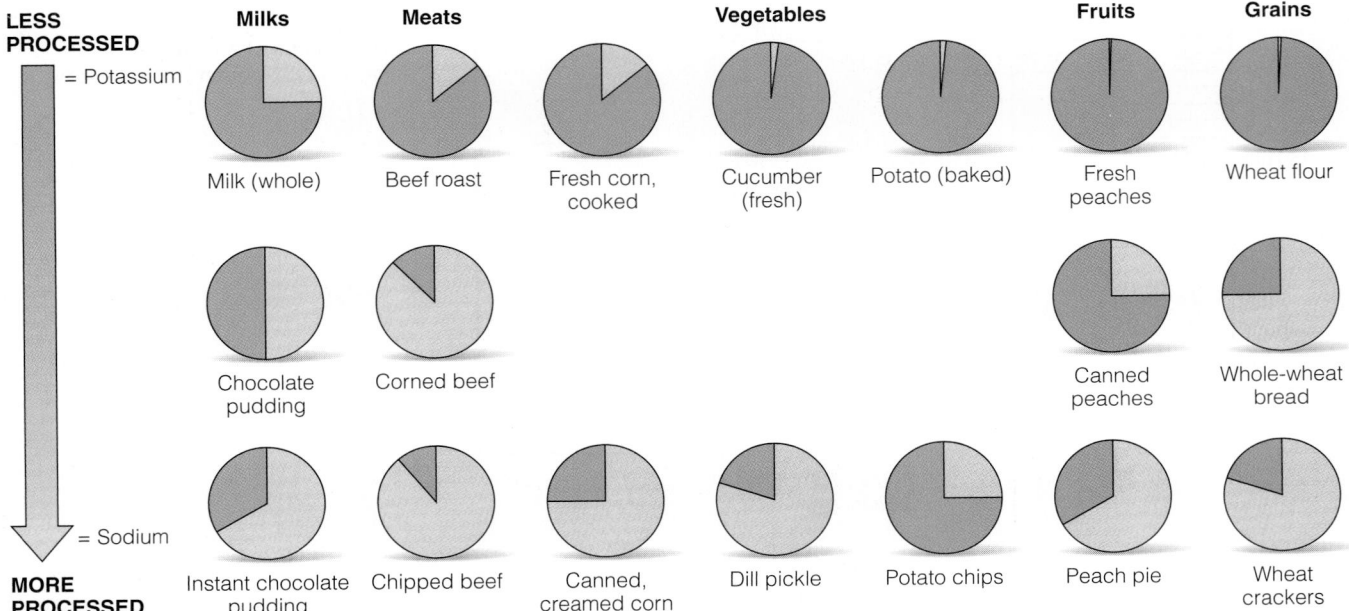

LESS PROCESSED
= Potassium

MORE PROCESSED
= Sodium

Milks	Meats		Vegetables		Fruits	Grains
Milk (whole)	Beef roast	Fresh corn, cooked	Cucumber (fresh)	Potato (baked)	Fresh peaches	Wheat flour
Chocolate pudding	Corned beef				Canned peaches	Whole-wheat bread
Instant chocolate pudding	Chipped beef	Canned, creamed corn	Dill pickle	Potato chips	Peach pie	Wheat crackers

Figure 9–2

What Processing Does to the Sodium and Potassium Contents of Foods
Note how potassium is lost and sodium is gained as foods become more processed.

Estimated minimum requirement for potassium: 2000 mg/day.

diuretics (dye-yoo-RET-ics): medications that promote the excretion of water through the kidneys. Not all diuretics increase the urinary loss of potassium. Some, called potassium-sparing diuretics, are less likely to result in a potassium deficiency (see Chapter 25).

steroids (STARE-oids): medications used to reduce tissue inflammation, to suppress the immune response, or to replace certain steroid hormones in people who cannot synthesize them.

cathartics (ca-THART-ics): strong laxatives.

Potassium

Potassium is the principal positively charged ion inside the body cells. It plays a major role in maintaining fluid and electrolyte balance and cell integrity. Potassium is also critical to keeping the heartbeat steady. The sudden deaths that occur in severe diarrhea and in children with kwashiorkor are likely due to heart failure caused by potassium loss. Potassium also assists in carbohydrate and protein metabolism.

Potassium Deficiency and Toxicity Potassium deficiency results more often from excessive losses than from deficient intakes. Deficiency arises in abnormal conditions such as diabetic acidosis, dehydration, or prolonged vomiting or diarrhea; potassium deficiency can also result from the regular use of certain medications, including **diuretics, steroids,** and **cathartics.** One of the earliest symptoms of deficiency is muscle weakness. Inadequate potassium intakes are possible with diets low in fresh fruits and vegetables, but out-and-out deficiencies of potassium are unlikely in healthy people.

In healthy people, potassium toxicity from foods is not a problem because the kidneys excrete excess potassium. Intakes from potassium supplements can reach toxic levels, however, and can cause death.

Potassium Recommendations and Food Sources As with sodium, an estimated minimum requirement for potassium has been determined. Surveys show wide variations in potassium intakes in the United States; people who emphasize fresh fruits and vegetables in their diets have high intakes. Potassium is abundant inside all living cells, both plant and animal, and because cells remain intact until foods are processed, the richest sources of potassium are *fresh* foods of all kinds—especially fruits and vegetables.

Potassium and Blood Pressure Diets low in potassium seem to play an important role in the development of high blood pressure. Research suggests that increasing potassium intakes may both prevent and help to correct hypertension.[4]

Calcium

Calcium occupies more space in this chapter than any other major mineral. Other minerals are revisited later in this book, where they play key roles in heart disease and kidney disease. Calcium, though, deserves emphasis here in the normal nutrition part of the book because an adequate intake of calcium early in life helps grow a healthy skeleton and prevent bone disease in later life.

Calcium Roles in the Body Calcium owns the distinction of being the most abundant mineral in the body. Ninety-nine percent of the body's calcium is stored in the bones, where it plays two important roles. First, it is an integral part of bone structure. Second, it serves as a calcium bank available to the body fluids should a drop in blood calcium occur.

Calcium in Bone As bones begin to form, calcium salts form crystals on a matrix of the protein collagen. As the crystals become denser, they give strength and rigidity to the maturing bones. As a result, the long leg bones of children can support their weight by the time they have learned to walk. Figure 9–3 (p. 214) shows the lacy network of calcium-containing crystals in the bone.

Many people have the idea that bones are inert, like rocks. Not so. Bones continuously gain and lose minerals in an ongoing process of remodeling. Growing children gain more bone than they lose, and healthy adults maintain a reasonable balance. When withdrawals substantially exceed deposits, however, problems such as osteoporosis develop.

Figure 9–4 (p. 214) shows the three phases of bone development throughout life. From birth to approximately age 20, the bones are actively growing by modifying their length, width, and shape. This rapid growth phase overlaps with the next period of peak bone mass development, which occurs between the ages of 12 and 30. During this period, skeletal mass increases. Bones grow thicker and denser by remodeling, a maintenance and repair process involving the loss of existing bone and the deposition of new bone. In the final phase, which begins between 30 and 40 years of age and continues throughout the remainder of life, bone loss exceeds new bone formation.

Calcium in Body Fluids The 1 percent of the body's calcium that circulates in the fluids as ionized calcium is vital to life. It helps regulate muscle contractions, transmit nerve impulses, clot blood, and secrete hormones, digestive enzymes, and neurotransmitters. Calcium also helps convey signals received at the cell surface to the inside of the cell. Calcium is a **cofactor** for several enzymes as well.

Calcium Balance Blood calcium concentration is tightly controlled. Whenever blood calcium rises too high, a system of hormones and vitamin D promotes its deposit into bone. Whenever blood calcium falls too low, the regulatory system acts in three locations to raise it:

1. The intestine absorbs more calcium.

2. The bones release more calcium.

3. The kidneys excrete less calcium.

Thus blood calcium rises to normal.

The calcium stored in bone provides a nearly inexhaustible source of calcium for the blood. Even in a calcium deficiency, blood calcium remains normal. Blood calcium changes only in response to abnormal regulatory control, not to diet. Blood calcium above normal causes **calcium rigor:** the muscles contract and cannot relax. Blood calcium below normal causes **calcium tetany**—also characterized by uncontrolled muscle contraction. These conditions are caused by a lack of vitamin D or by abnormal concentrations of the hormones that regulate calcium homeostasis.

Fresh fruits and vegetables provide potassium in abundance.

© Polara Studios, Inc.

The regulators are hormones from the thyroid and parathyroid glands, as well as vitamin D. One hormone, **parathormone,** raises blood calcium. Others, **calcitonin** and **thyrocalcitonin,** lower blood calcium by inhibiting release of calcium from bone. The hormonelike **vitamin D** raises blood calcium by acting at the three sites listed.

cofactor: a mineral element that, like a coenzyme, works with an enzyme to facilitate a chemical reaction.

calcium rigor: hardness or stiffness of the muscles caused by high blood calcium.

calcium tetany: intermittent spasms of the extremities due to nervous and muscular excitability caused by low blood calcium.

Water and the Minerals

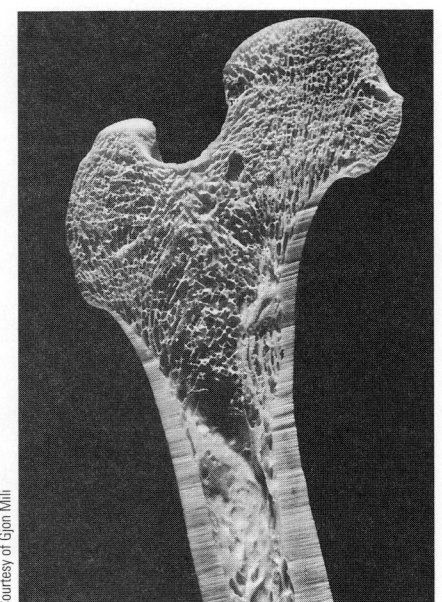

Figure 9–3

Cross Section of Bone
The lacy structural elements are *trabeculae* (tra-BECK-you-lee), which can be drawn on to replenish blood calcium.

osteoporosis (oss-tee-oh-pore-OH-sis): literally, porous bones; reduced density of the bones. Also known as *adult bone loss*, it is a condition in which the bones become porous and fragile. The causes of osteoporosis are multiple.
osteo = bone

Figure 9–4

Phases of Bone Development throughout Life
The active growth phase occurs from birth to approximately age 20. The next phase of peak bone mass development occurs between the ages of 12 and 30. The final phase, when bone loss exceeds formation, begins between the ages of 30 and 40 and continues throughout the remainder of life.

Although a chronic *dietary* deficiency of calcium or a chronic deficiency due to poor absorption does not change blood calcium, it does deplete the savings account in the bones. Because this is an important concept, we repeat: it is the bones, not the blood, that are robbed by calcium deficiency.

Calcium and Osteoporosis   Bone mass peaks at the time of skeletal maturity (about age 30), and a high peak bone mass is the best protection against later age-related bone loss and fracture. Adequate calcium nutrition during the growing years is essential to achieving optimal peak bone mass.[5] Following menopause, women lose about 15 percent of their bone mass, as do middle-aged and older men. When bone loss has reached such an extreme that bones fracture under even common, everyday stresses, the condition is known as **osteoporosis.** Osteoporosis afflicts as many as 28 million people, mostly women 45 years of age or older.[6] Men, however, are not immune to osteoporosis. Each year, a million and a half people—a quarter of them men—suffer broken hips, pelvis, legs, arms, hands, and ankles attributable to osteoporosis.[7]

Both genetic and environmental factors contribute to osteoporosis. Table 9–3 summarizes risk and protective factors for osteoporosis. Osteoporosis is more prevalent in women than men for several reasons. First, women consume only about half as much dietary calcium as men do. Second, at all ages, women's bone mass is lower than men's because women generally have smaller bodies. Finally, bone loss begins earlier in women than in men, and it accelerates after menopause.

In addition to calcium, many other minerals and vitamins, including phosphorus, magnesium, fluoride, and vitamin D, help to form and stabilize the structure of bones. Any or all of these elements are needed to prevent bone loss. The first, most obvious lines of defense, however, are to maintain a lifelong adequate intake of calcium and to "exercise it into place." Active bones are denser than sedentary bones.[8] Weight-bearing physical activity, such as walking, running, dancing, and weight training, prompts the bones to deposit minerals. It has long been known that when people are confined to bed, both their muscles and their bones lose strength. Muscle strength and bone strength go together: when muscles work, they pull on the bones, and both are stimulated to grow stronger.

Calcium and Hypertension   Some evidence suggests that calcium helps prevent hypertension.[9] Studies of populations prone to developing hypertension show that low dietary calcium correlates with high blood pressure.[10] Reports indicate that calcium supplements can sometimes lower blood pressure.[11] Some researchers speculate that calcium's effect on blood pressure is related to its action on the smooth muscle surrounding blood vessels.

Calcium Recommendations   As mentioned earlier, blood calcium concentration does not reflect calcium status. Calcium recommendations are therefore based on balance studies, which measure daily intake and excretion. An optimal calcium intake reflects the amount needed to retain the most calcium. The more

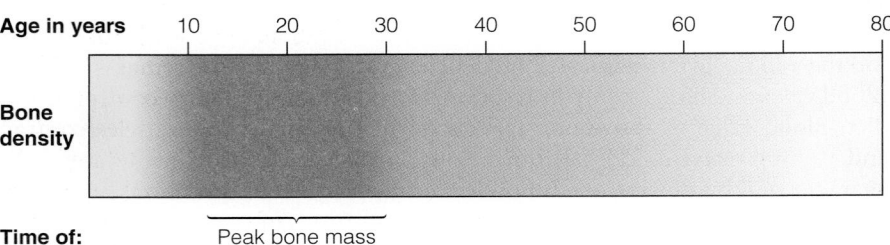

Age in years	10	20	30	40	50	60	70	80

Bone density

Time of: Peak bone mass

Table 9–3 Risk and Protective Factors That Correlate with Osteoporosis

Risk Factors	Protective Factors
High Correlation	
Advanced age; postmenopausal	African American
Alcohol abuse	Estrogens, long-term use
Anorexia nervosa	
Caucasian	
Chronic steroid use	
Female gender	
Rheumatoid arthritis	
Surgical removal of ovaries	
Thinness	
Moderate Correlation	
Chronic thyroid hormone use	Having given birth
Cigarette smoking	High body weight
Diabetes (type 1)	High-calcium diet
Early menopause	Regular physical activity
Excessive antacid use	
Family history of osteoporosis	
Low-calcium diet	
Sedentary lifestyle or immobility	
Vitamin D deficiency	
Probably Important But Not Yet Proved	
Alcohol taken in moderation	
Caffeine intake	Adequate vitamin K intake
High-fiber diet	Low-sodium diet (later years)
High-protein diet	
High-sodium diet	

SOURCE: Adapted from C. D. Arnaud and S. D. Sanchez, The role of calcium in osteoporosis, *Annual Review of Nutrition* 10 (1990): 397–414.

Table 9–4 Suggested Fluid Milk Intakes

Age	Suggested Intake
Children	2 c
Adolescents and young adults	3 c
Adults	2 c
Pregnant and lactating women	3 c
Pregnant and lactating adolescents	4 c
Women past menopause	3 c

calcium retained, the greater the bone density (within genetic limits) and, potentially, the lower the risk of osteoporosis. Calcium recommendations during adolescence are set high (1300 milligrams) to help ensure that the skeleton will be strong and dense. Between the ages of 19 and 50, recommendations are lowered to 1000 milligrams a day. For those over 50, recommendations are raised again, to 1200 milligrams, to minimize bone loss. Some authorities advocate calcium recommendations as high as 1500 milligrams per day for women over 50. Many women have intakes well below recommendations.

Calcium in Foods Calcium is found almost exclusively in a single food group—milk and milk products. For this reason, dietary recommendations advise daily consumption of reduced-fat, low-fat, or fat-free milk products. A cup of milk offers about 300 milligrams of calcium, so an adult who drinks 2 to 3 cups of milk a day is well on the way to meeting daily calcium needs. Pregnant and lactating teenagers need more (see Table 9–4). The other dairy food that contains comparable amounts of calcium is cheese. One slice of cheese (1 ounce) contains about two-thirds as much calcium as a cup of milk. Cottage cheese, however, contains much less. Snapshot 9–1 (p. 216) shows foods that are rich in calcium, and the "How to" box (p. 217) suggests ways of adding calcium to meals.

Some foods offer large amounts of calcium because of fortification. Calcium-fortified juice, high-calcium milk (milk with extra calcium added), and calcium-fortified cereals are examples. Some calcium-rich mineral waters provide as much as 500 milligrams of calcium per liter, while others provide slightly less.[12]

Calcium AI:
- *Adults (19–50 yr): 1000 mg/day.*
- *Adults (51 and older): 1200 mg/day.*

Snapshot 9–1

Calcium

All of these foods except the broccoli provide 10 percent or more of the calcium Daily Value (DV = 1000 mg/day).[a]

Milk: 30% DV per cup

Almonds: 10% DV per ⅓ c

Cheddar cheese: 31% DV per 1½ oz

Black-eyed peas: 11% DV per ½ c

Broccoli:[b] 9% DV per 1 c cooked

Sardines: 32% DV per 3 oz

[a]The Daily Values are based on a 2000-kcalorie diet.
[b]Although broccoli, kale, and some other cooked green leafy vegetables fall short of supplying 10 percent of the calcium DV in a serving, these foods are important sources of bioavailable calcium. Other greens such as spinach and chard contain calcium in an unabsorbable form. Note that the serving size for broccoli is a full cup, double the ½ cup serving of the Daily Food Guide.

Apparently, all fibers in plant foods—cellulose, hemicellulose, pectin, and others—bind calcium to some extent, as do **phytate** and **oxalate.** Phytate and oxalate are *binders* that combine with minerals to form complexes that the body cannot absorb.

Phosphorus RDA for adults: 700 mg/day.

Drinking calcium-rich water throughout the day offers a convenient way to meet both calcium and water needs.[13]

Among the vegetables, mustard greens, kale, parsley, watercress, and broccoli are good sources of available calcium. Some dark green, leafy vegetables—notably, spinach and Swiss chard—appear to be calcium-rich but actually provide very little, if any, calcium to the body. These foods contain binders that prevent calcium absorption.

People may think that taking a calcium supplement is preferable to getting calcium from food, but foods offer important fringe benefits. For example, drinking 2 cups of milk fortified with vitamins A and D will supply substantial amounts of other nutrients. Furthermore, the vitamin D, lactose, fat, and possibly other nutrients in the milk enhance calcium absorption. A calcium supplement supplies only calcium, and in a less absorbable form.

Aided by vitamin D, the body is able to regulate its absorption of calcium by altering its production of the calcium-binding protein. More of this protein is made if more calcium is needed. Infants and children absorb up to 60 percent of the calcium they ingest and pregnant women, about 50 percent. Other adults, who are not growing, absorb about 25 percent.

Phosphorus

Phosphorus is the second most abundant mineral in the body. About 85 percent of it is found combined with calcium in the crystals of the bones and teeth. As part of one of the body's buffer systems (phosphoric acid), phosphorus is also found in all body cells. Phosphorus is a part of DNA and RNA, the genetic code material present in every cell. Thus phosphorus is necessary for all growth. Phosphorus also plays many key roles in the transfer of energy that occurs during cellular metabolism. Phosphorus-containing lipids (phospholipids) help transport other lipids in the blood. Phospholipids are also principal components of cell membranes.

Animal protein is the best source of phosphorus because the mineral is so abundant in the cells of animals. Diets that provide adequate energy and protein also supply adequate phosphorus. Dietary deficiencies are unknown. A summary of facts about phosphorus appears in Table 9–5 (p. 219).

Magnesium

Magnesium barely qualifies as a major mineral. Only about 1 ounce of magnesium is present in the body of a 130-pound person, over half of it in the bones. Most of the rest is in the muscles, heart, liver, and other soft tissues,

How to

Add Calcium to Daily Meals

*M*any people cannot or will not drink enough milk to meet recommendations. To help deliver calcium, try the following suggestions:

1. Powdered fat-free milk is an excellent and inexpensive source of calcium and can be added to many foods (such as baked products and meat loaf) during preparation.

2. Yogurt and kefir (fermented dairy products) are acceptable substitutes for regular milk. Buttermilk and cheese (especially the low-fat or fat-free varieties) are also recommended substitutes. Cottage cheese and frozen yogurt contain about half the calcium of milk, while sherbet contains even less.

3. Puddings, custards, and baked goods can be prepared using appreciable amounts of milk.

4. People who adhere to a plant-based diet and those people who are either allergic to milk or lactose intolerant can use calcium-rich milk and cheese substitutes such as calcium-fortified soy milk or tofu (bean curd) and calcium-fortified juices.

5. Small fish, such as canned sardines, and other canned fish prepared with their bones, such as canned salmon, are also rich in calcium.

with only 1 percent in the body fluids. Bone magnesium seems to be a reservoir to ensure that some will be on hand for vital reactions regardless of recent dietary intake.

Magnesium is critical to the operation of hundreds of enzymes. Magnesium acts in all the cells of the soft tissues, where it forms part of the protein-making machinery and is necessary for the release of energy. Magnesium also helps muscles to relax after contraction and promotes resistance to tooth decay by holding calcium in tooth enamel.

Magnesium Deficiency Magnesium deficiency can result from vomiting, diarrhea, alcohol abuse, or protein malnutrition; in people who have been fed incomplete fluids intravenously for too long after surgery; or in people using diuretics. A severe magnesium deficiency causes tetany, an extreme and prolonged contraction of the muscles similar to the calcium tetany described earlier. Magnesium deficiency may also be related to cardiovascular disease and hypertension.[14] Magnesium deficiency is thought to cause the hallucinations commonly experienced during withdrawal from alcohol intoxication.

Magnesium Toxicity Magnesium toxicity is most often reported in older adults who abuse magnesium-containing laxatives, antacids, and other medications. The consequences can be severe: lack of coordination, confusion, coma, and, in extreme cases, death.

Magnesium Intakes and Food Sources Dietary intakes of magnesium average about three-quarters of the recommended intake for both men and women in the United States. Dietary intake data, however, do not assess the magnesium contribution of water. In various parts of the country, the water contains both calcium and magnesium and is known as "hard" water. Hard water can contribute significantly to magnesium intakes.

Magnesium-rich food sources include dark green, leafy vegetables, nuts, legumes, whole-grain breads and cereals, seafood, chocolate, and cocoa (see Snapshot 9–2 on p. 218). Magnesium is easily lost from foods during processing, so unprocessed foods are the best choices.

Magnesium RDA:
- *Men (19–30 yr): 400 mg/day. (31 and older): 420 mg/day.*
- *Women (19–30 yr): 310 mg/day. (31 and older): 320 mg/day.*

Snapshot 9–2
Magnesium
All of these foods provide 10 percent or more of the magnesium Daily Value (DV = 400 mg/day).[a]

Yogurt (plain) 11% DV per 1 c

Black-eyed peas: 11% DV per ½ c cooked

Baked potato: 14% DV per whole small potato

Spinach: 19% DV per ½ c cooked

Dried figs: 14% DV per ¼ c

Oysters: 13% DV per 3 oz steamed

[a]The Daily Values are based on a 2000-kcalorie diet.

Sulfur

The sulfur-containing amino acids are methionine and cysteine. Cysteine in one part of a protein chain can bind to cysteine in another part of the chain by way of a sulfur-sulfur bridge, thus helping to stabilize the protein structure.

The body does not use sulfur by itself as a nutrient. Sulfur is included here because it occurs in essential nutrients that the body does use, such as thiamin and certain amino acids. Sulfur is present in all proteins and plays its most important role in shaping strands of protein. The particular shape of a protein enables it to do its specific job, such as enzyme work. Skin, hair, and nails contain some of the body's more rigid proteins, and they have a high sulfur content.

There is no recommended intake for sulfur, and no deficiencies are known. Only a person who lacks protein to the point of severe deficiency will lack the sulfur-containing amino acids.

In Summary Table 9–5 offers a summary of the major minerals and their functions.

The Trace Minerals

Figure 9–1, earlier in this chapter, shows how tiny the quantities of trace minerals in the human body are. If you could remove all of them from your body, you would have only a bit of dust, hardly enough to fill a teaspoon. Yet each of the trace minerals performs some vital role for which no substitute will do. A deficiency of any of them can be fatal, and an excess of many can be equally deadly.

Recommendations have been established for nine of the trace minerals. Others are recognized as essential nutrients for some animals, but have not been proved to be required for human beings (see Table 9–6 on p. 220). Still others are under study to determine whether they, too, perform indispensable roles in the body.

Iron

hemoglobin: the oxygen-carrying protein of the red blood cells.
hemo = blood
globin = globular protein

myoglobin: the oxygen-carrying protein of the muscle cells.
myo = muscle

Every living cell—both plant and animal—contains iron. Most of the iron in the body is a component of the proteins **hemoglobin** in red blood cells and **myoglobin** in muscle cells. The iron in both hemoglobin and myoglobin helps them carry and hold oxygen and then release it. Hemoglobin in the blood carries oxygen from the lungs to tissues throughout the body. Myoglobin holds oxygen for the muscles to use when they contract. As part of many enzymes, iron is vital to the processes by which cells generate energy. Iron is also needed to make new cells, amino acids, hormones, and neurotransmitters.

Table 9–5 The Major Minerals—A Summary

Mineral Name	Chief Functions in the Body	Deficiency Symptoms	Toxicity Symptoms	Significant Sources
Sodium	With chloride and potassium (electrolytes), maintains cells' normal fluid balance and acid-base balance in the body. Also critical to nerve impulse transmission.	Muscle cramps, mental apathy, loss of appetite.	Hypertension.	Salt, soy sauce, processed foods.
Chloride	Part of the hydrochloric acid found in the stomach and necessary for proper digestion.	Growth failure in children; muscle cramps, mental apathy, loss of appetite; can cause death (uncommon).	Normally harmless (the gas chlorine is a poison but evaporates from water); can cause vomiting.	Salt, soy sauce; moderate quantities in whole, unprocessed foods, large amounts in processed foods.
Potassium	Facilitates reactions, including the making of protein; the maintenance of fluid and electrolyte balance; the support of cell integrity; the transmission of nerve impulses; and the contraction of muscles, including the heart.	Deficiency accompanies dehydration; causes muscular weakness, paralysis, and confusion; can cause death.	Causes muscular weakness; triggers vomiting; if given into a vein, can stop the heart.	All whole foods: meats, milk, fruits, vegetables, grains, legumes.
Calcium	The principal mineral of bones and teeth. Also acts in normal muscle contraction and relaxation, nerve functioning, blood clotting, blood pressure, and immune defenses.	Stunted growth in children; adult bone loss (osteoporosis).	Constipation; increased risk of urinary stone formation and kidney dysfunction; interference with absorption of other minerals.	Milk and milk products, oysters, small fish (with bones), tofu (bean curd), greens, legumes.
Phosphorus	Important in cells' genetic material, in cell membranes as phospholipids, in energy transfer, and in buffering systems.	Phosphorus deficiency unknown.	Excess phosphorus may cause calcium excretion.	All animal tissues.
Magnesium	Another factor involved in bone mineralization, the building of protein, enzyme action, normal muscular contraction, transmission of nerve impulses, and maintenance of teeth.	Weakness; confusion; if extreme, convulsions, bizarre movements (especially of eyes and face), hallucinations, and difficulty in swallowing. In children, growth failure.[a]	Large doses taken in the form of the laxative Epsom salts cause diarrhea.	Nuts, legumes, whole grains, dark green vegetables, seafoods, chocolate, cocoa.
Sulfur	A component of certain amino acids; part of the vitamins biotin and thiamin and the hormone insulin; combines with toxic substances to form harmless compounds; stabilizes protein shape by forming sulfur-sulfur bridges.	None known; protein deficiency would occur first.	Would occur only if sulfur amino acids were eaten in excess; this (in animals) depresses growth.	All protein-containing foods.

[a] A still more severe deficiency causes tetany, an extreme, prolonged contraction of the muscles similar to that caused by low blood calcium.

The special provisions the body makes for iron's handling show that it is a precious mineral to be tightly hoarded. For example, when a red blood cell dies, the liver saves the iron and returns it to the bone marrow, which uses it to build new red blood cells. Thus only tiny amounts of iron are lost, principally in urine, sweat, shed skin, and blood (if bleeding occurs).

Normally, only about 10 to 15 percent of dietary iron is absorbed; but if the body's supply is diminished or if the need increases for any reason (such as pregnancy), absorption increases. The body makes several provisions for absorbing

Table 9–6 Trace Minerals

RDA

Copper
Iodine
Iron
Molybdenum
Selenium
Zinc

Adequate Intake (AI)

Chromium
Fluoride
Manganese

Known Essential for Animals; Human Requirements under Study

Arsenic
Boron
Nickel
Silicon
Vanadium

Known Essential for Some Animals; No Evidence That Intake by Humans Is Ever Limiting; No Recommendation Necessary

Cobalt

NOTE: The evidence for requirements and essentiality is weak for the trace minerals cadmium, lead, lithium, and tin.

The storage proteins are **ferritin** (FERR-i-tin) and **hemosiderin** (heem-oh-SID-er-in).

One common test for anemia measures the **hemoglobin concentration** of blood.

- *Norms for adults:*
  Men: ≥13.5 g/100 ml.
  Women: ≥12 g/100 ml.
- *Norms for children:*
  Ages 2–5: ≥11 g/100 ml.
  Ages 6–12: ≥11.5 g/100 ml.

Note that hemoglobin is measured in grams per 100 ml, but often just the number of grams alone is used in speaking of it: "hemoglobin, 14."

transferrin (trans-FERR-in): the body's iron-carrying protein.

iron deficiency: having depleted iron stores.

iron-deficiency anemia: a blood iron deficiency characterized by small, pale red blood cells; also called **microcytic hypochromic anemia.**

micro = small; *cytic* = cells; *hypo* = too little; *chrom* = color

iron. A special protein in the intestinal cells captures iron and holds it in reserve for release into the body as needed; another protein transfers the iron to a special iron-carrier in the blood. The blood protein (**transferrin**) carries the iron to tissues throughout the body. When more iron is needed, more of these special proteins are produced so that more than the usual amount of iron can be absorbed and carried. If there is a surplus of iron, special storage proteins in the liver, bone marrow, and other organs store it.

Iron Deficiency Worldwide, **iron deficiency** is the most common nutrient deficiency, affecting more than one billion people. In developing countries, one-third of the children and women of childbearing age suffer from **iron-deficiency anemia.**[15] In the United States, iron deficiency is less prevalent, but still affects about 10 percent of toddlers, adolescent girls, and women of childbearing age.[16]

Women are especially prone to iron deficiency during their reproductive years because of blood losses during menstruation. Pregnancy places further iron demands on women. Iron is needed to support the added blood volume, the growth of the fetus, and blood loss during childbirth. Infants (six months or older) and young children receive little iron from their high-milk diets, yet need extra iron to support growth. The rapid growth of adolescence and, for females, the blood losses of menstruation also demand extra iron that a typical teen diet may not provide.

Causes of Iron Deficiency The cause of iron deficiency is usually inadequate intake from ignorance of what foods to choose, from sheer lack of food altogether, or from high consumption of iron-poor foods. In the Western world, high sugar and fat intakes are often responsible for low iron intakes. Blood loss is the primary nonnutritional cause, especially in poor regions of the world where parasitic infections of the GI tract may lead to blood loss.

Tests for Iron Deficiency The most common tests for iron-deficiency anemia measure the number and size of the red blood cells and the cells' hemoglobin content. Although these tests are easy, quick, and relatively inexpensive, they are late indicators of iron deficiency. Earlier stages of an iron deficiency can be detected by measuring the amount of transferrin in the blood, the amount of iron transferrin is carrying, and the amount of iron in storage.

Iron Deficiency and Anemia The distinction between iron deficiency and anemia is important. They often go hand in hand, but people can be anemic without being iron deficient and iron deficient without being anemic. Anemia is a symptom of a wide variety of disorders, some unrelated to nutrition, and some related to nutrients other than iron, such as folate and vitamin B_{12}. (Appendix D lists tests useful in identifying anemia and distinguishing between the major types of nutritional anemias.) In iron-deficiency anemia, new red blood cells are smaller and lighter red than normal (see Figure 9–5). The depleted cells cannot carry enough oxygen from the lungs to the tissues, so energy metabolism in all the cells is hindered. The entire body feels the effect.

Anemia is a clinical sign of severe iron deficiency. Other classic symptoms include fatigue, weakness, headaches, apathy, pallor, and poor tolerance to cold. One way the body accelerates heat production when the environmental temperature falls involves the neurotransmitter norepinephrine and the thyroid hormones, which speed up the metabolic rate. Iron deficiency impairs temperature regulation in both animals and human beings, probably by interfering with the normal production of these compounds.

Less severe iron deficiency produces symptoms, too. Long before the red blood cells are affected and anemia is diagnosed, a developing iron deficiency affects behavior. Even at slightly lowered iron levels, the complete oxidation of pyruvate is impaired, reducing physical work capacity and productivity. Children deprived of iron become irritable, restless, and unable to pay attention. These symptoms are among the first to appear when the body's iron begins to fall and among the first to disappear when iron status is restored.

Iron Deficiency and Pica A curious symptom sometimes seen in iron-deficient individuals is an appetite for ice, clay, paste, or other nonnutritious substances. Some people have been known to eat as many as eight trays of ice in a day, for example. This behavior, known as **pica,** has been observed for years, especially in women and children of low-income groups who are deficient in either iron or zinc. After iron is given, pica clears up dramatically within days, long before the red blood cells respond.

Caution on Self-Diagnosis Low hemoglobin may reflect an inadequate iron intake, and if it does, the physician may prescribe iron supplements. However, any nutrient deficiency or disease or agent that interferes with hemoglobin synthesis, disrupts hemoglobin function, or causes a loss of red blood cells can precipitate anemia.

Feeling fatigued, weak, and apathetic is thus a sign that something is wrong, but it does not indicate that a person should take iron supplements; it means that the person should consult a physician. In fact, taking iron supplements may be the worst possible thing a person can do, because such supplements can mask a serious medical condition, such as hidden bleeding from cancer or an ulcer. Furthermore, a person can waste precious time in not seeking treatment. Remember, don't self-diagnose.

Iron Overload Normally, the body protects itself against absorbing too much iron by setting up a block in the intestinal cells. The system can be overwhelmed, however, resulting in **iron overload.** Once considered rare, iron overload has increased in frequency over the last few decades.

Iron overload, known as **hemochromatosis,** is usually caused by a genetic disorder that enhances iron absorption. Hereditary hemochromatosis is the most common genetic disorder in the United States, affecting about 1.5 million people.[17] Other causes of iron overload include repeated blood transfusions, massive doses of supplementary iron, and other rare metabolic disorders. Long-term overconsumption of iron may cause **hemosiderosis,** a condition characterized by large deposits of the iron-storage protein hemosiderin in the liver and other tissues.

Some of the signs and symptoms of iron overload are similar to those of iron deficiency: apathy, lethargy, and fatigue. Therefore, taking iron supplements before assessing iron status is clearly unwise; hemoglobin tests alone would fail to make the distinction.

Iron overload is characterized by tissue damage, especially in iron-storing organs such as the liver. Infections are likely because bacteria thrive on iron-rich blood. Symptoms are most severe in alcohol abusers because alcohol damages the intestine, further impairing its defenses against absorbing excess iron. Untreated hemochromatosis aggravates the risk of diabetes, liver cancer, heart disease, and arthritis.

Iron overload is more common in men than in women and is twice as prevalent among men as iron deficiency. The fortification of many foods with iron makes it difficult to follow an iron-restricted diet.

Iron Poisoning The rapid ingestion of massive amounts of iron can cause sudden death. The most common cause of accidental poisoning in small children is ingestion of iron supplements or vitamins with iron.[18] The American Academy of Pediatrics has urged the Food and Drug Administration (FDA) to improve the labeling and packaging of iron-containing drugs and supplements. As few as 6 to 12 tablets have caused death in a child. A child suspected of iron poisoning should be rushed to the hospital to have the stomach pumped. Thirty minutes can make a crucial difference.

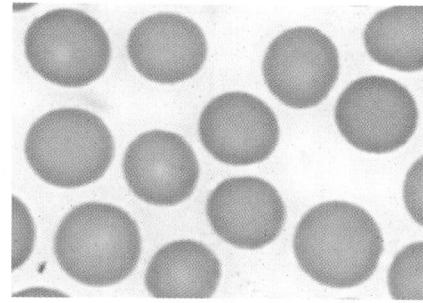

Normal red blood cells. Both size and color are normal.

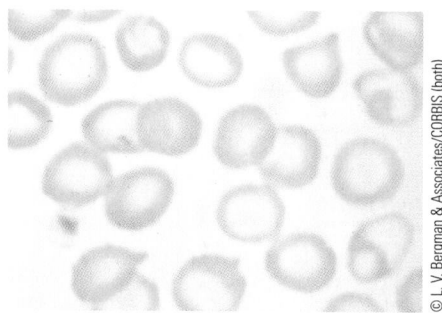

Blood cells in microcytic hypochromic anemia such as that caused by iron deficiency. These cells are small and pale because they contain less hemoglobin.

Figure 9–5
Normal and Anemic Blood Cells

Another common test, the **hematocrit,** represents the percentage of red blood cells in a whole blood sample.
- *Norms for adults:*
  Men: ≥41%.
  Women: ≥36%.
- *Norms for children:*
  Ages 2–5: ≥34%.
  Ages 6–12: ≥35%.

Transferrin can be measured directly or estimated by measuring the **total iron-binding capacity (TIBC)** and the **transferrin saturation.**

The skin of a fair person who is anemic may be noticeably pale, but in all people (including those who are dark skinned), the eye lining, normally pink, will be very pale, even white.

The effects of iron deficiency on children's behavior are revisited in Chapter 13.

Binding proteins in the intestinal cells (*mucosal ferritin* and *mucosal transferrin*) capture and hold unneeded iron to be shed with the cells, thereby forming a **mucosal block** to iron absorption.

Water and the Minerals

Iron RDA:

- Men (19 and older): 8 mg/day.
- Women (19–50 yr): 18 mg/day.
        (>50 yr): 8 mg/day.

About 40% of the iron in meat, fish, and poultry is bound into molecules of **heme** (HEEM), the iron-holding part of the hemoglobin and myoglobin proteins. This **heme iron** is much more absorbable than **nonheme iron.**

pica (PIE-ka): a craving for nonfood substances; also known as *geophagia* (jee-oh-FAY-jee-uh) when referring to clay-eating behavior.

picus = woodpecker or magpie

geo = earth

phagein = to eat

iron overload: toxicity from excess iron.

hemochromatosis (heem-oh-crome-a-TOE-sis): iron overload characterized by deposits of iron-containing pigment in many tissues, with tissue damage. Hemochromatosis is a hereditary defect in iron metabolism.

hemosiderosis (heem-oh-sid-er-OH-sis): a condition characterized by the deposition of the iron-storage protein hemosiderin in the liver and other tissues.

tannins: compounds in tea (especially black tea) and coffee that bind iron.

phytates: nonnutrient components of grains, legumes, and seeds. Phytates can bind minerals such as iron, zinc, calcium, and magnesium in insoluble complexes in the intestine, and the body excretes them unused.

Iron Recommendations The average diet in the United States provides only about 6 to 7 milligrams of iron in every 1000 kcalories. Men need 8 milligrams of iron each day; most men easily eat more than 2000 kcalories, so a man can meet his iron needs without special effort. The recommendation for women during childbearing years, however, is 18 milligrams. Because women have higher iron needs and typically consume fewer than 2000 kcalories per day, they have trouble achieving appropriate iron intakes. On the average, women receive only 10 to 11 milligrams of iron per day. A woman who wants to meet her iron needs from foods must emphasize the most iron-rich foods in every food group.

Iron in Foods Iron occurs in two forms in foods, one of which is up to ten times more absorbable than the other. The most absorbable form is heme iron, which is bound into the iron-carrying proteins hemoglobin and myoglobin in meats, poultry, and fish. Heme iron contributes a small portion of the iron consumed by most people, but it is absorbed at a fairly constant rate of about 23 percent. The less absorbable form is nonheme iron, found in meats and also in plant foods. People absorb nonheme iron at a lower rate (2 to 20 percent); its absorption depends on several dietary factors and iron stores. Most of the iron people consume is nonheme iron from vegetables, grains, eggs, meat, fish, and poultry. Snapshot 9–3 shows the iron found in usual serving sizes of different foods.

Iron absorption from foods can be maximized by two substances that enhance iron absorption: MFP factor and vitamin C. Meat, fish, and poultry contain a factor (MFP factor) other than heme that promotes the absorption of iron. MFP factor even enhances the absorption of nonheme iron from other foods eaten at the same time. Vitamin C eaten in the same meal also doubles or triples nonheme iron absorption. Additionally, cooking with iron skillets can contribute iron to the diet. Some substances impair iron absorption; they include the **tannins** of tea and coffee, the calcium in milk, and the **phytates** that accompany fiber in legumes and whole-grain cereals.[19] The accompanying "How to" box offers suggestions on obtaining adequate iron.

Zinc

Zinc is a versatile trace mineral required as a cofactor by more than 100 enzymes. These zinc-requiring enzymes perform tasks in the eyes, liver, kidneys, muscles, skin, bones, and male reproductive organs. Zinc works with the enzymes that make genetic material; manufacture heme; digest food; metabolize carbohydrate, protein, and fat; liberate vitamin A from storage in the liver; and dispose of damaging free radicals. Zinc also interacts with platelets in blood clotting, affects thy-

How to

Add Iron to Daily Meals

*T*he following set of guidelines can be used for planning an iron-rich diet:

- *Breads and cereals.* Use only whole-grain, enriched, and fortified products (iron is one of the enrichment nutrients).
- *Vegetables.* The dark green, leafy vegetables are good sources of vitamin C and iron. Eat vitamin C–rich vegetables often to enhance absorption of the iron from foods eaten with them.
- *Fruits.* Dried fruits, such as raisins, apricots, peaches, and prunes, are high in iron. Eat vitamin C–rich fruits often with iron-containing foods.

- *Milk and cheese.* Don't overdo foods from the milk group; they are poor sources of iron. But don't omit them either, because they are rich in calcium. Drink fat-free milk to free kcalories to be invested in iron-rich foods.
- *Meats.* Meat, fish, and poultry are excellent iron sources.
- *Meat alternates.* Include legumes frequently. A cup of peas or beans can supply up to 7 milligrams of iron.

---

roid hormone function, assists in immune function, and affects behavior and learning performance. Zinc is needed to produce the active form of vitamin A in visual pigments and is essential to wound healing, taste perception, the making of sperm, and fetal development. When zinc deficiency occurs, it impairs all these and other functions.

The body's handling of zinc differs from that of iron, but with some interesting similarities. For example, like iron, extra zinc that enters the body is held within the intestinal cells, and only the amount needed is released into the bloodstream. As with iron, zinc status influences the percentage of zinc absorbed from the diet; if more is needed, more is absorbed.

Zinc's main transport vehicle in the blood is the protein albumin. Research suggests that circulating albumin is a main determinant of zinc absorption. This may account for observations that zinc absorption declines in conditions that lower plasma albumin concentrations—for example, pregnancy and malnutrition.

Zinc Deficiency Zinc deficiency in human beings was first reported in the 1960s from studies of growing children and male adolescents in Egypt, Iran, and Turkey. Their diets were typically low in zinc and high in fiber and phytates (which impair zinc absorption). The zinc deficiency was marked by dwarfism, or severe growth retardation, and arrested sexual maturation—symptoms that were responsive to zinc supplementation.

Since that time, zinc deficiency has been recognized elsewhere and is known to affect more than growth. It drastically impairs immune function, causes loss of appetite, and, during pregnancy, may lead to developmental disorders. A detailed list of symptoms of zinc deficiency is presented later in Table 9–7. Conditions other than poor diet that contribute to the development of zinc deficiency include loss of blood due to parasitic infections, climates that increase sweat losses, and the practice of clay eating.

Pronounced zinc deficiency is not widespread in developed countries, but deficiencies do occur in the most vulnerable groups of the U.S. population—pregnant women, young children, the elderly, and the poor. Even mild zinc deficiency can result in metabolic changes such as impaired immune response, abnormal taste, and abnormal dark adaptation (zinc is required to produce the active form of vitamin A, retinal, in visual pigments).

This chili dinner provides iron and MFP factor from meat, iron from legumes, and vitamin C from tomatoes. The combination of heme iron, nonheme iron, MFP factor, and vitamin C helps to achieve maximum iron absorption.

Clay eating is a form of pica (see p. 221).

Water and the Minerals

Snapshot 9–4

Zinc

All of these foods provide 10 percent or more of the zinc Daily Value (DV = 15 mg/day).[a]

Enriched cereal (ready-to-eat): 21% DV per ¾ c

Crabmeat: 43% DV per 3 oz steamed

Yogurt: 15% DV per cup

Oysters: 187% DV per 3 oz steamed

Sirloin steak: 37% DV per 3 oz cooked

[a]The Daily Values are based on a 2000-kcalorie diet.

Pregnant teenagers are particularly vulnerable because they need zinc for their own growth, as well as for the developing fetus. Vegetarians, especially pregnant vegetarians, who consume large amounts of fiber, phytate, and dairy foods or low levels of protein need to scrutinize their diets for possible zinc deficiency. Research shows that both zinc intake and zinc absorption are reduced in people eating vegetarian diets that include milk and eggs compared with those whose diets include meat, fish, and poultry.[20] The researchers note, however, that despite a greater risk of zinc deficiency in people consuming vegetarian diets, zinc balance can be maintained with the inclusion of zinc-rich whole-grain breads and cereals and legumes.

Zinc Toxicity Zinc can be toxic if consumed in large enough quantities. A high zinc intake is known to produce copper-deficiency anemia by inducing the intestinal cells to synthesize large amounts of a protein that captures copper in a nonabsorbable form. Accidental consumption of high levels of zinc can cause vomiting, diarrhea, fever, exhaustion, and a host of other symptoms (see Table 9–7, later in the chapter). Large doses can even be fatal. The Tolerable Upper Intake Level for zinc for adults is 40 milligrams per day.

Zinc Recommendations and Food Sources The zinc recommendation for men is 11 milligrams per day; for women, 8 milligrams. Most people in the United States have zinc intakes that approximate recommendations.[21]

Zinc is most abundant in foods high in protein, such as shellfish (especially oysters), meats, and liver. In general, two ordinary servings a day of animal protein provide most of the zinc a healthy person needs. Milk, eggs, and whole-grain products are good sources of zinc if eaten in large quantities. For infants, breast milk is a good source of zinc, which is more efficiently absorbed from human milk than from cow's milk. Commercial infant formulas are fortified with zinc, of course. Snapshot 9–4 shows zinc-rich foods.

Zinc supplements are not recommended except for an accurately diagnosed zinc deficiency or when needed for use as a medication to displace other ions in unusual medical circumstances. Normally, it should be possible to obtain enough zinc from the diet.

Zinc RDA:
- *Men: 11 mg/day.*
- *Women: 8 mg/day.*

Selenium

Selenium is an essential trace mineral that functions as part of a group of antioxidant enzymes called glutathione peroxidases. These enzymes prevent free-radical formation, thus blocking the damaging chain reaction before it begins. Glutathione peroxidases and vitamin E work in concert. If free radicals do form, and a chain reaction starts, vitamin E halts it. Selenium also plays roles in converting thyroid hormone to its active form.

Selenium and Cancer The question of whether selenium protects against the development of some cancers is under investigation. Some research suggests that selenium supplements may reduce the incidence of some types of cancers, but given the potential for harm and the lack of additional evidence, recommendations to take selenium supplements would be premature.[22]

Selenium Deficiency Selenium deficiency is associated with heart disease in children and young women living in regions of China where the soil and foods lack selenium. The heart disease is named *Keshan disease* for one of the provinces of China where it was studied.

Selenium Toxicity High doses of selenium are toxic. Selenium toxicity causes vomiting, diarrhea, loss of hair and nails, and lesions of the skin and nervous system.

Selenium Recommendations and Intakes Anyone who eats a normal diet composed mostly of unprocessed foods need not worry about meeting selenium recommendations. Selenium is widely distributed in foods such as meats and shellfish and in vegetables and grains grown on selenium-rich soil. Some regions in the United States and Canada produce crops on selenium-poor soil, but people are protected from deficiency because they eat selenium-rich meat and supermarket foods transported from other regions.

Iodine

Iodine occurs in the body in minuscule amounts, but its principal role in human nutrition is well known, and the amount needed is well established. Iodine is an integral part of the thyroid hormones, which regulate body temperature, metabolic rate, reproduction, growth, the making of blood cells, nerve and muscle function, and more.

Iodine Deficiency When the iodine concentration in the blood is low, the cells of the thyroid gland enlarge in an attempt to trap as many particles of iodine as possible. If the gland enlarges until it is visible, the swelling is called a simple goiter. As many as 1 billion people border on iodine deficiency, and 200 million people worldwide have **goiter**.[23] In all but 4 percent of these cases, the cause is iodine deficiency. As for the 4 percent (8 million), those people have goiter because they overconsume plants of the cabbage family and others that contain an antithyroid substance whose effect is not counteracted by dietary iodine.

In addition to causing sluggishness and weight gain, an iodine deficiency may have serious effects on fetal development. Severe thyroid undersecretion during pregnancy causes the extreme and irreversible mental and physical retardation known as **cretinism**. A child with cretinism may have an IQ as low as 20 (100 is normal) and a face and body with many abnormalities. Iodine deficiency is one of the world's most common preventable causes of mental retardation and can be averted if the pregnant woman's deficiency is detected and treated in time.[24]

Iodine Toxicity Excessive intakes of iodine can enlarge the thyroid gland, just as deficiencies can. Intakes in the United States are slightly above the recommended intake of 150 micrograms, but still below the Tolerable Upper Intake Level of 1100 micrograms per day for an adult.[25]

Iodine Sources The ocean is the world's major source of iodine. In coastal areas, seafood, water, and even iodine-containing sea mist are important iodine sources. Further inland, the amount of iodine in the diet is variable and generally

Selenium RDA:
- *Adults: 55 µg/day.*

A thyroid antagonist that is found in food and causes *toxic goiter* is called a **goitrogen.**

Iodine RDA: 150 µg/day.

goiter (GOY-ter): an enlargement of the thyroid gland due to an iodine deficiency, malfunction of the gland, or overconsumption of a thyroid antagonist. Goiter caused by iodine deficiency is *simple goiter.*

cretinism (CREE-tin-ism): an iodine-deficiency disease characterized by mental and physical retardation.

Water and the Minerals

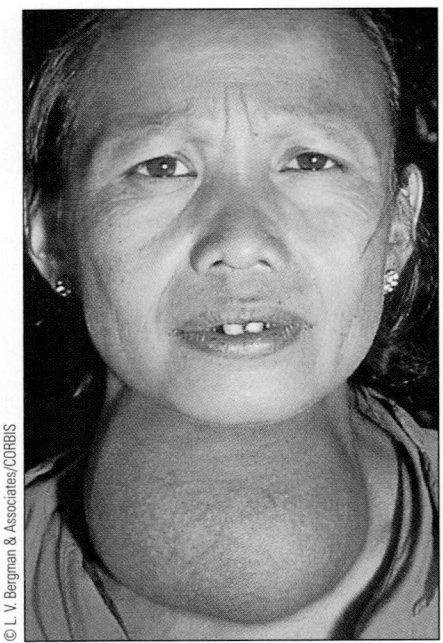

In iodine deficiency, the thyroid gland enlarges—a condition known as simple goiter.

reflects the amount present in the soil in which plants are grown or on which animals graze. In the United States and Canada, the use of iodized salt has largely wiped out the iodine deficiency that once was widespread.

The need for iodine is easily met by consuming seafood, vegetables grown in iodine-rich soil, and iodized salt. In the United States, you have to read the label to find out whether salt is iodized; in Canada, all table salt is iodized.

Iodine Used as a Medication An iodine-containing medication, **potassium iodide,** effectively blocks damage to the thyroid gland that could be caused during radiation emergencies, such as hostile attacks or malfunction of nuclear power plants. When given in the correct dosage within a certain timeframe relative to radiation exposure, potassium iodide can greatly reduce the likelihood of thyroid cancer development.[26] Given in the wrong dosage or with faulty timing, potassium iodide is useless or toxic. For this reason, concerned people who live near power plants are urged to rely on health professionals for guidance.

Copper

The body contains about 100 milligrams of copper. About one-fourth is in the muscles, one-fourth is in the liver, brain, and blood, and the rest is in the bones, kidneys, and other tissues. The primary function of copper in the body is to serve as a constituent of enzymes.[27] The copper-containing enzymes have diverse metabolic roles: they catalyze the formation of hemoglobin, help manufacture the protein collagen, assist in the healing of wounds, and help maintain the sheaths around nerve fibers. One of copper's most vital roles is to help cells use iron. Like iron, copper is needed in many reactions related to respiration and energy metabolism.

Copper Deficiency Copper deficiency is rare but not unknown. It has been seen in premature infants and malnourished infants. High intakes of zinc interfere with copper absorption and can lead to deficiency.[28]

Copper Toxicity Some genetic disorders create a copper toxicity. Copper toxicity from foods, however, is unlikely.

Copper Recommendations and Food Sources The RDA for copper is 900 micrograms per day, which is slightly below the average intake for adults in the United States.[29] The best food sources of copper are legumes, whole grains, seafood, nuts, and seeds.

Manganese

The human body contains a tiny 20 milligrams of manganese, mostly in the bones and glands. Manganese is a cofactor for many enzymes, helping to facilitate dozens of different metabolic processes. Deficiencies of manganese have not been noted in people, but toxicity may be severe. Miners who inhale large quantities of manganese dust on the job over prolonged periods show many symptoms of a brain disease, along with abnormalities in appearance and behavior.

Manganese requirements are low, and plant foods such as nuts, whole grains, and leafy green vegetables contain significant amounts of this trace mineral. Deficiencies are therefore unlikely.

Fluoride

Only a trace of fluoride occurs in the human body, but research demonstrates that where diets are high in fluoride during the growing years, crystalline deposits in bones and teeth are larger and more perfectly formed. When bones

Manganese AI:
- *Men: 2.3 mg/day*
- *Women: 1.6 mg/day*

potassium iodide: a medication approved by the FDA as safe and effective for the prevention of thyroid cancer caused by radioactive iodine known to be released during radiation emergencies.

Chapter Nine

and teeth become mineralized, first a crystal called hydroxyapatite forms from calcium and phosphorus. Then fluoride replaces the hydroxy portion of hydroxyapatite, forming **fluorapatite,** which makes the bones stronger and the teeth more resistant to decay. Once the teeth have erupted, the topical application of fluoride by way of toothpaste or mouth rinse continues to exert a caries-reducing effect.

Fluoride Deficiency Where fluoride is lacking in the water supply, the incidence of dental decay is high. Fluoridation of water to raise its fluoride concentration to 1 part per million is recommended as an important public health measure. Those fortunate enough to have had sufficient fluoride during the tooth-forming years of infancy and childhood are protected throughout life from dental decay. Dental problems are of great concern because they can lead to a multitude of other health problems affecting the whole body. Despite fluoride's value, the introduction of fluoride to a community may encounter opposition. Based on the accumulated evidence of its beneficial effects, water fluoridation has been endorsed by the National Institute of Dental Health, the American Dietetic Association, the American Medical Association, the National Cancer Institute, and the Centers for Disease Control and Prevention. Figure 9–6 shows the extent of fluoridation nationwide.

Fluoride Sources All normal diets include some fluoride, but drinking water, processed soft drinks and fruit juice made with fluoridated water, and fluoride toothpaste are the most common fluoride sources in the United States.[30] Fish and tea may supply substantial amounts as well.

In some areas, the natural fluoride concentration in water is high, and too much fluoride can damage teeth, causing **fluorosis.** In mild cases, the teeth develop small white specks; in severe cases, the enamel becomes pitted and permanently stained. Fluorosis occurs only during tooth development and cannot be reversed, making its prevention a high priority.

Chromium

Chromium is an essential mineral that participates in carbohydrate and lipid metabolism. Chromium enhances the activity of the hormone insulin.[31] Consequently, less insulin is needed to control blood glucose. When chromium is lacking, a diabetes-like condition may develop.

Chromium deficiency is unlikely, given the small amount of chromium required and its presence in a variety of foods. The more refined foods people eat, however, the less chromium they obtain from their diets. Unrefined foods such as liver, brewer's yeast, whole grains, nuts, and cheeses are the best sources.

Other Trace Minerals

An RDA has been established for one other trace mineral, molybdenum. **Molybdenum** functions as a working part of several metal-containing enzymes, some of which are giant proteins. Deficiencies or toxicities of molybdenum are unknown.

Nickel is recognized as important for the health of many body tissues. Nickel deficiencies harm the liver and other organs. Silicon participates in bone calcification, at least in animals. Tin is necessary for growth in animals and probably in people as well. Cobalt is found in the large vitamin B_{12} molecule. The future may reveal that other trace minerals also play key roles. Even arsenic—famous as the death potion in many murder mysteries and known to be a carcinogen—may turn out to be an essential nutrient in tiny quantities.

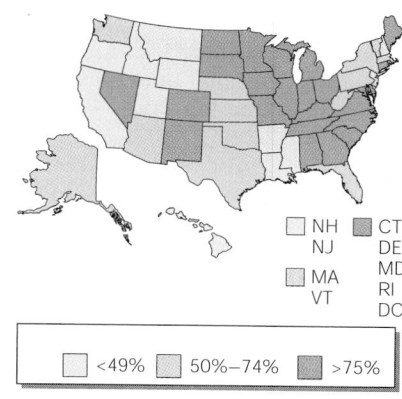

NH
NJ
MA
VT
CT
DE
MD
RI
DC

<49%   50%–74%   >75%

Figure 9–6

Percentage of State Populations with Access to Fluoridated Water through Public Water Systems

SOURCE: Centers for Disease Control and Prevention, Recommendations for using fluoride to prevent and control dental caries in the United States, *Morbidity and Mortality Weekly Report* (supplement), 17 August 2001, p. 10.

Fluoride AI:
- *Men: 4 mg/day.*
- *Women: 3 mg/day.*

To prevent fluorosis:
- *Monitor the fluoride content of the local water supply.*
- *Supervise children younger than six when they brush their teeth and use only a pea-size amount of toothpaste.*
- *Use fluoride supplements only as prescribed by a physician.*

Small organic compounds that enhance insulin's activity are called **glucose tolerance factors (GTF).** Some glucose tolerance factors contain chromium.

Chromium picolinate supplements are discussed in Nutrition in Practice 10.

Chromium AI:
- *Men (19–50 yr): 35 µg/day.
  (51 and older): 30 µg/day.*
- *Women (19–50 yr): 25 µg/day.
  (51 and older): 20 µg/day.*

Molybdenum RDA: *45 µg/day.*

fluorapatite (floor-APP-uh-tite): the stabilized form of bone and tooth crystal, in which fluoride has replaced the hydroxy portion of hydroxyapatite.

fluorosis (floor-OH-sis): mottling of the tooth enamel from ingestion of too much fluoride during tooth development.

Water and the Minerals

Table 9–7 The Trace Minerals—A Summary

Mineral Name	Chief Functions in the Body	Deficiency Symptoms	Toxicity Symptoms	Significant Sources
Iron	Part of the protein hemoglobin, which carries oxygen in the blood; part of the protein myoglobin in muscles, which makes oxygen available for muscle contraction; necessary for the utilization of energy.	Anemia: weakness, pallor, headaches, reduced work productivity, inability to concentrate, impaired cognitive function (chidren), lowered cold tolerance.	Iron overload: infections, liver injury, possible increased risk of heart attack, acidosis, bloody stools, shock.	Red meats, fish, poultry, shellfish, eggs, legumes, dried fruits.
Zinc	Part of the hormone insulin and many enzymes; involved in making genetic material and proteins, immune reactions, transport of vitamin A, taste perception, wound healing, the making of sperm, and normal fetal development.	Growth failure in children, sexual retardation, loss of taste, poor wound healing.	Fever, nausea, vomiting, diarrhea, muscle incoordination, dizziness, anemia, accelerated atherosclerosis, kidney failure.	Protein-containing foods: meats, fish, shellfish, poultry, grains, vegetables.
Selenium	Assists a group of enzymes that break down reactive chemicals that harm cells.	Predisposition to heart disease characterized by cardiac tissue becoming fibrous (uncommon).	Nausea, abdominal pain, nail and hair changes, nerve damage.	Seafoods, organ meats, other meats, whole grains and vegetables depending on soil content.
Iodine	A component of two thyroid hormones, which help to regulate growth, development, and metabolic rate.	Goiter, cretinism.	Depressed thyroid activity; goiterlike thyroid enlargement.	Iodized salt; seafood; bread; plants grown in most parts of the country and animals fed those plants.
Copper	Necessary for the absorption and use of iron in the formation of hemoglobin; part of several enzymes.	Anemia, bone abnormalities (rare in human beings).	Vomiting, diarrhea, liver damage.	Organ meats, seafood, nuts, seeds, whole grains, drinking water.
Manganese	Facilitator, with enzymes, of many cell processes.	(In experimental animals): poor growth, nervous system disorders, reproductive abnormalities.	Nervous system disorders.	Widely distributed in foods.
Fluoride	An element involved in the formation of bones and teeth; helps to make teeth resistant to decay.	Susceptibility to tooth decay.	Fluorosis (discoloration of teeth), nausea, diarrhea, chest pain, itching, vomiting.	Drinking water (if fluoride containing or fluoridated), tea, seafood.
Chromium	Associated with insulin and required for the release of energy from glucose.	Diabetes-like condition marked by an inability to use glucose normally.	None reported.	Meat, unrefined foods.

molybdenum (mo-LIB-duh-num): a trace element.

In Summary The body requires trace minerals in tiny amounts, and they function in similar ways—assisting enzymes all over the body. Eating a diet that consists of a variety of foods is the best way to ensure an adequate intake of these important nutrients. Many dietary factors, including the trace minerals themselves, affect the absorption and availability of these nutrients. Table 9–7 offers a summary of facts about trace minerals in the body.

Self Study

WATER AND MINERALS

The two minerals most likely to fall short in the diet are iron and calcium. Interestingly, both are found in protein-rich foods, but not in the same foods. Meats, fish, and poultry are rich in iron but poor in calcium. Conversely, milk and milk products are rich in calcium but poor in iron. Including meat or meat alternatives for iron and milk and milk products for calcium can help defend against iron deficiency and osteoporosis, respectively. Determine whether these food choices are typical of your diet.

- Do you eat a variety of foods, including some meats, seafood, poultry, or legumes daily?
- Do you drink at least 3 glasses of milk—or get the equivalent in calcium—every day?

Food Choices	Frequency per Week
Calcium-fortified foods (such as corn tortillas, tofu, cereals, or juices)	
Dark green vegetables (such as broccoli)	
Iron-fortified foods (such as breads or cereals)	
Legumes (such as pinto beans)	
Meats, fish, poultry, or eggs	
Milk or milk products	
Nuts (such as almonds) or seeds (such as sesame seeds)	
Small fish (such as sardines) or fish canned with bones (such as canned salmon)	
Whole or enriched grain products	

Self Check

1. Water excretion is governed primarily by the:
   a. liver.
   b. kidneys
   c. gallbladder.
   d. GI tract.

2. People's fluid needs are best met by:
   a. water.
   b. milk.
   c. juice.
   d. sodas.

3. Two situations in which a person may experience fluid and electrolyte imbalances are:
   a. vomiting and burns.
   b. diarrhea and cuts.
   c. broken bones and fever.
   d. heavy sweating and excessive carbohydrate intake.

4. Three-fourths of the sodium in people's diets comes from:
   a. fresh meats.
   b. home-cooked foods.
   c. frozen vegetables and meats.
   d. salt added to food by manufacturers.

5. Which mineral is critical to keeping the heartbeat steady and plays a major role in maintaining fluid and electrolyte balance?
   a. sodium
   b. calcium
   c. potassium
   d. magnesium

6. The two best ways to prevent age-related bone loss and fracture are to:
   a. take calcium supplements and estrogen.
   b. participate in aerobic activity and drink 8 glasses of milk daily.
   c. eat a diet low in fat and salt and refrain from smoking.
   d. maintain a lifelong adequate calcium intake and engage in weight-bearing physical activity.

7. Three good food sources of calcium are:
   a. milk, sardines, and broccoli.
   b. spinach, yogurt, and sardines.
   c. cottage cheese, spinach, and tofu.
   d. Swiss chard, mustard greens, and broccoli.

8. Foods high in iron that help prevent or treat anemia include:
   a. green peas and cheese.
   b. dairy foods and fresh fruits.
   c. homemade breads and most fresh vegetables.
   d. meat and dark green, leafy vegetables.

9. Two groups of people who are especially at risk for zinc deficiency are:
   a. Asians and children.
   b. infants and the elderly.
   c. smokers and athletes.
   d. pregnant adolescents and vegetarians.

10. A deficiency of ____ is one of the world's most common preventable causes of mental retardation.
    a. zinc
    b. iodine
    c. selenium
    d. magnesium

Critical Thinking

1. Which food would provide the most potassium?
   a. bologna
   b. potatoes
   c. pickles
   d. whole-wheat bread

2. Which of these people is *least* likely to develop an iron deficiency?
   a. 3-year-old boy
   b. 52-year-old man
   c. 17-year-old girl
   d. 25-year-old woman

Answers to these questions can be found in Appendix G.

Clinical Applications

1. What advice would you give a client who regularly relies on caffeine-containing beverages to quench thirst? Consider the effects of caffeine on both water and mineral balance.

2. Pull together information from Chapter 1 about the different food groups and the significant sources of minerals shown or discussed in this chapter. Consider which minerals might be lacking (or excessive) in the diet of a client who reports the following:

   ■ Relies on highly processed foods, snack foods, and fast foods as mainstays of the diet.

   ■ Never uses milk, milk products, or cheese.
   ■ Dislikes leafy green vegetables.
   ■ Never eats meat, fish, poultry, or even meat alternates such as legumes.

What additional information would help you pinpoint problems with mineral intake?

Nutrition on the Net

For further study of the topics of this chapter, access these websites.

Find updates and quick links to these and other nutrition-related sites at our website:
www.wadsworth.com/nutrition

Find out more about the importance of water from the International Food Information Council:
ificinfo.health.org/insight/waterref.htm

Find information about mineral supplements:
http://dietary-supplements.info.nih.gov

Search for minerals at the American Dietetic Association site:
www.eatright.org

To learn about the quality of drinking water in your area, visit:
www.epa.gov/ebtpages

Learn about sodium in foods and on food labels from the Food and Drug Administration:
www.fda.gov/fdac/foodlabel/sodium.html

Find tips and recipes for including more milk in the diet:
www.whymilk.com

Learn about the benefits of calcium from the National Dairy Council:
www.nationaldairycouncil.org

Find information about U.S. intakes of minerals and other nutrients:
www.cdc.gov/nchs/fastats/diet.htm

Search for the individual minerals by name at the U.S. Government health information site:
www.healthfinder.org

Learn more about iron overload from the Iron Overload Diseases Association:
www.ironoverload.org

Learn more about iodine deficiency and thyroid disease from the American Thyroid Association:
www.thyroid.org

Notes

1. S. M. Kleiner, Water: An essential but overlooked nutrient, *Journal of the American Dietetic Association* 99 (1999): 200–206.

2. USDA Center for Nutrition Policy and Promotion, Dietary guidance on sodium: Should we take it with a grain of salt? *Nutrition Today* 32 (1997): 250.

3. A. W. Cowley, Jr., Genetic and nongenetic determinants of salt sensitivity and blood pressure, *American Journal of Clinical Nutrition* 65 (1997): 587S–593S.

4. P. K. Whelton and coauthors, Effects of oral potassium on blood pressure: Meta-analysis of randomized controlled clinical trial,

Journal of the American Medical Association 277 (1997): 1624–1632.

5. E. A. Krall and B. Dawson-Hughes, Osteoporosis, in *Modern Nutrition in Health and Disease,* 9th ed., ed. M. E. Shils and coeditors (Baltimore, Md.: Williams & Wilkins, 1999), pp. 1353–1364.

6. NIH Consensus Development Panel, Osteoporosis prevention, diagnosis, and therapy, *Journal of the American Medical Association* 285 (2001): 785–795.

7. K. Delvaux and coauthors, Bone mass and lifetime physical activity in Flemish males: A 27-year follow-up study, *Medicine and Science in Sports and Exercise* 33 (2001): 1868–1875; National Center for Injury Prevention and Control, Falls and hip fractures among the elderly, *Unintentional Injury Fact Sheet,* 1998, available from **www.cdc.gov**; Standing Committee on the Scientific Evaluation of Dietary Reference Intakes, Food and Nutrition Board, Institute of Health, *Dietary Reference Intakes for Calcium, Phosphorus, Magnesium, Vitamin D, and Fluoride* (Washington, D.C.: National Academy Press, 1997), p. 83.

8. J. J. B. Anderson, Calcium requirements during adolescence to maximize bone health, *Journal of the American College of Nutrition* 20 (2001): 186S–191S; G. F. Maddalozzo and C. M. Snow, High intensity resistance training: Effects on bone in older men and women, *Calcified Tissue International* 66 (2000): 399–404; E. Ernst, Exercise for female osteoporosis: A systematic review of randomized clinical trials, *Sports Medicine* 25 (1998): 359–368; I. Vuori, Peak bone mass and physical activity: A short review, *Nutrition Reviews* 54 (1996): S11–S14.

9. T. A. Kotchen and J. M. Kotchen, Nutrition, diet, and hypertension, in *Modern Nutrition in Health and Disease,* 9th ed., ed. M. E. Shils and coeditors (Baltimore, Md.: Williams & Wilkins, 1999), pp. 1217–1227; D. A. McCarron, Role of adequate dietary calcium intake in the prevention and management of salt sensitive hypertension, *American Journal of Clinical Nutrition* 65 (1997): 712S–716S.

10. R. Jorde and K. H. Bonaa, Calcium from dairy products, vitamin D intake, and blood pressure: The Tromso study, *American Journal of Clinical Nutrition* 71 (2000): 1530–1535; C. G. Osborne and coauthors, Evidence for the relationship of calcium to blood pressure, *Nutrition Reviews* 54 (1996): 365–381; D. A. McCarron and D. Hatton, Dietary calcium and lower blood pressure: We can all benefit, *Journal of the American Medical Association* 275 (1996): 1128–1129.

11. J. H. Dwyer and coauthors, Dietary calcium, calcium supplementation, and blood pressure in African American adolescents, *American Journal of Clinical Nutrition* 68 (1998): 648–655; H. C. Bucher and coauthors, Effects of dietary calcium supplementation on blood pressure: A meta-analysis of randomized controlled trials, *Journal of the American Medical Association* 275 (1996): 1016–1022.

12. A. Azoulay, P. Garzon, and M. J. Eisenberg, Comparison of the mineral content of tap water and bottled waters, *Journal of General Internal Medicine* 16 (2001): 168–175.

13. J. Guillemant and coauthors, Mineral water as a source of dietary calcium: Acute effects on parathyroid function and bone resorption in young men, *American Journal of Clinical Nutrition* 71 (2000): 999–1002.

14. M. E. Shils, Magnesium, in *Modern Nutrition in Health and Disease,* 9th ed., ed. M. E. Shils and coeditors (Baltimore, Md.: Williams & Wilkins, 1999), pp. 169–189.

15. C. E. West, Strategies to control nutritional anemia, *American Journal of Clinical Nutrition* 64 (1996): 789–790.

16. A. C. Looker and coauthors, Prevalence of iron deficiency in the United States, *Journal of the American Medical Association* 277 (1997): 973–976.

17. R. E. Fleming and W. S. Sly, Mechanisms of iron accumulation in hereditary hemochromatosis, *Annual Review of Physiology* 64 (2002): 633–680; S. Tavill, Clinical implications of the hemochromatosis gene, *New England Journal of Medicine* 341 (1999): 755–756; Iron overload disorders among Hispanics—San Diego, California, 1995, *Morbidity and Mortality Weekly Report* 45 (1996): 991–993; D. H. G. Crawford and coauthors, Factors influencing disease expression in hemochromatosis, *Annual Review of Nutrition* 16 (1996): 139–160.

18. B. C. Morris, Pediatric iron poisonings in the United States, *Southern Medical Journal* 93 (2000): 352–358; J. P. Pestaner and coauthors, Ferrous sulfate toxicity: A review of autopsy findings, *Biological Trace Element Research* 69 (1999): 191–198.

19. Standing Committee on the Scientific Evaluation of Dietary Reference Intakes, Food and Nutrition Board, Institute of Health, *Dietary Reference Intakes for Vitamin A, Vitamin K, Arsenic, Boron, Chromium, Copper, Iodine, Iron, Manganese, Molybdenum, Nickel, Silicon, Vanadium, and Zinc* (Washington, D.C.: National Academy Press, 2001), pp. 9-16–9-17.

20. J. R. Hunt, L. A. Matthys, and L. K. Johnson, Zinc absorption, mineral balance, and blood lipids in women consuming controlled lactovegetarian and omnivorous diets for 8 wk, *American Journal of Clinical Nutrition* 67 (1998): 421–430.

21. Standing Committee on the Scientific Evaluation of Dietary Reference Intakes, 2001, p. 12-1.

22. Letters from V. Herbert, L. H. Kuller, J. S. Parker, and L. C. Clark, Selenium supplementation and cancer rates, *Journal of the American Medical Association* 277 (1997): 880–881; L. C. Clark and coauthors, Effects of selenium supplementation for cancer prevention in patients with carcinoma of the skin—A randomized controlled trial, *Journal of the American Medical Association* 276 (1996): 1984–1985.

23. A. Elnour and coauthors, Endemic goiter with iodine sufficiency: A possible role for the consumption of pearl millet in the etiology of endemic goiter, *American Journal of Clinical Nutrition* 71 (2000): 59–66.

24. N. Bleichrodt and coauthors, The benefits of adequate iodine intake, *Nutrition Reviews* 54 (1996): S72–S78.

25. Standing Committee on the Scientific Evaluation of Dietary Reference Intakes, 2001, p. 8-1.

26. Food and Drug Administration, Center for Drug Evaluation and Research, Guidance: Potassium iodide as a thyroid blocking agent in radiation emergencies, November 2001, available at **www.fda.gov/cder/guidance/index.htm**.

27. R. Uauy, M. Olivares, and M. Gonzalez, Essentiality of copper in humans, *American Journal of Clinical Nutrition* 67 (1998): 952S–959S.

28. B. Lonnerdal, Copper nutrition during infancy and childhood, *American Journal of Clinical Nutrition* 67 (1998): 1046S–1053S.

29. Standing Committee on the Scientific Evaluation of Dietary Reference Intakes, 2001, p. 7-1.

30. Centers for Disease Control and Prevention, Recommendations for using fluoride to prevent and control dental caries in the United States, *Morbidity and Mortality Weekly Report,* August 17, 2001, pp. 8–9.

31. S. Fairweather-Tait and R. F. Hurrell, Bioavailability of minerals and trace elements, *Nutrition Research Reviews* 9 (1996): 295–324.

Nutrition in Practice

■ VITAMIN AND MINERAL SUPPLEMENTS ■

About 50 percent of U.S. adults collectively spend close to $8 billion a year on vitamin and mineral supplements.[1] This trend is accelerating as scientists discover more and more links between nutrition and disease prevention. As discussed in Chapter 8, women of childbearing age need folate supplements to reduce the risk of neural tube defects. Many recent reports also indicate that the antioxidant nutrients beta-carotene, vitamin C, and vitamin E may be potent protectors against both cancer and heart disease. Before you race out to buy bottles of antioxidant supplements, however, it is important to know that the role of antioxidants in preventing disease remains to be confirmed by hundreds of scientific studies that are currently under way. Some findings so far are promising, but nevertheless tentative. The main message of this Nutrition in Practice is that most healthy people can get the nutrients they need from foods. Supplements cannot substitute for a healthy diet. For some people, however, certain nutrient supplements may be desirable.

Do foods really contain enough vitamins and minerals to supply all that most people need?

Emphatically, yes, for both healthy adults and children who choose a variety of foods. The Daily Food Guide and the Food Guide Pyramid described in Chapter 1 are the guides to follow to achieve adequate intakes. People who meet their nutrient needs from foods, rather than supplements, have little risk of deficiency or toxicity.

Do some people need supplements?

Yes, some people may suffer marginal nutrient deficiencies due to illness, alcohol or drug addiction, or other conditions that limit food intake. People who may benefit from nutrient supplements in amounts consistent with the RDA include:

- People with nutrient deficiencies.
- People with low food energy intakes (less than 1200 kcalories per day), such as habitual dieters, need a multivitamin and mineral supplement.
- People with illnesses that take away the appetite.
- People with illnesses that impair nutrient absorption, such as diseases of the gallbladder, pancreas, and digestive system.

- People taking medications that interfere with nutrient metabolism.
- People who are lactose intolerant, have milk allergies, or otherwise do not consume enough dairy products to forestall extensive bone loss need calcium.
- People with limited milk intake and sun exposure need vitamin D.
- Elderly people who may have difficulty chewing or swallowing and so do not eat enough food to meet nutrient needs.
- Women who bleed excessively during menstruation need iron supplements.
- People in certain stages of the life cycle who have increased nutrient needs (for example, infants need iron and fluoride, and some may need vitamin D; women of childbearing age need folate; pregnant women need iron; and the elderly need vitamin D).
- People who eat all-plant diets (vegans) and those with atrophic gastritis need vitamin B_{12}.
- Newborn infants need a single dose of vitamin K at birth under the direction of a physician.
- People who have infections or injuries or who have undergone major surgery. (The increased metabolic needs associated with these severe stresses are discussed in Chapter 27.)

Most adults can get all the nutrients they need by eating a varied diet of nutrient-dense foods. Nutrients are potentially toxic when taken in large doses, and individual tolerances vary depending on health and age. Whenever a health care professional finds a person's diet inadequate, the right corrective step is to improve the person's food choices and eating patterns, not to begin supplementation.

Why do so many people take supplements?

People frequently take supplements for mistaken reasons, such as "They give me energy" or "They make me strong." Other invalid reasons why people may take supplements include:

- Their feeling of insecurity about the nutrient content of the food supply.

Nutrition in Practice

© Tom Carter/PhotoEdit

experts agree that the antioxidant vitamins in these foods are probably the most important protective factors, but they also note that other constituents of fruits and vegetables (see Table NP8–2, p. 203) have not been ruled out as contributing factors.

The way to apply this information is to eat nutritious foods. Before supplementation is recommended as a strategy to prevent cancer or other diseases, researchers must determine the optimal doses to reduce risk and the potential adverse effects of long-term supplementation.

When a person needs a vitamin-mineral supplement, what kind should be used?

Take your health care professional's advice, if it is offered. If you are selecting a supplement yourself, a single, balanced vitamin-mineral supplement should suffice. Choose the kind that provides all the nutrients in amounts equal to, or very close to, the RDA (remember, you get some nutrients from foods). Avoid individual nutrients, unless prescribed by a health care professional. Avoid preparations that contain items not needed in human nutrition such as inositol.

Can supplement labels help consumers make informed choices?

Yes. To enable consumers to make more informed choices about nutrient supplements, the Food and Drug Administration (FDA), with the encouragement of the American Dietetic Association (ADA), published labeling regulations for supplements.[4] The Dietary Supplement Health and Education Act subjects supplements to the same general labeling requirements that apply to foods. Specifically:

- Nutrition labeling for dietary supplements is required. The nutrition panel on supplements is called "Supplement Facts" (see Figure NP9–1 on p. 234). The Supplement Facts panel lists the quantity and the percentage of the Daily Value for each nutrient in the supplement. Ingredients that have no Daily Value—for example, sugars and gelatin— appear in a list below the Supplement Facts panel.

- Labels may make nutrient claims (as "high" or "low") according to specific criteria (for example, "an excellent source of vitamin C").

- The FDA authorizes health claims on supplement labels about the relationship between folate and the risk of neural tube defects, calcium and osteoporosis, soluble fiber from whole oats and psyllium husks and heart disease, and sugar alcohols and dental caries.

- Their belief that extra vitamins and minerals will help them cope with stress.

- Their belief that supplements can enhance athletic performance or build lean body tissue without physical work.

- Their desire to prevent, treat, or cure symptoms or diseases ranging from the common cold to cancer.

Ironically, supplement users eat more nutrient-dense diets than nonusers and therefore need supplements less. In addition, little relationship exists between the nutrients people need and the ones they take in supplements. In fact, an argument against supplements is that they may lull people into a false sense of security. A person might eat irresponsibly, thinking, "My supplement will cover my needs."

Do antioxidant supplements prevent cancer and heart disease?

Again, it is better advice to eat a very nutritious diet. Evidence from population studies shows a correlation between low intakes of antioxidant nutrients and a high incidence of disease. Many studies show that low intakes of vegetables and fruits are consistently linked with an increased incidence of cancer.[2]

More than 200 population studies have examined the effects of fruits and vegetables on cancer risk, and the vast majority show that people who eat more of these foods are less likely to develop cancer.[3] Many

Nutrition in Practice

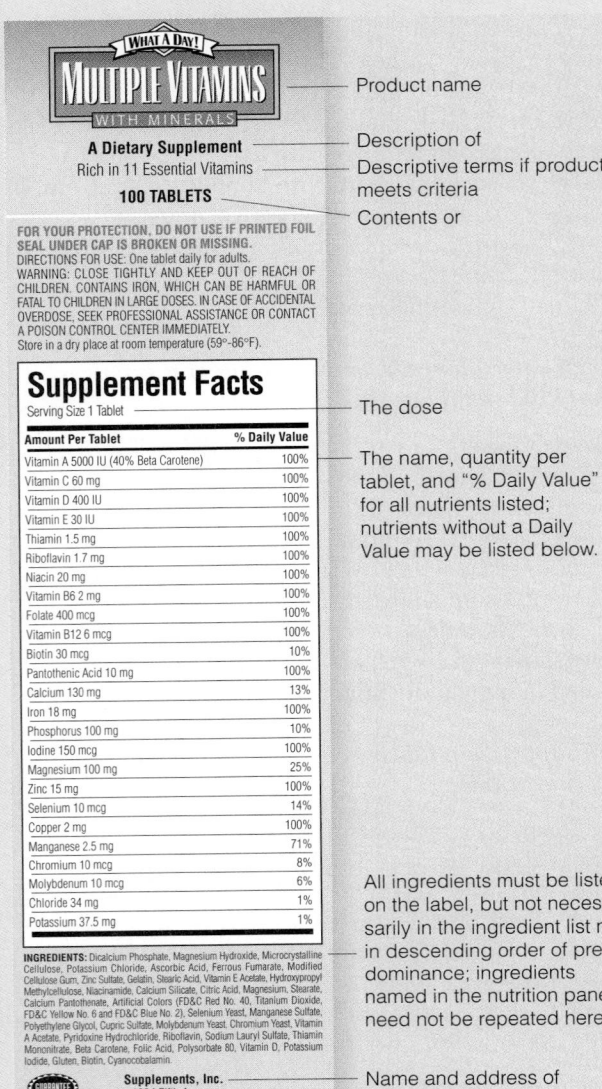

WHAT A DAY!

MULTIPLE VITAMINS
WITH MINERALS
——— Product name

A Dietary Supplement ——— Description of
Rich in 11 Essential Vitamins ——— Descriptive terms if product
meets criteria
100 TABLETS ——— Contents or

FOR YOUR PROTECTION, DO NOT USE IF PRINTED FOIL
SEAL UNDER CAP IS BROKEN OR MISSING.
DIRECTIONS FOR USE: One tablet daily for adults.
WARNING: CLOSE TIGHTLY AND KEEP OUT OF REACH OF
CHILDREN. CONTAINS IRON, WHICH CAN BE HARMFUL OR
FATAL TO CHILDREN IN LARGE DOSES. IN CASE OF ACCIDENTAL
OVERDOSE, SEEK PROFESSIONAL ASSISTANCE OR CONTACT
A POISON CONTROL CENTER IMMEDIATELY.
Store in a dry place at room temperature (59°–86°F).

Supplement Facts
Serving Size 1 Tablet ——— The dose

Amount Per Tablet	% Daily Value
Vitamin A 5000 IU (40% Beta Carotene)	100%
Vitamin C 60 mg	100%
Vitamin D 400 IU	100%
Vitamin E 30 IU	100%
Thiamin 1.5 mg	100%
Riboflavin 1.7 mg	100%
Niacin 20 mg	100%
Vitamin B6 2 mg	100%
Folate 400 mcg	100%
Vitamin B12 6 mcg	100%
Biotin 30 mcg	10%
Pantothenic Acid 10 mg	100%
Calcium 130 mg	13%
Iron 18 mg	100%
Phosphorus 100 mg	10%
Iodine 150 mcg	100%
Magnesium 100 mg	25%
Zinc 15 mg	100%
Selenium 10 mcg	14%
Copper 2 mg	100%
Manganese 2.5 mg	71%
Chromium 10 mcg	8%
Molybdenum 10 mcg	6%
Chloride 34 mg	1%
Potassium 37.5 mg	1%

The name, quantity per
tablet, and "% Daily Value"
for all nutrients listed;
nutrients without a Daily
Value may be listed below.

INGREDIENTS: Dicalcium Phosphate, Magnesium Hydroxide, Microcrystalline
Cellulose, Potassium Chloride, Ascorbic Acid, Ferrous Fumarate, Modified
Cellulose Gum, Zinc Sulfate, Gelatin, Stearic Acid, Vitamin E Acetate, Hydroxypropyl
Methylcellulose, Niacinamide, Calcium Silicate, Citric Acid, Magnesium, Stearate,
Calcium Pantothenate, Artificial Colors (FD&C Red No. 40, Titanium Dioxide,
FD&C Yellow No. 6 and FD&C Blue No. 2), Selenium Yeast, Manganese Sulfate,
Polyethylene Glycol, Cupric Sulfate, Molybdenum Yeast, Chromium Yeast, Vitamin
A Acetate, Pyridoxine Hydrochloride, Riboflavin, Sodium Lauryl Sulfate, Thiamin
Mononitrate, Beta Carotene, Folic Acid, Polysorbate 80, Vitamin D, Potassium
Iodide, Gluten, Biotin, Cyanocobalamin.

All ingredients must be listed
on the label, but not necessarily in the ingredient list nor
in descending order of predominance; ingredients
named in the nutrition panel
need not be repeated here.

GUARANTEE
Complete Satisfaction
or Your Money Back

Supplements, Inc. ———
1234 Fifth Avenue ——— Name and address of
Anywhere, USA manufacturer

Figure NP9–1
An Example of a Supplement Label

- Supplement labels are not allowed to include health claims on a number of other nutrient-disease relationships that have been approved for foods.

- Products may not bear claims to diagnose, treat, cure, or relieve a specific disease.

- Labels may make structure-function claims (see Chapter 1) about the role a nutrient plays in the body, explain how the nutrient performs its function, and indicate that consuming the nutrient is associated with general well-being.

- Labels may claim a substance benefits common complaints such as memory loss or menstrual cramps without proof of effectiveness.

In effect, the Dietary Supplement Health and Education Act resulted in the deregulation of the supplement industry. Unlike food additives or drugs, supplements do not need the FDA's approval before being marketed. Manufacturers alone decide whether their products are safe and effective. Should a problem arise, the burden falls to the FDA to prove that the supplement poses an unreasonable risk and should be removed from the market.

Notes

1. Position of The American Dietetic Association: Food fortification and dietary supplements, *Journal of the American Dietetic Association* 101 (2001): 115–125.

2. American Institute for Cancer Research, *Food, Nutrition and the Prevention of Cancer: A Global Perspective* (Washington, D.C.: American Institute for Cancer Research, 1997); K. A. Steinmetz and J. D. Potter, Vegetables, fruit, and cancer prevention: A review, *Journal of the American Dietetic Association* 96 (1996): 1027–1039.

3. T. P. Giovannucci and coauthors, Fruit, vegetables, dietary fiber, and risk of colorectal cancer, *Journal of the National Cancer Institute* 93 (2001): 525–533; American Institute for Cancer Research, 1997.

4. Commission on Dietary Supplement Labels issues final report, *Journal of the American Dietetic Association* 98 (1998): 270.

Jennie Oppenheimer/Studio Zocolo

Physical activity, or its lack, exerts a significant and pervasive influence on everyone's nutrition and overall health.

*P*erhaps you are already physically fit. If so, the following description applies to you. You are graceful and move with ease. You are strong and meet physical challenges without strain. You have endurance, and your energy lasts for hours. You meet daily physical challenges and have plenty of energy in reserve to handle emergencies. What's more, you are prepared to meet mental and emotional challenges, too—for physical fitness supports mental and emotional health and resilience as well.

Or perhaps you are not yet physically fit. Regardless, this chapter is written for "you," whoever you are and whatever your goals—whether you want to hone your athletic skills, improve your health, prepare for a career in health care, ensure your position on a sports team, lose weight, or become physically active.

This chapter begins by defining fitness and presenting its benefits. It goes on to explain how the body uses energy nutrients to fuel physical activity and concludes by describing how nutrition supports fitness.

Fitness

Fitness depends on a certain minimum amount of physical activity. Table 10–1 presents the American College of Sports Medicine (ACSM) guidelines on the quantity and quality of physical activity recommended for developing and maintaining fitness in healthy adults.[1] The main objective of these guidelines is to outline the types and amounts of physical activity needed to improve *physical fitness*. These familiar guidelines help adults develop programs to improve their cardiorespiratory endurance, body composition, strength, and flexibility.

The types and amounts of physical activity needed to promote *fitness*, however, may differ from those needed to obtain the *health* benefits of reduced disease risks. For health's sake, the ACSM and the *Dietary Guidelines for Americans* specify that people need to spend an accumulated minimum of 30 minutes in some sort of physical activity on most days of each week.[2] Eight minutes spent climbing up stairs, another 10 spent pulling weeds, and 12 more spent walking the dog all contribute to the day's total (see Figure 10–1). The guidelines for developing fitness are still optimal, though, because improving fitness provides additional health benefits (further reduction of cardiovascular disease risk and improved body composition, for example).[3]

Physical activity leads to fitness, and fitness, in turn, makes activity easy, a beneficial cycle. Activity and fitness are so closely connected that this chapter makes no distinction between them.

Definitions of Fitness

Narrowly defined, **fitness** refers to *the characteristics that enable the body to perform physical activity*. These characteristics include flexibility of the joints; strength and endurance of the muscles, including the heart muscle; and a healthy body composition. A broader definition of fitness is *the ability to meet routine physical demands with enough reserve energy to rise to a sudden challenge*. This definition shows how fitness relates to everyday life. Ordinary tasks such as carrying heavy suitcases, opening a stuck window, or climbing four flights of stairs, which might strain an unfit person, are easy for a fit person. Still another definition is *the body's ability to withstand stress*, meaning both physical and psychological stresses. These definitions do not contradict each other; all three describe the same wonderful condition of a healthy body.

The opposite of a physically active life is a **sedentary** life, which literally means "sitting down a lot." Today's world fosters inactivity by providing people with escalators, cars, and other labor-saving devices. As people go through life

fitness: the characteristics that enable the body to perform physical activity; more broadly, the ability to meet routine physical demands with enough reserve energy to rise to a physical challenge; or the body's ability to withstand stress of all kinds.

sedentary: physically inactive (literally, "sitting down a lot").

Table 10–1 Guidelines for Physical Fitness

	Cardiorespiratory	Strength	Flexibility
Type of Activity	Aerobic activity that uses large-muscle groups and can be maintained continuously	Resistance activity that is performed at a controlled speed and through a full range of motion	Stretching activity that uses the major muscle groups
Frequency	3 to 5 days per week	2 to 3 days per week	2 to 3 days per week
Intensity	55 to 90% of maximum heart rate	Enough to enhance muscle strength and improve body composition	Enough to develop and maintain a full range of motion
Duration	20 to 60 minutes	8 to 12 repetitions of 8 to 10 different exercises (minimum)	4 repetitions of 10 to 30 seconds per muscle group (minimum)

SOURCE: Adapted from American College of Sports Medicine, Position stand: The recommended quantity and quality of exercise for developing and maintaining cardiorespiratory and muscular fitness, and flexibility in healthy adults, *Medicine and Science in Sports and Exercise* 30 (1998): 975–991.

exerting minimal physical effort, they become weak and unfit and begin to feel unwell. In fact, a sedentary lifestyle fosters the development of several chronic diseases.

Benefits of Fitness

Extensive evidence confirms that regular physical activity promotes health and prevents disease.[4] Still, despite an increasing awareness of the health benefits that physical activity confers, more than 60 percent of adults in the United States are either irregularly active or completely inactive.[5] Physical inactivity is linked to

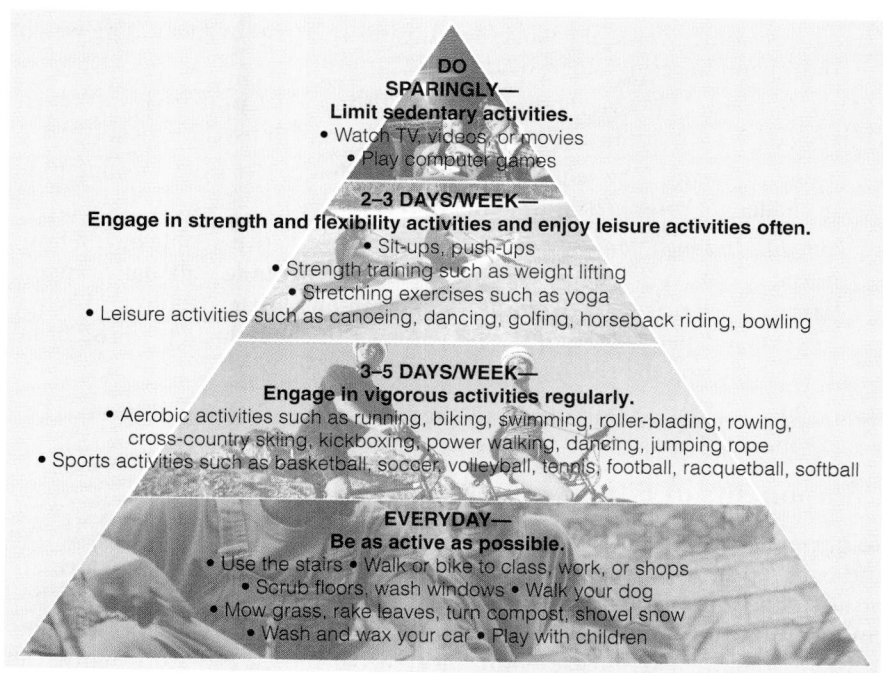

Figure 10–1
Physical Activity Pyramid

DO SPARINGLY—
Limit sedentary activities.
• Watch TV, videos, or movies
• Play computer games

2–3 DAYS/WEEK—
Engage in strength and flexibility activities and enjoy leisure activities often.
• Sit-ups, push-ups
• Strength training such as weight lifting
• Stretching exercises such as yoga
• Leisure activities such as canoeing, dancing, golfing, horseback riding, bowling

3–5 DAYS/WEEK—
Engage in vigorous activities regularly.
• Aerobic activities such as running, biking, swimming, roller-blading, rowing, cross-country skiing, kickboxing, power walking, dancing, jumping rope
• Sports activities such as basketball, soccer, volleyball, tennis, football, racquetball, softball

EVERYDAY—
Be as active as possible.
• Use the stairs • Walk or bike to class, work, or shops
• Scrub floors, wash windows • Walk your dog
• Mow grass, rake leaves, turn compost, shovel snow
• Wash and wax your car • Play with children

The following comparisons reflect similar differences in the risks associated with chronic disease and death:

- *Vigorous exercise vs. minimal exercise.*
- *Ideal weight vs. 20% overweight.*
- *Nonsmoking vs. smoking (one pack a day).*

the major degenerative diseases—heart disease, cancer, stroke, diabetes, and hypertension—that are the primary killers of adults in developed countries.[6]

People don't have to run marathons to reap the health rewards of physical activity.[7] In fact, people who regularly engage in just moderate physical activity live longer on average than those who are physically inactive.[8] The ACSM advises that public health efforts focus on "getting more people more active more of the time" rather than dictating a specific activity level for people to attain.

Activities that promote fitness are themselves enjoyable, and they quickly lead to rewards in terms of physical improvements. In general, physically fit people enjoy:

- *Restful sleep.* Rest and sleep occur naturally after periods of physical activity. During rest, the body repairs injuries, disposes of wastes generated during activity, and builds new physical structures.

- *Nutritional health.* Physical activity spends energy and thus allows people to eat more food. If they choose wisely, active people will consume more nutrients and be less likely to develop nutrient deficiencies.

- *Optimal body composition.* A balanced program of physical activity limits body fat and maintains lean tissue. Physically active people have relatively less body fat than sedentary people at the same body weight.[9]

- *Optimal bone density.* Weight-bearing physical activity builds bone strength and protects against osteoporosis.[10]

- *Resistance to colds and other infectious diseases.* Fitness enhances immunity.[11]

- *Low risks of some types of cancers.* Lifelong physical activity may help to protect against colon cancer, breast cancer, and others.[12]

- *Strong circulation and lung function.* Physical activity that challenges the heart and lungs slows the aging of the circulatory system.

- *Low risk of cardiovascular disease.* Physical activity lowers blood pressure, slows resting pulse rate, and lowers blood cholesterol, thus reducing the risks of heart attack and strokes.[13] Some research suggests that physical activity may reduce the risk of cardiovascular disease in another way as well—by reducing intra-abdominal fat stores.[14]

- *Low risk of type 2 diabetes.* Physical activity normalizes glucose tolerance, especially via the secretion of insulin.[15]

- *Reduced risk of gallbladder disease in women.* Regular physical activity reduces women's risk of gallbladder disease—perhaps by facilitating weight control and lowering blood lipid levels.[16]

- *Low incidence and severity of anxiety and depression.* Compared with sedentary people, physically active people deal better with psychological stress.

- *Strong self-image.* The sense of achievement that comes from meeting physical challenges promotes self-confidence.

- *Long life and high quality of life in the later years.* Active people have a lower mortality rate than sedentary people.[17] Even a two-mile walk daily can add years to a person's life.[18] In addition to extending longevity, physical activity supports independence and mobility in later life by reducing the risk of falls and minimizing the risk of injury should a fall occur.[19]

As a person becomes physically fit, the health of the entire body improves.

Components of Fitness

To be physically fit, a person needs to develop enough flexibility, muscle strength and endurance, and cardiorespiratory endurance to meet the everyday demands of life with some to spare and to achieve a reasonable body weight and **body composition. Flexibility** allows the joints to move with less chance of injury. **Muscle strength** and **muscle endurance** enable muscles to work harder and

body composition: the proportions of muscle, bone, fat, and other tissue that make up a person's total body weight.

flexibility: the capacity of the joints to move through a full range of motion; the ability to bend and recover without injury.

muscle strength: the ability of muscles to work against resistance.

muscle endurance: the ability of a muscle to contract repeatedly within a given time without becoming exhausted.

longer without fatigue. **Cardiorespiratory endurance** supports the ongoing action of the heart and lungs. Physical activity supports desirable lean body tissue and reduces excess body fat. A person who practices a physical activity *adapts* by becoming better able to perform it after each session—with more flexibility, more strength, and more endurance.

Conditioning by Training

The principles of **conditioning** apply to each component of fitness—flexibility, strength, and endurance. During conditioning, the body adapts microscopically to perform the work asked of it. The way to achieve conditioning is by **training,** primarily by applying the **progressive overload principle**—that is, by asking a little more of the body in each training session.

Physical activity helps you look good, feel good, and have fun, and it brings many long-term health benefits as well.

The Overload Principle You can apply the progressive overload principle in several different ways. You can perform the activity more often—that is, increase its **frequency.** You can perform it more strenuously—that is, increase its **intensity.** Or you can do it for longer times—that is, increase its **duration.** All three strategies, individually or in combination, work well. The rate of progression depends on individual characteristics such as fitness level, health status, age, and preference. If you enjoy your workout, do it more often. If you do not have much time, increase intensity. If you hate hard work, take it easy and go longer. If you want continuous improvements, remember to overload progressively as you reach higher levels of fitness.

Applying Overload When increasing the frequency, intensity, or duration of a workout, exercise to a point that only *slightly* exceeds the comfortable capacity to work. It is better to progress too slowly than to risk serious injury by overexertion. Other tips include:

- Be active all week, not just on the weekends.
- Use proper equipment and attire.
- Perform approved exercises using proper form.
- Train hard enough to challenge your strength or endurance a few times each week, not every time you work out. Between challenges, do moderate workouts and include at least one day of rest each week.
- Pay attention to body signals. Symptoms such as abnormal heartbeats, dizziness, lightheadedness, cold sweat, confusion, or pain or pressure in the middle of the chest, teeth, jaw, neck, or arm demand immediate medical attention.

Work out wisely. Do not start with activities so demanding that pain stops you within two days. Learn to enjoy small steps toward improvement. Fitness builds slowly.

Cautions on Starting Before beginning a fitness program, make sure it is safe for you to do so. The ACSM classifies individuals into three groups based on major coronary risk factors: "low risk" individuals have no more than one of the risk factors listed in the margin; "individuals at moderate risk" have two or more of the risk factors and/or symptoms suggestive of disease; and "individuals at high risk" are known to have cardiac, pulmonary, or metabolic disease.[20] Most apparently healthy people can begin **moderate exercise** programs such as walking or increasing daily activities without a medical examination, but people in either of the other two classifications need medical advice.

Warm-Up and Cool-Down Include **warm-up** and **cool-down** activities in each session. Warming up helps to prepare muscles, ligaments, and tendons for the upcoming activity and mobilizes fuels to support strength and endurance

Major coronary risk factors:

- *Age (men >45 yr; women >55 yr).*
- *Cigarette smoking.*
- *Diabetes.*
- *Family history of heart disease.*
- *Hypertension.*
- *Obesity (BMI ⩾ 30).*
- *Sedentary lifestyle.*
- *Serum cholesterol >200 mg/dL.*

cardiorespiratory endurance: the ability to perform large-muscle dynamic exercise of moderate-to-high intensity for prolonged periods.

conditioning: the physical effect of training; improved flexibility, strength, and endurance.

training: practicing an activity regularly, which leads to conditioning. (Training is what you do; conditioning is what you get.)

progressive overload principle: the training principle that a body system, in order to improve, must be worked at frequencies, durations, or intensities that gradually increase physical demands.

frequency: the number of occurrences per unit of time (for example, the number of activity sessions per week).

intensity: the degree of exertion while exercising (for example, the amount of weight lifted or the speed of running).

duration: length of time (for example, the time spent in each activity session).

Fitness and Nutrition

People's bodies are shaped by the activities they perform.

activities. Cool-down activity eases the transition from exercising to normal functioning. A few minutes of light activity facilitate the relaxation of tight muscles and enhance the circulation of blood through them. The circulation, in turn, brings accumulated heat from the body's core to the surface, where it can radiate away. As you approach the end of your workout, gradually ease up on the intensity of the activity (for example, if you are running, begin to slow to a light jog), reaching a minimum intensity over five to ten minutes. Stretching the muscles to promote flexibility is particularly well suited to the end of the cool-down.

The Body's Response to Physical Activity Fitness develops in response to demand and wanes when demand ceases. Muscles gain size and strength after being made to work repeatedly, a response called **hypertrophy.** Conversely, without activity, muscles diminish in size, a response called **atrophy.** As muscles atrophy, they lose strength and endurance.

Hypertrophy and atrophy are adaptive responses to the muscles' greater and lesser work demands, respectively. Thus cyclists often have strong, well-developed legs but less arm or chest strength; a tennis player may have one superbly strong arm, while the other is just average. For balanced muscular development, people should work different muscle groups from day to day. This strategy provides a day or two of rest for different muscle groups, giving them time to replenish nutrients and to repair any minor damage incurred by the activity.

Weight Training **Weight training** has long been recognized as a method to build and maintain muscle strength and endurance, but its benefits to health have emerged only recently. Once considered stressful to the heart, weight training has gained new respect among researchers and physicians. The American Heart Association's science advisory now endorses a program of progressive weight training to increase muscle strength and endurance; prevent and manage several chronic diseases, including cardiovascular disease; and enhance psychological well-being.[21]

By promoting strong muscles in the back and abdomen, weight training also improves posture and reduces the risk of back injury. Weight training can also help prevent the decline in physical mobility that often accompanies aging.[22] Older adults, even those in their eighties, who participate in weight training programs not only gain muscle strength, but also improve their muscle endurance, which enables them to walk significantly longer before exhaustion. Leg strength and walking endurance are powerful indicators of an older adult's physical abilities.

Weight training to improve muscle strength and endurance can also help to maximize and maintain bone mass.[23] Research shows that even in women past menopause (when most women are losing bone), a one-year program of weight training improved bone density.[24]

Depending on the technique, weight training can emphasize either muscle strength or muscle endurance. To emphasize muscle strength, combine high resistance (heavy weight) with a low number of repetitions. To emphasize muscle endurance, combine less resistance (lighter weight) with more repetitions.

Weight training enhances performance in other sports, too. Swimmers can develop a more efficient stroke and tennis players a more powerful serve, when they train with weights.

Cardiorespiratory Endurance

Although weight training provides some cardiovascular benefits, the kind of physical activity most beneficial to the health of the heart is cardiorespiratory endurance training. Everyone has felt the heartbeat pick up its pace during physical activity. The length of time a person can remain active with an elevated heart rate—that is, the ability of the heart and lungs to sustain a given physical demand—defines a person's cardiorespiratory endurance. Cardiorespiratory endurance reflects the health of the heart and circulatory system, on which all other body systems depend.

moderate exercise: activity that can be sustained comfortably for 60 minutes or so.

warm-up: five to ten minutes of light activity, such as easy jogging or cycling, prior to a workout to prepare the body for more vigorous activity.

cool-down: five to ten minutes of light activity, such as walking or stretching, following a vigorous workout to return the body's core gradually to near-normal temperature.

hypertrophy (high-PURR-tro-fee): an increase in size (for example, of a muscle) in response to use.

atrophy (AT-tro-fee): a decrease in size (for example, of a muscle) because of disuse.

weight training: the use of free weights or weight machines to provide resistance for developing muscle strength and endurance; also called **resistance training.** A person's own body weight may also be used to provide resistance as when a person does push-ups, pull-ups, or sit-ups.

Cardiorespiratory endurance training improves a person's ability to sustain vigorous activities such as running, brisk walking, or swimming. Such training enhances the ability of the heart, lungs, and blood to deliver oxygen to, and remove waste from, the body's cells. Thus cardiorespiratory endurance training is **aerobic.** As the cardiorespiratory system gradually adapts to the demands of aerobic activity, the body delivers oxygen more efficiently. In fact, the accepted measure of a person's cardiorespiratory fitness is maximal oxygen uptake (**VO₂max**). The benefits of cardiorespiratory training are not just physical, though, because all the body's cells, including the brain cells, require oxygen to function. When the cells receive more oxygen more readily, both the body and the mind benefit. Figure 10–2 on p. 242 shows the major relationships among the heart, circulatory system, lungs, and muscles.

Sustained muscular efforts as in a long-distance rowing event or a cross-country run involve *aerobic* work.

Benefits of Cardiorespiratory Conditioning The changes brought about by cardiorespiratory endurance training are called **cardiorespiratory conditioning.** Among its benefits, the total blood volume and number of red blood cells increase, so the blood can carry more oxygen. The heart muscle becomes stronger and larger, and each beat empties the heart's chambers more completely, so the heart pumps more blood per beat. Both **cardiac output** and **stroke volume** increase. As a result, fewer beats are necessary, so the pulse rate falls. The average resting pulse rate for adults is around 70 beats per minute, but people who achieve cardiorespiratory conditioning may have resting pulse rates of 50 or even lower. The muscles that inflate and deflate the lungs gain strength and endurance, so breathing becomes more efficient. Blood moves easily through the blood vessels because the muscles of the heart contract powerfully, and contraction of the skeletal muscles pushes the blood through the veins. Such improvements keep resting blood pressure normal. The improvements that come with cardiorespiratory endurance also raise blood HDL, the lipoprotein associated with lower heart disease risk.[25]

Which activities produce these beneficial changes? Effective activities elevate the heart rate, are sustained for longer than 20 minutes, and use most of the large-muscle groups of the body (legs, buttocks, and abdomen). Examples are swimming, cross-country skiing, rowing, fast walking, jogging, fast bicycling, soccer, hockey, basketball, water polo, lacrosse, and rugby.

A person's own perceived effort is usually a reliable indicator of the intensity of an activity. In general, when you're working out, do so at an intensity that raises your heart rate, but still leaves you able to talk comfortably. If you are more competitive and want to work to your limit on some days, a treadmill test can reveal your maximum heart rate. You can work out safely at up to 90 percent of that rate. The ACSM guidelines for developing and maintaining cardiorespiratory fitness were given in Table 10–1 on p. 237.

Muscle Conditioning A fringe benefit of cardiorespiratory training is that fit muscles use oxygen efficiently, reducing the heart's workload. An added bonus is that muscles that use oxygen efficiently can burn fat longer—a plus for body composition and weight control.

A Balanced Fitness Program In a balanced fitness program, aerobic activity improves cardiorespiratory fitness, stretching enhances flexibility, and weight training develops muscle strength and endurance. Table 10–2 on p. 242 provides an example of a balanced fitness program.

In Summary Physical activity brings positive rewards: good health and long life. To develop fitness—whose components are flexibility, muscle strength and endurance, and cardiorespiratory endurance—a person must condition the body, through training, to adapt to the activity performed.

Cardiorespiratory conditioning:
- *Increases cardiac output and oxygen delivery.*
- *Increases heart strength and stroke volume.*
- *Slows resting pulse.*
- *Increases breathing efficiency.*
- *Improves circulation.*
- *Reduces blood pressure.*

The importance of HDL to heart health is a topic of Chapter 25.

aerobic (air-ROE-bic): requiring oxygen. Aerobic activity strengthens the heart and lungs by requiring them to work harder than normal to deliver oxygen to the tissues.

VO₂ max: the maximum rate of oxygen consumption by an individual (measured at sea level).

cardiorespiratory conditioning: improvements in heart and lung function and increased blood volume, brought about by aerobic training.

cardiac output: the volume of blood discharged by the heart each minute.

stroke volume: the amount of oxygenated blood ejected from the heart toward body tissues at each beat.

Fitness and Nutrition

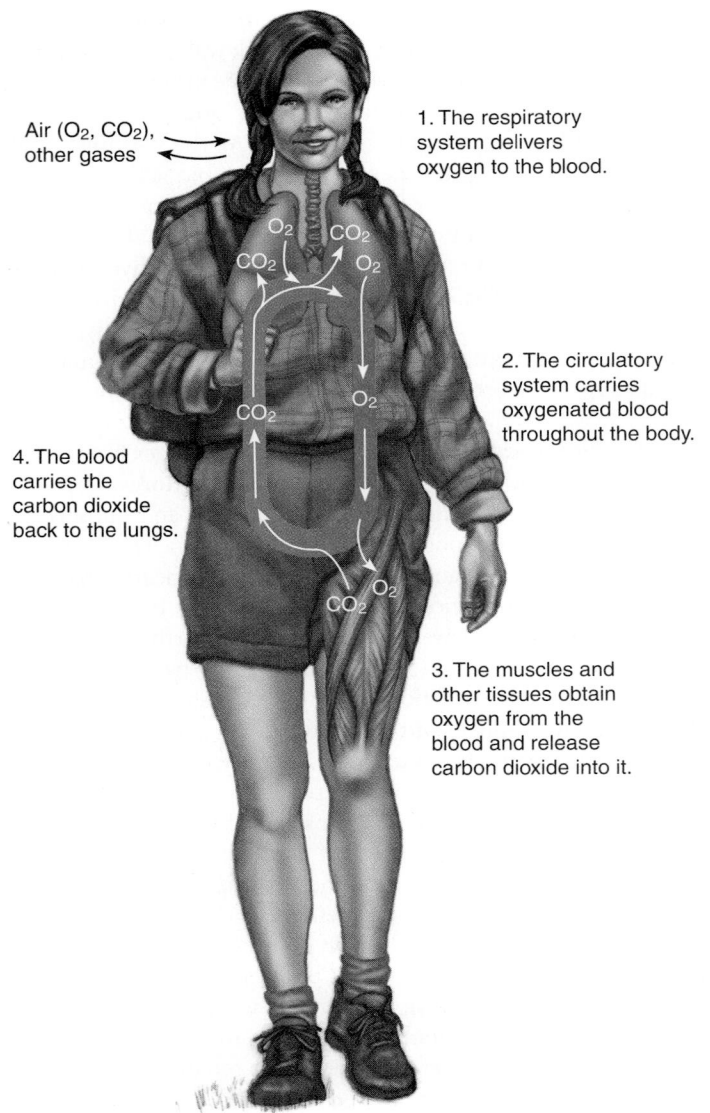

Figure 10–2

Delivery of Oxygen by the Heart and Lungs to the Muscles

The cardiorespiratory system responds to the muscles' demand for oxygen by building up its capacity to deliver oxygen. Researchers can measure cardiorespiratory fitness by measuring the maximum amount of oxygen a person consumes per minute while working out, a measure called VO_2max.

Air (O_2, CO_2), other gases

1. The respiratory system delivers oxygen to the blood.

2. The circulatory system carries oxygenated blood throughout the body.

4. The blood carries the carbon dioxide back to the lungs.

3. The muscles and other tissues obtain oxygen from the blood and release carbon dioxide into it.

Table 10–2   A Sample Balanced Fitness Program (45 minutes a day)
Monday, Wednesday, Friday
■ 5 minutes of warm-up activity
■ 30 minutes of aerobic activity
■ 10 minutes of cool-down activity and stretching
Tuesday, Thursday:
■ 5 minutes of warm-up activity
■ 30 minutes of weight training
■ 10 minutes of cool-down activity and stretching
Saturday or Sunday:
■ Sports, walking, hiking, biking, or swimming

The rest of this chapter describes the interactions between nutrients and physical activity. Nutrition alone cannot endow you with fitness or athletic ability, but along with the right mental attitude, it can complement the effort you put forth to obtain them. Conversely, unwise food selections can stand in your way.

The Active Body's Use of Fuels

The fuels that support physical activity are glucose (from carbohydrate), fatty acids (from fat), and, to a small extent, amino acids (from protein). The body uses different mixtures of fuels at different times depending on the intensity and duration of its activities and also depending on its own prior training.

During rest, the body derives a little more than half of its energy from fatty acids, most of the rest from glucose, and a little from amino acids. During physical activity, the body adjusts its fuel mix. The stored glucose of muscle glycogen is a major fuel for physical activity. In the early minutes of an activity, muscle glycogen provides the majority of energy the muscles use to go into action. As activity continues, messenger molecules, including the hormone epinephrine, flow into the bloodstream to signal the liver and fat cells to liberate their stored energy nutrients, primarily glucose and fatty acids. Thus hormones set the table for the muscles' energy feast, and the muscles help themselves to the fuels passing by in the blood.

Glucose Use during Physical Activity

Both the liver and muscles store glucose as glycogen; the liver can also make glucose from fragments of other nutrients. It has been said that muscles hoard their glycogen stores—they do not release their glucose into the bloodstream to share with other body tissues, as the liver does. This is fortunate. A muscle that shared its glycogen reserves with other tissues might lack glucose at a critical time such as when running from danger. A muscle that conserves its glycogen is prepared to act in emergencies because muscle glucose is the fuel for quick action. Later on, as activity continues, glucose from the liver's stored glycogen and dietary glucose absorbed from the digestive tract also become important sources of fuel for muscle activity.

Diet Affects Glycogen Storage and Use   The body constantly uses and replenishes its glycogen. How much carbohydrate a person eats influences how much glycogen is stored, which in turn influences performance. When glycogen is depleted, the muscles become fatigued.

A classic report compared fuel use during physical activity among three groups of runners on different diets. For several days before testing, one of the groups ate a normal mixed diet (55 percent of calories from carbohydrate); a second group ate a high-carbohydrate diet (83 percent of calories from carbohydrate); and the third group ate a high-fat diet (94 percent of calories from fat). As Figure 10–3 on p. 244 shows, the high-carbohydrate diet enabled the athletes to work longer before exhaustion. This study and many others that followed established that a high-carbohydrate diet enhances an athlete's endurance by ensuring ample glycogen stores.

Intensity of Activity Affects Glycogen Use   The body's glycogen stores are much more limited than its fat. A person with 30 pounds of body fat to spare may have only a pound or so of muscle and liver glycogen to draw on. How long an exercising person's glycogen will last depends not only on diet but also on the intensity of the activity. The most intense activities—the kind that make it difficult "to catch your breath," such as a quarter-mile run—use glycogen quickly. Less intense activities, such as jogging, during which breathing is steady and easy, use glycogen more slowly. Thus competitive athletes demand much more from their glycogen stores than do casual joggers. Joggers still use glycogen, however, and eventually they can run out of it. Glycogen depletion usually occurs within two hours from the onset of vigorous activity.[*]

During *moderate* physical activity, the lungs and circulatory system have no trouble keeping up with the muscles' need for oxygen. The individual breathes easily, and the heart beats at a faster pace than at rest but steadily—the activity is *aerobic*. During aerobic activity, muscles extract their energy from both glucose and fatty acids. By depending partly on fatty acids, moderate aerobic activity conserves glycogen stores.

Intense activity presents a different picture. When muscle exertion is so great that the demand for energy exceeds the capacity of the heart and lungs to supply oxygen, aerobic metabolism cannot sufficiently meet energy needs. This means that fat cannot be used because oxygen is required for its breakdown. The muscles must instead begin to rely more heavily on glucose, which can be partially broken down by **anaerobic** metabolism. Thus the muscles begin drawing more heavily on their limited glycogen supply.

anaerobic (AN-air-ROE-bic): not requiring oxygen. Anaerobic activity may require strength but does not work the heart and lungs very hard for a sustained period.

---

[*]Here "vigorous activity" means activity at 75 percent of VO$_2$max.

Figure 10–3

The Effect of Diet on Physical Endurance

A high-carbohydrate diet can increase an athlete's endurance. In this study, the high-fat diet provided 94 percent of kcalories from fat and 6 percent from protein; the normal mixed diet provided 55 percent of kcalories from carbohydrate; and the high-carbohydrate diet provided 83 percent of kcalories from carbohydrate.

High-fat diet

Normal mixed diet

High-carbohydrate diet

Maximum endurance time:

57 min

114 min

167 min

© Bonnie Kamin/PhotoEdit

Lactic Acid Anaerobic breakdown of glucose produces **lactic acid,** fragments of glucose molecules that accumulate in the tissues and blood. When the nervous and hormonal systems detect these fragments in the blood, they respond by speeding up the heart and lungs to draw in more oxygen and break down the fragments. At some point, however, the heart and lungs are no longer able to keep up, and lactic acid accumulates. If you exercise intensely, you may have to slow down or even stop to "catch your breath" (replenish your oxygen supply). Then your body begins relying on aerobic metabolism once more. At this point, lactic acid is burned for fuel or used by the liver to regenerate glucose.

Duration of Activity Affects Glycogen Use Glycogen use during physical activity depends on the *duration* of the activity as well as its *intensity.* In the first 10 minutes or so of an activity, the active muscles rely almost completely on their own stores of glycogen. Within the first 20 minutes or so of moderate activity, a person uses up about one-fifth of the available glycogen. As the muscles devour their own glycogen, they become ravenous for more glucose and increase their uptake of blood glucose dramatically.[26] If you test a person's blood glucose during moderate activity, you will find that it declines slightly, reflecting the muscles' use of blood glucose.

A person who continues exercising moderately for longer than 20 minutes begins to use less glucose and more fat for fuel. Still, glucose use continues, and if the activity goes on for long enough and at a high enough intensity, muscle and liver glycogen stores will run out almost completely. Physical activity can continue for a short time thereafter only because the liver scrambles to produce some glucose from available lactic acid and certain amino acids.

Glucose Depletion After a couple of hours of strenuous activity, glucose stores are depleted. When depletion hits, it brings nervous system function almost to a halt, making continued activity impossible. This is what marathon runners call "hitting the wall." To postpone exhaustion, endurance athletes must try to maintain their blood glucose concentrations for as long as they can. To maximize glucose supply, endurance athletes:

- Eat a high-carbohydrate diet (approximately 8 grams of carbohydrate per kilogram of body weight) regularly.
- Take glucose (usually in sports drinks, diluted fruit juice, or other sweet beverages) during endurance activity.
- Eat carbohydrate-rich foods after performance.
- Train the muscles to maximize glycogen stores.

lactic acid: a product of the incomplete breakdown of glucose during anaerobic metabolism. When oxygen becomes available, lactic acid can be completely broken down for energy or converted back to glucose.

How to

Maximize Glycogen Stores: Carbohydrate Loading

*S*ome athletes use a technique called **carbohydrate loading** to trick their muscles into storing extra glycogen before a competition. In general, the athlete tapers training during the week before the competition and then eats a high-carbohydrate diet during the three days just prior to the event.[a] Specifically, the athlete would follow the plan in the accompanying table.

Extra glycogen gained this way can benefit an athlete who must keep going for 90 minutes or longer. Those who exercise for shorter times simply need a regular high-carbohydrate diet. In a hot climate, extra glycogen confers an additional advantage: as glycogen breaks down, it releases water, which helps to meet the athlete's fluid needs.

Before the Event	Training Intensity	Training Duration	Dietary Carbohydrate
6 days	Moderate (70% VO$_2$max)	90 min	Normal (5 g/kg body weight)
5 days 4 days	Moderate (70% VO$_2$max)	40 min	Normal (5 g/kg body weight)
3 days 2 days	Moderate (70% VO$_2$max)	20 min	High-carbohydrate (10 g/kg body weight)
1 day	Rest	—	High-carbohydrate (10 g/kg body weight)

[a]E. Coleman, Carbohydrate and exercise, in *Sports Nutrition: A Guide for the Professional Working with Active People*, 3rd ed., ed. C. A. Rosenbloom (Chicago: The American Dietetic Association, 2000), pp. 13–31.

carbohydrate loading: a regimen of moderate exercise followed by the consumption of a high-carbohydrate diet that enables muscles to store glycogen beyond their normal capacities; also called **glycogen loading** or **glycogen super compensation.**

The last section of this chapter discusses how to design a high-carbohydrate diet for performance, and the accompanying "How to" box describes how to maximize glycogen stores for long endurance competitions.

Glucose during Activity Glucose ingested before or during a long-duration competition makes its way from the digestive tract to the working muscles. This external source of glucose augments dwindling internal glucose supplies from the muscle and liver glycogen stores.[27] Especially during games such as soccer or hockey, which last for hours and demand repeated bursts of intense activity, athletes may benefit from carbohydrate-containing drinks taken during the activity.

Before concluding that sugar might be good for your own performance, consider first whether you engage in *endurance* activity. Do you run, swim, bike, or ski nonstop at a rapid pace for more than an hour at a time or compete in games lasting for hours? If not, the sugar picture changes. For an everyday jog or swim lasting less than 60 minutes, sugar probably won't help performance, though it may do no harm, either. Even in athletes, extra carbohydrate does not benefit those who engage in sports in which fatigue is unrelated to blood glucose, such as 100-meter sprinting, baseball, casual basketball, and weight lifting.

Glucose after Activity Another method for gaining some extra glycogen involves eating carbohydrate after exercise. The athlete trains normally and then, within two hours after physical activity, consumes a high-carbohydrate meal, such as a glass of orange juice and some graham crackers, toast, or cereal. This method accelerates the rate of glycogen storage by 300 percent for a while.[28] Timing is important—eating the meal after two hours have passed reduces the glycogen synthesis rate by almost half.

Chapter 2 introduced the *glycemic effect* and discussed some possible health benefits of eating a diet ranking low on the glycemic index. For athletes wishing to

To make glycogen, muscles need carbohydrate, but they also need rest. Vary daily activity routines to work different muscles on different days.

Factors that affect glucose use during physical activity:
- *Carbohydrate intake.*
- *Intensity and duration of the activity.*
- *Degree of training.*

Fitness and Nutrition

Those who compete in endurance activities require fluid and carbohydrate fuel.

Foods with a high glycemic index:
- *Cornflakes.*
- *Mashed potatoes.*
- *Short-grain rice.*
- *Waffles.*
- *Watermelon.*
- *White bread.*

maximize muscle glycogen synthesis after strenuous training, however, eating foods with a high glycemic index (see the margin) may restore glycogen most rapidly.[29]

Training Affects Glycogen Use Training, too, affects how much glycogen muscles will store. Muscles that deplete their glycogen stores through work adapt to store greater amounts of glycogen to support that work. As you know, the more glycogen muscles store, the longer the stores last during physical activity.

Muscles make still another adaptation to training that affects glycogen use during activity. Trained muscles burn more fat, and at higher intensities, than untrained muscles, so they require less glucose to perform the same amount of work.[30] A person attempting an activity for the first time uses up much more glucose per minute than an athlete who is trained to perform it. A trained person can work at high intensities for longer periods than an untrained person while using the same amount of glycogen.

People with diabetes should know that the moderating effect of physical training on glucose metabolism may have implications for them. Those who must take insulin or insulin-eliciting medications sometimes find that as their muscles adapt to physical activity, they can reduce their daily drug doses. Physical activity may also improve type 2 diabetes by helping the body lose excess fat. For those with type 1 diabetes, physical activity has been shown to lower the risk of cardiovascular disease, increase insulin sensitivity, lower blood pressure, and improve blood lipids.[31]

Fat Use during Physical Activity

An active person who eats a fat-rich diet with little carbohydrate will burn more fat during activity but will sacrifice endurance, as Figure 10–3 showed. The importance of a high-carbohydrate diet for endurance has long been recognized. Some research suggests, however, that medium- and high-fat diets may benefit ultra-endurance performance and that severe dietary fat restriction may be detrimental for some athletes.[32] High-fat diets, however, carry risks of heart disease. Physical activity offers some protection against cardiovascular disease, but even physically active people and athletes can suffer heart attacks and strokes. Most nutrition experts agree that the potential for adverse health effects from prolonged high-fat diets makes them an unwise choice.

Athletes and active people who restrict fat below 20 percent of total energy intake may fail to consume adequate energy and nutrients. Sports nutrition experts recommend that endurance athletes consume 20 to 30 percent of their energy from fat in order to meet nutrient and energy needs.[33] One expert says the message is "not that high-fat diets improve performance, but rather that very low-fat diets inhibit performance."*

Body fat stores are more important as fuel for activity than is fat in the diet. Unlike the body's glycogen stores, which are limited, fat stores can fuel hours of activity without running out; body fat is (theoretically) an unlimited source of energy. Even the lean bodies of elite runners carry enough fat to fuel several marathon runs.

Early in activity, muscles begin to draw on fatty acids from two sources—fats stored within the working muscles and fats from fat deposits such as the fat under the skin. Areas that have the most fat to spare donate the greatest amounts of fatty acids to the blood (although they may not be the areas that you would choose to lose fat from). This is why "spot reducing" doesn't work: muscles do not own the fat that surrounds them. Fat cells release fatty acids into the blood for all the muscles to share. Proof of this is found in a tennis player's arms: the fatfolds measure the same in both arms, even though one arm has better-developed muscles than the other.

---

*The quotation is attributed to David R. Pendergast in, Cutting fat may crimp performance in endurance athletes, *Nutrition and the M.D.,* December 2000, pp. 3–4.

Intensity and Duration Affect Fat Use   The *intensity* of physical activity also affects the percentage of energy contributed by fat. As mentioned, fat can be broken down for energy in only one way—by aerobic metabolism. When the intensity of activity becomes so great that energy demands surpass the ability to provide energy aerobically, the body cannot burn more fat. Instead, it burns more glucose.

The *duration* of activity affects fat use as well. At the start of activity, the blood fatty acid concentration falls, but a few minutes into an activity, the hormone epinephrine signals the fat cells to break apart their stored triglycerides and to liberate fatty acids into the blood. After about 20 minutes of activity, the blood fatty acid concentration rises above the normal resting concentration. It is only during this phase of sustained, moderate activity, after the first 20 minutes, that the fat cells begin to shrink in size as they empty out their fat stores.

Training Affects Fat Use   Training—repeated aerobic activity—stimulates the muscles to develop more fat-burning enzymes. Aerobically trained muscles burn fat more readily than untrained muscles. With aerobic training, the heart and lungs also become stronger and better able to deliver oxygen to the muscles during high-intensity activities. This improved oxygen supply also enables the muscles to burn more fat.

Recommended Intensities and Durations   Health care professionals frequently advise people who want to control their body weight and lose fat to engage in activities of low-to-moderate intensity for a long duration, such as an hour-long, fast-paced walk. The reasoning behind such advice is that people exercising at low-to-moderate intensity are likely to stick with their activity for a longer time and are less likely to injure themselves. A person who stays with an activity routine long enough to enjoy the rewards will be less inclined to give it up and will, over the long term, reap many health benefits. Activity of low-to-moderate intensity that spends 1500 to 2000 kcalories per week is especially helpful for weight management.[34] Low-intensity physical activity has also been shown to reduce intra-abdominal fat and lower blood glucose in older, obese adults.[35] Walking is almost always the most appropriate type of physical activity for overweight or obese individuals.

Vigorous, high-intensity physical activity offers benefits as well, especially in terms of energy expenditure.[36] As activity intensity increases, the efficiency of performing the activity decreases, so a person spends more energy at a high intensity than at a low intensity. For example, it takes twice as much energy to cycle the same amount of time at a high intensity (21 miles per hour) as at a low intensity (13 miles per hour). High-intensity activity also enhances resting energy expenditure for hours after the activity is performed. Finally, the conditioned body that is adapted to strenuous and prolonged aerobic activity uses more fat all day long, not just during activity.[37] The bottom line on physical activity and weight and/or fat loss seems to be that total energy expenditure is the main factor, regardless of how you do it.[38]

Choosing an Activity   The intensity and type of physical activities that are best for one person may not be good for another. The intensity to choose depends on your present fitness: work so as to breathe hard, but not so hard as to incur an oxygen debt. As a general rule, you should be breathing easily enough to talk but not sing. If you can sing, pick up the pace; if you have to huff and puff to talk, slow down. If you have been sedentary for the past few years, the activity intensity that will initially make you short of breath will differ dramatically from the intensity of a fit person.

The type of physical activity that is best for you depends, too, on what you want to achieve and what you enjoy doing. If you are looking for health benefits, such as reducing your disease risks and lowering your blood cholesterol, then you might want to spend at least 30 minutes each day doing some kind of physical

Factors that affect fat use during physical activity:

- *Fat intake.*
- *Intensity and duration of the activity.*
- *Degree of training.*

To spend 1500 kcal/week:[*]

A 175 lb man might:

- *Run 30 min/day, 5 days.*
- *Walk 45 min/day, 5 days.*

A 125 lb woman might:

- *Run 45 min/day, 5 days.*
- *Walk 60 min/day, 5 days.*

[*]The man, because he weighs more, uses more energy per minute running or walking than the lighter-weight woman does.

Table 10–3 Fuels Used for Activities of Different Intensities and Durations				
Activity Intensity	**Activity Duration**	**Preferred Fuel Source**	**Oxygen Needed?**	**Activity Example**
Very high	20 sec to 3 min	Carbohydrate (lactic acid)	No (anaerobic)	¼-mile run at maximal speed
High	3 min to 20 min	Carbohydrate	Yes (aerobic)	Cycling, swimming, running
Moderate	More than 20 min	Fat	Yes (aerobic)	Hiking, jogging

The key to regular physical activity is finding an activity that you enjoy.

activity. If you are looking to lose weight and improve body composition, then choose an activity that you can sustain for 45 minutes or more at least three days a week. Choose an activity you enjoy: some people love walking, while others prefer to dance or ride a bike. If you want to be stronger and firmer, lift weights. And remember, muscle is more metabolically active than body fat, so the more muscle you have, the more energy you'll burn.

Table 10–3 summarizes fuel use during physical activity as discussed so far. You may wonder why the third energy-yielding nutrient, protein, is not listed in the table. The reason is that protein donates only a little energy to physical activity. It does, however, produce the structural material of muscle tissue, so it is important to active people.

Protein and Physical Activity

Physically active people use protein to build and maintain muscle and other lean tissue structures and, to a small extent, to fuel activity. The body handles protein differently during activity than during rest, however.

Protein for Building Muscle Tissue   In the hours of rest that follow physical activity, muscles speed up their rate of protein synthesis—they build more of the proteins they need to perform the activity. Additionally, whenever the body rebuilds a part of itself, it must tear down the old structures to make way for the new ones. Physical activity, with just a slight overload, calls into action both the protein-dismantling and the protein-synthesizing equipment of individual muscle cells that work together to remodel muscles.

Dietary protein provides the needed amino acids for synthesis of new muscle proteins. As Chapter 4 pointed out, however, the true director of synthesis of muscle protein is physical activity itself. Repeated activity signals the muscle cells' genetic material to begin producing more of the proteins needed to perform the work at hand.

The genetic protein-making equipment inside the nuclei of muscle cells seems to "know" when proteins are needed. Furthermore, it knows *which* proteins are needed to support each type of physical activity. Apparently, the intensity and pattern of muscle contractions initiate signals that direct the muscles' genetic material to make particular proteins. For example, a weight lifter's workout sends the information that muscle fibers need added bulk for strength and more enzymes for making and using glycogen. A jogger's workout stimulates production of proteins needed for aerobic oxidation of fat and glucose. Muscle cells are exquisitely responsive to the need for proteins and build them conservatively only as needed.

Finally, after muscle cells have made all the decisions about which proteins to build and when, protein nutrition comes into play. During active muscle-building phases of training, a weight lifter might add to existing muscle mass between ¼ ounce and 1 ounce (between 7 and 28 grams) of protein each day. This extra protein comes from ordinary food.

Protein Used for Fuel   Not only do athletes retain more protein, but they also use a little more protein as fuel.[39] Studies of nitrogen balance show that the body speeds up its use of amino acids for energy during physical activity, just as it speeds up its use of glucose and fatty acids. Protein contributes about 10 percent of the total fuel used, both during activity and during rest.

Diet Affects Protein Use during Activity The factors that regulate how much protein is used during activity seem to be the same three that regulate the use of glucose and fat. One factor is diet—a carbohydrate-rich diet spares protein from being used as fuel. Some amino acids can be converted into glucose when needed. Others, the **branched-chain amino acids,** can stand in for glucose in energy pathways. If the diet is low in carbohydrate, much more protein will be used in place of glucose.

Intensity and Duration Affect Protein Use Second, the intensity and duration of the activity also affect protein use.[40] Endurance athletes who train for over an hour a day, engaging in aerobic activity of moderate intensity and long duration, may deplete their glycogen stores by the end of their training and become more dependent on body protein for energy. The protein needs of bodybuilders and weight lifters are higher than those of sedentary people, but not as high as the protein intakes many bodybuilders consume.

Training Affects Protein Use Finally, the degree of training also affects the use of protein. Particularly in strength athletes such as bodybuilders, the higher the degree of training, the less protein a person uses.

Protein Recommendations for Active People As mentioned earlier, all active people, and especially those athletes in training, probably need somewhat more protein than do sedentary people. In the United States, however, average protein intakes are high enough to cover even the needs of most athletes. Therefore, athletes in training should attend to protein needs, but should back up the protein with ample carbohydrate. Otherwise, they will burn off as fuel the very protein they wish to retain in muscle.

A joint position paper from the American Dietetic Association (ADA) and the Dietitians of Canada (DC) recommends protein intakes somewhat higher than the 0.8 gram of protein per kilogram of body weight recommended for sedentary people.[41] Table 10–4 lists some recommendations and translates them into daily intakes for athletes.

After considering these recommendations, athletes may wonder whether their diets provide the protein they need. A later section translates protein recommendations into a diet plan and shows that no one needs protein supplements, or even large servings of meat, to obtain the highest protein intakes.

Physical activity itself triggers the building of muscle proteins.

Factors that affect protein use during physical activity:
- *Carbohydrate intake.*
- *Intensity and duration of the activity.*
- *Degree of training.*

branched-chain amino acids: amino acids that, unlike the others, can provide energy directly to muscle tissue: leucine, isoleucine, and valine.

Table 10–4 Recommended Protein Intakes for Athletes			
	Recommendations (g/kg/day)	Protein Intakes (g/day)	
		Males	Females
RDA for adults	0.8	56	44
Recommended intake for power (strength or speed) athletes	1.6–1.7	112–119	88–94
Recommended intake for endurance athletes	1.2–1.6	84–112	66–88
U.S. average intake		95	65

NOTE: Daily protein intakes are based on a 70-kilogram (154-pound) man and 55-kilogram (121-pound) woman.

SOURCES: Committee on Dietary Allowances, *Recommended Dietary Allowances,* 10th ed. (Washington, D.C.: National Academy Press, 1989); Position of The American Dietetic Association, Dietitians of Canada, and the American College of Sports Medicine: Nutrition and athletic performance, *Journal of the American Dietetic Association* 100 (2000): 1543–1556.

Fitness and Nutrition

Foods like these are packed with the nutrients that active people need.

In Summary The mixture of fuels the body uses during physical activity depends on diet, the intensity and duration of the activity, and training. During intense activity, the muscles use glucose primarily; during less intense, moderate activity, fat makes a greater energy contribution, and glycogen use is slower.

Vitamins and Minerals

Many of the vitamins and minerals assist in releasing energy from fuels and in transporting oxygen. This knowledge has led many people to believe, mistakenly, that vitamin and mineral *supplements* offer physically active people both health benefits and athletic advantages. (Nutrition in Practice 9 focuses on vitamin and mineral supplements, and Nutrition in Practice 10 explores supplements and other products people use in the hope of enhancing athletic performance.)

Supplements

Nutrient supplements do not enhance the performance of well-nourished people. Deficiencies of vitamins and minerals, however, do impede performance. In general, active people who eat enough nutrient-dense foods to meet energy needs also meet their vitamin and mineral needs. After all, active people eat more food; it stands to reason that with the right choices, they'll get more nutrients.

Some athletes mistakenly believe that taking vitamin or mineral supplements directly before competition will enhance performance. These beliefs are contrary to scientific reality. Most vitamins and minerals function as small parts of larger working units. After entering the blood, they have to wait for the cells to combine them with their appropriate other parts so that they can do their work. This takes time—hours or days. Vitamins or minerals taken right before an event are useless for improving performance, even if the person is actually suffering deficiencies of them.

In general, then, active people who eat well-balanced meals need no vitamins or minerals in supplement form with the possible exceptions of vitamin E and iron. During prolonged, high-intensity physical activity, the muscles' consumption of oxygen increases tenfold or more, enhancing the production of damaging free radicals in the body.[42] Vitamin E is a potent fat-soluble antioxidant that vigorously defends cell membranes against oxidative damage. Many athletes are taking vitamin E supplements in hopes of preventing oxidative damage to muscles. The results of some research lend support to this practice.[43] Research suggests that vitamin E supplements offer protection against exercise-induced oxidative stress.[44] Supplement doses in these studies varied considerably, however, and no one yet knows the precise dose that will offer the greatest benefits with the least risk of toxicity. Furthermore, although research suggests that antioxidant supplements such as vitamin E may offer protection against oxidative stress, there is little evidence that these supplements can improve performance.[45] Clearly, more research is needed before drawing conclusions about antioxidant supplements for active people. As the next section explains, however, some people may need iron supplements; iron helps deliver oxygen to the muscles, and iron deficiency can impair performance.

Iron Deficiency in Female Athletes

Endurance athletes, and especially female athletes, are prone to iron deficiency.[46] Physical activity may impair iron status in any of several ways. For one, iron may be excreted in sweat.[47] For another, iron may be lost through red blood cell destruction; blood cells are squashed when body tissues (such as the soles of the feet) make high-impact contact with an unyielding surface (such as the ground). In addition, physical activity may cause small blood losses through the digestive tract, at least in some athletes. Perhaps more significant than losses are the high iron demands by muscles to make the iron-containing molecules of aerobic metabolism. Habitually low intakes of iron-rich foods as well as increased losses and extra demands may contribute to iron deficiency in young female athletes.

Vegetarian female athletes may be especially vulnerable to iron insufficiency.[48] The bioavailability of iron is often poor in plant-based diets because such diets are high in fiber and phytic acid. Also, the nonheme iron in plant foods is not absorbed as well as the heme iron in animal-derived foods. Vegetarian diets are usually rich in vitamin C, however, which enhances iron absorption. To protect against iron deficiency, vegetarian athletes need to pay close attention to their intake of good dietary sources of iron (fortified cereals, legumes, nuts, and seeds) and include vitamin C–rich foods with each meal.[49] As long as vegetarian athletes, like all athletes, consume enough nutrient-dense foods, they can perform as well as anyone.

Iron deficiency impairs performance because iron helps deliver the muscles' oxygen. Insufficient oxygen delivery reduces aerobic work capacity, so the person tires easily. Whether marginal deficiency without clinical signs of anemia hinders physical performance is a point of debate among researchers.[50]

Sports Anemia   Early in training, athletes may develop low blood hemoglobin for a while. This condition, sometimes called **"sports anemia,"** is not a true iron deficiency condition. Strenuous training promotes destruction of the more fragile, older red blood cells, and the resulting cleanup work reduces the blood's iron content temporarily. Strenuous activity also promotes increases in the fluid of the blood; with more fluid, the red blood cell count in a unit of blood drops. Most researchers view sports anemia as an *adaptive,* temporary response to endurance training. True iron-deficiency anemia requires treatment with prescribed iron supplements, but sports anemia goes away by itself, even with continued training.

Iron Supplements May Be Needed   The best strategy concerning iron may be to determine individual needs. Many menstruating women probably border on iron deficiency even without the additional iron demand and losses incurred by physical activity. Teens of both sexes, because they are growing, have high iron needs, too. Especially for women and teens, then, prescribed supplements may be needed to correct a deficiency of iron that is confirmed by tests. (Medical testing is needed to eliminate nondietary causes of anemia, such as internal bleeding or cancer.)

In Summary   With the possible exceptions of vitamin E and iron, well-nourished active people do not need nutrient supplements.

Fluids and Electrolytes in Physical Activity

The body's need for water far surpasses its need for any other nutrient. If the body loses too much water, as in dehydration, its life-supporting chemistry becomes compromised.

The exercising body loses water primarily via sweat; second to that, breathing costs water, exhaled as vapor. During physical activity, both routes can be significant, and dehydration is a real threat. The first symptom of dehydration is fatigue. A water loss of even 1 to 2 percent of body weight can reduce a person's

Female athletes may be at special risk of iron deficiency.

sports anemia: a transient condition of low hemoglobin in the blood, associated with the early stages of sports training or other strenuous activity.

Fitness and Nutrition

Active people need extra fluid, even in cold weather.

Note: 10 degrees on the Celsius scale is about 18 degrees on the Fahrenheit scale.

Symptoms of heat stroke:
- *Headache.*
- *Nausea.*
- *Dizziness.*
- *Clumsiness.*
- *Stumbling.*
- *Sudden cessation of sweating (hot, dry skin).*
- *Internal (rectal) temperature above 104° Fahrenheit.*
- *Confusion or loss of consciousness.*

heat stroke: an acute and life-threatening reaction to heat buildup in the body.

hypothermia: a below-normal body temperature.

capacity to do muscular work.[51] A person with a water loss of about 7 percent is likely to collapse. The athlete who arrives at an event even slightly dehydrated starts out at a competitive disadvantage.

Fluid Losses

Working muscles produce heat. During intense activity, muscle heat production can be 15 to 20 times greater than at rest.[52] The body cools itself by sweating. Each liter of sweat dissipates almost 600 kcalories of heat, preventing a rise in body temperature of almost 10 degrees on the Celsius scale. The body routes its blood supply through the capillaries just under the skin, and the skin secretes sweat to evaporate and cool the skin and the underlying blood. The blood then flows back to cool the deeper body chambers.

Hyperthermia In hot, humid weather, sweat may fail to evaporate because the surrounding air is already laden with water. Body heat builds up and triggers maximum sweating, but without sweat evaporation, little cooling takes place. In such conditions, active people must take precautions to avoid **heat stroke.** Heat stroke is an especially dangerous accumulation of body heat with accompanying loss of body fluid. To reduce the risk of heat stroke, drink enough fluid before and during the activity, rest in the shade when tired, and wear lightweight clothing that encourages evaporation.[53] The rubber or heavy suits sold with promises of weight loss during physical activity are dangerous because they promote profuse sweating, prevent sweat evaporation, and invite heat stroke. If you experience any of the symptoms of heat stroke listed in the margin, stop your activity, sip cold fluids, seek shade, and ask for help. The condition demands medical attention; it can kill.

Hypothermia In cold weather, **hypothermia,** or loss of body heat, can pose as serious a threat as heat stroke does in hot weather. Inexperienced runners participating in long races on cold or wet, chilly days are especially vulnerable to hypothermia. Slow runners can produce too little heat to keep warm, especially if their clothing is inadequate. Early symptoms of hypothermia include shivering and euphoria. As body temperature continues to fall, shivering may stop, and weakness, disorientation, and apathy may set in. People with these symptoms soon become helpless to protect themselves from further body heat losses. Even in cold weather, the body still sweats and needs fluids, but the fluids should be warm or at room temperature, not cold, to help prevent hypothermia.

Fluid Needs during Physical Activity

Endurance athletes can lose 2 or more quarts of fluid in every hour of activity, but the digestive system can absorb only about a quart or so an hour. To prepare for fluid losses, a person must hydrate before activity. To replace fluid losses, the person must rehydrate during and after activity. Even then, in hot weather the digestive tract may not be able to absorb enough water fast enough to keep up with an athlete's sweat losses, and some degree of dehydration becomes inevitable. Athletes preparing for competition drink extra fluids in the last few days of training before the event. The extra fluid is not stored in the body, but drinking extra ensures maximum tissue hydration at the start of the event. Any coach or athlete who withholds fluids during practice for any reason takes a great risk and is subject to sanctions by the American College of Sports Medicine.

Athletes who rely on thirst to govern fluid intake can easily become dehydrated. During activity, thirst becomes detectable only *after* fluid stores are depleted. Don't wait to feel thirsty before drinking. Table 10–5 presents one schedule of hydration for physical activity. To find out how much water you need to replenish losses, weigh yourself before and after the activity. The difference is all water. Two cups (16 ounces) of fluid weigh about a pound.

Table 10–5 Hydration Schedule for Physical Activity	
When to Drink	**Approximate Amount of Fluid**
2 hr before activity	2 to 3 c
15 min before activity	1 to 2 c
Every 15 min during activity	½ to 1 c
After activity	At least 2 c for each pound of body weight lost

SOURCE: R. Murray, Fluid and electrolytes, in *Sports Nutrition: A Guide for the Professional Working with Active People*, 3rd ed., ed. C.A. Rosenbloom (Chicago: The American Dietetic Association, 2000), pp. 95–106.

Although water is the best fluid for most physically active people, many good-tasting sports drinks are available.

Water What is the best fluid to support physical activity? Surprisingly, the best drink for most active bodies is just plain cool water, for two reasons: (1) water rapidly leaves the digestive tract to enter the tissues, and (2) it cools the body from the inside out. As mentioned earlier, however, endurance athletes are an exception: they need more from their fluids than water alone. The first priority for endurance athletes should always be to replace fluids to prevent life-threatening heat stroke. But endurance athletes also need carbohydrate to supplement their limited glycogen stores, so glucose is important, too. The "How to" box on p. 254 compares water and sports drinks.

Electrolyte Losses and Replacement When a person sweats, small amounts of electrolytes—the electrically charged minerals sodium, potassium, chloride, and magnesium—are lost from the body along with water. Losses are greatest in beginners; training improves electrolyte retention.

To replenish lost electrolytes, a person ordinarily needs only to eat a regular diet that meets energy and nutrient needs. In events lasting more than one hour, sports drinks may be needed to replace fluids and electrolytes. Salt tablets should be avoided. As the "How to" box on p. 254 explains, they can worsen dehydration and impair performance.

Poor Beverage Choices

Carbonated beverages are not the best choice for meeting an athlete's fluid needs. Although they are composed largely of water, the air bubbles from the carbonation take up room in the stomach that might otherwise be filled with fluid that the athlete can absorb. Many carbonated beverages also contain caffeine. Caffeine's effects on athletic performance are discussed in this chapter's Nutrition in Practice.

Athletes, like others, sometimes drink beverages that contain alcohol, but these beverages are inappropriate as fluid replacements. Like caffeine, alcohol is a diuretic. Both substances promote the excretion of water; of vitamins such as thiamin, riboflavin, and folate; and of minerals such as calcium, magnesium, and potassium—exactly the wrong effects for fluid balance and nutrition. It is hard to overstate alcohol's detrimental effects on physical activity. It impairs temperature regulation, making hypothermia or heat stroke much more likely. It alters perceptions and slows reaction time. It depletes strength and endurance and deprives people of their judgment, thereby compromising their safety in sports. Many sports-related fatalities and injuries each year involve alcohol or other drugs.

Fluid replacement tips:[*]

- *To ensure adequate fluid intake without being distracted during an event: before the event, fill a 32 oz (4 c) water bottle and place two colored rubber bands about equal distance from the top. Finish off the first segment of the bottle in the first 30 minutes of activity; finish the next segment in the next 30 minutes; and the remainder in the next. Have someone refill the bottle if activity lasts longer than 90 minutes.*
- *The urine of a person who is adequately hydrated is the color of pale lemonade. Urine the color of apple juice indicates slight dehydration.*

[*]Ideas from J. Berning, nutrition professor and sports nutrition consultant, personal communication, 1999.

In Summary Evaporation of sweat cools the body. Heat stroke can be a threat to physically active people in hot, humid weather. Hypothermia threatens those who exercise in the cold. Physically active people lose fluids and must replace them to avoid dehydration. Carbonated beverages and those that contain alcohol are poor choices for fluid replacement.

Fitness and Nutrition

How to

Evaluate Sports Drinks

Many good-tasting drinks are marketed for athletes and active people. More than 20 sports drinks compete for their share of the $1 billion market. What do sports drinks have to offer? First, sports drinks offer fluids to help offset the loss of fluids during physical activity. As discussed earlier, however, plain water can do this, too.

Second, sports drinks supply glucose. A beverage that supplies glucose in some form can be useful during endurance activity lasting 60 minutes or more or during prolonged competitive games that demand repeated intermittent activity.[a] Not just any sweet beverage can meet this need, however, because a carbohydrate concentration greater than 8 percent can delay the emptying of fluid from the stomach and thereby slow down the delivery of water to the tissues. Most sports drinks contain an appropriate amount to ensure water absorption— about 7 percent glucose (about half the sugar of ordinary soft drinks, or about 5 teaspoons in each 12 ounces).

Third, sports drinks offer a little sodium and other electrolytes to help replace those lost during physical activity. Sodium in sports drinks also helps to improve palatability and fluid retention and maintains the osmotic drive for drinking fluid. This makes sense, physiologically, because the sensation of thirst is a function of changes in blood sodium concentration.[b]

In strenuous world-class competitions lasting for four hours or more, heavy sweating coupled with drinking large amounts of plain water has been reported to dangerously dilute blood sodium. Ultra-endurance athletes, therefore, may especially need to replace sodium.[c] They should obtain it from sports drinks and not from electrolyte or salt tablets, which can increase potassium losses, irritate the stomach, cause vomiting, and always pull water out of the tissues into the digestive tract at first. Athletes should avoid these products.

Unlike ultra-endurance athletes, most athletes do not need to replace the minerals lost in sweat immediately; a meal eaten within hours of competition replaces these minerals soon enough. Most sports drinks are relatively low in sodium, however, so healthy people who choose to use these beverages run little risk of excessive intake.

Most sports drinks also taste good. Manufacturers reason that if a drink tastes good, people will drink more, thereby ensuring adequate hydration. Research backs up such reasoning, too. Fluids that are flavored, sweetened, and cool stimulate fluid intake.[d] Finally, sports drinks can also provide a psychological edge to people who associate them with success in sports. Thus, for athletes who exercise for an hour or more, sports drinks offer some advantages over water.

[a]R. S. Welsh and coauthors, carbohydrates and physical/mental performance during intermittent exercise to fatigue, *Medicine and Science* in Sports and Exercise 34 (2002): 723–731.

[b]R. Murray, Fluid and electrolytes, in *Sports Nutrition: A Guide for the Professional Working with Active People,* 3rd ed. C. A. Rosenbloom (Chicago: The American Dietetic Association, 2000), pp. 95–106.

[c]American College of Sports Medicine, Position stand: Heat and cold illness during distance running, *Medicine and Science in Sports and Exercise* 28 (1996): i–x.

[d]J. H. Wilmore and coauthors, Role of taste preference on fluid intake during and after 90 minutes of running at 60% of VO$_2$max in the heat, *Medicine and Science in Sports and Exercise* 30 (1998): 587–595.

Food for Fitness

No particular diet best supports physical performance. Many different diets can be excellent for active people. To support health and athletic performance, however, food choices must comply with the rules for diet planning.

Choosing a Diet to Support Physical Activity

A physically active person needs a diet composed mostly of nutrient-dense foods, the kind that supply a maximum of vitamins and minerals for the energy they provide. When active people eat mostly refined, processed foods that have suffered nutrient losses and contain added sugar and fat, nutrition status suffers. Even if foods are fortified or enriched, manufacturers cannot replace the whole range of nutrients and nonnutrients lost in refining. Consider, for example, that manufacturers mill out much of a food's original magnesium and chromium but do not replace them. This doesn't mean that active people can never choose a white bread, bologna, and mayonnaise sandwich but only that later they should eat a large salad or big portions of vegetables and whole grains and drink a glass of milk to compensate. The nutrient-dense foods will provide the magnesium and chromium; the bologna sandwich provided extra energy, mostly from fat.

Balance Active people need to eat both for nutrient density and for energy. Physically active people, like most people, need to limit fat and saturated fat. Simply stated, a diet that is high in carbohydrate (60 to 70 percent of total calories), low in fat (20 to 30 percent), and adequate in protein (12 to 15 percent) is best for all these purposes. Even if the athlete does not compete in glycogen-depleting events, such a diet will provide adequate fiber while supplying abundant nutrients and energy.

With these principles in mind, compare the two 500-kcalorie sandwich meals in the margin. To get enough carbohydrate energy, just reduce the amount of fat and meat in a meal, and let carbohydrate-rich foods fill in for them.

Adding carbohydrate-rich foods is a sound and reasonable option for increasing energy intake, up to a point. It becomes unreasonable when the person cannot eat enough food to meet energy needs. At that point, the person can cram more food energy into the diet only by using refined sugars and fats or liquid meals. Still, these energy-rich additions must be superimposed on nutrient-rich choices; energy alone is not enough.

Some athletes use commercial high-carbohydrate liquid supplements to obtain the carbohydrate and energy needed for heavy training and top performance. Most of these products contain **glucose polymers** and about 18 to 24 percent carbohydrate. These supplements do not *replace* regular food; they are meant to be used in *addition* to it. Unlike the sports beverages discussed in the "How to" box, these high-carbohydrate supplements are too concentrated in carbohydrate to be used for fluid replacement.

Protein In addition to carbohydrate and some fat, active people need protein. What quantities of what kinds of foods supply enough protein to meet their needs? Meats and milk products head the list of protein-rich foods, but to recommend that active people eat more than the recommended servings of meat would be shortsighted advice for many reasons. People must protect themselves from heart disease, and even lean meats contain fat, much of it saturated fat. Besides, the extra servings of carbohydrate-rich foods such as legumes, grains, and vegetables that active people need to meet energy requirements also boost protein intakes.

Earlier in this chapter, Table 10–4 showed recommended protein intakes for a 55-kilogram female athlete and a 70-kilogram male athlete. It is likely that a person weighing 70 kilograms who engages in vigorous physical activity on a daily basis could require 3000 kcalories or more per day. As a general rule, endurance athletes should aim for an average intake of 50 kcalories per kilogram (2.2 pounds) of body weight (23 kcalories per pound of body weight). Others may need more. To meet such an energy requirement, an athlete should select from a variety of nutrient-dense foods. Figure 10–4 on p. 256 shows an example of a day's meals that provide 3300 kcalories and the extra nutrients athletes need. These meals supply over 130 grams of protein, more than the highest recommended intake for an athlete. For those with reasonable diets, protein is rarely a problem.

The meals in Figure 10–4 provide 63 percent of their kcalories from carbohydrate. Athletes who train exhaustively for endurance events may want to aim for somewhat higher carbohydrate levels—from 65 to 75 percent. Notice that breakfast, though light in fat, is filling and hearty. Current thinking supports the idea that athletes benefit from such a morning start. If you train early in the morning, try splitting breakfast into two parts. An hour or so before training, eat some toast, juice, and fruit. Later, after your workout, come back for the cereal and milk.

Small daily choices, when made consistently, enhance nutritional health.

Compare and decide which better meets your needs:

- *1 sandwich of 2 slices bologna, 2 slices white bread, 2 tbs mayonnaise (525 kcal, 9% protein, 23% carbohydrate, 68% fat).*

or

- *2 sandwiches of 2 slices lean ham, 4 slices whole-wheat bread, 2 tsp mayonnaise (503 kcal, 20% protein, 51% carbohydrate, 29% fat).*

Looking for an amino acid supplement that rates a perfect score of 100 for protein quality? Try 1 oz of chicken breast—it provides almost 10,000 mg of amino acids in perfect complement for use by the human body.

glucose polymers: compounds that supply glucose, not as single molecules, but linked in chains somewhat like starch. The objective is to attract less water from the body into the digestive tract.

Fitness and Nutrition

Breakfast
1 c shredded wheat.
1 c 1% low-fat milk.
1 small banana.
2 slices whole-wheat toast.
4 tsp jelly.
1½ c orange juice.

Snack
3 c plain popcorn.
A smoothie made from:
    1½ c apple juice.
    1½ frozen banana.

Lunch
2 turkey sandwiches.
1½ c 1% low-fat milk.
Large bunch of grapes.

Dinner
Salad: 1 c spinach, carrots,
and mushrooms.
    ½ c garbanzo beans.
    1 tbs sunflower seeds.
    1 tbs ranch salad
    dressing.
1 c spaghetti with
meat sauce.
1 c green beans.
1 corn on the cob.
2 slices Italian bread.
4 tsp butter.
1 piece angel food cake.
1¼ c fresh strawberries.
1 tbs whipping cream.
1 c 1% low-fat milk.

Total kcal: 3300

63% kcal from carbohydrate
22% kcal from fat
15% kcal from protein

All vitamin and mineral intakes exceed
the RDA for both men and women.

Figure 10–4
An Athlete's Meals

Good choices for pregame meals:
- *Apricot nectar, pineapple juice, grape juice, banana, toast with jam or jelly, pancakes with syrup, baked white or sweet potatoes, pasta with steamed vegetables, lentils or other peas, raisins, figs, dates, frozen yogurt, graham crackers, sponge cake, angel food cake.*

Not recommended:
- *Stuffing, muffins, biscuits, croissants, french fries, onion rings, potato chips, meats, cheese, pies, ice cream, eggnog, creams, nuts, butter, gravy, mayonnaise, salad dressing, frosted cakes.*

pregame meal: a meal eaten three to four hours before athletic competition.

Planning an Athlete's Meals

Table 10–6 shows some sample food patterns at various high-energy and high-carbohydrate intakes. These plans are effective only if the user chooses foods that provide nutrients as well as energy: extra milk for calcium and riboflavin; many servings of fruit for folate and vitamin C; energy-rich vegetables such as sweet potatoes, peas, and legumes; modest portions of lean meat for iron and other vitamins and minerals; and whole grains for B vitamins, magnesium, zinc, and chromium. In addition, these foods provide plenty of electrolytes.

Pregame Meals Science has recommendations for the **pregame meal.** The foods should be carbohydrate-rich and the meal light (300 to 800 kcalories). It should be easy to digest and should contain fluids. Breads, potatoes, pasta, and fruit juices—carbohydrate-rich foods low in fat, protein, and fiber—form the basis of the pregame meal. Bulky, fiber-rich foods such as raw vegetables or high-fiber cereals, although usually desirable, are best avoided just before competition. Such foods can cause stomach discomfort during performance. The competitor should finish eating three to four hours before competition to allow time for the stomach to empty before exertion.

What about drinks or candylike sport bars claiming to provide "complete" nutrition? These mixtures of carbohydrate, protein (usually amino acids), fat, some fiber, and certain vitamins and minerals usually taste good and provide additional food energy before a game or for those needing to gain weight. They fall short of providing "complete" nutrition, however, since they lack many of real food's nutrients and the nonnutrients that benefit health. These products

Table 10–6 High-Carbohydrate Food Patterns

Food Group	Number of Servings for a Daily Energy Intake of:					
	1500 kcal	2000 kcal	2500 kcal	3000 kcal	3500 kcal	4000[a] kcal
Milk	3	3	4	4	4	4
Fruit	5	6	7	9	10	12
Vegetable	3	3	3	5	6	7
Grain	7	11	16	18	20	24
Fat[b]	2	3	5	6	8	10
Meat (ounces)	5	5	5	5	6	6
Percent carbohydrate:	58%	58%	63%	64%	60%	62%

[a]A way to add more energy to the diet without adding much bulk is to snack on milkshakes or "complete meal" liquid supplements (see the text).

[b]A fat serving is 1 teaspoon of butter, margarine, oil, or the equivalent.

provide no special advantage for active people except one—they are easy to eat in the hours before competition. They are expensive, however.

Postgame Meals As mentioned earlier, eating high-carbohydrate foods *after* physical activity enhances glycogen storage. Since people usually are not hungry immediately following physical activity, carbohydrate-containing beverages such as sports drinks or fruit juices may be preferred. If an active person does feel hungry after an event, then foods high in carbohydrate and low in protein, fat, and fiber are the ones to choose—the same ones recommended prior to competition. Foods high in protein and fat should be avoided during the first few hours after activity as these foods may suppress hunger and thus limit carbohydrate intake.

In Summary The person who wants to excel physically will apply the most accurate nutrition knowledge along with dedication to rigorous training. A diet that provides ample fluid and consists of a variety of nutrient-dense foods in quantities to meet energy needs will enhance not only athletic performance but overall health as well. Training and genetics being equal, who would win a competition—the person who habitually consumes less than the amounts of nutrients needed or the person who arrives at the event with a long history of full nutrient stores and well-met metabolic needs?

Self Study

FITNESS AND NUTRITION

Fitness depends on a certain minimum amount of physical activity. Ideally, the quantity and quality of the physical activity you select will improve your cardiorespiratory endurance, body composition, strength, and flexibility. Examine your activity choices by keeping an activity diary for one week. For each physical activity, be sure to record the type of activity, the level of intensity, and the duration. In addition, record the times and places of beverage consumption and the types and amounts of beverages consumed. Now compare the choices you made in your one-week activity diary to the guidelines for physical fitness (see Table 10–1).

- How often were you engaged in aerobic activity to improve cardiorespiratory endurance? Was the intensity of aerobic activity between 55 and 90 percent of your maximum heart rate? Did each session last at least 20 minutes?

- How often did you participate in resistance activities to develop strength? Was the intensity enough to enhance muscle strength and improve body composition? Did you perform 8 to 10 different exercises, repeating each one 8 to 12 times?

- How often did you stretch to improve your flexibility? Was the intensity enough to develop and maintain a full range of motion? Did you hold each stretch 10 to 30 seconds and repeat each stretch at least four times?

- Do you drink plenty of fluids daily, especially water, before, during, and after physical activity?

- What changes could you make to improve your fitness?

Self Check

1. Regular physical activity helps protect against:
   a. backaches, cancer, and emphysema.
   b. cancer, diabetes, and heart disease.
   c. obesity, kidney disease, and anemia.
   d. high blood pressure, cancer, and osteopenia.

2. Fitness benefits health by:
   a. increasing lean body tissue and enhancing resistance to colds and other infectious diseases.
   b. lowering the risk of heart disease, decreasing muscle mass, and improving nutritional health.
   c. building bone strength, lowering the risk of some cancers, and increasing anxiety.
   d. reducing diabetes risk, compromising lung function, and promoting a strong self-image.

3. Which of the following characteristics is not a component of fitness?
   a. muscle endurance
   b. conditioning
   c. flexibility
   d. muscle strength

4. The progressive overload principle can be applied by performing:
   a. an activity less often.
   b. an activity for a shorter time.
   c. an activity with more intensity.
   d. a different activity each day of the week.

5. Cool-down activities after physical activity:
   a. facilitate the relaxation of tight muscles.
   b. enhance blood circulation through the muscles.
   c. ease the transition from activity to normal function.
   d. all of the above.

6. "Hitting the wall" is a term runners sometimes use to describe:
   a. dehydration.
   b. competition.
   c. indigestion.
   d. glucose depletion

7. Conditioned muscles rely less on _____ and more on _____ for energy.
   a. protein; fat
   b. fat; protein
   c. glycogen; fat
   d. fat; glycogen

8. Physically active young women, especially those who are endurance athletes, are prone to:
   a. energy excess.
   b. iron deficiency.
   c. protein overload.
   d. iodine deficiency.

9. Plain, cool water is the best fluid for everyday active people because it:
   a. rapidly leaves the digestive tract to enter the tissues and cool the body.
   b. tastes good.
   c. provides carbohydrate.
   d. leaves the digestive tract slowly.

10. A recommended pregame meal includes plenty of fluids and provides between:
    a. 300 and 800 kcalories, mostly from fat-rich foods.
    b. 50 and 100 kcalories, mostly from fiber-rich foods.
    c. 1000 and 2000 kcalories, mostly from protein-rich foods.
    d. 300 and 800 kcalories, mostly from carbohydrate-rich foods.

Critical Thinking

1. Some of the benefits of cardiorespiratory conditioning include:
    a. the blood carries less oxygen, the pulse rate slows down, and blood pressure increases.
    b. the blood carries less oxygen, the pulse rate increases, and blood pressure increases.
    c. the blood carries more oxygen, the blood moves more easily, and blood pressure increases.
    d. the blood carries more oxygen, the pulse rate slows down, and blood pressure falls.

2. Which of the following compounds is produced when muscles break down glucose anaerobically?
    a. ascorbic acid
    b. fatty acid
    c. lactic acid
    d. phytic acid

3. Which of the following is a characteristic of water metabolism during physical activity?
    a. The maximum loss of fluid per hour of exercise is about 0.5 quart.
    b. In cold weather, the need for water decreases dramatically because the body does not sweat.
    c. Sweat losses can exceed the capacity of the GI tract to absorb water, resulting in some degree of dehydration.
    d. Heavy sweating leads to a marked rise in the thirst sensation to stimulate water intake, which delays the onset of dehydration.

Answers to these questions appear in Appendix G.

Clinical Applications

1. During her freshman year in college, Kim spent much of her free time bonding with new friends over pizzas, burgers, and fried chicken. Now in her sophomore year, she has decided that it is time to shed the "freshman 15" she gained and get fit quickly.* She is replacing her pizzas and burgers with protein shakes and diet sodas; she avoids carbohydrates; and she takes vitamin supplements because she's heard they'll give her energy. Kim's demanding sophomore academic load leaves little time during the week to do anything but go to class, study, and work on assignments. She has therefore decided to spend much of her weekend time running, working out with weights, swimming, and playing pick-up soccer on Sunday afternoons. After a few weeks of her new eating and fitness plan, however, Kim is feeling tired and run down much of the time, her muscles ache, and she falls asleep when she's trying to study.

   ▪ What dietary advice would you suggest to help Kim feel healthier and more energetic?
   ▪ What fitness strategies would you offer Kim to prevent her fatigue and sore muscles?

2. Zak, a junior in college, has been weight training four days a week for the past year to gain strength and bulk up his muscles. Along with his training, he has changed his diet from a normal mixed diet to one that emphasizes large servings of meat, eggs, and milk. Zak also takes amino acid supplements in hopes of building more muscle, faster. What advice would you offer Zak about his dietary habits? What would you tell him about the supplements he is taking?

*Idea borrowed from Frances Sizer and Eleanor Whitney, *Nutrition Concepts and Controversies*, 9th ed. (Belmont, Calif.: Wadsworth/Thomson Learning, 2002).

Nutrition on the Net

For further study of the topics of this chapter, access these websites.

Find updates and quick links to these and other nutrition-related sites at our website:
www.wadsworth.com/nutrition/

Search the American College of Sports Medicine site for information on physical fitness:
www.acsm.org

Explore the many resources offered on the Nutrition and Physical Activity site from the Centers for Disease Control and Prevention:
www.cdc.gov/nccdphp/dnpa/

Visit the U.S. Government site for the Surgeon General's Report on Physical Activity:
www.cdc.gov/nccdphp/sgr/sgr.htm

To learn more about the President's Council on Physical Fitness and Sports, visit:
www.whitehouse.gov/WH/PCPFS/html/fitnet.html

Visit the Shape Up America site at:
www.shapeup.org

For information on sports drinks, visit the Gatorade Sports Science Institute site at:
www.gssiweb.com

Search among thousands of current scientific and medical abstracts for any topic related to exercise physiology at:
www.ncbi.nlm.nih.gov/PubMed/

Notes

1. American College of Sports Medicine, Position stand: The recommended quantity and quality of exercise for developing and maintaining cardiorespiratory and muscular fitness, and flexibility in healthy adults, *Medicine and Science in Sports and Exercise* 30 (1998): 975–991.

2. U.S. Department of Agriculture and U.S. Department of Health and Human Services, *Dietary Guidelines for Americans* (Washington, D.C.: Government Printing Office, 2000).

3. D. A. Leaf, D. L. Parker, and D. Schaad, Changes in VO_2max, physical activity, and body fat with chronic exercise: Effects on plasma lipids, *Medicine and Science in Sports and Exercise* 29 (1997): 1152–1159; P. T. Williams, Relationship of distance run per week to coronary heart disease risk factors in 8283 male runners, *Archives of Internal Medicine* 157 (1997): 191–198.

4. F. B. Hu and coauthors, Walking compared with vigorous physical activity and risk of type 2 diabetes in women, *Journal of the American Medical Association* 282 (1999): 1433–1439; S. W. Farrell and coauthors, Influences of cardiorespiratory fitness levels and other predictors on cardiovascular disease mortality in men, *Medicine and Science in Sports and Exercise* 30 (1998): 899–905; U.S. Department of Health and Human Services, *Physical Activity and Health: A Report of the Surgeon General Executive Summary* (Washington, D.C.: Government Printing Office, 1996).

5. U.S. Department of Health and Human Services, 1996.

6. F. B. Hu and coauthors, Physical activity and risk of stroke in women, *Journal of the American Medical Association* 283 (2000): 2961–2967; S. D. Hsieh and coauthors, Regular physical activity and coronary risk factors in Japanese men, *Circulation* 97 (1998): 661–665; I. Thune and coauthors, Physical activity and the risk of breast cancer, *New England Journal of Medicine* 336 (1997): 1269–1275; S. N. Blair, Physical inactivity and cardiovascular disease risk in women, *Medicine and Science in Sports and Exercise* 28 (1996): 9–10; NIH Consensus Development Panel on Physical Activity and Cardiovascular Health, Physical activity and cardiovascular health, *Journal of the American Medical Association* 276 (1996): 241–246.

7. A. L. Dunn and coauthors, Comparison of lifestyle and structured interventions to increase physical activity and cardiorespiratory fitness, *Journal of the American Medical Association* 281 (1999): 327–334.

8. U. M. Kujala and coauthors, Relationship of leisure-time physical activity and mortality: The Finnish Twin Cohort, *Journal of the American Medical Association* 279 (1998): 440–444.

9. J. H. Wilmore, Increasing physical activity: Alterations in body mass and composition, *American Journal of Clinical Nutrition* 63 (1996): 456S–460S.

10. R. D. Lewis, Nutrition, physical activity, and bone health in women, *International Journal of Sports Nutrition* 8 (1998): 250–284; E. Ernst, Exercise for female osteoporosis: A systematic review of randomised clinical trials, *Sports Medicine* 6 (1998): 359–368; D. Teegarden and coauthors, Previous physical activity relates to bone mineral measures in young women, *Medicine and Science in Sports and Exercise* 28 (1996): 105–113.

11. L. Hoffman-Goetz, Immunocompetence in physical activity, in *Nutrition in Exercise and Sport*, 3rd ed., I. Wolinsky (Boca Raton, Fla.: CRC Press, 1998), pp. 645–657.

12. M. E. Martinez and coauthors, Physical activity, body mass index, and prostaglandin E_2 levels in rectal mucosa, *Journal of the National Cancer Institute* 91 (1999): 950-953; Thune and coauthors, 1997; M. M. Kramer and C. L. Wells, Does physical activity reduce risk of estrogen-dependent cancer in women? *Medicine and Science in Sports and Exercise* 28 (1996): 322–334.

13. Hu and coauthors, 2000; J. E. Manson and coauthors, A prospective study of walking as compared with vigorous exercise in the prevention of coronary heart disease in women, *New England Journal of Medicine* 341 (1999): 650–658; Hsieh and coauthors, 1998; C. D. Lee, A. S. Jackson, and S. N. Blair, US weight guidelines: Is it also important to consider cardiorespiratory fitness? *International Journal of Obesity and Related Metabolic Disorders* 22 (1998): S2–S7; A. R. Folsom and coauthors, Physical activity and incidence of coronary heart disease in middle-aged women and men, *Medicine and Science in Sports and Exercise* 29 (1997): 901–909; G. B. M. Mensink and coauthors, Intensity, duration, and frequency of physical activity and coronary risk factors, *Medicine and Science in Sports and Exercise* 29 (1997): 1192–1198; Blair, 1996; NIH Consensus Development Panel on Physical Activity and Cardiovascular Health, 1996; P. T. Williams, High-density lipoprotein cholesterol and other risk factors for coronary heart disease in female runners, *New England Journal of Medicine* 334 (1996): 1298–1303.

14. G. R. Hunter and coauthors, Fat distribution, physical activity, and cardiovascular risk factors, *Medicine and Science in Sports and Exercise* 29 (1997): 362–369; A. Goulding and coauthors, More exercise, less central fat distribution in women, *Journal of the American Medical Association* 276 (1996): 193–194.

15. G. Perseghin and coauthors, Increased glucose transport-phosphorylation and muscle glycogen synthesis after exercise training in insulin-resistant subjects, *New England Journal of Medicine* 335 (1996): 1357–1362; S. N. Blair and coauthors, Physical activity, nutrition, and chronic disease, *Medicine and Science in Sports and Exercise* 28 (1996): 335–349.

16. M. F. Leitzmann and coauthors, Recreational physical acitvity and the risk of cholecystectomy in women, *New England Journal of Medicine* 341 (1999): 777–784.

17. Kujala and coauthors, 1998; L. H. Kushi and coauthors, Physical activity and mortality in postmenopausal women, *Journal of the American Medical Association* 277 (1997): 1287–1292.

18. A. A. Hakim and coauthors, Effects of walking on mortality among nonsmoking retired men, *New England Journal of Medicine* 338 (1998): 94–99.

19. L. DiPietro, The epidemiology of physical activity and physical function in older people, *Medicine and Science in Sports and Exercise* 28 (1996): 596–600.

20. American College of Sports Medicine, *ACSM's Guidelines for Exercise Testing and Prescription*, 6th ed. (Philadelphia: Lippincott, Willliams, & Wilkins, 2000), pp. 22–32; American College of Sports Medicine and American Heart Association, Joint position statement: Recommendations for cardiovascular screening, staffing, and emergency policies at health/fitness facilities, *Medicine and Science in Sports and Exercise* 30 (1998): 1009–1018.

21. M. L. Pollock and coauthors, AHA Science Advisory: Resistance exercise in individuals with and without cardiovascular disease:

Benefits, rationale safety, and prescription, *Circulation* 101 (2000): 828–833.

22. American College of Sports Medicine, Position stand: Exercise and physical activity for older adults, *Medicine and Science in Sports and Exercise* 30 (1998): 992–1008; P. A. Ades and coauthors, Weight training improves walking endurance in healthy elderly persons, *Annals of Internal Medicine* 124 (1996): 568–572.

23. J. E. Layne and M. E. Nelson, The effects of progressive resistance training on bone density: A review, *Medicine and Science in Sports and Exercise* 31 (1999): 25–30.

24. L. Metcalfe and coauthors, Postmenopausal women and exercise for prevention of osteoporosis: The bone, estrogen, strength training (BEST) study, *ACSM's Health and Fitness Journal,* May/June 2001.

25. Williams, 1996.

26. C. M. Donovan and K. D. Sumida, Training enhanced hepatic gluconeogenesis: The importance for glucose homeostasis during exercise, *Medicine and Science in Sports and Exercise* 29 (1997): 628–634.

27. C. Williams and C. Chryssanthopoulos, Pre-exercise food intake and performance, in *Nutrition and Fitness: Metabolic and Behavioral Aspects in Health and Disease,* ed. A. P. Simopoulos and K. N. Pavlou (New York: Karger, 1997), pp. 33–45.

28. J. A. M. Parkin and coauthors, Muscle glycogen storage following prolonged exercise: Effect of timing of ingestion of high glycemic index food, *Medicine and Science in Sports and Exercise* 29 (1997): 220–224.

29. E. Coleman, Carbohydrate and exercise, in *Sports Nutrition; A Guide for the Professional Working with Active People,* ed. C. A. Rosenbloom (Chicago: The American Dietetic Association, 2000, pp. 13–31; L. M. Burke, G. R. Collier, and M. Hargreaves, Glycemic index—A new tool in sports nutrition? *International Journal of Sports Nutrition* 8 (1998): 401–415.

30. A. R. Coggan, Plasma glucose metabolism during exercise: Effect of endurance training in humans, *Medicine and Science in Sports and Exercise* 29 (1997): 620–627.

31. Diabetes in the elite athlete, *Sports Medicine Digest* 18 (1998): 109–111.

32. P. J. Horvath and coauthors, The effects of varying dietary fat on performance and metabolism in trained male and female runners, *Journal of the American College of Nutrition* 19 (2000): 52–60.

33. E. Coleman, Does a low-fat diet impair nutrition and performance? *Sports Medicine Digest* 22 (2000): 41; Position of The American Dietetic Association, Dietitians of Canada, and the American College of Sports Medicine: Nutrition and athletic performance, *Journal of the American Dietetic Association* 100 (2000): 1543–1556.

34. M. Klem and coauthors, A descriptive study of individuals successful at long-term maintenance of substantial weight loss, *American Journal of Clinical Nutrition* 66 (1997): 239–246.

35. L. Pescatello and D. Murphy, Lower intensity physical activity is advantageous for fat distribution and blood glucose among viscerally obese older adults, *Medicine and Science in Sports and Exercise* 30 (1998): 1408–1413.

36. G. R. Hunter and coauthors, A role for high intensity exercise on energy balance and weight control, *International Journal of Obesity and Related Metabolic Disorders* 22 (1998): 489–493.

37. T. J. Horton and C. A. Geissler, Effect of habitual exercise on daily energy expenditure and metabolic rate during standardized activity, *American Journal of Clinical Nutrition* 59 (1994): 13–19.

38. M. A. Grediagin and coauthors, Exercise intensity does not affect body composition change in untrained, moderately overfat women, *Journal of the American Dietetic Association* 95 (1995): 661–665.

39. P. W. R. Lemon, Is increased dietary protein necessary or beneficial for individuals with a physically active lifestyle? *Nutrition Reviews* 54 (1996): S169–S175.

40. Lemon, 1996.

41. Position of The American Dietetic Association, Dietitians of Canada, and the American College of Sports Medicine, 2000.

42. J. M. McBride and coauthors, Effect of resistance exercise on free radical production, *Medicine and Science in Sports and Exercise* 30 (1998): 67–72; D. A. Leaf and coauthors, The effect of exercise intensity on lipid peroxidation, *Medicine and Science in Sports and Exercise* 29 (1997): 1036–1039; R. A. Fielding and M. Meydani, Exercise, free radical generation, and aging, *Aging: Clinical and Experimental Research* 9 (1997): 12–18.

43. P. M. Clarkson and H. S. Thompson, Antioxidants: What role do they play in physical activity and health? *American Journal of Clinical Nutrition* 72 (2000): 637S–646S; K. V. Reddy and coauthors, Pulmonary lipid peroxidation and antioxidant defenses during exhaustive physical exercise: The role of vitamin E and selenium, *Nutrition* 14 (1998): 448–451.

44. McBride and coauthors, 1998; L. Grievink and coauthors, Acute effects of ozone on pulmonary function of cyclists receiving antioxidant supplements, *Occupational and Environmental Medicine* 55 (1998): 13–17; Reddy and coauthors, 1998; M. Kanter, Free radicals, exercise and antioxidant supplementation, *Proceedings of the Nutrition Society* 57 (1998): 9–13.

45. W. J. Evans, Vitamin E, vitamin C, and exercise, *American Journal of Clinical Nutrition* 72 (2000): 647S–652S; L. Packer, Oxidants, antioxidant nutrients, and the athlete, *Journal of Sports Science* 15 (1997): 353–363.

46. J. Beard and B. Tobin, Iron status and exercise, *American Journal of Clinical Nutrition* 72 (2000): 594S–597S.

47. M. F. Waller and E. M. Haymes, The effects of heat and exercise on sweat iron loss, *Medicine and Science in Sports and Exercise* 28 (1996): 197–203.

48. Beard and Tobin, 2000; E. Coleman, Nutritional concerns of vegetarian athletes, *Sports Medicine Digest* 20 (1998): 22–23.

49. D. C. Nieman, Physical fitness and vegetarian diets: Is there a relation? *American Journal of Clinical Nutrition* 70 (1999): 570S–575S.

50. E. R. Eichner, Anemia in female athletes, *Sports Medicine Digest* 22 (2000): 42–43; Z. Y. Haas, Iron depletion without anemia and physical performance in young women, *American Journal of Clinical Nutrition* 66 (1997): 334–341.

51. M. N. Sawka and S. J. Montain, Fluid and electrolyte supplementation for exercise heat stress, *American Journal of Clinical Nutrition* 72 (2000): 564S–572S; C. V. Gisolfi, Fluid balance for optimal performance, *Nutrition Reviews* 54 (1996): S159–S168.

52. Gisolfi, 1996.

53. American College of Sports Medicine, Position stand: Heat and cold illness during distance running, *Medicine and Science in Sports and Exercise* 28 (1996): i–x.

CHAPTER 10

Nutrition in Practice

■ SUPPLEMENTS AND ERGOGENIC AIDS ATHLETES USE ■

In a world where body conditioning and skill are hard won, athletes gravitate to promises that they can easily improve their performance by taking pills, powders, or potions. Athletes often hear well-intended, but unsubstantiated, advice from their coaches and peers recommending that they use special nutrients, drugs, or procedures to enhance performance. The wish to win is strong but no amount of wishing can change the fact that a large majority of supplements sold for athletes are frauds. If the products that are tried have no effect and are harmless, they are only a waste of money; but some products are harmful or actually impair performance, and these are a waste of athletic potential as well. This Nutrition in Practice looks at scientific evidence for and against some of the most common dietary supplements and hormonal preparations available to athletes, and the glossary on p. 263 defines them.

What does ergogenic mean?

Ergogenic means work enhancing or work producing. In connection with athletic performance, **ergogenic aids** are substances or treatments that purportedly improve athletic performance above and beyond what is possible through training alone. Research findings do not, for the most part, support the claims made for ergogenic aids. When you hear a claim that a product is ergogenic, remember to consider the source of the claim and ask who may gain from the sale.

My coach told me to take protein supplements. Should I take them?

Protein powders and amino acid supplements are among the most common supplements athletes use.[1] The supplements are advocated to improve both strength and endurance, but as discussed in Chapter 10, well-nourished active people and athletes do not need them. Extra protein cannot be forced into the muscles to make them grow. Muscle cells accept nutrients only when they are needed. The cells "decide" what they need, based on the messages they receive from the hormones that regulate them and from demands put upon them. The way to make muscle cells grow, therefore, is to make them work. The only role

for diet in this process is to make protein available, and good diets always do. Although the protein needs of some endurance and strength athletes are higher than those of sedentary people, the additional protein is already present in a well-chosen diet as Chapter 10 described.

Whey protein appears to be particularly popular among athletes hoping to achieve greater muscle gains. A waste product of cheese manufacturing, whey protein is a common ingredient in many low-cost protein powders. Athletes and active people who want bigger muscles should know that whey protein does not increase muscle mass. To build bigger muscles, they need to eat food with adequate energy and protein to support the weight-training work that does increase muscle mass. Those who still think they need more whey should pour a glass of milk; one cup provides 1.5 grams of whey.

Most healthy athletes eating well-balanced diets do not need amino acid supplements either. Advertisers point to research that identifies the **branched-chain amino acids** as the main ones used as fuel by exercising muscles. What the ads leave out is that compared to glucose and fatty acids, branched-chain amino acids provide almost no fuel and that, anyway, ordinary foods provide them in abundance.

Large doses of branched-chain amino acids can raise plasma ammonia concentrations, which can be toxic to the brain.[2] Branched-chain amino acid supplements are neither effective nor safe and are not recommended.

I know that some athletes, especially endurance athletes, are taking carnitine supplements. What is carnitine?

Carnitine is a nonessential nutrient. Endurance athletes believe carnitine will help them burn more fat, thereby sparing glycogen during endurance events. Carnitine is also promoted to bodybuilders as a "fat burner."

In the body, carnitine facilitates the transfer of fatty acids across the mitochrondrial membrane. Supplement manufacturers suggest that with more carnitine available, fat oxidation will be enhanced, but this does

Nutrition in Practice

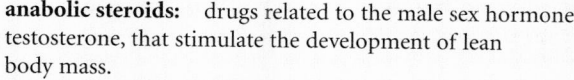

Glossary

anabolic steroids: drugs related to the male sex hormone, testosterone, that stimulate the development of lean body mass.

anabolic = promoting growth

sterols = compounds chemically related to cholesterol

branched-chain amino acids: the amino acids leucine, isoleucine, and valine, which are present in large amounts in skeletal muscle tissue; falsely promoted as fuel for exercising muscles.

caffeine: a natural stimulant found in many common foods and beverages, including coffee, tea, and chocolate; may enhance endurance by stimulating fatty acid release but also causes fluid losses. High doses cause headaches, trembling, rapid heart rate, and other undesirable side effects.

carnitine (CAR-ne-teen): a non-protein amino acid made in the body from lysine that helps transport fatty acids across the mitochondrial membrane. Carnitine supposedly "burns" fat and spares glycogen during endurance events, but in reality it does neither.

chromium (CROW-mee-um) **picolinate:** a trace mineral supplement; falsely promoted as building muscle, enhancing

energy, and burning fat. **Picolinate** (pick-oh-LYN-ate) is a derivative of the amino acid tryptophan that seems to enhance chromium absorption.

creatine (KREE-ah-tin): a nitrogen-containing compound that combines with phosphate to form the high-energy compound creatine phosphate (or phosphocreatine) in muscles. Claims that creatine enhances energy use and muscle strength need further confirmation.

DHEA (dehydroepiandrosterone) and **androstenedione:** hormones made in the adrenal glands that serve as precursors to the male hormone testosterone; falsely promoted as burning fat, building muscle, and slowing aging. Side effects include acne, aggressiveness, and liver enlargement.

ergogenic (ER-go-JEN-ick) **aids:** substances or techniques used in an attempt to enhance physical performance.

ergo = work

genic = gives rise to

whey protein: a by-product of cheese production; falsely promoted as increasing muscle mass.

---

not seem to be the case. Carnitine supplementation for 7 to 14 days neither raised muscle carnitine concentrations nor influenced fat or carbohydrate oxidation. It did, however, produce diarrhea in half of the men tested. Milk and meat products are good sources of carnitine, and supplements are not needed.

My friend who is a bodybuilder takes a supplement called chromium picolinate. Advertisements in magazines and health-food stores make all kinds of impressive claims about it. Are any of the claims true?

Chapter 9 introduced chromium as an essential trace mineral involved in carbohydrate and lipid metabolism. Advertisements in bodybuilding magazines claim that **chromium picolinate,** which is supposed to be more easily absorbed than chromium alone, builds muscle, enhances energy, and burns fat. Such claims derive from one or two initial studies on this mineral. Most studies of chromium picolinate and strength training that have followed, however, show no effects of chromium picolinate supplementation on strength, lean body mass, or body fat.[3]

The safety record of chromium picolinate is not unblemished. One athlete who ingested 1200 micrograms of chromium picolinate over two days' time

developed a dangerous condition of muscle degeneration, with the supplement strongly suspected as the cause.[4] Chromium-sensitive people may respond to chromium picolinate supplements with allergic reactions.[5]

A lot of my friends are taking creatine. Why is it so popular?

Interest in, and use of, creatine monohydrate supplements to enhance energy production during intense activity has grown dramatically in the last few years.[6] Power athletes such as weight lifters use **creatine** supplements to enhance stores of the high-energy compound creatine phosphate (or phosphocreatine) in muscles. Theoretically, the more creatine phosphate in muscles, the higher the intensity at which an athlete can train. High-intensity training stimulates the muscles to adapt, which, in turn, improves performance.

The results of some studies suggest creatine supplementation enhances performance of high-intensity strength activity such as weight lifting or repeated sprinting.[7] Other research findings conflict with reports that creatine supplements improve strength performance.[8]

Nutrition in Practice

Some medical and fitness experts voice concern that, like many performance enhancement supplements before it, creatine is being taken in huge doses (5 to 30 grams per day) before evidence of its value or safety has been ascertained.[9] Even people who eat red meat, which is a creatine-rich food, do not consume near the amount athletes are taking. Athletes who take megadoses of creatine risk possible long-term side effects such as organ and muscle damage. Despite the uncertainties, creatine supplements are not illegal in international competition.

What about caffeine? I've heard that it can improve endurance performance.

Many athletes believe that, just as **caffeine** provides mental stimulation during late-night study sessions, it can provide physical stimulation that might improve performance in endurance sports. As reasonable as this may sound, research findings are mixed on this point.[10] For example, the rowing performance of both men and women seems to be improved by taking caffeine (6 milligrams per kilogram of body weight).[11] Sprinters, though, gain little in terms of a performance edge from caffeine.[12] Any potential benefits of caffeine use must be weighed against the adverse effects—stomach upset, nervousness, irritability, headaches, dehydration, and diarrhea. Caffeine also constricts the arteries and raises blood pressure above normal, making the heart work harder to pump blood to the working muscles, an effect detrimental to sports performance.

Competitors should be aware that college, national, and international athletic competitions prohibit the use of caffeine in amounts greater than the equivalent of 5 or 6 cups of coffee consumed in a two-hour period prior to competition. Athletes are disqualified if urine tests detect more than this amount.

A theory holds that caffeine may assist endurance by stimulating the release of fatty acids from storage, thereby reducing the demand on glycogen stores and helping to make carbohydrate fuel available for exercise longer. Recent studies have all but reversed this line of thinking, however.[13] One double-blind study tested glycogen in the thigh muscle and free fatty acid concentrations in the blood in 20 bicyclers taking caffeine or a placebo. There was no difference in depletion of glycogen or use of fatty acids between the caffeine group and the placebo group, leading researchers to conclude that while athletes may enjoy a "wake-up" effect from caffeine, it does not alter energy fuel use.[14]

I have heard that anabolic steroids are dangerous, but one of my friends takes them. His mother is a doctor, and she constantly monitors his blood pressure when he is taking steroids. Are they safe in his case?

Anabolic steroids are not safe in your friend's case or in any case; they have dangerous side effects and are illegal. Technically called androgenic-anabolic steroid drugs, they are derivatives of the male sex hormone testosterone. Testosterone promotes the development of male characteristics (androgenic) and lean body mass (anabolic). Athletes take steroids to stimulate muscle bulking. The American College of Sports Medicine and the American Academy of Pediatrics condemn the use of steroids by athletes, and the International Olympic Committee has banned their use.[15] In support of its position, the committee cites the known toxic side effects and maintains that steroid use is a form of cheating. Competitors who use the drugs put other athletes in the difficult position of either conceding an unfair advantage to abusing competitors or taking steroids and accepting the risk of untoward side effects.

The list of hazards and adverse reactions from steroids continues to grow amid only a slight decline in use of the drugs. Among the side effects and adverse reactions that steroids produce are cancerous liver tumors that impair liver function, causing it to rupture and hemorrhage; testicular shrinkage in men and masculinization of women; cardiovascular problems; and sterility.

Your friend is sure to develop side effects no matter how closely a trainer or doctor monitors him. Table NP10–1 lists side effects and adverse reactions to anabolic steroids.

The dangers of steroid use cannot be overemphasized. Health care professionals are obligated to warn athletes of these dangers. Speak simply and emphatically: the price for the potential competitive edge that steroids confer is damaged health and sometimes life itself. The safest effective way to build muscle has always been through hard training, and always will be.

What are DHEA and androstenedione, and why do some athletes use them?

Some athletes use **DHEA** (dehydroepiandrosterone) and **androstenedione** as alternatives to anabolic steroids. Androstenedione, or "andro," made headlines in the late 1990s when the media reported its use by baseball great Mark McGwire.

DHEA and androstenedione are hormones made in the adrenal glands that serve as precursors to the male

Nutrition in Practice

Table NP10–1 Anabolic Steroids: Side Effects and Adverse Reactions

Mind

- Extreme aggression with hostility ("steroid rage"); mood swings; anxiety; dizziness; drowsiness; unpredictability; insomnia; psychotic depression; personality changes, suicidal thoughts

Face and Hair

- Swollen appearance; greasy skin; severe, scarring acne; mouth and tongue soreness; yellowing of whites of eyes (jaundice)
- In females, male-pattern hair loss and increased growth of face and body hair

Voice

- In females, irreversible deepening of voice

Chest

- In males, breathing difficulty, breast development
- In females, breast atrophy

Heart

- Heart disease; elevated or reduced heart rate; heart attack; stroke; hypertension; increased LDL; reduced HDL

Abdominal Organs

- Nausea; vomiting; bloody diarrhea; pain; edema; liver tumors (possibly cancerous); liver damage, disease, or rupture leading to fatal liver failure; kidney stones and damage; gallstones; frequent urination; possible rupture of aneurysm or hemorrhage

Blood

- Blood clots; high risk of blood poisoning; those who share needles risk contracting HIV (the AIDS virus) or other disease-causing organisms; septic shock (from injections)

Reproductive System

- In males, permanent shrinkage of testes; prostate enlargement with increased risk of cancer; sexual dysfunction; loss of fertility; excessive and painful erections
- In females, loss of menstruation and fertility; permanent enlargement of external genitalia; fetal damage, if pregnant

Muscles, Bones, and Connective Tissues

- Increased susceptibility to injury with delayed recovery times; cramps; tremors; seizurelike movements; injury at injection site
- In adolescents, failure to grow to normal height

Other

- Fatigue; increased risk of cancer

OK, protein powders and amino acid supplements are ineffective performance enhancers. Results of studies on chromium picolinate are inconsistent, and experts are concerned about long-term effects of creatine. Caffeine may or may not be effective, but can have adverse side effects and is illegal. Steroids pose serious health risks and are illegal. Do any of the substances athletes use to boost performance work?

For the most part, no. Many of these substances have been studied and found to be worthless. The glossary on p. 266 lists and describes many more substances promoted as ergogenic aids.

Health professionals can positively influence athletes and others interested in boosting athletic performance by stressing the measures that do help to enhance performance. They are, of course, regular training and sound nutrition.

hormone testosterone. Advertisements claim the hormones "burn fat," "build muscle," and "slow aging," but evidence to support such claims is lacking.

Short-term side effects of DHEA and androstenedione include oily skin, acne, body hair growth, liver enlargement, and aggressive behavior.[16] Long-term effects of DHEA and androstenedione use remain to be seen and may take years to become evident. The potential for harm from these supplements is great, and athletes, as well as others, should avoid them. DHEA and androstenedione are banned by the International Olympic Committee and the National Collegiate Athletic Association.

Nutrition in Practice

Glossary of Substances Promoted as Ergogenic Aids

arginine: a nonessential amino acid falsely promoted as enhancing the secretion of human growth hormone, the breakdown of fat, and the development of muscle.

bee pollen: a product consisting of bee saliva, plant nectar, and pollen that supposedly aids in weight loss and boosts athletic performance. It does neither and may cause an allergic reaction in individuals sensitive to it.

boron: a nonessential mineral that is promoted as a "natural" steroid replacement.

brewer's yeast: a preparation of yeast cells, containing a concentrated amount of B vitamins and some minerals; falsely promoted as an energy booster.

cell salts: a preparation of minerals supposedly harvested from living cells, sold as a health-promoting supplement.

coenzyme Q10: a lipid found in cells (mitochondria) shown to improve exercise performance in heart disease patients, but not effective in improving the performance of healthy athletes.

desiccated liver: dehydrated liver powder that supposedly contains all the nutrients found in liver in concentrated form; possibly not dangerous, but has no particular nutritional merit and is considerably more expensive than fresh liver.

DNA (deoxyribonucleic acid): the genetic material of cells necessary in protein synthesis; falsely promoted as an energy booster.

epoetin (eh-poy-EE-tin): a drug derived from the human hormone erythropoietin and marketed under the trade name Epogen; illegally used to increase oxygen capacity.

gelatin: a soluble form of the protein collagen, used to thicken foods; sometimes falsely promoted as a strength enhancer.

ginseng: a plant whose extract supposedly boosts energy. Side effects of chronic use include nervousness, confusion, and depression.

glycine: a nonessential amino acid, promoted as an ergogenic aid because it is a precursor of the high-energy compound creatine phosphate. Other amino acids commonly packaged for athletes that are equally useless include tryptophan, ornithine, arginine, lysine, and the branched-chain amino acids.

growth hormone releasers: herbs or pills that supposedly regulate hormones; falsely promoted as enhancing athletic performance.

guarana: a reddish berry found in Brazil's Amazon valley that is used as an ingredient in carbonated sodas and taken in powder or tablet form. Guarana is marketed as an ergogenic aid to enhance speed and endurance, an aphrodisiac, a "cardiac tonic," an "intestinal disinfectant," and a smart drug that supposedly improves memory and concentration and wards off senility. Because guarana contains seven times as much caffeine as its relative the coffee bean, there are concerns that high doses can stress the heart and cause panic attacks.

herbal steroids or **plant sterols:** curious mixtures of herbs, "adaptogens," and "aphrodisiacs" that supposedly enhance hormone activity. Products marketed as herbal steroids include astragalus, damiana, dong quai, fo ti teng, ginseng root, licorice root, palmetto berries, sarsaparilla, schizardra, unicorn root, yohimbe bark, and yucca.

HMB (beta-hydroxy-beta-methylbutyrate): a metabolite of the branched-chain amino acid leucine. Claims that HMB increases muscle mass and strength are based on the results of two studies from the lab that developed HMB as a supplement.

inosine: an organic chemical that is falsely said to "activate cells, produce energy, and facilitate exercise," but actually has been shown to reduce the endurance of runners.

ma huang: an evergreen plant derivative that supposedly boosts energy and helps with weight control. Ma huang contains ephedrine, a cardiac stimulant, and has been associated with high blood pressure, rapid heart rate, nerve damage, muscle injury, psychosis, stroke, and memory loss.

niacin: a B vitamin that when taken in excess rushes blood to the skin, producing vascularity and a red tint—physical attributes bodybuilders strive to attain prior to performance. These attributes do not enhance performance, and excess niacin can cause headaches and nausea.

octacosanol: an alcohol isolated from wheat germ; often falsely promoted as enhancing athletic performance.

ornithine: a nonessential amino acid falsely promoted as enhancing the secretion of human growth hormone, the breakdown of fat, and the development of muscle.

oryzanol: a plant sterol that supposedly provides the same physical responses as anabolic steroids without the adverse side effects; also known as *ferulic acid, ferulate,* or *FRAC.*

pangamic acid: also called vitamin B_{15} (but not a vitamin, nor even a specific compound—it can be anything with that label); falsely claimed to speed oxygen delivery.

phosphate pills: a product demonstrated to increase the levels of a metabolically important phosphate compound (diphosphoglycerate) in red blood cells and the potential of the cells to deliver oxygen to the body's muscle cells. However, it does not extend endurance or increase efficiency of aerobic metabolism and may cause calcium losses from the bones if taken in excess.

pyruvate: a 3-carbon compound derived during the metabolism of glucose, certain amino acids, and glycerol. Claims that pyruvate burns fat and enhances endurance are based on two studies of untrained individuals by the same author. Common side effects include intestinal gas and diarrhea.

RNA (ribonucleic acid): the genetic material of cells necessary for protein synthesis; falsely promoted as enhancing athletic performance.

(Continued on next page)

Nutrition in Practice

Glossary of Substances Promoted as Ergogenic Aids (continued)

royal jelly: the substance produced by worker bees and fed to the queen bee; falsely promoted as increasing strength and enhancing performance.

sodium bicarbonate: baking soda; an alkaline salt believed to neutralize blood lactic acid and thereby to reduce pain and enhance possible workload. "Soda loading" may cause intestinal bloating and diarrhea.

spirulina: a kind of alga ("blue-green manna") that supposedly contains large amounts of protein and vitamin B_{12}, sup-

presses appetite, and improves athletic performance. It does none of these things and is potentially toxic.

succinate: a compound synthesized in the body and involved in the TCA cycle; falsely promoted as a metabolic enhancer.

superoxide dismutase (SOD): an enzyme that protects cells from oxidation. When it is taken orally, the body digests and inactivates this protein; it is useless to athletes.

wheat germ oil: the oil from the wheat kernel; often falsely promoted as an energy aid.

Notes

1. E. A. Applegate and L. E. Grivetti, Search for the competitive edge: A history of dietary fads and supplements, *Journal of Nutrition* 127 (1997): 869S–873S.

2. E. Coleman, Branched-chain amino acids and fatigue, *Sports Medicine Digest* 18 (1996): 44.

3. J. M. Davis, R. S. Welsh, and N. A. Alerson, Effects of carbohydrate and chromium ingestion during intermittent high-intensity exercise to fatigue, *International Journal of Sports Nutrition and Exercise Metabolism* 10 (2000): 476–485; H. C. Lukaski and coauthors, Chromium supplementation and resistance training: Effects on body composition, strength, and trace element status of men, *American Journal of Clinical Nutrition* 63 (1996): 954–965; M. A. Hallmark and coauthors, Effects of chromium and resistance training on muscle strength and body composition, *Medicine and Science in Sports and Exercise* 28 (1995): 139–144.

4. W. R. Fuller, Suspected chromium picolinate–induced rhabdomyolysis, *Pharmacotherapy* 18 (1998): 860–862.

5. J. F. Fowler, Systemic contact dermatitis caused by oral chromium picolinate, *Cutis* 65 (2000): 116.

6. Applegate and Grivetti, 1997.

7. D. Preen and coauthors, Effect of creatine loading on long-term sprint exercise performance and metabolism, *Medicine and Science in Sports and Exercise* 33 (2001): 814–821; R. B. Kreider and coauthors, Effects of creatine supplementation on body composition, strength, and sprint performance, *Medicine and Science in Sports and Exercise* 30 (1998): 73–82; J. S. Volek and coauthors, Creatine supplementation enhances muscular performance during high-intensity resistance exercise, *Journal of the American Dietetic Association* 97 (1997): 765–770; S. M. Tolar, Creatine is an ergogen for anaerobic exercise, *Nutrition Reviews* 55 (1997): 21–23.

8. L. M. Odland and coauthors, Effect of oral creatine supplementation on muscle (PCr) and short-term maximum power output, *Medicine and Science in Sports and Exercise* 29 (1997): 216–219.

9. T. Noakes, as quoted in M. Gaie, Olympic athletes face heat, other health hurdles, *Journal of the American Medical Association* 276 (1996): 178–180.

10. C. D. Paton, W. G. Hopkins, and L. Vollebregt, Little effect of caffeine ingestion on repeated sprints in team-sport athletes, *Medicine and Science in Sports and Exercise* 33 (2001): 822–825; C. J. Sinclair and J. D. Geiger, Caffeine use in sports: A pharmacological review, *Journal of Sports Medicine and Physical Fitness* 40 (2000): 71–79; L. R. Bucci, Dietary supplements as ergogenic aids, in *Nutrition in Exercise and Sport*, 3rd ed., ed. I. Wolinsky (Boca Raton, Fla.: CRC Press, 1998), pp. 315–368; T. E. Graham and L. L. Spriet, Caffeine and exercise performance, *Sport Science Exchange* 1 (1996): 9.

11. C. R. Bruce and coauthors, Enhancement of 2000-m rowing performance after caffeine ingestion, *Medicine and Science in Sports and Exercise* 32 (2000): 1958–1963; M. E. Anderson and coauthors, Improved 2000-meter rowing performance in competitive oarswomen after caffeine ingestion, *International Journal of Sports Nutrition and Exercise Metabolism* 10 (2000): 464–475.

12. Paton, Hopkins, and Vollebregt, 2001.

13. T. E. Graham and coauthors, Caffeine ingestion does not alter carbohydrate or fat metabolism in human skeletal muscle during exercise, *Journal of Physiology* 529 (2000): 837–847.

14. D. Laurent and coauthors, Effects of caffeine on muscle glycogen utilization and the neuroendocrine axis during exercise, *Journal of Clinical Endocrinology and Metabolism* 85 (2000): 2170–2175.

15. D. H. Catlin and T. H. Murray, Performance-enhancing drugs, fair competition and Olympic sport, *Journal of the American Medical Association* 276 (1996): 231–237.

16. R. Skinner, E. Coleman, and C. A. Rosenbloom, Ergogenic aids, in *Sports Nutrition: A Guide for the Professional Working with Active People*, 3rd ed., ed. C. A. Rosenbloom (Chicago: The American Dietetic Association, 2000), pp. 107–146; E. Coleman, DHEA—An anabolic aid? *Sports Medicine Digest* 18 (1996): 140–141.

Jennie Oppenheimer/Studio Zocolo

CHAPTER 11 FOOD SAFETY

Food safety concerns:
- *Food-borne illness.*
- *Natural toxins.*
- *Environmental contaminants.*
- *Pesticide residues.*
- *Food additives.*

Warning signs of botulism:
- *Double vision.*
- *Weakening muscles.*
- *Difficulty swallowing.*
- *Difficulty breathing.*
- *Slurred speech.*

food-borne illness: an illness transmitted to human beings through food or water; caused by a poisonous substance *(food intoxication)* or an infectious agent *(food-borne infection)*; also called *food poisoning.*

toxins: poisons. Toxins produced by bacteria come in two varieties: enterotoxins, which act in the GI tract, and neurotoxins, which act on the nervous system.

Food and Drug Administration (FDA): an agency of the Public Health Service within the Department of Health and Human Services. The FDA is responsible for ensuring the safety and wholesomeness of all foods processed and sold in interstate commerce except meat, poultry, and eggs (which are under the jurisdiction of the USDA); inspecting food plants and imported foods; and setting standards for food composition.

pathogens: disease-causing microorganisms.

botulism: an often-fatal food poisoning caused by botulinum toxin, a toxin produced by the *Clostridium botulinum* bacterium that grows without oxygen in nonacidic canned foods.

*C*onsumers have concerns about the safety of their food. They want to know what causes **food-borne illness,** commonly called *food poisoning.* Other food safety concerns include the **toxins** that occur naturally in some foods as part of their normal composition and the contaminants (including pesticides) that can get into foods before they are harvested. Consumers also want to know whether food additives are safe or should be avoided. The next sections take up these issues in the order just mentioned, which is the order of concern. Food-borne illness is far and away the most important issue; food additives are the matter of least concern. The **Food and Drug Administration (FDA)** is the major agency in charge of monitoring the food supply in the United States, but other agencies are also involved.

Food-Borne Illness

In the United States, an estimated 76 million cases of GI distress (nausea, vomiting, and diarrhea) are caused yearly by food-borne illness.[1] Many cases of "the flu" may actually be episodes of food-borne illness. Each year, 325,000 people are hospitalized from food-borne illness and 5000 people die. Most vulnerable are pregnant women; very young, very old, sick, or malnourished people; and those with a weakened immune system (as in AIDS). The *Dietary Guidelines for Americans* advise people to take preventive steps to minimize their chances of contracting food-borne illnesses. Food-borne illness refers to either food-borne infection or food intoxication. Table 11–1 summarizes the most common and the most severe food-borne illnesses, their symptoms, and prevention methods.

Food-Borne Infections and Intoxications

A food-borne infection is an illness caused by eating foods contaminated by infectious microorganisms. Two of the most common food-borne **pathogens** are *Campylobacter jejuni* and *Salmonella,* which enter the GI tract in contaminated foods such as undercooked poultry and unpasteurized milk. Symptoms generally include abdominal cramps, fever, and diarrhea. If a person experiences these symptoms as the major or only symptoms of a bout of "flu," chances are excellent that what the person really has is a food-borne infection.*

Food intoxication is caused by toxins produced by microorganisms in food or within the digestive tract. The symptoms of one toxin stand alone as severe and commonly fatal—those of **botulism,** caused by the toxin of a microbe that grows inside improperly canned, home-canned, or vacuum-packed foods, or in homemade garlic- or herb-flavored oils stored at room temperature.[2] Botulism has also been associated with improperly stored foil-wrapped potatoes. When a restaurant served an appetizer made with baked potatoes that had been wrapped in aluminum foil and then stored at room temperature for several days, 30 people became deathly ill.[3] Botulism danger signs constitute a true medical emergency (see the margin). Even with medical assistance, survivors can suffer the effects for months, years, or a lifetime. So potent is the botulinum toxin that an amount as tiny as a single grain of salt can kill several people within an hour. The botulinum toxin is destroyed by heat, so canned foods that have been boiled for ten minutes are generally safe from this threat. Home-canned foods are safe if prepared by following proper canning techniques to the letter.† To prepare herb-

---

*Some viruses do cause intestinal distress, and those that do are usually transmitted via food; true influenza viruses cause symptoms primarily in the upper respiratory tract.
†The USDA's meat and poultry hotline answers questions about meat and poultry safety: (800) 535–455.

Table 11–1   Food-Borne Illnesses

Disease and Organism That Causes It	Estimated Yearly Occurrence[a]	Most Frequent Food Sources	Onset and General Symptoms	Prevention Methods
Food-Borne Infections				
Campylobacteriosis (KAM-pee-loh-BAK-ter-ee-OH-sis) *Campylobacter jejuni* bacterium	2 million; 100 deaths	Raw poultry, beef, lamb, unpasteurized milk (foods of animal origin eaten raw or undercooked or recontaminated after cooking).	Onset: 2 to 5 days. Diarrhea, nausea, vomiting, abdominal cramps, fever; sometimes bloody stools; lasts 7 to 10 days.	Cook foods (especially poultry) thoroughly; use pasteurized milk; use sanitary food-handling methods.
Cryptosporidiosis *Cryptosporidium parvum* microscopic parasite	30,000; 7 deaths. (In 1993, a Wisconsin water-borne outbreak affected more than 400,000 people)	Commonly, swimming or drinking contaminated water, even from treated sources. Highly chlorine-resistant. Contaminated raw produce and unpasteurized juices and cider.	Onset: 2 to 10 days. Diarrhea, loose or watery stools, stomach cramps, upset stomach, slight fever. Symptomless sufferers can pass the infection to others.	Wash all raw vegetables and fruits with uncontaminated water before peeling. Do not swallow drops of water while using pools, hot tubs, ponds, lakes, rivers, or streams for recreation.
Cyclosporiasis *Cyclospora cayetanensis* single-cell parasite	15,000; 8 deaths	Contaminated water; contaminated fresh produce.	Onset: average, 7 days. Watery diarrhea, loss of appetite, weight loss, stomach cramps, nausea, vomiting, muscle aches, low-grade fever, fatigue. Symptomless sufferers can spread the infection.	In areas of uncertain sanitation, drink only treated or boiled water, and eat only cooked hot foods or fruits you peel yourself.
Hemolytic-uremic syndrome *Escherichia coli (E. coli)* 0157:H7 bacterium	62,500; 50 deaths	Undercooked ground beef, unpasteurized milk and milk products, contaminated water, unpasteurized juices or cider, contaminated produce (especially alfalfa sprouts), and person-to-person contact.	Onset: 12 to 72 hr. Severe bloody diarrhea, abdominal cramps, acute kidney failure; death. Survivors may face kidney problems, hypertension, blindness, paralysis, and colon problems.	Cook ground beef thoroughly; avoid unpasteurized milk and juice products; use sanitary food-handling methods; use treated, boiled, or bottled water; susceptible people should avoid alfalfa sprouts.
Hepatitis (HEP-ah-TIE-tis) Hepatitis A virus	4,200; 4 deaths	Undercooked or raw shellfish; baked goods or other foods contaminated by infected food handlers.	Onset: 15 to 50 days (28 to 30 days average). Inflammation of the liver; fatigue; nausea, vomiting, or indigestion; jaundice (yellowed skin and eyes from buildup of wastes); muscle pain.	Cook foods thoroughly.
Listeriosis (lis-TER-ee-OH-sis) *Listeria monocytogenes* bacterium	2,500; 500 deaths	Raw meat and seafood, raw milk, and soft cheeses	Onset: 7 to 30 days. Mimics flu; blood poisoning; miscarriage of pregnancy; severe illness or death of newborn; meningitis (stiff neck, severe headache, and fever).	Use sanitary food-handling methods; cook foods thoroughly; use pasteurized milk.
Salmonellosis (sal-moh-neh-LOH-sis) *Salmonella* bacteria	1.34 million; 600 deaths	Raw or undercooked eggs, meats, poultry, milk and other dairy products, shrimp, frog legs, yeast, coconut, pasta, and chocolate.	Onset: 6 to 48 hr. Nausea, fever, chills, vomiting, abdominal cramps, diarrhea, and headache; death.	Use sanitary food-handling methods; use pasteurized milk; cook foods thoroughly; refrigerate foods promptly and properly.

(Continued on next page)

[a] Data from Diseases and pathogens under surveillance, Centers for Disease Control and Prevention, available at **www.cdc.gov/footnet/pus.htm**; P. S. Mead and coauthors, Food-related illness and death in the United States, *Emerging Infectious Diseases* 5 (1999), available at **www.cdc.gov/ncidod/eid/vol5no5/mead.htm**.

Table 11–1   Food-Borne Illnesses—continued

Disease and Organism That Causes It	Estimated Yearly Occurrence[a]	Most Frequent Food Sources	Onset and General Symptoms	Prevention Methods
Food-Borne Infections				
Shigellosis *Shigella* bacteria varieties	90,000; 14 deaths	Contaminated food (may look and smell normal); produce and other foods contaminated by poor sanitation practices of infected farmworkers or food handlers, sewage fertilizer in growing fields, or exposure to flies or other insects. Contaminated drinking or swimming water.	Onset: 1 to 2 days. Diarrhea, fever, stomach cramps. The diarrhea is often bloody. In young children, high fever, seizures. Symptomless sufferers can spread the bacteria to others.	Frequent and careful hand washing with soap. Those with shigellosis should not prepare food or beverages for others. In areas of uncertain sanitation, drink only treated or boiled water, and eat only cooked hot foods or fruits you peel yourself.
"Stomach flu"[b] (mistakenly called) Norwalk-type viruses	9.2 million; 124 deaths	Foods, such as sandwiches and salads, contaminated by infected food handlers. Contaminated produce. Oysters from waters contaminated with human sewage, such as boat bilge.	Onset: 18 to 72 hr. Acute digestive illness, with pain, vomiting, possibly diarrhea, headache, and low-grade fever.	Choose restaurants that pass health department inspections and enforce worker sanitation. If uncertain, order cooked foods served steaming hot. Avoid raw oysters.
"Stomach flu"[c] (mistakenly called) *Vibrio parahaemolyticus* and other *Vibrio* bacteria	5000; 31 deaths	Raw or undercooked shellfish, often oysters. Less commonly, skin infection when an open wound is exposed to warm seawater.	Onset: 24 hr. Watery diarrhea, abdominal cramping, nausea, vomiting, fever and chills.	Cook shellfish well, especially oysters. Purchase shellfish from a reputable dealer. Avoid exposing wounds to warm seawater.
Traveler's diarrhea A variety of microorganisms including *Giardia* and other protozoa	10 million international travelers affected	Contaminated water, undercooked ground beef, raw foods, imported unpasteurized soft cheeses.	Onset: 12 hr to several days. Loose and watery stools, nausea, vomiting, bloating, abdominal cramps.	Cook foods thoroughly; use safe, treated water and pasteurized milk; wash raw fruits and vegetables or avoid them in areas of uncertain sanitation.
Trichinosis *Trichinella spiralis* parasite	50; no deaths	Raw or undercooked pork or wild game (bear). Worms burrow through the body tissues to reach muscle tissue where they remain alive.	Onset: 24 hr. Abdominal pain, nausea, vomiting, diarrhea, and fever. One to two weeks later, muscle pain, low-grade fever, pain on breathing, edema (swelling), skin eruptions, loss of appetite, and weight loss. Drug therapy kills the worms, and deaths are rare.	Cook foods thoroughly.
Food Intoxications				
Botulism (BOT-chew-lizm) Botulinum toxin produced by *Clostridium botulinum* bacterium, which grows without oxygen, in low-acid foods, and at temperatures between 40° and 120°F; the **botulinum** (BOT-chew-line-um) **toxin** responsible for botulism is called **botulin** (BOT-chew-lin)]	Anaerobic environment of low acidity (canned corn, peppers, green beans, soups, beets, asparagus, mushrooms, ripe olives, spinach, tuna, chicken, chicken liver, liver pâté, luncheon meats, ham, sausage, stuffed eggplant, lobster, and smoked and salted fish).		Onset: 4 to 36 hr. Nervous system symptoms, including double vision, inability to swallow, speech difficulty, and progressive paralysis of the respiratory system; often fatal; leaves prolonged symptoms in survivors.	Use proper canning methods for low-acid foods; refrigerate homemade garlic and herb oils; avoid commercially prepared foods with leaky seals or with bent, bulging, or broken cans.
Staphylococcal (STAF-il-oh-KOK-al) **food poisoning** Staphylococcal toxin (produced by *Staphylococcus aureus* bacterium)	Toxin produced in meats, poultry, egg products, tuna, potato and macaroni salads, and cream-filled pastries.		Onset: ½ to 8 hr. Diarrhea, nausea, vomiting, abdominal cramps, and fatigue; mimics flu; lasts 24 to 48 hr; rarely fatal.	Use sanitary food-handling methods; cook food thoroughly; refrigerate foods promptly and properly; use proper home-canning methods.

[b] Though popularly called "stomach flu," the digestive disturbances caused by Norwalk-type viruses are unrelated to influenza.

[c] Though popularly called "stomach flu," the digestive disturbances caused by *Vibrio* organisms are unrelated to influenza.

flavored oils safely, wash and dry the herbs before adding to the oil. Keep the oil refrigerated, and throw out leftovers at the end of the week or whenever the oil has been left unrefrigerated for more than a few minutes. Commercially prepared oils are safer still because they have added acid and processing that cannot be duplicated at home.

Food Safety in the Marketplace

Transmission of food-borne illness is changing as the food supply changes. In the past, food-borne illness was caused by one person's error in a small setting, such as improperly refrigerated potato salad at a family picnic, and affected only a few victims. Today, people are eating more foods prepared and packaged by others. Consequently, when a food manufacturer or restaurant chef makes an error, food-borne illness can be epidemic. An estimated 80 percent of reported food-borne illnesses are caused by errors in a commercial setting, such as the improper **pasteurization** of milk at a large dairy.

In the mid-1990s, a fast-food restaurant served undercooked burgers tainted with a particularly dangerous *E. coli* strain known as 0157:H7. As a result, hundreds of people became ill, and at least three people died. This incident and others focused the national spotlight on two important safety issues: live, disease-causing organisms are commonly found in raw meats, and thorough cooking is necessary to make animal-derived foods safe.

Infections from *E. coli* 0157:H7 cause severe illness, with bloody diarrhea, severe intestinal cramps, and dehydration setting in a few days after eating bad meat, raw milk, or even fresh berries or organic produce that has become contaminated. It is important to obtain medical help when such symptoms appear and to be aware that antibiotics may worsen the condition.[4] Children given antibiotics to treat *E. coli* 0157:H7 infection were seven times more likely than untreated children to develop a kidney disease, *hemolytic-uremic syndrome,* that can lead to permanent kidney failure and death. The medications seem to make the bacterial toxin more available for absorption and thus enable it to reach higher tissue concentrations.

Although consumers came to fear undercooked commercial hamburgers after the incident at the fast-food restaurant, alfalfa sprouts are a more common source of disease from *E. coli* 0157:H7.[5] When the bacterium is present on the seed, it may multiply during sprouting, even when sprouts are grown at home using clean equipment and fresh water or in commercial sprout facilities with high sanitation standards. For this reason, some people, such as the very young or old or those with liver disease or compromised immunity, should probably avoid raw sprouts altogether.[6] Most healthy people can eat sprouts safely, however, if they select fresh-looking, crisp, green, living sprouts that have been kept under refrigeration and rinse them thoroughly with water before use.[7]

Industry Controls One result of the media reports and public concern about food-borne illness is a law requiring that producers of meat, poultry, seafood, and fresh fruit juices, and vegetable juices employ an effective prevention method known as a **Hazard Analysis Critical Control Points (HACCP)** plan.[8] Each producer must review its processes to identify "critical control points" where the risk of food contamination is high and then develop and implement a HACCP plan to prevent loss of control at those critical points. The results of the HACCP system seem promising. Since its implementation, *Salmonella* contamination of poultry, ground beef, and pork has decreased by almost 50 percent, 40 percent, and 25 percent, respectively.[9]

Consumer Awareness Canned and packaged foods sold in the grocery stores are almost invariably safe, but rare accidents do happen. Batch numbering makes it possible to recall contaminated foods through public announcements via

Cook hamburgers to 160°F; color alone cannot determine doneness. Some burgers will turn brown before reaching 160°F, while others may retain some pink color, even when cooked to 175°F.

pasteurization: the treatment of milk with heat sufficient to kill certain pathogens (disease-causing microbes) commonly transmitted through milk; not a sterilization process. Pasteurized milk retains bacteria that cause milk spoilage. Raw milk, even if labeled "certified," transmits many food-borne diseases to people each year and should be avoided.

Hazard Analysis Critical Control Points (HACCP): a systematic plan for identifying and correcting potential microbial hazards in the manufacturing, distribution, and commercial use of food products; commonly referred to as "HAASS-ip."

Figure 11–1
Four Steps to Fight Bac!

Thermy™

"IT'S SAFE TO BITE WHEN THE TEMPERATURE IS RIGHT!"

newspapers, television, and radio. In the grocery store, these guidelines can help consumers avoid buying foods that are contaminated:

■ Avoid packages with defective seals and wrappers.

■ Reject leaking or bulging cans.

■ Check safety "buttons" on jars to make sure the seal is intact.

■ Avoid partially frozen foods; those in chest-type freezers should be stored below the frost line.

■ Choose packages that have not been damaged, soiled, or punctured.

Improper handling of foods can occur anywhere along the line, from commercial manufacturers to large supermarkets to small restaurants to private homes. Maintaining a safe food supply requires everyone's efforts.

Food Safety in the Kitchen

Whether bacteria multiply and cause illness depends, in part, on what happens in the kitchen—whether the kitchen is in your home, a school cafeteria, a gourmet restaurant, or a canning plant. Foods can provide ideal conditions for bacteria to thrive and produce their toxins. Disease-causing bacteria require:

■ Warmth (40° to 140°F).

■ Moisture.

■ Nutrients.

To prevent bacterial growth, people who prepare foods can do these things: cook foods thoroughly, keep hot foods hot, keep cold foods cold, keep raw foods (meat, poultry, seafood) separate, and keep hands, utensils, surfaces, and the kitchen clean. To focus greater consumer attention on the importance of safe food handling, a partnership of government agencies, industry, and consumer groups developed the Fight Bac! campaign. Fight Bac! emphasizes the four steps people can take to fight food-borne bacteria and food-borne illness (see Figure 11–1).

Keep Hot Foods Hot   Cook foods long enough to reach an internal temperature that will kill microbes. To help consumers accomplish this, the United States Department of Agriculture (USDA) created "Thermy," the cartoon character in the margin who urges the use of a thermometer to test the temperatures of cooked foods. Figure 11–2 shows safe temperatures for cooked foods, and Table 11–2 describes different kinds of thermometers. To prevent bacterial growth when holding cooked food, keep it at 140°F or higher until it is served. Refrigerate leftover food immediately after serving.

The cardinal rule to protect yourself is to remember that food-borne illness is always a possibility when food is not kept hot. For example, the meatballs in a warming tray at a lovely buffet may be warm but not hot. Despite the beautiful setting, their low temperature is a warning flag. Food at 140°F feels hot, not just warm. The likelihood of illness is strong when food is not hot enough, and the pleasure of eating meatballs isn't worth the risk.

Keep Cold Foods Cold   Keeping cold foods cold starts when you leave the grocery store. If you are running errands, shop last, so the groceries will not stay in the car too long. (If the ice cream has begun to melt, it has been too long.) Upon arrival home, load foods into the refrigerator or freezer immediately. When serving food cold, let it stay at cool room temperature (about 68°F) for no more than two hours. If the room is warm (about 80°F), refrigerate the food after just one hour. Table 11–3 on p. 276 lists some safe keeping times for foods stored in the refrigerator at 40°F.

Keeping foods cold applies to defrosting foods before use, too. Bacterial growth begins on thawed portions of food even while the inner core is solidly frozen, so thaw meats or poultry in the refrigerator, not at room temperature. If

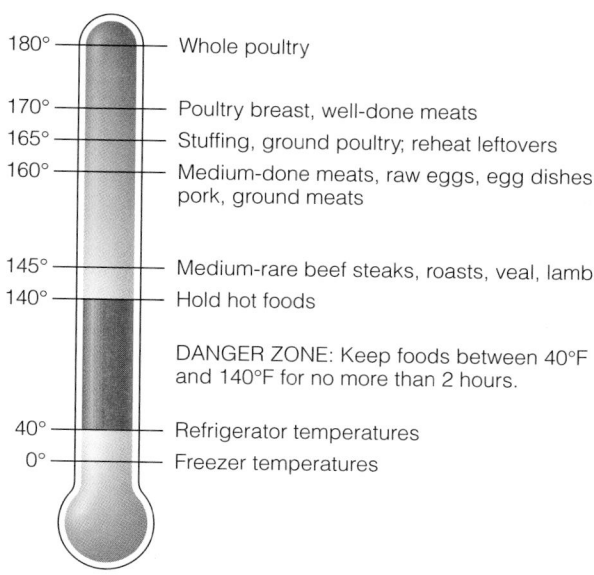

Figure 11–2

Recommended Safe Temperatures (Fahrenheit)

Bacteria multiply rapidly at temperatures between 40° and 140°F. Cook foods to the temperatures shown on this thermometer and hold them at 140°F or higher.

you must hasten thawing, use cool running water or a microwave oven set to defrost. Marinate foods in the refrigerator, not at room temperature.

Keep Raw Foods Separate Keeping raw foods separate means preventing **cross-contamination** of foods. Raw foods, especially meats, eggs, and seafood, are likely to contain bacteria. To prevent the bacteria from spreading, keep the raw foods and their juices away from ready-to-eat foods. For example, if you take burgers out to the grill on a plate, wash that plate in hot, soapy water before using it to hold the cooked burgers. If you use a cutting board to cut raw meat, wash the board, the knife, and your hands thoroughly before using the utensils to make a salad or other foods that are eaten raw.

Keep the Kitchen Clean Keeping the kitchen clean requires using freshly washed utensils and laundered towels and washing your hands with warm water and soap for a minimum of 20 seconds before and during food handling. If you are ill or have open sores, stay away from food. Clean equipment frequently and effectively. Microbes love to nestle down in small, damp spaces such as the inner

cross-contamination: the contamination of a food through exposure to utensils, hands, or other surfaces that were previously in contact with a contaminated food.

Table 11–2   Types of Thermometers
■ **appliance thermometer:** a thermometer that verifies the temperature of an appliance. An *oven thermometer* verifies that the oven is heating properly; a *refrigerator/freezer thermometer* tests for the proper refrigerator (<40°F) or freezer temperature (0°F).
■ **fork thermometer:** a utensil combining a meat fork and an instant-read food thermometer.
■ **instant-read thermometer:** a thermometer that, when inserted into food, measures its temperature in seconds; designed to test temperature of food at intervals, and not to be left in food during cooking.
■ **oven-safe thermometer:** a thermometer designed to remain in the food to give constant readings during cooking.
■ **pop-up thermometer:** a disposable timing device commonly used in turkeys. The center of the device contains a stainless steel spring that "pops up" when the food reaches the right temperature.
■ **single-use temperature indicator:** a type of instant-read thermometer that changes color to indicate that the food has reached the desired temperature. Discarded after one use, such thermometers are often used in retail food markets to eliminate cross-contamination.

Table 11–3 Safe Refrigerator Storage Times (≤40°F)

1 to 2 Days

Raw ground meats, breakfast or other raw sausages, raw fish or poultry; gravies

3 to 5 Days

Raw steaks, roasts, or chops; cooked meats, vegetables, and mixed dishes; ham slices; mayonnaise salads (chicken, egg, pasta, tuna)

1 Week

Hard-cooked eggs, bacon or hot dogs (opened packages); smoked sausages

2 to 4 Weeks

Raw eggs (in shells); bacon or hot dogs (packages unopened); dry sausages (pepperoni, hard salami); most aged and processed cheeses (swiss, brick)

2 Months

Mayonnaise (opened jar); most dry cheeses (parmesan, romano)

---

cells of sponges or the pores between the fibers of wooden cutting boards. Antibacterial sponges, cloths, boards, and utensils possess a chemical additive intended to prevent rapid bacterial growth, but the protection is not perfect—these products still need special handling to keep them safe. You can ensure the safety of cutting boards and sponges by washing them in a dishwasher or by treating them as suggested below. Alternatively, save sponges for car washing and other heavy cleaning chores, and clean the kitchen with washable dishcloths that can be laundered often.

To eliminate microbes in your kitchen, you have three choices, each with benefits and drawbacks. One is to poison the microbes on cutting boards, sponges, and other equipment with toxic chemicals such as bleach (one capful per gallon of water). The benefit is that chlorine can kill even the hardiest organism. The drawback is that chlorine is toxic to handle, can ruin clothing, and washes down household drains into the water supply and forms chemicals that can harm waterways and fish.

A second option is to treat kitchen equipment with heat. Soapy water heated to 140°F kills most harmful organisms and washes most others away. This method takes effort, though, since you have to use truly scalding water heated well beyond the temperature of the tap. Thirdly, an automatic dishwasher can combine both methods: it washes in water hotter than hands can tolerate, and a chlorine-containing dishwasher detergent can be used but, of course, with the environmental disadvantage that chlorine entails. Whichever strategy you use, you can have truly safe implements to prepare your food.

Troublesome Foods

Some foods are more hospitable to microbial growth than others. Especially vulnerable are moist foods, nutrient-rich foods, and foods that are chopped or ground.

Meats and Poultry Raw meats and poultry require special handling. Their packages bear labels to instruct consumers on meat safety (see Figure 11-3). Meats and poultry may contain bacteria, and they provide a moist, nutritious environment that is ideal for microbial growth. Ground meat is handled more than other kinds of meat and exposes much more surface area for bacteria to land on, so experts advise cooking it well done. Use a thermometer to test the internal temperature of poultry and meats, even hamburgers, to be certain they are done. Burgers often turn brown and appear cooked before their internal temperature is high enough to kill harmful bacteria.[10] Do not use or even taste a food with an "off" appearance or odor. Don't trust your senses of smell and sight alone to tell you that foods are safe, however. Most contamination is not detectable by odor, taste, or appearance. Even hot cooked food, if handled improperly prior to cooking, can cause illness.

Other Meat Concerns Reports from Europe on *bovine spongiform encephalopathy (BSE)*, commonly called **mad cow disease,** have sparked concerns in the United States, but so far, no cases of the disease have been identified.[11] Mad cow disease is a fatal condition that affects the central nervous system of cattle. A similar disease, called new variant Creutzfeldt-Jakob disease (nvCJD), thought to be the human form of mad cow disease, develops in people who have eaten beef from infected cattle. Almost 100 cases of nvCJD have been reported worldwide, almost all in the United Kingdom. Government agencies have taken numerous steps to prevent mad cow disease from entering the United States.

Eggs Concerns about food-borne illness from eating eggs contaminated with *Salmonella* prompted the FDA to take action to improve the food safety of eggs. Food preparation establishments such as supermarkets, restaurants, deli-

mad cow disease: an often-fatal illness of cattle affecting the nerves and brain; also called *bovine spongiform encephalopathy (BSE).*

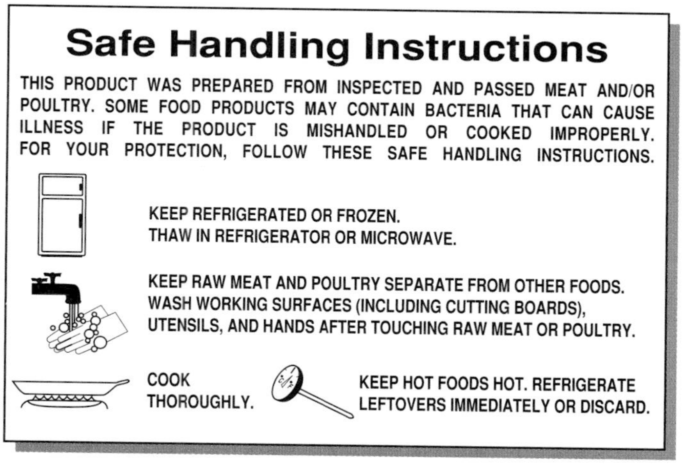

Figure 11-3
Safe Handling Instructions for Meat and Poultry

Safe Handling Instructions

THIS PRODUCT WAS PREPARED FROM INSPECTED AND PASSED MEAT AND/OR POULTRY. SOME FOOD PRODUCTS MAY CONTAIN BACTERIA THAT CAN CAUSE ILLNESS IF THE PRODUCT IS MISHANDLED OR COOKED IMPROPERLY. FOR YOUR PROTECTION, FOLLOW THESE SAFE HANDLING INSTRUCTIONS.

KEEP REFRIGERATED OR FROZEN. THAW IN REFRIGERATOR OR MICROWAVE.

KEEP RAW MEAT AND POULTRY SEPARATE FROM OTHER FOODS. WASH WORKING SURFACES (INCLUDING CUTTING BOARDS), UTENSILS, AND HANDS AFTER TOUCHING RAW MEAT OR POULTRY.

COOK THOROUGHLY.

KEEP HOT FOODS HOT. REFRIGERATE LEFTOVERS IMMEDIATELY OR DISCARD.

catessens, caterers, vending operations, hospitals, nursing homes, and schools are now required to refrigerate eggs at 45° F promptly upon delivery.[12] In addition, cartons for eggs in the shell must bear the statement shown in the margin.

Healthy people can still safely enjoy classic foods that call for raw or under-cooked eggs, such as Caesar salad dressing and hollandaise sauce, by preparing them with pasteurized egg substitutes, sold in cartons in the dairy or freezer case. These products may contain a few bacteria that escape the pasteurization process, however, so the elderly, the very young, and people who have weakened immune systems should avoid all uncooked or undercooked eggs, including pasteurized ones.[13]

Seafood As population density increases along the shores where seafood is harvested, pollution inevitably invades those waters and the seafood living there. Watchdog agencies monitor commercial fishing waters and try to keep harvesters out of unsafe areas. Nevertheless, chemical pollution and microbial contamination can originate in the boats and warehouses where seafood is cleaned, prepared, and refrigerated, as well as in the water. To help ensure safe seafood products, the FDA requires seafood marketers to adopt food safety practices based on the HACCP system mentioned earlier. Still, unwholesome foods can reach the market. In one season alone, black-market dealers may sell millions of dollars worth of clams and oysters taken illegally from polluted harvesting areas.

To keep seafood as fresh as possible, people in the industry "keep it cold, keep it clean, and keep it moving." Wise consumers eat it cooked. Experts are unanimous in saying that the risks of eating raw or lightly cooked seafood have become unacceptably high due to environmental contamination. The microorganisms that lurk in seafood are undetectable by an expert.* Eating raw oysters, for example, can be dangerous for anyone, but people with liver disease and those with suppressed immune systems are most vulnerable. Some seafood processors are now pasteurizing raw oysters in the shell to kill pathogens and prevent foodborne illness. According to consumers, this process does not change the texture or the raw flavor.

For adults and children alike, eating raw or lightly steamed seafood is a risky proposition even when it is prepared as **sushi** by a master Japanese chef. People who like sushi know that not all varieties are made from raw fish. Many types are made with cooked crabmeat and vegetables, avocado, or other delicacies and are perfectly safe to enjoy. Also, rumor has it that freezing fish will make it safe to eat raw, but this is only partly true. Freezing fish will kill mature parasitic worms, but only cooking can kill all worm eggs and other microorganisms that can cause illness.

SAFE HANDLING INSTRUCTIONS: To prevent illness from bacteria: keep eggs refrigerated, cook eggs until yolks are firm, and cook foods containing eggs thoroughly.

In Florida, containers of raw oysters must bear a warning that a risk is associated with consuming raw oysters or any raw animal protein. The risk is greatest for people who are ill or elderly or have weakened immune systems.

sushi: a Japanese dish consisting of vinegar-flavored rice, seafood, and colorful vegetables, typically wrapped in seaweed. Some sushi is wrapped in raw fish; other sushi contains only cooked ingredients.

* To speak with an expert about seafood safety, call the FDA seafood hotline: (800) FDA-4010.

Honey Honey has been found to contain dormant bacterial spores that can awaken in the human body to produce the deadly botulinum toxin mentioned earlier. Adults are big and strong enough to withstand the doses usually encountered, but infants under one year of age should never be fed honey. (It can also be contaminated with environmental pollutants picked up by the bees.) Honey has been implicated in several cases of sudden infant death.

Picnics

Picnics are fun and can be safe, too. Choose foods that last without refrigeration, such as fresh fruits and vegetables, breads and crackers, and canned spreads and cheeses that you can open and use on the spot. Aged cheeses, such as cheddar and swiss, do well for an hour or two, but for longer periods, carry them in an ice chest. Mayonnaise resists spoilage because of its acid content, but when mixed with chopped ingredients, such as pasta, meat, or vegetable salads, it spoils quickly. The chopped ingredients offer an extensive surface area for bacteria to invade, and the foods have been in contact with cutting boards, hands, and kitchen utensils that have transmitted bacteria to them. Chill chopped salads well before, during, and after the picnic. Keep mayonnaise itself cold.

If Illness Occurs

Local health departments and the USDA extension service can provide further information about food safety. Should efforts fail and mild food-borne illness develop, drink clear liquids to replace fluids lost through vomiting and diarrhea. If serious food-borne illness is suspected, first call a physician. Then, wrap the remainder of the suspected food and label its container so that it cannot be mistakenly eaten, place it in the refrigerator, and hold it for possible inspection by health authorities. The margin identifies foods most commonly implicated in food-borne illnesses.

Travel

Special food safety concerns arise when people travel. In many parts of the world, food-borne illness is likely to strike tourists even though the local people, eating exactly the same foods prepared the same way, remain healthy. That is because the locals have developed immunity to local disease-causing organisms, while tourists have no such protection. The accompanying "How to" box offers tips to travelers on avoiding food-borne infection.

Water

Foods are not alone in transmitting food-borne diseases; water is guilty, too. A glass of "water" is more than just water. Some diseases found on fresh fruits and vegetables and in raw oysters are transmitted through contaminated water.[14] In addition to microorganisms, water may contain many of the same impurities that foods do: environmental contaminants, pesticides, and additives such as chlorine used to kill pathogenic microorganisms.

Contamination Contamination can occur as water travels from the main water supply to homes. Lead or asbestos from old, corroded pipes can contaminate drinking water, as can bacteria and dirt from leaking pipes. People who suspect contamination of their water should have it tested where it flows out, at the tap.

Public water systems treat water to remove contaminants that have been detected above acceptable levels. Private well water is usually not treated or cleansed, so people who consume water from private wells are responsible for its safety and should test the water periodically.

Frequently unsafe:
- *Raw milk and milk products.*
- *Raw or undercooked seafood, meat, poultry, or eggs.*

Occasionally unsafe:
- *Airline food.*
- *Hamburgers.*
- *Salad bar items.*
- *Sandwiches.*
- *Soft cheeses (Mexican style, feta, brie, camembert, blue-veined).*
- *Sprouts.*
- *Unpasteurized fruit juices and ciders.*
- *Unwashed berries and grapes.*

Rarely unsafe:
- *Peeled fruit.*
- *Steaming-hot foods.*
- *High-sugar foods.*

How to

Achieve Food Safety while Traveling

*F*ood-borne illnesses contracted while traveling are colloquially known as traveler's diarrhea. A bout of this ailment can ruin the most enthusiastic tourist's trip. To avoid food-borne illness while traveling:

- Wash your hands often with soap and hot water, especially before handling food or eating.

- Eat only cooked food and canned foods. Eat raw fruits or vegetables only if you have washed them in boiled water and peeled them yourself. Skip salads, raw fish, and shellfish.

- Be aware that water, and ice made from it, may be unsafe, too. Take along disinfecting tablets or an element that boils water in a cup.

- Drink no beverages made with tap water. Drink only treated, boiled, canned, or bottled beverages, and

drink them without ice (ice may be made from contaminated water), even if they are not chilled to your liking. Refuse dairy products unless they have been properly pasteurized and refrigerated.

- Do not use the local water, even to brush your teeth, unless you boil or disinfect it first.

- Before you leave on the trip, ask your physician to recommend medicines to take with you in case your efforts to avoid illness fail.

One journalist succinctly sums up these recommendations, "Boil it, cook it, peel it, or forget it."[a] Chances are excellent that if you follow these rules, you will remain well.

[a]R. D. Williams, Boil it, cook it, peel it, or forget it, *FDA Consumer*, September 1991, p. 17.

Safe drinking water is a concern for everyone and must be protected to ensure continued health. To learn more about the water supply in your area, call the local public health agency.*

Bottled Water Some people turn to bottled water as an alternative to tap water. Bottled water is classified as a food, so it is regulated by the FDA and must meet safety standards similar to those set for public water systems. Bottled water must also be processed and labeled according to FDA regulations. Some bottled waters may have minerals or carbonation added. "Carbonated," "seltzer," and "tonic" waters are not considered waters, however, but soft drinks. The FDA requires labels to disclose the sources of bottled waters and to use legally defined descriptive terms.[15]

Advances in Food Technology and Safety

New advances in technology offer promise for the future purity of foods. Someday their use may dramatically improve the safety of foods for sale on the market.

Irradiation The FDA has approved the use of **irradiation** on certain foods to improve food safety. The American Dietetic Association makes this statement concerning irradiation: "Food irradiation enhances the safety and quality of the food supply and helps protect consumers from food-borne illness."[16] Irradiation kills microorganisms and insects on wheat; flour; spices; fresh and frozen beef, lamb, pork, and poultry; and some fresh fruits and vegetables. In addition, irradiation inhibits growth of sprouts on potatoes and onions and delays ripening in some fruits such as strawberries and mangoes. Milk products change flavor when irradiated and so are not candidates for the treatment. (Incidentally, the milk in those boxes kept at room temperature on grocery-store shelves is not irradiated, but processed with an **ultrahigh temperature treatment** for just long enough to sterilize it.)

irradiation: the application of ionizing radiation to foods to reduce insect infestation or microbial contamination or to slow the ripening or sprouting process; also called *cold pasteurization.*

ultrahigh temperature treatment (UHT): a process of sterilizing food by exposing it for a short time to temperatures above those normally used in processing.

*For information on safe drinking water in general, call the Environmental Protection Agency's hotline: (800) 426-4791.

This international symbol, called the *radura*, identifies retail foods that have been irradiated. The words "Treated by irradiation" or "Treated with irradiation" must accompany the symbol. The irradiation label is not required on commercially prepared foods that contain irradiated ingredients, such as spices.

The use of irradiation on food has been extensively evaluated and is supported by the American Dietetic Association, the World Health Organization, the American Medical Association, and other health agencies.[17] The process substantially reduces food-borne pathogens associated with fresh fruits and vegetables and is also effective in eliminating the *Salmonella* bacterium from poultry.

Consumer Concerns   Some consumers, associating radiation with cancer, birth defects, and mutations, have negative emotions about the use of irradiation on foods. Despite consumers' concerns, irradiation does not make the food radioactive. Vitamin loss is minimal and comparable to amounts lost in other food-processing methods. Irradiation cuts down on food spoilage and can replace some costly pesticides, thus reducing pesticide residues in food. Irradiation does change the taste of food slightly, and most consumers consider flavor to be important. Whether this effect on taste will influence consumer purchases of irradiated foods remains to be seen.

Regulation of Irradiation   The FDA has established regulations governing the uses of irradiation and allowed doses. Each food that has been treated with irradiation must say so on its label.

High-Intensity Pulsed Light   A new technology called high-intensity pulsed light has been approved by the FDA to enhance food safety. High-intensity pulsed light uses an intense flash of light to kill microorganisms on the surface of foods, packaging materials, and water. It extends the shelf life of foods without changing their nutritional properties. These new technologies show promise in the battle against food-borne pathogens. In combination with safe food handling by consumers, these processes will decrease the number of food-borne illnesses contracted each year and increase the safety of our food supply.

In Summary   Millions of people suffer mild to life-threatening symptoms caused by food-borne illnesses (review Table 11–1). Most of these illnesses can be prevented by storing and cooking foods at their proper temperatures and by preparing them in sanitary conditions. Like foods, water may contain infectious microorganisms, environmental contaminants, pesticide residues, and additives. The Environmental Protection Agency monitors the safety of the public water system, but many consumers choose bottled water instead of tap water.

Natural Toxins in Foods

Consumers concerned about food contamination may think that they can eliminate all poisons from their diets by eating only "natural" foods. On the contrary, nature has provided natural foods with the natural poisons they need to fend off diseases, insects, and other predators. Nevertheless, although the potential for harm exists, actual harm rarely occurs.

Most people would recognize the names belladonna and hemlock—both classic deadly poisons in the form of natural herbs. Few people know, however, that the herb sassafras contains a cancer-causing agent and is banned from use as an additive in commercially produced foods and beverages. Equally surprising is that cabbage, turnips, mustard greens, and radishes all contain small quantities of harmful goitrogens—compounds that can enlarge the thyroid gland and aggravate thyroid problems.

The cyanogens are another natural poison. These precursors to the deadly poison cyanide are found in *raw* lima beans and fruit seeds such as apricot pits. Many countries allow commercial growers to grow only those varieties of lima beans with the lowest cyanogen contents. As for fruit seeds, they are seldom deliberately eaten. An occasional swallowed seed or two presents no danger, but a couple of dozen seeds could be fatal to a small child.

Potatoes contain many natural poisons. One is solanine, a bitter, powerful, narcotic-like substance. The small amounts of solanine normally found in potatoes are harmless, but if potatoes are stored in the light, the solanine in them can build up to toxic levels. Cooking does not destroy solanine, but because most of a potato's solanine is in the green layer that develops just beneath the skin, it can be peeled off, making the potato safe to eat. If the potato tastes bitter, however, discard it.

In Summary Natural toxicants include the goitrogens in cabbage, cyanogens in lima beans, and solanine in potatoes. These examples of naturally occurring toxicants illustrate two familiar principles. First, any substance can be toxic when consumed in excess. Second, poisons are poisons, whether made by people or by nature. Remember: it is not the source of a chemical that makes it hazardous, but its chemical structure and the quantity consumed.

Environmental Contaminants in Foods

A justifiably high-ranking concern about our food supply is environmental contamination of foods. As populations increase worldwide and nations become more industrialized, this problem looms ever larger. A food **contaminant** is anything that does not belong there.

Harmfulness of Environmental Contaminants

The potential harmfulness of a contaminant depends in part on the extent to which it lingers in the environment or in the human body—how **persistent** it is. Some contaminants are short-lived because microorganisms or agents such as sunlight or oxygen can break them down. Some contaminants linger in the body for only a short time because the body can rapidly excrete them or metabolize them to harmless compounds. These contaminants present little cause for concern. Other contaminants, however, resist breakdown and interact with the body's systems without being metabolized or excreted. These can pass unchanged from food to the eater, and if the same food is eaten every day, larger and larger quantities of the contaminant may accumulate in the eater's body. Figure 11–4 on p. 282 shows how toxins accumulate at higher concentrations at each level of the food chain (**bioaccumulation**).

How much of a threat do environmental contaminants pose to the food supply? It depends on the contaminant. In general, the threat remains small because the FDA monitors the presence of contaminants in foods and issues warnings when contaminated foods appear in the market. In the event of an accidental industrial spill or one caused by a natural event, such as a volcano, however, the hazard can suddenly become great.

Mercury

Some contaminants build insidiously in the food supply. For example, increasing levels of the **heavy metal** mercury expelled from industrial sites have been detected in U.S. lakes, rivers, and ocean fisheries. More and more, the FDA has detected unacceptably high levels of mercury in fish and other wildlife. This finding prompted a recent advisory to all pregnant women, women who may become pregnant, nursing mothers, and young children against eating large commercially

contaminant: a substance that makes a food impure and unsuitable for ingestion.

persistent: of a stubborn or enduring nature; with respect to food contaminants, the quality of remaining unaltered and unexcreted in plant foods or in the bodies of animals and human beings.

bioaccumulation: the accumulation of a contaminant in the tissues of living things at higher and higher concentrations along the food chain.

heavy metal: any of a number of mineral ions such as mercury and lead, so called because they are of relatively high atomic weight. Many heavy metals are poisonous.

Figure 11–4

Bioaccumulation of Toxins in the Food Chain

This example features fish as the food for human consumption, but bioaccumulation of toxins occurs on land as well when cows, pigs, and chickens eat or drink contaminated foods or water.

◆ If none of the chemicals are lost along the way, people ultimately receive all of the toxic chemicals that were present in the original plants and plankton.

◆ Contaminants become further concentrated in larger fish that eat the small fish from the lower part of the food chain.

◆ Contaminants become more concentrated in small fish that eat the plants and plankton.

◆ Plants and plankton at the bottom of the food chain become contaminated with toxic chemicals, such as methylmercury (shown as red dots).

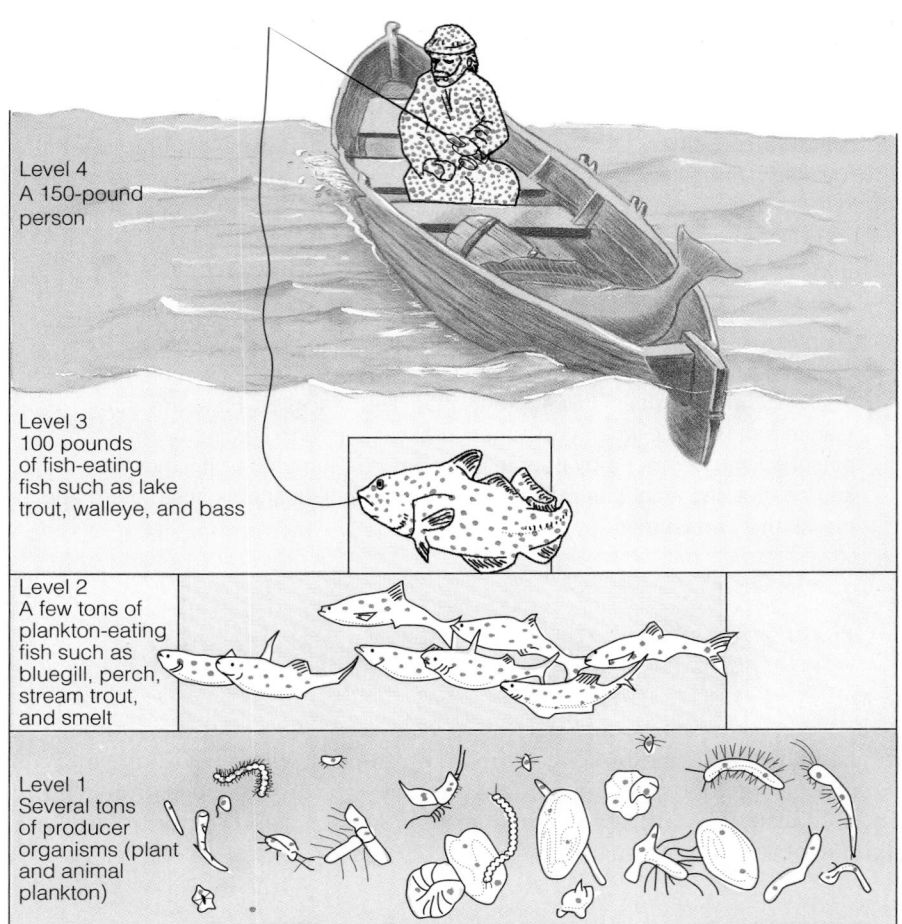

Level 4
A 150-pound person

Level 3
100 pounds of fish-eating fish such as lake trout, walleye, and bass

Level 2
A few tons of plankton-eating fish such as bluegill, perch, stream trout, and smelt

Level 1
Several tons of producer organisms (plant and animal plankton)

available predatory fish, such as king mackerel, swordfish, shark, and tilefish.[18] Further, the Environmental Protection Agency (EPA) has warned the same groups to limit their intakes of freshwater fish caught by family and friends in lakes, rivers, and streams to one fish meal per week.[19] Fish with mercury contents below the danger zone—and thus safe to consume—include canned and farm-raised fish, shellfish, and small ocean fish. Mercury is persistent in the environment, so efforts begun today to clean up U.S. waters will take years to diminish the threat to health.

In Summary Mercury has been used here to show the potential seriousness of food and water contamination by persistent contaminants and the best defenses against them. Thousands of other contaminants exist, but in all cases two principles apply. First, remain alert to the possibility of contamination of foods, and keep an ear open for public health announcements and advice. Second, do not eat any one food too often; vary your diet. Switching from food to food is an effective defensive strategy against the accumulation of toxins in your body. This is the principle of dilution: each food eaten dilutes contaminants that may be present in other components of the diet.

Pesticides

pesticides: chemicals used to control insects, diseases, weeds, fungi, and other pests on crops and around animals. Used broadly, the term includes *herbicides* (to kill weeds), *insecticides* (to kill insects), and *fungicides* (to kill fungi).

Pesticides are a special category of contaminants, differing from those just discussed in that they are applied to foods on purpose and in a manner that is regulated and controlled. Their use is controversial.

Hazards and Regulation of Pesticides

Pesticides do help to ensure the survival of some crops, but the damage they do to the environment is considerable and increasing. There is also some question about whether the widespread use of pesticides has really improved the overall yield of food. Pesticides and their use are monitored by government agencies.

Hazards of Pesticides   Many pesticides are poisons that can damage all living cells, not just those of pests. Their use, therefore, is hazardous to those who work with them: manufacturers, field workers, truck drivers, and anyone else who is exposed to them. The danger of misuses or spills is ever present, and serious accidents may not always be prevented despite safety regulations and precautions regarding pesticide use.

Consumers of produce in the marketplace have reason to be concerned about pesticides, too, because they may still linger in the foods to which they were applied in the field. Risks to health from pesticide exposure are probably small for healthy adults, but children, elderly people, and people with weakened immune systems may be vulnerable to some types of pesticide poisoning.

Regulation of Pesticides   The FDA and the EPA set legal limits on the types and amounts of pesticides permitted in foods. A pesticide's **tolerance level** is based on children. The government agencies set tolerance levels by first identifying foods that children commonly eat in large amounts and then considering the effects of pesticide exposure during each developmental stage.[20] Once tolerances are set, foods and livestock feeds are monitored for pesticides.

Pesticides from Other Countries   Today, approximately 70 percent of the fruits and vegetables consumed in the United States are imported from other countries, which do not have the same pesticide regulations as the United States and Canada.[21] Indeed, a loophole in federal law allows U.S. companies to produce pesticides that are banned here and sell them in other countries. Those countries then use the banned pesticides on their foods and ship the foods back to U.S. consumers—the so-called circle of poison. Federal inspectors monitor incoming foods, however, and refuse to let them enter the country if they are found to contain illegal residues. The United States, Mexico, and Canada are currently working to establish a pesticide policy for all of North America.* In addition, plans are being developed to allow the FDA to inspect foreign farms and ban any produce that does not meet U.S. food safety standards.[22]

Monitoring Pesticides

The FDA analyzes foods using methods that can detect **residues** well below tolerances. If the FDA finds violative levels, it can seize the products or order them destroyed. Four times a year, FDA surveyors buy over 200 foods in U.S. grocery stores in several cities, prepare the foods table ready, and then analyze them, not only for pesticides but for essential minerals, industrial chemicals, heavy metals, and radioactive materials. Food preparation often reduces levels of contaminants in foods, so the inspectors look for levels at least five times lower than permitted limits. Findings confirm that the bulk of the U.S. food supply is safe from excessive pesticide residues.

A problem, though, is that budget restraints limit the FDA's testing capacity. The FDA does not sample all food shipments or test for all pesticides. Fewer than 700 inspectors and scientists test food samples from the multitude of farms, groves, docks, airports, warehouses, and processing plants the agency oversees. The FDA cannot (nor can it be expected to) guarantee 100 percent safety in the

tolerance level: the maximum amount of a residue permitted in a food when a pesticide is used according to label directions.

residues: whatever remains. In the case of pesticides, those amounts that remain on or in foods when people buy and use them.

---

*These pesticide agreements are under the auspices of the North American Free Trade Agreement (NAFTA).

How to

Rid Foods of Pesticide Residues

*N*o matter where your food comes from, it is wise to remove parts that might contain pesticide residues:

- Trim the fat from meat and remove the skin from poultry and fish; discard fats and oils in broths and pan drippings. Avoid fish oil capsules. (Pesticide residues concentrate in the animal's fat.)
- Wash fresh produce in warm water. Use a scrub brush, and rinse thoroughly.
- Use a knife to peel an orange or grapefruit; do not bite into the peel.

- Discard the outer leaves of leafy vegetables such as cabbage and lettuce.
- Peel waxed fruit and vegetables; waxes don't wash off and can seal in pesticide residues.
- Peel vegetables such as carrots and fruits such as apples when appropriate. (Peeling removes pesticides that remain in or on the peel, but also removes fibers, vitamins, and minerals.)

Information about pesticides is available from the EPA's national pesticide hotline (800) 858-PEST) anytime day or night, 365 days a year.

food supply. Instead, it sets conditions so that substances do not become a hazard and acts promptly when problems or suspicions arise.

Avoiding Pesticides Consumers, therefore, have some responsibility for their own health and safety with respect to pesticides. They can learn about the potential benefits and dangers of pesticide use, discuss regulations and alternatives with others, advise their government representatives about their findings, and apply pressure wherever it will help change inappropriate procedures. Meanwhile, people can minimize their risks by following the guidelines offered in the accompanying "How to" box. In addition to the suggestions in the box, consumers can buy fresh foods grown locally, especially when they can confirm that the produce has been grown using responsible methods.

Organically Grown Crops Some farmers produce and market **organically grown crops:** crops grown and processed according to USDA regulations defining the use of synthetic fertilizers, herbicides, insecticides, fungicides, preservatives, and other chemical ingredients.[23] Similarly, meat and dairy products may be called organic if the livestock has been raised according to USDA regulations defining the grazing conditions and the use of organic feed, hormones, and antibiotics. In addition, producers may *not* claim products are organic if they have been irradiated, genetically engineered, or grown with fertilizer made from sewer sludge.

In addition to benefiting from reduced costs of farming, increased soil quality, and decreased chemical impact on the environment, farmers of organic crops stand to increase their share of the market. Many consumers are willing to pay more for organic foods, and sales are increasing in record numbers.

Implied in the definition of organic is that organic products are healthier for consumers than those grown using other methods, which may not be the case. Using unprocessed animal manure as an organic fertilizer, for example, may transmit bacteria, such as *E. coli*, to human beings. Both organic and conventional methods have advantages and disadvantages, and consumers must remain informed.

Organically grown foods that have met USDA standards may bear this seal on their labels.

organically grown crops: crops grown and processed according to USDA regulations defining the use of fertilizers, herbicides, insecticides, fungicides, preservatives, and other chemical ingredients.

Chapter Eleven
284

In Summary Pesticides can safely improve crop yields when used according to regulations, but can be hazardous when used inappropriately. The FDA tests both domestic and imported foods for pesticide residues in the fields and in market basket surveys of foods prepared table ready. Consumers can minimize their ingestion of pesticide residues on foods by following the guidelines in the accompanying box. Alternative farming methods may allow farmers to grow crops with few or no pesticides.

Incidental Food Additives

Indirect or **incidental additives** are contaminants that find their way into food as a result of some phase of production, processing, storage, or packaging. Examples of incidental additives include tiny bits of plastic, glass, paper, tin, and other substances from packages, as well as chemicals from processing, such as the solvent used to decaffeinate coffee.

Microwave Packaging

Some microwave products are sold in "active packaging" that participates in cooking the food. Pizza, for example, may rest on a cardboard pan coated with a thin film of metal that absorbs microwave energy and may heat up to 500°F. When exposed to the intense heat, some particles of the packaging components migrate into the food. Regular microwave packages heat up less, but particles still migrate. The materials from both kinds of packaging are under study to determine their safety for consumption. Until more is known, a wise choice is to use only glass or ceramic containers designed for microwaving and to avoid reusing disposable containers, such as margarine tubs, for heating foods.

Dioxins

Coffee filters, milk cartons, paper plates, and frozen food packages can all be made of bleached paper and so can contaminate foods with trace amounts of compounds known as **dioxins.** Dioxins form during the chlorination step in making bleached paper. They can migrate into foods that come in contact with the paper, but the amounts entering food are infinitesimally small—one part per trillion, or the equivalent of one second in 32,000 years. Such amounts do not appear to present a health risk to people, and drinking milk from bleached cartons appears to be safe. Dioxins are persistent, however, and they leach into the environment by way of both paper-mill effluent and discarded paper products in landfills. Thus, like the other contaminants described earlier (see Figure 11–4), dioxins bioaccumulate, becoming more and more concentrated in land, water, and animals until they build up to hazardous levels.

In Summary Incidental additives sometimes find their way into foods, but adverse effects are rare. These additives are well regulated. All food packagers are required to perform specific tests to discover whether materials from packages are migrating into foods; if they are, their safety must be confirmed by strict procedures similar to those governing intentional additives, discussed next.

Food Additives

Of all consumer concerns about food safety, preventing food poisoning is the most important. Next in importance are natural toxins, environmental contaminants, and pesticides. Last are **additives.** Manufacturers use food additives to give foods desirable characteristics: color, flavor, texture, stability, higher nutrient content, or resistance to spoilage. Consumers see additives mentioned on labels

© Polara Studios, Inc.

Many consumers are willing to pay a little more for pesticide-free produce.

Here's a way to tell if glass or other containers are made of microwave-safe materials. Microwave the empty container for one minute and carefully touch it.
- *Warm: Unsafe for microwave.*
- *Lukewarm: Safe for short reheating use.*
- *Cool: Safe for long microwave cooking times.*

incidental additives: substances that can get into food not through intentional introduction but as a result of contact with the food during growing, processing, packaging, storing, or some other stage before the food is consumed; also called *accidental* or *indirect additives.*

dioxins: toxic organic compounds containing chlorine that arise in industry as (among other things) by-products of the bleaching process.

additives: substances that are added to foods, but normally are not consumed by themselves as foods.

and wonder or worry about them, but additives are of very little concern compared to the other aspects of foods already covered. Nevertheless, for the sake of completeness, this section will first describe the regulations and procedures governing food additives and then will look at some individual additives.

Regulations Governing Additives

To get permission to use new additives in food products, manufacturers have to go through special procedures that can take many years. The manufacturer must test each new additive to satisfy the FDA of the following:

- The additive is effective (it does what it is supposed to do).
- It can be detected and measured in the final food product.

Then the manufacturer must study the effects of feeding the additive in large doses to animals under strictly controlled conditions to prove that:

- It is safe (it does not cause cancer, birth defects, or other injury).

Finally, the manufacturer must submit all test results to the FDA. Public hearings follow, where consumers are invited to participate and experts present testimony for and against granting permission to use the additive. Thus consumers' rights and responsibilities are written into the provisions for deeming additives safe.

When the FDA approves an additive, it writes a regulation stating in what amounts, for what purposes, and in what foods the additive may be used. No additives are permanently approved; all are periodically reviewed.

GRAS List Additives Many substances such as salt, sugar, caffeine, and herbs were exempted from complying with this procedure at the time it was first instituted because they had been used for a long time and their use entailed no known hazards. Some 700 substances in all were put on the **generally recognized as safe (GRAS) list.** When substantial scientific evidence or public outcry has questioned the safety of a substance on the GRAS list, however, its safety has been reevaluated. Whenever a legitimate question has been raised about a substance, it has been removed or reclassified.

Toxicity versus Hazard An important distinction governs decisions about an additive's safety—the distinction between **toxicity** as a property of substances and **hazard** associated with substances. Toxicity is a general property of all substances; hazard is the capacity of a chemical to produce injury under conditions of its use. All substances can be toxic at some level of consumption, but they are called hazardous only if they are actually consumed in sufficiently large quantities to cause harm. An additive is not considered to be a hazard if some immense amount that people never consume is toxic. The additive is a hazard only if it is toxic under the conditions of its actual use. A food additive is supposed to have a wide **margin of safety.**

Testing Procedures Most additives that involve risk are allowed in foods only at levels 100 times below those at which the risk is still known to be zero. Experiments to determine the extent of risk involve feeding test animals the substance at different concentrations throughout their lifetimes. The additive is then permitted in foods at 1/100 the level that causes no harmful effect whatever in the animals. In many foods, naturally occurring toxins appear at levels that bring their margins of safety closer to 1/10. Even nutrients, as you have seen, involve risks at high dosage levels.

The margin-of-safety concept also applies to nutrients when they are used as additives. Iodine is added to salt to prevent iodine deficiency, but it has to be added with care because it is a deadly poison in excess. Similarly, iron is added to refined bread and other grains (enrichment) and has doubtless helped prevent

generally recognized as safe (GRAS) list: a list, established by the FDA, of food additives long in use and believed safe.

toxicity: the ability of a substance to harm living organisms. All substances are toxic if the concentration is high enough.

hazard: a state of danger; used to refer to any circumstance in which harm is possible under normal conditions of use.

margin of safety: in reference to food additives, a zone between the concentration normally used and that at which a hazard exists. For common table salt, for example, the margin of safety is 1/5 (five times the concentration normally used would be hazardous).

many cases of iron-deficiency anemia in women and children who are prone to that disease. But the addition of too much iron could put men (who usually have enough iron in their bodies) at risk for iron overload. The Tolerable Upper Intake Level has to be remembered.

Two long-used preservatives.

Benefits versus Risks   Most additives used in foods are there because they offer benefits that outweigh their risks or that make the risks worth taking. In the case of color additives that only enhance the appearance of foods and do not improve their health value or safety, no amount of risk may be deemed worth taking. The FDA still approves only a few artificial colors for use in foods, and screening of these continues.

Furthermore, manufacturers must use only the amounts of additives necessary to get the needed effects, not more. Additives may not be used:

- To disguise faulty or inferior products.
- To deceive the consumer.
- Where they significantly destroy nutrients.
- Where their effects can be achieved by economical, sound manufacturing processes.

The regulations in force governing the management of intentional additives are well conceived and have been effective on the whole. Funding shortages limit the capabilities of watchdog agencies such as the FDA, however, and some mistakes and cases of false reporting are bound to slip by.

Some Intentional Additives

The next few paragraphs focus on a few individual food additives—notably, those that have received the most publicity because people ask questions about them most often. The order is alphabetical; it does not imply an order of importance.

Antimicrobial Agents   Foods can spoil in two ways: one dangerous, one not. The dangerous way is by becoming hazardous to health; the other way is by losing their flavor and attractiveness. An example of the dangerous way: bacteria, yeasts, and molds and other fungi growing in foods can cause food-borne illness. **Preservatives** known as **antimicrobial agents** protect foods from these microbes.

The best-known, most widely used antimicrobial agents are two common substances—salt and sugar. Salt preserves meat and fish; sugar preserves canned and frozen fruits, jams, and jellies. Both salt and sugar work by withdrawing water from the food; microbes cannot grow without water. Today, other additives such as potassium sorbate and sodium propionate are also used to extend the shelf life of baked goods, cheese, beverages, mayonnaise, margarine, and many other products.

Another group of antimicrobial agents, the **nitrites,** are added to foods for three main purposes: to preserve their color (especially the pink color of hot dogs and other cured meats); to enhance their flavor by inhibiting rancidity (especially in cured meats); and to protect against bacterial growth. In particular, nitrites prevent the growth of the botulinum bacterium that produces the deadly toxin described earlier.

Nitrites clearly perform important jobs, but they have been the object of controversy because in the human body they can be converted to **nitrosamines,** which cause cancer in animals. Some cured meats are available without nitrites, but reducing nitrites consumed in meats would hardly make a difference in a person's overall exposure to nitrosamine-related compounds. For example, the average cigarette smoker inhales 100 times the nitrosamines that the average bacon eater ingests. Likewise, a beer drinker imbibes up to roughly five times the

preservatives: antimicrobial agents, antioxidants, chelating agents, radiation, and other additives that retard spoilage or preserve desired qualities, such as softness in baked goods.

antimicrobial agents: substances used as food additives that prevent the growth of illness-causing microorganisms in foods.

nitrites: salts added to food to prevent botulism. An example is sodium nitrite.

nitrosamines (nigh-TROHS-uh-meens): derivatives of nitrites that may form when nitrites combine with amines.

Raw grapes may legally be treated with sulfites. Wash grapes thoroughly before eating.

Common examples of color additives:
- *Carotenoids.*
- *Blue #1 and #2 (brilliant blue and ingotine).*
- *Green #3 (fast green).*
- *Red #40 and #3 (allura red and erythrosine).*
- *Yellow #5 and #6 (tartrazine and sunset yellow).*

Foods containing tartrazine:
- *Orange drinks (Tang, Daybreak, Awake).*
- *Gatorade (lime flavored).*
- *Gelatin desserts (Jell-O, Royal).*
- *Golden Blend Italian dressing (Kraft).*
- *Some cake mixes and icings (Duncan Hines, Pillsbury, Cake Mate).*
- *Imitation banana or pineapple extract (McCormick).*
- *Seasoning salt (French's).*
- *Macaroni and cheese dinner (Kraft).*
- *"Cheez" curls and balls (Planter's).*
- *Fruit chews (Skittles).*
- *Butterscotch squares and candy corn (Brach's).*

antioxidants: compounds that protect other compounds from oxygen by reacting with oxygen themselves (first defined in Chapter 8). Antioxidants are used to prevent rancidity of fats in foods and other damage to food caused by oxygen. Examples are vitamins E and C, BHA, BHT, propyl gallate, and sulfites.

BHA, BHT: preservatives commonly used to slow the development of off-flavors, odors, and color changes caused by oxidation.

Chapter Eleven

amount that the bacon eater receives. Cosmetics deliver via absorption through the skin about twice the amount delivered from bacon. Even the air inside automobiles delivers measurable amounts of nitrosamines.

Antioxidants The other way foods can go bad is by undergoing changes in color and flavor caused by exposure to oxygen in the air (oxidation). Often these changes involve little hazard to health, but they damage the food's appearance, taste, and nutritional quality. Familiar examples of these changes are sliced apples or potatoes turning brown and oils going rancid. **Antioxidants** protect foods from this kind of spoilage. Among the antioxidants approved for use in foods are vitamin C (ascorbate) and vitamin E (tocopherol). Two other antioxidants in wide use are **BHA** and **BHT,** which prevent rancidity in baked goods and snack foods.[*]

The **sulfites** are another group of antioxidants. They are used to prevent oxidation in many processed foods, alcoholic beverages (especially wine), and drugs. They used to be popular with restaurant owners for use on salad bars because they keep raw fruits and vegetables looking fresh, but some people experience allergic reactions to sulfites—reactions that are sometimes dangerous and, for a few, deadly. The FDA now prohibits sulfite use on foods intended to be consumed raw, with the exception of grapes, and it requires sulfite-containing foods and drugs to include a warning on their labels. Restaurants, however, are not required to disclose whether they have used sulfites in food preparation. Therefore, concerned consumers must ask if sulfites have been used.[24] For most people, sulfites do not pose a hazard in the amounts used in products.

Artificial Colors As mentioned, the GRAS list still includes only about ten **artificial colors,** a highly select group that has survived considerable screening. They are among the most intensively investigated of all additives. In fact, they are much better known than the natural pigments of plants, and the limits on the safety of their use can be stated with greater certainty.

Still, the food colors have been more heavily criticized than almost any other group of additives. The reason, simply stated, is that they only make foods attractive, whereas other additives, such as preservatives, make foods safe. Hence, with food colors, we can afford to require that their use entail no risk, whereas with other additives we may have to compromise between the risks of using them and the risks of not using them.

The food color tartrazine (yellow dye number 5) causes an allergic reaction in susceptible people. Symptoms include hives, itching, and nasal congestion, sometimes severe enough to require medical treatment. It is not a common problem; only 1 or 2 in 10,000 individuals may have the reaction. Still, that is more than 20,000 individuals in the nation as a whole. These people rightly demand to know what foods contain the dye so that they can avoid it. It is not enough to avoid yellow-colored foods because tartrazine is used to confer turquoise, green, and maroon colors in foods and drugs as well. Legislation is now in force requiring that tartrazine be listed on all labels of foods that contain it.

Artificial Flavors and Flavor Enhancers While only a few artificial colors are currently permitted in foods, close to 2000 **artificial flavors** and **flavor enhancers** are approved, making them the largest single group of food additives. One of the best-known members of this group is monosodium glutamate, or MSG (trade name, Accent)—the monosodium salt of the amino acid glutamic acid. MSG is used widely in restaurants, especially Asian restaurants, as a flavor enhancer. Research indicates that in addition to enhancing other flavors, MSG may itself possess a basic taste independent of the well-known sweet, salty, bitter, and sour tastes.[†]

---

[*]BHA is butylated hydroxyanisole; BHT is butylated hydroxytolvene.
[†]The taste produced by MSG is termed *umami.*

MSG has received publicity because it may produce an adverse reaction called the **MSG symptom complex** in 1 to 2 percent of the population. MSG has been investigated extensively enough to be deemed safe for adults to use (except people who react adversely to it, of course), but it is kept out of foods for infants. Food labels require ingredient lists to itemize all additives, including MSG.

Nutrient Additives   Another class of additives includes nutrients added to improve or to maintain the nutritional value of foods.[25] Among the nutrient additives are the nutrients added to refined grains to enrich them, the iodine added to salt, vitamins A and D added to dairy products, and the nutrients added to fortified breakfast cereals. Nutrients are sometimes also added for other purposes. The use of vitamins C and E as antioxidants has already been mentioned.

Common examples of nutrient additives:
- *Thiamin, niacin, riboflavin, folate, and iron in grain products.*
- *Iodine in salt.*
- *Vitamins A and D in milk.*
- *Vitamin C in fruit drinks.*

Texture and Stability   **Thickening** and **stabilizing agents** may be added to foods during processing to maintain emulsions, foams, or suspensions or to lend a desirable thick consistency to foods. Dextrins (short chains of glucose formed as a breakdown product of starch), starch, and pectin are examples. Gums, such as carrageenan, guar, locust bean, agar, and gum arabic, are also added for thickening and stabilizing.

As this section has shown, no two additives are alike, and therefore generalizations about them are meaningless. No valid statement can be made that applies to the 3000-odd different substances commonly added to foods. Questions about which additives are safe and under what conditions of use must be asked and answered on an item-by-item basis.

The U.S. food supply is well monitored and well protected against hazards that might threaten people's health. Provided that consumers apply common sense in selecting and preparing their foods, they can enjoy the great benefits of an abundant and safe food supply.

In Summary   On the whole, the benefits of food additives seem to justify the risks associated with their use. The FDA regulates the use of the following intentional additives: antimicrobial agents (such as nitrites) to prevent microbial spoilage; antioxidants (such as vitamins C and E, sulfites, and BHA and BHT) to prevent oxidative changes; colors (such as tartrazine) and flavor enhancers (such as MSG) to appeal to senses; and nutrients (such as iodine in salt) to enrich or fortify foods.

Food Biotechnology

Hope for the future purity of foods comes on the crest of new advances in **biotechnology.** Biotechnology promises to produce greater crop yields, leaner meats, longer shelf lives, better nutrient composition, and fewer pesticides. Overall, biotechnology offers opportunities to enhance the quality, nutritional value, and variety of foods.[26]

Genetic Engineering

For centuries farmers have manipulated the genetics of plants and animals to shape the characteristics of their crops and livestock. Consider corn, for example. Wild, native corn bears only two or three kernels on a cob, but many years of patient selective breeding have produced the large, full, sweet ears people enjoy today. Many types of wild corn are now all but extinct. Half of the increases in U.S. crop yields in the twentieth century were due to such genetic improvements; the use of irrigation, fertilizers, and pesticides has also contributed. Farmers still

sulfites: salts containing sulfur that are added to fresh and frozen fruits and vegetables to prevent changes in color and texture due to oxidation. Sulfites appear on food labels as:
- Sulfur dioxide.
- Sodium sulfite.
- Sodium bisulfite.
- Potassium bisulfite.
- Sodium metabisulfite.
- Potassium metabisulfite.

artificial colors: certified food colors, added to enhance appearance (*certified* means approved by the FDA).

artificial flavors, flavor enhancers: chemicals that mimic natural flavors and those that enhance flavor.

MSG symptom complex: the acute, temporary, and self-limiting reactions experienced by sensitive people upon ingesting a large dose of MSG. The name *MSG symptom complex*, given by the FDA, replaces the former *Chinese restaurant syndrome.*

thickening and **stabilizing agents:** ingredients that maintain emulsions, foams, or suspensions, or lend a desirable thick consistency to foods.

biotechnology: the use of biological systems or organisms to create or modify products; also called *biogenetic engineering.*

With the benefits of a safe and abundant food supply comes the responsibility to select, prepare, and store foods safely.

use selective breeding to provide consumers with low-fat meats, high-yield grains, and a seemingly endless variety of fruits and vegetables.

Scientists can now speed up the process of genetic change through biotechnology. Farmers need no longer wait patiently for breeding to yield improved crops and animals, nor must they even respect natural lines of reproduction among species. Laboratory scientists can now select desirable traits from any of a number of species and insert those traits into the genetic material of crops and animals.

Among the new products of biotechnology are tomatoes that stay fresh much longer than others and so promise less waste and higher profits. Normally, tomatoes produce a protein that softens them after they have been picked. Scientists introduce into a tomato plant a gene that is a mirror image of the one that codes for the "softening" enzyme. This gene fastens itself to the RNA of the native gene and blocks its action. A vine-ripe tomato with this special gene rots more slowly than a normal tomato, allowing growers to harvest at the most flavorful and nutritious red stage. The red tomatoes will still last much longer during shipping and marketing than regular tomatoes harvested when green.

Similarly, soybeans may be implanted with a gene that will upgrade soy protein to a quality approaching that of milk. Corn may be modified to contain lysine and tryptophan, its two limiting amino acids. Fats and oils with a predetermined fatty acid composition may be possible within the decade. Crops that produce their own insecticides upon receiving genes from bacteria may render pesticides unnecessary. Shrimp may soon fight diseases with genetic ammunition borrowed from sea urchins. Livestock may receive growth-promoting hormones from bacteria. The possibilities seem unlimited, and though they sound fantastic, many are waiting on laboratory shelves for the time when they will be fully employed in agriculture.

While food industrialists hail biotechnology as a miracle, some other people fear that tampering with genetics may change organisms in ways not yet fully understood, even by the scientists who developed the techniques. They wonder what unknown changes take place when the genes of living things are manipulated and what the long-term consequences might be.

Regulations and Labeling

The FDA has taken the position that foods produced through biotechnology are not substantially different from others and require no special safety testing or labeling.[27] A product, such as the tomato described earlier, need not be tested because its new genes prevent synthesis of a protein and add nothing but a tiny fragment of genetic material. On the other hand, any substances introduced into a food by way of bioengineering must meet the same safety standards applied to all additives. Some people object to genetic tampering and want labels to help them identify "old-fashioned" tomatoes. They may not realize that most foods available today have already been altered genetically by selective breeding. The vegetable broccoflower, a product of sophisticated cross-breeding of broccoli with cauliflower, met no testing or approval barriers on its way to the dinner plate. Only after the vegetable became popular with consumers did scientists study its nutrient contents (see Appendix A for their findings).

In Summary Scientists are continuing to study the effects not only of biotechnology but of all sorts of new food-processing techniques. Their efforts to enhance food production will help meet the challenge of feeding an ever-increasing world population.

1. Which of the following is the major food source for transmission of *Campylobacter jejuni?*
   a. raw poultry
   b. uncooked seafood
   c. contaminated water
   d. imported soft cheeses

2. A method the food industry uses to identify points of contamination and implement controls to ensure food safety is called:
   a. margin of safety.
   b. generally recognized as safe.
   c. North American Free Trade Agreement.
   d. Hazard Analysis Critical Control Points.

3. The temperature danger zone for foods ranges from:
   a. −20°F to 120°F.
   b. 0°F to 100°F.
   c. 20°F to 120°F.
   d. 40°F to 140°F.

4. Examples of foods that frequently cause food-borne illness are:
   a. canned foods.
   b. steaming-hot foods.
   c. fresh fruits and vegetables.
   d. raw milk, seafood, meat, and eggs.

5. Irradiation can help improve the food supply by:
   a. minimizing the use of preservatives.
   b. improving the nutrient content of foods.
   c. cooking foods quickly.
   d. killing microorganisms.

6. A natural toxin present in foods that can cause illness when consumed in excess is:
   a. alar.
   b. solanine.
   c. botulinum toxin.
   d. *Salmonella.*

7. Additives approved for use must be:
   a. intentional and undetectable in the final product.
   b. indirect and quality enhancing.
   c. intentional and on the GRAS list.
   d. detectable and measurable in the final product.

8. Common antimicrobial additives include:
   a. salt and nitrites.
   b. carrageenan and MSG.
   c. dioxins and sulfites.
   d. vitamin C and vitamin E.

9. Common antioxidants include:
   a. BHA and BHT.
   b. tartrazine and MSG.
   c. sugar and vitamin E.
   d. nitrosamines and salt.

10. Biotechnological advances that have improved the food supply include:
    a. potatoes with solanine and cabbage without goitrogens.
    b. seafood without methylmercury and onions without an odor.
    c. larger cattle with a higher fat content and rennin produced by sheep.
    d. tomatoes with a longer shelf life and soybeans with a higher-quality protein.

Critical Thinking

1. Whenever Jodie's mom bakes a cake or cookies, Jodie loves to eat the batter left on the mixing spoon or in the bowl. Which of the following food-borne illnesses might Jodie become ill with if she keeps on eating raw batter?
   a. salmonellosis
   b. botulism
   c. *E. coli* infection
   d. none of the above

Answers to these questions appear in Appendix G.

Clinical Applications

1. Jack and Joann are hosting a cookout for several friends. The ground beef thawing in the refrigerator is still a little frozen, so Joann sets it on the counter to finish thawing. In the meantime, Jack peels and skewers raw shrimp to grill for those who don't eat burgers. He makes a marinade for the shrimp, places the skewered shrimp in the marinade, and sets the pan next to the thawing ground beef on the counter. Once the beef has thawed, Joann portions it out on a cutting board and makes individual burgers with her hands. She puts the raw burgers on a plate and sets them in the refrigerator until time to grill them. Joann then uses the cutting board and a knife to slice tomato, onion, and lettuce to serve with the burgers. Jack has the grill going outside and comes in and gets the burgers and shrimp. Joann pours the leftover marinade into a container for those who want to add some to their cooked shrimp. When the shrimp are done, Jack puts them on the plate he brought them out on and takes them inside. The burgers are now looking brown on the inside, so Jack removes them from the grill and brings them in on the same plate he carried them out on. Dinner is ready now, so everyone sits down to eat. Consider the four steps people can take to fight food-borne bacteria and make a list of the mistakes Jack and Joann made.

2. Describe what you would do to correct the mistakes Jack and Joann made.

Nutrition on the Net

For further study of the topics of this chapter, access these websites.

Find updates and quick links to these and other nutrition-related sites at our website:
www.wadsworth.com/nutrition

Get food safety tips from the Food and Drug Administration and the Center for Food Safety and Applied Nutrition or from the Fight BAC! Campaign of the Partnership for Food Safety Education:
www.foodsafety.gov or **www.fightbac.org**

Learn more about food-borne illnesses and livestock diseases such as BSE from the National Center for Infectious Diseases at the Centers for Disease Control and Prevention:
www.cdc.gov/ncidod

Learn about the various types of food thermometers and how and when to use them from the USDA Thermy Campaign:
www.fsis.usda.gov/thermy

Find commonsense health tips for travelers at the Centers for Disease Control and Prevention:
www.cdc.gov/travel

Learn more about food irradiation from the Foundation for Food Irradiation Education and the

International Consultative Group on Food Irradiation:
www.food-irradiation.com and
www.iaea.org/icgfi

Read a backgrounder on irradiation from the International Food Information Council:
ificinfo.health.org/foodirradiation/backgrounder.htm

Report adverse reactions to the FDA's MedWatch program at (800) 332-1088 or:
www.fda.gov/medwatch

Learn more about safe drinking water from the Environmental Protection Agency:
www.epa.gov/safewater

Visit the Environmental Protection Agency to review tips for food buying and preparation that will help minimize pesticide exposure:
www.epa.gov/pesticides/food

Get fish advisories from the Environmental Protection Agency:
www.epa.gov/ost/fish

Visit the Canadian Food Inspection Agency (CFIA):
www.cfia-acia.agr.ca

Learn more about food safety in the marketplace from the Food Safety Inspection Service:
www.usda.gov/fsis

Search for meat, poultry, seafood, fruits, vegetables, food additives, and food safety at the U.S. Government health information site:
www.healthfinder.gov

Learn more about organic foods and national organic food standards:
www.ams.usda.gov/nop

Visit the USDA Biotechnology Information Center:
www.nal.usda.gov/bic

Get a "pro" biotechnology perspective from the Council for Biotechnology Information:
www.whybiotech.com

Search for biotechnology at the International Food Information Council to find another "for" view:
ificinfo.health.org

Get a "con" biotechnology perspective from the Genetic Engineering section of Greenpeace, USA:
www.greenpeaceusa.org

Another "against" view is available from the Transgenic Café of the Union of Concerned Scientists:
www.ucsusa.org

Notes

1. Centers for Disease Control and Prevention, Preliminary Food-Net data on the incidence of foodborne illnesses—Selected sites, United States, *Morbidity and Mortality Weekly Report* 50 (2001): 241–246.

2. C. J. Lackey, Oil, herb, and garlic flavored, www.foodsafety.org, site visited on February 5, 1998.

3. F. J. Angulo and coauthors, A large outbreak of botulism: The hazardous baked potato, *Journal of Infectious Diseases* 178 (1998): 172–177.

4. L. B. Zimmerhackl, *E. coli*, antibiotics, and the hemolytic-uremic syndrome, *New England Journal of Medicine* 342 (2000): 1990–1991.

5. Sprouts lead beef in *E. coli* cases, *FDA Consumer*, November/December 1998, p. 2.

6. Sprout safety, high-risk groups warned: Don't eat alfalfa sprouts, *FDA Consumer*, November/December 1998, p. 2.

7. P. Kurtzweil, Questions keep sprouting about sprouts, *FDA Consumer*, January/February 1999, pp. 18–22.

8. National Advisory Committee on Microbiological Criteria for Foods, Hazard analysis and critical control point principles and application guidelines, *Journal of Food Protection* 61 (1998): 762–775; HHS News, www.fda.gov/bbs/topics/NEWS/2001

9. U.S. Department of Agriculture, Second progress report on *Salmonella* testing for raw meat and poultry products, *FSIS Backgrounder*, January 21, 1999.

10. J. Henkel, "Thermy" promotes thorough food cooking, *FDA Consumer*, September/October 2000, p. 35.

11. L. Bren, Trying to keep "mad cow disease" out of U.S. herds, *FDA Consumer*, March/April 2001, pp. 12–14.

12. R. J. Formanek, Highlights of FDA food safety efforts: Fruit juice, mercury in fish, *FDA Consumer*, March/April, 2001, pp. 15–17.

13. Even pasteurized eggs may contain harmful bacteria, *Tufts University Health and Nutrition Letter*, July 1998, p. 6.

14. Outbreaks of cyclosporiasis—United States, 1997, *Morbidity and Mortality Weekly Report* 46 (1997): 461–462; *Vibrio vulnificus* infections associated with eating raw oysters—Los Angeles, 1996, *Journal of the American Medical Association* 276 (1996): 937–938; J. W. Besser-Wiek and coauthors, Foodborne outbreak of diarrheal illness associated with *Cryptosporidium parvum*—Minnesota, 1995, *Morbidity and Mortality Weekly Report* 45 (1996): 783–784.

15. New bottled water standards, *FDA Consumer*, April 1996, p. 2.

16. Position of The American Dietetic Association: Food irradiation, *Journal of the American Dietetic Association* 100 (2000): 246–253.

17. S. L. Nightingale, Irradiation of meat approved for pathogen control, *Journal of the American Medical Association* 279 (1998): 9; M. T. Olsterholm, Cyclosporiasis and raspberries—Lessons for the future, *New England Journal of Medicine* 336 (1997): 1597–1598.

18. FDA announces advisory on methylmercury in fish, *FDA Talk Paper*, 2001, available from www.cfsan.fda.gov.

19. EPA national advice on mercury in fish caught by family and friends for women who are or may become pregnant, nursing

mothers, and young children, *EPA National Advisory,* 2001, available from www.epa.gov/ost/fish.

20. Food Quality Protection Act of 1996, Public Law No. 104-170, 110 Statute 1489; C. Marwick, New focus on children's environmental health, *Journal of the American Medical Association* 277 (1997): 871–872.

21. Olsterholm, 1997.

22. C. Marwick, "Fresh Produce Initiative" for imports, *Journal of the American Medical Association* 278 (1997): 1481.

23. www.ams.usda.gov/nop, site visited on October 5, 2001.

24. R. Papazian, Sulfites: Safe for most, dangerous for some, *FDA Consumer,* December 1996, pp. 11–14.

25. W. Mertz, Food fortification in the United States, *Nutrition Reviews* 55 (1997): 44–49.

26. Position of The American Dietetic Association: Biotechnology and the future of food, *Journal of the American Dietetic Association* 95 (1995): 1429–1432.

27. J. Henkel, Genetic engineering: Fast forwarding to future foods, *FDA Consumer,* April 1997, pp. 6–11.

Nutrition in Practice

■ ENVIRONMENTALLY CONSCIOUS FOODWAYS ■

The preceding chapter offered strategies and information for people to protect themselves from food-borne illness and contaminants. Earlier chapters suggested ways to achieve dietary goals such as nutritional adequacy and balance. People who follow the advice given can be satisfied that they have answers to the questions "How can I keep my foods safe?" and "How can I get the best health benefits from my foods?"

Some people want to achieve another goal when they shop for, prepare, and cook foods. They recognize that they will spend thousands of dollars on foods and will cook thousands of meals in a lifetime, and they perceive that their money and actions exert effects in the world outside their own personal lives. Increasingly, people today are asking, "What are the environmental impacts of my food choices? How do my actions in the kitchen affect the environment?" They want to make environmentally responsible choices.

What kinds of environmental impacts do people's food choices have?

Among the global resources involved in producing food are irrigation water, fertilizers, pesticides, fuel, and land and fisheries. In the U.S. market, tons of packaging materials and a massive transportation network burning immense quantities of **fossil fuels** are used to convey foods to consumers (see the glossary on p. 295 for fossil fuels). Each truckload of food produced in this country travels, on average, more than a thousand miles to reach the market. It costs 800 kcalories in fuel to make a can of diet soda that contains 1 kcalorie of food energy, and more water is used to make the can than to make the soda.

More environmentally benign choices are available. In place of vegetables shipped in from far away, people might choose to eat vegetables grown in their own home states, at least during the growing seasons. In place of several sodas in aluminum cans, a soda drinker might use one large recyclable bottle. Environmentally responsible food choices can also be made in the realms of food shopping, preparation, cooking, and cleanup.

What are some examples of environmentally responsible food shopping?

Food shopping involves going to the store, selecting foods once there, choosing among the packages in which those foods are sold, and choosing bags in which to carry the foods home. All of these actions exert impacts on the environment, and consumers can choose to minimize those impacts. Consider the shopping trips first. The environmentally conscious shopper knows that motor vehicles are the world's single largest source of air pollution and so tries to minimize car mileage spent on trips to and from the store. Strategies are to shop "carless," shop nearby, and shop only once a week. Whenever possible, ride a bicycle with baskets on the back, or walk and carry your own canvas bags or use a small cart. To shop for a week's meals at a time, a shopper can plan to buy foods with various shelf lives and to eat the most perishable ones first. For example, buy lettuce, cabbage, squash, and carrots. Use up the lettuce first, then the squash. The carrots and cabbage keep longer, so eat these later. Buy fruits of differing ripeness—for example, six bananas: two ripe, two nearly ripe, and two green. Use the ripe ones right away and the others as they become ripe. These strategies save time and money as well as fossil fuels.

Give some examples of environmentally responsible food choices.

Perhaps the advice most frequently given to save fuel and resources is to "eat low on the food chain." It takes much less land and fuel, and costs much less in pollution, to produce most plant foods than to produce most meats (see Figure NP11–1 on p. 296). Following this advice benefits nutritional health, too: recall from Chapter 1 that the Daily Food Guide recommends that adults eat 11 or more servings of plant foods (especially vegetables and grains) daily and only 4 or 5 servings of milk products and meats combined. This is also a good strategy for avoiding food contaminants. Another guideline is to eat foods that are processed as little as possible. It takes more energy to produce canned or frozen corn than fresh corn. Energy costs mean fuel costs, and fuel costs mean pollution.

Other environmentally aware food-shopping practices include buying products whose production benefits the land, or at least harms it minimally, and boycotting products whose production damages the land. Food produced locally requires less transportation, packaging, and refrigeration than shipped-in foods. It also makes sense to avoid buying canned beef

Nutrition in Practice

Courtesy of NASA

"We do not inherit the earth from our ancestors, we borrow it from our children." Ascribed to Chief Seattle, a nineteenth-century Native American leader.

products of any kind, including soups, chili, stews, corned beef, and even beef-flavored pet food. Some of these foods come at the expense of cleared rain forest land. About 200 square feet of rain forest are lost *permanently* for each pound of beef produced from cattle raised on the cleared land. Rain forest beef is not labeled, so the only way consumers can be sure they are not buying it is to buy no canned beef at all. Choose chicken and small fish more often. Chickens are often grown locally and require fewer resources to produce; small fish eat low on the food chain.

The task for consumers is to become aware of their impacts and to find **sustainable** ways of doing things. These are ways of living that use up resources at a rate that nature, forestry, or agriculture can replace. They are ways of living that pollute the earth at a rate that nature or human cleanup efforts can keep up with.

Do the packages foods come in exert impacts on the environment?

Yes. It costs energy and resources to make each can, foam tray, waxed or clay-coated cardboard container,

plastic bottle, or glass jar, and it costs land or pollution to dispose of them. In general, no packages at all are best for the environment; next best are minimal, reusable, or recyclable ones. Grocery bags represent a huge drain on energy and resources. Paper factories use chemicals such as toxic forms of chlorine bleach, which are released into waterways in such large quantities that the chemicals can destroy whole bays and fisheries. Some consumers are demanding alternatives to throw-away bags. Many shoppers carry reusable shopping bags to the store or ask for recyclable plastic bags—and then take care to recycle them. The third choice would be paper bags, and last would be nonrecyclable plastic.

What are the best ways to prepare and cook food from the environmental standpoint?

Many kitchens today are equipped with a vast array of electrical gadgets and small appliances designed to save time and human energy but at the expense of the earth's resources. The single largest contribution to air pollution comes from the generation and use of energy to run our homes and vehicles. Try to use fewer gadgets; mix batters, chop vegetables, and open cans by hand.

Fast cooking saves fuel and so pollutes less. Asian meals exemplify this principle: they are made of precut, bite-sized pieces of food, stir-fried fast in small amounts of oil. This cooking style both saves energy and preserves nutrients. The pressure cooker or the microwave can also cook food quickly. The pressure cooker can do the occasional big piece of meat the cook wants to serve whole; the microwave can cook vegetables, casseroles, or leftovers. Both may save using several burners on the stove, and both preserve nutrients better than most stove-top methods.

The oven, in contrast, can be a fuel waster. Efficient oven use is possible if the cook bakes or roasts a lot of food at one time. To use the stove top efficiently, a cook can use flat-bottomed pots with close-fitting lids that completely cover the burners. That way, each burner will donate all its heat to cooking something, not just heating the kitchen (and the planet). One can also turn

Glossary

fossil fuels: coal, oil, and natural gas, which all come from the fossilized remains of plant life of earlier times. These are nonrenewable fuels that pollute. (Renewable, or alternative, fuels, such as solar and wind energy, pollute less or not at all.)

sustainable: able to continue indefinitely; using resources at such a rate that the earth can keep on replacing them and producing pollutants at a rate with which the environment and human cleanup efforts can keep pace so that no net accumulation of pollution occurs.

Nutrition in Practice

Figure NP11–1

Eating Low on the Food Chain Saves Resources

It takes ten times as much land and fuel to feed people meat as to feed them plants.

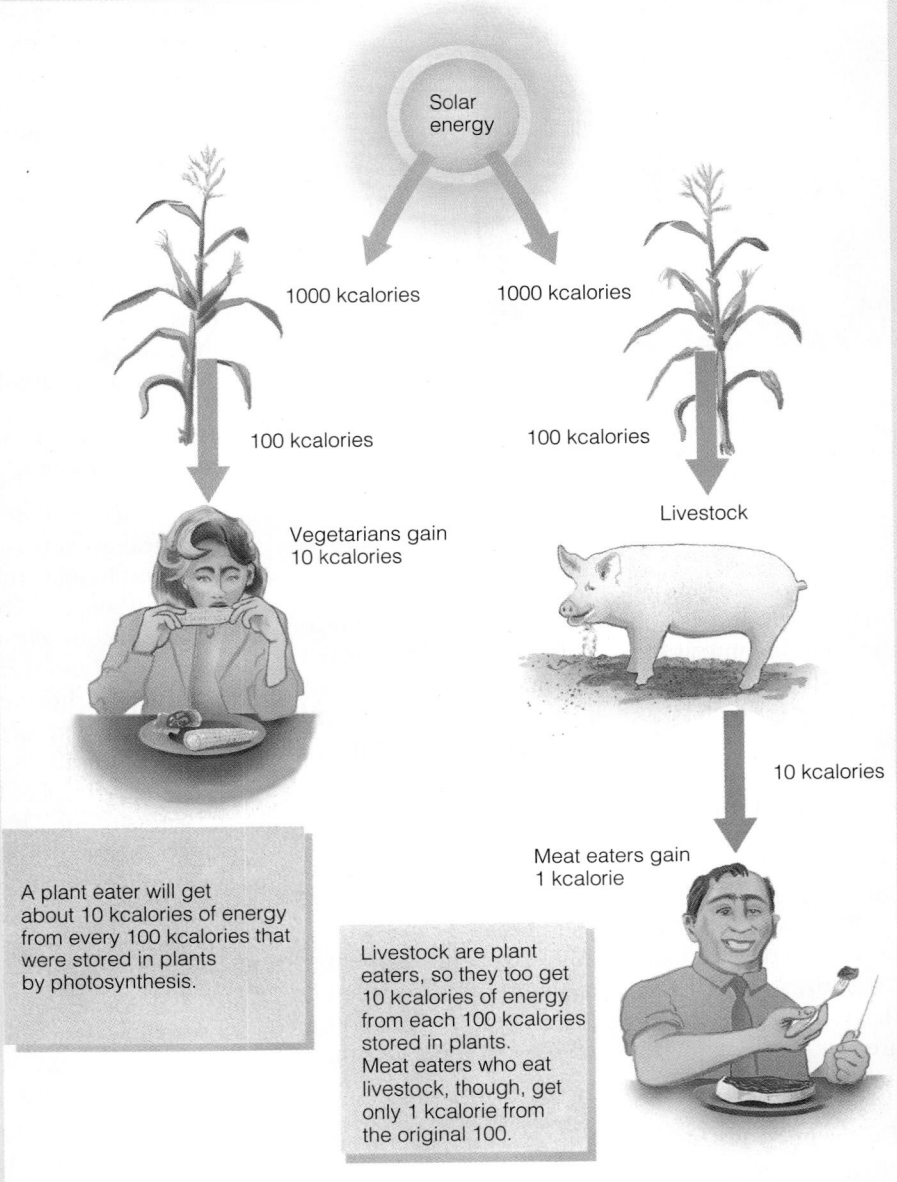

Solar energy

1000 kcalories

1000 kcalories

100 kcalories

100 kcalories

Vegetarians gain 10 kcalories

Livestock

10 kcalories

A plant eater will get about 10 kcalories of energy from every 100 kcalories that were stored in plants by photosynthesis.

Meat eaters gain 1 kcalorie

Livestock are plant eaters, so they too get 10 kcalories of energy from each 100 kcalories stored in plants. Meat eaters who eat livestock, though, get only 1 kcalorie from the original 100.

electric burners and ovens off before the food is fully cooked and let the cooking finish as the stove cools.

What about cleanup after a meal?

A big energy user associated with food preparation and cleanup is the water heater. The less hot water used, the less fuel must be burned to heat the water. People can save water-heating energy in many ways. They can set the water heater between 120 and 130°F, not hotter. They can put it on a timer so that each day it heats just enough water to meet that day's needs. They can wrap the water heater in insulation to keep it from losing heat to the surroundings, and they can wrap the hot-water pipes all the way to the points of use. When replacing a water heater or installing a new one, consumers can purchase a small, instantaneous-type water heater that heats the water only at the point of use, and only when needed. A consumer can choose a gas water heater, rather than an electric one; natural gas is a cleaner fossil fuel than the coal or oil usually burned to make electricity. Solar water heaters work well in sunny

Nutrition in Practice

regions. Water-saving faucets and showerheads can also save energy.

Someone who washes many dishes at a time should consider using a dishwasher, if it is affordable. People may think that the cost of the water, heat, and soap would be higher than the cost of washing by hand, but this is not the case. One school found that a normal machine cycle used on full loads consumed less than two-thirds the water used in hand washing. Using less hot water also means using less electricity to heat it. The savings are greatest if the dishes are not prerinsed and are allowed to air dry.

Consumers can purchase dishwashers, refrigerators, and other home appliances that bear the U.S. government's Energy Star logo for the highest energy efficiency (see Figure NP11–2). If everyone chose energy-efficient appliances and products with the Energy Star label, and did nothing else, over the next ten years, the nation's utility bill would be reduced by about $100 billion.

Are there "right" and "wrong" ways to throw garbage away, too?

An average American household of four people produces about 100 pounds of trash a week, much of it from the kitchen. To reduce your own trash burden, use reusable pans and dishes, rather than disposable ones that are used once and thrown away. Reduce your use of aluminum foil, paper towels, plastic storage bags, and other disposables. Find permanent replacements such as reusable storage containers and washable cloths.

National concern has focused on the issue of trash because the nation is running out of landfill space in which to dispose of all the trash. Landfill space is only one of many problems associated with trash, however. Every item thrown away is a resource lost: an aluminum can could be used to make a new aluminum can; a cereal box, but for its clay coating, could become recycled paper; a plastic bottle could become part of a beautiful carpet. Trash need not become an undesirable mess; recycled trash could be viewed as a usable resource. Yet much of the metal mined in the United States is used only once and then discarded. The aluminum thrown away every three months could rebuild the entire U.S. air fleet.

Garbage is a special case of a resource generated in the kitchen. Vegetable scraps, fruit peelings, and leftover plant foods are organic and biodegradable, like the leaves and grass cuttings people rake up in their yards. All of these materials can be piled up together with some soil and allowed to decompose naturally, forming compost, a rich, crumbly material that can be used as in nature to fertilize growing things. Some communities, recognizing this, conduct composting programs to recycle people's organic debris; some homeowners maintain their own composting piles. College campuses can use composted kitchen waste, mixed with grass cuttings and weeds, to mulch, fertilize, and enrich the soil they use in landscaping. Composting can even be done indoors in small odor-free bins and the resulting material used to pot plants.

As you can see, environmental awareness can permeate every aspect of food management from start to finish. And each choice made to save resources, cut energy use, or refrain from polluting helps to preserve and protect the environment.

Dietitians and foodservice managers have a special role to play, and their efforts can make an impressive difference.[1] The American Dietetic Association (ADA) urges members to conserve resources and minimize waste in both their professional and their personal lives.[2]

Notes

1. N. I. Hahn, The greening of a school district: How school foodservice led a recycling revolution, *Journal of the American Dietetic Association* 97 (1997): 371.
2. Position of The American Dietetic Association: Dietetics professionals can implement practices to conserve natural resources and protect the environment, *Journal of the American Dietetic Association* 101 (2001): 1221–1227.

Money Isn't All You're Saving

Figure NP11–2
Energy Star

Jennie Oppenheimer/Studio Zocolo

CHAPTER 12 | # NUTRITION THROUGH THE LIFE SPAN: PREGNANCY AND INFANCY

Both parents can prepare in advance for a healthy pregnancy.

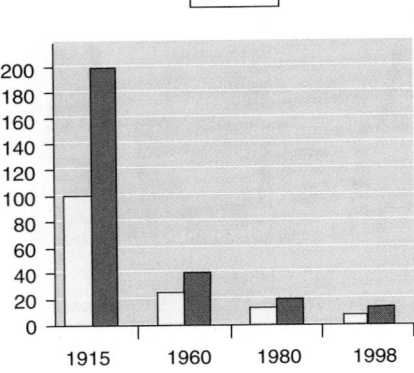

Figure 12–1

Infant Mortality Decline in the Twentieth Century
The graph shows infant deaths per 1000 live births.

SOURCE: Data from B. Guyer and coauthors, Annual summary of vital statistics: Trends in the health of Americans during the 20th century, *Pediatrics,* 106 (2000): 1307–1317.

low birthweight (LBW): a birthweight less than 5½ lb (2500 g); indicates probable poor health in the newborn and poor nutrition status of the mother during pregnancy. Normal birthweight for a full-term baby is 6½ to 8½ lb (about 3000 to 4000 g).

Low-birthweight infants are of two different types. Some are **premature;** they are born early and are of a weight **appropriate for gestational age (AGA).** Others have suffered growth failure in the uterus; they may or may not be born early, but they are **small for gestational age (SGA).**

*A*ll people need the same nutrients, but the amounts they need vary depending on their stage of life. This chapter focuses on nutrition in preparation for, and support of, pregnancy, lactation, and infancy. The next two chapters address the needs of children, adolescents, and older adults.

Pregnancy: The Impact of Nutrition on the Future

The woman who enters pregnancy with full nutrient stores, sound eating habits, and a healthy body weight has done much to ensure an optimal pregnancy. Then, if she eats a variety of nutrient-dense foods during her pregnancy itself, her own and her infant's health will benefit further.

Preparing for Pregnancy

Full nutrient stores *before* pregnancy are essential both to conception and to healthy infant development during pregnancy. In the early weeks of pregnancy, before many women are even aware that they are pregnant, significant developmental changes occur that depend on a woman's nutrient stores.

Fathers-to-be are also wise to consider their eating and other habits. Limited evidence suggests that men who consume too few fruits and vegetables containing vitamin C or who drink too much alcohol in the weeks before conception may sustain damage to their sperm's genetic material. This damage can cause birth defects in future children.

Prepregnancy Weight Appropriate weight prior to pregnancy also benefits pregnancy outcome. Being either underweight or overweight (see p. 306) presents medical risks during pregnancy and childbirth. Underweight women are therefore advised to gain weight before becoming pregnant, and overweight women to lose excess weight. Guidelines for weight gain and loss were offered in Chapter 7.

Infant Birthweight Infant birthweight correlates with prepregnancy weight and weight gain during pregnancy and is the most potent single predictor of the infant's future health and survival. An underweight woman has a high risk of having a **low-birthweight** baby, especially if she is unable to gain sufficient weight during pregnancy. Compared with normal-weight babies, low-birthweight babies are more likely to contract diseases and nearly 40 times more likely to die in the first month of life. Impaired growth and development during pregnancy may have long-term health effects as well. Research suggests that when nutrient supplies fail to meet demands, permanent adaptations take place that may predispose the infant to chronic diseases such as type 2 diabetes, hypertension, and heart disease in later life.[1] Other hazards of low birthweight may include lower adult IQ and other brain and sensory impairments, short stature, and educational disadvantages.[2] Underweight women are therefore advised to try to gain weight before becoming pregnant or to strive to gain adequately during pregnancy.

Nutritional deficiency, coupled with low birthweight, is the underlying cause of more than half of all the deaths worldwide of children under five years of age. In 1998, the U.S. infant mortality rate was among the lowest the nation has ever recorded: 7.2 deaths per 1000 live births.[3] This rate is still higher than that of some other developed countries, but as part of a significant steady decline for two decades, it stands as a tribute to public health efforts aimed at reducing infant deaths (see Figure 12–1).

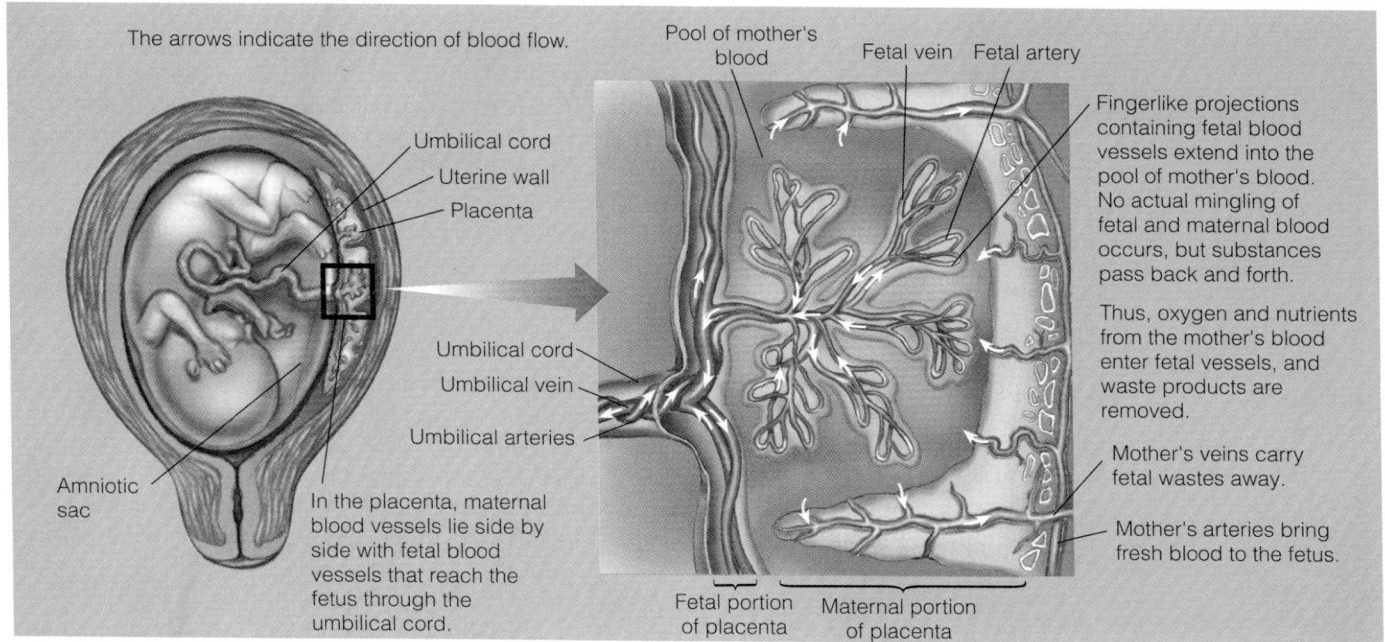

The arrows indicate the direction of blood flow.

Pool of mother's blood

Fetal vein   Fetal artery

Umbilical cord
Uterine wall
Placenta

Fingerlike projections containing fetal blood vessels extend into the pool of mother's blood. No actual mingling of fetal and maternal blood occurs, but substances pass back and forth.

Thus, oxygen and nutrients from the mother's blood enter fetal vessels, and waste products are removed.

Umbilical cord
Umbilical vein
Umbilical arteries

Amniotic sac

In the placenta, maternal blood vessels lie side by side with fetal blood vessels that reach the fetus through the umbilical cord.

Mother's veins carry fetal wastes away.

Mother's arteries bring fresh blood to the fetus.

Fetal portion of placenta   Maternal portion of placenta

Figure 12–2
The Placenta

Not all cases of low birthweight reflect poor nutrition. Other factors associated with low birthweight include heredity, disease conditions, smoking, and drug (including alcohol) use during pregnancy. Even with optimal nutrition and health during pregnancy, some women give birth to small infants for reasons unknown. Still, poor nutrition is the major factor in low birthweight, and generally, it is an avoidable one as later sections make clear.

Healthy Support Tissues   The mother's prepregnancy nutrition is crucial to a healthy pregnancy because it determines whether she will be able to grow healthy support tissues: the **placenta,** the **amniotic sac,** the **umbilical cord,** and the expanding **uterus** (see Figure 12–2). Malnutrition prior to and around conception keeps these tissues from developing fully.[4] If the placenta fails to develop properly, the **fetus** will not receive optimal nourishment.

In Summary   Maternal nutrition before and during pregnancy affects both the mother's health and the infant's growth and development. Appropriate weight before pregnancy and appropriate weight gain during pregnancy also benefit pregnancy outcome, and can reduce the risk of having a low-birthweight baby.

Nutrient Needs during Pregnancy

Between the moment of conception and the moment of birth, innumerable events determine the course and outcome of fetal development and, ultimately, the health of the newborn infant. With respect to nutrition, each organ needs nutrients most during its own intensive growth and development period. A nutrient deficiency during one stage of development might affect the heart and, during another stage, the developing limbs.

Developments that occur during a **critical period** can take place only at that time and at no other. Whatever nutrients and other environmental conditions are necessary during this period must be supplied on time if the organ is to reach its full potential. If the development of an organ is limited during a critical period, recovery is impossible.[5] Thus early malnutrition often does irreversible

placenta (pla-SEN-tuh): an organ that develops inside the uterus early in pregnancy, in which maternal and fetal blood circulate in close proximity and exchange materials. The fetus receives nutrients and oxygen across the placenta; the mother's blood picks up carbon dioxide and other waste materials to be excreted via her lungs and kidneys.

amniotic (am-nee-OTT-ic) **sac:** the "bag of waters" in the uterus, in which the fetus floats.

umbilical (um-BIL-ih-cul) **cord:** the rope-like structure through which the fetus's veins and arteries reach the placenta; the route of nourishment and oxygen into the fetus and the route of waste disposal from the fetus.

uterus (YOO-ter-us): the womb, the muscular organ within which the infant develops before birth.

fetus (FEET-us): the developing infant from eight weeks after conception until its birth.

critical period: a finite period during development in which certain events occur that will have irreversible effects on later developmental stages; usually a period of rapid cell division.

Nutrition Through the Life Span: Pregnancy and Infancy

Table 12–1 Factors Placing Pregnant Women at Nutritional Risk

Women likely to develop nutrient deficiencies include those who:

- Are young (adolescents).
- Have had many previous pregnancies (3 or more to mothers under age 20; 4 or more to mothers age 20 or older).
- Have short intervals between pregnancies (<18 months).
- Lack nutrition knowledge, have too little money to purchase adequate food, or have too little family support.
- Ordinarily consume an inadequate diet due to food faddism, preferences, weight-loss "dieting," uninformed vegetarianism, or eating disorders.
- Smoke cigarettes or use alcohol or drugs.
- Are lactose intolerant or suffer chronic health conditions requiring special diets.
- Are underweight or overweight at conception.
- Are carrying twins or triplets.
- Gain insufficient or excessive weight during pregnancy.
- Have a low level of education.

Pregnancy is often divided into thirds called *trimesters.*

Energy RDA during pregnancy (2nd and 3rd trimesters): +300 kcal/day.

Protein RDA during pregnancy: +10 g/day.

Recommended carbohydrate intake: about 50% of energy intake. In a 2000 kcal/day intake, this represents 1000 kcal of carbohydrate, or about 250 g. Four cups of milk will contribute about 50 g carbohydrate. An apple provides 12 g carbohydrate, and a slice of bread provides 12 g, so generous intakes of fruit and grain products are clearly beneficial.

Folate RDA during pregnancy: 600 μg/day.

damage, although this may not become fully apparent until the person has matured, so the problem may never be attributed to events of pregnancy. When people having such a poor start on life reach adulthood, they may be vulnerable to infections, and some may have high risks of diabetes, hypertension, stroke, or heart disease.[6] Table 12–1 provides a list of factors that make nutrient deficiencies likely during pregnancy. Notice that young age heads the list; a later section explains why pregnant adolescents are especially prone to malnutrition.

A woman's nutrient needs during pregnancy and lactation are higher than at any other time in her adult life and are greater for certain nutrients than for others. Figure 12–3 (p. 303) compares the nutrient needs of nonpregnant, pregnant, and lactating women. A study of the figure reveals some of the key needs.

Energy A pregnant woman needs extra food energy, but only a little extra—300 kcalories above the allowance for nonpregnant women—and only during the second and third trimesters. A woman can easily obtain 300 kcalories from just one extra serving from each of the five food groups—a slice of bread, a serving of vegetables, an ounce of lean meat, a piece of fruit, and a cup of fat-free milk. Pregnant teenagers, underweight women, or physically active women may require more.

Protein The RDA suggests an added 10 grams of protein per day throughout pregnancy. Many women in the United States exceed the recommended protein intake each day, so they already receive the 10 grams of additional daily protein recommended for pregnancy. In fact, pregnant women in the United States—even those with low incomes who are not participating in food assistance programs—generally receive between 75 and 110 grams of protein a day.

Obtaining enough protein need not pose a problem, even if the diet excludes all foods of animal origin. Pregnant vegetarian women who meet their energy needs by eating ample servings of protein-containing plant foods such as legumes, whole grains, nuts, and seeds meet their protein needs as well. Use of high-protein supplements during pregnancy can be harmful and is discouraged.

Carbohydrate Pregnant women need generous amounts of carbohydrate to spare the protein they eat. If added energy is needed, it is best obtained from carbohydrate.

Essential Fatty Acids The high nutrient requirements of pregnancy leave little room in the diet for excess energy from added purified fats such as oil, margarine, and butter. The essential fatty acids, however, are important to the growth of the fetus and are regarded by some as essential nutrients in early human development.[7] The brain is largely made of lipid material, and it depends heavily on products of both omega-3 and omega-6 fatty acids for its growth, function, and structure. If a pregnant woman eats a diet that regularly includes seafood, she receives a balance of the essential fatty acids and their derivatives. This benefits her pregnancy and later her infant by way of her milk. Supplements of fish oil are not recommended, however, both because they may carry concentrated toxins and because their effects on pregnancy remain unknown.

Vitamins The vitamins required for rapid cell proliferation—folate and vitamin B_{12}—are needed in large amounts during pregnancy. New cells are laid down at a tremendous pace as the fetus grows and develops. At the same time, the mother's red blood cells increase in number so the recommendation for folate during pregnancy rises from 400 to 600 micrograms a day.

As described in Chapter 8, folate plays an important role in preventing neural tube defects. Folate supplements (400 micrograms daily) taken one month before conception and continued throughout the first trimester can prevent neural tube defects.[8] Neural tube defects arise early in pregnancy, however, so the

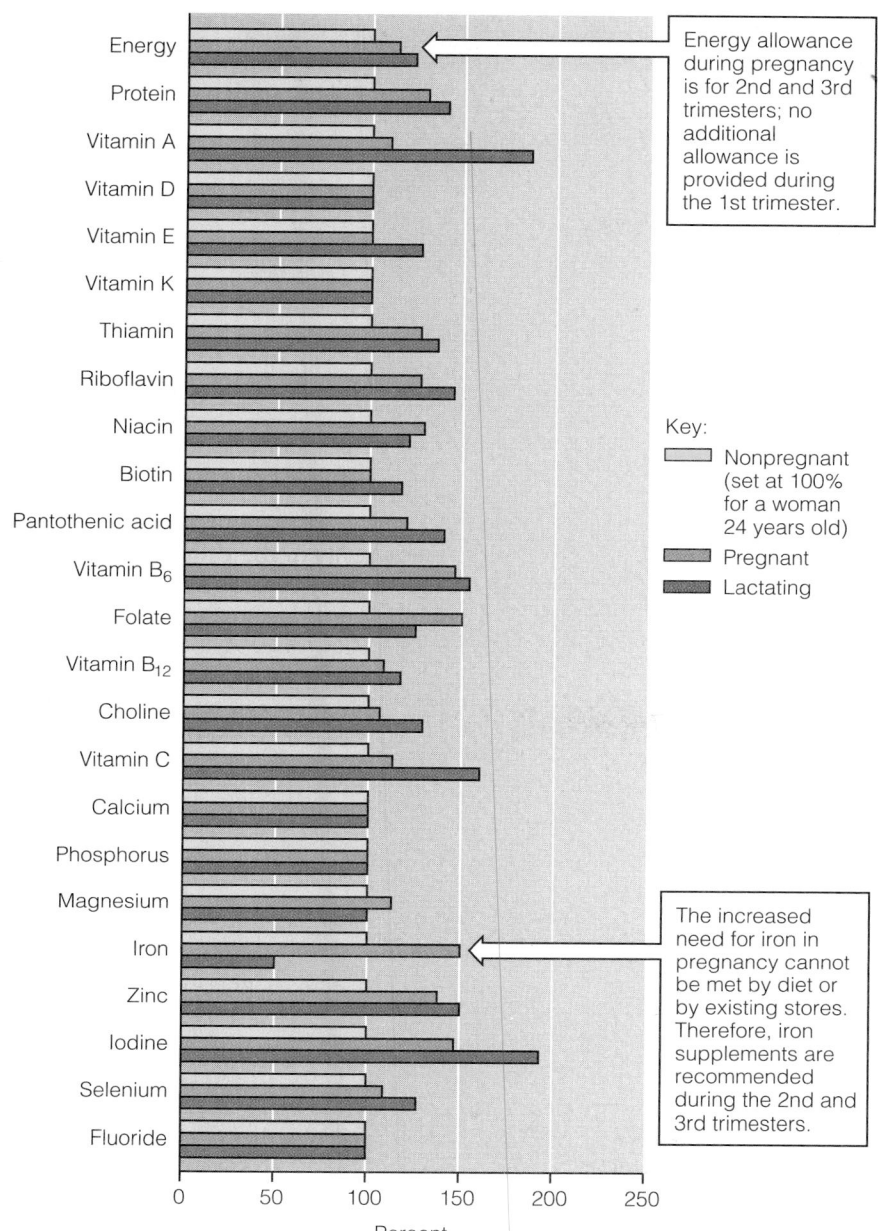

Figure 12–3

Comparison of Nutrient Recommendations for Nonpregnant, Pregnant, and Lactating Women
For actual values, turn to the table on the inside front cover.

American Academy of Pediatrics, the Public Health Services, the Institute of Medicine, and the March of Dimes recommend that all women capable of becoming pregnant take 400 micrograms of folate daily from supplements or fortified foods or a combination of the two, in addition to eating a variety of foods naturally rich in folate (see Table 12–2 on p. 304).[9]

Supplements and fortified foods offer women a convenient way to ensure sufficient folate regularly and continuously enough to benefit pregnancy.[10] Furthermore, the synthetic form of folate, *folic acid*, in supplements and fortified foods is better absorbed than the naturally occurring folate in foods. Thus folate status improves more with intakes of supplements or fortified foods than with intakes of only natural sources of folate.[11] The foods that naturally contain folate are still important, however, because they are rich sources of other vitamins, minerals, fiber, and the phytochemicals thought to protect against heart disease, cancer, and other diseases.

Table 12–2 Rich Folate Sources[a]	
Natural Folate Sources	**Fortified Folate Sources**
Liver (3 oz) 185 µg	Multi-Grain Cheerios Plus cereal (1 c) 400 µg[b]
Lentils (½ c) 180 µg	Product 19 cereal (1 c) 400 µg[b]
Chickpeas or pinto beans (½ c) 145 µg	Total cereal (1 c) 400 µg[b]
Asparagus (½ c) 125 µg	Pasta, cooked (1 c) 110 µg
Spinach (1 c raw) 115 µg	Rice, cooked (1 c) 80 µg
Avocado (½ c) 70 µg	Bagel (1 small whole) 50 µg
Orange juice (1 c) 60 µg	Waffles, frozen (2) 40 µg
Beets (½ c) 46 µg	Bread, white (1 slice) 20 µg

[a]Folate amounts for these and 2000 other foods are listed in the Table of Food Composition in Appendix A.
[b]Folate in cereals varies; read the Nutrition Facts panel of the label.

As of 1999, all refined grain products (cereal, pasta, flour, bread, rolls, buns, farina, grits, cornmeal, and rice) sold commercially in the United States are fortified with folic acid.* Folate fortification is expected to prevent half of all neural tube defects that occur each year. Research has already shown that folate fortification improves folate status.[12] Folate fortification does raise some safety concerns, however. A pregnant woman needs a greater amount of vitamin B_{12} to assist folate in the manufacture of new cells. High intakes of folate complicate the diagnosis of a vitamin B_{12} deficiency. For this reason, folate intakes should not exceed 1 milligram per day.[13]

Generally, even modest amounts of meat, fish, eggs, or milk products together with body stores easily meet the need for vitamin B_{12}. Those who exclude all foods of animal origin, however, may need daily supplements to prevent deficiency.

Vitamin D and Calcium for Bones   Vitamin D and the minerals involved in building the skeleton—calcium, phosphorus, and magnesium—are in great demand during pregnancy. Insufficient intakes may result in abnormal fetal bone development.

Intestinal absorption of calcium doubles early in pregnancy, when the mother's bones store the mineral. Later, as the fetal bones begin to calcify, there is a dramatic shift of calcium across the placenta. Whether the calcium added to the mother's bones early in pregnancy is withdrawn to build the fetus's bones later is unclear. In the final weeks of pregnancy, more than 300 milligrams a day are transferred to the fetus. Recommendations to ensure an adequate calcium intake during pregnancy are aimed at conserving the mother's bone mass while supplying fetal needs.

For women whose prepregnancy calcium intakes are below recommendations, as most are, increased calcium intakes may be especially important. Milk products offer many advantages over supplements, as emphasized in earlier chapters. Because bones are still actively depositing minerals until about age 25, adequate calcium is especially important for young women. Pregnant women under age 25 who consume less than 600 milligrams of calcium a day need to increase their intakes of milk, cheese, yogurt, and other calcium-rich foods. Alternatively, and less preferably, they may need a daily supplement of 600 milligrams of calcium.

Women who exclude milk products need calcium-fortified foods such as soy milk. It is worth noting that not all soy milk is fortified with calcium; products fortified with calcium and vitamin D are recommended. Calcium–fortified orange juice offers folate and vitamin C as well as calcium.

Vitamin B_{12} RDA during pregnancy: 2.6 µg/day.

Calcium AI during pregnancy:
  1300 mg/day (14 to 18 yr).
  1000 mg/day (19 to 50 yr).

Phosphorus RDA during pregnancy:
  1250 mg/day (14 to 18 yr).
  700 mg/day (19 to 50 yr).

Magnesium RDA during pregnancy:
  400 mg/day (14 to 18 yr) .
  350 mg/day (19 to 30 yr).
  360 mg/day (31 to 50 yr).

Four cups of milk a day will supply 1200 mg calcium. For other food sources of calcium, see Chapter 9.

*These products must be fortified with 1.4 milligrams of folate per 100 grams of food.

Fluoride Mineralization of the fetus's teeth begins in the fifth month after conception. For this and for bone development, fluoride may be needed. Fluoride crosses the placenta, and whether the placenta can defend against excess intakes is questionable. Therefore, fluoride supplements are not recommended for pregnant women who drink fluoridated water. For women who live in communities without fluoridated water, a fluoride supplement may protect fetal teeth.

Iron The body conserves iron especially well during pregnancy: menstruation ceases and absorption of iron increases up to threefold due to a rise in the blood's iron-absorbing and iron-carrying protein transferrin. Still, iron needs are so high that stores dwindle during pregnancy.

The developing fetus draws on the mother's iron stores to create stores of its own to last through the first four to six months of life. Iron losses also occur with the bleeding that is inevitable at birth.

Few women enter pregnancy with adequate iron stores. Women who enter pregnancy with iron-deficiency anemia have a greater-than-normal risk of delivering low-birthweight or preterm infants.[14] For all women not taking supplements containing iron, a daily iron supplement containing 30 milligrams is recommended during the second and third trimesters of pregnancy.

Zinc Zinc is required for DNA and RNA synthesis and thus for protein synthesis. Severe zinc deficiency predicts a low infant birthweight.[15] Zinc is most abundant in foods of high-protein content, such as shellfish, meat, and nuts, but the presence of other trace elements and fiber in foods may adversely affect zinc absorption. For example, iron interferes with the body's absorption and use of zinc, so women taking iron supplements (more than 30 milligrams per day) may also need zinc supplements.

Nutrient Supplements Women who make wise food choices during pregnancy can meet most of their nutrient needs except for iron. As already discussed, iron supplements are recommended during the second and third trimesters for all pregnant women. Daily multivitamin-mineral supplements are also recommended for women who do not eat adequately and for those in high-risk groups: women carrying multiple fetuses, cigarette smokers, and alcohol and drug abusers. The use of prenatal supplements may help to reduce the risks of preterm delivery, low infant birthweights, and birth defects.[16] Table 12–3 lists recommended amounts of supplements for pregnant women at nutritional risk.

Food Choices Because food energy needs increase less than nutrient needs, the pregnant woman must select foods of high nutrient density. For most women, appropriate choices include foods such as fat-free milk, fat-free plain yogurt, lean meats, eggs, dark green vegetables, vitamin C–rich fruits, legumes, and whole-grain breads and cereals. Table 12–4 provides a suggested food pattern.

Table 12–3 Nutrient Supplements during Pregnancy[a]	
Nutrient	**Amount**
Folate	400 ug
Vitamin B_6	2 mg
Vitamin C	50 mg
Vitamin D	5 ug
Calcium	600 mg
Copper	2 mg
Iron	30 mg
Zinc	15 mg

[a]For pregnant women at nutritional risk (see Table 12–1 on p. 302).

SOURCE: Reprinted with permission from *Nutrition during Pregnancy* © by the National Academy of Sciences. Published by the National Academy Press, Washington, D.C., 1990.

Fluoride AI during pregnancy: 3.0 mg/day.

Iron RDA during pregnancy: 27 mg/day.

In pregnancy, hemoglobin values of 12 g are not unusual, and 11 g is where the line defining "too low" is often drawn. Appendix E discusses more sensitive measures of iron status.

Food sources of iron:
- *Liver, oysters.*
- *Red meat, fish, other meat.*
- *Dried fruits (raisins, prunes).*
- *Legumes (dried beans, peas, lima beans).*
- *Dark green vegetables.*

Reminder: Vitamin C–rich foods enhance iron absorption from foods.

Zinc RDA during pregnancy:
13 mg/day (<18 yr).
11 mg/day (19 to 50 yr).

Table 12–4 Daily Food Choices for Pregnant and Lactating Women		
Food Group	**Number of Servings**	
	Adults	**Pregnant or Lactating Women**
Breads/cereals	6 to 11	7 to 11
Vegetables	3 to 5	4 to 5
Fruits	2 to 4	3 to 4
Meat/meat alternates	2 to 3	3
Milk/milk products	2	3 to 4

NOTE: Figure 1–5 in Chapter 1 provides a detailed summary of foods in each group with serving sizes.

Table 12–5 Recommended Weight Gains for Pregnancy[a]

- Underweight women (BMI <19.8): 28 to 40 lb
- Normal-weight women (BMI 19.8 to 26): 25 to 35 lb
- Overweight women (BMI 26 to 29): 15 to 25 lb
- Obese women (BMI >29): 13 lb minimum

[a]The BMI cutoff points in this table were established by the Subcommittee on Nutritional Status and Weight Gain during Pregnancy. The cutoff points defining underweight, normal weight, overweight, and obesity differ slightly from those established by the National Heart, Lung, and Blood Institute Expert Panel on the Identification, Evaluation, and Treatment of Overweight and Obesity in Adults.

SOURCE: Committee on Nutritional Status during Pregnancy and Lactation, Food and Nutrition Board, *Nutrition during Pregnancy* (Washington, D.C.: National Academy Press, 1990), pp.10, 12.

WIC, pronounced *WICK*, is the acronym for the federal Special Supplemental Food Program for **W**omen, **I**nfants, and **C**hildren. The U.S. Department of Agriculture (USDA) funds WIC, and state health departments administer the program.

Cravings for nonfood items such as clay, ice, laundry starch, and cornstarch are known as *pica*.

A prenatal weight-gain grid (see Appendix D) plots the rate of weight gain during pregnancy.

food cravings: deep longings for particular foods.

food aversions: strong desires to avoid particular foods.

Chapter Twelve

306

WIC   A woman of limited financial means may need help in obtaining needed food and information. At the federal level, the WIC program provides nutrition education and vouchers redeemable for nutritious foods to low-income pregnant women and their children. WIC provides food vouchers for eggs, milk, cereal, juice, cheese, legumes, and peanut butter to infants, children up to age five, and pregnant and breastfeeding women who qualify financially and are at medical or nutritional risk. For infants given formula, WIC also provides iron-fortified formulas. Studies of the nutrition and health effects of WIC have found that participation in the program benefits both the iron status and the growth and development of infants and children. WIC participation during pregnancy reduces the risks of delivering preterm or low-birthweight infants.[17]

Food Cravings and Aversions   Some women develop cravings for, or aversions to, certain foods and beverages during pregnancy. Individual **food cravings** during pregnancy do not seem to reflect real physiological needs. In other words, a woman who craves pickles does not necessarily need salt, nor does a chocolate craving indicate a need for sugar, caffeine, or fat. Similarly, cravings for ice cream are common during pregnancy, but do not signify a calcium deficiency. **Food aversions** and cravings that arise during pregnancy are probably due to hormone-induced changes in taste and sensitivities to smells.

In Summary   Energy and nutrient needs are high during pregnancy. A balanced diet that includes an extra serving from each of the five food groups can usually meet these needs with the exception of iron (supplements are recommended). Food cravings do not typically reflect physiological needs.

Weight Gain

All pregnant women must gain weight: fetal growth and maternal health depend on it. A pregnancy weight gain of 25 to 35 pounds is recommended for women who begin pregnancy at a healthy weight and are carrying a single fetus; for others, see Table 12–5. Some women, notably adolescents who are still growing themselves, should strive for gains at the upper ends of the target ranges. Short women (5 feet 2 inches and under) should strive for lower gains. Figure 12–4 shows the components of a weight gain of 30 pounds.

The ideal pattern of weight gain during pregnancy is thought to be about 3½ pounds during the first three months and a pound per week thereafter. Women lose some of the weight gained during pregnancy at delivery and most of the remainder within the following few weeks or months, as blood volume returns to normal and accumulated fluids are lost.

If a woman has gained more than the expected amount of weight early in pregnancy, she should not try to diet in the last weeks. A sudden large weight gain, however, is a danger signal that may indicate the onset of hypertension (see p. 308).

Physical Activity

Physical activity is important to the pregnant woman, not only to help her carry the extra weight of pregnancy without strain, but also to help ease her upcoming childbirth. Staying active during the course of a normal, healthy pregnancy can improve the fitness of the mother-to-be, facilitate labor, and reduce psychological stress. Women who remain active during pregnancy report fewer discomforts throughout their pregnancies and gain less weight than those who are not physically active.[18] Pregnant women should take care in choosing their physical activities, however. They should participate in "low-impact" activities and avoid

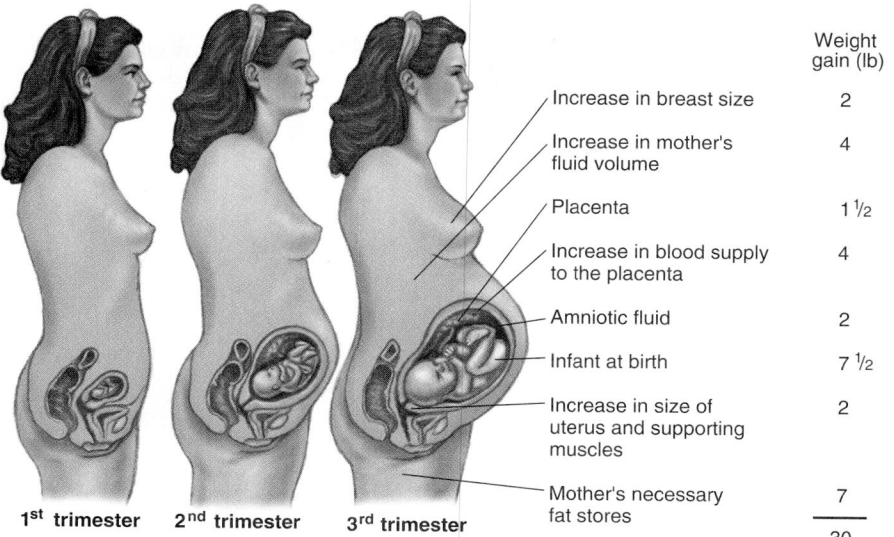

	Weight gain (lb)
Increase in breast size	2
Increase in mother's fluid volume	4
Placenta	1 ½
Increase in blood supply to the placenta	4
Amniotic fluid	2
Infant at birth	7 ½
Increase in size of uterus and supporting muscles	2
Mother's necessary fat stores	7
	30

1st trimester   2nd trimester   3rd trimester

Figure 12–4

Components of Weight Gain during Pregnancy

sports in which they might fall or be hit by other people or objects. As is true for everyone, the frequency, duration, and intensity of the activity affect the likelihood of the benefits or risks.[19] A pregnant woman should consult her health care provider before taking up additional activity. A few guidelines are offered in Table 12–6 on p. 308.

In Summary   A healthy pregnancy depends on a sufficient weight gain. Women who begin their pregnancies at a healthy weight need to gain about 30 pounds, which covers the growth and development of the placenta, uterus, blood, breasts, and infant. By remaining active throughout pregnancy, a woman can develop the strength she needs to carry the extra weight and maintain habits that will help her lose it after the birth.

Problems in Pregnancy

Just as adequate nutrition and normal weight gain support the health of the mother and growth of the fetus, maternal diseases detract. If discovered early, many diseases can be controlled—another reason why early prenatal care is recommended. Some nutrition measures can help alleviate the most common problems encountered during pregnancy.

Gestational Diabetes   Pregnancy precipitates the onset of diabetes in some women because placental hormones alter the way insulin works. This condition is known as **gestational diabetes.** In many cases, blood glucose becomes abnormal during pregnancy but usually returns to normal after the infant is born. Almost one-third of all women with gestational diabetes, however, develop type 2 diabetes later in life, especially if they are overweight. To ensure that the problems of diabetes are dealt with promptly, the American Diabetes Association recommends that most women receive glucose tolerance tests (see Chapter 24) during the sixth month of pregnancy.[20]

Without proper management, gestational diabetes can lead to complications such as a high-birthweight infant and difficult delivery; rates of **cesarean section** delivery are also exceptionally high (almost 30 percent).[21] An important aspect of nutrition management in gestational diabetes is prevention of excessive weight gain. Regular physical activity may help to prevent excessive weight gain, but it

Pregnant women can enjoy the benefits of physical activity.

gestational diabetes: the detection of abnormal glucose tolerance during pregnancy.

cesarean section: surgical childbirth, in which the infant is delivered through an incision in the woman's abdomen.

Nutrition Through the Life Span: Pregnancy and Infancy

Table 12–6   Guidelines for Physical Activity during Pregnancy

- Be physically active on a regular basis (at least three times a week), not intermittently.
- Warm up with 5 to 10 minues of light activity.
- Stop exercising if you feel overheated.
- Drink plenty of fluids before, during, and after physical activity.
- Avoid exerting yourself in hot, humid weather; avoid overheating.
- Avoid jarring or jerky motions.
- Avoid any activity that has the potential to cause even mild abdominal trauma.
- Discontinue any activity that causes discomfort.
- Do not exercise while lying on your back after the fourth month.
- Do not allow your heart rate to exceed 150 beats per minute.
- Cool down with 5 to 10 minutes of slow activity and gentle stretching.
- Eat enough to support the energy needs of pregnancy and physical activity.

must be individualized and medically monitored to ensure safety and effectiveness.[22] Insulin therapy may be required if blood glucose fails to normalize. Chapter 24 provides information about medical nutrition therapy for gestational diabetes.

Hypertension Hypertension complicates pregnancy and affects its outcome in different ways, depending on when the hypertension first develops and on how severe it becomes.[23] Hypertension can be a preexisting chronic condition that develops before a woman becomes pregnant or a transient condition that develops during the pregnancy and subsides after childbirth. In some cases, hypertension that develops during pregnancy warns of the ominous disorder preeclampsia.

Preexisting Chronic Hypertension In addition to the health risks normally imposed by hypertension (heart attack and stroke), high blood pressure increases the risks of a low-birthweight infant or the separation of the placenta from the wall of the uterus before the birth, resulting in stillbirth. Ideally, before a woman with hypertension becomes pregnant, her blood pressure will be under control.

The normal edema of pregnancy responds to gravity: blood pools in the ankles. The edema of preeclampsia is a generalized edema. The distinction helps with diagnosis.

Warning signs of preeclampsia:
- *Hypertension.*
- *Protein in the urine.*
- *Upper abdominal pain.*
- *Severe and constant headaches.*
- *Swelling, especially of the face.*
- *Dizziness.*
- *Blurred vision.*
- *Sudden weight gain (1 lb/day).*

Transient Hypertension of Pregnancy Some women first develop hypertension during the second half of pregnancy. Most often, the rise in blood pressure is mild and does not affect the pregnancy adversely. Blood pressure usually returns to normal during the first few weeks after childbirth. This transient hypertension of pregnancy differs from the life-threatening hypertension diseases of pregnancy—preeclampsia and eclampsia.

Preeclampsia Hypertension may signal the onset of **preeclampsia,** a condition characterized not only by high blood pressure but by protein in the urine and fluid retention (edema). Preeclampsia, which affects less than 10 percent of pregnant women, usually occurs with first pregnancies and almost always after 20 weeks' gestation. Symptoms typically regress within 48 hours of delivery. The edema of preeclampsia is a whole-body edema, distinct from the localized fluid retention women normally experience late in pregnancy.

Preeclampsia affects almost all of the woman's organs—the circulatory system, liver, kidneys, and brain. If it progresses, she may experience convulsions; when this occurs, the condition is called **eclampsia.** Maternal mortality during pregnancy is rare in developed countries, but eclampsia is the most common cause. Preeclampsia demands prompt medical attention. Treatment focuses on regulating blood pressure and preventing convulsions.

preeclampsia: a condition characterized by hypertension, fluid retention, and protein in the urine.

eclampsia: a severe stage of preeclampsia in which convulsions occur.

Morning Sickness Unlike the conditions just discussed, the nausea of "morning" (actually, anytime) sickness is usually benign, although it is distressing to some women. It arises from the hormonal changes taking place early in pregnancy, ranges from mild queasiness to debilitating nausea, and afflicts more than half of all pregnant women. Many women complain that smells, especially cooking smells, make them sick. Thus minimizing odors is a key to alleviating morning sickness.

Traditional strategies for alleviating nausea are listed in the margin, but many women benefit most from simply eating the foods they want when they feel like eating. A "How to" box in Chapter 28 provides additional suggestions that may help to combat nausea.

Heartburn Heartburn, a burning sensation in the lower esophagus near the heart, is common during pregnancy and is also benign. As the growing fetus puts increasing pressure on the woman's stomach, acid may back up and create a burning sensation in her throat. Tips to relieve heartburn are listed in the margin.

Constipation As the hormones of pregnancy alter muscle tone and the thriving infant crowds intestinal organs, an expectant mother may complain of constipation, another harmless but annoying condition. A high-fiber diet, physical activity, and a plentiful fluid intake will help relieve this condition. Also, responding promptly to the urge to defecate can help. Laxatives should be used only as prescribed by the physician. Mineral oil should not be used because it interferes with the absorption of fat-soluble vitamins.

In Summary Conditions such as gestational diabetes, hypertension, and preeclampsia can threaten the health and life of both mother and infant. Such conditions require medical and nutrition treatment. The nausea, heartburn, and constipation that sometimes accompany pregnancy can usually be alleviated with a few simple strategies.

Practices to Avoid

A general guideline for the pregnant woman is to eat a normal, healthy diet and practice moderation. A woman's daily choices during pregnancy take on enormous importance. Forewarned, pregnant women can choose to abstain from or avoid potentially harmful practices.

Cigarette Smoking Cigarette (and cigar) smoking is clearly a harmful practice. Smoking adversely affects a pregnant woman's nutrition status and thereby impairs fetal nutrition. Smokers tend to have lower intakes of dietary fiber, vitamin A, beta-carotene, folate, and vitamin C. Oxidants in cigarette smoke accelerate vitamin C metabolism and deplete smokers' body stores of this antioxidant, further compromising smokers' vitamin C status. Smoking also restricts the blood supply to the growing fetus and so limits the delivery of oxygen and nutrients and the removal of wastes. It slows growth, thus retarding physical development of the fetus, and it may cause behavioral or intellectual problems later on.[24] In addition, sudden infant death syndrome (SIDS), the unexplained deaths that sometimes occur in otherwise healthy infants, has been positively linked to the mother's cigarette smoking during pregnancy.[25]

A mother who smokes is more likely to have a complicated birth, and her infant is more likely to be of low birthweight.[26] The more a mother smokes, the smaller her baby will be. Of all preventable causes of low birthweight in the United States, smoking has the greatest impact.

Research suggests that even in women who do not smoke, exposure to **environmental tobacco smoke (ETS,** or secondhand smoke) during pregnancy

To alleviate the nausea of pregnancy:
- *On waking, arise slowly.*
- *Eat dry toast or crackers before getting out of bed.*
- *Chew gum or suck hard candies.*
- *Eat frequent small meals.*
- *Avoid foods with offensive odors.*
- *When nauseated, do not drink citrus juice, water, milk, coffee, or tea.*
- *Take prenatal vitamin and iron supplements on a full stomach or at a time of day when you feel well.*

To prevent or relieve heartburn:
- *Eat frequent small meals.*
- *Drink liquids between meals.*
- *Avoid spicy or greasy foods.*
- *Sit up while eating.*
- *Wait an hour after eating before lying down.*
- *Wait 2 hours after eating before exercising.*

environmental tobacco smoke (ETS): the combination of exhaled smoke (mainstream smoke) and smoke from lighted cigarettes, pipes, or cigars (sidestream smoke) that enters the air and may be inhaled by other people.

increases the risk of low birthweight and the likelihood of sudden infant death syndrome.[27] Constituents of cigarette smoke such as nicotine, cyanide, and others are directly toxic to a fetus and to the infant later on. With great urgency, the surgeon general has warned that parental smoking can kill an otherwise healthy fetus or newborn.

Caffeine Caffeine crosses the placenta, and the fetus has only a limited ability to metabolize it. No firm limit for caffeine intake is yet available. So far, research studies have not proved that caffeine causes birth defects in human beings (as it does in animals). Some evidence suggests that moderate-to-heavy use of caffeine (more than 300 milligrams per day—the equivalent of about 2 to 3 cups of coffee a day) may lower infant birthweight.[28] Another study of caffeine's effects on pregnancy, however, found that daily caffeine intake is not a risk factor for growth retardation.[29] Studies do show another possible danger of caffeine intake during pregnancy, however. Pregnant women who drink more than 3 cups of coffee a day may increase their risk of spontaneous abortion.[30]

In light of the evidence thus far, it seems most sensible to limit caffeine consumption to the equivalent of one cup of coffee or two 12-ounce cola beverages a day. Caffeine amounts in food and beverages are listed in Appendix A.

Medications Medications taken during pregnancy can cause complications and serious birth defects. The use of medications not prescribed by a physician, even over-the-counter drugs, herbal preparations, or high-dose vitamin supplements, is inadvisable.[31] Women are advised to take medications only if their physicians deem it necessary to protect their life and health. Drug labels warn: "As with any drug, if you are pregnant or nursing a baby, seek the advice of a health professional before using this product." For aspirin and ibuprofen, an additional warning immediately follows: "It is especially important not to use aspirin (or ibuprofen) during the last three months of pregnancy unless specifically directed to do so by a doctor because it may cause problems in the unborn child or excessive bleeding during delivery." Such warnings should be taken seriously.

Illicit Drugs The recommendation to avoid drugs during pregnancy includes illicit drugs, of course. Unfortunately, some pregnant women do use illicit drugs such as marijuana and cocaine. Marijuana or cocaine use during pregnancy adversely affects fetal growth and development.[32] Such drugs of abuse pass easily through the placenta and impair fetal development. Moreover, infants born to drug users face low birthweight, cardiovascular problems, and increased risk of death. If they survive, their cries, sleep, and behavior at birth are abnormal.[33] They may be hypersensitive or underaroused; those who test positive for drugs suffer the greatest effects of toxicity and withdrawal.

Environmental Contaminants Infants and young children of pregnant women exposed to environmental contaminants such as lead and mercury show signs of impaired cognitive development. During pregnancy, lead and mercury readily move across the placenta, inflicting severe damage on the developing fetal nervous system.

Unacceptably high concentrations of mercury in fish have prompted the Food and Drug Administration (FDA) to issue an advisory to all pregnant women, women who may become pregnant, lactating mothers, and young children against eating large ocean fish such as king mackerel, swordfish, shark, tuna, and tilefish.[34] Pregnant and lactating women are also advised to limit their consumption of canned tuna to one can per week, and young children are advised to eat less than a can per *month*. Furthermore, the Environmental Protection Agency (EPA) has warned the same groups of people to limit intakes of freshwater fish to one fish meal per week. Chapter 11 offers more details on contaminants in foods.

Vitamin-Mineral Megadoses   Many vitamins are toxic when taken in excess, and the minerals are even more so. Among vitamins, a single massive dose of preformed vitamin A (100 times the recommended intake) has caused birth defects. Chronic use of lower doses of vitamin A supplements (three to four times the recommended intake) may also cause birth defects. Intakes before the seventh week of pregnancy appear to be the most damaging. For this reason, vitamin A is not given as a supplement in the first trimester of pregnancy unless there is evidence of deficiency, which is rare. Women taking supplements should take heed—experts urge pregnant women not to exceed three times the recommended daily intake of vitamin A.

Dieting   Weight-loss dieting, even for short periods, is hazardous during pregnancy. Low-carbohydrate diets or fasts that cause ketosis deprive the growing brain of needed glucose and may impair its development. Energy restriction during pregnancy is dangerous for all women, regardless of their prepregnancy weights.

Alcohol   Drinking alcohol during pregnancy threatens the fetus with irreversible brain damage, growth retardation, mental retardation, facial abnormalities, vision abnormalities, low **Apgar scores,** and more than 40 identifiable health problems—a cluster of symptoms known as **fetal alcohol syndrome** or **FAS.**[35] The fetal brain is extremely vulnerable to a glucose or oxygen deficit, and alcohol causes both by disrupting placental functioning. In addition, alcohol itself crosses the placenta freely and is directly toxic to the defenseless fetal brain and nervous system.[36] The result is permanent brain damage and lifelong mental retardation. FAS is preventable by abstaining from drinking alcohol during pregnancy, but once present, it is incurable. FAS is known to occur with as few as two drinks a day.

About a fifth of women continue drinking alcohol after they learn that they are pregnant. In fact, one out of every 29 pregnant women reports "frequent" drinking (seven or more drinks per week or five or more drinks on one occasion).[37]

The pattern of a woman's drinking may be as important as her average alcohol intake. For example, a woman with an average intake of only 1 ounce of alcohol a day might not drink at all during the week, but then have 14 drinks each weekend. Thus the fetus might be intermittently exposed to high alcohol doses. No matter what the intake or pattern, the most severe impact is likely to occur in the first two months, before the woman may be aware that she is pregnant.

Research using animals shows that one-fifth of the amount of alcohol needed to produce major, outwardly visible defects will surely produce learning impairment or other defects in the offspring. The term *fetal alcohol effects,* originally used to describe this damage, has been replaced with two more descriptive terms, **alcohol-related neurodevelopmental disorder (ARND)** and **alcohol-related birth defects (ARBD).**[38] The terms describe conditions where there is a history of maternal alcohol intake and evidence of abnormalities related to alcohol.

Some children show no outward sign of impairment, but the damage is there on the inside.[39] Others may be short in stature or display subtle facial abnormalities. Most perform poorly in school and in social interactions and suffer a subtle form of brain damage. Anyone exposed to alcohol before birth may always respond differently to it, and also to certain drugs, than if no exposure had occurred. Even before fertilization, alcohol may damage the ovum or sperm in the mother- or father-to-be, and so lead to abnormalities in the child.

The children born with alcohol damage remain damaged. They may live, but they never fully recover. The facial abnormalities of FAS are apparent. A visual picture of the internal harm is impossible, however, but it is that damage that virtually seals the fate of the child for life. An estimated 5 to 30 of every 10,000 children are victims of this preventable damage, making FAS the leading known cause of mental retardation in the world. Moreover, for every baby diagnosed

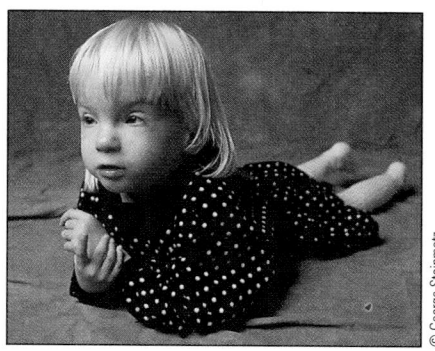

These facial traits are typical of fetal alcohol syndrome, caused by drinking during pregnancy—low nasal bridge, short eyelid opening, underdeveloped groove in center of the upper lip, small midface, short nose, and small head circumference.

Apgar scores: a system of scoring an infant's physical condition right after birth. Heart rate, respiration, muscle tone, response to stimuli, and color are ranked 0, 1, or 2. A low total score indicates that medical attention is required to facilitate survival.

fetal alcohol syndrome (FAS): the cluster of symptoms seen in an infant or child whose mother consumed excessive alcohol during her pregnancy. FAS includes, but is not limited to, brain damage, growth retardation, mental retardation, and facial abnormalities.

alcohol-related neurodevelopmental disorder (ARND): a condition caused by prenatal alcohol exposure. ARND is diagnosed when there is a confirmed history of substantial, regular maternal alcohol intake or heavy episodic drinking and behavioral, cognitive, or central nervous system abnormalities known to be associated with alcohol exposure.

alcohol-related birth defects (ARBD): a condition caused by prenatal alcohol exposure. ARBD is diagnosed when there is a history of substantial, regular maternal alcohol intake or heavy episodic drinking and birth defects known to be associated with alcohol exposure.

Nutrition Through the Life Span: Pregnancy and Infancy

with FAS, three or four with ARND may go undiagnosed until problems develop later in the preschool years. Upon reaching adulthood, such children are ill equipped for employment, relationships, and the other facets of life most adults take for granted.

The American Academy of Pediatrics takes the position that women should stop drinking as soon as they *plan* to become pregnant.[40] As mentioned, this step is important for fathers-to-be as well. It is important to know, though, that a woman who has drunk heavily during the first two-thirds of her pregnancy can still prevent some organ damage by stopping heavy drinking during the third trimester.

Experts have not always agreed that women need to abstain totally from using alcohol during pregnancy. Nevertheless, researchers looking for a "safe" intake limit have come full circle to concede that abstinence from alcohol is the best policy for pregnant women.

In Summary Abstinence from smoking, drugs, and alcohol greatly improves the outcome of pregnancy.

Adolescent Pregnancy

Each year in the United States, about one million adolescent girls between the ages of 12 and 19 become pregnant.[41] Of these, about half choose to continue their pregnancies. Many teenage women, especially the youngest ones, have not had time to store the nutrients needed to support their own rapid growth and development, much less nutrients needed to support pregnancy and the developing fetus. Nutrient shortages place both mother and infant at risk. Adolescents have more miscarriages, premature births, stillbirths, and low-birthweight infants than do adult women.[42] Their greatest risk, though, is death of the infant: mothers under 16 bear more babies who die within the first year than do women in any other age group. Clearly, teenage pregnancy is a major public health problem.

Pregnant teenagers suffer many illnesses. Rates of preeclampsia are 50 percent higher in teens than in older women. Other common problems of teen pregnancies are iron-deficiency anemia and prolonged labor.

To support the needs of both mother and fetus, a pregnant teenager with a BMI in the normal range is encouraged to gain about 30 pounds or so. Teenagers who gain less have smaller infants with associated risks. Adequate nutrition can substantially improve the health of the mother and infant; it is an indispensable component of prenatal care. Pregnant and lactating teenagers can use the Daily Food Guide presented in Table 12–4 (on p. 305), making sure to select at least 4 servings of milk or milk products daily.

In Summary Proper nutrition and adequate weight gain are especially important in reducing the risk of poor pregnancy outcome in pregnant adolescents.

Breastfeeding

The American Academy of Pediatrics (AAP) recommends that infants receive breast milk for at least the first 12 months of life.[43] The American Dietetic Association (ADA) advocates breastfeeding for the nutritional health it confers on the infant as well as for the physiological, social, economic, and other benefits it offers the mother.[44] Breast milk's unique nutrient composition and protective factors promote optimal infant health and development. The only acceptable alternative to breast milk is iron-fortified formula. Adequate nutrition of the mother supports successful **lactation,** and without it, lactation is likely to falter or fail.

lactation: production and secretion of breast milk for the purpose of nourishing an infant.

Nutrition during Lactation

By continuing to eat nutrient-dense foods, not restricting weight gain unduly, and enjoying ample food and fluid at frequent intervals throughout lactation, the mother who chooses to breastfeed her infant will be nutritionally prepared to do so. An inadequate diet does not support the stamina, patience, and self-confidence that nursing an infant demands. Figure 12–3 (on p. 303) shows how a lactating woman's nutrient needs differ from those of a nonpregnant woman, and Table 12–4 (on p. 305) presents a food pattern that meets those needs.

Energy A nursing mother produces about 25 ounces of milk a day, more or less, depending primarily on the infant's demand for milk. To produce milk, a woman needs extra energy—almost 650 kcalories per day above her regular need during the first six months of lactation. To meet this energy need, the woman is advised to eat an extra 500 kcalories of food each day. The other 150 kcalories may be drawn from the fat stores she accumulated during pregnancy. Energy needs of women who are breastfeeding exclusively range from 2500 to 3300 kcalories a day, depending on physical activity.[45] Severe energy restriction hinders milk production and can compromise the mother's health.

Weight Loss A question often raised is whether breastfeeding promotes a more rapid loss of the extra body fat accumulated during pregnancy. Results of studies about the relationship between feeding method and loss of body fat and body weight are inconsistent. In most studies where breastfeeding duration was three months or longer, researchers found that lactation accelerated a woman's weight loss.[46] This does not mean that a breastfeeding woman can eat unlimited food and still effortlessly return to prepregnancy weight. Breastfeeding costs energy, true, but carefully chosen programs of diet and physical activity are still the cornerstones of weight control. Physical activity in particular helps to reduce body fatness and improve fitness while having little effect on a woman's milk production or her infant's weight gain. A gradual weight loss (1 pound per week) is safe and does not reduce milk output.[47] Too large an energy deficit, however, especially soon after birth, will inhibit lactation.

Vitamins and Minerals Another question often raised is whether a mother's milk may lack a nutrient if she fails to get enough in her diet. The answer differs from one nutrient to the next, but in general, nutritional deprivation of the mother reduces the *quantity,* not the *quality,* of her milk. For protein, carbohydrate, fat, folate, and most minerals, the milk of a healthy mother has a fairly constant composition. Any excess water-soluble vitamins the mother takes in are excreted in the urine; the body does not release them into the milk. The amounts of fat-soluble vitamins in human milk, however, are affected by the mother's excessive or deficient intakes. For example, large doses of vitamin A correspondingly raise the concentration of this vitamin in breast milk. Vitamin supplementation of undernourished women appears to help normalize the vitamin concentrations in their milk and may be beneficial.

Water The volume of breast milk produced depends on how much milk the baby demands, not on how much fluid the mother drinks. The nursing mother is nevertheless advised to drink at least 2 quarts of liquids each day to protect herself from dehydration. To help themselves remember to drink enough liquid, many women make a habit of drinking a glass of milk, juice, or water each time the baby nurses as well as at mealtimes.

Particular Foods Some infants may be sensitive to foods such as cow's milk, onions, or garlic in the mother's diet and become uncomfortable when she eats them. Nursing mothers should not automatically avoid such foods, however. A

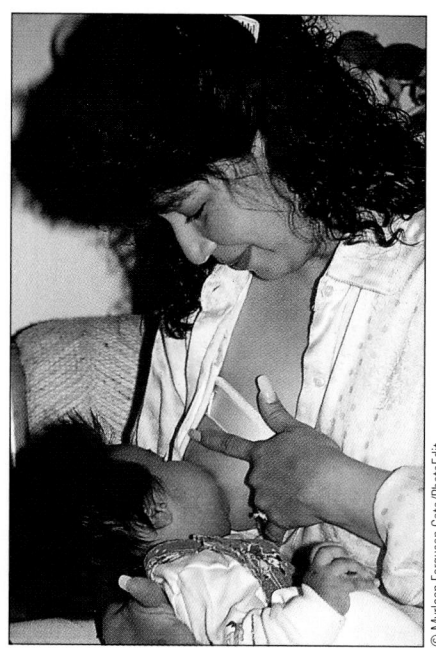

Breastfeeding is a natural extension of pregnancy—the mother's body continues to nourish the infant.

*Nutrition Through the
Life Span: Pregnancy and Infancy*

mother who is nursing her baby is advised to eat whatever nutritious foods she chooses. Then, if a particular food seems to cause the infant discomfort, she can try eliminating that food from her diet for a few days and see if the problem goes away.

Contraindications to Breastfeeding

Some substances impair maternal milk production or enter the breast milk and interfere with infant development. Some medical conditions prohibit breastfeeding.

Alcohol   Alcohol easily enters breast milk and can adversely affect the production, volume, composition, and ejection of breast milk as well as overwhelm an infant's immature alcohol-degrading system.[48] Alcohol concentration peaks within one hour after ingestion of even moderate amounts (equivalent to a can of beer). This amount may alter the taste of the milk to the disapproval of the nursing infant, who may, in protest, drink less milk than normal.

Caffeine   Caffeine can make a baby jittery and wakeful. As during pregnancy, caffeine consumption should be moderate.

Cigarette Smoke   Health care professionals should actively discourage smoking by lactating women. Research shows that lactating women who smoke produce less milk, and milk with a lower fat content, than mothers who do not smoke. Consequently, their infants gain less weight than infants of nonsmokers.

A lactating woman who smokes not only exposes her infant to nicotine and other chemicals via her breast milk, but may also expose the infant to sidestream smoke. Babies who are "smoked over" experience a wide array of health problems—poor growth, hearing impairment, vomiting, breathing difficulties, and even unexplained death.[49]

Medications and Illicit Drugs   If a nursing mother must take medication that is secreted in breast milk and is known to affect the infant, then breastfeeding must be put off for the duration of treatment. Meanwhile, the flow of milk can be sustained by pumping the breasts and discarding the milk. Many prescription medications do not reach nursing babies in sufficient quantities to affect them adversely and so have no impact on breastfeeding. Other drugs are not at all compatible with breastfeeding either because they are secreted into the milk and can harm the infant or because they suppress lactation. A nursing mother should consult with the prescribing physician before taking medicines. Breastfeeding is also contraindicated if the mother uses illicit drugs. Drug addicts, including alcohol abusers, are capable of taking such high doses that their infants can become addicts by way of breast milk.

Many women wonder about using oral contraceptives during lactation. One type that combines the hormones estrogen and progestin seems to suppress milk output, lower the nitrogen content of the milk, and shorten the duration of breastfeeding. In contrast, progestin-only pills have no effect on breast milk or breastfeeding and are considered appropriate for lactating women.

Maternal Illness   If a woman has an ordinary cold, she can go on nursing without worry. If susceptible, the infant will catch it from her anyway, and thanks to immunological protection, a breastfed baby may be less susceptible than a formula-fed baby would be. If a woman has a serious communicable disease such as tuberculosis or hepatitis, then mother and baby have to be separated. Breastfeeding may be continued by pumping the mother's breasts several times a day and letting the baby drink the milk from a bottle (see the margin for tips for safe handling).

The human immunodeficiency virus (HIV), responsible for causing AIDS, can be passed from an infected mother to her infant during pregnancy, at birth, or through breast milk, especially during the early months of breastfeeding.[50]

For safe breast milk storage:
- *Wash hands thoroughly before pumping.*
- *Clean pumping equipment according to the manufacturer's directions.*
- *Sterilize bottles, nipples, and rings before using.*
- *Refrigerate milk to be fed within 48 hours.*
- *Freeze milk to be stored longer than 48 hours.*
- *Thaw milk gently on defrost cycle of microwave or in refrigerator.*
- *Do not refreeze thawed milk.*

Thus, women in developed countries who have tested positive for HIV should not breastfeed if the infant is not infected. They should choose a safe alternative feeding method, such as breast milk from a milk bank.[51] Milk banks in the United States pasteurize donated human milk and make it available to infants who lack access to milk from their own mothers. Pasteurization destroys harmful organisms, such as HIV, but leaves intact most of the beneficial constituents of the milk.[52]

Throughout the world, breastfeeding prevents millions of infant deaths each year. In developing countries, where the feeding of inappropriate or contaminated formulas causes 1.5 million infant deaths each year, breastfeeding can be critical to infant survival. This advantage, however, must be weighed against the following: In 1999, 200,000 to 300,000 infants became infected with HIV by way of breastfeeding.[53] Thus the question whether HIV-infected women in developing countries should breastfeed comes down to a delicate balance between risks and benefits. For HIV-positive women in developing countries who are literate, have access to safe water, and have an uninterrupted supply of infant formula, replacement feeding may reduce the risk of infant illness and death by AIDS. For those mothers without safe water and with minimal education, the risk of replacement feeding may be substantial in terms of infant mortality. WHO and UNICEF, in acknowledging the transmission of HIV by way of breast milk, recommend formula feeding for babies of HIV-positive mothers in developing countries if they can be ensured uninterrupted access to safely prepared, nutritionally adequate breast milk substitutes.

In Summary The lactating woman needs enough energy and nutrients to produce about 25 ounces of milk a day. She also needs extra fluid. Alcohol, caffeine, smoking, and drugs may reduce milk production or enter breast milk and impair infant development. Some maternal illnesses are incompatible with breastfeeding.

Nutrition of the Infant

Early nutrition affects later development, and early feeding sets the stage for eating habits that will influence nutrition status for a lifetime. Trends change, and experts argue about the fine points, but properly nourishing an infant is relatively simple, overall. Common sense in the selection of infant foods and a nurturing, relaxed environment go far to promote an infant's health and well-being.

Nutrient Needs during Infancy

An infant grows faster during the first year than ever again, as Figure 12–5 shows. The growth of infants and children directly reflects their nutritional well-being and is an important parameter in assessing their nutrition status. Health care professionals use growth charts to evaluate the growth and development of children from birth to 20 years of age (see Appendix D).

Nutrients to Support Growth An infant's birthweight doubles by about four to five months of age, and it triples by the age of one year. (Consider that if an adult, starting at 120 pounds, were to do this, the person's weight would increase to 360 pounds in a single year.) By the end of the first year, the growth rate slows considerably. Between the first and second birthdays, the weight gained amounts to less than 10 pounds.

A newborn infant requires only about 650 kcalories per day, whereas most adults require about 2000 kcalories per day. In comparison to body weight, however, the difference is remarkable. Infants require about 100 kcalories per

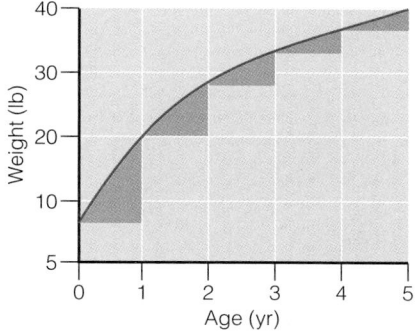

Figure 12–5

Weight Gain of Human Infants in Their First Five Years of Life
In the first year, an infant's birthweight may triple, but over the following several years, the rate of weight gain gradually diminishes.

Nutrition Through the Life Span: Pregnancy and Infancy

Figure 12–6

Recommended Intakes of a Five-Month-Old Infant and an Adult Male Compared on the Basis of Body Weight

Because infants are small, they need smaller total amounts of the nutrients than adults do, but when comparisons are based on body weight, infants need over twice as much of many nutrients. Infants use large amounts of energy and nutrients, in proportion to their body size, to keep all their metabolic processes going.

	Infants	Adults
Heart rate (beats/minute)	120 to 140	70 to 80
Respiration rate (breaths/minute)	20 to 40	15 to 20
Energy needs (kcal/body weight)	45/lb (100/kg)	<18/lb (<40/kg)

Energy
Protein
Vitamin A
Vitamin D
Vitamin E
Vitamin C
Folate
Niacin
Riboflavin
Thiamin
Vitamin B₆
Vitamin B₁₂
Calcium
Phosphorus
Magnesium
Iodine
Iron
Zinc

Vitamin D recommendations for an infant are 10 times greater *per pound of body weight* than those for an adult male.

Pound for pound, niacin recommendations for an infant and an adult male are similar.

Key:
- 20-year-old male (160 lb)
- 5-month-old infant (16 lb)

Recommendations for a male 20 years old

5 times as much per pound as an adult male

10 times as much per pound

After six months, the energy saved by slower growth is spent in increased activity.

kilogram of body weight per day; most adults require fewer than 40. A 170-pound adult who tried to eat like an infant would have to ingest over 7000 kcalories a day! Figure 12–6 compares a five-month-old infant's needs per kilogram of body weight with those of an adult male; as you can see, some of the differences are extraordinary. After six months, energy needs increase less rapidly as the growth rate begins to slow, but some of the energy saved by slower growth is spent in increased activity.

Water The most important nutrient of all, for infants as for everyone, is the one easiest to forget: water. Conditions that cause fluid loss, such as vomiting, diarrhea, sweating, or obligatory urinary loss without replacement, can rapidly propel an infant into life-threatening dehydration. In early infancy, breast milk or formula normally provides enough water for a healthy infant to replace water losses from the skin, lungs, feces, and urine.[54] An infant who is exposed to hot weather, has diarrhea, or vomits repeatedly, however, needs supplemental water to prevent dehydration.

In developed countries with well-nourished populations, such as the United States and Canada, the dietary practices that have the most influence on an infant's nutrition status are the type of milk the infant receives and the age at which solid foods are introduced. The remainder of this discussion is devoted to feeding the infant and identifying the nutrients most often deficient in infant diets.

Breast Milk

Breast milk excels as a source of nutrients for the young infant. With the possible exception of vitamin D, breast milk provides all the nutrients a healthy infant needs for the first four to six months of life. It provides many other health benefits as well.

Energy Nutrients   The energy nutrient balance of breast milk differs dramatically from the balance recommended for adults (see Figure 12–7). Yet, for infants, breast milk is the most nearly perfect food, proving that people at different stages of life really do have different nutrient needs.

The carbohydrate in breast milk (and infant formula) is lactose. In addition to being easily digested, lactose enhances calcium absorption.

The lipids in breast milk—and infant formula—provide the main source of energy in the infant's diet. Breast milk contains a generous proportion of the essential fatty acids linoleic acid and linolenic acid, as well as their longer-chain derivatives arachidonic acid and docosahexaenoic acid (DHA), which are found abundantly in both the retina of the eye and the brain; infant formula contains only linoleic acid and linolenic acid. Because breastfed infants receive more DHA than formula-fed infants, research is under way to determine the physiological significance of this difference.[55] One apparent benefit is that young children who were breastfed as infants have sharper vision than those who were fed formulas; this enhanced visual development is attributed to the DHA in breast milk.[56]

The protein in breast milk is largely **alpha-lactalbumin,** a protein the human infant can easily digest. Another breast milk protein, **lactoferrin,** indirectly benefits the baby's iron nutrition and also acts as an antibacterial agent. Lactoferrin is an iron-gathering compound that helps absorb iron into the infant's bloodstream, keeps intestinal bacteria from getting enough iron to grow out of control, and also works directly to kill some bacteria.

Vitamins and Minerals   The vitamin content of the breast milk of a well-nourished mother is ample. Even vitamin C, for which cow's milk is a poor source, is supplied generously by the breast milk of such a mother. The concentration of vitamin D in breast milk is low, but this is not a threat to light-skinned infants who are taken out into the sunshine regularly. A dark-skinned infant, or one who has little exposure to sunlight, however, may not make enough vitamin D to prevent rickets. Because so many variables exist regarding vitamin D and sunlight exposure, the AAP recommends vitamin D supplementation beginning at birth for breastfed babies who do not receive sufficient exposure to sunlight.

As for minerals, the 2-to-1 calcium-to-phosphorus ratio of breast milk is ideal for calcium absorption, and both of these minerals, along with magnesium, support the rate of growth expected in a human infant. Breast milk is also low in sodium. The limited amount of iron in breast milk is highly absorbable, and its zinc, too, is absorbed better than from cow's milk, thanks to the presence of a zinc-binding protein.

Supplements for Infants   Pediatricians may prescribe supplements containing vitamin D, iron, and fluoride (after six months of age). Table 12–7 (p. 318) offers a schedule of supplements during infancy.

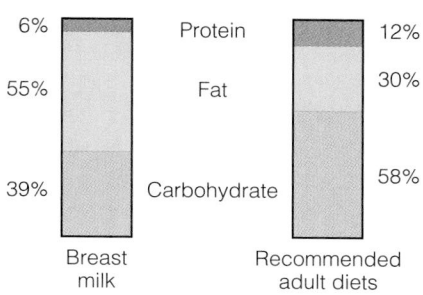

Figure 12–7

Percentages of Energy-Yielding Nutrients in Breast Milk and in Recommended Adult Diets

The balance of energy-yielding nutrients in human breast milk is ideal for infants and does not resemble the balance recommended for adults.

alpha-lactalbumin (lackt-AL-byoo-min): the chief protein in human breast milk, as **casein** (CAY-seen) is the chief protein in cow's milk.

lactoferrin (lack-toe-FERR-in): a factor in breast milk that binds iron and keeps it from supporting the growth of the infant's intestinal bacteria.

Table 12–7   Supplements for Full-Term Infants

	Vitamin D[a]	Iron[b]	Fluoride[c]
Breastfed infants:			
Birth to six months of age	√		
Six months to one year	√	√	√
Formula-fed infants:			
Birth to six months of age			
Six months to one year		√	√

[a]Vitamin D supplements are recommended for infants whose mothers are vitamin D deficient and for those who do not receive adequate exposure to sunlight.
[b]Infants four to six months of age need additional iron, preferably in the form of iron-fortified cereal for both breastfed and formula-fed infants and iron-fortified infant formula for formula-fed infants.
[c]The Committee on Nutrition of the American Academy of Pediatrics recommends initiating fluoride supplements at six months of age for breastfed infants, formula-fed infants who receive ready-to-use formulas (these are prepared with water low in fluoride), and those who receive formula mixed with water that contains little or no fluoride (less than 0.3 ppm).

SOURCES: Adapted from American Academy of Pediatrics, Committee on Nutrition, Nutrition and oral health, in *Pediatric Nutrition Handbook,* 4th ed., ed. R. E. Kleinman (Elk Grove Village, Ill.: American Academy of Pediatrics, 1998), pp. 523–529; American Academy of Pediatrics, Committee on Nutrition, Fluoride supplementation for children: Interim policy recommendations, Pediatrics 95 (1995): 777.

In addition, as discussed in Chapter 8, the AAP recommends a single dose of vitamin K at birth.[57] In many states, this preventive dose of vitamin K is required by law.

Immunological Protection   Breast milk offers the infant unsurpassed protection against infection.[58] Protective factors include antiviral agents, antibacterial agents, and other infection inhibitors.

During the first two or three days of lactation, the breasts produce **colostrum,** a premilk substance containing antibodies and white cells from the mother's blood. Colostrum is relatively sterile as it leaves the breast, and the infant cannot contract a bacterial infection from it even if the mother has one. Colostrum contains maternal immune factors that inactivate harmful bacteria within the digestive tract. Later, breast milk also delivers immune factors, although not as many as colostrum. Among them are **bifidus factors** and lactoferrin.

Breast milk also contains several enzymes, several hormones (including thyroid hormone and prostaglandins), and lipids, all of which protect the infant against infection. Research suggests that breastfeeding offers better protection against wheezing during the first few months of life than formula feeding does.[59] It seems, too, that breastfed babies are less prone to develop stomach and intestinal disorders during the first few months of life and so experience less vomiting and diarrhea than formula-fed infants do.[60] In fact, research shows that breast milk contains not only antibodies against the most common cause of diarrhea in infants and young children but also another factor that binds to, and inhibits replication of, the infective agent.[61]* Breastfeeding reduces the severity and duration of symptoms associated with this infection. Breastfeeding also protects against other common illnesses of infancy such as middle ear infection and respiratory illness.[62]

Some research shows that children with diabetes (type 1) almost always have antibodies to cow's milk protein in their pancreatic tissues.[63] This finding leads some to suspect that early feeding of cow's milk formula may set the stage for abnormal immune functioning that causes diabetes later on. Other researchers disagree.[64] In any case, breastfeeding prevents early exposure to cow's milk protein. Much remains to be learned about the composition and characteristics of human milk, but clearly it is a very special substance. Nutrition in Practice 12 offers suggestions for successful breastfeeding.

colostrum (co-LAHS-trum): a milklike secretion from the breast that is rich in protective factors. Colostrum is present during the first day or so after delivery, before milk appears.

bifidus (BIFF-id-us, by-FEED-us) **factors:** factors in colostrum and breast milk that favor the growth of the "friendly" bacterium *Lactobacillus* (lack-toe-ba-SILL-us) *bifidus* in the infant's intestinal tract. These bacteria prevent other, less desirable intestinal inhabitants from flourishing.

*The most common cause of diarrhea in the United States is rotavirus. More children are hospitalized for rotavirus infection than for any other single cause.

Case Study

PREGNANT WOMAN WITH A WEIGHT PROBLEM

Ellen is a 24-year-old woman who is four months pregnant. This is her first pregnancy, and she is eager to learn how to feed herself during pregnancy as well as her infant after birth. She is 5 feet 3 inches tall and currently weighs 150 pounds. Her prepregnancy weight was 148 pounds. Ellen is very concerned about her 2-pound weight gain.

■ Consult the BMI table (inside back cover), and using the "Healthy Weight" section, find a healthy weight in the middle of the range appropriate for a woman of Ellen's height.

■ Do you think that Ellen's weight at the start of her pregnancy was appropriate for her height? Why or why not?

■ Should Ellen be concerned about her 2-pound weight gain? Why or why not?

■ What advice should you give Ellen about her weight gain during pregnancy?

■ What other dietary advice would you give her?

■ Discuss methods of infant feeding with Ellen and describe some of the advantages breastfeeding would offer her.

■ What are the advantages of formula feeding?

■ What advice will you give Ellen if she decides to breastfeed?

■ What information should Ellen have about formula feeding?

The case study above presents a woman who is four months pregnant. Answering the questions offers practice in thinking through some of the issues related to pregnancy and breastfeeding.

Infant Formula

Breastfeeding offers many benefits to both mother and infant, and it should be encouraged whenever possible. The mother who has decided to use formula, however, should be supported in her choice just as the breastfeeding mother should be. She can offer the same closeness, warmth, and stimulation during feedings as the breastfeeding mother can.

Many mothers choose to breastfeed at first but wean their children within the first 1 to 12 months. Before infants reach a year of age, mothers must wean them onto *infant formula*, not onto plain cow's milk of any kind—whole, reduced fat, low fat, or fat-free.

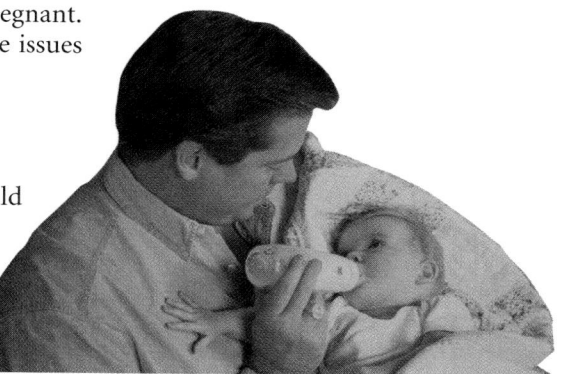

The infant thrives on infant formula offered with affection.

© Myrleen Cate/Index Stock Imagery

Infant Formula Composition Manufacturers can prepare formulas from cow's milk in such a way that they do not differ significantly from human milk in nutrient content. Figure 12–8 (on p. 320) illustrates the energy nutrient balance of both. Formulas contain no protective antibodies for human babies, but preventive medical care (vaccinations) and reliable public health measures (clean water) help minimize this disadvantage. The educated mother whose water supply is reliable can prepare safe, sanitary formulas. Lead-contaminated water, however, is a major source of lead poisoning in infants.

Infant Formula Standards National and international standards have been set for the nutrient contents of infant formulas. U.S. standards are based on AAP recommendations, and the FDA mandates quality control procedures to ensure that these standards are met. All standard formulas are therefore nutritionally similar. Small differences in nutrient content are sometimes confusing but usually are not important.

Formula preparation:

■ *Liquid concentrate (moderately expensive, relatively easy)—mix with equal part water.*

■ *Powdered formula (least expensive, lightest for travel)—read label directions.*

■ *Ready-to-feed (easiest, most expensive)—pour directly into clean bottles.*

Nutrition Through the Life Span: Pregnancy and Infancy

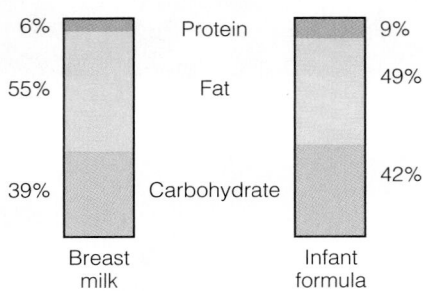

6%	Protein	9%
55%	Fat	49%
39%	Carbohydrate	42%
Breast milk		Infant formula

Figure 12–8

Percentages of Energy-Yielding Nutrients in Breast Milk and in Infant Formula

The proportions of energy-yielding nutrients in human breast milk and formula differ slightly.

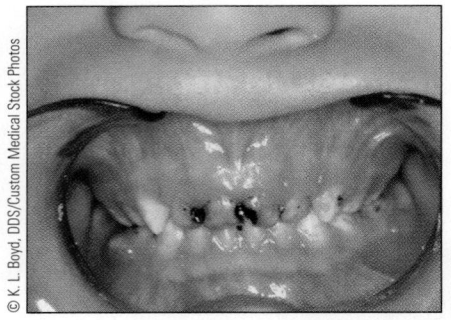

Nursing bottle tooth decay, an extreme example.

hypoallergenic formulas: clinically tested infant formulas that do not provoke reactions in 90% of infants or children with confirmed cow's milk allergy. Like all infant formulas, hypoallergenic formulas must demonstrate nutritional suitability to support infant growth and development. Extensively hydrolyzed and free amino acid–based formulas are examples.

nursing bottle tooth decay: extensive tooth decay due to prolonged tooth contact with formula, milk, fruit juice, or other carbohydrate-rich liquid offered to an infant in a bottle.

Special Formulas Standard formulas are inappropriate for some infants (see Figure 12–9). For example, premature babies require special formulas. Infants allergic to milk protein can drink special **hypoallergenic formulas** or formulas based on soy protein.[65] Soy formulas are lactose-free and so can be used for infants with lactose intolerance as well. They are also useful as an alternative to milk-based formulas for vegetarian families. For infants with other special needs, many other variations are available.

Risks of Formula Feeding In developing countries and in poor areas of the United States, formula may be unavailable, overdiluted in an attempt to save money, or prepared with contaminated water. Overdilution of formula can cause malnutrition and growth failure. Contaminated formula often causes infections leading to diarrhea, dehydration, and failure to absorb nutrients. Wherever sanitation is poor, breastfeeding should take priority over feeding formula. Breast milk is sterile, and its antibodies enhance an infant's resistance to disease.

Iron in Formula The AAP recommends iron-fortified formulas for all formula-fed infants.[66] Low-iron formulas have no role in infant feeding. Use of iron-fortified formulas has risen in recent decades and is credited with the decline of iron-deficiency anemia in U.S. infants.[67]

Nursing Bottle Tooth Decay Dentists advise against putting an infant to bed with a bottle. Salivary flow, which normally cleanses the mouth, diminishes as the baby falls asleep. Sucking for long times pushes the jawline out of shape and causes a bucktoothed profile (protruding upper and receding lower teeth). Furthermore, prolonged sucking on a bottle of formula, milk, or juice bathes the upper teeth in a carbohydrate-rich fluid that nourishes decay-producing bacteria. (The tongue covers and protects most of the lower teeth, but they, too, may be affected.) The result is extensive and rapid tooth decay. To prevent **nursing bottle tooth decay,** no child should be put to bed with a bottle as a pacifier.

The Transition to Cow's Milk

During an infant's first six months, formula must supply the nutrients of human milk in similar forms and proportions. Ordinary milk is an inappropriate replacement—primarily because cow's milk provides too little vitamin C and iron and too much sodium and protein. After the first year, the exact formulation of the milk selected is less critical, but milk or a suitable substitute still occupies a place in the diet that no other type of food can fill. Children one to two years of age should not drink reduced-fat, low-fat, or fat-free milk routinely; they need the fat of whole milk.

Introducing First Foods

Changes in the body organs during the first year affect the infant's readiness to accept solid foods. The immature stomach and intestines can digest milk sugar (lactose), but not starch until they are several months old. This is one of the many reasons why breast milk and formula are such good foods for an infant; they provide simple, easily digested carbohydrate that supplies energy for the infant's growth and activity.

The Need for Water An infant's kidneys are unable to concentrate waste efficiently, so the infant must excrete relatively more water than an adult to carry off a comparable amount of waste. Foods high in protein or electrolytes such as meat and eggs can cause dehydration if offered without water. Water should be offered to infants regularly once they are eating solid food. Water also provides fluid without additional food energy.

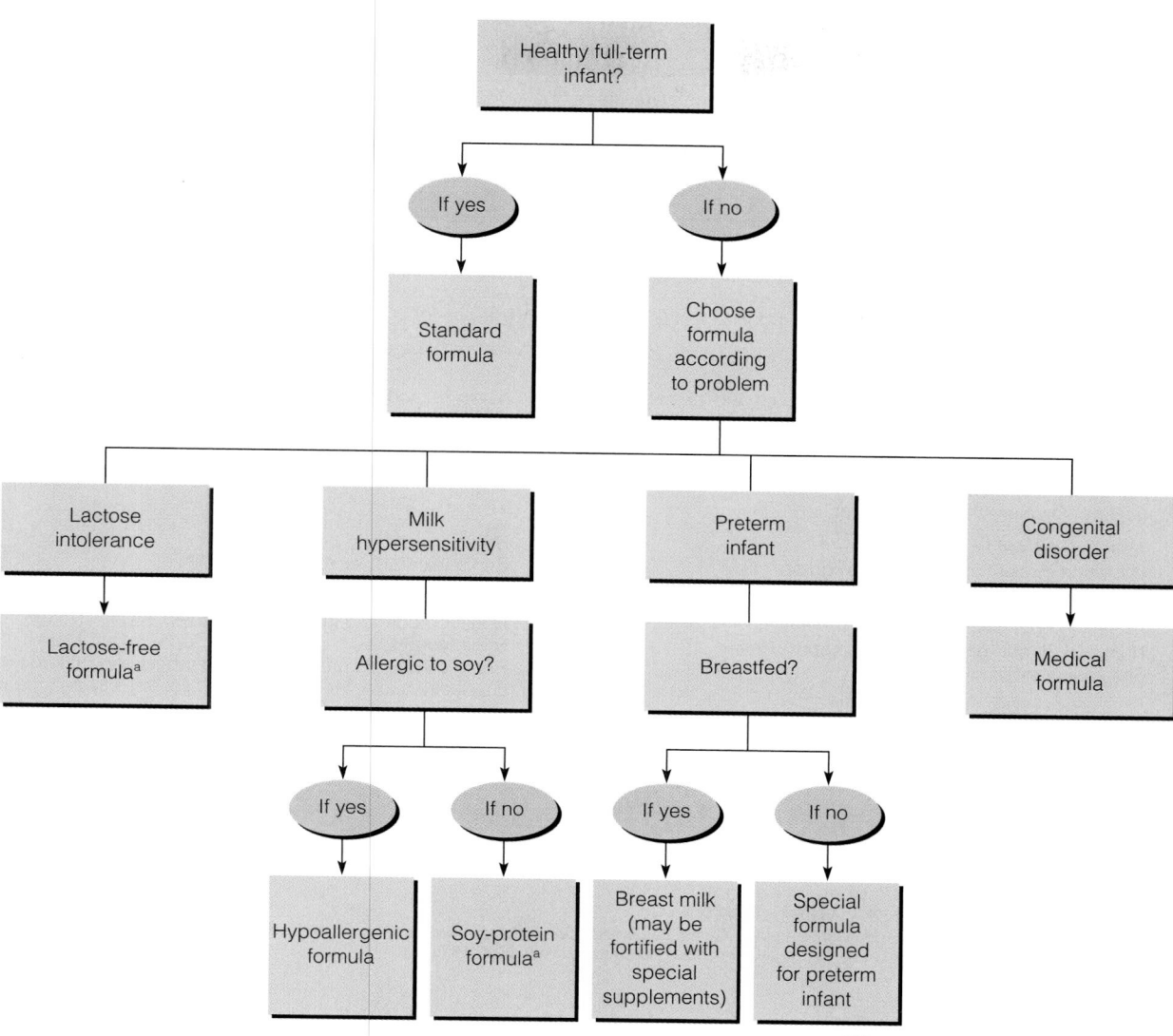

Figure 12–9
Choosing a Formula

ªManufacturers design soy-based formulas for infants with milk sensitivities—whether lactose intolerance or milk allergy. These formulas use corn syrup or sucrose in place of lactose.

When to Introduce Solid Food For an infant receiving formula or breast milk from a healthy, well-nourished mother, additions to the diet are not needed until the infant is four to six months old. Foods may be started gradually beginning sometime between four and six months, depending on the infant's readiness. Indications of readiness for solid foods include:

- The infant can sit with support and control head movements.
- The infant is six months old.

Infants vary; and the program of introducing solid foods depends on the individual infant's developmental readiness, not on any rigid schedule. Table 12–8 (on p. 322) presents a suggested sequence for introducing new foods.

The addition of foods to an infant's diet should be governed by three considerations: the infant's nutrient needs, the infant's physical readiness to handle different forms of foods, and the need to detect and control allergic reactions. With respect to nutrient needs, the nutrient needed earliest is iron, then vitamin C.

Foods to Provide Iron and Vitamin C Iron deficiency is common in young children throughout the world, especially between the ages of six months and three years when they are growing fast and milk, which is a poor source of iron,

Table 12–8 Infant Feeding Skills and Recommended Foods

Note: Because each stage of development builds on the previous stage, the foods from an earlier stage continue to be included in all later stages.

Age (mo)	Feeding Skill	Appropriate Foods Added to the Diet
0–4	Turns head toward any object that brushes cheek. Initially swallows using back of tongue; gradually begins to swallow using front of tongue as well. Strong reflex (extrusion) to push food out during first 2 to 3 months.	Feed breast milk or infant formula.
4–6	Extrusion reflex diminishes, and the ability to swallow nonliquid foods develops. Indicates desire for food by opening mouth and leaning forward. Indicates satiety or disinterest by turning away and leaning back. Sits erect with support at 6 months. Begins chewing action. Brings hand to mouth. Grasps objects with palm of hand.	Begin iron-fortified cereal mixed with breast milk, formula, or water. Begin pureed vegetables and fruits.
6–8	Able to feed self finger foods. Develops pincer (finger to thumb) grasp. Begins to drink from cup.	Begin breads and other cereals. Begin textured vegetables and fruits. Begin plain, unsweetened fruit juices from cup.
8–10	Begins to hold own bottle. Reaches for and grabs food and spoon. Sits unsupported.	Begin breads and cereals from table. Begin yogurt. Begin pieces of soft, cooked vegetables and fruit from table. Gradually begin finely cut meats, fish, casseroles, cheese, eggs, and legumes.
10–12	Begins to master spoon, but still spills some.	Include at least 4 servings of breads and cereals from table, in addition to infant cereal.[a] Include at least 2 servings of fruits and 3 servings of vegetables.[a] Include 2 servings of meat, fish, poultry, eggs, or legumes.[a]

[a]Serving sizes for infants and young children are smaller than those for an adult. For example, a serving might be ½ slice of bread instead of 1 slice, or ¼ cup rice instead of ½ cup.

SOURCE: Adapted in part from American Academy of Pediatrics, Committee on Nutrition, *Pediatric Nutrition Handbook*, 4th ed., ed. R. E. Kleinman (Elk Grove Village, Ill.: American Academy of Pediatrics, 1998), pp. 43–53.

has a large place in their diets. The iron an infant has stored from before birth typically runs out after the birthweight doubles, long before the end of the first year. This is why cow's milk should not be offered during the first year: it not only displaces iron-fortified formula but also causes GI blood loss in many infants. Infants can derive adequate iron first from breast milk or formula with iron, then from iron-fortified cereals, and later from meat or meat alternates such as legumes. Once infants are consuming iron-fortified cereals, parents or caregivers should begin selecting vitamin C–rich foods to go with meals to enhance iron absorption. The best sources of vitamin C are fruits and vegetables (see p. 192).

Fruit juice is a source of vitamin C, but some research shows that infants and young children may fail to grow and thrive when they drink so much juice that other, more nutrient- and energy-dense foods are displaced from their diets.[68] Although other research shows no relationship between juice consumption and children's growth, the AAP has nevertheless issued recommendations setting limits on juice consumption for infants and children.[69] Fruit juices should be served in a cup, not a bottle, and not before the infant is six months of age. Juices should be used moderately (4 to 6 ounces per day), so as not to displace other foods.

Physical Readiness for Solid Foods The ability to swallow solid food develops at around four to six months, and food offered by spoon helps to develop swallowing ability. At eight months to a year, an infant can sit up, can handle finger foods, and begins to teethe. At that time, hard crackers and other hard finger foods may be introduced to promote the development of manual dexterity and control of the jaw muscles. These feedings must occur under the watchful eye of an adult because the infant can also choke on such foods.

Some parents want to feed solids at an earlier age, on the theory that "stuffing the baby" at bedtime promotes sleeping through the night. There is no proof for this theory. On the average, infants start to sleep through the night at about the same age (three to four months) regardless of when solid foods are introduced.

Allergy-Causing Foods   New foods should be introduced singly and at intervals spaced to permit detection of allergies. For example, when cereals are introduced, rice cereal is offered first for several days; it causes allergy least often. Wheat cereal is offered last; it is the most common offender. If a cereal causes an allergic reaction (irritability due to skin rash, digestive upset, or respiratory discomfort), its use should be discontinued before going on to the next food.

Choice of Infant Foods   Baby foods commercially prepared in the United States and Canada are safe, and except for mixed dinners and heavily sweetened desserts, they generally have high nutrient density. An alternative for the parent who wants the infant to have family foods is to "blenderize" a small portion of the table food (cooked without salt) at each meal.

Foods such as iron-fortified cereals and formulas, mashed legumes, and strained meats provide iron.

Foods to Omit   Sweets of any kind (including baby food "desserts") have no place in an infant's diet. The added food energy conveys few if any nutrients to support growth and contributes to obesity. Canned vegetables are also inappropriate for infants; they often contain too much sodium. Honey should never be fed to infants because of the risk of botulism. Infants and even young children have difficulty chewing and swallowing foods such as popcorn, whole grapes, whole beans, hot dog slices, and nuts; consequently, they can easily choke on these foods. Also, an infant's caregiver must be on guard against food poisoning and take precautions against it as described in Chapter 11.

Foods at One Year   Whole milk is the best food to supply most of the nutrients the infant needs at one year of age; 2 to 3½ cups a day meet those needs sufficiently. More milk than this displaces iron-rich foods and can lead to the iron-deficiency anemia known as milk anemia. Other foods—meat and meat alternates, iron-fortified cereal, enriched or whole-grain bread, fruits, and vegetables—should be supplied in variety and in amounts sufficient to round out total energy needs. Ideally, a one-year-old will sit at the table, eat many of the same foods everyone else eats, and drink liquids from a cup—not a bottle. The sample menu on p. 324 shows a meal plan that meets a one-year-old's requirements.

Reminder: Milk anemia develops when an excessive milk intake displaces iron-rich foods from the diet.

Looking Ahead

Probably the most important single measure to undertake during the first year is to encourage eating habits that will support continued normal weight as the child grows. This means introducing a variety of nutritious foods in an inviting way, not forcing the infant to finish the bottle or the baby food jar, avoiding concentrated sweets and empty-kcalorie foods, and encouraging physical activity. Parents should avoid teaching infants to seek food as a reward, to expect food as comfort for unhappiness, or to associate food deprivation with punishment. If infants cry from thirst, give them water, not milk or juice. Infants seem to have no internal "kcalorie counter," and they stop eating when their stomachs feel full. Nutrient-dense, low-kcalorie foods will satisfy as long as they provide bulk.

Normal dental development is also promoted by supplying nutritious foods, avoiding sweets, and discouraging the association of food with reward or comfort. Dental health is the subject of Nutrition in Practice 2.

Ideally, a one-year-old eats many of the same healthy foods as the rest of the family.

Nutrition Through the Life Span: Pregnancy and Infancy

Menu for a One-Year-Old

Breakfast	Morning Snack	Lunch	Afternoon Snack	Dinner
½ c whole milk	½ c yogurt	1 c whole milk	½ c whole milk	1 c whole milk
½ c cereal	1 to 2 tbs fruit	2 to 3 tbs vegetables	½ slice toast	1 egg or ¼ c tofu
1 to 2 tbs fruit	Teething crackers	2 tbs chopped meat or well-cooked, mashed legumes	1 tbs peanut butter	½ c potato, rice or pasta
				2 to 3 tbs vegetables
				2 to 3 tbs fruit

NOTE: Fruit choices should include citrus fruits, melons, and berries, and vegetable choices should include dark green, leafy and deep yellow vegetables.

The AAP recommends against a fat-modified diet during infancy. The available evidence does not warrant dietary manipulation to lower serum cholesterol in infants.

Mealtimes

The wise parent of a one-year-old offers nutrition and love together. Both promote growth. Children "fed with love" grow more in both weight and height than children fed the same food in an emotionally negative climate.

The person feeding a one-year-old should be aware that exploring and experimenting are normal and desirable behaviors at this time in a child's life. The child is developing a sense of autonomy that, if allowed to develop, will lay the foundation for later confidence and effectiveness as an individual. The child's impulses, if consistently denied, can turn to shame and self-doubt. In light of the developmental and nutrient needs of one-year-olds, and in the face of their often contrary and willful behavior, a few feeding guidelines may be helpful:

- *Discourage unacceptable behavior (such as standing at the table or throwing food) by removing the child from the table to wait until later to eat.* Be consistent and firm, not punitive. The child will soon learn to sit and eat.
- *Let the child explore and enjoy food.* This may mean the child eats with fingers for a while. Use of the spoon will come in time.
- *Don't force food on children.* Provide children with nutritious foods, and let them choose which ones and how much they will eat. Gradually, they will acquire a taste for different foods. If children refuse milk, provide cheese, cream soups, and yogurt.
- *Limit sweets strictly.* Infants have little room in their 1000-kcalorie daily energy allowance for empty-kcalorie sweets, except occasionally.

These recommendations reflect a spirit of tolerance that serves the best interests of the child emotionally as well as physically. The Nutrition Assessment Checklist helps to identify nutrition-related factors that may help prevent or correct potential problems in pregnant women and infants. Details of nutrition assessment are presented in Chapter 16.

In Summary The primary food for infants during the first 12 months is either breast milk or iron-fortified formula. In addition to nutrients, breast milk also offers immunological protection. At about four to six months, infants should gradually begin eating solid foods. By one year, they are drinking from a cup and eating many of the same foods as the rest of the family.

Nutrition Assessment Checklist

FOR PREGNANT WOMEN AND INFANTS

Health Problems, Signs, and Symptoms

Check the medical record for:

☐ Gestational diabetes

☐ Hypertension

☐ Preeclampsia

☐ Neural tube defect in an infant born previously

☐ Alcohol or illicit drug abuse

☐ Chronic diseases

☐ History of previous pregnancies (number, intervals, outcomes, multiple births, and gestational age birthweights)

Note risk factors for complications during pregnancy, including:

☐ Cigarette smoking

☐ Low socioeconomic status

☐ Food faddism

☐ Weight-loss dieting

☐ Very young or old age

☐ Lactose intolerance

☐ Significant or prolonged vomiting

Note any complaints of:

☐ Morning sickness

☐ Heartburn

☐ Constipation

Medications

For pregnant women who are using drug therapy for medical conditions, note:

☐ Potential for contraindication to breastfeeding

☐ GI tract side effects that might reduce food intake or change nutrient needs

Food/Nutrient Intake

For all pregnant women, especially those considered at risk nutritionally, assess the diet for:

☐ Total energy

☐ Protein

☐ Calcium, phosphorus, magnesium, iron, and zinc

☐ Folate and vitamin B_{12}

☐ Vitamin D

For infants, note:

☐ Method of feeding (breastfeeding, formula, or both)

☐ Frequency and duration of breastfeeding

☐ Amount of infant formula

☐ Practice of putting infant to bed with bottle

☐ Solid foods the infant is fed, if any

☐ Amount of food the infant is fed

Height and Weight

Measure baseline height and weight:

☐ Prepregnancy weight

☐ Infant birthweight

Reassess weight at each medical checkup and determine whether gains are appropriate. Note:

☐ Weight gain during pregnancy

☐ Gestational age

☐ Weight, length, and head circumference of infants

Laboratory Tests

Monitor the following laboratory tests for pregnant women:

☐ Hemoglobin, hematocrit, or other tests of iron status

☐ Blood glucose

Monitor the following laboratory tests for infants:

☐ Blood glucose of infants born to mothers with gestational diabetes

☐ Results of tests for inborn errors

Physical Signs

Blood pressure measurement is a routine measurement in physical exams, but is especially important for pregnant women.

Look for physical signs of:

☐ Iron deficiency

☐ Edema

☐ Protein-energy malnutrition

☐ Folate deficiency

1. The most important single predictor of an infant's future health and survival is:
   a. the infant's birthweight.
   b. the infant's iron status at birth.
   c. the mother's weight at delivery.
   d. the mother's prepregnancy weight.

2. A mother's prepregnancy nutrition is important to a healthy pregnancy because it determines the development of:
   a. the largest baby possible.
   b. adequate maternal iron stores.
   c. an adequate fat supply for the mother.
   d. healthy support tissues—the placenta, amniotic sac, unbilical cord, and uterus.

3. Two nutrients needed in large amount during pregnancy for rapid cell proliferation are:
   a. vitamin B_{12} and vitamin C.
   b. calcium and vitamin B_6.
   c. folate and vitamin B_{12}.
   d. copper and zinc

4. For a woman who is at the appropriate weight for height and is carrying a single fetus, the recommended weight gain during pregnancy is:
   a. 40 to 60 pounds.
   b. 25 to 35 pounds.
   c. 10 to 20 pounds.
   d. 20 to 40 pounds.

5. Rewards of physical activity during pregnancy may include:
   a. weight loss.
   b. decreased incidence of pica.
   c. relief from morning sickness.
   d. reduced stress and easier labor.

6. A woman's need for _____ is greater during lactation than during pregnancy.
   a. vitamin D
   b. energy
   c. calcium
   d. folate

7. Breast milk is recommended for the first 12 months of life because it offers complete nutrition and _____ to the infant.
   a. fluoride
   b. fructose

   c. immunological protection
   d. pica

8. An acceptable substitute for breast milk during the first year is:
   a. low-fat cow's milk.
   b. apple juice.
   c. water.
   d. iron-fortified infant formula.

9. Indications of readiness for solid foods include:
   a. the infant cries a lot.
   b. the infant is two months old.
   c. the infant is not sleeping through the night.
   d. the infant can sit with support and can control head movements.

10. During the first year of life, the most important step to undertake to encourage healthy eating is to:
    a. give food as a reward or for comfort.
    b. give sweets as a reward for eating vegetables.
    c. introduce a variety of nutritious foods in an inviting way.
    d. restrict fat to less than 30 percent of energy intake.

Critical Thinking

1. Which of the following foods would a woman thinking about becoming pregnant need to eat to help prevent a neural tube defect in her infant?
   a. dairy products
   b. fish and shellfish
   c. potatoes
   d. liver, lentils, and orange juice

2. Which of the following foods should be offered to infants with iron-rich foods to enhance iron absorption?
   a. dairy products
   b. peanut butter
   c. breads and cereals
   d. fruits and vegetables

Answers to these questions can be found in Appendix G.

Clinical Applications

1. Consider the different factors in a pregnant woman's history that can affect her nutrition status and the outcome of her pregnancy. Describe what steps you would take to remedy potential problems for the following clients:
   a. A 15-year-old adolescent of low socioeconomic status in her first trimester of pregnancy. She began the pregnancy at a normal, healthy weight, but her weight gain during pregnancy so far has been less than expected. Her favorite

   beverages are soft drinks; her favorite foods are french fries and boxed macaroni and cheese.
   b. A lactose-intolerant, 22-year-old pregnant woman who has been eating a vegan diet for the past year or so. She began the pregnancy slightly underweight (BMI = 19.5), but her weight gain has been adequate and consistent during the four months of her pregnancy. She complains of feeling tired all the time.

2. What information would you give to a pregnant woman who is considering breastfeeding her infant, but isn't quite sure why she should?

3. The parents of a two-month-old infant have been told by the child's grandparents that they should introduce solid foods to help the baby sleep through the night. What advice would you give the parents?

Nutrition on the Net

For further study of the topics of this chapter, access these websites.

Find updates and quick links to these and other nutrition-related sites at our website:
www.wadsworth.com/nutrition

Visit the pregnancy and child health center of the Mayo Clinic:
www.mayohealth.org

Learn more about having a healthy infant and about birth defects from the March of Dimes:
www.modimes.org

Learn more about neural tube defects from the Spina Bifida Association of America:
www.sbaa.org

Search for birth defects, pregnancy, adolescent pregnancy, maternal and infant health, and breast-feeding at the U.S. Government health information site:
www.healthfinder.gov

Search for pregnancy at the American Dietetic Association site:
www.eatright.org

Learn more about the WIC program:
www.usda.gov/fns

Visit the American College of Obstetricians and Gynecologists:
www.acog.org

Learn more about gestational diabetes from the American Diabetes Association:
www.diabetes.org

Learn more about breastfeeding from LaLeche League International:
www.lalecheleague.org

Read *A Woman's Guide to Breastfeeding* at the American Academy of Pediatrics site:
www.aap.org

Notes

1. K. M. Godfrey and D. J. P. Barker, Fetal nutrition and adult disease, *American Journal of Clinical Nutrition* 71 (2000): 1344S–1352S.

2. M. Hack and coauthors, Outcomes in young adulthood for very-low-birth-weight infants, *New England Journal of Medicine* 346 (2002): 149–157.

3. B. Guyer and coauthors, Annual summary of vital statistics: Trends in the health of Americans during the 20th century, *Pediatrics* 106 (2000): 1307–1317.

4. W. W. Hay and coauthors, Workshop summary: Fetal growth: Its regulation and disorders, *Pediatrics* 99 (1997): 585–591.

5. D. J. Barker and P. M. Clark, Fetal undernutrition and disease in later life, *Reviews of Reproduction* 2 (1997): 105–112.

6. Godfrey and Barker, 2000; J. Newnham, Consequences of fetal growth restriction, *Current Opinion in Obstetrics and Gynecology* 10 (1998): 145–149; W. P. T. James, Long-term fetal programming of body composition and longevity, *Nutrition Reviews* 55 (1997): S31–S43.

7. S. J. Otto and coauthors, Changes in the maternal essential fatty acid profile during early pregnancy and the relation of the profile to diet, *American Journal of Clinical Nutrition* 73 (2001): 302–307; G. Hornstra, Essential fatty acids in mothers and their neonates, *American Journal of Clinical Nutrition* 71 (2000): 1262S–1269S; R. Uauy and coauthors, Role of essential fatty acids in the function of the developing nervous system, *Lipids* 31 (1996): S167–S176.

8. American Academy of Pediatrics, Folic acid for the prevention of neural tube defects, *Pediatrics* 104 (1999): 325–327.

9. American Academy of Pediatrics, 1999; March of Dimes Resource Center, Folic acid (Wilkes-Barre, Pa.: March of Dimes Birth Defects Foundation, 1999); Standing Committee on the Scientific Evaluation of Dietary Reference Intakes, Food and Nutrition Board, Institute of Medicine, *Dietary Reference Intakes for Thiamin, Riboflavin, Niacin, Vitamin B_6, Folate, Vitamin B_{12}, Pantothenic Acid, Biotin, and Choline* (Washington, D.C.: National Academy Press, 1998), pp. 8-1–8-68.

10. J. E. Brown and coauthors, Predictors of red cell folate level in women attempting pregnancy, *Journal of the American Medical Association* 277 (1997): 548–552.

11. H. McNulty, G. J. Cuskelly, and M. Ward, Response of red blood cell folate to intervention: Implications for folate recommendations for the prevention of neural tube defects, *American Journal of Clinical Nutrition* 71 (2000): 1308S–1311S.

12. P. F. Jacques and coauthors, The effect of folic acid fortification on plasma folate and total homocysteine concentrations, *New England Journal of Medicine* 340 (1999): 1449–1454.

13. Standing Committee on the Scientific Evaluation of Dietary Reference Intakes, 1998, pp. 196–305.

14. J. L. Beard, Effectiveness and strategies of iron supplementation during pregnancy, *American Journal of Clinical Nutrition* 71 (2000): 1288S–1294S.

15. J. C. King, Determinants of maternal zinc status during pregnancy, *American Journal of Clinical Nutrition* 71 (2000): 1334S–1343S.

16. M. M. Werler and coauthors, Multivitamin supplementation and risk of birth defects, *American Journal of Epidemiology* 150 (1999): 675–682; T. O. Scholl and coauthors, Use of multivitamin/mineral prenatal supplements: Influence on the outcome of pregnancy, *American Journal of Epidemiology* 146 (1997): 134–141.

17. A. L. Owen and G. M. Owen, Twenty years of WIC: A review of some effects of the program, *Journal of the American Dietetic Association* 97 (1997): 777–782.

18. J. F. Clapp, The effect of continuing regular endurance exercise on the physiologic adaptations to pregnancy and pregnancy outcome, *American Journal of Sports Medicine* 24 (1996): S28–S29.

19. J. M. Pivarnik, Potential effects of maternal physical activity on birth weight: Brief review, *Medicine and Science in Sports and Exercise* 30 (1998): 400–406.

20. L. C. Tolstoi and J. B. Josimovich, Gestational diabetes mellitus: Etiology and management, *Nutrition Today* 34 (1999): 178–188; The Expert Committee on the Diagnosis and Classification of Diabetes Mellitus, Report of the Expert Committee on the diagnosis and classification of diabetes mellitus, *Diabetes Care* (supplement 1) 21 (1998): 5–19.

21. C. D. Naylor and coauthors, Cesarean delivery in relation to birth weight and gestational glucose tolerance: Pathophysiology or practice style? *Journal of the American Medical Association* 275 (1996): 1165–1170.

22. Tolstoi and Josimovich, 1999.

23. B. M. Sibai, Treatment of hypertension in pregnant women, *New England Journal of Medicine* 335 (1996): 257–265.

24. C. D. Drews and coauthors, The relationship between idiopathic mental retardation and maternal smoking during pregnancy, *Pediatrics* 97 (1996): 547–553.

25. American Academy of Pediatrics, Task Force on Infant Sleep Position and Sudden Infant Death Syndrome, Changing concepts of sudden infant death syndrome: Implication for infant sleeping environment and sleep position, *Pediatrics* 105 (2000): 650–656.

26. J. M. Lightwood, C. S. Phibbs, and S. A. Glantz, Short-term health and economic benefits of smoking cessation: Low birth weight, *Pediatrics* 104 (1999): 1312–1320; B. Guyer and coauthors, Annual summary of vital statistics 1997, *Pediatrics* 102 (1998): 1333–1349.

27. American Academy of Pediatrics, 2000; Environmental tobacco smoke affects birth weight, *Journal of the American Medical Association* 279 (1998): 739; E. Cutz and coauthors, Maternal smoking and pulmonary neuroendocrine cells in sudden infant death syndrome, *Pediatrics* 98 (1996): 668–672.

28. T. S. Hinds and coauthors, The effect of caffeine on pregnancy variables, *Nutrition Reviews* 54 (1996): 203–207.

29. I. S. Santos and coauthors, Caffeine intake and low birth weight: A population-based, case-control study, *American Journal of Epidemiology* 147 (1998): 620–627.

30. S. Cnattingius and coauthors, Caffeine intake and the risk of first-trimester spontaneous abortion, *New England Journal of Medicine* 343 (2000): 1839–1845; M. A. Klebanoff and coauthors, Maternal serum paraxanthine, a caffeine metabolite, and the risk of spontaneous abortion, *New England Journal of Medicine* 341 (1999): 1639–1644.

31. G. Koren, A. Pastuszak, and S. Ito, Drugs in pregnancy, *New England Journal of Medicine* 338 (1998): 1128–1137.

32. F. D. Eyler and coauthors, Birth outcome from a prospective, matched study of prenatal crack/cocaine use I: Interactive and dose effects on health and growth, *Pediatrics* 101 (1998): 229–237; F. D. Eyler and coauthors, Birth outcome from a prospective, matched study of prenatal crack/cocaine use II: Interactive and dose effects on neurobehavioral assessment, *Pediatrics* 101 (1998): 237–241.

33. M. S. Scher, G. A. Richardson, and N. L. Day, Effects of prenatal cocaine/crack and other drug exposure on electroencephalographic sleep studies at birth and one year, *Pediatrics* 105 (2000): 39–48; V. Delaney-Black and coauthors, Prenatal cocaine exposure and child behavior, *Pediatrics* 102 (1998): 945–950; Eyler and coauthors, Birth outcome from a prospective, matched study of prenatal crack/cocaine use II, 1998.

34. FDA announces advisory on methylmercury in fish, *FDA Talk Paper*, 2001, available from **www.cfsan.fda.gov**.

35. National Institutes of Health News Release, Long-chain alcohol found to block mechanism of fetal alcohol syndrome, www.nih.gov/news/pr/may2001/niaaa-18.htm, website visited May 23, 2001; K. Strömand and A. Hellström, Fetal alcohol syndrome: An ophthalmological and socioeducational prospective study, *Pediatrics* 97 (1996): 845–850.

36. C. Ikonomidou and coauthors, Ethanol-induced apoptotic neurodegeneration and fetal alcohol syndrome, *Science* 287 (5455) (2000): 947–948.

37. Centers for Disease Control, Division of Birth Defects, Child Development, and Disability and Health, Fetal alcohol syndrome, **www.cdc.gov/nceh/cddh/fas/fasfact.htm**, website visited July 13, 2000.

38. Committee on Substance Abuse and Committee on Children with Disabilities, American Academy of Pediatrics, Fetal alcohol syndrome and alcohol-related neurodevelopmental disorders, *Pediatrics* 106 (2000): 358–361.

39. S. N. Mattson and coauthors, Heavy prenatal alcohol exposure with or without physical features of fetal alcohol syndrome leads to IQ deficits, *Journal of Pediatrics* 131 (1997): 718–721.

40. Committee on Substance Abuse and Committee on Children with Disabilities, 2000.

41. Committee on Adolescence, American Academy of Pediatrics, Adolescent pregnancy—Current trends and issues: 1998, *Pediatrics* 103 (1999): 516–520.

42. R. J. Trissler, The child within: A guide to nutrition counseling for pregnant teens, *Journal of the American Dietetic Association* 99 (1999): 516–520.

43. American Academy of Pediatrics, Work Group on Breastfeeding, Breastfeeding and the use of human milk, *Pediatrics* 100 (1997): 1035–1039.

44. Position of The American Dietetic Association: Breaking the barriers to breastfeeding, *Journal of the American Dietetic Association* 101 (2001): 1213–1220.

45. K. G. Dewey, Energy and protein requirements during lactation, *Annual Review of Nutrition* 17 (1997): 19–36.

46. M. J. Heinig and K. G. Dewey, Health effects of breast feeding for mothers: A critical review, *Nutrition Research Reviews* 10 (1997): 35–56.

47. M. A. McCrory, Does dieting during lactation put infant growth at risk? *Nutrition Reviews* 59 (2001): 18–27; C. A. Lovelady and coauthors, The effect of weight loss in overweight, lactating women on the growth of their infants, *New England Journal of Medicine* 342 (2000): 449–453.

48. J. L. B. Pharm, Breastfeeding and the use of recreational drugs—Alcohol, caffeine, nicotine, and marijuana, *Breastfeeding Reviews* 2 (1998): 27–30.

49. Pharm, 1998.

50. R. Nduati and coauthors, Effect of breastfeeding and formula feeding on transmission of HIV-1: A randomized clinical trial, *Journal of the American Medical Association* 283 (2000): 1167–1174; P. G. Miotti and coauthors, HIV transmission through breastfeeding: A study in Malawi, *Journal of the American Medical Association* 282 (1999): 744–749.

51. R. F. Black, Transmission of HIV-1 in the breast-feeding process, *Journal of the American Dietetic Association* 96 (1996): 267–274.

52. Black, 1996.

53. J. Humphrey and P. Iliff, Is breast not best? Feeding babies born to HIV-positive mothers: Bringing balance to a complex issue, *Nutrition Reviews* 59 (2001): 119–127.

54. American Academy of Pediatrics, Committee on Nutrition, *Pediatric Nutrition Handbook*, 4th ed., ed. R. E. Kleinman (Elk Grove Village, Ill.: American Academy of Pediatrics, 1998), pp. 43–53.

55. R. A. Gibson and M. Makrides, n-3 Polyunsaturated fatty acid requirements of term infants, *American Journal of Clinical Nutrition* 71 (2000): 251S–255S; M. Neuringer, Infant vision and retinal function in studies of dietary long-chain polyunsaturated fatty acids: Methods, results, and implications, *American Journal of Clinical Nutrition* 71 (2000): 256S–267S.

56. C. Williams and coauthors, Stereoacuity at age 3.5 y in children born full-term is associated with prenatal and postnatal dietary factors: A report from a population-based cohort study, *American Journal of Clinical Nutrition* 73 (2001): 316–322.

57. American Academy of Pediatrics, Committee on Nutrition, 1998, pp. 3–20.

58. A. L. Wright and coauthors, Increasing breastfeeding rates to reduce infant illness at the community level, *Pediatrics* 101 (1998): 837–844.

59. American Academy of Pediatrics, Work Group on Breastfeeding, 1997.

60. J. Raisler and coauthors, Breast-feeding and infant illness: A dose-response relationship? *American Journal of Public Health* 89 (1999): 25–30.

61. D. S. Newburg and coauthors, Role of human-milk lactadherin in protection against symptomatic rotavirus infection, *Lancet* 351 (1998): 1160–1164.

62. A. H. Cushing and coauthors, Breastfeeding reduces risk of respiratory illness in infants, *American Journal of Epidemiology* 147 (1998): 863–870; D. S. Newburg and J. M. Street, Bioactive materials in human milk, *Nutrition Today* 32 (1997): 191–201.

63. M. A. Atkinson and T. M. Ellis, Infants diets and insulin-dependent diabetes: Evaluating the "cows' milk hypothesis" and a role for antibovine serum albumin immunity, *Journal of the American College of Nutrition* 16 (1997): 334–340.

64. J. M. Norris and coauthors, Lack of association between early exposure to cow's milk protein and β-cell autoimmunity, *Journal of the American Medical Association* 276 (1996): 609–614.

65. Committee on Nutrition, American Academy of Pediatrics, Hypoallergenic infant formulas, *Pediatrics* 106 (2000): 346–349.

66. Committee on Nutrition, American Academy of Pediatrics, Iron fortification of infant formulas, *Pediatrics* 104 (1999): 119–123.

67. Committee on Nutrition, 1999.

68. B. A. Dennison, H. L. Rockwell, and S. L. Baker, Excess fruit juice consumption by preschool-aged children is associated with short stature and obesity, *Pediatrics* 99 (1997): 15–22.

69. J. D. Skinner and B. R. Carruth, A longitudinal study of children's juice intake and growth: The juice controversy revisited, *Journal of the American Dietetic Association* 101 (2001): 432–437; Committee on Nutrition, American Academy of Pediatrics, The use and misuse of fruit juice in pediatrics, *Pediatrics* 107 (2001): 1210–1213.

Nutrition in Practice

■ ENCOURAGING SUCCESSFUL BREASTFEEDING ■

As discussed in Chapter 12, breastfeeding offers benefits to both mother and infant. The American Academy of Pediatrics (AAP), the American Dietetic Association, and the U.S. Department of Health and Human Services all advocate breastfeeding as the preferred means of infant feeding.[1] Promotion of breastfeeding is an integral part of the WIC program's nutrition education component.[2] During the late 1980s, breastfeeding was on the decline after reaching a high of about 60 percent of mothers in 1984. National efforts to promote breastfeeding seem to be working, at least to some extent: an encouraging trend of breastfeeding is emerging, with 64 percent of mothers initiating breastfeeding in 1998.[3] Nevertheless, only about one in five infants is still being breastfed at five to six months of age. Many of the benefits of breastfeeding are enhanced by breastfeeding for at least four months. The AAP recommends that breastfeeding continue for at least a year and thereafter for as long as mutually desired.[4] Increasing the rates of breastfeeding initiation and duration is one of the goals of *Healthy People 2010:*

> Increase the proportion of mothers who breastfeed immediately after birth, for the first six months, and preferably, through the infant's first year of life. Increase the proportion of mothers who breastfeed exclusively.[5]

Despite the trend toward increasing breastfeeding, the percentage of mothers choosing to breastfeed their infants and continuing to do so still falls short of goals.

Why don't more women choose to breastfeed their infants?

Many experts cite two major deterrents: public advertising of infant formula, and the medical community's failure to encourage breastfeeding. As an example of the medical lack of encouragement, some hospitals routinely separate mother and infant soon after birth. The child's first feeding then comes from the bottle rather than the breast. Furthermore, many hospitals send new mothers home with free samples of infant formula. The World Health Organization opposes this practice because it sends a misleading message that medical authorities favor infant formula over breast milk for infants. Even in hospitals where women are encouraged to breastfeed

and are supported in doing so, little if any assistance is available after hospital discharge when many breastfeeding women still need assistance. Up to half of mothers who initially breastfeed their infants stop within a month—seemingly due to lack of knowledge.

Women who receive early and repeated breastfeeding information and support breastfeed their infants longer than other women do. Information and instruction are especially important during the *prenatal* period when most women decide whether to breastfeed or to feed formula. Nurses and other health care professionals can play a crucial role in encouraging successful breastfeeding by offering women adequate, accurate information about breastfeeding that permits them to make informed choices. Table NP12–1 lists ten steps maternity facilities and health care professionals can take to promote successful breastfeeding among new mothers.

If breastfeeding is a natural process, what do mothers need to learn?

Although lactation is an automatic physiological process, breastfeeding requires some learning. This learning is most successful in a supportive environment. It begins with preparatory steps taken before the baby is born.

What are these preparatory steps?

Toward the end of pregnancy and throughout lactation, a woman who intends to breastfeed should stop using soap and lotions on her breasts. The natural secretions of the breasts themselves lubricate the nipple area best. A few weeks before the baby is due, the woman should rub her breasts with a towel or allow them to rub against her outer clothing for a little while each day to toughen the nipples somewhat in preparation for the infant's sucking. A woman who plans to breastfeed should also acquire at least two nursing bras before her infant is born. The bras should provide good support and have drop-flaps so that either breast can be freed for nursing.

How soon after birth should breastfeeding start?

As soon as possible. Immediately after the delivery, for a short period, the infant is intensely alert and intent

Nutrition in Practice

Table NP12–1   Ten Steps to Successful Breastfeeding

To promote breastfeeding, every maternity facility should:

- Develop a written breastfeeding policy that is routinely communicated to all health care staff.
- Train all health care staff in the skills necessary to implement the breastfeeding policy.
- Inform all pregnant women about the benefits and management of breastfeeding.
- Help mothers initiate breastfeeding within ½ hour of birth.
- Show mothers how to breastfeed and how to maintain lactation, even if they need to be separated from their infants.
- Give newborn infants no food or drink other than breast milk, unless medically indicated.
- Practice rooming-in, allowing mothers and infants to remain together 24 hours a day.
- Encourage breastfeeding on demand.
- Give no artificial nipples or pacifiers to breastfeeding infants.[a]
- Foster the establishment of breastfeeding support groups and refer mothers to them at discharge from the facility.

[a]Compared with nonusers, infants who use pacifiers breastfeed less frequently and stop breastfeeding at a younger age. C. G. Victora and coauthors, Pacifier use and short breastfeeding duration: Cause, consequence, or coincidence? *Pediatrics* 99 (1997): 445–453.

SOURCE: U.S. Department of Health and Human Services, Office on Women's Health, *Breastfeeding: HHS Blueprint for Action on Breastfeeding* (Washington, D.C.: U.S. Department of Health and Human Services, 2000); United Nations Children's Fund and World Health Organization. *The UNICEF/Baby-Friendly Hospital Initiative: Ten Steps to Successful Breastfeeding* (New York: UNICEF, 1992).

on suckling. This is the ideal time for the first breastfeeding and facilitates successful lactation.

What does the new mother need to know in order to continue breastfeeding her infant successfully?

She needs to learn how to relax and position herself so that she and the infant will be comfortable and the infant can breathe freely while nursing. She also needs to understand that infants have a **rooting reflex** that makes them turn toward any touch on the face. (The accompanying glossary defines this and other relevant terms.) Consequently, she should touch the infant's cheek to her nipple so that the infant will turn the right way and start to nurse. The mother can then squeeze her areola, the colored ring around the nipple, between two fingers and slip enough of it into the infant's mouth to permit a good hold and strong pumping action (see Figure NP12–1, p. 332). The nipple must rest well back on the infant's tongue so that the infant's gums will squeeze on the glands that release the milk and swallowing will be effortless. To break the suction, if necessary, the mother can slip a finger between the infant's mouth and her breast.

Does it hurt to have the infant sucking so hard on the breast?

No, because the mother has a **letdown reflex** that forces milk to the front of her breast when the infant begins to nurse, virtually propelling the milk into the infant's mouth. Letdown is necessary for the infant to obtain milk easily, and the mother needs to relax for letdown to occur. The mother who assumes a comfortable position in an environment without interruptions will find it easiest to relax.

How long should the infant be allowed to nurse at each feeding?

Although the infant sucks half the milk from the breast within the first 2 minutes, and 80 to 90 percent of it within 4 minutes, sucking on each breast for 10 to 15 minutes is encouraged. The sucking itself, as well as the complete removal of milk from the breast, stimulates the mammary glands to produce milk for the next nursing session. Successive sessions should start on alternate breasts to ensure that each breast is emptied regularly. This pattern maintains the same

Glossary of Breastfeeding Terms

engorgement: overfilling of the breasts with milk.

letdown reflex: the reflex that forces milk to the front of the breast when the infant begins to nurse.

mastitis: infection of a breast.

rooting reflex: a reflex that causes an infant to turn toward whichever cheek is touched, in search of a nipple.

Nutrition in Practice

Figure NP12–1

Infant's Grasp on Mother's Breast
The mother squeezes the areola, slipping enough of it into the infant's mouth to promote good pumping action. The infant's lips and gums pump the areola, releasing milk from the mammary glands into the milk ducts that lie beneath the areola.

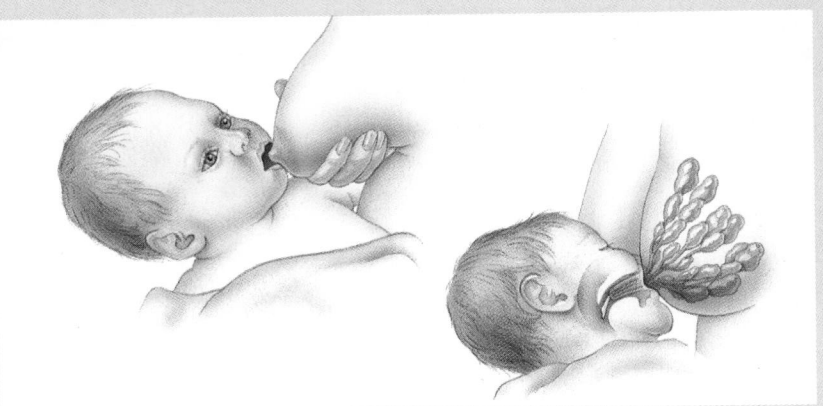

supply and demand for each breast and thus prevents either breast from overfilling.

Infants should be fed "on demand" and not held to a rigid schedule. The breastfed infant may average 8 to 12 feedings per 24-hour period during the first month or so. Once the mother's milk supply is well established and the infant's capacity has increased, the intervals between feedings will become longer.

What if a mother wants to skip one or two feedings daily—for example, because she works outside the home?

The mother can express breast milk into a bottle ahead of time, freeze the breast milk, and, when needed, substitute the expressed breast milk for a nursing session. Breast milk can be kept refrigerated for 48 hours or frozen (at a freezer temperature below 0°F) for several months.

The mother can hand express her breast milk or use one of several different breast pumps available. The bicycle-horn type of manual breast pump is difficult to keep clean and is not recommended. Cylinder-type manual pumps or electric breast pumps are safer and are also more efficient. Alternatively, a mother can substitute formula for those missed feedings and continue to breastfeed at other times.

What about problems associated with breastfeeding such as sore nipples or infection of the breast?

Most problems associated with breastfeeding can be resolved. Many mothers experience sore nipples during the initial days of breastfeeding. Sore nipples need to be treated kindly, but nursing can continue. Improper feeding position is a frequent cause of sore nipples: the mother should make sure the infant is taking the entire nipple and part of the areola onto the tongue. She should nurse on the less sore breast first to get letdown going while the infant is sucking hardest; then she can switch to the sore breast. Between times, she should expose her nipples to light and air to heal them.

Before lactation is well established, when the schedule changes, or when a feeding is missed, the breasts may become full and hard—an uncomfortable condition known as **engorgement.** The infant cannot grasp an engorged nipple and so cannot provide relief by nursing. A gentle massage or warming the breasts with a heating pad or in a shower helps to initiate letdown and to release some of the accumulated milk; then the mother can pump out some of her milk and allow the infant to nurse.

Infection of the breast, known as **mastitis,** is best managed by continuing to breastfeed. By drawing off the milk, the infant helps to relieve pressure in the infected area. The infant is safe because the infection is between the milk-producing glands, not inside them.

Even if everything is going smoothly, the nursing mother should ideally have enough help and support so that she can rest in bed a few hours each day for the first week or so. Successful breastfeeding requires the support of all those who care. This, plus adequate nutrition, ample fluids, fresh air, and physical activity, will do much to enhance the well-being of mother and infant.

Notes

1. Department of Health and Human Services, Office on Women's Health, *Breastfeeding: HHS Blueprint for Action on Breastfeeding* (Washington, D.C.: U.S. Department of Health and Human Services, 2000); American Academy of Pediatrics, Work Group on Breastfeeding, Breastfeeding and the use of human milk, *Pediatrics* 100 (1997): 1035–1039; Position of The American Dietetic Association: Breaking the barriers to breastfeeding, *Journal of the American Dietetic Association* 101 (2001): 1213–1220.

2. A. L. Owen and G. M. Owen, Twenty years of WIC: A review of some effects of the program, *Journal of the American Dietetic Association* 97 (1997): 777–782.

3. Department of Health and Human Services, Office on Women's Health, 2000.

4. American Academy of Pediatrics, 1997.

5. *Healthy People 2010: National Health Promotion and Disease Prevention Objectives* (Washington, D.C.: U.S. Department of Health and Human Services, 2000).

Jennie Oppenheimer/Studio Zocolo

NUTRITION THROUGH THE LIFE SPAN: CHILDHOOD AND ADOLESCENCE

333

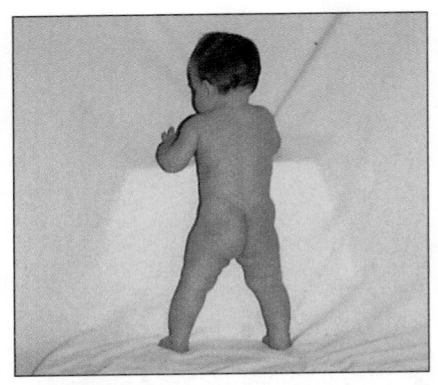

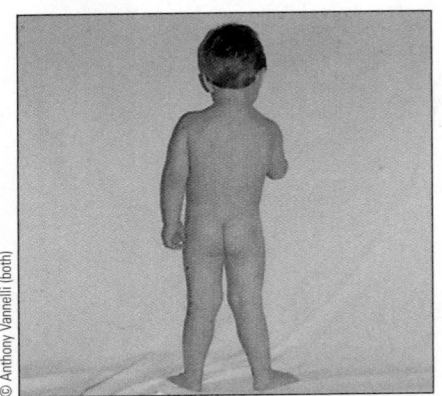

The body shape of a one-year-old (above) changes dramatically by age two (below). The two-year-old has lost much of the baby fat; the muscles (especially in the back, buttocks, and legs) have firmed and strengthened; and the leg bones have lengthened.

Nutrient needs change throughout life, depending on rates of growth, activity, and many other factors. Nutrient needs also vary from individual to individual, but generalizations are possible and useful. Sound nutrition throughout childhood promotes normal growth and development; facilitates academic and physical performance; and helps prevent obesity, diabetes, heart disease, cancer, and other degenerative diseases in adulthood. As children enter the teen years, a foundation built by years of eating nutritious foods best prepares them to meet the upcoming demands of rapid growth.

Early and Middle Childhood

After the age of one, growth rate slows, but the body continues to change dramatically. At one, infants have just learned to stand and toddle; by two, they walk confidently and are learning to run, jump, and climb. Nutrition and physical activity have helped them prepare for these new accomplishments by adding to the mass and density of their bone and muscle tissue. Thereafter, their bones continue to grow longer and their muscles to gain size and strength, though unevenly and more slowly, until adolescence.

Energy and Nutrient Needs

An infant's appetite declines markedly around the first birthday, consistent with the slowed growth rate. Thereafter, the appetite fluctuates. At times children seem to be insatiable, and at other times they seem to live on air and water. Parents and other caregivers need not worry about this—a child will need and demand much more food during periods of rapid growth than during slow periods. The perfect regulation of appetite in children of normal weight guarantees that their food energy intakes will be right for each stage of growth.

Children's Appetites Many people mistakenly believe that they must "make" their children eat the right amounts of food, and children's erratic appetites often reinforce this belief. Although children's food energy intakes vary widely from meal to meal, total daily energy intake remains remarkably constant. If children eat less at one meal, they eat more at the next, and vice versa.

Parents do, however, need to help children choose the right foods, and with overweight children, they may need to help more, as described later. Overweight children may not adjust their energy intakes appropriately, but may disregard appetite-regulation signals and eat in response to external cues, such as television commercials.

Energy Individual children's energy needs vary widely, depending on their growth and physical activity. A one-year-old child needs approximately 1000 kcalories a day; a three-year-old needs slightly more, about 1300 kcalories. By age ten, a child needs about 2000 kcalories a day. Total energy needs increase gradually with age, but energy needs per kilogram of body weight actually decline. Physically active children of any age need more energy because they spend more, and inactive children can become obese even when they eat less food than the average. Unfortunately, the prevalence of overweight in children has doubled over the past 20 years. Among preschool children as well as older children, overweight seems to be associated more with too little physical activity than with too much food intake.[1] Young girls, especially, show a marked reduction in their physical activity.[2]

Nutrients Steady growth during childhood necessitates a gradual increase of most nutrients. Nutrient recommendations cluster children into age groupings that reflect similarities in growth rate, biological changes, and hormone status (see inside front cover).

Ideally, children accumulate stores of nutrients before adolescence. Then, when they take off on the adolescent growth spurt and their nutrient intakes cannot keep pace with the demands of rapid growth, they can draw on the nutrient stores accumulated earlier. This is especially true of calcium; the denser the bones are in childhood, the better prepared they will be to support teen growth and still withstand the inevitable bone losses of later life.[3] Consequently, the way children eat influences their nutritional health during childhood, during their teen years, and for the rest of their lives.

Food Patterns for Children To provide all the needed nutrients, a child's meals and snacks should include a variety of foods from each food group—in amounts suited to the child's appetite and needs. Figure 13–1 shows the Food Guide Pyramid for young children. Notice that this pyramid differs slightly from the one for adults (see Chapter 1) in that a set number of servings, rather than a range, is suggested from each food group. Children 2 to 6 years old need at least the specified number of servings from each food group to meet their nutrient needs, but the serving sizes should vary according to age. Generally, children in the 4- to 6-year age group can eat the serving sizes recommended for adults. Caregivers of children 2 to 3 years old should offer servings that are about two-thirds the size of an adult serving. To satisfy the larger appetites of older children and adolescents, caregivers should provide additional servings of the same nutritious foods for the additional energy and nutrients growing children need. Unfortunately, few children's diets follow this nutritious pattern. Consequently, intakes of several nutrients, notably calcium, iron, and zinc, fall far below recommendations.[4]

Children's Food Choices Parents and other caregivers can do much to foster the development of healthy eating habits in a child. The challenge is to deliver nutrients in the form of meals and snacks that are both nutritious and appealing so that children will learn to enjoy a variety of nutritious foods.

Candy, cola, and other concentrated sweets must be limited in children's diets. If such foods are permitted in large quantities, the only possible outcomes are nutrient deficiencies, obesity, or both. Children can't be trusted to choose nutritious foods on the basis of taste alone; the preference for sweets is innate, and children naturally gravitate to them. Overweight children, especially, need help in selecting nutrient-dense foods that will meet their nutrient needs within their energy allowances. Underweight children or active, normal-weight children can enjoy higher-kcalorie foods, but these should still be nutritious. Examples are ice cream and pudding in the milk group and whole-grain or enriched pancakes and crackers in the bread group.

Malnutrition in Children

Most children in the United States and Canada are adequately nourished. Hunger and malnutrition are prevalent among some groups, however. Children in very low income families, for example, are more likely to be hungry and malnourished. About one in every nine U.S. households has one or more members who experiences pain from hunger caused by lack of food. In these households, well over 2.5 million children are hungry at least some of the time.[5] More than 9 million U.S. children fall under the broad definition of **food insecurity**—they do not know where their next meal is coming from or when it will come.

Effects of Hunger Both short-term and long-term hunger exert negative effects on behavior and health. Short-term hunger, such as when a child misses a meal, impairs the child's ability to pay attention and to be productive. Hungry children are irritable, apathetic, and uninterested in their environment. Long-term

food insecurity: limited or uncertain access to foods of sufficient quality or quantity to sustain a healthy and active life.

Nutrition Through the Life Span: Childhood and Adolescence

Figure 13–1
Food Guide Pyramid for Young Children

SOURCE: USDA Center for Nutrition and Policy Promotion, March 1999, Program AID 1649.

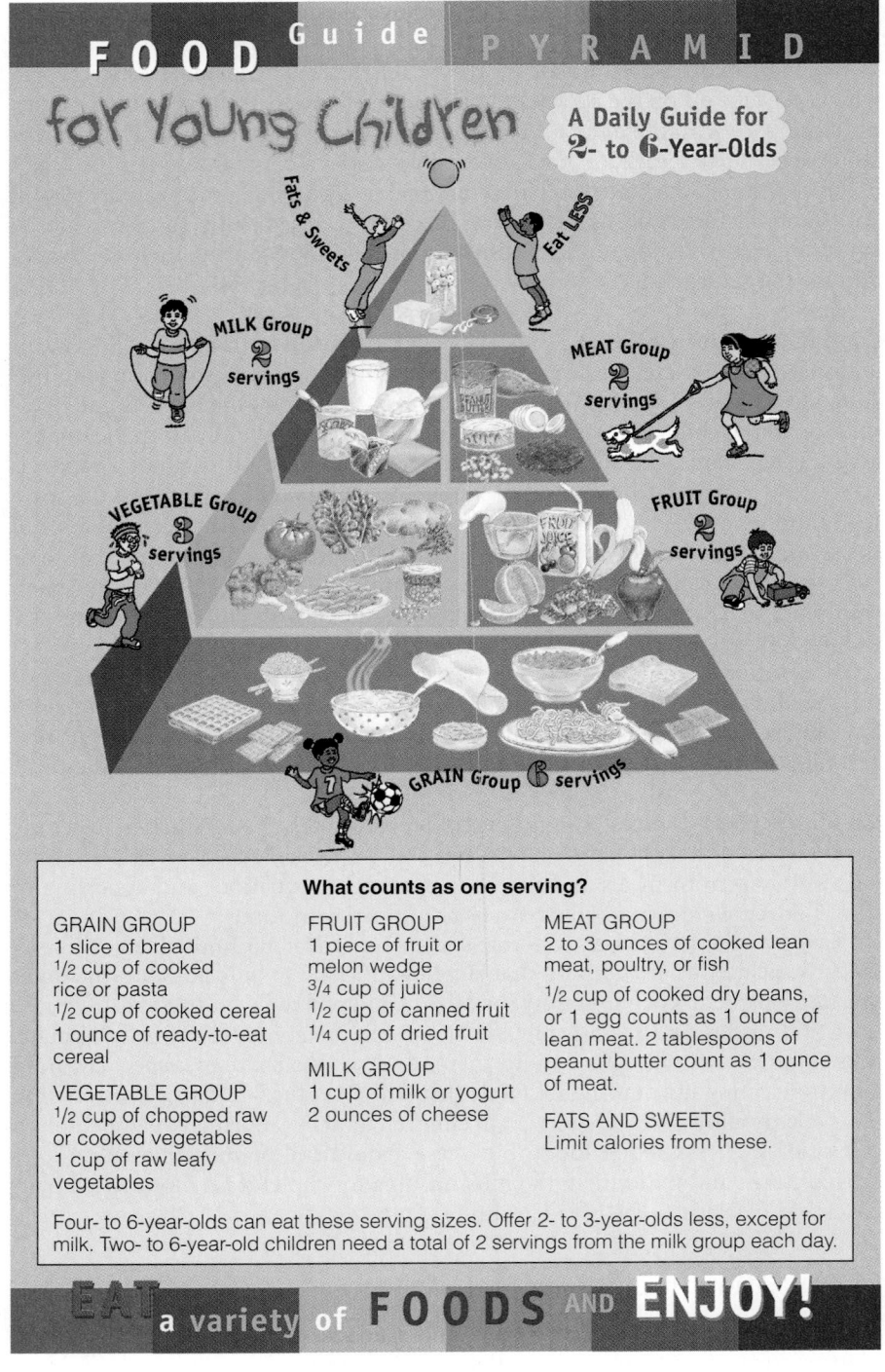

hunger impairs growth and immune defenses. Food assistance programs such as the WIC program and the School Breakfast and National School Lunch Programs are designed to protect against hunger and improve the health of children.

Hunger and School Performance   Children who eat nutritious breakfasts function better than their peers who do not. A nutritious breakfast that includes cereal is a central feature of a diet that meets the needs of children and supports their healthy growth and development.[6] Young children who participate in the School Breakfast Program improve their scores on achievement tests and are tardy or absent significantly less often than children who qualify for the program but do not participate. Common sense dictates that it is unreasonable to expect anyone to learn and perform work when no fuel has been provided. By the late morning, discomfort from hunger may become distracting even if a child has eaten breakfast. Chronically underfed children suffer all the more.[7]

The problem children face when attempting morning schoolwork on an empty stomach appears to be at least partly due to low blood glucose. The average child up to the age of ten or so needs to eat every four to six hours to maintain a blood glucose concentration high enough to support the activity of the brain and nervous system. The brain is the body's chief glucose consumer, and a child's brain is as big as an adult's is. A child's liver is considerably smaller than an adult's, however—and the liver is the organ responsible for storing glucose (as glycogen) and releasing it into the blood as needed. A child's liver cannot store more than about four hours' worth of glycogen—hence the need to eat fairly often. Teachers aware of the late-morning slump in their classrooms wisely request that a midmorning snack be provided; it improves classroom performance all the way to lunchtime.

Iron Deficiency and Behavior In U.S. children and adolescents, as in infants after six months, iron-deficiency anemia is the most prevalent nutrient deficiency. The best-known and most widespread effects of iron deficiency are its impacts on behavior. Most people are familiar with the role of iron in carrying oxygen in the blood. Another important function of iron is transporting oxygen within cells, where it is used to help produce energy. A lack of iron not only causes an energy crisis but also directly affects behavior, mood, attention span, and learning ability. Iron is also involved in the function of many molecules in the brain and nervous system. Much of the research on iron and behavior has focused on the proposal that even in the early stages of iron deficiency, an iron-dependent neurotransmitter is altered, and this change, in turn, impairs learning ability and behavior.

Iron deficiency is usually diagnosed by a deficit of iron in the *blood,* after the deficiency has progressed all the way to anemia. A child's *brain,* however, is sensitive to slightly lowered iron concentrations long before the blood deficits appear. Iron's effects are hard to distinguish from the effects of other factors in children's lives, but it is likely that iron deficiency manifests itself in a lowering of "motivation to persist in intellectually challenging tasks," a shortening of the attention span, and a reduction of overall intellectual performance.[8] Anemic children perform less well on tests and have more conduct disturbances than their nonanemic classmates. The effects of iron-deficiency anemia are especially detrimental to learning when combined with other nutrient deficiencies.[9]

Preventing Iron Deficiency To avert iron-deficiency anemia, children's foods must deliver about 10 milligrams of iron per day. To achieve this goal, milk intakes must be limited after infancy because milk is a poor source of iron. Children should receive enough milk products to ensure adequate calcium and riboflavin intakes, but no more. That means 2 cups of milk per day up to age 6, increasing to 3 cups per day from age 6 to 12. After age two, if reduced-fat milk is used instead of whole milk, the saved kcalories can be invested in iron-rich foods such as lean meats, fish, poultry, eggs, and legumes. Whole-grain or enriched breads and cereals also contribute iron. Table 13–1 lists iron-rich foods children like.

The iron status of children in the United States is improving. Infant feeding practices such as breastfeeding and the use of iron-fortified formulas may be improving iron status in later childhood.[10] Among children from low-income families, the WIC program, which provides supplemental iron-rich foods during infancy and early childhood, appears to be playing a role in improving iron status.

World Focus on Iron Deficiency The prevalence of iron deficiency among children throughout the world has become the focus of major public health organizations. The U.S. Public Health Service lists the reduction of iron deficiency

Table 13–1 Iron-Rich Foods Children Like[a]

Breads, cereals, and grains
Canned macaroni (½ c)
Canned spaghetti (½ c)
Cream of wheat (¼ c)
Fortified dry cereals (1 oz)[b]
Noodles, rice, or barley (½ c)
Tortillas (1 flour, 2 corn)
Whole-wheat, enriched, or fortified bread (1 slice)

Vegetables
Baked flavored potato skins (½ skin)
Cooked mushrooms (½ c)
Cooked mung bean sprouts or snow peas (½ c)
Green peas (½ c)
Mixed vegetable juice (1 c)

Fruits
Apple juice (1 c)
Canned plums (3 plums)
Cooked dried apricots (½ c)
Dried peaches (4 halves)
Raisins (1 tbs)

Meats and legumes
Bean dip (¼ c)
Canned pork and beans (⅓ c)
Mild chili or other bean/meat dishes (¼ c)
Meat casseroles (½ c)
Peanut butter and jelly sandwich (½ sandwich)
Lean chopped roast beef or cooked ground beef (1 oz)
Sloppy joes (½ sandwich)

[a]Each serving provides at least 1 milligram iron, or one-tenth of a child's RDA for iron. Vitamin C–rich foods included with these snacks increase iron absorption.
[b]Some fortified breakfast cereals contain more than 10 milligrams iron per half-cup serving (read the labels).

Table 13–2   Signs of Health and Malnutrition in Children

	Healthy	Malnourished
Hair	Shiny, firm in the scalp	Dull, brittle, dry, loose; falls out
Eyes	Bright, clear pink membranes; adjust easily to darkness	Pale membranes; spots; redness; adjust slowly to darkness
Teeth and gums	No pain or cavities, gums firm, teeth bright	Missing, discolored, decayed teeth; gums bleed easily and are swollen and spongy
Face	Good complexion	Off-color, scaly, flaky, cracked skin
Glands	No lumps	Swollen at front of neck and cheeks
Tongue	Red, bumpy, rough	Sore, smooth, purplish, swollen
Skin	Smooth, firm, good color	Dry, rough, spotty; "sandpaper" feel or sores; lack of fat under skin
Nails	Firm, pink	Spoon-shaped, brittle, ridged
Behavior	Alert, attentive, cheerful	Irritable, apathetic, inattentive, hyperactive
Internal systems	Heart rate, heart rhythm, and blood pressure normal; normal digestive function; reflexes and psychological development normal	Heart rate, heart rhythm, or blood pressure abnormal; liver and spleen enlarged; abnormal digestion; mental irritability, confusion; burning, tingling of hands and feet; poor balance and coordination
Muscles and bones	Good muscle tone and posture; long bones straight	"Wasted" appearance of muscles; swollen bumps on skull or ends of bones; small bumps on ribs; bowed legs or knock-knees

NOTE: The signs here are consistent with malnutrition but not diagnostic of it.

Paint is the primary source of lead in children's lives.

among young children as one of the nation's foremost health priorities. The World Health Organization and a United Nations subcommittee on nutrition have developed a ten-year plan to eliminate iron deficiency.

Other Nutrient Deficiencies   Iron is only one of several dozen nutrients that can be displaced by a diet of nutrient-poor foods. Any of the other nutrients may be lacking as well, and the deficiencies of those nutrients may also cause both behavioral and physical symptoms (see Table 13–2). A child with behavioral symptoms of nutrient deficiencies may be irritable, aggressive, disagreeable, or sad and withdrawn. Such a child may be labeled "hyperactive," "depressed," or "unlikable," but in fact these traits may arise from simple, albeit marginal, malnutrition. A study of school-aged children in a U.S. correctional institution found that children receiving a low-dose supplement of vitamins and minerals required discipline for fighting, vandalism, disrespect, refusal to work, and other infractions about half as often as those receiving a placebo.[11] Should suspicion of dietary inadequacies be raised, *no matter what other causes may be implicated,* the people responsible for feeding the child should take steps to correct those inadequacies promptly.

Lead Poisoning in Children

The damage caused by malnutrition may be compounded by environmental factors such as lead poisoning. A two-way interaction is typical: lead poisoning can cause an iron deficiency, and an iron deficiency can impair the body's defenses against lead absorption.[12] Adequate calcium may slow lead's absorption or interfere with its toxic effects in the body.[13] Like iron deficiency, mild lead toxicity has nonspecific effects, including diarrhea, irritability, reduced ability of the blood to carry oxygen, and fatigue. The symptoms may be reversible if exposure stops soon enough. With higher levels of lead, the signs become more pronounced, yet pinpointing a cause may still be difficult. Children lose their general cognitive, verbal, and perceptual abilities and develop learning disabilities and behavior problems. Still more severe lead toxicity can cause irreversible nerve damage, paralysis, mental retardation, and death. Table 13–3 lists the symptoms of lead toxicity.

Lead toxicity is most prevalent among children under age six—as many as 1 million may have blood concentrations high enough to cause mental, behavioral, and other health problems.[14] Lead aggressively attacks fetuses, infants, and children because the body absorbs lead most efficiently during times of rapid growth. Blood concentrations of lead generally reach a peak in two-year-old children; age two is the typical time of exploring surroundings "hand to mouth" and ingesting lead-tainted dirt and debris. Children's behaviors and activities—putting their hands in their mouths, playing in dirt, and eating nonfood items—favor their chances of exposure to lead. Of all the sources of lead for children, paint remains the most important. Three million young children live in homes where leaded surfaces are peeling and deteriorating. Each of these children is either already poisoned or in danger of becoming so.

A ban on leaded gasoline and reductions in other uses of lead have dramatically reduced the amount of lead in the environment.[15] The result has been a gratifying decline in children's blood lead concentrations since the 1970s. The decline follows exactly the reduction in the nation's use of leaded gasoline, leaded house paint, and lead-soldered food cans. A nationwide lead-monitoring system is now in place, and aggressive community programs are testing and treating children for lead poisoning. Despite such safeguards, the problem of exposure to lead still pervades children's lives (see Figure 13–2). The "How to" box on p. 341 suggests strategies to protect children from lead poisoning.

Food Allergies

Food allergies are frequently blamed for physical and behavioral abnormalities in children. In fact, only about 5 percent of children are diagnosed with true food allergies.[16] Food allergies diminish with age, until in adulthood they affect about 1 or 2 percent of the population.[17] A true food allergy occurs when a whole food protein or other large molecule enters the body and elicits an immunologic response. (Recall that large molecules of food are normally dismantled in the digestive tract to smaller ones before absorption.) The body's immune system reacts to a food protein or other large molecule as it does to an antigen—by producing antibodies or other defensive agents. A problem that results from exposure to food substances, but does not involve the immune system, is known as **food intolerance.**

Asymptomatic and Symptomatic Allergies   Allergies may have one or two components. They always involve antibodies; they may or may not involve symptoms. A person may produce antibodies without having any symptoms (known as asymptomatic allergy) or may produce antibodies and have symptoms (known as symptomatic allergy). A person who experiences symptoms without producing antibodies, however, does not have an allergy. This means that allergies have to be diagnosed by testing for antibodies.

Allergy Symptoms   A symptomatic allergy will exhibit different symptoms depending on the location of the reaction. In the digestive tract, the allergy may cause nausea or vomiting; in the skin, it may cause rashes; and in the nasal passages and lungs, it may cause inflammation or asthma. A dangerous, generalized, all-systems shock reaction known as **anaphylactic shock** can also occur.

Immediate and Delayed Reactions   Allergic reactions to food can occur with different timings, simply classified as immediate and delayed. In both, the antigen interacts immediately with the immune system, but symptoms may appear

Table 13–3   Symptoms of Lead Toxicity[a]
■ Learning disabilities
■ Low IQ
■ Behavior problems
■ Slow growth
■ Iron-deficiency anemia
■ Nervous system disorders
■ Impaired concentration
■ Reduced short-term memory
■ Slow reaction time
■ Seizures
■ Impaired hearing
■ Poor coordination

Reminder: *Antigens* are substances foreign to the body that elicit the formation of antibodies or an inflammation reaction from immune system cells. Food antigens are usually glycoproteins (large proteins with glucose molecules attached).

Reminder: *Antibodies* are large proteins that are produced in response to antigens and then inactivate the antigens.

food allergies: adverse reactions to foods that involve an immune response; also called *food-hypersensitivity reactions.*

food intolerance: an adverse response to a food or food additive that does not involve the immune system.

anaphylactic (an-AFF-ill-LAC-tic) **shock:** a life-threatening whole-body allergic reaction to an offending substance.

Figure 13–2

Sources of Lead Exposure

Lead finds its way into the bodies of children when they ingest lead-containing foods, water, dust, or paint chips, or when they breathe lead-laden air.

Lead in air

Lead in water

Lead solder in cans

Factory pollution

Car exhaust

Lead in air

Lead in pipes

Lead in old or imported pottery

Lead in food

Lead in old paint

Lead in soil

Lead dust on toys

Lead dust on pets

These symptoms can occur in minutes or hours after ingesting an allergen:

- *Tingling sensation in the mouth.*
- *Swelling of tongue and throat.*
- *Irritated, reddened eyes.*
- *Difficulty breathing.*
- *Hives, swelling, rashes.*
- *Vomiting, abdominal cramps, diarrhea.*
- *Drop in blood pressure.*
- *Loss of consciousness.*
- *Death.*

within minutes or after several (up to 24) hours. Identifying the food that causes an immediate allergic reaction is easy because symptoms correlate closely with the time of eating the food. Identifying the food that caused a delayed reaction is more difficult because the symptoms may not appear until a day after the offending food was eaten; by this time, many other foods will have been eaten, too, complicating the picture. Allergic reactions to single foods are common. Reactions to multiple foods are the exception, not the rule.

Anaphylactic Shock Peanuts, tree nuts, milk, eggs, wheat, soybeans, fish, and shellfish most often cause the life-threatening food allergy reaction of anaphylactic shock (see Table 13–4).[18] In anaphylactic shock, the airways in the lungs constrict, the blood pressure lowers, and the tongue and throat swell and cause suffocation.

Children (and adults) must stay alert for threatening symptoms of impending anaphylactic shock, such as a tingling of the tongue, throat, or skin, or difficulty breathing (see the margin). Any person with food allergies severe enough

How to

Protect Children from Lead Exposure

*D*efensive strategies to protect infants and children from lead exposure include:

- Test children for lead poisoning; effective screening with an appropriate questionnaire is essential to identifying high-risk children and thus treating the devastating effects.[a] About half the pediatricians surveyed reported universal screening.[b]

[a]S. J. Schaffer and coauthors, Lead poisoning risk determination in a rural setting, *Pediatrics* 97 (1996): 84–90; M. N. Haan, M. Gerson, and B. A. Zishka, Identification of children at risk for lead poisoning: An evaluation of routine pediatric blood lead screening in an HMO-insured population, *Pediatrics* 97 (1996): 79–83; D. C. Snyder and coauthors, Development of a population-specific risk assessment to predict elevated blood lead levels in Santa Clara County, California, *Pediatrics* 96 (1995): 643–648; Committee on Environmental Health, American Academy of Pediatrics, Lead poisoning: From screening to primary prevention, *Pediatrics* 92 (1993): 176–183.
[b]J. R. Campbell and coauthors, Blood lead screening practices among U.S. pediatricians, *Pediatrics* 97 (1996): 372–377.

- In contaminated environments, keep small children from putting dirty or old painted objects in their mouths, and make sure children wash their hands before eating. Similarly, keep small children from eating any nonfood items. Lead poisoning has been reported in young children who have eaten pool cue chalk.[c]
- Be aware that other countries do not have the same regulations protecting consumers against lead. Children have been poisoned by eating crayons made in China and drinking fruit juice canned in Mexico.
- Make infant formula from lead-free ingredients. Do not use lead-contaminated water.
- Feed children nutritious meals regularly.

[c]Pool cue chalk: A source of environmental lead, *Pediatrics* 97 (1996): 916–917.

---

to cause anaphylactic shock should wear a medical alert bracelet or necklace and carry a syringe of life-saving **epinephrine.**

Food Labeling To protect people with allergies, food manufacturers and consumer groups are working with the Food and Drug Administration (FDA) to ensure that allergens are appropriately labeled on food products. Voluntary efforts to list the eight most common food allergens in "plain language" are being initiated. For example, manufacturers are including the word *milk* in the ingredient list as well as the less familiar term *caseinate.* Figure 13–3 (p. 342) offers examples of appropriate labeling of food allergens. Food manufacturers and processors are also taking steps to avoid cross-contamination during production. For example, by processing an allergen-containing food such as egg noodles *after* plain noodles, rather than before, the risk of cross-contamination is significantly reduced.[19]

Other Adverse Reactions to Foods Adverse reactions to foods that are not true food allergies include:

- A reaction specific to the flavor enhancer monosodium glutamate, or MSG.
- Reactions to chemicals in foods, such as the natural laxative in prunes.
- Symptoms of digestive diseases, such as hernias and ulcers, that are aggravated by eating any food.
- Enzyme deficiencies, such as lactose intolerance, that cause symptoms superficially indistinguishable from those of food allergy.
- Psychological reactions based on the belief that certain foods cause certain symptoms.

The simple dislike of a food may be a clue to allergy or to any of these reactions.

Food Dislikes Parents are advised to watch for signs of food dislikes and take them seriously. Children's food aversions may be the result of nature's efforts to protect them from allergic or other adverse reactions. Test for allergies, and then

Table 13–4   Foods That Most Often Cause Serious Allergic Reactions[a]
- Eggs
- Fish
- Milk
- Peanuts
- Shellfish
- Soybeans
- Walnuts and other tree nuts
- Wheat

[a]Children often outgrow their allergies to milk, egg, soy, and wheat.

epinephrine: one of the stress hormones secreted whenever emergency action is needed; prescribed therapeutically to relax the bronchioles during allergy or asthma attacks.

Figure 13–3

Examples of Appropriate Labeling of Food Allergens

Food manufacturers are taking steps to ensure that consumers are aware of the presence of food allergens in packaged products.

ALLERGY INFORMATION: MANUFACTURED ON EQUIPMENT THAT PROCESSES PRODUCTS CONTAINING PEANUTS AND OTHER NUTS.

Cereal Giant

FOOD ALLERGIC CONSUMERS *See ingredient list*

...ICARBONATE, MALT ...XTRIN ...ND EXTRACT, **ALMOND FLOUR, PEANUT FLOUR.** FRESHNESS PRESERVED BY BHT.

CONTAINS WHEAT, MILK, SOY, PEANUT AND ALMOND INGREDIENTS.

Creme FILLED CAKES

apply nutrition knowledge conscientiously in deciding how to alter the diet. Don't risk feeding the child an unbalanced diet, which could lead to nutrient deficiencies. Whenever a food is excluded from the diet, care must be taken to include other foods that provide the same nutrients as the omitted food. Remember that children who must avoid certain foods need all their nutrients, just as other children do.

Hyperactivity

Hyperactivity affects behavior and learning in about 5 to 10 percent of young school-aged children.[20] Left untreated, it can interfere with a child's social development and ability to learn. Treatment focuses on relieving the symptoms and controlling the associated problems; there is no cure. Physicians often manage hyperactivity through behavior modification, special educational techniques, psychological counseling, and, in some cases, drug therapy.

Parents of hyperactive children sometimes seek help from alternative therapies, including special diets. They mistakenly believe a solution may lie in manipulating the diet—most commonly, by excluding sugar or food additives. Adding carrots or eliminating candy is such a simple solution that many parents eagerly give such diet advice a try. These dietary changes will not solve the problem of true hyperactivity. Studies have consistently found no convincing evidence that sugar causes hyperactivity or worsens behavior.[21] Recommendations to restrict sugar in children's diets to prevent or treat behavior problems are groundless. The accompanying case study offers an opportunity to think about these issues in relation to a specific child.

Children can become excitable, rambunctious, and unruly as a result of a desire for attention, lack of sleep, overstimulation, watching too much television or playing too many video games, too much caffeine from colas or chocolate, or a lack of physical activity. Such behaviors may suggest that more consistent care is needed. It helps to insist on regular hours of sleep, regular mealtimes, and regular outdoor activity.

hyperactivity: inattentive and impulsive behavior that is more frequent and severe than is typical of others a similar age; professionally called **attention-deficit/hyperactivity disorder (ADHD).**

In Summary Children's appetites and nutrient needs reflect their stage of growth. Long-term hunger and malnutrition impair growth and health. Short-term hunger exerts more subtle effects on children's health and behavior—such as poor academic performance. Iron deficiency is widespread and has many physical and behavioral consequences. Lead toxicity is prevalent among young children and can have irreversible effects on health and

Case Study

BOY WITH DISRUPTIVE BEHAVIOR

Freddie is a six-year-old boy who seldom sits still, often misbehaves, and is frequently sick. Freddie's eating habits are erratic and poor, as is his appetite. He often misses breakfast because he is too tired to get up in time to eat before school. By midmorning, Freddie is irritable and disruptive in the classroom. At lunchtime he trades the peanut butter and banana sandwich his mother packed in his lunchbox for a piece of cake. After school he hurries home to watch television while he eats his favorite snack—cola and potato chips. At dinnertime Freddie picks at his food because he isn't very hungry. Later on, when it's time for bed, Freddie complains that he's hungry. His parents let him stay up to have a bowl of cereal (the kind with marshmallows) before he finally falls asleep.

- What factors in Freddie's daily routine might be contributing to his restless behavior?
- Discuss some changes in diet that might improve Freddie's health and disposition.

behavior. True food allergies are somewhat rare in children, and children can outgrow some food allergies. Some allergies, however, can cause dangerous, life-threatening reactions in both children and adults. "Hyper" behavior is not caused by poor nutrition; misbehavior may reflect inconsistent care.

Food Choices and Eating Habits of Children

The childhood years are the parents' last chance to influence their children's food choices. Parents are **gatekeepers,** controlling the availability of foods in their children's environments. Gatekeepers who want to promote nutritious choices and healthful habits provide access to nutrient-dense, delicious foods and opportunities for active play at home. Food choices and regular physical activity can not only promote healthy growth but, as mentioned earlier, also can help prevent the degenerative diseases of later life. Many experts agree that early childhood is the time to put into effect practices that, until recently, were recommended only for adults. "Childhood Obesity and the Early Development of Chronic Diseases" is the title of the Nutrition in Practice that follows this chapter.

Mealtimes at Home

Feeding children requires not only providing a variety of nutritious foods but also nurturing the children's self-esteem and well-being. Parents face a number of challenges in preparing meals that both appeal to their children's tastes and provide needed nutrients. Because the interactions between parents and children can set the stage for lifelong attitudes and habits, a child's preferences should be treated with respect, even when nutrient needs must take precedence.

Honoring Children's Preferences Researchers attempting to explain children's food preferences encounter many contradictions. Children say they like colorful foods, yet most often reject green and yellow vegetables while favoring brown peanut butter and white potatoes, apple wedges, and bread. They do like raw vegetables better than cooked ones, though, so it is wise to offer vegetables that are raw or slightly undercooked and crunchy and bright in color. They should be warm, not hot, because a child's mouth is much more sensitive than an adult's. The flavor should be mild (a child has more taste buds), and smooth foods such

gatekeepers: with respect to nutrition, key persons who control other people's access to foods and thereby exert a profound impact on their nutrition. Examples are the spouse who buys and cooks the food, the parent who feeds the children, and the caregiver in a day-care center.

Little children like little tables and little portions.

as mashed potatoes or pea soup should have no lumps (a child wonders, with some disgust, what the lumps might be).

Young children like to eat at little tables and to be served little portions of food. They also love to eat with other children and have been observed to stay at the table longer and eat more food when in the company of their peers. Parents who serve food in a relaxed and casual manner, without anxiety, provide an environment in which a child's negative emotions will be minimized.

Avoiding Power Struggles It is not surprising that problems over food often arise during the second or third year, when children begin asserting their independence. Many of these problems stem from the conflict between children's developmental stages and capabilities and parents who, in attempting to do what they think is best for their children, try to control every aspect of eating. Such conflicts can disrupt children's abilities to regulate their own food intakes or to determine their own likes and dislikes. For example, many people share the misconception that children must be persuaded or coerced to try new foods. In fact, the opposite is true. When children are forced to try new foods, especially when offered rewards for eating a particular food (if you eat your vegetables, you can play video games), they are less likely to try those foods again than are children who are left to decide for themselves.[22] Similarly, when children are restricted from eating their favorite foods, they are more likely to want those foods.[23] As noted by dietitian and family therapist Ellyn Satter, the parent is responsible for *what* the child is offered to eat, but the child is responsible for *how much* and even *whether* to eat.[24]

When introducing new foods at the table, parents are advised to offer them one at a time and only in small amounts at first. The more often a food is presented to a young child, the more likely the child will like that food. Research shows that between five and ten exposures to a new food are necessary before a toddler shows an enhanced preference for the food.[25] Whenever possible, the new food should be presented at the beginning of the meal, when the child is hungry, but the child should make the decision to accept or reject it. Parents have their own inclinations and dislikes; so do children. It is best never to make an issue of food acceptance. A power struggle almost invariably sets a firm pattern of resistance and permanently closes the child's mind.

Television's Influence Watching television adversely affects children's nutritional health. As Chapter 7 pointed out, watching television contributes to obesity. Children who watch television for more than four hours a day, or during meals, are least likely to eat fruits and vegetables and most likely to be obese.[26] Not only are they inactive, but they often snack on the fattening foods that are advertised.

The average child sees about 10,000 commercials a year, and almost all of them urge viewers to purchase sugarcoated breakfast cereals, candy bars, chips, fast foods, and carbonated beverages. Those foods add sugar, fat, and salt to the diet and displace foods that provide needed nutrients. Many parents and pediatricians believe that food ads aimed at children should be banned because they support corporate profits rather than children's health. Alternatively, parents can teach their children how to evaluate food ads and make healthful choices.

Preventing Choking When feeding children, parents must always be alert to the dangers of choking. A choking child is a silent child—an adult should be present whenever a child is eating. Make sure the child sits when eating; choking is more likely when a child is running or falling. Round foods such as grapes, nuts, hard candies, marshmallows, and hot dog pieces are hard to control in a mouth with few teeth, and they can easily become lodged in the small opening of a child's trachea. Other potentially dangerous foods include tough meat, raw carrots and celery, popcorn, chips, and peanut butter eaten by the spoonful.

Play First Ideally, each meal is preceded, not followed, by the activity the child looks forward to the most. A number of schools have discovered that children eat a much better lunch if it is served after, rather than before, recess. Otherwise children "hurry up and eat" so that they can go play.

Child Participation Allowing children to help plan and prepare the family's meals provides enjoyable learning experiences and encourages children to eat the foods they have prepared. Vegetables are pretty, especially when fresh, and provide opportunities for children to learn about color, growing things and their seeds, and shapes and textures—all of which are fascinating to young children. Measuring, stirring, decorating, and arranging foods are skills that even a very young child can practice with enjoyment and pride (see Table 13–5).

Snacks Parents may find that their children often snack so much that they aren't hungry at mealtimes. Instead of teaching children *not* to snack, teach them *how* to snack. Provide snacks that are as nutritious as the foods served at mealtime. Snacks can even be mealtime foods that are served individually over time, instead of all at once on one plate. When providing snacks to children, think of the food groups and offer such snacks as pieces of cheese, tangerine slices, carrot sticks, and peanut butter on whole-wheat crackers (see Table 13–6, p. 346). Snacks that are easy to prepare should be readily available to children, especially if they arrive home after school before their parents.

Preventing Dental Caries Children frequently snack on sticky, sugary foods that stay on the teeth and provide an ideal environment for the growth of bacteria that cause dental caries. Teach children to brush and floss after meals, to brush or rinse after eating snacks, to avoid sticky foods, and to select crisp or fibrous foods frequently.

Serving as Role Models In an effort to practice these many tips, parents may overlook perhaps the single most important influence on their children's food habits—themselves. Parents who do not eat carrots should not be surprised when their children refuse to eat carrots. Likewise, parents who dislike the smell of brussels sprouts may not be able to persuade children to try them. Children learn much through imitation. Parents, older siblings, and other caregivers set an irresistible example by sitting with younger children, eating the same foods, and having pleasant conversations during mealtime.[27]

Nutrition at School

While parents are doing what they can to establish good eating habits in their children at home, others are preparing and serving foods to their children at day-care centers and schools. In addition, children begin learning about food and nutrition in the classroom. Meeting the nutrition and education needs of children is critical to supporting their healthy growth and development.[28]

The U.S. government funds several programs to provide nutritious, high-quality meals for children at school. Both the School Breakfast Program and the National School Lunch Program provide meals at a reasonable cost to children from families with the financial means to pay. Meals are available free or at reduced cost to children from low-income families.

School Breakfast The School Breakfast Program is available in slightly more than half of the nation's schools, and about 5 million children participate in it. Surveys show that the majority of children who eat school breakfasts are from low-income families. As research results continue to emphasize the positive impact breakfast has on school performance and health, campaigns to expand school breakfast programs are under way.

Table 13–5 Food Skills of Preschoolers[a]

Age 1–2 years, when large muscles develop, the child:
- uses short-shanked spoon.
- helps feed self.
- lifts and drinks from cup.
- helps scrub, tear, break, or dip foods.

Age 3 years, when medium hand muscles develop, the child:
- spears food with fork.
- feeds self independently.
- helps wrap, pour, mix, shake, or spread foods.
- helps crack nuts with supervision.

Age 4 years, when small finger muscles develop, the child:
- uses all utensils and napkin.
- helps roll, juice, mash, or peel foods.
- cracks egg shells.

Age 5 years, when fine coordination of fingers and hands develops, the child:
- helps measure, grind, cut, and grate.
- uses hand-cranked egg beater with supervision.

[a]These ages are approximate. Healthy, normal children develop at their own pace.

The school breakfast must contain at a minimum:
- *One serving of fluid milk.*
- *One serving of fruit or vegetable or full-strength juice.*
- *Two servings of bread or bread alternates; or two servings of meat or meat alternates; or one of each.*

Nutrition Through the Life Span: Childhood and Adolescence

Table 13–6 Healthful Snack Ideas—Think Food Groups, Alone and in Combination

Selecting two or more foods from different food groups adds variety and nutrient balance to snacks. The combinations are endless, so be creative.

Grain Products

Grain products are filling snacks, especially when combined with other foods:
- Cereal with fruit and milk.
- Crackers and cheese.
- Wheat toast with peanut butter.
- Popcorn with grated cheese.
- Oatmeal raisin cookies with milk.

Vegetables

Cut-up fresh, raw vegetables make great snacks alone or in combination with foods from other food groups:
- Celery with peanut butter.
- Broccoli, cauliflower, and carrot sticks with a flavored yogurt or cottage cheese dip.

Fruits

Fruits are delicious snacks and can be eaten alone—fresh, dried, or juiced—or combined with other foods:
- Apples and cheese.
- Bananas and peanut butter.
- Peaches with yogurt.
- Raisins mixed with sunflower seeds or nuts.

Meats and Meat Alternates

Meat and meat alternates add protein to snacks:
- Refried beans with nachos and cheese.
- Tuna on crackers.
- Luncheon meat on wheat bread.

Milk and Milk Products

Milk can be used as a beverage with any snack, and many other milk products, such as yogurt and cheese, can be eaten alone or with other foods as listed above.

School Lunch More than 25 million children receive lunches through the National School Lunch Program—half of them free or at a reduced price. School lunches are designed to provide at least a third of the recommendation for energy, protein, vitamin A, vitamin C, iron, and calcium. They must also include specified numbers of servings from each food group. Table 13–7 shows school lunch patterns for children of different ages.

Parents often rely on school lunches to meet a significant part of their children's nutrient needs on school days. Indeed, students who regularly eat school lunches have higher intakes of energy and nutrients than students who do not. Children don't always like what they are served, however, and school lunch programs must strike a balance between what children want to eat and what will nourish them and guard their health.

Many schoolchildren in the United States have significant risk factors for developing cardiovascular disease.[29] In an effort to help reduce their risk, the U.S. Department of Agriculture (USDA) has ruled that all government-funded meals served at schools must follow the *Dietary Guidelines for Americans*. This admirable attitude leaves many schools with a problem, however. Though a

Table 13–7 School Lunch Patterns for Different Ages

Food Group	Preschool (Age)		Grade School through High School (Grade)[a]		
	1 to 2	3 to 4	K to 3	4 to 6	7 to 12
Milk					
1 serving of fluid milk[b]	¾ c	¾ c	1 c	1 c	1 c
Meat or meat alternate 1 serving:					
Lean meat, poultry, or fish	1 oz	1½ oz	1½ oz	2 oz	3 oz
Cheese	1 oz	1½ oz	1½ oz	2 oz	3 oz
Large egg(s)	½	¾	¾	1	1½
Cooked dry beans or peas	¼ c	⅜ c	⅜ c	½ c	¾ c
Peanut butter	2 tbs	3 tbs	3 tbs	4 tbs	6 tbs
Peanuts, soynuts, tree nuts, or seeds[c]	½ oz	¾ oz	¾ oz	1 oz	1½ oz
Vegetable and/or fruit					
2 or more servings, both to total	½ c	½ c	½ c	¾ c (plus ½ c extra over a week)	¾ c
Bread or bread alternate					
Servings[d]	5 per week (minimum ½ per day)	8 per week (minimum 1 per day)	8 per week (minimum 1 per day)	8 per week (minimum 1 per day)	10 per week (minimum 1 per day)

[a]These patterns may be used so long as the meals served meet the *Dietary Guidelines for Americans* and provide one-third of the child's recommendations for nutrients.

[b]Whole milk and unflavored low-fat milk must be offered; flavored milks or fat-free milk may also be offered.

[c]These foods may meet no more than one-half a serving of meat and must be accompanied by other meat or alternate in the meal.

[d]A serving is 1 slice of whole-grain or enriched bread; a whole-grain or enriched biscuit, roll, muffin, or the like; or ½ cup cooked rice, pasta, or other grain.

SOURCE: U.S. Department of Agriculture, 1998.

school's own cafeteria may serve such meals, private vendors offer other, unregulated meals, even fast foods, side-by-side with the school lunches. Children receive a mixed message when they are left on their own to choose between the health-supporting school lunch and the high-fat, high-salt, low–nutrient density foods that their taste buds may prefer.

In Summary Adults at home and at school need to provide children with nutrient-dense foods and teach them how to make healthful choices. Adults also need to provide ample opportunity for children to be physically active.

The Teen Years

As children pass through **adolescence** on their way to becoming adults, they change in many ways. Their physical changes make their nutrient needs high, and their emotional, intellectual, and social changes make meeting those needs a challenge.

Teenagers make many more choices for themselves than they did as children. They are not fed, they eat; they are not sent out to play, they choose to go. At the same time, social pressures thrust choices at them: whether to drink alcoholic beverages and whether to develop their bodies to meet extreme ideals of slimness or athletic prowess. Their interest in nutrition derives from personal, immediate experiences. They are concerned with how diet can improve their lives now—

The American Dietetic Association has set nutrition standards for child-care programs. Among them, meal plans should:

- *Be nutritionally adequate.*
- *Involve parents in planning.*
- *Meet the* Dietary Guidelines for Americans.
- *Follow recommended meal patterns while respecting cultural and ethnic differences.*
- *Minimize added fat, sugar, and sodium.*
- *Emphasize fresh fruit, fresh and frozen vegetables, and whole grains.*
- *Respect children's small appetites.*

SOURCE: Position of The American Dietetic Association: Nutrition standards for child-care programs, *Journal of the American Dietetic Association* 99 (1999): 981–988.

adolescence: the period of growth from the beginning of puberty until full maturity. Timing of adolescence varies from person to person.

Nutrition Through the Life Span: Childhood and Adolescence

they engage in crash dieting in order to fit into a new bathing suit, avoid greasy foods in an effort to clear acne, or eat a plate of pasta to prepare for a big sporting event. In presenting information on the nutrition and health of adolescents, this chapter includes these many topics of interest to teens.

Growth and Development during Adolescence

With the onset of adolescence, the steady growth of childhood speeds up abruptly and dramatically, and the growth patterns of females and males become distinct. Hormones direct the intensity and duration of the adolescent growth spurt, profoundly affecting every organ of the body, including the brain. After two to three years of intense growth and a few more at a slower pace, physically mature adults emerge.

In general, a female's adolescent growth spurt begins at age 10 or 11 and a male's at 12 or 13. The spurt's duration is about two and a half years. Before **puberty,** the differences between male and female body composition are minimal. During the adolescent spurt, gender differences become apparent in the skeletal system, lean body mass, and fat stores. In males, the lean body mass—muscle and bone—becomes much greater, and in females, fat becomes a larger percentage of the total body weight. On average, males grow 8 inches taller, and females, 6 inches taller. Males gain approximately 45 pounds, and females, about 35 pounds.

Growth charts used for children must be abandoned when the signs of puberty begin to appear. Age in years indicates little about development. One way to monitor teen growth is to compare height and weight with previous measures taken at intervals. Rating scales based on stages of adolescent development are available and widely used to record developmental changes during puberty.

Energy and Nutrient Needs

The energy needs of teenagers vary greatly, depending on the current rate of growth, body size, and physical activity. Boys' energy needs may be especially high; they experience a more intense growth spurt and, as mentioned, develop more lean body mass than girls do. An active teenage boy of 15 may need 4000 kcalories or more a day just to maintain his weight. In general, because girls enter their growth spurts earlier and grow less than boys, their energy needs peak sooner and decline earlier than those of their male peers. An inactive girl of 15 whose growth is nearly at a standstill may need fewer than 2000 kcalories a day if she is to avoid excessive weight gain. Thus teenage girls need to pay special attention to being physically active and selecting foods of high nutrient density so that they will meet their nutrient needs without exceeding their energy needs.

Obesity The insidious problem of obesity becomes ever more apparent in adolescence and often continues into adulthood. One in every nine teens is overweight. The problem is most evident in females, especially those of African American descent. Without intervention, overweight teens will face numerous physical and socioeconomic consequences for years to come. The consequences of obesity are so dramatic and our society's attitude toward obese people is so negative that even teens of normal weight perceive a need to control their weight. When taken to the extremes, restrictive diets bring dramatic physical consequences of their own, as Nutrition in Practice 7 explains.

Vitamins Recommendations for most vitamins increase during the teen years (see the tables on the inside front cover). Several of the vitamin recommendations for adolescents are similar to those for adults, including the new recommendation for vitamin D. During puberty, both the activation of vitamin D and

puberty: the period in life in which a person becomes physically capable of reproduction.

the absorption of calcium are enhanced, thus supporting the intense skeletal growth of the adolescent years without additional vitamin D.

Iron The need for iron during adolescence differs for males and females. Iron needs increase in girls as they start to menstruate and in boys as their lean body mass develops. Iron intakes often fail to keep pace with increasing needs, especially for adolescent girls, who typically consume less iron-rich foods such as meat and fewer total kcalories than boys. For females, the RDA rises at adolescence and remains high into middle age. For males, the RDA returns to preadolescent values in early adulthood.

Calcium Adolescence is a crucial time for bone development, and the requirement for calcium reaches its peak during these years.[30] Unfortunately, many adolescents have calcium intakes below the current recommendations of 1300 milligrams per day through age 18. Low calcium intakes during the adolescent growth spurt, especially if paired with physical inactivity, may compromise the development of peak bone mass.[31] In contrast, increasing milk products in the diet to meet calcium recommendations greatly increases bone density. The attainment of maximal bone mass is considered the best protection against age-related bone loss and fractures. Once again, teenage girls are most vulnerable, for their milk—and therefore calcium—intakes begin to decline at the time when their calcium needs are greatest. Furthermore, women experience much greater bone losses than men in later life. In addition to dietary calcium, sports activities during adolescence build strong bones.

Nutritious snacks play an important role in an active teen's diet.

Food Choices and Health Habits

Teenagers like the freedom to come and go as they choose and eat what they want when they have time. With a multitude of afterschool, social, and job activities, they almost inevitably fall into irregular eating habits. Teens may begin to skip breakfast and to choose less milk, fruits and juices, and vegetables, and many more soft drinks each day.[32] Consequently, teenagers often miss out on the nutrients they need.

Wise gatekeepers provide access to nutritious foods that are low in sugar and fat. They make sure that their teenage sons and daughters and friends find plenty of nutritious, easy-to-grab food in the refrigerator (meats and low-fat cheeses for sandwiches, raw vegetables, milk, fruit, and fruit juices) and more in the pantry (breads, peanut butter, nuts, popcorn, cereals).

Snacks Snacks typically provide at least a fourth of the average teenager's daily food energy intake. Most often, favorite snacks are high in fat and sodium and low in calcium, iron, vitamin A, vitamin C, and folate. Most adolescents need to eat a greater variety of foods to obtain these nutrients. Table 13-6 on p. 346 shows how to combine foods from different food groups to create healthy snacks. Unfortunately, vending machines rarely offer nutrient-dense options, and nutrition information alone does not convince people to make healthy choices.

Beverages Teenagers frequently drink soft drinks with lunch, supper, and snacks (see Table 13-8 on p. 350). About the only time they select fruit juices is at breakfast. When they drink milk, they are more likely to consume it with a meal (especially breakfast) than as a snack. Soft drinks, when chosen as the primary beverage, may affect the density of the bones because they displace milk from the diet.[33] Because of their greater food intakes, boys are more likely to drink enough milk to meet their calcium needs, whereas girls typically fall short of calcium recommendations.

Soft drinks present another problem, too, when caffeine intake becomes excessive. Caffeine is a stimulant added during the manufacture of many soft drinks; on the average, caffeine-containing soft drinks deliver between 30 and 55

Table 13–8 Soft Drink Consumption by U.S. Adolescents

In general, soft drink consumption is inversely associated with the consumption of the nutrients in milk and fruit juices.

This Percentage of Adolescents:	Consumes This Many Soft Drinks Each Day:
22%	More than 26 oz (more than 3¼ c or 2 cans)
28%	13 to 26 oz (1¾ c to 3¼ c or 2 cans)
32%	Up to 13 oz (about 1¾ c or 1 can)
18%	None

SOURCE: Data from L. Harnack, J. Stang, and M. Story, Soft drink consumption among U.S. children and adolescents: Nutritional consequences, *Journal of the American Dietetic Association* 99 (1999): 436–441.

Appendix A provides a table of the caffeine contents of beverages, foods, and medications.

The dangers of steroid use are presented in Nutrition in Practice 10.

milligrams of caffeine per 12-ounce can. Caffeine increases the respiration rate, heart rate, blood pressure, and secretion of stress and other hormones. Caffeine seems to be relatively harmless, however, when used in moderate doses (the equivalent of fewer than, say, three 12-ounce cola beverages a day). In greater amounts, it can cause the symptoms associated with anxiety—sweating, tenseness, and inability to concentrate.

Eating Away from Home Adolescents eat about one-third of their meals away from home, and their nutritional welfare is enhanced or hindered by the choices they make.[34] A lunch of a hamburger, a chocolate shake, and french fries supplies substantial quantities of many nutrients at a kcalorie cost of about 800, an energy intake many adolescents can afford. When they eat this sort of lunch, teens can adjust their breakfast and dinner choices to include fruits and vegetables for vitamin A, vitamin C, folate, and fiber, and lean meats for iron and zinc.

Peer Influence Teenagers are intensely engaged in day-to-day life with their peers and preparing for their future lives as adults. Adults need to remember that teenagers have the right to make their own decisions—even if those decisions are not in line with the adults' own views. Gatekeepers can set up the environment so that nutritious foods are available and can stand by with reliable nutrition information and advice, but the rest is up to the teenagers. Ultimately, they make the choices.

Problems Adolescents Face

Physical maturity and growing independence present adolescents with new choices to make. The consequences of those choices will influence their nutritional health both today and throughout life. Some teenagers begin using drugs, alcohol, and tobacco; others wisely refrain. Information about the use of these substances is presented here because most people are first exposed to them during adolescence, but it actually applies to people of all ages.

Marijuana Almost half of the high school students in the United States report having at least tried marijuana.[35] When inhaled by smoking, the active chemicals in marijuana are rapidly and almost completely absorbed from the lungs.* They then travel in the blood to the various body tissues that metabolize them. The active ingredients from a single marijuana cigarette can linger in the body's fat a month or more before being excreted in urine.

Marijuana is unique among drugs in that it seems to enhance the enjoyment of eating, especially of sweets, a phenomenon commonly known as "the munchies." Why or how this effect occurs is not known; it may be a social effect induced by suggestibility, or perhaps the drug stimulates appetite. Whatever the reason, prolonged use of the drug does not seem to bring about weight gain.

Cocaine One in 12 high school seniors reports having used cocaine at least once.[36] Cocaine stimulates the nervous system and elicits the stress response—constricted blood vessels, raised blood pressure, widened pupils of the eyes, and increased body temperature. It also drives away feelings of fatigue. Cocaine occasionally causes immediate death—usually by heart attack, stroke, or seizure in an already damaged body system.

Weight loss is common, and cocaine abusers often develop eating disorders. Notably, the craving for cocaine replaces hunger; rats given unlimited cocaine will choose it over food until they starve to death. Thus, unlike marijuana use, cocaine use has major nutritional consequences.

*The active ingredient of marijuana, which is primarily responsible for its intoxicating effects, is delta-9-tetrahydrocannabinol, or THC.

Ecstasy The **club drug** known as **ecstasy** has become alarmingly popular in recent years. One in 12 high school seniors has tried ecstasy at least once. Ecstasy can damage brain cells, impair memory, increase the heart rate, and dangerously raise body temperature. Furthermore, uncertainties about the sources of ecstasy and other drugs or contaminants added during manufacture exacerbate the dangers of ecstasy use.[37] People who use ecstasy regularly lose weight for many of the same reasons listed in the margin.

Drug Abuse, in General The effects of other addictive drugs vary in degree but are similar to those caused by cocaine. Drug abusers face the multiple nutrition problems listed in the margin. During withdrawal from drugs, an important part of treatment is to identify and correct these nutrition problems.

Alcohol Abuse Sooner or later all teenagers face the decision whether to drink alcohol. The law forbids the sale of alcohol to people under 21, but most adolescents who seek alcohol can obtain it. Four out of five high school students have had at least one alcoholic beverage; about half drink regularly; and one in three students drinks heavily (defined as five or more drinks on at least one occasion in the previous month).[38]

Nutrition in Practice 23 describes how alcohol affects nutrition status. To sum it up, alcohol provides energy but no nutrients, and it can displace nutritious foods from the diet. Alcohol alters nutrient absorption and metabolism, so imbalances develop.

Smoking The prevalence of cigarette smoking among U.S. adolescents is on the decline, but every day, 6000 young people start smoking.[39] Cigarette smoking is a pervasive health problem causing thousands of people to suffer from cancer and diseases of the cardiovascular, digestive, and respiratory systems. These effects are beyond the scope of nutrition, but smoking cigarettes does influence hunger, body weight, and nutrient status.[40]

Smoking a cigarette eases feelings of hunger. When smokers receive a hunger signal, they can quiet it with cigarettes instead of food. Such behavior ignores body signals and postpones energy and nutrient intake. In rats, nicotine reduces food intake and increases the rate of energy expenditure, causing weight loss.

Indeed, smokers tend to weigh less than nonsmokers and to gain weight when they stop smoking. Weight gain is often a concern for people contemplating giving up cigarettes. They should know that the average person who quits smoking gains less than 10 pounds. Smokers wanting to quit need to prepare for this possibility and adjust their diet and activity habits so as to maintain weight during and after quitting. Smoking cessation programs need to include strategies for weight management.

Nutrient intakes of smokers and nonsmokers differ. Smokers tend to have lower intakes of dietary fiber, vitamin A, beta-carotene, folate, and vitamin C. The association between smoking and low vitamin C intake may be noteworthy, considering the altered metabolism of vitamin C in smokers and the protective effect of foods rich in this vitamin against some types of cancer. Research shows that compared to nonsmokers, smokers require almost twice as much vitamin C to maintain steady body pools. Oxidants in cigarette smoke accelerate vitamin C metabolism and deplete smokers' body stores of this antioxidant.[41] This depletion is even evident to some degree in nonsmokers who are exposed to passive smoke.

Beta-carotene enhances the immune response and protects against some cancer activity. Specifically, the risk of lung cancer is greatest for smokers who have the lowest intakes of carotene. Of course, such evidence should not be misinterpreted. It does not mean that as long as people eat their carrots, they can safely use tobacco. Nor does it mean that beta-carotene supplements would be beneficial. As mentioned in Chapter 8, some research shows beta-carotene supplements may have adverse effects in smokers.[42] Smokers were ten times more likely to get

Nutrition problems of drug abusers:
- *They buy drugs with money that could be spent on food.*
- *They lose interest in food during "highs."*
- *They use drugs that depress appetite.*
- *Their lifestyle fails to promote good eating habits.*
- *They use intravenous (IV) drugs. They may contract AIDS, hepatitis, or other infectious diseases, which increase their nutrient needs. Hepatitis also causes taste changes and loss of appetite.*
- *Medicines used to treat drug abuse may alter nutrition status.*

The vitamin C recommendation for people who regularly smoke cigarettes is an additional 35 mg/day.

club drug: any of a wide variety of drugs used by young adults at all-night dance parties such as "raves."

ecstasy: a street or slang term used for the drug methylenedioxymethamphetamine (MDMA). Ecstasy is chemically similar to amphetamine (speed) and mescaline (a hallucinogen).

Nutrition Through the Life Span: Childhood and Adolescence

lung cancer than nonsmokers. Both smokers and nonsmokers can, however, reduce their cancer risks by eating fruits and vegetables rich in carotene.

Smokeless Tobacco Nationwide, one in ten high school students reports having used smokeless tobacco products.[43] Like cigarettes, smokeless tobacco use is linked to many health problems, from minor mouth sores to tumors in the nasal cavities, cheeks, gums, and throat. The risk of mouth and throat cancers is even greater than for smoking tobacco. Other drawbacks to tobacco chewing and snuff dipping include bad breath, stained teeth, and blunted senses of smell and taste. Tobacco chewing also damages the gums, tooth surfaces, and jawbones, making it likely that users will lose their teeth in later life.

In Summary Nutrient needs rise dramatically as children enter the rapid growth phase of the teen years. The busy lifestyles of teenagers add to the challenge of meeting their nutrient needs—especially for iron and calcium. In addition to making wise food choices, adolescents need to refrain from using substances that will impair their health—including illicit drugs, alcohol, and tobacco.

Chapter 16 offers details about nutrition assessment.

Assessment of nutrition status in healthy children and teenagers can confirm that development is normal or can catch potential problems early. The Nutrition Assessment Checklist highlights problems to look for when working with children and teenagers.

Nutrition Assessment Checklist

FOR CHILDREN AND ADOLESCENTS

Health Problems, Signs, and Symptoms

Check the medical record for:

- ☐ Food allergies
- ☐ Attention-deficit/hyperactivity disorder (ADHD)
- ☐ Diabetes or other chronic disorders
- ☐ Eating disorders
- ☐ Lactose intolerance
- ☐ Alcohol, tobacco, or illicit drug abuse
- ☐ Obesity
- ☐ Pregnancy

Medications

For children or adolescents being treated with drug therapy for medical conditions, note:

- ☐ Side effects that might reduce food intake or change nutrient needs
- ☐ Proper administration of medication with respect to food intake

Food/Nutrient Intake

For all children and teens, especially those considered at risk nutritionally, assess the diet for:

- ☐ Total energy
- ☐ Protein
- ☐ Calcium and iron
- ☐ Vitamin A, vitamin C, and folate
- ☐ Fiber

Note the following:

- ☐ Number of days each week a nutritious breakfast is eaten
- ☐ Number of hours the child or teen sleeps each day
- ☐ Number of soft drinks the child or teen drinks each day
- ☐ Number of fast-food meals eaten each day
- ☐ Number and type of snacks eaten each day
- ☐ Type and amount of physical activity
- ☐ Amount of caffeine consumed

Height and Weight

Measure baseline height and weight.

- ☐ Reassess height, weight, and growth patterns at each medical checkup.
- ☐ Note significant obesity or underweight and intervention strategies employed.

Laboratory Tests

Monitor the following laboratory tests for children and adolescents:

- ☐ Hemoglobin, hematocrit, or other tests of iron status
- ☐ Blood glucose for children or teens with diabetes
- ☐ Blood lead concentrations

Physical Signs

Look for physical signs of:

- ☐ Protein-energy malnutrition
- ☐ Iron deficiency
- ☐ Vitamin A deficiency
- ☐ Vitamin C deficiency
- ☐ Folate deficiency

Self Check

1. The Food Guide Pyramid for young children differs from the adult Food Guide Pyramid in that:
   a. there are more food groups to choose from.
   b. there are fewer food groups to choose from.
   c. a set number of servings, rather than a range, is suggested.
   d. a range of servings, rather than a set number, is suggested.

2. Children who are hungry may be irritable or apathetic because:
   a. they need vitamin D.
   b. their blood glucose is low.
   c. their blood glucose is high.
   d. they have had too many sweets.

3. Two infant feeding practices that have improved the iron status of older children in the United States are:
   a. feeding infants liver and starches.
   b. breastfeeding and feeding iron-fortified formulas.
   c. giving iron supplements and increasing milk intake.
   d. feeding solids at an early age and giving iron-fortified milk.

4. Three symptoms of lead toxicity are:
   a. diarrhea, irritability, and fatigue.
   b. low blood sugar, hair loss, and skin rash.
   c. increased heart rate, hyperactivity, and dry skin.
   d. bleeding gums, brittle fingernails, and swollen glands.

5. Allergic reactions to foods are most often caused by:
   a. corn, rice, or meats.
   b. eggs, peanuts, or milk.
   c. red meats, milk, or MSG.
   d. seafood, dark greens, or lactose.

6. Which of the following is **not** true? Children who watch a lot of television are likely to:
   a. become obese.
   b. spend less time being physically active.
   c. learn healthy eating tips from programs.
   d. eat the foods most often advertised on television.

7. When introducing new foods to children:
   a. reward children as they try new foods.
   b. offer many choices to encourage variety.
   c. offer one new food at the end of the meal.
   d. offer one new food at the beginning of the meal.

8. During the growth spurt of adolescence:
   a. females gain more weight than males.
   b. males gain more fat, proportionately, than females.
   c. differences in body composition between males and females become apparent.
   d. similarities in body composition between males and females become apparent.

9. Two nutrients that are usually lacking in adolescents' diets are:
   a. zinc and fat.
   b. iron and calcium.

   c. protein and thiamin.
   d. vitamin A and riboflavin.

10. Smoking increases the need for:
    a. iron.   b. folate.
    c. vitamin C.   d. vitamin E.

Critical Thinking

Gina is usually a busy, playful three-year-old who loves to learn new tasks and catches on quickly. Lately, however, she is tired and irritable much of the time, is uninterested in new challenges, and cannot pay attention to learning something new for more than a few minutes at a time. Gina's mother wonders if her daughter's diet may have something to do with her changed behavior. As an infant, Gina ate the foods she was offered—her baby cereal, breast milk, vegetables, fruits, and table foods as they were introduced. In the last year and a half, however, it has become apparent that the toddler's favorite foods are milk and milk products such as cheese and yogurt. When she is hungry, this is what she wants to eat, and her mother, knowing that milk is a healthy food, obliges her. In fact, Gina drinks milk with every meal and snack, and frequently, her snacks are cheese bites with some fruit or crackers. Gina's mother knows she must offer more foods from each of the other food groups and fewer from the milk group, but which of the following foods would she be wise to emphasize immediately in hopes of improving Gina's behavior?
   a. roast beef, tuna fish, and baked beans
   b. orange juice, pineapple, and cantaloupe
   c. sweet potatoes, carrots, and broccoli
   d. bananas, yellow squash, and mashed potatoes

Answers to these questions appear in Appendix G.

Clinical Applications

1. At two and a half years old, Travis is healthy, though slightly underweight, and headstrong. Travis's mother hovers over him at every meal and insists that he take several bites of every food on his plate, even if he dislikes the food or it is unfamiliar to him. Even though Travis is hungry when he sits down to a meal, his mother's constant urging to get him to take bites quickly quells any interest the child had in eating. Travis simply folds his arms across his chest, closes his mouth tightly, and refuses to eat any more food. After more begging, pleading, and nagging, Travis's mother becomes angry and sends him away from the table. Travis is not allowed to snack between meals because his mother is concerned that snacks will ruin his appetite.

   ■ What factors might be contributing to Travis's refusal to eat?

   ■ Travis's mother is concerned about her son's underweight. What strategies would you suggest to help Travis gain weight?

   ■ What advice would you offer Travis's mother to help her improve mealtimes with her son?

2. Loni is a physically inactive, slightly overweight 15-year-old who started smoking cigarettes at the age of 13 to help her lose weight. She planned to quit smoking as soon as she dropped a few pounds. Smoking reduced her appetite somewhat, but when she did eat, she chose high-fat snack foods, cola drinks, and fast foods such as chicken nuggets and burgers. At 15, Loni is still overweight and is smoking more than a pack of cigarettes a day. She uses her lunch money to fund her cigarette habit, which leaves only a little change for crackers or chips from the vending machine for lunch. She's noticed that her complexion looks dry and off-color, her teeth are less white than they used to be, she's easily fatigued, and she gets sick more often than ever before.

   ■ What would you tell Loni to motivate her to stop smoking?

   ■ What nutrients might be affected by Loni's smoking, and how might her diet be adjusted to meet these needs?

   ■ What dietary advice would you suggest to help Loni look and feel healthier?

Nutrition on the Net

For further study of the topics of this chapter, access these websites.

Find updates and quick links to these and other nutrition-related sites at our website:
www.wadsworth.com/nutrition

Learn how to care for children and adolescents from the American Academy of Pediatrics and the Canadian Paediatric Society:
www.aap.org and **www.cps.ca**

Download the current growth charts and learn about their most recent revision:
www.cdc.gov/growthcharts

Get information on the Food Guide Pyramid for young children from the USDA:
www.usda.gov/cnpp

Get tips for feeding children from the American Dietetic Association, the I Am Your Child Program, and the Kids Food Cyber Club:
www.eatright.org, www.iamyourchild.org, and **www.kidsfood.org**

Get tips for keeping children healthy from the Nemours Foundation:
www.kidshealth.org

Visit the National Center for Education in Maternal & Child Health and the National Institute of Child Health and Development:
www.ncemch.org and **www.nih.gov/nichd**

Learn about the Child Nutrition Programs:
www.fns.usda.gov/fns

Learn how to reduce lead exposure in your home from the U.S. Department of Housing and Urban Development's Office of Lead Hazard Control:
www.hud.gov/lead

Visit the Lead Program of the Centers for Disease Control:
www.cdc.gov/nceh/programs/lead

Learn more about food allergies from the American Academy of Allergy, Asthma, and Immunology; the Food Allergy Network; and the International Food Information Council:
www.aaaai.org, www.foodallergy.org, and ificinfo.health.org

Learn more about hyperactivity from Children and Adults with Attention Deficit Disorders:
www.chadd.org

Visit the Public Health Service's Office of Women's Health site for messages on positive self-images, good nutrition, and fitness for girls between the ages of 9 and 14:
www.health.org/gpower/girlarea/bodywise

Visit the Milk Matters section of the National Institute of Child Health and Development (NICHD):
www.nichd.nih.gov

Learn more about caffeine from the International Food Information Council:
ificinfo.health.org

Get weight-loss tips for children and adolescents:
www.shapedown.com

Read the message for parents and teens on the risks of tobacco use from the American Academy of Pediatrics:
www.aap.org

Visit the Tobacco Information and Prevention Source (TIPS) of the Centers for Disease Control and Prevention:
www.cdc.gov/tobacco

Notes

1. Center for Nutrition Policy and Promotion, Facts about childhood obesity and overweightness, *Family Economics and Nutrition Review* 12 (1999): 52–53.

2. M. I. Goran and coauthors, Development changes in energy expenditure and physical activity in children: Evidence for a decline in physical activity in girls before puberty, *Pediatrics* 101 (1998): 887–891.

3. Committee on Nutrition, American Academy of Pediatrics, Calcium requirements of infants, children, and adolescents, *Pediatrics* 104 (1999): 1152–1157.

4. S. B. Roberts and M. B. Heyman, Micronutrient shortfalls in young children's diets: Common, and owing to inadequate intakes both at home and at child care centers, *Nutrition Reviews* 58 (2000): 27–29.

5. Federal Interagency Forum on Child and Family Statistics, *America's Children: Key National Indicators of Well-Being, 2001* (Washington, D.C.: Government Printing Office, 2001), available at **www.childstats.gov.**

6. A. F. Suber and coauthors, Dietary sources of nutrients among US children, 1989–1991, *Pediatrics* 102 (1998): 913–923; C. H. Ruxton and T. R. Kirk, Breakfast: A review of associations with measures of dietary intake, physiology, and biochemistry, *British Journal of Nutrition* 78 (1997): 199–214.

7. J. M. Murphy and coauthors, Relationship between hunger and psychosocial functioning in low-income American children, *Journal of the American Academy of Child and Adolescent Psychiatry* 37 (1998): 163–170.

8. J. S. Halterman and coauthors, Iron deficiency and cognitive achievement among school-aged children and adolescents in the United States, *Pediatrics* 107 (2001): 1381–1386.

9. E. Pollitt, Iron deficiency and educational deficiency, *Nutrition Reviews* 55 (1997): 133–141.

10. M. C. Holst, Developmental and behavioral effects of iron deficiency anemia in infants, *Nutrition Today* 33 (1998): 27–36.

11. S. J. Schoenthaler and I. D. Bier, The effect of vitamin-mineral supplementation on juvenile delinquency among American schoolchildren: A randomized, double-blind placebo-controlled trial, *Journal of Alternative and Complementary Medicine* 6 (2000): 7–17.

12. R. A. Goyer, Toxic and essential metal interactions, *Annual Review of Nutrition* 17 (1997): 37–50; P. Mushak and A. F. Crocetti, Lead and nutrition: Biologic interactions of lead with nutrients, *Nutrition Today* 31 (1996): 12–17.

13. T. D. Matte, Reducing blood levels: Benefits and strategies, *Journal of the American Medical Association* 281 (1999): 2340–2342; Mushak and Crocetti, 1996.

14. Matte, 1999; D. Farley, Dangers of lead still linger, *FDA Consumer,* January/February 1998, pp. 16–21.

15. Federal Interagency Forum on Child and Family Statistics, *America's Children: Key National Indicators of Well-Being, 1999*

(Washington, D.C.: Government Printing Office, 1999). Yearly updates available from **www.childstats.gov.**

16. R. Formanek, Food allergies: When food becomes the enemy, *FDA Consumer,* July/August 2001, pp. 10–16; J. M. Yeung, R. S. Applebaum, and R. Hildwine, Criteria to determine food allergen priority, *Journal of Food Protection* 63 (2000): 982–986.

17. H. Skolnick and coauthors, The natural history of peanut allergy, *Journal of Allergy and Clinical Immunology* 107 (2001): 367–374; Formanek, 2001.

18. K. J. Falci, K. L. Gombas, and E. L. Elliot, Food Allergen awareness: An FDA priority, *Food Safety Magazine,* February/March 2001, available at **www.cfsan.fda.gov/~dms/.**

19. K. Deibel and coauthors, A comprehensive approach to reducing the risk of allergens in foods, *Journal of Food Protection* 60 (1997): 436–441.

20. Committee on Quality Improvement, Subcommittee on Attention-Deficit/Hyperactivity Disorder, American Academy of Pediatrics, Clinical practice guideline: Diagnosis and evaluation of the child with attention-deficit/hyperactivity disorder, *Pediatrics* 105 (2000): 1158–1170.

21. J. W. White and M. Wolraich, Effect of sugar and mental performance, *American Journal of Clinical Nutrition* 62 (1995): 242S–249S; M. L. Wolraich, D. B. Wilson, and J. W. White, The effect of sugar on the behavior or cognition in children, *Journal of the American Medical Association* 274 (1995): 1617–1621.

22. L. L. Birch, Development of food preferences, *Annual Review of Nutrition* 19 (1999): 41–62.

23. J. O. Fisher and L. L. Birch, Restricting access to palatable foods affects children's behavioral response, food selection, and intake, *American Journal of Clinical Nutrition* 69 (1999): 1264–1272.

24. C. Evers, Empower children to healthful eating habits, *Journal of the American Dietetic Association* 97 (1997): S116.

25. Birch, 1999.

26. K. A. Coon and coauthors, Relationships between use of television during meals and children's food consumption patterns, *Pediatrics* 107 (2001): 167 [**www.pediatrics.org/cgi/content/full/107/l/37**]; C. J. Crespo and coauthors, Television watching, energy intake, and obesity in U.S. children: Results from the Third National Health and Nutrition Examination Survey, 1988–1994, *Archives of Pediatrics and Adolescent Medicine* 155 (2001): 360–365.

27. M. Nahikian-Nelms, Influential factors of caregiver behavior at mealtime: A study of child-care programs, *Journal of the American Dietetic Association* 97 (1997): 505–509.

28. Position of The American Dietetic Association: Nutrition standards for child-care programs, *Journal of the American Dietetic Association* 99 (1999): 981–988.

29. J. Anding and coauthors, Blood lipids, cardiovascular fitness, obesity, and blood pressure: The presence of potential coronary heart disease risk factors in adolescents, *Journal of the American Dietetic Association* 96 (1996): 238–242.

30. A. D. Martin and coauthors, Bone mineral and calcium accretion during puberty, *American Journal of Clinical Nutrition* 66 (1997): 611–615.

31. V. C. Lysen and R. Walker, Osteoporosis risk factors in eighth grade students, *Journal of School Health* 67 (1997): 317–321.

32. L. A. Lytle and coauthors, How do children's eating patterns and food choices change over time? Results from a cohort study, *American Journal of Health Promotion* 14 (2000): 222–228.

33. B. A. Spear, Adolescent growth and development, Journal of the American *Dietetic Association* 102 (2002): S23–S29.

34. B. H. Lin, J. Guthrie, and J. R. Blaylock, *The Diets of America's Children—Influences of Dining Out, Household Characteristics, and Nutrition Knowledge* (Washington, D.C.: U.S. Department of Agriculture, December 1996).

35. L. Kann and coauthors, Youth risk behavior surveillance— United States, 1997, *Journal of School Health* 68 (1998): 355–369.

36. Kann and coauthors, 1998.

37. National Institute on Drug Abuse, Club drugs: Community drug alert bulletin, available at www.clubdrugs.org, site visited on February 26, 2002.

38. Kann and coauthors, 1998.

39. Committee on Substance Abuse, American Academy of Pediatrics, Tobacco's toll: Implications for the pediatrician, *Pediatrics* 107 (2001): 794–798; Centers for Disease Control and Prevention, Trends in cigarette smoking among high school students—United States, 1991–1999, *Journal of the American Medical Association* 284 (2000): 1507–1508.

40. J. S. Hampl and N. M. Betts, Cigarette use during adolescence: Effects on nutritional status, *Nutrition Reviews* 57 (1999): 215–221.

41. J. Lykkesfeldt and coauthors, Ascorbate is depleted by smoking and repleted by moderate supplementation: A study in male smokers and nonsmokers with matched dietary antioxidant intakes, *American Journal of Clinical Nutrition* 71 (2000): 530–536.

42. K. Smigel, Beta-carotene fails to prevent cancer in two major studies: CARET intervention stopped, *Journal of the National Cancer Institute* 88 (1996): 145; G. S. Omenn and coauthors, Effects of a combination of beta carotene and vitamin A on lung cancer and cardiovascular disease, *New England Journal of Medicine* 334 (1996): 1150–1155.

43. Kann and coauthors, 1998.

Nutrition in Practice

■ CHILDHOOD OBESITY AND THE EARLY DEVELOPMENT OF CHRONIC DISEASES ■

Disease of the heart and blood vessels, known as **cardiovascular disease,** or **CVD,** is the number one killer of adults in the United States and Canada, but CVD begins in childhood. Over the past three decades, researchers have been observing how changes in body weight, blood lipids, blood pressure, and individual behaviors correlate with the development of CVD over time—from infancy to childhood through adolescence and into young adulthood. Some major findings have emerged from this research:

- Changes inside the arteries—changes predictive of CVD—are evident in childhood.

- Obesity in children affects these changes.

- Behaviors that influence the development of obesity and of CVD are learned and begin early in life. These behaviors include overeating, eating high-fat foods, physical inactivity, and cigarette smoking.

This Nutrition in Practice focuses on efforts to prevent childhood obesity and CVD, but the benefits extend to cancer, diabetes, and other chronic diseases as well. Unfortunately, type 2 diabetes, a chronic disease closely linked with obesity, has been on the rise among children as the prevalence of childhood obesity in the United States has increased.[1] The years of childhood are emphasized here, for the earlier in life health-promoting habits become established, the better they will stick.

What about genetics? Don't some people inherit the tendency to develop CVD regardless of the lifestyle habits they adopt?

Genetics does not appear to play a *determining* role in CVD; that is, a person is not simply destined at birth to develop CVD. Instead, genetics appears to play a *permissive* role—the potential is inherited, and then CVD will develop, if given a push by poor health choices such as excessive weight gain, poor diet, sedentary lifestyle, and cigarette smoking.

How does CVD develop, and when does its development begin?

Most CVD involves **atherosclerosis**—the accumulation of cholesterol and other blood lipids along the walls of the arteries (see the glossary on p. 358 for atherosclerosis and related terms). Frequently, atherosclerosis alters the flow of blood to the heart and can lead to hypertension and coronary heart disease (CHD), which, in turn, raises the likelihood of a heart attack. When atherosclerosis alters blood flow to the brain, a stroke can result. Infants are born with healthy, smooth, clear arteries, but within the first decade of life, **fatty streaks** may begin to appear. During adolescence, these fatty streaks may begin to turn to **fibrous plaques** (Figure 25–1 in Chapter 25 shows the formation of plaques in atherosclerosis). By early adulthood, the fibrous plaques may begin to calcify and become raised lesions, especially in boys and young men. As the lesions grow more numerous and enlarge, the heart disease rate begins to rise, and the rise becomes dramatic at about age 45 in men and 55 in women. From this point on, arterial damage and blockage progress rapidly, and heart attacks and strokes threaten life. In short, the consequences of atherosclerosis, which become apparent only in adulthood, have their beginnings in the first decades of life.[2]

Atherosclerosis is not inevitable; people can grow old with relatively clear arteries. Early lesions may either progress or regress, depending on several factors, many of which reflect lifestyle behaviors. Smoking, for example, is strongly associated with the prevalence of raised lesions, even in young adults.[3]

Parents don't need to worry about their children's blood cholesterol, do they?

Atherosclerotic lesions reflect blood cholesterol: as blood cholesterol increases, lesion coverage increases. Cholesterol values at birth are similar in all populations; differences emerge in early childhood. In countries where the adults have high blood cholesterol and high rates of CVD, the children also tend to have high blood cholesterol. Conversely, in countries where the adults have low blood cholesterol and low rates of CVD, the children tend to have low blood cholesterol, suggesting that adult heart disease tracks early and that early preventive efforts might reduce the incidence of later CVD.

Nutrition in Practice

Glossary

atherosclerosis (ath-er-oh-scler-OH-sis): a type of artery disease characterized by accumulations of lipid-containing material on the inner walls of the arteries.

athero = porridge or soft

scleros = hard

osis = condition

cardiovascular disease (CVD): a general term for all diseases of the heart and blood vessels. Atherosclerosis is the main cause of CVD. When the arteries that carry blood to the heart muscle become occluded, the heart suffers damage known as **coronary heart disease (CHD).**

cardio = heart

vascular = blood vessels

fatty streaks: accumulations of cholesterol and other lipids along the walls of the arteries.

fibrous plaques: mounds of lipid material, mixed with smooth muscle cells and calcium, which develop in the artery walls in atherosclerosis.

Such is the case among populations, but individual cholesterol status also becomes established in childhood, as early as one year. The best predictors of a person's blood cholesterol are that person's earlier baseline values: childhood values correlate with values in young adulthood. Quite simply, if you want to know a child's future cholesterol, measure it now. Standard values for cholesterol screening in children and adolescents are listed in Table NP13–1.[4]

Blood cholesterol also correlates with childhood obesity, especially central obesity.[5] In obese children, the LDL cholesterol value is often too high, while the HDL value is too low for health. These relationships are apparent throughout childhood, and their magnitude increases with age.

Is hypertension a concern for children and teenagers?

Pediatricians routinely monitor blood pressure in children and adolescents.[6] High blood pressure may signal an underlying disease or the early onset of hypertension. Hypertension accelerates the development of atherosclerosis. Standard values for hypertension screening in children and adolescents are given in Table NP13–2.[7] Like atherosclerosis and high blood cholesterol, hypertension may develop in the first decades of life, especially in obese children.[8] Children can control their hypertension by participating in regular aerobic activity and by losing weight or maintaining their weight as they grow taller. No evidence suggests a benefit of restricting sodium to lower blood pressure in children and adolescents.[9]

Is obesity a problem among children today?

Yes. Many experts agree that preventing or treating obesity in childhood will reduce the rate of CVD in adulthood. Without intervention, overweight children become overweight adolescents who become overweight adults, and being overweight exacerbates every chronic disease that adults face.[10]

Children are heavier today than they were 20 or so years ago. Since the late 1970s, the prevalence of overweight has doubled for children—and more than doubled for adolescents.[11] This pattern is a secular trend—that is, one that cannot be explained by genetics. Diet and physical activity must be responsible. Figure NP 13–1 (on p. 360) presents the BMI for children and adolescents, indicating cutoff points for overweight and obese.

Is it true that children are eating more food and more fat than ever before?

No. Children's energy intakes have remained relatively stable over the past 15 years. There has even been a slight decline in fat intake, from 38 to 33 percent of kcalories from fat daily.

Children who prefer high-fat foods, however, tend to be more overweight than their peers. Particularly noteworthy is the finding that children's fat preferences and consumption correlate with their parents' obesity as well. Such findings confirm the significant roles par-

Table NP13–1 Cholesterol Values for Children and Adolescents		
Disease Risk	Total Cholesterol (mg/dL)	LDL Cholesterol (mg/dL)
Acceptable	<170	<110
Borderline	170–199	110–129
High	≥200	≥130

NOTE: Adult values appear in Chapter 25. A deciliter (dL) is one-tenth of a liter or 100 milliliters.

Nutrition in Practice

Table NP13–2 Hypertension Standards for Children and Adolescents: Systolic over Diastolic Pressure (mm Hg)				
	6 to 9 yr	**10 to 12 yr**	**13 to 15 yr**	**16 to 18 yr**
High normal	111–121 over 70–77	117–125 over 75–81	124–135 over 77–85	127–141 over 80–91
Significant hypertension	122–129 over 70–85	126–133 over 82–89	136–143 over 86–91	142–149 over 92–97
Severe hypertension	>129 over >85	>133 over >89	>143 over >91	>149 over >97

ents play—teaching children about healthy food choices, providing children with low-fat selections, and serving as role models.

If diet is not to blame for the increasing prevalence of obesity among the young, then what is?

Most likely, children have grown more overweight because of their lack of physical activity. An inactive child can become obese and develop unhealthy blood lipids even while eating less food than an active child. Today's children are more sedentary and less physically fit than children were 20 years ago.

Watching television accounts for some 24 hours a week of sedentary behavior. Beyond these 24 hours, children spend additional sedentary time sitting at computers and playing video games. Children who watch television for more than four hours a day are most likely to be obese.[12]

Just as blood cholesterol and obesity track over the years, so does a person's level of physical activity.[13] Researchers studying almost 1000 teenagers found that over half of those who were initially described as inactive remained inactive six years later. Similarly, almost half of those who were physically active remained so. Compared with inactive teens, those who were physically active weighed less, smoked less, ate a diet lower in saturated fats, and had a better blood lipid profile. The message is clear: physical activity offers numerous health benefits, and children who are active today are most likely to be active for years to come.

What else can concerned adults do to help prevent childhood obesity?

In light of all these findings, parents and teachers are encouraged to make major efforts to prevent childhood obesity. Suggestions include the following: encourage children to eat slowly, to pause and enjoy their table companions, and to stop eating when they are full. Teach them how to select low-fat snacks and to serve themselves appropriate portions. Never force children to clean their plates.

Above all, be sensitive in teaching children nutrition principles that can help to prevent obesity. Children can easily get the idea that their worth is tied to their body weight. Some parents fail to realize that society's ideal of slimness can be perilously close to starvation, and that a child encouraged to "diet" cannot obtain the energy and nutrients required for normal growth and development. Even healthy children without eating disorders have been observed to limit their growth through "dieting." Weight gain in truly overweight children can be controlled safely without compromising growth, but should be overseen by a health care professional.

Do pediatricians check cholesterol on a routine visit?

Many children in the United States are not only overweight but also have high blood cholesterol. These children are quite likely to have parents who developed CVD early.[14] For this reason, selective screening is recommended for children and adolescents whose parents or grandparents have CVD, those whose parents have elevated blood cholesterol, and those whose family history is unavailable, especially if other risk factors are evident.[15] Since blood cholesterol in children is a good predictor of adult values, some experts recommend universal screening to identify all children with high blood cholesterol.[16] They note that many children who have high blood cholesterol would be missed under current screening criteria.

Nutrition in Practice

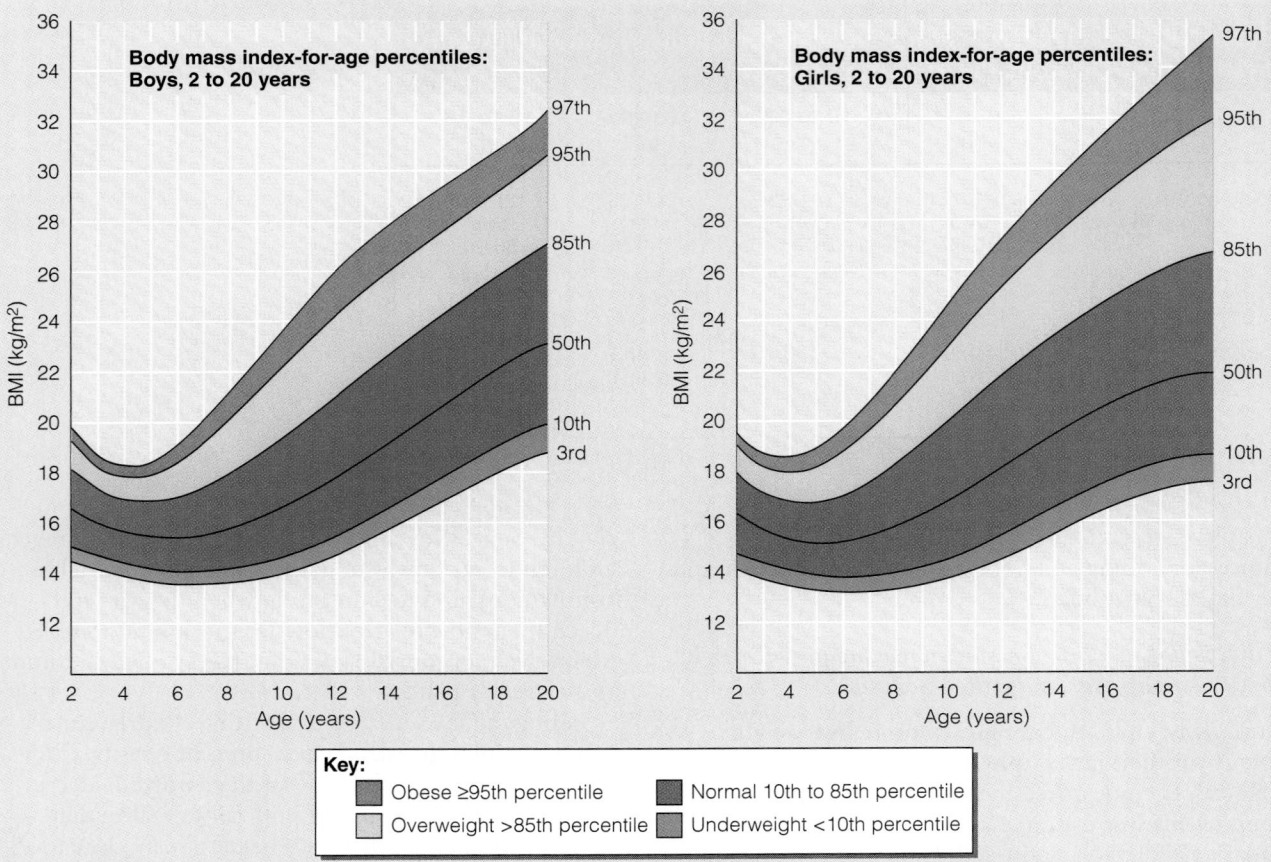

Figure NP13–1

**Body Mass Index-for-Age Percentiles:
Boys and Girls, Age 2 to 20.**

Among those children who may have high blood cholesterol, but may not meet screening criteria, are those who are overweight. The incidence of high blood cholesterol in obese children with no other criteria is similar to that in nonobese children with family histories of CVD. In addition to overweight, health care professionals should consider whether children consume a high-fat diet.[17]

Early—but not advanced—atherosclerotic lesions are reversible, making screening and education a high priority. Both those with family histories of CVD and those with multiple risk factors need intervention. Children with the highest risks of developing CVD are sedentary and obese, with high blood pressure and high blood cholesterol. In contrast, children with the lowest risks of heart disease are physically active and of normal weight, with low blood pressure and favorable lipid profiles. Routine pediatric care should identify these known risk factors and provide intervention when needed.

Are adult dietary recommendations appropriate for children?

Regardless of family history, all children over age two should eat a variety of foods and maintain desirable weight. Children should receive 20 to 30 percent of total energy from fat, less than 10 percent from saturated fat, and less than 300 milligrams of cholesterol per day.[18]

Recommendations limiting fat and cholesterol are not intended for infants or children under two years old. Infants and toddlers need a higher percentage of fat to support their rapid growth.

Healthy children over age two can begin the transition to eating according to recommendations by eating fewer high-fat foods, replacing some high-fat foods with low-fat choices, and selecting more fruits and vegetables.[19] All high-fat foods need not be eliminated, though. Healthy meals can still include moderate amounts of a child's favorite foods, even if they are

Nutrition in Practice

high-fat selections such as french fries and ice cream.[20] Without such additions, diets might be too low in fat, not to mention unappetizing and boring.

Balanced meals need to provide lean meat, poultry, fish, and vegetable sources of protein; fruits and vegetables; whole grains; and low-fat milk products. Such meals can provide enough food energy and nutrients to support growth and maintain blood cholesterol within a healthy range.

Pediatricians warn parents to avoid extremes; they caution that while intentions may be good, excessive food restriction may create nutrient deficiencies and impair growth. Furthermore, parental control over eating may instigate battles and foster attitudes about foods that can lead to inappropriate eating behaviors.

Is there anything else parents or caregivers can do to help children reduce their risks of CVD?

Even though the focus of this text is nutrition, another risk factor for CVD that starts in childhood and carries over into adulthood must also be addressed—cigarette smoking. Each day 5000 children light up for the first time—typically, in grade school. Among high school students, seven out of ten have tried smoking, and one in six smokes regularly.[21] Approximately 90 percent of all adult smokers began smoking before the age of 18.[22]

Efforts to teach children about the dangers of smoking need to be aggressive. Children are not likely to consider the long-term health consequences of tobacco use. They are more likely to be struck by the immediate health consequences, such as shortness of breath when playing sports, or social consequences, such as having bad breath. Whatever the context, the message to all children and teens should be clear: don't start smoking. If you've already started, quit.

In conclusion, adult CVD is a major pediatric problem.[23] Without intervention, some 60 million children are destined to suffer its consequences within the next 30 years. Optimal prevention efforts focus on children, especially on those who are overweight. Just as young children receive vaccinations against infectious diseases, they need screening for, and education about, CVD. Many health education programs have been implemented in schools around the country.[24] These programs are most effective when they include education in the classroom, heart-healthy meals in the cafeteria, fitness activities on the playground, and parental involvement at home.

Notes

1. American Diabetes Association, Type 2 diabetes in children and adolescents, *Pediatrics* 105 (2000): 671–680.

2. H. C. McGill and coauthors, Origin of atherosclerosis in childhood and adolescence, *American Journal of Clinical Nutrition* 72 (2000): 1307S–1315S; H. C. McGill, Childhood nutrition and adult cardiovascular disease, *Nutrition Reviews* 55 (1997): S2–S11.

3. McGill and coauthors, 2000.

4. Committee on Nutrition, American Academy of Pediatrics, Cholesterol in childhood, *Pediatrics* 101 (1998): 141–147.

5. N.-F. Chu and coauthors, Clustering of cardiovascular disease risk factors among obese schoolchildren: The Taipei Children Heart Study, *American Journal of Clinical Nutrition* 67 (1998): 1141–1146; F. J. van Lenthe and coauthors, Association of a central pattern of body fat with blood pressure and lipoproteins from adolescence into adulthood: The Amsterdam Growth and Health Study, *American Journal of Epidemiology* 147 (1998): 686–693.

6. National High Blood Pressure Education Program Working Group on Hypertension Control in Children and Adolescents, Update on the 1987 Task Force Report on High Blood Pressure in Children and Adolescents: A working group report from the National High Blood Pressure Education Program, *Pediatrics* 98 (1996): 649–658.

7. Committee on Sports Medicine and Fitness, American Academy of Pediatrics, Athletic participation by children and adolescents who have systemic hypertension, *Pediatrics* 99 (1997): 637–638.

8. A. R. Sinaiko, Hypertension in children, *New England Journal of Medicine* 335 (1996): 1968–1973.

9. B. Falkner and S. Michel, Blood pressure response to sodium in children and adolescents, *American Journal of Clinical Nutrition* 65 (1997): 618S–621S.

10. D. S. Freedman and coauthors, Relationship of childhood obesity to coronary heart disease risk factors in adulthood: The Bogalusa Heart Study, *Pediatrics* 108 (2001): 712–718; D. J. Gunnell and coauthors, Childhood obesity and adult cardiovascular mortality: A 57-y follow-up study based on the Boyd Orr cohort, *American Journal of Clinical Nutrition* 67 (1998): 1111–1118; R. C. Whitaker and coauthors, Predicting obesity in young adulthood from childhood and parental obesity, *New England Journal of Medicine* 337 (1997): 869–873.

11. Update: Prevalence of overweight among children, adolescents, and adults—United States, 1988–1994, *Morbidity and Mortality Weekly Report* 46 (1997): 199–202.

12. C. J. Crespo and coauthors, Television watching, energy intake, and obesity in U.S. children: Results from the Third National Health and Nutrition Examination Survey, 1988–1994,

Nutrition in Practice

Archives of Pediatrics and Adolescent Medicine 155 (2001): 360–365.

13. S. J. Marshall and coauthors, Tracking of health-related fitness components in youth age 9 to 12, *Medicine and Science in Sports and Exercise* 30 (1998): 910–916.

14. W. Bao and coauthors, Longitudinal changes in cardiovascular risk from childhood to young adulthood in offspring of parents with coronary artery disease: The Bogalusa Heart Study, *Journal of the American Medical Association* 278 (1997): 1749–1754.

15. Committee on Nutrition, 1998.

16. L. Van Horn and P. Greenland, Prevention of coronary artery disease is a pediatric problem, *Journal of the American Medical Association* 278 (1997): 1779–1780.

17. Committee on Nutrition, 1998.

18. Committee on Nutrition, 1998.

19. L. B. Dixon and coauthors, The effect of changes in dietary fat on the food group and nutrient intake of 4- to 10-year old children, *Pediatrics* 100 (1997): 863–872.

20. E. Satter, A moderate view on fat restriction for young children, *Journal of the American Dietetic Association* 100 (2000): 32–35.

21. Tobacco use among high school students—United States, 1997, *Morbidity and Mortality Weekly Report* 47 (1998): 229–233.

22. Tobacco use and usual source of cigarettes among high school students—United States, 1995, *Journal of the American Medical Association* 276 (1996): 184–185.

23. Van Horn and Greenland, 1997.

24. L. Calderon, Promoting a healthy lifestyle and encouraging advocacy among university and high school students, *Journal of the American Dietetic Association* 102 (2002): S71–S72; D. M. Hoelscher and coauthor, Designing effective nutrition interventions for adolescents, *Journal of the American Dietetic Association* 102 (2002): S52–S63.

Jennie Oppenheimer/Studio Zocolo

CHAPTER 14

NUTRITION THROUGH THE LIFE SPAN: LATER ADULTHOOD

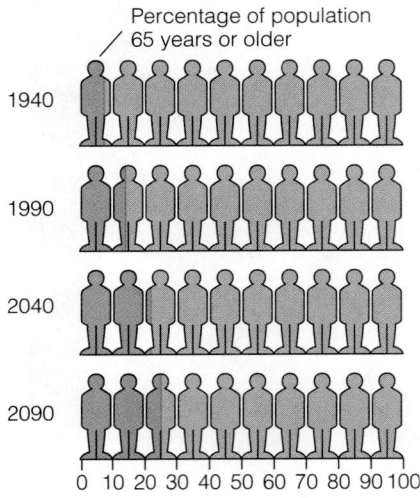

Percentage of population 65 years or older

1940

1990

2040

2090

0 10 20 30 40 50 60 70 80 90 100
Percentage of total population

Figure 14–1

The Aging of the U.S. Population
In 1940, 6.8 percent of the population was 65 or older. In 1990, 12.7 percent of us had reached age 65; by 2040, 21.7 percent will have reached age 65; and by 2090, nearly one out of four Americans will be 65 or older. An estimated 25,000 Americans now living are 100 years old or older.

*T*he last two chapters were devoted to stages of the life cycle that require special nutrition attention: pregnancy, lactation, infancy, childhood, and adolescence. Much of the text before that focused on nutrition to support wellness during adulthood. This chapter describes the special nutrition needs of the later adult years.

The most urgent nutrition need of older people, however, is to have made good food choices in the past! All of life's nutrition choices incur health consequences for the better or for the worse. A single day's intakes of nutrients may exert only a minute effect on body organs and their functions, but over years and decades, the repeated effects accumulate to have major impacts. This being the case, it is of great importance for everyone, of every age, to pay close attention today to nutrition.

The U.S. population is graying. The majority of citizens are now middle-aged, and the ratio of old people to young is increasing, as Figure 14–1 shows. Our society uses the arbitrary age of 65 to define the transition point between middle age and old age, but growing "old" happens day by day, with changes occurring gradually over time. Since 1950 the population of people over 65 has more than doubled. Remarkably, the fastest-growing age group is people over 85 years (see Figure 14–2).[1] The U.S. Bureau of the Census projects that by the year 2040 more than a million Americans will be 100 years old or older.

Life expectancy in the United States is 77 years, up from about 47 years in 1900.[2] Women today live about 6 years longer than men. Advances in medical science—antibiotics and other treatments—are largely responsible for almost doubling the life expectancy since 1900. Improved nutrition and an abundant food supply have also contributed to lengthening life expectancy. The human **life span,** currently estimated at 130 years, is the upper limit of human **longevity,** even given optimal nutrition.[3] With work progressing in medical and genetic technologies, however, the human life span may be extended significantly.[4]

The study of the aging process is among the youngest of the scientific disciplines. Not until the twentieth century did human beings achieve a life expectancy worthy of a science devoted to studying it. The idea that nutrition can influence the way the human body ages is particularly appealing because diet is a factor that people can control and change.

Nutrition and Longevity

What has been learned so far about the effects of nutrition and environment on longevity provides incentive for researchers to keep asking questions about how and why human beings age. Among their questions are:

- To what extent is aging inevitable, and can it be slowed through changes in lifestyle and environment?
- What roles does nutrition play in aging, and what roles can it play in retarding aging?

With respect to the first question, aging is a natural process, programmed into the genes at conception. People can, however, slow the process within the natural limits set by heredity. They can adopt healthy lifestyle habits such as eating nutritious food and engaging in physical activity. With respect to the second question, clearly, good nutrition can retard and ease the aging process in many significant ways.

Slowing the Aging Process

One approach researchers use to search out the secret of long life has been to study older people. Some people are young for their ages, others old for their ages. What makes the difference?

life expectancy: the average number of years lived by people in a given society.

life span: the maximum number of years of life attainable by a member of a species.

longevity: long duration of life.

Healthy Habits Six lifestyle habits seem to have a profound influence on people's health and therefore on their **physiological age:**

- Sleeping regularly and adequately.
- Eating regular meals, including breakfast.
- Keeping weight under control.
- Engaging in regular physical activity.
- Not smoking.
- Not using alcohol, or using it in moderation.

Over the years, the effects of these lifestyle choices accumulate—that is, those who follow all of these practices are in better health, even if older in **chronological age,** than people who fail to do so. In fact, the physical health of people who report all six positive health practices is comparable to that of people *30 years younger* who follow few or none. Other research confirms that these health habits both extend longevity and support independence in later life.[5] These findings suggest that even though people cannot alter the years of their births, they can alter the probable lengths and quality of their lives. Physical activity seems to be most influential in preventing or slowing the many changes that many people seem to accept as an inevitable consequence of old age. In other words, physical activity and long life seem to go together.[6]

Physical Activity The many and remarkable benefits of regular physical activity are not limited to the young: older adults who are active weigh less and have greater flexibility, more endurance, and better balance than those who are inactive.[7] They reap additional benefits as well; for example, evening activity helps to eliminate late night trips to the bathroom, moderate endurance activities improve the quality of sleep, and strength training significantly improves mobility and resistance to injury.[8] In fact, regular physical activity is a powerful predictor of a person's mobility in the later years.

Muscle mass and muscle strength tend to decline with aging, making older people vulnerable to falls and immobility. Falls are a major cause of fear, injury, disability, dependence, and even death among older adults. Regular physical activity tones, firms, and strengthens muscles, helping to improve confidence, reduce the risk of falling, and minimize the risk of injury should a fall occur. Strength training, even in frail, elderly people over 85 years of age, has been shown not only to improve muscle strength and mobility but to increase energy expenditure and energy intake, thereby enhancing nutrient intakes. This finding highlights another reason to be physically active: a person spending energy on physical activity can afford to eat more food and with it, more nutrients. People who are committed to an ongoing fitness program have higher energy and nutrient intakes than more sedentary people.

Activities of all kinds are recommended to maintain and promote health. Strength training improves muscle strength, which enhances a person's ability to perform many of life's daily tasks such as climbing stairs and carrying packages.[9] In fact, research shows that muscle strength during midlife (between the ages of 45 and 65) predicts health and disability 25 years later.[10] Among healthy, middle-aged men, those who initially had greater hand grip strength remained stronger and more physically able than their peers who had weaker hand grip strength. Hand grip strength correlates with strength of other muscles and is therefore an indicator of overall strength. The researchers speculate that greater muscle strength during midlife acts as a strength *reserve* later on, protecting older adults from disability even when chronic conditions develop. In short, improving overall strength during early and middle adulthood could potentially lower the risk of later physical disability. Aerobic activity improves cardiorespiratory endurance and lowers blood pressure and blood lipid concentrations.[11] Although aging affects both speed and endurance to some degree, older adults can still train and achieve exceptional performances.

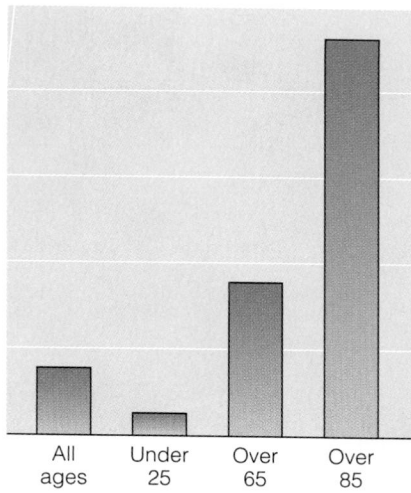

Figure 14–2

U.S. Population Growth, 1960 to Present
The "oldest old"—those over 85 years—are the fastest-growing age group in the United States. Since 1960, the population of those over 85 has more than doubled, increasing at more than five times the rate of the general population.

physiological age: a person's age as estimated from her or his body's health and probable life expectancy.

chronological age: a person's age in years from his or her date of birth.

Nutrition Through the Life Span: Later Adulthood

Strength training promotes strong muscles and bones and healthy appetites.

Ideally, physical activity should be part of each day's schedule and should be intense enough to prevent muscle atrophy and to speed up the heartbeat and respiration rate. Healthy older adults who have not been active can ease into a suitable routine. They can start by walking short distances until they can walk at least a mile three times a week, and then they can gradually increase their pace to achieve a 20- to 25-minute mile. With persistence, people can achieve great improvements at any age. Training not only benefits physical health but also increases the blood flow to the brain, thereby enhancing both mental ability and mood.[12]

Restriction of kCalories Another approach in aging research has been to manipulate animals' diets and look for effects on longevity. These studies have produced some interesting and suggestive findings. For example, rats live longer when their food intake is restricted in the early weeks of their lives, or even when it is restricted after they are mature. Research shows that it is the restriction of food energy rather than restriction of a specific nutrient that exerts the anti-aging action.[13]

Several mechanisms to explain how energy restriction prolongs life in animals have been proposed but not proved. Research suggests that food restriction may extend the life span by delaying age-related diseases and preventing damaging lipid oxidation.[14]

Experiments with food restriction and longevity in animals have *not* suggested any direct applications to human nutrition, though, and the animals given restricted feedings have suffered some distinct disadvantages. For example, the food restriction was so severe that half of the restricted animals died *very* early. The average length of life for the restricted rats was long because the few survivors lived a long time. Extreme starvation to extend life, like any extreme, is not worth the price.

Moderate energy restriction (80 percent of usual intake) in human beings may be valuable. When people restrict energy intake moderately, body weight, body fat, and blood pressure drop, and HDL cholesterol rises—favorable changes for preventing chronic diseases. Moderate energy restriction has no adverse effects on mental or physical performance.

Nutrition and Disease Prevention

Nutrition alone, even if ideal, cannot ensure a long and robust life. Nevertheless, nutrition clearly affects aging and longevity in human beings by way of its role in disease prevention. Among the better-known relationships between nutrition and disease are the following:

- Appropriate energy intake helps prevent *obesity, diabetes,* and related *cardiovascular diseases* such as atherosclerosis and hypertension (Chapters 7, 24, and 25) and may influence the development of some forms of *cancer* (Chapter 28).

- Adequate intakes of essential nutrients prevent *deficiency diseases* such as scurvy, goiter, anemia, and the like (Chapters 8 and 9).

- Variety in food intake, as well as ample intakes of certain fruits and vegetables, may be protective against certain types of *cancer* (Chapter 28).

- Moderation in sugar intake helps prevent *dental caries* (Nutrition in Practice 2).

- Appropriate fiber intakes help prevent disorders of the digestive tract such as *constipation, diverticulosis,* and possibly *colon cancer* (Chapters 2, 19, and 28).

- Moderate sodium intake and adequate intakes of potassium, calcium, and other minerals help prevent *hypertension,* at least in people who are genetically predisposed to it (Chapters 9 and 25).

- An adequate calcium intake throughout life helps protect against *osteoporosis* (Chapter 9).

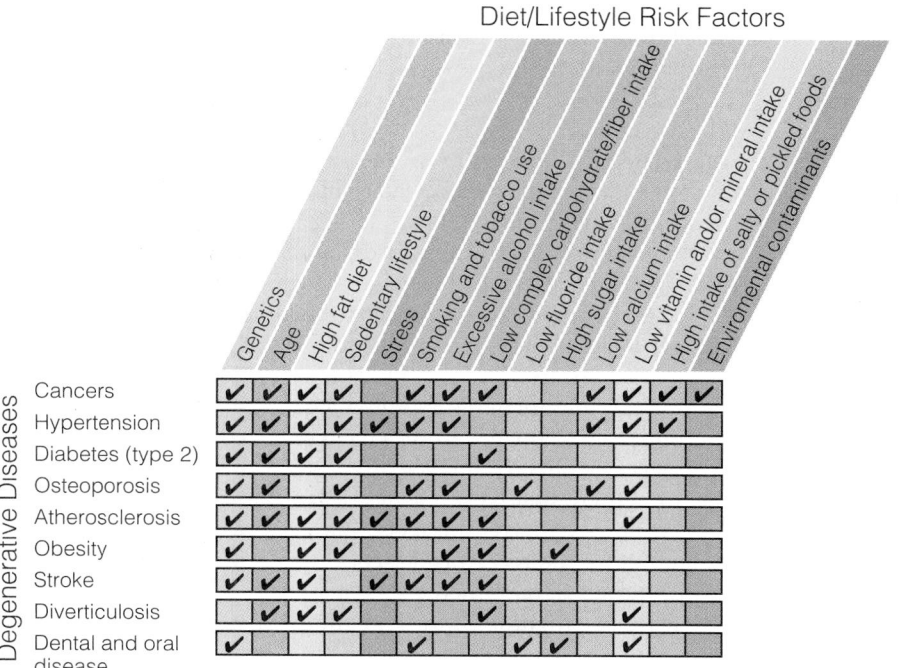

Diet/Lifestyle Risk Factors

Figure 14–3
Risk Factors and Degenerative Diseases
The chart at the top shows that the same risk factor can affect many chronic diseases. Notice, for example, how many diseases have been linked to a high-fat diet. The chart also shows that a particular disease, such as atherosclerosis, may have several risk factors.

The flow chart at the bottom shows that many of these conditions are themselves risk factors for other chronic diseases. For example, a person with diabetes is likely to develop atherosclerosis and hypertension. These two conditions, in turn, worsen each other. Notice how all these diseases are linked to obesity.

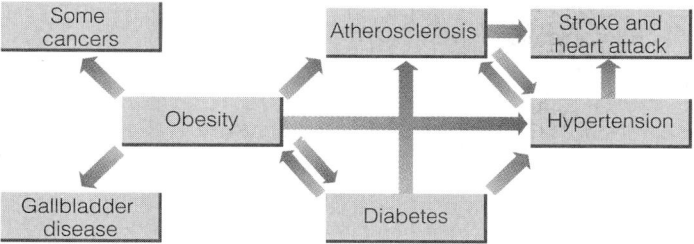

Figure 14–3 illustrates these relationships between diet and degenerative diseases. Note that the figure includes some risk factors such as heredity and age that cannot be modified, but notice how many can. Other, less well-established links between nutrition and disease are being discovered each day. Research that focuses on how life factors affect aging and disease processes is vital to ensuring that more and more people can look forward to long, healthy lives.

In Summary Life expectancy in the United States increased dramatically in the twentieth century. Factors that enhance longevity include limited or no alcohol use, regular balanced meals, weight control, adequate sleep, abstinence from smoking, and regular physical activity. Nutrition alone, even if ideal, cannot guarantee a long and robust life. At the very least, however, nutrition—especially when combined with regular physical activity—can influence aging and longevity in human beings by supporting good health and preventing disease.

Nutrition-Related Concerns during Late Adulthood

Nutrition through the prime years may play a greater role than has been realized in preventing many changes once thought to be inevitable consequences of growing older. The following discussions of cataracts, arthritis, and the aging brain show that nutrition may provide at least some protection against some of the conditions commonly associated with aging.

Cataracts

Cataracts are age-related thickenings in the lenses of the eye that impair vision. If not surgically removed, they ultimately lead to blindness. Cataracts occur even in well-nourished individuals due to ultraviolet light exposure, oxidative damage, injury, viral infections, toxic substances, and genetic disorders. Many cataracts, however, are vaguely called senile cataracts—meaning "caused by aging." In the United States, only about 5 percent of people younger than 50 years have cataracts; by age 65, the percentage jumps to over 50 percent.

Oxidative stress appears to play a significant role in the development of cataracts, and the antioxidant nutrients and phytochemicals in fruits and vegetables may help minimize the damage. Studies have reported an inverse relationship between cataracts and dietary intakes of carotenoids.[15] Specifically, foods rich in the carotenoid lutein, such as spinach, kale, and broccoli, were most consistently linked with a lower risk of cataracts. Taking supplements of vitamins C and E also seems to reduce the risk of cataracts.[16] One other diet-related factor may play a role in cataract development: overweight. In a study of more than 1700 men, those with the highest BMIs had the greatest risk of cataracts.[17] Abdominal fatness was also found to be a risk factor for cataracts. The role of obesity and abdominal fat in cataracts is not known, but researchers speculate that conditions that typically accompany obesity such as glucose intolerance and insulin resistance may provide clues.

Arthritis

The most common type of **arthritis** that disables older people is **osteoarthritis,** a painful swelling of the joints. During movement, the ends of bones are normally protected from wear by cartilage and by small sacs of fluid that lubricate the joint. With age, bones sometimes disintegrate, and the joints become malformed and painful to move. Osteoarthritis afflicts millions of people around the world, especially the elderly. Nutrition quackery to treat arthritis is abundant, but no known food or supplement prevents, relieves, or cures it. Table 14–1 presents some of the many *non*effective dietary treatments for osteoarthritis.

A known connection between osteoarthritis and nutrition is overweight. Weight loss can help overweight people with osteoarthritis, partly because the joints affected are often weight-bearing joints that are stressed and irritated by having to carry excess poundage. Interestingly, though, weight loss often relieves the worst pain of osteoarthritis in the hands as well, even though they are not weight-bearing joints. Jogging and other weight-bearing activities do not worsen osteoarthritis. In fact, both aerobic activity and weight training offer modest improvements in physical performance and pain relief.[18]

Currently, two supplements for treating osteoarthritis—glucosamine and chondroitin—are being acclaimed for their effectiveness in relieving symptoms. Indeed, the supplements seem to alleviate pain and improve mobility as well as over-the-counter pain relievers, but additional studies are needed to confirm reports.[19]

Another type of arthritis, known as **rheumatoid arthritis,** has a possible link to diet through the immune system. In rheumatoid arthritis, the immune system mistakenly attacks the bone coverings as if they were made of foreign tissue. The integrity of the immune system depends on adequate nutrition, and a poor diet may worsen arthritis. It is also possible that in some individuals, certain foods may stimulate the immune system to attack. For example, milk and milk products seem to aggravate arthritis in some people.

Another nutrient linked to rheumatoid arthritis is the omega-3 fatty acid found in fish oil, eicosapentaenoic acid (EPA). Research shows that the same diet recom-

cataracts: thickenings of the eye lenses that impair vision and can lead to blindness.

arthritis: inflammation of a joint, usually accompanied by pain, swelling, and structural changes.

osteoarthritis: a painful, chronic disease of the joints caused when the cushioning cartilage in a joint breaks down; joint structure is usually altered, with loss of function; also called *degenerative arthritis.*

rheumatoid arthritis: a disease of the immune system involving painful inflammation of the joints and related structures.

Table 14–1	Noneffective Dietary Strategies for Arthritis	
- Alfalfa tea	- Honey
- Aloe vera liquid	- Inositol
- Amino acid supplements	- Kelp
- Blackstrap molasses	- Lecithin
- Burdock root	- Para-amino benzoic acid (PABA)
- Calcium	- Raw liver
- Celery juice	- Superoxide dismutase (SOD)
- Cod liver oil	- Vitamin D
- Copper supplements	- Vitamin megadoses
- Dimethyl sulfoxide (DMSO)	- Watercress
- Fasting	- Yeast
- Fresh fruit	
- Garlic	

mended for heart health—one low in saturated fat from meats and milk products and high in oils from fish—helps prevent or reduce the inflammation in the joints that makes arthritis so painful. Researchers theorize that EPA probably interferes with the action of prostaglandins, compounds involved in inflammation.

When researchers investigate whether EPA *supplements* benefit sufferers of rheumatoid arthritis, their results are promising—fewer tender joints and less morning stiffness.[20] The researchers note that, overall, the response to the supplements is modest but nevertheless consistent.

Another possible link between nutrition and rheumatoid arthritis involves lipid peroxidation. Lipid peroxidation of the membranes within joints causes inflammation and swelling. Vitamin E helps prevent peroxidation, but it has not improved active cases of rheumatoid arthritis. This is not surprising, though, because the vitamin's role in lipid peroxidation is preventive, not restorative.

The Aging Brain

The brain, like all of the body's organs, responds to both inherited and environmental factors that can enhance or diminish its amazing capacities. One of the challenges researchers face when studying the aging of the brain in human beings is to distinguish among changes caused by normal, age-related, physiological processes; changes caused by diseases; and changes caused by cumulative, extrinsic factors such as diet.

The brain normally changes in some characteristic ways as it ages. For one thing, its blood supply decreases. For another, the number of **neurons,** the brain cells that specialize in transmitting information, diminishes as people age. When the number of nerve cells in one part of the **cerebral cortex** diminishes, hearing and speech are affected. Losses of neurons in other parts of the cortex can impair memory and cognitive function. When the number of neurons in the hindbrain diminishes, balance and posture are affected. Losses of neurons in other parts of the brain affect still other functions.

Clinicians now recognize that much of the cognitive loss and forgetfulness generally attributed to aging is due in part to extrinsic, and therefore controllable, factors such as nutrient deficiencies. In some instances, the degree of cognitive loss is extensive and attributable to a specific disorder such as a brain tumor. In cases such as Alzheimer's disease, deterioration may be genetically determined and will not yield to external approaches.

Alzheimer's Disease   In **Alzheimer's disease,** the most prevalent form of **senile dementia,** brain cell death occurs in the areas of the brain that coordinate memory and cognition. Dementia from conditions such as Alzheimer's afflicts 6 to 10

neurons: nerve cells; the structural and functional units of the nervous system. Neurons initiate and conduct nerve transmissions.

cerebral cortex: the outer surface of the cerebrum, which is the largest part of the brain.

Alzheimer's disease: a progressive, degenerative disease that attacks the brain and impairs thinking, behavior, and memory.

senile dementia: the loss of brain function beyond the normal loss of physical adeptness and memory that occurs with aging.

percent of U.S. adults 65 years of age and older—and the rate doubles when milder cases are included.[21] Diagnosis of Alzheimer's depends on its characteristic symptoms: the victim gradually loses memory and reasoning, the ability to communicate, physical capabilities, and eventually life itself.

Researchers are closing in on the cause of Alzheimer's disease. Clearly, genetic factors are involved.[22] Ultimately, researchers hope to use the genetic findings to develop early detection tests and a cure for this devastating disease.*

Treatment involves providing care to clients and support to their families. One medication, Aricept, relieves some of the memory impairment of Alzheimer's and is the most used Alzheimer's medication. Aricept helps only those with mild-to-moderate symptoms of Alzheimer's and does not slow or stop the progression of the disease.[23] Other medications may be used to control depression or behavior problems.

Most people have heard of an association between aluminum and the development of Alzheimer's, although a causal connection seems unlikely. Brain concentrations of aluminum in people with Alzheimer's exceed normal brain concentrations by some 10 to 30 times, but blood and hair aluminum remains normal, indicating that the accumulation is caused by something in the brain itself, not by an overload of aluminum in the body. Thus the high brain aluminum must be at least partly a result, rather than a cause, of Alzheimer's.

Free radicals may also be involved in Alzheimer's.[24] Researchers often find elevated levels of copper, iron, and zinc in the brain tissues of those with Alzheimer's and theorize that these metals may accelerate the progression of the disease, possibly by increasing oxidative stress.[25] Some research casts doubt on the idea that supplements of zinc or other trace minerals can worsen Alzheimer's, but other research lends support to the theory. To err on the side of safety, food sources, not concentrated supplements of trace minerals, may be advisable for people with the disease.[26]

Some preliminary but interesting research raises the possibility of a connection between elevated levels of the amino acid homocysteine and Alzheimer's.[27] As noted in Chapter 8, research shows a correlation between elevated blood levels of homocysteine and atherosclerosis.[28] Researchers are now beginning to recognize an association between atherosclerosis and Alzheimer's, which lends support to the hypothesis that elevated blood levels of homocysteine may play a role in Alzheimer's disease.[29] The possible link between elevated homocysteine and Alzheimer's, in turn, raises the question whether adequate amounts of B vitamins such as vitamin B_6, folate, and vitamin B_{12} may help prevent Alzheimer's. Increased intakes (slightly above the RDA) of vitamin B_6, folate, and vitamin B_{12} help to lower homocysteine.[30] The possibility that these B vitamins may be related to the development of Alzheimer's deserves a closer look.

Indeed, researchers have looked closer. In one study, the participants ate food from the same kitchen and shared similar environmental and lifestyle factors because they were all nuns living in the same convent. In this unique longitudinal study of 30 nuns, researchers measured the blood concentrations of nutrients, lipoproteins, total cholesterol, and other nutritional indicators.[31] Tests of brain function were also administered. The participants later died when they were 78 to 101 years of age. Upon their deaths, researchers examined their brains for signs of Alzheimer's disease such as atrophy of the cerebral cortex and number of plaques. The researchers' findings revealed a strong correlation between low blood folate and atrophy of the cortex, especially in those with large numbers of plaques. None of the other nutrients studied, including vitamin B_6 and vitamin B_{12}, correlated with atrophy of the cortex or other indicators of Alzheimer's.

In another study, researchers measured folate and vitamin B_{12} in 370 people 75 years of age or older. The participants did not have dementia. During three years

---

*A report on the genetic and other aspects of Alzheimer's is available from Alzheimer's Disease Education and Referral Center, P.O. Box 8250, Silver Springs, MD 20907-8250.

Table 14–2 Summary of Nutrient-Brain Relationships	
Brain Function	**Nutrient Deficiency**
Short-term memory loss	Folate, vitamin B_{12}, vitamin C
Poor performance in problem-solving tests	Riboflavin, folate, vitamin B_{12}, vitamin C
Dementia	Thiamin, niacin, folate, vitamin B_{12}, zinc
Cognition	Folate, vitamin B_6, vitamin B_{12}, iron
Degeneration of brain tissue	Vitamin B_6

Both foods and mental challenges nourish the brain.

of follow-up, participants with low concentrations of either folate or vitamin B_{12}, as well as those with low levels of both vitamins, had twice the risk of developing Alzheimer's disease as those with normal vitamin concentrations.[32] Thus the relationship between Alzheimer's and folate and vitamin B_{12} is gaining strength, but more research is needed to ascertain the exact nature of the relationship.

Maintaining appropriate body weight may be the most important nutrition concern for the person with Alzheimer's. Depression and forgetfulness can lead to poor food intake, and restlessness may increase energy needs. Perhaps the best that a caregiver can do nutritionally for a person with Alzheimer's is to supervise food planning and mealtimes. Providing well-liked and well-balanced meals and snacks in a cheerful atmosphere encourages food consumption. To minimize confusion, offer a few ready-to-eat foods, in bite-size pieces, with seasonings and sauces. To avoid mealtime disruptions, control distractions such as television, children, and the telephone.

Nutrient Deficiencies and Brain Function Poor nutrition in general can affect the brain in other ways. Severe dietary deficiencies of the B vitamins impair mental ability, including memory. Moderate, long-term nutrient deficiencies may contribute to the loss of memory and cognition that some older adults experience.[33] Table 14–2 summarizes some of the better-known connections between impaired brain function and severe nutrient deficiencies. If long-term, moderate nutrient deficiencies influence the loss of cognitive function that accompanies aging, then the loss may be preventable or at least diminished or delayed through diet.

In Summary Cataracts and arthritis afflict millions of older adults, while others face senile dementia and other losses of brain function. Some of these problems may be inevitable, but others are preventable, and good nutrition may play a key preventive role.

Energy and Nutrient Needs during Late Adulthood

Knowledge about the nutrient needs and nutrition status of older adults has grown considerably in the last decade or so. The Dietary Reference Intakes (DRI) cluster people over the age of 50 into two age categories—51 to 70 years old, and 71 years and older. It makes sense, and research is showing, that the nutrition needs of people 50 to 60 years old may be very different from those of people over 80.

Setting standards for older people is difficult, though, because individual differences become more pronounced as people grow older. One person may tend to omit vegetables from his diet, and by the time he is old, he will have an associated set of nutrition problems. Another may have omitted milk and milk products all her life—her nutrition problems will be different. Also, as people age,

Table 14–3	Examples of Physical Changes of Aging That Affect Nutrition
Mouth	Tooth loss, gum disease, and reduced salivary output impede chewing and swallowing. Swallowing disorders and choking may become likely. Discomfort and pain associated with eating may reduce food intake.
Digestive tract	Intestines lose muscle strength resulting in sluggish motility that leads to constipation (see Chapter 19). Stomach inflammation, abnormal bacterial growth, and greatly reduced acid output impair digestion and absorption. Pain may cause food avoidance or reduced intake.
Hormones	For example, the pancreas secretes less insulin and cells become less responsive, causing abnormal glucose metabolism.
Sensory organs	Diminished senses of smell and taste can reduce appetite; diminished sight can make food shopping and preparation difficult.
Body composition	Weight loss and decline in lean body mass lead to lowered energy requirements. May be preventable or reversible through physical activity.
Urinary tract	Increased frequency of urination may limit fluid intake.

they may suffer different chronic diseases and take different medications—both having impacts on nutrient needs. Table 14–3 lists some changes of aging that can affect nutrition. Even before all this, people start out with different genetic predispositions and ways of handling nutrients, and the effects of these become magnified over the years. Researchers have difficulty even defining "healthy aging," a prerequisite to developing recommendations that are designed to meet the "needs of practically all healthy persons." Still, some generalizations are valid. The next sections give special attention to a few nutrients of concern.

Energy and Energy Nutrients

Energy needs decline with advancing age. As a general rule, adult energy needs decline an estimated 5 percent per decade. For one thing, as people age, they usually reduce their physical activity, although they need not do so. For another, lean body mass diminishes, slowing the basal metabolic rate. Declining lean body mass and energy needs may not be entirely unavoidable. Physical activity to maintain lean body mass may be a key, as may a nutrient-dense diet. Loss of muscle mass, known as **sarcopenia,** can be significant in the later years (its prevalence is more than 50 percent among those older than 80), and its consequences, dramatic.[34] As skeletal muscle mass diminishes, people lose their ability to move and to maintain balance, making falls likely. As mentioned earlier, research shows that moderate weight-training exercise, even in the frail elderly, can have profound positive effects on strength and muscle mass. Research also shows a positive correlation between muscle mass and energy intake.[35] Physical activity not only increases energy expenditure but, along with sound nutrition, enhances bone density and supports many body functions as well.[36]

The lower energy expenditures of many older adults require that they eat less food energy to maintain their weights. Accordingly, the energy RDA for adults decreases slightly, beginning at age 51. Energy intakes typically decline in parallel with needs. Still, many older adults are overweight, indicating that their food intakes do not decline enough to compensate for their reduced energy expenditures. Overweight and obesity in older adults increase the risk for many diseases and for disabilities as well.[37]

On limited energy allowances, people must select mostly nutrient-dense foods. There is little leeway for sugars, fats, oils, or alcohol. Older adults can follow the Daily Food Guide (see pp. 16–17), making sure to get at least the minimum number of servings from each food group daily.

sarcopenia (SAR-koh-PEE-nee-ah): loss of skeletal muscle mass, strength, and quality.

Protein   The protein needs of older adults appear to be about the same as, or even greater than, those of younger people.[38] Protein is especially important for older adults to support a healthy immune system and to prevent muscle wasting. Since energy needs decrease, however, the protein has to be obtained from low-kcalorie sources of high-quality protein, such as lean meats, poultry, fish, and eggs; fat-free and low-fat milk products; legumes; and grains.

Carbohydrate and Fiber   As always, abundant carbohydrate is needed to protect protein from being used as an energy source. Complex carbohydrate foods such as whole grains, vegetables, and fruits are also rich in fiber and essential vitamins and minerals.

Average fiber intakes among older adults fall short of current recommendations. The combination of high-fiber foods and ample water can alleviate constipation—a condition common among older adults, and especially among nursing home residents. Physical inactivity and medications also contribute to the high incidence of constipation.

Many nursing home residents are malnourished and underweight. For these people, a diet that emphasizes fiber-rich foods such as whole grains, fruits, and vegetables may be too low in concentrated protein and energy. Protein- and energy-dense snacks such as hard-boiled eggs, tuna fish and crackers, peanut butter on graham crackers, and homemade soups are valuable additions to the diets of underweight or malnourished older adults. Enteral formulas served between meals can also add energy and kcalories to a day's intake.

Fat   As is true for people of all ages, keeping fat intake to 30 percent of total energy or less continues to be important. Not only are the foods lowest in fat often richest in vitamins, minerals, and phytochemicals, but limiting fat may help retard the development of cancer, atherosclerosis, and other degenerative diseases. For some older adults, though, limiting fat intake too severely may lead to nutrient deficiencies and weight loss—two problems that carry greater health risks in the elderly than overweight.

Water

Dehydration is a risk for older adults, who may not notice or pay attention to their thirst, or who find it difficult and bothersome to get a drink or to get to a bathroom. Older adults who have lost bladder control may be afraid to drink too much water. Despite real fluid needs, older people do not seem to feel thirsty or notice mouth dryness.[39] Many nursing home employees say it is hard to persuade their elderly clients to drink enough water and fruit juices.

Total body water decreases as people age, so even mild stresses such as fever or hot weather can precipitate rapid dehydration in older adults. Dehydrated older adults seem to be more susceptible to urinary tract infections, pneumonia, pressure ulcers, confusion, and disorientation.[40] An intake of 6 to 8 glasses of water a day is recommended. Milk and juices may replace some of this water, but beverages containing alcohol or caffeine are not as effective because of their diuretic effect.

Water recommendation for adults: 1 to 1½ oz/kg actual body weight, with a minimum of 50 oz (6¼ c) daily.

Vitamins and Minerals

As research reveals more about how specific vitamins and minerals influence disease prevention and how age-related physiological changes affect nutrient metabolism, optimal intakes of vitamins and minerals for different groups of older adults are being defined. This section highlights the vitamins and minerals of greatest concern to older adults.

Vitamin D   Older adults face a greater risk of vitamin D deficiency than younger people do. Only vitamin D–fortified milk provides significant vitamin D, and many older adults drink little or no milk. Consequently, many older adults have vitamin

Vitamin D AI during late adulthood:
10 μg/day (51–70 yr).
15 μg/day (>70 yr).

Vitamin B$_6$ RDA during late adulthood:
1.7 mg/day (men).
1.5 mg/day (women).

Vitamin B$_{12}$ RDA during late adulthood:
2.4 μg/day.

Folate RDA during late adulthood:
400 μg/day.

Iron RDA during late adulthood:
8 mg/day.

Zinc RDA during late adulthood:
11 mg/day (men).
8 mg/day (women).

Calcium AI during late adulthood:
1200 mg/day.

atrophic gastritis (a-TRO-fik gas-TRI-tis): a condition characterized by chronic inflammation of the stomach accompanied by a diminished size and functioning of the mucosa and glands.

D intakes of less than half of recommendations. Further compromising the vitamin D status of many older people, especially those in nursing homes, is their limited exposure to sunlight. Finally, aging reduces the skin's capacity to make vitamin D and the kidneys' ability to convert it to its active form. To prevent bone loss and to maintain vitamin D status in older people, especially in those who engage in minimal outdoor activity, recommendations for vitamin D were recently raised.[41]

Vitamin B$_6$ Studies on vitamin B$_6$ reveal that its metabolism is altered with age, resulting in a higher requirement. Many older adults consume far less than the RDA for vitamin B$_6$. Research suggests that immune response is impaired with vitamin B$_6$ deficiency. Such findings may have important implications for older adults because the aging process itself seems to be accompanied by a decline in immune function.

Another approach to determining the vitamin B$_6$ requirements of older adults, as well as the requirements for vitamin B$_{12}$ and folate, focuses on the amino acid homocysteine discussed earlier in connection with Alzheimer's disease. An elevated homocysteine level is recognized as an independent risk factor for heart disease and stroke in the United States.[42] Homocysteine concentrations rise with vitamin B$_6$, vitamin B$_{12}$, or folate deficiencies (see Chapter 8).

Vitamin B$_{12}$ The committee on DRI recommends that adults aged 51 years and older obtain 2.4 micrograms of vitamin B$_{12}$ daily *and* that vitamin B$_{12}$–fortified foods (such as fortified cereals) or supplements be used to meet much of the DRI recommended intake.[43] The committee's recommendation reflects the finding that between 10 and 30 percent of people older than 50 years lose the ability to produce enough stomach acid to make the protein-bound form of vitamin B$_{12}$ available for absorption. Synthetic vitamin B$_{12}$ is reliably absorbed, however.

One cause of this malabsorption of protein-bound vitamin B$_{12}$ is a condition known as **atrophic gastritis.** The prevalence of atrophic gastritis among those 60 years of age and older is high.

Folate   As is true of vitamin B$_6$ and vitamin B$_{12}$, folate intakes of older adults typically fall short of recommendations. The elderly are also more likely to have medical conditions or to take medications that can compromise folate status.

Iron   Among the minerals, iron deserves first mention. Iron-deficiency anemia is less common in older adults than in younger people, but it still occurs in some, especially in those with low food energy intakes. Aside from diet, other factors in many older people's lives make iron deficiency likely: chronic blood loss from disease conditions and medicines, and poor iron absorption due to reduced stomach acid secretion and antacid use. Anyone concerned with older people's nutrition should keep these possibilities in mind.

Zinc   Zinc intake is commonly low in older people. Zinc deficiency can depress the appetite and blunt the sense of taste, thereby leading to low food intakes and worsening of zinc status. Many medications that older adults commonly use can impair zinc absorption or enhance its excretion and thus lead to deficiency.

Calcium   The importance of abundant dietary calcium throughout life to protect against osteoporosis has been emphasized throughout this book. The calcium intakes of many people, especially women, in the United States are well below the recommended amount. If milk causes stomach discomfort, as many older adults report, then lactose-modified milk or other calcium-rich foods should take its place.

Nutrient Supplements for Older Adults

People judge for themselves how to manage their nutrition, and some turn to supplements. Advertisers target older people with appeals to take supplements and eat "health" foods, claiming that these products prevent disease and promote

Table 14–4 Strategies for Growing Old Gracefully

- Choose nutrient-dense foods.
- Maintain appropriate body weight.
- Reduce stress by identifying priorities and setting limits.
- For women, consult a physician about planning a strategy to protect against osteoporosis.
- For people who smoke, quit.
- Expect to enjoy sex, and learn new ways of enhancing it.
- Use alcohol only moderately, if at all; use medications only as prescribed.
- Take care to prevent accidents.
- Expect good vision and hearing throughout life; obtain glasses and hearing aids if necessary.
- Be alert to confusion as a disease symptom, and seek diagnosis.
- Control depression through activities and friendships.
- Drink 8 glasses of water every day.
- Practice mental skills. Keep on solving math problems and crossword puzzles, playing cards or other games, reading, writing, imagining, and creating.
- Make financial plans early to ensure security.
- Accept change. Work at recovering from losses; make new friends.
- Cultivate spiritual health. Cherish personal values. Make life meaningful.
- Go outside for sunshine and fresh air as often as possible.
- Be physically active. Walk, run, dance, swim, bike, row, or climb for aerobic activity. Lift weights, do calisthenics, or pursue some other activity to tone, firm, and strengthen muscles. Change activities to suit changing abilities and tastes.
- Be socially active—play bridge, join an exercise group, take a class, teach a class, eat with friends, volunteer time to help others.
- Stay interested in life—pursue a hobby, spend time with grandchildren, take a trip, read, cultivate a garden, or go to the movies.
- Enjoy life.

longevity. About half of all women over 65 years of age take some type of nutrient supplement, while about one-fifth of older men do. Quite often those who take supplements are not deficient in the nutrients being supplemented.

Elderly people often benefit from a balanced low-dose vitamin and mineral supplement, however. Such supplements supply many of the needed minerals along with the vitamins often lacking in older people's diets without providing too much of any one nutrient.[44] Many times, those taking such supplements suffer fewer infectious diseases.[45]

Food is still the best source of nutrients for everybody, however. Supplements are just that—supplements to foods, not substitutes for them. For anyone who is motivated to obtain the best possible health, it is never too late to learn to eat well, become physically active, and adopt other lifestyle changes such as quitting smoking, moderating alcohol use, and the like. Table 14–4 offers strategies for growing old gracefully. Table 14–5 (p. 376) summarizes the nutrient concerns of aging.

The Effects of Drugs on Nutrients

As people grow older, the use of medicines—from over-the-counter types such as aspirin and laxatives to prescription medications of all kinds—becomes commonplace and accounts for about 25 percent of all medications sold. Most medications interact with one or more nutrients in several ways, usually resulting in greater-than-normal needs for these nutrients. Chapter 15 discusses diet-medication interactions and describes the many reasons why elderly people are vulnerable to such interactions.

The most common drug that can affect nutrition in older people is alcohol. A recent estimate sets the incidence of alcoholism in people over 60 in our

Table 14–5 Summary of Nutrient Concerns in Aging

Nutrient	Effect of Aging	Comments
Water	Lack of thirst and decreased total body water make dehydration likely.	Mild dehydration is a common cause of confusion. Difficulty obtaining water or getting to the bathroom may compound the problem.
Energy	Need decreases.	Physical activity moderates the decline.
Fiber	Likelihood of constipation increases with low intakes and changes in the GI tract.	Problems chewing fibrous foods, inadequate water intakes and lack of physical activity, along with some medications, compound the problem.
Protein	Needs may stay the same or increase slightly.	Low-fat, high-fiber legumes and grains meet both protein and other nutrient needs.
Vitamin B_6	Intakes may be low; needs may increase.	Deficiency may hinder immune response.
Vitamin B_{12}	Atrophic gastritis is common.	Deficiency causes neurological damage.
Vitamin D	Increased likelihood of inadequate intake; less likely to go outdoors; skin synthesis declines.	Daily limited sunlight exposure may be of benefit.
Iron	In women, status improves after menopause; deficiencies are linked to chronic blood losses and low stomach acid output.	Adequate stomach acid is required for absorption; antacid or other medicine use may aggravate iron deficiency; vitamin C and meat increase absorption.
Zinc	Intakes may be low.	Medications interfere with absorption; deficiency may depress appetite and sense of taste.
Calcium	Intakes may be low; osteoporosis common.	Stomach discomfort commonly limits milk intake; calcium substitutes are needed.

society at 2 to 10 percent. The effects of alcohol on people of all ages are explained in Nutrition in Practice 23.

In Summary Table 14–5 summarizes the nutrient concerns of aging. The ever-growing number of older people creates an urgent need to know more about how their nutrient requirements differ from those of others and how such knowledge can enhance their health.

Food Choices and Eating Habits of Older Adults

To provide any benefit, strategies and interventions to improve a person's nutrition status must be based on knowledge of food preferences and eating patterns. Menus and feeding programs for older adults must take into consideration not only the food likes and dislikes but also the living conditions, economic status, and medical conditions of this diverse group of people. If nutrition intervention is to be successful, it is essential to know what foods people will eat, in what settings they like to eat these foods, and whether they can buy and prepare meals.

Older people are, for the most part, independent, socially sophisticated, mentally lucid, fully participating members of society who report themselves to be happy and healthy. In fact, chronic disabilities among the elderly have declined dramatically in recent years.[46] Older people spend more money per person on foods to eat at home than other age groups and less money on foods away from home. Manufacturers would be wise to cater to the preferences of older adults by providing good-tasting, nutritious foods in easy-to-open, single-serving packages with labels that are easy to read. Such services enable older adults to maintain their independence; most of them want to take care of themselves and need to feel a sense of control and involvement in their own lives. As discussed earlier, another way older adults can take care of themselves is by remaining or becoming physically active. Physical activity helps preserve one's ability to perform daily tasks and so promotes independence.[47]

Individual Preferences Familiarity, taste, and health beliefs are most influential on older people's food choices. Eating foods that are familiar, especially those that recall family meals and pleasant times, can be comforting. Older adults are choosing low-fat poultry and fish, low-fat milk and milk products, and high-fiber breads and grains, indicating their belief in the importance of diet in supporting good health. Few older adults, however, consume the recommended amounts of milk products.

Meal Setting The food choices and eating habits of older adults are also affected by the changes in lifestyle that often accompany aging in this society. Whether people live alone, with others, or in institutions affects the way they eat. For example, men living alone are most likely to be poorly nourished. Older adults who live alone do not make poorer food choices than those who live with companions; rather, they consume too little food: loneliness is directly related to inadequacies, especially of energy intakes.

Depression Another factor affecting food intake and appetite in older people is depression. Though not an inevitable component of aging, depression is more common with advancing age. Loss of appetite and motivation to cook or even to eat frequently accompanies depression. An overwhelming feeling of grief and sadness at the death of a spouse, friend, or family member may leave many people, particularly elderly people, with a feeling of powerlessness to overcome the depression. The support and companionship of family and friends, especially at mealtimes, can help overcome depression and enhance appetite. The accompanying case study presents a man who has several of these problems. Use the suggestions here, and in the Nutrition in Practice that follows this chapter, to help develop solutions. The Nutrition Assessment Checklist helps to pinpoint nutrition-related factors to look for when working with older adults. To *determine* the risk of malnutrition in older clients, health care providers can keep in mind the characteristics listed in the margin.

In Summary Food choices of older adults are affected by health status and changed life circumstances.

Taking time to nourish your body well is a gift you give yourself.

Shared meals can brighten the day and enhance the appetite.

Risk factors for malnutrition in older adults:

- *Disease.*
- *Eating poorly.*
- *Tooth loss or oral pain.*
- *Economic hardship.*
- *Reduced social contact.*
- *Multiple medications.*
- *Involuntary weight loss or gain.*
- *Needs assistance with self-care.*
- *Elderly person older than 80 years.*

Case Study

ELDERLY MAN WITH A POOR DIET

Mr. Brezenoff is a 75-year-old man who lives alone. He has been losing weight slowly since his wife died a year ago. At 5 feet 8 inches tall, he currently weighs 124 pounds. His previous weight was 150 pounds. In talking with Mr. Brezenoff, you realize that he doesn't even like to talk about food, let alone eat it. "My wife always did the cooking before, and I ate well. Now I just don't feel like eating." You manage to find out that he skips breakfast, has soup and bread for lunch, and sometimes eats a cold-cut sandwich or a frozen dinner for supper. He seldom sees friends or relatives. Mr. Brezenoff has also lost several teeth and doesn't eat any raw fruits or vegetables because he finds them hard to chew. He lives on a meager but adequate income.

- Consult the BMI table (inside back cover), and judge whether Mr. Brezenoff is at a healthy weight. What other assessments might you use to back up your judgment?

- Is his weight loss significant?

- What factors are contributing to his poor food intake?

- What nutrients are probably deficient in his diet?

- Look at Mr. Brezenoff as an individual and suggest ways he can improve his diet and his lifestyle.

- What other aspects of Mr. Brezenoff's physical and mental health should you consider in helping him to improve his food intake?

Nutrition Assessment Checklist

FOR OLDER ADULTS

Health Problems, Signs, and Symptoms

Check the medical record for:

- ☐ Chronic diseases (cancer, heart disease, hypertension, diabetes)
- ☐ Dehydration
- ☐ Arthritis
- ☐ Cataracts
- ☐ Alzheimer's disease or other dementia or confusion
- ☐ Alcohol abuse
- ☐ Depression
- ☐ Dental disease or tooth loss
- ☐ Cigarette, cigar, or pipe smoking; use of other tobacco products
- ☐ Swallowing disorders
- ☐ Constipation
- ☐ Inflammation of the stomach (gastritis)

Medications

For older adults being treated with drug therapy for medical conditions, note:

- ☐ Use of multiple medications—prescription and/or over-the-counter medications such as laxatives and pain relievers
- ☐ Side effects that might reduce food intake or change nutrient needs
- ☐ Proper administration of medication with respect to food intake
- ☐ Malnutrition—is the person's nutrition status questionable even before considering side effects of medications that worsen nutrition status?
- ☐ Diminished mental capacity that might interfere with taking correct medications and doses
- ☐ Dehydration (can alter effects of medications)

Food/Nutrient Intake

For all older adults, especially those at risk nutritionally, assess the diet for:

- ☐ Total energy
- ☐ Protein
- ☐ Calcium, iron, and zinc
- ☐ Vitamin B_6, vitamin B_{12}, folate, and vitamin D

Note the following:

- ☐ Number of meals eaten each day
- ☐ Number and ages of people in household
- ☐ Amount of milk consumed each day
- ☐ Type and frequency of outdoor activity
- ☐ Type and frequency of physical activity
- ☐ Financial resources
- ☐ Transportation resources
- ☐ Physical disabilities
- ☐ Mental alertness

Height and Weight

Measure baseline height and weight.

- ☐ Reassess height and weight at each medical checkup.
- ☐ Note significant overweight or underweight, which warrants intervention.
- ☐ Use fatfold measures to reveal altered body composition that may indicate malnutrition and loss of lean tissue.

Laboratory Tests

- ☐ Hemoglobin, hematocrit, or other tests of iron status
- ☐ Serum albumin or other measures of protein status
- ☐ Serum folate
- ☐ Serum B_{12}

Physical Signs

Look for physical signs of:

- ☐ Protein-energy malnutrition
- ☐ Iron and zinc deficiency
- ☐ Folate deficiency

1. The fastest-growing age group in the United States is:
   a. 21 years of age.
   b. 30 to 45 years of age.
   c. 50 to 70 years of age.
   d. over 85 years of age.

2. Which of the following lifestyle habits can enhance the length and quality of people's lives?
   a. moderate smoking
   b. 6 hours of sleep daily
   c. regular physical activity
   d. moderate alcohol intake

3. Among the better-known relationships between nutrition and disease prevention are:
   a. appropriate fiber intake helps prevent goiter.
   b. moderate sodium intake helps prevent obesity.
   c. moderate sugar intake helps prevent hypertension.
   d. appropriate energy intake helps prevent diabetes and cardiovascular disease.

4. A disease of the immune system that involves painful inflammation of the joints is:
   a. sarcopenia.
   b. osteoarthritis.
   c. senile dementia.
   d. rheumatoid arthritis.

5. Examples of low-kcalorie, high-quality protein foods include:
   a. cottage cheese, sour cream, and eggs.
   b. green and yellow vegetables and citrus fruits.
   c. potatoes, rice, pasta, and whole-grain breads.
   d. lean meats, poultry, fish, legumes, fat-free milk, and eggs.

6. For malnourished and underweight people, protein- and energy-dense snacks include:
   a. fresh fruits and vegetables.
   b. yogurt and cottage cheese.
   c. whole grains and high-fiber legumes.
   d. hard-boiled eggs and peanut butter and crackers.

7. Which of the following does **not** explain why dehydration is a risk for older adults?
   a. They do not seem to feel thirsty.
   b. Total body water increases with age.
   c. They may find it difficult to get a drink.
   d. They may have difficulty swallowing liquids.

8. Inadequate milk intake and limited exposure to sunlight contribute to older adults' risk of:
   a. vitamin A deficiency.
   b. vitamin D deficiency.
   c. riboflavin deficiency.
   d. vitamin B_6 deficiency.

9. Two risk factors for malnutrition in older adults are:
   a. loneliness and multiple medication use.
   b. increased energy needs and lack of fiber.
   c. decreased mineral absorption and antioxidant intake.
   d. high carbohydrate intake and lack of physical activity.

10. Two strategies to improve nutrition status when growing old include:
    a. increase vitamin A intake and exercise 30 minutes daily.
    b. choose nutrient-dense foods and maintain appropriate weight.
    c. avoid high-fiber foods and take a daily vitamin-mineral supplement.
    d. eat at least one big meal per day and drink at least 10 glasses of water daily.

Critical Thinking

1. When caring for a person with Alzheimer's disease, which of the following foods would be a priority?
   a. fresh fruits
   b. fat-free cheeses and yogurt
   c. meats, milk, and eggs
   d. steamed vegetables

Answers to these questions can be found in Appendix G.

Clinical Applications

Ms. Hamilton is an 80-year-old woman in excellent health who lives alone, eats a well-balanced diet, enjoys an active social life, and walks every day. Consider the ways Ms. Hamilton's health and nutrition status might be affected by the following situations:

- Many of Ms. Hamilton's friends pass away or move into extended care facilities.

- Ms. Hamilton falls and breaks her hip.
- Ms. Hamilton begins to feel isolated and depressed.

Describe interventions the health care professional can take to help Ms. Hamilton deal with each situation to prevent her from falling into a downward spiral.

Nutrition on the Net

For further study of the topics of this chapter, access these websites.

Find updates and quick links to these and other nutrition-related sites at our website: **www.wadsworth.com/nutrition**

Search for aging, arthritis, and Alzheimer's on the U.S. Government health information site: **www.healthfinder.gov**

Visit the National Aging Information Center of the Administration on Aging: **www.aoa.dhhs.gov/naic**

Visit the American Geriatrics Society:
www.americangeriatrics.org

Visit the National Institution on Aging:
www.nih.gov/nia

Visit the American Association of Retired Persons:
www.aarp.org

Get nutrition tips for growing older in good health from the American Dietetic Association:
www.eatright.org

Learn more about cataracts from the National Eye Institute and the American Society of Cataract and Refractive Surgery:
www.nei.nih.gov and **www.ascrs.org**

Learn more about arthritis from the Arthritis Foundation and the National Institute of Arthritis and Musculoskeletal and Skin Diseases Information Clearinghouse of the National Institutes of Health:
www.arthritis.org and **www.nih.gov/niams**

Learn more about Alzheimer's disease from the NIA Alzheimer's Disease Education and Referral Center and the Alzheimer's Association:
www.alzheimers.org and **www.alz.org**

Learn more about malnutrition in older adults and the Nutrition Screening Initiative from the American Academy of Family Physicians:
www.aafp.org

Find out about federal government programs designed to help senior citizens maintain good health:
www.seniors.gov

Notes

1. Position of The American Dietetic Association: Nutrition, aging, and the continuum of care, *Journal of the American Dietetic Association* 100 (2000): 580–595.

2. D. L. Hoyert and coauthors, Annual summary of vital statistics: 2000, *Pediatrics* 108 (2001): 1241–1255; Centers for Disease Control and Prevention, National Center for Health Statistics, *Monthly Vital Statistics Report,* October 1996, p. 4.

3. K. G. Manton and E. Stallard, Longevity in the United States: Age and sex-specific evidence on life span limits from mortality patterns 1960–1990, *Journal of Gerontology* 51A (1996): B362–B375.

4. L. Guarente, G. Ruvkun, and R. Amasino, Aging, life span, and senescence, *Proceedings of the National Academy of Sciences* 95 (1998): 11-34–11-36; D. A. Banks and M. Fossel, Telomeres, cancer, and aging: Altering the human life span, *Journal of the American Medical Association* 278 (1997): 1345–1348.

5. A. J. Vita and coauthors, Aging, health risks, and cumulative disability, *New England Journal of Medicine* 338 (1998): 1035–1041.

6. B. A. Franklin, Improved fitness = increased longevity, *ACSM'S Health & Fitness Journal,* March/April 2001, pp. 32–33; I. M. Lee and R. S. Paffenbarger, Associations of light, moderate, and vigorous intensity physical activity with longevity: The Harvard Alumni Health Study, *American Journal of Epidemiology* 151 (2000): 293–299; U. M. Kujala and coauthors, Relationship of leisure-time, physical activity, and mortality: The Finnish Twin Cohort, *Journal of the American Medical Association* 279 (1998): 440–444; S. N. Blair and coauthors, Changes in physical fitness and all-cause mortality: A prospective study of healthy and unhealthy men, *Journal of the American Medical Association* 273 (1995): 1093–1098.

7. American College of Sports Medicine, Position stand, Exercise and physical activity for older adults, *Medicine and Science in Sports and Exercise* 30 (1998): 992–1008.

8. A. C. King and coauthors, Moderate-intensity exercise and self-rated quality of sleep in older adults, *Journal of the American Medical Association* 277 (1997): 32–37; W. Evans, Functional and metabolic consequences of sarcopenia, *Journal of Nutrition* 127 (1997): S998–S1003.

9. W. J. Evans and D. Cyr-Campbell, Nutrition, exercise, and healthy aging, *Journal of the American Dietetic Association* 97 (1997): 632–638.

10. T. Rantanen and coauthors, Midlife hand grip strength as a predictor of old age disability, *Journal of the American Medical Association* 281 (1999): 558–560.

11. J. W. R. Twisk, H. C. G. Kemper, and W. van Mechelen, Tracking of activity and fitness and the relationship with cardiovascular disease factors, *Medicine and Science in Sports and Exercise* 32 (2000): 1455–1461; C. D. Lee, S. N. Blair, and A. S. Jackson, Cardiorespiratory fitness, body composition, and all-cause and cardiovascular disease mortality in men, *American Journal of Clinical Nutrition* 69 (1999): 373–380; J. S. Green and S. F. Crouse, The effects of endurance training on functional capacity in the elderly: A meta-analysis, *Medicine and Science in Sports and Exercise* 27 (1995): 920–926.

12. S. M. Avent and D. M. Landers, The effects of exercise on mood in the elderly: A meta-analysis, presented at the annual meeting of the American College of Sports Medicine, Seattle; as yet unpublished, as cited in *Sports Medicine Digest* 21 (1999): 72.

13. R. Weindruch and R. S. Sohal, Caloric intake and aging, *New England Journal of Medicine* 337 (1997): 986–994.

14. J. Wanagat, D. B. Allison, and R. Weindruch, Caloric intake and aging: Mechanisms in rodents and a study in nonhuman primates, *Toxicological Sciences* (supplement) 52 (1999): 35–40; R. B. Verdery and coauthors, caloric restriction increases HDL2 levels in rhesus monkeys (Macaca mulatta), *American Journal of Physiology* 273 (1997): E714–719; R. Weindruch, Caloric restriction and aging, *Scientific American,* January 1997, pp. 46-52.

15. L. Chasan-Taber and coauthors, A prospective study of carotenoid and vitamin A intakes and risk of cataract extraction in US women, *American Journal of Clinical Nutrition* 70 (1999): 509–516; L. Brown and coauthors, A prospective study of carotenoid intake and risk of cataract extraction in US men, *American Journal of Clinical Nutrition* 70 (1999): 517–524.

16. M. C. Leske and coauthors, Antioxidant vitamins and nuclear opacities: The longitudinal study of cataract, *Ophthalmology* 105 (1998): 831–836; P. F. Jacques and coauthors, Long-term vitamin C supplement use and prevalence of early age-related lens opacities, *American Journal of Clinical Nutrition* 66 (1997): 911–916.

17. D. A. Schaumberg and coauthors Relations of body fat distribution and height with cataract in men, *American Journal of Clinical Nutrition* 72 (2000): 1495–1502.

18. W. H. Ettinger and coauthors, A randomized trial comparing aerobic exercise and resistance exercise with a health education program in older adults with knee osteoarthritis: The Fitness Arthritis and Seniors Trial (FAST), *Journal of the American Medical Association* 277 (1997): 25–31.

19. T. E. McAlindon and coauthors, Glucosamine and chondroitin

for treatment of osteoarthritis: A systematic quality assessment and meta-analysis, *Journal of the American Medical Association* 283 (2000): 1469–1475.

20. J. M. Kremer, n-3 fatty acid supplements in rheumatoid arthritis, *American Journal of Clinical Nutrition* 71 (2000): 349S–351S.

21. H. C. Hendrie, Epidemiology of dementia and Alzheimer's disease, *American Journal of Geriatric Psychiatry* 6 (1998): S3–S18.

22. E. Rogaeva and coauthors, Evidence for an Alzheimer disease susceptibility locus on chromosome 12 and further locus heterogeneity, *Journal of the American Medical Association* 280 (1998): 614–618; W. S. Wu and coauthors, Genetic studies on chromosome 12 in late-onset Alzheimer disease, *Journal of the American Medical Association* 280 (1998): 619–622.

23. A. T. Hingley, Alzheimer's: Few clues on the mysteries of memory, *FDA Consumer,* May/June 1998, pp. 27–31.

24. Y. Christen, Oxidative stress and Alzheimer disease, *American Journal of Clinical Nutrition* 71 (2000): 621S–629S.

25. M. A. Lovely and coauthors, Copper, iron and zinc in Alzheimer's disease senile plaques, *Journal of the Neurological Sciences* 158 (1998): 47–52; C. R. Cornett, W. R. Markesbery, and W. D. Ehmann, Imbalances of trace elements related to oxidative damage in Alzheimer's disease brain, *Neurotoxicology* 19 (1998): 339–345.

26. M. A. Lovely, C. Xie, and W. R. Markesbery, Protections against amyloid beta peptide toxicity by zinc, *Brain Research* 823 (1999): 88–95; Cornett, Markesbery, and Ehmann, 1998; Lovely and coauthors, 1998; M. P. Cuajungco and G. J. Lees, Zinc metabolism in the brain, *Neurobiology of Disease* 4 (1997): 137–169; F. C. Potocnik and coauthors, Zinc and platelet membrane microviscosity in Alzheimer's disease: The in vivo effect of zinc on platelet membranes and cognition, *South African Medical Journal* 87 (1997): 1116-1119.

27. J. W. Miller, Homocysteine and Alzheimer's disease, *Nutrition Reviews* 57 (1999): 126–129.

28. P. M. Ridker and coauthors, Homocysteine and risk of cardiovascular disease among postmenopausal women, *Journal of the American Medical Association* 281 (1999): 1817–1821; K. Robinson and coauthors, Low circulating folate and vitamin B6 concentrations: Risk factors for stroke, peripheral vascular disease, and coronary artery disease, European COMAC Group, *Circulation* 97 (1998): 437–443.

29. A. Hoffman and coauthors, Atherosclerosis, apolipoprotein E, and prevalence of dementia and Alzheimer's disease in the Rotterdam Study, *Lancet* 349 (1997): 151–154, as cited in Miller, 1999.

30. A. Chait and coauthors, Increased dietary micronutrients decrease serum homocysteine concentrations in patients at high risk of cardiovascular disease, *American Journal of Clinical Nutrition* 70 (1999): 881–887.

31. D. A. Snowdon and coauthors, Serum folate and the severity of atrophy of the neocortex in Alzheimer disease: Findings from the Nun Study, *American Journal of Clinical Nutrition* 71 (2000): 993–998.

32. H. X. Wang and coauthors, Vitamin B$_{12}$ and folate in relation to the development of Alzheimer's disease, *Neurology* 56 (2001): 1188–1194.

33. I. H. Rosenberg, B vitamins, homocysteine, and neurocognitive function, *Nutrition Reviews* 59 (2001): S69–S74; A. La Rue and coauthors, Nutritional status and cognitive functioning in a nor-

mally aging sample: A 6-y reassessment, *American Journal of Clinical Nutrition* 65 (1997): 20–29.

34. A. S. Nicolas and coauthors, Successful aging and nutrition, *Nutrition Reviews* 59 (2001): S88–S90; R. Roubenoff and C. Castaneda, Sarcopenia—Understanding the dynamics of aging muscle, *Journal of the American Medical Association* 286 (2001): 1230–1231.

35. R. Baumgartner and coauthors, Predictors of skeletal muscle mass in elderly men and women, *Mechanisms of Ageing and Development* 107 (1999): 123–136.

36. L. DiPietro, The epidemiology of physical activity and physical function in older people, *Medicine and Science in Sports and Exercise* 28 (1996): 596–660.

37. Position of The American Dietetic Association, 2000; M. Visser and coauthors, High body fatness, but not low fat-free mass, predicts disability in older men and women: The Cardiovascular Health Study, *American Journal of Clinical Nutrition* 68 (1998): 584–590.

38. Nicolas and coauthors, 2001; W. D. Campbell, Dietary protein requirements of older people: Is the RDA adequate? *Nutrition Today* 31 (1996): 192–197.

39. S. M. Kleiner, Water: An essential but overlooked nutrient, *Journal of the American Dietetic Association* 99 (1999): 200–206.

40. J. C. Chidester and A. A. Spangler, Fluid intake in the institutionalized elderly, *Journal of the American Dietetic Association* 97 (1997): 23–28; S. A. Gilmore and coauthors, Clinical indicators associated with unintentional weight loss and pressure ulcers in elderly residents of nursing facilities, *Journal of the American Dietetic Association* 95 (1995): 984–992.

41. Standing Committee on the Scientific Evaluation of Dietary Reference Intakes, Food and Nutrition Board, Institute of Medicine, *Dietary Reference Intakes for Calcium, Phosphorus, Magnesium, Vitamin D, and Fluoride* (Washington, D.C.: National Academy Press, 1997).

42. J. Selhub and coauthors, Association between plasma homocysteine concentrations and extracranial carotid-artery stenosis, *New England Journal of Medicine* 332 (1995): 286–291.

43. C. Ho and coauthors, Practitioners' guide to meeting the vitamin B$_{12}$ Recommended Dietary Allowance for people aged 51 years and older, *Journal of the American Dietetic Association* 99 (1999): 725–727; Standing Committee on the Scientific Evaluation of Dietary Reference Intakes, Food and Nutrition Board, Institite of Medicine, *Dietary Reference Intakes for Thiamin, Riboflavin, Niacin, Vitamin B$_6$, Folate, Vitamin B$_{12}$, Pantothenic Acid, Biotin, and Choline* (Washington, D.C.: National Academy Press, 1998), pp. 7-1–7-27.

44. R. D. Chandra, Graying of the immune system: Can nutrient supplements improve immunity in the elderly? *Journal of the American Medical Association* 277 (1997): 1398–1399.

45. M. A. Johnson and K. H. Porter, Micronutrient supplementation and infection in institutionalized elders, *Nutrition Reviews* 55 (1997): 400–404.

46. Y. Liao and coauthors, Quality of the last year of life of older adults: 1986 vs. 1993, *Journal of the American Medical Association* 283 (2000): 512–518; K. G. Manton, L. Corder, and E. Stallard, Chronic disability trends in elderly United States population: 1982–1994, *Proceedings of the National Academy of Sciences of the USA* 94 (1997): 2593–2598.

47. DiPietro, 1996.

Nutrition in Practice

■ FOOD FOR SINGLES ■

Singles of all ages face difficulties in purchasing, storing, and preparing food. Large packages of meat and vegetables are often intended for families of four or more, and even a head of lettuce can spoil before one person can use it all. Many singles live in small dwellings and have little storage space for foods. A limited income presents additional obstacles. The following ideas can help to solve some of these problems.

What advice can help a person on a fixed income acquire groceries?

Once the rent, utilities, and other bills are paid, most people on a fixed income don't have much money for groceries. First, make sure these people know what food assistance programs are available. Such programs are available to older adults who need help obtaining nourishing meals because of financial or other difficulties. Table NP14–1 (p. 384) summarizes food assistance programs for the elderly.

People who have the means to shop and cook for themselves can cut their food bills just by being wise shoppers. The first decision a person with a tight grocery budget must make is where to shop. Large supermarkets are usually less expensive than convenience stores, but the cost of transportation to the market is a consideration. Once a person decides where to shop, a grocery list that includes specials and coupons will save money and help reduce impulse buying. Specials and coupons are a bargain only when the items featured are those that the shopper needs and uses. Foods that are almost always good buys include rice and fat-free dry milk, which can be stored on a shelf for months at room temperature; fresh produce in season; and dried beans and peas.

All foods, whether featured as specials or not, are bargains only when they are available in quantities that can be used without waste or spoilage. For example, turkey may be on sale and is one of the most economical meats for the nutrients it offers, but only a person with ample storage space for leftovers would benefit from buying a whole turkey.

How can older people and others who live alone buy small quantities when so many foods are packaged for families?

All singles face this problem. Packages of meat and fresh vegetables often come already wrapped in large

servings. Even milk is often available only in gallons or half-gallons and can spoil before one person can use it all. People living alone can try these hints. First, in the dairy case, buy fresh milk in the size best suited for you. If your grocer doesn't carry pints or half-pints, try a nearby service station or convenience store. Pint-size and even cup-size boxes of heat-treated milk are also available and can be stored unopened on a shelf for up to three months without refrigeration. Dry powdered milk can be stored for months before it is reconstituted. You can use only the amount you need and store the rest.

Next, among the meats, buy only what you will use. Ask the grocer to break open a package of wrapped meat and rewrap the portion that you need. Alternatively, if you have ample freezer space, you can buy large packages of meat, such as pork chops, ground beef, or chicken, when they are on sale. Then, immediately divide the package into individual servings. Wrap them in aluminum foil, not freezer paper: the foil can become the liner for the pan in which you bake or broil the meat, thus saving work over the sink. Don't label these individually; just put them all in a brown bag marked "hamburger" or "chicken thighs" or whatever, along with the date. The bag is easy to locate in the freezer, and you'll know when your supply is running low.

As for meat alternates, buy eggs by the half-dozen—break the carton of a dozen eggs in half. Eggs do keep for long periods, though, if stored in the refrigerator and are such a good source of high-quality protein that you will probably use a dozen before they lose their freshness. Dried beans and peas offer high-quality protein, fiber, and many other nutrients for practically pennies and have a long shelf life.

Among fruits and vegetables, purchase fresh ones individually. Buy only three pieces of each kind of fresh fruit: a ripe one, a semiripe one, and a green one. Eat the first right away, the second soon after, and let the last one ripen on the windowsill. If vegetables are packaged in large quantities, ask the grocer to break open the package so that you can buy what you need. Buy small cans of fruits and vegetables even though they are more expensive per unit. Remember that it is expensive to buy a regular-size can and let the unused portion spoil in the refrigerator. If you have space in your freezer, buy frozen vegetables in large bags rather

Nutrition in Practice

Table NP14–1 Food Assistance Programs for Older Adults

Elderly Nutrition Program

- *Services:* Congregate and home-delivered meals to improve older people's nutrition status. Supportive services include transportation to congregate meal sites; shopping assistance; information and referral; and, to some extent, nutrition counseling and education.
- *Impact:* The Elderly Nutrition Program improves the nutrient content of high-risk older adults' diets and offers socialization and recreation. Many of the nutrition programs around the country go above and beyond federal requirements of congregate and home meals by offering lunch clubs, ethnic meals, and meals for older homeless people. An estimated 25 percent of our nation's elderly poor benefit from these meals.[a]

Food Stamps

- *Services:* Income supplement for low-income households in the form of a card similar to a credit card that can be used to purchase food.
- *Impact:* The Food Stamp program serves more as an income supplement for some elderly participants than as a device to improve nutrition status. For other elderly food stamp participants, nutrient intakes are higher than those of nonparticipants with similar incomes.

Meals on Wheels

- *Services:* Direct meal delivery to the homebound elderly, integrated into the meal delivery services provided by the Elderly Nutrition Program.
- *Impact:* Meals on Wheels focuses on filling the need for weekend and holiday meals for homebound elderly people, a service that is limited in the Elderly Nutrition Program.

[a]Federal program nourishes poor elderly, *Journal of the American Medical Association* 278 (1997): 1301.

than in small boxes. You can take out the exact amount you need and close the bag tightly with a rubber band or twist tie. If you return the package quickly to the freezer each time, the vegetables will stay fresh for a long time.

Finally, breads and cereals usually must be purchased in large quantities. When you buy bread, take out the amount you will use in a few days and store the rest in the freezer. Cereal grains (rice, barley, oatmeal) and pastas, like the dried beans mentioned earlier, generally have a long shelf life if you keep them sealed in jars.

What are some suggestions for when a person just has to buy more food than he or she can use?

One suggestion is to make mixtures of leftovers that are already on hand. A thick stew prepared from leftover green beans, carrots, cauliflower, broccoli, and any meat with added onion, pepper, celery, and potatoes makes a complete and balanced meal—except for milk, but then powdered milk can be added to the stew.

Another suggestion is to set aside a shelf in the kitchen for rows of glass jars containing staple items that a person usually can't buy in single-serving quantities—rice, tapioca, lentils and other dry beans, flour,

cornmeal, fat-free dry milk, pasta, cereal, or coconut, to name only a few possibilities. Freeze each filled jar for one night first to kill any insect eggs that might be present. The jars will then keep bugs out of the food indefinitely. The jars make an attractive display and are reminders of different choices on hand to vary menus. Cut the directions-for-use labels from the packages and tape them on the jars.

Creative chefs think of various ways to use foods when only large amounts are available. For example, a head of cauliflower can be divided into thirds. Cook one-third and eat it as a hot vegetable. Put the other two-thirds into a vinegar and oil marinade for use as an appetizer or in a salad. Keep half a package of frozen vegetables with other vegetables to be used in soup or stew.

Sometimes single people, especially older people, don't feel like cooking a meal and may not get the nutrients they need.

People need to take advantage of times when they do feel like cooking and cook several meals at a time. For example, boil three potatoes with skins. Eat one hot with margarine and chives. When the others have

Nutrition in Practice

cooled, use one to make a potato-cheese casserole ready to be put into the oven for the next evening's meal. Then slice the third one into a covered bowl and pour the juice from pickles over it. The pickled potato will keep several days in the refrigerator and can be used later in a salad.

Depending on freezer space, make double or triple portions of a dish that takes time to prepare: a casserole, vegetable pie, or meat loaf. Freeze individual portions in containers that can be heated later. Be sure to date these so you will use the oldest first. The work will seem worthwhile when several meals are prepared at once.

Frozen dinners can also be useful when you don't feel like cooking. Many such dinners that are now available are low in kcalories and nutritious. Adding a fresh salad, a whole-wheat roll, and a glass of milk can make a nice meal.

More convenient than frozen foods are take-out delicatessen-style foods from grocery stores. Such foods can be purchased while shopping for other items. Convenience doesn't come without a price, of course, but the foods cost less than similar foods from restaurants.

A bonus for singles is that you specify the amount you need and portion it onto your plate at home—you need not buy more than you will eat.

People need to socialize, too. Sometimes, the loneliness that results from isolation can impair a single person's appetite and motivation to cook appetizing meals. To combat this isolation, a person can cook a lot of food and invite someone to share it. Before long, the guest will probably invite the host or hostess back, and both people will get to enjoy a meal they wouldn't have thought to cook for themselves.

One more suggestion for those who are alone at mealtime is this: make it a special occasion. One way to do this is to set the table with a tablecloth, a napkin, a full set of utensils, and fresh flowers. Set a pot of stew or homemade soup with vegetables and fresh herbs on low heat to cook, and make a salad. Get comfortable in a stuffed chair, and enjoy a book or your favorite music until the rich aroma of your simmering dinner beckons. After serving your plate, light a candle, dim the lights, savor the food, and relish some of the best company you will ever have—your own.

Jennie Oppenheimer/Studio Zocolo

CHAPTER 15 | # HEALTH PROBLEMS, MEDICATIONS, AND COMPLEMENTARY THERAPIES

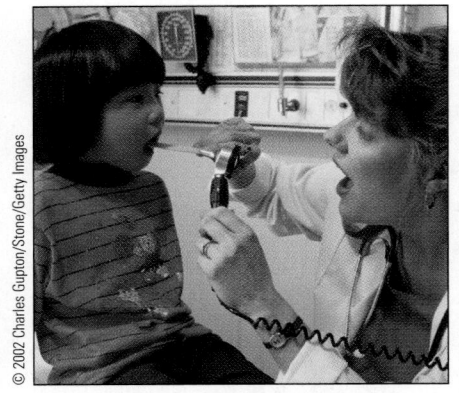

A nutritious diet protects against illness and supports recovery from illnesses that do develop.

For the purposes of this discussion, health problems include:

- *Ilnesses, such as diabetes and cancer.*
- *Injuries, such as gunshot wounds and burns.*
- *Symptoms (subjective findings), such as nausea and fatigue.*
- *Clinical signs (objective findings), such as high blood pressure and elevated blood lipids.*

Reminder: *Free radicals* are highly reactive chemicals that can set off chain reactions that damage lipids and proteins. In addition to free radicals derived from natural body processes, environmental factors such as ultraviolet radiation, ozone, tobacco smoke, and pollution, as well as tissue damage, generate free radicals.

Reminder: *Phytochemicals* are nonnutrient compounds found in plant foods that have biological activity in the body (see Nutrition in Practice 8). Chapters 8 and 9 provide more information about antioxidant nutrients.

acute: developing rapidly and associated with symptoms that are severe, but limited in duration.

chronic: developing gradually and associated with symptoms that progressively worsen with time.

oxidative stress: a condition in which the production of oxidants and free radicals exceeds the body's ability to neutralize them.

A balance of energy and nutrients supports a physically fit body and an alert mind throughout life. Nutrient imbalances, on the other hand, contribute to the development of some health problems and complicate the treatment of many more. Turning now to clinical nutrition, this chapter describes the potential effects of illnesses and their treatments on nutrition status and nutrient needs and defines the responsibilities of various health care professionals in preventing malnutrition.

What Causes Health Problems?

Some causes of health problems are readily identifiable. An infant may "catch" a bacterial infection from a family member; an adult may break a leg during a fall. In many other cases, no single cause can be identified. Many health problems appear to result from a combination of factors including genetic, environmental, and dietary factors.

The immune system plays a pivotal role in defending the body against infectious diseases, ridding the body of toxic substances and worn-out cells, and repairing damaged tissues. A growing body of evidence suggests that within the body, two processes related to immune system functions—the production of free radicals and inflammation—play important roles in the development of both **acute** and **chronic** health problems. A nutritious diet may exert its health-protective effects, in part, by minimizing the damage that may ensue following free-radical production and inflammation. Nutrition in Practice 15 provides more information about the immune system and the ways energy and nutrients affect its performance.

Free Radicals and Disease

Although the body uses most of its oxygen in energy-producing metabolic pathways, oxygen sometimes reacts with other body compounds, forming free radicals (see Chapter 8). Most free radicals are unstable and highly reactive. To regain stability, they alter the chemical properties of vulnerable molecules in the area. Now the vulnerable molecule becomes a free radical itself, and a chain reaction is set in motion. Free radicals frequently target the fatty acids in lipoproteins and the fatty acids and protein in cell membranes. When the chain reaction continues unchecked, tissue damage and loss of cell function can follow. Free radicals can also attack DNA and lead to mutations.

Protection from Free Radicals Survival would be impossible without defense mechanisms to halt the formation of free radicals. A system of enzymes provides one line of defense; these enzymes intercept free radicals and convert them to less toxic substances. The action of these enzymes depends on nutrient cofactors, including the minerals copper, zinc, manganese, and selenium. Antioxidant nutrients, including vitamins C and E, provide a second line of defense; these nutrients stop the free-radical chain reaction. A variety of phytochemicals, including carotenoids and flavonoids, also serve as antioxidants. As earlier chapters mention, the antioxidant properties of the nutrients and phytochemicals found in fruits and vegetables may help explain why eating a variety of these foods appears to provide an edge in preventing chronic diseases.

Damage from Free Radicals Occasionally, the production of free radicals overwhelms the body's ability to neutralize them—a condition called **oxidative stress.** Prolonged or severe oxidative stress results in tissue damage. The accumulation of this damage over time is believed to contribute to aging and the development of many chronic diseases including heart disease, arthritis, and cataracts. Free-radical-induced mutations in DNA may give rise to cancers.

Inflammation

Inflammation is the immune system's response to tissue injury. The chapters that follow provide many examples of health problems associated with inflammation. As you read through these chapters, keep in mind that tissues can be damaged from bodily injuries (blows and cuts), heat (burns), infectious agents (bacteria and viruses), inadequate blood flow (hemorrhages), stress on blood vessels (high blood pressure), or chemicals (free radicals). In addition to free radicals, other chemicals that can lead to inflammation include some drugs and herbs, alcohol, and even blood glucose and blood lipids when they remain elevated over time.

Inflammatory Responses During the initial response, blood clots form to seal off injured blood vessels and stop any bleeding. Chemical signals from immune system cells direct changes that cause blood-clotting, infection-fighting, and wound-healing factors to adhere to the injured area. The walls of the blood vessels swell, allowing infection-fighting factors to enter the injured area. Once bleeding and infection are under control, tissue repair begins. A scar is a visible reminder of tissue repair.

Inflammation and Health Problems Although inflammation serves a vital role in preventing blood loss and infection and repairing tissue damage, chronic inflammation or exaggerated inflammatory responses have negative effects because they:

- Slow blood flow in the injured area and enhance the permeability of blood vessels, which can alter organ function.

- Induce chemical changes that favor the formation of blood clots, which can obstruct blood vessels, lead to tissue death, and alter organ function.

- Generate free radicals and lead to oxidative stress, which can damage healthy tissue and induce further inflammation.

- Replace healthy tissue with scar tissue, which can grow out of control (**fibrosis**) and interfere with organ function.

Inflammation appears to be an important contributor to many diseases including heart disease, diabetes, arthritis, Alzheimer's disease, and possibly osteoporosis. Chapters 24 and 25 provide more information about inflammation and diabetes and heart disease. Chapter 27 describes how severe injuries, such as extensive burns, produce heightened inflammatory responses and oxidative stress and rapidly raise the body's demands for nutrients. Chapter 28 explains how inflammation may contribute to malnutrition in the end stages of many chronic diseases.

> *In Summary* Health problems arise from different causes, and often several factors contribute to disease development. The production of free radicals and inflammation may contribute to the negative effects of many diseases. Energy and nutrients that support immune function and those with antioxidant properties may play important roles in preventing these processes from leading to chronic health problems.

Nutrition in Health Care

Nutrient imbalances, both excesses and deficiencies, can lead to health problems and make existing problems worse. Once a person becomes ill, energy and nutrient deficiencies can follow, regardless of whether a person is well-nourished,

© BSIP Agency/Index Stock Imagery

Plant foods contain nutrients and phytochemicals with antioxidant properties, which may help explain the association between generous consumption of these foods and lower incidences of many chronic diseases.

inflammation: the immune system's response to any type of tissue injury. The symptoms of inflammation include redness, swelling, and pain.

fibrosis (figh-BROH-sis): abnormal growth of scar tissue.

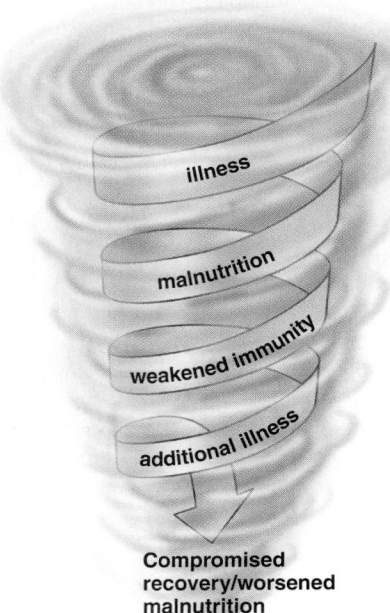

Regardless of where a person enters the spiral, the effects of illness, malnutrition, and impaired immunity can interact to compromise recovery and worsen malnutrition.

Figure 15–1
Illness, Malnutrition, and Immunity

Reminder: During surgery, tissues are damaged by surgical incisions and sometimes by the removal of tumors or dysfunctional tissues, and inflammatory responses are called into action.

Pressure sores trigger inflammatory responses. In the case of pressure sores, tissues are damaged by lack of oxygen to the skin.

pressure sores: the breakdown of skin and underlying tissues due to constant pressure and lack of oxygen to the affected area; also called **decubitus** (dee-CUE-bih-tus) **ulcers** or **bedsores.**

under-nourished, or over-nourished at the onset of the illness. Without adequate energy, protein, vitamins, and minerals, the body may be unable to maintain immune defenses, heal wounds, utilize medications, and support organ function. Then the compromised person may fall victim to additional complications, and a downward spiral is set in motion (see Figure 15–1).

Health Problems and Treatments: Effects on Nutrition Status

As the remaining chapters of this book show, health problems, symptoms, and treatments can lead to malnutrition by impairing a person's ability to eat, interfering with digestion and absorption, and altering metabolism and excretion. Figure 15–2 provides examples of ways illnesses can affect each of these processes. Regardless of which function is initially affected, once a problem intensifies, all functions are affected. When problems with food intake progress unchecked, for example, the body's ability to digest and absorb nutrients declines, and the body alters its metabolism in an effort to maintain homeostasis.

Effects of Treatments Medications (described later in this chapter) and other treatments can significantly affect nutrition status. Preparation for diagnostic tests and surgeries, as well as the treatment of some disorders, may require that a person not eat. For the person who needs many such tests and procedures or many days of bowel rest, food intake may be very poor at a time when nutrient needs may be especially high due to illness. Consequences of some surgeries, especially those of the GI tract, often include limited food intake or altered digestion and absorption of nutrients. The location of a wound (burns affecting the hands, for example) or surgical incision may make eating difficult or painful. When treatment of a medical condition includes a highly restrictive diet, the client's motivation to eat may be impaired by limited food choices. Often people with the health problems most likely to affect nutrition status also require many treatments that compound the problem. A person with malabsorption, for example, may require a battery of tests necessitating bowel rest and then may later require surgery and a restrictive diet.

Immobility and Pressure Sores Bed rest (a treatment for some conditions) and immobility (a consequence of some medical conditions) further compromise nutrition status. Without physical activity, the muscles and bones begin to lose nitrogen and calcium, respectively, which can further contribute to malnutrition.

Immobility, poor food intake, and protein-energy malnutrition (PEM) are all associated with the development of **pressure sores.** Pressure sores can form wherever there is constant pressure on the skin. The elderly, people who are unable to respond to pain or change body positions, and people who are malnourished are most likely to develop pressure sores. Once pressure sores develop, energy and nutrient needs increase, especially if the pressure sore becomes infected.

Indirect Effects In addition to the direct effects of medical conditions on nutrition status, the costs of health care can seriously affect economic status and lifestyle choices, limiting a person's ability to obtain adequate amounts of foods and the space and equipment necessary to store and prepare them. Health problems, especially chronic diseases or terminal illnesses, can have significant effects on mental health, which can also affect nutrition status (see Nutrition in Practice 16). With chronic disorders that demand a substantial commitment of time and financial resources for management, nutrition problems are quite likely.

Responsibility for Nutrition Care

With the many potential effects of illness on nutrition status, how can health care professionals truly make a difference? For busy health care professionals with many responsibilities, it may be easy to put clients' nutrition needs on the back burner.

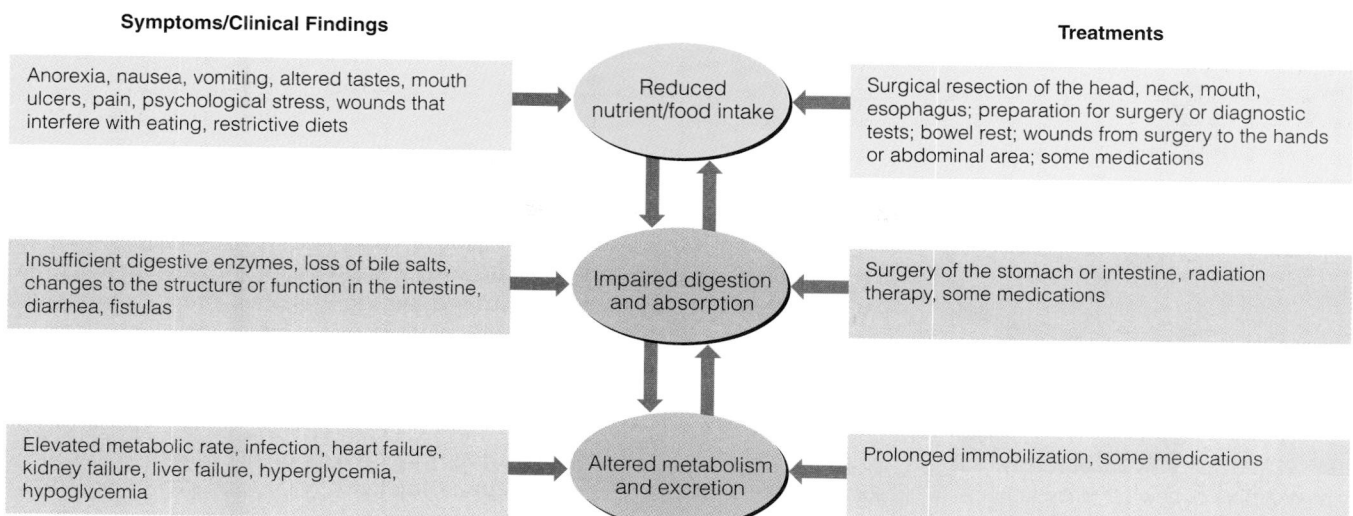

Symptoms/Clinical Findings

Anorexia, nausea, vomiting, altered tastes, mouth ulcers, pain, psychological stress, wounds that interfere with eating, restrictive diets

Insufficient digestive enzymes, loss of bile salts, changes to the structure or function in the intestine, diarrhea, fistulas

Elevated metabolic rate, infection, heart failure, kidney failure, liver failure, hyperglycemia, hypoglycemia

Reduced nutrient/food intake

Impaired digestion and absorption

Altered metabolism and excretion

Treatments

Surgical resection of the head, neck, mouth, esophagus; preparation for surgery or diagnostic tests; bowel rest; wounds from surgery to the hands or abdominal area; some medications

Surgery of the stomach or intestine, radiation therapy, some medications

Prolonged immobilization, some medications

Figure 15–2

Examples of Conditions Associated with Illnesses that Can Affect Nutrition Status

After all, the effects of nutrition therapy are not always as immediate or apparent as the effects of other treatments. Yet poor nutrition can significantly impair clients' health, affect their responses to other treatments, and delay recovery.

Health care professionals who routinely "think nutrition" can make a remarkable difference in helping clients maintain or improve health and quality of life. As an added bonus, attention to nutrition can save time and minimize health care costs by preventing, forestalling, and efficiently treating health problems and their complications.

Physicians Physicians hold the ultimate responsibility for ensuring that all the client's medical needs, including nutrition, are met. The physician prescribes the client's diet and writes the **diet order** in the medical record. The physician may also write orders that relate to nutrition care, including orders for **nutrition assessment,** diet counseling, and evaluation of the client's food intake. Physicians, in turn, rely on registered dietitians, nurses, and other health care professionals to alert them to nutrition problems, suggest strategies for handling nutrition problems, and provide requested nutrition services.

Clinical Dietitians Clinical dietitians have the primary responsibility for ensuring that clients receive optimal nutrition care. Clinical dietitians conduct nutrition assessments and provide **medical nutrition therapy.** Not all facilities employ dietitians or employ them on a regular basis, however, and even in those that do, the dietitian cannot see every client. Most often, the dietitian will routinely monitor clients with written orders for special diets or other nutrition services. Sometimes nutrition problems are overlooked when a client seeks health care, or nutrition status gradually deteriorates over the course of care; in these circumstances, the physician's written orders may not reflect the need for a dietitian's services. Consequently, nutrition problems may not be addressed in a timely manner—or at all. Thus dietitians depend on other health care professionals to alert them to clients who are having nutrition-related problems. Dietitians also rely on other health care professionals to provide information about clients' lifestyles and beliefs that affect their abilities to follow nutrition care plans.

Dietetic Technicians Dietetic technicians may take diet histories, collect information for nutrition screenings and assessments, calculate special diets, work directly with clients who are having problems with foods, and help with nutrition education. Dietetic technicians may work independently or assist dietitians.

diet order: a statement of the client's diet prescription that the physician writes in the health record.

nutrition assessment: a comprehensive evaluation of a person's nutrition status, completed by a registered dietitian. Chapter 16 provides more information.

medical nutrition therapy: the provision of a client's nutrient, dietary, and nutrition education needs based on a comprehensive nutrition assessment.

Health Problems, Medications, and Complementary Therapies

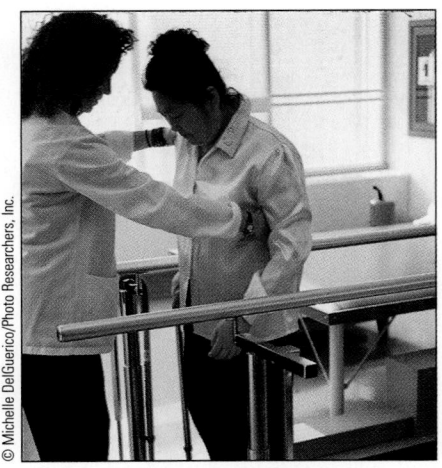

As an example of how attention to nutrition eases the tasks of health care professionals, consider the physical therapist teaching a person to walk. Optimal muscle strength, supported by an optimal intake of nutrients, makes the job easier for both the physical therapist and the client.

Nurses Nurses, with their frequent and intimate contact with clients, are in a unique position to identify clients who need nutrition services and to provide information that can ensure successful nutrition care. Nurses often use information from nursing assessments to identify clients who need complete nutrition assessments. Nurses and dietitians share responsibility for identifying diet-drug interactions, which can alter a medication's effectiveness or affect a client's nutrition status. (Diet-drug interactions receive additional attention later in this chapter.)

Nurses also provide direct nutrition services, such as encouraging clients to eat, finding practical solutions to food-related problems, measuring height and weight, recording information about a client's food intake, and answering questions about special diets. In facilities that do not employ dietitians, the nurse often assumes the primary responsibility for nutrition care.

Other Health Care Professionals Other health care professionals may also assist with nutrition care. Pharmacists, physical therapists, occupational therapists, social workers, nursing assistants, and home health care aides can be instrumental in alerting dietitians or nurses to nutrition-related problems and sharing relevant information about clients' health status or personal histories or concerns.

In Summary Illnesses and their treatments have both direct and indirect effects on nutrition status. Either or both can interfere with the intake, digestion, absorption, metabolism, and excretion of nutrients. Although physicians and dietitians shoulder the most responsibility for nutrition care, the efforts of many health care professionals ensure that clients receive optimal nutrition care.

Medications and Complementary Therapies

People today increasingly rely on **medications** and dietary supplements to prevent and treat health problems. As the population expands and people live longer, the number of people taking medications is growing rapidly. At the same time, people are using more and more complementary therapies, especially dietary supplements. Whether a person uses a vitamin C supplement to prevent a cold, an herbal remedy to treat indigestion, an over-the-counter medication to relieve pain, or a prescription medication to lower blood pressure, the potential for adverse side effects exists. The chemicals that compose medications and dietary supplements can affect metabolism and body processes just as nutrients (which are also chemicals) from foods can.

Reminder: *Dietary supplements* are chemicals (drugs) taken by mouth that contain ingredients such as vitamins, minerals, herbs, amino acids, enzymes, organ tissues, metabolites, extracts, or concentrates. Examples include vitamin-mineral supplements, creatine, melatonin, tryptophan, shark cartilage, and coenzyme Q.

Conventional and Complementary Therapies

What distinguishes conventional from complementary therapies? Conventional therapies:

- Receive the greatest emphasis in required courses in medical schools.
- Are routinely prescribed by physicians in health care facilities.
- Are generally supported by research or clinical practice.

Conventional therapies include treatments that most physicians routinely prescribe such as medications, surgery, and medical nutrition therapy. Most physicians do not routinely recommend complementary therapies, and they often do not even know when clients are using them.

medications: chemicals (drugs) that alter one or more body functions that are marketed only with approval of the Food and Drug Administration and only after research shows that they are safe and effective.

Table 15–1   Fields of Alternative Medicine and Selected Examples

Mind-body interventions	Alternative systems of medical practice	Pharmacological and biological treatments
Biofeedback	Acupuncture	Cartilage therapy
Faith healing	Ayurveda	Chelation therapy
Hypnotherapy	Homeopathic medicine	Ozone therapy
Imagery	Naturopathic medicine	Herbal medicine
Meditation	Manual healing methods	Diet and nutrition in the prevention and treatment of chronic disease
Bioelectromagnetic applications in medicine	Biofield therapeutics	Macrobiotic diets
Electroacupuncture	Chiropractic	Orthomolecular medicine
Microwave resonance therapy	Massage therapy	

In contrast, complementary therapies:

- Receive far less emphasis in medical school courses. (Many older practicing health care professionals have no formal training in complementary therapies; more recent graduates often have some training.)
- Often derive from folklore, tradition, and testimonial accounts rather than research.

Complementary therapies include herbs and other dietary supplements; other examples are listed in Table 15–1.

The Gray Areas   Distinguishing between conventional and complementary therapies is not always easy. For example, when clinical trials of complementary therapies have been conducted, they often include only small numbers of people, have not been repeated, or lack the controls necessary to draw conclusions. In fairness, however, even some conventional therapies derive largely from logic and clinical experience rather than research. An example is the clear-liquid diet traditionally provided following surgery. Such a diet is believed to be easier to tolerate after surgery than solid foods. Some clinicians question this assumption and advocate the use of easy-to-digest solid foods instead. Research to back either position is minimal.

Conversely, just because a complementary therapy is based on hearsay or folklore does not mean it is not effective. A complementary therapy may become a conventional therapy if enough research and clinical experience become available to support its use.

The Black and White Areas   One thing is clear—people are using complementary therapies regardless of whether conventional health care professionals prescribe them. In 1992, the National Institutes of Health established its National Center for Complementary and Alternative Medicine to study complementary therapies and determine which therapies are effective. More and more health care professionals are learning about complementary therapies and helping clients to use them along with conventional therapy. Medical schools today are more likely to offer elective courses in complementary therapies or discuss these therapies in required courses than in the past. Some major hospitals now have complementary medicine centers, and some health plans cover the costs of complementary therapies.

Alternative versus Complementary Therapy   The term *alternative therapy* is frequently used interchangeably with the term *complementary therapy*. We prefer the term *complementary therapy* because most people use alternative therapies in

Health care professionals face a challenge in helping clients prevent and recognize adverse side effects and interactions associated with medications and dietary supplements.

Health Problems, Medications, and Complementary Therapies

While over-the-counter medications contain less potent ingredients than prescription medications, clients may take them for the wrong reasons, in the wrong amounts, and for longer-than-recommended periods of time.

addition to, rather than in place of, conventional therapies. The remainder of this chapter discusses medications, which are conventional treatments, and dietary supplements, which are complementary treatments. Other chemicals people ingest, including alcohol (see Nutrition and Practice 23) and illegal drugs (see Chapter 13), can also affect nutrition status and interact with drugs.

Medications

Over-the-counter and prescription medications fall within conventional medical therapies. Medications make a claim to affect a health problem, and that claim must be backed up by research. Medications reach the market only after they have been extensively studied and approved by the Food and Drug Administration (FDA). Thus the benefits and risks of medications are relatively clear, and information about proper dosing is available. Medication labels, even for medicated throat lozenges, must carry general precautions and dosing information. The amount of active ingredients in each dose is also carefully controlled. Even with these safeguards in place, however, serious side effects may become evident after a product is marketed, and a previously approved medication may be withdrawn from the market.

Prescription Medications Prescription medications are often more potent and used for longer periods of time than over-the-counter medications, and they also carry a greater risk of serious side effects. Thus they require a prescription, which is intended to ensure that a physician evaluates the client's health status and determines that the benefits from using a prescription medication outweigh the risks of potential or actual side effects. Physicians, pharmacists, nurses, and dietitians share responsibility for educating clients so that they know how to use prescription medications properly and how to prevent, recognize, and handle side effects.

Over-the-Counter Medications Many medications, once available only by prescription, become available over-the-counter. Over-the-counter medications are most often intended for temporary use to relieve common but relatively minor and short-term health problems. Examples include aspirin to treat headaches or pain, decongestants to relieve stuffy noses, and antacids to combat indigestion. Although they are less potent and are used for shorter periods of time than prescription medications, over-the-counter medications may present a hidden danger by masking symptoms that require a physician's care. Because people often self-medicate with over-the-counter medications, they may use these medications for the wrong reasons, in the wrong amounts, or for longer-than-recommended periods of time. People who self-medicate may also be unaware of potential interactions with prescription medications and dietary supplements.

Dietary Supplements

Before medications became widely available, people often relied on herbs and plants to cure aches and ills with varying degrees of success. Upon scientific scrutiny, some of these folk remedies reveal ingredients that explain their value in treating health problems. The herb valerian, for example, which has long been used as a tranquilizer, contains oils that have a sedative effect. The compounds that some plants make are so beneficial, in fact, that they are isolated and used in many modern medicines. The quantities of active ingredients in the plants, however, are often much lower than in the medications, where the ingredients are isolated. Other dietary supplements have failed to show benefits; many more have simply not been studied. Many of these dietary supplements are safe, even though they may not be effective.

Nutrition in Practices 9 and 10 provided information about dietary supplements, including examples of risks associated with their use. When nutrient supplements, such as antioxidant supplements, are taken at levels that greatly exceed recommended intakes, the supplement's action is pharmacological rather than physiological; that is, the supplement acts more like a drug than a nutrient or other food component.

Marketing Unlike medications, dietary supplements can be marketed without studies to document their effectiveness and safety and without the prior approval of the FDA. Although labels of dietary supplements can make no direct claim to affect a health problem, manufacturers creatively circumvent this restriction. The label of an herbal product cannot claim that an herb alleviates insomnia. It can claim that the herb promotes restful sleep (see "Structure—Function Claims in Chapter 1). Such a claim may be made without research to support the claim provided that the label carries the disclaimer: "Has not been evaluated by the Food and Drug Administration."

Although many dietary supplements are harmless, others can be dangerous. In contrast to the rules for medications, the FDA has the burden of proving that a dietary supplement is not safe. And such proof often must be substantial. Thus some products remain on the market despite warnings from the FDA. Such is the case with ephedra (commonly known as ma huang), a product promoted as an energy booster and a weight-loss aid, even though its use has been associated with many adverse effects and even death.[1]

Active Ingredients Not only do the active ingredients in a dietary supplement not have to be proved to be safe and effective, but the amounts stated on the label can vary widely from the amounts actually present in the product. A study of the supplement DHEA, for example, found that only 7 out of 16 products contained from 90 to 110 percent of the label claim.[2] One product contained no DHEA, and two others contained only trace amounts.

Even when the active ingredients in a dietary supplement have been shown to be safe and effective, the dose suggested on the label may not provide the amount of active ingredients found to be effective. Preliminary clinical testing of the herb saw palmetto, for example, has shown some benefits over the placebo for treating some types of prostate-related problems. In a recent test, however, only 8 of 13 brands of saw palmetto provided effective amounts of the active ingredient when used as suggested on the label; the cost ranged from 44 cents to $1.44 per day. Of the remaining 5 brands, the label-recommended doses supplied as little as 1 percent of the effective dose. Thus people may spend a great deal of money on products that may or may not be safe, effective, or reliably labeled.

Other Problems Associated with Dietary Supplements Health care professionals and clients face a countless array of herbs and dietary supplements, which come in a multitude of brand names, strengths, and formulations of the same product. Although some clients consult traditional health care professionals or licensed, alternative practitioners, others simply self-medicate or ask the advice of store clerks and friends. In so doing, clients may delay or discontinue effective conventional treatments for remedies that may have little merit. The consequences can sometimes be serious and irreversible.

Furthermore, clients who use herbs and dietary supplements often fail to tell health care professionals about them. Some clients may simply forget to mention the products, especially if health care professionals don't ask. Others mistakenly believe that herbs and dietary supplements are natural and safe and pose no health risks. Still others may withhold information about herb or dietary supplement use because they feel health care professionals will disapprove or have little knowledge about these products, which is true in many cases.

People who use herbs often use them along with conventional therapies to treat or prevent minor health problems.

The regulations that govern the labeling and marketing of dietary supplements are contained in the *Dietary Supplement Health and Education Act (DSHEA)*. Nutrition in Practice 9 provides more information.

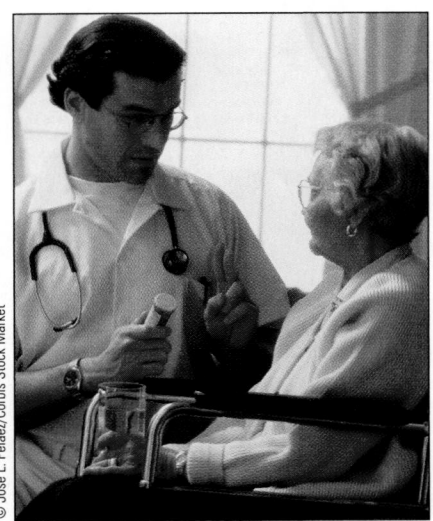

Remember to find out about all drugs clients take, including prescription and over-the-counter medications, herbs, and other dietary supplements.

Clients may also be unaware that dietary supplements can interact with medications—sometimes with potentially serious consequences. Several herbs can lead to complications during and after surgery:[3]

- Garlic, ginkgo, and ginseng raise the likelihood of bleeding.
- Ginseng can cause blood glucose to fall too low.
- Ephedra raises the likelihood of cardiovascular problems.
- Kava and valerian enhance the sedative effects of drugs used for anesthesia.
- St. John's wort increases the metabolism of many drugs.

Because research on dietary supplements is often lacking, assessing potential interactions is difficult for health care professionals and clients alike.[4] Even less is known about the effects of herbs and other dietary supplements on nutrition status. In the United States, about 44 percent of people take prescription drugs, and of these, more than 18 percent say they also use an herbal product or a megadose of a nutrient supplement.[5] Of the 18 percent who use herbal products or dietary supplements, 46 percent do so with no medical supervision.

In Summary Clearly, people use a wide variety of chemicals to prevent and treat health problems, and all of these pose risks, some known, some unknown. Unlike medications, dietary supplements can be marketed before they have been proved safe and effective. Their active ingredients may vary considerably between products. The sheer number of medications, herbs, and dietary supplements makes it difficult for health care professionals and consumers to evaluate risks, and undoubtedly, new problems will surface in the future.

Weighing Risks and Benefits

Physicians decide what medications to prescribe based on the client's health problems, any drugs currently being taken, and the potential benefits of the medication weighed against the risks. When clients self-medicate with over-the-counter medications or dietary supplements, they may not have all the information needed to make educated decisions.

The remainder of this chapter specifically addresses interactions between medications, dietary supplements, and nutrient needs or nutrition status. Keep in mind that many of the risks for diet-drug interactions also apply to medication-medication interactions. For simplicity, we use the term *diet-drug interactions* to include interactions among prescription medications, over-the-counter medications, dietary supplements, foods, individual nutrients, and other food components, such as caffeine and fiber.

Learning about Benefits and Risks

With hundreds of prescription medications, over-the-counter medications, herbs, and dietary supplements available, how do health care professionals learn about potential diet-drug interactions? This chapter provides a basic introduction and some examples, but is by no means complete.

Getting Started Learning about diet-drug interactions is difficult at first, but becomes easier with experience. Initially, looking up and learning about medications and dietary supplements encountered in clinical practice is an inescapable task. Drug guides that are available for purchase or use in health care facilities or libraries provide information about prescription and over-the-counter medications; many current guides also include information about dietary supplements.

Table 15–2 Popular Herbs, Their Common Uses, and Risks

Common Name	Claims and Uses[a]	Risks[b]
Echinacea (root)	Boosts immunity; promotes wound healing; shortens duration of colds and flus	Generally considered safe; may potentiate the effects of warfarin and interfere with immunosuppressants
Evening primrose (oil)	Relieves symptoms of premenstrual syndrome	Few studies to support safety and effectiveness
Garlic (bulb)	Lowers blood lipids and prevents atherosclerosis; lowers blood pressure	Generally considered safe; may cause mouth odors, gas, and heartburn; may potentiate the effects of warfarin, antiplatelet agents, and antidiabetic agents
Ginkgo (leaf)	Enhances memory; improves peripheral circulation	Generally considered safe; may cause headaches and GI distress; may potentiate the effects of warfarin and antiplatelet agents
Ginseng (root)	Combats fatigue; restores stamina and impaired concentration; improves sexual performance	May cause hypertension, insomnia, headaches, euphoria, diarrhea, and edema; may interact with monoamine oxidase (MAO) inhibitors, aspirin, caffeine, warfarin, heparin, anti-inflammatory agents, and antidiabetic agents
Kava (root)	Promotes sleep; relaxes muscles; reduces stress and anxiety	Generally considered safe; may interfere with the actions of central nervous system (CNS) stimulants and potentiate the effects of CNS depressants (including alcohol)
Milk thistle (fruit)	Protects liver tissue	Generally considered safe; may cause diarrhea
Saw palmetto (fruit)	Relieves symptoms of enlarged prostate; enhances sexual performance; enlarges mammary glands	Generally considered safe; may cause GI distress; may interact with medications used to treat prostate problems and steroid hormones (testosterone, oral contraceptives)
St. John's wort (leaf and flower)	Relieves depression and anxiety	Considered to be safer than prescription antidepressants; may cause GI distress, dizziness, confusion, and sedation; may interact with antidepressants, cyclosporine, digoxin, oral contraceptives, antiviral agents, theophylline, warfarin, and calcium channel blockers (anti-hypertensive agents)
Valerian (root)	Promotes sleep; relieves anxiety	Generally considered safe; may cause headaches, insomnia, and restlessness; long-term use not recommended; may potentiate the effects of CNS depressants and limit the effects of CNS stimulants

[a]Reminder: Although some dietary supplements have undergone clinical studies to support health-related claims, research is limited.
[b]Allergies are always a possible side effect.

SOURCES: J. E. Robbers and V. E. Tyler, *Herbs of Choice: The Therapeutic Use of Phytochemicals* (New York: The Haworth Herbal Press, 1999); P. Shah and K. L. Grant, An overview of common herbal supplements, *Support Line,* October 2000, pp. 3–7; S. M. Riddle, Drug interactions: Examining the impact of botanicals and dietary supplements, *Support Line,* October 2000, pp. 9–13, 16–18.

It is vital that the guide be current—many new medications reach the market each year, and new information about side effects and precautions also becomes available. Several Internet sources for information about drugs and complementary therapies are listed at the end of this chapter. Table 15–2 lists popular herbs and their common uses and risks. Remember, however, that information about the safety and effectiveness of herbs and dietary supplements is often scarce.

Evaluating Risks Health care professionals should be aware that some clients are at high risk for adverse nutrition-related side effects from medications, herbs, and dietary supplements. These include women who are pregnant or nursing and clients who:

- Take medications and dietary supplements for long periods of time.
- Regularly take several medications or dietary supplements.

Consider elderly people to be at high risk for diet-drug interactions.

- Regularly use alcohol or illegal drugs, either of which can affect nutrition status or interact with other drugs.
- Fail to use medications or other remedies as directed.
- Are in poor health, especially if they have altered organ function.
- Are in poor nutrition status.
- Have medical conditions that markedly raise nutrient needs.

Several of these conditions may coexist, especially in the elderly. Consider that elderly people are more likely than others to have:

- Chronic diseases that require the use of multiple medications over long periods of time.
- Difficulty taking medications as prescribed, either due to changes in cognitive function or due to physical or financial problems that affect their ability to acquire and pay for medications.
- Malnutrition due to chronic diseases, altered organ function, or limited food intake as a consequence of poor mental and emotional health, physical disabilities, or financial problems.

Not surprisingly, studies of institutionalized elderly people show that multiple medication use may significantly affect their nutrition status.[6]

Limiting Risks

The most important steps health care professionals can take to limit the risks of drug-related side effects is to know what and how much prescription medications, over-the-counter medications, dietary supplements, alcohol, and illegal drugs each client takes. Always ask! Clients may be taking medications prescribed by different physicians; taking health remedies recommended by alternative medicine practitioners, store clerks, trainers, or friends; or simply trying a remedy they read about. Health care professionals who use a nonjudgmental, culturally sensitive, respectful, and open-minded approach are most likely to learn the facts about what drugs clients take and why they are taking them. For clients with limited knowledge of the products they use, ask them to bring in their medications and dietary supplements.

Physician-Recommended Drugs Once health care professionals know what products the client uses, the next step is to evaluate why the client is taking them and limit the total number of products or doses the client uses whenever possible. Ideally, the client's primary physician evaluates all the client's prescription and other physician-recommended medications and makes recommendations when some of these can be eliminated or switched to a more appropriate dose or drug.

Self-Prescribed Medications For clients who regularly use over-the-counter medications and dietary supplements, find out the reasons why. Perhaps a problem needs to be investigated, or the client may be able to make lifestyle changes that eliminate the problem without the use of a drug. A client who regularly uses an over-the counter medication or herb to relieve indigestion, for example, may benefit from nutrition counseling that helps to identify foods and eating habits that contribute to indigestion (see Chapter 18). In some cases, a client who takes a vitamin or mineral supplement may be able to include more food sources of the nutrient and eliminate the need for the supplement.

For clients contemplating the use of a dietary supplement, discuss what is known about the therapy's safety and effectiveness. Recommend trying only one new supplement at a time so that the client can determine if it provides a benefit. Suggest that the client keep a symptom record, logging the severity and frequency of symptoms as well as factors that make the symptoms worse or better.[7]

Additional Recommendations Encourage clients to purchase all medications at the same pharmacy and to tell the pharmacist about their use of over-the-counter medications, dietary supplements, and prescriptions from other pharmacies. Many pharmacists track potential drug interactions using computer programs. With a complete account of clients' medication use, the pharmacist will be able to alert physicians and clients to potential problems.

In Summary Health care professionals limit the likelihood of serious side effects from medications and dietary supplements by evaluating their clients' risks for serious side effects, knowing what drugs and how much of each clients take, and eliminating the use of unnecessary drugs. Health care professionals share responsibility for explaining what is known about the safety and effectiveness, precautions, side effects, and interactions of each drug or dietary supplement.

Diet-Drug Interrelationships

Current information about diet-drug interactions involving dietary supplements is very limited, and this discussion, therefore, focuses mainly on diet-drug interactions. As more and more people use dietary supplements, however, the likelihood of additional interactions will grow. Health care professionals who understand the ways nutrients and drugs can interact will be best prepared to understand and recognize such interactions. Clinicians frequently overlook or fail to recognize diet-drug interactions, yet these interactions can raise health care costs and result in serious, and even fatal, complications.[8] With hundreds of diet-drug interactions known to exist and many more to be identified in the future, health care professionals must learn how to prevent, recognize, and handle them. The "How to" box on p. 400 offers practical suggestions.

Foods and food components, including nutrients and other components, can alter the absorption, metabolism, and excretion of medications. Likewise, medications can alter food intake and the absorption, metabolism, and excretion of nutrients. Table 15–3 on p. 401 lists the general classes of medications notable for their interactions with foods and food components. The following sections provide examples of different types of interactions.

Medications and Food Intake

Medications can reduce food intake by directly suppressing the appetite or by causing complications that make eating difficult (see Table 15–4 on p. 401). Conversely, some medications heighten the appetite and lead to weight gain. Still other medications enable the person to eat by providing relief from complications that interfere with the appetite.

Altering the Appetite Most medications prescribed for obesity work by intentionally suppressing the appetite. Amphetamines, sibutramine (Meridia), and the herb ephedra are examples. Sometimes, however, appetite suppression is unintentional. Such is the case when amphetamines are prescribed to treat attention-deficit/hyperactivity disorder. In this case, amphetamines are used to improve concentration and behavior, and appetite suppression and weight loss are undesirable side effects.

Medications such as megestrol acetate, dronabinol, growth hormone, and testosterone can improve the appetite and help people gain weight. Unintentional weight gains can result from the use of some antianxiety agents, antidepressants, and antipsychotics.

Dronabinol, a prescription medication, contains THC, the active ingredient in marijuana.

Health Problems, Medications, and Complementary Therapies

How to

Manage Diet-Drug Interactions

*B*egin with a list of medications prescribed for a particular client. Ask the client about the types and amounts of over-the-counter medications and dietary supplements used on a regular basis. Using a drug guide, look up each medication and make a note of:

- The appropriate method of administering the medication (twice daily or at bedtime, for example).
- How the medication should be given with respect to foods or specific nutrients (give on an empty stomach, give with food, do not give with milk, give iron supplements at least two hours apart from the medication dose, for example).
- How the medication should be given with respect to other medications.
- Side effects that can affect food intake (nausea, vomiting, or sedation, for example), nutrient needs (hypokalemia or hyperglycemia, for example), or dietary recommendations (constipation or flatulence, for example).

Standards set by the Joint Committee on Accreditation of Healthcare Organizations (JCAHO) include the provision that all clients be educated about potential diet-drug interactions. The physician, nurse, pharmacist, or dietitian should review all precautions related to medications and dietary supplements with the client, explain signs of potential nutrient deficiencies or nutrition-related problems, and advise the client of actions to take if problems arise.

Use a similar process for any dietary supplements the client is taking. Using a reliable reference, study what is known about the dietary supplements, how they should be taken, and their potential interactions and side effects.

Clients, particularly those who must take multiple medications, may need help figuring out when to take each medication to avoid medication-medication or diet-drug interactions. Chapter 16 describes ways to uncover information about a client's usual eating habits. Use this information to coordinate medications that must be administered with regard to food intake or specific dietary components. For additional information about a medication or potential interaction, remember to ask the pharmacist.

For medications that can lead to nutrient deficiencies or alter nutrient needs, remain alert for signs of nutrient imbalances, especially when:

- Imbalances are commonly noted with the use of the medication.
- The adverse effect persists over time.
- The client is in a high-risk group (see pp. 397–398).
- The client will need to take the medication for a long period of time.

Remember to alert the dietitian if you suspect a problem or think that the client can benefit from nutrition counseling.

Causing or Alleviating Complications Medications can lead to symptoms and complications that make it difficult to eat. Sedatives, for example, can make a person too tired to eat. Many medications including some antibiotics, many medications used in the treatment of cancer, garlic, ginkgo, saw palmetto, St. John's wort, and iron supplements can lead to indigestion and nausea and thus limit food intake. Complications that limit food intake are significant only when they persist over time. Almost all medications, for example, can cause nausea for some people. Often nausea subsides after the first few doses of the medication. If nausea persists, however, weight loss and malnutrition can follow, especially if the medication must be taken over a long period of time.

Some medications treat symptoms and complications that can reduce food intake. Antinauseants and antiemetics, for example, help reduce nausea and vomiting. The herbs peppermint and chamomile are promoted as aids to alleviate indigestion.

Contributing Energy Medications can sometimes add energy to the diet. The energy from medications that contain sugar is generally minimal. One intravenous sedative (propofol), however, is packaged in lipid. This sedative can provide a substantial number of kcalories.

Table 15–3 Medications That Can Affect Nutrition Status

Classification	Possible Side Effects That Can Affect Nutrition Status
Analgesics, narcotic	Sedation, nausea and vomiting, reduced motility of GI tract
Antacids	Constipation, diarrhea
Antibiotics	Nausea, vomiting, diarrhea
Anticonvulsants	Nausea, vomiting, GI distress
Antidepressants	Weight changes, dry mouth, nausea and vomiting, diarrhea, constipation
Antidiabetic agents	GI distress, diarrhea
Antidiarrheals	Nausea, constipation
Antifungal agents	Depressed appetite, nausea and vomiting, GI distress, diarrhea
Antihypertensives	Nausea, drowsiness, dry mouth, constipation, dizziness
Antilipemics	Nausea, GI distress, constipation
Antineoplastics	Depressed appetite, nausea and vomiting, dry mouth, taste alterations, mouth ulcers, mouth inflammation, fatigue, diarrhea, fever
Antiulcer agents	Reduced absorption of iron and vitamin B_{12}
Antiviral agents	Depressed appetite, nausea and vomiting, GI distress
Central nervous system stimulants	Depressed appetite, dry mouth, taste alterations
Corticosteroids	Nausea and vomiting, insulin resistance, altered calcium and vitamin D metabolism, negative nitrogen balance, sodium and fluid retention
Diuretics	Altered excretion of sodium, potassium, magnesium, phosphorus, calcium, and zinc
Hormonal agents	Appetite and weight changes; various other side effects depending on agent
Immunosuppressants	Nausea and vomiting, diarrhea, constipation, impaired renal function
Laxatives	GI gas, laxative dependency, nutrient malabsorption

Table 15–4 Examples of Drug-Induced Side Effects That Can Limit Food Intake

- Altered tastes
- Anorexia
- Belching
- Bloating
- Blurred vision
- Chest pain
- Confusion
- Congestion, nasal
- Constipation
- Coughing
- Cramps, abdominal
- Diarrhea
- Dizziness
- Dry mouth
- Epigastric pain
- Fatigue
- GI distress
- Indigestion
- Inflammation of mouth tissue
- Intestinal gas
- Mouth ulcers or lesions
- Nausea
- Pain
- Sedation
- Shortness of breath
- Throat irritation

Diet-Drug Interactions and Absorption

Some foods and food components affect the ways medications are absorbed and consequently how much of a medication becomes available to the body. Medications can also affect the ways some nutrients are absorbed.

Bioavailability is the term used to describe the amount of a drug that is available for use in the body.

Diet Effects on Medication Absorption Foods frequently affect medication absorption. Foods reduce the absorption of one antihypertensive drug, captopril, and improve the absorption of another, hydralazine. Thus captopril should be taken on an empty stomach, while hydralazine should be taken with food. In some cases, foods delay, but do not reduce, a medication's absorption. Aspirin works faster when taken on an empty stomach than when given with food, but because aspirin can irritate the GI tract, taking it with food can reduce nausea.

Individual nutrients and nonnutrients in foods can also affect drug absorption. Among the more common substances in foods that can bind with medications and reduce their absorption are minerals, fiber, phytates, and oxalates. Note that antacids contain a variety of minerals, which may include aluminum, calcium, magnesium, and sodium. Thus, when any of these minerals interferes with the absorption of a medication, antacids or other medications (including mineral supplements) that contain the offending mineral must be taken at another time.

Reminder: *Phytates* are nonnutrient components of food found in the husks of grains, legumes, and seeds. *Oxalates* are also nonnutrients: they are found in significant amounts in rhubarb, spinach, beets, nuts, chocolate, tea, wheat bran, and strawberries.

Medication Effects on Nutrient Absorption Laxatives provide an example of how medications can interfere with nutrient absorption. Some laxatives move foods rapidly through the intestine, reducing the time available for nutrient absorption. Other laxatives reduce nutrient absorption for different reasons. For

The minerals in dairy products interfere with the absorption of some medications, including the antibiotic tetracycline.

example, fat-soluble vitamins (notably, vitamin D) dissolve in and are excreted along with mineral oil, an indigestible oil that is sometimes used as a laxative. Calcium, too, is excreted. An added danger with all laxatives is that a person who uses them daily for a long time may find that the intestines can no longer function without them. The more often laxatives are used, the more likely that nutrient deficiencies will develop.

Other medications improve nutrient absorption. Enzyme replacements, for example, help clients with insufficient digestive enzymes to absorb protein, fat, and carbohydrate. Lactase enzyme replacements help clients with lactose intolerance to absorb lactose.

Other Absorption-Related Interactions Nutrients and medications can also interact and reduce the absorption of both. A classic example is the interaction between the antibiotic tetracycline and the minerals calcium and iron. When either of these minerals and tetracycline are taken at the same time, the mineral binds to the tetracycline, and both are excreted. To circumvent this problem, clients are instructed to take tetracycline on an empty stomach at least one hour before or two hours after meals or after using milk, milk products, calcium-containing antacids, and mineral supplements.

Diet-Drug Interactions and Metabolism

Medications taken orally and absorbed through the GI tract or those delivered directly into the circulation can interact with the components from foods that enter the bloodstream or the components from intravenous nutrients. The alterations in metabolism can affect the ways medications work, affect the availability of nutrients, or in other ways exert negative effects on body processes.

Interactions Affecting Medications Vitamin K and the anticlotting medication warfarin (Coumadin) provide an example of how a nutrient can affect the way a medication works. The chemical structure of warfarin resembles vitamin K, and it is this property that makes it an effective anticlotting agent. Warfarin interferes with the synthesis of clotting factors that require vitamin K. The prescribed warfarin dose depends, in part, on how much vitamin K is in the diet. If a person's vitamin K intake changes, as may happen in summer when lettuce and greens are in season, then the physician has to alter the medication dose. To avoid potential problems, people taking warfarin should try to eat a consistent intake of vitamin K every day, and that amount should meet dietary recommendations. A note of caution: many herbs and dietary supplements can interact with warfarin and either reduce or potentiate its effectiveness. In either case, clients should be advised to avoid these herbs and dietary supplements or to use them consistently so that the proper warfarin dose can be determined. Herbs that can interact with warfarin include danshen, don quai, echinacea, feverfew, garlic, ginger, ginkgo, ginseng, and St. John's wort.[9] Other dietary supplements that can interact with warfarin include coenzyme Q, omega-3 fatty acids, and vitamin E (doses greater than 400 IU per day).

In some cases, nutrients must be available to maximize a medication's effectiveness. The medication aldendronate sodium (Fosamax), used to increase bone mass and prevent osteoporosis, for example, depends on an adequate supply of vitamin D and calcium, either from the diet or from supplements, for maximum effectiveness. Erythropoeitin, a medication that helps correct anemia, works best when the diet provides nutrients necessary to synthesize hemoglobin, including iron, folate, and vitamin B_{12}.

Among the notable foods and food components that can affect medication metabolism are grapefruit juice (but not other citrus juices), caffeine, and natural licorice. An area currently generating a great deal of interest and research is the interaction between grapefruit juice and a variety of medications. So far,

Chemicals in grapefruit enhance the bioavailability of many medications including some used to lower blood pressure and blood lipids.

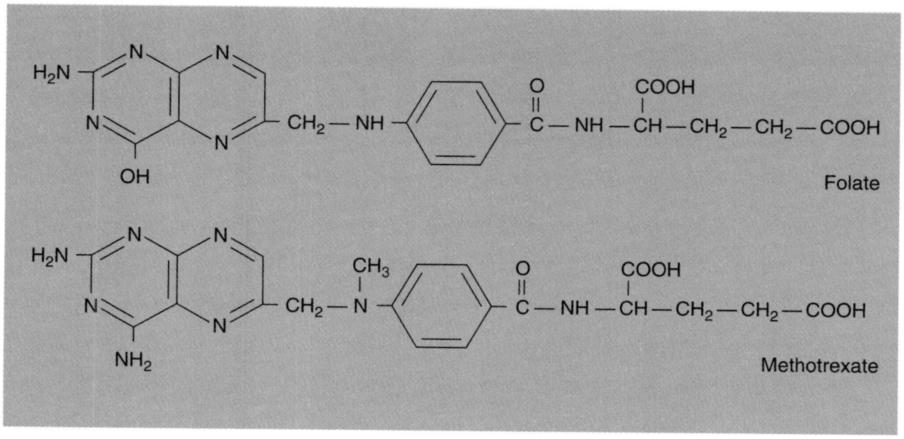

Figure 15-3
Folate and Methotrexate
By competing for the enzyme that activates folate, methotrexate prevents cancer cells from obtaining the folate they need to multiply. In the process, normal cells are also deprived of the folate they need.

grapefruit juice has been shown to raise blood levels of a variety of medications; some examples include calcium channel blockers used to treat hypertension, cholesterol-lowering medications (statins), the antianxiety agent buspirone, and the immunosuppressant cyclosporine.

Caffeine, which acts as a central nervous system stimulant, can enhance the actions of other central nervous stimulants, such as amphetamines or ephedra, and limit the effectiveness of some central nervous system depressants, such as barbituates. Natural licorice and the herb licorice root can complicate drug therapy that includes diuretics and antihypertensive agents because they promote sodium retention and potassium excretion. Most licorice sold as candy or breath fresheners in the United States is not natural licorice, however, but a flavored substitute that does not interact with medications.

Interactions Affecting Nutrients Corticosteroids, which act as anti-inflammatory agents and immunosuppressants, provide an example of how a medication can affect nutrient metabolism. Corticosteroids alter hormones that affect the way the body uses calcium and vitamin D, and in so doing, they raise the risk of osteoporosis.

Methotrexate, a medication used to treat certain cancers, provides another example. Methotrexate resembles folate in structure (see Figure 15-3). Because of this similarity, methotrexate prevents the conversion of folate to its active form, and signs of folate deficiency develop. Methotrexate may also be used to treat rheumatoid arthritis, psoriasis, and inflammatory bowel diseases, but in these cases, lower doses of the medication are prescribed, and signs of folate deficiencies are less common.

Aspirin can also alter folate metabolism but in a different way. Aspirin competes with folate for its protein carrier, thus hindering the body's use of the vitamin. When aspirin is used over long periods of time, health care professionals should ensure that either the diet or supplements are supplying sufficient folate to meet the added needs.

Other Interactions Tyramine, a substance found in some foods, and monoamine oxidase (MAO) inhibitors, medications prescribed to treat certain forms of severe depression, provide an example of how foods and medications can alter metabolism and lead to a potentially fatal outcome. MAO inhibitors block the action of the enzyme in the brain that normally inactivates tyramine. When people who take MAO inhibitors consume large amounts of tyramine, tyramine remains active and stimulates the release of the neurotransmitter norepinephrine. Severe headaches and hypertension can result, and if blood pressure rises high enough, it can be fatal. For this reason, people taking MAO inhibitors are advised to restrict their intakes of foods rich in tyramine (see Table 15-5 on p. 404).

Table 15–5	Foods Restricted in a Tyramine-Controlled Diet
Beverages	Red wines including chianti, sherry[a]
Cheeses	Aged cheeses, American, camembert, cheddar, gouda, gruyère, mozzarella, parmesan, provolone, romano, roquefort, stilton[b]
Meats	Liver; dried, salted, smoked, or pickled fish; sausage; pepperoni; salami; dried meats
Vegetables	Fava beans; Italian broad beans; sauerkraut; snow peas; fermented pickles and olives
Other	Brewer's yeast;[c] all aged and fermented products; soy sauce in large amounts; cheese-filled breads, crackers, and desserts; salad dressings containing cheese

NOTE: The tyramine contents of foods vary from product to product depending on the methods used to prepare, process, and store the food. In some cases, as little as 1 ounce of cheese can cause a severe hypertensive reaction in people taking monoamine oxidase inhibitors. In general, the following foods contain small enough amounts of tyramine that they can be consumed in small quantities: ripe avocado, banana, yogurt, sour cream, acidophilus milk, buttermilk, raspberries, and peanuts.
[a]Most wine and domestic beer can be consumed in small quantities.
[b]Unfermented cheeses, such as ricotta, cottage cheese, and cream cheese, are allowed.
[c]Products made with baker's yeast are allowed.

Diet-Drug Interactions and Excretion

Nutrients and medications can also interact and affect the way one or the other is excreted. When diet-drug interactions cause nutrients to be excreted in greater-than-normal amounts, deficiencies can develop. When diet-drug interactions cause medications to be excreted in greater-than-normal amounts, the medication may not remain available to the body for as long as intended.

Nutrient Effects on Medication Excretion Nutrients can alter urinary acidity, which, in turn, can affect the reabsorption of a medication from the kidneys back into the blood. An acidic urine limits the excretion of acidic drugs like aspirin, and megadoses of vitamin C contribute to urinary acidity. Consequently, when a person takes aspirin along with megadoses of vitamin C, the aspirin remains available to the body for longer periods of time. Dietary sodium intake greatly affects the reabsorption of the medication lithium, used to treat certain psychiatric disorders. People taking lithium are advised to maintain a consistent intake of sodium from day to day in order to maintain a stable blood level of lithium.

Medication Effects on Nutrient Excretion Medications can also alter the urinary excretion of nutrients. For example, some diuretics accelerate the excretion of calcium, potassium, magnesium, and zinc. Use of the antifungal agent amphotericin B leads to loss of potassium and magnesium in the urine.

Other Ingredients in Medications

Besides the active ingredients, medications may contain other substances such as sugar, sorbitol, lactose, sodium, and caffeine. For most people who use medications on occasion and in small amounts, such ingredients pose no problems. When medications are taken regularly or in large doses, however, people with specific problems may need to be aware of these additional ingredients and their effects.

Many liquid preparations contain sugar or sorbitol to make them taste better. People who must regulate their intakes of carbohydrates, such as people with diabetes, need to consider the amount of sugar these medications contribute to their diets. Large doses of liquids containing sorbitol may result in diarrhea. This can be a problem for adults who must use a pediatric liquid formulation to take a medication. Because they must use more of the liquid than a child would use, they may develop diarrhea with repeated use. The lactose added as a filler to some medications may cause problems for people who cannot digest lactose or those who cannot metabolize galactose.

Antibiotics and antacids often contain sodium. People who take Alka-Seltzer may not realize that a single two-tablet dose may exceed their safe sodium intakes for a whole day. Medications given by vein provide water and frequently provide sodium, potassium, and other electrolytes, or dextrose (the name for glucose in intravenous solutions). The contributions these nutrients make must be considered when a client's diet must be restricted in any of these nutrients. Interactions between medications, tube feedings, and intravenous feedings can also occur; these will be described in Chapters 20 and 21.

Reminder: Lactose is composed of two simple sugars, glucose and galactose.

In Summary Considering the many ways that drugs and nutrients can interact and the number of medications and dietary supplements available to clients, it is no wonder that serious side effects are increasingly being recognized. Health care professionals are challenged to understand the mechanisms of diet-drug interactions, identify them when they occur, and prevent them whenever possible.

Self Check

1. Nutrients with antioxidant properties may protect against some diseases by:
   a. creating mutations in DNA.
   b. halting free-radical chain reactions.
   c. stimulating free-radical production.
   d. ensuring that the body uses all of its oxygen to produce energy.

2. The term that describes the state in which free-radical production exceeds the body's ability to halt it is:
   a. inflammation.
   b. mutation.
   c. oxidative stress.
   d. free-radical hypermetabolism.

3. Inflammation occurs in response to:
   a. tissue damage.
   b. nutrient deficiencies.
   c. antioxidant nutrients.
   d. production of free radicals.

4. Nutrient imbalances:
   a. do not contribute to the development of an illness or recovery from illness.
   b. may contribute to both the development of an illness and recovery from illness.
   c. may contribute to the development of some illnesses, but seldom affect recovery.
   d. seldom contribute to the development of illnesses, but may affect recovery.

5. Health care professionals can assist dietitians in the nutrition care of clients by:
   a. calculating clients' nutrient needs.
   b. conducting nutrition assessments.
   c. providing medical nutrition therapy.
   d. alerting dietitians to clients who are having nutrition-related problems.

6. Over-the-counter and prescription medications:
   a. cannot make health claims.
   b. must be used under the supervision of a physician.

   c. do not require approval of the FDA prior to reaching the market.
   d. require research to support their safety and effectiveness prior to reaching the market.

7. Which of the following have the greatest risk of serious side effects if used inappropriately?
   a. prescription medications
   b. over-the-counter medications
   c. herbs
   d. dietary supplements

8. Over-the-counter medications:
   a. are not required to list precautions on their labels.
   b. are often used for longer periods than prescription medications.
   c. may be used for the wrong reasons and in the wrong amounts.
   d. are unlikely to react with prescription medications.

9. Compared to medications, dietary supplements are:
   a. more likely to contain a specific amount of active ingredient per dose.
   b. more likely to make health claims.
   c. less likely to have proof of safety and effectiveness.
   d. less likely to be used.

10. Adverse diet-drug interactions are more likely to occur if:
    a. the person taking the medication is malnourished.
    b. one medication or herb is taken exclusively.
    c. all organ systems are fully functional.
    d. medications or dietary supplements are taken for a few days.

11. The most important step health care professionals can take to limit the risk of drug-related side effects is to:
    a. encourage clients to take as many medications as possible.
    b. encourage clients to communicate openly about what medications, herbs, and dietary supplements they take and how much of each they are taking.

c. encourage clients to use more over-the-counter medications, herbs, and dietary supplements and fewer prescription medications.

d. encourage clients who want to try using herbs and dietary supplements to begin by using only three or four products at a time.

12. A client is taking the anticoagulant warfarin. The health care professional should periodically evaluate the client's intake of:
a. folate.
b. vitamin K and dietary supplements.
c. vitamin D, caffeine, oxalates, and phytates.
d. vitamin D, calcium, caffeine, and licorice.

Critical Thinking

1. Mr. Asmov has experienced a loss of appetite, difficulty swallowing, and mouth pain as a consequence of an illness. In this case, malnutrition may develop due to:
a. altered metabolism.
b. reduced food intake.
c. altered excretion of nutrients.
d. altered digestion and absorption.

2. The health care professional recognizes that a client who exercises daily, adheres to a low-fat diet, and uses a garlic extract to prevent heart disease is practicing:
a. alternative therapies.
b. medical nutrition therapy.
c. complementary therapies.
d. conventional medical therapies.

3. Like elderly people, premature infants are more likely than others to develop diet-drug interactions. Factors that may contribute to this increased risk include all of the following **except:**
a. premature infants are more likely to be malnourished.
b. premature infants have organ systems that are immature.
c. premature infants are more likely to be given multiple medications.
d. premature infants are more likely to grow faster and have lower intelligence.

Answers to these questions can be found in Appendix G.

Clinical Applications

1. An elderly person uses mineral oil (an over-the-counter medication) to treat chronic constipation. Drawing on the discussion of laxatives on pp. 000, discuss why elderly people are at particular risk for developing both diet-drug interactions and serious consequences from taking mineral oil. Consider the reasons and implications for the following statements to guide your thinking: elderly people are more likely to use laxatives to treat chronic constipation, elderly people are more likely to have chronic diseases and take many medications, elderly people are more likely to be malnourished, and elderly people are more likely to develop osteoporosis.

2. A client states that she has a hard time sleeping even though she is using three different herbs and a dietary supplement to help her sleep. She says she drinks 8 to 10 cups of coffee each day, the last of which she drinks after dinner. The client's job often requires her to work into the evening, and she seldom engages in physical activity. After she finishes working, she "unwinds" for a few hours by finding interesting chat rooms and surfing the Net. What dietary and lifestyle factors might contribute to the client's inability to fall asleep? Can you suggest lifestyle changes she might try? What might you suggest that she do to determine if any of the herbs or the dietary supplement she uses is helping her sleep?

Nutrition on the Net

For further study of the topics of this chapter, access these websites and search for the phrases or words in quotation marks:

Find updates and quick links to these and other nutrition-related sites at our website: **www.wadsworth.com/nutrition**

Look for information about side effects of drugs at RxList or Medscape DrugInfo: **www.rxlist.com** or **promini.medscape.com/drugdb/search.asp**

Investigate dietary supplements by visiting the Food and Drug Administration, the National

Center for Complementary and Alternative Medicine, the National Institutes of Health's Office of Dietary Supplements, and The Natural Pharmacist: **www.fda.gov** or **nccam.nih.gov** or **odp.od.nih.gov/ods** or **www.TNP.com**

Notes

1. J. B. Leikin and L. Klein, Ephedra causes myocardites, *Clinical Toxicology* 38 (2000): 353–354.

2. J. Parasrampuria, K. Schwartz, and R. Petesch, Quality control of dehydroepiandrosterone dietary supplement products, *Journal of the American Medical Association* 280 (1998): 1565.

3. M. K. Ang-Lee, J. Moss, and C. S. Yuan, Herbal medicines and perioperative care, *Journal of the American Medical Association* 286 (2001): 208–216.

4. S. M. Riddle, Drug interactions: Examining the impact of botanicals and dietary supplements, *Support Line,* October 2000, pp. 8–18.

5. D. M. Eisenberg, Trends in alternative medicine use in the United States: 1990–1997, *Journal of the American Medical Association* 280 (1998): 1569–1575.

6. C. W. Lewis, E. A. Frogillo, and D. A. Roe, Drug-nutrient interactions in long-term care facilities, *Journal of the American Dietetic Association* 95 (1994): 192–194.

7. D. M. Eisenberg, Advising patients who seek alternative medical therapies, *Annals of Internal Medicine* 127 (1997): 61–69.

8. L. Chan, Redefining drug-nutrient interactions, *Nutrition in Clinical Practice* 15 (2000): 249–252.

9. Riddle, 2000.

Nutrition in Practice

▪ A CLOSER LOOK AT THE IMMUNE SYSTEM ▪

Most people appreciate the ways that medications and dietary supplements serve to prevent and treat health problems, but they often know far less about their body's defenses for preventing health problems and limiting damage from those problems that do manage to make them sick. The **immune system** protects the body from infectious agents and toxins, enables the body to repair damaged cells, and rids the body of worn-out cells. This Nutrition in Practice presents a simplified description of the immune system and examines its relationships to nutrition. The accompanying glossary defines terms used in this discussion.

What are the central organs of the immune system?

The immune system depends on a complex interplay of many organs and cells scattered strategically throughout the body. The skin provides a physical barrier that protects the body from the countless infectious agents that

teem in the world around us. The skin is thick and coated with protective waxes that thwart attempts by infectious agents to penetrate it. Glands found within the skin also secrete chemicals that destroy some microbes.

The mucous membranes line body passages that interface with the outside environment. Mucous membranes line the eyes, nose, mouth, lungs, GI tract, and genitourinary tract. Mucous membranes produce mucus—a secretion containing chemicals and enzymes that destroy invading organisms. Mucus also forms a sticky coat that traps microbes and allows them to be safely excreted from the body.

The GI tract provides additional protection against invading organisms. Microbes that reach the stomach face destruction from highly acidic gastric juices and enzymes. Those that do survive enter the intestine, where further defenses are in place. Recall from Chapter 5 that the surface of the intestine is lined with fin-

Glossary

allergen: any substance that triggers an inappropriate immune response to a substance not normally harmful to the body.

antibodies: proteins produced by B-cells in response to invasion of the body by specific antigens.

antigen: any substance that triggers an immune system response, including bacteria, viruses, fungi, parasites, worn-out cells, and malignant cells.

asthma: a respiratory disease in which lung tissue becomes irritated and inflamed, interfering with breathing.

autoimmune disorders: disorders that result from immune system defenses attacking the body's own cells.

B-cells: lymphocytes that produce antibodies.

cell-mediated immunity: immunity conferred by T-cells traveling to the invasion site to fight specific antigens.

cytokines: special proteins that direct immune and inflammatory responses.

humoral immunity: immunity conferred by B-cells, which produce antibodies that travel through the blood to the invasion site.

immune system: the body's system of defense against harmful substances.

lymphocytes: cells made in lymph tissues that travel throughout the lymphatic and circulatory systems.

lymph tissues: tissues that contain lymphocytes.

natural killer cells: lymphocytes that confer nonspecific immunity. Natural killer cells destroy viruses and tumor cells.

nonspecific immunity: immunity directed at foreign substances in general, rather than specific antigens.

phagocytes: large white blood cells that confer nonspecific immunity. Phagocytes engulf and destroy foreign substances. Phagocytes that travel in the blood are called *monocytes;* when monocytes embed themselves in tissues, they grow larger and are called *macrophages.* Other types of phagocytes include *neutrophils, polymorphonuclear leukocytes,* and *basophils.*

specific immunity: immunity directed at specific organisms. The B-cells and T-cells confer this type of immunity.

T-cells: lymphocytes that react to specific antigens by traveling directly to the invasion site. Some T-cells (cytotoxic T-cells) kill invaders; others (helper/inducer T-cells) activate immune responses; still others (suppressor T-cells) turn off immune responses.

translocation: the movement of bacteria into the body through the intestine.

Nutrition in Practice

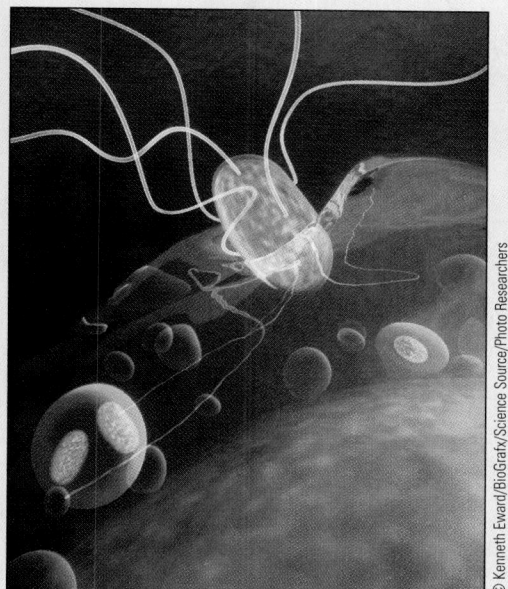

Invaders that enter the body trigger an array of immune system responses. Here a phagocyte engulfs and destroys a bacterium.

gerlike projections called villi. Healthy villi are crowded close together, forming a physical barrier that prevents microbes from passing between the villi. Interspersed among the villi are mucus-secreting cells and lymph tissue that house immune cells to fend off invaders (see Figure NP15–1 on p. 410). Consequently, substances can pass from the intestine to the inside of the body only by crossing the cells' membranes, and the cells are remarkably efficient at keeping intruders out.

The immune system allows some species of bacteria to flourish within the large intestine. These harmless bacteria help prevent the growth of harmful bacteria by competing with them for nutrients and space. They also produce short-chain fatty acids that prevent harmful bacteria from sticking to the intestinal surface.

The cells of the immune system—the **lymphocytes**—are housed in **lymph tissues.** Lymph tissues include the bone marrow, thymus, lymph nodes, spleen, tonsils, adenoids, appendix, and Peyer's patches (clumps of lymph tissue in the intestine). Lymph tissues can also be found interspersed in the mucous membranes.

How do immune system cells fit into the body's defense system?

Immune system cells continuously circulate throughout the body by way of the blood and lymphatic system, where they vigilantly search for foreign substances. They react when organisms evade the physical and chemical barriers just described and trigger an immune response. Any substance that triggers an immune response is called an **antigen.** Antigens include infectious agents (including bacteria, viruses, fungi, and parasites), worn-out and malignant cells, and tissues or cells from another person. Harmless proteins, such as milk protein, can trigger immune responses if they enter the circulation as a protein rather than amino acids. In such a case, the antigen is called an **allergen,** and the immune response is called an allergic response. In some medical disorders, immune responses are mounted against the body's own cells. Such **autoimmune disorders** include rheumatoid arthritis and lupus and some cases of diabetes mellitus.

What roles do cells play in immune responses?

Four types of white blood cells, the **phagocytes** and three types of lymphocytes, shoulder much of the responsibility for immune responses. The three types of lymphocytes are **natural killer cells, T-cells,** and **B-cells.** Figure NP15–2 on p. 411 describes immune cells, their actions, and the results of their actions. Phagocytes and natural killer cells direct their actions toward any type of antigen and confer **nonspecific immunity.** T-cells and B-cells, on the other hand, direct their actions toward specific antigens and confer **specific immunity.**

What are the differences between phagocytes and natural killer cells?

Phagocytes, large white blood cells, act as scavengers in the blood and in tissues. Phagocytes rid the body of worn-out cells and debris. When a phagocyte encounters an antigen, it engulfs and digests it in a process called *phagocytosis.* Some phagocytes play a crucial role in initiating immune responses. As these phagocytes engulf a foreign particle, they display a portion of the antigen on their own cell surface. This display triggers lymphocytes into action.

Still other phagocytes contain granules filled with potent chemicals that destroy microbes and trigger inflammatory responses (see p. 389). Inflammatory responses help limit the spread of infectious agents and promote tissue healing.

Phagocytes also secrete enzymes and special proteins called **cytokines.** Cytokines bind to receptors on target cells and direct the actions of many other cells and substances. They stimulate cell growth, activate cells, and destroy target cells, to name a few of their actions. Cytokines also cause many clinical findings associated with infections including fever and appetite suppression.

Nutrition in Practice

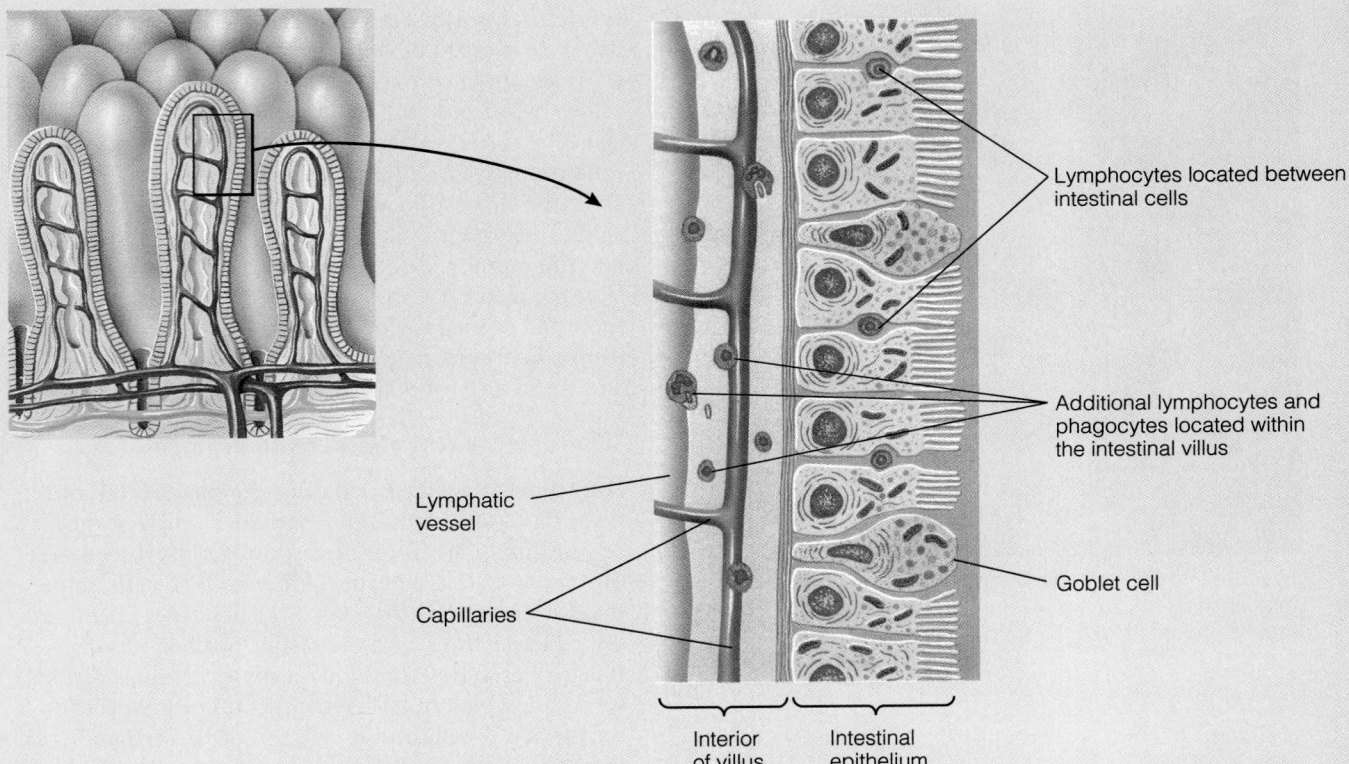

Lymphocytes located between intestinal cells

Additional lymphocytes and phagocytes located within the intestinal villus

Goblet cell

Lymphatic vessel

Capillaries

Interior of villus

Intestinal epithelium

Figure NP15–1
Immune Cells of the Intestinal Villi

Unlike phagocytes, which engulf and digest antigens, natural killer cells bind to their target and secrete chemicals that puncture the antigen's cell membrane. Natural killer cells recognize and destroy abnormal cells and many infectious agents, especially viruses. Like phagocytes, natural killer cells may also help direct immune responses by secreting cytokines.

How does specific immunity work?

Unlike phagocytes and natural killer cells, the T-cells and B-cells recognize specific antigens. A single T-cell or B-cell can attack only one type of antigen. After the body makes enough cells to destroy a particular antigen, some of the cells retain the necessary information to serve as memory cells so that the immune system can rapidly respond if the same infection should recur.

What functions do T-cells perform?

T-cells participate in **cell-mediated immunity,** so named because the cells themselves travel to the invasion site to wage a battle against foreign antigens. Like some phagocytes, some T-cells secrete cytokines that regulate immune responses. Some of the regulatory T-cells activate immune responses, while others suppress responses once an infection is under control. Still other T-cells destroy specific antigens. They rid the body of virus-infected cells or cells transformed by cancer. T-cells also lead to the rejection of newly transplanted tissue, which is why physicians prescribe drugs (immunosuppressants) to inactivate them when an organ transplant has been performed.

Do B-cells work in a different way?

Yes. The B-cells confer **humoral immunity,** so named because the cells' secretions, not the cells themselves, mount the defensive effort. B-cells respond to antigens by rapidly making cells that produce **antibodies.** Each cell derived from a given B-cell synthesizes millions of identical antibodies and pours them into the bloodstream. Some antibodies directly inactivate microbes. Others stick to the surfaces of antigens and make them easy prey for attack by phagocytes. Still others block viruses from entering cells.

Nutrition in Practice

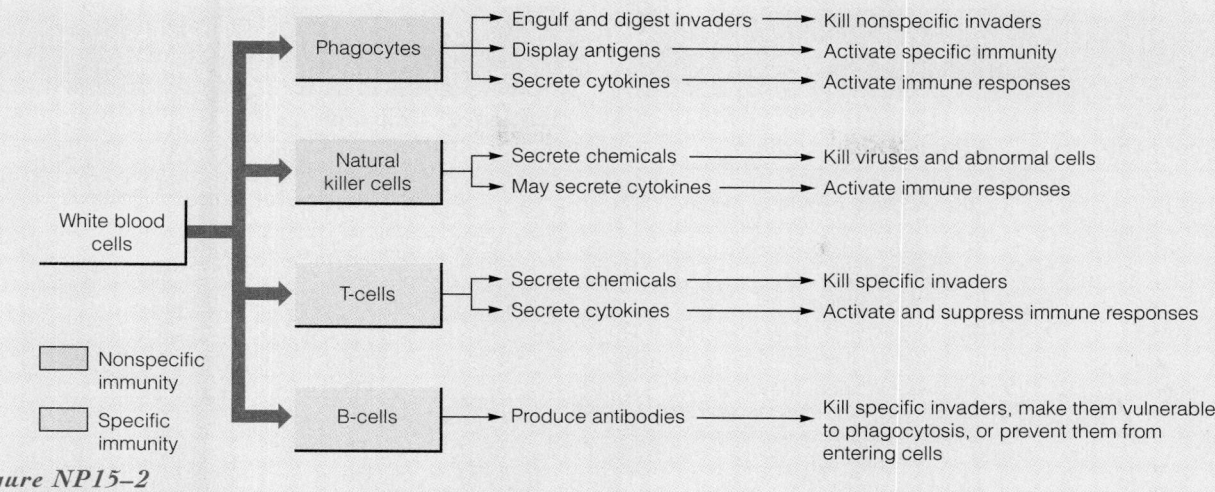

Figure NP15–2
Immune System Cells

Does nutrition affect the body's defense systems?

Nutrition has many interrelationships with the body's defense systems. As Chapter 15 described, nutrients with antioxidant properties may help limit tissue damage during inflammation by neutralizing free radicals. Protein-energy malnutrition (PEM) profoundly affects the body's ability to fight infectious agents. PEM compromises the body's physical barriers to infectious agents. The skin, for example, loses collagen and connective tissue and becomes thinner as a consequence of malnutrition. In another effect of malnutrition, the microvilli of the intestines flatten and shrink, and these physical changes may allow pathogens to cross the intestinal barrier—a process called **translocation.** Translocation may be a cause of serious infections and multiple organ failure in critically ill people (see Nutrition in Practice 27). In these cases, using the GI tract to provide nutrients (as opposed to feeding by vein) and providing specific amino acids (glutamine and arginine) may help bolster immune responses (see Chapter 27).

PEM also compromises cellular immune responses. Even moderate PEM impairs all components of immune defenses, especially the T-cells and cell-mediated immunity. The effects on T-cell function are especially important because regulatory T-cells direct further immune responses. PEM may also reduce the quantity of antibodies in mucus. People with PEM frequently develop infections, and PEM raises the risk of death from infection for children in developing countries.[1] Table NP15–1 on p. 000 summarizes the effects of malnutrition on the immune system.

More recently, researchers have found that obesity and the low-kcalorie diets often used in its treatment can also affect immune defenses. In addition, many micronutrients can affect an individual's ability to respond to antigens.

How does obesity affect the body's defense systems?

Although far less research has been conducted on the effects of obesity on immune function than the effects of PEM, studies suggest that defense mechanisms are impaired in obese individuals.[2] Specifically, the responses of T-cells and B-cells to antigens may be reduced. Low-kcalorie diets may also have these effects, although it is unclear whether it is the low-kcalorie diet, the weight loss it incurs, the amount of fat or type of fat consumed, or an inadequate intake of essential nutrients that is responsible for the changes.

How can the type of fat affect immune responses?

Both the total amount of fat in the diet and the type of fat appear to influence immune system functions. High-fat diets impair immune responses, and lowering the total fat content of the diet may improve them.[3] Specific fatty acids exert their effects on immune responses by altering the fluidity of cell membranes. The degree of cell membrane fluidity, in turn, affects the degree to which structures on cell surfaces, such as receptors, respond to stimulation, and receptors play an important role in immune responses. Diets rich in omega-3 fatty acids (such as fish oils) help modulate

Table NP15–1   Effects of Malnutrition on the Immune System

Immune System Component	Effects of Malnutrition
Lymph tissues	Thymus gland atrophied; lymph nodes and spleen smaller; T-cell areas depleted of lymphocytes
Skin	Thinner, with less connective tissue and collagen
Mucus	Reduced in quantity and amount of antibodies it contains
Intestine	Microvilli flattened; reduced number of T-cells in lymph tissue, reduced quantity of mucus with fewer antibodies
Phagocytosis	Kill time delayed
Cell-mediated immunity	Circulating T-cells and responsiveness of T-cells reduced
Humoral immunity	Responsiveness of B-cells reduced

Table NP15–2   Effects of Selected Micronutrient Deficiencies on Immune Function

Deficiency	Impairments
Vitamin A	T-cell and antibody production; response of lymphocytes, resistance to infections
Vitamin B_6	Antibody production and response of lymphocytes
Vitamin E	Phagocytosis, antibody production, response of lymphocytes, increased activity of viruses
Zinc	T-cell production, response of lymphocytes, resistance to infection

inflammatory responses and limit tissue damage—a benefit not conferred by diets rich in omega-6 fatty acids (such as corn and safflower oils).

A recent study suggests that diets rich in omega-6 fatty acids may raise the risk of **asthma** in young children.[4] In some cases of asthma, allergens from the air, foods, or drugs irritate lung tissue, triggering inflammation, which, in turn, obstructs the airways in the lungs and makes it difficult for the person to breathe. Diets rich in omega-6 fatty acids may prolong inflammation and raise the likelihood of asthma.

Do other nutrients affect immune functions?

Among the micronutrients, vitamin A appears to have a particularly strong relationship with immunity. All body surfaces, both inside and out, maintain their integrity with the help of vitamin A. These body surfaces include the skin and mucous membranes. All lymphocytes require vitamin A to develop and function properly. Vitamin A deficiency can alter the response of some antibodies to antigens and may also exert effects on the network of cytokines secreted during immune responses.

Other nutrient deficiencies that may impair immune functions include those of vitamin E, vitamin C, vitamin

B_6, and zinc. Table NP15–2 summarizes the effects of micronutrient deficiencies on immune functions.

The healthy immune system is both fascinating and remarkable in its ability to distinguish the body's own cells from microbes and malignant cells. Occasionally, the immune system is defective or fails, and serious consequences follow. Some of the many medical conditions and treatments with direct connections to the immune system that you will encounter in the remaining chapters of this book include inflammatory bowel diseases, severe stress, atherosclerosis, cancer, HIV infections, and organ transplants. This Nutrition in Practice will prepare you for a deeper understanding of these disorders and treatments and their connections to nutrition.

Notes

1. D. L. Pelletier and coauthors, The effects of malnutrition on child mortality in developing countries, *Bulletin of the World Health Organization* 73 (1995): 443–448.

2. L. Langseth, Dietary factors which alter immune responses, in *Nutrition and Immunity in Man* (Brussels, Belgium: ILSI Europe Concise Monographs, International Life Sciences Institute Press, 1999).

3. P. C. Calder, Dietary fatty acids and the immune system, *Nutrition Reviews* (supplement) 56 (1998): 70–73.

4. M. M. Haby and coauthors, Asthma in preschool children: Prevalence and risk factors, *Thorax* 56 (2001): 589–595.

Jennie Oppenheimer/Studio Zocolo

CHAPTER 16 | NUTRITION SCREENING AND ASSESSMENT

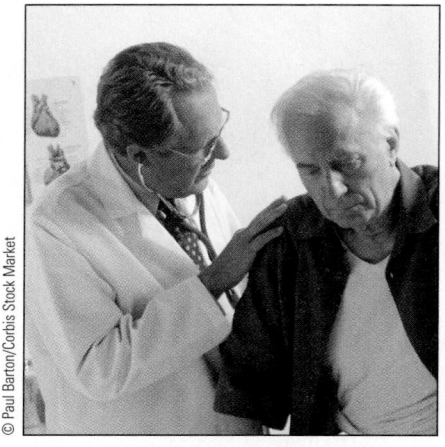

Illnesses and treatments can threaten nutrition status and quickly lead to malnutrition.

Reminder: A *nutrition assessment* is a comprehensive evaluation of a person's nutrition status.

*A*s Chapter 15 showed, illnesses and treatments can have profound effects on nutrition status. Poor nutrition status, in turn, complicates recovery from disease. Health care professionals face a challenge in identifying clients with malnutrition or those likely to develop malnutrition. This chapter briefly describes the process used to ensure that clients receive optimal nutrition care. It emphasizes techniques health care professionals use to spot clients who are malnourished or likely to become malnourished so that corrective measures can be implemented.

Identifying Nutrition Needs

Nutrition screening and nutrition assessment both identify nutrition needs. Each serves a different purpose, however. Nutrition screenings are, in a sense, mini-nutrition assessments. They identify clients with problems that often lead to malnutrition.

Nutrition Screening Nutrition screenings can be adapted for use in any health care setting. Formal procedures for nutrition screening are most likely to be used routinely in hospitals and long-term care facilities. In these settings, clients are more likely to be malnourished, but time constraints make it difficult to conduct complete assessments for all clients.

Nurses, nursing assistants, and dietetic technicians most often conduct nutrition screenings. The Joint Commission on Accreditation of Healthcare Organizations (JCAHO) recommends that a nutrition screening be completed on each client shortly after admission, ideally within 24 hours, and at regular intervals thereafter. A person may be admitted to a health care facility in good nutrition status, but nutrient stores may decline as a consequence of the client's medical condition and treatments. Declines in nutrition status during the course of hospitalization increase the likelihood of complications and the cost of care.[1]

The exact criteria that determine nutrition risk vary from facility to facility and are based, in part, on the client population. The screening criteria in a hospital that serves a pediatric population, for example, differ from those in a long-term care facility that serves a geriatric population. In general, people most likely to develop malnutrition include:

- Those with compromised nutrition status prior to developing a health problem.
- Those with one or more health problems associated with protein-energy malnutrition (PEM—described later).
- Those with no appetites or poor appetites for more than a few days.
- Those with gastrointestinal symptoms that interfere with eating or those with severe, persistent diarrhea or vomiting.
- Those who are significantly underweight or who have lost significant amounts of weight (described later in this chapter).

If a nutrition screening reveals nutrition risk, the client is referred to the dietitian for a nutrition assessment. Figure 16–1 provides an example of a nutrition screening tool.

nutrition screening: a tool for quickly identifying clients at risk for malnutrition so that they can receive complete nutrition assessments.

Nutrition Assessment Nutrition assessment provides the foundation upon which to evaluate a client's current nutrition status, determine nutrient and nutrition education needs, and plan realistic and effective interventions to meet those needs. Whereas a nutrition screening looks for conditions that place a client at risk for nutrition problems, a nutrition assessment requires a professional judgment about assessment findings and their relevance to the client's

Nutrition Screening

A. Diagnosis

If the patient has at least ONE of the following diagnoses, check the box and proceed to section E to consider the patient AT NUTRITIONAL RISK and stop here.

- ❑ Anorexia nervosa/bulimia nervosa
- ❑ Malabsorption (celiac sprue, ulcerative colitis, Crohn's disease, short bowel syndrome)
- ❑ Multiple trauma (closed-head injury, penetrating trauma, multiple fractures)
- ❑ Decubitus ulcers
- ❑ Major gastrointestinal surgery within the past year
- ❑ Cachexia (temporal wasting, muscle wasting, cancer, cardiac)
- ❑ Coma
- ❑ Diabetes
- ❑ End-stage liver disease
- ❑ End-stage renal disease
- ❑ Nonhealing wounds

B. Nutrition intake history

If the patient has at least ONE of the following symptoms, check the box and proceed to section E to consider the patient AT NUTRITIONAL RISK and stop here.

- ❑ Diarrhea (>500 mL x 2 days)
- ❑ Vomiting (>5 days)
- ❑ Reduced intake (<$\frac{1}{2}$ normal intake for >5 days)

C. Ideal body weight standards

Compare the patient's current weight for height to the ideal body weight chart on the back of this form. If at <80% of ideal body weight, proceed to section E to consider the patient AT NUTRITIONAL RISK and stop here.

D. Weight history

Any recent unplanned weight loss? ❑ No ❑ Yes Amount (lb or kg) _____

If yes, within the _____ weeks or _____ months

Current weight (lb or kg) _____

Usual weight (lb or kg) _____

Height (ft, in, or cm) _____

Find percentage of weight lost: $\frac{\text{usual wt} - \text{current wt}}{\text{usual wt}} \times 100 = $ _____ % wt loss

Compare the % wt loss with the chart values and check appropriate value

Length of time	Significant (%)	Severe (%)
❑ 1 week	1–2	>2
❑ 2–3 weeks	2–3	>3
❑ 1 month	4–5	>5
❑ 3 months	7–8	>8
❑ 5+ months	10	>10

If the patient has experienced a significant or severe weight loss, proceed to section E and consider the patient AT NUTRITIONAL RISK

E. Nurse assessment

Using the above criteria, what is this patient's nutritional risk? (check one)

- ❑ LOW NUTRITIONAL RISK

SOURCE: From Kovacevich, et al., *Nutrition in Clinical Practice*, Vol. 12, no. 1, February 1997. Reprinted with permission.

Figure 16–1
Nutrition Screening Tool

Nutrition Screening and Assessment

nutrition-related needs. Follow-up assessments uncover changes in nutrition status and help measure the success of the implemented strategies.

Registered dietitians most often conduct nutrition assessments, sometimes with the assistance of dietetic technicians. In some cases, nurses and physicians skilled in clinical nutrition may also perform complete assessments, which draw on many sources of information including:

- Health, drug, personal, and diet histories.
- Anthropometric measurements.
- Laboratory tests.
- Physical examinations.

A meaningful assessment depends on both accurate information and the careful interpretation of each finding in relation to the others. The remainder of this chapter describes the most common techniques and tests that provide clues about nutrition status. Many other measures can help evaluate nutrition status; their use varies from facility to facility.

In Summary Nutrition screenings, conducted when clients are admitted to health care facilities and at regular intervals thereafter, help identify clients who are most likely to develop malnutrition and therefore need complete nutrition assessments. Dietitians conduct complete nutrition assessments to evaluate a client's nutrition status, determine nutrient and nutrition education needs, and plan strategies to meet those needs.

Histories

Table 16–1 on p. 000 summarizes the types of information from a client's history that provide clues about nutrition status, the client's nutrient needs, and the many lifestyle and personal factors that shape the **nutrition care plan.** As Figure 16–2 on p. 000 shows, health, medications, nutrition status, and personal factors are interrelated; all must be considered to extract meaningful information. A thorough history points out potential problems requiring further investigation.

Health History A review of a client's **health history** identifies past, current, and potential health problems. Table 16–2 lists health problems that often lead to PEM. Clients found to have any of these conditions during nutrition screening receive complete nutrition assessments. Besides identifying risk for PEM, a review of the health history uncovers other nutrition-related information including:

- Current health problems that require diet modifications, including overweight.
- Potential health problems that modified diets might help to prevent or delay.
- Symptoms and clinical findings that can affect food intake or alter nutrient needs.
- Disorders and treatments that demand a great deal of time, motivation, or financial resources.
- Physical disabilities that interfere with a person's ability to purchase, prepare, and/or eat adequate amounts of food.

A review of health problems helps direct the assessment process. For example, when a person reports problems sleeping at night, the astute assessor checks the client's diet for sources of caffeine and the times of day they are consumed.

nutrition care plan: a strategy for meeting nutrient and nutrition education needs identified through nutrition assessment.

health history: an account of the client's current and past health status and risk factors for disease. Traditionally, the health history has been called the *medical history.* The term *health history* now seems more appropriate, however, because the contents describe the client's health status, and the goal of medical care is health promotion and disease prevention.

Table 16–1　Histories and Nutrition Assessments

Type of History and Significant Information	What It Identifies
Health history ■ Current health problem(s) ■ Past health problem(s) ■ Family health history ■ Previous surgeries ■ Potential health problem(s)	Health factors that affect nutrient or nutrition education needs or place the client at risk for poor nutrition status.
Medication history ■ Prescription medications ■ Over-the-counter medications ■ Herbal supplements ■ Dietary supplements ■ Illegal drugs	Medications, alternative therapies, and illegal drug use that can affect nutrient needs or alter nutrition status.
Personal history ■ Age ■ Gender ■ Cultural/ethnic identity ■ Occupation ■ Role in family ■ Educational level ■ Motivational level ■ Economic status	Factors that affect nutrient needs, influence food choices, or limit diet therapy options.
Diet history ■ Food intake ■ Eating habits and patterns ■ Lifestyle patterns	Nutrient intake and imbalances, reasons for potential nutrition problems, and dietary factors important to shaping a nutrition care plan.

Table 16–2　Health Problems Associated with Protein-Energy Malnutrition (PEM)[a]

- Acquired immune deficiency syndrome (AIDS)
- Alcoholism
- Anorexia nervosa
- Bone marrow transplants
- Burns, second or third degree covering more than 20% of body surface area
- Cancer, some types
- Chronic obstructive pulmonary disease, later stages
- Chronic pancreatitis
- Congestive heart failure, later stages
- Crohn's disease
- Cystic fibrosis
- Depression, especially in the elderly
- Diarrhea, prolonged, severe
- Dysphagia
- Feeding disabilities
- Infections
- Kidney failure, later stages
- Liver failure, later stages
- Malabsorption
- Nonhealing wounds
- Pressure sores
- Septicemia
- Surgical resection of the mouth, esophagus, and small intestine
- Trauma (tissue damage)
- Vomiting, prolonged, severe

[a]The conditions listed here frequently lead to PEM. Many other conditions can lead to malnutrition, especially if nutrition-related problems are not addressed and corrected in a timely manner.

Drug History As Chapter 15 discussed, the medications (prescription and over-the-counter) and dietary supplements the client takes can affect nutrition status. A **drug history** includes all medications and dietary supplements the client uses, the dose, and how often each is taken. Any illegal drugs used are included as well.

Personal History To determine nutrient needs and develop realistic interventions that will help clients attain their nutrition goals, the dietitian considers intellectual, spiritual, social, psychological, and financial factors to help identify influences on food choices, resources for dealing with health and nutrition problems, and constraints that may interfere with prescribed interventions. A **personal history** includes factors such as age and gender, number of people in the household, education level, occupation, income, ethnic background, cultural orientation, and religious affiliation, as well as perception of health status, level of psychological stress, and coping skills.

Personal factors help direct the assessment process. Age, for example, helps pinpoint nutrient needs and also cues the assessor to look for nutrition problems common to a specific age. For another example, if the assessor discovers that a person depends on a significant other to make health care decisions, then the assessor includes that person in the assessment process. If the assessor suspects that a client's limited financial resources may be placing a nutritious diet out of reach, then the assessor remains alert to the possibility of malnutrition throughout the remainder of the assessment.

drug history: an account of all prescription and over-the-counter medications, dietary supplements, and illegal drugs a client uses as well as the amounts the client takes and the length of time the client has taken the drugs.

personal history: an account of socioeconomic and psychosocial factors that affect a person's nutrient needs or ability or willingness to follow nutrition advice.

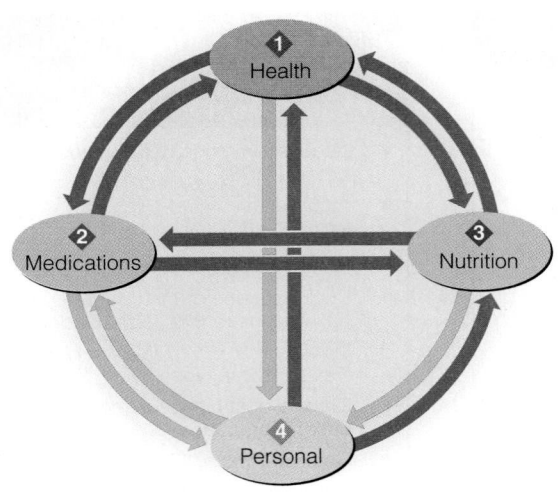

1. Health status can alter nutrient requirements, alter nutrition status, and directly affect the need for medications. Health care costs and the effects of health status on quality of life can alter lifestyle and seriously affect economic status.

2. Medications can alter nutrient requirements and affect nutrition status. Medications can have positive effects, negative effects, or both positive and negative effects on health. The costs of medications and their side effects can affect economic status, lifestyle, and quality of life.

3. Nutrition status can affect a client's health status, which, in turn, can affect the need for medications. Nutrition status and food intake can also alter the way medications are digested, absorbed, metabolized, and excreted. Although many personal factors are independent of nutrition status, poor nutrition and health status can affect a person's quality of life and have an impact on lifestyle. The costs of some nutrition therapies as well as a person's quality of life can interfere with a person's ability to maintain an optimal economic status.

4. Personal factors can have an impact on nutrient needs, food availability, and food choices. Personal factors can affect health status by altering nutrition status. Socioeconomic factors can also increase the predisposition to some diseases, such as hypertension and diabetes, and can affect people's ability to pay for health care, as well as their attitudes and beliefs about seeking health care and/or following health care recommendations.

Figure 16–2
Interrelationships among Health, Medications, Nutrition Status, and Personal Factors

Diet History Dietitians or dietetic technicians help clients complete diet histories, which include information about food intake, eating patterns, and lifestyle habits. The **diet history** uncovers current or potential nutrient imbalances and patterns and habits that support health or contribute to health problems. Table 16–3 shows examples of questions included in diet histories. An important component of the diet history is an account of how much and what a person is eating, as described in the next section.

In Summary Health, drug, personal, and diet histories provide a wealth of information about a person's nutrient needs and socioeconomic and psychosocial factors that influence the person's ability to follow nutrition advice. All of these factors assist the health care professional in understanding the person's needs so that the nutrition interventions can be realistic and attainable.

Food Intake Data

Finding out how much food a person eats and how an illness has affected food intake is important for both nutrition screenings and nutrition assessments. For people who need to make changes in diet, finding out what a person currently eats helps the dietitian evaluate the general quality of the diet and provides information necessary to devise a nutrition care plan.

Identifying Potential Food Intake Problems

Although exact guidelines vary, clients who eat less than half of their usual intakes for five or more days risk malnutrition. Assessment of weight, rate of weight loss (described later in this chapter), and symptoms and clinical findings help the dietitian clarify the severity of food intake problems and assist in identifying appropriate interventions.

Information from Clients

Nutrition screenings generally rely on clients' perceptions and accounts of their food intakes. For example, the nurse conducting a nutrition screening might ask the client, "Have you noticed any changes in the amount of food you are eating?" If the response indicates that the client has not been eating very well, the nurse then asks, "How long have you had the prob-

diet history: a comprehensive record of eating-related behaviors and the foods a person eats.

lem?" The nurse might also inquire about how much less the client has been eating. If the nurse determines that the problem is significant, the client is referred to a dietitian for a nutrition assessment.

Observation of Food Intake in Health Care Facilities For clients admitted to facilities that serve foods, health care professionals can use direct observation to identify problems with food intake by taking these steps:

- Regularly check the client's tray to be sure that foods are being eaten. Remember that appetites can change during the course of an illness, and catching problems early can help prevent serious problems later.

- If the client is to receive no food or is unable to eat, find out how long it has been since the client has eaten. Ask if the client is expected to be able to eat soon.

- If the client is unable to eat, check to see if nutrients are being delivered by tube (Chapter 20) or by vein (Chapter 21).

Communicate problems to the dietitian or physician, record the problems in the medical record, and follow up to make sure the problems are being addressed.

For some clients who are having problems with appetite, the physician may order a **kcalorie count**—a procedure that evaluates what and how much a client is eating. To conduct a kcalorie count, nurses or nursing assistants (or, in some cases, the client) write down all the foods and beverages, as well as the amounts of each, the client receives. After the client finishes the food, beverage, or meal, the amount that is left is also recorded. Dietitians or dietetic technicians retrieve the record, deduce what has been eaten, and use the information to estimate energy and nutrient intake.

Other Tools for Gathering Food Intake Data

When health care professionals need information about what and how much a client eats at home, they can use simple tools to get a picture of the client's diet. The tools described here are the 24-hour recall, the usual intake method, food frequency questionnaires, and food records. The best tool to use depends on the purpose of the record.

The 24-Hour Recall and Usual Intake Method The **24-hour recall** asks the client to recount everything eaten or drunk during the previous day. The **usual intake method** is similar, but asks the client to recount everything eaten or drunk in a typical day. For both methods, the assessor asks about the times when meals or snacks are eaten, the amounts of foods eaten, and the ways foods are prepared.

To obtain food intake data using a 24-hour recall or usual intake method, the assessor might begin by asking, "What is the first thing you ate or drank yesterday?" (Or "What is the first thing you usually eat or drink during the day?") After the client responds, the next question might be, "What time was that?" For each food the client recounts, the health care professional asks how it was prepared or eaten. Similar questions follow until the intake for the previous day or typical day is complete. When using the 24-hour recall, health care professionals ask clients if their intake that day was fairly typical. If not, the assessor finds out how it varied from the usual intake. Another approach might be to gather information from three or more 24-hour periods, being sure to include at least one weekend day.

Clients may forget to mention the beverages they drink and the fats, sweeteners, condiments, and other foods they may use, unless specifically prompted to do so. Clients may state they ate 2 slices of toast, for example, and may not think to mention that they spread 4 pats of butter and 2 tablespoons of jelly on the toast. For another example, they may mention they drank a cup of coffee, but not mention that the "coffee" was half milk or cream.

Table 16–3 Examples of Questions to Uncover Dietary Patterns or Habits

- Do you have any favorite foods?
- Do you dislike any foods?
- Are there any foods that you do not eat for any reason?
- Are you allergic or intolerant to any foods?
- Are you on a special diet?
- How many times a day do you typically eat meals and snacks?
- Where do you typically eat?
- How many times a week do you typically eat out?
- How is your appetite?
- Who does the shopping in your household?

NOTE: In addition to questions such as these, a food intake tool would be used to evaluate actual food intake.

To obtain accurate information about food intake, health care professionals must conduct interviews and ask questions in a nonjudgmental manner.

kcalorie count: a determination of a client's food intake from a direct observation of how much the client eats.

24-hour recall: a record of foods eaten by a person in the previous 24 hours.

usual intake method: a record of the foods eaten by a person in a typical day.

Nutrition Screening and Assessment

The Food Frequency Questionnaire A **food frequency questionnaire** is typically used to compare a client's usual food intake with the Daily Food Guide and thus helps to identify energy intakes, type and amount of fat in the diet, and specific nutrient imbalances. Clients may be asked how many servings of breads, cereals, or grain products; vegetables; fruits; meat, poultry, fish, and alternatives; milk, cheese, yogurt, and milk products; fats, oils, and sweets they eat in a typical day, week, or month. Clients are often also asked specific questions about the type of fat in their diets. For example, "How often do you eat red meat in a typical week?" or "What type of milk do you drink?" Food frequency questionnaires can also be used in addition to a 24-hour recall or usual intake method to verify the accuracy of a client's recollection of food intake.

The Food Record **Food records** maintained over several days provide valuable information about a client's food intake as well as the person's response to and compliance with medical nutrition therapy or tolerance for foods. The client writes down all foods and beverages consumed, times foods are eaten, amounts consumed, and methods of preparation. Depending on the purpose of the food record, the person may be asked to record other information as well. If the record is to be used to help a person change eating behaviors and lose weight, it might also include information about the person's mood, the occasion (party, holiday, family meal), behaviors associated with eating food (watching TV, driving in the car, sitting at the table with the family), and physical activity. When the purpose of the food record is to establish blood glucose control (see Chapter 24), records include details of medication administration, physical activity, illness, and the results of blood glucose monitoring. When the purpose of the record is to establish food tolerances (such as the amount of lactose a person can handle), food records also include symptoms associated with eating (for example, cramps, diarrhea, nausea, or hives).

Figure 7–7 on p. 154 provides an example of a food and activity record.

Food records help pinpoint problem food patterns so that solutions can be implemented. Unfortunately, they require a great deal of time to complete, and clients must be motivated to keep them accurately. Another drawback is that clients may either consciously or unconsciously change their eating behaviors while keeping the records. If clients understand their diets but are not following them, the clients may record what they believe they should be eating, rather than what they are actually eating.

Applying Food Intake Information In addition to helping evaluate a client's appetite and usual eating habits, information from 24-hour recalls, usual intakes, food frequencies, and food records can help identify nutrition-related problems, eating patterns that must be considered when nutrition care plans are devised, and problems the client may have with a prescribed diet. As Chapter 15 described, dietitians or nurses can also use information from food intake records to help clients plan when and how to take their medications relative to food intake.

To estimate the energy and energy nutrients in a client's diet, dietitians often use the exchange system described in Chapter 24. To check for nutrition-related problems, dietitians compare the client's intake to the Daily Food Guide and other dietary guidelines. Are all food groups included in appropriate amounts? If not, which nutrient may be excessive or deficient (see Figure 1–5 on pp. 16–17)? Are food choices varied, or is the diet monotonous? Does the diet provide adequate fiber or excessive amounts of refined sugars? Are the total amount of fat and types of fat appropriate? Is caffeine, alcohol, or salt use excessive? Dietitians may also use a computerized nutrient analysis program to evaluate the energy and nutrient content of the client's diet.

In assessing food intake data, the dietitian keeps the client's health problems in mind. Consider a client who reports problems with constipation, for example. The problem alerts the dietitian to look for sources of fiber and the volume of fluid in the client's diet.

food frequency questionnaire: a tool for gathering food intake data that asks clients about the types and amounts of foods they routinely eat.

food records: logs that list all the foods eaten over a period of time and that may also include records of behaviors and symptoms, physical activity, and medications; also called **eating** or **food diaries.**

Table 16–4   Anthropometric and Other Body Measurements

Measurement	What It Reflects
Abdominal girth measurement	Abdominal fluid retention and abdominal organ size
Hand grip strength	Ability of muscles to perform work (protein status)
Height-weight	Overnutrition and undernutrition; growth in children
%IBW, %UBW,[a] recent weight change	Overnutrition and undernutrition
Head circumference	Brain growth and development in infants and children under age two
Fatfold	Subcutaneous and total body fat
Midarm muscle circumference	Skeletal muscle mass (protein status)
Size of reaction to skin test	Immune function (protein status)
Waist circumference	Fat distribution

[a]%IBW = percent ideal body weight; %UBW = percent usual body weight.

In Summary Finding out about a client's food intake is an important component of both nutrition screenings and nutrition assessments. The client's description of food intake is a good starting point for identifying nutrition-related problems, and several tools can be used to clarify what and how much a person is eating. These tools include the 24-hour recall, the usual intake method, food frequency questionnaires, and food records.

Anthropometric Measurements

Anthropometric measurements, introduced in Chapter 1, measure physical characteristics of the body. Height and weight are the most common anthropometric measurements, and both are important components of nutrition screenings and assessments. As Table 16–4 shows, other anthropometric measurements include head circumference (described in this chapter), abdominal girth (see Chapter 23), and waist circumference (see Chapter 7). Other body measurements serve to evaluate the body's composition or ability to perform functional tasks. These include fatfold measurements, hydrodensitometry (underwater weighing), and bioelectrical impedance (see Figure 16–3 on p. 422).

Length or height, weight, and head circumference are routinely measured in children, as are height and weight in adults. Dietitians use anthropometric measurements to evaluate current nutrition status and to look for patterns—obviously high or low measurements and upward or downward trends that occur in the individual over time. They also use height and weight measurements to estimate energy and energy nutrient needs and then use the measurements again to see whether their estimates require further refinements.

Length, Height, and Weight Length measurements for infants and children up to age two or three and height measurements for older children help evaluate growth, which depends on adequate nutrition. Poor growth in children is an important indicator of malnutrition. For adults, height measurements alone do not reflect current nutrition status but help to estimate desirable weight and energy needs. Length may also be measured in adults and children who are unable to stand for physical or medical reasons. Recumbent length measurements can be significantly greater than height measurements, however, and when length measurements are used to estimate desirable weight or energy needs, the estimates must be carefully interpreted. The "How to" boxes on pp. 423 and 424 describe the proper techniques for measuring length, height, and weight.

Appendix D provides more information about waist circumference and fatfold measurements.

Measure *length* for children under age three when using growth charts from birth to 36 months to plot growth data; measure *height* for children two or older when using growth charts for children over two years.

anthropometric (an-throw-poe-MEH-trick) **measurements:** measurements of the physical characteristics of the body, such as height and weight.
- *anthropos* = human
- *metric* = measure

Figure 16–3
Methods for Assessing Body Fat

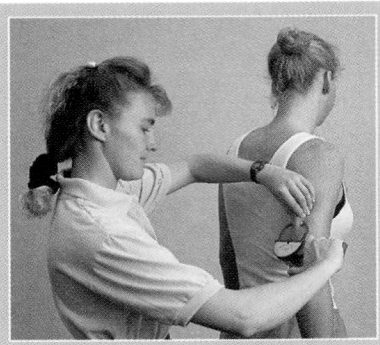

Fatfold measures: The assessor measures body fat by using a caliper to gauge the thickness of a fold of skin on the back of the arm (over the triceps), below the shoulder blade (subscapular), and in other places (including lower-body sites) and then compares these measurements with standards.

Hydrodensitometry: The assessor measures body density by weighing the person first on land and then again while submerged in water. The difference between the person's actual weight and underwater weight provides a measure of the body's volume. A mathematical equation using the two measurements (volume and actual weight) allows the assessor to calculate body density, from which the percentage of body fat can be estimated.

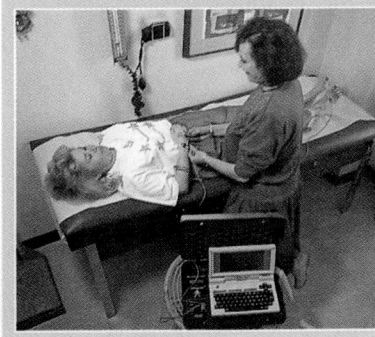

Bioelectrical impedance: The assessor measures body fat by using a low-intensity electrical current. Because electrolyte-containing fluids, which readily conduct an electrical current, are found primarily in lean body tissues, the leaner the person, the less resistance to the current. The measurement of electrical resistance is then used in a mathematical equation to estimate the total body water, lean body mass, and body fat.

© David Young-Wolff/PhotoEdit (all)

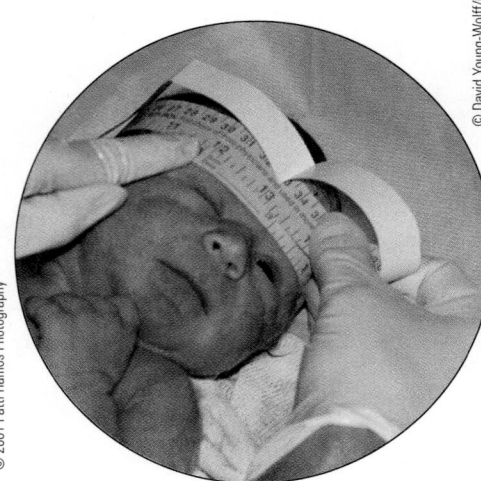

© 2001 Patti Ramos Photography

Head circumference measurements reflect brain growth.

Reminder: *The body mass index (BMI) is an index of a person's weight in relation to height, determined as follows:*

$$BMI = \frac{weight\ (kg)}{height\ (m)^2}.$$

Head Circumference Health professionals may also measure head circumference to confirm that an infant's growth is proceeding normally. To measure head circumference, the assessor places a nonstretchable tape so that it encircles the largest part of the infant's head: just above the eyebrows, just above the point where the ears attach, and around the occipital prominence at the back of the head. The measurement is recorded to the nearest ⅛ inch or 0.5 centimeter.

Analysis of Measures in Infants and Children To evaluate physical development, health professionals periodically measure length or height, weight, and head circumference (as appropriate), plot the results on growth charts (see Appendix D), and analyze the results. For children age two and over, height and weight can also be used to calculate the body mass index (BMI). Growth charts consist of percentiles, which are used to compare weight to age, length or height to age, weight to length or height, and BMI to age. Although individual growth patterns vary, ideally a child's length or height, weight, and BMI should fall roughly in the same percentile and remain at about the same percentile throughout childhood.

In addition to nutrition, genetic factors influence height; therefore, when analyzing the measurements, the health care professional considers the heights of family members and disorders that affect growth independently of nutrition.

How to

Measure Length and Height

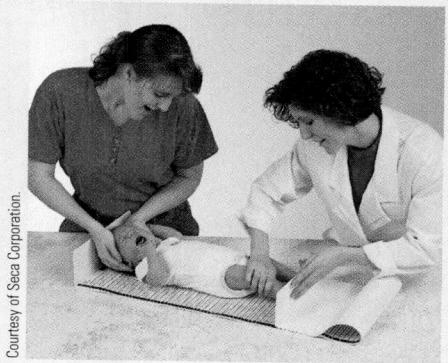

Lying with legs straight for a length measurement can be a trying experience for an infant.

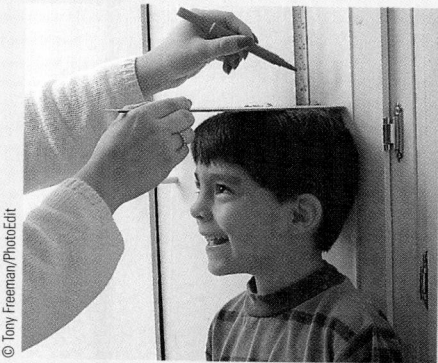

Health care professionals obtain accurate height measurements using a measuring tape or a board fixed to a wall.

*T*ips for measuring length and height include:

- Always measure—never ask! Self-reported heights are often greater than measured heights. If height is not measured, document that the height is self-reported.

- Measure length for infants and young children using a measuring board with a fixed headboard and movable footboard. It often takes two people to measure length. One person gently holds the infant's head against the headboard; the other straightens the infant's legs and moves the footboard to the bottom of the infant's feet.

  Using a measuring tape is a less exacting way to measure length in infants and is the method used to measure length for others who cannot stand erect. To use this method, straighten out the infant's or person's body, make a mark at the top of the head, make another mark at the bottom of the heel, and then measure the distance between the two marks.

- Measure height against a wall to which a nonstretchable measuring tape or a board has been fixed. Ask the person to stand erect without shoes and with heels together. The person's line of sight should be horizontal, with the heels, buttocks, shoulders, and head touching the wall. Place a ruler or other flat, stiff object on the top of the head at a right angle to the wall and carefully note the height measurement. Although less accurate, the measuring rod of a scale can also be used to measure height. To use this method, follow the same general procedure, asking the person to face away from the scale. Take extra care to ensure that the client is standing erect and that the line of sight is horizontal.

- Immediately record length and measurements to the nearest ¼ inch or 0.5 centimeter.

State of hydration influences weight measurements. Children who are retaining fluids may have weights that are deceptively high; children who are dehydrated may have weights that are deceptively low.

Length or height and weight below the 25th percentile or a sudden drop in a previously steady growth pattern suggests growth retardation, an important sign of malnutrition. A weight percentile significantly lower than the height percentile—for example, weight in the 25th percentile and height in the 95th percentile—or a BMI-for-age below the 5th percentile indicates that the child is underweight. Conversely, a weight percentile significantly higher than the height percentile—for example, weight in the 95th percentile and height in the 25th percentile—or a BMI in the 95th percentile or above indicates that the child is overweight. A BMI at or greater than the 85th percentile indicates risk for overweight.

Measure Weight

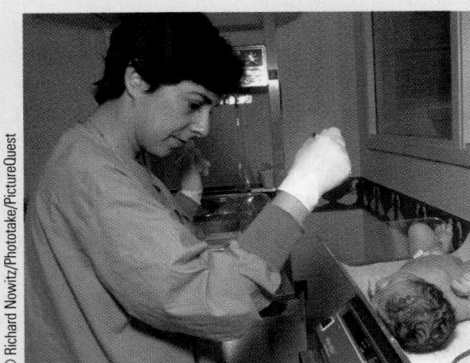

Infants are weighed on scales equipped with platforms that allow them to sit or lie down during the measurement. The scale shown here is an electronic scale.

To improve the accuracy of weight measurements, keep in mind these tips:

- Always measure—never ask! Self-reported weights are often inaccurate and are frequently lower than measured weights.
- Use the right equipment. Beam balance or electronic scales that have been calibrated and checked for accuracy at regular intervals provide the most reliable weight measurements. Infant scales are equipped with platforms that allow infants to sit or lie down while being weighed. Once children can stand erect, regular scales provide accurate weight measurements. Special scales and hospital beds with built-in scales assist in weighing clients who are unable to stand on regular scales.
- Follow standard procedures. For children under two or three (depending on the growth chart used), weigh the child without clothing or diapers. For all others, try to use the same scale and take weight measurements at about the same time of the day (preferably before breakfast), in about the same amount of clothing (without shoes), and after the person has voided.
- Record the measurement immediately either to the nearest ¼ pound or 0.1 kilogram.

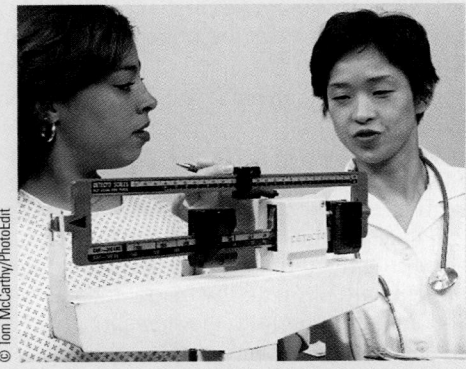

Beam balance or electronic scales provide accurate weight measurements for older children and adults.

In children whose growth has been retarded, nutrition rehabilitation will ideally induce height and weight to increase to higher percentiles. In overweight children, the goal is for weight to remain stable as height increases, until weight becomes appropriate for height.

Head circumference reflects brain growth, which occurs rapidly before birth and during early infancy and is nearly complete by age seven. Malnutrition before birth and during early childhood can impair brain development. In addition to nutrition factors, genetic variations and certain disorders can also influence head circumference.

Table 7–1 on p. 141 shows the relationship of BMI values to health risks in adults.

Analysis of Measures for Adults   For people who are healthy or whose weights have not been affected by illness, the BMI can be used to assess health risks in relation to height and weight. For people with illnesses that lead to weight loss

ow to

Estimate %IBW and %UBW

*T*o estimate %IBW, compare the client's current weight with the desirable body weight from standard weight-for-height tables:

$$\%IBW = \frac{\text{actual body weight}}{\text{desirable (ideal) weight}} \times 100.$$

For example, suppose you wish to calculate %IBW for a man who is 5 feet 8 inches tall and weighs 123 pounds. Use the Healthy Weights for Adults Table on the inside back cover to estimate desirable (ideal) weight. The desirable weight range for someone of this height is 125 to 164 pounds. Because the client is a man, begin with the upper half of the weight range (144.5 to 164 pounds). Use the midpoint (154 pounds) as the desirable weight estimate. In this example, the desirable weight is 154 pounds.

$$\%IBW = \frac{123 \text{ lb}}{154 \text{ lb}} \times 100 = 80\%.$$

The man in this example is at 80 percent of his ideal body weight. Table 16–5 on page 000 indicates that at 80 percent IBW he is mildly underweight.

This man has lost 15 pounds in the last month. To calculate his %UBW, use this formula:

$$\%UBW = \frac{\text{actual weight}}{\text{usual weight}} \times 100.$$

Calculate the usual weight (138 pounds) by adding the weight loss (15 pounds) to the current body weight (123 pounds).

$$\%UBW = \frac{123 \text{ lb}}{138 \text{ lb}} \times 100 = 89\%.$$

The man is at 89 percent of his usual body weight. A look at Table 16–5 on p. 000 reveals that a person at 89 percent UBW is mildly underweight. Based on %UBW, the degree of underweight is less severe than the %IBW implied, because this man has consistently weighed less than standard weight. Nevertheless, his recent rate of weight change is very significant: he has lost almost 4 pounds per week.

---

practitioners often use weight-for-height tables (see inside back cover) and compare a person's actual weight with desirable weight, a figure commonly known as the percent ideal body weight (%IBW). Actual weight can also be compared with usual body weight to generate the percent usual body weight (%UBW), which considers what is normal for a particular individual. Both nutrition screenings and nutrition assessments generally include an evaluation of the %IBW and %UBW.

The %UBW is particularly useful for evaluating the degree of nutrition risk associated with illness. In cases where an overweight individual becomes acutely ill and is rapidly losing weight, the health care professional relying on the %IBW may inadvertently overlook significant weight loss. Conversely, for clients who have been underweight throughout life, %IBW may overstate the degree of weight loss. The "How to" box and Table 16–5 on p. 426 show how to calculate and evaluate the %IBW and the %UBW.

In assessing weight loss, consider not only the amount of loss, but also the rate. A person who loses less than 5 percent UBW over a six-month period is at minimal risk for poor nutrition status. A loss of 5 to 10 percent UBW over six months is significant, and a loss of greater than 10 percent UBW is highly significant.

Keep in mind that a person's state of hydration can significantly influence body weight measurements. Disorders or medications that cause fluid retention can mask significant weight loss. Conversely, a person who is dehydrated may have a body weight that is deceptively low.

Reminder: Weight-for-height tables suggest a weight range, rather than pinpoint one ideal weight—a healthful reminder that several weights at any given height can be consistent with good health.

Although the term *ideal body weight* is a misnomer, it is the term most likely to be used in health care settings, and so it is used here.

Table 16–5 Weight Parameters and Nutrition Status		
%IBW	%UBW	Nutrition Status
>130	—	Obese
110–129	—	Overweight
90–109	—	Adequate
80–89	85–95	Mildly underweight
70–79	75–84	Moderately underweight
<70	<75	Severely underweight

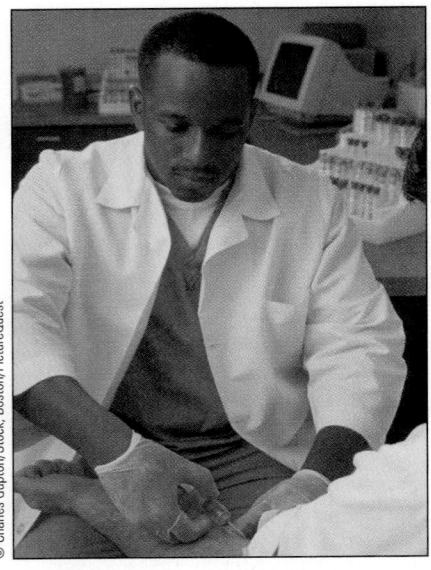

Lab tests are indirect body measurements—they help determine what is happening inside the body.

The *serum* is the watery portion of the blood that remains after removal of the cells and clot-forming material; *plasma* is the fluid that remains when unclotted blood is centrifuged. Usually, serum and plasma concentrations are similar, but plasma samples are more likely to clog mechanical blood analyzers, so serum samples are preferred.

Albumin transports many nutrients through the blood to the tissues. When albumin levels are low, the tissues may be unable to receive the nutrients they need, creating secondary deficiencies of these nutrients.

Laboratory Tests of Nutrition Status

Laboratory tests measure levels of metabolites to evaluate the body's state of health or its response to various treatments. Laboratory tests can provide information about protein-energy balance, vitamin-mineral status, fluid balance, body composition, organ function, and metabolic status. They can also help determine if nutrition therapy is appropriate or if a person is complying with a special diet. Though laboratory tests provide valuable information for nutrition assessments, lab test results may be unavailable for nutrition screenings.

This section describes some tests useful in assessing nutrition status and response to diet therapy. Other tests, such as serum glucose, cholesterol, and tests that define fluid and electrolyte balance, acid-base balance, and organ function, help pinpoint disorders or problems with nutrition implications. Table 16–6 on p. 427 lists some common laboratory tests with nutrition implications and shows what these tests reflect. Tests relevant to specific disorders will be discussed in the appropriate chapters.

Interpreting Laboratory Tests To interpret laboratory tests that fall outside the normal range requires consideration of all the factors that influence the test results. For laboratory tests that suggest malnutrition, the person's state of health and hydration can influence test results independently of nutrition factors. With dehydration, lab results may be deceptively high; with fluid retention, lab results may be deceptively low.

The body's way of handling nutrients also affects lab tests. The low blood concentration of a nutrient, for example, may reflect a primary deficiency of that nutrient, but it may also be secondary to the deficiency of one or several other nutrients involved in the transport or metabolism of that nutrient. Nutrient concentrations in the blood and urine sometimes reflect recent intakes rather than long-term intakes. Thus blood concentrations of a nutrient may be normal, even when tissue levels are deficient.

Serum Proteins and the Total Lymphocyte Count It is beyond the scope of this text to describe all lab tests used to assess nutrition status, define organ function, and develop nutrition care plans. Instead, the emphasis is on lab tests commonly used to evaluate protein status—serum proteins and the total lymphocyte count. Appendix D provides information about nitrogen balance studies and tables of lab tests useful in detecting various vitamin and mineral deficiencies, including those associated with nutrition-related anemias. Table 16–7 on p. 428 shows how serum proteins and the total lymphocyte count relate to nutrition status.

Factors Affecting Serum Protein Levels Serum protein levels reflect protein intake (the availability of amino acids) and the body's metabolism, degradation, and distribution of the specific protein. The larger the body's pool of a serum protein, and the slower its rate of degradation, the longer it takes for the protein to be affected by changes in diet. During acute stresses that cause significant inflammation, such as an extensive burn injury, some proteins shift from the blood to the extracellular and intracellular fluid. Thus lab tests of these serum proteins will reveal low levels, even though the body's pool may not be low. Furthermore, serum proteins are synthesized in the liver, so low levels can also reflect altered liver function. As Table 16–8 on p. 428 shows, a variety of medical conditions and treatments can alter levels of serum proteins and the total lymphocyte count independently of nutrition factors.

Albumin Albumin is the most abundant serum protein, and its serum levels are inexpensive to measure. Serum albumin is affected by many medical conditions and is slow to reflect changes in nutrition status because of its large body pools

Table 16-6 Examples of Routine Laboratory Tests with Nutrition Implications

Test	Uses
Hematology	
Hemoglobin (Hg)	To detect anemia and determine state of hydration.
Hematocrit (Hct)	To detect anemia and determine state of hydration.
White blood cells (WBC)	To detect infection and determine total lymphocyte count.
Mean corpuscular volume (MCV)	To detect anemia and determine its causes.
Mean corpuscular hemoglobin (MCH)	To detect anemia and determine its causes.
Mean corpuscular hemoglobin concentration (MCHC)	To detect anemia and determine its causes.
Blood Chemistry	
Proteins	
Total protein[a]	To detect PEM and various nutrient imbalances.
Albumin	To detect PEM and determine state of hydration.
Transferrin	To detect PEM and iron status, and monitor response to feeding.
Prealbumin (transthyretin)	To detect PEM and monitor response to feeding.
Electrolytes	
Sodium	To check state of hydration.
Potassium	To monitor acid-base balance and renal function and detect imbalances.
Chloride	To monitor acid-base balance and detect GI losses of chloride (from vomiting or nasogastric suctioning).
Carbon dioxide	To monitor acid-base balance.
Other	
Glucose	To detect diabetes mellitus, pancreatic tumors, and hypoglycemia and monitor glucose intolerance.
Glycated hemoglobin	To monitor blood glucose control over the past 2–3 months.
Blood urea nitrogen	To monitor renal function and determine state of hydration.
Calcium	To detect hormonal imbalances, certain malignancies, and calcium imbalances.
Phosphorus	To detect imbalances and PEM and monitor renal function and response to feeding.
Magnesium	To monitor renal function and response to feeding and detect PEM.
Cholesterol	To assess risk of heart disease and detect possibility of obstructive jaundice.
Uric acid	To detect gout and determine state of hydration.
Serum creatinine	To monitor renal function and determine state of hydration.
Serum enzymes	
Creatinine phosphokinase (CPK)	To monitor heart function and muscle damage.
Lactic dehydrogenase (LDH)	To monitor heart and renal function.
Alanine transaminase (ALT, formerly SGPT)	To monitor heart and liver function.
Aspartate transaminase (AST, formerly SGOT)	To monitor heart and liver function.
Alkaline phosphatase	To monitor liver function.
Serum amylase	To monitor pancreatic function.
Serum lipase	To monitor pancreatic function.

NOTE: This table presents a partial listing of the major uses of certain commonly performed lab tests that have implications for nutrition.
[a]More than half of the total protein is albumin.

and slow rate of degradation. In people with chronic PEM, serum albumin levels remain normal for long periods of time despite a depletion of body proteins; the levels fall only after prolonged malnutrition. Likewise, albumin concentrations increase slowly with appropriate nutrition support, so albumin is not a sensitive indicator of response to nutrition therapy.

Serum Transferrin Transferrin transports iron, and its concentrations reflect both protein-energy and iron status. As with albumin, markedly reduced transferrin levels indicate severe PEM. Transferrin breaks down in the body more

Table 16–7 Serum Proteins, the Total Lymphocyte Count, and Nutrition Status

Indicator	Degree of Depletion			
	Normal	**Mild**	**Moderate**	**Severe**
Albumin (g/100 ml)	3.5–5.4	2.8–3.4	2.1–2.7	<2.1
Transferrin (mg/100 ml)	200–400	150–200	100–149	<100
Prealbumin (mg/100 ml)	23–43	10–15	5–9	<5
Retinol-binding protein[a] (mg/100 ml)	3–7	—	—	—
Total lymphocyte count (mm³)	2500	<1500	<1200	<800

NOTE: To convert albumin (g/100 ml) to standard international units (g/L), multiply by 100. To convert transferrin (mg/100 ml) to standard international units (g/L), multiply by 0.01.
[a]Levels less than normal suggest compromised protein status. The actual degree of depletion (mild, moderate, and severe) has not been defined.

rapidly than albumin, but is still relatively slow to respond to nutrition therapy. Thus it may not be a sensitive indicator of response to nutrition therapy. Furthermore, interpreting transferrin levels as an indicator of protein-energy status is difficult when iron deficiency is present. Transferrin rises as iron deficiency grows worse and falls as iron status improves.

Prealbumin is also known as transthyretin or thyroxin-binding protein.

Prealbumin and Retinol-Binding Protein Prealbumin and retinol-binding protein levels decrease rapidly during PEM and respond quickly to changes in protein intake. Thus lab tests of prealbumin and retinol-binding protein are more sensitive tests of protein status than serum albumin, but they are also more expensive and less commonly measured in clients. Measurement of prealbumin and retinol-binding protein is often reserved for clients with disorders that markedly raise metabolic rates because such disorders can rapidly lead to malnutrition.

Total lymphocyte count (mm³) = white blood cells (mm³) × percent lymphocytes.

Total Lymphocyte Count PEM compromises the immune system, reducing the number of some white blood cells (lymphocytes), which are important in resisting and fighting infections and directing immune responses. Two laboratory

Table 16–8 Factors Affecting Serum Protein Levels and the Total Lymphocyte Count

Lab Test	Factors That Influence Values
Albumin	Chronic PEM, metabolic stress, liver disease, kidney disease (nephrotic syndrome), and eclampsia may lower serum levels.
Transferrin	Chronic PEM, metabolic stress, liver disease, kidney disease (nephrotic syndrome), and the use of some antibiotics lower serum levels; pregnancy, iron deficiency, and use of oral contraceptives raise serum levels.
Prealbumin	PEM, metabolic stress, hemodialysis, and hypothyroidism lower serum levels; kidney disease and the use of corticosteroids may raise serum levels.
Retinol-binding protein	PEM, metabolic stress, liver disease, hyperthyroidism, vitamin A deficiency, and cystic fibrosis may lower serum levels; kidney disease may raise serum levels.
Total lymphocyte count	PEM, metabolic stress, and the use of chemotherapy, immunosuppressants, and corticosteroids lower levels; infections raise levels.

tests, the number of white blood cells and the percentage of lymphocytes, are used to derive the total lymphocyte count. The total lymphocyte count is inexpensive and easy to obtain, but its value in nutrition assessment is limited because so many variables affect its level.

In Summary Anthropometric measurements and laboratory tests provide valuable information regarding nutrition status. Single measurements allow for a comparison with norms. Repeated measurements allow health care professionals to evaluate how nutrition status is affected by illness or by medical nutrition therapy. Body measurements, together with historical information and physical findings, described next, help define nutrition status and nutrient needs.

Physical signs of B vitamin deficiencies include dry, cracked lips and sores in the corners of the lips.

Physical Signs of Malnutrition

One clinician astutely summarized the role of the physical examination in nutrition assessment this way: "To me, physical examination proves the saying 'A picture is worth a thousand words.'"[2] Indeed, health care professionals can simply look at people to see if they are overweight, underweight, lethargic, confused, or unable to feed themselves, to give a few examples.

With closer examination, a skilled health care professional can learn to detect physical signs of nutrient deficiencies and excesses. Signs of malnutrition appear most rapidly in parts of the body where cell replacement occurs at a rapid rate, such as the hair, skin, and digestive tract (including the mouth and tongue). The summary tables in Chapters 8 and 9 include physical signs of specific vitamin and mineral imbalances. Table 16–9 lists physical signs of PEM and of vitamin and mineral malnutrition.

For the purposes of this discussion, physical signs of malnutrition include clinical signs such as elevated blood pressure and changes in taste perception.

Fluid Balance Among the most useful physical signs of nutrition status are those that reflect dehydration and fluid retention (see Table 16–10). Along with laboratory tests, physical signs of hydration help guide medical and nutrition

Table 16–9 Physical Signs of Nutrient Imbalances

Body System	Acceptable	Signs of Malnutrition	Other Possible Causes
Hair	Shiny, firm in scalp	Dull, brittle, dry, loose; falls out (PEM) corkscrew hair (copper)	Excessive hair bleaching; hair loss from aging, chemotherapy, or radiation therapy
Eyes	Bright, clear pink membranes; adjust easily to light	Pale membranes (iron); spots, dryness, night blindness (vitamin A); redness at corners of eyes (B vitamins)	Anemia, unrelated to nutrition; eye disorders; allergies
Lips	Smooth	Dry, cracked, or with sores in the corner of the lips (B vitamins)	Sunburn, windburn, excessive salivation from ill-fitting dentures or other disorders
Mouth and gums	Red tongue without swelling, normal sense of taste; teeth without caries; gums without bleeding, swelling, or pain	Smooth or magenta tongue (B vitamins), decreased taste sensations (zinc); swollen, bleeding gums (vitamin C)	Medications, periodontal disease (poor oral hygiene)
Skin	Smooth, firm, good color	Poor wound healing, (PEM, vitamin C, zinc); dry, rough, lack of fat under skin (essential fatty acids, PEM, B vitamins); bruising, bleeding under skin (vitamins C and K)	Poor skin care, diabetes mellitus, aging, medications
Nails	Smooth, firm, pink	Ridged (PEM), spoon shaped, pale (iron)	
Other		Dementia, peripheral neuropathy (B vitamins); swollen glands at front of neck (PEM, iodine); bowed legs (vitamin D)	Disorders of aging (dementia), diabetes mellitus (peripheral neuropathy)

Table 16–10 Physical Signs of Dehydration and Fluid Retention

Dehydration
- Sunken eyes
- Hollow cheekbones
- Dry mucous membranes
- Loss of skin turgor (elasticity)[a]
- Weak cry[b]
- Depression of the anterior fontanel[b]
- Deep, gasping respirations
- Weak, rapid pulse
- Thirst
- Reduced urinary output
- Weight loss

Fluid retention
- Edema
- Ascites (abdominal fluid retention)
- Elevated blood pressure
- Increased urinary output
- Weight gain

[a]May not be a useful parameter in the elderly.
[b]Findings specific to infants.

therapy and are important in the interpretation of weight measurements. Various medical conditions and treatments (including medications) can upset fluid balances, and signs of fluid imbalance can vary as well.

Limitations of Physical Findings Identifying and interpreting physical signs of malnutrition require knowledge, skill, and clinical judgment. Many physical signs are nonspecific; they can reflect any of several nutrient deficiencies as well as conditions not related to nutrition (see Table 16–9). For example, cracked lips may be caused by any of several B vitamin deficiencies or sunburn, windburn, or dehydration. Food intake information and laboratory tests can provide further evidence to support a suspected nutrient deficiency.

In Summary The information gathered from nutrition assessments conducted at regular intervals allows health care professionals to construct nutrition care plans and see how well they are working. Nutrition assessments include historical information, anthropometric measurements, laboratory tests, and physical examinations. The case study that follows provides practice in identifying and evaluating factors that suggest malnutrition or shape future nutrition therapy.

Case Study

NUTRITION SCREENING AND ASSESSMENT

Mrs. Genosa is an 85-year-old retired schoolteacher who has been a widow for 15 years. She has been admitted to the hospital with pneumonia and has congestive heart failure and diabetes. She routinely takes several medications, and, in addition to these, the physician has ordered antibiotics to treat the pneumonia. During an initial nutrition screening, Mrs. Genosa states that she has been eating very poorly over the past two weeks. She says that she usually weighs about 125 pounds, and that was her weight at her last physician's visit one month ago. Although she feels she has been losing weight, she doesn't know how much weight she has lost or when she started losing weight. Mrs. Genosa currently weighs 115 pounds and is 5 feet, 2 inches tall. A physical exam reveals edema, and laboratory tests confirm that she is retaining fluid. As a result of the nutrition screening, Mrs. Genosa has been referred to the dietitian for a complete nutrition assessment.

1. From the brief description provided, what factors in Mrs. Genosa's health, medication, personal, and diet histories might alert the nurse to risk for malnutrition?

2. Identify a desirable body weight for Mrs. Genosa and calculate her %IBW and %UBW. What do the results reveal? What effect does fluid retention have on Mrs. Genosa's weight?

3. How might fluid retention alter Mrs. Genosa's serum protein levels?

4. What tools might the dietitian conducting a complete nutrition assessment use to estimate what and how much Mrs. Genosa has been eating?

5. Describe some other types of assessment information the dietitian would need to develop a nutrition care plan.

Self Check

1. The purpose of nutrition screening is to:
   a. define nutrition status.
   b. determine nutrient needs.
   c. evaluate blood levels of nutrients.
   d. identify clients who need nutrition assessments.

2. All of the following factors place a client at risk for poor nutrition status **except:**
   a. a health problem frequently associated with PEM.
   b. use of several prescription medications that may affect nutrient needs.
   c. a personal history that reveals that the client lives with a spouse in a middle-income neighborhood.
   d. a significant decrease in appetite that lasts for five or more days.

3. Nutrition assessments rely on information from all of the following **except:**
   a. histories.
   b. cost of care.
   c. laboratory tests.
   d. anthropometric measurements.

4. Factors such as a person's age, education, and ethnic identity are part of the _____ history.
   a. personal
   b. health
   c. diet
   d. drug

5. Health care professionals can use anthropometric measurements to:
   a. predict laboratory test results.
   b. detect undernutrition and overnutrition.
   c. monitor disease progression.
   d. uncover health problems that interfere with eating.

6. Both height and weight measurements:
   a. are affected by fluid status.
   b. are laboratory measurements.
   c. cannot be taken on bedridden clients.
   d. are routine measurements in health care facilities.

7. The %IBW of a person who weighs 185 pounds and has a desirable body weight of 150 pounds is:
   a. 23 percent.
   b. 50 percent.
   c. 123 percent.
   d. 150 percent.

8. Laboratory tests:
   a. measure levels of metabolites.
   b. are not influenced by state of hydration.
   c. cannot be influenced by the way the body handles metabolites such as nutrients.
   d. measure metabolites in all fluid compartments so that test results reveal the body's total pool of metabolites.

9. Prealbumin and retinol-binding protein:
   a. decrease slowly as a consequence of PEM.
   b. respond slowly to changes in protein intake.
   c. are more expensive to measure than albumin.
   d. are not useful for clients with disorders that markedly raise the metabolic rate.

10. Which of the following is true about physical signs of nutrient imbalances?
   a. Physical signs are highly specific.
   b. Physical signs alone can confirm nutrient toxicities.
   c. Physical signs alone can confirm nutrient deficiencies.
   d. Physical signs can reflect both nutrient imbalances and conditions unrelated to nutrition.

Critical Thinking

1. After a nutrition screening revealed that a client had been eating very poorly at home during the past several weeks and had lost a considerable amount of weight, the client was referred to the dietitian for a complete nutrition assessment. Which food intake tool would the dietitian most likely use to get a clearer picture of how much the client has been eating?
   a. food record
   b. kcalorie count
   c. food frequency questionnaire
   d. 24-hour recall or usual intake method

2. A nutrition screening reveals that a client who weighed 160 pounds six months ago currently weighs 140 pounds. The rate of weight loss is:
   a. not significant.
   b. significant.
   c. highly significant.
   d. difficult to determine without additional information.

3. A client has just begun to eat after days without significant amounts of food. Which of the following laboratory tests would be expected to respond most quickly to changes in energy and protein intakes?
   a. albumin
   b. prealbumin
   c. transferrin
   d. total lymphocyte count

4. Physical signs of PEM might include all of the following **except:**
   a. low serum albumin.
   b. dull, brittle hair.
   c. lack of fat under the skin.
   d. wasted appearance.

Answers to these questions can be found in Appendix G.

Clinical Applications

1. Describe the possible nutrition implications of these findings from a client's history and physical examination: age 73, lives alone, recently lost spouse, uses a walker, no teeth, pale skin, lack of energy, history of hypertension and diabetes, several medications prescribed.

2. Nurses and nurse's aides frequently shoulder much of the responsibility for collecting food intake data for kcalorie counts because they often deliver food trays and snacks and later retrieve them. Why is it important for a nurse or aide to verify and record both what the client receives (both the foods and the amounts) and the foods that remain uneaten? When might clients be enlisted to aid in the collection of food intake data, and when might such a course be unwise?

3. Calculate the %IBW and %UBW for a man who is 5 feet 11 inches tall with a current weight of 160 pounds and a usual body weight of 180 pounds. Use the desirable body weight table on p. 000 to estimate an ideal body weight. What additional information will be important for you to find out about this man's weight loss?

Nutrition on the Net

For further study of the topics of this chapter, access these websites and search for the phrases or words in quotation marks:

Find updates and quick links to these and other nutrition-related sites at our website:
www.wadsworth.com/nutrition

Locate a food frequency questionnaire and assessment tool for children at the National Network for Child Care. Go to the site map, click on "nutrition," and then click on "How's Your Nutritional Health":
www.nncc.org/Nutrition/health.quiz.html

Calculate your nutrient intake using a nutrient analysis program at NAT Tools for good health:
www.nat.umc.edu

Complete a health assessment that includes questions about nutrition at Ask the Dietitian:
www.dietitian.com/ibw/ibw.html

Find out more about body composition analysis, including underwater weighing and bioelectrical impedance, at the "Virtual" Nutrition Center:
www.sci.lib.ucl.edu/HSG/Nutrition.html HYDRO

Notes

1. C. Braunschweig, S. Gomez, and P. M. Sheean, Impact of declines in nutritional status on outcomes in adult patients hospitalized for more than 7 days, *Journal of the American Dietetic Association* 100 (2000): 1316–1322.

2. K. Hammond, Nutrition focused physical assessement, *Support Line,* August 1996, pp. 1–4.

Nutrition in Practice

■ NUTRITION AND MENTAL HEALTH ■

Mental and nutritional health go together. Mentally healthy people have the capacity to feed themselves well. Well-nourished people experience none of the nutrient deficiencies that might contribute to poor mental health. By the same token, when either type of health is impaired, both may be affected. People who frequently experience strong emotions or those with mental health problems often have poor diets. Conversely, people who are malnourished may experience apathy and are often in poor mental health. It is important to understand these connections, for they can affect everyone from the man or woman on the street to the person institutionalized with a severe psychiatric illness. The professional who recognizes the relationships between nutrition and mental health is in a better position to offer effective care. The accompanying glossary defines terms associated with mental health.

How can emotions affect the nutritional health of ordinary people?

To understand the connections between emotions and nutritional health, consider what ordinary anxiety does to your own eating habits. Do you lose your appetite? Overeat? Eat "junk" foods instead of balanced meals? Your reaction depends on your personality type. Temporary emotional stresses may have little effect on nutrition, but if the stress persists over time or occurs frequently, the resulting changes in eating habits can lead to underweight, overweight, or nutrient imbalances. People who are unable to cope with emotional and physical stresses frequently develop depression, and depression carries with it an even greater risk of malnutrition.

How can depression lead to nutrition problems?

Quite often people who are depressed lose interest in caring for themselves and in participating in usual activities such as eating, socializing, or pursuing hobbies. When people fail to care for themselves and cut themselves off from pleasurable activities and friendships, depression deepens and becomes a self-aggravating condition. People with depression may feel worthless, hopeless, drained of energy, and unable to sleep and concentrate. People who are depressed are likely to have little interest in preparing and eating food.

As Chapter 15 described, physical illnesses can also lead to depression. Not surprisingly, many health problems raise the likelihood of both depression and malnutrition. The pain, loss of physical independence, and economic hardships imposed by serious illness or deteriorating health, as well as by some medications, can lead to both depression and malnutrition. The emotional adjustments necessary to accept and handle a terminal illness, such as an HIV infection, can also lead to both depression and malnutrition. Once people who become ill lose significant amounts of weight, depression may deepen as their physical appearance deteriorates, and they become unable to perform routine tasks and maintain quality of life.

Although depression and its consequences affect people of all ages, depression is pervasive among the elderly and is a common cause of weight loss,

Glossary

delusions (dee-LOO-shuns): inappropriate beliefs not consistent with the individual's own knowledge and experience.

dementia (dee-MEN-she-ah): irreversible loss of mental function.

emotions: mental states such as love or hate that arise from subjective experiences rather than conscious efforts.

mood disorders: mental illness characterized by episodes of severe depression or excessive excitement (mania) or both.

paranoia (PAR-ah-NOY-ah): mental illness characterized by delusions of persecution.

schizophrenia (SKITZ-oh-FREN-ee-ah): mental illness characterized by an altered concept of reality and, in some cases, delusions and hallucinations.

senility (see-NIL-ih-tee): mental or physical weakness associated with old age.

Nutrition in Practice

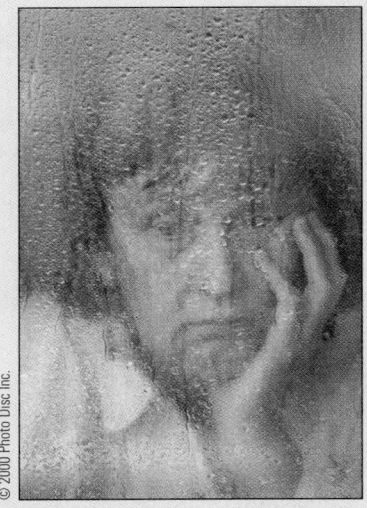

Depression and loneliness can profoundly affect nutrition status.

especially for those in nursing homes.[1] Depression may be overlooked in the elderly because health care professionals may mistakenly regard depression as a normal consequence of aging.[2] Thus depression may progress unrecognized and lead to serious problems.

Why is depression more prevalent in the elderly?

The elderly often experience the serious mental and physical stresses that can lead to depression. Depression in the elderly often results from loneliness associated with social isolation and the loss of loved ones, mobility, or sense of purpose. Health problems, physical disabilities, and the use of multiple medications may also contribute to depression in the elderly. Many authorities believe that among the elderly, loneliness is particularly relevant to depression and malnutrition. For human beings, eating is as much a social and psychological event as a biological one. Without companionship, appetite diminishes. Some 6 million adults over age 65 live alone. Their most pressing need seems to be for companionship; food takes second place. Social interaction is important to mental health, and elderly people of all classes in our society, both the financially secure and the poverty-stricken, tend to become isolated. Jack Weinberg, professor of psychiatry at the University of Illinois, wrote perceptively of this problem:

> In our efforts to provide the aged with a proper diet, we often fail to perceive it is not what the older person eats but with whom that will be the deciding factor in proper care for him. The oft-repeated complaint of the older patient that he has little incentive to prepare food for only himself is not merely a statement of fact but also a rebuke to the questioner for failing to perceive his isolation and aloneness and to realize that food . . . for one's self lacks the condiment of another's presence which can transform the simplest fare to the ceremonial act with all its shared meaning.[3]

A sad spiral can set in when a lonely person begins to neglect to eat well. Malnutrition worsens the apathy felt due to loneliness. Then the person has even less energy and less desire to secure nourishment.

Can something be done to prevent such a spiral?

Caring health care professionals remain alert for signs of depression in people of all ages, especially elderly people who live alone and those who have recently lost a spouse or loved one. When depression is recognized early, appropriate treatment can help prevent a decline in health and nutrition status. Counselors and social workers can help elderly clients work through depression and find solutions to their loneliness. Health care professionals can help clients understand how depression affects food intake and health and how eating a well-balanced diet can prevent health and nutrition problems. Meal plans that emphasize foods that are easy to prepare and eat can help some clients meet their nutrient needs when they lack the motivation to eat. Encouraging clients to eat with family or friends or at congregate meal sites can help combat loneliness.

In what ways can nutrient deficiencies affect mental health?

Among all age groups, nutrient deficiencies can affect mental health. Protein-energy malnutrition during pregnancy, for example, may lead to mental retardation in the infant. Children malnourished early in life often exhibit behavioral and social deficits as well as physical retardation. Conversely, children who are neglected early in life show a greater tendency to suffer from severe malnutrition than children who receive love and attention.

Individual nutrient deficiencies can also affect brain function. Table 14–2 on p. 371 lists some brain functions affected by nutrient deficiencies. People with B vitamin deficiencies often exhibit symptoms ranging from confusion, apathy, fatigue, memory deficits, and irritability to delirium and psychoses. Severe niacin deficiencies, for example, can lead to **dementia**.[4] Deficiencies of folate and vitamin B_{12} are also associated with memory loss, depression, and dementia.[5] Depression is a common manifestation of folate deficiency,

Nutrition in Practice

and conversely, people diagnosed with depressive disorders frequently have low serum or red blood cell folate levels.[6] In the latter case, it is not clear whether folate deficiency leads to depression or depression leads to low folate levels by altering food intake.

Is senility in the elderly related to malnutrition?

Sometimes the confusion caused by nutrient deficiencies is incorrectly diagnosed as **senility.** An elderly person may even be wrongly confined to a nursing home. The story is told of a woman who exhibited the classic signs of senility—mental confusion, inability to make decisions, and forgetting to perform important tasks, such as turning off a stove burner. The woman's family decided to move her into a nursing home. While she was waiting for a place there, her family took her into their own home. After several weeks of eating good meals and enjoying social stimulation, the woman became her old self again and was able to return to her home. This story has been repeated with many variations and serves to remind us to think about loneliness and nutrition before concluding that a person is senile and needs institutional care. What harm could there be in first trying good, balanced meals served with tender, loving care?

Suppose a person is truly mentally ill. Then what considerations apply?

Nutrition in psychiatric care is a specialty all its own because there are so many connections. Mental illnesses characterized by depression, illogical thinking, dementia, **paranoia, delusions,** and inappropriate eating habits can alter food intake and thus interfere with nutrition status. Some of these disorders include **schizophrenia,** Alzheimer's disease (see p. 369), **mood disorders,** substance abuse, and eating disorders.

People with mental disorders characterized by depression risk poor nutrition status for reasons already described. People with mental illnesses charac-terized by illogical thinking or dementia may have little interest in food or may be unable to make appropriate food choices. Those who are paranoid may believe that foods are being used to poison them. People suffering from delusions may attribute magical powers to certain foods and insist on eating only those foods. Medications used in the treatment of mental illnesses can also interact with nutrients and alter nutrition status.

What mental disorders affecting nutrition are especially common?

Among the most common mental disorders with considerable effects on nutrition status are alcoholism and eating disorders. The relationships between alcoholism and nutrition are treated in Nutrition in Practice 23, and eating disorders are discussed in Nutrition in Practice 7.

Nutrition affects the brain and the mind, and the brain and the mind affect the way people eat. All are interrelated, and the wise health care professional will keep these interrelationships in focus.

Notes

1. G. K. Kennedy, The geriatric syndrome of late-life depression, *Psychiatric Services* 46 (1995): 43–48.

2. C. Ryan and M. E. Shea, Recognizing depression in older adults: The role of the dietitian, *Journal of the American Dietetic Association* 96 (1996): 1042–1044.

3. J. Weinberg, Psychological implications of the nutritional needs of the elderly, *Journal of the American Dietetic Association* 60 (1972): 293–296.

4. J. E. Morley, Nutritional modulation of behavior and immunocompetence, *Nutrition Reviews* (supplement 2) 52 (1994): 6–8.

5. T. Bottiglieri, Folate, vitamin B_{12}, and neuropsychiatric disorders, *Nutrition Reviews* 54 (1996): 382–390.

6. J. E. Alpert and M. Fava, Nutrition and depression: The role of folate, *Nutrition Reviews* 55 (1997): 145–149.

Jennie Oppenheimer/Studio Zocolo

*C*hapter 15 showed how illnesses and treatments affect nutrient needs, and Chapter 16 described some tools and techniques health care professionals use to evaluate clients' nutrition status. This chapter describes the ways dietitians estimate nutrient needs and complete nutrition care plans. Assuring that a person's nutrient needs are met is a key part of this process, so this chapter also describes modified diets and looks at how the client's needs are communicated among health care team members.

Identifying Nutrient Needs

The client's age, health problems, treatments, current nutrition status, and activity level provide the information the dietitian needs to estimate energy and nutrient needs. The earlier chapters of this book described methods for estimating the energy and protein needs of healthy people. These estimates serve as reasonable estimates for people with stable chronic disorders, such as diabetes and heart disease. As the chapters that follow show, many acute health problems and end-stage chronic diseases alter energy and nutrient needs.

Chapter 6 describes methods for calculating energy needs. Chapter 4 discusses protein needs. Chapters 12 through 14 describe the energy and nutrient needs for infants, children, pregnant women, and older adults.

Energy   In hospitals, other methods and estimates may be used to determine energy needs. The "gold standard" for calculating energy needs—**indirect calorimetry**—provides an estimate of a person's resting energy expenditure by measuring the ratio of carbon dioxide expired to the amount of oxygen inspired. Indirect calorimetry requires special equipment and skilled clinicians to perform the measurements and interpret the results.

More often, clinicians rely on formulas to estimate energy needs. Table 17–1 shows one such formula—the Harris Benedict equation. This equation estimates resting or basal energy expenditure. Additional factors may be added to the basal energy expenditure to account for the added energy demands imposed by activ-

---

Table 17–1   The Harris-Benedict Equation for Estimating Basal Energy Expenditure (BEE)

Women:

$$BEE = 655 + (9.6 \times wt^a \text{ in kg}) + (1.8 \times ht \text{ in cm}) - (4.7 \times age \text{ in yr}).$$

Men:

$$BEE = 66.5 + (13.8 \times wt^a \text{ in kg}) + (5 \times ht \text{ in cm}) - (6.8 \times age \text{ in yr}).$$

Add to BEE for stress:

  0–20% for most stresses (elective surgery, minor infection, long bone fracture)

 30–50% for head injuries

20–110% for severe burns[b]

For people with a %IBW greater than 125, instead of actual weight, use an adjusted body weight in the BEE equation. To determine adjusted body weight, use this equation:[c]

  (actual body weight − IBW) × 25%[d] + IBW = adjusted body weight

NOTE: wt = weight; ht = height; yr = years; kg = kilograms; cm = centimeters.
[a]Use actual body weight, not ideal body weight.
[b]Many formulas are available for estimating energy needs for burn injuries. A variety of formulas are described and referenced in D. J. Rodriguez, Nutrition in major burn patients: State of the art, *Support Line,* August 1995, pp. 1–8. Chapter 27 provides more information.
[c]From J. M. Karkeck, Adjustment for obesity, *American Dietetic Association Renal Practice Group Newsletter,* Winter, 1984.
[d]Approximately 25 percent of body fat tissue is metabolically active.
SOURCES: M. M. McMahon, Nutrition support of hospitalized patients, presented at the Eighth Annual Advances and Controversies in Clinical Nutrition, Mayo Clinic Foundation, April 5–7, 1998; E. Weeks and M. Elia, Observations on the patterns of 24-hour energy expenditure changes in body composition and gastric emptying in head-injured patients receiving nasogastric tube feeding, *Journal of Parenteral and Enteral Nutrition* 20 (1996): 31–37.

indirect calorimetry (kal-oh-RIM-eh-tree): an indirect estimate of resting energy needs made by measuring the ratio of carbon dioxide expired to the amount of oxygen inspired and using the results in a mathematical equation.

How to

Estimate Energy and Protein Needs

*B*ernadette is a 39-year-old female, who is 5 feet 3 inches tall and weighs 130 pounds. She recently underwent elective surgery. Her energy needs can be estimated using the Harris-Benedict equation as follows:

Weight in kilograms =
130 lb ÷ 2.2 kg/lb = 59 kg.

Height in centimeters =
63 in × 2.54 cm/in = 160 cm.

BEE = 655 + (9.6 × wt in kg)
+ (1.8 × ht in cm) − (4.7 × age in yr).
655 + (9.6 × 59 kg) + (1.8 × 160 cm)
− (4.7 × 39) = 655 + 566 + 288 − 183 = 1326 kcal.

Next add 20% × BEE for surgery:[a]

1326 kcal × 20% = 262 kcal.
1326 + 265 = 1591 kcal.

[a]Alternatively, multiply the BEE by 120% (× 1.20), which is the mathematical equivalent.

Bernadette needs about 1591 kcalories to meet her BEE and additional energy needs due to surgery; clinicians monitor weight changes to determine if actual needs are higher or lower. Her energy needs will change as stress resolves.

Protein needs for Bernadette can be estimated at 1.0 to 1.5 grams of protein per kilogram of body weight per day. Use her weight of 59 kilograms to make the calculation:

59 kg × 1.0 g/kg = 59 g protein.
59 kg × 1.5 g/kg = 89 g protein.

Bernadette needs an estimated 59 to 89 grams of protein daily.

ity and stress. Many clinicians consider only the added demands of stress, however. One reason is because the Harris Benedict equation overestimates resting energy expenditure. Furthermore, most people in the hospital setting participate in minimal physical activity. The chapters that follow provide examples of stress factors.

Alternatively, body weight provides a simpler method of estimating body weight. For people who are not obese, energy needs range from 25 to 35 kcalories per kilogram of body weight. For those who are obese, energy needs are lower—21 kcalories per kilogram of body weight.

Regardless of what method is used to estimate energy needs, the dietitian reevaluates the client's needs periodically by assessing weight changes. Actual energy needs may be higher or lower than original estimates, or energy needs may change as clients' reach their weight goals.

Protein Healthy adults need 0.8 gram of protein per kilogram of body weight per day. For the elderly, protein needs range from 0.8 to 1.0 gram of protein per kilogram of body weight per day. As the chapters that follow describe, some health problems raise protein needs to as high as 2.5 grams per kilogram per day. The margin provides some examples. The "How to" box shows how to calculate energy and protein needs.

Protein needs can also be estimated by conducting nitrogen balance studies, but such studies require considerable time and are most likely to be used in cases where clients are critically ill. Dietitians monitor the client's serum protein levels to help determine if the client's protein needs are being met.

Micronutrients Exact vitamin and mineral needs during illness are largely unknown. Later chapters point out specific nutrients that may need to be supplemented as a result of a health problem.

Health condition	Protein needs (g/kg body weight/day)
Following surgery	*1.0 to 1.5*
Sepsis	*1.2 to 1.5*
Severe burns	*2.0 to 2.5*

Reminder: *Nitrogen balance* is the amount of nitrogen consumed compared to the amount of nitrogen excreted. Appendix D provides additional information about nitrogen balance studies.

Nutrition Intervention

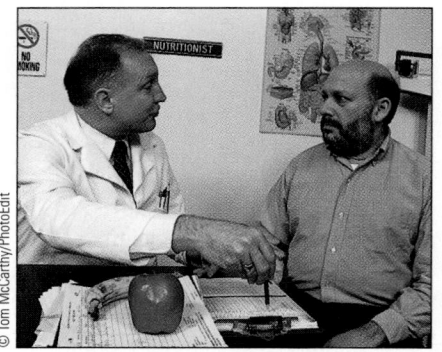

Optimal communication requires sensitivity to cultural orientation, education, and motivation.

Reminder: The *nutrition care plan* is a plan that translates nutrition assessment data into a strategy for meeting a client's nutrition and nutrition education needs.

nursing diagnoses: identification of the client's actual or potential health problems that require intervention by the nurse.

clinical pathways, critical pathways, or **care maps:** charts or tables that outline a plan of care for a specific diagnosis, treatment, or procedure, with a goal of providing the best possible outcome at the lowest cost. The plan, developed by the health care team after a careful study of each facility's unique client population, is regularly reassessed and improved.

outcome measures: indicators that describe an observable change and used to evaluate the effects of interventions.

In Summary Information gathered during nutrition assessment provides the dietitian with the information necessary to determine the client's nutrient needs. Dietitians often rely on formulas to estimate energy and protein needs and reassess nutrition status periodically to make adjustments.

Medical Nutrition Therapy

After completing a nutrition assessment, the dietitian identifies actual and potential nutrition problems. Does the client's history indicate the need for changes in diet? Is body weight appropriate? Do laboratory tests or physical signs suggest malnutrition? Are nutrient needs altered due to growth, illness, or medications? Will long-term dietary adjustments be necessary? Answers to questions like these enable the dietitian to determine what nutrition problems need to be addressed.

Goals and Interventions The nutrition care plan sets goals and identifies interventions that aim to resolve the client's immediate and long-term nutrition problems. Medical nutrition therapy has the greatest chance for success when clients help set the goals and take an active role in determining the interventions they feel they can follow. To achieve the desired outcome, care plans must take into account the client's food habits, lifestyle, motivation, and other personal factors. As Nutrition in Practice 17 describes, a person's genetic makeup may also influence medical nutrition therapy.

The dietitian should also ensure that the care plans developed by other health care team members coordinate with the nutrition care plan. Nursing care plans, for example, should include **nursing diagnoses** that address nutrition-related problems and nursing interventions relevant to the diagnoses. Table 17–2 lists examples of nursing diagnoses with nutrition implications. In some cases, plans for addressing nutrition problems are incorporated into a total medical care plan developed by the health care team. Such plans, called **clinical pathways, critical pathways,** or **care maps,** are charts or tables that outline a coordinated plan of care for a specific medical diagnosis, treatment, or procedure.

Addressing Nutrition-Related Problems Goals for nutrition care plans are stated in terms of **outcome measures,** such as target ranges for body weight or blood glucose levels. Outcome measures provide a means for evaluating the success of therapy. Consider, for example, a problem of diarrhea. If a medication is causing the diarrhea, an appropriate nutrition strategy might be to prevent dehydration and electrolyte imbalances by encouraging the client to consume ample fluids and electrolytes while the physician determines if another medication might be appropriate. Measurable goals might include a target range for serum electrolytes, urinary output, and blood pressure, as well as reduced frequency of bowel elimination and the production of stools of normal volume and consistency. If the diarrhea results from lactose intolerance, the goals might be the same, but the problem-solving strategy would include the temporary elimination of milk and milk products.

Addressing Nutrition Education Needs Nutrition education is a key part of the nutrition care plan. Goals for nutrition education can also be defined in terms of measurable outcomes. For the client experiencing diarrhea as a result of lactose intolerance, for example, the dietitian plans counseling sessions to teach the client how to plan a nutritionally adequate diet that limits lactose from milk and milk products. Examples of measurable outcomes for nutrition education might include the client's ability to verbally identify foods containing lactose and plan a day's menus appropriate for the restrictions. Keep in mind that the nutrition education plan must be flexible to accommodate the client's personal goals, understanding of the information presented, and motivation to practice the suggestions offered.

Nurses, thanks to their frequent daily contact with clients, can offer important support in the educational aspect of client care. Clients often think of questions long after the dietitian has left, and they most often ask the nurse. The nurse who is confident of the answers should provide them. If not sure of the answers, the nurse should express honest uncertainty and reassure the client that the information will be provided. The nurse should then inform the dietitian that the client needs a follow-up visit.

Ongoing Evaluation As the planned strategies are implemented, the dietitian regularly evaluates their effectiveness in achieving goals. If, for example, a client on a weight-reduction diet fails to lose weight, a change may be needed. Is the client eating too much? Is the client too inactive? Perhaps the client should keep a food and activity record to help identify problems with the weight-loss plan.

If a client's situation changes, so may nutrition status and nutrient needs. For example, when a pregnant woman delivers her baby, she will need instructions on how to feed her infant. She will also need information on how to revise her diet to support lactation (if she is breastfeeding) and return to a healthy weight.

A care plan may appear to be ideal, but still fall short of meeting goals if a client is unable or unwilling to comply with it. If the client is unable to comply, reassessing communication techniques may help. Perhaps the level of instruction needs to be simplified or cultural differences addressed. If the client is unwilling, despite the best efforts of health care professionals, little can be done except to try later when the client may be more receptive.

In Summary Once nutrient and nutrition education needs have been identified, the dietitian develops a nutrition care plan to meet those needs. The plan includes measurable goals and interventions to meet those goals. Once implemented, the plan is regularly evaluated to see how well it is working. Often a nutrition care plan includes diet modifications, described in the next section, to help clients meet their needs.

Table 17–2 Examples of Nursing Diagnoses with Nutrition Implications[a]
■ Altered nutrition: Less than body requirements
■ Altered nutrition: More than body requirements
■ Altered nutrition: Risk for more than body requirements
■ Altered growth and development
■ Altered oral mucous membrane
■ Body image disturbance
■ Chronic confusion
■ Chronic pain
■ Diarrhea
■ Feeding self-care deficit
■ Hopelessness
■ Impaired memory
■ Impaired physical mobility
■ Impaired skin integrity
■ Impaired social interaction
■ Impaired swallowing
■ Ineffective breastfeeding
■ Ineffective individual coping
■ Ineffective infant feeding pattern
■ Powerlessness
■ Risk for loneliness
■ Self-esteem disturbance
■ Social isolation
■ Unilateral neglect

[a]North American Nursing Diagnosis Association–approved nursing diagnoses.

Standard and Modified Diets

An essential part of medical nutrition therapy is to determine the appropriate amounts of energy, protein, carbohydrate, fat, vitamins, major minerals, trace elements, and water in the form that best meets the client's needs. For **standard** or **regular diets,** any food can serve as a source of energy or nutrients. **Modified diets** vary from standard diets in the types or amounts of nutrients, foods, or food components they provide.

Diet Modifications

Modifying the standard diet is much like tailoring a suit. A tailored suit is the same suit after alterations—only it fits better. In the case of a modified diet, the tailoring may involve changing the consistency; adjusting the amounts of energy, individual nutrients, or fluid; altering the number of meals; or including or eliminating certain foods. For example, a person who cannot chew easily needs soft foods; a person who is retaining fluid may need to limit sodium; a person with diabetes mellitus needs a diet that provides consistent amounts of carbohydrate at regular intervals.

standard or **regular diets:** diets that include all foods and meet the nutrient needs of healthy people.

modified diets: diets adjusted to meet medical needs. Such diets may be adjusted in consistency, level of energy and nutrients, amount of fluid, or number of meals, or by the inclusion or elimination of certain foods.

Sometimes modified diets are temporary. An example is the liquid diet shown here.

Health care professionals consult diet manuals to clarify what foods are included on or excluded from different diets.

A physician's order to give a client nothing orally (including food, beverages, and medications) reads **NPO,** the abbreviation for *non per os,* which means "nothing by mouth." **PO** stands for *per os,* which means "by mouth" or "orally."

house diet: a menu preselected by the dietary department.

diet manual: a book that describes the foods allowed and restricted on a diet, outlines the rationale and indications for use of each diet, and provides sample menus.

Treating Symptoms and Clinical Findings   It is helpful to think about modified diets in terms of the symptoms or conditions they prevent or relieve rather than in terms of disorders. Two people with the same disorder may need two different diets. Consider two people with an HIV (human immunodeficiency virus) infection, for example: one may need a diet that will help control nausea; the other may need a fat-modified diet to control elevated blood lipids. Conversely, people with two different disorders may develop similar problems, and both may benefit from the same diet. Many neurological disorders, for example, can lead to problems with swallowing (dysphagia). Whatever the cause, the diet aims to ease the task of swallowing.

Diet Orders   As Chapter 15 described, the physician prescribes the client's diet and writes the diet order in the medical record. The physician often relies on the dietitian or nurse to suggest a diet prescription or make recommendations when changes in the diet order appear warranted or when clarity is lacking.

To avoid confusion, physicians should provide clear and precise diet orders. For example, a "low-sodium diet" order should specify the amount of sodium; otherwise, "low-sodium" could be interpreted to mean any amount from 500 to 4000 milligrams. For uncomplicated diets, such as low-sodium diets, the foodservice department often sends a preselected diet (**house diet**) to the client until orders are clarified, and the level of restriction is recorded in the medical record. For more complicated diets, such as renal diets, meals will not be sent until the order is clarified. Occasionally, diet orders may be inappropriate. For example, the physician may describe an obese individual as "well-nourished" and order a regular diet. Unless the nurse identifies the problem, the client will not be referred to a dietitian and will receive an inappropriate diet and no nutrition advice. As another example, a physician may order that a client receive no food or fluids after midnight for a lab test to be conducted in the morning. The doctor assumes that the order will be discontinued after the test, but it may not be. The client may miss several meals before someone notices the error. These examples illustrate situations where communication between health care professionals can make a difference in client care. Whenever you notice inappropriate diet orders, contact the dietitian or alert the physician.

Diets in Facilities That Serve Foods   In an in-patient health care facility (such as a hospital, nursing home, or in-patient mental health care facility), providing an appropriate diet appears deceptively simple: appropriate foods are delivered to clients. Behind the scenes, however, much work goes into providing appropriate foods.

In large facilities, the staff of dietitians may compile a **diet manual,** subject to approval by the hospital administrator, several physicians, and representatives of the nursing service. Small facilities may adopt the diet manual of another hospital or an organization such as the state dietetic association. The diet manual describes the foods allowed and not allowed on each diet, outlines the rationale and indications for use of each diet, provides information on the nutritional adequacy of the diets, and offers sample menus. The dietary department uses the manual to design menus for each diet.

The exact foods excluded from or included on a specific modified diet, and even the name given to the diet, may differ among health care facilities, generally in minor ways. These variations reflect different schools of thought regarding diet. Health care professionals consult diet manuals as a standard of practice whenever they have questions about the foods provided for a modified diet.

Aside from planning appropriate diets, the foodservice department assures that the appropriate foods are carefully prepared and delivered. Nutrition in Practice 19 describes how foodservice systems operate. Equally important, once delivered, the food must be eaten by the client. As Chapter 15 described, many

How to

Help Clients Improve Their Food Intakes

1. Empathize. If the person is frightened, angry, or confused, show that you care and are there to help. Imagine feeling too sick to move or too tired to sit up. Show that you understand how difficult eating may be.

2. Motivate. Be sure the client understands how important nutrition is to recovery.

3. Help clients select foods they like and mark menus appropriately. Call the dietitian if the client needs extra help. When appropriate and permissible, let a friend or family member bring in favorite foods from outside the hospital. This may be especially helpful for clients with strong ethnic, religious, or personal food preferences.

4. Solve eating problems. Encourage clients who feel full after a short time to eat the most nutritious foods first and save liquids until after meals. For clients who are weak or tired, suggest foods that require little effort to eat. Eating a roast beef sandwich, for example, requires less effort than cutting and eating a steak; drinking soup is easier than eating it with a spoon. For clients who either fill up quickly when eating or are weak or tired, smaller meals combined with snacks, such as a sandwich at bedtime and milkshakes or instant breakfast drinks between meals, can improve intake considerably.

5. Help clients prepare for meals. Encourage clients to wash their hands and faces and to brush their teeth or rinse their mouths before eating. Help them get comfortable, either in bed or in a chair. Adjust the extension table to a comfortable distance and height, and make sure it is clean. A clean, odor-free room also helps. Take these steps before the tray arrives, so the meal can be served promptly and at the right temperature.

6. Check for accuracy and appearance. When the food cart arrives, check the client's tray. Confirm that the client is receiving the right diet, that the foods on the tray are the ones the client marked on the menu, and that the foods look appealing. Order a new tray if foods are not appropriate.

7. Help with eating. Help clients who need assistance in opening containers or cutting foods and those who are unable to feed themselves.

8. Take a positive attitude toward the hospital's food. Never say something like "I couldn't eat this stuff either." Instead, say, "The foodservice department really tries to make foods appetizing. I'm sure we can find a solution."

medical conditions and treatments can depress the appetite. Dietitians, nurses, and their assistants play a central role in helping clients to eat. The accompanying "How to" box offers suggestions for improving clients' food intakes. The chapters that follow provide many examples of modified diets, and the next section describes the ways nutrients can be delivered.

Feeding Routes

Nutrients can be delivered using the GI tract (**enteral nutrition**) or by bypassing the GI tract and supplying nutrients intravenously (**parenteral nutrition**). Enteral diets include both oral diets and tube feedings. Most often, people meet their nutrient needs by eating table foods. If their nutrient needs are high or their appetites are poor, liquid formulas taken orally can help meet some or all of their nutrient needs. Sometimes, however, a person's medical condition makes it difficult for the person to eat foods or drink formula supplements. Two options remain: **tube feedings** or **intravenous** feedings.

Tube Feedings   With a tube feeding, nutritionally complete formulas are delivered through a tube placed in the stomach or intestine (Chapter 20 provides the details). Tube feedings are preferred to intravenous feedings, but they can be used only when clients are able to digest and absorb enough nutrients to meet their needs. A person in a coma, for example, though unable to eat, may be able to digest foods and absorb nutrients. In such a case, a tube feeding would be an appropriate option.

enteral (EN-ter-all) **nutrition:** the provision of nutrients using the GI tract. Enteral nutrition includes both oral diets and tube feedings.
enteron = intestine

parenteral (par-EN-ter-all) **nutrition:** the provision of nutrients bypassing the intestine.
par = beside

tube feedings: liquid formulas delivered through a tube placed in the stomach or intestine.

intravenous: through a vein.

The medical record serves as a tool for communicating information about a client's health and responses to treatments.

Intravenous Feedings   In some cases, however, a person's medical condition prohibits the use of the GI tract to deliver nutrients. If the person is malnourished and the GI tract cannot be used for long periods of time, then intravenous feedings can supply nutrients (see Chapter 21).

In Summary Providing appropriate nutrients in an appropriate form is an important component of medical nutrition therapy. When standard diets fail to meet these needs, diet modifications are necessary. The physician prescribes each client's diet. The exact foods included on and excluded from modified diets, as well as serving sizes, are specified in each facility's diet manual. The GI tract most often serves as the conduit to deliver nutrients to the body, but in some cases nutrients must be delivered intravenously.

The Medical Record

Maintaining strong professional communication networks benefits both health care professionals and their clients. Conversely, miscommunication between professionals can result in inappropriate therapy or ineffective care with serious consequences for clients' health. Medical records are legal documents that record a client's health history; the assessment, diagnosis, and prognosis of medical problems; the measures being taken to treat those problems; and the results of tests and therapy. Writing in the medical records allows health care professionals to document the actions taken to comply with physicians' orders, the client's responses to those actions, and recommendations. This information helps determine if medical orders are being followed and directs future care.

Types of Medical Records   Medical records can be organized in many ways. In today's cost-conscious health care environment, time constraints compel health care professionals to minimize the time spent in documenting medical care.[1] Traditional problem-oriented medical records (POMR), which focus on a client's medical problems and the strategies being used to address those problems, are gradually being replaced by outcome- or goal-oriented medical records, which focus on observable medical goals. A problem-oriented medical record for a person with diabetes mellitus, for example, would list diabetes mellitus as a problem and then define the strategies used to control the diabetes, such as a 1600-kcalorie diet, oral hypoglycemic agents, and the like. A goal-oriented medical record for the same client would list the client's goal weight and target blood glucose measurements, and the health care team would document how these values have responded to therapy.

Recording Nutrition Information   Learn how to effectively use the record in the facility where you work. Regardless of the approach used, be sure the client's medical record includes important nutrition-related information. Examples of important information include:

- Documentation of nutrition screening.
- Assessment data relevant to nutrition status.
- Recommended medical nutrition therapy, including goals.
- Acceptance and tolerance of current diet.
- Problems with food intake.
- Documentation of diet counseling, including an assessment of the client's understanding of the diet.
- Any planned follow-up or referral to another person or agency.

Case Study

NUTRITION CARE PLANS

Sam is a nine-year-old Native American who was admitted to the hospital after he passed out while playing with friends. Tests confirm a diagnosis of diabetes mellitus. Sam will be in the hospital for several days until his blood glucose levels are under control. During this time, he and his family will begin learning about diabetes mellitus, the diet Sam will have to follow, how to use insulin, how to monitor blood glucose levels, and how to coordinate diet, insulin, and physical activity.

The details of diabetes mellitus are reserved for Chapter 24, but for now consider what steps will be necessary to develop a nutrition care plan.

1. What is the first step for the dietitian to take before a care plan can be developed?

2. From the limited information available, what personal factors will be important in devising a realistic care plan? Consider how these factors might affect the diet prescribed as well as nutrition education.

3. The extensive amount of information that Sam and his family will need to handle his diabetes will require the combined expertise of many health care professionals. Describe ways these health care professionals can communicate with each other to provide the most effective care.

4. Sam will need follow-up to learn more about diabetes and make the adjustments that will allow him to cope with diabetes. Why is it important that plans for follow-up be addressed before Sam leaves the hospital?

In Summary Effective nutrition care addresses the unique needs of the individual, framing nutrition needs in the context of the client's personal and medical needs. This chapter has described techniques for building effective nutrition care plans. The accompanying case study offers an opportunity to apply some of the information presented in this chapter. The remaining chapters describe how nutrition care can support recovery when health problems arise.

Self Check

1. Although indirect calorimetry is the "gold standard" for determining resting energy expenditure, its use is limited because:
   a. special computer software programs are required to complete the mathematical equations.
   b. special equipment and skilled clinicians are required to perform the measurements of oxygen and carbon dioxide and interpret the results.
   c. the technique can be used only on healthy adults.
   d. the results are invalid if the person is not meeting all energy and nutrient requirements.

2. All of the following statements about protein needs are true except:
   a. health problems do not affect protein needs.
   b. healthy adults need 0.8 gram of protein per kilogram of body weight per day.
   c. the elderly need from 0.8 to 1.0 gram of protein per kilogram of body weight per day.
   d. assessment of serum proteins can help clinicians determine whether protein needs are being met.

3. Health care professionals other than dietitians:
   a. should not include nutrition interventions in their respective care plans.

   b. should not be concerned with the dietitian's nutrition care plan.
   c. should ensure that their respective care plans include nutrition interventions that are consistent with the nutrition care plan as appropriate.
   d. are responsible for calculating a client's energy and protein needs.

4. The most important factors that affect the manner in which nutrition information is presented to a client are:
   a. the client's nutrient needs and nutrition status.
   b. the client's ability level and motivation.
   c. the client's health and drug histories.
   d. the medical record and nutrition screening results.

5. A dietitian instructs a client on a weight-reduction diet and exercise program and sets a goal for weight loss of 1 pound per week. The next step for the dietitian is to:
   a. assume the plan is working.
   b. formulate a new nutrition care plan.
   c. find a new strategy for meeting weight-loss goals.
   d. plan a follow-up meeting to weigh the client and see how well the plan is working.

6. All of the following statements describe diet orders **except:**
   a. they should never be questioned.
   b. they are prescribed by the physician.
   c. they should be written clearly and precisely.
   d. the physician may write an order that reads, "NPO."

7. Modified diets:
   a. treat one health problem exclusively.
   b. are not part of medical nutrition therapy.
   c. differ from standard diets only in consistency.
   d. treat conditions and symptoms that may occur as a consequence of several health problems.

8. A nurse checks the food on a client's tray and is not sure if the food is allowed on the client's diet. An appropriate action for the nurse to take would be to:
   a. check the care plan.
   b. check the diet order.
   c. check the diet manual.
   d. check the medical record.

9. Enteral nutrition includes all of the following **except:**
   a. standard diets.
   b. intravenous nutrients.
   c. modified diets composed of table foods.
   d. formula supplements taken orally or delivered by tube.

10. All of the following are true about medical records **except:**
   a. they are legal documents.
   b. they provide information about a client's medical problems and the measures being taken to address those problems.

   c. they all conform to the same format.
   d. they demand the time of health care professionals.

Critical Thinking

1. At a minimum, the nutrition problems of a client with a very poor appetite and weight of less than 80 percent IBW would include:
   a. fatigue.
   b. risk for malnutrition.
   c. fluid and electrolyte imbalance.
   d. difficulty chewing and swallowing foods.

2. Based solely on the information presented in question 1, which of the following interventions is the most appropriate for the client?
   a. referral for a complete nutrition assessment
   b. indirect calorimetry to determine resting energy expenditure
   c. nitrogen balance studies to calculate protein needs
   d. determination of serum protein levels

3. Measurable goals for a client who must lose weight might include:
   a. the client's food intake record.
   b. the type of diet the client is expected to follow.
   c. the family members' attitudes toward the client's weight.
   d. the number of pounds the client is expected to lose each week.

Answers to these questions can be found in Appendix G.

Clinical Applications

1. A client who recently suffered a heart attack is on a special diet. The client tells the nurse that the dietitian has talked to him about his diet and that he is totally confused. He confides that diet is the last thing on his mind right now. What actions should the nurse and/or dietitian take?

2. An initial nutrition screening of an elderly client admitted to the hospital for surgery revealed that the client was not at risk for poor nutrition status. The client was not on a modified diet and was not referred for nutrition assessment, and so has not seen a dietitian. Following surgery, ongoing assessment has revealed that the client has developed several complications, and recovery has been slower than expected. The

nurse working with the client notices that the client has eaten only minimal amounts of food for several days. Describe several steps that can be taken to uncover and address problems the client might be having with food.

3. Assume that the client in Clinical Application 2 loses a considerable amount of weight and the physician orders a kcalorie count. The physician's written order is recorded in the medical record, as is a note from the dietitian verifying the procedure. Considering the communication channels described in this chapter, what steps might the dietitian take to ensure that the client's intake will be recorded through all shifts?

Nutrition on the Net

For further study of the topics of this chapter, access these websites and search for the phrases or words in quotation marks:

Find updates and quick links to these and other nutrition-related sites at our website:
www.wadsworth.com/nutrition

Learn about medical nutrition therapy: visit Ask the Dietitian: **www.dietitian.commedtherapy.html**

Find out more about nursing diagnoses by visiting the North American Nursing Diagnosis Association:
www.nanda.org

Note

1. C. J. Klein, J. B. Bosworth, and C. E. Wiles, Physicians prefer goal-oriented note format more than three to one over the outcome-focused documentation, *Journal of the American Dietetic Association* 97 (1997): 1306–1310.

Nutrition in Practice

■ THE NEW GENETICS: IMPLICATIONS FOR NUTRITION INTERVENTION ■

The **Human Genome Project (HGP)**, a joint effort of the National Institutes of Health and the Department of Energy, began in 1990. This ambitious project seeks to identify and locate all the **genes** in the human body. To put the scope of the project in perspective, consider that researchers estimate that mapping and sequencing the human **genome** will require the reading of over 3 billion bits of information.[1] Unraveling the **genetic code** will help us understand the similarities and differences between people and the ways our heredity and environment affect our health. As a result of the knowledge and technologies generated from the HGP, medical practice is rapidly changing, and health care professionals are challenged to understand the new genetics (also called **genomics**) and its implications for various health care disciplines. This Nutrition in Practice describes some of these changes, emphasizing those with implications for nutrition intervention. It does not address the ethical, legal, and social issues emanating from the HGP. The glossary defines related terms.

What is a genome?

A genome is a complete set of instructions for making an organism and controlling its functions. The instructions reside in the chromosomes found in each cell of an organism. The human genome consists of 23 pairs of chromosomes. One chromosome of each pair is inherited from the mother and the other from the father. Each chromosome, in turn, contains numerous genes. Genes are composed of long stretches of **DNA**—complementary strands of chemical building blocks that form a double helix (see the photo on p. 448). Each gene carries the blueprint for a specific protein such as an enzyme or a muscle protein. Each gene also contains an area that responds to chemical signals and determines when, and to what extent, its protein will be made (**gene expression**). Through their regulation of protein synthesis, genes orchestrate all body functions; those functions that specifically relate to nutrition include digestion, absorption, metabolism, and excretion. Figure NP17–1 shows the relationship between cells, chromosomes, and genes.

What is the relationship between a person's genetic makeup and disease?

As Chapter 15 described, diseases often result from a combination of hereditary and environmental factors. All diseases have a hereditary component although the relative contributions of heredity and environment vary. Inherited metabolic disorders, such as Down syndrome, result from a gene defect—they are genetically determined. Diseases caused by infectious organisms are primarily environmentally determined.[2] Sometimes, however, altering environmental factors can prevent the consequences of an inherited disorder. Conversely, one's genetic makeup can alter susceptibility and response to environmental stress. A person's ability to fight off an infection or mend a broken bone, for example, depends, in part, on genes that regulate the immune system or mend damaged tissues.

The incidence and prevalence of many chronic diseases, including diabetes, hypertension, and heart diseases, vary among individuals, families, and nations. Multiple genetic and environmental factors interact and lead to the progression of these diseases. A gene may raise the likelihood that a person will develop a disease, but the gene does not actually cause the disease. Not all people who carry the gene will develop the disease. Instead, environmental factors, such as diet, tip the scales in determining who among the susceptible people develop the disease.

What are some examples of the ways genes interact with diet?

One example is phenylketonuria, or PKU (see Nutrition in Practice 20). A person with PKU inherits a gene that is unable to metabolize the amino acid phenylalanine. Without a special diet that is low in phenylalanine, mental retardation occurs as a consequence. Neither the defective gene nor a high-phenylalanine diet alone leads to mental retardation; both must be present for the consequences to occur.

Genetic diseases can alter nutrient requirements. In one genetic disease, hemochromatosis (see Chapter 23), excessive iron accumulates in the liver and causes

Nutrition in Practice

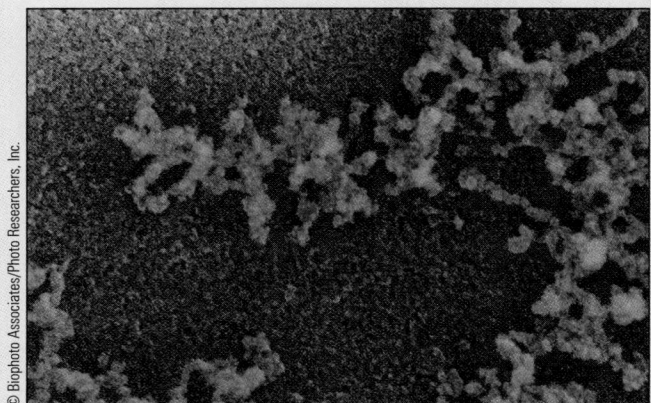

DNA consists of two strands of chemical bases strung along a backbone of sugar (deoxyribose). The strands coil around each other, and the entire molecule coils in a spiral to form a double helix.

tissue damage. The diet must limit iron to avoid further damage. No doubt, as more is learned about genetic variations, other significant differences in nutrient requirements among individuals will be recognized.

Other examples of how genes and diet interact involve enzyme systems that metabolize dietary compounds, alcohol, chemicals, and carcinogens so that they can be safely excreted from the body. Some phytochemicals, for example, may raise the production of enzymes that transform carcinogens into harmless compounds.

Genetics also determines how a person responds to different diets. Armed with an understanding of an individual's genetic makeup, it may one day be possible to tailor nutrition interventions to truly meet the needs of the individual.

Can you give some examples?

One example involves serum cholesterol levels. The diet recommendations to prevent cardiovascular disease focus on measures that lower the cholesterol carried on the low-density lipoproteins—LDL cholesterol for short. Altering the types of fat consumed is effective for some individuals, but not for others. Likewise, oat bran has proved effective in lowering LDL cholesterol for some people, but not for others. People's responses to diet therapy in these cases appear to be related to their **genotype.**

Another example involves variations in a person's genotype, calcium absorption, and dietary calcium. The genotype may determine whether a person needs a high-calcium diet to prevent the loss of bone density that leads to osteoporosis.[3]

In cases such as these, knowing the person's genotype will help dietitians tailor nutrition advice specifically for the individual. Clients will be able to avoid unnecessary restrictions and determine the strategies that will be most effective for them.

How is health care changing as a result of the HGP?

Only time can reveal the full implications of the HGP for medical practice. Even now, however, genetics is playing an increasing role in the diagnosis, monitoring, and treatment of diseases. As reported in the lay press, for example, genes that raise the likelihood of breast cancer have been identified. People with a family history of breast cancer may choose genetic testing to uncover whether they carry the gene. Those who carry the gene can then take measures to reduce their risk of developing breast cancer and receive regular checkups to find the disease early should it develop.

Among the most exciting possibilities offered by the new genetics is the prospect of better treatments and

Glossary

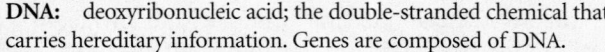

DNA: deoxyribonucleic acid; the double-stranded chemical that carries hereditary information. Genes are composed of DNA.

gene expression: the production of a protein by the gene that codes for it.

genes: the portion of a chromosome that carries the blueprint for making a protein.

gene therapy: treatment that replaces a defective gene with a normal gene.

genetic (gen-ET-ick) **code:** The hereditary information programmed into the DNA of a living organism.

genome (GEE-nome): the complete set of chromosomes, which contain instructions for making an organism and controlling its functions.

genomics (gee-NOM-icks): pertaining to the genes.

genotype (GEEN-oh-type): the complete genetic makeup of an individual.

Human Genome Project (HGP): an extensive research project, funded through the National Institutes of Health and the Department of Energy, that aims to identify and locate all the genes in the human body.

Nutrition in Practice

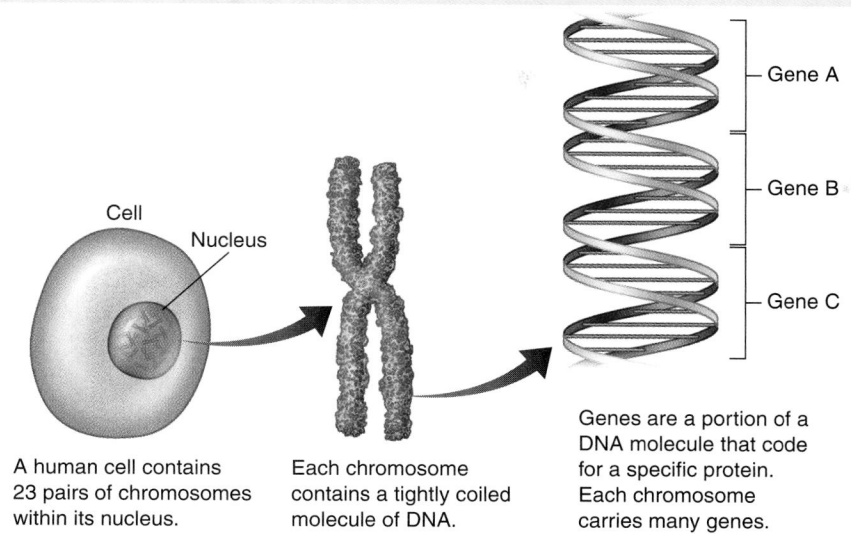

Figure NP17–1
Cells, Chromosomes, and Genes

Gene A

Gene B

Gene C

Cell

Nucleus

A human cell contains
23 pairs of chromosomes
within its nucleus.

Each chromosome
contains a tightly coiled
molecule of DNA.

Genes are a portion of a
DNA molecule that code
for a specific protein.
Each chromosome
carries many genes.

even cures for diseases. Consider that physicians typically treat illnesses with drugs derived from plants and synthetic compounds only after the person develops symptoms. With knowledge of a person's genotype and the implications of any genetic variations, physicians may be able to predict the consequences to body functions and prescribe drugs or other interventions to prevent disease development. Drug development will be highly specific—that is, drugs will be able to target specific proteins. Such drugs have the potential to work more efficiently and have fewer side effects.[4]

Gene therapy—the use of normal genes to replace defective genes—holds great promise as a treatment, or even cure, for inherited and acquired diseases. Replacing a defective gene that causes an inherited metabolic disorder, for example, may be a cure for that disorder. Adding a gene that suppresses tumor growth and treats or cures cancer is another example.

What can professionals do to prepare for the impact of the new genetics on health care?

Probably the best advice is to start preparing now! If you have never taken a course in genetics, learn the basics and become familiar with the terminology related to the field. Those who have already taken a course should brush up on what they learned and expand their knowledge base. Keep up with the current literature to stay abreast of new developments in the field. With respect to nutrition, it is important to specifically understand how people's genotypes affect their nutrient requirements and their responses to diet interventions.[5]

What will change in the years to come as a result of the HGP? If the predictions for the future of health care materialize, your health record will most likely include your complete genome and an accounting of all genetic variations that might affect your susceptibility to disease or predict your response to diet interventions, drug therapy, or other environmental treatments.[6] As a result, your health care will be highly individualized, and preventive measures and treatments will be far more precise and effective. Some negative consequences are likely as well. Your medical information may be available to insurers and employers, which, in some cases, may make it harder for you to get insurance coverage or may limit your employment opportunities.

Nutrition on the Net

For further study of the topics of Nutrition in Practice, access these websites and search for the phrases or words in quotation marks:

Find updates and quick links to these and other nutrition-related sites at our website:
www:wadsworth.com/nutrition

Find a wealth of information and many references related to the HGP and genetics by visiting the Human Genome Project Information website:
www.ornl.gov/hgmis/medicine/tnty.html

Review information about the implications of the HGP for various health care disciplines by visiting the Human Genome Education Model Project II:
http://gucdc.georgetown.edu/hugem/intoduc.html

Read an article about "genomic medicine" at Medscape Cardiology:
http://cardiology.medscape.com

Nutrition in Practice

Notes

1. Human Genome Education Model Project II, The human genome project, **http://gucdc.georgetown.edu/hugem/project.htm**, site visited September 17, 2001.

2. R. E. Patterson, D. L. Eaton, and J. D. Potter, The genetic revolution: Change and challenge for the dietetics profession, *Journal of the American Dietetic Association* 99 (1999): 1412–1420.

3. A. P. Simopoulos, Genetic variation and nutrition, *Nutrition Reviews* (supplement 2) 57 (1999): 10–19.

4. Human Genome Project Information, Medicine and the new genetics, **www.ornl.gov/hgmis/medicine/medicine.html**, site visited September 17, 2001.

5. Human Genome Education Model Project, Fact sheet 29: Genetics and dietetic practice, **gucdc.georgetown.edu/hugem/fs29.htm**, site visited September 17, 2001.

6. Human Genome Project Information, Fast forward to 2020: What to expect in molecular medicine, **www.ornl.gov/hgmis/medicine/tnty.html**, site visited September 17, 2001.

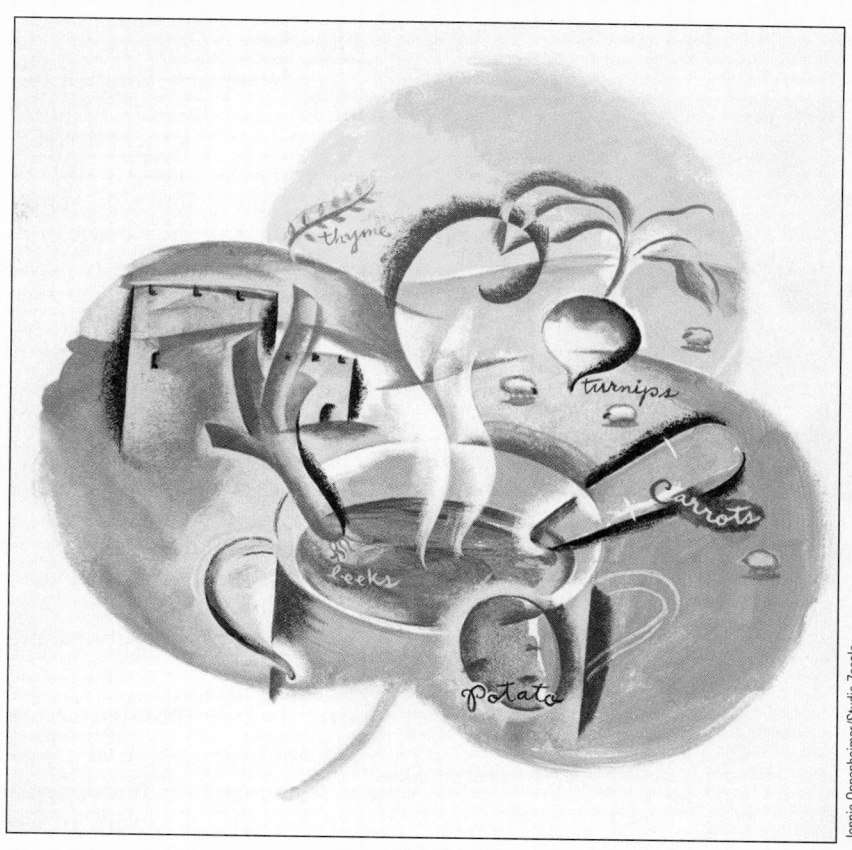

Jennie Oppenheimer/Studio Zocolo

CHAPTER 18

CONSISTENCY-MODIFIED AND OTHER DIETS FOR UPPER GI TRACT DISORDERS

The body depends on the upper GI tract, which includes the mouth, esophagus, and stomach, to deliver the food from which it extracts nutrients.

As Chapters 12 through 14 explain, nutrient needs fluctuate in response to a multitude of physical and physiological changes that occur throughout life. Illnesses also alter nutrient needs. The remaining chapters of this book explore the many ways that modified diets serve to treat illnesses and their symptoms.

Just as the assimilation of nutrients begins with physical processes—chewing, swallowing, and the passage of food through the GI tract—your study of therapeutic nutrition begins by discussing foods and diets that affect these physical processes. This chapter describes the consistency-modified and other diets that help treat conditions that primarily affect the upper GI tract. The next chapter describes how diets assist in the treatment of lower GI tract disorders. Table 18–1 lists conditions for which consistency-modified and other diets are indicated.

Consistency-Modified Diets

Consistency-modified diets provide foods modified in firmness and texture. Examples of consistency-modified foods include liquid, pureed, chopped, tender-cooked, or whole foods that are easy to eat. Routine progressive diets and mechanical soft diets are examples of diets that include consistency-modified foods.

Routine Progressive Diets

In health care facilities that serve meals, a common diet order reads, "progress diet from clear liquids to a regular diet as tolerated." Diets that progress from liquids to solids, or **routine progressive diets,** are typically used for clients after uncomplicated surgeries, for clients experiencing nausea or diarrhea, or for clients who have not been eating foods orally and are beginning to eat again. At each step of a progressive diet, other diet modifications may also apply. For example, caffeine may be restricted. Then the diet order might read, "progress diet from caffeine-restricted clear liquids to a caffeine-restricted, regular diet."

Clear-Liquid Diets Progressive diets frequently begin with clear-liquid diets. **Clear-liquid diets** provide energy and fluids and electrolytes and consist mainly of foods that are transparent and liquid at body temperature. Clear liquids provide minimal stimulation to the GI tract—they are easily and almost completely absorbed. Thus they help determine if the digestive system is working well enough to handle more complex foods. Table 18–2 lists the foods allowed on clear-liquid diets, and the sample menu shows an example of a day's meals. Once the person tolerates clear liquids, the diet may progress to full liquids.

Full-Liquid Diets   **Full-liquid diets** include both clear and opaque liquid foods and some semiliquid foods (see Table 18–2 and the sample menu). A full-liquid diet may serve as the second diet as a person progresses from clear liquids to regular foods, or as a diet for people who are unable to chew or swallow regular foods for medical reasons or because they are too ill to eat.

Note that many foods provided on full-liquid diets contain lactose. The same conditions for which liquid diets are prescribed often result in temporary lactose intolerance. When potential problems with lactose intolerance are suspected, lactose-free formulas can meet nutrient needs, or the diet may progress directly from clear liquids to solid foods.

Cautious Use of Liquid Diets Liquids are offered in small amounts at first to make sure the person can tolerate them. Many clients willingly accept liquid diets because they are too ill to eat solid foods. When left on the diet for any length of time, however, people may understandably find these foods unappetizing and

consistency-modified diets: diets that include foods modified in firmness, such as liquid, pureed, chopped, tender-cooked, or soft whole foods (such as bananas).

routine progressive diets: diets that advance from liquids to solids as a client's tolerances permit.

clear-liquid diets: diets consisting of foods that are mainly liquid and transparent at body temperature.

full-liquid diets: diets consisting of both clear and opaque liquid foods and near-liquid foods.

boring. More importantly, both clear-liquid and full-liquid diets are lacking in many nutrients and, therefore, should be used only temporarily. When clients are unable to tolerate solid foods for more than a day or two, liquid formulas (see Chapter 20) provide an option. Formulas are of a known nutrient composition, and they can be selected to meet nutrient needs and reduce the likelihood of GI problems, including lactose intolerance.

Solid Foods and Diet Progression   Once a person is tolerating liquids, the next step is solid foods. The exact diet used as the first solid diet depends on the client's medical condition and tolerances. Some clients begin receiving regular foods, although the first foods offered may be those least likely to irritate the GI tract (described later in this chapter) and produce gas (see Chapter 19).

Advancing the Diet   The nurse is often responsible for monitoring the client's tolerance and readiness to advance through the stages of a progressive diet. Indigestion, nausea, vomiting, diarrhea, cramping, or other GI upsets indicate intolerance. At each progressive step, a client may be intolerant to particular foods (orange juice or milk, for example), rather than to the diet itself. In such a case, the offending food is withheld temporarily.

Surgery and Routine Progressive Diets

Diet Order
Progress diet from clear liquids to full liquids to a solid diet as tolerated.

Ideally, all clients enter surgery at appropriate weight with full nutrient reserves. Such optimal status may be difficult to achieve, however. The illness that necessitates the surgery, the drug therapy used in its treatment, or the psychological stress associated with it may interfere with a person's intake of food or with the body's use of nutrients. Additionally, the person may have to fast or follow a nutritionally inadequate liquid diet before undergoing diagnostic and laboratory tests, further taxing nutrient status. The health care professional who appreciates the vital roles nutrition plays in resistance and recovery will watch for signs of nutrition risk and encourage clients to eat as well as possible, when possible.

Table 18–1   Indications for Consistency-Modified and Other Diets
Progressive Diets
■ Diarrhea
■ Following bowel rest
■ Following surgery
■ Myocardial infarction (heart attack)
■ Refeeding in protein-energy malnutrition
Mechanical Soft Diet
■ Broken jaw
■ Dry mouth
■ Dysphagia
■ Ill-fitting dentures
■ Missing or no teeth
■ Oral infections (thrush and herpes)
■ Periodontal disease
■ Severe dental caries
■ Stroke
■ Ulcers of the mouth or gums
Diets to Reduce Gastric Irritation
■ Esophageal varices
■ Gastritis
■ Gastroesophageal reflux disease (GERD)
■ Indigestion
■ Peptic ulcers
■ Reflux esophagitis

Table 18–2 Foods Included on Liquid Diets	
Clear-Liquid Diets	**Full-Liquid Diets**
Bouillon	All clear liquids
Broth, clear	Butter
Carbonated beverages	Commercially prepared liquid formulas (all)
Coffee, regular and decaffeinated	Cooked cereals, strained
Commercially prepared clear-liquid formulas	Cream
Fruit drinks	Custard
Fruit ices	Ice cream, plain
Fruit juices, strained	Instant breakfast drinks
Gelatin	Margarine
Hard candy	Milk, all types
Honey	Pudding
Lemonade	Sherbet
Popsicles	Soups, strained vegetable, meat, or cream
Salt	Sour cream
Salt substitutes	Vegetable juices, strained
Sugar	Vegetable purees, diluted in cream soups
Sugar substitutes	Yogurt
Tea, regular and decaffeinated	

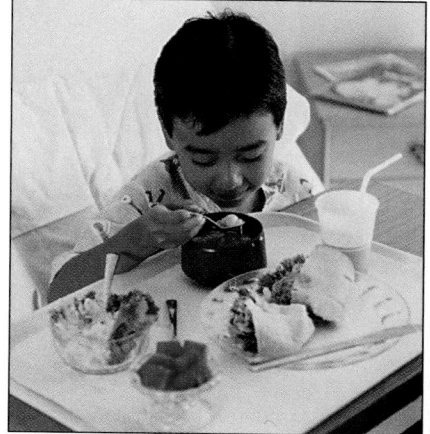

Clients often look forward to their first meal of solid foods after a meal or two of liquids only.

Consistency-Modified and Other Diets for Upper GI Tract Disorders

453

Clear-Liquid Diet Menu

Breakfast	Lunch	Supper	Between Meals
Strained orange juice	Bouillon	Bouillon	Soft drinks
Flavored gelatin	Apple juice	Cranberry juice	Gelatin
Ginger ale	Flavored gelatin	Fruit ice	Fruit juices
Coffee or tea	Coffee or tea	Flavored gelatin	
Sugar	Sugar	Coffee or tea	
		Sugar	

This section describes the nutrient needs of well-nourished clients who undergo uncomplicated surgery that requires general anesthesia and at least an overnight stay in the hospital. Malnourished clients or those who undergo extensive surgery are severely stressed and need the high-kcalorie, high-protein diets described in Chapter 27. Special considerations also apply to clients undergoing surgeries of the GI tract (Chapter 22 provides the details). Clients who undergo minor, outpatient surgeries often can begin eating solid foods soon after surgery, although health care professionals may advise them to try water first and mild foods for the first few meals.

Immediate Presurgery Diet Physicians generally order all foods and fluids withheld for at least eight hours before surgery. Ensuring that no food is in the upper GI tract helps prevent regurgitation and **aspiration,** which can occur during anesthesia or recovery. Aspiration can lead to serious, and sometimes fatal, respiratory infections. In the special case of GI surgery, people receive liquid or very-low-residue diets (see Chapter 19) for two or three days before surgery.

Immediate Postsurgery Support In the immediate postsurgical period, the most important nutrition-related task for the medical team is to maintain fluid and electrolyte balances. Clients lose blood, fluid, and electrolytes during surgery, and they may lose additional fluids thereafter from fever, draining wounds, vomiting, and diarrhea. Excessive losses without replacement can lead to dehydration and shock.

Immediately following surgery, GI motility slows, and the person is given nothing to eat or drink orally. Clients receive intravenous infusions (see Chapter 21) to supply fluid and electrolytes. Physicians determine fluid and electrolyte needs based on the person's blood pressure, pulse rate, urinary output, level of consciousness, breathing patterns, body temperature, and laboratory test results.

aspiration: the drawing of food, saliva, or gastric secretions into the lungs.

Full-Liquid Diet Menu

Breakfast	Lunch	Supper	Between Meals
Orange juice	Apricot nectar	Apple juice	Milkshakes (made with plain ice cream),
Strained oatmeal	Yogurt, plain	Creamed soup	
Milk	Pudding	Custard	plain ice cream, eggnog,
Sugar	Milk	Milk	pudding,
Margarine	Coffee or tea	Coffee or tea	custard
	Sugar	Sugar	or gelatin

Postsurgical Diets   When GI tract motility resumes, the postoperative client is ready to begin eating again. A routine progressive diet, beginning with a clear-liquid diet, is a frequent diet order. However, research to support the use of liquid diets following surgery is lacking.[1] Many people can eat regular foods by the second meal after surgery. People who experience nausea or distention may benefit from eating small servings of easy-to-digest foods. Clients undergoing certain GI surgeries may be advanced to low-fiber or low-residue diets (see Chapter 19) rather than regular diets. Clients undergoing surgery of the mouth or esophagus may advance from liquid diets to the mechanical soft diets described next.

Mechanical Soft Diets

Mechanical soft diets provide foods modified to make them easy to chew and swallow. In most cases, mechanical soft diets include all foods and seasonings. Individual tolerances determine if foods should be provided in liquid, pureed, chopped, or tender-cooked form. Foods that are naturally soft, such as ripe bananas, are provided as tolerated.

Difficulties Chewing   People may have difficulties chewing for many reasons ranging from sedation and pain to neurological disorders, facial injuries, mouth ulcers, reduced flow of saliva, missing teeth, or ill-fitting dentures (see Table 18–3). People with persistent problems with chewing may eat too little, lose too much weight, and suffer the consequences of a deteriorating nutrition status.

Diet Options   People's tolerances of food consistencies vary greatly. Following surgery to repair a broken jaw, for example, a person may be able to consume liquids only. With time, however, the person gradually progresses to solid foods. Nurses and dietitians work together with the client, family, and caregivers to identify which foods the client can safely handle. The goal is to provide a wide variety of foods that are as similar as possible to those of a regular diet. Such a strategy enhances the appetite and minimizes the likelihood of nutrient deficiencies. Therefore, soft natural foods, tender-cooked foods, and chopped foods are provided whenever possible; only when the person cannot chew or swallow adequate amounts of these foods are pureed foods provided.

The person recovering from a broken jaw, described above, may progress from liquids to pureed foods, then to ground or chopped foods, and then to regular foods. In some cases, however, a person may need to follow a mechanical soft diet permanently. Moist, soft-textured foods, such as foods prepared with sauces and gravies, often work well. Drinking liquids along with meals makes it easier to chew and swallow foods.

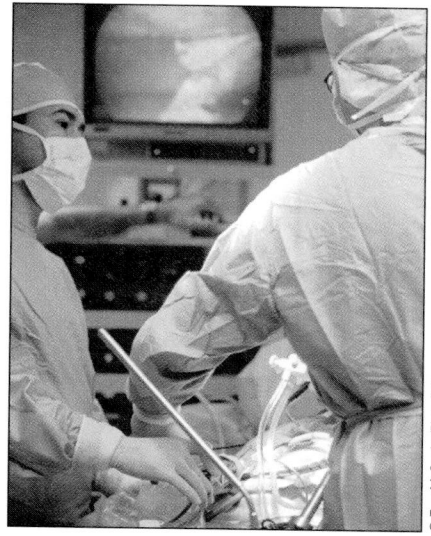

For clients who have undergone surgery and are beginning to eat again, progressive diets test their ability to digest food.

Reminder: The process of chewing is called **mastication.**

A person without teeth is **edentulous** (ee-DENT-you-lus).
e = without
dens = teeth

mechanical soft diets: diets that exclude only those foods that a person cannot chew or swallow; also called **dental soft diets** or **edentulous diets.**

Table 18–3 Conditions That May Interfere with Chewing and Swallowing	
Achalasia	Ill-fitting dentures
Alzheimer's disease	Missing teeth
Broken jaw	Multiple sclerosis
Cancer	Myasthenia gravis
Candida albicans (thrush) infection	Oral herpes infection
Chemotherapy	Parkinson's disease
Congenital defects of upper GI tract	Periodontal disease
Dental caries	Radiation therapy of the head and neck
Dryness of mouth	Sensitivity of mouth to hot or cold
Dysphagia	Strokes
Guillain-Barré syndrome	Surgery of the mouth, head, or neck
Head injury	Ulcers of mouth, gums, or esophagus
HIV Infection	

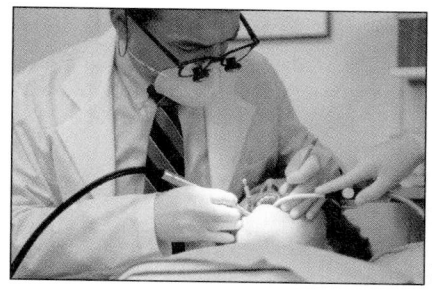

Mechanical soft diets may temporarily benefit clients who undergo some dental procedures, such as tooth extractions or root canals.

Consistency-Modified and Other Diets for Upper GI Tract Disorders

Improve Acceptance of Pureed Foods

Take a moment to think about a meal of pureed foods. A typical dinner of baked chicken, mashed potatoes, and green beans is blenderized to white mush, more white mush, and a green blob. The foods may taste great, but on seeing the plate, the person may be reluctant to try the first bite. To stimulate the appetite, use creative techniques such as these for preparing and serving food:

- Encourage clients and their caregivers to prepare a variety of favorite foods and blenderize them to a tolerable consistency. The smell of favorite foods and the thought of consuming them stimulate the appetite.

- Consider color when planning meals. The meal of baked chicken, mashed potatoes, and green beans, described above, can be made more appealing by substituting mashed sweet potatoes for the white potatoes. Arranging foods attractively on a plate with appropriate garnishes also adds color and eye appeal.

- Serve foods at the right temperature and puree them so that they are smooth and thick—not watery and thin. Commercially available thickeners add shape, texture, and even nutrients to food.

- Experiment with seasonings and spices to enliven food flavors, excluding only those that the person cannot tolerate for personal or medical reasons. Seasoning food to accommodate personal tastes adds flavor, allows for variety, and improves the appetite.

- Supplement the diet with nutritious liquids such as milk, instant breakfast drinks, or liquid formulas (see Chapter 20).

Efforts to improve the acceptance of pureed foods can go a long way toward helping people to eat and maintain or improve their weights. When efforts to improve a client's intake of pureed foods are unsuccessful, feeding the person by tube becomes an option (see Chapter 20).

Cautious Use of Pureed Foods Pureed foods can meet all nutrient needs and may taste good, but the monotony of eating foods of the same consistency for every meal, day after day, can create a psychological block to eating. Commercial products that thicken pureed foods and allow them to be shaped attractively can make a remarkable difference in enhancing the appetite. The accompanying "How to" box offers other suggestions for clients and caregivers to improve the acceptance of pureed foods.

Health care professionals employed in long-term care facilities such as nursing homes and rehabilitation centers face a formidable challenge in ensuring that clients receiving pureed foods over long periods of time continue to eat enough to meet their nutrient needs. In addition to the problems associated with pureed diets, clients in long-term care facilities lack the comforts of home and may be overwhelmed by their medical conditions, isolation, and loss of control. Addressing the client's emotional needs is critical to stimulating the appetite.

Examples of drugs that can lead to mouth dryness include some anticonvulsants, antidepressants, anticholinergics, and antihistamines.

mouth ulcers: lesions or sores in the lining of the mouth. Certain medications, radiation therapy, and some disorders, such as oral herpes infections, can cause mouth ulcers.

tonsillectomy (tawn-sill-ECK-tah-me): surgical removal of the tonsils.

Mouth and Throat Pain The mechanical soft diet for people with **mouth ulcers** or inflammation of the lips, mouth, or throat (following a **tonsillectomy,** for example) provides moist, soft-textured foods and eliminates spicy, salty, or acidic foods (such as citrus fruits and juices and tomato products) that may be painful to eat. In addition, nuts or seeds in foods (such as sesame or poppy seeds in breads) and sticky foods (like peanut butter) can cause discomfort. Heat may intensify pain, too. Many clients prefer cold foods or foods served no warmer than room temperature.

Reduced Flow of Saliva Some health problems and medications lead to mouth dryness. People with a reduced flow of saliva often prefer moist, soft-textured foods. Clients can moisten foods with sauces and gravies. Salty foods and snacks dry the mouth and should be avoided. Encourage clients to practice good oral hygiene; when salivary flow is reduced, the mouth is poorly defended against dental caries. Clients can stimulate salivary secretions by sucking on sugarless candy, chewing sugarless gum, or using drugs that stimulate the flow of saliva.

Dysphagia

Diet Order
Individualized mechanical soft diet.

Problems with swallowing, known as **dysphagia,** can arise from many causes, including aging, neurological disorders (including strokes), developmental disabilities, certain surgeries of the head or neck, and some treatments for cancer. Dysphagia may occur when the muscles of the mouth or esophagus fail to push foods to the back of the throat to initiate swallowing or to propel foods through the esophagus to the stomach. In **achalasia,** the lower esophageal sphincter fails to relax in response to the presence of foods in the esophagus, and foods and liquids accumulate in the esophagus until either the pressure of the food forces the sphincter to open or the food is regurgitated.

Subtle and Dangerous Dysphagia often goes undiagnosed, especially when associated with aging, because symptoms may be sporadic and mild at first. All people "catch food in their throat" at one time or another, so a person who experiences this feeling more frequently may consider the condition normal and ignore it. Eventually, the person may eat less, experience marked weight loss, and develop nutrient deficiencies.

Dysphagia can be dangerous for other reasons as well. People with dysphagia experience aspiration and respiratory infections more frequently than people without dysphagia. Some people with dysphagia fail to cough in response to the presence of saliva or food in the trachea. Saliva or food particles, along with bacteria, may then pass unnoticed into the lungs, and the bacteria may multiply. Such "silent" aspiration, which has been observed in people following strokes, carries with it the risk of serious pneumonia and death, underscoring the importance of early detection of dysphagia.

People with achalasia generally experience only mild dysphagia at first. Eventually, the esophagus may enlarge and regurgitation may occur frequently. The loss of nutrients through regurgitation as well as a fear of eating to avoid the discomfort of vomiting can markedly interfere with nutrition status.

Signs of Dysphagia Health care professionals should be alert to subtle symptoms of dysphagia including an unexplained decline in food intake or repeated bouts of pneumonia. Other symptoms include pain upon swallowing, weight loss, a fear of eating certain foods or any food at all, a feeling that food is sticking in the throat, a tendency to hold food in the mouth rather than swallowing it, coughing or choking during meals, frequent throat clearing, drooling, or a change in voice quality. Depending on the cause of dysphagia, the voice may be hoarse, or nasal, or have a "wet" sound. Diagnosis is based on extensive testing that may include cranial nerve assessment, X rays, fluoroscopy, and measurements of esophageal sphincter pressure and esophageal peristalsis.

Dietary Interventions for Dysphagia The diet for dysphagia leaves little room for error or experimentation. Speech therapists, dietitians, physicians, and nurses work together to assess the client's swallowing abilities and design a highly individualized diet that meets nutrient needs and maintains an adequate fluid balance. The diet plan considers both the cause of dysphagia and the types of solid and liquid foods a client can handle. Sticky foods and small pieces of food or foods that break into small pieces when eaten (such as rice or chopped meats) can be difficult for some clients to handle. Others may have problems swallowing foods with more than one texture, such as yogurt with fruit or dry cereal with milk. Figure 18–1 on p. 458 describes the characteristics of the four levels of diets for dysphagia.[2]

With respect to liquids, the thickness (viscosity) of liquids clients can safely swallow varies and ranges from true thin liquids to liquids the consistency of

Conditions associated with dysphagia include:
- *Acquired immune deficiency syndrome (AIDS).*
- *Aging.*
- *Alzheimer's disease.*
- *Brain tumors.*
- *Cancers of the head and neck.*
- *Developmental feeding disorders.*
- *Down syndrome.*
- *Guillain-Barré syndrome.*
- *Head injuries.*
- *Lou Gehrig's disease (amyotrophic lateral sclerosis).*
- *Multiple sclerosis.*
- *Myasthenia gravis.*
- *Parkinson's disease.*
- *Polio.*
- *Reflux esophagitis.*
- *Strokes.*

The **lower esophageal sphincter (LES)** is also called the **cardiac sphincter** or **gastroesophageal sphincter.**

The fear of eating is called **sitophobia** (SIGH-toe-FOE-bee-ah).
sitos = food
phobia = fear

dysphagia (dis-FAY-gee-ah): difficulty in swallowing.

achalasia (ack-ah-LAY-zee-ah): failure of the lower esophageal sphincter to relax and allow foods to pass from the esophagus to the stomach; formally called **achalasia of the cardia** or **cardiospasm.**
a = not
chalasis = relaxation

Stage I: All meat and meat alternates, breads and cereals, fruits and vegetables pureed to a pudding-like consistency. Sticky foods, such as melted cheese or peanut butter, are not allowed.

Stage II: Ground meats or soft, moist casseroles; soft poached or scrambled eggs; soft, bite-size pasta; rice; smooth cooked cereals; pancakes; mashed or minced soft fruits and vegetables.

Stage III: Moist, shaved, tender meats; poached or scrambled eggs; egg salad; soft breads and cookies without nuts or seeds; graham crackers; cooked or cold cereals with milk; fresh or canned fruits without seeds or tough skins; well-cooked or canned vegetables; sliced cucumber with no seeds or skin.

Stage IV: All of the above plus lightly toasted breads; all eggs; tender-cooked or soft raw vegetables.

Figure 18–1

Diets for Dysphagia

SOURCE: Adapted from American Dietetic Association and Dietitians of Canada, *Manual of Clinical Dietetics,* 6th ed. (Chicago: American Dietetic Association, 2000), pp. 680–685.

© Polara Studios, Inc. (all)

pudding or very thick milkshakes. Although some clients with dysphagia can swallow true liquids, many need thicker liquids because they flow slowly, allowing the client time to coordinate swallowing movements. Thin liquids can be thickened with commercial thickeners, baby rice cereal, baby apple flakes, and pureed fruits and vegetables.

Appropriate body positioning during meals minimizes the risk of choking. Advise the client to sit upright at a 90-degree angle with hips flexed, feet flat on the floor, and the head tilted slightly forward.

With time, swallowing function may change. The health care team continuously monitors the person and alters the diet to include appropriate foods. Ideally, the diet will progress to a diet of whole foods.

Tube Feedings   People who are unable to eat adequate amounts of foods orally, particularly those who are severely malnourished or those whose swallowing function continues to deteriorate, may need to be fed by tube (Chapter 20). Tube feedings delivered into the stomach may be contraindicated, however, due to the high risk of aspiration in people with dysphagia, and feeding by tube directly into the intestine provides a safer alternative.

In Summary   Routine progressive diets and mechanical soft diets include consistency-modified foods. Progressive diets test clients' tolerances for foods and begin with liquid foods and progress to solid foods. Many disorders that affect the mouth and esophagus can interfere with chewing and swallowing and may require a modification in food consistency. Dysphagia, a serious swallowing disorder that frequently goes undiagnosed, can lead to repeated bouts of pneumonia and even death. Medical nutrition therapy for chewing and swallowing disorders includes a highly individualized mechanical soft diet based on the client's individual tolerances.

Diets to Treat Gastric-Related Problems

Dietary recommendations to treat the gastric-related problems described in this section aim to minimize irritation to the esophagus and stomach (see Table 18–4). Such diets, which have traditionally been called **bland diets**, benefit clients with **gastroesophageal reflux disease (GERD)**, gastritis, or ulcers. The same suggestions may also benefit clients who suffer from temporary bouts of indigestion. The suggestions in Table 18–4 serve as a starting point for minimizing gastric discomforts. Health care professionals often begin with these suggestions and then individualize the diet based on each person's tolerances.

Gastroesophageal Reflux Disease (GERD)

Diet Order
Frequent small meals of bland, low-fat foods.

Many people suffer from GERD—acid indigestion and **heartburn** caused by the backflow of the stomach's acid fluids into the esophagus (acid indigestion). GERD occurs when the lower esophageal sphincter fails to close tightly enough to keep the stomach's contents from backing up into the esophagus.

Causes of GERD   Conditions and substances that weaken the lower esophageal sphincter itself or raise the pressure in the stomach increase the likelihood of GERD (see Table 18–5, on p. 460). GERD frequently occurs in people with asthma, peptic ulcers (described later in this chapter), irritable bowel

The viscosity of liquids that people with dysphagia can handle ranges. Viscosity level standards and examples include:
- *Thin: All liquids without modification.*
- *Nectarlike: Liquids with the thickness of nectar, tomato juice, and eggnog.*
- *Honeylike: Liquids with the thickness of honey or tomato sauce.*
- *Puddinglike: Liquids with the thickness of pudding or very thick milkshakes.*

Table 18–4   Dietary Recommendations for Minimizing Gastric Irritation
■ Avoid alcohol.
■ Most foods can be used in limited amounts, but avoid any food or spice that repeatedly causes discomfort.
■ Limit coffee (regular or decaffeinated), black pepper, garlic, cloves, and chili powder.

bland diets: diets that aim to minimize gastric acid secretion and limit gastric irritants.

gastroesophageal reflux disease (GERD): gastroesophageal reflux is the backflow of gastric juices into the esophagus. GERD is diagnosed when reflux occurs frequently or produces significant heartburn or indigestion.

heartburn: a burning sensation felt behind the sternum that is caused by the presence of gastric juices in the esophagus; also called **pyrosis** (pie-ROE-sis).

Table 18–5 Conditions and Substances Associated with GERD

Conditions

- Ascites (accumulation of fluid in the abdomen)
- Bending over after eating
- Delayed gastric emptying
- Eating especially large meals
- Hiatal hernias
- Lying flat within two hours of eating
- Obesity
- Pregnancy
- Wearing clothing that fits tightly across the waist or abdomen

Substances

- Alcohol
- Anticholinergic agents
- Caffeine
- High-fat foods
- Tobacco

Antisecretory agents are also called **anti-GERD agents.**

hiatal hernias: protrusions of a portion of the stomach through the esophageal hiatus of the diaphragm. There are several types of hiatal hernias, but the type most commonly associated with reflux is a **sliding hiatal hernia.**

reflux esophagitis (eh-sof-ah-JYE-tis): inflammation of the esophagus caused by the reflux of gastric juices.

   re = back
   fluxus = flow

epigastric: the region of the body just above the stomach.

   epi = above
   gastric = stomach

esophageal ulcers: lesions or sores in the lining of the esophagus.

esophageal stricture: narrowing of the inner diameter of the esophagus from inflammation and scarring.

Barrett's esophagus: changes in the cells of the esophagus associated with chronic reflux that raise the risk of cancer of the esophagus.

syndrome (Chapter 19), developmental disabilities (see Nutrtition in Practice 18), and sliding **hiatal hernias** (see Figure 18–2).

Consequences of GERD In some cases, reflux causes no symptoms or problems other than acid indigestion and heartburn. In other cases, however, the acidic gastric fluids irritate and erode the lining of the esophagus, leading to a painful inflammation called **reflux esophagitis.** In addition to **epigastric** pain, clients with reflux esophagitis may also report symptoms not associated with the GI tract, including chest pain, chronic coughing, hoarseness, and sore throat. Severe and chronic irritation may lead to **esophageal ulcers;** the consequent bleeding and scarring can narrow the inner diameter of the esophagus (**esophageal stricture**), and dysphagia may develop. Chronic blood loss can lead to anemia.

Chronic reflux esophagitis is also associated with **Barrett's esophagus,** a condition linked to an increased risk of cancer of the esophagus.[3] Chronic pulmonary disease can develop if the gastric contents are frequently aspirated into the lungs through the throat.

Prevention and Treatment Treatment for active GERD aims to alleviate reflux and prevent inflammation of the esophagus and its associated pain by reducing gastric acidity, reducing pressure in the stomach, and eliminating foods, substances, or activities that weaken the lower esophageal sphincter. The "How to" box on p. 462 offers tips for preventing reflux and treating its symptoms.

Drug Therapy Turn on the television set and quite likely you will see several advertisements touting antacids and other over-the-counter medications that are highly effective in both preventing and relieving acid indigestion. The ads imply that rather than eat sensibly and listen to the body's signals about what kinds and amounts of foods to eat, people can eat anything, anyway they like, and then simply take medication to avoid the discomforts of acid indigestion. All medications carry risks, however, and everyone should be encouraged to take steps to prevent acid indigestion without medication. People who must rely on medications to treat GERD or reflux esophagitis should do so with the advice of their physicians.

Physicians frequently prescribe antacids to neutralize gastric acidity and antisecretory agents to suppress or inhibit gastric acid secretion (see the Diet-Drug Interactions box on p. 464). Medications that strengthen lower esophageal sphincter pressure (cholinergics) may be used when other measures prove unsuccessful in controlling symptoms. Surgery may be necessary in some cases. The case study on p. 463 provides questions that review the nutrition needs of a client with GERD and reflux esophagitis.

Gastritis

Diet Order
Bland diet as tolerated.

Gastritis is an inflammation of the lining of the stomach that can develop suddenly (acute gastritis) or over time (chronic gastritis). Unresolved gastritis can lead to ulcers (described in the next section), hemorrhage, shock, obstruction, and perforation and raises the risk of gastric cancer.

Acute Gastritis Acute gastritis most often follows the repeated use of aspirin or other medications that irritate the gastric mucosa. *Helicobacter pylori* infections and other bacterial infections, alcohol abuse, food irritants, food allergies, food poisoning, radiation therapy, and metabolic stress can also

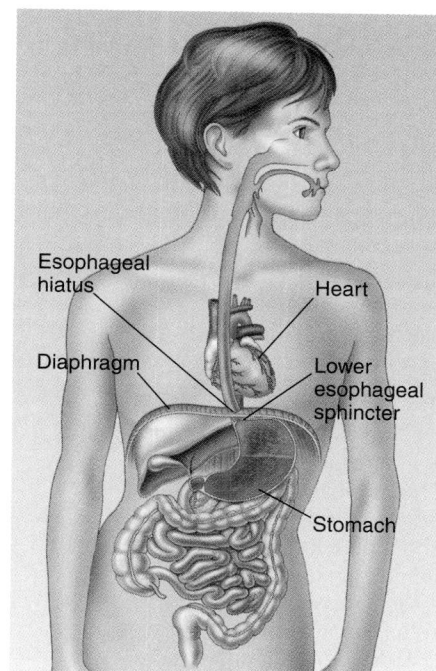

Esophageal hiatus
Diaphragm
Heart
Lower esophageal sphincter
Stomach

The stomach normally lies below the diaphragm, and the esophagus passes through the esophageal hiatus. The lower esophageal sphincter prevents reflux of stomach contents.

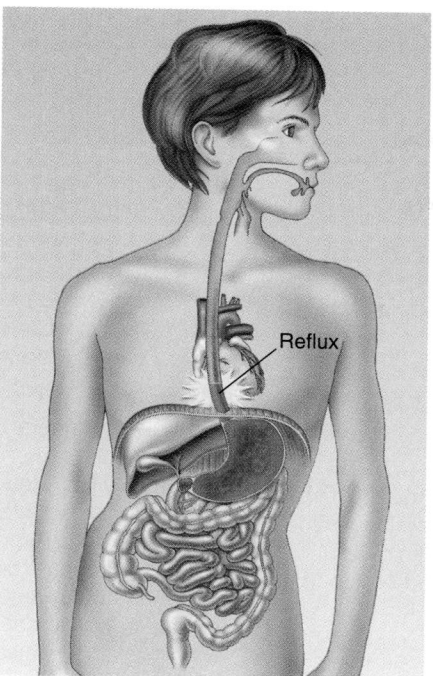

Reflux

Whenever the pressure in the stomach exceeds the pressure in the esophagus, as can occur with overeating and overdrinking, the chance of reflux increases. The resulting "heartburn" is so-named because it is felt in the area of the heart.

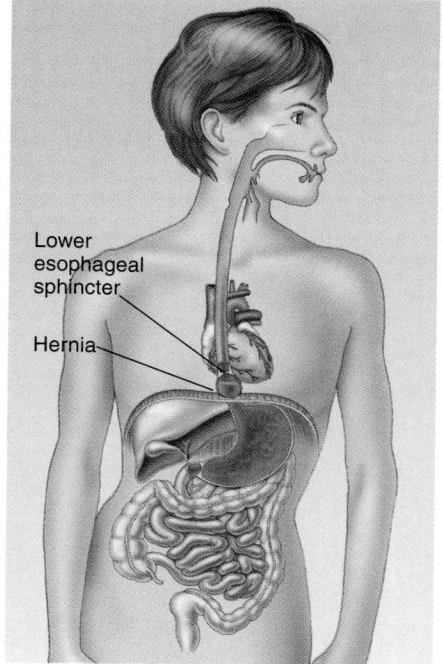

Lower esophageal sphincter
Hernia

Acid reflux may occur as a consequence of a hiatal hernia. A sliding hiatal hernia occurs when part of the stomach, with the lower esophageal sphincter, slips through the diaphragm.

Figure 18–2

The Upper GI Tract, Heartburn, and Hiatal Hernia

cause gastritis. Symptoms of acute gastritis vary and may include anorexia, nausea and vomiting, stomach pain, and fever.

Treatment for acute gastritis depends on the cause, but in all cases any food or substance that irritates the gastric mucosa is withheld (see Table 18–4). For the person with acute gastritis who cannot eat because of nausea or vomiting, foods may be withheld for a day or two. Then, the diet progresses from liquids to a bland diet until acute gastritis resolves.

Chronic Gastritis Chronic gastritis may be associated with aging, *Helicobacter pylori* and other infections, and conditions that cause the chronic reflux of basic fluids from the duodenum into the stomach. With time, gastric acid secretions diminish, and the production of intrinsic factor falters. Vitamin B_{12} malabsorption and pernicious anemia follow. People with chronic gastritis may have mild symptoms similar to those of acute gastritis or no symptoms at all. For this reason, a diagnosis of pernicious anemia may actually lead to the diagnosis of chronic gastritis. In the later stages of unresolved chronic gastritis, the gastric cells gradually shrink and lose their function (atrophic gastritis).

Chronic gastritis requires diagnosis and treatment before damage progresses too far. Interventions need to begin early enough to prevent complications such as dehydration, malnutrition, or damage to the esophagus. Health care professionals advise clients to avoid foods that irritate the gastric mucosa (see Table 18–4). For clients with pernicious anemia, vitamin B_{12} injections or prescription vitamin B_{12} nasal sprays provide vitamin B_{12} without the need for absorption. Physicians may also prescribe antacids, antisecretory agents, and antiulcer agents (see the Diet-Drug Interactions box on p. 464).

Reminder: **Pernicious anemia** is a reduced number of red blood cells caused by a lack of intrinsic factor and consequent vitamin B_{12} malabsorption.

Reminder: **Atrophic gastritis** is the severe form of gastritis in which chronic inflammation diminishes the size and function of the stomach's mucosal cells and glands.

gastritis: inflammation of the stomach lining.

Helicobacter pylori: a bacterium that may lead to gastritis and peptic ulcers and may raise the risk of cancer of the stomach.

Consistency-Modified and Other Diets for Upper GI Tract Disorders

How to

How to Prevent and Treat GERD

*T*o prevent and treat GERD and its associated discomfort, recommend that clients:

- Avoid large meals to avoid distending the stomach. A distended stomach exerts pressure on the lower esophageal sphincter.

- Relax during mealtimes, eat foods slowly, and chew them thoroughly to avoid swallowing air and distending the stomach.

- Limit foods that weaken lower esophageal sphincter pressure or increase gastric acid secretion, including fat, alcohol, coffee (regular and decaffeinated), chocolate, spearmint, peppermint, and any food that causes discomfort.

- Lose weight, if necessary. Overweight tends to increase abdominal pressure.

- Refrain from lying down, bending over, and wearing tight-fitting clothing or belts, particularly after eating, to avoid increasing pressure in the stomach.

- Elevate the head of the bed by 4 to 6 inches. Keeping the chest higher than the stomach helps prevent reflux.

- Refrain from smoking cigarettes. Smoking relaxes the lower esophageal sphincter.

- During periods of active reflux esophagitis, avoid foods and beverages that irritate the esophagus, such as citrus fruits and juices, tomatoes and tomato-based products, pepper, spices, and very hot or very cold foods, according to individual tolerances.

Individual tolerances of types and amounts of foods and spices vary markedly especially during periods of active esophagitis. Health care professionals can help clients pinpoint food intolerances by advising them to keep a record of the types and amounts of foods and beverages consumed, the time of consumption, GI symptoms, and time of occurrence. Assessment of the record by a dietitian or health care professional will help determine the types and amounts of food the client can handle without discomfort. If a client cannot tolerate citrus fruits and juices and tomatoes and tomato-based products, ensure that the diet provides adequate vitamin C from other foods or supplements.

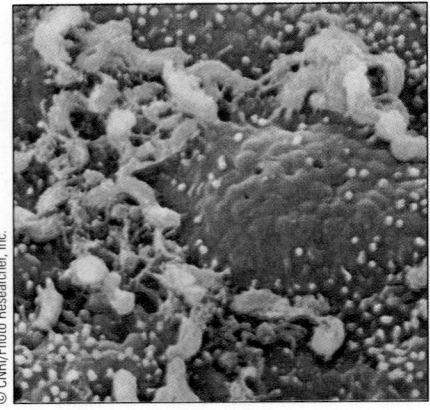

Heliobacter pylori, the bacterium implicated in gastritis, ulcers, and gastric cancer, protects itself from destruction by embedding itself in the lining of the stomach and producing bases to neutralize the stomach's acid. Inflammation ensues as the immune system responds to the invading organism.

peptic ulcer: a lesion or erosion of the cells of the mucosa of the lower esophagus, stomach, or small intestine. Ulcers may also develop in the mouth, upper esophagus, and large intestine, or on the skin.

Ulcers

Diet Order
Bland diet as tolerated.

Ulcers can develop both inside and outside the body, but the term *ulcer* used alone generally refers to a **peptic ulcer**—an erosion of the top layer of cells from the lining of the esophagus, stomach, or small intestine. This erosion leaves the underlying layers of cells exposed to gastric juices. When the erosion reaches the capillaries, the ulcer bleeds, and when gastric juices reach the nerves, they cause pain.

Causes of Ulcers Ulcers occur as a consequence of three major causes: *Helicobacter pylori* infections, the chronic use of certain anti-inflammatory drugs, and disorders that cause excessive gastric acid secretion. One such disorder, the **Zollinger-Ellison syndrome,** results from a tumor of the pancreas that produces gastrin, which, in turn, stimulates the production of gastric acid. Severe peptic ulcer disease follows.

A study of more than 45,000 men suggests that higher intakes of vitamin A, fruits and vegetables, and soluble dietary fibers are associated with a lower incidence of duodenal ulcers.[4] Further research is necessary to determine if other closely related nutrition (or other) factors may actually account for the protective effects the researchers observed.

Treatment for Ulcers Treatment for ulcers aims at relieving pain, healing the ulcer, and minimizing the likelihood of recurrence. Drug therapy plays the primary role in the treatment of ulcers (see the Diet-Drug Interactions box on

Case Study

ACCOUNTANT WITH GERD

Mrs. Scarlatti, a 49-year-old accountant, recently underwent a complete physical examination. She told her physician that she had been feeling fairly well until she began experiencing heartburn, which has become more frequent and more painful. The heartburn usually occurs after she has eaten a large meal, particularly when she lies down after eating. After a series of tests, the physician diagnosed GERD.

Mrs. Scarlatti's past health history shows no indication of significant health problems. During her last physical, the physician did advise her to stop smoking cigarettes and to lose 20 pounds, which she has yet to do. The nutrition assessment reveals that Mrs. Scarlatti is experiencing a great deal of stress because it is the middle of the tax season. She usually does not have time for breakfast, eats a lunch of fast foods hurriedly

while continuing to work, and eats a large dinner around 8:00 P.M. She generally drinks one or two alcoholic beverages before going to sleep. Her current height and weight are 5 feet 6 inches and 170 pounds.

- Explain to Mrs. Scarlatti what GERD is, and describe how it leads to heartburn. Explain the potential consequences of untreated GERD.

- From the brief history provided, list the factors and behaviors that increase Mrs. Scarlatti's chances of experiencing reflux and the pain of heartburn. What recommendations can you make to help her change these behaviors? What medications might the physician prescribe and why?

---

p. 464). Physicians prescribe antibiotics to treat bacterial infections when they are present. Antiulcer agents that protect the esophagus, stomach, or intestinal lining from acid erosion and antisecretory agents that suppress or inhibit gastric acid secretion may also be prescribed. As for diet, only foods that cause discomfort for the individual need to be eliminated completely. Suggestions that may reduce gastric acid secretion and speed healing include:

- Eat a variety of nutritious foods.

- Avoid overeating, frequent meals, bedtime snacks, alcohol, aspirin, nonsteroidal anti-inflammatory agents, and cigarette smoking.

- Limit foods that cause discomfort and caffeine, black pepper, garlic, cloves, and chili powder.

Surgery to sever the nerves that stimulate gastric acid production or to remove all or part of the stomach (see Chapter 22) may be indicated in some cases. For clients with Zollinger-Ellison syndrome, surgical removal of the tumor, when possible, reduces gastric acid production.

Zollinger-Ellison syndrome: marked hypersecretion of gastric acid and consequent peptic ulcers caused by a tumor of the pancreas.

In Summary People with GERD, gastritis, and ulcers, may benefit from dietary recommendations that reduce gastric irritation and gastric acid secretions. The Diet-Drug Interactions box on p. 464 describes the nutrition implications of medications used in the treatment of upper GI tract disorders. The Nutrition Assessment Checklist on p. 465 summarizes information that health care professionals can use to assess and monitor the nutrition status of clients with upper GI tract disorders.

Diet-Drug Interactions

Antacids

Antacids neutralize gastric acid.

Aluminum-containing antacids contribute aluminum to the diet and may cause constipation or lead to phosphorus deficiency. Long-term or inappropriate use can lead to aluminum toxicity.

Calcium-containing antacids contribute calcium to the diet and may cause constipation. Concurrent use with vitamin D supplements or foods containing large amounts of vitamin D can lead to elevated blood calcium levels (hypercalcemia).

Magnesium-containing antacids contribute magnesium to the diet and may cause diarrhea; long-term use may lead to magnesium toxicity.

Sodium bicarbonate, an antacid that contributes sodium to the diet, can alter serum electrolyte levels and raise the pH of the blood. Sodium bicarbonate should not be taken with milk—hypercalcemia can result.

Antibiotics

Antibiotics are used to treat *Helicobacter pylori* infections

When *amoxicillin* is given without regard to food, nausea and diarrhea are common side effects. Encourage clients to take with food to reduce nausea.

Clarithromycin may cause taste alterations, nausea, and diarrhea.

Metronidazole may cause taste alterations (metallic taste), and no alcohol should be used during treatment and for 24 hours afterward. Alcohol can react with metronidazole and result in nausea, vomiting, headache, cramps, and flushing of the skin.

Tetracycline and its classic interactions with nutrients were mentioned in Chapter 15. Calcium, magnesium, zinc, aluminum, antacids that contain any of these nutrients, and vitamin-mineral supplements should not be given from one hour before to two hours after a tetracycline dose. Iron supplements should not be given from two hours before to three hours after a tetracycline dose. Tetracycline can also cause nausea and diarrhea.

Antisecretory Agents

Antisecretory agents inhibit gastric acid secretion.

Cimetidine may increase the formation of toxic metabolites from pennyroyal.

When taken with tomato-based juices, the potency of *nizatidine* is reduced.

Esomeprazole, unsoprazole, omeprazole, and *rabeprazole* (proton pump inhibitors) may interfere with iron absorption. When iron supplements are necessary, they should be given two hours before or after taking proton pump inhibitors.

Nutrition Assessment Checklist

FOR PEOPLE WITH UPPER GI TRACT DISORDERS

Medical History

Does the client's health history reveal medical conditions or treatments that:

- ☐ Interfere with chewing and swallowing?
- ☐ Often lead to dysphagia?

Does the client have a medical diagnosis of:

- ☐ GERD?
- ☐ Gastritis?
- ☐ Pernicious anemia?
- ☐ Peptic ulcers?

Medications

Check the medication and dosing schedule for:

- ☐ Medications that cause nausea, taste alterations, or diarrhea.
- ☐ Clients receiving tetracycline for *Helicobacter pylori* infections. Do not give tetracycline one hour before or two hours after giving milk and milk products or antacids. Do not give tetracycline two hours before or three hours after giving iron supplements.
- ☐ Clients receiving antisecretory agents and iron supplements. Give the antisecretory agent two hours before or after the iron supplement.

Food/Nutrient Intake

Confer with clients and the dietitian to help pinpoint food intolerances for clients with:

- ☐ GERD
- ☐ Reflux esophagitis
- ☐ Gastritis
- ☐ Peptic ulcers

For clients on long-term mechanical soft diets, regularly note:

- ☐ Appetite
- ☐ Variety of food being offered and eaten
- ☐ Consistencies the client can handle

Anthropometrics

Measure baseline height and weight. Address weight loss early to prevent malnutrition for clients with:

- ☐ Any condition requiring mechanical soft diets for long periods of time
- ☐ Dysphagia
- ☐ Chronic gastritis

Laboratory Tests

Check laboratory tests for signs of dehydration for clients with:

- ☐ Dysphagia
- ☐ Chronic gastritis

Check laboratory tests for nutrition-related anemias (see Appendix D) for people with:

- ☐ Gastritis
- ☐ Conditions that require long-term use of antisecretory agents

Physical Signs

Look for physical signs of:

- ☐ Dehydration (especially for people with dysphagia)
- ☐ Iron deficiency (especially for people who take tetracycline or antisecretory agents)
- ☐ Vitamin B_{12} deficiency (especially for people with chronic gastritis)

Self Check

1. Which of the following statements describes both clear- and full-liquid diets?
   a. They can be used for long periods of time.
   b. They are deficient in nutrients and are temporary diets.
   c. They are contraindicated for people experiencing GI disturbances.
   d. They are contraindicated for people who are unable to chew and swallow whole foods.

2. A client's diet order reads, "progress diet from clear liquids to a regular diet as tolerated." What signs and symptoms suggest the client is not ready to advance to the next step of the diet progression?
   a. nausea, vomiting, diarrhea, and cramping
   b. dehydration or overhydration
   c. pain and confusion
   d. headache or fever

3. Immediately following surgery to remove the appendix (a procedure that requires general anesthesia and a hospital stay), a well-nourished client is most likely to:
   a. receive no foods or fluids by mouth.

b. receive a clear-liquid diet.

c. receive a mechanical soft diet as tolerated.

d. receive a regular diet.

4. The health care professional working with a client on a mechanical soft diet recognizes that:

a. highly seasoned foods are restricted.

b. only pureed foods should be given to minimize the risk of aspiration.

c. the diet cannot be planned to meet nutrient needs and supplements are necessary.

d. the client can have any food that can be comfortably and safely chewed and swallowed.

5. For people with dysphagia:

a. the diet requires a great deal of experimentation to uncover which foods a person can tolerate.

b. meals should be eaten in bed with the head tilted back as far as possible.

c. the diet is based on a health care team's assessment of a person's swallowing abilities and close monitoring of the person's ability to handle different foods.

d. coughing during meals indicates that the person is able to clear the throat and is not at risk for aspiration.

6. Diets to reduce gastric irritation limit foods that:

a. are difficult to digest.

b. are high in vitamin A and fiber.

c. are difficult to chew or swallow.

d. stimulate gastric acid production.

7. GERD is a common disorder of the upper GI tract characterized by:

a. an inability to swallow foods.

b. acid indigestion and heartburn.

c. erosion of the lining of the stomach.

d. erosion of the lining of the esophagus.

8. Reflux esophagitis is:

a. an inflammation of the esophagus caused by the backflow of acidic gastric fluids from the stomach.

b. a protuberance of a portion of the stomach above the lower esophageal sphincter.

c. an erosion of the lining of the stomach caused by excess gastric acid.

d. an obstruction of the lower esophagus that results in dysphagia.

9. Nutrition concerns most commonly associated with gastritis include:

a. malnutrition and pernicious anemia.

b. dysphagia and pernicious anemia.

c. dysphagia and electrolyte imbalances.

d. iron-deficiency anemia and malabsorption.

10. Causes of ulcers include all of the following **except:**

a. bacterial infections.

b. high pressure in the stomach.

c. nonsteroidal anti-inflammatory medications.

d. tumors of the pancreas.

Critical Thinking

1. Which of the following is a physical sign of dysphagia that might be revealed during a nutrition assessment?

a. repetitive throat clearing

b. history of repeated episodes of pneumonia

c. silent aspiration

d. diet history that reveals a reduced appetite over several months

2. The health care professional working with a client with reflux esophagitis recognizes that the client understands her diet when she says:

a. "I need to eat three meals a day and eat them as quickly as possible."

b. "I need to eat food slowly, relax during meals, and lie down after eating."

c. "I need to drink more citrus juices and tomato-based products at the first sign of heartburn."

d. "I need to limit my intake of fat, alcohol, and coffee."

3. A bland diet and antisecretory agents are recommended for a client with peptic ulcers. The client complies with the suggestions in Table 18–4, but continues to complain of indigestion and gastric pain. The most appropriate advice would be the recommendation to:

a. eliminate all highly seasoned foods.

b. take foods in frequent small meals.

c. keep a record of food intake, gastric symptoms, and when symptoms occur.

d. drink milk with each meal.

Answers to these questions can be found in Appendix G.

Clinical Applications

1. People on mechanical soft diets differ in the kinds of foods they can handle, in the lengths of time that diet modifications remain necessary, and in the help they need from health care professionals. Think about the difference between working with a person who has had no teeth for years and a person who recently had mouth surgery and is just beginning to eat again. Describe some nutrition-related concerns you might have for the person who has been following a mechanical soft diet for years. How would these concerns differ for a person who needs the mechanical soft diet only temporarily? Contrast the amounts of time a nurse or dietitian might need to spend with the two types of clients.

2. Many of the diets described in this chapter are highly individualized. A particular food may cause discomfort for one person and have no effect on another. Describe practical ways to keep track of food intolerances.

Nutrition on the Net

For further study of the topics of this chapter, access these websites.

Find updates and quick links to these and other nutrition-related sites at our website:
www.wadsworth.com/nutrition

Search for dysphagia, GERD, gastritis, and ulcers by visiting Gut Feelings and the American College of Gastroenterology:
www.gutfeelings.com and **www.acg.gi.org**

Find additional resources and information about dysphagia by visiting the Dysphagia Resource Center:
www.dysphagia.com

Discover additional facts about GERD by visiting the GERD Information Center:
www.gerd.com

Learn more about *Helicobacter pylori* infections and their consequences by visiting the Helicobacter Foundation:
www.helico.com

Learn more about ulcers and their treatments by visiting the U.S. Department of Health and Human Services:
www.hoptechno.com/book35.htm

Notes

1. K. M. Jeffery and coauthors, The clear liquid diet is no longer a necessity in the routine management of surgical patients, *American Journal of Surgery* 62 (1996): 167–170.

2. P. Felt, The National Dysphagia Diet Project: The science and practice, *Nutrition in Clinical Practice* (supplement) 14 (1999): 60–63.

3. J. Lagerhan and coauthors, Symptomatic gastroesophageal reflux as a risk factor for esophageal adenocarcinoma, *New England Journal of Medicine* 340 (1999): 825–831.

4. W. H. Aldoori and coauthors, Prospective study of diet and the risk of duodenal ulcer in men, *American Journal of Epidemiology* 145 (1997): 42–50.

Nutrition in Practice

■ HELPING PEOPLE WITH FEEDING DISABILITIES ■

Chapter 18 described problems encountered when people have difficulties with chewing and swallowing. This Nutrition in Practice discusses a broader problem faced by thousands of people who must cope with disabilities that interfere with the process of eating. These obstacles can arise at any time in a person's life and from any number of causes. An infant may be born with a physical impairment such as cleft palate; an adolescent may suffer nerve damage from injuries sustained in a car accident; a middle-aged adult may lose motor control following a stroke; an older adult may struggle with the pain of arthritis or the mental deterioration of dementia. Table NP18–1 lists some of the conditions that may lead to feeding problems.

How do disabilities impair eating?

To get food from the plate to the stomach requires an amazing number of individual coordinated movements. Consider the infant learning to feed himself. Every single step—sitting, grasping and bringing utensils or food to the mouth, biting, chewing, and swallowing—requires coordinated movements. Any injury or disability that interferes with these movements can lead to feeding problems. Other types of disabilities may also have nutrition-related consequences.

What are some examples?

An example of a nutrition-related effect unrelated to the process of eating involves energy needs. A disability may make it difficult for a person to engage in enough physical activity to support a healthy appetite or, conversely, to burn energy. Then the person's energy intake may either fail to meet needs or result in excessive weight gain. A person who has lost a limb to amputation also has altered energy needs. Energy needs are reduced in proportion to the weight and metabolism represented by the missing limb, but may be increased if extra energy is needed to perform tasks such as propelling a wheelchair. When people have disabilities that cause involuntary motor activity, energy needs may be exceedingly high.

As another example, a person who has problems with sight, or who cannot drive or walk or carry groceries, or who cannot plan meals and think through what to buy or cook, has a disability that affects eating. Disabilities of any type can cause people to have trouble maintaining adequate nutrition status.[1] Their number one problem is inadequate food intake, which leads to malnutrition, underweight, and, in children, poor growth.[2] Children and adolescents are especially vulnerable to these problems. Nutrition status may be further impaired by symptoms and medications that interfere with the appetite or alter nutrient needs.

What symptoms and medications can affect nutrition status?

Depending on the disability, many symptoms can interfere with eating. Some of these include nausea, frequent coughing, choking, GERD, and language and hearing disabilities. Frequent coughing and choking can lead to respiratory infections, which raise nutrient needs. People with speech and hearing problems may have a difficult time communicating with caregivers about thirst and hunger. People with disabilities that interefere with the ability to walk and move risk bone demineralization and pressure sores (see Chapter 15).

The disabilities that lead to feeding problems frequently require the use of multiple medications, which can have a significant impact on food intake and nutrient needs (see Chapter 15). Furthermore, people with feeding problems may encounter emotional and social problems. For example, children fail to receive the social training that mealtimes provide, and any person may miss the social stimulation that goes with eating in the company of others.

How can health care professionals promote independent eating for people with disabilities?

The evaluation and treatment of feeding problems ideally involve the joint efforts of several health care professionals, including a physician, a dietitian, a psychologist, an occupational therapist, a physical therapist, a speech-language pathologist, a dentist, and one or several nurses. Together, health care professionals assess each client's ability to perform eating-related tasks such as chewing, swallowing, grasping utensils,

Nutrition in Practice

Adaptive feeding equipment can help clients with feeding disabilities gain independence.

Table NP18–1 Conditions That May Lead to Feeding Problems

The following conditions may lead to feeding problems by interfering with a person's ability to suck, bite, chew, swallow, or coordinate hand-to-mouth movements.

- Accidents
- Amputations
- Arthritis
- Birth defects
- Cerebral palsy
- Cleft palate
- Down syndrome
- Head injuries
- Huntington's chorea
- Hydrocephalia

- Language, visual, or hearing impairment
- Microcephalia
- Multiple sclerosis
- Muscle weakness
- Muscular dystrophy
- Neuromotor dysfunction
- Parkinson's disease
- Polio
- Spinal cord injuries
- Stroke

using utensils to pick up foods, and bringing foods from the plate to the mouth.

What roles do the various team members play in dealing with feeding problems?

The physician diagnoses the client's medical problems, prescribes treatments, and coordinates the health care team. The nurse develops a nursing care plan to address the client's needs and educates the client and caregivers in accordance with the plan. The dietitian assesses the client's nutrition status, plans a diet, monitors medications for potential diet-drug interactions, and provides nutrition counseling.[3] Table NP 18–2 lists important nutrition assessment parameters.

A speech-language pathologist most often evaluates a client's ability to chew and swallow and trains the client to use lips, tongue, and throat for eating and speaking. The occupational therapist may work with the client to evaluate the need for special feeding devices and shows the client how to use them. A dentist may evaluate a client's dental health and provide instructions on maintaining oral hygiene. The health care team and client face many challenges in solving feeding problems, but the personal satisfaction of seeing someone gain control of her eating can be very rewarding.

Do clients receive behavioral training, such as practice eating sessions?

Yes, such training sessions are essential. Direct observation of the client during mealtimes allows the health care team to evaluate current eating behaviors, demonstrate feeding techniques, monitor the client's and the caregiver's understanding of the techniques, and

Table NP18–2 Nutrition Assessment for People with Feeding Disablities

Diet History

Feeding environment
Feeding position
Symptoms associated with eating (coughing, gagging)
Amount of food lost from feeding utensil or mouth during feeding
Person responsible for feeding
Client/caregiver attitudes toward feeding

Food Intake Data

Length of feeding
Types and amounts of foods consumed
Food aversions, intolerances, or allergies
Total fluid intake

Anthropometrics

Height
Weight
For children, plot data on growth charts

Laboratory Data

Hemoglobin
Hematocrit
Albumin
Prealbumin

Physical Findings

Physical signs of nutrient imbalances
Poor dental health

Nutrition in Practice

evaluate how well the care plan is working. To illustrate, consider an example of a child with a feeding problem caused by hypersensitivity to oral stimulation. The health care professional may start by teaching the caregiver to gently and playfully stroke the client's face with a hand, washcloth, or soft toy. Once the child tolerates touch on less sensitive areas of the face, the health care professional may encourage the caregiver to slowly begin to rub the child's lips, gums, palate, and tongue. With time, the child may develop a tolerance for the presence of food in the mouth.

What special feeding devices are available for helping people to eat?

Figure NP18–1 on p. 471 shows a few of many special feeding devices and describes their uses. These devices can make a remarkable difference in a person's ability to eat independently. For a person who cannot grasp an ordinary fork, for example, a special fork may be the key to future health.

   Sometimes, despite the best efforts of all involved, the client still can't eat enough food by mouth. In these cases, feeding by tube (see Chapter 20) can help improve nutrition status.[4]

With so many things to do, how do clients and caregivers cope?

You have identified an area of great concern regarding feeding disabilities. Eating is only one of many routine tasks that clients with disabilities face. The time and patience needed to learn and handle the many tasks required to care for a person with disabilities often result in a great deal of frustration and distress for both the client and the caregiver. A child with cerebral palsy, for example, may take ten times longer to eat than children without cerebral palsy. Some mothers have reported spending up to seven hours a day feeding these children. Besides feeding, the caregiver must

often help the disabled person with many other tasks, and all may require a great deal of time. Caregivers may feel overwhelmed with responsibility and have little time to care for themselves and other family members. Psychologists can offer counseling to clients or caregivers to help them adjust; all members of the health care team can offer emotional support and practical suggestions to ease caregivers' responsibilities and frustrations.

Successful therapy for feeding disabilities requires the involvement of many health care professionals and depends on the accurate identification of impaired feeding skills and appropriate interventions. Ideally, with training, people with disabilities attain total independence—they are able to prepare, serve, and eat nutritionally adequate food daily without help. In some cases, these goals can be attained with the help of caregivers. The combined efforts of the health care team can support both clients and caregivers in enhancing quality of life and in achieving independence to the greatest degree possible.

Notes

1. Position of The American Dietetic Association: Nutrition services for children with special health needs, *Journal of the American Dietetic Association* 95 (1995): 809–812.
2. R. D. Stevenson, Feeding and nutrition in children with developmental disabilities, *Pediatric Annals* 24 (1995): 225–260.
3. H. H. Cloud, Expanding roles for dietitians working with persons with developmental disabilities, *Journal of the American Dietetic Association* 97 (1997): 129–130.
4. R. Tawfik and coauthors, Caregivers' perceptions following gastrostomy in severely disabled children with feeding problems, *Developmental Medicine and Child Neurology* 39 (1997): 746–751.

Nutrition in Practice

Utensils

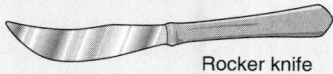

Rocker knife

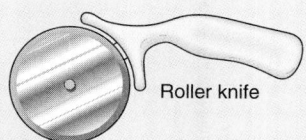

Roller knife

People with only one arm or hand may have difficulty cutting foods and may appreciate using a *rocker knife* or a *roller knife.*

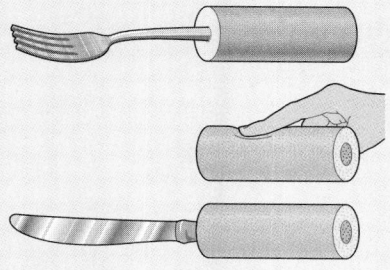

People with a limited range of motion can feed themselves better when they use *flatware with built-up handles.*

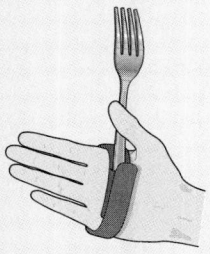

People with extreme muscle weakness may be able to eat with a *utensil holder.*

For people with tremors, spasticity, and uneven jerky movements, *weighted utensils* can aid the feeding process.

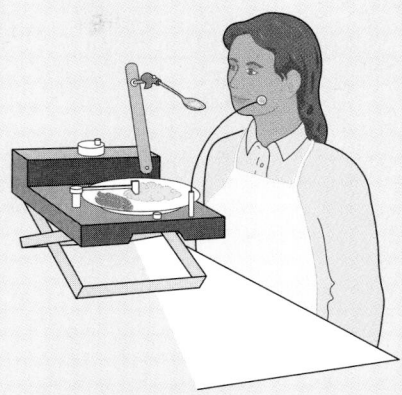

Battery-powered feeding machines enable people with severe limitations to eat with less assistance from others.

Plates

People who have limited dexterity and difficulty maneuvering food find *scoop dishes* or *food guards* useful.

People with uncontrolled or excessive movements might move dishes around while eating and may benefit from using *unbreakable dishes with suction cups.*

Cups

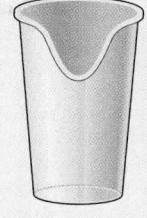

People with limited neck motion can use a *cutout plastic cup.*

Two-handed cups enable people with moderate muscle weakness to lift a cup with two hands.

People with uncontrolled or excessive movements might prefer to drink liquids from a *covered cup* or glass with a *slotted opening* or *spout.*

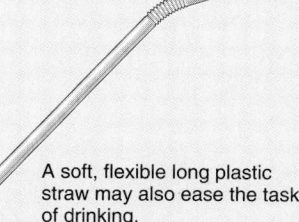

A soft, flexible long plastic straw may also ease the task of drinking.

Figure NP18–1
Examples of Adaptive Feeding Devices

Jennie Oppenheimer/Studio Zocolo

FIBER-MODIFIED DIETS FOR LOWER GI TRACT DISORDERS

High-fiber foods promote normal bowel movements.

© Peter Kane/The Stock Shop

Unlike the consistency-modified and other diets used to treat conditions of the upper GI tract, fiber- and residue-modified diets serve primarily to treat conditions of the lower GI tract. This chapter describes fiber- and residue-modified diets and shows their uses in conditions characterized by altered intestinal motility or those for which minimal fecal bulk aids recovery. Chapter 22 describes disorders that affect the absorption of nutrients.

Fiber-Modified Diets

Fiber-modified diets include both high-fiber and low-fiber diets. Table 19–1 lists indications for the use of fiber-modified diets. The amount of fiber in the diet affects the total fecal volume (**residue**) and the time it takes for the stool to pass through the body. As Chapter 2 described, insoluble fibers add to stool bulk, increase GI transit time, and stimulate bowel movements. Soluble fibers, on the other hand, contribute less to stool bulk, delay GI transit time, delay glucose absorption, and lower blood cholesterol.

High-Fiber Diets As Chapter 2 described, health authorities recommend that people adopt high-fiber diets (20 to 35 grams of fiber) by emphasizing the consumption of whole-grain breads and cereals, fruits and vegetables, and legumes. Figure 2–4 on p. 46 shows the fiber content of selected foods. As fibers pass through the intestine, they draw fluids from the intestinal contents along with them; therefore, health care professionals encourage people on high-fiber diets to drink plenty of fluids. The sample menu shows a day's meals for a high-fiber diet, and the "How to" box on p. 475 suggests ways to help clients raise their fiber intakes.

High-fiber diets help maintain intestinal health and are of particular benefit to people with constipation, irritable bowel syndrome, or diverticulosis. They also benefit people with inflammatory bowel diseases that are in remission. High-fiber diets may also help in weight-loss efforts (Chapter 7), in the regulation of blood glucose (Chapter 24) and blood lipids (Chapter 25), and in reducing the risk of cancer. Diets high in soluble fibers, but low in insoluble fibers, may also help treat diarrhea.

Intestinal Gas Excessive intestinal gas production, and the bloating and GI discomfort that accompany it, can be uncomfortable side effects of high-fiber diets. In the colon, bacteria metabolize undigested and unabsorbed dietary fibers and

High-Fiber Diet Menu

Breakfast	Lunch	Supper	Snack
1 c multigrain cereal	1 c black bean soup	3 oz baked fish	3 c popcorn
½ c strawberries	3 oz broiled chicken	½ c brown rice	1 c tomato juice
1 c fat-free milk	½ c steamed broccoli	½ c peas	
2 slices whole-wheat toast	½ c baked sweet potatoes	1 whole-wheat dinner roll	
2 tbs peanut butter	1 fresh pear	2 tsp margarine	
1 c coffee	1 whole-wheat dinner roll	1 piece carrot cake	
	1 tsp margarine	1 c fat-free milk	

How to

Help Clients Add Fiber to Their Diets

During the first few weeks on a high-fiber diet, a person may feel bloated, pass gas frequently, or experience heartburn. Provide these suggestions to help:

- *Go slow.* Add high-fiber foods gradually and in small portions at first. Increase portion sizes and add foods as tolerance improves.
- *Add fluids.* Fiber attracts water as it moves through the intestine. Aim to drink 64 ounces of water each day. If you have trouble remembering to drink fluids, set up a schedule of when you will drink fluids each day and follow the schedule until drinking becomes automatic.
- *Experiment.* Try small servings of various fiber-containing foods at first and adopt those that are most pleasing.
- *Mix high-fiber foods with other foods.* Sprinkle bran flakes, wheat germ, or raisins on salads or applesauce. Add bran or mashed legumes to meat loaf. Add legumes and other high-fiber vegetables to soups and salads.

---

produce gas in the process. People with excessive gas may feel bloated and experience abdominal pain, cramps, and **flatus.** In addition to fibers, any undigested and unabsorbed food can cause intestinal gas, and people with malabsorption often experience this problem. High-fat foods, fructose, and sugar alcohols (sorbitol, mannitol, xylitol, and maltitol) taken in large amounts may be incompletely absorbed and cause gas for some people (see Table 19–2). Unabsorbed lactose can cause gas for people with lactose intolerance (described in Chapter 22). People troubled by gas need to determine which foods bother them and then avoid those foods or eat them in moderation. Medications that help reduce intestinal gas production, called antiflatulents, are also available.

Low-Fiber Diets   Low-fiber diets serve to reduce the total fecal volume and intestinal transit time. All foods produce some residue, but residue consists primarily of dietary fibers. Some health care facilities serve both low-fiber and low-residue diets, although there is little research to support the use of low-residue diets, and low-fiber diets minimize intestinal residue to an acceptable level in most cases.[1] Table 19–3 lists low-fiber foods, and the sample menu on p. 476 shows a day's meals.

Low-fiber foods are least likely to obstruct an intestinal tract that is narrowed by inflammation or scarring or in which GI motility is slow. Low-fiber foods are also less likely than high-fiber foods to produce intestinal gas. Low-fiber diets may be prescribed before and after intestinal surgery. People at risk for small bowel obstructions, as well as those with delayed gastric emptying, active inflammatory bowel diseases, or diverticulitis, may also benefit from low-fiber diets.

Table 19–2 Foods or Substances That May Produce Gas	
Apples	Honey
Asparagus	Legumes
Beer	Maltitol
Bran	Mannitol
Broccoli	Milk and milk products
Brussels sprouts	Nuts
Cabbage	Onions
Carbonated beverages	Peppers, green
Cauliflower	Prunes
Corn	Radishes
Cream sauces	Raisins
Cucumber	Sorbitol
Fried foods	Soybeans
Fructose	Wheat
Gravy	Xylitol
High-fat meats	

Table 19–3 Low-Fiber Foods	
Meat	Tender meat, poultry, seafood, and eggs
Breads and Cereals	Refined breads, cereals, rice, and pasta
Fruits	Cooked or canned peeled apples, apricots, peaches, pineapple, plums; bananas, cherries, cranberry sauce, mandarin oranges, tangerines; fruit juice
Vegetables	Most well-cooked vegetables without seeds or husks: lettuce, tomato paste, tomato puree, vegetable juice
Other	Avocado, nuts

residue: the total amount of material in the colon; includes dietary fiber and undigested food, intestinal secretions, bacterial cell bodies, and cells shed from the intestinal mucosa.

flatus (FLAY-tuss): gas in the intestinal tract or the expelling of gas from the intestinal tract, especially through the anus.

Breakfast	Lunch	Supper	Snack
Orange juice	Baked fish	Roast beef	Applesauce
Soft-cooked egg	White rice	Mashed potatoes	Vanilla wafers
Puffed rice cereal	Green beans	Cooked carrots	
White bread toast	Small banana	Canned peaches	
Coffee or tea	Roll	Roll	
Milk for cereal	Margarine	Margarine	
Creamer	Coffee or tea	Coffee or tea	
	Sugar	Sugar	
	Creamer	Creamer	

In addition to restricting fiber, low-residue diets generally limit milk and milk products and meats with tough connective tissue because these foods may contribute to the intestinal residue. In health care facilities that serve both low-fiber and low-residue diets, menus for both these diets are the same.

Because low-fiber diets are often used to alleviate uncomfortable GI symptoms, dietary advice may also include recommendations to limit foods that stimulate gastric acid or cause gastric irritation.

In Summary The amount of fiber in foods affects stool volume and intestinal transit time. High-fiber diets assist in the treatment of constipation, some cases of irritable bowel syndrome, and inflammatory bowel diseases that are in remission. Diets high in soluble fibers also serve in the treatment of diarrhea. Low-fiber diets help minimize fecal volume and intestinal gas production. People undergoing intestinal surgery, those at risk for small bowel obstructions, and those with delayed gastric emptying, diverticulitis, or active inflammatory bowel diseases may benefit from low-fiber diets.

Motility Disorders

Health problems that alter the transit of foods through the GI tract include such problems as delayed gastric emptying, constipation, diarrhea, and irritable bowel syndrome. Because dietary fibers affect GI transit time as well as stool volume and composition, fiber-modified diets serve in the treatment of motility disorders.

Delayed Gastric Emptying

Although low-fiber diets generally serve to treat disorders of the lower GI tract, delayed gastric emptying is an exception. Both temporary conditions (following surgery, for example) and chronic conditions (as a complication of diabetes mellitus, for example) can delay the rate at which the stomach empties. Table 19–4 lists examples. In some disorders, delayed gastric emptying occurs because the pyloric sphincter fails to open completely and the narrowed opening prevents the passage of food into the intestine. Plant matter from food (fibers, leaves, and skins) may stagnate in the stomach and form a mass called a **phytobezoar**.

Diet Order
Low-fiber, low-fat diet provided in frequent, small meals.

phytobezoar (FIGH-toe-BEE-zor): a mass of plant matter (fibers, leaves, and skins) that forms a ball and may block the outlet from the stomach to the intestine. *Trichobezoars* contain hair and nails. People with some psychiatric disorders may chew and swallow their hair and nails.

Symptoms of Delayed Gastric Emptying People experiencing delayed gastric emptying feel full after eating small amounts of food and may experience nausea, vomiting, abdominal pain, and bloating. The discomfort of eating can lead to anorexia and weight loss.

Treatments for Delayed Gastric Emptying   Treatments aim to control the underlying medical disorder and may include drugs to stimulate gastric motility. Diet therapy aims to eliminate symptoms and provide adequate nutrition. Often a low-fiber, low-fat diet given in frequent, small meals is helpful. Fiber and fat slow the rate at which the stomach empties, and large volumes of food distend the stomach and increase the likelihood of GI upsets. In severe cases, the person tolerates only liquids and only in small amounts. In such cases, formula diets provide nutrients. If there is a complete obstruction, or if the person is unable to drink enough liquids, feeding by tube below the stomach or intravenous (IV) feedings can provide nutrients.

Constipation

Diet Order
High-fiber diet.
Encourage fluids.

Constipation describes a symptom, not a disease. People with constipation may pass stools that are difficult or painful to excrete or experience a reduced frequency of bowel movements compared to their typical pattern (see Nutrition in Practice 5). Abdominal discomfort, headaches, backaches, and passage of gas sometimes accompany constipation. In severe cases, hardened stool may form an obstruction (**fecal impaction**). Left untreated, pressure may build above the impaction, and the intestine may rupture, which can lead to **peritonitis.**

Causes of Constipation   Constipation may be associated with fluid and electrolyte and hormonal imbalances, many diseases of the GI tract, chronic laxative abuse, stress, lack of physical activity, and various medications, including narcotic analgesics, anticholinergics, aluminum- and magnesium-containing antacids, and some antihypertensives. In addition, dietary supplements, including chondroitin sulfate, glucosamine, iron, calcium, and magnesium supplements, can lead to constipation.[2] Expectant mothers may experience constipation as the growing fetus crowds intestinal organs and hormonal changes alter intestinal muscle tone. Constipation is a problem for many elderly people, particularly those in institutions. With aging, gastrointestinal motility slows, and elderly people are less likely than others to engage in regular physical activity, drink adequate amounts of fluids, and eat foods high in fiber. Elderly people may also use several medications that contribute to constipation. Some people, especially elderly people, report problems with constipation and take laxatives to treat it, although objective measurements of stool consistency and frequency fail to confirm constipation.[3]

Treatment of Constipation   When constipation occurs as a result of an underlying medical condition, treatment of that disorder may help alleviate constipation. Medical nutrition therapy includes a high-fiber diet, plenty of fluids, and regular physical activity. Laxatives, often **bulk-forming agents,** may be prescribed if other measures fail to relieve constipation (see the Diet-Drug Interactions box on p. 486).

Diarrhea and Dehydration

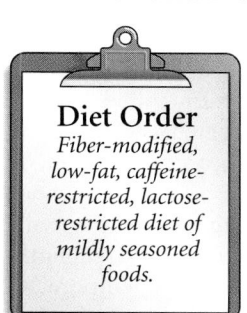

Diet Order
Fiber-modified, low-fat, caffeine-restricted, lactose-restricted diet of mildly seasoned foods.

Diarrhea is not a disease, but a complication of some medical conditions and a side effect of many medications and other treatments. It can be acute, lasting less than two weeks, or chronic, lasting longer. Mild diarrhea that remits in 24 to 48 hours is seldom a cause for concern unless the person is already dehydrated. A person with severe, persistent diarrhea may rapidly become dehydrated, lose weight, and develop multiple nutrient deficiencies. A child or infant can lose proportionately more fluid and weight than an adult and can develop severe

Table 19–4   Conditions Associated with Delayed Gastric Emptying

Temporary Delayed Gastric Emptying

- Diabetic ketoacidosis
- Electrolyte imbalances
- Hyperglycemia
- Infection
- Some drugs (examples: alcohol, nicotine, aluminum-containing antacids, opiates)
- Surgery

Chronic Delayed Gastric Emptying

- Connective tissue disorders
- Diabetic neuropathy
- Gastric surgery
- Gastroesophageal reflux disease (GERD)
- Neurological disorders
- Peptic ulcers
- Severe protein-energy malnutrition

Anticholinorganics are drugs that slow GI motility.

Reminder: Diarrhea is the frequent passage of watery bowel movements. Severe, chronic diarrhea that does not respond to treatment is **intractable diarrhea.**

fecal impaction: a dry, compacted mass of fecal matter in the colon or rectum.

peritonitis (pear-ih-toe-NIGH-tus): infection and inflammation of the membrane lining the abdominal cavity caused by leakage of infectious organisms through a perforation in an abdominal organ.

bulk-forming agents: laxatives composed of fibers that work like dietary fibers. They attract water in the intestine to form a bulky stool, which then stimulates peristalsis. Metamucil and Fiberall are examples.

Fiber-Modified Diets
for Lower GI Tract Disorders

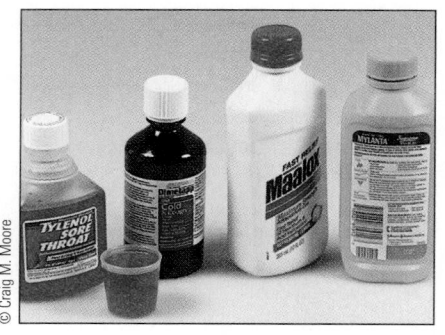

People who use high doses of sugar-free liquid medications, such as adults who have trouble swallowing pills, may develop diarrhea from the sorbitol these medications contain.

dehydration and malnutrition in a short time. Table 16–10 on p. 430 listed findings associated with dehydration.

Causes of Diarrhea   Acute diarrhea that occurs abruptly in a healthy person frequently results from viral, bacterial, or protozoal infections or as a side effect of medications or dietary supplements. It can also occur in the person who begins to eat foods or begins a tube feeding after a period of fasting, starvation, or IV nutrition. Infants may develop diarrhea when given formulas their immature GI tracts cannot tolerate or when they are ill. When used in large quantities, food ingredients such as sorbitol and olestra may cause diarrhea in some people.

Chronic diarrhea can occur as a result of disorders that alter GI tract motility, such as the irritable bowel syndrome. It can also develop as a consequence of malabsorption (see Chapter 22) and of some infections, such as human immunodeficiency virus (HIV).

Treatment of Diarrhea   The treatment of diarrhea requires treatment of the primary medical condition. If a food is responsible, that food must be omitted from the diet. If a medication is responsible, a different medication or medication form (injectable versus oral, for example) may alleviate the problem. Infections are treated with appropriate anti-infective agents. Antidiarrheals slow GI motility and are often prescribed along with other therapies to treat diarrhea (see the Diet-Drug Interactions box on p. 486).

Fluid and Electrolyte Replacement   Until the diarrhea resolves, treatment includes replacement of lost fluids and electrolytes to prevent dehydration. Fluids that provide both glucose and sodium help maximize the body's absorption of fluids and electrolytes.[4] Good choices for mild diarrhea include diluted fruit juices, sports drinks, and caffeine-free carbonated beverages. For mild-to-moderate cases of diarrhea, oral rehydration formulas—simple solutions of water, salts, and sugar—provide needed fluids and electrolytes. Salty broths, soups, bouillon, flavored ades, and jello can be used to supplement replacement fluids. For severe cases, especially if the client is dehydrated or experiencing persistent vomiting, IV fluids and electrolytes replace losses.

Yogurt contains *lactobacillus,* a bacterium that helps to establish a healthy bacterial flora and secretes lactase to help digest lactose. Nutrition in Practice 22 provides more information about *lactobacillus* and other probiotics.

Oral Diets   If eating aggravates diarrhea, clients may be advised to drink only clear liquids (including oral rehydration formulas). In severe cases, it may become necessary to stop placing demands on the GI tract by withholding all foods and beverages until the diarrhea remits, usually in a day or two. For people who can tolerate solid foods, the liberal use of foods with soluble fibers can often help control diarrhea (see Table 19–5). Foods with insoluble fibers should be avoided temporarily. Yogurt, which contains helpful bacteria, can help to control diarrhea in some cases (see Nutrition in Practice 22). The diet also temporarily excludes lactose, caffeine, highly seasoned foods, foods high in fat, foods that cause gas, and any food that aggravates the diarrhea. Frequent, small meals are easiest to tolerate at first. Permanent dietary changes may be necessary for diarrhea caused by food sensitivities or allergies.

irritable bowel syndrome: an intestinal disorder of unknown cause characterized by abdominal discomfort, cramping, diarrhea or constipation, or alternating diarrhea and constipation; also called *spastic colon.*

Diet Order
High-fiber, fat-restricted diet of mildly seasoned foods.

Irritable Bowel Syndrome

Irritable bowel syndrome, sometimes called spastic colon, is a common motility disorder characterized by abdominal pain associated with diarrhea, constipation, or alternating episodes of both.[5] The person may also experience bloating, flatulence, the passage of mucus with stools, indigestion, and nausea. Symptoms frequently occur shortly after a person eats and resolve temporarily following a bowel movement. For many, the symptoms are mild, but in some cases, the need to be close to a bathroom can interfere with work or prevent participation in social activities.

In the United States, irritable bowel syndrome affects about 5 million people annually and occurs more frequently in women.[6] Symptoms often begin in childhood or young adulthood and decrease with age. Even though symptoms persist for several months, people with irritable bowel syndrome and no other health problems generally remain in good health and seldom experience malnutrition.

Causes of Irritable Bowel Syndrome The causes of irritable bowel syndrome remain elusive, but stress and anxiety are believed to be contributing factors. Symptoms often worsen during periods of psychological stress or, for women, during menstruation. Other contributing factors may include abnormal GI tract motility and hypersensitivity to intestinal distention. Some studies suggest that bacterial overgrowth in the small intestine (see Chapter 22) may also play a role.[7] Foods and food components, including gas-forming foods (review Table 19–2), high-fat foods, fat substitutes, lactose, fructose, sorbitol, caffeine, and alcohol, may aggravate the symptoms of irritable bowel syndrome, but are not believed to cause it. Medications and herbal remedies can also aggravate diarrhea or lead to constipation.

Treatment of Irritable Bowel Syndrome Medical therapy for irritable bowel syndrome focuses on stress management and medical nutrition therapy. Drug therapy, often reserved for cases that do not respond to other treatments, may include antidepressants, anticholinergics, antidiarrheals, and laxatives (see the Diet-Drug Interactions box on p. 486).

Medical Nutrition Therapy To help minimize abdominal distention, improve digestion, and reduce the chance of swallowing air with foods (which can contribute to abdominal distention and gas in the intestine), advise clients with irritable bowel syndrome to avoid eating too much or too fast and to chew foods thoroughly. Clients respond uniquely to specific dietary interventions for irritable bowel syndrome, and no one diet is suitable for all clients. Dietitians often rely on food records that include food intake, fluid intake, stool consistencies, other GI symptoms, and stress levels to help uncover individual food intolerances.

Many clients benefit from a fat-restricted diet with a liberal fiber and fluid intake. The dietitian often adjusts types of fiber (soluble and insoluble) depending on whether the client primarily experiences constipation or diarrhea. For some clients, adding fiber to the diet aggravates symptoms because fiber-containing foods increase intestinal gas production. In these cases, fiber from bulk-forming agents may be preferable to dietary fibers. Other foods are restricted only when they cause discomfort to the individual. When lactose intolerance is a problem, a lactose-restricted diet (see Chapter 22) markedly improves symptoms.[8] The case study on p. 480 provides questions to help you apply information about irritable bowel syndrome to a clinical situation.

In Summary Motility disorders, including delayed gastric emptying, constipation, diarrhea, and irritable bowel syndrome, cause a great deal of discomfort for people who experience them. Fiber-modified diets help alleviate these problems in some cases and lessen uncomfortable symptoms in others.

Intestinal Obstructions

Ileus, a partial or complete blockage of the intestinal lumen, can occur in both the small and the large intestine. Inflammation, scar tissue, tumors, congenital malformations, protracted contraction of the intestinal muscles, or a fecal

> **Table 19–5 Dietary Suggestions for Treating Diarrhea**
>
> - Drink plenty of fluids including diluted fruit juices, sports drinks, or oral rehydration formulas.
> - Add salt to foods.
> - Use foods containing soluble fibers including bananas, peeled apples, applesauce, peeled pears, oranges, peeled white or sweet potatoes, oatmeal, barley, millet, couscous, noodles, rice, white bread, and saltine crackers.
> - Avoid foods with insoluble fibers including whole-wheat breads and cereals, bran, raw vegetables, corn, peas, nuts, seeds, and skins of fruits and vegetables.
> - Avoid caffeine and lactose, highly seasoned foods, high-fat foods, foods that cause gas, and any food that aggravates the diarrhea.
> - Eat frequent, small meals.

ileus (ILL-ee-us): an intestinal obstruction.

Fiber-Modified Diets for Lower GI Tract Disorders

Case Study

MARKETING PROFESSIONAL WITH IRRITABLE BOWEL SYNDROME

Sudah Patel is a 22-year-old recent college graduate who began her first professional job in a marketing firm one month ago. As a college student, she occasionally experienced abdominal pain and cramping after eating. She also experienced frequent bouts of diarrhea and noticed that she felt better for a while after a bowel movement. Once Sudah began her new job, she noticed that her symptoms were occurring more frequently. At first she attributed her symptoms to the stress related to her new job, but when the symptoms continued for several months, she decided to see her physician. Sudah is 5 feet 3 inches tall and weighs 118 pounds. After taking a careful history and conducting several tests to rule out other bowel disorders, the physician diagnosed irritable bowel syndrome. The physician prescribed bulk-forming agents and advised Sudah to keep a food intake and symptoms record for one week. Sudah was then referred to a dietitian to review the records and recommend appropriate dietary suggestions. In reviewing Sudah's food intake records,

the dietitian notices that Sudah eats many traditional Indian foods, which are highly seasoned and often include legumes. Sudah tells the dietitian that she began to drink coffee and caffeinated cola drinks in college so that she could study for longer periods of time. She has continued to drink coffee since that time.

- How would you explain irritable bowel syndrome to Sudah? What role does stress play in irritable bowel syndrome?

- Can Sudah's diet be responsible for causing irritable bowel syndrome? Can any of these foods or food components aggravate her symptoms? How will the dietitian use the food intake and symptoms record to devise medical nutrition therapy for Sudah?

- What type of diet benefits people with irritable bowel syndrome? What problem(s) can this diet cause?

Inflammation can occur as a consequence of many disorders of the intestine including diverticulitis, Crohn's disease, and ulcerative colitis (described later in this chapter).

Diet Order
Low-fiber diet in frequent, small meals.

impaction can physically block the movement of foods and fluids through the intestine (**mechanical ileus**). Sometimes the muscles of the intestine fail to function (**adynamic ileus**), often as a result of abdominal surgery.

Consequences of Intestinal Obstructions The severity of complications depends on the degree and type of obstruction. Adynamic ileus may resolve without special treatment in a few days, for example, but complete obstructions can result in life-threatening complications in a matter of hours.

When the intestine becomes obstructed, its contents stagnate above the blockage. Peristalsis increases as the body attempts to push the intestinal contents past the obstruction. Injury to the intestinal cells and intestinal distention follow. The distention cuts off blood flow to the area, which, in turn, can lead to malabsorption, acid-base imbalances, dehydration, and tissue death. A **perforation** may form in the damaged intestinal lining and can lead to peritonitis and **sepsis.**

People with ileus of the small intestine may develop severe pain, abdominal distention, nausea, and vomiting. Those with a complete mechanical ileus may vomit fecal material, while those with adynamic ileus may vomit gastric fluids and bile. People with mechanical obstructions of the large intestine often develop constipation and sudden, intermittent, and severe abdominal pain.

Treatments for Intestinal Obstructions In many cases, surgery is needed to treat complete obstructions or correct the underlying medical condition. Most often, people with complete obstructions cannot be fed by mouth or by tube. If they are malnourished or at risk for malnutrition, they may receive IV nutrition.

People with partial obstructions and those with adynamic ileus can sometimes be treated without surgery. In these cases, a tube is inserted from the nose to the stomach and is used to suction fluids and gas to help relieve pressure and

mechanical ileus: ileus caused by a physical obstruction.

adynamic or **paralytic ileus:** ileus caused by the failure of the intestinal muscles to function.

perforation (per-foe-RAY-shun): a hole or tear.

sepsis: the spread of an infection from a local area into the blood, which alters blood flow to vital organs and can lead to multiple organ failure and death (see Chapter 27).

reduce abdominal distention. Intravenous fluids help restore fluid and electrolyte and acid-base balances. Antibiotics are given to prevent and treat infections. Depending on the degree of obstruction, some people with partial or intermittent obstructions can eat an oral diet of low-fiber foods in frequent, small meals.

In Summary Intestinal obstructions arise from a variety of causes, and they can be either partial or complete. The underlying medical condition and the degree of obstruction determine what actions will be taken to correct the problem. Low-fiber foods served in frequent, small meals can benefit some clients with partial or intermittent obstructions.

Inflammatory Bowel Diseases

Diet Order
*Individualized,
fiber-modified diet
as tolerated.*

Two of the most prevalent **inflammatory bowel diseases (IBD)** are **Crohn's disease** and **ulcerative colitis.** IBD share some clinical features but are distinct disorders. Although the exact causes of IBD remain unknown, research suggests that genetic, immune system, and environmental factors play a role. Investigators have identified a genetic mutation that may increase susceptibility to Crohn's disease.[9] People with the mutation appear to mount an immune system response to bacteria that normally reside in the intestine and cause no such response in healthy people. Inflammation and tissue damage follow. IBD occur more frequently in people who have relatives with IBD and in American Jews of European descent. Adolescents and adults between the ages of 15 and 35 and, to a lesser degree, those older than 50 are most likely to develop IBD.

Crohn's disease most often affects the ileum and colon, but can occur throughout the GI tract. Crohn's disease affects all layers of the intestine, and healthy sections of intestine may be found between diseased portions. The person often experiences intermittent fatigue, abdominal pain, and diarrhea.

Ulcerative colitis develops only in the large intestine and affects only the inner layer of the intestine. During active episodes, ulcerative colitis causes almost continuous diarrhea with malabsorption and great losses of fluids and electrolytes. Bloody diarrhea, cramping, abdominal pain, anorexia, and weight loss are clinical manifestations of the disorder. Iron-deficiency anemia may develop as a consequence of blood loss.

Complications Associated with Inflammatory Bowel Diseases   As IBD progresses, fibrous tissue (scar tissue) may form in the intestine, reducing its absorptive ability, narrowing the intestinal lumen, and sometimes creating an obstruction. Localized infections can develop. The intestine may also rupture, which can lead to peritonitis.

Malabsorption, weight loss, and malnutrition frequently occur, especially with Crohn's disease. The area and extent of the intestine affected by the IBD determine the type of malabsorption and the degree to which it occurs. Malabsorption can lead to gas and bloating, which aggravate abdominal pain and further reduce food intake. Growth failure occurs in up to 50 percent of children and adolescents with Crohn's disease, and 90 percent in this age group are underweight.[10] Lactose intolerance is common. Iron deficiencies are more common in IBD affecting the colon because blood is lost regularly.

Fistulas may develop if an inflamed loop of intestine sticks to another loop of intestine, to another organ, or to the skin and gradually erodes. If a fistula forms between the stomach or the upper portion of the small intestine and the colon, ingested food is shunted directly into the colon, and malabsorption

Figure 22–1 in Chapter 22 shows where various nutrients are absorbed along the intestine and lists the consequences of disorders or surgeries that affect different portions of the intestine.

inflammatory bowel diseases (IBD): diseases characterized by inflammation of the bowel.

Crohn's disease: inflammation and ulceration along the length of the GI tract, often with granulomas.

ulcerative colitis (ko-LYE-tis): inflammation and ulceration of the colon.

fistulas (FIS-chew-lahs): abnormal openings formed between two organs or between an internal organ and the skin.
    fistula = pipe

*Fiber-Modified Diets
for Lower GI Tract Disorders*

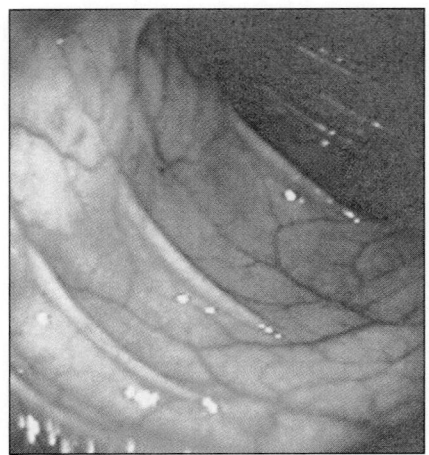

The normal colon has a smooth, shiny surface with a visible pattern of fine blood vessels.

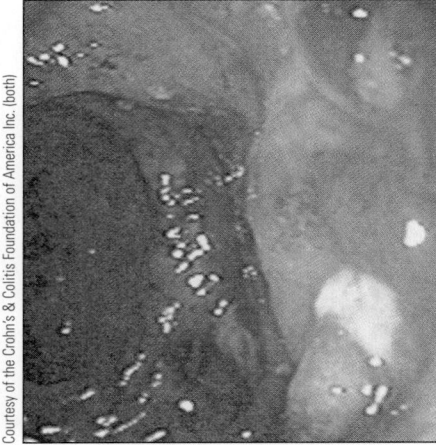

In ulcerative colitis, the colon appears inflamed and reddened, and ulcers are visible.

worsens. Bacteria from the colon can then invade the stomach or upper small intestine, contributing to further malabsorption (see "Bacterial Overgrowth" in Chapter 22), raising the risk of serious infections, and causing severe inflammation, nausea, and vomiting. If a fistula forms between the small intestine and the skin, significant malabsorption, dehydration, and sepsis can follow.

Treatments for Inflammatory Bowel Diseases   Treatments for IBD aim to reduce inflammation, control symptoms, and prevent malnutrition. People with IBD often rely on medications to reduce inflammation and control symptoms. They may remain symptom-free for long periods of time, but the symptoms tend to recur. Medications used in the treatment of IBD may include analgesics, antidiarrheals, anti-infective agents, and immune system suppressors and modulators; their potential nutrition-related side effects are numerous and are described in the Diet-Drug Interactions box on p. 486. Surgery may be necessary to remove a diseased or obstructed portion of the intestine or to repair a fistula. For people with ulcerative colitis, removal of the diseased portion of the colon (described later in this chapter) may cure the disorder. For people with Crohn's disease, surgery may reduce pain and alleviate some complication, but the disease tends to recur in other portions of the intestine.

Medical Nutrition Therapy for Inflammatory Bowel Diseases   Medical nutrition therapy for people with IBD is highly individualized. During periods of remission, health care professionals encourage clients to eat high-fiber foods to help maintain the health of the intestinal tract. Those who cannot maintain a desirable weight need high-kcalorie, high-protein diets to help prevent or treat malnutrition (see Chapters 27 and 28). Liquid formulas (Chapter 20) may be recommended, especially for children with IBD. Children sometimes eat table foods during the day and receive tube feedings at night.

During active periods of disease, eating often intensifies uncomfortable symptoms. Clients need only eliminate those foods that cause symptoms. The most common offenders include high-fiber foods, lactose-containing foods, and high-fat foods, and these foods are restricted to the degree necessary to control symptoms. Low-fiber diets may also be recommended for people with partial intestinal obstructions or those at high risk for obstructions. Sometimes eating smaller amounts of food more often is helpful. Vitamin and mineral supplements are generally recommended.

For people with active ulcerative colitis, no dietary interventions seem to lessen symptoms. People with severe abdominal pain and diarrhea need complete bowel rest.

Tube Feedings and Parenteral Nutrition   For people with active Crohn's disease who have intestinal obstructions or fistulas or for whom eating significantly worsens symptoms, foods and fluids may temporarily be withheld while fluids and electrolytes are replaced intravenously. For people who are already severely malnourished or for those who have gone for several days without eating, feeding by tube (Chapter 20) should be considered. In some cases, feeding tubes can be placed so as to bypass a fistula or partial obstruction. Intravenous nutrition (Chapter 21) can provide nutrients when oral or tube feedings significantly aggravate pain and diarrhea, when the bowel is obstructed, when complete bowel rest might help a fistula to close, or when oral or tube feedings cannot meet nutrient requirements.

In Summary   Malnutrition and fluid and electrolyte imbalances frequently occur as a consequence of IBD. The nutrition effects of IBD depend on the extent and portion of the intestine that are affected. Individual food tolerances vary greatly, and medical nutrition therapy is tailored to each person's unique responses. During periods of remission, high-fiber diets may help maintain intestinal health, if clients can tolerate them. During periods of active Crohn's disease, low-fat, low-fiber diets may be necessary.

Diverticular Disease

Diet Order
High-fiber diet.

Sometimes pouches of the intestinal wall (called **diverticula**) bulge out through the muscles surrounding the large intestine, often at points where blood vessels enter the muscles (see Figure 19–1). Most authorities believe that low-fiber diets and constipation lead to the development of diverticula (**diverticulosis**). Without adequate fiber, stools become hard and difficult to pass. The person must then strain the intestinal muscles to defecate, and this raises pressure in the colon. As pressure builds, it forces weakened areas of the intestinal membrane outward through the muscle layer.

People with diverticulosis are frequently symptom-free and unaware of the disorder. Others may develop cramps, bloating, and constipation. Diverticulosis often develops with age. About half of all people between the ages of 60 and 80 have diverticulosis; virtually all people over age 80 have it.[11]

Diverticulitis   In about 10 to 25 percent of people with diverticulosis, a localized area of inflammation and infection develops around a diverticulum, a condition called **diverticulitis.** Although the cause is unknown, it is believed that bacteria or fecal matter gets trapped in a diverticulum and leads to infection. People with diverticulitis may experience abdominal pain and distention, alternating episodes of diarrhea and constipation, indigestion, flatus, and fever. Occasionally, a diverticulum ruptures, causing a localized or sometimes life-threatening infection (peritonitis). Bleeding from a diverticulum may also occur. If the diverticula become inflamed repeatedly, the intestinal wall can form scar tissue and thicken (fibrosis), narrowing the intestinal lumen and creating an obstruction. An inflamed bowel segment can also stick to other pelvic organs, forming a fistula.

Treatments for Diverticular Disease   People with diverticulosis require no medications, but are advised to eat a high-fiber diet to reduce pressure in the colon and stimulate peristalsis. A study of over 45,000 people suggests that high-fiber, low-fat diets are associated with a lower incidence of symptomatic diverticular disease.[12] Traditionally, diet advice for people with diverticulosis included a recommendation to avoid foods with seeds such as okra and strawberries because seeds were believed to get trapped in the diverticula and cause irritation. However, evidence to support this theory is lacking.[13] During periods of active diverticulitis, however, a low-fiber restriction is appropriate. As diverticulitis resolves, a gradual reintroduction of fiber-containing foods is appropriate.

In Summary   Diverticular disease, a frequent consequence of aging, is believed to develop as a result of rising pressure in the colon. By adding fecal bulk, high-fiber foods reduce pressure in the colon and can help to prevent diverticular disease. If diverticulitis develops, low-fiber diets may be used temporarily to minimize uncomfortable GI symptoms.

Colostomies and Ileostomies

Treatment for some medical conditions and surgeries affecting the large intestine necessitates that the feces be diverted from all or portions of the colon. A **colostomy** is a surgical procedure that creates an opening (**stoma**) between the

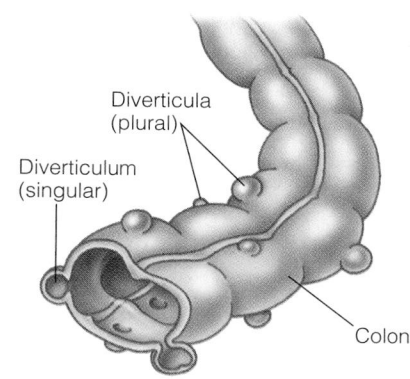

Figure 19–1
Diverticula

diverticula (dye-ver-TIC-you-lah): sacs or pouches that develop in the weakened areas of the intestinal wall (like bulges in an inner tube of a bike tire where the wall is weak).
  divert = to turn aside

diverticulosis (DYE-ver-tic-you-LOH-sis): the condition of having diverticula.
  osis = condition

diverticulitis (DYE-ver-tic-you-LYE-tis): infected or inflamed diverticula.
  itis = inflammation

colostomy (co-LOSS-toe-me): surgery that creates an opening from any portion of the colon through the abdominal wall and out through the skin.
  colo = colon

stoma (STOH-ma): a surgically formed opening. After a colostomy or ileostomy, a stoma is formed by bringing the cutoff end of the intestine through the abdominal wall, rerouting the excretion of wastes.
  stoma = window

Figure 19–2
Colostomy and Ileostomy

Colostomy

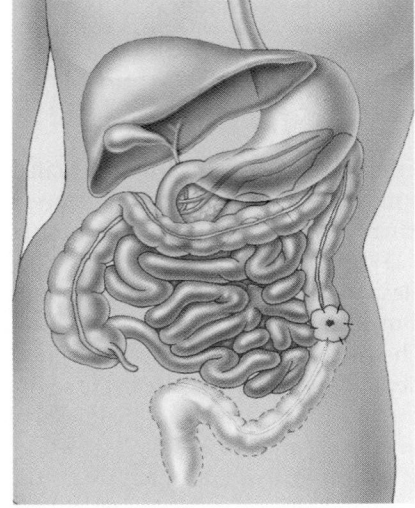

In a colostomy, the rectum and anus are removed, and the stoma is formed from the remaining colon.

Ileostomy

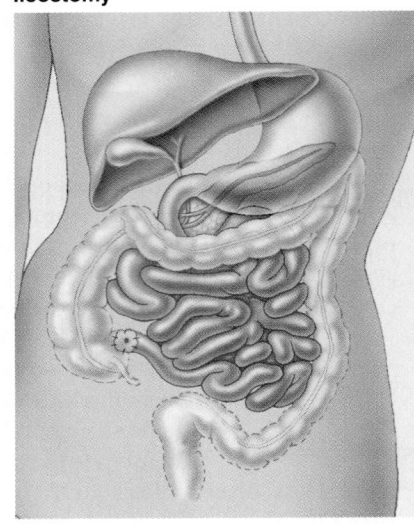

In an ileostomy, the entire colon, rectum, and anus are removed, and the stoma is formed from the ileum.

Inflammatory bowel diseases, diverticulitis, tumors of the colon and rectum, intestinal trauma, intestinal tissue death, and radiation enteritis are some conditions that may necessitate a colostomy or ileostomy. In some cases, a portion of the small intestine must also be resected, and in these cases, significant malabsorption can result. Chapter 22 provides more information about the nutrition effects of surgery involving the small intestine.

Diet Order
Progress diet from a low-fiber diet of moderately seasoned food to a diet as tolerated. Encourage fluids.

colon and the surface of the skin to allow for defecation when feces cannot pass through the rectum and anus. A pouch placed over the stoma collects the feces. Colostomies may be permanent or temporary.

In an **ileostomy,** the entire colon is bypassed, and the stoma is created from the ileum to the skin. In an alternative to an ileostomy, the ileal pouch/anal anastomosis, the surgeon removes the diseased colon and rectal tissue and connects the ileum to the anus. A temporary ileostomy allows time for the tissue to heal; thereafter, defecation occurs through the anus, rather than a stoma. Figure 19–2 shows examples of a colostomy and ileostomy. In this section, the word *ostomy* is used to refer to both colostomies and ileostomies.

Water is absorbed along the length of the colon, and when the colon is resected, additional fluid will be lost in the stool. Stool consistency will vary depending on both the length and the portion of the resected bowel. In general, the greater the length of colon left intact, the more fluid that will be absorbed and the more formed the stool.

Medical Nutrition Therapy following Surgery   Once fluid and electrolyte balances have been restored and solid foods are permitted following surgery, people who have undergone ostomies often receive low-fiber diets of mildly seasoned foods to prevent obstructing the ostomy, promote healing of the stoma, and prevent GI upsets. Clients are encouraged to judiciously include other foods as soon as possible, however. Foods should be added one at a time and in small amounts to assess their effects. If the added food presents problems, the person can try it again in a few weeks or months.

Preventing Obstructions   Some foods are more likely than others to be incompletely digested and obstruct the stoma. These include stringy foods such as celery, spinach, and bean sprouts; foods with tough skins such as dried fruits, raw apples, and corn; tough, fibrous meats; foods with seeds; mushrooms; nuts; coconut; and popcorn. Practitioners report that some of these foods can be used if the client cuts the food into small pieces and chews them thoroughly. An undi-

ileostomy (ILL-ee-OSS-toe-me): surgery that creates an opening from the ileum through the abdominal wall and out through the skin.
  ileo = ileum

gested portion of a mushroom, for example, may act as a plug and obstruct a stoma, but if it is cut into small pieces, it may be tolerated.

Encouraging Fluids   People with ostomies, especially those with ileostomies, need extra fluids because they are absorbing less fluid from the large intestine. They may tend to restrict their fluid intakes, however, for fear of increasing output from the stoma. Remind clients that extra fluids are essential and advise them to drink at least 8 to 10 cups of fluids per day.

Controlling Diarrhea   People with ostomies may benefit from foods rich in soluble fiber that thicken the stool and help control diarrhea. These foods include applesauce, bananas, cheese, creamy peanut butter, and starchy foods such as breads, rice, potatoes, and oatmeal. Foods that may aggravate diarrhea include apple, grape, and prune juices; highly seasoned foods; foods high in sugar; alcohol; and coffee. The foods listed here are suggestions only; what works for the individual is determined by trial and error.

Reducing Gas and Odors   People with ostomies are often concerned about gas and odors associated with foods. Gas-forming foods in general were listed in Table 19–2, but certain foods seem to cause gas and odors for people with ostomies: asparagus, legumes, beer, carbonated beverages, eggs, fish, garlic, onions, and strong cheeses. Foods thought to reduce odors include buttermilk, parsley, and yogurt.

In Summary   For people with ostomies, limiting some high-fiber foods can help minimize the likelihood of obstructing the stoma. Providing foods high in soluble fiber, however, can help minimize diarrhea.

This chapter has described fiber-modified diets and their uses in disorders affecting the transit of food through the GI tract or those that might lead to obstructions or uncomfortable symptoms. The Diet-Drug Interactions box on p. 486 provides examples of nutrition-related concerns related to medications used in the treatment of these disorders. The Nutrition Assessment Checklist on p. 487 highlights nutrition-related findings important to consider in clients with these disorders.

Analgesics

Analgesics, which relieve pain, can cause nausea, vomiting, and GI upsets. When oral intake is permitted, giving the medication along with food can minimize these side effects. *Narcotic analgesics* can lead to constipation; these medications can also lead to lethargy, which can contribute to reduced food intake.

Antianxiety Agents

Antianxiety agents reduce anxiety and can cause drowsiness, which can interfere with food intake. *Alprazolam* and *chlordiazepoxide* can stimulate the appetite and lead to weight gain. *Diazepam, lorazepam,* and *oxazepam* can cause constipation, nausea, and vomiting, and drowsiness. Caution clients on antianxiety agents to avoid alcohol.

Antidepressants

Antidepressants treat depression and frequently cause mouth dryness, nausea, vomiting, and constipation. *Nefazone* should be administered without food or with a consistent intake of food to ensure proper drug absorption. The herb *belladonna* can enhance the effects of antidepressants. People taking *St. John's wort* and antidepressants should do so only under medical supervision to minimize potential interactions.

Antidiarrheals

Antidiarrheals treat diarrhea. They seldom result in significant nutrition-related side effects, with the exception of *opium* and *paraegoric,* which can cause nausea and vomiting, constipation, and lethargy.

Antiemetics

Antiemetics, which relieve nausea and vomiting, frequently lead to drowsiness. *Prochlorperazine* and *thiethyperazine maleate* can also lead to mouth dryness and constipation.

Anti-Inefective Agents

Anti-infective agents combat infections. Numerous anti-infection agents with a variety of nutrition-related interactions may be used in the treatment of diarrhea, diverticulitis, and infections that arise in clients recovering from surgery of the intestine. Consult a drug guide for specific examples.

Immunosuppressants (for Inflammatory Bowel Diseases)

Immunosuppressants modulate immune responses. *Prednisone* can stimulate the appetite and lead to fluid retention—both can lead to weight gain. Long-term use also leads to negative nitrogen and calcium balances and osteoporosis. For clients who take prednisone, high-protein, high-calcium, high-potassium diets are recommended. If lactose intolerance is a problem, clients may need to use digestive aids or calcium supplements. Clients who are not losing fluids through diarrhea and malabsorption may need to limit sodium. The dietary supplement *melatonin* can antagonize the effects of prednisone. Two other immunosuppressants, *azathioprine* and *6-mercaptopurine,* may lead to nausea and vomiting.

Laxatives

Laxatives alleviate constipation. Common side effects include nausea and cramps. Mineral oils can reduce the absorption of fat-soluble vitamins. Clients on laxatives are encouraged to eat high-fiber foods and drink generous amounts of fluids.

Uncategorized Drugs (for Inflammatory Bowel Diseases)

Sulfasalazine may cause nausea, vomiting, and diarrhea, and it may also lead to folate deficiency. The drug should be taken with food at regular intervals during the day. *Mesalamine* may lead to anorexia and weight loss, but nutrition-related side effects are less common with mesalamine than sulfasalazine. Clients unable to eat enough folate from foods may need supplements. Encourage the client to drink fluids. *Infliximab* may cause nausea and abdominal pain and is given by injection.

Nutrition Assessment Checklist

FOR PEOPLE WITH LOWER GI TRACT DISORDERS

Medical History

Does the client have a medical diagnosis of:

- ☐ Irritable bowel syndrome?
- ☐ Crohn's disease?
- ☐ Ulcerative colitis?
- ☐ Diverticular disease?

Has the client had surgery that includes:

- ☐ Colostomy?
- ☐ Ileostomy?

Does the client have the following symptoms or complications:

- ☐ Constipation?
- ☐ Diarrhea/dehydration?
- ☐ Lactose intolerance?
- ☐ Malabsorption?
- ☐ Infection?
- ☐ Fistulas?
- ☐ Obstructions?

Medications

Check for medications or herbal remedies that may:

- ☐ Cause constipation or diarrhea
- ☐ Interfere with food intake including those that cause nausea, vomiting, cramps, dry mouth, or drowsiness.
- ☐ After nutrient needs, including prednisone, sulfasalazine, and antidepressants

Food/Nutrient Intake

Note the following conditions and contact the dietitian if you suspect a problem with:

- ☐ Poor appetite or limited food intake
- ☐ Food intolerances
- ☐ Inadequate fiber/fluid intake for those with constipation
- ☐ Fluid intake

Anthropometrics

Measure baseline height and weight. Address weight loss early to prevent malnutrition for clients with:

- ☐ Severe or persistent diarrhea
- ☐ Malabsorption

Laboratory Tests

Check laboratory tests for signs of dehydration for clients with:

- ☐ Severe or persistent diarrhea
- ☐ Ostomies

Check laboratory tests for signs of nutrition-related anemias and nutrient deficiencies (see Appendix D) for clients with:

- ☐ Severe or persistent diarrhea
- ☐ Inflammatory bowel disease

Physical Signs

Look for physical signs of:

- ☐ Dehydration (especially for people with severe or persistent diarrhea or malabsorption
- ☐ PEM
- ☐ Folate and vitamin B_{12} deficiencies (especially for people with Crohn's disease
- ☐ Mineral deficiencies (especially for people who have severe and persistent diarrhea, ulcerative colitis, or ostomies)

Self Check

1. A health care professional talking with a client about a high-fiber diet explains that all of the following foods contain fiber **except:**
   a. whole-grain breads and cereals.
   b. milk and milk products.
   c. legumes.
   d. fresh fruits and vegetables.

2. Why are high-fiber foods more likely than low-fiber foods to cause GI discomfort?
   a. High-fiber foods may delay gastric emptying and lead to excessive gas production.
   b. High-fiber foods often contain more fat and lactose.
   c. Low-fiber foods are mildly seasoned foods.
   d. Low-fiber diets are liquid diets.

3. The primary constituent of intestinal residue is:
   a. intestinal secretions.
   b. shed intestinal cells.
   c. undigested fibers.
   d. waste products of energy metabolism.

4. Dietary therapy for people with delayed gastric emptying most often includes:
   a. foods low in soluble fiber.
   b. foods low in insoluble fiber.
   c. high-fiber foods given in frequent, small meals.
   d. low-fiber, low-fat foods given in frequent, small meals.

5. The health care professional advising an elderly client with constipation encourages the client to eat a:
   a. low-fat diet rich in potassium.
   b. low-fiber diet rich in calcium.
   c. high-fiber, bland diet.
   d. high-fiber diet and drink plenty of fluids.

6. Which statement best describes the use of fiber-modified diets in the treatment of diarrhea?
   a. Low-fiber foods are recommended.
   b. High-fiber foods are recommended.
   c. Foods high in soluble fiber are recommended.
   d. Foods high in insoluble fiber are recommended.

7. A common disorder characterized by a disturbance in the motility of the GI tract with symptoms that frequently occur shortly after a person eats is called:
   a. irritable bowel syndrome.
   b. inflammatory bowel disease.
   c. diverticular disease.
   d. ulcerative colitis.

8. Treatments for partial intestinal obstructions may include all of the following **except**:
   a. insertion of a tube from the nose to the stomach to help relieve pressure and distention.
   b. laxatives to help propel the intestinal contents.
   c. low-fiber diets.
   d. antibiotics.

9. The health care professional working with a client with IBD recognizes that all of the following can affect dietary recommendations **except**:
   a. individual food tolerances.
   b. risk for intestinal obstructions.
   c. risk for diverticulitis.
   d. symptoms that are aggravated by eating.

10. Long-term management of diverticular disease includes a:
    a. high-fiber diet.
    b. low-fiber diet.
    c. high-fiber, lactose-free diet.
    d. low-fiber, lactose-free diet.

11. Clients undergoing ostomies may receive a low-fiber, bland diet to:
    a. prevent malabsorption.
    b. increase intestinal motility.
    c. prevent mineral deficiencies.
    d. help the stoma heal and prevent obstructions.

Critical Thinking

1. A health care professional working with a client following a low-fiber diet learns that the client eats a sandwich and fruit and drinks a glass of milk for lunch. Considering this meal only, what other information will help the health care professional assess the client's compliance with the diet?
   a. the type of bread and fruit the client ate
   b. the time the client ate in relation to other meals
   c. the type of milk and the condiments the clients used
   d. the length of time the client has been following the diet

2. Given that low-fiber diets are less likely to cause uncomfortable GI symptoms than high-fiber diets, why not continue to recommend low-fiber diets to people with IBD and those who have recovered from diverticulitis?
   a. Low-fiber diets are deficient in nutrients.
   b. Low-fiber diets are monotonous to follow.
   c. High-fiber diets reduce fluid requirements.
   d. High-fiber diets promote intestinal health.

Answers to these questions can be found in Appendix G.

Clinical Applications

1. Many symptoms and disorders of the GI tract described in this chapter and the last are associated with aging. Review both chapters and list these symptoms and disorders. Referring to Chapter 14, describe the effects of aging on the GI tract and describe how these changes might relate to the symptoms and disorders you find.

2. The last chapter described how an obstruction of the esophagus might occur as a consequence of chronic reflux esophagitis, and this chapter discussed how the stomach might become obstructed as a result of a defective pyloric sphincter function. Contrast complete obstructions of the upper GI tract with complete intestinal obstructions. In which case would enteral nutrition be more likely to be of benefit, and why?

Nutrition on the Net

For further study of the topics of this chapter, access these websites.

Find updates and quick links to these and other nutrition-related sites at our website: **www.wadsworth.com/nutrition**

Search for more information on the disorders described in this chapter by visiting the National Institute of Diabetes & Digestive & Kidney Diseases, the American College of Gastroenterology, Medscape Gastroenterology, and InteliHealth: **www.niddk.nih.gov/health/digest/digest.html, www.acg.gi.org, www.medscape.com/medscape/ gastro,** and **www.intelihealth.com**

Click on the Medical Library at the Crohn's and Colitis Foundation to find more information about IBD: **www.ccfa.org**

Learn more about colostomies and ileostomies by visiting the United Ostomy Foundation: **www.uof.org**

Notes

1. American Dietetic Association and Dietitians of Canada, *Manual of Clinical Dietetics,* 6th ed. (Chicago: The American Dietetic Association, 2000), p. 705.

2. J. E. Suneson, Irritable bowel syndrome: A practical approach to medical nutrition therapy, *Support Line,* February 1999, pp. 11–15.

3. D. Harari and coauthors, Bowel habit relation to age and gender: Findings from the National Health Interview Survey and clinical implications, *Archives of Internal Medicine* 156 (1996): 315–320; C. S. Probert and coauthors, Evidence for the ambiguity of the term constipation: The role of irritable bowel syndrome, *Gut* 35 (1994): 1455–1458.

4. W. McCray and B. Krevsky, Diarrhea in adults: When is intervention necessary? *Hospital Medicine* 35 (1999): 39–46.

5. L. R. Schiller, Irritable bowel syndrome: One physician's perspective, *Support Line,* October 1998, pp. 3–8.

6. Suneson, 1999.

7. M. Pimentel and coauthors, Eradication of small intestinal bacterial overgrowth reduces symptoms of irritable bowel syndrome, *American Journal of Gastroenterology* 95 (2000): 3503–3506.

8. C. J. Bohmer and H. A. Tuynam, The effect of a lactose-restricted diet in patients with a positive lactose tolerance test, earlier diagnosed as irritable bowel syndrome: A 5-year follow-up study, *European Journal of Gastroenterology and Hepatology* 13 (2001): 941–944.

9. J. P. Hugot and coauthors, Association of NOD2 leucine-rich repeat variants with susceptibility to Crohn's disease, *Nature* 411 (2001): 599–603.

10. Controlling childhood Crohn's disease requires a multipronged approach, *Drug and Therapy Perspectives* 17 (2001): 5–8, available online at **www.medscape.com**, site visited October 9, 2001.

11. National Institute of Diabetes & Digestive & Kidney Diseases, Diverticulosis and diverticulitits, November 1998, **www.niddk.nih.gov/health/digest/pubs/divert/divert.htm**, site visited October 3, 2001.

12. W. H. Aldoori and coauthors, A prospective study of diet and the risk of symptomatic diverticular disease in men, *American Journal of Clinical Nutrition* 60 (1994): 757–764.

13. National Digestive Diseases Information Clearinghouse, November 1998.

Nutrition in Practice

■ FOODS AND FOODSERVICE IN HEALTH CARE FACILITIES ■

Hospitals and long-term care facilities prepare and provide foods for clients with a variety of medical conditions. The facility's foodservice department faces a challenge in planning, producing, and delivering appetizing, nutritious meals designed to accommodate dozens of special diets and food preferences. Although this discussion focuses on foodservice in hospitals, much of the information applies to foodservice in any health care facility, including nursing homes, assisted living centers, rehabilitation centers, and residential mental health care facilities. An important difference between hospitals and long-term health care facilities deserves attention, however. When clients in hospitals eat poorly, they can make up for nutrient deficits by eating well when they return home. Residents of a long-term care facility do not have this option. For this reason, foodservice departments in long-term care facilities must make even greater efforts to ensure that their clients receive nutritious foods and eat them.

I don't know too many people who have been in a hospital, but those I do know all seem to have a comment about the food. Why?

In the hospital, eating offers clients familiarity in an otherwise strange environment. Most people generally look forward to eating, and for many people in the hospital, a healthy appetite signals a return to health. Eating may become even more enjoyable than usual. It is also one of the few hospital experiences where clients have a choice. Consider that clients usually cannot choose when they will receive tests, how many times blood will be drawn, what nurse will care for them, or what time they will have surgery. But they usually can select their meals, and they can use those meals to exercise control or express their feelings: they can eat or refuse to eat!

Many obvious and not-so-obvious factors shape a client's perceptions of hospital foods. Complaints may have little to do with the food itself, but instead serve as a way for clients to vent fear, frustration, anger, and physical pain. Clients need opportunities to express their feelings, and often you may find that a problem can be resolved simply by listening and providing emotional support.[1]

Aren't there valid complaints about food sometimes?

Yes, of course. Many disorders, medications, and treatments can dramatically alter taste perceptions and lead to complaints about food.[2] In addition, the hospital may not prepare foods in the same way the client does at home—sometimes a considerable problem for a person who must eat three meals a day for many days in the hospital. A client may be expecting to enjoy a favorite food for dinner, only to be disappointed with the way the food is prepared. Unfortunately, hot foods may not be hot and cold foods may not be cold by the time they arrive in the client's room, or foods may have been left in the room while the client was gone for tests or therapy. The client receives meals at specified times regardless of hunger and often must eat in bed without companionship, which can be more of a chore than a pleasurable experience. Meals may also be unwelcome if the person is in pain or has been sedated.

What can be done to help correct these problems?

The majority of people in the hospital will eat adequate amounts of foods, even though they complain about them. If they fail to eat enough food to meet their nutrient needs, the deficit will be easy to correct once they are at home eating familiar foods. Chapter 17 provides suggestions for helping people to eat. For people in pain, administering pain medications so that they will be effective during meals can be helpful. For people with altered taste perceptions, the dietitian can work closely with them to uncover the tastes and food preferences that they can tolerate and enjoy.

Some problems can be handled directly by the person caring for the client. For example, the nurse or assistant providing trays to clients should distribute the trays as soon as possible after the food carts arrive. That person can also make sure that foods and utensils are arranged attractively before they are served. At other times, the foodservice department must be contacted to solve food-related problems. For example, the foodservice department should be contacted if a client receives the wrong diet or consistently receives foods that differ

Nutrition in Practice

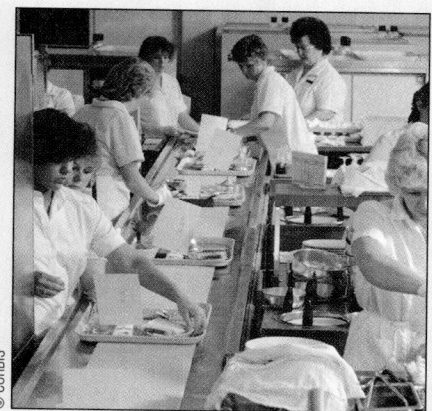

Foodservice departments prepare foods to accommodate dozens of special diets and hundreds of food preferences.

from those that were requested. Foodservice departments often conduct periodic surveys to uncover problems clients may have with foods or foodservice so that they can take steps to correct them.

Who is responsible for problems directly related to the foodservice department?

The responsibility of budgeting, planning, preparing, and serving appropriate meals rests with either a chief administrative dietitian or a foodservice manager. In some facilities, foodservice companies from outside the hospital perform these duties.

Clinical dietitians work directly with clients to assess their nutrition status, plan appropriate diets, and provide nutrition education. Clinical dietitians may also assist in menu planning, especially for special diets. In some facilities, dietetic technicians assume some administrative or clinical responsibilities. Other dietary employees include clerks, porters, and other assistants. Keep in mind that many dietary employees do not have formal education in nutrition, and their ability to interpret diet orders and provide accurate information is limited.

How does the foodservice department know what foods to serve each client?

Most hospitals provide menus from which clients can select their meals. A client who must follow a modified diet receives menus that include only foods specified in the hospital's diet manual for that particular diet (see Figure NP19–1 on p. 492). By allowing a choice, this system helps to ensure that clients receive foods they prefer and will eat. An added advantage for people on

special diets is that they become familiar with their diets by marking appropriate menus.

Although procedures vary between hospitals, generally dietary employees deliver menus to each client's room early in the day and pick them up again later in the day. Each menu shows the client's name and room number, as well as the name of the meal (breakfast, lunch, or supper), the type of diet, and the day the menu will be served. Often clients make selections for the next few days to give the foodservice department time to collect the menus and estimate the amount and type of food to prepare. Menus are usually color coded by diet. Color coding helps ensure that foodservice employees put the right types of foods on food trays and helps the person delivering the tray to quickly determine if the right diet has been delivered.

If clients make their own menu selections, how can they be receiving foods they don't like?

Several problems can occur. Clients typically select one or more items from each food category on the menu. Clients may not receive foods they enjoy if they fail to mark the menus or inadvertently make the wrong food selections. If menus are not marked or if a menu is lost, the client receives meals selected by the foodservice department. Consider these potential problems:

- Clients may have difficulty seeing, reading, understanding, or physically marking menus.
- Clients may not understand that their selections will be for the next (or another) day.
- Clients may be out of their rooms (for tests, procedures, or physical activity) or asleep when the menus arrive; when the clients return or wake up, they may not see the menus or may have missed the menu pickup time.
- Clients may be too ill or too disinterested in food to make menu selections.

Occasional problems with menu selections can usually be corrected simply by explaining the menu system to clients or taking time to help them mark menus. If clients continue to complain about food selections, contact the dietetic technician or dietitian.

What happens to menus once they are collected?

Once food selections have been made and menus collected, menus are often checked by a member of the foodservice staff (usually, a dietetic technician or dietitian) to ensure that selections are appropriate. Completed menus can provide valuable clues about a person's usual eating habits or understanding of a

Nutrition in Practice

REGULAR	SUNDAY
🌸〜 Lunch 〜🌸	

Meats

Baked chicken ♥	Fried fish

Starch

Cornbread dressing	Parsleyed potatoes ♥

Vegetables

Baby carrots ♥	Green beans ♥

Soup/Salad/Juice — **Dressings**

Coleslaw	French
Clam chowder	Thousand Island
Gelatin	Italian
Tossed salad ♥	Diet Thousand Island ♥

Desserts

Apple pie	Fresh fruit ♥

Breads

Dinner roll	Bran bread ♥
White or wheat bread	Crackers

Beverages & Condiments

Coffee	Sugar
Decaf. coffee	Sugar substitute
Hot tea	Herb seasoning
Decaf. hot tea	Creamer
Iced tea	Lemon
Whole milk	Mustard
Buttermilk	Mayonnaise
2% milk	Catsup
Skim milk ♥	Margarine
Chocolate milk	

Name _____ Room _____

BLAND/LOW FIBER	SUNDAY
🌸〜 Lunch 〜🌸	

Meats

Baked chicken	Baked fish (cod)

Starch

Rice	Boiled potatoes

Vegetables

Baby carrots	Green beans

Soup/Salad/Juice — **Dressings**

Gelatin	Mayonnaise
Chicken broth	

Desserts

Apple pie	Pears

Breads

Dinner roll	Crackers
White bread	

Beverages & Condiments

Decaf. coffee	Sugar
Decaf. hot tea	Sugar substitute
Decaf. iced tea	Creamer
Whole milk	Lemon
2% milk	Margarine
Buttermilk	Catsup
Skim milk	
Lemonade	

No pepper

Name _____ Room _____

KCALORIE RESTRICTED, DIABETIC	SUNDAY
1200 CALORIES	
🌸〜 Lunch 〜🌸	

LF = Low Fat   LSLF = Low Sodium, Low Fat

Meat Exchange (Select _1_)

LSLF Baked chicken	LSLF Baked fish
(2 oz)	(2 oz)

Starch Exchange (Select _1_)

Clam chowder (1 c)	LF dinner roll (1)
LSLF Rice (1/3 c)	White bread (1 slice)
LSLF Boiled potatoes	Wheat bread (1 slice)
(1/2 c)	Bran bread (1 slice)
	Crackers (6)

Vegetable Exchange (Select _2_)

LSLF Baby carrots	LSLF green beans
(1/2 c)	(1/2 c)

Fruit Exchange (Select _1_)

Diet pears (1/2 c)	Fresh fruit

Milk Exchange (Select _1_)

Whole milk (1 c) omit 2 fats	Buttermilk (1 c)
2% milk (1 c) omit 1 fat	Skim milk (1 c)

Fat Exchange (Select _1_)

Margarine (1 tsp)	Creamer (1 = 1/2 fat)
Diet mayonnaise (1/2 oz)	

Calorie-free foods

Coffee	LSLF Coleslaw (1/2 c)
Decaf. coffee	Tossed salad (1/2 c)
Hot tea	Diet gelatin (1/2 c)
Decaf. hot tea	Diet French
Iced tea	Diet Thousand Island
Sugar substitute	Diet Italian
Lemon	Mustard
Herb seasoning	Diet catsup

Name _____ Room _____

People on regular diets select the foods of their choice. The regular menu may also be used for high-kcalorie, high-protein diets. The menu items marked with a heart guide people in selecting foods that are lower in fat, cholesterol, sodium, and caffeine or higher in fiber than other menu choices.

Foods for bland/low-fiber diets are similar to those for regular diets. Foods from the regular menu that are not appropriate have been eliminated from the menu, and substitutes have been made. When permitted, regular coffee or tea can be added to the menu.

For kcalorie-restricted and diabetic diets, the number of exchanges allowed is written on the menu beforehand. (This example uses a 1200-kcalorie diet.) Note that the meat exchange is written in 2-ounce portions so that 1 serving = 2 exchanges.

Figure NP19–1
Sample Lunch Menus

---

modified diet. In checking menus, for example, the technician may notice that one person on a regular diet is selecting very little or that another is selecting too much. In another case, the technician may notice that a person on a kcalorie-restricted diet is not selecting the appropriate number of servings of allowed foods. Such problems suggest the need for further intervention.

Do all facilities offer selective menus?

Some hospitals and other health care facilities do not offer selective menus. Instead, they serve a standard house diet, adjusting the menu for individual food preferences. For example, clients can request simple changes, such as the substitution of one vegetable for another. Similarly if the regular menu offers fried fish,

a person on a low-fat or low-kcalorie diet would receive baked fish.

How do foodservice departments prepare foods for a variety of diets?

The logistics of preparing foods tailored to each modified diet can be overwhelming. For this reason, foodservice departments use systems designed to limit costs and minimize errors. Foods prepared for regular and bland/low-fiber diets are prepared with some fat and salt because these dietary components are not restricted on such diets. Note that the other diet menus shown in Figure NP19–1 provide a number of low-fat (LF) or low-sodium (LS) or low-sodium, low-fat (LSLF) foods.

Nutrition in Practice

LOW FAT/LOW CHOLESTEROL/CARDIAC SUNDAY

❀∼ Lunch ∼❀

LF = Low Fat   LSLF = Low Sodium, Low Fat

Meats

LSLF Baked chicken   LSLF Baked fish (cod)

Starch

LSLF Rice   LSLF Boiled potatoes

Vegetables

LSLF Baby carrots   LSLF Green beans

Soup/Salad/Juice   **Dressings**

LSLF Coleslaw	Diet French
Gelatin	Diet Thousand Island
Tomato soup	Diet Italian
Tossed salad	

Desserts

Pears   Fresh fruit

Breads

LF Dinner roll	Bran bread
White bread	LS Crackers
Wheat bread	

Beverages & Condiments

Coffee	Creamer
Decaf. coffee	Sugar
Hot tea	Sugar substitute
Decaf. hot tea	Herb seasoning
Iced tea	Lemon
Buttermilk	Margarine
Skim milk	Mustard
	Diet mayonnaise
	Catsup

Name _____ Room _____

People on low-fat, low-cholesterol diets who also need kcalorie restriction receive a kcalorie-restricted menu to control portion sizes and number of servings. Both menus provide low-fat, low-cholesterol foods. Foods not appropriate for a low-fat, low-cholesterol diet, such as whole milk, would be crossed off the menu beforehand.

LOW SODIUM   SUNDAY

❀∼ Lunch ∼❀

LF = Low Fat   LSLF = Low Sodium, Low Fat

Meats

LSLF Baked chicken   LSLF Baked fish (cod)

Starch

LSLF Rice   LSLF Boiled potatoes

Vegetables

LSLF Baby carrots   LSLF Green beans

Soup/Salad/Juice   **Dressings**

LSLF Coleslaw	Diet French
LS Chicken broth	Diet Thousand Island
Apple juice	Diet Italian
Tossed salad	

Desserts

Pears   Fresh fruit

Breads

Dinner roll	Bran bread
White bread	LS Crackers
Wheat bread	

Beverages & Condiments

Coffee	Sugar
Decaf. coffee	Sugar substitute
Hot tea	Creamer
Decaf. hot tea	Lemon
Iced tea	Herb seasoning
Whole milk	Margarine
2% milk	Diet mustard
Skim milk	Diet mayonnaise
No salt	Diet catsup

Name _____ Room _____

Low-sodium menus are similar to those provided for low-fat, low-cholesterol diets, but they eliminate high-sodium foods, such as tomato soup. The person on a low-sodium, low-fat, low-cholesterol diet selects foods from a low-fat menu with high-sodium foods crossed off beforehand. If the person is also on a low-kcalorie diet, foods would be selected from a low-kcalorie menu with high-sodium foods crossed off the menu beforehand.

RENAL   SUNDAY

❀∼ Lunch ∼❀

LF = Low Fat   LSLF = Low Sodium, Low Fat

Meats

LSLF Baked chicken   LSLF Baked fish

Starch

LSLF Rice   LSLF Dialyzed potatoes

Vegetables

LSLF Baby carrots   LSLF Green beans

Soup/Salad/Juice   **Dressings**

Lemonade	Diet French
LSLF Coleslaw	Diet Thousand Island
Tossed salad	Diet Italian
(no tomato)	

Desserts

Pears   Apple pie

Breads

Dinner roll	Bran bread
White bread	LS Crackers
Wheat bread	

Beverages & Condiments

Coffee	Sugar
Decaf. coffee	Sugar substitute
Hot tea	Creamer
Decaf. hot tea	Lemon
Iced tea	Margarine
	Diet mustard
	Mayonnaise
No salt	

Name _____ Room _____

Renal diets must be highly individualized, and the person checking the menu has to carefully consider the client's selections and make appropriate changes when necessary.

If the foodservice department were to prepare a food (green beans, for example) for each different diet, it would have to prepare green beans made with some fat and salt, green beans made without fat, green beans made without salt, and green beans made without fat or salt. Instead, only two types of green beans are prepared—one with some fat and salt, and the other without fat or salt. Clients can then add allowed ingredients to food. For example, the person on a low-salt diet could add margarine to the green beans; the person on a low-fat diet could add salt.

Keep in mind that green beans are only one of many menu items in a day, and you can see why preparing individual foods for each diet is not feasible.

However, you can also see how such a system can lead to complaints about food. A person on a low-fat diet, for example, may not realize that the food has been prepared without any salt. In such a case, the client should be advised to add other seasonings to enliven the flavor of the food.

How do foods get delivered to clients?

Sometimes foods are prepared in a main kitchen, assembled on trays, and delivered to the nursing unit; then foods intended to be eaten hot are heated in areas close to the clients' rooms. In other cases, foods are delivered directly from the main kitchen, using serving

Nutrition in Practice

equipment that helps keep hot foods hot and cold foods cold. In either case, foodservice personnel deliver food carts directly to the nursing unit; then nursing or foodservice personnel take a tray to each client. Efficient delivery of foods to the nursing unit and then to the client helps ensure that clients receive foods at the appropriate temperature.

Once the client is finished eating, the tray is returned to the food cart. Foodservice personnel pick up the carts and return them to the foodservice department.

How can I use this information to help clients eat?

Learning about the foodservice system in the health care facility where you work can help save both you and your clients needless aggravation. Ask to spend a few hours or a day working with different foodservice employees to see firsthand how the department operates and what problems they encounter. If that is not possible, learn the facility's procedures for ordering diets, making diet changes, reporting problems with a client's tray, or making special requests. Remember that requests are not simply made by one individual to another. Often many people are involved in processing a single request, and the number of requests made during any one meal can be considerable. Translating requests (for example, preparing another tray) takes time, and delays are unavoidable. The best strategy is prevention—make sure that clients mark menus and that you call in requests as early as possible to allow the

foodservice department the time to process the request.

One of the most important things to know about your facility's foodservice system is the time when meals are actually assembled so that you can call in requests well before this time. Once tray assembly begins, dietary employees are extremely busy, and requests will be difficult to process.

With so many people and steps involved in foodservice, and so many clients with individual dietary needs, food preferences, and emotional responses, it is easy to see many opportunities for problems to arise. Once you understand the many factors that affect clients' appetites and perceptions of hospital foods as well as how the foodservice department operates, you can begin to tackle problems efficiently and avoid needless frustration for your clients and yourself.

Notes

1. M. A. Hess, Taste: The neglected nutritional factor, *Journal of the American Dietetic Association* (supplement 2) 97 (1997): 205–207.

2. M. Bélanger and L. Dubé, The emotional experience of hospitalization: Its moderation and its role in patient satisfaction with foodservice, *Journal of the American Dietetic Association* 96 (1996): 354–360.

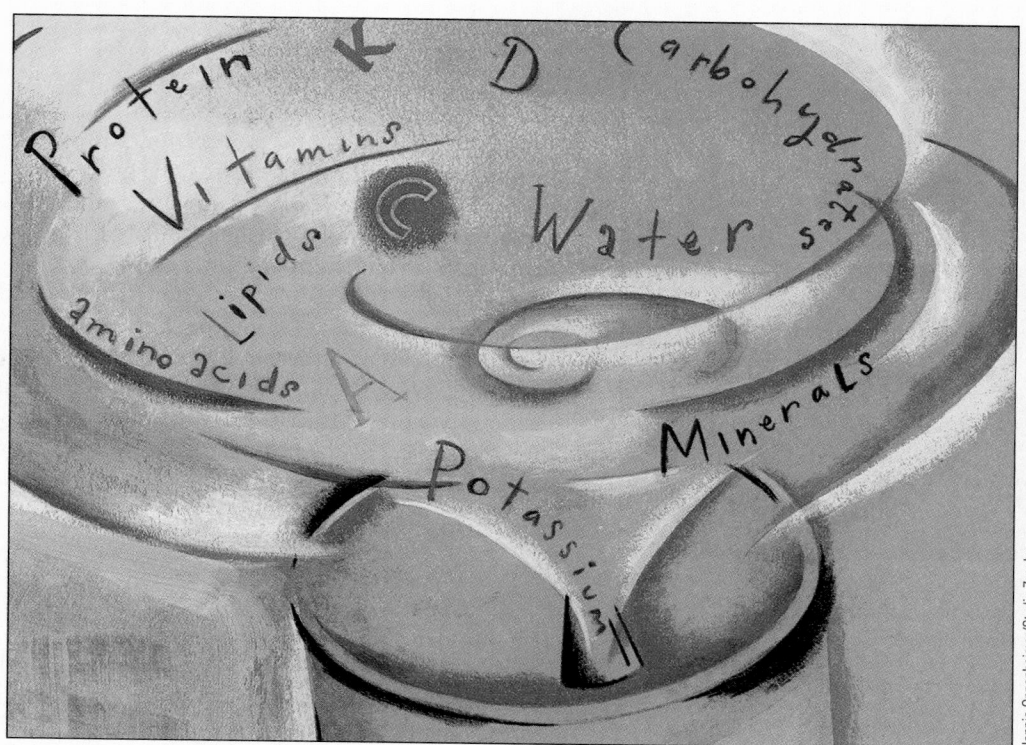

Jennie Oppenheimer/Studio Zocolo

ENTERAL FORMULAS

Pharmacies and grocery stores carry many enteral formulas intended primarily for oral use.

*T*o meet nutrient needs with conventional foods, a person must be able to chew and swallow and digest and absorb nutrients in the amounts necessary to satisfy metabolic demands. As Chapters 18 and 19 have shown, however, illnesses may interfere with eating to such a degree that conventional foods fail to deliver necessary nutrients. If a poor appetite is the primary nutrition problem, **enteral formulas** given orally can help clients meet nutrient needs, if the clients can drink them in sufficient amounts.

For clients who cannot eat enough food or drink enough formula to meet nutrient needs, however, it may be necessary to deliver nutrients by tube or by vein. Formulas provided either orally or by tube are enteral feedings, the subject of this chapter. Enteral feedings are possible whenever a client can digest and absorb nutrients via the GI tract. Otherwise, nutrients are given by vein as parenteral feedings, the subject of the next chapter. (Figure 21–1 in the next chapter summarizes some of the factors involved in deciding the most appropriate way to feed a client.)

Enteral Formulas: What Are They?

The number of enteral formulas on the market is staggering (some examples are listed in Appendix F). Television and magazine advertisements promote some of these products, which are available over-the-counter in pharmacies and grocery stores (Ensure® and Boost® are examples). Most formulas are available in ready-to-use form, although some come in powdered form that must be mixed with water. Formulas can meet a variety of medical and nutrition needs and can be used alone or given along with other foods. Thus a formula is simply a standard or modified diet provided in liquid form.

Whenever formula is the primary source of nutrients, **complete formulas** are necessary. Such is the case when a client is on a tube feeding or an oral liquid diet for more than a few days. Complete formulas, when consumed in appropriate amounts, supply all the nutrients a client needs. Complete formulas can also be (and often are) used in smaller quantities to supplement table foods.

Types of Formulas

Formulas are classified in many ways, but for purposes of this book, it is reasonable to think of two major kinds categorized by the type of protein they supply. Standard formulas contain complete proteins, whereas hydrolyzed formulas contain small fragments of proteins, which may include free amino acids, dipeptides, and tripeptides.

Standard Formulas To benefit from **standard formulas,** people must be able to digest and absorb nutrients without difficulty. Standard formulas contain whole proteins or one or a combination of **protein isolates** (purified proteins).

Hydrolyzed Formulas To simplify the body's digestive work, a complete protein can be hydrolyzed—that is, partially broken down to yield small peptides. Alternatively, a formula can be made from free amino acids. In this text, we call both types *hydrolyzed* for simplicity. **Hydrolyzed formulas** are often very low in fat or provide some fat from medium-chain triglycerides (MCT) to ease digestion and absorption. They also provide easy-to-absorb carbohydrates. People who cannot digest nutrients well may benefit from hydrolyzed formulas.

Modular Formulas **Modular formulas** serve as sources of a single nutrient (protein, carbohydrate, or fat) that can be added to enteral formulas to alter nutrient composition (for example, to add kcalories or protein). Modular for-

enteral formulas: liquid diets designed to be delivered through the GI tract, either orally or by tube.

complete formulas: liquid diets designed to supply all needed nutrients when consumed in sufficient volume.

standard formulas: liquid diets that contain complete molecules of protein; also called **intact** or **polymeric formulas.**

protein isolates: proteins that have been separated from a food. Examples include casein from milk and albumin from egg.

hydrolyzed formulas: liquid diets that contain broken-down molecules of protein, such as amino acids and short peptide chains; also called **monomeric formulas.**

modular formulas: formulas that provide a single nutrient and are designed to be added to other formulas to alter nutrient composition or combined to create a highly individualized formula.

mulas can also be combined with other modular formulas and liquid vitamin and mineral preparations to construct individualized formulas for clients with unique nutrient needs. Designing, preparing, and delivering such a formula is a challenge that requires in-depth nutrition knowledge and skills.

Distinguishing Characteristics

Formulas differ not only in the form of protein they contain but also in the amount of energy, the proportion of nutrients, and the sources of energy nutrients they provide. Most formulas are lactose-free and gluten-free. Although formulas differ, most fit into general categories and within each category can often be used interchangeably. For example, standard formulas provide similar amounts of energy and nutrients and require the same degree of digestive function. They may derive their nutrients from different sources, but the sources are all of similar quality. Thus the physician or dietitian has many choices in selecting a formula for an individual client. Table 20–1 lists examples of protein, carbohydrate, and fat sources in formulas.

Nutrient Density Standard formulas provide about 1.0 kcalorie per milliliter. Nutrient-dense formulas provide 1.2 to 2.0 kcalories per milliliter and meet energy and nutrient needs in a smaller volume. Thus nutrient-dense formulas often benefit clients with high nutrient needs or those with fluid restrictions. Clients just beginning a tube feeding or those with increased fluid requirements may benefit from a diluted formula or a formula containing less than 1.0 kcalorie per milliliter. Formulas also vary in the percentage of energy from protein, fat, and carbohydrate.

Residue and Fiber Many formulas are low in residue, while others are fiber enriched. Low- to moderate-residue formulas are least likely to cause gas and abdominal distention and thus are often well tolerated by many who need them: people with GI tract disorders (inflammatory bowel diseases or partial obstructions, for example), those who have undergone surgeries of the GI tract, or those beginning enteral nutrition after periods of GI tract disuse. People who depend on tube feedings for long periods of time, those with constipation, and some people with short-bowel syndrome (see Chapter 22) may benefit from fiber-enriched formulas. Formulas with soluble fibers may help control diarrhea in some people receiving enteral formulas.

Osmolality Osmolality is a measure of the concentration of molecular and ionic particles in a solution. A formula that approximates the osmolality of blood serum (about 300 milliosmoles per kilogram) is referred to as an **isotonic formula.** A **hypertonic formula** has a higher osmolality than serum. Most people tolerate both isotonic and hypertonic formulas without difficulty. When hypertonic formulas are delivered directly into the intestine, however, they can lead to diarrhea. To prevent this problem, hypertonic formulas delivered into the intestine are initially delivered at a slow, even rate and gradually increased as tolerated.

Costs Costs of individual products vary greatly. As a general rule, however, hydrolyzed formulas, products formulated for specific disorders (renal or respiratory failure, for example), and modular formulas are more expensive than standard formulas.

In Summary Enteral formulas are liquid mixtures of nutrients compounded to meet a variety of medical and nutrition needs. Standard formulas meet the nutrient needs of people who can digest and absorb nutrients without difficulty, whereas hydrolyzed formulas meet the nutrient needs of people who may have some difficulty with digestion and absorption.

Table 20–1 Examples of Protein, Carbohydrate, and Fat Sources in Enteral Formulas

Protein sources
- Casein
- Egg white
- Free amino acids
- Hydrolyzed casein, whey, or soy protein
- Soy
- Whey

Carbohydrate sources
- Cornstarch
- Corn syrup
- Fructose
- Guar gum
- Maltodextrin
- Sucrose

Fat sources
- Canola oil
- Fish oil
- MCT oil
- Safflower oil
- Soybean oil
- Sunflower oil

For practical purposes, 1 ml (milliliter) is equivalent to 1 cc (cubic centimeter).

Since hydrolyzed formulas are almost completely absorbed, they contribute little residue to the intestine.

osmolality (OZ-moh-LAL-eh-tee): a measure of the concentration of particles in a solution, expressed as the number of milliosmoles (mOsm) per kilogram.

isotonic formula: a formula with an osmolality similar to that of blood serum (300 mOsm/kg).
- *iso* = the same
- *ton* = tension

hypertonic formula: a formula with an osmolality greater than that of blood serum.
- *hyper* = greater, more

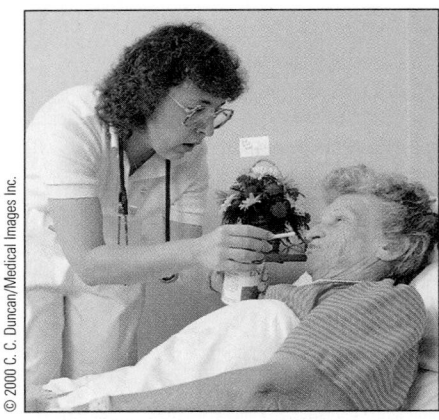

Formula supplements often help meet nutrient needs when clients cannot eat enough conventional foods.

Caution: Enteral formulas cannot be delivered by vein.

In addition to enteral formulas, oral supplements come in many other forms including fortified puddings and snack bars.

Formulas differ in the amount of energy they deliver, sources of energy nutrients, and percentages of energy nutrients. Some contain fiber, while others are designed to provide minimal residue. A formula's osmolality can also vary. In some cases, hypertonic formulas may cause diarrhea, which can often be prevented by delivering the formula at a slow, even rate. Like table foods, enteral formulas can meet a variety of dietary needs. With that in mind, it is important to identify which people benefit from such formulas.

Enteral Formulas: Who Needs What?

All people with functional GI tracts who cannot get the nutrients they need from table foods can potentially benefit from enteral formulas. They could also get nutrients from parenteral nutrition, but enteral nutrition is preferable whenever it is possible. Compared with parenteral nutrition, enteral nutrition helps maintain normal gut and immune function better, causes fewer complications, and is less costly. Enteral feedings also help stimulate intestinal adaptation following intestinal resections and long periods of GI tract disuse.

Similarly, oral feedings are preferred to tube feedings whenever a person can drink the formula and drink enough of it. In so doing, the client avoids the stress of the procedure to insert a feeding tube, and nurses save valuable time. Orally provided formulas are also less costly than feeding by tube and less likely to result in complications.

Enteral Formulas Provided Orally

For people who can tolerate only liquids for long periods and those who need hydrolyzed formulas, formulas provided orally can meet all nutrient needs in ways that foods cannot. If people can drink enough of the formula, they can avoid being fed by tube.

More often, formulas provided orally supplement a conventional diet. Some people can eat table foods, but not in the quantities they need to meet nutrient needs. Enteral formulas provide a reliable source of nutrients and work particularly well for adding energy and protein to the diet. Psychologically, liquids seem less filling than foods, and they are easier for debilitated, weak, or tired clients to handle. Nutrition in Practice 20 describes how enteral formulas provided orally meet the special needs of people with inborn errors of metabolism.

When a client uses an oral formula, taste becomes an important consideration. Allowing clients to sample different products and flavors and select the ones they like best helps promote acceptance.[1] The "How to" box offers suggestions for helping clients accept oral formulas.

Tube Feedings

Tube feedings are simply complete formulas delivered through a tube into the stomach or intestine. An individual who has a functional GI tract but is unable to ingest enough nutrients (or the appropriate type) by mouth to meet nutrient needs may need a tube feeding. Candidates for tube feedings may include:

- People with physical problems that seriously interfere with chewing and swallowing.
- People with no appetite for an extended time, especially if they are malnourished.
- People with a partial obstruction, some types of fistulas, or altered motility in the upper GI tract.

How to

Help Clients Accept Oral Formulas

People on enteral formulas are often quite ill and frequently have poor appetites. Even when a person enjoys a formula, palatability can become a problem after a while. Hydrolyzed formulas are often less palatable than standard formulas, and clients may find them difficult to accept. Caring professionals can help by using these suggestions:

- Let the client try both different flavors and different formulas appropriate for the client's needs; use those the client likes best.
- Serve formulas attractively and remind clients to drink them. Formulas offered in a glass are more appealing than those served from a can with an unfamiliar name. Some people find the smell of for-mulas unappealing. Covering the top of the glass with plastic wrap or a lid, leaving just enough room for a straw, can help.

- Provide easy access. Keep the formula close to the client's bed where little effort is required to reach it, and within sight to remind the client to drink it. Clients who are very ill may lack the motivation even to reach for formula, let alone drink it. In such cases, offer the formula ready-to-drink and in small amounts frequently through the day.
- Keep the formula cold so that it will be refreshing when the client drinks it.
- If the client stops drinking the formula after a while, recommend different flavors or another formula to help relieve boredom.

- People in a coma.
- People with high nutrient requirements.
- People who have undergone extensive intestinal resections and are just beginning enteral feedings.
- People who are unable to ingest a hydrolyzed formula orally.

Feeding Tube Placement Feeding tubes are inserted into different locations along the GI tract depending on the client's medical problems and the estimated length of time that the feeding will be required. Figure 20–1 shows feeding tube placement sites, and the glossary on p. 500 describes these sites. When clients are expected to be on tube feedings for less than about four weeks, feeding tubes are frequently inserted through the nose and passed into the stomach or intestine (**transnasal** placement). The client often remains fully alert during the procedure and helps pass the tube by swallowing. For an infant, the feeding tube may be inserted through the mouth and into the stomach (**orogastric** placement) before each feeding and removed immediately after the feeding to allow the infant to breathe easily and reduce the risk of aspiration.

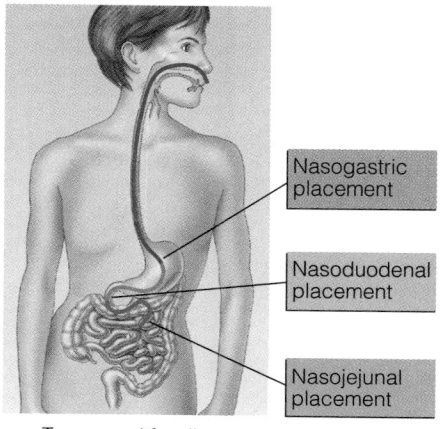

Transnasal feeding tube placements

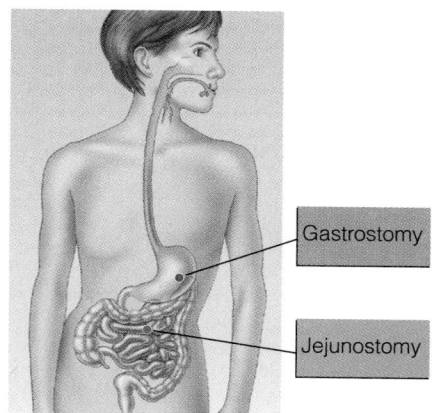

Enterostomies

Figure 20–1
Feeding Tube Placement Sites

Enteral Formulas

Glossary

These terms are listed in order from the nose to lower organs of the digestive system.

transnasal: through the nose. A **transnasal feeding tube** is one that is inserted through the nose.

> *naso* = nose

nasogastric (NG): from the nose to the stomach.

nasoenteric: from the nose to the stomach or intestine. Nasoenteric feedings include nasogastric, nasoduodenal, and nasojejunal feedings. Most clinicians use nasoenteric to refer to nasoduodenal and nasojejunal feedings only.

nasoduodenal (ND): from the nose to the duodenum.

nasojejunal (NJ): from the nose to the jejunum.

orogastric: from the mouth to the stomach. This method is often used to feed infants because they breathe through their noses, and a nasogastric tube can hinder the infant's breathing.

enterostomy (EN-ter-OSS-toe-mee):   an opening into the stomach or jejunum through which a feeding tube can be passed.

gastrostomy (gas-TROSS-toe-mee):   an opening in the stomach made surgically or under local anesthesia through which a feeding tube can be passed. The technique for creating a gastrostomy under local anesthesia is called **percutaneous endoscopic gastrostomy,** or **PEG** for short. When the feeding tube is guided from such an opening into the jejunum, the procedure is called **percutaneous endoscopic jejunostomy (PEJ),** a misnomer because the enterostomy is in the stomach rather than the jejunum.

jejunostomy (JEE-ju-NOSS-toe-mee):   an opening in the jejunum made surgically or under local anesthesia through which a feeding tube can be passed. The technique for creating a jejunostomy under local anesthesia is called a **direct endoscopic jejunostomy (DEJ).** Note: Some clinicians also refer to this procedure as a PEJ, which is a more accurate use of the term than the more common use described above.

When a client will be on a tube feeding for a longer period, or when a feeding tube cannot be passed through the nose, esophagus, or stomach due to an obstruction or for other medical reasons, an opening can be made into the stomach or jejunum. A tube **enterostomy** can be made either surgically or nonsurgically using local anesthesia. Table 20–2 compares some of the features of various tube feeding sites. The "How to" box on p. 501 suggests ways to reduce anxiety for clients beginning a tube feeding.

Table 20–2   Comparison of Feeding Tube Sites[a]

Insertion Method and Feeding Site	Advantages	Disadvantages
Transnasal	Does not require surgery or incisions for placement.	Easy to remove by disoriented clients; long-term use may irritate the nasal passages, throat, and esophagus.
Nasogastric	Easiest to insert and confirm placement; feedings can often be given intermittently and without an infusion pump.	Highest risk of aspiration in compromised clients.
Nasoduodenal and nasojejunal	Lower risk of aspiration in compromised clients; allow for enteral nutrition earlier than gastric feedings following severe stress; may allow for enteral feeding when partial obstructions, fistulas, or other medical conditions prevent gastric feeding.	More difficult to insert and confirm placement; feedings require an infusion pump for administration; may take longer to reach nutrition goals.
Tube enterostomies	Allow lower esophageal sphincter to remain closed, reducing the risk of aspiration; more comfortable than transnasal insertion for long-term use; site is not visible under clothing.	May require general anesthesia for insertion; require incisions; greater risk of complications from the insertion procedure; greater risk of infection; may cause skin irritation around the insertion site.
Gastrostomy	Feedings can often be given intermittently and without a pump; easier to insert than a jejunostomy.	Moderate risk of aspiration in high-risk clients.
Jejunostomy	Lowest risk of aspiration; allows for enteral nutrition earlier following severe stress; may allow for enteral feeding when partial obstructions, fistulas, or medical conditions prevent gastric feeding.	Most difficult to insert; feedings require an infusion pump for administration; may take longer to reach nutrition goals.

[a]Relative to each tube feeding site. The actual advantages and disadvantages of different insertion procedures depend on the person's medical condition.

How to

Help Clients Cope with Tube Feedings

The thought of being "force-fed" is frightening to many people. One person may envision a large feeding tube and fear that the procedure will be extremely painful. Another may have heard about tube feedings only from the popular press and associate them with irreversible comas. All clients benefit when they understand the insertion procedure, the expected duration of the tube feeding, and the strategic role that nutrition plays in recovery from disease. These pointers can help health care professionals prepare clients for transnasal tube feedings:

- Allow clients to see and touch the feeding tube. Seeing firsthand that the tube is soft and narrow (only about half the diameter of a pencil) often alleviates anxiety. Show clients how the feeding apparatus is attached to the feeding tube, and explain how the feeding will work. Use dolls or stuffed toys to demonstrate tube insertion and feeding procedures to young children.

- Explain that the client remains fully alert during the procedure and helps pass the tube by swallowing. A numbing solution sprayed on the back of the throat minimizes discomfort and prevents gagging during the procedure.

- Tell the client that once the tube has been inserted, most people become accustomed to its presence within a few hours. In most cases, the client can easily swallow foods and liquids with the tube in place. If permitted, favorite foods or beverages can still be enjoyed.

- Assure the client that the tube feeding will be temporary, if such assurance is appropriate.

Although a tube feeding may be frightening for some, for others, it is a relief. People who understand that they should eat, but cannot do so, may be relieved to receive sound nutrition without any effort. As they feel better and begin to eat again, the volume of the tube feeding can often be reduced and then discontinued when oral intake is adequate.

Some people feel a loss of control over their lives; others feel self-conscious about how the feeding tube looks or awkward about moving around with the equipment. A few simple measures can help:

- Involve older children, teens, and adults in the decision-making and care process whenever possible. Clients can help arrange daily feeding schedules, and some can also perform many of the feeding procedures themselves.

- Show clients how to manipulate the feeding equipment so that they can get out of bed and move around.

- Recommend that clients walk around and socialize with others, if permissible.

- Encourage clients to maintain contact with friends and keep busy with hobbies and activities they enjoy. This measure is especially important for children, teens, and those on long-term feedings.

- For infants and children, keep the developmental age of the child in mind and work with parents to ensure that appropriate feeding skills are mastered. For infants, providing a pacifier during feedings helps maintain the associations between sucking, swallowing, eating, and fullness. When possible, some of the tube feeding formula may be provided by bottle or by spoon to further develop skills.

The more complex the procedure, the easier it becomes for health care professionals to focus on the procedure and forget about the client's emotions. No matter how many technicalities you have to keep in mind, remember to stay focused on the person receiving your care.

Gastric Feedings When formulas are delivered into the stomach, either through a **nasogastric** tube or a **gastrostomy,** the digestive process begins in the stomach, just as it does with an oral diet. The stomach empties its contents at a controlled rate and delivers small volumes of nutrients into the intestine. Thus gastric feedings are often preferred whenever they are possible. Gastric feedings are not possible for people with gastric obstructions or conditions that significantly interfere with the stomach's ability to empty. Following a severe stress, such as intestinal surgery, for example, GI motility may be temporarily disrupted. Activity resumes more quickly in the small intestine than in the stomach, however. Thus, after severe stress, the delivery of formulas into the small intestine can be initiated earlier than gastric feedings.

Gastric feedings may also be a problem for people at risk for aspiration. In these clients, formula may reflux from the stomach into the esophagus, and the

The final location of the feeding tube tip determines the type of feeding. If a feeding tube is passed through a gastrostomy into the duodenum or jejunum, the feeding is intestinal rather than gastric.

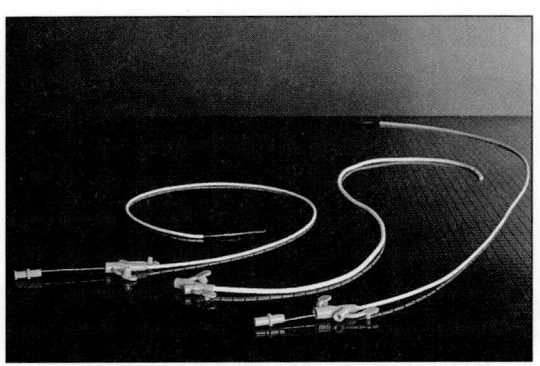

Feeding tubes come in many lengths and diameters. The thin wires protruding from the end of the feeding tubes are stylets, which stiffen the tube to ease insertion and are discarded thereafter. The Y connector (shown here in orange) provides a port for administering water or medications without disrupting the feeding.

The outer diameter of a feeding tube is measured using the French scale, where each unit is about one-third of a millimeter. Thus the outer diameter of a 10 French feeding tube is a little over 3 mm. The inner diameter varies depending on the thickness of the material used to construct the tube.

aspiration pneumonia: an infection of the lungs caused by inhaling fluids regurgitated from the stomach.

gastric decompression: the removal of pressure and gas from the stomach.

Chapter Twenty

client may aspirate the formula into the lungs. **Aspiration pneumonia,** a lung infection that can be fatal, may develop. To minimize the possibility of aspiration, clinicians may prefer a **nasoenteric** feeding for clients at risk. Alternatively, for a client with a very high risk of aspiration, some clinicians prefer a gastrostomy or **jejunostomy,** which allows the lower esophageal sphincter to remain tightly closed.

The major disadvantage of intestinal feedings is loss of the controlled emptying action of the stomach. Assuring passage of the feeding tube into the appropriate location is also more difficult, and placement of the tube must be confirmed before the feeding begins. Formulas must be carefully administered to avoid diarrhea and dehydration.

Feeding Tubes Feeding tubes are soft and flexible and come in a variety of diameters and lengths. Many have special characteristics that make them desirable for specific purposes. For example, tubes with double lumens allow for **gastric decompression** and intestinal feedings at the same time.

Which feeding tube is appropriate depends on the client's age and size, medical condition, how the tube will be placed (transnasally or through an enterostomy), how far the final placement will be from the insertion site (from the nose to the stomach or intestine, for example), and the tube's inner diameter. Once the appropriate length is selected, the smallest tube through which the formula will flow without clogging the tube is selected. Unclogging a tube is a difficult procedure that interrupts the feeding schedule and may require insertion of a new tube, causing stress and anxiety for the client and raising the cost of the feeding.

Formula Selection

To select an appropriate formula requires a logical approach. Figure 20–2 shows some considerations involved. In a nutshell, the formula that meets the client's medical and nutrient needs with the lowest risk of complications and at the lowest cost is the best choice. If no formula can be found that meets the client's needs, then modules can be used to create an appropriate formula.

Selecting a formula that meets the client's nutrient needs is paramount. Nutrient requirements are estimated based on a careful assessment of the client's age, nutrition status, medical condition, ability to digest and absorb nutrients, and metabolic rate. Standard formulas are appropriate for the vast majority of clients. However, the person with a functional, but impaired, GI tract may benefit from hydrolyzed formulas. Besides the client's ability to digest and absorb nutrients, other nutrition-related factors that affect the selection of a formula include:

- The client's energy, protein, fluid, and nutrient needs. High nutrient needs must be met in the volume of formula the client can tolerate. If nutrients (including water) must be restricted, the selected formula should deliver the prescribed amount of nutrients in the volume prescribed.
- The need for residue or fiber modifications. The choice of formulas is narrowed when a person needs a low-residue or a high-fiber diet.
- Individual tolerances (food allergies and sensitivities). Most formulas are lactose-free because temporary or permanent lactose intolerance is a common problem for people who need enteral formulas. Many formulas are also gluten-free and can accommodate the needs of people with celiac disease.
- The size of the inner lumen of the feeding tube. When a small-diameter tube has been inserted, the formula selected must flow readily through the tube to prevent it from becoming obstructed.

In addition, health care facilities cannot stock all formulas, so formula selection is limited by availability. In the final analysis, the dietitian or physician can make only an educated guess in selecting the best formula for an individual. The health

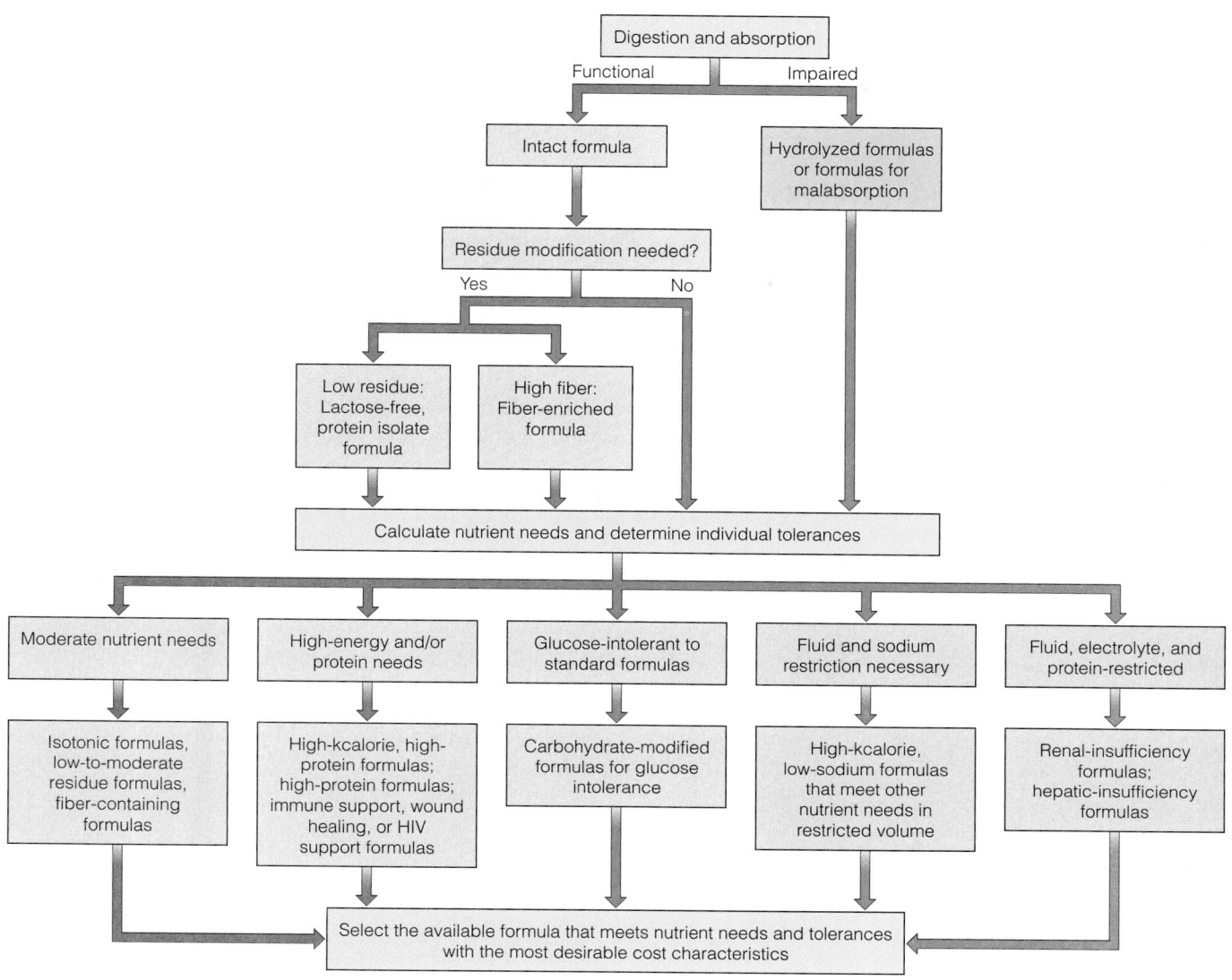

Figure 20–2
Selecting a Formula

care team monitors each person's nutrition status and responses to the formula to help ensure that individual needs are being met.

In Summary People who can digest and absorb foods but cannot get the nutrients they need from conventional foods can benefit from enteral formulas. Whenever possible, people are encouraged to drink the formula, but sometimes the formula must be delivered through feeding tubes inserted through the nose to the stomach or intestine, or through enterostomies made directly into the stomach or intestine. The health care team considers the client's medical and nutrient needs and selects a formula that meets those needs with the least chance of complications.

Tube Feedings: How Are They Given?

Once a feeding route has been selected and the feeding tube inserted, attention turns to delivering the formula. Many people beginning a tube feeding are seriously ill or malnourished. Using the safest methods of preparing and administering

Just as the actual foods provided on special diets vary from one health care facility to the next, the exact procedures for safely handling tube feedings and providing them to clients vary as well. The procedures presented in the following sections are guidelines only.

Enteral Formulas

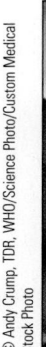

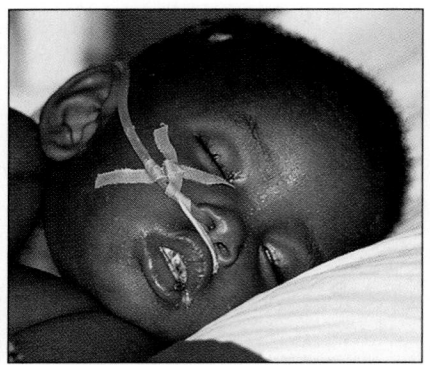

The safe delivery of a tube feeding helps clients meet their medical and nutrition goals with minimal discomfort.

the formula helps minimize the risk of complications, which can delay or prevent the attainment of medical and nutrition goals. Although the following sections specifically address the preparation and administration of tube feedings in health care institutions, the principles apply to people who receive tube feedings at home as well.

Safe Handling

People who are ill or malnourished may have suppressed immune systems that make them vulnerable to infection from food-borne illness. To prevent contamination, all personnel involved in preparing or delivering formulas should handle them only in clean environments, using clean equipment and clean hands. The foodservice department or pharmacy most often assumes responsibility for diluting formulas or mixing them from powders and for delivering formulas. Formulas are labeled with the client's name, room number, date, and time of preparation (if necessary) and then sent to the nursing station.

Open and Closed Feeding Systems Formulas are available in both **open feeding systems** and **closed feeding systems.** Open systems require that a formula be transferred from its original packaging to a feeding container. The feeding container is then connected to the feeding tube so that the formula can be delivered to the client. Examples include formulas that come in cans or standard bottles or those that must be diluted or mixed from powder.

Some formulas come prepackaged in containers that can be directly connected to a feeding tube. Closed feeding systems reduce the risk of bacterial contamination of the formula, save nursing time, and can hang for longer periods of time than open systems.[2] Although closed systems cost more initially, they may actually be less expensive in the long run by preventing bacterial contamination and thus avoiding the costs of treating infections.

At the Nursing Station Once a formula reaches the nursing station, the nursing staff assumes responsibility for its safe handling. Hands should be carefully washed before handling formulas and feeding containers. In some facilities, nonsterile gloves are worn whenever formulas are handled (see the photo on p. 505). The following steps reduce the likelihood of formula contamination when open feeding systems are in use:

- Before opening a can of formula, carefully clean the can opener and the lid. If you do not use the entire can at one feeding, label the can with the time it was opened.

- Store opened cans or mixed formulas in clean, closed containers. Refrigerate the unused portion of formula promptly.

- Discard unlabeled or improperly labeled containers and all opened containers of formula not used within 24 hours.

open feeding systems: enteral formula delivery systems that require formula to be transferred from its original packaging to a feeding container before it can be administered through a feeding tube.

closed feeding systems: enteral formula delivery systems in which the formula comes prepackaged in a container that is ready to be attached to a feeding tube for administration.

At Bedside The procedures that minimize the risk of bacterial infections during a feeding vary between open and closed feeding systems. For open systems:

- Hang no more than a 4- to 6-hour supply of mixed or diluted formula or an 8- to 12-hour supply of ready-to-use formula.

- Discard any formula that remains, rinse out the feeding bag and tubing, and add fresh formula to the feeding bag.

- Flush the tube feeding line with water before administering a fresh supply of formula.

For closed systems:

- Hang no more than a 24- to 48-hour supply of formula.

Exact hang times for closed systems vary, and the manufacturer should be consulted for an appropriate hang time. Regardless of whether an open or a closed feeding system is used, use a new feeding container and tubing (except the feeding tube itself) every 24 to 48 hours, as appropriate.

Initiating and Progressing a Tube Feeding

Two of the most serious complications associated with tube feedings are the inadvertent placement of a transnasal tube into the respiratory tract and the aspiration of formula from the stomach into the lungs. In both cases, potentially fatal infections can follow. To minimize the risk of such complications, most clinicians use X rays to verify the position of the feeding tube before a feeding is initiated, and nurses confirm the tube's position several times a day. Lying flat raises the risk of aspiration; therefore, the client's upper body is elevated to at least a 30- to 45-degree angle during the feeding and for 30 minutes after the feeding whenever possible. Although the practice is controversial, sterile blue food coloring is sometimes added to formulas so that the formula can easily be distinguished from pulmonary secretions.

Formula Delivery Techniques A day's volume of formula can be given in relatively large amounts at intervals or in relatively small amounts continuously throughout the day. A client may start on a continuous feeding and gradually be converted to an intermittent feeding. Each method has specific uses, advantages, and disadvantages.

Intermittent Feedings **Intermittent feedings** are intended for delivery into the stomach (not intestine), and no more than 250 to 400 milliliters of formula is given over 30 minutes or more using the gravity drip method or an infusion pump. The larger the volume of formula needed to meet nutrient needs, the more frequently feedings are delivered. (People who have very high nutrient needs benefit from either high-nutrient-density formulas or continuous feedings.) Rapid delivery (in 15 minutes or less) of a large volume of formula (300 to 400 milliliters)—called a **bolus feeding**—often leads to complaints of abdominal discomfort, nausea, fullness, and cramping, especially when a feeding is first initiated. This makes sense. After all, people do not gobble down a meal in just a few minutes, especially if they have not been eating for a while.

Intermittent feedings may be difficult for a client to tolerate, and because there is a greater volume of formula in the stomach at one time, the risk of aspiration is higher than with continuous feedings. Intermittent feedings work well for clients able to tolerate them, however. Often clients gradually adapt to larger volumes of formula given over shorter periods of time. Such feedings mimic the usual pattern of eating and allow the client freedom of movement between meals. They also require less time, making them less costly and easier for people to use at home.

Continuous Feedings **Continuous feedings** are delivered slowly and in constant amounts over a period of 8 to 24 hours. Such feedings benefit people who have received no food through the GI tract for a long time, those who are hypermetabolic, and those who are receiving intestinal feedings. Although continuous feedings are often well tolerated and are associated with a relatively low risk of aspiration, they require infusion pumps to help ensure accurate and constant flow rates; the pump makes the feedings more costly to provide and limits the client's freedom of movement. The "How to" box on p. 506 explains several ways to plan tube feeding schedules.

Formula Volume and Strength Almost all people can receive undiluted formula (either isotonic or hypertonic) at the start of a feeding. Formulas, especially hypertonic formulas, are given slowly at first, and the volume is gradually

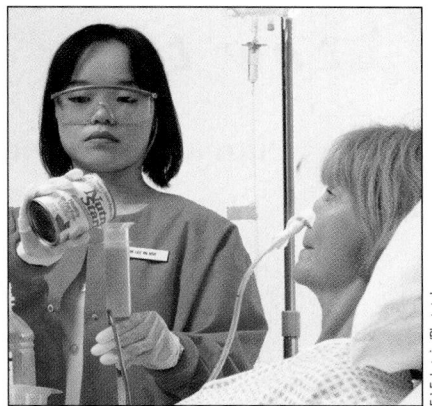

With time, clients often adapt to larger amounts of formula administered over shorter periods of time.

Caution: The bright lights, beeps, and controls of an infusion pump may make it irresistible to young children. Keep the infusion pump at a safe distance from children to prevent them from toppling the pump or IV pole and possibly injuring themselves or damaging the equipment.

intermittent feedings: delivery of about 250 to 400 ml of formula over 30 minutes or more.

bolus feeding: delivery of about 300 to 400 ml of formula over 15 minutes or less.

continuous feedings: slow delivery of formula in constant amounts over an 8- to 24-hour period.

Enteral Formulas

Plan a Tube Feeding Administration Schedule

*A*fter selecting a formula that meets the client's medical and nutrient needs, the planner, usually a dietitian, determines the volume of formula per day that meets those needs. Consider a client who needs 2000 milliliters of formula per day. If the client is to receive the formula intermittently six times a day, he needs about 330 milliliters of formula at each feeding (2000 ml ÷ 6 feedings = 333 ml/feeding). Alternatively, if he is to receive the same volume of formula eight times a day, then he needs 250 milliliters (or about one can of ready-to-feed formula) at each feeding (2000 ml ÷ 8 feedings = 250 ml/feeding). He will probably tolerate this volume of formula best if it is given to him over 30 minutes or more at each feeding. If the client is to receive the formula continuously over 24 hours, he needs about 85 milliliters of formula each hour (2000 ml ÷ 24 hr = 83 ml/hr).

As an example of a volume progression for a continuous tube feeding, start the feeding at 30 ml/hr at full strength and then progress as follows:
- *After 8 hours: 50 ml/hr.*
- *After 16 hours: 70 ml/hr.*
- *After 24 hours: 90 ml/hr.*

The following formula is one of several used to estimate fluid requirements for adults and children:
- *Allow 100 ml/kg for each of the first 10 kg of body weight.*
- *Allow 50 ml/kg for each of the next 10 kg of body weight.*
- *Allow 20 ml/kg for each kilogram of body weight above 20 kg.*

gastric residual: the volume of formula that remains in the stomach from a previous feeding.

increased. In rare cases, formulas may need to be diluted initially, the strength increased gradually, and then the rate advanced.

Intermittent feedings may start at about 100 to 150 milliliters at each feeding and then be increased by 50 to 100 milliliters daily until the goal volume is reached.[3] Continuous feedings may start at about 10 to 40 milliliters per hour. If the person tolerates the formula, the rate of feeding can be increased by about 10 to 20 milliliters per hour every 8 to 12 hours, depending on the location of the feeding tube (gastric or intestinal), the person's medical condition, and the nutrient density of the formula.[4] For both intermittent and continuous feedings, if the new rate is not tolerated, back up and proceed more slowly, giving the person additional time to adapt. If the person on an intermittent feeding cannot tolerate the feeding, a continuous feeding may be a better choice than intermittent feeding.

Supplemental Water Formulas contain a considerable amount of water, and they often meet a substantial portion of clients' water needs. Standard formulas contain about 850 milliliters of water per liter of formula. Nutrient-dense formulas provide less water. In addition to the formula itself, water provided through the feeding tube helps meet additional fluid needs and helps prevent clogged feeding tubes. Most adults require about 2000 milliliters (approximately 2 quarts) of water daily. For people with kidney, liver, and heart diseases, less water may be necessary. Clients with fever, excessive sweating, severe vomiting, diarrhea, fistula drainage, high-output ostomies, blood loss, or open wounds require additional water.

Attention to indicators of body water balance can help determine how much additional water an individual needs. In alert adults, thirst is a good indicator of water needs; a person complaining of thirst generally needs water. In the elderly, however, thirst may be slow to develop in response to dehydration. Health care professionals monitor clients' weight changes, record their intake and output, and measure urine specific gravity to evaluate hydration status. Table 16–10 on p. 430 listed physical and laboratory indices of dehydration.

Water is often given by syringe or gravity bag at room (for gastric feedings) or body (for intestinal feedings) temperature.[5] Automatic flush pumps that deliver a selected volume of water each hour are also available. For clients on continuous feedings, the feeding tube is generally flushed with water every four hours. For clients on intermittent feedings, the feeding tube is generally flushed with water before and after each feeding.

Gastric Residuals Nurses measure the **gastric residual** to ensure that the stomach is emptying properly and to prevent nausea, vomiting, and possible aspiration of formula into the lungs. The gastric residual is the volume of formula that remains in the stomach from a previous feeding. It is measured by gently withdrawing the gastric contents through the feeding tube using a syringe. The gastric residual is measured before each feeding for intermittent feedings and every four hours for continuous feedings. The gastric residual that is considered excessive varies among facilities, but ranges from about 100 to 200 milliliters. If the residual is excessive, the feeding is held for an hour or two, and then the residual is rechecked. If excessive residuals persist, the physician may withhold the feeding, reduce the rate of administration, or begin drug therapy to stimulate gastric emptying.

Medications Delivered through Feeding Tubes Clients receiving tube feedings are also likely to be receiving numerous medications. Often these medications are delivered through feeding tubes, and in some cases, complications can occur. Keep in mind that enteral formulas can interact with medications in the same ways that foods can. For example, the vitamin K in a formula can interact with warfarin (see Chapter 15) just as the vitamin K in foods can. The health care team considers the effects of drug therapy on nutrient requirements, the effects of the formula on medication absorption, the effects of medications on physical properties of formulas, and the development and prevention of complications.

Medication Forms A medication may come in any of several forms including tablets, liquid, injectable, and intravenous. The following guidelines may be helpful in preventing medication-medication interactions, medication-formula interactions, or clogged feeding tubes when medications are delivered through feeding tubes:

- Give medications by mouth instead of by tube whenever possible.
- Use the liquid form of medications, whenever available. Dilute hypertonic liquid medications with water before administering them through the feeding tube.
- Use injectable or intravenous forms of medications if they are not available in liquid form.
- Ideally, tablets should not be crushed and administered through feeding tubes. If using tablets is unavoidable, crush the tablets to a fine powder and mix with water before administering them.
- Never crush tablets intended to release their contents slowly (time-released tablets) or enteric-coated medications. In these cases, another medication form must be given.
- Unless the compatibility of multiple medications is known, do not mix medications together or mix medications with the formula hanging in the feeding container. Instead, give each medication individually using the separate port whenever available.
- Flush the feeding tube with warm (body temperature) water before and after administering each medication.
- Avoid medications known to be incompatible with formulas. Bulk-forming agents, for example, can clog the feeding tube. Other medications that may be incompatible with some formulas are listed in Table 20–3 on p. 508.

Like hypertonic formulas, hypertonic liquid medications can lead to diarrhea. Examples of hypertonic medications include potassium chloride elixir (a potassium supplement), liquid multivitamin preparations, Tagamet liquid, and theophylline elixir.

GI Side Effects Medications can trigger GI side effects, including nausea, vomiting, and diarrhea, that are also frequent complications of tube feedings. Medications are a frequent culprit when diarrhea is associated with tube feedings. In one study, the most common cause of diarrhea (responsible for about half the cases) in clients receiving tube feedings was sorbitol-containing medications.[6]

Additional Considerations The location of the feeding tube (whether gastric or intestinal) is also relevant to medication administration. Medications designed to dissolve in the mouth should not be given through a feeding tube.

Table 20–3   Examples of Medications That May Be Incompatible with Some Formulas

Aluminum and magnesium hydroxide	MCT oil
Chlorpromazine concentrate	Mellaril concentrate
Cibalith-S syrup	Mellaril oral solution
Cimetidine	Paregoric elixir
Dimetane elixir	Potassium chloride
Dimetapp elixir	Reglan syrup
Feosol elixir	Riopan
Fleet's phosphosoda	Robitussin expectorant
Gevrabon liquid	Sudafed syrup
Klorvess syrup	Thorazine concentrate
Mandelamine Forte suspension	Zinc sulfate capsules

NOTE: These substances may be compatible with some formulas and not others.

SOURCES: Z. M. Pronsky, *Food-Medication Interactions*, 11th ed. (Pottstown, Pa.: Food-Medication Interactions, 2000); F. C. Thompson, M. R. Naysmith, and A. Lindsay, Managing drug therapy in patients receiving enteral and parenteral nutrition, *Hospital Pharmacist* 7 (2000): 155–164.

Medications that depend on the stomach's acidic environment for absorption may be poorly absorbed if delivered directly into the duodenum or jejunum. Similarly, a medication that is optimally absorbed in the duodenum may be poorly absorbed in the jejunum. In such cases, oral, intravenous, or injectable forms of the medication should be provided.

Addressing Tube Feeding Complications   Table 20–4 summarizes complications associated with tube feedings and shows that many problems can be prevented or corrected by selecting the formula and feeding route wisely, preparing the formula correctly, and delivering it appropriately. Attention to the person's primary medical condition and medications is important as well. Table 20–5 (on p. 510) provides a monitoring schedule that helps ensure early detection of problems that may be encountered when clients are fed by tube.

Types of Complications   Failure to estimate nutrient needs correctly or to ensure that the selected formula meets these needs limits the client's ability to achieve or maintain adequate nutrition status. Mechanical problems, such as a clogged feeding tube, a malfunctioning feeding pump, or a tube that becomes dislodged from its appropriate location, can lead to complications and interrupt the feeding schedule. Other complications related to the formula or its administration can result in nausea, vomiting, diarrhea, cramps, constipation, delayed gastric emptying, abdominal distention, and aspiration.

Metabolic complications such as dehydration, fluid and electrolyte imbalances, and elevated blood glucose can also occur. When complications arise, they can place further stress on the client and make it more difficult to achieve medical and nutrition goals.

What to Chart   Chapter 17 emphasized the importance of the medical record as a legal document and communication tool. Before reimbursing for the cost of tube feedings, many types of insurance (including Medicare) and managed care organizations require that the physician appropriately document the client's need for a tube feeding as well as the justification for using an infusion pump or a special formula, when appropriate. Medicare does not cover the cost of tube feeding unless the physician believes the condition necessitating the tube feeding will last for at least three months. Furthermore, Medicare does not cover the cost of tube feeding for clients with functioning GI tracts who are unable to eat due to a lack of appetite.

Table 20–4 Causes and Prevention or Correction of Tube Feeding Complications

Complications	Possible Causes	Preventive/Corrective Measures
Aspiration of formula	Compromised lower esophageal sphincter, delayed gastric emptying	Use nasoenteric, gastrostomy, or jejunostomy feedings in high-risk clients; check tube placement; elevate head of bed during and for 45 minutes after feeding; check gastric residuals.
Clogged feeding tube	Formula too thick for tube	Select appropriate tube size; flush tubing with water before and after giving formula; use infusion pump to deliver thick formulas; remedies reported to help unclog feeding tubes include cola, cranberry juice, meat tenderizer, and pancreatic enzymes.
	Medications delivered through feeding tube	Use oral, liquid, or injectable medications whenever possible; dilute thick or sticky liquid medications with water before administering; crush tablets to a fine powder and mix with water; flush tubing with water before and after medications are given; give medications individually; do not add medications to the feeding container.
Constipation	Low-fiber formula	Provide additional fluids; use high-fiber formula.
	Lack of exercise	Encourage walking and other activities, if appropriate.
Dehydration and electrolyte imbalance	Excessive diarrhea	See items under *Diarrhea*.
	Inadequate fluid intake	Provide additional fluid.
	Carbohydrate intolerance	Use continuous drip administration of formula; monitor blood glucose; select a formula with a lower amount or different type of carbohydrate.
Diarrhea, cramps, abdominal distention	Excessive protein intake	Monitor blood electrolyte levels; reduce protein intake.
	Bacterial contamination	Use fresh formula every 24 hours; store opened or mixed formula in a refrigerator; rinse feeding bag and tubing before adding fresh formula; change feeding apparatus every 24 hours; prepare formula with clean hands using clean equipment in a clean environment.
	Lactose intolerance	Use lactose-free formula in lactose-intolerant and high-risk clients.
	Hypertonic formula	Use small volume of formula and increase volume gradually.
	Rapid formula administration	Slow administration rate or use continuous drip feedings.
	Malnutrition/low serum albumin	Use small volume of dilute formula and increase volume and concentration gradually.
Hyperglycemia	Diabetes, hypermetabolism, drug therapy	Check blood glucose; slow administration rate; provide adequate fluids; select a formula with a lower amount or different type of carbohydrate.
Nausea and vomiting	Obstruction	Discontinue tube feeding.
	Delayed gastric emptying	Check gastric residual; slow administration rate, use continuous drip feedings, or discontinue tube feeding.
	Intolerance to concentration or volume of formula	Use small volume of formula and increase volume and concentration gradually; use continuous drip feedings.
	Psychological reaction to tube feeding	Address client's concerns.
Skin irritation at enterostomy site	Leakage of GI secretions and friction caused by the tube	Keep site clean; inspect area for redness, tenderness, and drainage; use protective skin cream.

NOTE: Many of the complications presented here can be caused by the client's primary disorder or drug therapy rather than the tube feeding itself. In such a case, the corrective measure would include treatment of the disorder or a change in drug therapy. Additionally, other corrective measures that require a physician's order are not shown here.

The health care team should routinely document the following information for clients on tube feedings:

- The nutrition goals for the client.
- The condition necessitating the tube feeding.
- The placement—both where (gastric or intestinal) and how (transnasal or enterostomy) the feeding tube is placed—and the type and size of the feeding tube.
- The techniques and results of tests used to confirm the tube's placement.

Table 20–5 Suggested Guidelines for Monitoring Clients on Tube Feedings

Before starting a new feeding:	Complete a nutrition assessment.
	Check tube placement.
Before each intermittent feeding:	Check client position.
	Check gastric residual.
	Check tube placement.
After each intermittent feeding:	Flush feeding tube with water.
Every half hour:	Check gravity drip rate, when applicable.
Every hour:	Check pump drip rate, when applicable.
Every 4 hours:	Check vital signs, including blood pressure, temperature, pulse, and respiration.
Every 6 hours:	Check blood glucose; monitoring blood glucose can be discontinued after 48 hours if test results are consistently negative in a nondiabetic client.
Every 4 to 6 hours of continuous feeding:	Check client position.
	Check gastric residual.
	Flush feeding tube with water.
Every 8 hours:	Check intake and output.
	Check specific gravity of urine.
	Check tube placement.
	Chart client's total intake of, acceptance of, and tolerance to tube feeding.
Every 24 to 48 hours	Weigh client.
	Change feeding container and attached tubing.
	Clean feeding equipment.
Every 7 to 10 days:	Reassess nutrition status.
As needed:	Observe client for any undesirable responses to tube feeding; for example, delayed gastric emptying, nausea, vomiting, or diarrhea.
	Check nitrogen balance.
	Check laboratory data.
	Chart significant details.

- The formula selected to meet nutrition goals and its nutrient composition.
- The recommended administration schedule (concentration and rate) and method of delivery (intermittent or continuous, gravity drip or infusion pump).
- The client's education regarding the nutrition goals and the tube feeding procedure.
- The client's physical and emotional responses to the tube insertion and tube feeding procedure.
- The client's tolerances to the formula and administration schedule, complications (if any), and corrective actions recommended.
- The actual amount of formula the client receives.
- All substances delivered through the feeding tube including the formula, dye, additional water, medications, and any substances used to unclog the feeding tube (review Table 20–4).
- Confirmation that measures for monitoring the client have been taken.
- The reasons why a tube feeding was interrupted or could not be delivered, if necessary.

The medical record serves as both a legal record and an invaluable source of information for investigating problems so that corrective measures can be taken promptly.

Case Study

GRAPHICS DESIGNER REQUIRING ENTERAL NUTRITION

Mrs. Innis is a 24-year-old graphics designer who suffered multiple fractures when she fell from a cliff while hiking. She has been in the hospital for seven days and has no appetite. Mrs. Innis has lost 8 pounds over the course of her hospitalization. Due to the nature of her injuries, Mrs. Innis is in traction and is immobile, although the head of her bed can be elevated to 45 degrees. From the history, the dietitian determined that Mrs. Innis's nutrition status was adequate prior to hospitalization. The health care team agrees that a nasoduodenal tube feeding should be instituted before nutrition status deteriorates further. The standard formula selected for the feeding is lactose-free, and Mrs. Innis's nutrient requirements can be met with 2200 milliliters of the formula per day.

1. What steps can be taken to prepare Mrs. Innis for tube feeding? Why might nasoduodenal placement of the feeding tube be preferred to nasogastric placement?

2. The physician's orders specify that the feeding should be given continuously over 18 hours. Develop a tube feeding schedule.

3. What parameters should be monitored to ensure that Mrs. Innis's fluid needs are being met? How can additional fluids be given? Describe precautions that should be taken if Mrs. Innis is to receive medications through the feeding tube.

4. After three days of feeding, Mrs. Innis develops diarrhea. Look at Table 20–4 on p. 509 to determine the possible causes. What measures can be taken to correct the various causes of diarrhea?

5. What information should be charted in Mrs. Innis's medical record? When Mrs. Innis is ready to eat table foods again, what steps will the health care team take?

In Summary To maximize the benefits of tube feedings, formulas must be prepared and delivered using techniques that minimize the risk of complications. Formulas must be prepared safely to minimize the risk of bacterial contamination and its consequent effects on health. The delivery of a formula may be intermittent or continuous, depending on the client's medical condition. Formulas are given slowly at first and gradually increased until the client is meeting nutrition goals. When feeding tubes are used to deliver medications, special care must be taken to ensure that the client does not develop complications as a consequence. Throughout a tube feeding, health care professionals should chart the client's progress in the medical record to ensure that the client obtains the most benefits from the procedure.

From Tube Feedings to Table Foods

Once the medical condition that required a tube feeding resolves, the client can gradually shift to an oral diet as the volume of formula is tapered off. The client should be eating about two-thirds of the estimated nutrient needs by mouth before the tube feeding is discontinued.[7] In many cases, the person can begin to drink the same formula that is being delivered by tube. As clients begin to take more food or formula orally, they receive less of the formula by tube. Some people cannot make the transition to oral intake for medical reasons and go home on tube feedings. Chapter 21 discusses how home tube feedings work. The accompanying case study helps you to consider the many factors involved in tube feedings, and the Nutrition Assessment Checklist on p. 512 reviews key points for assessing nutrition status in people receiving tube feedings.

Nutrition Assessment Checklist

FOR PEOPLE RECEIVING TUBE FEEDINGS

Medical History

Check the medical record for medical conditions that:

- ☐ Alter nutrient needs (including malabsorption) and narrow the selection of formula
- ☐ Suggest how long the feeding will be required and how the tube will be inserted (transnasal versus enterostomy)
- ☐ Narrow the choice of placement sites (gastric versus intestinal)

Regularly review the medical record for complications that may suggest the need to alter the tube feeding formula or delivery technique, including:

- ☐ Aspiration of formula
- ☐ Constipation
- ☐ Dehydration/electrolyte imbalance
- ☐ Diarrhea
- ☐ Hyperglycemia
- ☐ Nausea and vomiting
- ☐ Skin irritation

Medications

Check medications for those that can cause side effects similar to those associated with tube feedings including:

- ☐ Nausea/vomiting
- ☐ Diarrhea
- ☐ Constipation
- ☐ GI discomfort/cramps

For medications being delivered through the feeding tube, check:

- ☐ Form of medication (avoid crushing tablets; use other drug forms, if possible)
- ☐ Consistency of liquid medications (dilute thick or sticky liquids)
- ☐ Compatibility with formulas (see Table 20–3 on p. 508)
- ☐ Type of medication (do not crush time-released tablets)

Food/Nutrient Intake

To assess the adequacy of a tube feeding, check to see if:

- ☐ Nutrition goals are consistent with nutrient needs
- ☐ Formula is being administered as prescribed
- ☐ Table foods are meeting nutrient needs before stopping the feeding
- ☐ Supplemental water is being given to meet needs

Anthropometrics

Measure baseline height and weight. If weight is not consistent with meeting nutrition goals:

- ☐ Reestimate energy needs and determine if energy needs have been correctly calculated
- ☐ Check to see if formula is being delivered as prescribed
- ☐ Check for physical and laboratory signs of dehydration or overhydration

Laboratory Tests

Check serum and urine tests for signs of:

- ☐ Dehydration
- ☐ Electrolyte imbalances
- ☐ Glucose intolerance
- ☐ Adequacy of protein intake (prealbumin and retinol-binding protein), when available
- ☐ Improvement or deterioration of medical condition

Physical Signs

Look for physical signs of:

- ☐ Dehydration
- ☐ Overhydration
- ☐ Delayed gastric emptying (gastric residuals)
- ☐ Malnutrition

In Summary Tube feeding is a practical solution to feeding a person who is unable to consume adequate nutrients by mouth. However, a person without a functional GI tract cannot benefit from a tube feeding. In such a case, intravenous nutrition (the subject of the next chapter) can be a lifesaving treatment option.

---

1. Which of the following statements is correct?
   a. Standard formulas contain whole proteins or protein isolates.
   b. Standard formulas contain free amino acids or small peptide chains.
   c. Hydrolyzed formulas are made from pureed meats.
   d. Hydrolyzed formulas may contain protein isolates or whole proteins.

2. When a client cannot meet nutrient needs from table foods due to a poor appetite:
   a. parenteral nutrition is preferred to tube feedings.
   b. parenteral nutrition is preferred to enteral formulas provided orally.
   c. enteral formulas are inappropriate.
   d. enteral formulas provided orally are preferred to formulas provided by tube.

3. For a client expected to be able to eat table foods in less than a month, but with a high risk of aspiration, an appropriate placement of a feeding tube would most likely be:
   a. nasogastric.
   b. nasoenteric.
   c. gastrostomy.
   d. jejunostomy.

4. In selecting an appropriate enteral formula for a client, the primary consideration is:
   a. the formula's osmolality.
   b. the client's nutrient needs.
   c. the availability of infusion pumps.
   d. the formula's cost and availability.

5. What step can health care professionals take to prevent bacterial contamination of tube feeding formulas?
   a. Deliver the formula continuously.
   b. Do not change the feeding bag and attached tubing.
   c. Discard all opened containers of formula not used in 24 hours.
   d. Add formula to the feeding container before it empties completely.

6. When compared to intermittent feedings, continuous feedings:
   a. require a pump for infusion.
   b. allow greater freedom of movement.
   c. are more like normal patterns of eating.
   d. are associated with more GI side effects.

7. The volume of formula that remains in the stomach from a previous feeding is called:
   a. residue.
   b. osmolar load.
   c. gastric residual.
   d. intermittent feeding.

8. The nurse using the feeding tube to deliver medications recognizes that:
   a. medications generally do not result in GI complaints.
   b. medications can be added directly to the feeding container.
   c. thick or sticky liquid medications and crushed tablets can clog feeding tubes.
   d. enteral formulas do not interact with medications in the same way that foods do.

9. Tube feedings can gradually be discontinued when:
   a. discharge planning begins.
   b. the client experiences hunger.
   c. the medical condition resolves.
   d. the client is able to eat foods or drink formula in sufficient amounts.

Critical Thinking

1. For which of the following health problems might a hydrolyzed formula be beneficial?
   a. delayed gastric emptying
   b. dysphagia
   c. diverticular disease
   d. Crohn's disease affecting the small intestine

2. If a person needs 2100 kcalories per day to meet his energy needs, about how many milliliters per hour would the person need if the formula delivers 1.5 kcalories per milliliter?
   a. 58
   b. 88
   c. 100
   d. 108

3. A client needs 1800 milliliters of formula a day. If the client is to receive the formula intermittently every 4 hours, she will need _____ milliliters of formula at each feeding.
   a. 225
   b. 300
   c. 400
   d. 425

4. If the client in question 3 above is to receive formula every 6 hours, she will need _____ milliliters of formula at each feeding:
   a. 200
   b. 350
   c. 450
   d. 500

Answers to these questions can be found in Appendix G.

Clinical Applications

1. Complex procedures, such as those necessary to insert feeding tubes and deliver tube feedings, require attention to many technical details, making it easy to focus on the procedure and forget about the client. Imagine that you need a transnasal tube feeding. How might you react to the news that you need the feeding and to the insertion procedure? What things would you miss most about eating table foods?

Think about ways a nurse might help you deal with these feelings.

2. Review Chapters 18 and 19 and note symptoms or disorders that may require the use of tube feedings. For each, consider when and why a tube feeding might be appropriate and which conditions, if any, might require hydrolyzed formulas.

Nutrition on the Net

For further study of the topics of this chapter, access these websites.

Find updates and quick links to these and other nutrition-related sites at our website: **www.wadsworth.com/nutrition**

Learn more about organizations that promote the appropriate use of enteral and parenteral nutrition by visiting the American Society for Parenteral and Enteral Nutrition, the Canadian Parenteral-Enteral Nutrition Association, and the European Society of

Parenteral and Enteral Nutrition at: **www.clinnutri.org, www.magi.com/~cpena**, and **www.espen99.org**

View illustrations of PEG tube insertions by visiting Healthgate: **www.healthgate. com/sym/surg68.shtml**

Visit the Children's Hospital Medical Center for information about caring for feeding tubes: **www.cincinnatichildrens.org/family/pep/homecare**

Find the nutrient composition and uses for specific enteral formulas and other oral supplements by visiting formula manufacturers including Mead Johnson Nutritionals, Nestlé Clinical Nutrition, and Ross Laboratories: **www.meadjohnson.com, www.nestleclinicalnutrition.com**, and **www.ross.com**

Notes

1. A. Skipper, C. Bohac, and M. B. Gregoire, Knowing brand name affects patient preferences for enteral supplements, *Journal of the American Dietetic Association* 99 (1999): 91–92.

2. V. Vanek, Closed *versus* open enteral delivery systems: A quality improvement study, *Nutrition in Clinical Practice* 15 (2000): 234–243; M. E. Rupp and coauthors, Evaluation of bacterial contamination of a sterile, non-air-dependent enteral feeding system in immunocompromised patients, *Nutrition in Clinical Practice* 14 (1999): 135–137.

3. M. H. Delegge and B. M. Rhodes, Continuous versus intermittent feedings: Slow and steady or fast and furious? *Support Line,* October 1998, pp. 11–15.

4. R. J. Merritt, ed., *The ASPEN Nutrition Support Manual* (Silver Springs, Md.: American Society for Parenteral and Enteral Nutrition, 1998).

5. J. Lipp, L. M. Lord, and L. H. Scholer, Fluid management in enteral nutrition, *Nutrition in Clinical Practice* 14 (1999): 232–237.

6. M. S. Williams and coauthors, Diarrhea management in enterally fed patients, *Nutrition in Clinical Practice* 13 (1998): 225–229.

7. T. Lykins, Nutrition support clinical pathways, *Nutrition in Clinical Practice* 11 (1996): 16–20.

Nutrition in Practice

▪ ENTERAL FORMULAS FOR INBORN ERRORS OF METABOLISM ▪

Chapter 20 described the use of enteral formulas for people who have nutrient needs that cannot be met using table foods. Such is the case for people with some inborn errors of metabolism, and in these cases, enteral formulas play a vital role in management. This Nutrition in Practice explains inborn errors of metabolism and discusses the role of diet in two of these disorders: phenylketonuria and galactosemia. The glossary on p. 516 defines terms related to inborn errors of metabolism.

What is an inborn error of metabolism?

An **inborn error of metabolism** is a genetic error (**mutation**) that alters the production of a protein (see also Nutrition in Practice 17). In many cases, the protein is an enzyme. When the body fails to make an enzyme, makes an enzyme in insufficient amounts, or makes an enzyme with an abnormal structure, body functions that depend on that enzyme cannot proceed as they normally do. For example, if an enzyme is missing or malfunctioning in the metabolic pathway that converts compound A to compound B, then compound A accumulates and compound B becomes deficient. Both the excess of compound A and the lack of compound B can lead to a variety of problems and are sometimes fatal. Furthermore, these imbalances affect other metabolic pathways, creating another array of problems.

What kinds of problems result from inborn errors?

Inborn errors can lead to many problems, although the problems are not always serious. In some instances, the compound that accumulates is not toxic, and the compound that is deficient is not essential, so the individual experiences no problem. The person most likely does not even know about the error. In other cases, however, inborn errors have severe consequences, including mental retardation and complications that limit life expectancy. Without prompt diagnosis and treatment, they can be lethal.

What are the treatments for inborn errors of metabolism?

Medical nutrition therapy comprises the primary treatment for many inborn errors of metabolism. With an understanding of the biochemical pathway involved, a clinician can sometimes manipulate the diet to compensate for excesses and inadequacies. Management involves restricting dietary precursors that occur prior to the error in the metabolic pathway, replacing needed products that fail to be produced, or both. The goal of therapy is to:

- Prevent the accumulation of toxic metabolites.
- Replace essential nutrients that are deficient as a result of the defective metabolic pathway.
- Provide a diet that supports normal growth, development, and maintenance.

Meeting these three objectives is a major challenge that was previously unattainable. New knowledge about the body's many biochemical pathways, coupled with current technology for synthesizing formulas with specific nutrient compositions, has greatly enhanced the treatment of inborn errors.

Why can't table foods meet nutrient needs?

In some cases, medical nutrition therapy for an inborn error of metabolism can be planned around table foods. In many cases, however, the compound that accumulates is found widely in many foods. In such cases, it is often impossible to obtain other needed nutrients. Enteral formulas play critical roles in supporting health under such circumstances.

Can you give me a specific example?

A classic example that illustrates the use of enteral formulas in treatment is the most common inborn error of metabolism: **phenylketonuria (PKU).** PKU is one of many inborn errors that affect amino acid metabolism. Other disorders can affect carbohydrate, lipid, vitamin, and mineral metabolism. PKU affects approximately 1 out of every 10,000 newborns in the United States each year. The ability to detect and treat PKU has saved and significantly improved the lives of many people; its example offers hope to those suffering from other inborn errors.

What causes PKU?

Classic PKU results from a deficiency of an enzyme that converts the essential amino acid phenylalanine to

Nutrition in Practice

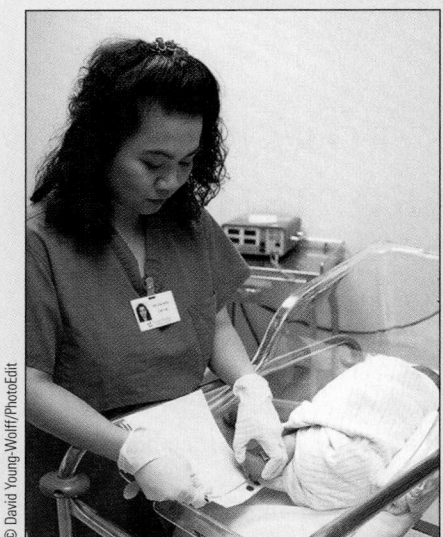

A simple blood test screens newborns for PKU—the most common inborn error of metabolism.

tyrosine (see Figure NP20–1). Without the enzyme, abnormally high concentrations of phenylalanine and other related compounds accumulate and damage the developing nervous system. Simultaneously, the body cannot make tyrosine or other compounds (such as the neurotransmitter epinephrine) that normally derive from tyrosine. Under these conditions, tyrosine becomes an essential amino acid; that is, the body cannot make it, and so the diet must supply it.

PKU is a hidden disease that cannot be seen at birth, yet diagnosis and treatment beginning in the first few days of life can prevent its devastating effects. At first, the only signs are a skin rash and light skin pigmentation. Without treatment, signs of developmental delay begin to appear between three and six months. By one year, irreversible brain damage is clearly evident. For these reasons, and because PKU is the most

common inborn error of metabolism, all newborns in the United States receive a screening test for PKU. The test must be conducted after the infant has consumed several meals containing protein (usually after 24 hours and before seven days).

What is the dietary treatment for PKU?

Essentially, the diet restricts phenylalanine and supplements tyrosine to maintain blood concentrations within a safe range. The diet's effectiveness is remarkable; in almost every case, it can prevent the devastating array of symptoms described (providing the disorder is identified soon after birth). As most dietitians can attest, though, the diet is more easily described than designed.

Because phenylalanine is an essential amino acid, the diet cannot exclude it completely. Children with PKU require as much phenylalanine as other children, but they cannot handle excesses without detrimental effects. If phenylalanine intake is too low, children suffer bone, skin, and blood disorders; growth and mental retardation; and death. Failure to eat the recommended amount of phenylalanine, along with inadequate energy intakes, can lead to negative nitrogen balance and a consequent *rise* in blood phenylalanine levels.[1] Therefore, the diet must strike a balance, providing enough phenylalanine to support normal growth and health, but not enough to cause harm. To ensure that blood phenylalanine and tyrosine concentrations remain within a safe range, children with PKU receive blood tests periodically and alter their diets when necessary.

How do formulas fit into the diet of people with PKU?

Phenylalanine is widespread in foods that contain protein, and therefore, the diet for PKU includes very little natural protein. The diet excludes high-protein foods such as meat, fish, poultry, cheese, eggs, milk, nuts, and legumes. Also excluded are commercial breads and pas-

Glossary

galactosemia (ga-LAK-toe-SEE-me-ah): an inborn error of metabolism in which enzymes that normally metabolize galactose to compounds the body can handle are missing and an alternative metabolite accumulates in the tissues, causing damage.

inborn error of metabolism: an inherited flaw evident as a metabolic disorder or disease present from birth.

mutation: an alteration in a gene such that an altered protein is produced.
muta = change

PKU, phenylketonuria (FEN-el-KEY-toe-NEW-ree-ah): an inborn error of metabolism in which phenylalanine, an essential amino acid, cannot be converted to tyrosine. Alternative metabolites of phenylalanine (phenylketones) accumulate in the tissues, causing damage, and overflow into the urine.

Nutrition in Practice

Normal:

Normally, the amino acid phenylalanine follows two pathways, one in the liver, the other in the kidneys. In the liver, the enzyme phenylalanine hydroxylase adds a hydroxyl group (OH) to produce the amino acid tyrosine. Tyrosine, in turn, produces melanin, the pigmented compound found in skin and brain cells; the neurotransmitters epinephrine and norepinephrine; and the hormone thyroxin. In the kidneys, enzymes convert phenylalanine to by-products that are excreted.

Figure NP20–1
Biochemical Alterations in PKU

In the liver:

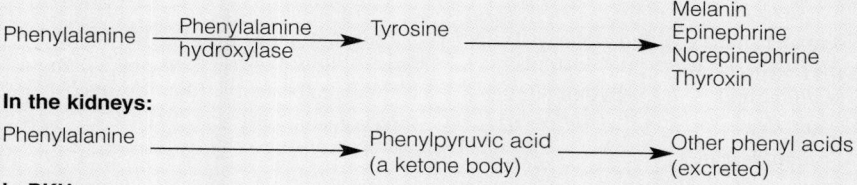

In the kidneys:

In PKU:

Individuals with PKU lack the liver enzyme phenylalanine hydroxylase, impairing conversion of phenylalanine to tyrosine. Phenylalanine accumulates in the liver and blood, reaching the kidneys in abnormally high concentrations. In the kidneys, an aminotransferase enzyme converts phenylalanine to the ketone body phenylpyruvic acid, which spills into the urine—thus the name phenylketonuria.

In the liver:

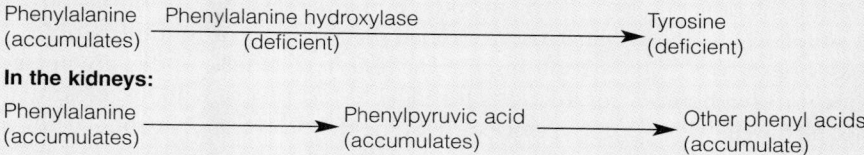

In the kidneys:

---

tries made from regular flour, which has a high phenylalanine content. Basically, the diet allows foods that contain some phenylalanine, such as fruits, vegetables, and cereals, and those that contain none, such as fats, sugars, jellies, and some candies. Clearly, it is impossible to create such a diet that is adequate in energy and all other amino acids and other nutrients using only whole, natural foods.

Formulas for people with PKU supply energy, protein, tyrosine, vitamins, and minerals. Some formulas are phenylalanine-free; others supply small amounts of phenylalanine. The less phenylalanine in the formula, the more phenylalanine the person can eat from foods. Health care professionals vigilantly monitor nutrition status to ensure that nutrient needs are met. The menu on p. 518 provides a sample phenylalanine-restricted diet for a child with PKU.

Do people follow the phenylalanine-restricted diet for life?

During the early years of central nervous system development, the diet for PKU is critical to preventing irreversible mental retardation. Until the late 1970s, researchers assumed that the child with PKU could abandon the special diet after the first few years of life when the central nervous system had completed its development. Unfortunately, however, elevated phenyl-

alanine concentrations in the older child do cause problems such as short attention span, poor short-term memory, and poor eye-to-hand coordination, although the damage is less severe than at an earlier age. A child with PKU who discontinues the controlled diet may experience problems in school performance, mood, and behavior. Clinicians generally encourage children to continue the low-phenylalanine diet indefinitely. As children reach their teen years, they begin to assume more responsibility for their food choices, and it becomes very important that they understand their disorder and its treatment.[2] This is especially important for girls, because they may eventually become pregnant.

How does PKU affect a pregnancy?

High blood phenylalanine in a pregnant woman with PKU presents a hostile environment to fetal development. The fetus's blood phenylalanine concentrations rise even higher than the mother's, and the mother may experience a spontaneous abortion; or her infant may suffer mental retardation, congenital heart disease, and low birthweight. For these reasons, women with elevated phenylalanine concentrations need counseling prior to pregnancy so that they understand the problems their condition may create for their children.

Dietary control of maternal PKU may protect the fetus, at least in part, if implemented early enough.

Nutrition in Practice

Phenylalanine-Restricted Menu for a Child with PKU

Breakfast	Snack	Lunch	Snack	Supper	Snack
2 tbs raisins	4 oz orange juice	½ small banana	Low-phenylalanine formula	2 tbs instant potatoes (without milk)	2 tbs raisins
5 tbs cream of rice		2 tbs tomato soup (without milk)	5 round butter crackers	3 tbs green beans	Low-phenylalanine formula
2 tsp sugar		3 tbs rice		4 tbs vegetable and beef broth	
Low-phenylalanine formula		1½ tsp margarine		1½ tsp margarine	
		Low-phenylalanine formula		¾ c sliced peaches	
				Low-phenylalanine formula	

Dietary control does not ensure a successful outcome of pregnancy, but the infants of women who follow a phenylalanine-restricted diet from at least one to two months prior to conception and continue it throughout pregnancy are more likely to have higher birthweights, larger head circumferences, fewer malformations, and higher scores on intelligence tests than the infants of women who begin diet therapy during their pregnancies or not at all. Again, women rely on formulas to help them meet the nutrient demands of pregnancy.

What is an example of an inborn error of metabolism that can be managed without a special formula?

Galactosemia is an example. Galactosemia occurs when any one of three enzymes that convert galactose to glucose is missing or defective. Growth failure, liver enlargement, cataracts, neurological abnormalities, coma, and death can result. Infants with galactosemia cannot tolerate breast milk or standard infant formulas because they contain lactose (and thus galactose). Instead they rely on lactose-free, soy-based infant formulas to meet nutrient needs.

Once a child is eating solid foods, the diet for galactosemia is simpler than the diet for PKU for two reasons. First, unlike phenylalanine, galactose is not an essential nutrient. Diets for galactosemia need only to restrict galactose, not to provide a perfectly calculated dose. Second, galactose occurs primarily in milk and milk products, unlike phenylalanine, which appears in all proteins. Fresh blueberries and honeydew melon contain enough free galactose to warrant their exclusion from the diet as well.[3] This is not to say that the diet is easy to follow; many commercially prepared products contain milk.

Early introduction of lactose-free formulas and galactose-restricted diets prevent or minimize some symptoms, but the long-term effects of diet are not as dramatic as for PKU.[4] Most people with galactosemia experience developmental delays, speech abnormalities, mental disabilities, cataracts, and, in women, ovarian dysfunction.

As scientific understanding of human genetics and biochemistry increases, more and more inborn errors affecting enzyme function are being recognized. Understanding the roles of enzymes in metabolism sometimes makes it possible to develop enteral formulas that compensate for these defects and to develop other treatments to prevent complications. In such cases, diet can make a dramatic difference in people's lives.

Notes

1. G. P. Duran and coauthors, Necessity of complete intake of phenylalanine-free mixture for metabolic control of phenylketonuria, *Journal of the American Dietetic Association* 99 (1999): 1559–1563.

2. R. Singh and coauthors, Impact of a camp experience on phenylalanine levels, knowledge, attitudes, and health beliefs relevant to nutrition management of phenylketonuria in adolescent girls, *Journal of the American Dietetic Association* 100 (2000): 797–803.

3. S. S. Gropper and coauthors, Free galactose content of fresh fruits and strained fruit and vegetable baby foods: More foods to consider for the galactose-restricted diet, *Journal of the American Dietetic Association* 100 (2000): 573–575.

4. K. Widhalm, B. D. O. M. da Cruz, and M. Koch, Diet does not ensure normal development in galactosemia, *Journal of the American College of Nutrition* 16 (1997): 204–208.

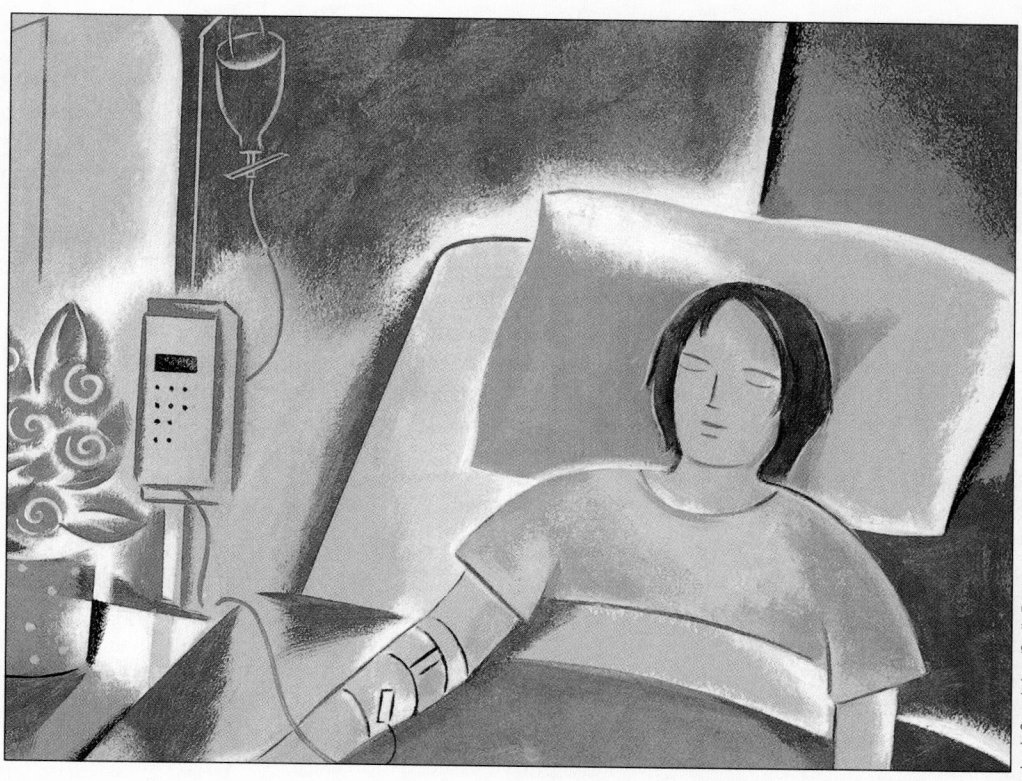

Jennie Oppenheimer/Studio Zocolo

CHAPTER 21 | PARENTERAL NUTRITION

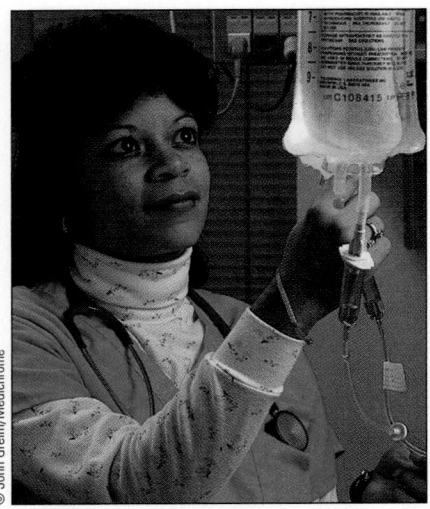

Many IV solutions look alike, and the only way to discover their contents is to read the label.

Reminder: *Sepsis* is the presence of disease-causing organisms in the blood—a life-threatening complication.

When parenteral nutrition is necessary to meet nutrient needs, some nutrients should be provided enterally (whenever possible) to help maintain the structure and function of the GI tract.

Reminder: Formulas intended for enteral use cannot be delivered intravenously.

Reminder: A conditionally essential amino acid is one that is normally nonessential but must be supplied by the diet when the body's need for it exceeds the body's ability to produce it in sufficient amounts.

dextrose monohydrate: a form of glucose that contains a molecule of water and is stable in IV solutions. IV dextrose solutions provide 3.4 kcal/g.

*T*he science of medical nutrition as we know it today was shaped tremendously by the demonstration in 1968 that all nutrient needs could be met by vein.[1] Practitioners now had the means to feed people who otherwise might have died from malnutrition. With time, clinicians learned more and more about the solutions and delivery techniques used to provide parenteral nutrition. They also discovered that although parenteral nutrition is a lifesaving procedure, it is also very costly and is associated with serious complications including sepsis, liver dysfunction, progressive kidney problems, bone disorders, and many nutrient deficiencies. These findings prompted a renewed appreciation for the GI tract, and clinicians today subscribe to the adage "If the GI tract works, use it." Only when people cannot meet their nutrient needs using the enteral route should they receive total parenteral (or intravenous) nutrition support. Figure 21–1 summarizes the decision-making process in selecting the most appropriate feeding method.

Intravenous Solutions: What Are They?

As is true of all medical nutrition therapy, the decision to use intravenous (IV) nutrition, the method of delivery, and the type and amount of nutrients to provide are based on a thorough assessment of the client's medical condition and nutrient needs. Infusion of IV nutrients immediately changes blood levels of fluids, electrolytes, and other nutrients and, therefore, requires vigilant attention to the individual's responses and adjustments to the formula, if appropriate.

Intravenous Nutrients

A variety of nutrients can be administered by vein in different combinations. These IV solutions may contain any or all of the essential nutrients: water, amino acids, carbohydrate, lipid, vitamins, and minerals. The "How to" box on p. 522 explains how to read IV solution abbreviations.

Amino Acids Protein from food reaches the bloodstream in the form of amino acids. Thus IV solutions supply amino acids rather than whole proteins. Standard IV solutions contain both essential and nonessential amino acids to meet the body's protein needs. Special solutions that contain only essential amino acids or have large amounts of certain amino acids and small amounts of others meet protein needs for specific medical conditions. Products designed for liver failure, for example, may contain more branched-chain amino acids and fewer aromatic amino acids (see Chapter 23).

Sometimes a nonessential amino acid may be omitted from standard parenteral solutions because it does not mix well or is unstable in the solution. For example, glutamine, which may be a conditionally essential amino acid for some clients, is not stable in IV solutions. Providing glutamine as a dipeptide solves the instability problem, and studies suggest that short-chain peptides can be digested to free amino acids by enzymes bound to cell membranes. Glutamine is not routinely added to parenteral solutions, however. In the United States, glutamine dipeptide solutions are not available. Although a powdered form of glutamine that can be added parenteral amino acid solutions is available, it requires special handling that limits its use.[2]

Carbohydrate Standard IV solutions provide carbohydrate as **dextrose monohydrate.** This form of dextrose is more suitable for IV solutions because it contains some water. Dextrose monohydrate provides 3.4 kcalories per gram, whereas glucose provides 4.0.

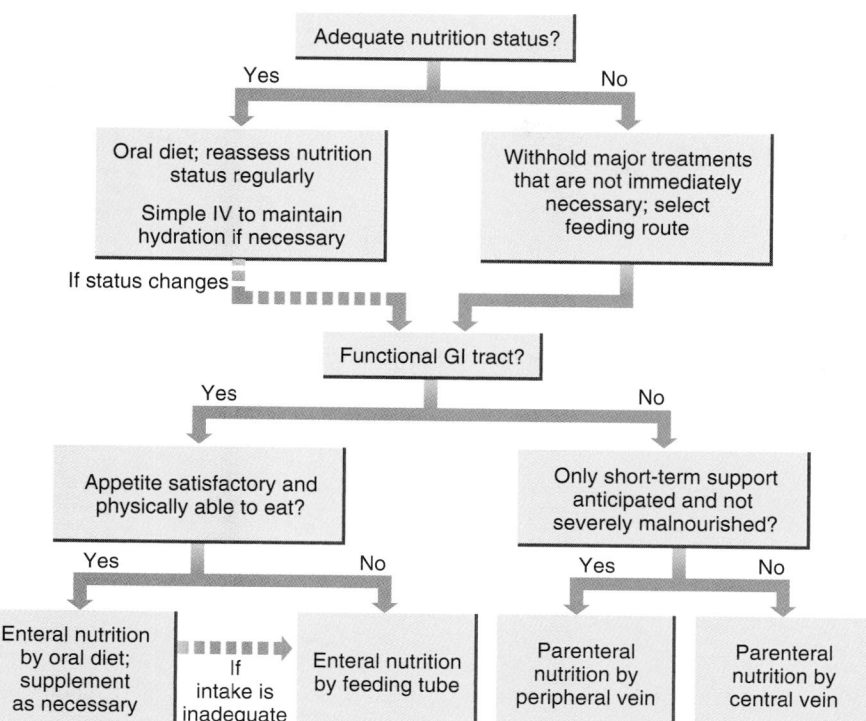

Figure 21–1

Selecting a Feeding Route

Adequate nutrition status?

Yes → **Oral diet; reassess nutrition status regularly** / **Simple IV to maintain hydration if necessary**

No → **Withhold major treatments that are not immediately necessary; select feeding route**

If status changes

Functional GI tract?

Yes → **Appetite satisfactory and physically able to eat?**

No → **Only short-term support anticipated and not severely malnourished?**

Appetite satisfactory and physically able to eat?
- Yes → **Enteral nutrition by oral diet; supplement as necessary** — If intake is inadequate → **Enteral nutrition by feeding tube**
- No → **Enteral nutrition by feeding tube**

Only short-term support anticipated and not severely malnourished?
- Yes → **Parenteral nutrition by peripheral vein**
- No → **Parenteral nutrition by central vein**

Lipid Lipid emulsions deliver fat in IV solutions. Lipid emulsions are provided either daily or periodically (once or twice a week). If provided daily, IV fat serves as a concentrated source of energy; if offered less often, it serves primarily as a source of essential fatty acids. A 500-milliliter bottle of a 10 percent fat emulsion provides 550 kcalories (1.1 kcalories per milliliter). The same volume of a 20 percent and 30 percent fat emulsion provides 1000 kcalories (2.0 kcalories per milliliter) and 1500 kcalories (3.0 kcalories per milliliter), respectively.

Lipid emulsions are contraindicated for some newborns with markedly elevated **bilirubin** levels, some people with elevated blood lipids, people with severe liver disease, and those with severe egg allergies. Cautious use of IV fats is recommended for severely stressed people with compromised immune systems and for people with atherosclerosis, moderate liver disease, blood coagulation disorders, pancreatitis, and some lung problems.

Currently, only lipid emulsions composed of long-chain triglycerides (LCT) derived from either soybean oil or a combination of soybean and safflower oils are available in the United States.[3] Around the world, however, products containing medium-chain triglycerides (MCT) and those that contain monounsaturated fatty acids from olive oil or omega-6 fatty acids from fish oil are also in use. Products containing **structured lipids** have also shown promising results in studies, but emulsions containing structured lipids are not commercially available.[4]

Micronutrients Vitamins, electrolytes (minerals), and trace elements may all be used in IV solutions. The Food and Drug Administration (FDA) recently issued new specifications for IV multivitamin formulations.[5] In the past, vitamin K was not added to adult IV multivitamin formulations, but once new formulations become available, vitamin K will be included. Pediatric multivitamin formulations, already contain vitamin K and will continue to do so.

Failure to include vitamins in parenteral nutrition solutions can result in serious nutrient deficiencies and even death.[6] During 1988, for example, a national shortage of parenteral multivitamin formulations led to the omission of

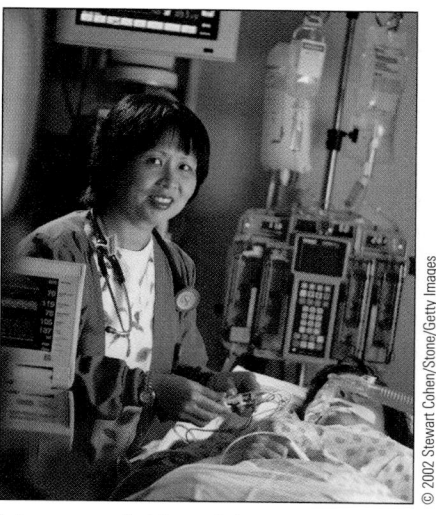

© 2002 Stewart Cohen/Stone/Getty Images

Intravenous lipid emulsions, made from plant-derived oils and egg phospholipids, provide energy and essential fatty acids and can be easily identified by their milky white color.

bilirubin: a pigment in the bile whose blood concentration may rise as a result of medical conditions that affect the liver.

structured lipids: triglycerides that have been artificially constructed to include both LCT and MCT fatty acids attached to the glycerol backbone.

vitamins in parenteral solutions and resulted in severe thiamin deficiencies and three deaths.[7] Conversely, long-term parenteral nutrition can sometimes lead to elevated manganese (a trace mineral) levels; these levels return to normal when manganese is removed from the IV solution.[8]

Some electrolytes (particularly, calcium and phosphorus) can precipitate with other IV solution components, potentially resulting in life-threatening complications. Pharmacists skilled in formulating IV solutions follow specific mixing procedures to minimize the risk of precipitation and protect the stability of IV solutions.

Other Additives   To avoid the need for a separate infusion site, IV medications are sometimes added directly to IV solutions or infused along with the solutions through a separate port (Y-connector). These practices are not recommended, however, because interactions between medications and IV solutions can and do occur. Only medications proved to be physically compatible with, and biochemically stable in, IV solutions are safe to administer together with the solution. Intravenous catheters with separate lumens allow medications to be delivered separately from the IV solution and limit the risks of adverse interactions.

Other problems may occur when medications are added directly to the IV solution. If the total volume of the solution is not infused, the client does not receive the full dose of the medication. Conversely, if the addition of a medication to an IV solution is not noted clearly in the medical record, the physician may inadvertently reorder the medication, and the client may suffer potentially severe consequences.

Types of Intravenous Formulations

Intravenous solutions can be compounded in many ways. The concentrations of energy nutrients and electrolytes are often expressed as percentages (see the accompanying "How to" box), although the recommended practice is for physicians' orders for IV solutions and labels for their containers to state the formula's composition in both grams per day and grams per liter.[9]

Simple Intravenous Solutions   Simple IV solutions typically contain 5 percent dextrose and/or normal saline. Other electrolytes or salts may be added as

Normal saline is a solution of sodium and chloride in water.

How to

Calculate the Energy Nutrient Content of IV Solutions Expressed as Percentages

*T*he percentage of a substance in an IV solution tells you how many grams of that substance are present in 100 milliliters.[a] For example, a 5 percent dextrose solution contains 5 grams of dextrose per 100 milliliters. A 0.9 percent normal saline solution contains 0.9 gram sodium per 100 milliliters.

Suppose a person is receiving 1800 milliliters of a complete nutrient IV solution containing 16 percent dextrose, 5 percent amino acids, and 2.5 percent lipid. First, determine the number of grams of each energy nutrient:

- For dextrose:

$$\frac{16 \text{ g dextrose}}{100 \text{ ml}} = \frac{X \text{ g dextrose}}{1800 \text{ ml}}$$

$$\frac{16 \text{ g} \times 1800 \text{ ml}}{100 \text{ ml}} = 288 \text{ g dextrose}.$$

- For amino acids:

$$\frac{5 \text{ g amino acids}}{100 \text{ ml}} = \frac{X \text{ g amino acids}}{1800 \text{ ml}}$$

$$\frac{5 \text{ g} \times 1800 \text{ ml}}{100 \text{ ml}} = 90 \text{ g amino acids}.$$

- For lipid:

$$\frac{2.5 \text{ g}}{100 \text{ ml}} = \frac{X \text{ g lipid}}{1800 \text{ ml}}$$

$$\frac{2.5 \text{ g} \times 1800 \text{ ml}}{100 \text{ ml}} = 45 \text{ g lipid}.$$

Next, calculate the total kcalories in 1800 milliliters of the solution by multiplying the kcalories per gram and then adding the totals:

$$
\begin{aligned}
288 \text{ g dextrose} \times 3.4 \text{ kcal/g} &= 979 \text{ kcal.} \\
90 \text{ g amino acids} \times 4.0 \text{ kcal/g} &= 360 \text{ kcal.} \\
45 \text{ g lipid} \times 9.0 \text{ kcal/g}^{b} &= 405 \text{ kcal.} \\
\text{Total} &= 1744 \text{ kcal.}
\end{aligned}
$$

Note that this solution provides 71 percent of the nonprotein kcalories from carbohydrate and 29 percent from fat.

[a]The preferred method for ordering energy nutrients in IV solutions and labeling their containers is to state the amount of each energy nutrient in grams per day and grams per liter. National Advisory Group on Standards of Practice, Guidelines for parenteral nutrition, *Journal of Parenteral and Enteral Nutrition* 22 (1998): 49–66.
[b]Intravenous lipid actually contains more than 9.0 kcalories per gram, but 9.0 kcalories per gram is an acceptable estimate.

needed. Three liters of the typical IV solution containing dextrose provide about 150 grams of dextrose or about 500 kcalories per day. Although simple IV solutions provide some energy, they fall far short of meeting energy and nutrient needs. The volume of simple IV solutions required to meet energy needs exceeds the volume of fluid the body can safely handle.

Complete Nutrient Solutions Complete nutrient solutions provide amino acids, dextrose, fatty acids, vitamins, minerals, and trace elements. Solutions delivered through the **peripheral veins** typically contain from 5 to 10 percent dextrose (like simple IVs) and 3 to 5 percent amino acids and provide about 1500 to 2000 kcalories per day; lipid emulsions contribute more than half of the total kcalories.

Complete nutrient solutions delivered through the **central veins** meet energy needs primarily from concentrated dextrose solutions. These solutions ideally provide about 70 to 85 percent of the nonprotein energy from dextrose and 15 to 30 percent from fat.[10] If lipids are not used as an energy source, however, essential fatty acid requirements may be met by giving IV lipids periodically (two to three times per week).

The dextrose concentrations of solutions infused through peripheral veins should not exceed 10 percent, and the osmolarity of the solution (a measure of the concentration of particles) should not exceed 600 milliosmoles per liter (mOsm/L).

In Summary Intravenous solutions contain all or a combination of nutrients and can include water, amino acids, dextrose, lipid, and micronutrients. The nutrients may be given as simple IV solutions that contain primarily water, dextrose, and minerals or complete nutrient solutions that meet total nutrient needs.

peripheral veins: the small-diameter veins that carry blood from the arms and legs.

central veins: the large-diameter veins located close to the heart.

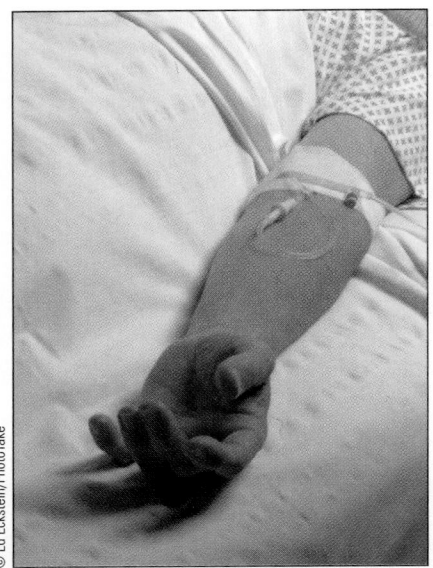
The peripheral veins can provide access to the blood for the delivery of simple IV solutions, PPN, and central TPN.

Intravenous Nutrition: Who Needs What?

The different types of IV solutions deliver nutrients for different purposes. The composition of the solution and the delivery method depend on the person's immediate medical and nutrient needs, nutrition status, and anticipated length of time on IV nutrition support. When enteral feeding is not possible and complete nutrient solutions are required, they should be given before nutrition status is severely compromised. It is much easier to maintain nutrition status than to try to replenish lost nutrient stores.

Simple Intravenous Infusions People with medical conditions that disrupt fluid and electrolyte or acid-base balances may need simple IV solutions to help them maintain or restore their body's normal homeostasis. Simple IV solutions are delivered through a peripheral **IV catheter** inserted into a small-diameter peripheral vein, such as those in the forearm and back of the hand. For people who are expected to be unable to eat for more than a few days, especially those who are malnourished or have high nutrient requirements, simple IV solutions are insufficient.

Peripheral Parenteral Nutrition (PPN) For some people, nutrient needs can be met using the peripheral veins. **Peripheral parenteral nutrition (PPN)** relies on IV lipid emulsions to provide a concentrated source of kcalories in a form that is isotonic and less irritating to the blood than concentrated dextrose solutions, which irritate the peripheral veins and cause them to collapse.

The relatively large volume of fluid necessary to deliver a limited number of kcalories and the need for accessible and strong peripheral veins limit the use of PPN. Peripheral parenteral nutrition best suits people with normal renal function who need only short-term nutrition support (about 7 to 14 days), people who need additional nutrients temporarily to supplement an oral diet or tube feeding, and those in whom inserting an IV catheter into a central vein is medically unsound. People who need fluid restrictions, people who have high energy requirements, and people for whom lipid emulsions are contraindicated are not candidates for PPN. In addition, the need for strong peripheral veins limits the use of PPN for people with weak veins that collapse easily.

Central Total Parenteral Nutrition (Central TPN) Another method used to meet all nutrient needs by vein is **central total parenteral nutrition,** or **central TPN** for short. In central TPN, the tip of a central venous catheter is either placed directly into a large-diameter central vein or threaded into a central vein through a peripheral vein (see Figure 21–2). The advantage of TPN is that it allows the infusion of concentrated IV solutions, which deliver larger amounts of energy and nutrients in smaller volumes than are possible with PPN solutions. The central veins lie close to the heart, where a large volume of blood rapidly dilutes the TPN solution. Consider that even under resting conditions the heart pumps more than a gallon of blood out of its chambers and into the arteries each minute. By the time the TPN solution reaches the peripheral veins, it is no longer concentrated enough to irritate the blood vessels.

Central TPN is indicated whenever parenteral nutrition will be required for long periods of time, when nutrient requirements are high, or when people are severely malnourished (see Table 22–1). A **peripherally inserted central catheter** is associated with fewer insertion- and infection-related complications than a catheter inserted directly into a central vein and is also less costly. Although access to a strong peripheral vein is still required, once inserted, a peripherally inserted central catheter generally remains usable longer than a

IV catheter: a thin tube inserted into a peripheral or central vein. Additional tubing connects the IV solution to the catheter.

peripheral parenteral nutrition (PPN): the provision of an IV solution that meets nutrient needs delivered into the peripheral veins.

central total parenteral nutrition (central TPN): the provision of an IV solution that meets nutrient needs delivered into a central vein.

peripherally inserted central catheter (PICC): a catheter inserted into a peripheral vein and advanced into a central vein.

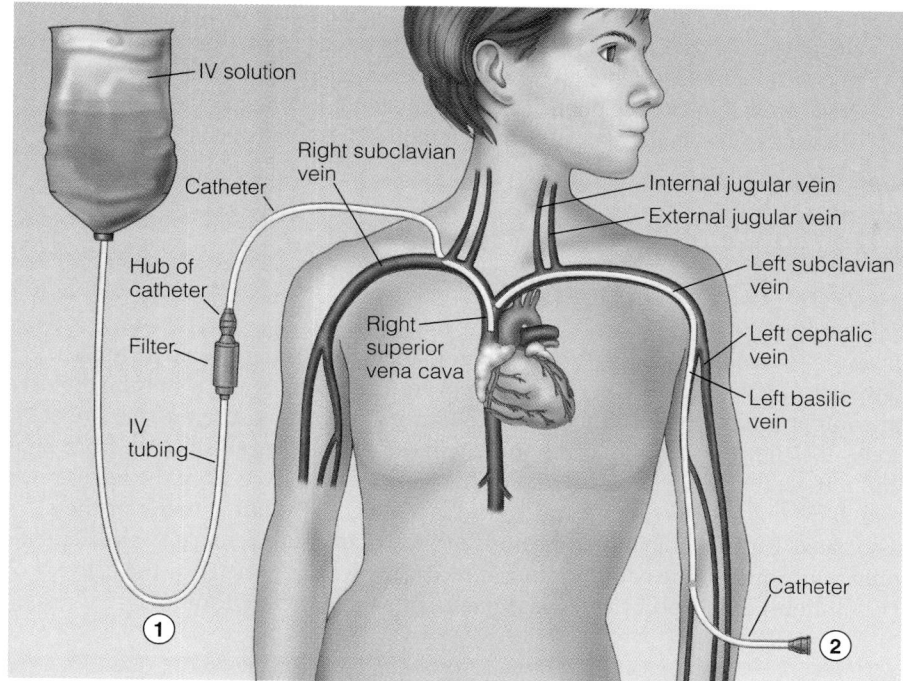

Figure 21-2

Central TPN
❶ Traditionally, central TPN catheters enter the circulation at the right subclavian vein and are threaded into the superior vena cava with the tip of the catheter lying close to the heart. Sometimes catheters are threaded into the superior vena cava from the left subclavian vein, the internal jugular vein, or the external jugular vein.
❷ Peripherally inserted central catheters usually enter the circulation at the basilic or cephalic vein and are guided up toward the heart so that the catheter tip rests in the superior vena cava.

standard peripheral IV catheter. Thus, when people need TPN to meet nutrient needs for weeks or months or when they need nutrition support but have few strong veins remaining, a peripherally inserted central catheter is an option.[11]

In Summary Simple IV solutions serve to maintain fluid, electrolyte, and acid-base balances, but they provide little energy and few nutrients. To meet all nutrient needs by vein, complete nutrient solutions can be delivered using both peripheral and central veins. PPN works best for people with strong peripheral veins who will not need parenteral nutrition for more than

---

Table 21-1 Possible Indications for Central TPN
■ Bone marrow transplants
■ Extensive small bowel resections
■ GI tract obstructions
■ High-output enterocutaneous fistulas
■ Hypermetabolic disorders, or major surgery, when it is anticipated that the GI tract will be unusable for more than 2 weeks
■ Intractable diarrhea
■ Intractable vomiting
■ Low birthweight with necrotizing enterocolitis (severe GI inflammatory disease) or bronchopulmonary dysplasia (chronic lung disease)
■ Severe acute pancreatitis
■ Severe malnutrition if surgical or intensive medical intervention is necessary
■ Severe nausea and vomiting associated with pregnancy (hyperemesis gravidarum) when they last more than 14 days
■ When it is anticipated that enteral nutrition cannot be established within 7 to 14 days of hospitalization

NOTE: Parenteral nutrition is indicated only when enteral nutrition is contraindicated. If short-term parenteral nutrition support is anticipated (less than 14 days), PPN might be preferred.

SOURCE: Adapted from A.S.P.E.N. Board of Directors and the Clinical Guidelines Task Force, Guidelines for the use of parenteral and enteral nutrition in adult and pediatric patients, *Journal of Parenteral and Enteral Nutrition* (supplement) 26 (2002): 1–138.

a week or two. They must be able to tolerate lipid emulsions and large amounts of fluids. Central TPN meets nutrient needs when clients need long-term parenteral nutrition, when they lack the strong veins necessary for PPN, when they are malnourished, when their nutrient needs are high, or when they have medical conditions that require fluid restrictions.

Intravenous Solutions: How Are They Given?

Intravenous feedings are like tube feedings in that careful attention to formula selection, preparation, and delivery helps support nutrition status and minimize the risks of complications. To prevent bacterial contamination and ensure the stability of IV solutions, they are carefully compounded in the pharmacy and shielded from light and refrigerated. Prior to infusion, the IV solution is removed from the refrigerator and allowed to reach room temperature. As Table 21–2 shows, many of the risks associated with IV nutrition are more serious than those associated with enteral nutrition. Some of the complications directly involve the catheter or the delivery of IV solutions; others occur as a consequence of altered metabolism.

Table 21–2 Complications Associated with Intravenous Nutrition

Catheter- or Care-Related Complications

Air leaking into catheter, obstructing blood flow (air embolism)
Air or gas in the chest (pneumothorax)[a]
Blood clot (thrombosis)
Blood in the chest (hemothorax)[a]
Catheter inadvertently placed in subclavian artery (arterial puncture)[a]
Catheter occlusion
Catheter tip broken off, obstructing blood flow (catheter embolism)[a]
Fluid in the chest (hydrothorax)[a]
Hole or tear in heart made by catheter tip (myocardial perforation)[a]
Improperly positioned catheter tip
Infection
Inflammation of vein (phlebitis)
Infusion pump malfunctions
Sepsis

Metabolic or Nutrition-Related Complications

Acid-base imbalances
Bone demineralization
Coma from excessive glucose load (hyperosmolar, hyperglycemic, nonketotic coma)[a]
Dehydration
Electrolyte imbalances
Elevated blood glucose (hyperglycemia)
Elevated liver enzymes
Essential fatty acid deficiency[a]
Fatty liver
Fluid overload
High blood ammonia levels (hyperammonemia)
Low blood glucose (hypoglycemia)
Trace element deficiencies
Vitamin, mineral and trace element deficiencies

[a]Central TPN.

Insertion and Care of Intravenous Catheters

Insertion of a peripheral catheter for PPN is the same as for simple IV solutions. Skilled nurses can place peripherally inserted catheters for simple IVs, PPN, or TPN, but a catheter for direct central vein access requires surgical insertion by a qualified physician. The client may be awake for the procedure, but is given a local anesthetic.

Vein Irritation The difficulty of maintaining the integrity of peripheral veins frequently interferes with the ability to sustain PPN. Veins frequently become inflamed (phlebitis) and sometimes infected. Often the catheter must be removed and reinserted at a new site; consequently, long-term feedings using peripheral veins are difficult and rarely indicated. Peripherally inserted central catheters are less irritating to the veins and can often remain in place longer than PPN catheters.

Catheter-Related Infections Infections can develop at the catheter site in both PPN and central TPN. Compared with peripherally inserted catheters (for either PPN or TPN), though, central TPN presents a greater risk of introducing disease-causing microorganisms into the bloodstream. Nurses inspect the catheter site regularly and use aseptic techniques when changing the dressing covering a catheter insertion site.

Administration of Complete Nutrient Solutions

Just as a tube feeding is started slowly to give the GI tract time to adapt to the formula, a complete nutrient solution is started slowly to allow time for the body to adapt to the glucose load and osmolarity (concentration) of the solution. People who need parenteral nutrition often have elevated blood glucose as a consequence of their medical conditions (see Chapter 27). They generally receive insulin to prevent blood glucose from rising to dangerously high levels. Insulin doses must be carefully adjusted when glucose solutions are introduced, increased, or decreased. Rapid changes in the infusion rate, especially for central TPN, can cause severe hyperglycemia or hypoglycemia, which can lead to coma, convulsions, and even death. An infusion pump ensures an accurate and steady delivery rate. When the administration of solution gets behind or ahead of schedule, the drip rate should be adjusted to the correct hourly infusion rate, but no attempt should be made to speed up or slow down the rate to meet the originally ordered volume. All changes to the infusion rate must be made gradually and cautiously, and infusion pumps must be checked regularly for malfunctions. Problems are more likely to occur in people with organ dysfunction or in infants with immature organ systems.

Electrolytes and blood glucose are monitored vigilantly initially and periodically thereafter. If tests indicate electrolyte imbalances or unacceptably high blood glucose, the causes are investigated and treated. Table 21–3 on p. 528 provides guidelines for monitoring clients on PPN and TPN.

Intravenous Lipid Infusion Traditionally, IV fat emulsions are infused separately from the base solution containing amino acids, dextrose, and micronutrients. People sometimes experience adverse reactions to IV lipid emulsions, most often when the IV lipids are administered too rapidly. Immediate reactions may include fever, warmth, chills, backache, chest pain, allergic reactions, palpitations, rapid breathing, wheezing, cyanosis, nausea, and an unpleasant taste in the mouth. To guard against adverse reactions, the client receives only small amounts of lipid emulsion over the first 15 to 30 minutes. After that time, the rate can be increased.

When IV lipid emulsions are used as an energy source, they are often added directly to the base solution and infused along with it. The use of **total nutrient admixtures** for clients in the hospital as well as at home has grown dramatically. Total nutrient admixtures must be compounded carefully, refrigerated prior to use, and mixed gently before they are infused.

Reminder: Keep IV poles and infusion pumps at a safe distance from young children so that they will not topple the pole and pump, potentially injuring themselves and damaging the equipment.

To prevent hyperglycemia, the final volume of a TPN solution should provide no more than 7 g of carbohydrate per kilogram of body weight per day.

total nutrient admixtures: intravenous solutions that contain all nutrients, including lipid emulsions; also called **three-in-one (3-in-1) admixtures** or **all-in-one admixtures.**

Table 21–3 Guidelines for Monitoring Clients Receiving Parenteral Nutrition

Before starting TPN:	Complete nutrition assessment.
	Record weight.
	Confirm placement of catheter tip by X ray.
	Check blood glucose, electrolytes, chemical profile, and complete blood count.
Every 4 to 6 hours:	Check blood glucose.[a]
	Monitor vital signs.
	Check pump infusion rate.
Daily:	Monitor weight changes.[b]
	Record intake and output.
	Inspect catheter site for signs of inflammation and infection.
Weekly:	Reassess nutrition status.
	Monitor serum proteins, ammonia, enzymes, triglycerides, cholesterol, and bleeding indices.
	Check the complete blood count.
As needed:	Monitor serum transferrin, electrolytes, calcium, magnesium, phosphorus, blood urea nitrogen, and creatinine.

[a]If blood glucose levels are stable, they can be checked once daily after the first week.
[b]After the first week, weight can usually be measured less often (about twice a week).

Cyclic Infusion A person on **cyclic parenteral nutrition** receives a TPN solution at a constant rate for 8 to 12 hours a day with breaks in the infusion during the rest of the day. The infusion can be given during the night to allow freedom for routine daytime activities. Consequently, cyclic parenteral nutrition is suited to long-term TPN, especially for clients who receive TPN at home. When a person receives a TPN solution continuously, insulin levels stay high. As a result, the person cannot mobilize fat stores for energy or for essential fatty acids; eventually, fat may be deposited in the liver. Cyclic TPN reverses these problems. Additionally, fewer kcalories seem to be effective in maintaining nitrogen balance, probably because the person uses body fat for energy. Some people, however, cannot tolerate the delivery of a day's volume of solution over a short period of time.

Discontinuing Intravenous Nutrition

Although some clients will need parenteral nutrition for the rest of their lives, most often the medical problem causing the need for IV nutrition resolves, and the client can gradually shift to an enteral diet. The transition requires careful planning, especially when the client has been receiving no nutrients enterally for a long time. During long periods of disuse, the intestinal villi shrink and lose some of their function. Reintroducing nutrients to the GI tract at the appropriate rate and volume will stimulate the progressive restoration of the villi's normal structure and function and prevent malabsorption and other GI discomforts.

To prevent hypoglycemia when a person is being taken off central TPN, the infusion rate of the solution is often tapered off gradually, although limited studies suggest that this step is unnecessary.[12] PPN can be discontinued without tapering.

Transitional Feedings The transition from IV feeding to an enteral diet can be accomplished in different ways and often involves a combination of feeding methods. One way is to start an oral diet while the person is still on IV nutrition. The diet is often progressive, beginning with liquids provided in small amounts. If the person cannot eat enough food to meet at least 50 percent of daily nutrient needs within a few days, and intake does not seem to be improving, a tube feeding may be considered. Whether a person is given a tube feeding or provided

cyclic parenteral nutrition: the continuous administration of a parenteral solution for 8 to 12 hours with time periods when no nutrients are infused.

an oral diet, the volume of the IV solution is reduced as the volume of enteral feeding is increased. The person who cannot tolerate enteral feedings can still rely on TPN to meet nutrient needs. Parenteral nutrition can be discontinued when at least 70 to 75 percent of estimated energy needs are being met by oral intake, tube feeding, or a combination of the two.[13] Chapter 20 described the transition from tube feedings to table foods.

Psychological Effects Returning to oral intake after having been fed either intravenously or by tube can have a variety of psychological effects. Some people may be extremely eager to eat again, and food can be an important morale booster. Others may be apprehensive about eating, particularly if they have had extensive GI problems. Appetite may be slow to return for some. In such circumstances, all members of the health care team can support the successful reintroduction of food. Recognize the person's concerns, and provide reassurances that help will be available throughout the process.

In Summary Complete nutrient solutions are given slowly at first to allow the body time to adapt to the glucose concentration and osmolarity of the solution. Changes are made carefully and slowly. Parenteral nutrition can be given either continuously for 24 hours or over 8 to 12 hours with breaks in the infusion during the rest of the day. As the need for parenteral nutrition resolves, clients are gradually shifted to an enteral diet (either an oral diet or tube feeding), while the volume of parenteral nutrition is gradually reduced.

Specialized Nutrition Support at Home

Occasionally, a client must continue to receive nutrition support (tube feedings or parenteral nutrition) after the primary medical condition has stabilized. In such a case, home nutrition support may be an option.

Since the first report of a person sent home successfully on TPN in 1969,[14] the use of home nutrition support has expanded rapidly. In 1992, about 40,000 people received parenteral nutrition at home, and more than 150,000 received enteral nutrition at home.[15] Since that time, clinicians report a reduction in the use of home parenteral nutrition and an increase in the use of home enteral nutrition.[16] As the number of people benefiting from home nutrition programs continues to grow, health care professionals who work with these programs are gaining valuable experience and improving the quality of home nutrition support. Medical supply companies provide the equipment, formulas, and service necessary to support home nutrition care.

The Basics of Home Programs

As with tube feedings and TPN in the hospital, the main objective of home nutrition support is to maintain or achieve adequate nutrition status. Nutrition support at home, however, has an added dimension: it permits a person to continue nutrition care in familiar surroundings. If you have ever been in the hospital, or taken a long trip for that matter, you probably remember the comfort you experienced when you returned to your own bed, knew where things were, and could get things when you needed them.

Candidates for Home Nutrition Support The nutrition support team most frequently determines whether a client is a candidate for a home nutrition program. In addition to medical considerations, the candidate for home nutrition support and those who care for that person must have rational, stable personalities so that they can successfully handle the problems that arise. They must be capable of learning the necessary techniques and of dealing with complications. They must be willing to comply with recommendations. They must have a home that is clean and has electricity, refrigeration, formula preparation and storage areas, a telephone for emergencies, and a safe water supply. They must also have adequate financial resources and access to the equipment, supplies, and professional support that are integral components of a successful home program.

Roles of Health Care Professionals The health care team members not only assess the client's medical and nutrition needs, but also evaluate the individual's home environment so that they can plan realistic feeding and care schedules and find practical solutions to the unique problems the client will face at home. Health care teams who work with home nutrition programs must also understand the regulations that govern payment for home nutrition programs by both government and private insurance. They must use this knowledge to document the need for services to ensure maximum reimbursement for the client.

Once a home nutrition program is initiated, the physician directs the program and monitors the client's care. Nurses train the client in the appropriate techniques; dietitians monitor nutrition status; pharmacists coordinate the delivery of formulas, medications, and supplies with the home care company; and social workers provide insurance assistance. All team members are expected to answer questions and provide emotional support. Home visits by a nurse, and sometimes the dietitian, help ensure that the client is able to comply with the demands of the home care program. A qualified nurse, dietitian, or physician must be available to answer questions and handle problems as they arise.

To compare the expense of receiving nutrients from parenteral or enteral solutions with the cost of table foods, consider that the average monthly "grocery bill" for parenteral and enteral nutrition is $5800 and $1500, respectively, per person.

Cost Considerations One recent analysis found that the average cost of maintaining a home parenteral nutrition program was $70,000 per year with a range of $15,000 to $169,000 per year.[17] Costs for enteral nutrition programs are lower (average $18,000 per year with a range of $5000 to $50,000 per year), but are still substantial. Insurance and Medicare often pay for a portion of the costs, but people on home programs are also more likely than others to have high medical costs for managing their diseases as well. It is easy to see why the economic impact of home nutrition programs is often the primary concern for people on these programs.

Adjustments for Clients Although home nutrition programs can extend and improve the quality of life, clients may struggle with the many lifestyle adjustments the programs entail. In addition to the economic impact of the therapy, clients may find the demands of the home schedule to be frustrating, inconvenient, and monotonous. Among physical complaints, clients cite frequent urination (due to high volumes of fluids), disturbed sleep (often as a consequence of waking up to go to the bathroom), and feeling weak as common problems.[18] Clients say that the lifestyle areas most affected by home nutrition programs are sleep, travel, exercise, and social life.[19] Despite the big impact a home program can have on lifestyle, clients report that their lives have improved with home nutrition therapy.[20] Many people who require nutrition support at home can also eat some foods by mouth, and some resume activities, such as going to work, driving, and playing sports.

How Home Enteral and Parenteral Nutrition Programs Work

People on home enteral nutrition programs commonly have neurological disorders that interfere with swallowing, digestive diseases, or cancer.[21] People on home TPN often have acquired immune deficiency syndrome (AIDS), cancer, Crohn's disease, short-bowel syndrome, or other intestinal disorders. Most often, home TPN is relatively short term, lasting for less than eight weeks, as opposed to lifelong.[22]

Case Study

ARTIST REQUIRING PARENTERAL NUTRITION

Adam Goldstein, a 25-year-old artist with Crohn's disease underwent an extensive small bowel resection five days ago. Adam had received central TPN prior to surgery and continues to receive it. After ten days, a tube feeding was begun in very small amounts.

- Review possible reasons for the use of central TPN in the management of active Crohn's disease. Why might Adam require central TPN following surgery? How would you explain the need for central TPN to Adam?

- Describe the components of a typical TPN solution. Calculate the energy content of 1 liter of a solution that provides 140 grams dextrose, 45 grams amino acids, and 20 grams lipid. If Adam's energy requirement is 2100 kcalories per day, how many liters of solution will he need each day?

- Why are enteral feedings important for Adam, even though he relies on parenteral nutrition to meet his nutrient needs?

Assuming that Adam will eventually be able to tolerate the tube feeding, how will the health care team help Adam make the transition from parenteral feedings to tube feeding? How will the health care team know when it is safe to take Adam off TPN? Consider some of the physiological and psychological problems Adam might face when he begins eating an oral diet.

- If Adam is unable to meet nutrient needs orally, he may need to continue a tube feeding or central TPN at home. Consider the factors health care professionals might look for in determining if Adam will be a candidate for a home nutrition program. Consider some of the benefits of a home program for Adam. What are some of the problems he might encounter on a home program?

Home Enteral Nutrition Both transnasal tubes and tube enterostomies provide access to the GI tract for home tube feedings (see Chapter 20). When possible, intermittent feeding schedules are arranged so that clients are free to move around between meals. Some people are able to meet some of their nutrient needs by eating during the day and using tube feedings at night. Clients on continuous feedings who must use pumps can obtain small, lightweight pumps that are easily concealed in carrying cases and allow freedom of movement.

Home TPN Different types of home TPN programs are currently in use. Ideally, clients are given as much responsibility for their own care as they can handle. For example, a client who is capable of changing the catheter dressings is trained to do so. Typically, caregivers also learn the techniques so that they can assist as needed.

Special catheters designed for long-term use are often inserted for home TPN. Many times these catheters are tunneled under the skin so that the catheter exit site lies in the abdominal area where the client can access and care for it easily. The day's volume of TPN solution is frequently delivered within 8 to 12 hours using an infusion pump. The client infuses the solution while sleeping or at some other convenient time and thus is free to move about unencumbered for much of the day. For those who need continuous feedings, a lightweight carrying case that holds a small pump and IV bags enables the client to move around freely.

Portable pumps and convenient carrying cases allow people who require nutrition support at home to move about freely.

In Summary Unquestionably, tube feedings and parenteral nutrition provide a lifesaving feeding alternative for people who cannot eat traditional diets either in the hospital or at home. The case study presents an example for review. The Nutrition Assessment Checklist on p. 532 reviews areas of concern for people on parenteral nutrition support.

Nutrition Assessment Checklist

FOR PEOPLE RECEIVING PARENTERAL NUTRITION

Medical History

Check the medical record for medical conditions that:

- ☐ Prevent the use of enteral nutrition
- ☐ Suggest how long parenteral nutrition will be required
- ☐ Indicate the appropriate feeding route (peripheral versus central)

Review the medical record for complications that may suggest the need to alter the IV solution or the administration technique:

- ☐ Hyperglycemia/hypoglycemia
- ☐ Dehydration/overhydration
- ☐ Electrolyte imbalances
- ☐ Acid-base imbalances
- ☐ Essential fatty acid deficiencies
- ☐ Vitamin, mineral, and trace element deficiencies
- ☐ Fatty liver

Medications

Check medications and TPN solution for:

- ☐ Physical compatibility with the TPN solution
- ☐ Effects of TPN solution on medication absorption and availability

When a medication is added directly to the TPN solution:

- ☐ Note times that the TPN solution is not delivered
- ☐ Note that the full amount of medication is not delivered when the feeding is stopped

Food/Nutrient Intake

If the client is not meeting nutrition goals, check to see if the:

- ☐ TPN solution is being delivered as prescribed
- ☐ Infusion pump is working correctly
- ☐ Client's nutrient needs were correctly determined

When the client is ready to begin an enteral feeding program:

- ☐ Record all amounts of foods or formulas eaten or provided enterally

- ☐ Reduce volume of TPN solution as enteral intake increases

Anthropometrics

Measure baseline height and weight and daily weights. For clients who are not meeting goals for weight:

- ☐ Reestimate energy needs and verify that the TPN solution is delivering an amount sufficient to meet those needs
- ☐ Check to see if formula is being delivered as prescribed
- ☐ For rapid weight changes, check for physical and laboratory signs of dehydration or overhydration

Laboratory Tests

Check serum and urine tests for signs of:

- ☐ Dehydration/overhydration
- ☐ Hyperglycemia/hypoglycemia
- ☐ Electrolyte imbalances
- ☐ Acid-base imbalances
- ☐ Vitamin and trace element deficiencies
- ☐ Essential fatty acid deficiencies
- ☐ Adequacy of protein intake (prealbumin and retinol-binding protein), when available
- ☐ Improvement or deterioration of medical condition, which can alter nutrient needs

Physical Signs

Check:

- ☐ Catheter insertion site for signs of infection or inflammation
- ☐ Blood pressure, temperature, pulse, and respiration for signs of fluid, electrolyte, acid, and base imbalances

Look for physical signs of:

- ☐ Dehydration/overhydration
- ☐ PEM
- ☐ Essential fatty acid deficiencies
- ☐ Vitamin, mineral, and trace element deficiencies

Self Check

1. Which of the following cannot be delivered intravenously?
   a. dextrose
   b. amino acids
   c. lipid emulsions
   d. hydrolyzed enteral formulas

2. A simple IV solution is most appropriate for people who need fluid and electrolytes and:
   a. are well nourished and are expected to eat in a few days.
   b. cannot eat for long periods of time.
   c. have high nutrient needs.
   d. are malnourished.

3. All of the following may benefit from PPN **except:**
   a. those with normal renal function who need short-term parenteral nutrition.
   b. those who need long-term parenteral nutrition.
   c. those in whom inserting an IV catheter into a central vein would be medically unsound.
   d. those on oral or tube feedings who need additional nutrients temporarily.

4. For a client receiving central TPN who also receives IV lipid emulsions two or three times a week, the lipid emulsions serve primarily as a source of:
   a. vitamin C.
   b. essential fatty acids.
   c. fat-soluble vitamins.
   d. concentrated energy.

5. Compared to solutions delivered by peripheral vein, solutions delivered by central vein provide:
   a. a lower osmolality.
   b. more fat, less dextrose.
   c. more dextrose, less fat.
   d. more vitamins and minerals.

6. The person who is a good candidate for central TPN rather than PPN:
   a. needs long-term parenteral nutrition support.
   b. does not have high nutrient requirements.
   c. is well-nourished and expected to be able to eat soon.
   d. has strong peripheral veins and moderate nutrient needs.

7. The solutions infused through peripherally inserted central catheters are the same as those infused for:
   a. PPN.
   b. simple IVs.
   c. central TPN.
   d. enteral feedings.

8. Which type of IV insertion procedure is associated with the highest risk of sepsis?
   a. central TPN catheter
   b. peripheral catheter for PPN
   c. peripheral catheter for simple IVs
   d. peripherally inserted central catheter

9. A gradual transition from an IV feeding to an oral diet is primarily designed to:
   a. improve the appetite.
   b. prevent hypoglycemia.
   c. prevent apprehension about eating.
   d. ensure that nutrient needs will continue to be met.

10. The health care team evaluating a client's ability to manage a home enteral or parenteral nutrition program would be least concerned with:
   a. the client's financial resources.
   b. the client's psychological status.
   c. the exact composition of the enteral formula or IV solution.
   d. the client and caregivers' abilities to learn the necessary techniques and follow medical instructions.

Critical Thinking

1. Ideally, nutrition support for a person recovering from an extensive small bowel resection includes:
   a. PPN with no enteral nutrition.
   b. central TPN with no enteral nutrition.
   c. central TPN with minimal use of IV fat emulsion.
   d. a combination of parenteral and enteral nutrition.

2. Rapid weight gain in a person who is dehydrated and is just beginning to receive IV fluids may be caused by:
   a. excessive energy from the IV fluids.
   b. lack of activity.
   c. fluid retention.
   d. vitamin deficiencies.

3. Which of the following disorders is most likely to interfere with the body's ability to use the GI tract to obtain food?
   a. dysphagia
   b. complete bowel obstruction
   c. delayed gastric emptying
   d. irritable bowel syndrome

Answers to these questions can be found in Appendix G.

Clinical Applications

1. One liter of a TPN solution contains 500 milliliters of 50 percent dextrose solution and 500 milliliters of 5 percent amino acid solution. Determine the daily kcalorie and protein intakes of a person who receives 2 liters of such a solution. Calculate the average daily energy intake if the person also receives 500 milliliters of a 20 percent fat emulsion three times a week.

2. Consider what it must be like to be on a home TPN program with no foods allowed by mouth. What would be the advantages of being at home instead of in the hospital? What would be the disadvantages? Think of how you would manage feedings. How would you feel about the time, costs, and commitment required to maintain this therapy? If you were not permitted to take any foods by mouth, how would you feel about not being able to eat after a long time? How would you handle holidays and special occasions that often center around food?

Nutrition on the Net

For further study of the topics of this chapter, access these websites.

Find updates and quick links to these and other nutrition-related sites at our website:
www.wadsworth.com/nutrition

Visit the websites of organizations that support the appropriate use of parenteral and enteral nutrition including the American Society for Parenteral and Enteral Nutrition, the Canadian Parenteral-Enteral Nutrition Association, and the European Society of Parenteral and Enteral Nutrition:
www.nutritioncare.org, www.magi.com/~cpena, and **www.espen99.org**

Explore resources available to clients on home nutrition programs by visiting the Oley Foundation site:
www.wizvax.net/oleyfdn

Discover resources for caregivers of children with tube feedings by visiting the Children's Hospital Medical Center:
www.cincinnatichildrens.org/family/pep/homecare

Notes

1. D. W. Wilmore and S. J. Dudrick, Growth and development of an infant receiving all nutrients exclusively by way of the vein, *Journal of the American Medical Association* 203 (1968): 860–864; C. T. Spencer and C. Compher, The development of TPN: An interview with pioneer surgical nutritionist Jonathan E. Rhoads, *Journal of the American Dietetic Association* 101 (2001): 747–750.

2. L. R. Kearns and coauthors, Update on parenteral amino acids, *Nutrition in Clinical Practice* 16 (2001): 219–225.

3. D. F. Driscoll, Intravenous lipid emulsions, *Nutrition in Clinical Practice* 16 (2001): 215–218.

4. J. W. Kruimel and coauthors, Parenteral structured triglyceride emulsion improves nitrogen balance and is cleared faster from the blood of moderately catabolic patients, *Journal of Parenteral and Enteral Nutrition* 25 (2001): 237–244.

5. U.S. Food and Drug Administration, Parenteral multivitamin products; Drugs for human use; Drug efficacy study implementation; Amendment, *Federal Register* 65 (no. 77), April 20, 2000, pp. 21200–21201.

6. G. S. Sacks, P. Trainor, and M. S. Gibson, Alternate dosing of multivitamins in parenteral nutrition: Economically savvy or clinically unsafe? *Nutrition in Clinical Practice* (supplement) 16 (2001): 1–4.

7. National Advisory Group on Standards of Practice, Guidelines for parenteral nutrition, *Journal of Parenteral and Enteral Nutrition* 22 (1998): 49–66.

8. K. Masumoto and coauthors, Manganese intoxication during intermittent parenteral nutrition: Report of two cases, *Journal of Parenteral and Enteral Nutrition* 25 (2001): 95–99.

9. National Advisory Group on Standards of Practice, 1998.

10. National Advisory Group on Standards of Practice, 1998.

11. J. Z. Rogers, K. McKee, and E. McDermott, Peripherally inserted central venous catheters, *Support Line,* October 1995, pp. 6–9; S. C. Loughran and M. Borzatta, Peripherally inserted central catheters: A report of 2506 catheter days, *Journal of Parenteral and Enteral Nutrition* 19 (1995): 133–136.

12. R. Nirula, K. Yamada, and K. Waxman, The effect of abrupt cessation of total parenteral nutrition on serum glucose: A randomized trial, *American Journal of Surgery* 66 (2000): 866–869.

13. T. C. Clark, Nutrition support clinical pathways, *Nutrition in Clinical Practice* 11 (1996): 16–20; The 1995 A.S.P.E.N. standards for nutrition support: Hospitalized patients, *Nutrition in Clinical Practice* 10 (1995): 206–207.

14. M. E. Shils and coauthors, Long-term parenteral nutrition through an external arterivenous shunt, *New England Journal of Medicine* 283 (1970): 314–343.

15. P. Reddy and M. Malone, Cost and outcome analysis of home parenteral and enteral nutrition, *Journal of Parenteral and Enteral Nutrition* 22 (1998): 302–310.

16. C. J. Rollins, Home care issues with multivitamin therapy, *Nutrition in Clinical Practice* (supplement) 16 (2001): 12–16; E. Steiger and C. Ireton-Jones, The evolution of home parenteral nutrition in the United States, *Nutrition in Clinical Practice* 16 (2001): 236–239.

17. Reddy and Malone, 1998.

18. M. F. Winkler, Nutrition support from hospital to home, Presented at the 82nd Annual Meeting of the American Dietetic Association, October 21, 1999.

19. Reddy and Malone, 1998.

20. M. Malone, Effect of home nutrition support on patient's lifestyle (abstract), *Journal of Parenteral and Enteral Nutrition* (supplement) 19 (1995): 23.

21. S. M. Schneider and coauthors, Outcome of patients treated with home enteral nutrition, *Journal of Parenteral and Enteral Nutrition* 25 (2001): 203–209.

22. Rollins, 2001.

Nutrition in Practice

■ NUTRITION IN THE CHANGING HEALTH CARE ENVIRONMENT ■

Decades of medical research have resulted in an explosion of knowledge and technologies to diagnose and treat diseases. Among the benefits spurred by this explosion are a greater understanding of the GI tract and the ability to safely feed people by tube and by vein. This astounding progress, however, has come at a tremendous financial cost. The United States spends more money on health care than any other nation, yet the high cost of health care has put medical care out of the reach of many citizens. Without attention to cost containment, the health status of the nation is threatened; sophisticated medical services do little good if people cannot use them. The medical community, which once embodied the idealistic approach of sparing no cost when it came to health care, has embraced the reality that cost is an element of quality. To examine all the ramifications of cost containment for health care delivery is beyond the scope of this discussion. Instead, this Nutrition in Practice identifies major trends affecting medical services, including nutrition services, and points out some ways nutrition can help reduce health care costs.

What measures can be taken to contain health care costs?

Among other things, efforts to control health care costs aim to eliminate duplication of services, reduce the number of hours spent in client care, maximize the use of health care professionals, and limit access to unnecessary services. The task is to cut costs without compromising the quality of care.

A major change in health care delivery that has evolved from efforts to control costs is the shift from the traditional **indemnity insurance** (fee-for-service) system to **managed care** (see the glossary on p. 536). A traditional fee-for-service health insurance plan allows clients to use the physicians and health care facilities of their choice and then reimburses the clients for a percentage of the services covered by the plan. Managed care organizations, which include **health maintenance organizations (HMOs)** and **preferred provider organizations (PPOs),** strive to provide high-quality, low-cost care by controlling the access to and cost of services. Each client selects a primary care physician,

who then determines what services the client needs. Managed care is moving toward a capitated system. **Capitation** is a system where health care providers agree to supply all the services a client needs for a set monthly fee. Because the providers assume the financial responsibility if they fail to operate within a predetermined budget, they have a strong incentive to control costs. Managed care organizations, which were rare just 15 years ago, are increasingly replacing traditional insurance plans. While managed care organizations have been effective at controlling costs, many consumers and health care professionals worry that quality of care has taken a backseat.

What exactly is quality of care?

Obviously, obtaining the best treatments to maintain optimal health and treat illness is an important part of quality health care. To deliver such care requires the services of competent, skilled, and caring health care professionals who attend to clients' physical, emotional, intellectual, social, and spiritual needs. In a cost-conscious health care environment, an added requirement is "at the least cost."

How can quality of care be measured?

In cost-conscious health care, quality of care is most often defined in terms of outcomes.[1] Examples of desirable outcomes include:

- Ability to function independently.
- Reduced number or length of hospital stays.
- Prevention of diseases and complications.
- Extended survival time.

Desirable outcomes improve a person's quality of life, extend its length, save money, or all of these. People who are able to function independently, for example, enjoy the freedom from dependence on others and also save the expense of having others care for them.

Health care services are cost justified when they have a reasonable chance of improving a client's outcome. Services that produce the best outcomes at the lowest costs are the most cost-effective. The more expensive the procedure or test, the more critical

Nutrition in Practice

Cost-cutting measures have redefined the responsibilities of health care professionals as well as the settings in which they work.

justifying its cost becomes. Thus it may not be cost-effective to perform a nutrition assessment on every client admitted to a hospital, whereas nutrition screening is cost-effective. For another example, providing enteral formulas orally to clients who can drink them is more cost-effective than providing the formula by tube, whenever possible.

In what ways can nutrition affect quality of care?

A review of the desirable outcomes listed above reveals many ways that nutrition can affect quality of care and reduce health care costs. By providing clients with the nutrients they need to maintain both physical and mental health, nutrition can significantly improve the client's ability to function independently, prevent or forestall diseases, limit complications, and maintain quality of life. All of these factors can help prevent the need for medical treatments, hospitalization, or lengthy hospital stays. Both timely identification of nutrition

problems and appropriate intervention are essential to improving outcomes and cutting costs. Malnutrition, a persistent and common problem in hospitalized clients, is associated with significantly longer hospital stays, higher costs, and the need for home health care.[2] Based on one analysis of data from several sources, researchers estimate that early attention to malnutrition in hospitalized clients can reduce length of stay and may result in cost savings of approximately $8300 per hospital bed per year.[3] A recent study found that clients whose nutrition status deteriorated during the course of hospitalization experienced higher rates of complications and incurred higher hospital costs than clients who maintained their nutrition status.[4]

Outside the hospital, attention to nutrition can be cost-effective as well. The authors of a review article estimate a potential savings of billions of health care dollars from an increased public awareness of the roles of nutrition in preventing several common disorders (type 2 diabetes, hypertension, coronary heart disease, stroke, cancer, osteoporosis, and depression) and consequent changes in eating habits.[5] Medical nutrition therapy can also have a significant impact on the costs of treating common chronic diseases such as diabetes and cardiovascular diseases.[6]

Medical nutrition therapy is a relatively inexpensive service, and, as the examples above indicate, it can also be cost-effective. (Note: In most managed care settings, medical nutrition therapy is treated as specialty care that requires referral by the primary care physician.) In representing nutrition professionals, the American Dietetic Association takes the position that managed care organizations and integrated health delivery systems should provide medical nutrition therapy as an essential component of health care and that it should be provided by qualified nutrition professionals.[7] The task for nutrition professionals is to document the cost savings of nutrition intervention and ensure that nutri-

Glossary of Health Care Insurance Terms

capitation: prepayment of a set fee per client in exchange for medical services.

enteral formulary: the formulas available for use in a health care facility.

health maintenance organizations (HMOs): managed care organizations that limit the subscriber's choice of health care professionals to those affiliated with the organization and control access to services by directing care through a primary care physician.

indemnity insurance: traditional fee-for-service insurance.

managed care: a health care delivery system that aims to provide cost-effective health care by coordinating services and limiting access to services.

preferred provider organizations (PPOs): managed care organizations that encourage subscribers to select health care providers from a group that has contracted with the organization to provide services at lower costs.

Nutrition in Practice

tion is addressed in standards that guide the care of clients.

What types of standards guide the care of clients?

Health care professionals and the organizations that represent them are working together to define outcome-oriented standards of practice and measures of care. The Joint Commission on Accreditation of Healthcare Organizations (JCAHO), an agency that oversees the accreditation of health care facilities, requires compliance with nutrition care standards to identify, address, and monitor each client's nutrition needs.[8] The standards promote coordination and communication among disciplines in an effort to improve quality of care in a cost-efficient manner. Hospitals and other health care facilities also develop protocols and standards that best serve their unique client populations with the goal of providing the best quality of care.

How have the responsibilities of health care professionals changed as a result of efforts to reduce costs while maintaining quality of care?

One of the trends spurred by the changes is the replacement of traditional health care delivery, which is discipline- or department-specific, with a multidisciplinary or team approach. By working together, health care professionals can coordinate their services, which enables them to eliminate unnecessary services and solve problems in a timely and efficient manner. Traditionally, for example, nurses performed nursing functions and left the responsibility for the nutrition care of clients to the dietitian. When a client was admitted to the hospital, the nurse would perform a nursing assessment, and the dietitian, a nutrition assessment. Yet much of the information from these assessments overlaps. As Chapter 16 described, a nurse can use the nursing assessment to determine the client's nutrition risk and make appropriate referrals to the dietitian. The dietitian, in turn, can rely on much of the information the nurse collects during the nursing assessment to complete a nutrition assessment. Teams, such as nutrition support teams, can also help reduce costs while guarding quality of care.

What are nutrition support teams?

Nutrition support teams often include a physician, nurse, dietitian, and pharmacist. Team members have a special interest in and in-depth knowledge of nutrition assessment, tube feedings, and parenteral nutrition. As Chapters 20 and 21 show, the appropriate use of these techniques requires skill and considerable amounts of time.

Team members develop standards and protocols for nutrition screening, nutrition assessment, tube feedings, and parenteral nutrition; identify and monitor the care of people who need nutrients from enteral formulas or parenteral nutrition; and serve as consultants to the rest of the institution's staff. Team members also identify and solve problems regarding individual clients, problems with procedures or equipment, or issues appropriate to their specialty area. The team members analyze new products and review current research to determine which products and services might benefit their clients or free valuable staff time. As an example, the nutrition support team might evaluate open and closed enteral feeding systems to see if the higher initial costs of a closed feeding system might be offset by a lower incidence of bacterial contamination of formulas and reduced hospital costs for treating infections and for nursing time spent with clients. Figure NP21–1 on p. 538 shows the responsibilities of individual team members as well as those shared by the team. Nutrition support teams play an important role in ensuring quality nutrition care while reducing health care costs.

Why can't just one health care professional perform these roles? Does it really take a team?

To understand how various disciplines contribute their unique skills to solve problems efficiently, consider this example. During nutrition support team rounds, the dietitian mentions a recurring problem with the collection of urine for nitrogen balance studies, which has invalidated several test results, wasted time (and money), and eliminated a source of information that could improve client care. The nurse recalls that many of the nurses are unfamiliar with the procedure and suggests an in-service session on the proper techniques for nitrogen balance studies. The nurse agrees to schedule and conduct the session with the help of the dietitian. In an alternative scenario, the pharmacist (reacting to the dietitian's comments) relays information he read about a cost-effective and efficient procedure for conducting nitrogen balance studies. The team agrees that each team member will review the article and decide at the next meeting whether to test the new procedure. Clients benefit from many eyes noting problems before they become serious and many brains searching for better ideas and solutions.

In a similar manner, the team relies on the expertise of its members to select the enteral formulas the hospital will stock (**enteral formulary**); maintaining the formulary allows the hospital to purchase formula in bulk and thus saves money. Team members also

Nutrition in Practice

Figure NP21–1
The Nutrition Support Team

The physician
- Diagnoses medical problems
- Performs medical procedures
- Coordinates and prescribes therapy
- Directs and supervises team
- Approves guidelines and protocols
- Consults with other physicians

The nurse
- Assesses nursing needs
- Performs direct client care
- Explains medical procedures and treatment plans
- Instructs clients regarding medical care
- Acts as a liaison between team and nursing staff
- Coordinates discharge plans

All team members
- Review current research
- Analyze new products
- Develop guidelines
- Provide in-service training
- Monitor clients
- Correct problems
- Educate clients
- Evaluate the outcome of the care provided and cost savings
- Promote the appropriate use of nutrition support
- Improve communications among team members and between the team and other health care professionals

The dietitian
- Assesses nutrition status
- Determines clients' nutrient needs
- Recommends appropriate diet therapy
- Reevaluates clients regularly
- Instructs clients about their diets
- Acts as a liaison between the team and the dietary department

The pharmacist
- Recommends appropriate drug therapy
- Identifies medication-medication and diet-medication interactions
- Identifies medication-related complications
- Educates clients about their medications
- Acts as a liaison between the team and the pharmacy

develop protocols for cost-effective, high-quality care. One example might be the development of a clinical pathway (see Chapter 17). For the nutrition support team, examples might include clinical pathways for enteral and parenteral nutrition. Such a plan defines a time frame for each intervention with a goal of providing the best outcome at the lowest cost. Once the plan is in place, each team member assists in studying unexpected outcomes or variances from the expected course of treatment and fine-tuning the plan as necessary to achieve the best outcomes. Specific examples of cost savings generated by nutrition support teams include:

- Preventing overuse of parenteral nutrition, which is costly. In one hospital that implemented procedures to reduce the inappropriate use of parenteral nutrition, quality nutrition care could be achieved with an estimated cost savings of over $225,000 for one year.[9] In another study, reducing the inappropriate use of parenteral nutrition generated potential cost savings estimated at $500,000 per year.[10]

- Selecting cost-effective enteral and parenteral formulas and supplies. In one hospital seeking to reduce formula waste and the incidence of clogged feeding tubes and encourage appropriate formula administration techniques, clinicians studied the costs and benefits of a new type of pump (automatic flush pump) that automatically delivers water through feeding tubes at intervals. The potential cost saving generated by reducing the number of clogged feeding tubes was estimated at $43,350 per month.[11] In addition, the cost saving realized because nurses spent less time manually flushing feeding tubes was estimated at $40,150 per year.

These are but a few of a growing number of examples that suggest nutrition support teams improve the quality of care and generate cost savings at the same time.

Can health care facilities cut costs in any other ways?

Health care facilities also save costs by maximizing the use of skilled assistants in health care. A dietetic techni-

cian assisting a clinical dietitian, for example, may collect the data for a nutrition assessment, and the dietitian may analyze the data. Because the dietitian's time is reserved for tasks that require a dietitian's unique skills, the dietitian can manage more clients, and money is saved. Assistants for physicians, nurses, and pharmacists perform similar roles.

Health care facilities also increasingly rely on clients and caregivers to perform as many health-related tasks as possible. An extension of this trend is a great reduction in the length of hospital stays and the growth of home health care. As a result, hospital staffs have been reduced.

Does the trend toward home health care have any effect on nutrition?

Early hospital discharges and limited hospital staffs raise the likelihood that nutrition problems will not be identified or not be corrected by the time clients leave the hospital. Clients may be discharged before they have regained health, and they may require additional medical and nutrition care in outpatient settings and at home. Estimates suggest that more than 20,000 providers meet the needs of the 8 million Americans who currently receive home care for either long- or short-term medical needs.[12] Nurses most often provide nutrition services that are part of a home care treatment plan, although the number of dietitians in home health care is expected to grow.[13] As Chapter 21 described, the use of home nutrition support, especially home enteral nutrition, has grown significantly and the growth is expected to continue. Health care professionals are challenged to ensure that home care agencies understand the cost-effectiveness of nutrition services in helping clients maintain health and quality of life.

The full implications of cost containment and its impact on the nation's health remain to be seen. In the midst of these changes, health care professionals have often been pushed to provide better outcomes with less assistance and time. Little wonder, then, that the changes are often met with resistance. Yet significant improvements in quality of care continue to be realized.[14] Health care professionals who react positively to the inevitable changes have the opportunity to make a significant difference in ensuring the success of the health care system for the future.

Notes

1. D. A. August, Outcomes research, nutrition care, and the provider-payer relationship, *Nutrition in Clinical Practice* (supplement) 13 (1998): 8–11.

2. C. S. Chima and coauthors, Relationship of nutritional status to length of stay, hospital costs, and discharge status of patients hospitalized in the medicine service, *Journal of the American Dietetic Association* 97 (1997): 975–978.

3. H. N. Tucker and S. G. Miguel, Cost containment through nutrition intervention, *Nutrition Reviews* 54 (1996): 975–978.

4. C. Braunschweig, S. Gomez, and P. M. Sheean, Impact of declines in nutritional status on outcomes in adult patients hospitalized for more than 7 days, *Journal of the American Dietetic Association* 2000 (100): 1316–1322.

5. As cited in D. J. Rodriquez, Managed care: Key concepts and nutritional implications, *Support Line*, October 1999, pp. 8–15.

6. G. Sikand and coauthors, Dietitian intervention improves lipid values and saves medication costs in men with combined hyperlipidemia and a history of niacin noncompliance, *Journal of the American Dietetic Association* 2000 (100): 218–224; J. Sheils, R. Rubin, and D. C. Stapleton, The estimated costs and savings of medical nutrition therapy: The Medicare population, *Journal of the American Dietetic Association* 99 (1999): 428–435.

7. Position of The American Dietetic Association: Nutrition services in managed care, *Journal of the American Dietetic Association* 96 (1996): 391–395.

8. D. Dougherty and coauthors, Nutrition care given new importance in JCAHO standards, *Nutrition in Clinical Practice* 10 (1995): 26–31.

9. N. M. Pace and coauthors, Performance model anchors successful nutrition support protocol, *Nutrition in Clinical Practice* 12 (1997): 274–279.

10. E. B. Trujillo and coauthors, Metabolic and monetary costs of avoidable parenteral and enteral nutrition use, *Journal of Parenteral and Enteral Nutrition* 23 (1999): 109–113.

11. M. P. Petnicki, Cost savings and improved patient care with the use of flush enteral feeding pump, *Nutrition in Clinical Practice* (supplement) 13 (1998): 29–41.

12. National Association of Home Care, Basic statistics about home care 2000, **www.nahc.org**, site visited on October 18, 2001.

13. P. S. Anthony, The business of home care, *Support Line*, October 1999, pp. 16–21; M. S. Schiller, M. B. Arensberg, and B. Kantor, Administrators' perceptions of nutrition services in home health care agencies, *Journal of the American Dietetic Association* 98 (1998): 56–61.

14. P. J. Schneider, A. Bothe, and M. Bisognago, Improving the nutrition support process: Assuring that more patients receive optimal nutrition support, *Nutrition in Clinical Practice* 14 (1999): 221–226.

Jennie Oppenheimer/Studio Zocolo

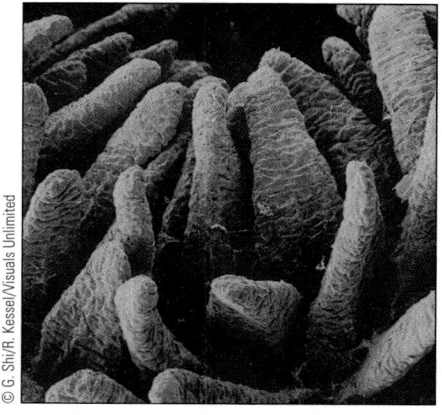

Health problems that damage the intestinal cells, speed the transit of nutrients through the intestine, or limit enzyme or bile salt activity can lead to malabsorption.

© G. Shi/R. Kessel/Visuals Unlimited

Table 22–1 Indications for Carbohydrate- and Fat-Modified Diets for Malabsorption Syndromes
Carbohydrate-Modified Diets
▪ Celiac disease
▪ Dumping syndrome
▪ Lactose intolerance
Fat-Modified Diets
▪ Acute pancreatitis
▪ Bacterial overgrowth
▪ Bile duct obstruction
▪ Blind loop syndrome
▪ Chronic pancreatitis[a]
▪ Crohn's disease
▪ Cystic fibrosis[a]
▪ HIV infection
▪ Liver disease
▪ Short-bowel syndrome

[a]For these disorders, limiting fat can help reduce steatorrhea, but the need for energy is an overriding concern. Rather than limiting the type or amount of fat, clients take enzyme replacements to minimize steatorrhea.

soaps: chemical compounds formed between a basic mineral (such as calcium) and unabsorbed fatty acids.

*M*alabsorption can occur for a variety of reasons and can involve one or many nutrients. The vitamin B_{12} malabsorption that accompanies chronic gastritis (see Chapter 18), for example, results from a reduced synthesis of intrinsic factor, which is necessary for vitamin B_{12} absorption. This chapter describes disorders that require carbohydrate- and fat-modified diets to treat malabsorption. Table 22–1 provides examples.

Malabsorption Syndromes

Malabsorption of several nutrients can occur when nutrients from food fail to enter the intestine at a controlled rate. Diseases that inflame or scar the intestine (such as the inflammatory bowel diseases described in Chapter 19) or those that necessitate the removal of a portion of the intestine can also lead to malabsorption. In these cases, the specific nutrients that may be malabsorbed and the degree of malabsorption depend on the amount and location of the intestine that is affected. Figure 22–1 shows where various nutrients are absorbed in the intestine and describes the possible consequences of diseases or surgeries that affect the different parts of the intestine.

As this chapter shows, other causes of malabsorption include enzyme deficiencies, food sensitivities, and bacterial overgrowth in the stomach and small intestine. As Chapter 23 explains, disorders of the pancreas, liver, and biliary tract can lead to malabsorption by disrupting the synthesis or secretion of bile and/or digestive enzymes. Table 22–2 provides examples of tests used to evaluate the absorption of various nutrients.

Malabsorption and Nutrition Status   Few disorders that lead to malabsorption affect only one nutrient. Depending on the cause of fat malabsorption, for example, protein, carbohydrate, vitamins, and minerals may also be affected. Disorders that cause malabsorption and their treatments can further tax nutrition status by leading to complications that alter food intake, incur further nutrient losses, and/or raise nutrient needs (see Table 22–3 on p. 544). Dietary restrictions may make it difficult to obtain other nutrients, and deficiencies of these nutrients may follow.

Fat Malabsorption   Disorders and treatments that lead to malabsorption often affect fat absorption more profoundly than carbohydrate or protein absorption because fat digestion and absorption are more complex. Fat malabsorption and the loss of fat in the stool are also more likely to severely compromise nutrition status. With fat malabsorption, food energy, essential fatty acids, and fat-soluble vitamins (which are normally absorbed along with fat) are also lost. Some minerals, including calcium and magnesium, form **soaps** with unabsorbed fatty acids, and they too are lost in the stools. The loss of vitamin D (a fat-soluble vitamin) aggravates calcium malabsorption, and deficiencies of vitamin D and calcium lead to bone diseases, which commonly accompany fat malabsorption syndromes.

Calcium malabsorption also raises the likelihood that kidney stones will form. Oxalate (see Chapter 26) normally binds with calcium in the gut and is excreted along with it. But when fatty acids form soaps with calcium, the oxalate remains. The colon absorbs the unbound oxalate, but the body cannot metabolize it, and so excretes it in the urine. High urinary oxalate, in turn, favors the formation of kidney stones. Figure 22–2 on p. 544 shows the consequences of fat malabsorption.

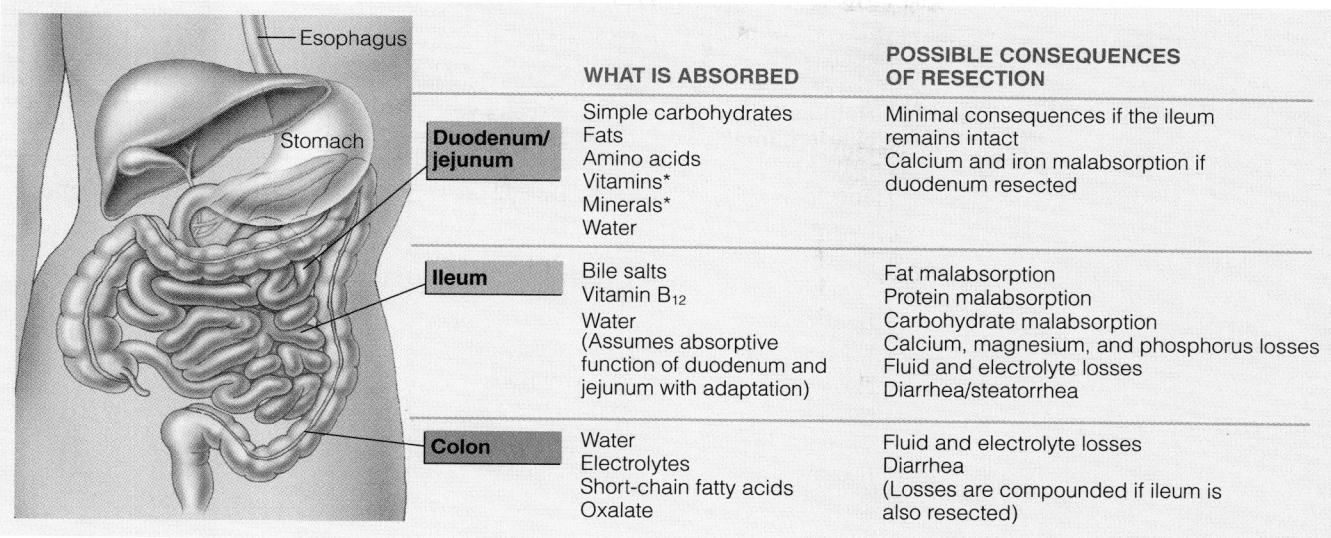

	WHAT IS ABSORBED	POSSIBLE CONSEQUENCES OF RESECTION
Duodenum/ jejunum	Simple carbohydrates Fats Amino acids Vitamins* Minerals* Water	Minimal consequences if the ileum remains intact Calcium and iron malabsorption if duodenum resected
Ileum	Bile salts Vitamin B_{12} Water (Assumes absorptive function of duodenum and jejunum with adaptation)	Fat malabsorption Protein malabsorption Carbohydrate malabsorption Calcium, magnesium, and phosphorus losses Fluid and electrolyte losses Diarrhea/steatorrhea
Colon	Water Electrolytes Short-chain fatty acids Oxalate	Fluid and electrolyte losses Diarrhea (Losses are compounded if ileum is also resected)

*The absorption of vitamins and minerals begins in the duodenum and continues throughout the length of the small intestine.

Figure 22–1

Absorption and Consequences of Intestinal Disorders or Surgeries
About 90 to 95 percent of nutrient absorption takes place in the first half of the small intestine.

In Summary Many disorders can lead to the malabsorption of one or more nutrients for a variety of reasons. Malnutrition threatens in disorders that cause malabsorption, not only because of the malabsorption of nutrients, but also because food intake and nutrient needs can also be seriously affected. Disorders that lead to fat malabsorption can seriously impair nutrition status. With fat malabsorption, needed energy, essential fatty acids, fat-soluble vitamins, and some minerals are lost as well. Malnutrition and bone diseases are frequent complications. Oxalate stones can also form.

Carbohydrate-Modified Diets for Malabsorption Syndromes

Carbohydrate-modified diets used in the treatment of malabsorption syndromes include the lactose-restricted diet, the postgastrectomy diet, and the gluten-free diet. The simplest of these is the lactose-restricted diet, which relieves the uncomfortable symptoms of lactose intolerance. The postgastrectomy diet

Table 22–2   Examples of Tests Used to Evaluate Absorption
■ Breath hydrogen: Breath sample checked for hydrogen before and after a standard lactose dose is consumed. A rise of hydrogen 10 to 20 times above baseline indicates lactose malabsorption.
■ Direct stool examinations: Stool checked for weight and fat (greater-than-normal weight or excess fat suggests malabsorption).
■ Chemical analysis of fecal fat: Fecal fat of greater than 7 grams/day when the diet includes 100 grams of fat/day indicates fat malabsorption.
■ Chemical analysis of fecal nitrogen: Normal fecal nitrogen is less than 2 grams/day.
■ D-xylose test: Test of carbohydrate absorption.
■ Serum calcium: Low levels seen in calcium or vitamin D malabsorption.
■ Serum carotene: Low levels accompany steatorrhea.
■ Schilling test: Identifies vitamin B_{12} malabsorption.

Carbohydrate- and Fat-Modified Diets for Malabsorption

Table 22–3 Possible Causes of Malnutrition in Malabsorption Syndromes

Reduced Nutrient Intake	Excessive Nutrient Losses	Raised Nutrient Needs
Abdominal pain	Blood loss	High basal energy expenditure
Anorexia	Diarrhea	Infection
Bowel rest	Fistulas	Inflammation
Emotional stress	General malabsorption	Medications
Food intolerance	Medications	Surgery
Indigestion	Steatorrhea	
Medications	Vomiting	
Nausea		
Obstructions		
Restrictive diets		

controls malabsorption that may occur following surgery of the stomach, and gluten-free diets reverse malabsorption caused by celiac disease.

Lactose-Restricted Diets

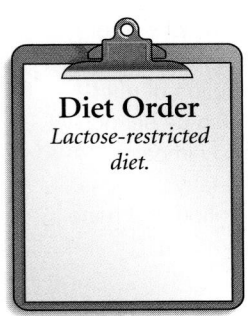

Diet Order
Lactose-restricted diet.

Lactose-restricted diets are highly individualized diets that most often limit, but do not exclude, milk and milk products. Lactose-restricted diets treat malabsorption caused by a deficiency of lactase, the enzyme that splits lactose to glucose and galactose.

Causes of Lactose Intolerance In rare cases, a person is born with a lactase deficiency; more often, lactase activity gradually diminishes with age. Lactose intolerance is prevalent among people in certain ethnic groups includ-

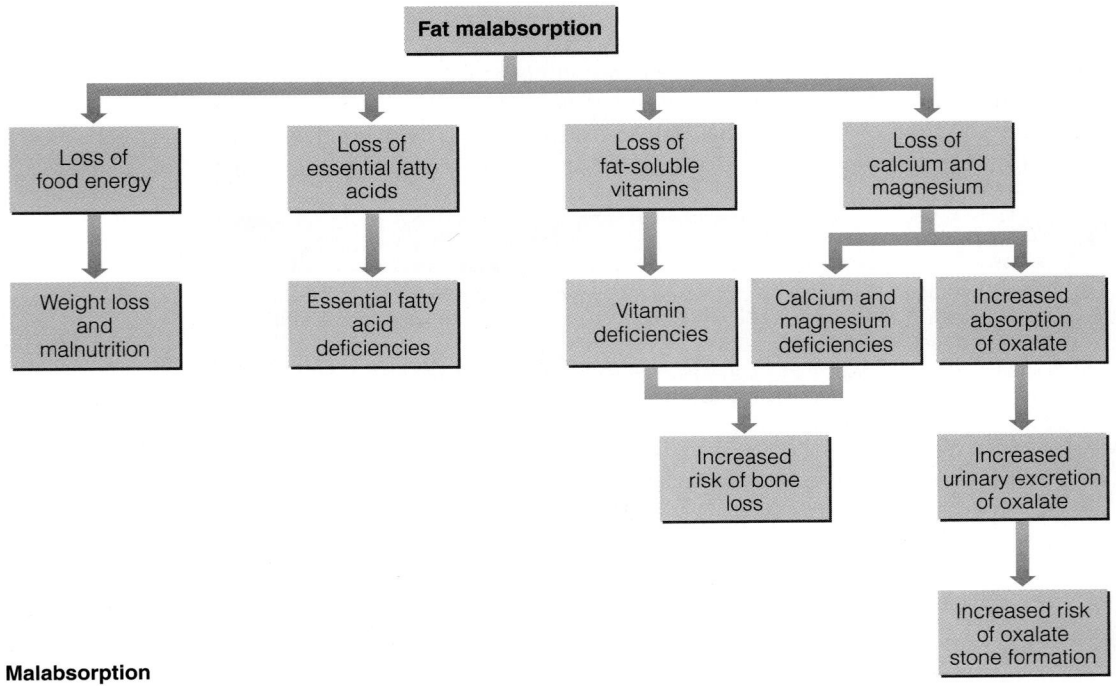

Figure 22–2
The Consequences of Fat Malabsorption

ing those of Mediterranean origin, African Americans, Asians, Jews, Mexicans, and Native Americans. Permanent or temporary lactase deficiencies can develop as a consequence of any disorder or condition that damages the delicate intestinal microvilli including malnutrition, Crohn's disease (Chapter 19), radiation therapy (Chapter 28), and many of the conditions described later in this chapter.

Vigorous advertising campaigns to promote products for lactose intolerance have led many people to believe they have lactose intolerance, when, in fact, they do not.[1] Some believe they are intolerant to even the smallest amounts of lactose. These people may unnecessarily eliminate milk and milk products from the diet and inadvertently develop calcium and vitamin D deficiencies in the process.

Medical Nutrition Therapy for Lactose Intolerance   Medical nutrition therapy includes an individualized lactose-restricted diet and advice on preventing calcium and vitamin D deficiencies. Dietitians advise clients to test their tolerance for lactose by gradually increasing consumption of lactose-containing foods to the point that precipitates symptoms of lactose intolerance—bloating, cramps, and diarrhea. People with lactose intolerance can often include up to 1 or 2 cups of milk a day, provided that the milk is taken with food and in small amounts at a time.[2] Many people can eat cheeses, particularly aged cheeses.[3] Some tolerate chocolate milk better than plain milk. Most tolerate yogurt well because yogurt contains bacteria that produce lactase and thus help digest lactose.

Lactase Enzyme Preparations   People can also add a lactase enzyme preparation to milk before they drink it or take enzyme tablets whenever they eat lactose-containing foods. Lactose-free milk and milk products treated with lactase are also available. Products to aid lactose digestion are unnecessary in many cases, however, because many people tolerate a fair amount of lactose.

Hidden Sources of Lactose   People who are highly sensitive to even the smallest amounts of lactose need to avoid regular milk and milk products altogether and look for hidden sources of lactose on food labels. Products that may contain hidden lactose include baked goods, such as cookies and cake, and processed foods, such as luncheon meats and commercially prepared sauces and gravies. People with temporary lactose intolerance are advised to temporarily restrict all milk and milk products and gradually reintroduce them in small amounts.

Preventing Calcium and Vitamin D Deficiencies   People who restrict milk and milk products because they have lactose intolerance, or believe they have it, risk calcium and vitamin D deficiencies. To help prevent such deficiencies, encourage people to include milk and milk products in their diets to the extent that they can tolerate these foods, with or without the use of enzyme preparations. People who fail to receive adequate amounts of calcium from milk and milk products should be encouraged to eat other food sources of calcium, including calcium-fortified juices, cereals, and snacks; broccoli; mustard greens; kale; and sardines. For people who cannot get enough calcium from conventional foods, calcium supplements are indicated, particularly for children, adolescents, and pregnant, lactating, or postmenopausal women. Vitamin D is not a problem if the person gets regular exposure to sunlight; otherwise, a supplement may be necessary.

Most people with lactose intolerance can drink milk, if they drink it along with other foods and limit the amount they have at any one time.

Four to 6 oz of milk contain about 6 to 9 g of lactose.

Clients unable to tolerate even small amounts of lactose need to read food labels and avoid foods (or medications) containing any of the following ingredients:

- *Buttermilk.*
- *Cheese flavors.*
- *Curds.*
- *Fat-free milk powder or solids.*
- *Lactose.*
- *Malted milk.*
- *Milk or milk solids.*
- *Sweet or sour cream.*
- *Whey*

Diet Order
Progress diet from a postgastrectomy diet to a regular diet as tolerated.

The Postgastrectomy Diet

The **postgastrectomy** diet aims to provide the energy and nutrients necessary to support recovery and minimize complications associated with gastric surgery. Following gastric surgery, malabsorption can occur if the stomach loses its ability to control the rate at which it releases food

postgastrectomy (post-gas-TREK-toe-me): following surgery that removes all (total gastrectomy) or part (subtotal or partial gastrectomy) of the stomach.
   post = after
   gastr = belly
   ectomy = excision

Carbohydrate- and Fat-Modified Diets for Malabsorption

Table 22–4 Postgastrectomy Diet[a]

Meat and Meat Alternates

Any type allowed. Fried meats and eggs and highly seasoned meats may cause discomfort.

Milk and Milk Products

Withheld initially and then gradually introduced as tolerated.

Grains and Starchy Vegetables

Allowed (up to 5 servings per day): Plain breads, crackers, rolls, unsweetened cereal, rice, pasta, corn, lima beans, parsnips, peas, white potatoes, sweet potatoes, pumpkin, yams, winter squash.

Excluded: Sweetened cereal; breads or cereal containing dates or raisins; pastries, doughnuts.

Nonstarchy Vegetables

Allowed (unlimited): Any type as tolerated.
Excluded: Vegetables prepared with sugar or creamed.

Fruits

Allowed (up to 3 servings per day): Unsweetened fruits and fruit juices.
Excluded: Sweetened fruits and fruit juices; dates, raisins, and other dried fruits.

Fats

Any type allowed except sweetened salad dressings.

Beverages

Allowed: Coffee, tea, artificially sweetened drinks. Coffee may cause discomfort.
Excluded: Alcohol; sweetened milk, beverages, and fruit drinks; cocoa.

[a]Clients with dumping syndrome who are unable to tolerate a sufficient variety or volume of foods over long periods of time often require nutrient supplements.

into the intestine. The postgastrectomy diet limits the total amount of carbohydrate as well as the amount of simple sugars. To provide energy, slow the passage of foods through the intestine, minimize diarrhea, and prevent irritation to the GI tract, the diet emphasizes foods high in protein and moderate in fat and limits foods that stimulate gastric acid or raise the likelihood of gastroesophageal reflux. Table 22–4 lists the foods allowed and excluded from postgastrectomy diets, and the sample menu below shows a day's meals.

Postgastrectomy Diet Menu

Breakfast	*Midmorning Snack*	*Lunch*	*Midafternoon Snack*	*Supper*	*Evening Snack*
1 poached egg	¼ c cottage cheese	2 oz hamburger patty	2 tbs peanut butter	2 oz boiled ham	¼ c tuna
1 slice toast	1 graham cracker	¼ c mashed potatoes	3 saltine crackers	¼ c rice	1 tsp mayonnaise
1 tsp butter		1 tsp margarine		½ c carrots	1 slice bread
Decaffeinated coffee (take 30–60 minutes after meal)		½ small banana		2 tsp butter	
		Iced tea (take 30–60 minutes after meal)		¼ c unsweetened peach slices	
				Tea (take 30–60 minutes after meal)	

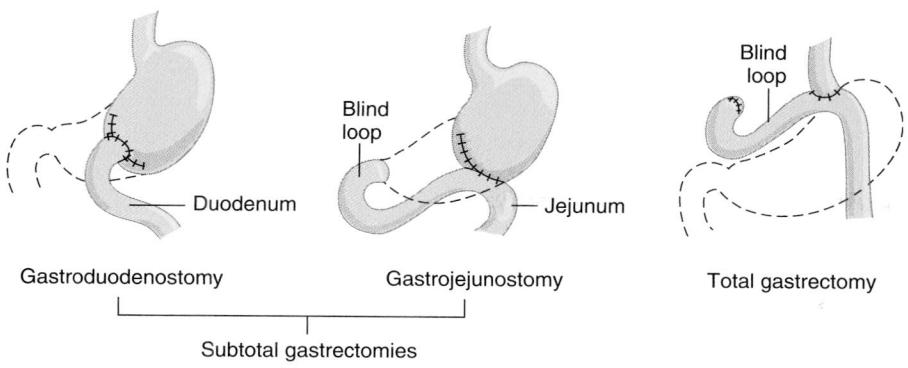

Figure 22–3
Typical Gastric Surgery Resections
In a gastric resection, part or all of the stomach is surgically removed. The dashed lines show the removed section.

Blind loop

Blind loop

Blind loop

Duodenum

Jejunum

Gastroduodenostomy

Gastrojejunostomy

Total gastrectomy

Subtotal gastrectomies

Gastric Surgery   Several surgical procedures affect the functions of the stomach. During a gastrectomy, the surgeon removes either a portion or all of the stomach. Figure 22–3 shows examples of gastrectomy procedures. Another type of gastric surgery, **pyloroplasty,** enlarges the pyloric sphincter (the sphincter that joins the stomach and small intestine) so that the intestinal fluids (with a higher pH) reflux into the stomach and neutralize gastric acidity. During a **vagotomy,** the surgeon severs the nerves that stimulate gastric acid production. A vagotomy may accompany either a gastrectomy or a pyloroplasty in some cases. In **gastric partitioning,** a treatment for severe obesity, the stomach remains intact, but all or a portion of the stomach is bypassed.

Postsurgical Care   Immediately following surgery, the client receives no foods or fluids by mouth. Health care professionals monitor fluid and electrolyte balances carefully, and the physician corrects any imbalances promptly. After several days, clear liquids and then solids are gradually introduced in very small amounts offered frequently. Many clients can tolerate solid foods by about the fourth or fifth day. Initially, the dietitian visits the client after each meal to check the client's tolerance for different foods and carefully tailors the diet to meet the client's needs. Gradually, most people begin to tolerate limited amounts of concentrated sweets, larger quantities of food, and some liquids with meals.

Dumping Syndrome   One complication that can arise when the portion of the stomach containing the pyloric sphincter has been removed, bypassed, or disrupted is the **dumping syndrome** (see Figure 22–4 on p. 548). Without a functional pyloric sphincter, the stomach loses control over the rate at which food enters the intestine. Instead, food gets "dumped" rapidly into the jejunum. (The duodenum is short, and even if it had not been bypassed surgically, foods pass quickly through it into the jejunum.) With digestion of the food mass, the intestinal contents rapidly become concentrated (hypertonic). As fluids from the body move into the intestinal lumen to dilute the concentration, the volume of circulating blood falls, causing weakness, dizziness, and a rapid heartbeat. The large volume of hypertonic fluid and unabsorbed material in the jejunum causes pain and **hyperperistalsis,** and diarrhea results.

Hypoglycemia   Two to three hours later, many of the same symptoms may appear again, and others occur: dizziness, fainting, nausea, and sweating. This time the cause is different. The body efficiently absorbed so much glucose from the meal that blood glucose rose quickly. The pancreas responded by overproducing insulin, which made blood glucose fall too low **(hypoglycemia).** Now, low blood glucose is causing the symptoms.

Diet Adjustment   Not all people experience the diarrhea of dumping syndrome following gastric surgery or vagotomies. Even fewer develop hypoglycemia. Most people who initially experience the dumping syndrome gradually adapt to a

Clear liquids appropriate for a postgastrectomy diet include items such as broth, bouillon, unsweetened gelatin, and diluted unsweetened fruit juices provided in small amounts.

Reminder: Something that is hypertonic has a concentration of particles greater than the concentration of particles in the blood. Fluid flows from the area of lesser concentration to the area of greater concentration.

pyloroplasty (pie-LOOR-oh-PLAS-tee): surgery that enlarges the pyloric sphincter.

vagotomy (vay-GOT-oh-mee): surgery that severs the nerves that stimulate gastric acid secretion.

gastric partitioning: surgery for severe obesity that limits the functional size of the stomach.

dumping syndrome: the symptoms that result from the rapid entry of undigested food into the jejunum: sweating, weakness, and diarrhea shortly after eating and hypoglycemia later.

hyperperistalsis (HY-per-pear-ih-STALL-sis): rapid movement through the intestine.

hypoglycemia (HY-poe-gly-SEE-me-ah): low blood sugar. The type of hypoglycemia that occurs following gastric surgery is called *reactive* or *postgastrectomy* hypoglycemia.

Carbohydrate- and Fat-Modified Diets for Malabsorption

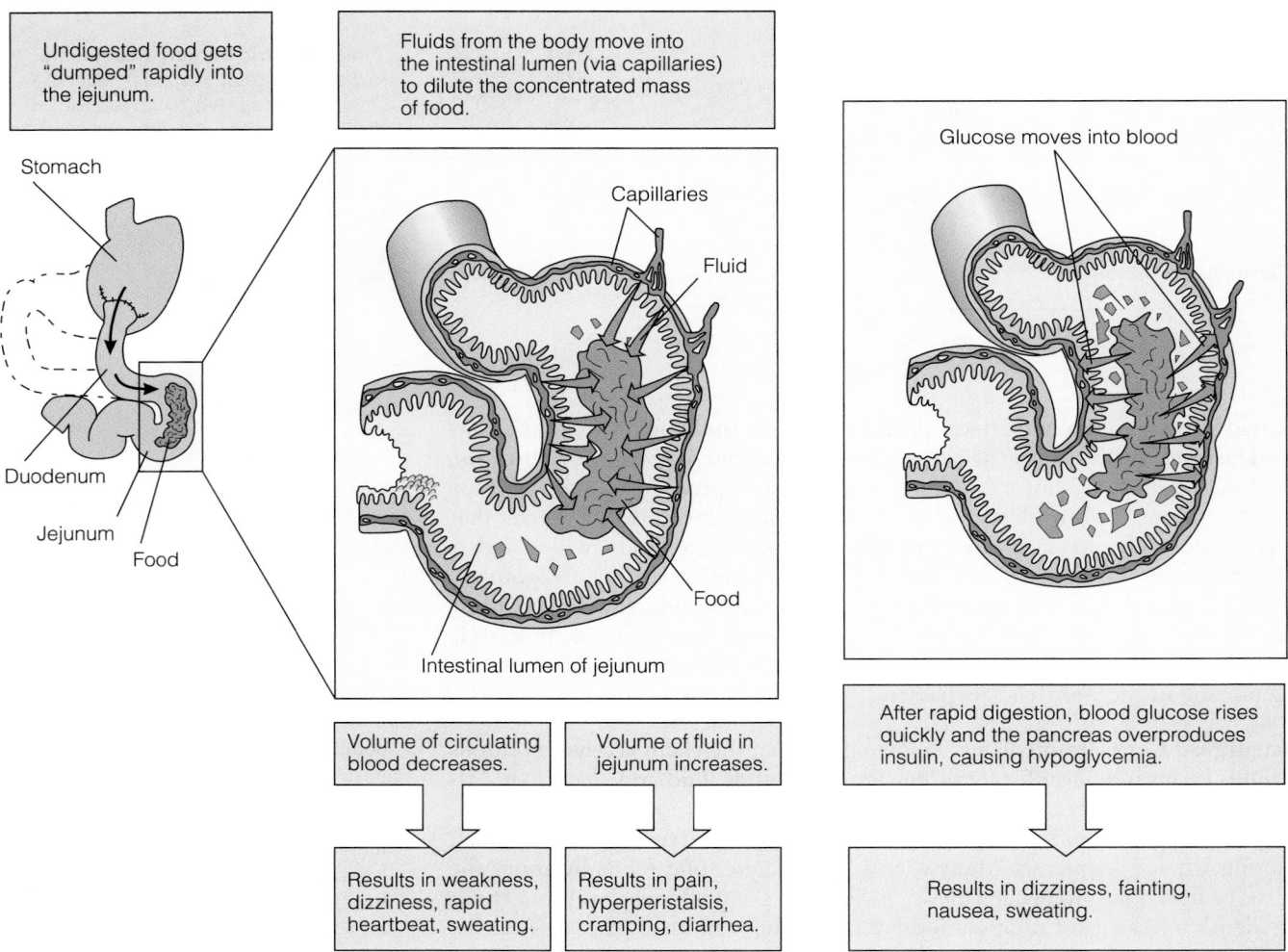

Undigested food gets "dumped" rapidly into the jejunum.

Fluids from the body move into the intestinal lumen (via capillaries) to dilute the concentrated mass of food.

Glucose moves into blood

Stomach

Duodenum

Jejunum

Food

Capillaries

Fluid

Food

Intestinal lumen of jejunum

Volume of circulating blood decreases.

Volume of fluid in jejunum increases.

After rapid digestion, blood glucose rises quickly and the pancreas overproduces insulin, causing hypoglycemia.

Results in weakness, dizziness, rapid heartbeat, sweating.

Results in pain, hyperperistalsis, cramping, diarrhea.

Results in dizziness, fainting, nausea, sweating.

Figure 22–4

The Dumping Syndrome

When partially digested food rapidly enters the jejunum, it is quickly digested and creates a concentrated mass. Fluid from the intestinal capillaries enters the jejunum, diminishing blood volume and stimulating peristalsis. The result: low blood pressure and diarrhea.

fairly regular diet. Nevertheless, a postgastrectomy diet, along with the measures described in the "How to" box, serves to control the dumping syndrome. With time, the symptoms of the dumping syndrome resolve or improve. Sometimes adding pectin (a type of dietary fiber) to the diet can help control the dumping syndrome. If dietary and medical management fails to resolve the problem, additional surgery may be necessary.

Weight Loss and Nutrient Deficiencies   Although weight loss is the goal for people undergoing gastric partitioning, people undergoing other gastric surgeries often lose significant amounts of weight before surgery, and they risk serious malnutrition following surgery. Unfortunately, many people continue to lose weight and develop nutrient deficiencies. In addition to malabsorption, food intake may be poor due to early satiety, postsurgical pain, fear that eating will bring on the symptoms of the dumping syndrome, reflux esophagitis and its associated epigastric pain that commonly occur following gastric surgery, and sometimes dysphagia.

Clients who have undergone some gastric surgeries experience fat malabsorption. Surgical procedures that divert food from the duodenum interfere with the release of the hormones that mediate the secretion of enzymes and bile to aid fat digestion. Surgeries that reduce gastric acid secretion or create a blind loop (see Figure 22–3) may allow bacteria to colonize in the stomach and small intes-

How to

Prevent the Dumping Syndrome

*P*eople who experience the dumping syndrome benefit from the following suggestions:

- Chew foods thoroughly and eat slowly.
- Avoid concentrated sweets (sugar, cookies, cakes, pies, or soft drinks) because the body digests and absorbs these carbohydrates rapidly and breaks them down into many particles that draw fluids from the body into the intestine.
- Eat frequent, small meals to fit the reduced storage capacity of the stomach.
- Drink liquids in small amounts about 45 minutes before or after meals, not with them. This precaution prevents overloading the stomach's reduced storage capacity and slows the transit of food from the stomach to the intestine.

- Lie down immediately after eating for 30 to 60 minutes to help slow the transit of food to the intestine. Clients who experience reflux, however, should not lie down after eating.
- Be aware that lactose intolerance may develop and add to the problem of diarrhea and abdominal pain. Enzyme-treated milk and milk products should also be avoided because the enzymes break down lactose to glucose and galactose—simple sugars that further concentrate the intestinal contents and promote dumping. Discontinue all types of milk and milk products until recovery is under way. Then try them gradually and in small amounts.

tine. In both cases, fat malabsorption can result. A significant number of people who undergo gastric surgery develop a bone disease, a consequence of fat, calcium, and vitamin D malabsorption.

Anemia   Iron-deficiency anemia frequently occurs following gastric surgery, although it may take several years to develop. The rapid transit of food through the stomach reduces iron's exposure to gastric acid, interfering with the conversion of iron to its absorbable form. In addition, iron absorption is normally more efficient in the duodenum than in the rest of the small intestine, and following gastric surgery, the food bolus either bypasses the duodenum altogether or moves rapidly through it. Inadequate intake of iron from foods and blood loss also contribute to the problem. An iron supplement helps correct iron deficiencies.

Inadequate intake and malabsorption can also lead to anemia caused by folate and, less often, vitamin B_{12} deficiencies. To correct deficiencies, clients receive supplements.

Although one might expect vitamin B_{12} deficiencies to be common after gastric surgery because intrinsic factor production could be impaired, surgeons often avert this problem by leaving intact a small area of the stomach where intrinsic factor production occurs. Vitamin B_{12} deficiencies are common following total gastrectomies, however.

Gastric Partitioning   Unlike with other gastric surgeries, weight loss is a goal following gastric partitioning. Most clients lose weight for about 12 to 18 weeks after surgery and lose about 50 to 60 percent of their excess body weight.[4]

The safety and effectiveness of gastric partitioning depend, in large part, on compliance with medical nutrition therapy, which aims to prevent nutrient deficiencies and promote eating and lifestyle habits that will enable the client to lose weight and then to maintain a desirable weight. Poor dietary habits may prevent weight loss, rupture staples, or obstruct the small opening left for food to pass into the lower stomach. Other potential complications include infections, nausea, vomiting, dehydration, dumping syndrome, gastroesophageal reflux, and, as a result of all this and more, depression.

Diet following Gastric Partitioning   After surgery, clients typically follow a mechanical soft diet beginning with small amounts of liquids. After about three months, clients can generally begin to eat solid foods. Clients are cautioned to chew foods thoroughly to reduce the risk of dislodging surgical staples and to prevent foods from obstructing the opening into the lower

Carbohydrate- and Fat-Modified Diets for Malabsorption

Table 22–5 Gluten-Free Diet[a]

Meat and Meat Alternates

Any allowed except those that are breaded, prepared with bread crumbs or ingredients that are not allowed, or creamed.

Milk and Milk Products

Any allowed if client is not intolerant to lactose except milk mixed with Ovaltine, commercial chocolate milk with a cereal additive, milk beverages flavored with malt, pudding thickened with wheat flour, ice cream or sherbet containing gluten stabilizers, processed cheese foods and spreads, Roquefort and ricotta cheese.

Fruits and Vegetables

Any allowed except those that are breaded, prepared with bread crumbs, or creamed.

Starches and Grains

Allowed: Bread, cereal, or dessert products made from buckwheat, cornmeal, soybean flour, rice flour, or potato flour; tapioca; popcorn and hominy; rice, cream of rice, puffed rice, rice flakes; sago; millet; flax; teff; sorghum; amaranth; quinoa.

Not allowed:[b] Bread, cereal, or dessert products made from wheat, rye, barley, and oats; wheat starch; low-gluten flours; commercially prepared mixes for biscuits, cornbread, muffins, pancakes, cakes, cookies, or waffles; bran; pasta, macaroni, and noodles; malt; pretzels; wheat germ; doughnuts; ice cream cones; matzo.

Other

Not allowed: Beer; ale; certain whiskeys (Canadian rye); alcohol-based extracts; cereal beverages (Postum); root beer; commercial salad dressings that contain gluten stabilizers; soy sauce; soups containing any ingredients not allowed (such as barley or noodles); products made with hydrolyzed vegetable protein.

[a]This is a partial list. Celiac organizations can provide additional information.
[b]Note that many special products made with allowed ingredients are available. Gluten-free pastas and macaroni, for example, are acceptable substitutes for regular pastas and macaroni.

---

stomach. Regardless of food consistency, clients tolerate only small amounts of food (about a cup of food at a time); overeating or overdrinking can cause nausea, reflux, and vomiting.

Food Selections Clients must understand that if they regularly drink high-kcalorie liquids or eat high-kcalorie foods, even if only in small quantities, they will not lose weight. Their selections also affect the nutritional quality of the diet. Nutrient deficiencies, particularly of vitamin B_{12}, iron, and vitamin D, are common. Careful planning, diligent compliance with the prescribed diet, and vitamin-mineral supplements help ensure that nutrient needs are met within a strictly limited energy intake.

Gluten-Free Diets

Diet Order
Gluten-free diet.

The gluten-free diet eliminates certain grains that inflame the intestine and lead to malabsorption for people with **celiac disease. Gluten** is a protein found in wheat, and **gliadin** is the fraction of gluten that causes sensitivity in celiac disease. Barley and rye have similar protein fractions that also cause problems in celiac disease. As Table 22–5 shows, many commonly used foods contain these grains in various forms, many of which are not obvious and may not be listed on food labels. Table 22–5 also lists starches and grains that are acceptable for gluten-free diets.

celiac (SEE-lee-ack) **disease:** a sensitivity to a part of the protein gluten that causes flattening of the intestinal villi and malabsorption; also called *gluten-sensitive enteropathy* or *celiac sprue.*

gluten (GLUE-ten): a protein found in wheat.

gliadin (GLY-ah-din): the fraction of gluten that causes the toxic effects in celiac disease. Corresponding protein fractions in barley, rye, and possibly oats also have these effects.

Causes of Celiac Disease Celiac disease, a hereditary intestinal disorder, affects about 1 in every 250 people.[5] Although it was once thought to be a childhood disease, it is now known that the disease can also appear in adults. In celiac disease, gliadin and corresponding protein fractions act as toxic substances that damage intestinal cells. The immune system responds by triggering inflammation that further damages the intestinal villi, and malabsorption results.

Consequences of Celiac Disease People with celiac disease may have difficulty absorbing many nutrients, notably, fat, fat-soluble vitamins, folate, vitamin B_{12}, iron, calcium, and magnesium. (Fat malabsorption and its consequences are described later in this chapter.) Lactose intolerance is common. Clinical manifestations of the disease vary and may range from simple iron and folate deficiencies to serious malnutrition, fatigue, and severe anemia. A vitamin K deficiency may precipitate clotting abnormalities, and the person may bleed easily. Reduced bone mineral density and bone diseases are common.[6] People with unexplainable bone disease may actually have celiac disease that has gone undiagnosed because GI symptoms are either mild or absent.[7]

In some people, sensitivity to gluten manifests itself primarily as a rash on the skin. Although some people with this form of gluten sensitivity also exhibit GI symptoms, most often GI problems are milder than in the other form of celiac disease.

Treatment of Celiac Disease Lifelong adherence to a gluten-free diet serves as the primary treatment for celiac disease. Once the person follows a gluten-free diet for a few weeks, the intestinal changes reverse almost completely. With early diagnosis and treatment, children and adolescents with celiac disease who adhere to a gluten-free diet can achieve normal bone mass and avoid bone diseases.[8] In adults, bone demineralization is not always reversible.[9] Lactose intolerance may be permanent. A gluten-free diet also helps control the skin rash characteristic of gluten sensitivity.

Offending Grains Although authorities agree that gluten-free diets should strictly exclude wheat, barley, and rye, controversies surround the use of oats and other grains and foods, including buckwheat, amaranth, quinoa.[10] In the United States, all celiac organizations oppose the inclusion of oats, which has traditionally been excluded from the gluten-free diet, but some organizations in Europe consider oats safe.[11] Studies suggest that moderate amounts of oats are probably safe to include, but some authorities believe that oats may be contaminated with wheat during processing.

Even the smallest amounts of offending grains can cause problems for people with celiac disease. Therefore, people with the disease must be vigilant in preparing and storing foods. They must be careful to avoid contaminating the foods they eat with bread crumbs from toasters, cutting boards, or grills or with crumbs that may be in margarine or jelly or in fats used to fry foods.

Food Choices If people on gluten-free diets do not take care to vary their food choices and select appropriate foods, malnutrition can be a problem.[12] Food manufacturers make many gluten-free products that serve as acceptable substitutes for foods not permitted on gluten-free diets. These products help people with celiac disease expand their food choices and enjoy foods that would otherwise be forbidden. Clients must select products carefully, however; some products contain lower levels of B vitamins, iron, and fiber than the foods they are intended to replace.[13]

Gluten-free products help people with celiac disease enjoy a wide variety of foods.

© Polara Studios, Inc.

In Summary Carbohydrate-modified diets used to treat malabsorption include the lactose-restricted diet, the postgastrectomy diet, and the gluten-free diet. Lactose-restricted diets, which may include moderate amounts of lactose, help control the diarrhea and pain associated with lactose intolerance.

Carbohydrate- and Fat-Modified Diets for Malabsorption

Postgastrectomy diets limit total carbohydrate and simple carbohydrates to minimize malabsorption following gastric surgery. Gluten-free diets eliminate wheat, barley, rye, and possibly oats. Strict adherence to a gluten-free diet reverses the intestinal damage that leads to malabsorption in celiac disease.

Fat-Modified Diets for Malabsorption Syndromes

As Chapter 19 described, dietary fats delay the rate at which the stomach empties, and for that reason, low-fat diets may be used to treat disorders that delay gastric emptying or for gastroesophageal reflux disease. Fat-modified diets can also help ease fat malabsorption and prevent its consequences.

Diet Plans

Individual tolerances determine the specific amount of fat prescribed: fat is restricted only as much as necessary to prevent **steatorrhea** (the loss of fat in the stools) because the person needs food energy. Typically, the person begins with a diet providing about 20 to 25 percent of total energy as fat (usually 50 grams or less of fat). Fat intake gradually increases if additional energy is needed and as tolerance permits. Foods are provided in frequent, small meals because fat is best tolerated in small amounts at a time. Steatorrhea or worsening steatorrhea indicates intolerance to fat. Table 22–6 provides instructions for a fat-restricted diet, and the sample fat-restricted diet menu shows a day's meals. In some cases, **pancreatic enzyme replacements** are provided with meals to aid digestion and absorption.

Adding kCalories   Whenever fat must be restricted to minimize malabsorption, additional fat may be given as **medium-chain triglycerides (MCT).** MCT are easier to digest and absorb than long-chain triglycerides (LCT). MCT supply almost as many kcalories as regular fats, but they do not provide essential fatty acids, so some LCT are required. The "How to" box on p. 554 offers suggestions for helping people accept fat-restricted diets and includes tips for using MCT.

Restriction of Oxalate   As described earlier, people with fat malabsorption syndromes risk kidney stones because the colon absorbs unbound oxalate, which must then be excreted in the urine. Although some of the oxalate the body excretes in the urine comes from food, even more oxalate is synthesized within the body. One way the body synthesizes oxalate is through a pathway that begins with vitamin C. Thus, to reduce the likelihood that a kidney stone will form, clients are advised to restrict foods high in oxalate (see Chapter 26) and avoid vitamin C supplements.

Bacterial Overgrowth

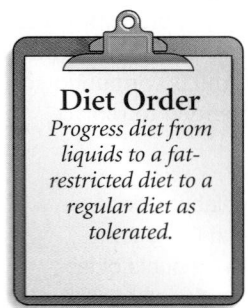

Diet Order
Progress diet from liquids to a fat-restricted diet to a regular diet as tolerated.

Although the colon normally houses a considerable bacterial population, the small intestine is protected from bacterial overgrowth by gastric acid, which kills bacteria, and peristalsis, which flushes GI secretions and microorganisms through the small intestine before they can multiply. Conditions that disrupt these protective mechanisms can allow bacteria to flourish in the stomach and small intestine. Malabsorption of fat and fat-soluble vitamins can occur as a consequence, because the bacteria alter the bile salts necessary for fat digestion and absorption. The

Examples of foods high in oxalate include spinach, rhubarb, beets, nuts, chocolate, tea, wheat bran, and strawberries.

steatorrhea (STEE-ah-toe-REE-ah): fatty diarrhea characterized by loose, foamy, foul-smelling stools. Soaps that form between unabsorbed fatty acids and minerals give steatorrhea its foamy appearance.

pancreatic enzyme replacements: extracts of pork or beef pancreatic enzymes that are taken as supplements to aid digestion.

medium-chain triglycerides (MCT): triglycerides containing fatty acids with 8 to 12 carbon atoms; they require minimal lipase and no bile for absorption.

Chapter Twenty-Two

Table 22–6　Fat-Restricted Diet (50 grams)

Use:

1. Fat-free milk, fat-free cheeses, fat-free yogurt, sherbet, and fruit ices.
2. Low-fat egg substitutes.
3. Up to 6 oz of lean meats and poultry without skin daily. One egg may be used instead of 1 oz of meat, three times a week.
4. Three to 5 servings of fat daily. One serving is any one of the following:

    1 tsp butter, margarine, shortening, oil, or mayonnaise

    1 tbs reduced-fat butter, margarine, or mayonnaise

    1 strip crisp bacon

    1 tbs salad dressing or 2 tbs reduced-fat salad dressing

    ⅛ avocado

    2 tbs cream (half and half)

    10 peanuts

    8 large black olives or 10 large green olives

    If fat is used to cook or season food, it must be taken from this allowance.
5. All vegetables prepared without fat.
6. All fruits prepared without fat.
7. Plain white or whole-grain bread; fat-free cereals, pasta, rice, noodles, and macaroni.
8. Clear soups.
9. Angel food cake and fruit whips made with gelatin, sugar, and egg-white meringues.
10. Jelly, jam, honey, gumdrops, jelly beans, and marshmallows.

Do not use:

1. Whole milk, chocolate milk, whole-milk cheeses, and ice cream.
2. Biscuits, french toast, waffles, muffins, pastries, cakes, pies, sweet rolls, breads, or vegetables made with fat.
3. More than one egg a day, fried or fatty meats (sausage, luncheon meats, spareribs, frankfurters), duck, goose, or tuna packed in oil (unless well drained).
4. More than allowed servings of fat.
5. Desserts, candy, or anything made with chocolate, nuts, or foods not allowed.
6. Creamed soups made with whole milk.

Suggestions:

1. To make the diet still lower in fat, reduce the fat and meat (and egg) servings.
2. To raise the fat content, give additional fat or meat servings.
3. To improve acceptance of the diet, check the fat content of a well-liked food and allow that food if possible. Use the exchange system fat list for alternate suggestions for fat servings (see Appendix B).

NOTE: A "How to" box on p. 67 provides additional tips for lowering fat in the diet.

Fat-Restricted Diet Menu

Breakfast	Lunch	Supper	Snack
Soft-cooked egg	3 oz broiled chicken	3 oz lean roast beef	Fruit ice
Dry cereal	Rice	Mashed potatoes	
Orange juice	Green beans	Peas	
Coffee, sugar	Tossed salad	Bread	
Whole-wheat toast	1 tbs low-fat French dressing	1 tsp margarine	
½ tsp margarine	Fresh apple	Peaches	
Fat-free milk	Iced tea, sugar	Fat-free milk	
	1 tsp margarine		

All foods are prepared without added fat.

How to

Improve Acceptance of Fat-Restricted Diets

*F*at-restricted diets can be difficult to follow. Fats give flavors, aromas, and textures to foods—characteristics that people may miss. Unlike some diets that can be introduced gradually, the fat-restricted diet for malabsorption must sometimes be implemented immediately without giving the person time to adapt to the changes. These suggestions may help:

- Provide clients with tips for making foods palatable while lowering fat intake, such as those found in the box on p. 67.
- Remind clients that new fat-free and low-fat products appear on market shelves daily, and most people find these products very acceptable. Caution clients to avoid products containing fat substitutes (such as olestra), however. While healthy people can use fat substitutes in appropriate amounts, people with

digestive problems and fat-soluble vitamin malabsorption may find that fat substitutes aggravate their conditions.

People who use MCT need additional advice:

- Advise clients to add MCT to the diet gradually. Nausea, vomiting, diarrhea, abdominal pain, and distention can result from using too much MCT all at once.
- Recommend that clients improve the palatability of MCT oil by substituting it for regular oil in salad dressing and for cooking and baking and by adding it to beverages, desserts, sauces, casseroles, and other dishes.
- Warn clients that MCT products are expensive, and explain that these products can be purchased at pharmacies and are sometimes covered by medical insurance.

bacteria also disrupt the absorption of vitamin B_{12} and can lead to vitamin B_{12} deficiencies.

Causes of Bacterial Overgrowth   As Figure 22–3 on p. 547 shows, in some types of gastric surgery a portion of the small intestine is bypassed, allowing bacteria to flourish. The bypassed portion of the intestine is called a blind loop, and the symptoms associated with bacterial overgrowth are called the **blind loop syndrome.**

Other conditions and treatments can lead to bacterial overgrowth by significantly reducing gastric acid secretions; these include chronic gastritis (Chapter 18), medications (antisecretory agents), chronic pancreatitis (Chapter 23), and HIV infections (Chapter 28). Small bowel obstructions (Chapter 19) and nerve dysfunction associated with diabetes (Chapter 24) can lead to bacterial overgrowth by altering peristalsis.

Consequences of Bacterial Overgrowth   Typical symptoms of bacterial overgrowth include chronic diarrhea, intestinal gas, malnutrition, and weakness. In some cases, signs of fat-soluble vitamin deficiencies may develop including dry, scaly skin (vitamin A deficiency), bruising (vitamin K deficiency), and numbness, tingling, muscle spasms, and bone pain (vitamin D and calcium deficiencies). Pernicious anemia can develop as a consequence of vitamin B_{12} deficiency.

blind loop syndrome: the problems of fat and vitamin B_{12} malabsorption that result from the overgrowth of bacteria in a bypassed segment of the intestine.

Treatments for Bacterial Overgrowth   The primary treatment for bacterial overgrowth is antibiotics. In some cases, antibiotics correct the problem, but symptoms recur weeks to months later. During active periods of bacterial overgrowth, clients benefit from fat-restricted diets. Oral vitamin and mineral supplements serve to correct nutrient deficiencies with the exception of vitamin B_{12}. Because bacterial overgrowth impairs the absorption of vitamin B_{12}, injections or prescription nasal sprays supply vitamin B_{12}.

Intestinal Surgeries and Short-Bowel Syndrome

Diet Order
Fat-modified and carbohydrate-modified diet as tolerated.

The treatment of inflammatory bowel diseases, cancer of the intestine, intestinal obstructions, fistulas, diverticulitis, trauma to the intestine, or impaired blood supply to the intestine may include surgery to remove a portion of the intestine. When portions of the small intestine must be removed, the length, location, and health of the remaining intestine determine if the person will be able to meet nutrient needs using the GI tract. (Review Figure 22–1 on p. 543 to see how absorption is affected by surgical resections.) Generally, up to 50 percent of the intestine can be resected without serious nutrition consequences. When the absorptive surface of the small intestine is significantly reduced, however, **short-bowel syndrome**—characterized by severe diarrhea and malabsorption following surgery—results. Gradually, diarrhea and malabsorption may lessen as adaptation occurs in the remaining bowel.

Adaptation   After an intestinal resection, the healthy portions of the intestine that remain get longer, thicker, and wider, and they absorb nutrients more efficiently. Adaptation begins soon after surgery and may continue for a year or more. The presence of nutrients in the remaining gut stimulates adaptation. Intestinal disease in the remaining portions of the intestine (such as active Crohn's disease or inflammation caused by radiation therapy) may impede adaptation. Following extensive small bowel resections, adaptation is most likely to allow clients to meet their nutrient needs enterally if the ileum, the ileocecal valve, and the colon remain intact.

The Ileum and Ileocecal Valve   An intact ileum reabsorbs bile salts, helping to maintain the body's bile salt pool and prevent fat malabsorption. The ileum is also the site where vitamin B_{12} is absorbed. When the ileum has been resected, vitamin B_{12} deficiencies and fat malabsorption and its consequences often follow. With surgical removal of the ileocecal valve, the contents of the small intestine empty rapidly into the colon, fluid and electrolyte losses are compounded, and bacteria from the colon can invade the small intestine, leading to bacterial overgrowth. Bacterial overgrowth further damages bile salts and worsens fat and vitamin B_{12} malabsorption.

The Colon   For the person with an extensive resection of the small intestine, the presence of an intact colon helps reduce fluid and electrolyte losses. Furthermore, bacteria in the colon can metabolize unabsorbed carbohydrate to short-chain fatty acids. Short-chain fatty acids, in turn, can provide energy the body can use and help promote adaptation in the colon.

As a guideline, enteral nutrition is possible with a functional (healthy) bowel length of greater than 40 inches (100 centimeters) with the colon removed, or greater than 24 inches (60 centimeters) with the colon intact.[14] When people cannot meet nutrient needs enterally, two options remain: permanent intravenous (IV) nutrition or intestinal transplantation. Due to the major risks associated with intestinal transplantation, this procedure is reserved for clients who require lifelong IV nutrition support, but who develop complications that make IV support extremely difficult or impossible.

Medical Nutrition Therapy for Short-Bowel Syndrome   Immediately following surgery, fluid and electrolyte balances are markedly upset, and IV fluids serve to correct these imbalances. For resections of less than 50 percent of the intestine, oral nutrition often begins after a few days. With extensive intestinal resections, severe diarrhea, malabsorption, and lack of GI motility prevent use of the

short-bowel or **short-gut syndrome:** severe malabsorption that may occur when the absorptive surface of the small bowel is reduced, resulting in diarrhea, weight loss, bone disease, hypocalcemia, hypomagnesemia, and anemia.

Carbohydrate- and Fat-Modified Diets for Malabsorption

Case Study

ARTIST WITH SHORT-BOWEL SYNDROME

Adam Goldstein is a 25-year-old artist with an 8-year history of Crohn's disease (see p. 531). Adam is 5 feet, 9 inches tall. Three years ago, Adam underwent a small bowel resection and remained free of active disease for two years. During that time, Adam's symptoms subsided, he gained weight, and with the exception of milk products that aggravate his lactose intolerance, he was able to eat many foods. Adam has been able to maintain a modest income from selling his paintings and working part-time in an art gallery. Several months ago, Adam experienced a severe flare-up of his Crohn's disease. He lost 10 pounds and currently weighs 158 pounds. He experienced fatigue and severe abdominal pain that persisted despite aggressive medical management that included intravenous nutrition. Adam underwent another resection five days ago, which left him with 36 percent of healthy small intestine. His colon is intact. He is experiencing extensive diarrhea.

- What is Crohn's disease, and why is surgery sometimes necessary in its treatment (see Chapter 19)?

- Calculate Adam's IBW. What nutrition concerns do his current weight and recent weight loss suggest? What other nutrition problems might Adam be experiencing as a consequence of Crohn's disease? Describe some complications and other factors that can affect nutrient needs in people with Crohn's disease.

- What factors determine how well a person will be able to meet nutrient needs enterally following an extensive intestinal resection?

- How are nutrient needs met following extensive intestinal resections? Why? How are enteral feedings usually delivered at first?

- What diet will be advised for Adam once he is able to eat table foods? How would this advice differ if Adam's medical condition had necessitated removal of his colon?

GI tract to provide nutrients. Once fluid and electrolyte balances are stabilized, IV nutrition is cautiously initiated. Once adaptation is under way, generally in about one to three weeks, diarrhea and malabsorption begin to lessen, and enteral nutrition (usually by tube feeding) can be started in limited amounts. If tolerance for enteral feeding improves, IV feedings are tapered and gradually discontinued. Some people may be unable to discontinue IV nutrition and will require lifelong IV feedings to supply all or some of their nutrient needs.

Once enteral diets are possible, clients whose colons remain intact benefit from high-carbohydrate (50 to 60 percent of the total kcalories and limited in simple sugars), low-fat (20 to 30 percent of total kcalories), oxalate-restricted diets provided in five or six meals a day.[15] Bacteria in the colon break down complex carbohydrates to short-chain fatty acids, which provide a source of energy for the body. For people without intact colons, no bacteria are available to provide additional energy from carbohydrate, and oxalate is not a problem because unbound oxalate is absorbed in the colon. Furthermore, carbohydrates, especially simple carbohydrate, may accelerate fluid losses when no colon is present, so an appropriate diet provides more fat (30 to 40 percent of total kcalories) and less carbohydrate (40 to 50 percent of total kcalories) compared to the client with an intact colon. In all cases, dietitians work closely with clients to help them identify individual food intolerances and to offer suggestions to help control the rate at which foods pass through the intestinal tract. Vitamin and mineral supplements are routinely recommended. For those with vitamin B_{12} malabsorption, vitamin B_{12} injections or prescription nasal sprays are necessary to correct deficiencies and maintain vitamin B_{12} status. Use the case study above to apply the concepts described in this section to a client with short-bowel syndrome.

In Summary In some cases, fat-modified diets help minimize fat malabsorption and its consequences. To help people with fat malabsorption get enough energy from foods, some of the fat can be supplied as MCT, which

Analgesics

See p. 486.

Antidiarrheals

Octreotide, sometimes used to treat the dumping syndrome, can cause nausea and abdominal pain. *See also p. 486.*

Anti-Infective Agents

Numerous *anti-infective agents* with a variety of nutrition-related interactions may be used in the treatment of disorders that lead to malabsorption, including those used to treat infections associated with gastric and intestinal surgeries and bacterial overgrowth. Consult a drug guide for specific examples.

Antisecretory Agents

See p. 486.

Enzyme Replacements

Pancreatic enzyme replacements sometimes cause nausea and may interfere with iron absorption. They must be taken with meals and snacks. Enteric-coated forms should not be crushed or chewed. Capsules containing enteric-coated microspheres (tiny spheres of medication) may be sprinkled on soft foods (such as applesauce), provided they can be swallowed without chewing and are followed by a glass of water or juice. *Lactase* tablets or capsules should be taken just before using a lactose-containing food.

Immunosuppressants (for Inflammatory Bowel Diseases)

See p. 486.

are easier to digest and absorb than LCT. Bacterial overgrowth leads to fat malabsorption by altering the bile salts necessary for their absorption. Fat malabsorption is more likely to occur as a consequence of intestinal surgery when the ileum and ileocecal valve must be resected. The Diet-Drug Interactions box shows examples of nutrition-related concerns associated with medications used to treat diseases that lead to malabsorption. The Nutrition Assessment Checklist on p. 558 highlights nutrition-related findings to consider for clients with malabsorption.

Nutrition Assessment Checklist

FOR PEOPLE WITH MALABSORPTION

Medical History

Does the client have a medical diagnosis of:

☐ Lactose intolerance?

☐ Crohn's disease?

☐ Ulcerative colitis?

☐ Short-bowel syndrome?

☐ Celiac disease?

Has the client had surgery that includes:

☐ Gastric resection?

☐ Gastric partitioning?

☐ Intestinal resection?

Does the client have the following symptoms or complications:

☐ Diarrhea/dehydration?

☐ Lactose intolerance?

☐ Malabsorption?

☐ Essential fatty acid deficiency?

☐ Bone disease

☐ Oxalate stones?

☐ Infection?

☐ Fistulas?

☐ Obstructions?

☐ Bacterial overgrowth?

Medications

Check for medications or herbal remedies that may:

☐ Cause constipation or diarrhea

☐ Interfere with food intake including those that cause nausea, vomiting, cramps, dry mouth, or drowsiness

Food/Nutrient Intake

Note the following conditions and contact the dietitian if you suspect a problem with:

☐ Poor appetite or limited food intake.

☐ Food intolerances (including amount of carbohydrate and fat for those with malabsorption, and amount of lactose for those with lactose intolerance)

☐ Inadequate calcium intake for those with lactose intolerance or malabsorption

☐ Fluid intake

Anthropometrics

Measure baseline height and weight. Address weight loss early to prevent malnutrition for clients with:

☐ Severe or persistent diarrhea

☐ Malabsorption

Laboratory Tests

Check laboratory tests for signs of dehydration for clients with:

☐ Severe or persistent diarrhea

☐ Malabsorption

☐ Gastric and intestinal resections

Check laboratory tests for signs of nutrition-related anemias and nutrient deficiencies (see Appendix D) for clients with:

☐ Severe or persistent diarrhea

☐ Malabsorption

Physical Signs

Look for physical signs of:

☐ Dehydration (especially for people with severe or persistent diarrhea or malabsorption)

☐ PEM

☐ Essential fatty acid deficiencies (especially for people with fat malabsorption)

☐ Folate and vitamin B_{12} deficiencies (especially for people with Crohn's disease, bacterial overgrowth, or resection of the ileum)

☐ Mineral deficiencies (especially for people with severe and persistent diarrhea, lactose intolerance, fat malabsorption, or resection of the ileum)

1. Problems associated with fat malabsorption include all of the following **except:**
   a. bone diseases.
   b. kidney stones.
   c. severe weight loss and PEM.
   d. essential amino acid deficiencies.

2. For people with permanent lactose intolerance, lactose-restricted diets:
   a. include lactose-containing foods according to individual tolerances.
   b. strictly limit lactose from all sources.
   c. require the use of lactase-containing digestive aids.
   d. require careful review of the client's intake of folate and vitamin B_{12}.

3. Compared to foods containing carbohydrate, foods containing protein and moderate amounts of fat are emphasized in the postgastrectomy diet because they:
   a. contain pectin.
   b. give a greater feeling of fullness.
   c. produce fewer particles when digested.
   d. attract fluid into the intestine more rapidly.

4. In celiac disease, a fraction of the protein gluten in wheat and corresponding protein fractions in some other grains:
   a. bind to fats and calcium so that they cannot be absorbed.
   b. are clearly marked on food labels when they are used in food products.
   c. allow oxalate to be absorbed and contribute to a higher incidence of kidney stones.
   d. act as toxic substances, inflame the intestine, and cause the intestinal villi to lose their absorptive capacity.

5. Which strategy is best for providing energy for people on fat-restricted diets?
   a. Provide 50 percent of energy as fat.
   b. Do not restrict fat any more than necessary.
   c. Restrict use of MCT, and provide additional fat as LCT.
   d. Encourage clients to eat three large meals and avoid food between meals.

6. Common nutrition problems associated with bacterial overgrowth in the stomach and small intestine include:
   a. sensitivity to gluten and gliaden.
   b. permanent loss of digestive enzymes.
   c. fat malabsorption and vitamin B_{12} deficiencies.
   d. increased absorption of bile salts and constipation.

7. A complex of symptoms that may occur whenever the absorptive surface of the small intestine is significantly reduced is called the:
   a. blind loop syndrome.
   b. dumping syndrome.
   c. short-bowel syndrome.
   d. Zollinger-Ellison syndrome.

8. Diets for people who have undergone extensive intestinal resections but whose colons remain intact should be:
   a. gluten-free.
   b. low in simple sugars, high in protein, and moderate in fat.
   c. low in carbohydrate to prevent lactose intolerance.
   d. high in complex carbohydrate and restricted in fat.

Critical Thinking

1. Which of the following snacks would be an appropriate choice for a person on a postgastrectomy diet?
   a. eggnog
   b. cookies and milk
   c. vanilla milkshake
   d. saltine crackers and peanut butter

2. The health care professional interviewing a client who has done well for three years following a gastrectomy, but who is now experiencing fatigue and looks pale, should be alert to the possibility of:
   a. dumping syndrome.
   b. iron-deficiency anemia.
   c. blind loop syndrome.
   d. fat malabsorption.

3. The health care professional recognizes that a client understands the basics of a gluten-free diet when the client states:
   a. "I must avoid all products containing wheat, barley, and rye."
   b. "I must avoid all products containing wheat, corn, and rice."
   c. "I must avoid all products containing barley, soybeans, and corn."
   d. "I must limit my use of products containing wheat, barley, rice, and oats."

4. (Hint: Review pp. 481–482 and Figure 22–1 on p. 543 before completing the following statement.) In comparing Crohn's disease affecting the small intestine and ulcerative colitis:
   a. ulcerative colitis is more likely to cause fat malabsorption.
   b. Crohn's disease is less likely to lead to nutrient deficiencies.
   c. ulcerative colitis is more likely to lead to fluid and electrolyte imbalances.
   d. Crohn's disease is more likely to lead to fluid and electrolyte imbalances.

Answers to these questions can be found in Appendix G.

Clinical Applications

1. Using Table 22–6 on p. 553 as a guide, plan a day's menus for a diet containing 50 grams of fat. Take care to make the menus both palatable and nutritious. How can these menus be improved using the suggestions in the box on p. 67?

2. As stated in this chapter, treatment of celiac disease is deceptively simple—eliminate wheat, barley, rye, and possibly oats. Randomly select ten of your favorite snack and convenience foods. Take a trip to the grocery store and check the labels of

the products you selected and see if they are allowed on gluten-free diets. (As you complete this part of the assignment, keep in mind that the labels may not list all offending ingredients.) Find acceptable substitutes for the products that are not allowed, either by substituting other foods or by checking for gluten-free products in the grocery store. (If you have access to a computer, you may want to look up sites that sell gluten-free products to get an idea of what's available.) No doubt, this will be a challenging assignment.

Nutrition on the Net

For further study of the topics of this chapter, access these websites.

Find updates and quick links to these and other nutrition-related sites at our website: **www.wadsworth.com/nutrition**

Search for more information about malabsorption and malabsorption syndromes by visiting the National Institute of Diabetes & Digestive & Kidney Diseases, the American College of Gastroen-terology, Medscape Gastroenterology, and InteliHealth: **www.niddk.nih.gov/health/digest/digest.html, www.acg.gi.org, www.medscape.com/medscape/gastro/,** and **www.intelihealth.com**

Learn more about gluten-free foods, gluten-free diets, and celiac disease by visiting the Celiac Diseases Foundation, the Canadian Celiac Association, the Celiac Sprue Association, and the Gluten Intolerance Group of North America: **www.celiac.org, www.celiac.ca, www.csaceliacs.org,** and **www.gluten.net**

Discover resources for health care professionals and people with inflammatory bowel diseases, including information about intestinal resections, by visiting the Crohn's and Colitis Foundation: **www.ccfa.org**

Notes

1. F. L. Suarez, D. A. Savaiano, and M. D. Levitt, A comparison of symptoms after the consumption of milk or lactose-hydrolyzed milk by people with self-reported severe lactose intolerance, *New England Journal of Medicine* 333 (1995): 1–4; F. L. Suarez and coauthors, Tolerance to the daily ingestion of two cups of milk by individuals claiming lactose intolerance, *American Journal of Clinical Nutrition* 65 (1997): 1502–1506.

2. Suarez and coauthors, 1997.

3. S. R. Hertzler and coauthors, How much lactose is low lactose? *Journal of the American Dietetic Association* 96 (1996): 243–246.

4. G. S. M. Cowan and A. Frank, Medical management of obesity, *Medscape Diabetes & Endocrinology Clinical Management Modules,* June 14, 2001, available at: **www.medscape.com/Medscape/ endocrinology,** site visited June 25, 2001.

5. R. P. Anderson and coauthors, In vivo antigen challenge in celiac disease identifies a single transglutaminase-modified peptide as the dominant A-gliadin T-cell epitope, *Nature Medicine* 6 (2000): 337–342; T. Not and coauthors, Celiac diseases risk in the USA: High prevalence of antiendomysium antibodies in healthy blood donors, *Scandinavian Journal of Gastroenterology* 33 (1998): 494–498.

6. S. Mora and coauthors, Reversal of low bone mineral density with a gluten-free diet in children and adolescents with celiac disease, *American Journal of Clinical Nutrition* 67 (1998): 477–481; G. R. Corazza and coauthors, Propeptide of type I procollagen is predicative of posttreatment bone mass gain in adult celiac disease, *Gastroenterology* 113 (1997): 67–71.

7. J. L. Shaker and coauthors, Hypocalcemia and skeletal disease as presenting features of celiac diseases, *Archives of Internal Medicine* 157 (1997): 1013–1016.

8. Mora and coauthors, 1998.

9. Corazza and coauthors, 1997.

10. A. Fasano, Celiac disease: The past, the present, the future, *Pediatrics* 107 (2001): 768–770; T. Thompson, Case problem: Questions regarding the acceptability of buckwheat, amaranth, quinoa, and oats from a patient with celiac disease, *Journal of the American Dietetic Association* 101 (2001): 586–587.

11. T. Thompson, Questionable foods and the gluten-free diet: Survey of current recommendations, *Journal of the American Dietetic Association* 100 (2000): 463–465.

12. M. T. Bardella and coauthors, Body composition and dietary intakes in adult celiac disease patients consuming a strict gluten-free diet, *American Journal of Clinical Nutrition* 72 (2001): 937–939.

13. T. Thompson, Folate, iron, and dietary fiber contents of the gluten-free diet, *Journal of the American Dietetic Association* 100 (2000): 1389–1396; T. Thompson, Thiamin, riboflavin, and niacin contents of the gluten-free diet: Is there a cause for concern? *Journal of the American Dietetic Association* 99 (1999): 858–862.

14. M. D. Johnson, Management of short bowel syndrome—A review, *Support Line,* December 2000, pp. 11–13, 16–23.

15. T. Byrne and coauthors, Beyond the prescription: Optimizing the diet of patients with short bowel syndrome, *Nutrition in Clinical Practice* 15 (2000): 306–311.

Nutrition in Practice

■ PROBIOTICS AND INTESTINAL HEALTH ■

Countless bacteria reside in the world around us, continuously attempting to enter the warm, nutrient-rich environment within the human body. Consider that bacterial cells associated with the skin, lungs, and gut outnumber human cells in the body by more than ten times.[1] The adult intestine harbors some 500 different species of bacteria, some of them harmless (**nonpathogenic**) and some of them potentially harmful (**pathogenic**).[2] Conditions that upset the balance of harmless and potentially harmful bacteria may predispose the host to health problems. **Probiotics,** foods and dietary supplements that provide live bacteria, may protect health by improving the balance between harmful and harmless bacteria in the gut. (The glossary on p. 562 defines terms related to this discussion.)

How do harmless bacteria protect health?

As Nutrition in Practice 15 described, harmless bacteria protect the body from microorganisms by competing with them for nutrients and space. Harmless bacteria also stimulate immune responses that protect the body from harmful microorganisms in the intestine, but they themselves do not stimulate inflammatory responses. In a sense, the harmless bacteria keep the intestinal immune defenses on high alert so that they can quickly respond to harmful microorganisms. Because they do not stimulate inflammatory responses, these bacteria do not interfere with normal intestinal functions like absorption.

Are the bacteria in the gut the same for all people?

No, many factors can affect the composition of intestinal bacteria (see Table NP22–1 on p. 562). Diet is one factor. The bacterial flora of a breastfed infant differs from that of a bottle-fed infant. The presence of certain nondigestible food ingredients favors the growth and activity of healthful bacteria—these ingredients are called **prebiotics.** Diet regulates the species and concentrations of intestinal microorganisms.[3] The composition of the bacterial flora, in turn, may affect both absorptive and immune functions. Researchers found that introducing one strain of harmless bacteria into mice with sterile GI tracts enhanced the production of genes important for nutrient digestion and absorption as well as genes that limit damage from inflammatory

responses.[4] Furthermore, different strains of bacteria resulted in variations in the genes produced, suggesting that optimal intestinal function depends, in part, on the composition of the bacterial flora.

What conditions upset the balance between harmless and harmful bacteria in the gut?

To put the problem in perspective, it is helpful to begin by looking briefly at the history of infectious disease. Infectious diseases like smallpox once claimed the lives of many children and limited the life expectancy of adults. The discovery that bacteria cause many infectious diseases spurred the development of antibiotics to kill microorganisms and stop them from growing. Unfortunately, antibiotics do not specifically target harmful bacteria—harmless bacteria may become innocent victims in the assault.

Other measures to control infectious diseases, such as the purification of water, the pasteurization of milk (and now eggs), and the processing of food, have changed the amounts and types of microorganisms that people ingest. Prior to the introduction of these measures, people's diets contained more fermented products (which contain probiotics) than today's diets, which often contain large quantities of processed and preserved foods with few probiotics.

Interestingly, during the period when infectious disease rates began to fall, the incidence of diseases related to inflammatory responses—allergies, asthma, inflammatory bowel diseases, diabetes, and arthritis—began to rise.[5] As Chapter 19 described, altered intestinal immune responses to normally harmless bacteria may play a role in inflammatory bowel diseases. Other studies have shown that probiotics can markedly improve signs of allergic responses in children and adults with food sensitivities.[6]

Have any other beneficial uses for probiotics been identified?

As Chapter 22 explained, probiotics in yogurt produce lactase, which helps people with lactose intolerance digest lactose. Although research is preliminary, evidence suggests that probiotics may prove valuable in the treatment of many intestinal disorders, including diarrhea (antibiotic-induced, traveler's, and childhood),

Nutrition in Practice

Glossary

nonpathogenic (non-path-oh-GEN-ick): not causing disease.
  patho = disease
  genic = producing

pathogenic: causing disease

prebiotics: nondigestible food ingredients that favor the growth and activity of harmless bacteria.
  pre = before
  bio = life

probiotics: foods and dietary supplements that contain live bacteria.
  pro = for

inflammatory bowel diseases, and irritable bowel syndrome.[7] Differences in bacterial flora between breast- and bottle-fed infants may explain the protection that breastfed infants have against diarrhea. Animal studies suggest that probiotics may be protective against colon cancer. For women, probiotics may help reduce infections of the urinary and genital tracts because these infections often arise from colonic bacteria.

Is it possible to get probiotics from foods today?

Yes, and the best sources are yogurt and milk to which probiotic bacteria have been added. Fermented milk products, such as kefir, are another source. Other food sources of probiotics include fermented plant products, such as brined olives, salted gherkins, and sauerkraut.[8] Dietary supplements containing probiotics are also available. The most common species of bacteria used in foods and dietary supplements are *Lactobacillus* and *Bifidobacterium*.

Probiotics include foods and dietary supplements that contain live bacteria that support GI health.

© Polara Studios, Inc.

Table NP22–1   Factors Affecting the Composition of Intestinal Bacteria
■ Age
■ Alcohol consumption
■ Diet
■ Genetics
■ Immune system health
■ Intestinal pH and transit time
■ Nutrient requirements
■ Stress

Are any problems associated with using probiotics?

Some experts have expressed concerns over the safety of probiotics.[9] Although probiotics have been used for many years and have generally been regarded as safe, clinicians report that bacteria used in probiotics have been cultured from infection sites in a number of cases. This suggests that probiotics may be harmful in some cases, especially for people with severely compromised immune systems.

Other concerns center around the concentrations and availability of probiotics in foods and supplements. Currently, there are no mandatory industry standards for probiotics in foods, and the concentration of probiotics in foods varies widely. It may be difficult to determine how much of a product must be consumed to achieve the desired effect.

In recent years, more and more information has surfaced regarding the role of the intestinal tract in immunity. Some of this information has sparked a renewed appreciation for the bacteria that flourish in the intestine and promote health. By restoring the bacterial balance that may have been disrupted by attempts to wipe out infectious diseases, probiotics show promise in the treatment of a variety of disorders, especially those of the intestinal tract.

Notes

1. E. Isolauri, Probiotics in human disease, *American Journal of Clinical Nutrition* (supplement) 73 (2001): 1142–1146; L. Kopp-Hoolihan, Prophylactic and therapeutic uses of probiotics: A review, *Journal of the American Dietetic Association* 101 (2001): 229–238, 241.

2. J. M. Saavedra, Clinical applications of probiotic agents, *American Journal of Clinical Nutrition* (supplement) 73 (2001): 1147–1151.

3. Kopp-Houlihan, 2001.

4. L. V. Hooper and coauthors, Molecular analysis of commensal host-microbial relationships in the intestine, *Science* 291 (2001): 881–884.

5. Isolauri, 2001.

6. E. Isolauri and coauthors, Probiotics in the management of atopic eczema, *Clinical and Experimental Allergy* 30 (2000): 1604–1610; L. Pelto and coauthors, Probiotic bacteria down-regulate the milk-induced inflammatory response in milk-hypersensitive subjects but have an immunostimulatory effect in healthy subjects, *Clinical and Experimental Allergy* 28 (1998): 1474–1479; H. Majamaa and E. Isolauri, Probiotics: A novel approach in the management of food allergy, *Journal of Allergies and Clinical Immunology* 99 (1997): 179–186.

7. Isolauri, 2001; Kopp-Houlihan, 2001; Saavedra, 2001; G. W. Elmer, Probiotics: "Living drugs," *American Journal of Health System Pharmacists* 58 (2001): 1101–1109.

8. G. Molin, Probiotics in foods not containing milk or milk constituents, with special reference to *Lactobacillus plantarum* 299v, *American Journal of Clinical Nutrition* (supplement) 73 (2001): 380–385.

9. N. Ishibashi and S. Yamazaki, Probiotics and safety, *American Journal of Clinical Nutrition* (supplement) 73 (2001): 465–470.

Jennie Oppenheimer/Studio Zocolo

CHAPTER 23

ENERGY-, FAT-, AND PROTEIN-MODIFIED DIETS FOR DISEASES OF THE GALLBLADDER, PANCREAS, AND LIVER

Most gallstones are made primarily of cholesterol; they can be as small as a grain of sand or as large as a ping-pong ball.

As Chapter 22 mentioned, disorders of the pancreas and liver can lead to malabsorption, especially fat malabsorption. The **biliary tract,** which includes the gallbladder and the bile ducts, conducts bile from the liver to the gallbladder and from the gallbladder to the intestine. For these reasons, disorders of the biliary tract can also lead to malabsorption. As Figure 23–1 shows, the pancreatic duct joins the common bile duct to form a common entry into the small intestine. When the bile ducts become obstructed, damage to the pancreas and liver can follow, and this damage can lead to malabsorption as well. This chapter begins with a discussion of gallstones and continues with a look at disorders of the pancreas and liver.

Gallstones

Diet Order
Safe, weight-loss diet.

The liver produces bile, which consists primarily of cholesterol but also contains bile salts, bile pigments (bilirubin), and water. Bile then travels through the bile ducts to the gallbladder, where it remains until it is needed for fat digestion. During bile's time in storage, water is slowly extracted from it. The hardened material may eventually clump together and form a **gallstone.** The formation or presence of stones in the gallbladder, or **cholelithiasis,** affects about 20 million Americans, or about 10 percent of the population.

Types of Gallstones

Two types of gallstones may form and are distinguished by their composition. The vast majority of gallstones (about 80 percent of all cases) contain primarily cholesterol; the rest (20 percent) contain mainly calcium salts and bilirubin.

Cholesterol Gallstones   Most clinicians believe that cholesterol gallstones form when bile contains too much cholesterol, when substances present in bile allow cholesterol to crystalize, or when the gallbladder does not contract forcibly enough to regularly empty its contents. Sluggish motility in the large intestine may also play a role.[1] The cholesterol in bile precipitates as small crystals, which gradually fuse together and grow. Stones can be tiny specks or as large as ping-pong balls, although most stones are less than an inch in diameter.

Pigment Stones   Pigment stones form when excess bile pigments crystalize and grow. Pigment stones are more likely to develop in people with cirrhosis (described later in this chapter), biliary tract infections, or some hereditary blood disorders, including sickle-cell anemia. The remainder of this discussion focuses on the most common type of stone—cholesterol stones.

Causes of Gallstones

Although the exact causes of gallstones are unknown, some people have a greater chance of developing gallstones than others. Native Americans and Mexican Americans, for example, have a high rate of gallstones. The three most important risk factors, however, are body weight, gender, and age.

Body Weight   Obesity, especially in women, raises the likelihood of gallstones. People who are obese have a three to seven times greater risk of developing gallstones than people of desirable weight. Obese people produce too much cholesterol in their livers and then excrete more in bile.

biliary tract: the gallbladder and bile ducts.

gallstone: a hard mass formed when crystals of cholesterol or bile pigments precipitate together.

cholelithiasis (KOH-lee-lih-THIGH-ah-sis): the formation or presence of stones in the gallbladder or common bile duct.
  chole = bile
  lith = stone
  iasis = condition

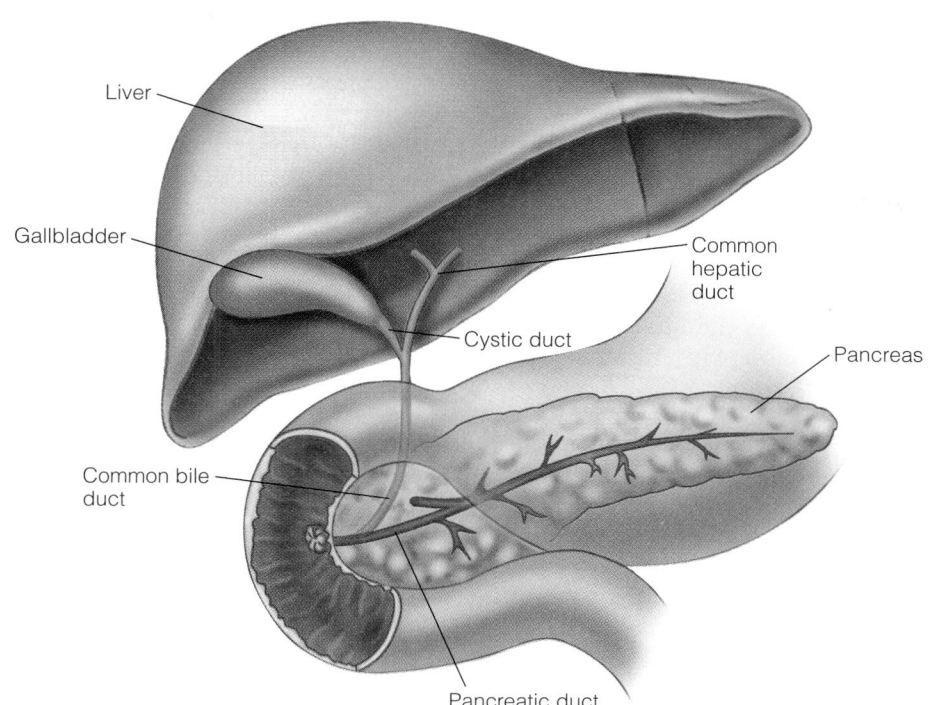

Figure 23–1
The Gallbladder, Pancreas, Liver, and Biliary Tract

Liver

Gallbladder

Common hepatic duct

Cystic duct

Pancreas

Common bile duct

Pancreatic duct

Weight loss can also raise the risk of gallstones, particularly if the person fasts or quickly loses a lot of weight. Gallstones are common following surgeries for obesity and also for people who follow diets of less than 800 kcalories per day. It may be that very-low-kcalorie diets fail to provide sufficient fat to stimulate efficient emptying of the gallbladder.[2] It is unclear whether rapid weight loss leads to gallstones or increases the likelihood that **silent stones** will become problematic. Dieting may shift the balance of the cholesterol and bile salts excreted in bile in a way that favors the formation of gallstones. When people go for long periods without eating (as many dieters do), the gallbladder may not contract enough to dispel its contents, allowing more time for gallstones to grow.

Gender   Women between the ages of 20 and 60 have twice the risk of developing gallstones as men. Women who are pregnant, use estrogen replacement therapy, or take birth control pills also have a higher risk of developing gallstones. Extra estrogen increases the amount of cholesterol in bile and may also disrupt the gallbladder's contractions. In women, but not in men, serum levels of ascorbic acid relate inversely to the prevalence of gallbladder disease.[3] (Ascorbic acid plays a role in the catabolism, or breakdown, of cholesterol.)

Age   The risk of gallstones increases with age for both men and women. By age 60, 10 percent of men and 20 percent of women have gallstones.[4]

Other Relationships   Gallstones are more frequent in people with diabetes, those who use cholesterol-lowering medications, and those on total parenteral nutrition. People with Crohn's disease, especially those who undergo multiple intestinal resections, are also more likely to develop gallstones.[5] People with short-bowel syndrome commonly experience gallstones; often the gallbladder is removed during extensive small bowel resections as a preventive measure.

Ironically, a high-cholesterol diet alone does not seem to increase the likelihood of gallstones. (Excessive kcalories, and consequently obesity, present the greatest risk.) Other dietary factors have also been implicated in the development of

silent stones: gallstones that cause no symptoms.

Energy-, Fat-, and Protein-Modified Diets for Diseases of the Gallbladder, Pancreas, and Liver

gallstones, but whether altering these factors can prevent gallstones is unknown. Among these factors, soluble dietary fibers may protect against gallstones by reducing the concentration of cholesterol in bile.[6] Simple sugars, alcohol, and, possibly, diets rich in animal fats and low in vegetable fats may favor the development of gallstones.[7] More research is needed to determine if specific dietary recommendations, other than avoiding obesity, can reduce the risk of gallbladder disease.

Consequences of Gallstones

Most often, gallstones cause no symptoms (silent stones) and require no treatment. Once symptoms develop, however, the symptoms persist and occur more frequently. The longer a stone remains in the gallbladder, the greater the likelihood that symptoms will develop.

Common Symptoms   When gallstones become symptomatic, they cause a steady, intense pain in the upper abdomen that increases in intensity and lasts from 30 minutes to several hours. Sometimes the pain spreads to the chest and shoulders and mimics the symptoms of a heart attack. Indigestion, nausea, vomiting, gas, and bloating may also occur. For some, symptoms begin following a fatty meal; for others, the pain occurs at night and awakens them from sleep. When a stone enters a bile duct, the pain that results is called **biliary colic.**

Immediate medical attention is required for anyone who has pain that does not resolve over time; develops fever, chills, severe nausea and vomiting, or jaundice; or excretes clay-colored stools. In such cases, the gallstones may be causing serious problems.

Serious Complications   If a gallstone lodges in the bile duct (**choledocholithiasis**) as it is expelled from the gallbladder during a contraction, it can obstruct the flow of bile and lead to inflammation of the gallbladder (**cholecystitis**) or the bile duct itself (**cholangitis**). If the stone obstructs the pancreatic duct, acute pancreatitis (described next) can follow. If the bile ducts are blocked over the course of years, the liver cells may be progressively destroyed, and liver failure (described later in this chapter) can follow. Elderly people and people with diabetes may develop serious complications without first experiencing the symptoms of gallstones.[8]

Prevention and Treatment of Gallstones

The health risks of obesity (see Chapter 7) often outweigh the risks of gallstones during weight loss. Nevertheless, it makes sense to avoid unnecessary risks. Although research is still needed to back up this recommendation, it seems prudent for those undertaking a weight-loss program to follow a plan that allows for a gradual weight loss, contains some fat, and provides meals throughout the day with no long periods of fasting.

As described earlier, no treatment is necessary for gallstones that do not produce symptoms. Medications and surgery can be used to treat gallstones once they become a problem.

Medications   In some cases, a medication called ursodiol, a natural bile salt, can dissolve a small gallstone by reducing the amount of cholesterol the liver adds to the bile (see also the Diet-Drug Interactions box on p. 584). Ursodiol can also help prevent the formation of gallstones for people on weight-loss diets. Other medications used in the treatment of gallstones include antiemetics to reduce nausea and vomiting, analgesics to relieve pain, and sometimes antibiotics.

Surgery   The treatment for most symptomatic gallstones is surgical removal of the gallbladder (**cholecystectomy**). Cholecystectomy is the most common surgery performed in the United States, with over 500,000 people undergoing the procedure each year.[9]

biliary colic: pain associated with gallstones that have entered the common bile duct.

choledocholithiasis (koh-LED-oh-koh-lih-THIGH-ah-sis): the presence of gallstones in the common bile duct.

cholecystitis (KOH-lee-sis-TIE-tis): inflammation of the gallbladder.

cholangitis (KOH-lan-JYE-tis): inflammation of the bile ducts.

cholecystectomy (KOH-lee-sis-TEK-toe-mee): surgical removal of the gallbladder.

Medical Nutrition Therapy for Gallstones For people who develop symptoms following a fatty meal, a low-fat diet is appropriate. Those who need to lose weight benefit from a safe weight-loss plan that provides for a gradual weight loss and encourages regular meals. Once the gallbladder is removed, bile flows directly from the liver into the intestine. Although most people experience no symptoms once they recover from surgery, others continue to experience nausea, bloating, gas, and abdominal pain; others develop diarrhea from the continuous secretion of bile into the small intestine. Certain foods, especially fatty or spicy foods, trigger pain for some people, and if so, these foods should be avoided.

> *In Summary* Gallstones are the most common disorder affecting the biliary tract, and surgery to remove the gallbladder is the most common surgery in the United States. Gallstones can become serious if a gallstone lodges in the bile duct and obstructs the flow of bile. Inflammation of the bile duct, gallbladder, pancreas, and liver is possible. Obesity and rapid weight loss are among the diet-related factors that increase the likelihood of gallstones.

The acinar cells of the pancreas produce digestive enzymes, which then travel through the pancreatic duct to the intestine.

Pancreatis

The pancreas secretes the enzymes necessary for the digestion of protein, fat, and carbohydrate, together with bicarbonate-rich juices that provide the optimal pH necessary to activate these enzymes. Normally, the pancreas stores digestive enzymes in an inactive form to protect itself from digestion. In **pancreatitis,** however, the pancreas becomes inflamed, digestive enzymes are activated within the pancreas, and the enzymes damage the pancreas itself. The blood picks up some of these enzymes; thus serum amylase and lipase rise and serve as indicators of pancreatitis.

Reminder: The pancreas has both exocrine (the secretion of digestive enzymes and pancreatic juices) and endocrine functions (the secretion of insulin and glucagon). Chapter 24 provides more information about the pancreas's endocrine functions.

Acute Pancreatitis

Diet Order
Progress diet from liquids to a low-fat diet to a regular diet as tolerated.

Acute pancreatitis most often develops as a consequence of gallstones or alcoholism (see Nutrition in Practice 23); sometimes, though, the reasons are unclear because a variety of medical conditions, including some infections, and some drugs can also precipitate pancreatitis. Sudden, severe abdominal pain, nausea, vomiting, and diarrhea often accompany acute pancreatitis. Mild cases may subside in a few days, but in other cases, impaired pancreatic function may persist for weeks or months. In severe cases, life-threatening complications including shock and multiple organ failure (see Chapter and Nutrition in Practice 27), pancreatic hemorrhages, fistulas, abscesses, peritonitis, and sepsis may develop. Malnutrition and chronic pancreatitis can also follow severe cases.

Medical Nutrition Therapy for Acute Pancreatitis Initially, food is withheld because foods stimulate pancreatic secretions. Fluids and electrolytes are provided intravenously. In some cases, a tube is inserted into the stomach to suction gastric secretions and help relieve pain and distention (gastric decompression). Oral intake begins when abdominal pain subsides and serum amylase returns to normal or near-normal levels. The diet progresses from liquids to a fat-restricted diet and, eventually, a regular diet as tolerated. Alcohol is restricted to prevent further damage to the pancreas.

Tube Feedings and Enteral Nutrition Evidence suggests that infections, shock, and multiple organ failure associated with severe acute pancreatitis sometimes arise from bacteria in the intestine and that delivering tube feedings of easy-to-

pancreatitis (PAN-cree-ah-TIE-tis): inflammation of the pancreas.

Energy-, Fat-, and Protein-Modified Diets for Diseases of the Gallbladder, Pancreas, and Liver

digest (hydrolyzed) formulas directly into the jejunum early after severe pancreatitis develops may limit the risks of infection and multiple organ failure.[10] In some cases, parenteral nutrition may be necessary, especially when fistulas develop or when the GI tract is immobile (adynamic ileus) for long periods of time.

Chronic Pancreatitis

When severe pancreatitis or repeated episodes of pancreatitis permanently damage the pancreas, absorption, especially of fat, becomes permanently impaired. Chronic pancreatitis is most commonly associated with alcoholism. Abdominal pain is often severe and unrelenting, vomiting is frequent, and severe weight loss and malnutrition are common.

Diet Order
High-kcalorie,
high-protein diet.

Medical Nutrition Therapy for Chronic Pancreatitis   Medical nutrition therapy aims to maintain optimal nutrition status, reduce malabsorption, and avoid subsequent attacks of acute pancreatitis. High-kcalorie, high-protein diets containing normal amounts of fat are frequently prescribed. Enzyme replacements taken with meals help the person digest and absorb protein and fat while minimizing steatorrhea. Clients are strongly cautioned to avoid alcohol.

Other Complications   Some clients with chronic pancreatitis also develop glucose intolerance, diabetes, or hypoglycemia. In these cases, clients' diets require additional adjustments (see Chapter 24). Furthermore, clients who develop chronic pancreatitis as a consequence of alcohol abuse may have multiple nutrient deficiencies and alterations in metabolism, and these require additional dietary adjustments (see Nutrition in Practice 23). The case study on p. 583 describes a client with chronic pancreatitis.

Cystic Fibrosis

Diet Order
High-kcalorie,
high-protein diet.

Until recently, few infants born with **cystic fibrosis,** the most common fatal genetic disorder of the white population, survived to adulthood. Now, with early detection and advances in medical therapy and nutrition care, the outlook is much brighter, with some surviving into their forties and even fifties.

Consequences of Cystic Fibrosis

People with cystic fibrosis produce secretions of thick, sticky mucus that may seriously impair the function of many organs, notably the lungs and pancreas. Many complications may result. Those with nutrition implications include chronic lung disease, malabsorption, and the loss of electrolytes in the sweat.

Respiratory Infections   Labored breathing and stagnation of the thick mucus in the bronchial tubes provide an ideal environment for bacteria to multiply. Lung infections occur frequently, and they further damage lung tissue and raise nutrient needs in people with cystic fibrosis.

Malabsorption   Thick mucus can damage the pancreas, bile ducts, and liver and interfere with the secretion of digestive enzymes, pancreatic juices, pancreatic hormones, and bile. The damage often becomes worse with time. Cystic

cystic fibrosis (SIS-tic fie-BRO-sis): a hereditary disorder characterized by the production of thick mucus that affects many organs including the pancreas, lungs, liver, heart, gallbladder, and small intestine.

fibrosis is believed to cause some degree of pancreatic insufficiency in all cases, with about 85 to 90 percent of cases serious enough to require enzyme replacement therapy. The endocrine cells of the pancreas may also be affected, and diabetes mellitus is a frequent complication.

Medical Nutrition Therapy for Cystic Fibrosis

Although the person with cystic fibrosis may have a hearty appetite, meeting the high energy and nutrient needs imposed by labored breathing, repeated infections, and the loss of nutrients through malabsorption presents a considerable challenge. With such high energy needs, fat restrictions are inappropriate. An unrestricted high-kcalorie, high-protein diet carefully tailored to individual needs and tolerances supports nutritional health, while enzyme replacements help control steatorrhea and relieve abdominal pain. Multivitamin and fat-soluble vitamin supplements are routinely recommended. The liberal use of table salt is encouraged to make up for losses of electrolytes in the sweat. Encouraging fluid intake is also important. Fluids help liquefy thick secretions and prevent dehydration.

Infants with Cystic Fibrosis Breast milk, standard infant formulas, and hydrolyzed infant formulas can all meet the nutrient needs of infants with cystic fibrosis, provided that enzyme replacements are given as well. Breast milk is low in sodium, so the breastfed infant is also given additional table salt mixed with water to replace electrolytes lost through sweat. Infant formulas contain sufficient sodium to prevent depletion.

Body Weight and Stature For all people with cystic fibrosis, every effort is made to maintain an appropriate weight for height. If weight falls to between 85 and 90 percent of desirable weight for height, the diet should also include high-kcalorie snacks and formula supplements to meet energy needs. A client whose weight falls below 75 percent of desirable weight for height may benefit from a home nutrition program that includes an oral diet during waking hours and tube feedings at night.

Malnutrition characterized by poor linear growth in children with cystic fibrosis is associated with reduced life expectancy.[11] Shorter people with cystic fibrosis are more likely to die before taller people with the disorder, suggesting that every effort should be made to ensure that children with cystic fibrosis are growing at an optimal rate.

In Summary Disorders of the pancreas, especially chronic pancreatitis and cystic fibrosis, frequently lead to malabsorption and malnutrition. In acute pancreatitis, food is withheld until inflammation subsides, and then the diet is progressed from liquids to a low-fat diet and then to a regular diet as tolerated. In chronic pancreatitis and cystic fibrosis, a fat-restricted diet is not appropriate. Instead enzyme replacements are given to reduce malabsorption.

Fatty Liver and Hepatitis

The liver is the metabolic crossroads of the body, and its health is crucial, not only to digestion and absorption, but to every body function. Table 23–1 on p. 572 summarizes some of these functions. **Fatty liver** and **hepatitis** are two of the more common signs of liver dysfunction. Both inflame and enlarge the liver. Dietary factors may play a role in the development of both disorders, although both may also arise from causes unrelated to diet.

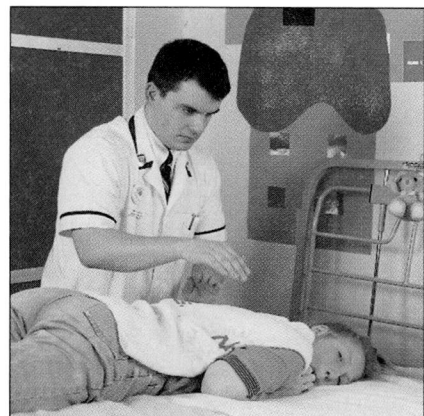

Chest physical therapy (postural drainage) promotes maximal lung function for people with cystic fibrosis.

fatty liver: an accumulation of triglycerides in the liver resulting from many disorders, including exposure to excessive alcohol, excessive weight gain, and diabetes mellitus; also called **hepatic steatosis, steatohepatitis,** and **fatty infiltration of the liver.**

hepatitis (hep-ah-TIE-tis): inflammation of the liver.

hepatic = liver

Energy-, Fat-, and Protein-Modified Diets for Diseases of the Gallbladder, Pancreas, and Liver

Table 23–1 Functions of the Liver

- Metabolizes carbohydrate, protein, and fat so that energy is available to body cells.
- Stores glycogen, most vitamins, and many minerals.
- Manufactures cholesterol, which serves as a precursor for steroid hormones.
- Packages lipids in lipoproteins for transport throughout the body.
- Manufactures bile to aid fat digestion.
- Makes nonessential amino acids and keeps amino acid composition in balance for energy use.
- Converts amino acids to glucose for energy use.
- Converts ammonia from the blood to urea so that it can be excreted by the kidneys.
- Makes plasma proteins including lipoproteins, clotting and immune system factors, and proteins that carry nutrients throughout the body.
- Activates many nutrients.
- Detoxifies drugs (including alcohol) and other substances that are harmful to the body.
- Inactivates hormones.
- Dismantles worn-out red blood cells and recycles the iron they contain.

Fatty Liver

Fatty liver is a clinical finding common to many conditions associated with liver dysfunction. Fatty liver can develop from the liver's exposure to toxic substances such as alcohol (see Nutrition in Practice 23) or as a consequence of excessive weight gain, insulin resistance, or diabetes mellitus. Fatty liver can also develop as a consequence of inadequate intake of protein (as in protein-energy malnutrition, or PEM), an infection or malignant disease, drug therapy (such as therapy with corticosteroids or tetracycline), long-term total parenteral nutrition (TPN), or small bowel bypass surgery. Fatty liver does not arise from eating too much fat alone; excess total kcalories (from either fat or carbohydrate) and high blood glucose levels (as found in insulin resistance and poorly controlled diabetes) appear to have a greater association with fatty liver.

The exact reasons why fats (triglycerides) accumulate in the liver are unknown, but the liver may either synthesize too much fat, use too little for energy, take up too much from the blood, release too little back to the blood, or make a combination of these errors. In severe cases, liver weight may increase from a typical weight of about 3 pounds to as much as 11 pounds, with triglycerides increasing from 5 percent to as much as 40 percent of liver weight. Laboratory findings associated with fatty liver may include elevated serum transaminases (ALT and AST), alkaline phosphatase, bilirubin, glucose, triglycerides, and lipids.

Consequences of Fatty Liver Temporary fatty liver usually causes no harm. Fatty liver associated with TPN, for example, may resolve as the feeding continues, when the feeding is changed to a cyclic infusion, when the energy provided in TPN is reduced, or when TPN is discontinued. In other cases, however, the liver's accumulation of fat can damage liver cells, lead to inflammation, and result in permanent liver damage.

Treatment of Fatty Liver The appropriate therapy for fatty liver focuses on eliminating the cause and reversing its effects. Fatty liver related to obesity and diabetes mellitus requires weight reduction and treatment of elevated blood glucose and blood lipids, as appropriate. Fatty liver caused by malnutrition requires a gradual introduction of a high-kcalorie, high-protein diet adequate in all other nutrients. Fatty liver due to alcohol abuse requires abstinence from alcohol and an adequate diet to replenish nutrient stores. Fatty liver caused by drug therapy requires a change in medications.

Nutrition in Practice 25 describes the development of fatty liver as a consequence of obesity and insulin resistance. This form of fatty liver disease is called **nonalcoholic fatty liver disease.**

The two transaminase enzymes that may be elevated in liver disease are **alanine transaminase (ALT)** and **aspartate transaminase (AST).**

Hepatitis

In hepatitis, inflammation and enlargement of the liver most often occur as a consequence of one of six viral infections. Of the three most common viruses known to cause hepatitis (A, B, C), hepatitis A virus (HAV) is the highly contagious form that can be spread through contaminated foods and water and between people living together, although it usually resolves and does not become chronic. Hepatitis B virus (HBV) and hepatitis C virus (HCV) can lead to serious, permanent liver damage. Some people with HCV also have human immunodeficiency virus (HIV) infections (see Chapter 28). An estimated 1.2 million Americans are believed to be carriers of HBV, and another 3.5 million are carriers of HCV. Other, less common viruses can also cause hepatitis, and hepatitis can occur as a consequence of damage to liver cells by chronic and excessive alcohol ingestion, certain medications and illicit drugs, or other chemical toxins. The dietary supplements chaparral, bee pollen, germander, jin bu huan, ma huang, skullcap, mistletoe, senna, and valerian root have also been reported to lead to hepatitis.[12]

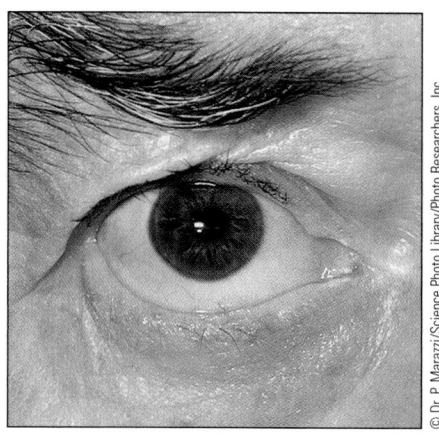

Jaundice is easy to see in the whites of the eyes.

Symptoms of Hepatitis For the person with hepatitis, the symptoms may be mild, and the disease may go undiagnosed. In other cases, flu-like symptoms including fatigue, anorexia, nausea, vomiting, diarrhea or constipation, fever, and pain in the area of the liver may develop. If hepatitis progresses to a severe stage, bilirubin accumulates in the inflamed liver and spills into the blood, causing **jaundice** and producing a dark urine. Serum transaminase levels (AST and ALT) rise, and the liver becomes enlarged and tender.

Consequences of Hepatitis The origin and type of hepatitis, the extent of liver damage, and the person's response to treatment all determine how seriously the disease will affect health. In many cases, liver cells gradually regenerate, and liver function recovers. Recovery from HAV generally occurs over several weeks. Ten percent of people with HBV and 85 percent of those with HCV develop chronic hepatitis, which increases the risk of cirrhosis (described next), liver cancer, and liver failure. Viral hepatitis can also lead to both acute and chronic renal failure (see Chapter 26).[13] Far less frequently, acute hepatitis progresses rapidly and leads to acute liver failure **(fulminant liver failure),** which is often fatal. This is more likely to occur when hepatitis results from medications or chemical agents that are toxic to liver cells.

Treatment of Hepatitis Vaccines to prevent HAV and HBV, but not HCV, infections are available. When hepatitis does occur, treatment aims to reduce further insult to the liver cells; thus the person must abstain from alcohol and avoid taking any unnecessary medications and dietary supplements. An HAV infection generally resolves without medications. HBV and HBC infections generally require treatment with antiviral agents (lamivudine and ribavirin, respectively) and interferon (an immune system modulator). Other cases of severe hepatitis may be treated with anti-inflammatory agents (see the Diet-Drug Interactions box on p. 584).

Medical Nutrition Therapy Liver cells need nutrients to help them recover from hepatitis. The person with hepatitis who is in good nutrition status receives a regular, well-balanced diet. The malnourished person with hepatitis receives a high-kcalorie, high-protein diet to replenish nutrient stores. For the person with mild anorexia, frequent small meals, formula supplements, or both may be helpful. For the person with vomiting, intravenous replacement of fluids is important. Parenteral nutrition is an alternative if the person is malnourished and vomiting persists.

In Summary Fatty liver develops when the liver is unable to process fats as it normally does. The treatment of fatty liver depends on the cause, and many of the causes are nutrition related. Hepatitis can develop as a consequence of several viral infections or from alcohol abuse or the

jaundice (JON-dis): a characteristic yellowing of the skin, whites of the eyes, mucous membranes, and body fluids resulting from the accumulation of bilirubin in the blood.

fulminant (FULL-mih-nant) **liver failure:** liver failure that rapidly progresses to a life-threatening stage.

Energy-, Fat-, and Protein-Modified Diets for Diseases of the Gallbladder, Pancreas, and Liver

ingestion of chemicals, including some medications and illicit drugs, that are toxic to liver cells. The goal of nutrition support for hepatitis is to provide the nutrients liver cells need to regenerate.

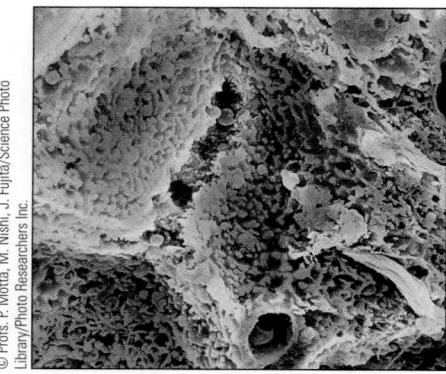

This micrograph of a normal liver shows several hepatocytes (dark reddish-brown), bile ducts (yellowish green), red blood cells (bright red), and a macrophage (yellow).

In a liver damaged by cirrhosis, the hepatocytes lose their structure and flexibility, obstructing the flow of blood and bile through the liver and ultimately preventing the liver from carrying out its functions.

Ingestion of chaparral tea has led to liver failure requiring transplantation. Two other herbs, pennyroyal and comfrey, may also be toxic to liver cells.

cirrhosis (sih-ROW-sis): an advanced form of liver disease in which scar tissue replaces liver cells that have permanently lost their function.

portal hypertension: elevated blood pressure in the portal vein caused by obstructed blood flow through the liver.

collaterals: small branches of a blood vessel that develop when blood flow through the liver is obstructed; also called **shunts.**

varices (VAIR-ih-seez): blood vessels that have become twisted and distended.

Cirrhosis

In **cirrhosis,** generally a consequence of serious chronic liver diseases, liver cells are damaged to the point that they cannot regenerate, and liver function falters and, finally, fails. Scar tissue (fibrosis) forms within the liver, altering the structure of the liver cells and disrupting the blood flow through it. Most often the damage occurs over time, but sometimes the liver can fail suddenly.

Chronic alcohol abuse is the most common cause of cirrhosis (**Laennec's cirrhosis**) in the United States, although not all people with cirrhosis are alcohol abusers, and not all alcohol abusers develop cirrhosis. Other causes of cirrhosis include:

- Infections, including those that lead to chronic hepatitis (**postnecrotic cirrhosis**).
- Progression from fatty liver in people who do not consume excessive alcohol (**nonalcoholic fatty cirrhosis**).
- Diseases of the bile ducts, including **biliary atresia** and **primary biliary cirrhosis,** or obstructions of the bile ducts due to a gallstone or tumor (**obstructive cirrhosis**).
- Later stages of chronic heart failure (**cardiac cirrhosis**).
- Severe reactions to medications and prolonged exposure to toxic chemicals (**toxic cirrhosis**).
- Inherited metabolic disorders, including those that cause excessive accumulation of iron (**hemochromatosis**) or copper (**Wilson's disease**) in the liver or alter the liver's ability to use glucose (**glycogen storage diseases**).

Sometimes cirrhosis develops for no identifiable cause (**idiopathic cirrhosis**). The accompanying glossary defines terms related to various types of cirrhosis.

Consequences of Cirrhosis

Unlike healthy liver tissue, which is soft and flexible, scar tissue is unyielding—a difference with major implications for health. Many people with cirrhosis have no symptoms at first, but as the disorder progresses, symptoms and clinical findings may include nausea, vomiting, weight loss, liver enlargement, edema, jaundice, mental disturbances, and itching from the accumulation of bile pigments under the skin. The liver progressively loses function and, in the end stages, fails. Figure 23–2 on p. 576 shows many of the consequences of cirrhosis described in the sections that follow.

Portal Hypertension The portal vein and the hepatic artery carry 11½ quarts of blood every minute to the miles of intermeshed blood vessels within the liver. This huge volume of blood cannot pulse easily through the scarred tissue of a cirrhotic liver. Consequently, blood flow to and through the liver decreases, blood backs up, and pressure in the portal vein rises sharply, causing **portal hypertension.**

Collaterals and Esophageal Varices With normal blood flow through the portal vein obstructed, pressure forces some of the blood to take a detour from the portal vein through smaller vessels around the liver. These **collaterals,** or **shunts,** can develop throughout the GI tract, but often occur in the area of the esophagus. As pressure builds, the collaterals become enlarged and twisted, forming **varices.** Esophageal varices bulge into the lumen of the esophagus, and they can

Glossary of Terms Related to Types of Cirrhosis

biliary (BILL-ee-air-ee) **atresia** (ah-TREE-zee-ah): a disorder of infants characterized by absent or injured bile ducts.

cardiac cirrhosis: severe liver damage associated with the later stages of congestive heart failure (see Chapter 25).

glycogen storage diseases: several inherited disorders in which a person lacks one of the enzymes that allow glycogen stores to be utilized efficiently. The accumulation of glycogen in the liver leads to severe liver damage.

hemochromatosis (HE-moe-CROW-mah-toe-sis): an inherited disorder in which a person absorbs too much iron from the intestine and stores too much iron in the liver. The accumulation of iron in the liver leads to severe liver damage.

idiopathic (id-ee-oh-PATH-ic) **cirrhosis:** severe liver damage for which the cause cannot be identified.

Laennec's (LAY-eh-necks) **cirrhosis:** severe liver damage related to alcohol abuse.

obstructive cirrhosis: severe liver damage caused by obstruction of the bile ducts due to a stone or tumor.

nonalcoholic fatty cirrhosis: severe liver damage that develops from fatty liver that is unrelated to alcohol abuse.

postnecrotic cirrhosis: severe liver damage that develops as a complication of chronic hepatitis.

primary biliary cirrhosis: severe liver damage from the gradual destruction of the bile ducts; can lead to chronic pancreatitis and fat malabsorption.

toxic cirrhosis: severe liver damage that results from toxic levels of chemicals.

Wilson's disease: an inherited disorder that can lead to severe liver damage. People with Wilson's disease absorb too much copper from the intestine and have too little of the protein that transports copper from the liver to the sites where it is needed.

---

rupture and lead to massive, and sometimes fatal, bleeding. Blood from ruptured varices anywhere in the GI tract can travel to the intestine and serve as a source of ammonia (described later).

Ascites and Edema Rising pressure in the portal vein forces plasma out of the liver's blood vessels and into the abdominal cavity, causing the abdomen to swell. This accumulation of fluid, or edema, in the abdominal cavity is called **ascites.** At the same time, the diseased liver is unable to synthesize adequate amounts of albumin, which normally creates a pressure that draws fluids from tissues back into the blood. As fluid accumulates in the abdomen, less blood reaches the kidneys, a sign the body interprets as fluid depletion. In response, the body makes more **aldosterone** and **antidiuretic hormone (ADH),** hormones that expand the body's fluid volume by triggering retention of sodium and water in the kidneys. As a result of sodium and water retention, ascites worsens, and edema may spread to all body compartments (peripheral edema). To make matters worse, the diseased liver cannot dispose of aldosterone as it usually does, so aldosterone levels remain high. Thus ascites is a self-aggravating condition. Ascites frequently causes early satiety and nausea and also raises the basal metabolic rate.

Elevated Blood Ammonia Levels Ammonia is a normal but toxic product of protein digestion and metabolism (see Figure 23–3 on p. 577). The healthy liver detoxifies ammonia by removing it from circulation and converting it to urea. As liver disease progresses, ammonia-laden blood bypasses the liver by way of the collaterals, the liver may be unable to detoxify the ammonia it does retrieve, and blood ammonia rises. **Hyperammonemia** disrupts central nervous system function.

Hepatic Encephalopathy and Hepatic Coma Liver diseases can alter mental function (**hepatic encephalopathy**) and lead to **hepatic coma**—a dangerous complication that may develop when liver function deteriorates significantly. The causes of encephalopathy and hepatic coma remain elusive, although elevated blood ammonia appears to play a role. As Figure 23–3 shows, most of the ammonia in the body comes from the GI tract. Thus events that contribute protein to the GI tract (such as GI bleeding) or lengthen the time that protein remains in the GI tract (such as constipation) may raise blood ammonia and increase the risk of coma. Infections, malnutrition, and any acute stress result in

Reminder: Normally, almost all the blood from the intestine passes through the liver; this blood is delivered from the *portal vein.* The *hepatic vein* returns blood from the liver to the heart. The *hepatic artery* delivers oxygen-rich blood from the heart back to the liver.

ascites (ah-SIGH-teez): a type of edema characterized by the accumulation of fluid in the abdominal cavity.

aldosterone (al-DOSS-ter-own): a hormone secreted by the adrenal gland that acts on the kidneys to increase sodium and fluid retention.

antidiuretic hormone (ADH): the hormone secreted by the pituitary gland that increases the reabsorption of water by the kidneys.

hyperammonemia (HIGH-per-AM-moe-KNEE-me-ah): elevated blood ammonia.

hepatic encephalopathy (en-SEF-ah-LOP-ah-thee): mental changes associated with liver disease that may include irritability, short-term memory loss, and an inability to concentrate.
  encephalo = brain
  pathy = disease

hepatic coma: a state of unconsciousness that results from severe liver disease.

Energy-, Fat-, and Protein-Modified Diets for Diseases of the Gallbladder, Pancreas, and Liver

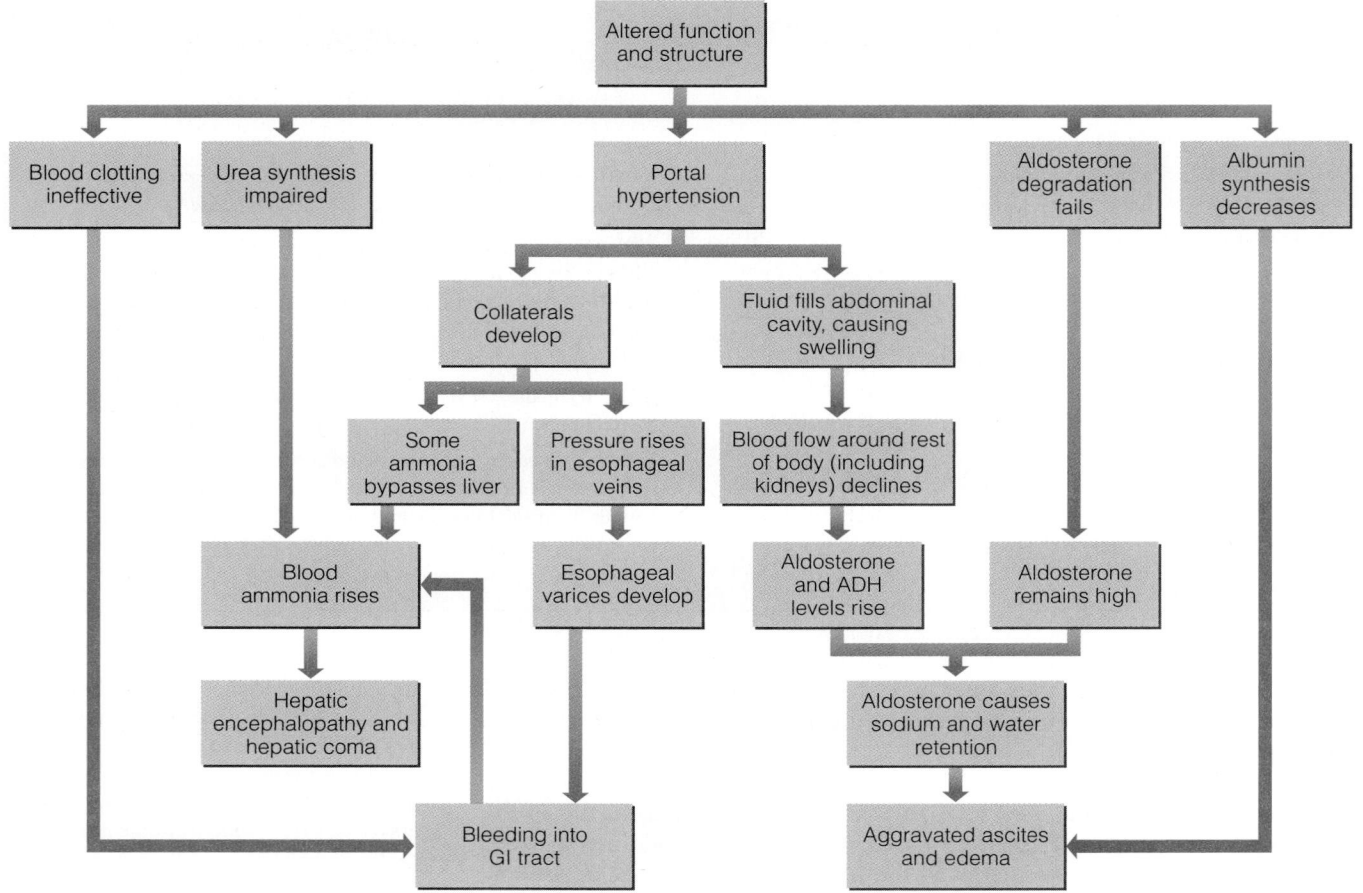

Figure 23–2

The Consequences of Cirrhosis

Sedatives can lead to constipation, and for this reason, sedative use can precipitate hepatic coma.

Reminder: Phenylalanine and tyrosine are aromatic amino acids; leucine, isoleucine, and valine are the branched-chain amino acids.

fetor hepaticus (FEE-tor he-PAT-eh-cuss): a pungent odor of the breath that may develop in people with impending hepatic coma.

flapping tremor: uncontrolled movement of a muscle group causing the outstretched arm and hand to flap like a wing; occurs in disorders that cause encephalopathy; also called **asterixis** (AS-ter-ICK-sis).

Chapter Twenty-Three

protein catabolism, and this too leads to increased ammonia production and risk of coma.

Elevated blood ammonia is associated with encephalopathy and hepatic coma, but the degree of ammonia elevation does not correlate with the severity of symptoms. A possible explanation for this poor correlation is that blood ammonia does not always parallel brain ammonia concentration. When brain ammonia is high, the body produces greater quantities of two substances (glutamine and alpha-ketoglutarate), and the degree of their elevation tends to correlate with the severity of symptoms.

Blood amino acid patterns also change in the later stages of liver disease: aromatic amino acid levels rise, and branched-chain amino acid levels fall. These changes may alter chemicals in the brain or add ammonia to the blood and contribute to encephalopathy and hepatic coma in some people.

People with hepatic encephalopathy may suffer from irritability, short-term memory loss, and an inability to concentrate. People with impending hepatic coma exhibit changes in judgment, personality, or mood. They may be unable to draw even simple shapes, such as a star. A sweet, musty, or pungent odor (**fetor hepaticus**) may develop on the breath. **Flapping tremor** may also develop in the precoma state. Just before passing into a coma, the person becomes very difficult to arouse.

Clotting Abnormalities People with cirrhosis bruise and bleed easily because the damaged liver fails to make an adequate supply of clotting factors. Problems with clotting raise the risk of a fatal hemorrhage if varices develop and bleed. Blood loss in the GI tract also contributes to rising blood ammonia.

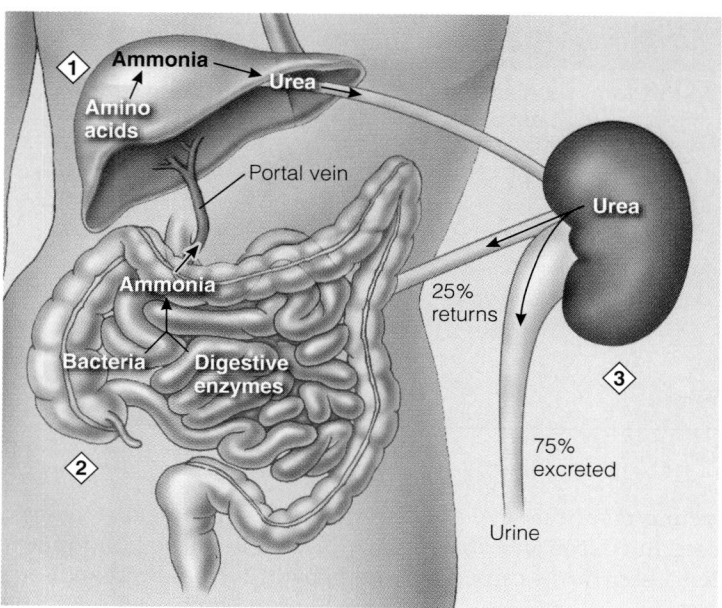

Figure 23-3

Ammonia Production in the Body

❶ The liver makes urea mainly from ammonia formed in the GI tract and also makes some during amino acid metabolism.

❷ Most ammonia in the body derives from the GI tract. Intestinal bacteria make ammonia from undigested protein and urea. Digestive enzymes make ammonia as they dismantle protein from the diet, blood in the GI tract, and shed intestinal cells. Some foods directly contribute to ammonia.

❸ The kidneys excrete about 75 percent of ammonia as urea, but about 25 percent of urea is returned to the intestine, where it can form ammonia again.

Insulin Resistance   Insulin resistance can lead to fatty liver, which may eventually progress to cirrhosis, and people with cirrhosis often develop insulin resistance, hyperinsulinemia, and hyperglycemia, although the reasons for this are unclear. In some cases, the pancreas is unable to produce enough insulin to overcome the insulin resistance, and diabetes mellitus (see Chapter 24) develops.

Malnutrition and Wasting   As liver deterioration progresses, malnutrition and wasting become evident. Table 23–2 on p. 578 summarizes possible causes of wasting in people with liver disorders, which can include anorexia and reduced food intake, altered metabolism, and malabsorption. Fat malabsorption occurs when a liver disorder interferes with bile salt production, when the flow of bile from the liver is obstructed, or when the person develops chronic pancreatitis.

Other Consequences   When liver function deteriorates significantly, the alterations in fluid and electrolyte balances, as well as the buildup of waste products in the blood, can precipitate kidney failure (see Chapter 26) and the **hepatorenal syndrome.** Bone disorders that are generally unresponsive to vitamin D and calcium supplementation can also become a problem, especially for people with cirrhosis and malabsorption (see Chapter 22).

Treatment of Cirrhosis

Treatment of cirrhosis aims to preserve remaining organ function, reversing the damage that has occurred to whatever extent is possible, and to prevent and control complications. Treatment depends, in part, on addressing the underlying medical condition leading to cirrhosis. When cirrhosis has progressed to a severe stage, liver transplantation becomes a treatment option in some cases.

Drug Therapy   One goal of drug therapy in cirrhosis is to control blood ammonia levels. About two-thirds of the body's ammonia comes from the intestine (see Figure 23–3), and drug therapy often includes antibiotics to limit the growth of intestinal bacteria and laxatives to speed intestinal transit time and provide less time for ammonia absorption.

Other medications may include antihypertensives to control portal hypertension and diuretics to reduce fluid retention and prevent ascites and edema. Oral antidiabetic agents may be used to control hyperglycemia. Cholestyramine,

hepatorenal syndrome: the combined symptoms of liver and renal failure that occur as a consequence of severe liver damage.

Energy-, Fat-, and Protein-Modified Diets for Diseases of the Gallbladder, Pancreas, and Liver

Table 23–2   Possible Causes of Wasting in Liver Disease

Reduced Nutrient Intake	Excessive Nutrient Losses	Raised Nutrient Needs
Abdominal pain	Blood loss through GI bleeding	Ascites (raises metabolic rate)
Anorexia	Diarrhea	Infections
Early satiety	Malabsorption	Inflammation
Esophageal varices	Medications	Malnutrition
Medications	Steatorrhea	Medications
Nausea	Vomiting	
Restrictive diets		
Vomiting		

Antidiabetic agents lower blood glucose and *antilipemics* lower blood lipids.

For people with hemochromatosis, treatment may include opening a vein to withdraw blood (phlebotomy), which reduces iron stores by stimulating red blood cell production.

an antipemic, may be used to alleviate the excessive itching that may accompany liver disease. Interferon and ribavirin may be used to improve immune responses in people with cirrhosis caused by viral hepatitis. People with Wilson's disease may be treated with penicillamine, which binds with copper and allows it to be excreted in the urine. The Diet-Drug Interactions box on p. 584 describes the nutrition-related concerns associated with these medications.

Promising treatments for liver disease include the dietary supplement milk thistle (silymarin), which may prove to have a protective effect on liver cells.[14] Milk thistle increases protein synthesis in liver cells and acts as an antioxidant and free-radical scavenger. Preliminary studies suggest that an extract from soybeans, polyenylphosphatidylcholine (PPC), may also protect the livers of people with liver disease associated with alcohol abuse.[15]

Medical Nutrition Therapy   Medical nutrition therapy for people with cirrhosis carefully considers each client's needs for energy, protein, sodium, and fluid (see Table 23–3). Needs vary considerably and depend on the complications present in each case. In all cases, to protect the liver from further injury, clients with cirrhosis must abstain completely from alcohol and avoid the use of unnecessary medications, herbal supplements, and other drugs.

Energy   The compromised liver needs energy to protect its functioning cells and to prevent protein catabolism. People with ascites, malabsorption, or infections have higher energy needs than people without these complications. For people with ascites, energy needs are based on the person's desirable or estimated **dry weight** (weight without ascites) to avoid overestimating energy needs. Energy needs range from 120 to 175 percent of the basal energy expenditure (see Table 17–1 on p. 438).

Protein   For people with cirrhosis who do not show signs of impending coma, protein restrictions are unnecessary, and a high-protein diet (1.0 to 1.5 grams of protein per kilogram of body weight per day) is appropriate.[16] For people who develop signs of impending coma, the cause is investigated, and the underlying condition is treated. In a small number of people (7 to 9 percent), dietary protein can precipitate hepatic encephalopathy.[17] For people who are protein sensitive, the diet plan is low in protein at first, and protein is gradually increased. The final amount of protein that is optimal depends on the person's nutrition status and individual tolerance to protein. For a person who is malnourished, up to 1.5 grams of protein per kilogram of body weight is the goal. The final level of protein the person tolerates may be less than this amount, however. Sometimes medications can be adjusted to control symptoms, allowing a higher protein intake. Vegetable proteins may be better tolerated than animal proteins, perhaps because vegetables contain fewer ammonia-forming constituents and aromatic amino

dry weight: weight after excess fluids have been removed from the body.

Table 23–3 Diet Guidelines for Liver Failure

Energy

- Without ascites, malnutrition, or infection: 120% of basal energy expenditure (BEE); see Table 17–1 on p. 438.
- With ascites, malnutrition, or infection: 150–175% of BEE.

Protein

- Not sensitive to protein: 1.0 to 1.5 g protein/kg body weight/day.
- Protein sensitive: Start with 0.5 to 0.7 g protein/kg body weight/day; increase gradually to level of tolerance with a goal of at least 1.0 g protein/kg body weight/day and up to 1.5 g protein/kg body weight/day if the person is malnourished.
- Very sensitive to protein (unable to tolerate at least 0.8 g protein/kg body weight/day): Consider vegetable-based diets or enteral supplements enriched with branched-chain amino acids and low in aromatic amino acids.

Carbohydrate

- Not restricted.
- For people with insulin resistance or diabetes, provide up to 50 to 60% of kcalories from carbohydrates (mainly complex carbohydrates) with a consistent carbohydrate intake from day to day and at each meal and snack.

Fat

- Not restricted unless fat malabsorption is present.
- With fat malabsorption: Restrict fat only as necessary to control steatorrhea (see Chapter 22); use medium-chain triglycerides (MCT) to increase kcalories as necessary.

Sodium

- Restrict only as necessary to control ascites, but not less than 2 g sodium/day in most cases.

Fluid

- Restrict to 1.0 to 1.5 liters per day for people with ascites who also have low blood sodium levels. In severe cases, limit fluids to 500–750 milliliters plus urinary output/day.

Vitamins and Minerals

- Ensure adequate intake from diet or supplements based on individual needs.

SOURCE: Adapted from J. Hasse and coauthors, Nutrition therapy for end-stage liver disease: A practical approach, *Support Line*, August 1997, pp. 8–15.

© Polara Studios Inc.

The foods shown here provide 60 grams of protein—a day's protein allowance for some people with liver disease.

acids and more branched-chain amino acids than meats do. In addition, the fiber from plant foods speeds intestinal transit time, reducing the amount of time for ammonia absorption from the intestine.

In some cases, the person who is protein sensitive can tolerate only low protein intakes (0.5 to 0.7 gram of protein per kilogram of body weight per day). For these clients, some clinicians recommend special enteral (or parenteral, when needed) formulas that are low in aromatic amino acids and high in branched-chain amino acids (see Appendix H).

If a person lapses into a coma, all protein is withheld temporarily until the cause is determined and corrective measures are taken. For protein-sensitive people, protein is then gradually reintroduced in small amounts and increased gradually as tolerated.

In advanced liver disorders, the liver's inability to activate nutrients can result in deficiencies despite an adequate intake. In that case, providing the nutrient in its active form may prevent deficiencies. One example involves methionine, an essential amino acid. To perform its functions, methionine needs to be activated to S-adenosylmethionine (SAMe) in the liver. Researchers have shown that providing methionine as SAMe can reduce mortality in some children with cirrhosis.[18]

Carbohydrate and Fat   In a diet for liver disease, carbohydrate generally provides 50 to 60 percent of energy needs. A high-carbohydrate (50 grams) snack at bedtime may help prevent excessive breakdown of fat and protein during an

Energy-, Fat-, and Protein-Modified Diets for Diseases of the Gallbladder, Pancreas, and Liver

overnight fast.[19] People with hypoglycemia or hyperglycemia should follow the guidelines for diets in diabetes; that is, eat mostly complex carbohydrates and eat them at consistent times throughout the day.

Because fat helps make foods appetizing and delivers energy efficiently, fat plays an important role in the diet of a person with cirrhosis. Fat is restricted only if the person develops fat malabsorption. In these cases, medium-chain triglycerides (MCT) can provide additional kcalories. People with liver disease related to the biliary tract or those in the later stages of cirrhosis are most likely to malabsorb fat.

Sodium and Fluid For people with ascites, the diet restricts sodium. The lower the sodium intake, the more quickly ascites resolves. Very-low-sodium diets are unpalatable to many people, however, so most clinicians recommend a diet that allows from 2 to 4 grams of sodium and depend on diuretics to mobilize excess fluids. Table 23–4 provides more information about a 2-gram sodium diet.

In people with ascites, the amount of sodium in the body is excessive, even though blood levels are sometimes low. In these cases, the low blood sodium levels (**hyponatremia**) signify fluid overload. Thus, when people with ascites have low blood sodium, adding sodium to the diet is inappropriate; additional sodium leads to further fluid retention. Instead, fluids may be restricted to about 1.0 to 1.5 liters per day. To assess changes in fluid balance, health care professionals monitor weight and measure abdominal girth. Rapid weight gain and an increasing abdominal girth indicate fluid retention; sudden weight loss and decreasing abdominal girth indicate successful fluid excretion.

Vitamins The liver's central role in the metabolism and storage of vitamins and minerals, combined with coexisting conditions (such as malabsorption, alcoholism, and malnutrition), explains why nutrient deficiencies commonly occur in people with liver diseases. Virtually all people with advanced liver disease require supplements of some vitamins and minerals.

The B vitamins serve as cofactors for the liver's many metabolic reactions and repair work; deficiencies of thiamin, vitamin B_6, riboflavin, and folate are common. Fat-soluble vitamins may need to be supplemented if fat malabsorption develops. If the diseased liver fails to synthesize adequate amounts of retinol-binding protein, body tissues may not receive the vitamin A they need. Vitamin D deficiencies may develop if the liver is unable to perform its roles in the activation of vitamin D. Although the bone diseases associated with liver disease are generally not responsive to vitamin D supplements, supplements are provided to ensure adequate vitamin D intakes. Vitamin K deficiencies can prolong the time it takes for blood to clot, a dangerous complication that increases the risk of massive bleeding from the GI tract. (Recall that liver damage itself can interfere with blood coagulation.)

Other Minerals Calcium deficiencies can develop from three causes in people with cirrhosis: steatorrhea, low serum albumin (albumin, which carries calcium in the blood, is manufactured in the liver), and impaired vitamin D metabolism. Thus it is important that the diet or supplements provide adequate calcium to prevent deficiencies. Magnesium deficiencies are also common. Zinc stores may also be depleted as a consequence of liver damage. Limited evidence suggests that zinc may help prevent the muscle cramps frequently experienced by people with liver disorders.[20]

Enteral and Parenteral Nutrition People with cirrhosis face many difficulties in eating enough food to maintain nutrition status. The "How to" box on p. 582 offers tips for maintaining an adequate intake. If the person with cirrhosis cannot take enough food or formula by mouth, tube feedings or TPN is indicated. As mentioned earlier, enteral and parenteral formulas designed for liver disease

Abdominal girth is measured by placing a tape measure around the back and over the person's umbilicus.

Vitamin D is supplied as 25-hydroxyvitamin D (ergocalciferol).

One laboratory test that measures the time it takes for blood to clot is called the **pro-thrombin time.** Both vitamin K deficiency and liver disease can prolong the pro-thrombin time.

hyponatremia (HIGH-poe-nay-TREE-mee-ah): low levels of sodium in the blood.

Table 23–4 Two-Gram Sodium-Restricted Diet[a]

General Guidelines

About 75 percent of sodium in the typical diet comes from processed foods, and about 10 percent comes from unprocessed natural foods. About 15 percent of sodium in a typical diet comes from table salt. With this in mind:

- Choose fresh foods and foods frozen or canned without added salt.
- Avoid adding salt to foods while cooking.
- Avoid adding salt to foods at the table.
- When eating out, ask that meals be prepared without salt.

Sodium in Foods

All foods contain sodium, but some contain more than others. Use the information about the average sodium contents of foods to tailor the diet to the individual's preferences.

Food Group	Serving Size	Sodium (mg) per Serving
Fresh meats, poultry, freshwater fish; low-sodium canned meats and fish; low-sodium peanut butter and cheese; unsalted cottage cheese, soybeans, and textured vegetable protein	1 oz	20
Regular fat-free, low-fat, and whole milk and yogurt	8 oz	120
Eggs	1	60
Fresh artichokes, beets, carrots, and celery; beet, collard, dandelion, mustard, and turnip greens	½ c	50
Regular canned vegetables	½ c	300
Regular white and whole-grain bread	1 slice	150
Butter and margarine	1 tsp	50
Salt	½ tsp	1000

Other Foods

These foods can be used freely or with some limits with respect to sodium, although some of these foods may need to be restricted for weight or blood lipid control:

- All fruits and fruit juices.
- Low-sodium canned or frozen vegetables without added salt except those listed above; low-sodium vegetable juices.
- Low-sodium bread and bread products; puffed rice and wheat and shredded wheat cereals; rice; pasta.
- Soups, casseroles, and recipes made with allowed foods and ingredients.
- Unsalted butter, margarine, nuts, and gravy; low-sodium mayonnaise and salad dressing; shortening.
- Low-sodium catsup, mustard, tabasco sauce, and other condiments; low-sodium baking powder.

These foods and dishes prepared with them are high in sodium and should be avoided:

- Cured, canned, salted, or smoked meats, poultry, and fish such as bacon, luncheon meats, corned beef, kosher meats, and canned tuna and salmon; imitation fish products, salted textured vegetable protein, peanut butter, and nuts.
- Buttermilk, regular cheeses.
- Maraschino cherries; crystallized or glazed fruits; dried fruits with sodium sulfite added.
- Pickles, pickled vegetables, sauerkraut, and regular vegetable juices.
- Instant and quick-cooking hot cereals; commercial bread products made from self-rising flour or cornmeal; salted snack foods.
- Salt pork and bacon; commercial salad dressing; olives; regular gravy; catsup, baking powder, soy sauce, bouillon.

A Sample Two-Gram Sodium-Restricted Diet

Using the information above, many diet plans that meet individual needs are possible. Using the guidelines for a heart-healthy diet, a typical plan for a day might look like this:

Food Group	Sodium (mg)
Meat, 6 oz (6 × 20 mg)	120
Milk, 2 c (2 × 120 mg)	240
Fruit, 3 servings	negligible
Vegetables, ½ c vegetables with some sodium (1 × 50)	50
Vegetables, other vegetables and legumes, 2 servings	negligible
Whole-grain bread, 6 slices (6 × 150 mg)	900
Salted margarine, 6 servings (6 × 50 mg)	300
Total	1610

Clients can use the remainder of the sodium allowance for whatever foods they choose. The sodium content of other foods can be determined by reading food labels or using food composition tables. One client may choose to use some table salt or a favorite food that contains sodium.

[a]The 2400-milligram sodium diet described in Chapters 24 and 25 is more liberal than the diet shown here. The 2400-milligram sodium diet requires that a person choose unprocessed foods as much as possible and avoid table salt and highly salted foods.

How to

Help Clients with Cirrhosis Eat Enough Food

*P*eople with cirrhosis often have difficulty eating enough food to prevent malnutrition and its consequences. Ascites and gastrointestinal symptoms such as nausea and vomiting can interfere with food intake. The person with encephalopathy may be confused about what to eat or have little interest in eating. People with sodium restrictions may have difficulty adjusting to a low-salt diet. To facilitate diet compliance:

■ Individualize the diet plan based on each person's symptoms and responses. The diet should restrict protein, fat, sodium, or fluid only when such restrictions are warranted.

■ Use the suggestions for improving food intake in the "How to" box in Chapter 17.

■ Advise clients to eat frequent small meals and to eat a high-carbohydrate bedtime snack. Eating frequent small meals can help with nausea and can also improve glucose tolerance. For people who are protein sensitive, eating small amounts of protein frequently throughout the day improves tolerance.

■ Point out that there is probably no food that cannot be incorporated into the diet on special occasions and in limited amounts. (The same advice does not apply to alcohol, which is a drug, not a food.) Advise clients to talk with the dietitian about how to change

their meal plans so that favorite foods can be incorporated from time to time.

Some suggestions that can help clients adhere to their sodium restrictions include:

■ Suggest that clients replace the salt they use for cooking and seasoning with herbs or spices like basil, bay leaves, curry, garlic, ginger, lemon, mint, oregano, rosemary, and thyme.

■ Suggest that clients experiment with low-sodium products to find the ones they like. (As Chapter 25 describes, people on potassium-sparing diuretics should be cautioned to avoid salt substitutes that replace sodium with potassium.)

■ Advise clients to check food labels to learn the sodium content of the foods they eat. They may be able to find similar products with lower sodium contents.

Continue to offer support and encouragement to the client with cirrhosis. Severe weight loss is less likely to occur if nutrition intervention is provided before problems progress too far. A box in Chapter 26 provides suggestions that may be helpful to clients with cirrhosis who must restrict fluids.

provide fewer aromatic and more branched-chain amino acids than standard formulas (see Appendix H). These formulas are costly and are generally reserved for people highly sensitive to protein.

People with bleeding esophageal varices will be unable to consume food by mouth and are often given simple intravenous solutions to maintain fluid and electrolyte balances. When they begin to eat, they are frequently given liquid, and then soft, foods. Parenteral nutrition should be considered if the person is malnourished or if the health care team anticipates that oral intake will not resume for an extended time. The case study on p. 583 asks you to use clinical knowledge and judgment in answering questions about cirrhosis.

In Summary In cirrhosis, a complication of a variety of disorders, liver cells undergo permanent changes, and function falters. Consequences of liver failure may include portal hypertension, varices in the esophagus and GI tract, ascites and edema, elevated blood ammonia levels, hepatic encephalopathy, hepatic coma, clotting abnormalities, insulin resistance, malnutrition, renal failure, and bone disease. Therapy often includes medications and a carefully tailored diet that considers the client's energy, protein, sodium, fluid, vitamin, and mineral needs. Once liver failure reaches its final stages, liver transplantation is necessary to sustain life.

Case Study

CARPENTER WITH CHRONIC PANCREATITIS AND CIRRHOSIS

Mr. Sloan, a 48-year-old carpenter, has been hospitalized many times and has been diagnosed with chronic pancreatitis and alcoholic cirrhosis. He recognizes his problem with alcohol abuse and has entered alcohol rehabilitation programs several times over the last few years. Nevertheless, he is still drinking. Mr. Sloan was recently admitted to the hospital with advanced liver disease and signs of impending hepatic coma. At 5 feet 7 inches tall, Mr. Sloan, who once weighed 150 pounds, now weighs 120 pounds. He is jaundiced and looks thin, although his abdomen is distended with ascites. Laboratory findings include elevated AST, ALT, alkaline phosphatase, and blood ammonia.

1. Can you explain to Mr. Sloan what chronic pancreatitis and cirrhosis are and what their consequences are? From the limited information available, what can you determine about Mr. Sloan's nutrition status? What medical problem makes it difficult to interpret Mr. Sloan's actual weight? How can his weight measurements help determine if his condition is improving? Calculate Mr. Sloan's energy needs. What factors influence protein needs for a person with impending hepatic coma? What signs suggest that a person is in a precoma state?

2. Why is Mr. Sloan's abdomen distended? Explain the development of ascites in liver disease and how diet is adjusted.

3. Would you expect Mr. Sloan's blood ammonia levels to be high? Why or why not?

4. Describe portal hypertension, jaundice, and esophageal varices. Describe a diet appropriate for Mr. Sloan. Should fat be restricted in his diet? Why or why not? How would Mr. Sloan's diet be changed if he were found to have bleeding esophageal varices?

Liver Transplantation

If liver failure progresses to a severe and irreversible state, liver transplantation becomes an option when an acceptable organ donor can be found. While a liver transplant candidate is awaiting a donor, nutrient imbalances should be identified and corrected to whatever extent possible.

Nutrition before Transplantation   In severe liver failure, malnutrition has often progressed for some time. Not only is malnutrition common in liver transplant candidates, but it also increases the risk of complications following a liver transplant. The person equipped with adequate nutrient stores faces the transplant better prepared to fight infections, heal wounds, and mount a stress response.

Difficulties arise in assessing nutrition status in liver transplant candidates because the metabolic effects of liver dysfunction and those of malnutrition are difficult to distinguish. Edema may mask weight loss and alter other anthropometric measurements. Fluid retention can also interfere with laboratory tests.

Nutrition following Transplantation   Following liver transplantation, liver function determines nutrient needs. All people are hypermetabolic after surgery, and energy needs must be met. In the immediate posttransplant period, energy needs range from about 1.2 to 2.0 times the basal energy expenditure (see Table 17–1 on p. 438) or 30 to 35 kcalories per kilogram of body weight.[21] Immunosuppressant drugs and catabolism following surgery raise protein needs to about 1.2 to 2.0 grams of protein per kilogram of body weight per day. Medications can contribute to nutrient imbalances by causing nausea, vomiting, diarrhea, and mouth sores. The nutrition support team may use indirect calorimetry to estimate energy needs and carefully monitors clinical and laboratory data so that specific nutrient recommendations can be made.

Although TPN has been the traditional source of nutrients following transplantation, tube feedings delivered directly into the intestine have also proved

Energy-, Fat-, and Protein-Modified Diets for Diseases of the Gallbladder, Pancreas, and Liver

Diet-Drug Interactions

Antidiabetic Agents

For *antidiabetic agents,* see Chapter 24.

Antihypertensives

For *antihypertensives,* see Chapter 25.

Anti-infectives

The antibiotic frequently used to control blood ammonia levels in liver disease is *neomycin.* Because neomycin kills bacteria in the intestine, the person may need a vitamin K supplement. The person may be prone to other intestinal infections that ordinarily are suppressed by normal bacterial flora. Neomycin can also be toxic to the kidneys. The antiviral agents *ribavirin* and *lamivudine* may be used in the treatment of liver diseases. Lamivudine can lead to nausea, vomiting, diarrhea, anorexia, and fatigue.

Anti-inflammatory Agents

For *anti-inflammatory agents,* see p. 486.

Antilipemics

For *antilipemics,* see Chapter 25.

Diuretics

Diuretics remove excess fluids from the body. *Spironolactone* is a potassium-sparing diuretic frequently used to treat ascites. People taking spironolactone should avoid foods high in potassium, potassium supplements, and salt substitutes that contain potassium. See also Chapter 25.

Gallstone Solubilizers

Ursodiol can lead to nausea, vomiting, GI distress, diarrhea, and constipation. It should not be taken along with aluminum-containing antacids.

Immunosuppressants

Immunosuppressants are used for people who undergo liver transplants. For *cyclosporine* and a general discussion of immuno-suppressants, see Chapter 26. Another immunosuppressant some-times used to prevent tissue rejection following a liver transplant is *tacrolimus.* Tacrolimus can be given with food to reduce nausea, vomiting, and GI upsets. Other nutrition-related side effects include anorexia, diarrhea, constipation, anemia, hyperglycemia, potassium imbalances, low blood magnesium levels, and ascites.

Interferon

Interferon can lead to nausea, vomiting, weight loss, fever, fatigue, and depression.

Laxatives

The laxative commonly used in the treatment of liver disease is *lactulose.* Lactulose can cause belching, cramps, and diarrhea. Encourage clients to replace fluid and electrolytes lost through diarrhea.

Pancreatic Enzyme Replacements

For *pancreatic enzyme replacements,* see p. 557.

Penicillamine

Penicillamine is used in the treatment of Wilson's disease. Foods decrease the absorption of penicillamine, and it should be taken on an empty stomach. Nutrition-related side effects may include altered taste perceptions, anorexia, nausea, vomiting, diarrhea, vitamin B_6 deficiency, iron deficiency, and loss of protein in the urine. The person should avoid foods and supplements that contain copper. Iron supplements should be taken at least two hours before or after penicillamine is given.

---

successful in meeting nutrient needs. Early enteral nutrition may reduce the incidence of infection, a particularly important consideration for people with suppressed immune systems.

In Summary Halting or delaying the progression of liver dysfunction depends in large part on attention to nutrition and nutrition assessment parameters. Once liver failure progresses, a liver transplant becomes a life-sustaining option. Attention to nutrition needs prior to and after a liver transplant provides the best assurance that a person will be able to heal wounds and fight infections. The Diet-Drug Interactions box describes the nutrition-related concerns associated with the use of the drugs mentioned in this chapter. The accompanying Nutrition Assessment Checklist reviews important points to keep in mind when assessing the nutrition status of people with liver disorders.

Nutrition Assessment Checklist

FOR PEOPLE WITH DISORDERS OF THE GALLBLADDER, PANCREAS, AND LIVER

Medical History

Does the client have a medical diagnosis of:

- ☐ Gallstones?
- ☐ Pancreatitis?
- ☐ Cystic fibrosis?
- ☐ Fatty liver?
- ☐ Hepatitis?
- ☐ Cirrhosis?

Has the client had surgery that includes:

- ☐ Cholecystectomy?
- ☐ Liver transplant?

Does the client have the following symptoms or complications that may alter medical nutrition therapy:

- ☐ Overweight?
- ☐ Intolerance to spicy or fatty foods?
- ☐ Malnutrition?
- ☐ Malabsorption?
- ☐ Infection?
- ☐ Fistulas?
- ☐ Esophageal varices?
- ☐ Ascites?
- ☐ Hepatic encephalopathy?
- ☐ Hepatic coma?
- ☐ Insulin resistance/diabetes mellitus?
- ☐ Renal failure?

Medications

For clients with liver dysfunction, note that the risk of diet-drug interactions is very high because many drugs are metabolized in the liver. The risk of interactions is intensified for clients with:

- ☐ Ascites (medications may take a long time to reach the liver)
- ☐ Renal failure (medications are often metabolized further in the kidneys and excreted in the urine)
- ☐ Malabsorption and malnutrition
- ☐ Multiple medication prescriptions and long-term medication use

Food/Nutrient Intake

For clients with gallstones who are overweight and dieting, note:

- ☐ The total energy in the diet
- ☐ The client's eating habits, especially fasting or skipping meals

For clients with fatty liver, pay special attention to the client's intake of:

- ☐ Energy, if the client is overweight or malnourished, has diabetes, or is receiving TPN
- ☐ Carbohydrate, if the client has diabetes or is receiving TPN
- ☐ Alcohol

For clients with pancreatitis, hepatitis, and cirrhosis, make a note of:

- ☐ Appetite
- ☐ Adequacy of energy and nutrient intake
- ☐ Alcohol use

For clients with protein-sensitive encephalopathy:

- ☐ Ensure that total energy intake is adequate.
- ☐ Find the level of protein restriction that is appropriate for the individual.
- ☐ Base energy needs on desirable or estimated dry weight to avoid overfeeding.

Anthropometrics

Take baseline height and weight measurements and monitor weight regularly. For clients with ascites and edema:

- ☐ Use weight and abdominal girth to monitor the degree of fluid retention.
- ☐ Remember that the client may be malnourished even though weight may be deceptively high.

Laboratory Tests

Check laboratory tests for signs of malnutrition and nutrient deficiencies. Note that albumin and serum proteins are often reduced in people with liver disease and are difficult to interpret as an indicator of nutrition status. Check laboratory test results for complications associated with liver failure including:

- ☐ Anemia
- ☐ Fluid retention
- ☐ Hypoglycemia and hyperglycemia
- ☐ Renal function tests

Physical Signs

Look for physical signs of:

- ☐ Dehydration (from malabsorption)
- ☐ Fluid retention (ascites and edema)
- ☐ PEM (muscle wasting and unintentional weight loss)
- ☐ B vitamin deficiencies
- ☐ Fat-soluble vitamin deficiencies
- ☐ Mineral deficiencies

1. Which of the following diet-related factors is most closely associated with the risk of gallstones?
   a. low-fiber diets
   b. overweight
   c. dietary cholesterol
   d. simple sugars and alcohol

2. The client with chronic pancreatitis benefits from:
   a. following a very-low-fat diet.
   b. moderately restricting alcohol.
   c. eating three meals a day and avoiding snacks.
   d. using enzyme replacements to improve fat absorption.

3. Dietary recommendations for people with cystic fibrosis include:
   a. limited use of table salt.
   b. a fat-restricted diet.
   c. a high-kcalorie, high-protein diet.
   d. fluid restrictions.

4. Which of the following diet strategies would be most appropriate for helping to reverse fatty liver associated with obesity?
   a. weight-loss diet
   b. low-protein diet
   c. fat-restricted diet
   d. fluid- and sodium-restricted diet

5. Which of the following statements about hepatitis is true?
   a. Chronic hepatitis can progress to cirrhosis.
   b. Regardless of the type of hepatitis, symptoms are severe.
   c. People with hepatitis always need high-kcalorie, high-protein diets.
   d. HCV infections are often mild and can be spread through contaminated foods and water.

6. The consequences of cirrhosis are primarily due to:
   a. chronic malnutrition.
   b. chronic alcohol abuse.
   c. liver cell damage and altered hepatic blood flow.
   d. elevated blood ammonia, amino acid, and sodium levels.

7. Esophageal varices are a dangerous complication of liver disease primarily because they:
   a. interfere with food intake.
   b. can lead to massive bleeding.
   c. divert blood flow from the GI tract.
   d. cause portal hypertension and collateral development.

8. For the person with cirrhosis, short-term memory loss and an inability to concentrate are signs of:
   a. coma.
   b. encephalopathy.
   c. hyperammonemia.
   d. hepatorenal syndrome.

9. Medical nutrition therapy for most people with cirrhosis includes diets that are:
   a. restricted in fat.
   b. high in kcalories and low in protein.
   c. based on symptoms and responses to treatment.
   d. restricted in protein, carbohydrate, sodium, and fluid.

10. With respect to vitamins and minerals, people with cirrhosis:
    a. may develop clotting abnormalities associated with vitamin A deficiency.
    b. frequently develop vitamin C deficiencies.
    c. seldom require nutrient supplements.
    d. may develop calcium deficiencies.

11. For the person undergoing a liver transplant:
    a. immunosuppressant drugs seldom alter nutrient needs.
    b. enteral nutrition is contraindicated following the transplant.
    c. attention to nutrition before a transplant provides few benefits.
    d. attention to nutrition before a transplant improves chances of recovery.

Critical Thinking

1. A 24-year-old woman with cystic fibrosis who weighs 96 pounds has a desirable weight of 130 pounds. For this woman, medical nutrition therapy would ideally include:
   a. an unsupplemented high-kcalorie, high-protein diet.
   b. a high-kcalorie, high-protein diet with formula supplements.
   c. a high-kcalorie, high-protein diet during the day, supplemented with tube feedings during the night.
   d. a high-kcalorie, high-protein diet during the day, with parenteral nutrition during the night.

2. A person with ascites is most likely to have which of the following complications associated with liver disease?
   a. portal hypertension
   b. elevated blood ammonia levels
   c. reduced whole-body and blood sodium
   d. insulin resistance and diabetes mellitus

3. All of the following statements are true about individualized diets for people with cirrhosis **except:**
   a. people with ascites need to restrict sodium.
   b. people with steatorrhea may need to restrict fat.
   c. people with ascites and low blood sodium levels need to restrict sodium and fluid.
   d. people who are protein sensitive always need to restrict protein to 0.5 to 0.7 gram of protein per kilogram of body weight per day.

Answers to these questions can be found in Appendix G.

Clinical Applications

1. Consider the ways that malabsorption caused by disorders of the intestines differs from the malabsorption caused by disorders of the pancreas, liver, and biliary tract. First, describe how each of the following disorders can lead to malabsorption: Crohn's disease of the small intestine, celiac disease, short-bowel syndrome, pancreatitis, cystic fibrosis, and cirrhosis. (Example: Crohn's disease of the small intestine inflames the intestinal mucosa and thus reduces the intestinal surface area for absorption.) What are the drawbacks of low-fat diets in the treatment of chronic pancreatitis and cystic fibrosis?

2. Using Table 23–3 on p. 579 and Table 17–1 on p. 438 as guides, calculate the energy and protein requirements of a man with cirrhosis and malnutrition who is not sensitive to protein and is 5 feet 11 inches tall and weighs 160 pounds. How many grams of protein would this man need if he is sensitive to protein and can tolerate between 0.8 and 1.0 gram of protein per kilogram of body weight per day?

Nutrition on the Net

For further study of the topics of this chapter, access these websites.

Find updates and quick links to these and other nutrition-related sites at our website: **www.wadsworth.com/nutrition**

Learn more about preventing liver disorders and obtain information about gallstones, fatty liver, hepatitis, cirrhosis, diseases related to these conditions, and links to other websites by visiting the American Liver Foundation and the Canadian Liver Foundation: **www. liverfoundation.org** and **www.liver.ca**

Check out the facts about all the disorders covered in this chapter, including pancreatitis and cystic fibrosis, by visiting the National Institute of Diabetes & Digestive & Kidney Diseases: **www.niddk.nih/gov**

Discover resources for people with cystic fibrosis by visiting the Cystic Fibrosis Foundation and the Canadian Cystic Fibrosis Foundation: **www.cff.org** and **www.ccff.org**

Uncover resources and support for children with liver diseases and liver transplants by visiting the Children's Liver Alliance: **www.livertx.org**

Review up-to-date information about hepatitis by visiting the Hepatitis Education Project, the Hepatitis Foundation International, and the Hepatitis B Foundation: **www.scn.org/ health/hepatitis/index.htm, www.hepfi.org**, and **www2.hepb.org**

Visit the Center Span Transplant News Network for updates about liver and other organ transplants: **www.centerspan.org**

Notes

1. R. H. Dowling, Review: Pathogenesis of gallstones, *Alimentary Pharmacology and Therapeutics* (supplement 2) 14 (2000): 39–47.

2. D. Festi and coauthors, Review: Low calorie intake and gallbladder motor function, *Alimentary Pharmacology and Therapeutics* (supplement 2) 14 (2000): 51–53.

3. J. A. Simon and E. S. Hudes, Serum ascorbic acid and gallbladder disease prevalence among US adults: The Third Health and Nutrition Examination Survey (NHANES III), *Archives of Internal Medicine* 160 (2000): 931–936.

4. Gallstones: A national health problem (1996), available at **www.liverfoundation.org,** site visited November 8, 2001.

5. M. Fraquelli and coauthors, Gallstone disease and related risk factors in patients with Crohn's disease, *Archives of Internal Medicine* 161 (2001): 2201–2204.

6. W. H. Schwesinger and coauthors, Soluble dietary fiber protects against cholesterol gallstone formation, *American Journal of Surgery* 177 (1999): 307–310.

7. M. Tseng, J. E. Everhart, and R. S. Sandler, Dietary intake and gallbladder disease, *Public Health Nutrition* 2 (1999): 161–172; G. Miscianga and coauthors, Diet, physical activity, and gallstones—A population-based case-control study in southern Italy, *American Journal of Clinical Nutrition* 69 (1999): 120–126.

8. S. Santen, Cholecystitis and biliary colic from emergency medicine, available at **www.emedicine.com/EMERG/topic98.htm,** updated January 23, 2001, site visited March 20, 2001.

9. Gallstones: A national health problem, 1996; National Diseases Information Clearinghouse, Gallstones, available at **www.niddk.nih.gov,** November 1998, site visited November 8, 2001.

10. P. Lehocky and M. G. Sarr, Early enteral feeding in severe acute pancreatitis: Can it prevent secondary pancreatic (super) infection? *Digestive Surgery* 17 (2000): 571–577; R. Anderson and X. D. Wang, Gut barrier dysfunction in experimental acute pancreatitis, *Annals of the Academy of Medicine,* Singapore 28 (1999): 141–146.

11. L. T. Beker, E. Russek-Cohen, and R. J. Fink, Stature as a prognostic factor in cystic fibrosis survival, *Journal of the American Dietetic Association* 101 (2001): 438–442.

12. J. A. Shad, C. G. Chinn, and O. S. Brann, Acute hepatitis after ingestion of herbs, *Southern Medical Journal* 92 (1999): 1095–1097.

13. N. K. Krane and P. Gaglio, Viral hepatitis as a cause of renal disease, *Southern Medical Journal* 92 (1999): 354–360.

14. J. E. Robbers and V. E. Tyler, *Tyler's Herbs of Choice* (Binghamton, N.Y.: Hawthorn Herbal Press, 1999), pp. 76–79.

15. C. S. Lieber, Alcohol: Its metabolism and interaction with

nutrients, *Annual Reviews of Nutrition* 20 (2000): 395–430; C. S. Lieber, Alcoholic liver disease: New insights in pathogenesis lead to new treatments, *Journal of Hepatology* (supplement 1) 32 (2000): 113–128.

16. G. L. Braden, Clinical Gastroenterology: Highlights of the Annual Postgraduate Course, Part II, 1999 American College of Gastroenterology Annual Scientific Meeting, available from Medscape at **http://gastroenterology.medscape.com,** site visited March 17, 2001.

17. Hepatic encephalopathy—Effective treatments available once acute precipitants have been eliminated, *Drug and Therapy Perspectives* 17 (2001): 8–11.

18. Lieber, Alcoholic liver disease: New insights in pathogenesis lead to new treatments, 2000.

19. W. Change and coauthors, Effects of extra-carbohydrate supplementation in the late evening on energy expenditure and substrate oxidation in patients with liver cirrhosis, *Journal of Parenteral and Enteral Nutrition* 21 (1997): 96–99.

20. M. Kugelman, Preliminary observation: Oral zinc sulfate replacement is effective in treating muscle cramps in cirrhotic patients, *Journal of the American College of Nutrition* 19 (2000): 13–15.

21. J. M. Hasse, Recovery after organ transplantation in adults: The role of postoperative nutrition therapy, *Topics in Clinical Nutrition* 13 (1998): 15–26.

Nutrition in Practice

■ ALCOHOL ABUSE: EFFECTS ON NUTRITION STATUS ■

As Chapter 23 described, alcohol abuse is a frequent cause of chronic pancreatitis and liver diseases. When used in excess, alcohol has a toxic effect on many organ systems, and its impacts on nutrition are so profound that they deserve special attention here. The glossary on p. 590 defines terms used in this Nutrition in Practice.

Like all drugs, alcohol—properly termed **ethanol,** the active ingredient of alcoholic beverages—offers both benefits and hazards. Wine, beer, and other alcoholic beverages have been associated with pleasure and relaxation for more than 5000 years. People have always known that these beverages affected their moods, sensations, and behavior. Taken in moderation, alcohol can relax people, reduce their inhibitions, encourage social interactions, and possibly reduce the risk of coronary heart disease (see Chapter 25). Taken in excess, alcohol can be devastatingly destructive. The key to unlocking the benefits of alcohol without unleashing its toxic effects lies with the concept of moderation.

What constitutes moderate alcohol use?

People's tolerances to alcohol differ, and it is impossible to name an exact amount of alcohol that is appropriate for everyone. The amount a person can drink safely depends on genetics, health, gender, body composition, age, and family history. To provide people with a guideline, authorities have attempted to set limits that are acceptable for most healthy adults. This amount is supposed to be enough to elevate mood with a minimum risk of long-term harm to health. An accepted definition of moderation is not more than two drinks a day for the average-sized man and not more than one drink a day for the average-sized woman or for any person age 65 and over. Women and elderly people have less total body water than men and, therefore, higher alcohol concentrations than men after consuming the same amount of alcohol.

A drink is any alcoholic beverage that delivers ½ ounce of pure ethanol:

- 4 to 5 ounces of wine.
- 10 ounces of wine cooler.
- 12 ounces of regular or **light beer.**
- 1½ ounces of hard liquor (80 **proof** whiskey, scotch, brandy, rum, gin, or vodka).

Alcoholic beverages are often served at celebrations and on special occasions.

Doubtless, some people could consume slightly more; others could not handle nearly so much without significant risk.

What happens to alcohol in the body?

Unlike the energy nutrients, which undergo digestion before they can be absorbed, 95 percent of the alcohol consumed is absorbed directly in the stomach or the small intestine. Once inside the bloodstream, alcohol quickly reaches and enters the body cells. Within minutes of ingestion, alcohol reaches the brain where it quickly depresses inhibitions and judgment and impairs motor skills and reaction times. In the liver, alcohol is detoxified so that it can be excreted from the body. The liver performs this function at a fixed rate—that is, the liver can metabolize only so much alcohol at a time, regardless of how much alcohol is present. Therefore, if drinking continues, more and more alcohol reaches the brain. Breathing and the heart rate slow. Death can follow if alcohol depresses breathing too much. Such an occurrence is unlikely, however. More often, the body reacts to heavy drinking by vomiting to expel additional alcohol before it can be absorbed, or people lapse into comas before they can drink a fatal dose.

Nutrition in Practice

What are the long-term effects of alcohol on the body?

Alcohol's rapid and easy access into the cells explains why so many organ systems can be affected by excessive alcohol intakes(see Figure NP23–1.) In addition to the effects on the brain and central nervous system just described, alcohol most often affects the pancreas, liver, GI tract, and cardiovascular system. (The profound consequences of alcohol on the developing fetus are described in Chapter 12.) As Chapter 23 described, chronic, excessive alcohol use can lead to pancreatitis, fatty liver, alcoholic hepatitis, and cirrhosis. Chronic alcohol abuse can also slow GI motility and inflame the cells of the GI tract. In the stomach, alcohol stimulates gastric acid secretions, and reflux esophagitis, nausea, gastritis, and ulcers are more common in people who drink too much alcohol. Alcohol consumption also increases the risks of cancer of the mouth, throat, and esophagus.[1]

Heavy drinkers experience a high incidence of hypertension, heart attacks, arrhythmias, and deterioration of the heart muscle (see Chapter 25). A recent study found that men who consumed more than 14 drinks of hard liquor per week had an 82 percent higher risk of developing type 2 diabetes than men who did not drink alcohol or those who drank it in moderation.[2] Those who consumed more than 21 drinks of any type of alcoholic beverage per week were 50 percent more likely to develop diabetes than others. Drinking less than 14 drinks of alcohol had no effect on the incidence of diabetes, and the relationship between alcohol consumption and type 2 diabetes was not observed in women.

Alcohol raises the likelihood of serious drug-drug and diet-drug interactions. The liver uses the same enzyme system to metabolize some alcohol and certain drugs. When the enzyme system is busy metabolizing alcohol, it cannot metabolize the drugs that depend on the same system. Thus, not only may a drug be slow to work, but it may also build up with the result that its effects are amplified once it gains access to the enzyme system. The more of a drug that is active at one time, the more likely the risk of interactions. Conversely, once a heavy drinker quits drinking, drugs are rapidly metabolized, so larger doses may be required; the change in metabolism make it hard for physicians to determine an appropriate dose.

Other effects of alcohol on the body include sexual dysfunction and an increased susceptibility to lung infections and lung cancer. Women who drink alcohol may have a greater risk of breast cancer than women who do not drink.[3] The more alcohol people drink, the greater their risks of long-term health consequences. For example, although even moderate alcohol use alters brain function, excessive alcohol use shrinks and destroys brain cells.

What is the difference between drinking alcohol regularly and alcohol abuse?

Alcohol abuse refers to patterns of drinking that result in health problems, social problems, or both. **Alcohol addiction,** often called alcoholism or alcohol dependence, refers to a disease that is characterized by abnormal alcohol-seeking behavior that leads to impaired control over drinking. Alcohol abusers and alcohol-addicted individuals experience many of the same harmful effects of alcohol consumption; the distinguishing characteristics of alcohol addiction are physical dependence on alcohol and an impaired ability to

Glossary

alcohol abuse: a pattern of drinking that includes failure to fulfill work, school, or home responsibilities; drinking in situations that are physically dangerous (as in driving while intoxicated); recurring alcohol-related legal problems (as in aggravated assault charges); or continued drinking despite ongoing health or social problems that are caused by or worsened by alcohol use.

alcohol addiction: a pattern of drinking that includes a strong craving for alcohol, a loss of control and an inability to stop drinking once begun, withdrawal symptoms (nausea, sweating, shakiness, and anxiety) after heavy drinking, and the need for increasing amounts of alcohol in order to feel "high"; also called *alcoholism.*

ethanol: the type of alcohol found in wine, beer, and other alcoholic drinks.

euphoria (you-FOR-ee-ah): a sense of happiness and physical well-being.

light beer: beer that contains the same amount of alcohol as regular beer, but with fewer kcalories.

proof: a description of the amount of alcohol in a product. Liquor that is 80 proof is 40% alcohol. Liquor that is 100 proof is 50% alcohol.

Wernicke's (VER-nih-keys) **encephalopathy:** a brain dysfunction and disturbances of motor coordination caused by severe thiamin deficiency; also called **Wernicke's-Korsakoff syndrome.**

Nutrition in Practice

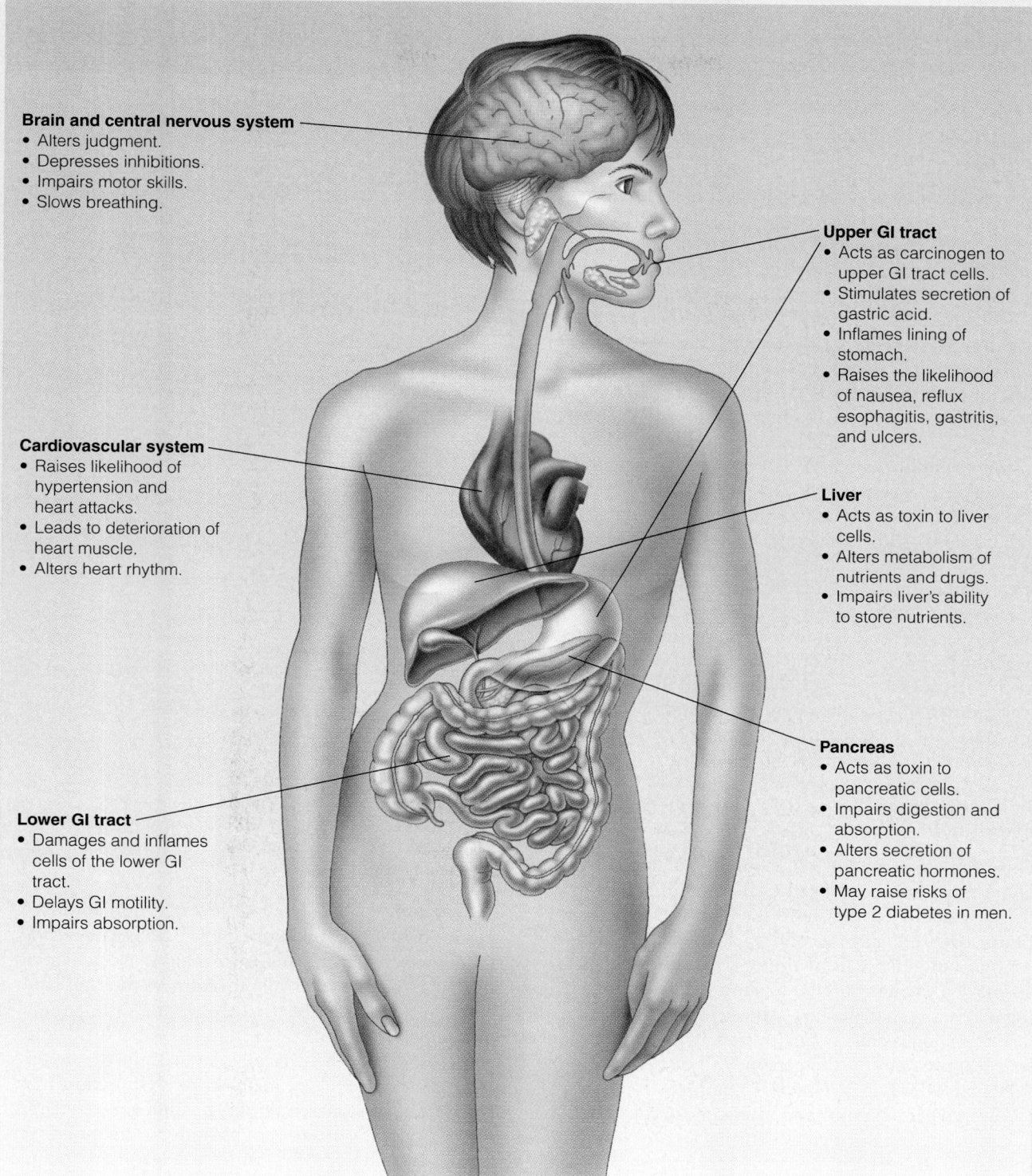

Brain and central nervous system
- Alters judgment.
- Depresses inhibitions.
- Impairs motor skills.
- Slows breathing.

Cardiovascular system
- Raises likelihood of hypertension and heart attacks.
- Leads to deterioration of heart muscle.
- Alters heart rhythm.

Lower GI tract
- Damages and inflames cells of the lower GI tract.
- Delays GI motility.
- Impairs absorption.

Upper GI tract
- Acts as carcinogen to upper GI tract cells.
- Stimulates secretion of gastric acid.
- Inflames lining of stomach.
- Raises the likelihood of nausea, reflux esophagitis, gastritis, and ulcers.

Liver
- Acts as toxin to liver cells.
- Alters metabolism of nutrients and drugs.
- Impairs liver's ability to store nutrients.

Pancreas
- Acts as toxin to pancreatic cells.
- Impairs digestion and absorption.
- Alters secretion of pancreatic hormones.
- May raise risks of type 2 diabetes in men.

Figure NP23–1
Alcohol's Effect on Organ Systems

Nutrition in Practice

control alcohol intake. Alcohol abuse and addiction exert a heavy toll on the health of the 18 million people in the United States who meet the criteria for alcohol abuse, alcohol addiction, or both.

In what ways can alcohol abuse or addiction affect nutrition status?

The effects of alcohol on nutrition and metabolism—both directly and as a consequence of the alcohol-related diseases described earlier—are significant. Every alcohol abuser and alcohol-addicted person should be considered at risk for poor nutrition status.

Alcohol is rich in energy (7 kcalories per gram), but like pure fat or sugar kcalories, the kcalories from alcohol are empty kcalories. People who use moderate amounts of alcohol regularly may gain weight as a result of the extra kcalories they consume. For heavy drinkers, however, weight loss is more likely. Alcohol produces **euphoria**, which depresses appetite, and the more alcohol a person drinks, the less likely that he or she will eat enough food to obtain adequate nutrients. Table NP23-1 shows the kcalories in typical alcoholic beverages. Nutrient deficiencies are an almost inevitable result of alcohol abuse, not only because the person who drinks heavily eats less and obtains fewer nutrients from food, but also because alcohol interferes with the body's absorption, metabolism, and excretion of nutrients. Once alcohol has damaged the liver, GI tract, or cardiovascular system, many additional factors are at work that can severely affect a person's nutrition status, as previous chapters have shown.

In what ways can alcohol affect the body's utilization of nutrients?

One example is alcohol's effect on blood glucose. Hypoglycemia is most likely to develop in people who drink alcohol after they have been fasting for several hours. By-products of alcohol metabolism inhibit the synthesis of glucose from amino acids (gluconeogenesis). When there is no food to supply glucose directly, hypoglycemia may develop. Although alcohol use can lead to hypoglycemia in any person, hypoglycemia can be a real problem for the person with diabetes mellitus (Chapter 24 provides more information).

Even when vitamins and minerals are consumed in adequate amounts, nutrient deficiencies are common in alcohol abusers. The excessive consumption of alcohol interferes with the availability and activation of virtually every vitamin and many minerals. Table NP23–2 lists examples of nutrients commonly affected by chronic, excessive alcohol use. One example is thiamin.

Table NP23–1 kCalories in Alcoholic Beverages and Mixers		
Beverage	**Amount (oz)**	**Energy (kcal)**
Beer		
Regular	12	150
Light	12	78–131
Nonalcoholic	12	32–82
Distilled liquor (gin, rum, vodka, whiskey)		
80 proof	1½	100
86 proof	1½	105
90 proof	1½	110
Liqueurs		
Coffee liqueur, 53 proof	1½	175
Coffee and cream liqueur, 34 proof	1½	155
Crème de menthe, 72 proof	1½	185
Mixers		
Club soda	6	0
Cola	6	75
Cranberry juice cocktail	6	110
Diet drinks	6	1
Ginger ale	6	65
Grapefruit juice	6	70
Orange juice	6	80
Tomato or vegetable juice	6	35
Tonic water	6	70
Wine		
Dessert	3½	110–135
Nonalcoholic	5	10
Red or rosé	5	110
White	5	100
Wine cooler	10	140

To metabolize alcohol, the body requires thiamin. If a person has been drinking heavily and has eaten very little, severe thiamin deficiencies can develop. In addition to a low intake of thiamin from the diet, alcohol abusers experience impaired absorption of thiamin, reduced activation of thiamin to its usable form, and altered ability of the liver to store thiamin. Severe thiamin deficiencies cause major brain dysfunction characterized by disordered thinking, confusion, impaired memory, and disturances of motor coordination (**Wernicke's encephalopathy**), which can lead to permanent brain damage and death if left untreated.

Alcohol's effect on folate is dramatic. When alcohol is present, the body behaves as if it were actively trying to expel folate from its sites of action and storage. The

Nutrition in Practice

Table NP23–2 Vitamin and Mineral Imbalances Associated with Chronic, Excessive Alcohol Consumption

Nutrient	Reasons for Imbalance
Fat-soluble vitamins	
Vitamin A	Limited intake; reduced absorption, activation, and storage; accelerated metabolism and excretion.
Vitamin D	Limited intake and exposure to sunlight; reduced absorption and activation; accelerated metabolism.
Vitamin K	Limited intake and impaired synthesis by bacterial flora; malabsorption.
Water-soluble vitamins	
Thiamin	Limited intake; reduced absorption, activation, and storage.
Folate	Limited intake; reduced absorption, activation, and availability of circulating folate; accelerated excretion.
Pyridoxine	Limited intake and accelerated excretion.
Minerals	
Iron (toxicity)	Excessive intake; enhanced absorption.
Iron (deficiency)	Loss of iron from GI hemorrhages or bleeding from trauma.
Magnesium	Limited intake; reduced absorption; accelerated excretion.
Zinc	Limited intake; accelerated excretion.

Alcohol also acts as a diuretic and promotes water excretion by the kidneys. Important minerals, such as zinc and magnesium, are lost as well. In short, alcohol directly and profoundly affects nutrition status.

Can an adequate diet protect the body from the harmful effects of alcohol?

No. Eating well and even taking supplements of protein, vitamins, and minerals do not protect the drinker. There is no set level of safe drinking where no adverse effects take place. Even just a couple of drinks set in motion the destructive processes described, but if the drinking has been moderate, the next day's abstinence can repair the damage. Someplace between total abstinence and the extreme of alcoholism, there may be alcohol intakes moderate enough not to harm health, but the more a person drinks, the closer to a dangerous extreme that person comes.

liver, which normally stores folate to meet the body's needs, instead releases folate into the blood. As the blood folate concentration rises, the kidneys react as if the body has an excess of folate and are deceived into excreting the excess. To make matter worse, alcohol also interferes with the action of what little folate is left, inhibiting the production of new cells, especially the rapidly dividing cells of the intestine and the blood. Damage to the intestine from both folate deficiency and alcohol toxicity impairs the intestine's ability to continuously release and retrieve folate as it normally does, and it also fails to absorb folate that may trickle in from food as well. Thus alcohol abuse causes a folate deficiency that is self-aggravating and seriously impairs digestive system function. The combination of poor folate status and alcohol consumption can also lead to anemia and has been implicated in promoting colorectal cancer.

Alcohol can lead directly to iron overload and indirectly to iron deficiency. People who drink some alcoholic beverages, especially wine, may actually increase their intake of iron. Alcohol also increases the secretion of gastric acid, which enhances iron absorption. As Chapter 23 described, excessive iron storage in the liver can further injure the liver. Iron deficiencies develop largely due to GI bleeding associated with alcohol abuse or due to accidents or other trauma that occur more commonly in people who drink alcohol excessively.

Notes

1. M. J. Thun and coauthors, Alcohol consumption and mortality among middle-aged and elderly U.S. adults, *New England Journal of Medicine* 337 (1997): 1705–1714; American Cancer Society, Cancer Resource Center, **www3.cancer.org**, site visited November 12, 2001.

2. W. H. L. Kao and coauthors, Alcohol consumption and the risk of type 2 diabetes mellitus: Atherosclerosis risk in communities study, *American Journal of Epidemiology* 154 (2001): 748–757.

3. American Cancer Society, Cancer Resource Center, **www.cancer.org**, site visited November 12, 2001; R. C. Ellison and coauthors, Exploring the relation of alcohol consumption to risk of breast cancer, *American Journal of Epidemiology* 154 (2001): 740–747; S. A. Smith-Warner and coauthors, Alcohol and breast cancer in women: A pooled analysis of cohort studies, *Journal of the American Medical Association* 279 (1998): 535–540.

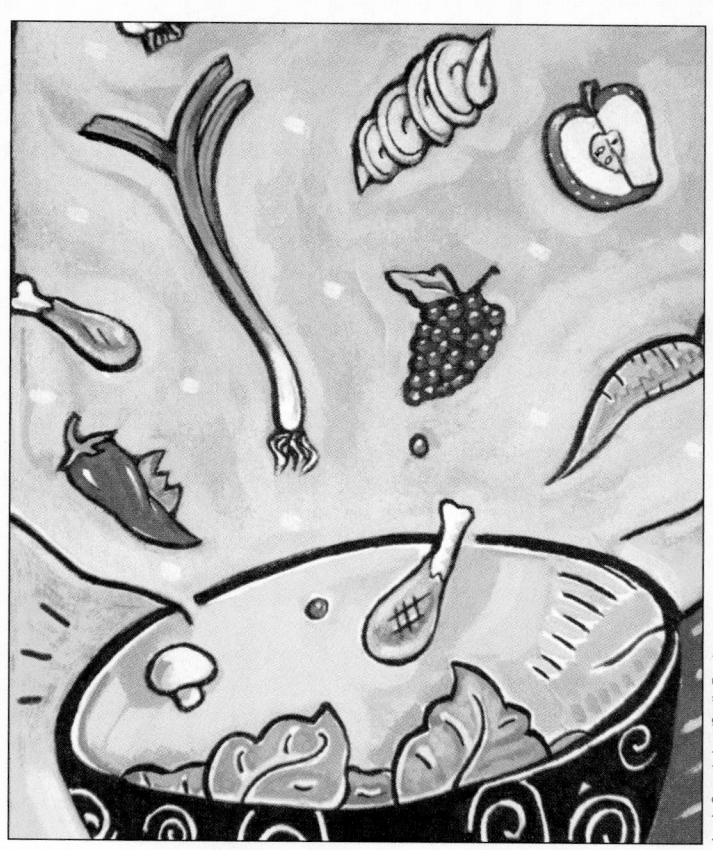

Jennie Oppenheimer/Studio Zocolo

CARBOHYDRATE-MODIFIED DIETS FOR DIABETES

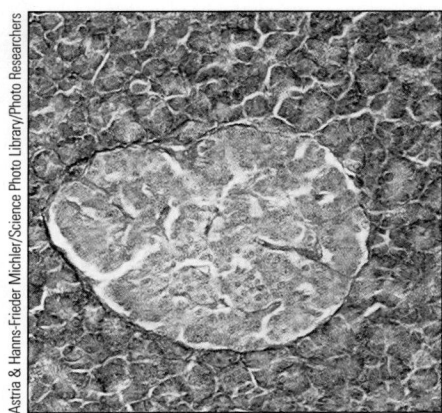

Astria & Hanns-Frieder Michler/Science Photo Library/Photo Researchers

Specialized endocrine cells in the pancreas, the islets of Langerhans, produce hormones. Among these cells are the alpha cells, which produce glucagon, and the beta cells, which produce insulin.

Reminder: Insulin is the hormone that, among other things, enables many cells to take up glucose from the blood and store energy fuels.

Immune system disorders in which the body destroys its own cells or tissues are called **autoimmune disorders.** A family history of type 1 diabetes and laboratory tests that detect antibodies to insulin help clinicians predict which individuals will develop type 1 diabetes and also help them distinguish between type 1 and type 2 diabetes.

Type 1 diabetes diagnosed in adults is sometimes called **latent autoimmune diabetes of adults (LADA).**

diabetes (DYE-uh-BEET-eez) **mellitus** (MELL-ih-tus or mell-EYE-tus): a group of metabolic disorders of glucose regulation and utilization.
 diabetes = passing through (the body)
 mellitus = honey-sweet (sugar)

type 1 diabetes: the less common type of diabetes in which the person produces no insulin at all.

*A*s Chapter 22 showed, diet planners use carbohydrate-modified diets to prevent malabsorption and complications caused by the dumping syndrome, lactose intolerance, and celiac disease. This chapter describes the carbohydrate-modified diets used to regulate blood glucose levels for people with **diabetes mellitus.**

What Is Diabetes Mellitus?

Diabetes mellitus describes a group of metabolic disorders characterized by elevated blood glucose and altered energy metabolism and caused by defective insulin secretion, defective insulin action, or a combination of the two. There are two major types of diabetes. Their distinguishing features are summarized in Table 24–1. A later section of this chapter describes the special case of gestational diabetes. Other types of diabetes can occur as a consequence of a variety of disorders.

Types of Diabetes

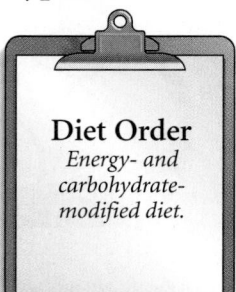

Diet Order
Energy- and carbohydrate-modified diet.

Over 16 million people in the United States have diabetes, and that number is expected to rise to 22 million by 2025. Of the 16 million people who currently have diabetes, about one-third do not know they have it.[1] Undiagnosed diabetes is especially dangerous because its damaging effects (described later) may begin to develop years before the symptoms of diabetes appear.[2]

Type 1 Diabetes   In **type 1 diabetes,** the less common of the two major types of diabetes (about 5 to 10 percent of diagnosed cases), the pancreas gradually loses its ability to synthesize insulin. In most cases, the individual inherits a defect in which immune cells mistakenly attack and destroy the insulin-producing pancreatic cells. By the time type 1 diabetes is diagnosed, the destruction of beta cells has often, but not always, reached a point where insulin must be supplied. Although type 1 diabetes has historically been considered a disease of childhood or early adolescence, the incidence of diagnosis during adulthood is considerable.[3] People with siblings or parents with type 1 diabetes have a high risk of developing the disease themselves. Studies are

Table 24–1 Features of Type 1 and Type 2 Diabetes		
	Type 1	**Type 2**
Age of Onset	<20 (mean age, 17)	10–19, >40
Associated Conditions	Viral infection, heredity	Obesity, heredity, aging
Insulin Required?	Yes	Sometimes
Cell Response to Insulin	Normal	Resistant
Clinical Findings (generally)	Hyperglycemia with ketoacidosis	Hyperglycemia without ketoacidosis
Prevalence in Diabetic Population	5 to 10%	90 to 95%
Other Names	Insulin-dependent diabetes mellitus (IDDM)	Noninsulin-dependent diabetes mellitus (NIDDM)
	Juvenile-onset diabetes	Adult-onset diabetes
	Ketosis-prone diabetes	Ketosis-resistant diabetes
	Brittle diabetes	Lipoplethoric diabetes
		Stable diabetes

currently under way to determine if the destruction of beta cells in type 1 diabetes can be delayed or prevented.[4]

Type 2 Diabetes The predominant type of diabetes mellitus (90 to 95 percent of cases), and the type most likely to go undiagnosed, is **type 2 diabetes.** In type 2 diabetes, the pancreas produces insulin, and the cells respond to it, but with less sensitivity (**insulin resistance**). As blood glucose rises, the pancreas makes more insulin, and blood insulin rises to abnormally high levels (hyperinsulinemia). During this period of **impaired glucose tolerance,** the body is able to maintain blood glucose within a fairly normal range but at a cost. The chronic demand for insulin gradually exhausts the beta cells of the pancreas, and finally insulin production falters.

People most likely to develop type 2 diabetes include:

- Those who are obese.
- Those who have first-degree relatives with diabetes.
- Members of high-risk ethnic populations (African Americans, Asian Americans, Pacific Islanders, Hispanic Americans, and Native Americans).
- Those who are over age 45.
- Women who have given birth to babies weighing over 9 pounds or who have been diagnosed with diabetes while pregnant.

The incidence of type 2 diabetes is reaching epidemic proportions. Impaired glucose tolerance and type 2 diabetes are associated with excess body fat, especially abdominal fat; physical inactivity; and aging. Obesity aggravates insulin resistance: as body fat increases, body tissues become less and less able to respond to insulin. Although type 2 diabetes frequently occurs in obese, middle-aged or older adults, children and adolescents represent a sizable and growing number of cases. Children and adolescents with type 2 diabetes are generally obese (body mass index of 27 or greater) and between 10 and 19 years old, and they have a strong family history of diabetes. In addition, about 20 percent of people over 65 have diabetes mellitus.[5] The incidence in people over 80 may be as high as 40 percent.[6]

The Diabetes Prevention Program (DPP), a nationwide multicenter study to determine whether type 2 diabetes can be prevented, found that people with impaired glucose tolerance can dramatically lower their risk of developing diabetes with a moderate program of physical activity and weight loss.[7] The risk of diabetes fell by 58 percent for people who lost an average of 5 to 7 percent of their body weight and exercised moderately for about 150 minutes per week. The same study also showed that people with impaired glucose tolerance can reduce their risk of diabetes by 31 percent if they take a medication called metformin (described later in this chapter). Additional studies are under way to determine whether these benefits will persist over time.

Acute Complications of Diabetes

Elevated blood glucose leads to immediate complications. Figure 24–1 on p. 598 presents an overview of these acute metabolic changes and clinical manifestations. Type 1 diabetes incurs more immediate and severe acute complications than type 2 diabetes does because people with type 1 diabetes produce little or no insulin, so no glucose enters the cells. The glossary on p. 599 defines diabetes-related symptoms and complications.

Hyperglycemia and Glycosuria When a person absorbs carbohydrate from a meal or snack, blood glucose rises. The rising blood glucose triggers the release of insulin, which allows glucose to enter the cells, and blood glucose falls. With insufficient or ineffective insulin, blood glucose remains high, and symptoms of **hyperglycemia** develop. Despite the high blood glucose, the cells are deprived of

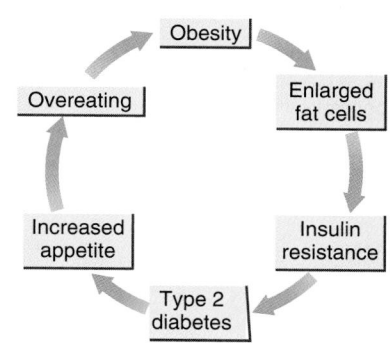

First-degree relatives are immediate family members—fathers, mothers, brothers, and sisters.

type 2 diabetes: the more common type of diabetes that develops gradually and is associated with obesity and insulin resistance.

insulin resistance: the condition in which the cells fail to respond to insulin as they do in healthy people.

impaired glucose tolerance: inability to maintain normal blood glucose levels without excessive insulin production. Some people with impaired glucose tolerance have fasting glucose levels somewhat higher than normal but not high enough to diagnose diabetes. Others have normal blood glucose levels most of the time, but when given a large amount of glucose, their blood glucose rises too high.

Carbohydrate-Modified Diets for Diabetes

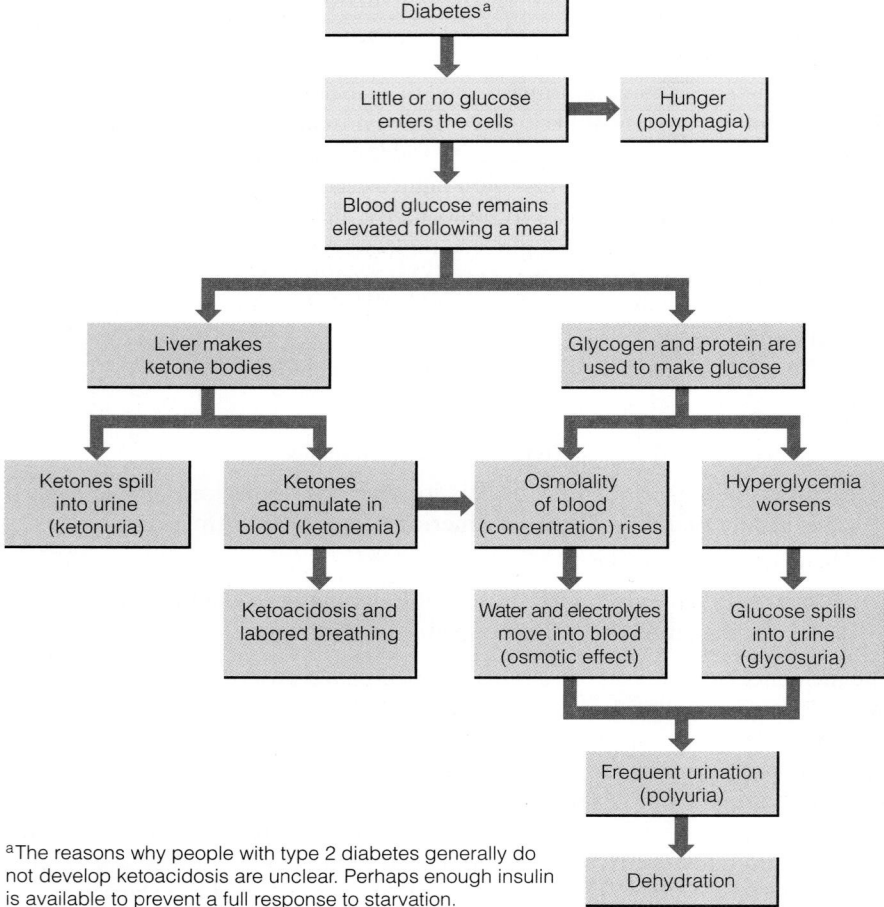

aThe reasons why people with type 2 diabetes generally do not develop ketoacidosis are unclear. Perhaps enough insulin is available to prevent a full response to starvation.

Symptoms of hyperglycemia:
- *Intense thirst and, sometimes, hunger.*
- *Increased urination.*
- *Blurred vision.*
- *Fatigue.*
- *Acetone breath.*
- *Labored breathing.*

When counterregulatory hormone levels rise, they cause insulin resistance and signal the liver to mobilize glycogen and protein to provide glucose.

Reminder: Ketone bodies are produced by the incomplete breakdown of fat when glucose is not available to the cells (see Chapter 4).

counterregulatory hormones: hormones that oppose insulin actions. The counterregulatory hormones include glucagon, epinephrine, norepinephrine, cortisol, and growth hormone.

energy. The body responds as it would to starvation—levels of **counterregulatory hormones** rise, insulin resistance develops or worsens, and the body mobilizes its glycogen stores and uses protein to make more glucose. Blood glucose rises higher still. Some people experience hunger and eat excessively (**polyphagia**), which aggravates hyperglycemia.

Under normal conditions, glucose is not excreted in the urine, but eventually the concentration of glucose in the blood exceeds the kidneys' ability to reabsorb it (the **renal threshold**), and some glucose is excreted in the urine (**glycosuria**) along with fluids and electrolytes. However, not enough glucose is excreted to reduce blood glucose to acceptable levels.

Ketosis and Acidosis When little or no insulin is available, even very high blood glucose levels fail to suppress the activity of the counterregulatory hormones. These hormones stimulate the production of glucose and ketone bodies (**ketosis**). Ketone bodies, which are acidic, begin to accumulate excessively in the blood (**ketoacidosis**) and spill into the urine (**ketonuria**). A fruity odor develops on the breath of a person with ketoacidosis (**acetone breath**). The lungs must work hard to help restore acid-base balance, and labored breathing and fatigue develop as a consequence.

Dehydration and Fluid and Electrolyte Imbalances High blood glucose and ketosis, either alone or in combination, raise the number of solutes in the blood, and water is drawn from the tissues to dilute the concentration (osmosis). At the same time, fluid is lost from the body through the kidneys, which must excrete the excess glucose and ketone bodies. Thus hyperglycemia and ketosis can lead to severe dehy-

Glossary of Diabetes-Related Symptoms and Complications

acetone breath: a distinctive fruity odor that can be detected on the breath of a person who is experiencing ketosis.

dawn phenomenon: early morning hyperglycemia that develops in response to elevated levels of counterregulatory hormones that act to raise blood glucose after an overnight fast.

diabetic coma: unconsciousness precipitated by hyperglycemia, dehydration, ketosis, and acidosis in people with diabetes.

gangrene: death of tissue due to a deficient blood supply and/or infection.

gastroparesis (GAS-troh-pah-REE-sis): delayed gastric emptying.

glycosuria (GLY-ko-SUE-ree-ah) or **glucosuria** (GLUE-ko-SUE-ree-ah): glucose in the urine, which generally occurs when blood glucose exceeds 180 mg/dL.

hyperglycemia: elevated blood glucose. Normal fasting blood glucose is less than 110 mg/dL. Fasting blood glucose between 110 and 125 mg/dL suggests impaired glucose tolerance; values of 126 mg/dL or higher suggest diabetes.

hyperosmolar, hyperglycemic coma: coma that occurs in uncontrolled type 2 diabetes precipitated by the presence of hypertonic blood and dehydration.

hypoglycemia: low blood glucose.

ketoacidosis: lowering of the blood's normal pH due to the accumulation of acidic ketones.

ketonuria: ketones in the urine.

ketosis: accelerated production of ketones.

nephropathy (neh-FROP-ah-thee): a disorder of the kidneys.

neuropathy (new-ROP-ah-thee): a disorder of the nerves.

nocturnal hypoglycemia: hypoglycemia that occurs while a person is sleeping.

polydipsia (POLL-ee-DIP-see-ah): excessive thirst.

polyphagia (POLL-ee-FAY-gee-ah): excessive eating.

polyuria (POLL-ee-YOU-ree-ah): excessive urine production.

rebound hyperglycemia: hyperglycemia resulting from excessive secretion of counterregulatory hormones in response to excessive insulin and consequent low blood glucose levels; also called the **Somogyi** (so-MOHG-yee) **effect.**

renal threshold: the point at which a blood constituent that is normally reabsorbed by the kidneys reaches a level so high the kidneys cannot reabsorb it. The renal threshold for glucose is generally reached when blood glucose rises above 180 mg/dL.

retinopathy: a disorder of the retina.

---

dration. This series of events explains why people with undiagnosed diabetes produce excessive urine (**polyuria**) and develop excessive thirst (**polydipsia**).

Electrolytes that normally reside in the cells, especially potassium and phosphorus, move into the blood as proteins are sacrificed to make glucose. In addition, when glucose is excreted in the urine, electrolytes are excreted along with it. Both of these factors lead to electrolyte imbalances and aggravate dehydration.

Coma in Diabetes Severe hyperglycemia with or without ketosis constitutes a medical emergency that is treated in the hospital by carefully administering insulin and correcting fluid and electrolyte and acid-base imbalances using intravenous (IV) fluids. Without treatment, severe hyperglycemia can lead to coma. **Hyperosmolar, hyperglycemic coma** can develop from extremely high blood glucose (greater than 600 milligrams per deciliter) and severe dehydration without ketoacidosis or with only minimal ketoacidosis. Coma that occurs when ketoacidosis is also present is called **diabetic coma.** In this type of coma, blood glucose does not rise as high as in hyperosmolar, hypoglycemic coma, but acidosis becomes a serious problem.

Hyperglycemia in Treated Diabetes Once diabetes has been diagnosed and treatment is under way, hyperglycemia can develop when a person eats too much carbohydrate or uses too little medication (when necessary). It can also occur during times when counterregulatory hormone levels rise. Such is the case when the person with diabetes experiences hyperglycemia after an overnight fast (**dawn phenomenon**); when the person with high blood glucose engages in strenuous physical activity; when the person injects too much insulin (**rebound hyperglycemia**); or when the person develops an illness or infection. When these problems occur, insulin doses must be adjusted.

The odor associated with *acetone breath* is typically described as "fruity" for lack of a better description, but it is actually a sharper, more distinctive smell similar to that of nail polish remover, which contains acetone.

Carbohydrate-Modified Diets for Diabetes

Even a minor illness such as a cold or flu can cause blood glucose to rise dramatically. Infections and injuries that evoke inflammatory responses are the most common causes of severe hyperglycemia in people with diabetes.

Hypoglycemia   **Hypoglycemia,** or low blood glucose, arises from the inappropriate management of diabetes, rather than from the disease itself. It can result from taking too much insulin or glucose-lowering medications, strenuous physical activity, skipped or delayed meals, inadequate food intake, vomiting, or severe diarrhea. Some people develop hypoglycemia while they are sleeping (**nocturnal hypoglycemia).** The symptoms of hypoglycemia include mental confusion and shakiness, which may make it difficult for the person to recognize the problem and take corrective actions. (Nutrition in Practice 24 provides more information about hypoglycemia.)

Left untreated, hypoglycemia can lead to coma and death. Children and people who have had diabetes for a long time may not notice the warning signs, however, and thus risk severe hypoglycemia. Repeated episodes of hypoglycemia may permanently impair cognitive function.[8] Avoiding hypoglycemia is especially important for children, particularly children under seven years old, because hypoglycemia can interfere with normal brain development.[9]

Notice that many of the symptoms of hypoglycemia are those of alcohol intoxication. If the true problem goes unrecognized, the person may die. To prevent such a tragic mistake, advise every person with diabetes to wear identification in the form of a bracelet or necklace.

Chronic Complications of Diabetes

Exposure of the tissues to high glucose concentrations over time results in the chronic complications of diabetes. When blood glucose concentrations remain high, glucose attaches to proteins in the blood and cells lining the blood vessels. These **glycated proteins** damage the structures of the blood vessels and nerves. Tissue damage elicits inflammatory responses, which thicken the walls of the blood vessels, alter blood flow, and bring about further tissue damage. As the cells lining the blood vessels thicken, circulation becomes poor, and nerve function falters. Infections are likely to occur due to poor circulation coupled with glucose-rich blood and urine. Infections often go undetected due to impaired nerve function; **gangrene** may follow. People with diabetes must pay special attention to hygiene and keep alert for early signs of infection.

Cardiovascular Diseases   Coronary heart disease and hypertension (see Chapter 25) and their complications—heart attacks and strokes—commonly develop in people with diabetes. Cardiovascular disease tends to develop early in people with type 2 diabetes, progress rapidly, and be more advanced at the time of diagnosis.[10] More than 80 percent of people with diabetes die as a consequence of cardiovascular diseases, especially heart attacks. If nerve function is also impaired, the person may suffer a heart attack and not even realize it.

As Chapter 25 will show, people with diabetes often have many risk factors for coronary heart disease including altered blood lipid levels, hypertension, and obesity. The combination of insulin resistance, secretion of more and more insulin to maintain blood glucose, obesity, hypertension, elevated LDL and triglycerides, and lowered HDL is called the **metabolic syndrome;** it frequently leads to type 2 diabetes and cardiovascular disease. Nutrition in Practice 25 describes the metabolic syndrome in more detail.

Nephropathy and Retinopathy   When hyperglycemia damages the blood vessels that supply blood, oxygen, and nutrients to the kidneys and eyes, loss of kidney function, retinal degeneration, and loss of vision may follow. About 85 percent of people with diabetes have **nephropathy, retinopathy,** or both. Consequently, diabetes is a leading cause of both kidney failure and blindness.

Hypoglycemia in a person who uses insulin is also called an **insulin reaction** or **insulin shock.**

Symptoms of hypoglycemia:
- *Hunger.*
- *Headache.*
- *Sweating.*
- *Shakiness.*
- *Nervousness.*
- *Confusion.*
- *Disorientation.*
- *Slurred speech.*

Reminder: The damaging effects of insulin resistance and impaired glucose tolerance on the blood vessels and nerves may begin before a diagnosis of diabetes is made.

A heart attack that goes unrecognized is called a **silent heart attack.**

glycated proteins: proteins that have glucose attached to them.

metabolic syndrome: the combination of insulin resistance, hyperinsulinemia, obesity, hypertension, elevated LDL and triglycerides, and reduced HDL that is frequently associated with type 2 diabetes and cardiovascular disease; also called *syndrome X* and *insulin-resistance syndrome.*

Neuropathy   Nerve tissues may also be damaged, resulting in **neuropathy.** At first, the person may experience a painful prickling sensation, often in the arms and legs, which progresses until the person loses sensations in the hands and feet. Injuries to these areas may go unnoticed, and infections can progress rapidly. If tissues die as a consequence, amputation of the affected limb (usually the toes, feet, or legs) may be necessary. Neuropathy can also delay gastric emptying **(gastroparesis).** When the stomach empties slowly after a meal, the person may experience a premature feeling of fullness, nausea, vomiting, weight loss, and poor blood glucose control due to irregular nutrient absorption. Although neuropathy has long been thought to be the cause of delayed gastric emptying in people with diabetes, after-meal blood glucose levels may also play a role.[11]

In Summary   Diabetes develops when little or no insulin is produced (type 1 diabetes) or when insulin is ineffective (type 2 diabetes). In either case, undiagnosed or uncontrolled diabetes can lead to acute complications, including hyperglycemia and hypoglycemia. If left untreated, both of these complications can lead to coma and death. The chronic complications of diabetes include repeated infections, cardiovascular diseases, kidney disease, loss of vision, and neuropathy.

Treatment of Diabetes Mellitus

A diagnosis of diabetes can be devastating. An elderly person with recently diagnosed diabetes may fear possible complications and be overwhelmed by the lifestyle changes necessary to control the disorder. Preteens and teenagers often have intense difficulty accepting a diagnosis of diabetes. As young adults strive for independence from their families, they become increasingly self-conscious and reject those things that set them apart from peer groups—including the tasks necessary to control their diabetes. Once children reach school age, they should be active participants in their treatment plans. Adolescents need to know they can manage their disease themselves and achieve the independence they strive for.

Effective treatment plans for diabetes must consider individual needs. A plan for a child, for example, must be flexible enough to accommodate a typical child's activity levels (and appetite), which often vary widely from day to day.

Treatment Goals

The goals of both medical and nutrition therapy for diabetes are to maintain blood glucose within a fairly normal range, achieve optimal blood lipid levels, control blood pressure, support health and well-being, and prevent and treat complications. The most important of these goals is blood glucose control. Table 24–2 on p. 602 compares two treatment plans for controlling blood glucose—traditional therapy and intensive therapy.

Benefits of Intensive Therapy   A major multicenter clinical trial (the Diabetes Control and Complications Trial, or DCCT) clearly showed that compared to traditional therapy, intensive therapy reduces the risks of onset and progression of nephropathy, retinopathy, and neuropathy by about 50 percent. The study included both adults and children age 13 and older and found comparable benefits for both groups. A follow-up study found that the benefits of intensive therapy persist for at least four years (the time period of the study), even though blood glucose levels tend to rise with time.[12] The United Kingdom Prospective Diabetes Study found similar benefits for people with type 2 diabetes.[13]

Treatment Plans   A treatment plan that includes diet, physical activity, and medication is formulated for each client after an accurate and thorough assessment of the client's medical condition, developmental stage, and psychosocial and economic needs. The plan is modified as needed based on subsequent monitoring

Chapter 25 discusses measures to improve blood lipid levels and control blood pressure.

Carbohydrate-Modified Diets for Diabetes

Physical activity plays an important role in the management of diabetes.

Table 24–2   Comparision of Traditional and Intensive Therapy for Diabetes

	Traditional	Intensive
Diet	Consistent intake of energy and carbohydrates at set times each day.	Consistent intake of energy and carbohydrates at set times with adjustments in medications to accommodate deviations from the meal plan.
Physical Activity	Regular schedule of physical activity at set times.	Regular schedule of physical activity at set times with adjustments in carbohydrate intake and medications when activity schedule or intensity changes.
Insulin	One or two injections at same time each day.	Multiple daily injections or insulin pump. Adjust insulin doses to accommodate deviations from meal plan or physical activity schedule.
Blood Glucose Monitoring	Once or twice a day.	At least three or four times a day. More frequent monitoring when meal plan or physical activity schedule changes.
Advantages	Easier to learn, requires less time to follow, fewer injections and finger pricks, less risk of hypoglycemia and weight gain.	Better blood glucose control, lower risk of neuropathy and nephropathy, greater flexibility.
Disadvantages/ Contraindications	Poorer blood glucose control (blood glucose stays higher), greater risk of complications, inflexible—deviations from diet/physical activity plan not recommended.	Greater risk of hypoglycemia and excessive weight gain; not recommended for children under age 2 or older people with significant cardiovascular disease.

of the client's condition. The fewer the disruptions to the client's lifestyle and family (or caregiver) relationships, the more successful the management plan.

Together with diet and insulin, physical activity plays an important role in the management of diabetes. The client with diabetes should be carefully evaluated to determine a safe type and amount of physical activity. For people with type 2 diabetes, a regular program of physical activity improves blood glucose control, contributes to weight loss, improves blood lipid levels, lowers blood pressure, and relieves emotional stress. Although physical activity has not been shown to aid blood glucose control in people with type 1 diabetes, its value lies in the benefits it confers on the cardiovascular system.

Evaluating Diabetes Control

The health care team routinely evaluates the client's blood glucose levels, blood lipid levels, blood pressure, weight, and reflexes to see how well the treatment plan is working and to check for signs of complications. Urine tests can help detect the early signs of kidney disease, eye exams help detect the early signs of retinopathy, and foot exams help detect early signs of infection.

Blood Glucose   Clients learn to monitor their blood glucose levels at home, often using computerized blood glucose meters. Clients who practice intensive therapy check their blood glucose at least four times a day, and those who practice traditional therapy check it once or twice a day. At medical appointments, the health care team reviews the client's test results and records to monitor the client's progress and look for patterns that suggest the need for adjustments in the treatment plan.

Traditionally, health care professionals guide diabetes therapy based on fasting blood glucose. Emerging therapies for diabetes place increasing emphasis on the control of after-meal (**postprandial**) blood glucose, however. Postprandial blood glucose shows a strong correlation with measures of long-term diabetes control and the development of cardiovascular complications.[14]

Urinary Ketones   Health care professionals advise clients to monitor ketones in the urine when they find that blood glucose levels are consistently high or during illness. As described earlier, high blood glucose may predispose the person to ketosis and coma.

Glycated Hemoglobin   Physicians periodically evaluate blood glucose control through laboratory measurement of the percentage of **glycated hemoglobin.** As blood glucose rises, glucose attaches to amino acids on hemoglobin molecules and remains there until the red blood cells that carry the hemoglobin die (about 120 days). Thus the percentage of glycated hemoglobin reflects diabetes control over the past two to three months, rather than just prior to the test. The type of glycated hemoglobin most commonly measured in the United States is **HbA$_{1c}$** (hemoglobin A$_{1c}$) or **A1C** for short. Table 24–3 lists normal and target levels for blood glucose and A1C.

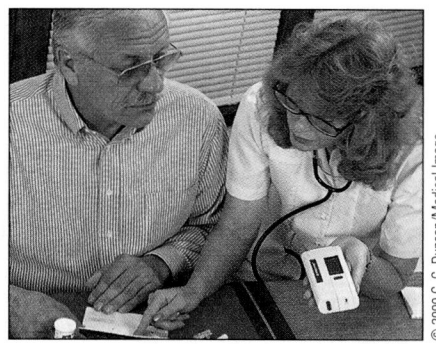

Blood glucose monitoring helps people with diabetes learn how their blood glucose responds to carbohydrate, physical activity, and illness and helps them maintain it in a safe range.

The American College of Endocrinology and the American Association of Clinical Endocrinologists recommend that HbA$_{1c}$ be called simply A1C and also recommend that clients know their A1C level and their target A1C level.

In Summary   The goals of diabetes management are to control blood glucose, blood lipids, and blood pressure; support health; and prevent and treat complications. Tight control of blood glucose levels reduces the complications associated with diabetes. Clients monitor blood glucose (and sometimes urinary ketones), and clinicians monitor A1C to evaluate treatment plans and determine when adjustments are necessary.

postprandial (post-PRAN-dee-all): following a meal.
   post = after
   prandial = breakfast

glycated (GLIGH-kate-id) **hemoglobin:** hemoglobin with glucose molecules attached to its amino acids; also called *glycosylated hemoglobin*. The type of glycated hemoglobin most commonly measured is hemoglobin A$_{1c}$ (**HbA$_{1c}$**), or **A1C** for short.

Table 24–3   Normal and Target Levels for Blood Glucose and A1C

Measurement	Normal Level	Target Level	Level Requiring Further Action
Plasma glucose,[a] fasting or before meals (mg/dL)	<110	90–130	<90/>150
Plasma glucose, 2-hour postprandial (mg/dL)[b]	<140	N/A	N/A
Plasma glucose, bedtime (mg/dL)	<120	110–150	<110/>180
Whole blood glucose, fasting or before meals (mg/dL)	<100	80–120	<80/>140
Whole blood glucose, 2-hour postprandial (mg/dL)[b,c]	<125	N/A	N/A
Whole blood glucose, bedtime (mg/dL)	<110	100–140	<100/>160
A1C (%)	<6	<7[d]	<8[d]

SOURCE: Adapted from American Diabetes Association, Standards of medical care for patients with diabetes mellitus, *Diabetes Care* (supplement 1) 25 (2002): 33–49.

NOTE: The values shown here are average values for nonpregnant adults; individual target levels may vary.

[a]Blood glucose readings obtained from glucose monitors measure whole blood; however, some monitors convert the measurement to plasma values. Plasma blood glucose is about 10 to 15 percent higher than whole blood glucose. Clients should be instructed to learn which value their monitor reports.

[b]From the American College of Endocrinology and the American Association of Clinical Endocrinologists, ACE Consensus Conference on Guidelines for Glycemic Control, August 20–21, 2001, available from **www.aace.com,** site visited October 26, 2001.

[c]Estimated by adding 10 percent to the postprandial whole blood glucose.

[d]The ACE Consensus Conference recommends a target level of <6.5 percent for A1C and recommends that further action be taken when values exceed 6.5 percent.

The diet for diabetes emphasizes a consistent intake of carbohydrate from well-balanced meals spaced evenly throughout the day.

Medical Nutrition Therapy for Diabetes

Medical nutrition therapy for diabetes is most appropriately designed and implemented by a skilled dietitian. A complete nutrition assessment provides the foundation. The dietitian uses the assessment to learn about the client's usual eating habits, analyzes the diet's composition and adequacy, and devises a plan that considers the client's age and daily routines that affect food intake. After the client tries the plan for a while, adjustments often need to be made. Sometimes nutrient needs change (as a child grows, for example), or the client may have problems with blood glucose management or simply find it hard to follow certain parts of the plan.

Energy and Nutrients

The diet for diabetes parallels a healthy diet for all people in both amounts and types of nutrients. Attention to all energy nutrients is important: controlling carbohydrate prevents hyper- and hypoglycemia; controlling protein may help preserve kidney function; controlling fat helps prevent cardiovascular complications.

Energy Medical nutrition therapy for diabetes first focuses on providing food energy in the amount necessary to achieve or maintain a healthy, realistic body weight and to support growth in children and pregnant women. The diet planner calculates the person's average daily energy intake and uses it to determine an appropriate energy level based on body weight, monitors weight periodically, and adjusts the diet as necessary. Protein should account for about 15 to 20 percent of the energy intake, carbohydrate and monounsaturated fat together should provide 60 to 70 percent of the total energy, and the remainder should come from polyunsaturated (about 10 percent) and saturated fats.[15]

People with newly diagnosed or poorly controlled type 1 diabetes are likely to be thin despite eating apparently adequate amounts of food. With treatment, especially intensive therapy, however, excessive weight gain can be a problem.[16] Teenagers with type 1 diabetes who practice intensive therapy are often heavier than teenagers without diabetes. Children and adults with type 2 diabetes are likely to be overweight and often benefit from weight loss. For overweight people with type 2 diabetes, even moderate weight loss (10 to 20 pounds) can help reverse insulin resistance, improve the blood lipid profile, and reduce blood pressure. A diet plan that moderately restricts energy intake (500 to 1000 fewer kcalories than necessary for weight maintenance) provides for a realistic and gradual weight loss.[17] Elderly people with diabetes, even those with type 2 diabetes, are often underweight.

An early sign of kidney disease is the excretion of small amounts of albumin in the urine or **microalbuminuria.**

Protein Clients with kidney disease may need to further restrict protein to less than 15 to 20 percent of the total kcalories (see Chapter 26). A small study suggests that people with type 2 diabetes may improve their blood glucose and blood lipid levels by adhering to a diet that limits protein from animal sources (one serving every other day).[18]

Carbohydrate To control blood glucose, the person with diabetes needs to have glucose available throughout the day, but not so much at one time that blood glucose levels rise too high, or so little that blood glucose falls too low. Of all the energy nutrients, carbohydrates have the greatest effect on blood glucose. Once eaten, carbohydrates raise blood glucose in about an hour. Some protein and fat from a meal may also raise blood glucose, but to a lesser degree and much more slowly than carbohydrates.

A person who is not ill and follows a regular physical activity program, takes a prescribed dose of medication at a set time, and then eats about the same amount of carbohydrate at about the same time each day is likely to have a safe amount of glucose available to the body when it is needed. Thus, for people with diabetes, consistent timing and composition of meals and snacks from day to day improve blood glucose control. Without medication adjustments, too much carbohydrate at a meal or snack can cause hyperglycemia. Skipping meals or eating too little can lead to hypoglycemia. An evening snack is especially important because it helps sustain the person's blood glucose through the night and helps prevent nocturnal hypoglycemia.

Carbohydrate Sources Encourage clients with diabetes to select whole-grain breads and cereals, legumes, fruits, and vegetables. In addition to carbohydrates, these foods provide fiber, vitamins, and minerals and offer many health benefits (see Chapter 2). Authorities recommend a daily dietary fiber intake of 20 to 35 grams for all people, including those with diabetes. Diets containing 50 grams of dietary fiber may improve blood glucose control, reduce the frequency of hypoglycemia, and lower blood lipids in people with diabetes, but this level of fiber may be difficult to consume without adverse GI side effects.[19]

Traditionally, concentrated sweets were strictly excluded from the diet for diabetes, but now they are restricted only to the same extent as they are for all people. The total amount of carbohydrate is of greater concern in diabetes than the type of carbohydrate. The person with diabetes can consume concentrated sweets and nutritive sweeteners as a limited part of a healthy diet, as long as they are counted as part of the carbohydrate allowance. Artificial sweeteners (such as saccharin and aspartame) and products made from them contain no carbohydrate and minimal kcalories and can be used in place of sugar.

Glycemic Index As Chapter 2 described, carbohydrates from different sources have varying effects on blood glucose after a meal. Foods with a high glycemic index raise blood glucose faster and to a greater extent than foods with a low glycemic index. In general, high-fiber foods tend to have a lower glycemic index. Controversy abounds regarding the value of the glycemic index of foods and the diet for diabetes.[20] Some feel enough evidence exists to support the use of glycemic indexing in the diet for diabetes. Others feel that, at present, too little information is available to warrant the imposition of additional restrictions and that recommendations to include high-fiber foods are sufficient. Although various health organizations around the world and many clinicians endorse the use of the glycemic index, the American Diabetes Association currently does not.

Fat People with diabetes who have acceptable blood lipid concentrations benefit from limiting saturated fat intake to less than 10 percent of kcalories and cholesterol to less than 300 milligrams. Those who have elevated LDL cholesterol and those who are overweight may need to restrict saturated fat intake to 7 percent or less of total kcalories and cholesterol to less than 200 milligrams daily. Those with elevated LDL and a desirable body weight can replace the energy from saturated fat with either carbohydrate or monounsaturated fat.

With a low intake of saturated fat (10 percent or less) and an appropriate energy intake, diets high in carbohydrate (about 55 to 60 percent) and low in total fat (30 percent or less) can help lower LDL and blood cholesterol. So can diets providing less carbohydrate (about 45 percent) and more total fat (about 35 percent), provided that the additional energy from fat comes from monounsaturated sources. Compared to the high-carbohydrate diet, the high monounsaturated fat diet results in a lower postprandial rise in blood glucose, insulin, and triglycerides, but does not improve fasting blood glucose or A1C.[21]

Reminder: Nutritive sweeteners are carbohydrates and include sucrose, fructose, sorbitol, mannitol, and xylitol.

The Canadian Diabetes Association recommends that no more than 10 percent of the total daily kcalories come from sugar and that the sugar used be spread throughout the day and consumed along with meals.

Reminder: Cholesterol is transported through the blood packaged with lipoproteins. LDL are low-density lipoproteins, and VLDL are very-low-density lipoproteins. See Chapter 25 for more information about the relationship of lipoproteins to cardiovascular disease.

People with very high triglycerides ($\geq$1000 mg/dL) need to restrict all types of dietary fat to less than 10% of kcalories.

To lower fat and cholesterol intakes, clients can use low-fat and fat-free milk and milk products and lean meats, among other strategies (see Chapter 25 for more information). People with diabetes who use reduced-fat products must use them cautiously.[22] People may overeat when they believe they are saving kcalories by eating reduced-fat products, and this is not always the case. Additionally, carbohydrate often replaces the fat in these products. The energy and carbohydrate that the reduced-fat products contribute to the person's food intake must be considered as part of the energy and carbohydrate allowance in the diet plan.

Reminder: The fat substitute olestra does not contribute kcalories or carbohydrate.

Sodium   People with diabetes frequently develop hypertension, and limiting sodium can help reduce blood pressure. An intake of 2400 milligrams of sodium (6 grams salt) or less per day is appropriate for all people, including those with diabetes. People with diabetes and hypertension or kidney disease may need to restrict sodium even more (2000 milligrams of sodium or less per day).

Alcohol   The person whose blood glucose is well controlled can usually include moderate amounts of alcoholic beverages (no more than one drink per day for women or two drinks per day for men), with the consent of the physician. Alcohol can cause hypoglycemia in any person, however, and people with diabetes who take insulin (described in a later section) are particularly likely to develop this complication. People who take insulin are advised to drink alcohol only with meals, and in addition to the usual meal plan. To protect against hypoglycemia, no foods should be omitted. (Remember, too, that the person with hypoglycemia may appear to be intoxicated and alcohol use can add confusion to a potentially dangerous situation.)

People with a history of alcohol abuse, pancreatitis, abnormal blood lipids, or neuropathy and women who are pregnant are strongly cautioned to avoid alcohol completely. Alcohol use is also discouraged for people who are overweight; if it is used, alcohol should be considered as part of the fat allowance. Drinks that contain simple sugars (mixers, sweet wines, and liqueurs) are best avoided. If they are used, the person must count their carbohydrate contents as part of the daily carbohydrate allowance. The combination of alcohol and some **oral antidiabetic agents** (described in a later section) may cause nausea, vomiting, headache, cramps, flushing of the skin, and a rapid heartbeat (**disulfiram-like reaction**).

Reminder: One alcoholic drink is defined as 1½ oz of distilled liquor, 12 oz of beer, or 5 oz of wine. Note: Light beer contains the same amount of alcohol as regular beer and less than half the carbohydrate. Regular beer (12 oz) contains about 13 g of carbohydrate, light beer about 5 g. Wine contains about 2 to 3 g of carbohydrate. The carbohydrate from beer and wine should be considered in the meal plan. For people who need to lose weight, 1 drink = 2 fat exchanges. A later section explains exchanges.

Other Micronutrients   Vitamin and mineral needs for people with diabetes are the same as for healthy people. Magnesium deficiencies have been linked to insulin resistance and hypertension, and chromium deficiencies have been linked to glucose intolerance. Supplementation of these nutrients confers no added benefits to people with diabetes, however, unless they have documented deficiencies. People who take some types of diuretics (see Chapter 25), though, may need to take potassium supplements.

Other Dietary Considerations

Physical activity, missed meals, and illnesses can cause special problems for people with diabetes. Maintaining normal blood glucose requires special attention to diet during these times.

Physical Activity and Food Intake   All persons, especially those with diabetes, need to make sure they are adequately hydrated before and during physical activity by drinking liquids throughout the day. The person who takes insulin or hypoglycemic agents may need to eat before, during, and after vigorous physical activity. Especially important is carbohydrate, which is readily available from fruits, fruit juices, yogurt, crackers, and other starches. The amount of carbohydrate the client needs depends on the type of activity, its duration, the client's individual responses, and the results of blood glucose tests.

oral antidiabetic agents: medications taken by mouth to lower blood glucose levels in people with type 2 diabetes.

disufiram-like reaction: nausea, vomiting, headache, cramps, flushing of the skin, and a rapid heartbeat that can occur when some medications are taken along with alcohol. The medication disulfiram produces these effects when combined with alcohol to discourage alcohol abusers from using alcohol.

Missed Meals and Illness When people with diabetes are ill, their blood glucose levels often rise dramatically. During illness, clients may be advised to change their doses of medication, reduce carbohydrate intakes somewhat, or a combination of both. Whenever they must miss a meal for any reason, however, they need some carbohydrate to forestall hypoglycemia. For adults, adjusting medications and providing about 45 to 50 grams of carbohydrate (3 or 4 carbohydrate exchanges) every three or four hours is generally sufficient to keep blood glucose in an acceptable range and prevent ketosis. If appetite is poor, people can use juice, flavored gelatin, soft drinks, or frozen juice bars to meet their carbohydrate needs.

Treating Hypoglycemia If a client develops hypoglycemia for any reason, a judicious approach prevents overtreatment and subsequent hyperglycemia. As soon as the symptoms are observed, the person needs 10 to 15 grams of carbohydrate. People who take oral agents that interfere with the digestion of sucrose and complex carbohydrates need to take glucose to treat hypoglycemia; for all others, glucose is the best choice, but any carbohydrate that is readily available and easy to eat is a good choice. Advise clients to avoid foods that also contain fat, which slows the absorption of carbohydrate. Blood glucose is then checked within 15 to 20 minutes to see if it has risen to an acceptable level. If not, an additional 10 to 15 grams of carbohydrate are given, and blood glucose is rechecked. The procedure continues until blood glucose returns to an acceptable range. Blood glucose should then be rechecked 60 minutes later. Advise clients to carry some convenient source of carbohydrates with them at all times, so they can act immediately when hypoglycemic symptoms occur. Repeated treatment of hypoglycemia can lead to weight gain.

If hypoglycemia becomes severe, the person may be disoriented, unable to recognize a hypoglycemic reaction, and unable to swallow safely. In such cases, the person needs to receive IV glucose or the hormone glucagon or both to counteract the insulin reaction. Without treatment, the person may lapse into shock and die.

Enteral and Parenteral Formulas The indications for enteral and parenteral nutrition for people with diabetes are the same as for other people (see Chapters 20 and 21). When people with diabetes require enteral or parenteral formulas, however, they may also be experiencing insulin resistance, so adjustments may need to be made. Often health care professionals provide or adjust insulin doses to cover the higher blood glucose levels. If insulin fails to control hyperglycemia, people on parenteral nutrition may need to receive less energy from dextrose and more from IV lipid emulsions. People who cannot handle the carbohydrate in standard enteral formulas may benefit from formulas that contain less total carbohydrate (see Appendix F).

Meal-Planning Strategies

No single approach to medical nutrition therapy meets everyone's needs, and dietitians use several approaches to help clients follow a consistent diet plan and maintain their target blood glucose levels. Some diet strategies use food guides or simple menus to teach clients to plan diets. Traditionally, however, diet planners use the exchange system to help people with diabetes plan their diets.

Exchange Lists The exchange system sorts foods into three main groups by their proportions of carbohydrate, fat, and protein. The foods in these three groups—the carbohydrate group, the fat group, and the meat and meat substitutes group (protein)—are then organized into several exchange lists. Figure 24–2 on pp. 608–609 shows examples of foods and portion sizes for each group and Table 24–4 on p. 610 shows the energy and energy nutrients content of each exchange list.

Easy-to-eat sources of carbohydrate (10 to 15 g per serving):
- *2 to 3 tsp honey.*
- *4 to 5 hard candies (such as Lifesavers).*
- *5 to 6 large jelly beans.*
- *4 oz orange or other fruit juice.*
- *1 tbs icing from a can or tube.*
- *Glucose gel or tablets (check label for amount).*

Starch

1 starch exchange is like:
1 slice bread.
¾ c ready-to-eat cereal.
½ c cooked pasta, rice noodles, or bulgur.
⅓ c cooked rice.
½ c cooked beans.
½ c corn, peas, or yams.
1 small (3 oz) potato.
½ bagel, English muffin, or bun.
1 tortilla, waffle, roll, taco, or matzoh.
(1 starch = 15 g carbohydrate, 3 g protein, 0–1 g fat, and 80 kcal.)

Vegetables

1 vegetable exchange is like:
½ c cooked carrots, greens, green beans, brussels sprouts, beets, broccoli, cauliflower, or spinach.
1 c raw carrots, radishes, or salad greens.
1 large tomato.
(1 vegetable = 5 g carbohydrate, 2 g protein, and 25 kcal.)

Fruits

1 fruit exchange is like:
1 small banana, nectarine, apple, or orange.
½ large grapefruit, pear, or papaya.
½ c orange, apple, or grapefruit juice.
17 small grapes.
⅓ cantaloupe (or 1 c cubes).
2 tbs raisins.
1½ dried figs.
3 dates.
1½ carambola (star fruit).
(1 fruit = 15 g carbohydrate and 60 kcal.)

© Polara Studios, Inc. (all)

Meat and substitutes (very lean)

1 very lean meat exchange is like:
1 oz chicken (white meat, no skin).
1 oz cod, flounder, or trout.
1 oz tuna (canned in water).
1 oz clams, crab, lobster, scallops, shrimp, or imitation seafood.
1 oz fat-free cheese.
½ c cooked beans, peas, or lentils.[a]
¼ c fat-free or low-fat cottage cheese.
2 egg whites (or ¼ c egg substitute).
(1 very lean meat = 7 g protein, 0–1 g fat, and 35 kcal.)

[a]½ c cooked beans = 1 very lean meat exchange *plus* 1 starch exchange.

Meats and substitutes (lean)

1 lean meat exchange is like:
1 oz beef or pork tenderloin.
1 oz chicken (dark meat, no skin).
1 oz herring or salmon.
1 oz tuna (canned in oil, drained).
1 oz low-fat cheese or luncheon meats.
(1 lean meat = 7 g protein, 3 g fat, and 55 kcal.)

Meats and substitutes (medium-fat)

1 medium-fat meat exchange is like:
1 oz ground beef.
1 oz pork chop.
1 egg.
¼ c ricotta.
4 oz tofu.
(1 medium-fat meat = 7 g protein, 5 g fat, and 75 kcal.)

Figure 24–2

The Exchange System: Example Foods, Portion Sizes, and Energy Nutrient Contributions

Appendix B includes the complete U.S. exchange system.

The carbohydrate group includes these exchange lists:

- Starch (cereals, grains, pasta, breads, crackers, snacks, starchy vegetables, and dried beans, peas, and lentils).
- Fruit.
- Milk and some milk products.
- Other carbohydrates (desserts and snacks with added sugars and fats).
- Vegetables (nonstarchy vegetables).

Some foods in the carbohydrate group also contain fat, and the fat must be counted in addition to the carbohydrate. Milk provides an example. The exchange system encourages users to think of fat-free milk as milk and of whole milk as milk with added fat.

Other carbohydrates

1 other carbohydrates exchange is like:
2 small cookies.
1 small brownie or slice of cake.
5 vanilla wafers.
1 granola bar or ½ fat-free granola bar
½ c ice cream.
(1 other carbohydrate = 15 g
carbohydrate and may be exchanged for
1 starch, 1 fruit, or 1 milk. Because many
items on this list contain added sugar
and fat, their fat and kcalorie values vary,
and their portion sizes are small.)

Meats and substitutes (high-fat)

1 high-fat meat exchange is like:
1 oz pork sausage.
1 oz luncheon meat (such as bologna).
1 oz regular cheese (such as cheddar or swiss).
1 small hot dog (turkey or chicken).[b]
2 tbs peanut butter.[c]
(1 high-fat meat = 7 g protein, 8 g fat, and 100 kcal.)

[b]A beef or pork hot dog counts as 1 high-fat meat exchange *plus* 1 fat exchange.
[c]Peanut butter counts as 1 high-fat meat exchange *plus* 1 fat exchange.

Milks (fat-free and low-fat)

1 fat-free milk exchange is like:
1 c fat-free or 1% milk.
¾ c fat-free yogurt, plain.
1 c fat-free or low-fat buttermilk.
½ c evaporated fat-free milk.
⅓ c dry fat-free milk.
(1 fat-free milk = 12 g carbohydrate,
 8 g protein, 0–3 g fat, and 90 kcal.)

Milks (reduced-fat)

1 reduced-fat milk exchange is like:
1 c 2% milk.
¾ c low-fat yogurt, plain.
(1 reduced-fat milk = 12 g carbohydrate,
8 g protein, 5 g fat, and 120 kcal.)

Milks (whole)

1 whole-milk exchange is like:
1 c whole milk.
½ c evaporated whole milk.
(1 whole milk = 12 g carbohydrate,
8 g protein, 8 g fat, and 150 kcal.)

THE FAT GROUP

Fats

1 fat exchange is like:
1 tsp butter.
1 tsp margarine or mayonnaise (1 tbs reduced fat).
1 tsp any oil.
1 tbs salad dressing (2 tbs reduced fat).
8 large black olives.
10 large peanuts.
⅛ medium avocado.
1 slice bacon.
2 tbs shredded coconut.
1 tbs cream cheese (2 tbs reduced fat).
(1 fat = 5 g fat and 45 kcal.)

NOTE: Health recommendations urge people to limit their intakes of saturated fats; butter, bacon, coconut, and cream cheese contain saturated fats.

The fat group includes typical fats such as butter, margarine, oil, and salad dressing as well as high-fat foods such as nuts, olives, bacon, avocado, coconut, and cream cheese. By including items like bacon and avocados on the fat list, the exchange system alerts users to foods that are unexpectedly high in fat. The fat list also shows which fats are monounsaturated, polyunsaturated, and saturated to help clients easily plan lipid-lowering diets.

The meat and meat substitutes group includes high-protein foods. The meat list is also separated into categories based on their fat contents. Very lean meats count as a meat exchange, and lean, medium-, and high-fat meats count as meat and some fat. Meat exchanges that are particularly high in cholesterol are noted.

Carbohydrate-Modified Diets for Diabetes

Table 24–4 The Exchange Lists: Energy Nutrients Per Serving				
Group/Lists	Carbohydrate (g)	Protein (g)	Fat (g)	Energy (kcal)
Carbohydrate Group				
Starch	15	3	1 or less	80
Fruit	15	—	—	60
Milk				
Fat-free and very low fat	12	8	0–3	90
Low-fat	12	8	5	120
Whole	12	8	8	150
Other carbohydrates	15	varies	varies	varies
Vegetable	5	2	—	25
Meat and Meat Substitutes Group				
Meat				
Very lean	—	7	0–1	35
Lean	—	7	3	55
Medium-fat	—	7	5	75
High-fat	—	7	8	100
Fat Group				
Fat	—	—	5	45

By strictly defining portion sizes, all of the foods on a given exchange list provide approximately the same amounts of energy (kcalories) and energy nutrients (carbohydrate, fat, and protein). Therefore, any food on a list can be exchanged, or traded, for any other food on that same list without affecting a plan's energy balance. Clients also learn how to use food labels to determine the number of exchanges for foods not shown on the exchange lists. Review the "How to" box on pp. 612–613 to learn the basics of using exchange lists to plan a diet for diabetes.

Carbohydrate Counting Carbohydrate counting is another strategy for planning diets for diabetes. Carbohydrate counting allows clients more flexibility in adjusting their diets while still maintaining blood glucose control. To learn carbohydrate counting, clients may first learn the exchange list system to help them establish healthy eating habits, understand portion sizes, control their intake of energy and energy nutrients, and learn to eat consistent amounts of carbohydrates at regular times. Clients must also learn to perform the mathematical operations necessary to calculate their carbohydrate intakes.

Once clients understand how to maintain a healthy diet and control their blood glucose levels, they can focus on the carbohydrate they eat to simplify meal planning and manage times when their intakes vary. Clients who monitor their blood glucose and keep records of their glucose levels, the time and amount of carbohydrate they eat at each meal, and the time and amount of insulin or oral antidiabetic agents they use gain the most benefits from carbohydrate counting. The "How to" box on p. 615 shows how carbohydrate counting works.

In Summary The diet for diabetes parallels a healthy diet for all people, but emphasizes a consistent intake of carbohydrate spaced evenly throughout the day. Clients need to adjust their diets when they engage in physical activity, miss meals, or become ill. Clients can use different strategies to adjust their diets to maintain blood glucose control. Learning to use exchange lists and carbohydrate counting are two examples.

Drug Therapy for Diabetes

People with type 1 diabetes need insulin to control blood glucose and obtain energy. People with type 2 diabetes can sometimes control blood glucose with diet and physical activity. When these measures fail, one or a combination of oral antidiabetic agents may be prescribed. If oral antidiabetic agents fail to control blood glucose, the physician may prescribe insulin or insulin analogs, alone or in combination with oral antidiabetic agents.

Actions of Medications

Medications do not replace diet and physical activity in the management of diabetes; advise clients to continue these therapies. The Diet-Drug Interactions box on p. 616 includes the agents described here. Medications used to treat cardiovascular and renal complications that may accompany diabetes are described in Chapters 25 and 26, respectively.

Oral Antidiabetic Agents   The number of oral agents available to treat people with type 2 diabetes has grown markedly in recent years. Some oral agents work by stimulating the release of insulin from the beta cells; these include sulfonylureas, replaglinide, and nateglinide. Sulfonylureas (mainly chlorpropamide) are more likely than other oral agents to lead to hypoglycemia. Sulfonylureas may also react with alcohol (see p. 606) and can cause a disulfiram-like reaction. Nateglinide stimulates insulin release for only short periods following meals to help minimize postprandial hyperglycemia without overtaxing the pancreas. Metformin and the thioglitazones (rosiglitazone and piolitazone) work primarily by lessening peripheral insulin resistance. The alpha-glucosidase inhibitors, acarbose and meglitol, block the digestion of starches and slow the digestion of disaccharides, limiting the rise in blood glucose after a meal. People who use alpha-glucosidase inhibitors must use glucose to treat episodes of hypoglycemia.

Insulin and Insulin Analogs   For people who need insulin, commercial insulin comes in different forms that act with different timings so that it can be delivered in a manner that mimics the body's normal insulin actions as closely as possible. As Figure 24–3 shows, insulin can be either rapid acting (regular), intermediate acting (NPH and lente), or long acting (ultralente). Insulin analog (lispro) is a rapid-acting insulin whose amino acid composition has been modified so that it works faster and has a shorter duration of action. As a result, lispro reduces

Reminder: Metformin is the oral agent that has been shown to reduce the incidence of type 2 diabetes in people with impaired glucose tolerance.

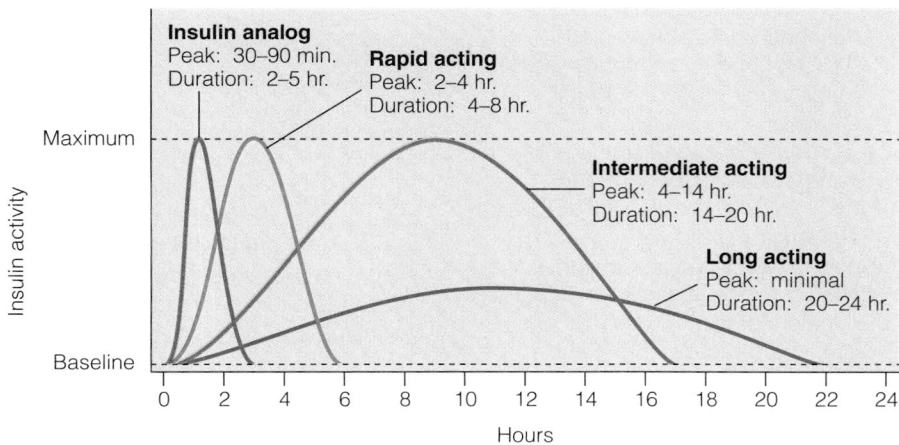

Figure 24–3
Actions of Insulin Types

Plan a Diet for Diabetes Using Exchange Lists

*U*sing exchange lists to plan diets takes time at first, but with practice, the planning process becomes routine. This box describes a simplified diet plan.

The first step is to assess each individual to determine a reasonable body weight and the energy intake necessary to achieve or maintain that body weight. For this example, we will use a 45-year-old woman who is 5 feet 2 inches tall and is comfortable with the weight of 120 pounds that she has maintained throughout her adult life. From an assessment of food intake, the dietitian estimates that the woman has maintained her weight on about 2000 kcalories per day with 25 percent of kcalories from protein, 45 percent from carbohydrate, and 30 percent from fat.

1. The next step is to compare the client's intake to the recommended diet for diabetes and make adjustments, if necessary. The dietitian begins by looking at protein. The client currently eats 25 percent of her kcalories from protein, somewhat more protein than the recommended amount (10 to 20 percent of the total kcalories). To minimize the changes the client will need to make, the dietitian plans the diet to include 20 percent of kcalories from protein. The recommended grams of protein is calculated as follows:

   ▪ .20 × 2000 kcal = 400 kcal.
     400 kcal ÷ 4 kcal/g = 100 grams.

2. Thus 80 percent or 1600 kcalories remain for carbohydrate and fat. After reviewing information about the woman's blood lipids, which are within acceptable limits, the dietitian plans the diet to keep fat at the current and acceptable level of 30 percent.

   ▪ .30 × 2000 kcal = 600 kcal.
     600 kcal ÷ 9 kcal/g = 67 g.

3. This means that 50 percent of the kcalories remain for carbohydrate.

   ▪ .50 × 2000 kcal = 1000 kcal.
     1000 kcal ÷ 4 kcal/g = 250 g.

   In summary, the diet will provide 2000 kcalories with a distribution of about 20 percent protein (100 grams or 400 kcalories), 30 percent fat (67 grams or 600 kcalories), and 50 percent carbohydrate (250 grams or 1000 kcalories).

4. Now it is time to translate the diet prescription into a meal plan. Table 24–4 on p. 610 shows the grams of carbohydrate, protein, and fat and the energy value in each serving on an exchange list. Using this table and the client's food intake record as a guide, the dietitian first plans servings of foods that contain carbohydrate, then protein, and finally fat, trying to match foods as closely as possible to the client's usual food

intake. This process takes practice and requires some adjusting based on trial and error. Most often, the final result does not fit the meal plan exactly, but comes close. Table 24–5 shows how the dietitian might plan a day's exchanges for the woman in this example. Note that the plan falls within the guidelines of the Daily Food Guide on pp. 16–17. Lower-fat foods are encouraged. The client's food intake record indicates that the client drinks 1% low-fat milk, so the plan uses very-low-fat milk for calculations. The plan is also based on lean meat exchanges. If the client occasionally chooses to use another type of milk or meat, the number of fat servings must be adjusted accordingly. For example, if the woman eats 2 ounces of a high-fat meat (16 grams of fat) instead of lean meat (6 grams of fat), she must then use about 2 fewer fat exchanges during the day (10 grams of fat). The plan shown in Table 24–5 does not include the

Table 24–5 A Day's Exchanges for a Sample 2000-kCalorie Diet				
Exchange Group/List	Number of Exchanges	Carbohydrate (g)	Protein (g)	Fat[a] (g)
Carbohydrate group[b]				
Starch	10	150	30	0
Fruit	3	45	—	—
Milk, very low fat	3	36	24	9
Vegetable	5	25	10	—
Meat and meat substitutes group				
Lean	5	—	35	15
Fat group	8	—	—	40
Total grams		256	99	64
Total kcalories		1024	396	576
% kcalories		51	20	29

[a]To ease calculation, exchanges from the carbohydrate groups are assumed to have 0 grams fat. If the client uses a fat-containing exchange, the fat can be deducted from the daily fat allowance.
[b]Foods from the "other carbohydrates" list can be substituted for a starch, fruit, or milk list exchange. Any fat in the selected food is then deducted from the daily fat allowance.

How to

Plan a Diet for Diabetes Using Exchange Lists—continued

Table 24–6 Dividing a Day's Exchanges between Meals and Snacks

Exchange Group/List	Number of Exchanges	Breakfast	Lunch	Supper	Bedtime Snack
Carbohydrate group					
Starch	10	2	3	3	2
Fruit	3	1		1	1
Milk, very low fat	3	1	1		1
Vegetable	5		2	3	
Meat and meat substitutes group					
Lean	5		2	2	1
Fat group	8	2	2	2	2

"other carbohydrates" list; starches and other carbohydrates (described later) can be substituted for foods on this list.

5. Distribute foods into meals that fit the client's usual eating patterns. Table 24–6 shows how the day's exchanges might be divided for the woman in this example. With this information in hand, the dietitian and client can begin to fill in the plan with real foods to create a sample menu such as the one shown in Table 24–7.

6. Teach clients how to tailor the diet to meet their own preferences. For example, foods from the starch, fruit, milk and other carbohydrate lists contain similar amounts of energy and carbohydrate and can be substituted for one another from time to time. Regular substitution is discouraged, however, because each list makes unique contributions to other nutrient needs. A client who regularly substitutes fruit for milk, for example, may not be getting enough calcium. The client who regularly substitutes milk for a starch or fruit may not be getting enough fiber.

7. Three servings of free foods can be included as long as they are spread throughout the day. Free foods contain up to 20 kcalories and 5 grams of carbohydrate per serving. A serving of food that contains a carbohydrate-based fat substitute that provides 5 grams or less carbohydrate counts as a free food.

Table 24–7 Translating a Day's Exchanges into a Day's Meals

Breakfast

½ c high-fiber cereal (1 starch)

1 c very-low-fat milk (1 milk)

½ bagel, 1 oz (1 starch)

1 tsp margarine (1 fat)

½ banana (1 fruit)

Coffee

Lunch

1 c spaghetti (2 starch) served with:

2 oz meatballs (2 medium-fat meat)[a]

½ c spaghetti sauce (1 starch, 1 fat)

1 c broccoli (2 vegetables)

1 c very-low-fat milk (1 milk)

Supper

2 oz grilled salmon (2 lean meat)

⅔ c brown rice (2 starch)

1 c cooked carrots (2 vegetables)

1 c mixed green salad (1 vegetable)

Salad dressing made with 2 tsp olive oil (2 fat) and red wine vinegar

1 whole-wheat roll (1 starch)

1 tsp margarine (1 fat)

1 c diced cantaloupe (1 fruit)

Iced tea

Bedtime Snack

1 fat-free granola bar (2 starch)

2 tbs peanut butter (1 high-fat meat plus 1 fat)[b] spread over slices of 1 small apple (1 fruit)

1 c very-low-fat milk (1 milk)

NOTE: Compared to the plan shown in Table 24–6, this menu provides one fewer fat at breakfast and one additional fat at supper.

[a]Two medium-fat meats = 2 lean meats plus 1 fat.

[b]Two tbs peanut butter = 1 high-fat meat (8 grams of fat) plus 1 fat exchange (5 grams fat) or the equivalent of one lean meat (3 grams fat) and 2 fats (10 grams of fat)

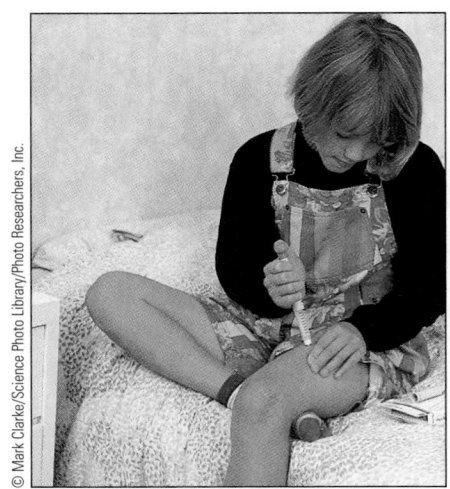

Injections are one option for delivering insulin to people with diabetes.

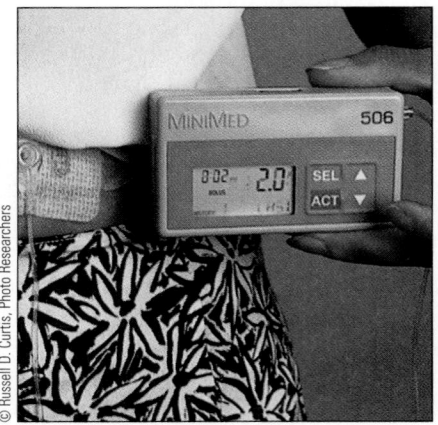

For people who use external insulin pumps, insulin passes from the pump through tubing that enters the body through the abdomen.

Human insulin, the most widely used insulin, is synthesized from bacteria or yeast using recombinant DNA (genetic engineering). Years ago, most insulins were derived from pork or beef sources or a combination of the two.

With exercise, counterregulatory hormone levels rise, which elevates blood glucose and provides additional fuel to the body. If blood glucose is already elevated at the onset of exercise, then it may rise to dangerously high levels.

multiple daily injections: delivery of a mixture of insulins by injection three or more times daily.

postprandial hyperglycemia to a greater extent than regular insulin and is also associated with a lower risk of hypoglycemia between meals and during the night. Some insulins come mixed together so that only one injection is required to provide both types of insulin.

Some people experience a temporary remission from diabetes after their initial treatment with insulin—a time referred to as the "honeymoon phase." Recall that some beta cell function may remain when type 1 diabetes is diagnosed. With insulin treatment and relief from hyperglycemia, the beta cells of the pancreas temporarily regain their function.

Insulin Administration

Normally, the body secretes a constant, baseline amount of insulin at all times and more after blood glucose rises following a meal. The person with type 1 diabetes often receives NPH (intermediate-acting) insulin to meet baseline needs and regular (rapid-acting) insulin and/or insulin analogs to process energy nutrients after meals.

Currently, people who need insulin must inject it or use external pumps to deliver the insulin they need. Personal preferences, motivational level, and financial considerations guide clients in deciding which delivery system works best for them. In the future, people may also have the option of inhaling insulin or using it in an oral form.[23] The chemical encapsulating oral insulin allows it to evade any digestion in the stomach and small intestine. Once in the intestine, the chemical creates small temporary openings between the tight junctions of the intestinal wall that allow insulin to pass into the bloodstream as a protein rather than as individual amino acids.

Insulin Injections   Clients who practice intensive therapy use **multiple daily injections** (a mixture of two or more types of insulin three to four times daily) or external pumps to meet their insulin needs. People who practice traditional therapy use insulin less often (one to two times daily). Single injections are seldom effective for people with type 1 diabetes. People with type 2 diabetes may be treated with insulin alone, or they may use it in combination with oral antidiabetic agents. Often a single injection of NPH (intermediate-acting) insulin is given at bedtime. Insulin analogs may also be used in the treatment of type 2 diabetes.

Insulin Pumps   For those who choose pumps to deliver insulin, external pumps, about the size of a beeper, hold enough insulin to meet needs for two or three days. From the pump, insulin enters the body through tubing and a needle inserted into the abdominal area. Implanted pumps with built-in glucose sensors that automatically dispense insulin in response to changing glucose levels may also be an option for insulin delivery in the future.[24]

Physical Activity and Insulin   People who use insulin should check their blood glucose before and after engaging in physical activity to help them maintain their blood glucose levels. Physical activity should not be undertaken if blood glucose is too low (less than 100 milligrams per deciliter) because hypoglycemia can quickly develop. If blood glucose is too high (greater than 300 milligrams per deciliter), exercising can cause blood glucose levels to rise even higher.

Generally, clients should take insulin more than an hour before physical activity. Vigorous physical activity and warm temperatures speed blood flow, increase the rate of insulin absorption, and set the stage for a hypoglycemic reaction, which may occur after several hours. Reducing the insulin dose before and after the activity by up to 30, or even 50, percent can help to prevent this sequence of events. People prone to nocturnal hypoglycemia who engage in strenuous physical activity are often advised to undertake the activity early in the day or to reduce the insulin dose when the activity is undertaken late in the day.

Help Clients Count Carbohydrates

1. Begin by calculating how much carbohydrate the person usually eats at each meal and snack. For example, assume that the client is an elderly man who maintains his weight and controls his blood glucose levels on an intake of 1700 kcalories. Use exchanges to determine his usual carbohydrate intake. In this example, the client and dietitian determined that his usual eating pattern includes:

   - Breakfast: 2 starch, 1 fruit, 1 milk.
   - Lunch: 2 starch, 1 fruit, 2 vegetables.
   - Midafternoon snack: 1 fruit, 1 milk.
   - Supper: 2 starch, 1 fruit, 2 vegetables.
   - Bedtime snack: 1 starch, 1 fruit.

2. Next, use Table 24–4 on p. 610 to determine the grams of carbohydrate in each meal. Breakfast, for example, contains:

   $$2 \text{ starch} = 2 \times 15 \text{ g carbohydrate} = 30 \text{ g}$$
   $$1 \text{ fruit} = 1 \times 15 \text{ g carbohydrate} = 15 \text{ g}$$
   $$1 \text{ milk} = 1 \times 12 \text{ g carbohydrate} = 12 \text{ g}$$
   $$\text{Total} = 57 \text{ g carbohydrate.}$$

   Using the same method to calculate the carbohydrate from other meals and snacks, the carbohydrate contents are as follows:

   - Lunch: 55 g.
   - Midafternoon snack: 27 g.
   - Supper: 55 g.
   - Bedtime snack: 30 g.

3. Translate the grams of carbohydrate into carbohydrate choices. Each carbohydrate choice equals about 15 grams of carbohydrate. To simplify carbohydrate counting, servings from the starch, fruit, and milk lists are considered to be one carbohydrate choice. To determine the number of carbohydrate choices, divide the carbohydrate at each meal and snack by 15 and round the results:

   - Breakfast:

     57 g carbohydrate ÷ 15 = 3.8 or 3½ to 4 carbohydrate choices.

   - Lunch:

     55 g carbohydrate ÷ 15 = 3.7 or 3½ to 4 carbohydrate choices.

   - Midafternoon snack:

     27 g carbohydrate ÷ 15 = 1.8 or 1½ to 2 carbohydrate choices.

   - Supper:

     55 g carbohydrate ÷ 15 = 3.7 or 3½ to 4 carbohydrate choices.

   - Bedtime snack:

     30 g carbohydrate ÷ 15 = 2 carbohydrate choices.

4. Advise the client to include these same amounts of carbohydrate consistently at each meal and snack. Clients can use exchanges, food labels, and food composition tables to determine the carbohydrate contents of the foods they eat.

5. Encourage clients to carefully weigh and measure all foods initially. Once clients are familiar with portion sizes, they can weigh and measure foods less often, but should still check portion sizes occasionally.

6. Remind clients that they still need to be aware of the total amount of food they eat as well as the type and amount of fat they are eating.

Clients who use insulin can learn how much insulin they need to cover their carbohydrate intakes below. Then, if they change their carbohydrate intakes on occasion, they can adjust their insulin dose accordingly.

Intentional Misuse of Insulin As described earlier, adolescents with diabetes are often heavier than adolescents without diabetes, and they are also very self-conscious about their weights. Although the incidence of overt eating disorders is no greater among teenagers with diabetes than among their peers, some overweight teens with diabetes, particularly girls, may intentionally use less than recommended amounts of insulin or omit insulin doses as a means of weight control (recall that hyperglycemia leads to weight loss).[25] Such practices can lead to both acute and chronic complications.

Carbohydrate-to-Insulin Ratios Clients who practice intensive therapy can work with health care professionals to determine the unique way their body uses insulin in response to carbohydrate. To uncover their **carbohydrate-to-insulin ratio,** clients eat a consistent amount of carbohydrate, monitor their blood glucose, and keep records of their food intake, insulin use, physical activity, and

carbohydrate-to-insulin ratio: the grams of carbohydrate covered by one unit of insulin. The lower the ratio, the more insulin is needed to cover carbohydrate intake.

Carbohydrate-Modified Diets for Diabetes

Alpha-glucosidase Inhibitors (acarbose and meglitol)

Alpha-glucosidase inhibitors are taken at the start of each meal. Clients using these agents must use glucose to treat hypoglycemia. Nutrition-related side effects include abdominal pain, gas, and diarrhea.

Insulin

The timing of *insulin* administration and meals varies, depending on the action of the insulin prescribed. Insulin may cause hypoglycemia.

Lispro (insulin analog)

Lispro is taken 5 to 10 minutes before meals. Lispro lowers the risk of hypoglycemia compared to insulin.

Metformin

Metformin is taken once or twice a day before meals (breakfast or breakfast and dinner). GI side effects are uncommon, but metformin may cause diarrhea or leave a metallic taste in the mouth.

Nateglinide

Nateglinide is taken immediately before eating. Nateglinide can cause hypoglycemia and sometimes diarrhea.

Repaglinide

Repaglinide is taken before meals. Repaglinide can cause hypoglycemia and diarrhea.

Sulfonylureas (chlorpropamide, glipzide, glyburide, glimepiride)

Sulfonylureas are generally taken one or two times a day, before meals. Disulfiram-like reactions can occur when large amounts of alcohol are taken (especially with *chlorpropamide*). These medications can also cause hypoglycemia.

Thioglitazones (rosiglitazone, piolitazone)

Thioglitazones are taken once or twice a day before meals. Nutrition-related side effects are uncommon. One medication of this type, *troglitazone,* was taken off the market due to concerns that its use might be a cause of liver failure.

NOTE: Ask clients who take insulin or antidiabetic agents about their use of dietary supplements or herbal teas that contain the following ingredients, which may affect blood glucose: chromium, ephedra, garlic, ginger, ginseng, and niacin.

---

illness. From these records, health care professionals calculate the number of units of insulin a person needs per gram of carbohydrate. Then clients can use their carbohydrate-to-insulin ratio to adjust their insulin doses when they eat more or less carbohydrate than their meal plan provides.

Pancreas and Beta Cell Transplants For people with type 1 diabetes who encounter serious problems managing their diseases with insulin, pancreas transplants have been successful in providing functional, insulin-producing beta cells. Most often, a pancreas transplant is combined with a kidney transplant because the person often has significant kidney problems as well. A combination pancreas-kidney transplant can eliminate the need for insulin and for dialysis (see Chapter 26) and greatly enhance the person's quality of life.

Very promising clinical trials have shown that transplanted beta cells infused into the portal vein can secrete insulin. Early studies show that beta cell transplantation can eliminate the need for insulin injections or infusions in people with type 1 diabetes, providing hope that a cure for type 1 diabetes may be on the horizon.[26] Researchers have also successfully generated insulin-producing cells from embryonic stem cells, a technique that could provide a vast pool of beta cells and minimize the need for organ donors as a source of beta cells.[27]

Stem cells are undifferentiated cells found in bone marrow and lymph tissues that may give rise to a variety of cells.

Coordinating Medications, Diet, and Physical Activity

Once a client implements a diet, medication, and physical activity plan, adjustments are necessary from time to time. A child may grow, an adult may change jobs and need to adjust the plan, or the treatment plan may not work when it is put into practice. Clients and health care professionals can use blood glucose,

Table 24–8 Strategies for Managing Hyperglycemia and Hypoglycemia

Problem	Possible Solutions[a]
Hyperglycemia	
Before breakfast	▪ Adjust dose of intermediate-acting insulin at bedtime.[b]
Before lunch	▪ Adjust morning dose of rapid-acting insulin.[b]
	▪ Reduce amount of carbohydrate at breakfast.
	▪ Reduce or omit midmorning snack.
	▪ Change time of breakfast or midmorning snack.
	▪ Add physical activity after breakfast.
Before dinner	▪ Adjust afternoon dose of rapid-acting insulin.[b]
	▪ Reduce carbohydrate at lunch.
	▪ Reduce or omit midafternoon snack.
	▪ Change time of lunch or midafternoon snack.
	▪ Add physical activity between lunch and dinner.
At bedtime	▪ Adjust insulin dose before dinner.[b]
	▪ Reduce amount of carbohydrate at dinner.
	▪ Reduce or omit evening snack.
	▪ Add physical activity after dinner.
Hypoglycemia	
Before breakfast	▪ Adjust dose of intermediate- or long-acting insulin at bedtime.[b]
	▪ Add carbohydrate at evening snack.
	▪ Avoid strenuous activity late in the day.
Before lunch	▪ Adjust morning dose of rapid-acting insulin.[b]
	▪ Add carbohydrate at breakfast.
	▪ Add a morning snack.
	▪ Change time of breakfast, lunch, or morning snack.
	▪ Adjust physical activity schedule.
Before dinner	▪ Adjust afternoon dose of rapid-acting insulin.[b]
	▪ Add carbohydrate at lunch.
	▪ Add an afternoon snack.
	▪ Change time of lunch, dinner, or afternoon snack.
	▪ Adjust physical activity schedule.
At bedtime	▪ Adjust insulin dose before dinner.[b]
	▪ Add carbohydrate at dinner.
	▪ Add an evening snack.
	▪ Change time of dinner or evening snack.

[a]Skilled health care professionals gather additional data to find the best solution for problems with blood glucose control. Is the problem an isolated occurrence or a pattern? Has food intake changed? If yes, why? Has the activity level changed? Has illness been a problem? Whenever altering the diet or physical activity plan is difficult for the client, insulin is adjusted to correct problems, if possible.
[b]Insulin doses can be adjusted in amount or timing or both.

diet, medication, and physical activity records to look for solutions to problems clients may have with blood glucose control. Table 24–8 summarizes examples of strategies for correcting problems with hyperglycemia and hypoglycemia.

In Summary Although all people with type 1 diabetes eventually need insulin to treat diabetes, some people with type 2 diabetes can manage their blood glucose with diet and physical activity alone. Others with type 2 diabetes can be managed with oral antidiabetic agents, insulin, insulin analogs, or a combination of these. Currently, insulin must be administered by injection or by using an insulin pump. Pancreas transplants are options for treating diabetes. Beta cell transplants hold a great deal of promise as a possible cure.

Carbohydrate-Modified Diets for Diabetes

Careful control of blood glucose during pregnancy offers the best chance of a safe delivery and healthy infant for women with diabetes.

The hormones that oppose the actions of insulin during late pregnancy include *placental lactogen, cortisol, prolactin,* and *progesterone.*

The term that describes infants that are large for gestational age is **macrosomia.**

Chapter 12 provides more information about gestational diabetes.

Diabetes Management in Pregnancy

All women who become pregnant face new challenges that include altering their diets to meet the demands of the growing fetus and addressing complications that occur as a consequence of pregnancy. Pregnancy elevates blood insulin and alters insulin resistance in all women. Because pregnancy in women with diabetes stresses an already altered glucose regulatory system, they should expect control to become more difficult during pregnancy.

Diabetes during Pregnancy

Diet Order
Energy- and carbohydrate-modified diet.

Women with either type 1 or type 2 diabetes who are contemplating pregnancy should know that uncontrolled diabetes in early pregnancy raises the risk of spontaneous abortions.[28] A fetus exposed to high blood glucose and ketones may develop birth defects. The extra glucose also means that the fetus must make extra insulin to handle the load, which may lead to severe hypoglycemia in the infant after birth. High blood glucose levels also "overfeed" the growing fetus, resulting in large infants that are difficult to deliver. As a consequence, more pregnant women with diabetes require a cesarean delivery. Women with diabetes are more likely than other women to develop pregnancy-related hypertension (see Chapter 12).

Preexisting Diabetes   Women with diabetes who are contemplating pregnancy need to receive preconception care, which aims to achieve the best possible blood glucose control before pregnancy, and continued care to maintain control during pregnancy. Women in the Diabetes Control and Complications Trial who tightly controlled their blood glucose levels and became pregnant experienced rates of spontaneous abortion and birth defects similar to those of women without diabetes.[29]

Gestational Diabetes   Women who have never had diabetes or never knew that they had it may be diagnosed with diabetes during pregnancy (gestational diabetes). Gestational diabetes is the most common medical complication of pregnancy. The American Diabetes Association recommends that, with few exceptions, women be screened for diabetes between 24 and 28 weeks of gestation.

Treatment of Diabetes during Pregnancy

Obstetricians recommend blood glucose monitoring for all pregnant women with any type of diabetes. Establishing blood glucose control is important to the health of both mother and infant. For pregnant women with diabetes, individualized diets tailored to meet the added nutrient demands of pregnancy and carefully coordinated with insulin therapy (when necessary) and physical activity are central to therapy.

Medical Nutrition Therapy   The diet plan aims to maintain blood glucose levels and prevent complications. It should provide adequate but not excessive kcalories to support appropriate weight gain (see the weight-gain recommendations on p. 306). The margin summarizes guidelines for energy needs for pregnant women with diabetes. Nutrient needs are highly individualized. Careful and ongoing nutrition assessments help dietitians determine if the diet should be adjusted.

Consistency in meals and snacks helps assure an ongoing supply of glucose without inducing hyperglycemia. A careful distribution of carbohydrate throughout the day and the use of low–glycemic index foods may help limit postprandial hyperglycemia.[30] Frequent small meals and snacks often benefit women who do not use insulin. For women who use insulin, the dietitian adjusts the carbohydrate distribution to accommodate each woman's lifestyle and insulin administration schedule. A bedtime snack is often recommended to prevent nocturnal hypoglycemia and ketosis in the mother and to provide fuel to the developing fetus.

Some women with gestational diabetes experience high blood glucose in the morning; in such cases, limiting carbohydrate (to less than 20 percent of the day's total carbohydrate) at breakfast helps maintain morning blood glucose levels in an acceptable range until counterregulatory hormone levels diminish. Concentrated sweets are discouraged because they quickly raise blood glucose.

Preventive Measures after Gestational Diabetes   For most women with gestational diabetes, glucose tolerance returns to normal after pregnancy, but women with gestational diabetes and their offspring risk developing permanent diabetes (usually type 2 diabetes) later in life, especially if they are overweight.[31] The offspring of women with gestational diabetes are also likely to develop obesity and impaired glucose tolerance. For these reasons, health care professionals closely monitor women who have experienced gestational diabetes and their offspring and recommend that they seek medical attention if they develop symptoms suggestive of diabetes. Education focuses on strategies to achieve and maintain a healthy weight and to develop a regular program of physical activity. Use the case study on p. 620 to review the connections between gestational and type 2 diabetes.

In Summary   Careful management of blood glucose levels before and during pregnancy helps women with diabetes limit the risks of complications to mother and infant. The diet plan for diabetes during pregnancy carefully controls energy intakes and provides for a consistent intake of carbohydrate intake throughout the day. Women with gestational diabetes and their offspring may develop diabetes at a later time, and health care professionals check for diabetes at regular intervals following the pregnancy. The Nutrition Assessment Checklist highlights areas of concern for people with diabetes.

Energy needs for diabetes during pregnancy based on body weight:

- *For women with a desirable prepregnancy weight (BMI 20–26), 30 kcal/kg body weight.*
- *For women who are obese prior to pregnancy (BMI >30), 24 kcal/kg body weight.*

Case Study

SCHOOL COUNSELOR WITH TYPE 2 DIABETES

Mrs. Lopez is a 41-year-old Hispanic American woman recently diagnosed with type 2 diabetes. Mrs. Lopez developed gestational diabetes while she was pregnant with her second child. Her blood glucose returned to normal following pregnancy, and she was advised to get regular checkups, maintain a desirable weight, and engage in regular physical activity. Although she visits her physician at least once a year, she has been unable to maintain a healthy weight. At 5 feet 3 inches tall, Mrs. Lopez currently weighs 155 pounds. She is determined to lose weight and begin an activity plan because she fears that she might need to use insulin injections. She is also concerned about her husband and children because they are overweight as well. The physician refers Mrs. Lopez to a dietitian to help her plan a diet.

■ What factors in Mrs. Lopez's history increase her risk for diabetes? Are her husband and children also at risk?

■ Describe the general characteristics of the diet that will be appropriate for Mrs. Lopez to follow. In what ways can weight loss and physical activity benefit Mrs. Lopez? How will the dietitian determine what diet-planning strategy will work best for Mrs. Lopez?

■ If Mrs. Lopez is unable to control her blood glucose with diet and physical activity, what type of medication would the physician most likely prescribe? Can you explain to Mrs. Lopez why she would probably not require insulin at this time?

■ Why might the dietitian suggest nutrition counseling for the entire family?

Nutrition Assessment Checklist

FOR PEOPLE WITH DIABETES

Medical History

Check the medical record to determine:

- ☐ Type of diabetes
- ☐ Duration of diabetes
- ☐ Acute and chronic complications
- ☐ Other medical conditions, including pregnancy, that alter nutrient needs

Medications

For clients with preexisting diabetes who use antidiabetic agents, insulin, or both, note:

- ☐ Type(s) of antidiabetic agent or insulin
- ☐ Administration schedule

Check for other medications and note possible diet-drug interactions, including:

- ☐ Antilipemics (to lower blood lipids, see Chapter 25)
- ☐ Antihypertensive agents (to reduce blood pressure, see Chapter 25)
- ☐ Diuretics (to reduce blood pressure, see Chapter 25)
- ☐ Dietary supplements

Food/Nutrient Intake

To devise an acceptable meal plan and coordinate medications, obtain:

- ☐ An accurate and thorough record of food intake and usual eating habits
- ☐ An account of usual physical activities

At medical checkups, reassess the client's ability to:

- ☐ Maintain an appropriate energy intake
- ☐ Maintain a consistent intake of carbohydrate
- ☐ Adjust the diet for missed meals due to illness
- ☐ Use appropriate amounts and types of foods to treat hypoglycemia

Anthropometrics

Take accurate baseline height and weight measurements as a basis for:

- ☐ An appropriate energy intake
- ☐ Initial insulin therapy

Reassess height and weight for children and weight for adults periodically to ensure that the meal plan provides an appropriate energy intake.

Laboratory Tests

Check the following tests to monitor the success of diabetes therapy:

- ☐ Results of home blood glucose monitoring
- ☐ Glycated hemoglobin
- ☐ Blood lipids
- ☐ Ketones in urine
- ☐ Microalbuminemia, when available

Physical Signs

Look for signs of:

- ☐ Dehydration, especially in the elderly
- ☐ Nutrient deficiencies and excesses

Self Check

1. Diabetes mellitus describes a group of disorders characterized primarily by:
   a. altered fat metabolism and hyperglycemia.
   b. altered fluid and electrolyte balance.
   c. altered energy metabolism and hyperglycemia.
   d. insulin resistance.

2. Which of the following is the major characteristic of type 1 diabetes?
   a. The pancreas makes little or no insulin.
   b. It frequently goes undiagnosed.
   c. It is the predominant form of diabetes.
   d. Insulin secretion is ineffective in preventing hyperglycemia.

3. Which of the following describes type 2 diabetes?
   a. Autoimmunity is the primary cause.
   b. The pancreas makes little or no insulin.
   c. Hyperglycemia with ketoacidosis is a common complication.
   d. Chronic complications may have begun to develop before it is diagnosed.

4. Sudden hyperglycemia in a person who has consistently maintained good blood glucose control can be precipitated by:
   a. infections or illnesses.
   b. chronic alcohol ingestion.

c. undertreatment of hypoglycemia.

d. conditions that lower levels of counterregulatory hormones.

5. Repeated episodes of hypoglycemia in the person with diabetes can result in all of the following **except:**
   a. weight gain.
   b. retinopathy.
   c. shock and death.
   d. cognitive impairment.

6. The chronic complications associated with diabetes result from:
   a. alterations in kidney function.
   b. weight gain and hypertension.
   c. damage to blood vessels and nerves.
   d. infections that deplete nutrient reserves.

7. All of the following statements are true regarding physical activity and diabetes **except:**
   a. people with type 2 diabetes can improve blood glucose control through a regular program of physical activity.
   b. people with type 1 diabetes can improve blood glucose control through a regular program of physical activity.
   c. physical activity helps control weight, blood pressure, and blood lipids.
   d. people with type 1 diabetes can improve cardiovascular health through a program of regular physical activity.

8. The health care professional working with a client with diabetes emphasizes that the diet should provide:
   a. a very low intake of fat.
   b. more protein than regular diets.
   c. a restricted intake of simple sugars and concentrated sweets.
   d. a consistent carbohydrate intake from day to day and at each meal and snack.

9. Which of the following is true regarding the use of alcohol in a diet for diabetes?
   a. A serving of alcohol is always considered part of the carbohydrate allowance.
   b. Alcohol can cause hypoglycemia in all people, including those with diabetes.
   c. People with well-controlled blood glucose levels should refrain from alcohol use.
   d. In combination with alcohol, some types of insulin can cause a disulfiram-like reaction in people with type 2 diabetes.

10. The meal-planning strategy that is most effective for the person with diabetes is:
    a. the one that best helps the client control blood glucose levels.
    b. food guides and sample menus.

c. carbohydrate counting.

d. the exchange list system.

Critical Thinking

1. Which of the following describes a measurable goal of diabetes therapy?
   a. support of health and well-being
   b. blood glucose control
   c. emotional encouragement
   d. weight-loss diet

2. Which of the following best describes insulin therapy in a person receiving both NPH insulin and an insulin analog?
   a. Since NPH insulin is of long duration, there is no need for the insulin analog.
   b. Since glucose is not available from food between meals, the insulin analog alone would cover the person's insulin needs.
   c. NPH insulin covers basal insulin needs while the insulin analog covers the carbohydrate from meals.
   d. NPH insulin covers the carbohydrate from meals while the insulin analog covers basal insulin needs.

3. Which of the following describes the treatment plan for a person with type 1 diabetes who practices intensive therapy?
   a. The person uses multiple daily injections, monitors blood glucose several times a day, and adjusts insulin to accommodate meals and snacks.
   b. The person uses a single injection of insulin, monitors blood glucose once a day, and eats a specified amount of food at the same time each day.
   c. The person learns to control blood glucose and can then quit blood glucose monitoring as long as a consistent amount of carbohydrate is eaten at each meal and snack.
   d. The health care team assumes responsibility for monitoring the client's blood glucose and A1C to ensure that blood glucose remains tightly controlled.

4. Women with pregnancies complicated by diabetes:
   a. should not eat a snack at bedtime.
   b. need more carbohydrate than healthy women.
   c. need more carbohydrate than women with diabetes who are not pregnant.
   d. need the same number of kcalories to support the pregnancy as women without diabetes.

Answers to these questions appear in Appendix G.

Clinical Applications

1. Using the box on pp. 612–613, plan a diet using the exchange lists for a sedentary woman with type 1 diabetes who is 5 feet 9 inches tall and weighs 160 pounds. Assume that the distribution of kcalories will be 55 percent from carbohydrate, 20 percent from protein, and 25 percent from fat. Develop a sample menu.

2. An important part of learning is being able to apply knowledge and guidelines to real-life situations. Using Table 24–8 on p. 617 as a guide, think about the possible remedies for either hyperglycemia or hypoglycemia. Describe at least one situation when it might be preferable to alter the insulin dose and one situation when it might be preferable to alter the carbohydrate intake.

3. Take a trip to a pharmacy and price these items: blood glucose meter, test strips appropriate for the glucose meter you select, lancets, insulin, and syringes. Determine the approximate cost of insulin for a person who uses 14 units of regular insulin and 26 units of NPH insulin in three injections daily.

Then estimate the cost of testing blood glucose four times daily. Consider how an external pump might affect the total cost of managing diabetes. How does the need for a well-balanced diet influence the cost of diabetes care?

Nutrition on the Net

For further study of the topics of this chapter, access these websites.

Find updates and quick links to these and other nutrition-related sites at our website:
www.wadsworth.com/nutrition

Visit the American Diabetes Association and the Joslin Diabetes Center to find information on a wide range of topics related to diabetes:
www.diabetes.org and **www.joslin.org**

Review the qualifications of diabetes educators by visiting the American Association of Diabetes Educators site:
www.aadenet.org

Learn more about alternative therapies for diabetes by visiting The Natural Pharmacist encyclopedia:
www.alternativediabetes.com

Read about the National Institute of Diabetes & Digestive & Kidney Diseases study of the long-term effects of weight loss on type 2 diabetes by visiting Look AHEAD (Action for Health and Diabetes):
www.LookAHEADstudy.org

Notes

1. A. H. Mokdad and coauthors, The continuing increase of diabetes in the U.S., *Diabetes Care* 24 (2001): 1278–1283.

2. The Expert Committee on the Diagnosis and Classification of Diabetes Mellitus, Report of the Expert Committee on the diagnosis and classification of diabetes mellitus, *Diabetes Care* (supplement 1) 25 (2002): 5–20.

3. M. A. Atkinson and G. S. Eisenbarth, Type 1 diabetes: New perspectives on disease pathogenesis and treatment, *Lancet* 358 (2001): 221–229.

4. E. A. Simone, D. R. Wegmann, and G. S. Eisenbarth, Immunologic "vaccination" for the prevention of autoimmune diabetes (type 1a), *Diabetes Care* (supplement 2) 22 (1999): 7–15.

5. A. D. Mooradian and coauthors, Diabetes care for older adults, *Diabetes Spectrum* 12 (1999): 70–77.

6. S. Saffel-Shrier, Carbohydrate loading for older patients, *Diabetes Spectrum* 13 (2000): 158–162.

7. National Institute of Diabetes & Digestive & Kidney Diseases, Diet and exercise dramatically delay type 2 diabetes: Diabetes medication metformin also effective, available at **www.niddk.nih.gov,** site visited October 19, 2001.

8. I. J. Deary, Hypoglycemia-induced cognitive decrements in adults with type 1: A case to answer? *Diabetes Spectrum* 10 (1997): 13–15.

9. American Diabetes Association, Implications of the Diabetes Control and Complications Trial, *Diabetes Care* (supplement 1) 25 (2002): 25–27.

10. B. Janand-Delenne and coauthors, Silent myocardial ischemia in patients with diabetes, *Diabetes Care* 22 (1999): 1396–1400.

11. P. Bytzer and coauthors, Prevalence of gastrointestinal symptoms associated with diabetes mellitus: A population-based survey of 15,000 adults, *Archives of Internal Medicine* 161 (2001): 1989–1996; C. K. Rayner and coauthors, Relationships of upper gastrointestinal motor and sensory function with glycemic control, *Diabetes Care* 24 (2001): 371–381.

12. The Diabetes Control and Complications Trial/Epidemiology of Diabetes Interventions and Complications Research Group, Retinopathy and nephropathy in patients with type 1 diabetes four years after a trial of intensive therapy, *New England Journal of Medicine* 342 (2000): 381–389.

13. American Diabetes Association, Implications of the United Kingdom Prospective Diabetes Study, *Diabetes Care* (supplement 1) 25 (2002): 28–32.

14. D. S. Bell, Importance of postprandial glucose control, *Southern Medical Journal* 94 (2001): 804–809; E. Bastyr and coauthors, Therapy focused on lowering postprandial glucose, not fasting glucose, may be superior for lowering HbA1c, *Diabetes Care* 23 (2000): 1236–1241.

15. American Diabetes Association, Evidence-based nutrition principles and recommendations for the treatment and prevention of diabetes and related complications, *Diabetes Care* (supplement 1) 25 (2002): 56–60.

16. The Diabetes Control and Complications Trial Research Group, Influence of intensive diabetes treatment on body weight and composition of adults with type 1 diabetes in the Diabetes Control and Complications Trial, *Diabetes Care* 24 (2001): 1711–1721.

17. American Diabetes Association, Evidence-based nutrition principles and recommendations for the treatment and prevention of diabetes and related complications, 2002.

18. G. Arsenis and D. Goettelman, Alternative treatments of type 2 diabetes mellitus (abstract), presented at the 83rd Annual Meeting of the Endocrine Society, **www.endo-society.org,** site visited October 23, 2001.

19. R. Giacco and coauthors, Long-term dietary treatment with increased amounts of fiber-rich low-glycemic index natural foods improves blood glucose control and reduces the number of hypoglycemic events in type 1 diabetic patients, *Diabetes Care* 23 (2000): 1461–1466; M. Chandalia and coauthors, Beneficial effects of high dietary fiber intake in patients with type 2 diabetes, *New England Journal of Medicine* 342 (2000): 1440–1441.

20. S. Carden, The glycemic index in diabetes meal planning, 60th Scientific Sessions of the American Diabetes Association, June 13, 2000, **www.medscape.com/medscape/cno/2000/ADA/Story.cfm?story_id=1393,** site visited on February 14, 2001.

21. J. Connolly-Schoonen, Ask the experts on . . . overweight type 2 diabetic with elevated triglycerides, **http://diabetes/medscape.com,** site visited August 13, 2001; A Garg, High-monounsaturated-fat diets for patients with diabetes mellitus: A meta-analysis, *American Journal of Clinical Nutrition* (supplement) 67 (1998): 577–582.

22. American Diabetes Association, Evidence-based nutrition principles and recommendations for the treatment and prevention of diabetes and related complications, 2002.

23. B. L. Laube, Treating diabetes with aerolized insulin, *Chest* (supplement 3) 120 (2001): 99–106; M. Hung, Phase II trials of oral insulin progressing, Medscape News, **http://diabetes.medscape.com,** site visited July 2, 2001; Reuters Medical News, Oral insulin delivery may soon be possible, **http://pharmacists.medscape.com,** site visited August 23, 2001.

24. Z. T. Bloomgarden, New technologies for diabetes: Continuous glucose monitoring and new strategies for insulin administration, **www.medscape.com,** site visited April 9, 2001.

25. K. S. Bryden and coauthors, Eating habits, body weight, and insulin misuse, *Diabetes Care* 22 (1999): 1956–1960.

26. G. L. Warnock, Frontiers in tranplantation of insulin-secreting tissue for diabetes mellitus, *Canadian Journal of Surgery* 42 (1999): 421–426.

27. American Diabetes Association News, Diabetes breakthrough: Unlimited source of healthy beta cells, **www.diabetes.org/ada/isletcell.asp,** site visited April 18, 2001.

28. American Diabetes Association, Position statement: Preconception care of women with diabetes, *Diabetes Care* (supplement 1) 25 (2002): 82–89.

29. The Diabetes Control and Complications Trial Research Group, Pregnancy outcomes in the Diabetes Control and Complications Trial, *American Journal of Obstetrics and Gynecology* 174 (1996): 1343–1353.

30. M. Romon and coauthors, Higher carbohydrate intake is associated with decreased incidence of newborn macrosomia in women with gestational diabetes, *Journal of the American Dietetic Association* 101 (2001): 897–902.

31. American Diabetes Association, Gestational diabetes mellitus, *Diabetes Care* (supplement 1) 25 (2002): 94–96.

Nutrition in Practice

■ HYPE OR HYPOGLYCEMIA? ■

As Chapter 24 described, *hypoglycemia* simply means "low blood glucose" and refers not to a disease, but to a clinical sign of altered carbohydrate metabolism. In diabetes, hypoglycemia can result from taking too much medication, missing or delaying a meal, eating too little food, or exercising too strenuously. Chapter 22 explained that hypoglycemia may also follow gastric surgery. Indeed, there are real cases of hypoglycemia, but they are uncommon. Nevertheless, popular books and articles claim that many people suffer from hypoglycemia and that hypoglycemia explains many vague symptoms people experience but for which no medical explanation is found. The situation is analogous to other "diseases of the month" such as lactose intolerance or chronic fatigue syndrome. Any type of abdominal pain or indigestion must be lactose intolerance, and any type of exhaustion must be chronic fatigue syndrome. This Nutrition in Practice explores the facts and myths about hypoglycemia. The glossary defines terms related to hypoglycemia.

What are the true symptoms of hypoglycemia?

The symptoms of hypoglycemia depend on the type of hypoglycemia and its cause. The two major types are **reactive** (or **postprandial**) **hypoglycemia** and **fasting hypoglcemia.** Examples of reactive hypoglycemia include the hypoglycemia that sometimes follows gastric surgery (see Chapter 22) or the abrupt disruption of TPN (Chapter 21). The hypoglycemia associated with diabetes mellitus is an example of fasting hypoglycemia. Table NP24–1 on p. 627 shows the differences between reactive and fasting hypoglycemia.

What is reactive hypoglycemia?

Reactive hypoglycemia occurs within two to five hours after eating. Ordinarily, blood glucose initially rises and then falls after eating; in healthy people, the decline is gradual, glucose remains in the normal range, and the transition occurs without notice. In people with reactive hypoglycemia, however, blood glucose falls too low. The body "reacts" to low blood glucose by triggering the release of the hormones glucagon and epinephrine, which oppose insulin's actions. The release of epinephrine, the "emergency hormone," causes symptoms that mimic an anxiety attack—weakness, rapid heartbeat, sweating, anxiety, hunger, and trembling.

What causes blood glucose to fall too low in reactive hypoglycemia?

The rapid absorption of simple carbohydrates and the consequent release of a large amount of insulin can cause blood glucose to fall too low as a consequence of the dumping syndrome. Similarly, a sudden disruption in the delivery of a concentrated dextrose solution (as in TPN) can lead to hypoglycemia. Reactive hypoglycemia may also occur in early type 2 diabetes, when hyperinsulinemia is more likely than in late type 2 diabetes. In other rare cases, the reason for the rapid decline in blood glucose is unknown (**idiopathic** reactive hypoglycemia). This is the type of hypoglycemia people may believe they have even if they do not.

Why do some people believe they have hypoglycemia?

Searching the Internet for websites about hypoglycemia provides valuable insights. Consider that one site implies that a person with these symptoms and clinical findings may have hypoglycemia:

- Exhaustion.
- Headaches.
- Temper outbursts.

Glossary

fasting hypoglycemia: low blood glucose that develops gradually, does not occur in response to food intake, and primarily affects the brain and central nervous system.

idiopathic (ID-ee-oh-PATH-ick): arising from an unknown cause.

reactive hypoglycemia: low blood glucose that develops two to five hours following a meal. Also called **postprandial hypoglycemia.**

Nutrition in Practice

People who experience hypoglycemia benefit from eating regularly throughout the day and avoiding refined carbohydrates.

- Irritability.
- Sleeping problems.
- Indecisiveness.
- Nervousness.
- Depression.
- Overweight.

If you think about these symptoms and problems, you will probably agree that they are common and can occur for many reasons, often with a psychological component. A person bothered by these symptoms who visits a physician and is given a clean bill of health may have difficulty accepting that nothing is physically wrong. In other cases, ill-informed practitioners "diagnose" hypoglycemia on the basis of their clients' verbal reports alone.

What tests determine if a person really has hypoglycemia?

A diagnosis of hypoglycemia generally depends on three conditions:

- The client's symptoms.
- A measurement of blood glucose that is 45 mg/dL or less in a woman or 55 mg/dL or less in a man at the same time that the person is experiencing the symptoms of hypoglycemia.
- The prompt relief of symptoms when a simple carbohydrate is given.

In the past, glucose tolerance tests (see Chapter 24) were used to diagnosis hypoglycemia, but these tests may actually trigger hypoglycemia because a fairly large dose of simple carbohydrate is given all at once. Therefore, glucose tolerance tests may show that hypoglycemia is present, even though the person would not actually have a problem when provided typical servings of carbohydrate.

How is reactive hypoglycemia treated?

For people with reactive hypoglycemia with no known cause, a judicious carbohydrate-modified diet often brings relief. These suggestions can help:

- Eat regular meals and avoid long periods without any food.
- Avoid low-carbohydrate dieting and eating sudden large amounts of carbohydrate, especially high–glycemic index foods.
- Include protein and fat along with carbohydrate for meals and snacks to help delay the absorption of carbohydrate.
- If several average-size meals fail to relieve symptoms, try smaller meals eaten more frequently.

The remedy, then, is similar to the diet for diabetes: eat a consistent amount of carbohydrate from balanced meals at regular times. Two other measures may also help:

- Limit alcohol, which can lead to hypoglycemia (see Nutrition in Practice 23 and Chapter 24).
- Limit caffeine, which may disrupt blood flow and the delivery of glucose to the brain.

Those changes seem beneficial for everyone. So what is the harm in people thinking they have hypoglycemia when they don't?

For most people, there is no harm. For some people, however, there is a danger that they may self-diagnose hypoglycemia and fail to visit a doctor when they actually have another disorder that requires treatment.

How does fasting hypoglycemia differ from reactive hypoglycemia?

In reactive hypoglycemia, blood glucose falls too low following a meal or carbohydrate load. In fasting hypoglycemia, the drop in blood glucose is not related to food intake alone. Fasting hypoglycemia is most often associated with diabetes. In diabetes, for example, taking too much insulin or some oral antidiabetic medications (sulfonylureas) for the amount of food eaten can

Nutrition in Practice

Table NP24–1	Characteristics of Reactive and Fasting Hypoglycemia	
	Reactive Type	**Fasting Type**
Onset of Symptoms	Sudden; occurs 2 to 5 hours after meals	Gradual
Type of Symptoms	Anxiety, weakness, sweating, rapid heartbeat, hunger, trembling	Headache, mental dullness, fatigue, confusion, amnesia, seizures, unconsciousness
Duration of Symptoms	Transient	Persistent
Possible Causes	Early type 2 diabetes, gastric surgery, TPN	Hormonal imbalance, diabetes, medications, tumors
Clinical Course	Less serious; treat with diet	Can be serious; treat underlying problems

drive too much glucose into the cells. Besides diabetes, fasting hypoglycemia can arise from medically diverse disorders, including certain tumors of the pancreas or liver that interfere with normal blood glucose regulation.

Are the symptoms of fasting hypoglycemia different from those of reactive hypoglycemia?

The symptoms of fasting hypoglycemia differ from those of reactive hypoglycemia because they are not related to epinephrine release. Instead, blood glucose falls slowly, and the major effect is on the brain and central nervous system. The symptoms include headaches, blurred vision, mental dullness, fatigue, confusion, amnesia, and even seizures and unconsciousness.

What are the treatments for fasting hypoglycemia?

Chapter 24 described how an evenly spaced, consistent carbohydrate intake can help prevent fasting hypo-

glycemia in people with diabetes and also discussed how carbohydrates are used to treat hypoglycemia. Surgery is the primary treatment for tumors that precipitate fasting hypoglycemia, although carbohydrate-modified diets (as described for reactive hypoglycemia) may be used temporarily.

Some people have reactive hypoglycemia, although it is a rare condition. Many more people believe they have it. Because the treatment for reactive hypoglycemia is basically a nutritious, balanced diet, no harm can come from the misbelief in most cases. In some cases, however, clients may have true medical problems that go undiagnosed.

CHAPTER 25

FAT-CONTROLLED, MINERAL-MODIFIED DIETS FOR CARDIOVASCULAR DISEASES

Table 25–1 Indications for the Use of Fat-Restricted, Fat-Modified Diets

- Atherosclerosis
- Chronic heart failure
- Chronic renal disease
- Diabetes mellitus
- HIV infections
- Hyperlipidemia
- Hypertension
- Myocardial infarction
- Nephrotic syndrome
- Stroke

Reminder: Cholesterol is carried in several lipoproteins, including chylomicrons, very-low-density lipoproteins (VLDL), low-density lipoproteins (LDL), and high-density lipoproteins (HDL).

The technical term for abnormal blood lipids is **dyslipemia;** elevated blood lipids may also be called **hyperlipidemia.** The **blood lipid profile** measures each type of lipoprotein.

cardiovascular diseases (CVD): diseases of the heart and blood vessels.

atherosclerosis (ATH-er-oh-skler-OH-sis): the buildup of plaque along the inner walls of the arteries, which narrows the lumen of the artery and restricts blood flow to the tissue it supplies.

coronary heart disease (CHD): heart damage that results from an inadequate supply of blood to the heart.

plaques (PLACKS): mounds of lipid material (mostly cholesterol) covered with fibrous connective tissue and embedded in artery walls that may harden with time.

angina (an-JYE-nah or AN-ji-nah): a painful feeling of tightness or pressure in and around the heart, often radiating to the back, neck, and arms; caused by a lack of oxygen to an area of heart muscle.

*T*he fat-modified diets described in this chapter limit both the amount and type of fat (see Table 25–1). Such diets aim to prevent and treat **cardiovascular diseases (CVD)**—the leading cause of death in the United States and around the world. Cardiovascular diseases claim over 10,000 more lives in the United States than the next six leading causes of death combined.[1] Men have a greater risk of heart disease and are at risk at an earlier age than women, but this gap closes with age. Most cardiovascular diseases involve **atherosclerosis** and hypertension. Coronary heart disease and strokes are the most frequent consequences.

Atherosclerosis

Atherosclerosis or "hardening of the arteries" occurs when fibrous **plaques,** composed primarily of cholesterol, build up in the arteries, especially at branch points (see Figure 25–1). The first sign of atherosclerosis is soft fatty streaks visible along the walls of the arteries. Often these fatty streaks gradually enlarge and harden as they fill with lipids and minerals (especially calcium) and become encased in fibrous connective tissue (scar tissue). Plaque stiffens an artery and narrows its diameter. Atherosclerosis may obstruct blood flow through any blood vessel, although the coronary arteries are frequently affected.

Causes of Atherosclerosis   What causes plaque to develop is unknown, but many scientists believe that tissue damage and inflammatory responses play a role. The cells lining the blood vessels may incur damage from high cholesterol levels, hypertension, toxins from tobacco products, or some viral and bacterial infections. Inflammatory responses increase the permeability of the blood vessels and allow immune system cells (macrophages) and LDL cholesterol to deposit in the blood vessel walls. Free radicals produced during inflammatory responses oxidize LDL cholesterol, and the macrophages engulf it. Macrophages swell with large quantities of oxidized LDL cholesterol and eventually become the cells that comprise plaque. Inflammatory responses also direct changes that favor the formation of blood clots and allow minerals to harden plaque and form the fibrous connective tissue that encapsulates it.

Blood Lipids and Atherosclerosis   The blood cholesterol most clearly linked to atherosclerosis is LDL cholesterol. If excess LDL cholesterol remains after the body's cells take up the amount they need, then the excess becomes available for oxidation. HDL also carry cholesterol in the blood, but it is believed they carry cholesterol away from the arteries and back to the liver. Although the mechanisms are unclear, high levels of HDL protect against the development of plaque. Conversely, low HDL may favor the development of plaque.

Elevated triglycerides are also linked to atherosclerosis. Elevated triglycerides often occur together with elevated LDL, low HDL, and other conditions that favor plaque development (overweight, metabolic syndrome, diabetes).

Consequences of Atherosclerosis   When progression of atherosclerosis in the coronary arteries restricts blood flow and damages the heart muscle, **coronary heart disease,** or **CHD,** develops. The person with CHD often experiences pain and pressure in and around the area of the heart (**angina**). If blood flow to the heart is cut off, that area of heart muscle dies, and a myocardial infarction (heart attack) results.

A sudden spasm or surge in blood pressure in an artery can tear away part of the fibrous coat covering a plaque. When this happens, the body responds to the damage as it would to other tissue injuries. **Platelets,** tiny disc-shaped bodies,

Figure 25–1
**The Formation of Plaques
in Atherosclerosis**

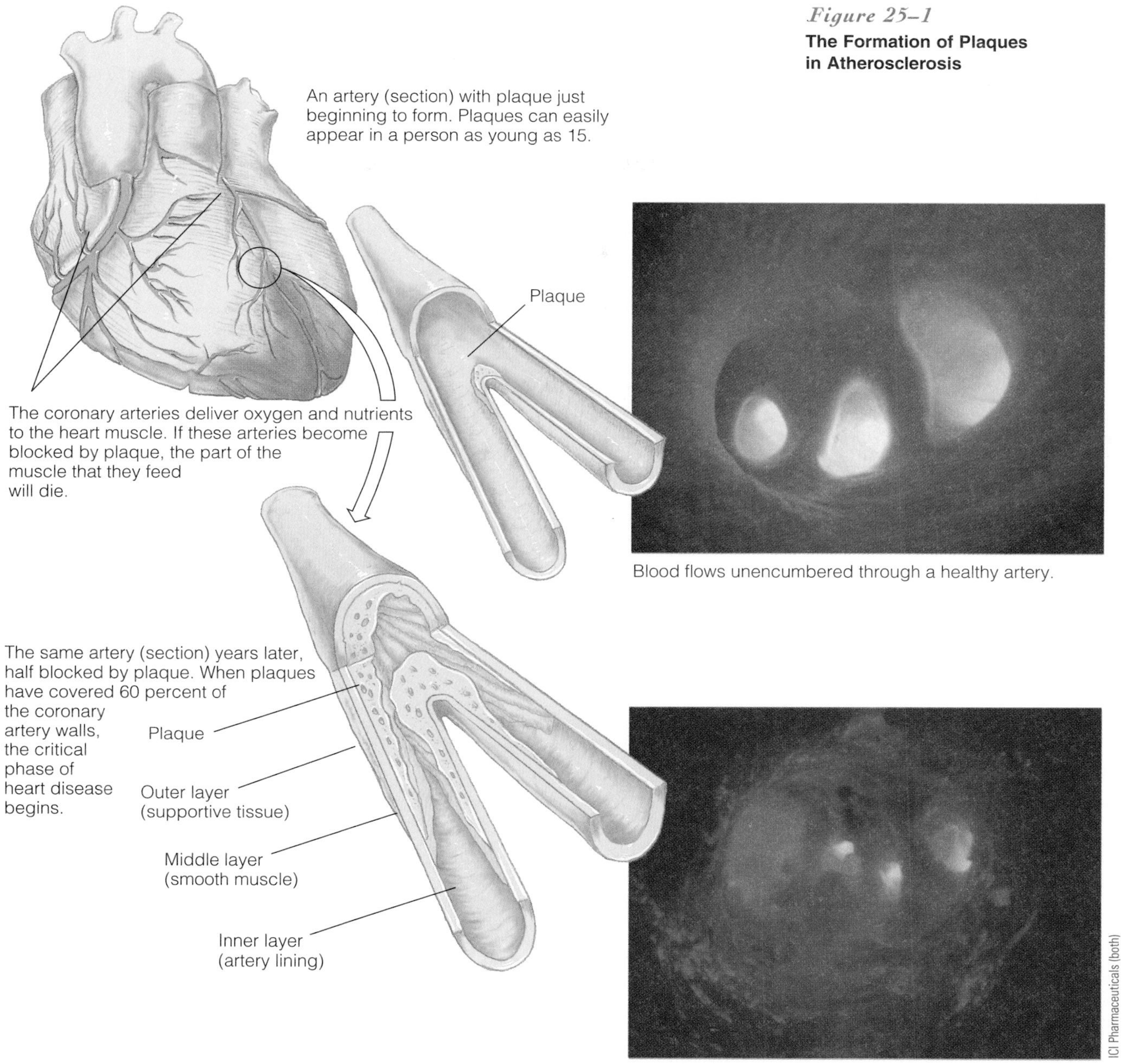

An artery (section) with plaque just beginning to form. Plaques can easily appear in a person as young as 15.

Plaque

The coronary arteries deliver oxygen and nutrients to the heart muscle. If these arteries become blocked by plaque, the part of the muscle that they feed will die.

Blood flows unencumbered through a healthy artery.

The same artery (section) years later, half blocked by plaque. When plaques have covered 60 percent of the coronary artery walls, the critical phase of heart disease begins.

Plaque

Outer layer (supportive tissue)

Middle layer (smooth muscle)

Inner layer (artery lining)

ICI Pharmaceuticals (both)

Plaques along an artery narrow its diameter and obstruct blood flow. Clots can form, aggravating the problem.

cover the damaged area, and together with other factors, they form a clot (**thrombosis**). A blood clot (**thrombus**) on a plaque may grow large enough to close off a blood vessel. A portion of a clot may also break free from the plaque (**embolus**) and travel through the circulatory system until it lodges in a small artery and suddenly shuts off the blood flow to that area (**embolism**). The gradual or sudden loss of blood flow to the portion of tissue supplied by the clotted artery robs the tissue of oxygen and nutrients, and the tissue may eventually die.

Sometimes atherosclerosis obstructs blood flow to the brain; the result may include a transient ischemic attack or stroke (described later in this chapter). Blood flow to the kidneys can also be affected. Renal failure is described in Chapter 26.

platelets: tiny, disc-shaped bodies in the blood that are important in clot formation.

thrombosis (throm-BOH-sis): the formation or presence of a blood clot in the vascular system. A *coronary thrombosis* occurs in a coronary artery, and a *cerebral thrombosis* occurs in an artery that feeds the brain.

thrombus (THROM-bus): a blood clot that blocks a blood vessel or a cavity of the heart.

*Fat-Controlled, Mineral Modified
Diets for Cardiovascular Diseases*

Technically, transient ischemic attacks and strokes are **cerebrovascular diseases** rather than cardiovascular diseases. In this book we consider both to be cardiovascular diseases for simplicity.

Optimal resting blood pressure for adults averages about 120 over 80 millimeters of mercury (mmHg). At a blood pressure of 140 over 90 mmHG or higher, the risks of heart attacks and strokes increase in direct proportion to rising blood pressure (see Table 25–3 later in the chapter).

The resistance to the flow of blood caused by the reduced diameter of the smallest arteries and capillaries at the periphery of the body is called **peripheral resistance.**

hypertension: elevated blood pressure.

primary prevention: efforts aimed at preventing CHD.

secondary prevention: efforts aimed at preventing complications associated with CHD or CHD risk equivalents.**embolus** (EM-boh-lus): a traveling blood clot.

embolism (EM-boh-lizm): the obstruction of a blood vessel by an embolus, causing sudden tissue death.

aneurysm (AN-you-riz-um): a ballooning out of a portion of a blood vessel (usually an artery) due to weakness of the vessel's wall.

Hypertension

Chronic elevated blood pressure, or **hypertension,** affects some 50 million people in the United States.[2] Although one out of four adults in the United States may have high blood pressure, almost a third of these people do not know they have it. Thus hypertension is sometimes called the "silent killer"; it may go undetected until the person experiences a potentially fatal complication such as a heart attack or stroke. Sometimes hypertension develops as a consequence of another disorder, but most often the cause is unknown. The higher the blood pressure above normal, the greater the risk of heart disease.

Blood Pressure Regulation The body's ability to maintain blood pressure is vital to life. The heart's pumping action must create enough force to push the blood through the major arteries into the smaller arteries and finally into tiny capillaries, whose thin porous walls permit fluid exchange between the blood and tissues (see Figure 25–2). The nervous system helps maintain blood pressure by adjusting the size of the blood vessels and by influencing the heart's pumping action. The kidneys also help regulate blood pressure by setting in motion mechanisms that change the blood volume. The narrower the blood vessels or the greater the volume of blood in the circulatory system, the harder the heart must pump (and the more pressure the heart must create) to feed the tissues.

Consequences of Hypertension Hypertension stiffens the arteries and restricts blood flow through them. Thus hypertension can lead to many of the same complications as atherosclerosis—heart attacks, transient ischemic attacks, strokes, and renal failure. Constant high pressure damages the arterial walls and may eventually cause them to weaken and balloon out (**aneurysm**). Aneurysms that go undetected can burst and lead to massive bleeding and death, particularly when a large vessel such as the aorta is affected. In the small arteries of the brain, an aneurysm may lead to stroke, and in the eye, it may lead to blindness. Likewise the kidneys may be damaged (kidney disease) when the heart is unable to adequately pump blood through them. Strain on the heart's pump, the left ventricle, can enlarge and weaken it until finally it fails (heart failure is described later).

Interrelationships between Hypertension and Atherosclerosis Hypertension injures the arterial walls, and plaques and clots are especially likely to form at damage points. Thus hypertension sets the stage for atherosclerosis or makes it worse. Once plaques or clots develop, they may reduce the diameter of an artery and raise blood pressure even higher. Hypertension and atherosclerosis are mutually aggravating conditions, and both contribute to acute and chronic heart failure and strokes.

In Summary Most CHD involves atherosclerosis, hypertension, or a combination of the two. In atherosclerosis, cholesterol-filled plaques develop on the inner walls of the arteries. The plaques gradually enlarge and harden and may block the flow of blood to organs. A plaque may also rupture, and the clot that forms to repair the damage may further restrict blood flow or may break away and cut off blood flow through a smaller artery. Hypertension obstructs blood flow and strains the heart. Atherosclerosis and hypertension often occur together, and both are self-aggravating conditions that can lead to heart failure, transient ischemic attacks, strokes, and renal disease.

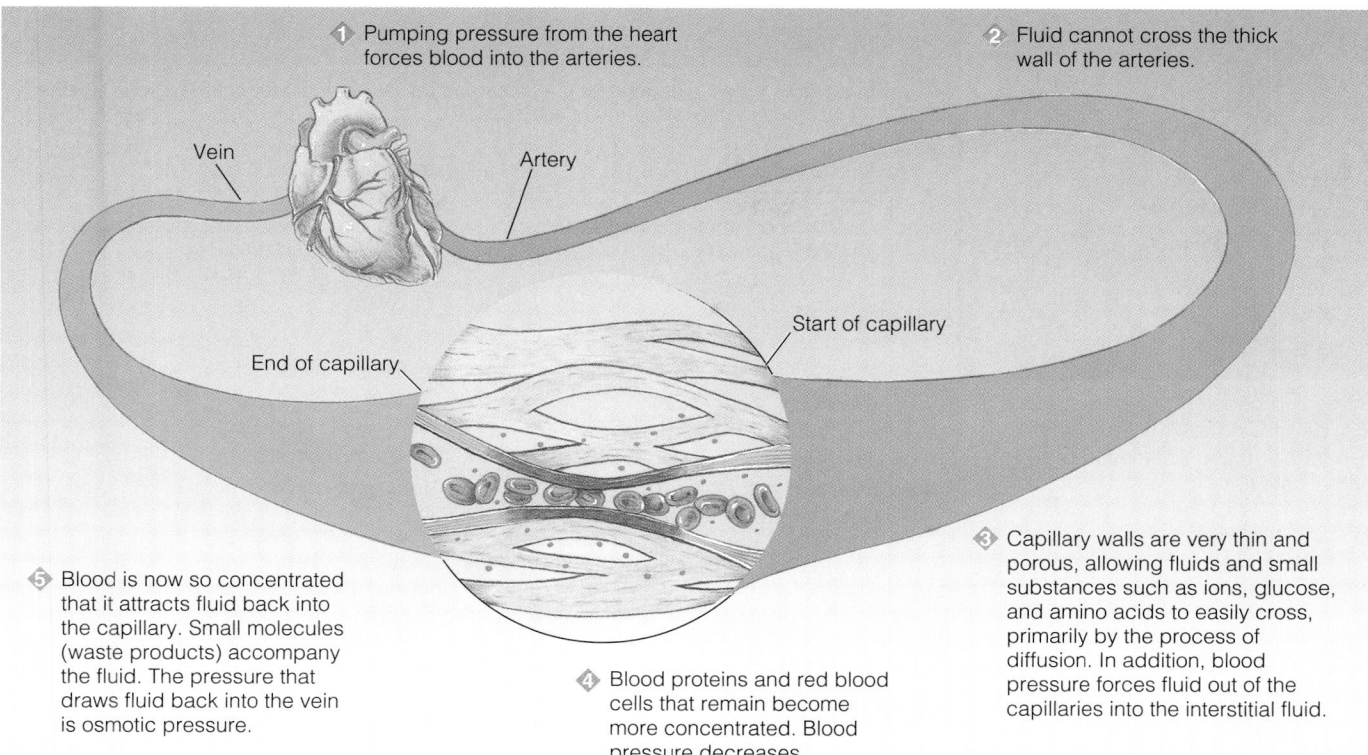

① Pumping pressure from the heart forces blood into the arteries.

② Fluid cannot cross the thick wall of the arteries.

Vein

Artery

Start of capillary

End of capillary

③ Capillary walls are very thin and porous, allowing fluids and small substances such as ions, glucose, and amino acids to easily cross, primarily by the process of diffusion. In addition, blood pressure forces fluid out of the capillaries into the interstitial fluid.

⑤ Blood is now so concentrated that it attracts fluid back into the capillary. Small molecules (waste products) accompany the fluid. The pressure that draws fluid back into the vein is osmotic pressure.

④ Blood proteins and red blood cells that remain become more concentrated. Blood pressure decreases.

Figure 25–2

Blood Pressure and Fluid Exchange
At the same time that the heart pushes blood into an artery, the small-diameter arteries and capillaries at its other end resist the blood's flow (peripheral resistance). Both actions contribute to the pressure inside the artery. Another determining factor is the volume of fluid in the circulatory system, which depends in turn on the number of dissolved particles in that fluid.

Major Risk Factors and Protective Factors

Diet Order
Energy- and fat-controlled, low-salt diet.

Efforts to prevent CHD (**primary prevention**) and its complications in those who already have the disease (**secondary prevention**) focus on normalizing blood lipids and blood pressure. The margin lists both the major risk factors for CHD and the factors that protect against CHD.[3] Earlier sections of this chapter have described the relationships of LDL, HDL, and hypertension to CHD. Many risk factors are interrelated—lack of physical activity, overweight, hypertension, elevated blood cholesterol, and type 2 diabetes frequently occur together, for example. Emotional stress and drinking too much alcohol may also raise the likelihood that CHD will develop. A host of other factors may contribute to the risk of CHD or offer protection against it, but these factors have not been studied thoroughly or studies have found conflicting results, so no concrete recommendations can be made at this time. Table 25–2 on p. 634 summarizes many of the diet-related factors that may offer protection against CHD, and the following sections provide more information.

Besides diabetes, kidney diseases (see Chapter 26) frequently lead to CHD. People with HIV infections who are treated with certain medications (see Chapter 28) often develop central obesity and the metabolic syndrome.

Preventive Efforts By middle age, most adults have at least one risk factor for CHD, and many have more than one.[4] Both the United States and Canada recommend screening to identify individuals at risk so as to offer preventive advice

Risk factors for CHD that cannot be modified:
- *Increasing age.*
- *Male gender.*
- *Heredity.*

Modifiable risk factors for CHD:
- *Tobacco smoke.*
- ***Elevated blood cholesterol (LDL).***
- ***Hypertension.***
- ***Excess body weight and body fat, especially abdominal fat.***
- ***Diabetes mellitus.***

Factors that protect against CHD:
- ***High HDL (>60 mg/dL).***
- ***Physical activity.***

NOTE: Risk and protective factors in bold type have a relationship with diet.

Fat-Controlled, Mineral Modified Diets for Cardiovascular Diseases

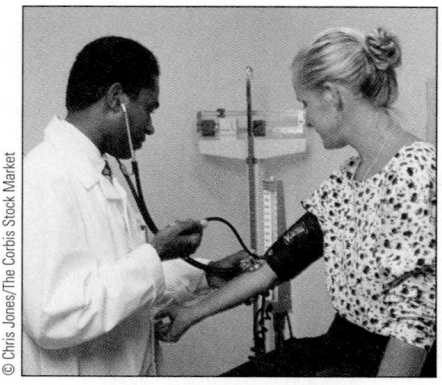

Screening people for hypertension is a first step toward early detection of hypertension and prevention of complications.

Table 25–2   Diet-Related Factors That May Protect against CHD

In addition to reducing total fat and saturated fat (described later in this chapter), these dietary factors may also protect against CHD.

Dietary Factor	Protection against CHD
Soluble fiber (apples and other fruits, soy, barley, legumes)	■ Lowers blood cholesterol, especially in oats, those with high cholesterol ■ Lowers risk of heart attack ■ Improves LDL-to-HDL ratio
Omega-3 fatty acids (fish and some plant oils)	■ Limit clot formation ■ Prevent irregular heartbeats ■ Lower risk of heart attack
Alcohol (in moderation)	■ Raises HDL ■ Prevents clot formation
Folate, vitamin B_6, vitamin B_{12}	■ Reduce homocysteine
Vitamin E (vegetable oils and margarines, some nuts, wheat germ)	■ Slows progression of plaque formation ■ Lowers risk of heart attack in people with CHD ■ Limits LDL oxidation
Soy (protein and isoflavones)	■ Lowers blood cholesterol ■ Raises HDL cholesterol ■ Improves LDL-to-HDL ratio

and treatment. Such public health programs are proving successful: since 1960, both blood cholesterol levels and deaths from cardiovascular disease among U.S. adults have shown a continuous and substantial downward trend.[5]

With respect to hypertension, the most effective single step people can take is to find out whether they have it. A major national effort to identify and treat hypertension is currently under way. Even mild hypertension can be serious; early treatment promotes health and a higher-quality, longer life. Table 25–3 lists the criteria for defining blood lipids, blood pressure, and measures of excess body weight and body fat in relation to CHD risk.

Age, Gender, and Heredity   Three of the major risk factors for CHD cannot be modified by diet or otherwise: age (greater than 45 years for men, greater than 55 years for women), gender, and heredity. With increasing age, the risk of CHD and its complications increases as well. Men generally have higher blood cholesterol and a greater risk of CHD and heart attacks at an earlier age than women. Men's blood cholesterol early in adulthood strongly correlates with their risk of developing heart disease later in life.[6]

Cardiovascular disease occurs about 10 to 12 years later in women than in men. Women younger than 45 tend to have lower LDL cholesterol than men of the same age, but a woman's blood cholesterol typically begins to rise between ages 45 and 55. Women who use birth control pills and also smoke or have hypertension are at greater risk of CHD than other women.

Risks of CHD are higher for children of parents with heart disease and for people belonging to certain ethnic groups. These groups include African Americans, Mexican Americans, Native Americans, Native Hawaiians, and some Asian Americans.

The Metabolic Syndrome   Individually, abnormal blood lipids, hypertension, impaired glucose tolerance, and obesity all raise the likelihood of developing CHD, but when they occur together (the metabolic syndrome), they synergistically raise the risk. The most recent guidelines issued by the Expert Panel of the

Table 25–3 Standards for CHD Risk Factors[a]

	Desirable	Borderline	High
Lipids (mg/dL)[b]			
Total cholesterol	<200	200–239	≥240
LDL cholesterol	<130	130–159	≥160
Triglycerides	<150	150–199	≥200
Body Weight and Distribution			
Body mass index (BMI)	18.5–24.9	25–29.9	≥30
Waist circumference (in)			
Men			>40
Women			>35

	Systolic	Diastolic
Blood Pressure (mm/Hg)		
Optimal	<120	<80
Normal	120–129	80–84
High-normal	130–139	85–89
Mild	140–159	90–99
Moderate	160–179	100–109
Severe	≥180	≥110

[a]See Nutrition in Practice 13 for lipid and blood pressure standards for children.
[b]HDL cholesterol is considered to be protective against CHD when it is ≥60 mg/dL and low when it is <40 mg/dL. An LDL-to-HDL ratio of >5 in men and >4.5 in women indicates risk for CHD.
SOURCES: Blood lipid standards adapted from The Expert Panel, Executive Summary of the third report of the National Cholesterol Education Program (NCEP) Expert Panel on Detection, Evaluation and Treatment of High Blood Cholesterol in Adults (Adult Treatment Panel III), available from the National Heart, Lung and Blood Institute, **www.nhlbi.nih.gov/guidelines/cholesterol/index.htm**; Hypertension standards adapted from the Sixth Report of the Joint National Committee on Prevention, Detection, Evaluation, and Treatment of High Blood Pressure, National High Blood Pressure Education Program, National Heart, Lung, and Blood Institute, National Institutes of Health, November 1997, p. 11.

National Cholesterol Education Program recognize the impact of the metabolic syndrome on cardiovascular health and recommend early diagnosis and treatment.[7] Nutrition in Practice 25 presents more information about the metabolic syndrome and its relationship to cardiovascular disease.

Tobacco Smoke   Smoking stresses the cardiovascular system by depriving the heart of oxygen and raising the blood pressure. It also damages platelets, making blood clot formation more likely. Toxins in tobacco smoke damage the blood vessels and lead to inflammation, setting the stage for plaque formation. People who smoke cigars and pipes have an increased risk of CHD, but not as high as the risk for people who smoke cigarettes. Finally, even nonsmokers have a higher risk of CHD if they are regularly exposed to other people's smoke (secondhand or environmental tobacco smoke).

Protective Effects of Physical Activity   Regular exercise confers many benefits to the cardiovascular system and helps control other risk factors for CHD. Aerobic activities, undertaken faithfully for 30 minutes or more as a daily or every-other-day routine, provide the most benefits. Such a physical activity program can strengthen the heart and blood vessels, expand the volume of oxygen the heart can deliver to the tissues at each beat and so reduce the heart's workload, and bring about a redistribution of body water that eases the transit of blood through the peripheral arteries.[8] In addition, regular aerobic activity lowers LDL, raises HDL, lowers blood pressure, speeds loss of weight and body fat, improves blood glucose control in people with type 2 diabetes, and reduces emotional stress. These

A person with any three of the following clinical findings is considered to have the metabolic syndrome:

- *Abdominal obesity (waist circumference of >40 in for men or >35 in for women).*
- *Triglycerides of ≥150 mg/dL.*
- *HDL of <40 in men or <50 in women.*
- *Blood pressure of ≥130/85 mmHg.*
- *Fasting glucose of ≥110 mg/dL.*

Regular aerobic exercise can help to defend against heart disease by strengthening the cardiovascular system, promoting weight loss, reducing blood pressure, and improving blood lipid and blood glucose levels.

Examples of aerobic physical activities that can improve cardiovascular fitness include brisk walking, jogging, running, cycling, and rowing.

CHD risk equivalents: disorders that raise the risk of heart attacks, strokes, and other complications associated with cardiovascular diseases to the same degree as existing CHD. These disorders include symptomatic carotid artery disease, peripheral arterial disease, abdominal aortic aneurysm, and diabetes mellitus.

changes are so beneficial that some experts believe that physical activity should be *the* primary focus of cardiovascular disease prevention efforts.[9]

Some researchers wonder if physical activity itself raises blood HDL or if the weight loss that often accompanies exercise is the real protective factor. For women, weight loss through diet alone appears to *lower* HDL, but when diet is combined with moderate aerobic activity, HDL do not decline. In fact, HDL increase substantially in women who exercise regularly.[10] In men, diet raises HDL, and the combination of activity and diet results in a significantly greater rise in HDL than diet alone.

If heart and artery disease has already set in, a monitored program of physical activity may actually help to reverse it. Activity may stimulate development of new arteries to feed the heart muscle, which may account for the excellent recovery seen in some heart attack victims who exercise regularly.

Diet-Related Risk Factors   It befits a nutrition book to focus on dietary strategies to prevent heart disease, and the following sections describe those strategies and their relationship to CHD risk. It should be noted, however, that only some of the major risk factors shown in the margin on p. 633 can be modified by diet. As appealing as the solution may sound, diet may not reduce the risk of heart disease and stroke as successfully as other interventions do. Treatment with statins (one type of medication used to lower LDL cholesterol) significantly reduces the risk of cardiovascular complications in both men and women; people with diabetes, peripheral vascular disease, and those who have previously suffered a stroke; and people over age 75, even when LDL cholesterol is normal or below normal.[11] Dietary changes, however, confer additional benefits when they are combined with other strategies, including medications, regular physical activity, and smoking cessation.

In Summary   Some risk factors for CHD, such as age, gender, and heredity, cannot be modified, but many other risk factors, including smoking and exposure to tobacco smoke, elevated LDL cholesterol, hypertension, physical inactivity, obesity and overweight, and diabetes mellitus, can be modified or controlled. Diet can play a role in lowering blood lipids, blood pressure, and weight and also in controlling blood glucose levels. Diet strategies for reducing the risk of CHD are most effective when combined with other strategies.

Prevention and Treatment of Coronary Heart Disease

Plans for preventing and treating CHD, including medical nutrition therapy, aim to normalize blood lipids and blood pressure, alter modifiable risk factors, and prevent complications. Plans for preventing CHD focus on lifestyle changes; clients are encouraged to increase physical activity, lose weight (if necessary), implement dietary changes, and reduce exposure to tobacco smoke either by quitting smoking or by avoiding secondhand smoke. Treatment plans for people with existing CHD or conditions that place them at high risk for heart attacks and strokes (**CHD risk equivalents**) also focus on lifestyle changes first, but the target level for LDL is lower (see Table 25–4). If lifestyle changes fail to reduce lipid levels or to reduce blood pressure to an acceptable level, then lipid-lowering and/or antihypertensive medications are added to the treatment plan. Many physicians routinely advise clients to take low doses of aspirin daily. Estimates suggest that a staggering 65 million people in the United States should be targeted for lifestyle changes to prevent CHD and that another 36 million need medication as well.[12]

Table 25–4 Target and Treatment Levels for LDL Cholesterol Based on Risk Factors

Risk Factors	Target LDL Level (mg/dL)	LDL Level (mg/dL) at Which to Implement Lifestyle Changes	LDL Level (mg/dL) Suggesting the Need for Drug Therapy)
0–1	<160	≥160	≥190
2 or more	<130	≥130	≥160; ≥130[a]
CHD or CHD risk equivalent	<100	≥100	≥130[b]

[a]The Expert Panel recommends that physicians perform a risk assessment on clients who have two or more risk factors for CHD. Those with a lower risk of complications (less than 10 percent) begin drug therapy at higher levels of LDL than those with a higher risk of complications (10 to 20 percent).
[b]Some physicians prefer to begin drug therapy when LDL cholesterol ranges from 100 to 129 mg/dL if lifestyle changes are unsuccessful in reducing LDL cholesterol to less than 100 mg/dL.
SOURCE: Adapted from National Heart, Lung, and Blood Institute, National Cholesterol Education Program, ATPIII Guidelines At-A-Glance Quick Desk Reference, available at **www.nhlbi.hih.gov,** sited visited October 13, 2001.

A varied, nutritious diet confers many benefits to support the health of the heart and blood vessels.

For people with diabetes, controlling blood glucose levels is an important strategy. As Chapter 24 mentioned, postprandial (after a meal) blood glucose may have a closer relationship with CHD risk than fasting blood glucose. For people with advanced atherosclerosis, therapy may also include surgery to restore blood flow to the affected organ.

Diet Strategies

Diet strategies to both prevent and treat CHD focus on four main goals: a healthy eating pattern, a healthy body weight, a desirable blood cholesterol and lipoprotein profile, and a desirable blood pressure.[13] The specific strategies for achieving a desirable blood cholesterol and lipoprotein profile vary somewhat, depending on whether the objective is to prevent CHD in a healthy person with desirable LDL cholesterol levels or to improve blood lipid levels in high-risk groups—people with elevated LDL cholesterol, preexisting CHD, insulin resistance, or diabetes mellitus. Table 25–5 on p. 638 summarizes diet strategies for preventing CHD and reducing elevated LDL cholesterol; the "How to" box on p. 639 offers practical suggestions, and the sample menu on p. 640 shows a day's meals. Dietitians often use the exchange system described in Chapter 24 or similar lists to help clients plan the diet.

A Healthy Eating Pattern In the past, the American Heart Association's dietary recommendations for preventing and treating CHD centered specifically on the total energy and the amount and type of fat in the diet. Although guidelines for energy and fat remain pivotal, the current recommendations have a broader focus on eating patterns that foster both general and cardiovascular health.

Besides limiting fats (described later), the heart-healthy diet encourages consumption of complex carbohydrate–rich foods that provide nutrients, phytochemicals, and fiber that benefit health and may protect against CHD. The soluble fiber found in oats, barley, and pectin-rich fruits and vegetables, for example, helps to reduce blood lipids.[14] Carbohydrate-rich foods also provide minerals, which may help control blood pressure (described later); antioxidant nutrients, which may protect against LDL oxidation (see Chapters 8 and 15); and B vitamins, which may help lower blood homocysteine levels. (Studies have shown a positive association between elevated blood homocysteine and the risk of CHD.)[15] Although an adequate supply of folate, vitamin B_6, and vitamin B_{12} lowers homocysteine levels, whether such a diet-induced reduction also reduces CHD risk remains to be answered.

Fat-Controlled, Mineral Modified Diets for Cardiovascular Diseases

Reminder: Unlike most fatty acids in which the hydrogens next to double bonds occur on the same side of the carbon chain, the hydrogens next to double bonds in *trans*-fatty acids fall on opposite sides of the carbon chain (see p. 59).

Table 25–5 Diet Strategies for Meeting American Heart Association Dietary Goals

Goals:

A healthy eating pattern, a healthy body weight, a desirable blood cholesterol and lipoprotein profile, and a desirable blood pressure.

Diet Strategies:

- Balance energy intake with energy needs to maintain weight, or limit energy to lose weight.
- Achieve a level of physical activity that balances with energy intake (to maintain weight) or exceeds energy intake (to lose weight).
- Limit foods high in saturated fatty acids (<10 percent of total energy intake) *trans*-fatty acids, and cholesterol (< 300 milligrams).[a,b]
- Limit total fat to <30 percent of total energy.
- Replace saturated fats with carbohydrate from grains, legumes, fruits, and vegetables or with unsaturated fats (both long-chain omega-3 polyunsaturated fats and monounsaturated fats) from fish, vegetable oils, and nuts.
- Eat 5 or more servings of a variety of fruits and vegetables each day.
- Include 6 or more servings of a variety of starchy vegetables and grain products, including legumes and whole grains, each day.
- Use 6 ounces or less of lean meat, skinless poultry, or fish each day. Limit the use of whole eggs to 4 or less per week (including those used in cooking).[c]
- Include fish in at least two meals per week.
- Include at least 2 servings of fat-free or low-fat milk products each day for children 1 to 8 years old; 3 servings a day for adults aged 19 to 50 years; 3 to 4 servings for women who are pregnant or breastfeeding; and 4 servings for children and adolescents 9 to 18 years old and adults 51 years and older.
- Limit the intake of salt (sodium chloride) to <6 grams (2400 milligrams sodium) per day.
- Limit alcohol consumption (no more than 1 drink a day for women or 2 drinks a day for men).

[a]For high-risk individuals—those with elevated LDL cholesterol, CHD, or CHD risk equivalents—the saturated fat and cholesterol should be lower: <7 percent of total energy intake and <200 milligrams, respectively.

[b]See p. 67 in Chapter 3 for ways to lower saturated fat and cholesterol intake and the margin on p. 63 for a list of foods high in *trans*-fatty acids.

[c]High-risk individuals are advised to eat fewer than 5 ounces of lean meat, skinless poultry, and fish per day and fewer than 2 eggs per week.

SOURCES: Adapted from R. M. Krauss and coauthors, AHA dietary guidelines revision 2000: A statement for healthcare professionals from the nutrition committee of the American Heart Association; American Heart Association, An eating plan for a healthy America: The new 2000 food guidelines, available at **www.americanheart.org**, site visited February 21, 2001; National Heart, Lung, and Blood Institute, Tipsheet: Step II diet daily food guide, available at **www.nhlbi.nih.gov/chd/Tipsheets/daily.htm**, site visited February 21, 2001

Chapter 7 provides more information about central body fat and its effects on the body.

A Healthy Body Weight   The dietary guidelines recommend regular physical activity (described earlier) and an appropriate energy intake to achieve and maintain a desirable weight. Obesity, especially central obesity, is associated with elevated blood lipids, hypertension, insulin resistance, and diabetes. For people who are overweight, weight loss reduces LDL cholesterol, triglycerides, blood pressure, and insulin resistance. Even a modest weight loss can help reduce the risk of CHD and stroke.

A Desirable Blood Cholesterol and Lipoprotein Profile   Too much total fat, saturated fatty acids, *trans*-fatty acids, and, to a lesser extent, dietary cholesterol raise blood LDL cholesterol levels. To achieve a healthy blood cholesterol and lipoprotein profile, the dietary guidelines recommend that all people limit total fat to less than 30 percent of daily kcalories and avoid *trans*-fatty acids. For people without CHD, CHD risk equivalents, or elevated LDL cholesterol, saturated fat should comprise less than 10 percent of the total kcalories, and dietary cholesterol should be less than 300 milligrams a day.* In most cases, foods high in

*These recommendations for fat and cholesterol correspond to the Step I diet that has been recommended in the past.

Help Clients Implement Heart-Healthy Diets

*F*or many people, following a heart-healthy diet translates into major changes in the foods they eat. They may find it easier to adapt to the diet if changes are made gradually. The following suggestions can help:

- Gradually increase servings of fruits, vegetables, and grains to reduce gas and bloating that may accompany a higher-than-normal fiber intake. Advise clients to drink plenty of water as well.

- For a client who usually eats only one or two servings of fruits or vegetables a day, start by including at least one serving of a fruit or vegetable at each meal (3 servings). Then add an additional serving at two meals or have one fruit and one vegetable as a snack (total 5 servings). Use a similar approach to add starchy vegetables and grains and include at least 6 servings daily. Choose extra servings of fruits, vegetables, starchy vegetables and grains if additional kcalories are needed to round out the meal plan.

- Cut back on servings of meat, poultry, and fish by eating one-third to one-half the amount usually eaten at each meal. The goal is to include no more than 5 to 6 ounces of meat, poultry, and fish a day.

- For a client who doesn't drink milk or has only one serving, start by adding a glass of fat-free or very-low-fat (1%) milk at one meal; then add a glass at another meal. A third glass of milk can be used as a snack or taken with a third meal.

- Reduce salt intake by gradually cutting down on the salt added to food and used in cooking. The goal is to reduce salt intake to about 1 teaspoon a day, including the salt in foods and added as a seasoning.

To help clients implement their diet, recommend they use:

- Whole-grain breads and cereals instead of refined breads and cereals.
- Whole fruits and vegetables, rather than juice, as often as possible.
- Fresh, frozen, dried, or canned fruit packed in its own juice.
- Fresh, frozen, or canned vegetables that contain no added salt.
- Fat-free, ½%, or 1% fat milk and low-fat and fat-free yogurt.
- Casseroles, pasta meals, and stir-fry dishes to limit meat and increase servings of vegetables and grains.
- Two or more vegetarian meals per week.
- Fats that contain 2 grams or less of saturated fat per serving including liquid and tub margarines and canola, corn, olive, safflower, sesame, soybean, and sunflower oils.

- Liquid or tub margarines with no *trans*-fatty acids.
- Low-salt or salt-free products.
- Products containing fat substitutes in moderate amounts.
- Sodium-free spices such as basil, bay leaves, curry, garlic, ginger, lemon, mint, oregano, pepper, rosemary, and thyme to season foods.

(The box on p. 67 offered many tips for reducing fat and the type of fat in the diet.)

Offer clients these suggestions for snacks:

- Frozen low-fat or fat-free yogurt.
- Fruit.
- Angel food cake.
- Low-fat ice cream (no more than 3 grams fat per ½ cup), fruit ices, sherbets, or sorbets.
- Raw vegetables.
- Unsalted pretzels and nuts.
- Plain popcorn without butter, margarine, or salt.

Suggest that clients limit or avoid these foods:

- Foods of low nutrient density including high-fat foods, foods high in sugar, and alcohol.
- Foods prepared with hydrogenated fat, which contains trans-fatty acids, including doughnuts, crackers, commercially prepared baked goods, and fried foods in restaurants.

Recommend that clients also avoid these high-salt foods:

- Avoid processed foods to the greatest degree possible.
- Pickles, olives, and sauerkraut.
- Cured or smoked meats, such as beef jerky, bologna, corned or chipped beef, frankfurters, ham, luncheon meats, salt pork, and sausage. Many of these foods are also high in fat.
- Salty or smoked fish, such as anchovies, caviar, salted or dried cod, herring, sardines, and smoked salmon.
- Salted snack foods like potato chips, pretzels, popcorn, nuts, and crackers. Many of these foods are also high in fat.
- Bouillon cubes, seasoned salts, and soy, steak, Worcestershire, and barbecue sauces.
- Cheeses, especially processed cheeses. Many cheeses are also high in fat.
- Salted canned and instant soups.
- Prepared horseradish, catsup, and mustard.

Clients with lactose intolerance unable to tolerate the recommended servings of dairy foods can use the digestive aids described in Chapter 22 (see p. 545).

Fat-Controlled, Low-Salt Diet Menu

Breakfast
¾ c cantaloupe
½ c all-bran cereal
Fat-free milk
2 slices whole-wheat toast
2 tsp diet margarine
½ c orange juice
Coffee or tea
Sugar (optional)

Lunch
Broiled chicken breast,
3 oz
Baked beans
Carrots
Tossed salad
1 slice multigrain bread
Low-fat dressing
Fat-free milk
1 small apple

Snack
Popcorn
Grape juice

Supper
Baked flounder, 3 oz
Brown rice
Broccoli
Hard roll
Margarine
1 medium orange
Fat-free milk

Foods for this menu are prepared with little or no salt and a minimal amount of fat. The fats used for cooking or for flavor would be monounsaturated and polyunsaturated.

cholesterol are also high in saturated fat. Eggs and some vegetable oils (coconut, palm, and palm kernel) are exceptions. Eggs are low in saturated fats and high in cholesterol; coconut, palm, and palm kernel oils are high in saturated fats and low in cholesterol.

For people with elevated LDL cholesterol, CHD, or CHD risk equivalents, the recommended diet further reduces saturated fat to 7 percent of daily kcalories and cholesterol to less than 200 milligrams per day.* Compared to the diet for healthy people, the diet for high-risk groups provides less lean meat, skinless poultry, fish, and eggs (see Table 25–5) and strictly limits high-fat meats, especially organ meats (liver, brain, and kidney). Dietitians can help plan the diet to ensure that saturated fats and cholesterol are reduced without sacrificing the nutritional quality of the diet.

Polyunsaturated fatty acids, monounsaturated fatty acids, and, to a lesser extent, soluble fibers lower LDL cholesterol. Either unsaturated fat or carbohydrate can replace saturated fats in a heart-healthy diet. As Chapter 24 described, however, high-carbohydrate diets can elevate blood triglycerides and reduce HDL cholesterol, which may be a problem for people with impaired glucose tolerance or type 2 diabetes. In these cases, replacing saturated fats with unsaturated fats instead of carbohydrate may help improve the lipoprotein profile.

Fish oils, rich in omega-3 polyunsaturated fatty acids, improve blood lipids (primarily by lowering triglycerides), prevent blood clots, and may reduce the risk of sudden death associated with CHD.[16] For these reasons, the dietary guidelines recommend at least two servings of fish per week. Plant sources of omega-3 fatty acids, which include flaxseed and flaxseed oil, canola oil, soybean oil, and nuts, may also confer benefits.[17] For people who have suffered a heart attack, long-term use of omega-3 fatty acids may reduce overall mortality as well as sudden death from CHD.[18]

Achieve and Maintain a Normal Blood Pressure Diet strategies to reduce hypertension have traditionally included recommendations to control weight, reduce salt intake, increase potassium intake, and limit alcohol. Although health authorities continue to recommend these modifications, they also recommend a diet that contains adequate amounts of calcium and magnesium and that limits

People with elevated triglycerides should avoid simple sugars, which often cause triglycerides to rise.

*These recommendations for fat and cholesterol correspond to the Step II diet that has been recommended in the past.

saturated fats and cholesterol. Results of the Dietary Approaches to Stop Hypertension (DASH) trial show that a diet rich in fruits, vegetables, and low-fat dairy products and with reduced total fat and saturated fat can lower blood pressure to a significant degree.[19] When the DASH diet is combined with a limited intake of sodium, the effects on blood pressure are greater still.[20] The DASH diet also lowers total cholesterol and LDL cholesterol.[21] Thus the heart-healthy dietary guidelines embrace these strategies in an overall diet to prevent and treat CHD. (The box on p. 639 included suggestions for limiting sodium.)

For many years, controversy surrounded recommendations to limit salt to prevent or reduce hypertension. The second study of the DASH diet, which looked at the effects of sodium restriction on blood pressure control, however, provides strong evidence that sodium restriction plays an important role. Lowering sodium intakes reduced blood pressure regardless of gender, race, presence or absence of preexisting hypertension, or whether people followed the DASH diet or a typical American diet. The most benefits to blood pressure, however, occurred in the groups who followed the DASH diet and lowered their sodium intakes. Furthermore, the lower the sodium intake, the greater the drop in blood pressure.

Alcohol Moderate consumption of alcohol may reduce the risk of CHD by raising HDL cholesterol and preventing clot formation.[22] Beneficial effects of alcohol are most apparent in people over age 50, those with other risk factors, and those with high LDL. For others, the benefits may not be apparent. At least one recent study reports that abstainers and moderate drinkers shared similar risks of dying from heart disease.[23]

These findings pose a dilemma for health care professionals who are well aware of the potentially damaging effects of alcohol on many body systems (see Nutrition in Practice 23). Too much alcohol can raise blood pressure, and, for people who drink more-than-moderate amounts of alcohol, limiting alcohol intake can reduce blood pressure.[24] Alcohol can also raise triglycerides, and people with elevated triglycerides are advised to restrict alcohol. Most authorities do not advise clients who do not drink alcohol to begin to do so to reduce their risk of CHD. If clients do drink alcohol, clinicians stress that moderation is the key.

Alcohol from any source—red or white wine, beer, or distilled liquor—appears to protect against CHD.[25] However, wine may provide additional protections.[26] In addition to alcohol, wine contains phytochemicals that may act as antioxidants and limit LDL oxidation and may alter prostaglandin metabolism and reduce blood clot formation.[27] Other studies suggest that the beneficial effects of wine over other sources of alcohol may actually reflect higher intelligence quotients (IQ), higher parental educational level, and higher socioeconomic status among wine drinkers compared to those who do not drink wine.[28]

Plant Sterols Margarines made from plant sterols are being marketed as aids to lower cholesterol. Plant sterols, which are found in oil derived from some complex carbohydrates (including soy), resemble cholesterol in structure; they reduce the absorption of dietary cholesterol and lower blood cholesterol levels.[29] Even vegetarians, however, may find it difficult to consume enough plant sterols from food to lower blood cholesterol. Thus the margarines serve to supplement plant sterols in the diet. The long-term safety of these products is unclear, however, and most authorities recommend that margarines made from plant sterols be used only by people with elevated LDL cholesterol or preexisting CHD or CHD risk equivalents.

Soy Protein and Isoflavones When soy proteins are substituted for animal proteins, blood levels of LDL cholesterol and triglycerides fall, and HDL levels do not.[30] The cholesterol-lowering effect may be due to soy isoflavones—phytochemicals that have an estrogen-like effect.[31] For people who have elevated cholesterol levels and are following low-fat, low–saturated fat diets, adding soy

Like other low-fat diets, the DASH diet also lowers HDL—an undesirable outcome. Whether this outcome raises the risk of CHD is unknown, although some studies suggest that people with both low LDL and low HDL do not have an increased risk of CHD.

The DASH diet trial studied the effects of three levels of sodium restriction: 3300 mg, 2400 mg, and 1500 mg.

Fat-Controlled, Mineral Modified Diets for Cardiovascular Diseases

protein daily can significantly lower LDL cholesterol.[32] The Food and Drug Administration (FDA) allows foods that contain 6.25 grams of soy protein per serving to carry a health claim for reduced risk of heart disease.

Drug Therapy

Drug therapy can effectively lower blood lipids and reduce the risk of complications, especially when used together with diet therapy and a physical activity program. As Table 25-4 showed, the client's risk for CHD and its complications and the LDL cholesterol level guide clinicians in determining whether to recommend lifestyle changes alone or to add a lipid-lowering drug as well.

Medications   Besides lipid-lowering medications, medications used in the treatment of cardiovascular diseases may include antihypertensives and diuretics to reduce blood pressure, aspirin and anticoagulants to prevent clot formation, and nitroglycerin to alleviate angina. Some antihypertensive agents can lead to potassium deficiencies; others can cause potassium to rise dangerously high. (Clients with diabetes may also be taking oral antidiabetic agents or insulin.) All of these medications are associated with significant risks and nutrition-related side effects (see the Diet-Drug Interactions box on p. 647), a problem compounded by the fact that drug therapy often includes multiple medications and continues for many years or even life. Elderly clients especially risk diet-medication interactions.

Dietary Supplements   People may use dietary supplements including antioxidant nutrients and folate (described earlier in this chapter) and garlic, black tea, and coenyzme Q to protect against cardiovascular diseases. Garlic is a relatively safe remedy that, when used along with dietary changes, may counter the tendency of blood to clot and modestly lower serum cholesterol and blood pressure. Black teas provide antioxidants that may play a role in CHD prevention. Coenzyme Q has antioxidant properties and may help lower blood pressure. Although all of these products appear to be safe, they have not been thoroughly tested, and they cannot take the place of conventional treatments for cardiovascular disease. The Diet-Drug Interactions box describes herbs used for other reasons that may interact with medications prescribed for cardiovascular diseases.

In Summary   Lifestyle changes that involve diet, physical activity, and avoidance of tobacco smoke comprise the first line of defense against CHD. A heart-healthy diet focuses on an overall healthy eating pattern, an appropriate body weight, a desirable cholesterol and blood lipid profile, and a desirable blood pressure. Among the many medications that may be used in the treatment of CHD, lipid-lowering agents (antilipemics) and antihypertensives are commonly prescribed.

Heart Failure and Strokes

When atherosclerosis and hypertension run their course, the consequences can be fatal. Most often, these consequences include heart failure or strokes.

Heart Attacks

The heart receives nutrients and oxygen, not from inside its chambers, but from arteries on its surface (see Figure 25–1 on p. 631). As mentioned earlier, a **heart attack,** or **myocardial infarction (MI),** occurs when CHD robs the

heart attack or **myocardial** (my-oh-CAR-dee-al) **infarction** (in-FARK-shun) or **MI:** sudden tissue death caused by blockages of vessels that feed the heart muscle; also called **cardiac arrest** or **acute heart failure.**

> *myo* = muscle
> *cardial* = heart
> *infarct* = tissue death

Diet Order
Energy- and fat-controlled, low-salt diet.

Manage Diets after a Myocardial Infarction (MI)

- Offer nothing by mouth until shock resolves.
- After several hours, give a 1000- to 1200-kcalorie diet that progresses from liquids to low-sodium soft foods of moderate temperature in frequent, small feedings.
- After five to ten days, adjust the diet to meet individual needs, generally to three meals a day.
- Although somewhat controversial, many practitioners restrict caffeine completely during the first few days

after an MI and generally recommend a moderate restriction (no more than 3 cups of a caffeine-containing beverage per day) thereafter.
- Gradually progress the diet to a heart-healthy diet with less than 7 percent saturated fat and less than 200 milligrams cholesterol with three meals a day.

heart muscle of its blood supply. Treatment aims to relieve pain, stabilize the heart rhythm, and reduce the heart's workload.

Immediate Care   Immediately following a heart attack, blood flow is disrupted, and **shock** results. During this critical period, food is withheld. After several hours of observation, the person can usually begin to eat again, as outlined in the "How to" box above. Medical nutrition therapy aims to reduce the work of the heart and therefore restricts total energy, the amount of food or drink at each feeding, sodium, and caffeine. Because nausea is a common problem following an MI, liquids are offered first. Low-sodium foods prevent fluid retention, and low-fiber, moderately seasoned foods can help prevent nausea, gas, and abdominal distention. Gas and abdominal distention can push the diaphragm up toward the heart and stress the heart muscle. Temperature extremes can stimulate nerves that slow the heart rate, so foods are offered at moderate temperatures. Caffeine stimulates the metabolic rate and is often restricted.

Long-Term Diet Therapy   After the person is out of immediate danger (in about five to ten days), the diet is tailored to meet individual needs and to deal with conditions such as hyperlipidemia, hypertension, obesity, and diabetes. Because the person has CHD, the heart-healthy diet providing less than 7 percent of total kcalories as saturated fat and less than 200 milligrams of cholesterol is generally appropriate. Such a diet can be planned to provide three meals a day, but people who still have chest pain after a heart attack may continue to benefit from eating frequent, small meals. Advise clients to eat slowly and to avoid strenuous physical activity before and after meals.

Encouraging Lifestyle Changes   Often a person who has experienced an MI is eager to apply strategies to reduce the risks of further CHD, and health care professionals can use the opportunity to offer sound and useful counsel. Continue to offer support and encouragement for as long as possible. Clients may readily return to their old habits when their symptoms disappear if they have not fully incorporated new healthful behaviors into their lives.

Chronic Heart Failure

Heart attacks represent acute heart failure, but heart failure can also occur gradually. Coronary heart disease and hypertension are common causes. A person may develop chronic heart failure following a heart attack or a severe stress. The conditions that lead to **chronic,** or **congestive, heart failure (CHF)** cause the

shock: a clinical condition characterized by disrupted circulation of blood from the peripheral circulation to the heart. Every type of injury is accompanied by some degree of shock.

chronic or **congestive heart failure (CHF):** a syndrome in which the heart gradually weakens and can no longer adequately pump blood through the circulatory system.

Fat-Controlled, Mineral Modified Diets for Cardiovascular Diseases

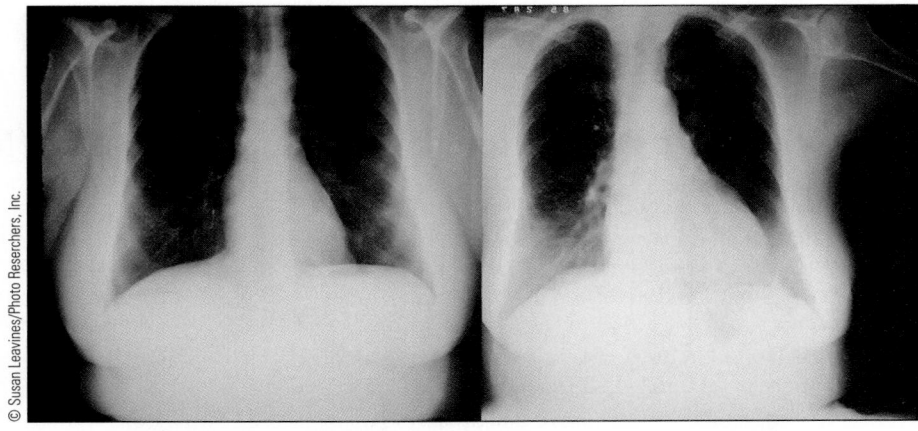

An overburdened heart enlarges in an effort to supply blood to the body's tissues.

Diet Order
Energy- and fat-controlled, low-salt diet.

heart muscle to work unusually hard. As a result, the heart muscle enlarges (**cardiomegaly**) and gradually weakens as it strains to supply adequate blood to the tissues.

Consequences of CHF   As heart failure progresses, reduced blood flow impairs the function of all organs. Reduced blood flow to the kidneys triggers the retention of fluid, further stressing the heart and compounding the stagnation of fluid in the body. Peripheral, pulmonary, and hepatic edema may develop as the person becomes increasingly "congested" with excess fluids. Pulmonary edema increases the likelihood of respiratory infections, which can further stress the heart and lungs.

CHF and Nutrition Status   As CHF progresses, energy needs increase because organ systems, particularly the heart and lungs, must work extra hard to maintain their functions. At the same time, the disrupted blood flow limits the supply of nutrients and oxygen to the organs and tissues. Repeated respiratory infections can further tax nutrition status. People in the later stages of CHF are often unable to eat enough to meet energy and protein demands; oral intake may be limited due to anorexia, altered taste sensitivity, intolerance to food odors, physical exhaustion, the low-sodium diet used for treatment (described later), and medications. Weight loss in people with CHF may go unnoticed until it has progressed considerably because edema masks their underweight condition. Thus protein-energy malnutrition (PEM) can occur as a consequence of CHF (**cardiac cachexia**), particularly when the disorder progresses to the later stages. Malnutrition then further contributes to the weakness of the heart muscle and raises the likelihood of respiratory infections.

Treatment of CHF   Drug therapy for CHF includes diuretics to reduce the fluid volume and drugs (cardiac glycosides) to increase the strength of heart muscle contractions. People taking thiazide or loop diuretics and cardiac glycosides together are at high risk for potassium deficiency and may be prescribed a potassium supplement. Stool softeners may be prescribed, particularly for elderly clients who frequently experience constipation, because straining to empty the bowels can stress the heart. Initially, bed rest helps reduce the heart's workload. Once recovery is under way, the person must rest frequently and avoid overexertion.

Medical Nutrition Therapy   The person newly diagnosed with CHF who is overweight benefits from a safe weight-loss program. Medical nutrition therapy for the wasting associated with later stages of CHF aims to preserve or restore

cardiomegaly (CAR-dee-oh-MEG-ah-lee): enlargement of the heart.
  mega = large

cardiac cachexia (ka-KEKS-ee-ah): chronic PEM that develops as a consequence of heart failure. Research suggests that cytokines play a role in the development of PEM in the late stages of CHF.
  kakos = bad
  hexia = condition

nutrition status and reduce the work of the heart. Providing enough energy to maintain body weight is vital, but providing too much energy increases the body's metabolic rate, stressing the heart. Likewise, giving too much fluid and sodium expands the body's fluid volume, which also taxes the heart. When clients are unable to eat adequate amounts of table foods, formulas of high nutrient density, which provide energy and protein with less fluid, may help clients meet their nutrient needs.

The heart-healthy diet for people with existing CHD is also appropriate. Restricting sodium is important. In mild heart failure, the level of sodium restriction recommended for heart health (2400 milligrams) is appropriate. As CHF progresses, sodium may need to be further restricted to 2 grams or less per day. In most cases, diuretics serve to eliminate excess fluids from the body. In severe cases of heart failure, however, people may need to restrict their fluid intake to 1000 to 2000 milliliters per day. Dietary fiber is carefully adjusted: the goal is to provide some fiber to prevent constipation, but to avoid amounts and types of fibers that produce gas and abdominal distention.

Energy needs for people with severe heart failure may be as high as 20 to 30% above basal energy needs (see Table 17–1 on p. 438)

Like a 2400-milligram sodium diet, a 2000-milligram (2-gram) sodium diet restricts high-sodium processed foods such as salad dressings; smoked, salted, kosher, and luncheon meats; salted snack foods; and soy sauce. Table 23–4 on p. 581 provides more information about a 2-gram sodium diet.

Chapter 26 includes a box (see p. 672) with suggestions for helping clients handle fluid restrictions.

Strokes

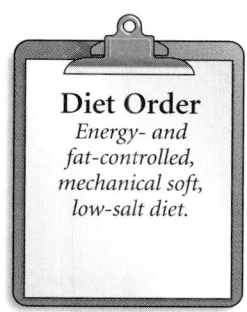

Diet Order
Energy- and fat-controlled, mechanical soft, low-salt diet.

Temporary interference with blood flow to the brain may result in a **transient ischemic attack (TIA),** a condition that causes changes in mental status that may last for a few minutes or a few hours. People who experience a TIA may or may not develop a total blockage of blood flow to a portion of the brain that results in a **stroke, or cerebral vascular accident (CVA).** Most strokes occur as a consequence of atherosclerosis, hypertension, or a combination of the two.

Complications Affecting Food Intake   For some stroke victims, recovery is unremarkable; others may require months of rehabilitative therapy. Victims may suffer from temporary or permanent problems that interfere with the ability to communicate and sometimes develop dysphagia (see Chapter 18). The inability to communicate effectively makes it difficult for them to tell health care professionals about foods they can or would like to eat or about problems they may be having with swallowing foods.

Following a stroke, some people develop physical problems that interfere with their ability to prepare foods or to handle the physical process of eating. Such problems can include an inability to grasp utensils or coordinate movements that bring foods or liquids from the table to the mouth. Nutrition in Practice 18 described some ways to handle such problems.

Medical Nutrition Therapy   Clients with significant problems with chewing and swallowing may require tube feedings until they have recovered from the stroke and a speech pathologist determines which foods they can safely chew and swallow. Long-term diet therapy for stroke victims depends on the underlying medical condition, but often includes measures to help lower blood lipids and blood pressure. For people with limited activity, food energy may need to be limited. A client who must relearn to walk or use other muscle groups can best do so if not overweight. Underweight can also hinder physical rehabilitation—another reason to ensure that the person does not become malnourished.

transient ischemic attack (TIA): a temporary reduction in blood flow to the brain, which causes temporary symptoms that vary depending on the part of the brain that is affected. Common symptoms include light-headedness, visual disturbances, paralysis, staggering, numbness, or dysphagia.

stroke: an event in which the blood flow to a part of the brain is cut off; also called a **cerebral vascular accident (CVA).**

Case Study

HISTORY PROFESSOR WITH CARDIOVASCULAR DISEASE

Mr. Jablonski, a 48-year-old history professor, has a blood lipid profile that includes elevated LDL cholesterol. He is 5 feet 7 inches tall and weighs 200 pounds. Mr. Jablonski has a family history of CHD. His diet history shows excessive intakes of food energy, cholesterol, total fat, saturated fat, and salt. He smokes a pack of cigarettes a day, and his lifestyle leaves him little time for physical activity. Mr. Jablonski also has hypertension, for which antihypertensive agents have been prescribed. He frequently forgets to take his pills, though, and his blood pressure is often quite high.

- Name the risk factors for CHD and hypertension in Mr. Jablonski's history. Which of them can he control? Which can be helped by diet? What complications might you expect if his condition goes untreated?

- What type of diet, if any, would you recommend to treat Mr. Jablonski's high LDL cholesterol and hypertension?

Explain the rationale for each diet change. How will his current diet change? Prepare a day's menus for Mr. Jablonski. What suggestions might you offer to help him make the necessary diet changes?

- What laboratory and clinical tests would you expect to see monitored regularly? Why?

- Name at least three ways in which Mr. Jablonski could benefit from losing weight. How does physical activity fit into a weight-loss plan? Describe other ways Mr. Jablonski might benefit from a physical activity program.

- Discuss nutrition considerations if Mr. Jablonski should suffer a heart attack or stroke. Describe the relationships of these disorders to elevated blood lipids and hypertension.

In Summary Some of the major consequences of atherosclerosis and hypertension include heart failure and strokes. The heart can fail suddenly (myocardial infarction) or gradually (congestive heart failure). For both types of heart failure, heart-healthy diets are appropriate once clients are out of immediate danger. In the later stages of congestive heart failure, clients may need to restrict sodium, and health care professionals must remain alert for signs of PEM. The dietary needs of people who experience a stroke vary, depending on the extent and the area of the brain affected. The accompanying case study presents a client with CHD. Carefully consider the questions posed to review the information presented in this chapter. The Diet-Drug Interactions box lists the nutrition-related concerns of the drugs described in this chapter. The Nutrition Assessment Checklist helps to identify nutrition-related factors that may help to prevent or treat disorders of the heart and blood vessels.

Diet-Drug Interactions

Anticoagulants (including aspirin)

Aspirin can be given with food to reduce nausea and GI distress. Long-term use of aspirin may lead to folate and vitamin C deficiencies. *Ticlopidine* is best absorbed along with foods. Ticlopidine may cause nausea, GI distress, and diarrhea. For people taking *warfarin,* maintaining a consistent vitamin K intake from day to day helps assure the medication's effectiveness. Warfarin may also cause diarrhea. The herbs *danshen, dong quai, ginkgo, cayenne, feverfew, echinacea, ginger, St. John's wort, garlic* and *coenzyme Q, omega-3 fatty acids,* and high doses of *vitamin E* can affect clotting times and must be used cautiously by people who are also taking anticoagulants.

Antihypertensives

All clients on *antihypertensives* should limit alcohol and avoid *natural licorice* and the herbs *ephedra, yohimbe,* and *hawthorn. Ginger* should be used cautiously. People on *ACE inhibitors* should avoid salt substitutes containing potassium and use potassium supplements (if prescribed) cautiously because these agents can raise blood potassium levels. One ACE inhibitor, *fosinopril,* should not be given along with calcium or magnesium supplements or calcium- or magnesium-containing antacids because calcium and magnesium reduce the drug's absorption. *Alpha-adrenergic blockers* can lead to weight gain and fatigue. *Beta-blockers,* which also act as antianginals, should be given with foods to reduce nausea and GI distress. Grapefruit juice should not be given along with *calcium channel blockers* because it can potentiate the effects of the drugs. *Clonidine* may dry the mouth and lead to constipation and drowsiness.

Antilipemics

Cholestyramine can cause nausea, GI distress, and constipation and can lead to fat-soluble vitamin deficiencies because it reduces their absorption. *Colestipol* is less likely than cholestyramine to cause nausea and lead to fat-soluble vitamin deficiencies, but it can lead to constipation. *Gemfibrozil* can lead to nausea and GI distress. The *statins (lovastatin, pravastatin, simvastin)* should not be given with grapefruit juice. People taking statins should limit alcohol use and tell their physicians if they take *niacin* supplements.

Cardiac Glycosides

Digitoxin, digoxin, and *digitalis* can lead to anorexia and nausea. Low blood potassium and elevated blood calcium levels can lead to drug toxicity, and clients are advised to eat a high-potassium diet. If they use calcium and vitamin D supplements, they must be used cautiously. Magnesium-containing antacids or supplements given along with *cardiac glycosides* can reduce drug absorption. The herb *hawthorn* can potentiate the effects of cardiac glycosides. *Licorice root* can lead to digoxin toxicity. *Ephedra* and *Indian snakeroot* should be avoided completely.

Diuretics

Diuretics can cause potassium imbalances and lead to muscle weakness, unexplained numbness and tingling sensations, irregular heartbeats, and cardiac arrest. *Thiazide* and *loop diuretics* increase the urinary excretion of potassium, and clients on these diuretics are encouraged to include rich sources of potassium daily and may be prescribed potassium supplements. When taken internally, *aloe vera* can potentiate the loss of potassium in people taking potassium-wasting diuretics. *Indian snakeroot* can potentiate the effects of thiazide and loop diuretics. Potassium-sparing diuretics *(amiloride, spironolactone,* and *triamterene),* on the other hand, lead to potassium retention. Clients on potassium-sparing diuretics should avoid excessive potassium intakes, potassium supplements, and salt substitutes that contain potassium. Sometimes a combination of both types of diuretics is used to avoid potassium imbalances.

Nutrition Assessment Checklist

FOR PEOPLE WITH CARDIOVASCULAR DISORDERS

Medical History

Check the medical record for a diagnosis of:

☐ Coronary heart disease

☐ Hypertension

☐ Congestive heart failure

Review the medical record for complications related to cardiovascular disease:

☐ Heart attacks

☐ Stroke

☐ Weight loss related to chronic heart failure

Note risk factors for CHD and strokes related to diet, including:

☐ Elevated LDL cholesterol

☐ Obesity or overweight

☐ Diabetes

Medications

For people who are using drug therapy for cardiovascular diseases, note:

☐ Side effects that may alter food intake

☐ Potential for folate and vitamin C deficiencies for people taking aspirin

☐ Consistency of vitamin K intake from day to day for people taking warfarin

☐ Intake of potassium (including potassium from salt substitutes) for people taking ACE inhibitors, cardiac glycosides, and diuretics

☐ Intake of fat-soluble vitamins for people taking cholestyramine

☐ Use of alcohol and natural licorice for people taking antihypertensive agents

☐ Use of grapefruit juice for people taking calcium channel blockers and statins

☐ Use of calcium- and magnesium-containing antacids and supplements for people taking fosinopril and cardiac glycosides

☐ Use of herbs and their potential effects on medications

Food/Nutrient Intake

For all clients, especially those with CHD or hypertension, or those with risk factors for CHD or hypertension, assess the diet for:

☐ Total energy

☐ Total fat and sources of saturated fat, monounsaturated fat, polyunsaturated fat, and cholesterol

☐ *Trans*-fatty acids

☐ Fiber

☐ Salt, potassium, calcium, and magnesium

☐ Folate and other B vitamins

☐ Alcohol

Considerations for clients with complications of the heart and blood vessels include:

☐ Energy and nutrient intake for clients in the later stages of chronic heart failure

☐ Physical disabilities that interfere with clients' abilities to prepare and eat foods and/or dysphagia following a stroke

Anthropometrics

Measure baseline height and weight and reassess weight at each medical checkup.

Note whether clients are meeting weight goals, including:

☐ Weight loss for clients who are overweight, especially those who have or are at risk for CHD, hypertension, or chronic heart failure

☐ Weight maintenance for people in the later stages of chronic heart failure

Remember that weight may be deceptively high in people who are retaining fluids, especially those in the later stages of chronic heart failure.

Laboratory Tests

Monitor the following laboratory tests for people with cardiovascular diseases or those at risk for CHD and hypertension:

☐ Blood lipids

☐ Blood glucose for people with diabetes

☐ Blood potassium for people taking diuretics and/or cardiac glycosides

☐ Indicators of fluid retention for people with heart failure

Physical Signs

Blood pressure measurement is routine in physical exams, but is especially important for people:

☐ With cardiovascular diseases or following a heart attack or stroke

☐ At risk for CHD or hypertension

Look for physical signs of:

☐ Folate and vitamin C deficiencies in people taking aspirin

☐ Potassium imbalances (muscle weakness, numbness and tingling sensations, irregular heartbeats) in people taking diuretics or cardiac glycosides

☐ Fat-soluble vitamin deficiencies in people taking cholestyramine

☐ Fluid overload in people in the later stages of chronic heart failure

1. Which of the following is the leading cause of death around the world?
   a. cardiovascular diseases
   b. inflammatory bowel diseases
   c. diabetes mellitus
   d. gastroesophageal reflux disease

2. The form of blood cholesterol most clearly associated with an increased risk of atherosclerosis is:
   a. triglycerides.
   b. chylomicrons.
   c. LDL cholesterol.
   d. HDL cholesterol.

3. Which of the following statements about hypertension is true?
   a. Low blood pressure reduces life expectancy.
   b. The narrower the diameter of the arteries, the greater the blood pressure.
   c. People can tell when their blood pressure rises.
   d. Hypertension aggravates atherosclerosis, but atherosclerosis does not aggravate hypertension.

4. Modifiable risk factors for CHD include:
   a. age and heredity.
   b. heredity, smoking, and salt intake.
   c. obesity, diabetes, and elevated LDL cholesterol.
   d. physical activity, hypertension, gender, and heredity.

5. Compared to the heart-healthy diet recommended to prevent CHD, the diet recommended for people with elevated LDL cholesterol, preexisting CHD, and diabetes is:
   a. lower in kcalories.
   b. higher in complex carbohydrates.
   c. lower in total fat and cholesterol.
   d. lower in saturated fat and cholesterol.

6. The benefits of adhering to the DASH diet include all of the following **except:**
   a. lower blood pressure.
   b. lower LDL.
   c. lower HDL.
   d. lower total cholesterol.

7. Diet-related components that may be protective against CHD include all of the following **except:**
   a. fiber.
   b. B vitamins.
   c. vitamin K.
   d. antioxidant nutrients.

8. For people receiving drug therapy for either hyperlipidemia or hypertension:
   a. physical activity must be restricted.
   b. the risk of diet-drug interactions is high.
   c. diet restrictions and physical activity programs are no longer effective.
   d. the risk of potassium imbalances is great for people taking anticoagulants.

9. Diet strategies to help a person recover from a myocardial infarction include:
   a. low-fiber foods to prevent constipation.
   b. foods of moderate temperatures to avoid slowing the heart rate.
   c. three larger meals to allow the heart more time to rest between meals.
   d. high-kcalorie, high-protein, low-sodium liquids at first to speed repair of the heart and prevent fluid retention.

10. Which of the following are associated with malnutrition in chronic heart failure?
    a. anorexia, medications, and respiratory infections
    b. dehydration, fat malabsorption, and lactose intolerance
    c. essential fatty acid deficiencies and calcium imbalances
    d. vitamin K deficiencies, dysphagia, and low-fiber, low-potassium diets

Critical Thinking

1. Which of the following most closely meets the recommendations to lower the risk of CHD?
   a. sandwich made with freshly baked skinless chicken on whole-wheat bread
   b. canned lentil soup with saltine crackers
   c. fresh strawberries on a slice of pound cake
   d. soft pretzels topped with mustard

2. Which of the following statements about alcohol and cardiovascular diseases might a health care professional relay to a client who wishes to start using alcohol to lower the risks of CHD?
   a. Alcohol lowers blood pressure.
   b. Alcohol intake should not exceed more than 4 drinks a day.
   c. A reduced risk of CHD is observed at all levels of alcohol intake.
   d. Risks associated with alcohol use may outweigh its benefits in protecting against CHD.

3. Since a client suffered a stroke, she has experienced repeated bouts of pneumonia. The health care professional should be alert for clinical signs that suggest that the client is experiencing:
   a. congestive heart failure.
   b. dysphagia and silent aspiration.
   c. a silent heart attack.
   d. the metabolic syndrome.

4. A client with CHD and hypertension who is taking a cholesterol-lowering medication and an antihypertensive mentions that he is also taking several dietary supplements. The health care professional should:
   a. check a drug guide or other resource to make sure that the dietary supplements and medications have no known interactions.
   b. tell the client not to take the dietary supplements because the risk of drug-drug interactions is too high.
   c. tell the client the dietary supplements are ineffective and will provide no benefits.
   d. tell the client not to take the diet supplements because the risk of diet-drug interactions is too high.

Answers to these questions can be found in Appendix G.

1. Consider the risk factors for CHD. Describe possible interrelationships among the factors. For example, a woman over age 55 is also at higher risk of diabetes; a person with diabetes is more likely to have hypertension.

2. Plan a week's menus that incorporate the diet strategies listed in Table 25–5. Begin by calculating how much energy you need. If you are at a desirable weight, you need about 15 kcalories per pound of body weight to maintain your weight if you are active and 13 kcalories if you are less active. If you are overweight, you can use 10 kcalories per pound of body weight to estimate your energy needs. Use the exchange lists to determine the number of servings of fats, starchy vegetables and grains, lean meat, and alcohol that you can include in your diet.

Nutrition on the Net

For further study of the topics of this chapter, access these websites.

Find updates and quick links to these and other nutrition-related sites at our website: **www.wadsworth.com/nutrition**

Visit the American Heart Association to find more information about CHD and their consequences, including the complete American Heart Association dietary guidelines, planning guides to implement the dietary guidelines, the DASH diet, tools for assessing your personal risk of heart disease, links to related websites, and much more: **www.americanheart.org**

For other CHD risk assessment tools, more information about CHD and its consequences, suggestions for implementing heart-healthy diets, and the summary of the Expert Panel of the National Cholesterol Education Program, visit the National Heart, Lung, and Blood Institute: **www.nhlbi.nih.gov/nhilbi/nhlbi.htm**

Discover resources on CHD and strokes by visiting the Heart and Stroke Foundation of Canada: **www.hsf.ca**

Review the risks and benefits of alternative therapies for heart disease by visiting The Natural Pharmacist: **www.TNP.com**

Notes

1. American Heart Association, 2001 Heart and Strokes Statistical Update, **www.americanheart.org/statistics/cvd.html**, site visited on February 9, 2001.

2. American Heart Association, 2001 Heart and Strokes Statistical Update, 2001.

3. American Heart Association, Risk factors for coronary heart disease: AHA Scientific position, 2000, **www.americanheart.org,** site visited February 20, 2001.

4. G. S. Berenson and coauthors, Association between multiple cardiovascular risk factors and atherosclerosis in children and young adults, *New England Journal of Medicine* 339 (1998): 1650–1656.

5. W. D. Rosamond and coauthors, Trends in the incidence of myocardial infarction and in mortality due to coronary heart disease, 1987 to 1994, *New England Journal of Medicine* 339 (1998): 861–867.

6. J. Stamler and coauthors, Relationship of baseline serum cholesterol levels in 3 large cohorts of younger men to long-term coronary, cardiovascular, and all-cause mortality and to longevity, *Journal of the American Medical Association* 284 (2000): 311–318; D. Steinberg and A. M. Gotto, Preventing coronary artery disease by lowering cholesterol levels, *Journal of the American Medical Association* 282 (1999): 2043–2050.

7. The Expert Panel, Executive summary of the third report of the National Cholesterol Education Program (NCEP) Expert Panel on Detection, Evaluation and Treatment of High Blood Cholesterol in Adults (Adult Treatment Panel III), available from the National Heart, Lung and Blood Institute, **www.nhlbi.nih.gov/guidelines/cholesterol/index.htm,** site visited October 30, 2001.

8. G. F. Fletcher and coauthors, Statement on exercise: Benefits and recommendations for physical activity programs for all Americans, *Circulation* 94 (1996): 857–862.

9. F. W. Booth and coauthors, Waging war on modern chronic diseases: Primary prevention through exercise biology, *Journal of Applied Physiology* 88 (2000): 774–787; F. B. Hu and coauthors, Physical activity and risk of stroke in women, *Journal of the American Medical Association* 283 (2000): 2961–2967; F. W. Farrell and coauthors, Influences of cardiorespiratory fitness levels and other predictors on cardiovascular disease mortality in men, *Medicine and Science in Sports and Exercise* 30 (1998): 899–905.

10. P. T. Williams, High-density lipoprotein cholesterol and other risk factors for coronary heart disease in female runners, *New England Journal of Medicine* 334 (1996): 1298–1303.

11. J. Plutzky, The Heart Protection Study and other developments in atherosclerosis, American Heart Association Scientific Session 2001, November 15, 2001, available at **www.medscape.com,** site visited December 3, 2001.

12. B. J. Goldstein, Insulin resistance: The core defect in type 2 diabetes, in Insulin Resistance: Implications for Metabolic and Cardiovascular Diseases, a Medscape CME program, available from **www.medscape.com,** site visited October 13, 2001.

13. R. M. Krauss and coauthors, AHA dietary guidelines, revision 2000: A statement for healthcare professionals from the nutrition committee of the American Heart Association, *Circulation* 102 (2000): 2284–2299.

14. L. Van Horn, Fiber, lipids, and coronary heart diseases, *Circulation* 95 (1997): 2701–2704.

15. E. Arnesen and coauthors, Serum total homocysteine and coronary heart disease, *International Journal of Epidemiology* 24 (1995): 704–709; M. J. Stampfer and coauthors, A prospective study of plasma homocyste(e)ine and risk of myocardial infarction in US physicians, *Journal of the American Medical Association* 268 (1992): 887–880.

16. W. S. Harris, Cardioprotective effects of σ-3 fatty acids, *Nutrition in Clinical Practice* 16 (2001): 6–12; C. R. Harper and T. A. Jacobson, The role of omega-3 fatty acids in the prevention of coronary heart disease, *Archives of Internal Medicine* 161 (2001): 2185–2192; C. M. Albert and coauthors, Fish consumption and risk of sudden cardiac death, *Journal of the American Medical Association* 279 (1998): 23–28; W. S. Harris, σ-3 Fatty acids and serum lipoproteins: Human studies, *American Journal of Clinical Nutrition* (supplement) 65 (1997): 1645–1654; T. A. Mori and coauthors, Interactions between dietary fat, fish, and fish oils and their effects on platelet functions in men at risk of cardiovascular disease, *Thrombosis and Vascular Biology* 17 (1997): 279–286.

17. C. Von Shacky and coauthors, The effect of dietary omega-3 fatty acids on coronary atherosclerosis: A randomized, double-blind, placebo-controlled trial, *Annals of Internal Medicine* 130 (1999): 554–562.

18. GISSI-Prevenzione Investigators, Dietary supplementation with σ-3 polyunsaturated fatty acids and vitamin E, after myocardial infarction: Results of the GISSI-Prevenzione trial, *Lancet* 9177 (1999): 447–455.

19. L. J. Appel and coauthors, A clinical trial of the effects of dietary patterns on blood pressure, *New England Journal of Medicine* 339 (1997): 1117–1124.

20. F. M. Sacks and coauthors, Effects on blood pressure of reduced sodium and the Dietary Approaches to Stop Hypertension (DASH) diet: DASH-Sodium Collaborative Research Group, *New England Journal of Medicine* 344 (2001): 3–10.

21. E. Obarzanek and coauthors, Effects on blood lipids of a blood pressure–lowering diet: The Dietary Approaches to Stop Hypertension (DASH) Trial, *American Journal of Clinical Nutrition* 74 (2001): 80–89.

22. P. R. Ridker and coauthors, Association of moderate alcohol consumption and plasma concentrations of endogenous tissue-type plasminogen activator, *Journal of the American Medical Association* 272 (1994): 929–933; J. M. Gaziano and coauthors, Moderate alcohol intake, increased levels of high-density lipoprotein and its subfractions, and decreased risk of myocardial infarction, *New England Journal of Medicine* 329 (1993): 1829–1834;

23. C. L. Hart and coauthors, Alcohol consumption and mortality from all causes, coronary heart disease, and stroke: Results from a prospective cohort study of Scottish men with 21 years of followup, *British Medical Journal* 318 (1999): 1725–1729.

24. X. Xin and coauthors, Effects of alcohol reduction on blood pressure, *Hypertension* 38 (2001): 1112–1117.

25. E. M. Rimm and coauthors, Review of moderate alcohol consumption and reduced risk of coronary heart disease: Is the effect due to beer, wine, or spirits? *British Medical Journal* 312 (1996): 731–736.

26. M. Grønbaek and coauthors, Type of alcohol consumed and mortality from all causes, coronary heart disease, and cancer, *Annals of Internal Medicine* 133 (2000): 411–419.

27. S. V. Nigdikar and coauthors, Consumption of red wine polyphenols reduces the susceptibility of low-density lipoproteins to oxidation in vivo, *American Journal of Clinical Nutrition* 68 (1998): 258–265.

28. E. L. Mortensen and coauthors, Better psychological functioning and higher social status may largely explain apparent health benefits of wine: A study of wine and beer drinking in young Danish adults, *Archives of Internal Medicine* 161 (2001): 1844–1848.

29. H. Gylling, Reduction of serum cholesterol in postmenopausal women with previous myocardial infarction and cholesterol malabsorption induced by dietary sitostanol ester margarine, *Circulation* 95 (1997): 4226–4231; B. V. Howard and D. Kritchevsky, Phytochemicals and cardiovascular disease: A statement for healthcare professionals from the American Heart Association, *Circulation* 95 (1997): 2591–2593.

30. J. W. Anderson, B. M. Johnstone, and M. E. Cook-Newell, Meta-analysis of the effects of soy protein intake on serum lipids, *New England Journal of Medicine* 333 (1995): 276–282.

31. J. R. Crouse and coauthors, A randomized trial comparing the effect of casein with that of soy protein containing varying amounts of isoflavones on plasma concentrations of lipids and lipoproteins, *Archives of Internal Medicine* 159 (1999): 2070–2072.

32. S. R. Teixeira and coauthors, Effects of feeding 4 levels of soy protein for 3 and 6 weeks on blood lipids and apolipoproteins in moderately hypercholesterolemic men, *American Journal of Clinical Nutrition* 71 (2000): 1077–1084.

Nutrition in Practice

■ MORE ABOUT THE METABOLIC SYNDROME ■

As Chapters 24 and 25 described, several risk factors for type 2 diabetes, cardiovascular diseases (CVD), and their complications frequently occur together. These risk factors—insulin resistance, high triglycerides, low HDL cholesterol, and hypertension—collectively comprise the metabolic syndrome, which is sometimes called syndrome X or the insulin resistance syndrome. Often obesity, especially central obesity, triggers insulin resistance, which then leads to a cascade of metabolic events that impair the health of the cardiovascular system. Conversely, because the risk factors associated with the metabolic syndrome are modifiable, it might be possible to reduce the incidence of diabetes and CVD. Researchers continue to unravel the mechanisms of the metabolic syndrome, but many pieces of the puzzle have already emerged. This Nutrition in Practice summarizes some of the most important findings. The accompanying glossary defines terms related to the metabolic syndrome.

What exactly is happening in a person who is insulin resistant?

In a person who is insulin resistant, insulin fails to lower blood glucose as it would in a healthy, insulin-sensitive person. The liver responds by making more glucose to provide energy for the cells. At first, the beta cells of the pancreas secrete more insulin in response. At this stage, people with insulin resistance have elevated levels of both glucose and insulin (hyperinsulinemia). While chronic elevated blood glucose levels are associated with the acute and **microvascular** complications of diabetes (neuropathy, nephropathy, and retinopathy), insulin resistance may play a bigger role in the cardiovascular complications. Figure NP25–1 on p. 654 shows how insulin resistance leads to problems typical of the metabolic syndrome.

Most, but not all, people with type 2 diabetes have insulin resistance, especially those who are overweight. Not all people with insulin resistance have type 2 diabetes, however, or develop it later. For some people, the extra insulin secretion compensates for insulin resistance, and they are able to maintain normal blood glucose levels. People who may be insulin resistant and not have diabetes include some who have hypertension or have survived a heart attack and women who have **polycystic ovary syndrome.** In addition, some people

with HIV infections who take certain medications (protease inhibitors) develop central obesity and insulin resistance.[1]

What causes insulin resistance?

Although the exact causes remain unknown, a great deal of information is emerging. The accumulation of fatty acids in muscle cells plays an important role, although the reasons why fatty acids accumulate is subject to debate. One explanation is that fatty acids accumulate because they are released excessively from fat cells and taken up excessively by muscle cells. Another is that fatty acids accumulate because with insulin resistance, the muscle cells cannot metabolize them properly.

Normally, skeletal muscle uses both glucose and fatty acids for fuel. After a meal, insulin levels rise, and glucose enters the muscle cells, which use it to produce energy. During a fast, insulin levels fall, and adipose cells release free fatty acids. Skeletal muscle cells use some of these free fatty acids for energy during the fast. Most of the rest reach the liver, where they may be used for energy or packaged in lipoproteins and returned to the circulation.

The adipose cells of people who are insulin resistant, however, may continue to release free fatty acids even when insulin rises. This condition is called **insulin resistance** of adipose tissue.[2] People who are insulin resistant tend to have high levels of circulating fatty acids despite high levels of insulin. In these people, the high fatty acid levels can overload skeletal muscle and the liver. An overload of lipid in skeletal muscle impairs the uptake of glucose in the muscle cells and results in **insulin resistance of skeletal muscle.** Some evidence suggests that a reduced use of fatty acids for fuel during fasting in people who are insulin resistant may also contribute to the excess storage of fatty acids in skeletal muscle.[3]

What is the relationship of obesity to insulin resistance?

Obesity, especially central obesity, leads to insulin resistance and hyperinsulinemia and is a risk factor for both type 2 diabetes and CVD. Many people with the metabolic syndrome are obese, but others are not. The reasons some people tend to develop central obesity

Nutrition in Practice

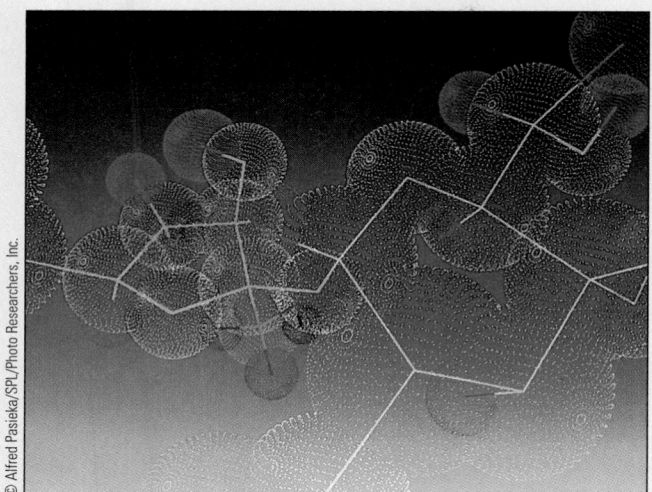

Adverse effects on many body functions occur when fat and muscle cells fail to respond to insulin, shown here.

are largely unknown, although heredity and aging are believed to play roles.

All fat tissues can release fatty acids into the bloodstream, but **visceral** (central) **fat** releases fatty acids at a faster rate than **peripheral** (subcutaneous) **fat.**[4] As described above, excess fatty acids may then accumulate in muscle cells. Free fatty acids in the blood also stimulate the liver to produce glucose (gluconeogenesis). Both of these effects raise blood glucose and insulin levels.

What is the connection between insulin resistance and blood lipids?

The hyperinsulinemia that accompanies insulin resistance ultimately leads to altered blood lipid patterns. Triglyceride levels rise, and with this rise, the size of the LDL particles is reduced. Among the LDL, small LDL have a particularly strong correlation with CHD risk.[5] Compared to regular-size LDL, small LDL may cross through the arterial wall and be oxidized more readily. In addition to elevated triglycerides and small LDL, HDL levels fall.

When excess fatty acids reach the liver, they may accumulate excessively and lead to fatty liver (see Chapter 23). Although the reasons are unclear, fatty liver results in various changes in normal fat metabolism. Some of these changes include:

- Increased production of VLDL (triglyceride-rich lipoproteins). Serum triglycerides and VLDL rise, and small LDL rise as well.

- Impaired action of the enzyme (lipoprotein lipase) that allows fat cells to take up fatty acids from lipoprotein. Triglycerides and triglyceride-rich lipoproteins remain high.

- Overactivity of the enzyme that degrades HDL. HDL levels fall.

Thus, once insulin resistance and hyperinsulinemia are set in motion, blood glucose rises higher (less is taken up by skeletal muscle), and lipid abnormalities that favor the development of CVD ensue. Insulin resistance and hyperinsulinemia can also affect the volume of blood, impair the ability of the blood vessels to dilate and constrict, and raise the likelihood that clots will form. All of these effects also add to the risk of CVD.

How can insulin resistance alter the blood volume?

Insulin influences sodium reabsorption in the kidneys. People who are insulin resistant reabsorb more sodium and fluid along with it, and the increased blood volume contributes to hypertension.

Glossary

fibrinogen (fie-BRIN-oh-jen): a protein produced by the liver that is essential to blood clotting.

insulin resistance of adipose tissue: failure of the enzyme that releases free fatty acids from fat cells to adequately reduce its activity in the presence of insulin.

insulin resistance of skeletal muscle: failure of the muscle cells to take up glucose from the blood in response to insulin as they normally would.

microvascular: pertaining to the capillaries. Retinopathy, nephropathy, and neuropathy are microvascular complications.

nitric oxide (NO): a substance produced by the vascular endothelium that causes blood vessels to dilate and inhibits clot formation.

peripheral fat: located directly beneath the skin.

plasminogen (plaz-MIN-oh-jen) **activator inhibitor-1 (PAL-1):** a substance important in blood clotting.

polycystic (POL-ee-SIS-tik) **ovary syndrome:** a disorder of women characterized by ovaries enlarged with fluid-filled sacs and elevated levels of male hormones (androgens).

visceral fat: fat located between abdominal organs.

Nutrition in Practice

Figure NP25–1
Insulin Resistance and the Metabolic Syndrome

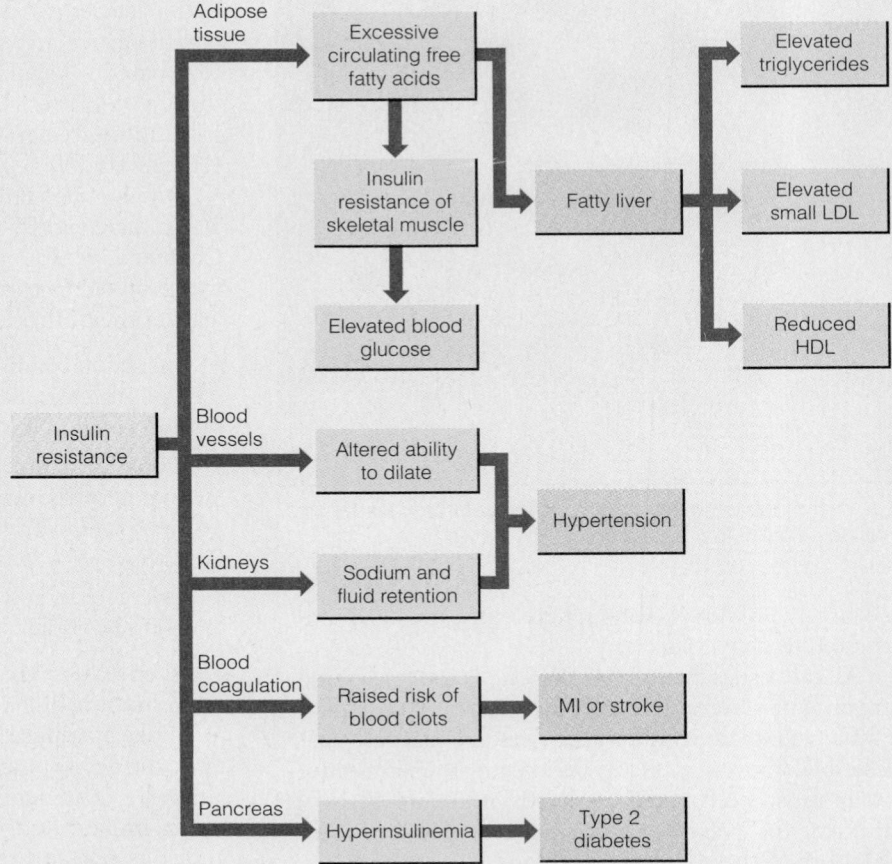

In what ways are blood vessels affected by insulin resistance?

Like many other cells, the cells lining the blood vessels respond to insulin. For healthy people, the rise in insulin following a meal relaxes the blood vessels, causing them to dilate. Once a blood vessel is dilated, blood flow through the vessel increases. Insulin resistance changes the ability of the blood vessels to relax so that blood flow can increase. In insulin-sensitive individuals, blood flow to muscle cells increases rapidly and significantly in response to insulin. In obese people, blood flow takes longer to increase to the same level, and in people with type 2 diabetes, blood flow takes even longer to increase, and it never rises to the same level as it does in insulin-sensitive and obese people.[6] These changes may also contribute to the hypertension that frequently accompanies insulin resistance and diabetes.[7]

What effect does insulin resistance have on blood clotting?

Insulin resistance alters the body's coagulation system in a way that favors the formation of clots. Among the blood factors that favor the formation of clots are **fibrinogen** and **plasminogen activator inhibitor-1,** and both of these are elevated in people with insulin resistance.[8] Improved blood glucose control may lower plasma fibrinogen, suggesting that prompt treatment of insulin resistance may help reduce the progression of the syndrome and the development of cardiovascular complications.[9]

An interesting hypothesis that may help to explain how insulin resistance affects both the ability of the blood vessels to dilate and blood coagulation suggests that insulin resistance interferes with the production of **nitric oxide (NO).**[10] NO is continuously produced from arginine (a nonessential amino acid) by the cells lining the blood vessels. NO relaxes the smooth muscle cells within the blood vessels and also inhibits factors that favor clot formation. In people with insulin resistance, NO production appears to falter.[11] Interestingly, limited studies have shown that the oral administration of arginine can help to improve insulin sensitivity in both the peripheral tissues and the liver.[12] Although the results appear promising, the studies were conducted on lean individuals with type 2 diabetes who

Nutrition in Practice

were in good metabolic control. It remains to be seen whether the same effects will be observed in larger groups of people with insulin resistance or diabetes with varying weights and degrees of metabolic control.

Can the metabolic syndrome be prevented?

Although the answer is unknown, many studies suggest that prevention may be possible. As Chapter 24 described, interventions that included weight loss, a low-fat diet, and a moderate program of physical activity had a dramatic effect on preventing type 2 diabetes in people with insulin resistance. In addition, the drug metformin, which works primarily by reducing the production of glucose by the liver, has also been shown to be a successful treatment. The thioglitazones may also have important applications in reversing the metabolic syndrome. These medications work primarily by reducing peripheral insulin resistance. Thioglitazones appear to alter fatty acid metabolism through mechanisms that remain unknown. They may help shift fat from visceral fat stores to peripheral fat stores.[13]

A number of contributing factors raise the risk for CVD in people who develop insulin resistance and the clinical findings associated with the metabolic syndrome. Abnormal blood lipids, changes in the blood vessels, hypertension, and an increased risk of clot formation are some of these clinical findings. Equally as important as determining how insulin resistance exerts its negative effects on the cardiovascular system is learning if the cascade of events can be prevented with lifestyle changes or other interventions.

Notes

1. L. A. Kosmiski and coauthors, Fat distribution and metabolic changes are strongly correlated and energy expenditure is increased in the HIV lipodystrophy syndrome, *AIDS* 15 (2001): 1993–2000.

2. S. M. Grundy, Pathogenesis of atherogenic dyslipidemia, *Drug Benefit Trends* 15 (2000): 22–27.

3. D. E. Kelley and B. H. Goodpaster, Skeletal muscle triglyceride: An aspect of regional adiposity and insulin resistance, *Diabetes Care* 24 (2001): 933–941.

4. B. J. Goldstein, Insulin resistance: The core defect in type 2 diabetes, in Insulin Resistance: Implications for Metabolic and Cardiovascular Diseases, a Medscape CME program, available at **www.medscape.com,** site visited October 13, 2001.

5. R. M. Krauss, Triglycerides and atherogenic lipoproteins: Rationale for lipid management, *American Journal of Medicine* (supplement) 105 (1998): 58–62.

6. A. D. Baron, Insulin and the vasculature—Old actors, new roles, *Journal of Investigative Medicine* 44 (1996): 406–412.

7. D. B. Corry and M. L. Tuck, Pathogenesis of hypertension in diabetes, *Journal of Cardiovascular Pharmacology* (supplement 4) 28 (1996): 6–15.

8. Grundy, 2000; G. Imperatore and coauthors, Plasma fibrinogen: A new factor of the metabolic syndrome, *Diabetes Care* 21 (1998): 649–654.

9. G. Bruno and coauthors, Association of fibrinogen with glycemia control and albumin excretion rate in patients with non-insulin-dependent diabetes mellitus, *Annals of Internal Medicine* 125 (1996): 653–657.

10. Type 2 diabetes mellitus: New perspectives of an old problem, *Clinical Courier,* November 1999, pp. 1–7.

11. Baron, 1996.

12. P. Piatti and coauthors, Long-term oral L-arginine administration improves peripheral and hepatic insulin sensitivity in type 2 diabetic patients, *Diabetes Care* 24 (2001): 875–880.

13. Goldstein, 2001.

Jennie Oppenheimer/Studio Zocolo

*D*iets modified in protein, minerals, and fluids serve in the treatment of kidney stones and altered **renal** function. This chapter begins by describing kidney stones and their dietary management. It then describes the management of disorders characterized by alterations in renal function—the nephrotic syndrome and renal failure.

The kidneys produce urine, which travels through the ureters to the bladder where it is stored temporarily and then excreted from the body (see Figure 26–1 on p. 659). The **nephrons** are the kidneys' functional units. Within each nephron, the **glomerulus** serves as a gate through which fluids and other blood components enter the nephron and form **filtrate**. As the filtrate passes through the **tubule,** some of its components are returned to the body, and others are passed on through the ureter to be excreted in the urine. Healthy kidneys, by continuously filtering the blood, maintain the body's fluid and electrolyte and acid-base balances and eliminate metabolic waste products. In addition, the kidneys:

- Help regulate blood pressure by secreting the enzyme **renin.** Renin triggers the release of the hormone **aldosterone,** which, in turn, signals the kidneys to retain sodium and water and raises blood pressure.
- Produce the hormone **erythropoietin,** which stimulates red blood cell production.
- Convert vitamin D to its most active form (**1, 25-dihydroxy vitamin D**) and play an important role in maintaining healthy bone tissue.

The kidneys' functions are so vital to life that if the kidneys fail, life cannot continue for more than a few days without medical intervention. The glossary below defines terms related to the kidneys and their functions.

Kidney Stones

Of the disorders that affect the kidneys and urinary tract, **kidney stones** are the most common. About 500,000 people in the United States develop kidney stones each year. Most of these people are men over 20 years old. Kidney stones are often quite painful, although sometimes they produce no symptoms. Stones often recur, but recurrences may be preventable. No single diet serves in the treatment

Kidney stones that have passed from the kidney to the ureter are technically ureteral stones, but most people use the term *kidney stone* to describe stones anywhere in the urinary tract.

renal: pertaining to the kidneys.

kidney stones: crystals of salts or other components that form a concentrated mass in the kidney; also called **renal calculi.**

calculi = pebbles

Glossary of Kidney-Related Terms

aldosterone (al-DOS-ter-own or AL-dough-STEER-own): a hormone secreted from the adrenal glands that signals the kidneys to retain sodium and fluid.

erythropoietin (eh-RITH-row-POY-eh-tin): a hormone, secreted by the kidneys in response to oxygen depletion or anemia, that stimulates the bone marrow to produce red blood cells.
erthro = red (blood cell)
poi = to make

filtrate: in the kidneys, the fluid that passes from the blood through the capillary walls of the glomeruli, eventually forming urine.

glomerulus (glow-MARE-you-lus): a cup-shaped membrane enclosing a tuft of capillaries within a nephron. (The plural is *glomeruli.*)

nephrons (NEF-rons): the working units of the kidneys. Each nephron consists of a glomerulus and a tubule.

1,25-dihydroxy vitamin D: the active form of vitamin D.

renin (REN-in): an enzyme, secreted by the kidneys in response to a reduced blood flow, that triggers the release of the hormone aldosterone.

tubule: a tubelike structure that surrounds the glomerulus and descends through the nephron. A pressure gradient between the glomerular capillaries and the tubule returns needed materials to the blood and moves wastes into the tubule to be excreted in the urine.

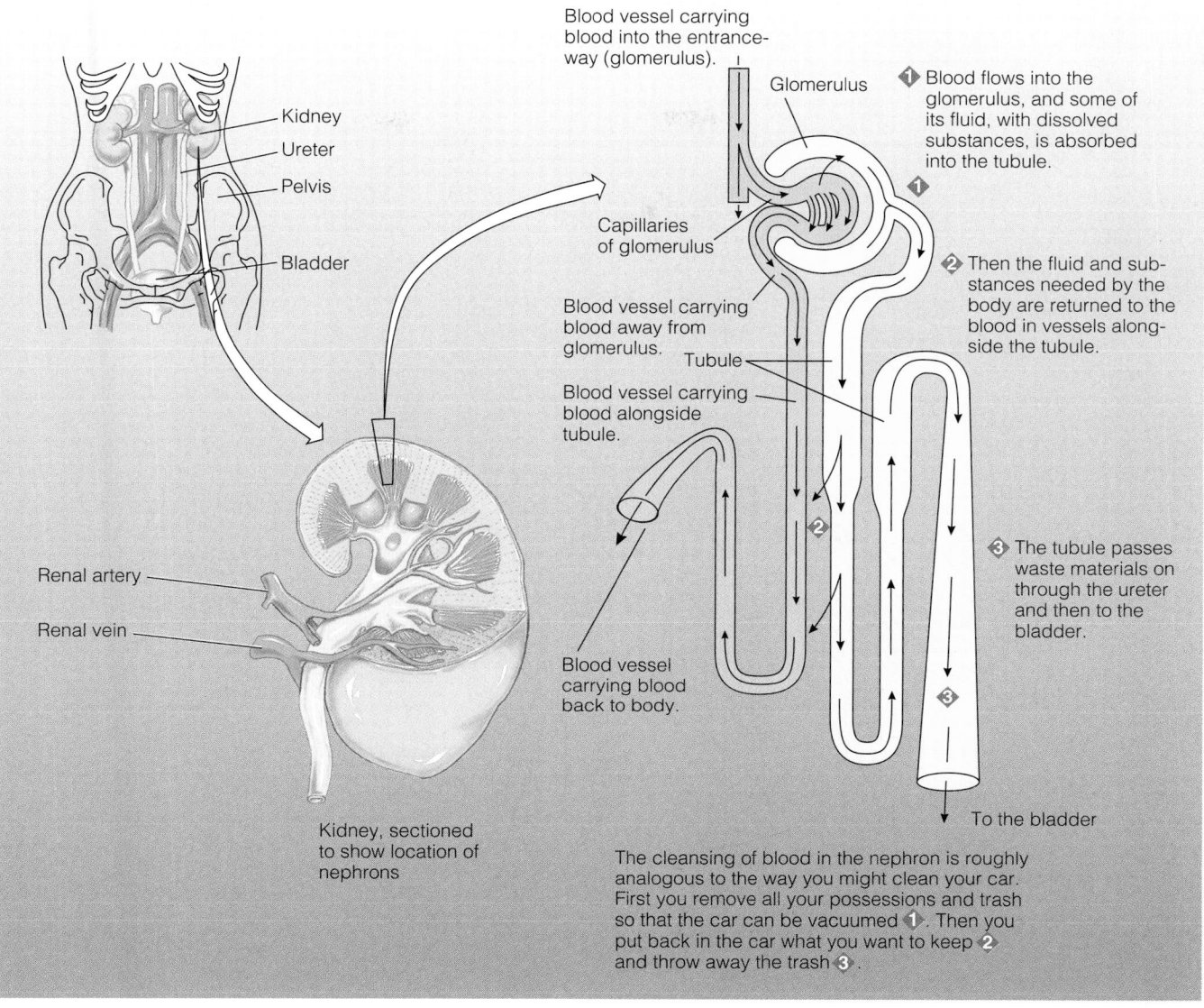

A nephron (a working unit of the kidney)

Blood vessel carrying blood into the entrance-way (glomerulus).

Glomerulus

Kidney
Ureter
Pelvis
Bladder

Capillaries of glomerulus

Blood vessel carrying blood away from glomerulus.

Tubule

Blood vessel carrying blood alongside tubule.

Renal artery
Renal vein

Blood vessel carrying blood back to body.

Kidney, sectioned to show location of nephrons

❶ Blood flows into the glomerulus, and some of its fluid, with dissolved substances, is absorbed into the tubule.

❷ Then the fluid and substances needed by the body are returned to the blood in vessels alongside the tubule.

❸ The tubule passes waste materials on through the ureter and then to the bladder.

To the bladder

The cleansing of blood in the nephron is roughly analogous to the way you might clean your car. First you remove all your possessions and trash so that the car can be vacuumed ❶. Then you put back in the car what you want to keep ❷ and throw away the trash ❸.

Figure 26–1

The Urinary Tract, a Kidney, and a Nephron

of kidney stones because they vary in composition and in their responses to dietary adjustments.

Causes of Kidney Stones

Kidney stones develop when stone constituents become concentrated in the urine and form crystals that grow. The stone constituents vary, but more than three-fourths of them contain calcium as calcium oxalate, calcium phosphate, or a combination of calcium, oxalate, and phosphate. Less commonly, stones are composed of uric acid, the amino acid cystine, or magnesium ammonium phosphate. Table 26–1 on p. 660 lists some conditions associated with stone formation.

Calcium Stones   Some people with calcium stones excrete normal amounts of calcium in the urine; others excrete excessive calcium (**hypercalciuria**). The reason that people with normal amounts of calcium in their urine form calcium

Reminder: Chapters 15 and 22 described how immobilization and fat malabsorption, respectively can lead to kidney stones.

hypercalciuria (HIGH-per-kal-see-YOU-ree-ah): excessive urinary excretion of calcium.

Protein-, Mineral-, and Fluid-Modified Diets for Disorders of the Kidneys and Urinary Tract

The most common type of kidney stone is composed of calcium oxalate crystals, shown here.

The most common type of kidney stone is composed of calcium oxalate crystals, shown here.

Table 26–1	Conditions Associated with Kidney Stones
■ Cystinuria	■ Osteoporosis
■ Fat malabsorption	■ Paget's disease
■ Glucocorticoid excess	■ Recurrent urinary tract infections
■ Gout	■ Renal tubular acidosis
■ Hyperparathyroidism	■ Use of indinavir (antiviral agent)
■ Hyperthyroidism	■ Vitamin D toxicity
■ Immobilization (prolonged bed rest or paralysis)	
■ Malignancies (some types)	

gout: an inherited metabolic disorder that results in excessive uric acid in the blood and urine and the deposition of uric acid in and around the joints, which causes acute arthritis and joint inflammation.

cystinuria (SIS-tin-NEW-ree-ah): an inherited metabolic disorder that is characterized by the excessive urinary excretion of cystine, lysine, arginine, and ornithine and commonly leads to kidney stone formation.

struvite (STREW-vite): crystals of magnesium ammonium phosphate formed by the action of bacterial enzymes.

renal colic: the severe pain that accompanies the movement of a kidney stone through the ureter to the bladder.

dysuria (dis-YOU-ree-ah): painful or difficult urination.

hematuria (HE-mah-TOO-ree-ah): blood in the urine.

stones is unclear, although they may lack factors that normally inhibit stone formation in most people. People with hypercalciuria are either more efficient at absorbing calcium from the intestine or more wasteful in their excretion of calcium than most people.

Uric Acid Stones   Uric acid stones are frequently associated with **gout,** a metabolic disorder characterized by elevated levels of uric acid in the blood and urine. Uric acid stones form when the urine becomes persistently acid, contains excessive uric acid, or both.

Cystine Stones   Cystine stones form as a consequence of an inherited disorder of amino acid metabolism called **cystinuria.** As the name implies, cystinuria causes the abnormal excretion of cystine in the urine.

Magnesium Stones   Stones composed of magnesium ammonium phosphate, or **struvite,** are associated with recurring urinary tract infections. Urinary tract infections and struvite stones are more common in women than men.

Consequences of Kidney Stones

In most cases, kidney stones pose no serious medical problems, especially when they are few and small. Small stones (less than one-fifth of an inch in diameter) may readily pass through the ureters (see Figure 26–1) and out of the body via the urine with minimal treatment.

Renal Colic   Large stones cannot pass easily through the ureters. When a large stone enters a ureter, it produces a sharp, stabbing pain, called **renal colic.** Typically, the pain starts suddenly in the back and intensifies as the stone follows the ureter's course down the abdomen toward the groin. The intense pain often occurs when the person moves about and may be accompanied by nausea and vomiting. When the stone reaches the bladder, the pain subsides abruptly.

Urinary Tract Complications   Symptoms associated with kidney stones may include frequent urination, urgency of urination, **dysuria,** and **hematuria.** Stones that cannot pass through the ureter may cause a urinary tract obstruction or infection and serious bleeding.

The New England Journal of Medicine October, 1992, p. 1142 with permission

<caption>The New England Journal of Medicine October, 1992, p. 1142 with permission</caption>

Prevention and Treatment of Kidney Stones

Treatment of the underlying medical condition is necessary to help prevent recurrences of kidney stones. Specific treatment measures vary according to the composition of the stone, but advice to prevent stones always includes this recommendation: increase fluid intake to dilute the urine. People who have had kidney stones need to drink enough fluid (mostly water) to maintain a urine volume of at least 2 liters per day. Generally, this requires a total intake of about 3 to 4 liters of fluid throughout the day. People who are physically active or live in warm climates may need additional fluids. People with fevers, diarrhea, or vomiting also need additional fluids until these conditions resolve. Once a stone has formed, drinking plenty of fluids (more than 3 liters a day) can sometimes help a small kidney stone to pass through the ureter.

Water should comprise at least 50 percent of the fluid intake. The choice of other beverages may also be important. Coffee (both regular and decaffeinated), tea, and wine may lower the likelihood of stone formation; grapefruit juice may raise the likelihood.[1]

Other Preventive Measures   For people who have never had a kidney stone, a high intake of calcium from foods may actually reduce the likelihood that a kidney stone will form.[2] Using calcium supplements, especially if they are taken without meals or with foods low in oxalate, may raise the risk of kidney stones. Whether it is the calcium from foods or some other constituent in the calcium-containing food that provides protection remains to be determined.

Preventing urinary tract infections is an important strategy for preventing struvite stones. Limited studies suggest that cranberry juice may help prevent urinary tract infections in women by preventing bacteria from adhering to the inner lining of the urinary tract.[3] People with urinary tract infections may also need to take anti-infective agents.

Drug Therapy   For people who excrete too much calcium, thiazide diuretics help to reduce the amount of calcium excreted in the urine. People with uric acid stones are often treated with allopurinol, a medication that reduces urinary uric acid concentrations. Other medications may include agents that reduce urinary acidity. The Diet-Drug Interactions box on p. 674 includes more information about medications used to treat kidney stones. Large stones that block the flow of urine or cause an infection require removal either surgically or, more commonly, by using shock waves to break the stone into pieces small enough to pass through the urinary tract (lithotripsy).

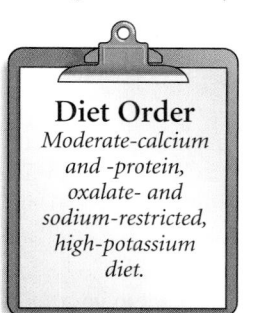

Diet Order
Moderate-calcium and -protein, oxalate- and sodium-restricted, high-potassium diet.

Medical Nutrition Therapy for Calcium and Oxalate Stones   Medical nutrition therapy for people with hypercalciuria includes the recommendation to avoid excessive calcium intakes, but not to let calcium intakes fall below recommended intakes. People with hypercalciuria who follow a low-calcium diet generally excrete more calcium than they ingest, indicating that they are losing calcium from their bones. Calcium from food sources should be encouraged. A person who cannot eat sufficient dietary calcium should be advised to use appropriate amounts of calcium supplements cautiously and to take them with meals. (Calcium restriction is not appropriate in the treatment of calcium stones unrelated to hypercalciuria, including stones related to fat malabsorption.)

People with calcium oxalate stones are advised to limit their intakes of foods high in oxalate (see Table 26–2 on p. 662). **Hyperoxaluria** increases the likelihood of calcium oxalate stone formation more than hypercalciuria does. Megadoses of vitamin C over long times raise urinary oxalate concentrations, so people at risk for oxalate stones are advised to avoid vitamin C in excess of recommended amounts.

Drinking plenty of water regularly throughout the day is the most important measure a person can take to prevent kidney stones.

For people with hypercalciuria who absorb too much calcium, calcium is restricted as follows:
- *800 mg/day for most adults.*
- *1200 mg/day for pregnant and lactating women.*
- *1200 to 1500 mg/day for postmenopausal women.*

For people with hypercalciuria and normal calcium absorption, calcium is restricted to 1000 mg/day.

hyperoxaluria (HIGH-per-OX-all-YOU-ree-ah): excessive urinary excretion of oxalate.

Protein-, Mineral-, and Fluid-Modified Diets for Disorders of the Kidneys and Urinary Tract

Table 26-2	Foods High in Oxalate	
Vegetables	**Fruits**	**Other**
Beans, green and wax	Blackberries	Chocolate and chocolate beverages*
Beets*	Blueberries	Cocoa
Celery	Currants, red	Coffee
Chard, swiss	Gooseberries	Draft beer
Collard greens	Grapes, Concord	Fruit cake
Dandelion greens	Lemon peel	Grits
Eggplant	Lime peel	Nuts, nut butters*
Endive	Orange peel	Peanut butter*
Escarole	Raspberries	Pepper
Leeks	Rhubarb*	Soybean crackers
Legumes	Strawberries*	Tea*
Okra		Tofu
Parsley		Wheat bran*
Potatoes, sweet		Wheat germ
Spinach*		
Squash, summer		

NOTE: The oxalate content of many foods has not been analyzed, and even fewer studies have been conducted to determine which foods raise urinary oxalate. The foods marked with an asterisk have been documented to raise urinary oxalate and should be avoided by people who form calcium stones.

In addition to avoiding excessive calcium and restricting oxalate, limiting salt and including foods high in potassium may also be important in preventing calcium oxalate stones. Excessive salt increases urinary calcium excretion in all people, but causes a proportionately greater amount of calcium to be excreted in people with hypercalciuria.[4] For people taking thiazide diuretics, moderate salt restriction and inclusion of high-potassium foods help limit the urinary excretion of both calcium and potassium. Some clinicians recommend a moderate protein intake (0.8 to 1.0 gram protein per kilogram of body weight per day) from either animal or plant sources to minimize calcium excretion.[5]

Uric Acid Stones   Diets restricted in **purine** are commonly prescribed to prevent uric acid stones. A purine-restricted diet limits red meats, particularly organ meats, and anchovies, sardines, and scallops. The benefits of such a diet are unproven, but avoiding excessive protein may be useful, and many foods high in purine are also high in protein. Alcohol intake is also limited.

Cystine Stones   The body synthesizes cystine, a nonessential amino acid, from methionine, an essential amino acid. Therefore, people with cystinuria benefit from a diet that provides enough methionine to meet the body's needs without providing too much. Medications to reduce urinary acidity may also be beneficial.

In Summary   Kidney stones form when stone constituents become very concentrated in the urine. The majority of kidney stones contain calcium, usually as calcium oxalate. Drinking plenty of water throughout the day is the most important dietary measure for preventing kidney stones of any type. People who form calcium oxalate stones may need to moderately limit calcium and protein, avoid foods high in oxalate and sodium, and include high-potassium foods. Treatment of uric acid stones includes a purine-restricted or moderate-protein diet, while the treatment of cystine stones includes a methionine-restricted diet.

purine (PU-reen): an end product of nucleotide metabolism that eventually breaks down to form uric acid.

The Nephrotic Syndrome

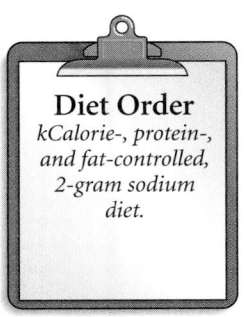

Diet Order
kCalorie-, protein-, and fat-controlled, 2-gram sodium diet.

The **nephrotic syndrome** is not a disease, but rather a distinct cluster of symptoms caused by damage to the glomerular capillaries. The damage may occur as a consequence of diabetes mellitus, hypertension, infections (either of the kidneys or elsewhere in the body), immunological and hereditary disorders, chemicals (medications, illegal drugs, or contaminants), and some cancers to name a few. The damage alters the permeability of the glomerular capillaries and allows plasma proteins to escape in the urine (**proteinuria**). Along with proteinuria, low serum albumin, edema, and elevated blood lipids are typical clinical findings. The nephrotic syndrome is often an early sign of deteriorating renal function, especially in people with diabetes (see Chapter 24). In some cases, treatment of the underlying condition corrects the disorder.

Consequences of the Nephrotic Syndrome

The consequences of the nephrotic syndrome include protein-energy malnutrition (PEM), anemia, infection, blood coagulation disorders, and accelerated atherosclerosis. Many of the consequences of this disorder are the same as those of malnutrition, which is not surprising because both alter protein status. If the nephrotic syndrome progresses to renal failure, the person develops uremia and other manifestations, as a later section describes.

Loss of Blood Proteins   As plasma proteins escape through the urine, blood proteins plummet. Albumin, the major plasma protein, is also the major protein lost in the urine, and its blood level is markedly reduced. Because albumin acts as a carrier for a variety of nutrients, hormones, and medications, blood levels of these may drop as well. Other proteins that are lost in the urine include immunoglobulins, transferrin, and vitamin D–binding protein. Losses of immunoglobulins raise the likelihood of infections. Loss of transferrin, the iron-carrying protein, may lead to anemia. The loss of vitamin D–binding protein may lead to vitamin D deficiency and impaired calcium absorption. Some calcium may also be lost in the urine along with the albumin that carries it. Consequently, rickets may develop, especially in children. If protein loss continues without replacement, lean body tissues break down, and PEM and general malnutrition follow. Figure 26–2 on p. 664 shows the consequences of falling levels of various proteins.

Edema   For many years, clinicians believed that edema occurred in the nephrotic syndrome due to the loss of albumin, which helps maintain the pressure that draws fluid from tissues into the bloodstream. However, although plasma proteins are lost as a consequence of the nephrotic syndrome, the nephrons reabsorb more sodium than normal, and this change is believed to be the primary cause of edema.[6] Fluid retention can lead to hypertension, which can further damage the kidneys.

Altered Blood Lipids   Elevated cholesterol, triglycerides, LDL, and VLDL are characteristic of the nephrotic syndrome. Hyperlipidemia may further impair renal function. Blood coagulation disorders and blood clots are frequent complications. Hypertension, hyperlipidemia, and blood coagulation disorders combine to raise the risk of cardiovascular diseases and stroke. Blood clots can also form in the renal veins and further injure the kidneys.

nephrotic (neh-FRAUT-ic) **syndrome:** the cluster of clinical findings that occur when glomerular function falters, including proteinuria, low serum albumin, edema, and elevated blood lipids.

proteinuria (pro-teen-YOUR-ee-ah): the loss of protein in the urine. The loss of albumin in the urine is *albuminuria*.

Protein-, Mineral-, and Fluid-Modified Diets for Disorders of the Kidneys and Urinary Tract

663

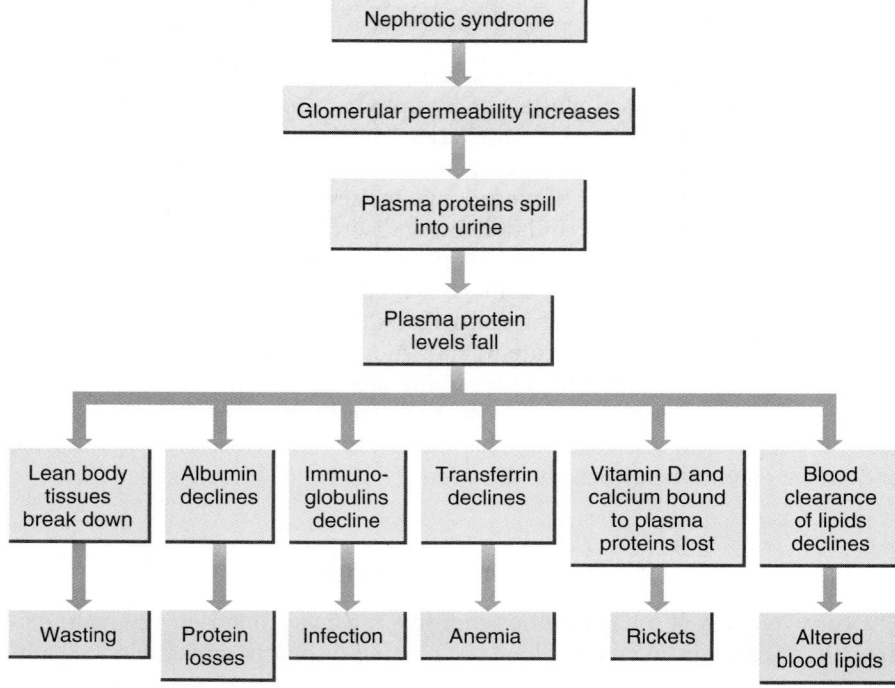

Treatment of the Nephrotic Syndrome

Medical treatment of the nephrotic syndrome first requires treatment of the underlying disorder and then drug and medical nutrition therapy to resolve symptoms. Treatments aim to minimize urinary protein losses, reduce blood lipids and hypertension, and prevent deterioration of renal function and malnutrition. Medications may include anti-infective agents, anticoagulants, antihypertensives, anti-inflammatory agents, antilipemics, and diuretics (see the Diet-Drug Interactions box on p. 674). Diet is central to preventing PEM and alleviating edema.

Energy A diet adequate in energy (35 kcalories per kilogram of body weight per day) sustains weight and spares protein. Weight loss or infections signal the need for additional kcalories. People who are overweight benefit from weight-loss diets to help control blood lipids and blood glucose (when necessary).

Protein In the past, high-protein diets (about 120 grams per day) were often prescribed for people with the nephrotic syndrome, stemming from the belief that high dietary protein would compensate for protein losses. High-protein diets accelerate protein synthesis, but they also incur greater urinary protein losses.[7] Furthermore, extra dietary protein may accelerate deterioration of renal function. For these reasons, protein is provided at about the RDA (0.8 to 1.0 gram per kilogram of body weight per day).

Fat A diet low in saturated fat, and cholesterol, such as the plans described in Chapter 25, can help control the elevated blood lipids associated with the nephrotic syndrome. Often, however, people with the nephrotic syndrome are unable to control blood lipids adequately using diet alone, and physicians may prescribe antilipemic agents as well. The use of fish oils may delay the progression of renal disease in people with the nephrotic syndrome associated with the abnormal accumulation of immunoglobulin A complexes in the kidneys, although more research is necessary to confirm such a benefit.[8]

Reminder: If blood lipids are elevated, an appropriate diet provides less than 30% of the total energy from fat, with less than 7% saturated fat and less than 200 mg of cholesterol per day.

Immunoglobulin A (IgA) nephropathy is the most common cause of the nephrotic syndrome worldwide.

Sodium A person with the nephrotic syndrome avidly retains sodium, and the resulting fluid retention can lead to edema and hypertension. Although the level of sodium restriction varies depending on the individual's response to diuretics, the diet often provides about 2 grams of sodium (see Table 23–4 on p. 581). Clients placed on thiazide or loop diuretics should be encouraged to select foods rich in potassium (see p. 662).

> *In Summary* The nephrotic syndrome describes the symptoms of proteinuria, edema, altered blood lipids, and altered blood clotting associated with damage to the glomerular capillaries. Losses of protein in the urine can lead to infections, anemia, rickets, PEM, and general malnutrition. Medications and a diet adequate in energy, moderate in protein, and low in sodium are the primary treatments.

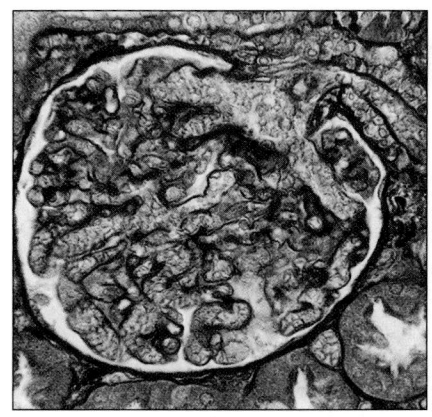

Renal failure follows the loss of a significant number of healthy nephrons (shown here).

Renal Failure

Diet Order
Protein-, mineral-, and fluid-controlled diet.

In renal failure, the rate at which the kidneys form filtrate, the **glomerular filtration rate (GFR),** declines to the point that the body can no longer maintain fluid and electrolyte and acid-base balances. The decline may occur suddenly (**acute renal failure**) or gradually (**chronic renal failure**).

Causes of Renal Failure

Acute renal failure most often occurs because blood flow to the kidneys is suddenly disrupted. Most often the disruption is a consequence of a severe stress such as a heart attack, shock, or severe blood loss during surgery or from trauma. Less commonly, urinary tract obstructions, infections, or some drugs directly damage the kidneys and lead to acute renal failure.

Chronic renal failure may develop in people with the nephrotic syndrome as renal function eventually deteriorates. Thus diabetic nephropathy, hypertension, certain infections, and immunological and hereditary diseases can lead to chronic renal failure. Other causes include renal artery obstructions, atherosclerosis of the renal arteries, and chronic heart failure. In a few cases, renal function is permanently damaged following acute renal failure.

Consequences of Renal Failure

The consequences of acute renal failure are immediate and dramatic. Nevertheless, acute renal failure is often reversible. In chronic renal failure, the consequences develop gradually, and the deterioration in renal function is irreversible.

Uremia As the nephrons fail, the body's principal nitrogen-containing metabolic waste products—blood urea nitrogen (BUN), creatinine, and uric acid—accumulate in the blood (**uremia**), and urine output slows. The symptoms and clinical findings associated with uremia are collectively called the **uremic syndrome.** The skin becomes dry and scaly, and the person may itch uncomfortably. Skin hemorrhages may be visible. In the later stages, urea (which can be excreted through the sweat) may crystallize on the skin, a symptom known as **uremic frost.** Nausea, vomiting, diarrhea, gastritis, and GI bleeding frequently occur.

Blood Chemistry Alterations In addition to retaining nitrogen-containing waste products, the body retains excess fluids, electrolytes, and acids produced during metabolism and normally excreted in the urine. As a consequence, the person develops edema, electrolyte imbalances, and acidosis, which stress the

Upper respiratory tract infections, especially those caused by streptococci, and viral infections, including hepatitis B virus, hepatitis C virus, and human immunodeficiency virus (HIV), can lead to chronic renal failure.

Normal blood urea nitrogen (BUN) is 10 to 20 mg/dL. A BUN of 50 to 150 mg/dL indicates serious impairment of renal function. BUN may rise as high as 150 to 250 mg/dL in the severe stages of renal disease.

glomerular filtration rate (GFR): the rate at which the kidneys form filtrate, usually measured by determining the amount of creatinine excreted in 24 hours. The normal GFR is about 130 ml/min for males and 120 ml/min for females.

acute renal failure: the sudden loss of the kidneys' ability to function.

chronic renal failure: the gradual and irreversible deterioration of kidney function.

uremia (you-REE-me-ah): abnormal accumulation of nitrogen-containing substances, especially urea, in the blood; also called *azotemia* (AZE-oh-TEE-me-ah).

uremic syndrome: the cluster of clinical findings associated with the buildup of nitrogen-containing waste products in the blood; may include fatigue, diminished mental alertness, agitation, muscle twitches, cramps, anorexia, nausea, vomiting, inflammation of the membranes of the mouth, unpleasant taste in the mouth, itchy skin, skin hemorrhages, gastritis, GI bleeding, and diarrhea.

uremic frost: the appearance of urea crystals on the skin.

Protein-, Mineral-, and Fluid-Modified Diets for Disorders of the Kidneys and Urinary Tract

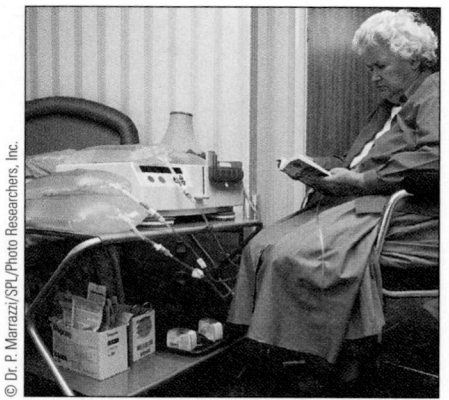

Dialysis provides an artificial means of maintaining the body's chemical balance when the kidneys fail.

cardiovascular and pulmonary systems. Elevated blood potassium can trigger irregular heartbeats (arrhythmias) and heart failure.

Treatment of Renal Failure

Treatments for renal failure aim to correct the underlying cause. In acute renal failure, effective treatment can often prevent permanent kidney damage. In chronic renal failure, the focus of treatment is to delay the progression of renal failure.

Medical Nutrition Therapy   Medical nutrition therapy plays an important role in the treatment of both acute and chronic renal failure. For acute renal failure, energy and protein needs are often the same as for severe stress—these will be described in Chapter 27. A later section of this chapter describes medical nutrition therapy for chronic renal failure.

Medications   Medications are prescribed to restore fluid and electrolyte balances and reduce complications. They may include insulin, antihypertensives, antilipemics, and medications to bind potassium and phosphorus and minimize their absorption. Other medications may include antiemetics, antisecretory agents, cardiac glycosides, and diuretics.

Dialysis   **Dialysis** is generally undertaken to restore fluid and electrolyte balances in acute renal failure and in the end stages of chronic renal failure. Nutrition in Practice 26 answers questions about dialysis and continuous renal replacement therapy (CRRT) and the effects of these procedures on nutrient needs. A kidney transplant is an option for chronic end-stage renal failure if an acceptable organ donor can be found.

In Summary   When the kidneys fail, the body is no longer able to maintain its chemical balance. The kidneys may fail suddenly, or renal function may decline over time. Some of the consequences of renal failure include uremia, the retention of fluid and electrolytes, and acidosis.

Chronic Renal Failure

In the early stages of chronic renal failure, the body compensates for the loss of some nephron function by enlarging the remaining functional nephrons. The hypertrophied nephrons work so efficiently that the GFR may fall to 75 percent of its normal rate before the symptoms of renal failure appear.

Consequences of Chronic Renal Failure

Eventually, the body exhausts the overworked nephrons, and renal function deteriorates. (The nephrons' effort is similar to that of the pancreatic beta cells in type 2 diabetes, which at first produce more and more insulin in response to high blood glucose and later become exhausted and unable to produce adequate insulin.) In **end-stage renal disease (ESRD),** the GFR drops below 20 percent of normal. **Renal insufficiency** describes the period in which kidney function has deteriorated but not to the point of ESRD. Uremia and changes in blood chemistry develop gradually and are pronounced in ESRD.

Cardiovascular Diseases   Accelerated atherosclerosis and cardiovascular diseases frequently accompany renal failure, and they can aggravate hypertension and increase the risk of congestive heart failure, heart attacks, and pulmonary edema.

The capacity of the kidneys to function despite loss of some nephrons is referred to as **renal reserve.** Normal values for GFR in males and females are about 130 and 120 ml/min, respectively. In either gender, a GFR of 56 to 100 ml/min constitutes mild renal failure, 25 to 55 ml/min constitutes moderate renal failure, and 24 ml/min or less constitutes severe renal failure.

dialysis (dye-AL-ih-sis): removal of waste from the blood through a semipermeable membrane using the principles of simple diffusion and osmosis. The two main types are **hemodialysis** and **peritoneal dialysis** (see Nutrition in Practice 26).

end-stage renal disease (ESRD): the severe stage of chronic renal failure in which dialysis or a kidney transplant is necessary to sustain life. In ESRD, the GRF falls to less than about 25 ml/min, and the BUN may rise as high as 150 to 250 mg/dL.

renal insufficiency: the stage of renal failure in which renal function is reduced but not to a life-threatening degree.

Table 26–3	Possible Causes of Wasting in Renal Failure	
Reduced Nutrient Intake	**Excessive Nutrient Losses**	**Raised Nutrient Needs**
Anorexia	Dialysis	Hormonal alterations
Fatigue	Diarrhea	Infection
Medications	GI bleeding	Inflammation
Nausea	Medications	Medications
Pain	Numerous blood tests	
Restrictive diet	Poor absorption	
Taste alterations	Vomiting	

Altered hormone levels together with the retention of waste products lead to hypertension, hyperglycemia, and elevated blood lipids (especially triglycerides).

Other Complications Anemia may result from depressed erythropoietin synthesis by the damaged kidneys, restrictive diets, nausea and vomiting, GI blood losses, and blood losses through dialysis and from frequent blood testing. The inability of the kidneys to excrete phosphorus elevates blood phosphorus and upsets the body's balance of phosphorus and calcium, often leading to bone diseases. The kidneys' inability to activate vitamin D, coupled with poor intakes of calcium, also contributes to bone diseases.

Bone disorders resulting from calcium and phosphorus imbalances in renal disease are called **renal osteodystrophies** (OS-tee-oh-DIS-tro-fees).

Growth Failure and Wasting Wasting and PEM frequently complicate chronic renal disease in both children and adults. Children with renal disease need nutrition intervention before the end of puberty if they are to make up growth deficits. By attending to diet, adults with renal disease can maintain or restore their nutrition status, avoid complications, and improve their quality of life. Nutrition status becomes more difficult to maintain as renal disease progresses, however. Table 26–3 summarizes the causes of wasting associated with renal failure.

Medical Nutrition Therapy for Chronic Renal Disease

Table 26–4 on p. 668 summarizes the changes in nutrient needs as renal insufficiency progresses to ESRD. The sample menu on p. 669 shows a day's meals for a renal diet. Note that the two dialysis procedures (hemodialysis and peritoneal dialysis) affect nutrient needs differently. The physician monitors the client's renal function and medical status to determine the appropriate diet prescription.

The complexity of the renal diet, as well as its critical role in the treatment of renal disease, underscores the need for a specialist, a renal dietitian, to educate clients and provide diet plans. Other health care professionals must understand the general concepts in order to communicate effectively with clients.

Energy All people with renal failure need adequate energy to maintain a desirable weight and prevent protein catabolism. As renal failure progresses, growth failure and wasting become more likely, and consuming adequate energy becomes more difficult. Nausea, vomiting, anorexia, taste alterations, and fatigue that may occur as a consequence of the uremic syndrome can all reduce food intake. As additional dietary restrictions are imposed, people are challenged to find a variety and quantity of foods they can eat and enjoy.

Notice from Table 26–4 that energy needs are slightly lower once dialysis, especially peritoneal dialysis, begins. This is because the person obtains some glucose from fluids (**dialysate**) used to remove waste products (Nutrition in

dialysate (dye-AL-ih-SATE): a solution used during dialysis to draw wastes and fluids from the blood.

Protein-, Mineral-, and Fluid-Modified Diets for Disorders of the Kidneys and Urinary Tract

Table 26–4 Nutrient Needs in Chronic Renal Failure

Nutrients[a]	Renal Insufficiency (Predialysis)	Hemodialysis	Peritoneal Dialysis
Energy (kcal/kg)[b]	30–40	30–35	25–35
Protein (g/kg)	0.6–0.8	1.2–1.4	1.2–1.5
Fluid (ml)	Typically not restricted	500–750 plus daily urine output, or 1000 if anuric	≥2000
Sodium (g)	2–4	2–3	3–4
Potassium (mg/kg)	Typically not restricted	40	Typically not restricted
Phosphorus (mg)	<1200[c]	800–1200[c]	1200[c]
Supplements			
Calcium (mg)	1000–1500	1000–1500	1000–1500
Folate (mg)	1	1	1
Vitamin B6 (mg)	5	10	10
Vitamin D	As appropriate	As appropriate	As appropriate

NOTE: The actual amounts of these nutrients in the diet must be highly individualized based on each person's responses. For example, energy needs in renal insufficiency may be lower for people who are overweight or higher for people who are underweight.

[a]Besides the specific nutrients listed, all others should meet recommended amounts.

[b]For children, an intake of 100 kcalories per kilogram of body weight is desirable, but 80 kcalories per kilogram of body weight is good. At a minimum, energy intake should not fall below the RDA.

[c]The extent of phosphorus restriction depends on serum phosphorus. The goal is to maintain serum phosphorus between 4.5 and 6.0 milligrams per deciliter. Often, phosphate binders are useful for this purpose.

SOURCES: Adapted from Meeting the challenge of the renal diet: A preview of the "National Renal Diet" educational series, *Journal of the American Dietetic Association* 93 (1993): 637–639; J. A. Beto, Which diet for which renal failure: Making sense of the options, *Journal of the American Dietetic Association* 95 (1995): 898–903.

Practice 26 provides the details). Energy needs are based on body weight, and people on dialysis retain fluid between treatments. For people on dialysis, body weight immediately following a dialysis treatment (**dry weight**) provides the best estimate of true body weight.

Protein Controlling protein (nitrogen) intake is a primary goal in the treatment of chronic renal failure. Providing the right amount of protein to the person in renal failure, however, is like walking a tightrope. Too little protein, and the person develops malnutrition. Too much protein, and blood urea (the toxic waste product of protein metabolism) rises. For people with renal insufficiency, restricting protein may help protect the remaining nephrons, although human studies have not clearly demonstrated a beneficial effect. As renal insufficiency progresses, dietary protein restrictions tighten. When protein intakes fall below the RDA recommendation, consuming adequate energy to spare protein and eating high-quality protein become vitally important. Clinicians generally recommend a protein-restricted diet that derives at least half of its protein from sources such as eggs, milk, meat, poultry, and fish. The remaining protein is derived from plant sources. A diet that includes protein from both animal and plant sources may be best because it combines the high quality of animal proteins with the low–saturated fat, low-cholesterol nature of plant proteins.

As renal failure progresses, some clinicians prescribe very-low-protein diets (0.3 gram protein per kilogram of body weight per day) supplemented with essential amino acids or essential amino acid precursors (keto acids), although this method has not been proved to be beneficial. Once the client begins dialysis, protein restrictions can be relaxed because dialysis removes nitrogenous waste products and incurs protein losses.

For people with a GFR less than 55 but greater than 25 ml/min, protein is restricted to 0.8 g/kg body weight/day (the RDA). For people with more advanced renal insufficiency (GFR less than 25 ml/min), protein is restricted to 0.6 g/kg body weight/day until dialysis begins.

dry weight: weight after excess fluids are removed from the body.

Renal Diet Menu

Breakfast
1 egg, fried with margarine
1 slice toast
Margarine
Jelly
½ c grape juice
Coffee
Nondairy creamer
Sugar

Lunch
Sandwich with 2 oz turkey
2 slices bread
Mayonnaise
Lettuce leaf
1 c green beans
Margarine
½ c strawberries
Whipping cream
Sugar
½ c milk

Supper
3 oz roast beef
½ c rice
Margarine
½ c mushrooms sautéed in olive oil and
seasonings
½ c applesauce
Iced tea
Sugar

Snacks
Hard candy, gumdrops, marshmallows
Carbonated beverages

This diet menu provides 60 grams of protein and controls phosphorus, potassium, and sodium intake. To increase the kcalories in this diet, prepare food with oil or unsalted margarine (regular margarine, if allowed) and use additional fat, sugar, or syrup whenever possible. For example, canned fruit packed in heavy syrup, rather than juice, adds kcalories.

Lipid and Carbohydrate   The ideal renal diet restricts total fat, saturated fat, and cholesterol to help control elevated blood lipid levels. A diet rich in complex carbohydrates helps minimize elevated blood glucose and triglycerides. Total carbohydrate, and especially simple carbohydrates, needs to be restricted further for people on peritoneal dialysis (see Nutrition in Practice 26) because they absorb considerable amounts of glucose as a consequence of the dialysis procedure.

Sodium and Fluids   As renal failure progresses, the person excretes less urine and cannot handle normal amounts of sodium and fluids. At this point, limiting sodium and fluids helps to prevent hypertension, edema, and heart failure. Individual needs for sodium and fluids are determined by carefully monitoring each person's weight, blood pressure, urine output, and blood electrolyte levels. A rapid rise in body weight and blood pressure suggests that the person is retaining sodium and fluid; conversely, a rapid decline in body weight and blood pressure (a desirable outcome of dialysis) indicates fluid loss.

Fluids are not restricted in renal insufficiency until urine output decreases. After that time, fluids are restricted. For the person who is neither dehydrated nor overhydrated, daily fluid needs amount to the daily urine output plus 500 to 750 milliliters to provide for obligatory water losses. Once a person is on dialysis, sodium and fluid intakes are controlled to allow a weight gain of about 2 pounds (of fluid) between dialysis treatments, although larger weight gains are common.

Potassium   Many people with renal insufficiency and those on peritoneal dialysis can handle typical, but not excessive, intakes of potassium (about 2.5 to 3.5 grams per day) until urine output falls below one liter per day. Some people with diabetic nephropathy may experience **hyperkalemia** earlier in the course of renal failure. People on hemodialysis may develop hyperkalemia between dialysis treatments. In such cases, potassium may be moderately restricted to about 1.5 to 3.0 grams per day. Remember, however, that individual needs vary. People taking thiazide and loop diuretics may need to adjust their potassium intakes accordingly.

Phosphorus, Calcium, and Vitamin D   Controlling blood phosphorus and calcium in renal failure may help to slow its progression and prevent bone diseases. As renal function declines, blood phosphorus levels rise and levels of active

© Craig M. Moore

When a diet must be restricted in protein, it is important that the protein that is consumed be of high quality, including protein from eggs, milk, meat, poultry, and fish.

hyperkalemia (HIGH-per-kay-LEE-me-ah): excessive potassium in the blood.

Protein-, Mineral-, and Fluid-Modified Diets for Disorders of the Kidneys and Urinary Tract

669

In limited amounts, the fruits and vegetables pictured here provide a level of potassium acceptable for renal diets.

Phosphorus appears in many high-protein foods including milk and milk products, eggs, peanut butter, sardines, and legumes. Bran cereal is also high in phosphorus.

The active form of supplemental vitamin D, or **calcitriol**, can be provided orally or intravenously.

vitamin D fall. The low amount of active vitamin D limits calcium absorption, and the combination of high blood phosphorus and limited availability of calcium leads to the loss of calcium from the bones and consequent bone disorders.

Dietary phosphorus restrictions are instituted early in the course of renal failure to help control rising blood phosphorus. Fortunately, protein-restricted diets are restricted in phosphorus as well. As renal failure progresses and especially when protein is liberalized, physicians often prescribe medications that bind phosphorus in the GI tract and make it unavailable for absorption. Calcium salts, and most recently a binder that does not contain calcium, can be used to bind phosphorus. Aluminum salts can also bind phosphorus, but the use of these salts is discouraged because aluminum toxicity can be a problem for people on dialysis.

Most people with renal disease need calcium supplements. Supplemental vitamin D in its active form can help maintain blood calcium and prevent bone disease. Some people, however, develop **hypercalcemia.** The renal team monitors serum calcium closely to prevent both low and high blood calcium and prescribes calcium and vitamin D supplements accordingly. Calcium-containing phosphate binders provide some calcium, but the absorption of calcium from these products varies widely. For people who tend to develop hypercalcemia, the phosphate binders that do not contain calcium may be useful.

Other Vitamins   People with renal failure frequently develop folate and vitamin B_6 deficiencies because of restrictive diets, loss of vitamins during dialysis, drug therapy, and altered metabolism. Medical nutrition therapy often includes folate and vitamin B_6 in generous amounts, along with the recommended amounts of the remaining water-soluble vitamins.[9] Adequate amounts of folate and vitamins B_6 and B_{12} may also help to lower serum homocysteine levels (see Chapter 25), which may help to protect against cardiovascular diseases in people with renal failure. Furthermore, these vitamins are important in preventing anemia.

Thiamin deficiencies mimic many of the complications associated with uremia, including changes in mental function (encephalopathy) that sometimes occur in people with end-stage renal disease. A recent investigation showed that providing thiamin can dramatically reverse unexplained encephalopathy when thiamin deficiency is the cause.[10]

Because many people with renal failure have high blood oxalate levels, intakes of vitamin C from both the diet and supplements are often limited to less than 100 milligrams per day. When oxalate levels are elevated, oxalate crystals can become deposited in soft tissues and result in complications including kidney stones and heart attacks. Supplementation of fat-soluble vitamins other than vitamin D usually is not necessary.

Trace Minerals   The administration of human erythropoietin along with the B vitamins and iron needed to synthesize hemoglobin is effective in treating iron-deficiency anemia, once a common and debilitating problem for people with chronic renal disease. Clients should be cautioned to avoid iron supplements that also contain vitamin C.

For people with renal failure, aluminum and magnesium can reach toxic levels. Therefore, clients should avoid aluminum- and magnesium-containing antacids as well as supplements, laxatives, and enemas containing magnesium.

People on dialysis frequently complain of anorexia, altered taste perceptions (dysgeusia), and a decreased sexual drive, symptoms typical of zinc deficiency. Clients who have these symptoms may need zinc supplements if their serum zinc levels are inadequate. Recall that children have particularly high needs for zinc, and zinc deficiencies can contribute to growth retardation.

Enteral and Parenteral Nutrition   Enteral formulas designed for people with renal insufficiency who have not begun dialysis can meet energy needs in smaller volumes. Such formulas provide less protein than other formulas with high

hypercalcemia (HIGH-per-cal-SEE-me-ah): excessive calcium in the blood.

nutrient density. Enteral and parenteral formulas designed for people on dialysis are also available. Compared with standard enteral formulas, renal formulas have fewer electrolytes and more kcalories per milliliter (see Appendix F). Renal TPN formulas are compounded with lower concentrations of both nonessential and essential amino acids and higher concentrations of dextrose than in standard TPN solutions. Electrolytes are added in appropriate amounts.

Diet Planning   The ideal renal diet presents a challenge: provide adequate energy, but restrict protein, fat, and, sometimes, simple carbohydrate. Complex carbohydrates, which might appear to be the ideal energy source, are often also rich sources of potassium. Because potassium must be restricted, complex carbohydrates must be carefully selected. Diet planners must accept that under such circumstances, no diet is truly ideal. They must recognize that the need to adjust protein and electrolytes outweighs the need to restrict fat and simple carbohydrate.

To help meet energy needs, clients include as many complex carbohydrates as their diet plans allow. They supplement their meals with formulas high in kcalories but restricted in protein and electrolytes and use foods such as sugars (hard candy and jelly) and fats (margarine and oil) freely. The person with elevated blood lipids is advised to restrict fat and modify the type of fat to whatever extent is possible. The person with diabetes or hyperglycemia is advised to eat a consistent carbohydrate intake at regular intervals and to adjust insulin to cover carbohydrate intake.

To help individuals on renal diets find foods they will accept and enjoy, food lists similar to the exchange system are available. Whereas the exchange system for diabetes groups foods by their energy, carbohydrate, protein, and fat contents, renal food lists group foods by their energy, protein, sodium, potassium, and phosphorus contents. The "How to" box on p. 672 provides suggestions to ease the task of complying with a renal diet.

Diet Compliance   The challenges dietitians face in designing renal diets pale in comparison with those encountered by clients and caregivers who must follow a complicated medical plan of which diet is only a part. To help clients understand their medical plans and support their efforts, the members of the health care team must effectively communicate with their clients and, just as important, listen to them. Within a tangled web of grossly altered metabolic processes, dialysis lines, and toxic waste products is a person, often a frightened or discouraged one. All members of the health care team need to understand the plan if they are to offer the most effective support.

A nutrition education approach that emphasizes self-management appears to be a useful strategy for helping clients comply with a renal diet.[11] This approach guides clients in identifying goals, selecting strategies to meet goals, and evaluating progress.

Diets following Kidney Transplants

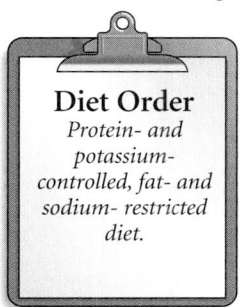

Diet Order
Protein- and potassium-controlled, fat- and sodium- restricted diet.

A preferable alternative to dialysis in end-stage renal disease is a kidney transplant. Kidney transplants can successfully restore kidney function and promote normal growth. For this reason, transplants are particularly desirable in children. Given a choice, many would prefer transplants, but suitable kidney donors cannot always be found. About 12,500 people received kidney transplants in 1999, while about 48,000 others remained on waiting lists.[12]

Immunosuppressive Drug Therapy   After receiving a new kidney, the person must take large doses of immunosuppressive medications to prevent tissue rejection (see the Diet-Drug Interactions box). Infections and increased susceptibility

How to

Help Clients Comply with a Renal Diet

*T*he following suggestions can assist clients in complying with the renal diet:

1. To keep track of fluid intake:
   - Fill a container with an amount of water equal to your total fluid allowance. Each time you use a liquid food or beverage, discard an equivalent amount of water from the container. The amount remaining in the container will show you how much fluid you have left for the day.
   - Be sure to save enough fluid to take medications.

2. To help control thirst:
   - Chew gum or suck hard candy.
   - Freeze fluids so they take longer to consume.
   - Add lemon juice to water to make it more refreshing.
   - Gargle with refrigerated mouthwash.

3. To prevent the diet from becoming monotonous:
   - Experiment with new combinations of allowed foods.
   - Use favorite foods whenever possible.
   - Substitute nondairy products for regular dairy products. Nondairy products are lower in protein, phosphorus, and potassium than regular dairy foods, and they can substitute for milk and add energy to the diet.
   - Add zest to foods by seasoning with garlic, onion, chili, curry powder, oregano, pepper, or lemon juice.
   - Consult a dietitian when you want to eat restricted foods. Many restricted foods can be used occasionally and in small amounts if the diet is carefully adjusted.

Table 26–5 Dietary Guidelines following Kidney Transplants

- **Energy** Adequate to achieve or maintain desirable body weight.
- **Protein** 1 gram per kilogram of body weight. (Adjust based on renal function tests.)
- **Fat** ≤30 percent of total kcalories; ≤7 percent saturated fat; ≤200 milligrams cholesterol.
- **Sodium** 3 to 4 grams per day.
- **Potassium** Adjust according to individual needs.

SOURCE: Adapted from J. A. Beto, Which diet for which renal failure: Making sense of the options, *Journal of the American Dietetic Association* 95 (1995): 898–903.

---

to malignant tumors are common adverse effects of these medications. Muscular weakness, GI bleeding, protein catabolism, carbohydrate intolerance, sodium retention, fluid retention, hypertension, weight gain, and a characteristic puffy-faced appearance commonly accompany immunosuppressant therapy. Diuretics are frequently prescribed to promote the excretion of fluid and sodium and to prevent hypertension.

Diet Interventions   Once recovery is under way, the degree of renal function guides diet therapy. Typical post-transplant diet modifications appear in Table 26–5. Protein is provided in amounts adequate to prevent the protein catabolism that immunosuppressants may incur, but not so much as to tax renal function. Because blood lipids are frequently elevated, clients are advised to follow a fat-modified diet. Sodium restrictions help prevent fluid retention and hypertension. Depending on the type of diuretic prescribed and the person's responses, potassium intake may have to be adjusted.

A person with a kidney transplant may reject the new kidney either temporarily or permanently. During these times, dialysis must be reinstituted, and the person must return to the prescribed diet for renal failure. Clients may find this regression difficult to accept and should be prepared for the possibility before it occurs. The accompanying case study helps you apply information about chronic renal disease and kidney transplants.

In Summary   Chronic renal failure develops gradually from a variety of disorders that permanently damage the kidneys. Diet plans may control energy, protein, fluid, sodium, potassium, phosphorus, and calcium. When end-stage renal disease develops, dialysis or a kidney transplant is necessary to sustain life. The Diet-Drug Interactions box describes nutrition-related concerns of medications used in the treatment of disorders of the kidneys and urinary tract. Use the Nutrition Assessment Checklist to review assessment findings relevant to people with disorders of the urinary tract and kidneys.

Case Study

CHILD WITH CHRONIC RENAL FAILURE

Jason is a nine-year-old child who developed chronic renal failure after suffering from a streptococcal infection (strep throat). Jason's renal function has declined steadily over the last three years, and he is currently receiving hemodialysis three times a week. A search is on for a suitable kidney donor. Jason is 4 feet 3 inches tall and weighs 55 pounds. He follows a fluid-, sodium-, potassium-, and phosphorus-restricted diet. His typical energy intake is about 1000 kcalories per day.

- Describe chronic renal failure. What happens to the GFR and BUN as renal function declines? What determines when renal failure reaches its end stage?

- Look at Jason's height and weight. Use a growth chart to determine how Jason's height and weight compare to those of other children of the same age (see Appendix D). Discuss

reasons why growth may be compromised in a child with renal failure.

- Describe the reasons for each of Jason's dietary restrictions. Calculate Jason's energy needs and compare them to his typical energy intake. Consider the effect of Jason's energy intake on his growth. What other nutrients are important to consider in Jason's diet?

- Consider the ways Jason's diet will change if he undergoes a successful kidney transplant.

- Discuss some diet strategies you can suggest to Jason and his parents to help him comply with his diet. Consider the impact of renal disease on Jason, his family, and his interactions with friends.

Anticoagulants

For *anticoagulants,* see p. 647.

Antiemetics

For *antiemetics,* see p. 486.

Antigout

Allopurinol and *colchicine* should both be given with meals. Nutrition-related side effects of allopurinol are uncommon, but colchicine can cause nausea, vomiting, GI distress, and diarrhea. Colchicine reduces the absorption of vitamin B_{12}, and the person may need to take vitamin B_{12} supplements.

Antihypertensives

For *antihypertensives,* see p. 647.

Anti-infective Agents

Numerous *anti-infective agents* may be used in the treatment of kidney stones, nephrotic syndrome, and acute or chronic renal failure. Consult a current drug guide for specific interactions.

Antilipemics

For *antilipemics,* see p. 647.

Antisecretory Agents

For *antisecretory agents,* see p. 464.

Cardiac Glycosides

For *cardiac glycosides,* see p. 647.

Diuretics

For *diuretics,* see p. 647.

Exchange Resins

Cellulose sodium phosphate (prescribed for people with hypercalciuria who absorb too much calcium) can cause altered tastes, GI distress, and diarrhea and can lead to low blood levels of magnesium. Clients should take the medication with meals and should not take a magnesium supplement for at least one hour before or after taking the resin. *Sodium polystyrene* (prescribed to reduce blood potassium levels) can cause anorexia, nausea, vomiting, constipation, and low blood levels of potassium (intentional) and calcium. The powder can be mixed with sorbitol-containing syrup to combat constipation. Calcium-containing antacids or calcium supplements should not be taken within several hours of taking sodium polystyrene.

Immunosuppressants

Immunosuppressants have multiple effects on many organ systems, may be toxic to the kidneys and liver, and can lead to many problems including nausea, vomiting, and a high risk of infections. Cyclosporine can elevate blood potassium, and it should not be given with grapefruit or grapefruit juice (unless prescribed). The person taking cyclosporine should not use potassium supplements or salt substitutes containing potassium. *St. John's wort* reduces bioavailability of cyclosporine. *Azathioprine* can also lead to pancreatitis. Lymphocyte immune globulin and muromonab-cd3 can lead to pulmonary edema.
See also *anti-inflammatory agents* on p. 486.

Phosphate Binders

For *calcium acetate, calcium carbonate, calcium citrate, aluminum carbonate,* and *aluminum hydroxide,* see calcium- and aluminum-containing antacids on p. 464. *Sevelamer hydrochloride* does not contain calcium or aluminum. It is given with meals and can cause nausea, flatulence, GI distress, diarrhea, and, less frequently, constipation.

Nutrition Assessment Checklist

FOR PEOPLE WITH RENAL AND URINARY TRACT DISORDERS

Medical History

Check the medical record to determine:

- ☐ Type of kidney stone
- ☐ Degree of renal function
- ☐ Cause of nephrotic syndrome or renal failure
- ☐ Type of dialysis, if appropriate
- ☐ If client has received a kidney transplant

Review medical record for complications that may alter nutrient needs:

- ☐ PEM
- ☐ Severe stress
- ☐ Infection
- ☐ Anemia
- ☐ Diabetes mellitus
- ☐ Hyperlipidemia
- ☐ Hypertension
- ☐ Heart failure (acute or chronic)

Medications

For clients with kidney stones, note diet-medication interactions for medications taken by the client, including:

- ☐ Anti-infective agents
- ☐ Diuretics
- ☐ Exchange resins (cellulose sodium phosphate)
- ☐ Antigout agents

Note that clients with nephrotic syndrome, renal insufficiency, or renal failure and those who have had a kidney transplant risk medication-related malnutrition for many reasons, including:

- ☐ Long-term use of medications
- ☐ Multiple medication use with many of the medications having significant effects on nutrition status
- ☐ Altered renal function, which compounds nutrition risks
- ☐ Preexisting malnutrition due to the disorder itself and complications of the disorder
- ☐ Reduced food intake, altered digestion and absorption, altered metabolism, and the altered excretion of nutrients due to the medications as well as the disorders themselves

For all clients with kidney diseases, note:

- ☐ Use of over-the-counter medications and supplements that may contain electrolytes that must be controlled
- ☐ Use of herbs and other remedies, which can have a significant impact on clients who suffer from malnutrition and renal insufficiency or renal failure and who use multiple medications

- ☐ Use of fluids to take medications

Food/Nutrient Intake

For people with kidney stones or a past history of kidney stones:

- ☐ Stress the importance of drinking plenty of fluids regularly throughout the day.
- ☐ Assess intake of calcium, oxalate, salt, and protein as appropriate for type of stone.

For people who wish to try cranberry juice cocktail to prevent urinary tract infections, explain that:

- ☐ Studies are limited and they were conducted on older women; but, at worst, cranberry juice cocktail is not harmful.
- ☐ The amount of cranberry juice cocktail used by study participants was 10 ounces per day.
- ☐ A 10-ounce serving of regular cranberry juice cocktail has about 180 kcalories, while the same serving of low-kcalorie cranberry juice has about 60 kcalories.

For clients with the nephrotic syndrome, renal insufficiency, or renal failure, or those who have undergone kidney transplants, regularly assess intakes of:

- ☐ Energy
- ☐ Protein
- ☐ Sodium
- ☐ Potassium

In addition, for clients with renal insufficiency or renal failure, assess intakes of:

- ☐ Fluid
- ☐ Phosphorus
- ☐ Calcium
- ☐ Vitamins
- ☐ Minerals

Anthropometrics

Take accurate baseline height and weight measurements. Keep in mind that:

- ☐ Fluid retention in people with the nephrotic syndrome or renal failure can mask malnutrition.
- ☐ For people on dialysis, the weight measured immediately following a dialysis treatment, called the "dry weight," most accurately reflects the person's true weight.
- ☐ Rapid weight gain between dialysis treatments often reflects fluid retention. For clients who regularly have problems with fluid retention, review fluid intake to ensure that the client understands and is complying with diet recommendations.

(Continued on next page)

Nutrition Assessment Checklist

For People with Renal and Urinary Tract Disorders—continued

☐ Weight loss is expected and intentional following a dialysis treatment.

Laboratory Tests

Note that serum protein levels are often low in people with the nephrotic syndrome or renal failure. Review the following laboratory test results to assess degree of renal function and response to treatments:

☐ Glomerular filtration rate (GFR)

☐ Creatinine

☐ Blood urea nitrogen (BUN)

☐ Electrolytes

Check laboratory test results for complications associated with renal disease including:

☐ Anemia

☐ Hyperglycemia

☐ Hyperlipidemia

☐ Hyperparathyroidism (bone diseases)

Physical Signs

For clients with the nephrotic syndrome and renal insufficiency or renal failure, look for physical signs of:

☐ Dehydration and fluid retention

☐ Iron deficiency

☐ Uremia

☐ Bone diseases

☐ Hyperkalemia

☐ Zinc deficiencies

Self Check

1. Which of the following is not a function of the kidneys?
   a. activation of vitamin K
   b. maintenance of acid-base balance
   c. elimination of metabolic waste products
   d. maintenance of fluid and electrolyte balance

2. Treatment for all kidney stones includes a:
   a. protein-restricted diet.
   b. methionine-restricted diet.
   c. calcium intake that meets but does not exceed the DRI.
   d. fluid intake to maintain a urine volume of at least 2 liters a day.

3. People with calcium oxalate stones may benefit from diets that restrict:
   a. oxalate and sodium.
   b. calcium and potassium.
   c. protein and methionine.
   d. calcium and phosphorus.

4. A person with the nephrotic syndrome is prone to infections due to losses of:
   a. albumin.
   b. transferrin.
   c. lean body mass.
   d. immunoglobulins.

5. Diet recommendations for the nephrotic syndrome include:
   a. no diet restrictions.
   b. sodium restrictions.

   c. protein intakes that are less than the RDA.
   d. protein intakes that are 1.5 to 2.0 times the RDA.

6. Renal failure is characterized by:
   a. the loss of plasma proteins.
   b. declining levels of BUN, creatinine, and uric acid.
   c. uremia and decreasing urine output.
   d. a characteristic puffy-faced appearance.

7. Complications commonly associated with chronic renal failure may include:
   a. growth failure and renal colic.
   b. nausea, vomiting, and reflux esophagitis.
   c. anemia, bone disease, cardiovascular diseases, and malnutrition.
   d. anemia, edema, potassium deficiencies, and organ rejection.

8. Which of the following nutrients may be unintentionally restricted when a person follows a renal diet?
   a. fluid
   b. calcium
   c. potassium
   d. phosphorus

9. The health care professional recognizes that compared to the diet of a person with renal failure who is not on dialysis, the diet of a person on dialysis is:
   a. lower in protein.
   b. higher in protein.

    c. lower in potassium and phosphorus.
    d. higher in potassium and phosphorus.
10. The diet following a kidney transplant is characterized by all of the following **except:**
    a. low in sodium.
    b. low in fat.
    c. potassium controlled.
    d. high in protein.

Critical Thinking

1. A person who has renal insufficiency with a GFR of 40 and weighs 125 pounds needs about _____ grams of protein per day.
    a. 20
    b. 25
    c. 45
    d. 57

2. A person with renal insufficiency who excretes 500 milliliters of urine each day needs to drink about _____ milliliters of fluid each day.
    a. 500–750
    b. 750–1000

    c. 1000–1250
    d. 1300–1500

3. A likely cause of dry, scaly, itchy skin in a person with chronic renal failure is:
    a. uremic frost.
    b. uremic syndrome.
    c. nephrotic syndrome.
    d. essential fatty acid deficiency.

4. A person with chronic renal failure and a poor appetite complains that "food just doesn't taste right anymore." What might be a possible cause?
    a. zinc deficiency
    b. iron deficiency
    c. aluminum toxicity
    d. magnesium toxicity

Answers to these questions can be found in Appendix G.

Clinical Applications

1. Consider that a person with chronic renal failure may need multiple medications to control disease progression and treat symptoms and complications. For people with diabetes and hyperlipidemia who develop renal failure, medications might include insulin, antihypertensives, diuretics, antilipemics, antiemetics (for nausea), antisecretory agents, and phosphate binders. Review the nutrition-related side effects of these medications. Describe the ways these medications can make it harder for people to maintain nutrition status.

2. Think about the case of a person with type 2 diabetes, elevated triglycerides and LDL cholesterol, and diabetic nephropathy. First consider the recommended diet for type 2 diabetes and heart health. In what ways might careful adherence to such a diet prevent or delay the development of diabetic nephropathy? How would the diet change if the person developed the nephrotic syndrome and eventually kidney failure? Using the box on p. 672 as a guide, offer practical suggestions for helping the person adjust to a renal diet.

3. Referring to Figure 23–3 on p. 577 and determining the relationship of ammonia to urea, explain why you would expect blood ammonia levels to rise in liver disease and blood urea nitrogen levels to rise in renal disease.

Nutrition on the Net

For further study of the topics of this chapter, access these websites.

Find updates and quick links to these and other nutrition-related sites at our website:
www.wadsworth.com/nutrition

Learn more about kidney stones and their treatments at the Oxalosis and Hyperoxaluria Foundation and the American Foundation for Urologic Disease:
www.ohf.org and **www.afud.org**

Search for specific topics related to kidney diseases, dialysis, and kidney transplants, including materials for clients with kidney diseases, at the National Institute of Diabetes & Digestive & Kidney Diseases, the National Kidney Foundation, the Kidney Foundation of Canada, and Renalnet Kidney Information Clearinghouse:
www.niddk.nih.gov, www.kidney.org, www.kidney.ca, and
www.renalnet.org/renalnet/renalnet.cfm

Discover resources to support people with kidney diseases by visiting the American Association of Kidney Patients:
www.aakp.org

Learn more about dialysis by visiting the Kidney Dialysis Foundation:
www.kdf.sg

Notes

1. G. C. Curhan and coauthors, Beverage use and risk of kidney stones in women, *Annals of Internal Medicine* 128 (1998): 534–540.

2. G. C. Curhan and coauthors, Comparison of dietary calcium with supplemental calcium and other nutrients as factors affecting the risk for kidney stones in women, *Annals of Internal Medicine* 126 (1997): 497–504.

3. S. Ahuja, B. Kaack, and J. Roberts, Loss of fimbrial adhesion with the addition of *Vaccinum macrocarpon* to the growth medium of P-fimbriated *Escheria coli*, *Journal of Urology* 159 (1998): 559–562; J. Avorn and coauthors, Reduction of bacteriuria and pyuria after ingestion of cranberry juice, *Journal of the American Medical Association* 271 (1994): 751–754.

4. W. J. Burtis and coauthors, Dietary hypercalciuria in patients with calcium oxalate stones, *American Journal of Clinical Nutrition* 60 (1994): 424–429.

5. L. K. Massey and S. A. Kynast-Gales, Diets with either beef or plant proteins reduce risk of calcium oxalate precipitation in patients with a history of calcium kidney stones, *Journal of the American Dietetic Association* 101 (2001): 326–331.

6. S. R. Orth and E. Ritz, The nephrotic syndrome, *New England Journal of Medicine* 338 (1998): 1202–1211.

7. R. Rodrigo and M. Pino, Proteinuria and albumin homeostasis in the nephrotic syndrome: Effect of dietary protein intake, *Nutrition Reviews* 54 (1996): 337–347.

8. J. V. Donadio, σ-3 Fatty acids and their role in nephrologic practice, *Current Opinions in Nephrology and Hypertension* 10 (2001): 639–642; J. V. Donadio, Use of fish oil to treat patients with immunoglobulin A nephropathy, *American Journal of Clinical Nutrition* (supplement) 71 (2000): 373–375.

9. R. Makoff and H. Gonick, Renal failure and concomitant derangement of micronutrient metabolism, *Nutrition in Clinical Practice* 14 (1999): 238–246.

10. S. Hung and coauthors, Thiamine deficiency and unexplained encephalopathy in hemodialysis patients, *American Journal of Kidney Diseases* 38 (2001): 941–947.

11. B. P. Gillis and coauthors, Nutrition intervention program of the Modification of Diet in Renal Disease Study: A self-management approach, *Journal of the American Dietetic Association* 95 (1995): 1288–1294.

12. National Kidney Foundation, 25 facts about organ donation and transplantation, **www.kidney.org**, website visited November 6, 2001.

Nutrition in Practice

■ DIALYSIS AND NUTRITION ■

Although there is no perfect substitute for one's own kidneys, dialysis offers a life-sustaining treatment option for people with end-stage renal disease. Dialysis can serve as a permanent treatment for kidney failure or as a temporary measure to sustain life until a suitable kidney donor can be found. Dialysis also benefits the person with acute renal failure who needs immediate help in restoring normal blood balances. All health care professionals need to understand that renal diseases alter nutrient needs and that dialysis affects those needs. Those who routinely work with clients with renal diseases, however, need to understand more about dialysis procedures. This Nutrition in Practice discusses the various types of dialysis procedures and explains why different procedures affect nutrient needs in different ways.

What is dialysis?

Dialysis removes excess fluids and wastes from the blood, in part, by employing the principles of simple **diffusion** and **osmosis** across a **semipermeable membrane** (see the glossary on p. 680). For **hemodialysis** and **peritoneal** dialysis, a solution similar in composition to normal blood plasma, called the dialysate, is placed on one side of the semipermeable membrane; the person's blood flows by on the other side.

How do the semipermeable membrane and dialysate work together to remove excess wastes?

A semipermeable membrane acts like a filter. Different types of membranes have pores of different sizes, and the size of the pores determines which molecules will be able to pass through the membrane and which will not. Small molecules like urea and electrolytes cross the membrane freely; large molecules like proteins are less likely to cross the membrane.

Wastes are removed from the blood by altering the composition of the dialysate. When the dialysate contains a lower concentration of a substance than the blood, that substance—provided it can cross the membrane—will diffuse out of the blood. To maximally remove waste products like urea from the blood, the dialysate contains no urea. In some cases, however, the dialysate must be adjusted so that only excesses will be removed. Thus potassium can be removed from the blood, for example, by providing a dialysate with a lower concentration of potassium than the person's blood. Some potassium must remain in the dialysate, however, or blood potassium will fall too low.

Dialysis can filter excess substances only from the blood.[1] Thus electrolytes like potassium and phosphorus that reside mainly in the intracellular fluid are more difficult to control with dialysis.

The dialysate can also be used to add needed components back into the blood. For a person with acidosis, for example, bases such as bicarbonate can be added through the dialysate. The base moves by diffusion into the person's blood to ease acidosis.

How are excess fluids removed?

The pressure created by proteins that cannot cross the membrane tends to hold excess fluid in the blood. Thus, to remove excess fluid (and sodium along with it), pressure gradients must be created between the blood and the dialysate in such a way that water and small molecules are "pushed" through the pores of the membrane (**ultrafiltration**). Figure NP26–1 on p. 681 illustrates the different forces that work to remove wastes and fluids during dialysis.

What is hemodialysis?

In hemodialysis, the blood enters a machine called a **dialyzer** or artificial kidney. The dialyzer houses a series of tubes made from synthetic semipermeable membranes. Blood is pumped out of the body and into the dialyzer where it flows between the tubes that carry the dialysate. The dialysis machine applies a negative pressure to the dialysate side of the membrane so that the excess water and sodium can be removed from the blood. The dialysate extracts wastes from the blood, and then the blood is returned to the body.

Hemodialysis can take place in a hospital or clinic or in the home. Often the person receives three treatments a week, and each treatment takes from two to four hours.

How does peritoneal dialysis work?

In peritoneal dialysis, the peritoneal membrane (the membrane that covers the abdominal organs) serves as the semipermeable membrane. The dialysate is infused

Nutrition in Practice

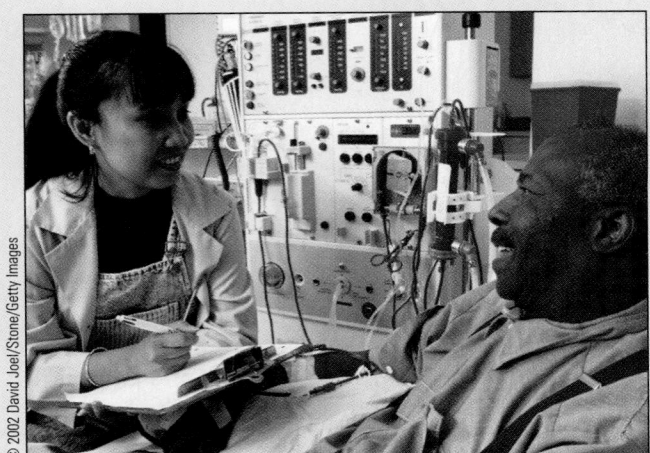

During hemodialysis, shown here, blood passes through a dialyzer where wastes are extracted. The cleansed blood is then returned to the body.

directly through a tube into the peritoneal space—the space within the person's abdomen that overlays the intestine. There are several different peritoneal dialysis techniques, and the technique used determines how often treatments are given and how long fluid remains in the abdomen. With the most common technique, **continuous ambulatory peritoneal dialysis (CAPD),** the dialysate is placed in the abdomen where it remains for four to six hours. The dialysate is then drained from the abdomen through a tube and replaced with fresh dialysate. It takes about 30 to 40 minutes to drain the dialysate and replace it, and generally the solution is changed four times a day.

Because negative pressure cannot be created in the peritoneal cavity as it can in a dialyzer, glucose (and sometimes amino acids) is added to the dialysate. Increasing the concentration of glucose creates osmotic pressure so that fluids and sodium can be removed through osmosis.

Unlike hemodialysis, peritoneal dialysis does not require that blood exit the body. In addition, the removal of fluids and wastes is more gradual.

Are there any other ways to remove fluids and wastes from the body?

For people in acute renal failure, another procedure, called **continuous renal replacement therapy (CRRT),** allows removal of fluids or wastes.[2] CRRT is usually reserved for people with acute renal insufficiency or renal failure who cannot tolerate hemodialysis or peritoneal dialysis for medical reasons.

Several different methods can be used to perform CRRT, depending on the individual's needs. If the purpose is to remove fluids only, no dialysate is used. Instead blood is circulated through a filter where pressure gradients force all blood components small enough to pass through the pores of a semipermeable membrane out of the blood (ultrafiltration). The final composition of the filtrate is similar to that of the blood but without plasma proteins and blood cells, which are too large to pass through the membrane. Thus excess electrolytes and waste products are not removed, but the total volume is reduced. In some cases, intravenous fluids replace components needed by the client that were unavoidably removed during ultrafiltration (**hemofiltration**). If blood containing electrolytes and wastes is

Glossary

continuous ambulatory peritoneal dialysis (CAPD): the most common type of peritoneal dialysis that people use at home.

continuous renal replacement therapy (CRRT): a slow and continuous type of dialysis used in the treatment of acute renal failure.

dialyzer (dye-ah-LYES-er): the machine used for hemodialysis; also called an **artificial kidney.**

diffusion: movement of solutes from an area of high concentration to one of low concentration.

hemodialysis: removal of fluids and wastes from the blood by passing it through a dialyzer.

hemofiltration: removal of fluids and wastes from the blood by using ultrafiltration and fluid replacement.

osmosis: movement of water from an area of low solute concentration to one of high solute concentration.

peritoneal dialysis: removal of wastes and fluids from the body by using the peritoneal membrane as a semipermeable membrane.

semipermeable membrane: a membrane with pores that allow some particles to pass through the membrane but not others.

ultrafiltration: removal of fluids and small- to medium-size molecules from the blood by using pressure to transfer the blood across a semipermeable membrane.

Nutrition in Practice

Diffusion

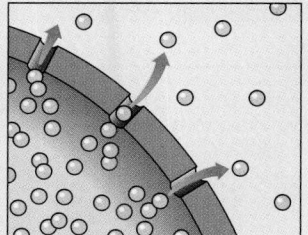

Small molecules (electrolytes and waste products) move from an area of high concentration to an area of low concentration by diffusion.

Osmosis

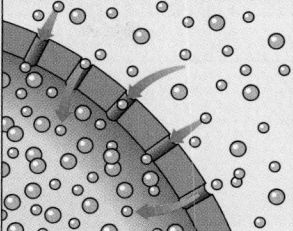

Water moves from an area of low concentration to an area of high concentration by osmosis. In other words, water moves from an area that has fewer particles dissolved in it to an area with more particles dissolved in it.

Ultrafiltration

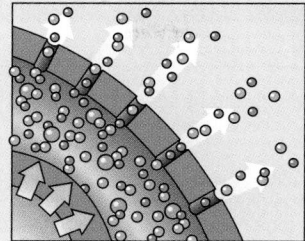

Pressure squeezes water and small molecules through the pores of a semipermeable membrane during ultrafiltration.

Figure NP26–1

Diffusion, Osmosis, and Ultrafiltration

removed and some of the fluid is replaced with a very dilute dextrose solution, for example, gradually the concentration of electrolytes and wastes in the blood is lowered, as is the amount of fluids. The filter can also be adjusted to perform dialysis as well as filtration, making it possible to provide slow, continuous dialysis. The methods can also be combined so that ultrafiltration and dialysis can take place together. Table NP26–1 on p. 682 lists the advantages and disadvantages of hemodialysis, peritoneal dialysis, and CRRT.

Do the different forms of dialysis affect energy needs?

For people on dialysis, dextrose (glucose) is added to the dialysate. Glucose helps to draw fluid into the dialysate through osmosis, but the person undergoing dialysis may also absorb some of it. During hemodialysis, glucose absorption is minimal. In peritoneal dialysis, however, glucose gradually crosses the peritoneal membrane. This makes the dialysate less efficient at drawing fluid out of the blood and also means that the person may absorb significant amounts of glucose. The glucose may provide as many as 600 to 800 kcalories per day.

For people on CRRT, the absorption of glucose is also considerable and can range from about 35 to 45 percent to as high as 60 to 70 percent, depending on the type of CRRT and many other factors.[3] Because people undergoing CRRT are often catabolic, they frequently experience insulin resistance, and the carbohydrate provided must be carefully adjusted to avoid serious problems with hyperglycemia.

Why do protein needs change when people begin dialysis?

Compared to people with renal failure who are not on dialysis, people on dialysis receive diets higher in pro-

tein. In part, this is because dialysis can remove the nitrogen-containing waste products of protein metabolism. In addition, some amino acids and smaller proteins may be lost in the dialysate. In hemodialysis, protein losses average about 5 to 8 grams per treatment. Protein losses from peritoneal dialysis are slightly higher (about 10 to 20 grams per day) because more blood proteins pass into the dialysate through the peritoneal membrane than through the synthetic membrane used for hemodialysis. Clients with acute renal failure often receive TPN solutions during the time that they are undergoing CRRT. People undergoing CRRT retain about 90 percent of infused amino acids.[4]

Are other nutrient needs affected by different dialysis procedures?

Another difference between hemodialysis and peritoneal dialysis with respect to nutrition concerns potassium. In hemodialysis, the dialysate contains potassium; because a relatively large volume of blood is pumped through the dialyzer, potassium levels can drop rapidly, and hypokalemia can follow, disrupting heart function. In peritoneal dialysis and CRRT, the exchange of potassium between the blood and the dialysate occurs more slowly, and blood potassium does not change rapidly. In peritoneal dialysis and CRRT, potassium may be added in varying amounts or left out of the dialysate altogether, depending on the person's blood potassium levels.

Since peritoneal dialysis can remove more potassium than hemodialysis, the person on peritoneal dialysis generally does not have to restrict dietary sources of potassium. The person on hemodialysis, however, does need to moderately restrict potassium between dialysis treatments.

Nutrition in Practice

Table NP26–1   Advantages and Disadvantages of Various Dialysis Procedures

	Advantages	Disadvantages
Hemodialysis	Acts quickly; shorter treatment time; minimal absorption of glucose; can be done at home or in clinic.	Requires access to blood; risk of hypotension, hypokalemia, arrhythmias, blood clots, and anemia; requires a dialyzer.
Peritoneal Dialysis	Gradual removal of fluids and wastes; low risk of hypotension, hypokalemia, and blood clots; fewer dietary restrictions; requires little equipment; easy to do at home.	Requires access to peritoneal cavity; takes a great deal of time; risk of peritonitis; greater loss of protein.
Continuous Renal Replacement Therapy (CRRT)	Gradual removal of fluids and wastes; useful only for acute renal failure; less risk of hypotension; done only in hospital.	Requires direct access to blood.

Dialysis and CRRT help remove wastes and fluids normally removed by functional kidneys. Although these procedures cannot restore the hormonal functions of the kidneys, they provide a lifesaving means of alleviating the symptoms of uremia, hypertension, and edema and limiting the risk of heart failure.

Notes

1. D. I. Charney, Medical treatment in renal disease: Basic concepts in dialysis, *Support Line*, February 1998, pp. 3–7.

2. D. J. Rodriguez and W. M. Sandoval, Nutrition support in acute renal failure patients: Current perspectives, *Support Line*, December 1997, pp. 3–7.

3. As cited in D. C. Kaufman and coauthors, Adjustment of nutrition support with continuous hemodiafiltration in a critically ill patient, *Nutrition in Clinical Practice* 14 (1999): 120–123.

4. Kaufman and coauthors, 1999; Rodriguez and Sandoval, 1997.

Jennie Oppenheimer/Studio Zocolo

CHAPTER 27 | ENERGY- AND PROTEIN-MODIFIED DIETS FOR ACUTE STRESS

Although all illnesses threaten the body's internal balances, the healthy body adapts to minor stresses quickly and efficiently.

Examples of stressors that can lead to acute stress include infections; surgery; burns; fractures; deep, penetrating wounds (gunshot wounds, fistulas, or surgical incisions); and hemorrhaging. These stressors can lead to shock or tissue death (necrosis) and many other complications.

The cells of the immune system—white blood cells—include a variety of phagocytes and lymphocytes (see Nutrition in Practice 15).

metabolic stress: the state in which a body's internal balance (homeostasis) is upset by a threat to a person's physical well-being (**stressor**). The terms **acute stress** and **severe stress** are used in this chapter to refer to pathological stresses that rapidly and markedly raise the body's metabolic rate and significantly upset its normal internal balance.

physiological stresses: disruptions to the body's internal balance caused by processes necessary to sustain life.

pathological stresses: disruptions to the body's internal balance that lie beyond its normal and healthy functioning.

cytokines (SIGH-toe-kynes): proteins that help regulate immune system responses. Cytokines trigger hypermetabolism and cause anorexia, fever, and discomfort.

*P*revious chapters have provided many examples of disorders that benefit from energy-modified diets, which include both low-kcalorie and high-kcalorie diets. This chapter describes the use of energy-modified, high-protein diets to treat medical conditions that create **metabolic stress**—a disruption of the body's internal balance. (Nutrition in Practice 16 describes emotional stress and its connections to nutrition.)

Types of Stress

The body routinely experiences some degree of stress and makes adjustments to restore balance. These stresses fall within the body's normal and healthy functioning and are known as **physiological stresses.** Examples include the release of insulin in response to a surge in blood glucose following a meal or retention of fluid in the kidneys in response to dehydration. In contrast, **pathological stresses** push the body beyond these limits. The amount of stress the body tolerates without significant disruption varies from individual to individual and depends on many factors, including health and nutrition status.

Conditions that cause acute, severe stress lead to major disruptions of the body's normal internal balance and rapidly and markedly raise its metabolic rate. These conditions include uncontrolled infections and extensive tissue damage, such as deep, penetrating wounds or multiple broken bones. When confronted with such a severe stress, the body employs complicated mechanisms to reestablish its balance. Once in play, these mechanisms have both helpful and harmful effects.

In Summary Although the body is regularly confronted with physiological stress, the body is adept at regaining balance. Severe pathological stresses, however, greatly upset the body's homeostasis and require more complicated responses than everyday stresses.

The Body's Responses to Acute Stress

Diet Order
Energy-modified, high-protein diet.

To regain homeostasis during acute stress, the body speeds up its metabolic rate (hypermetabolism) to mobilize glucose and amino acids. From these nutrients the body can synthesize the special factors it needs to limit and repair damage. The exact factors and the amount the body makes depend, in part, on what condition precipitates the stress and the severity of the stress. Mending a broken bone requires different factors than does healing a wound or fighting an infection. Healing a blister produces less stress than mending a ruptured organ.

Researchers are working to elucidate the body's complex and interrelated responses to stress, which ultimately affect many body systems. Much of that research focuses on **cytokines**—proteins that direct changes in the cardiovascular and nervous systems and stimulate the production of the many immune system cells and chemicals necessary to recover from stress (see Nutrition in Practice 15).[1]

Immune System Responses

The body's natural system of defense against pathogens—the immune system— enables the body to fight off infectious agents and repair tissue damage. The immune system defends the body so alertly and silently that most healthy people

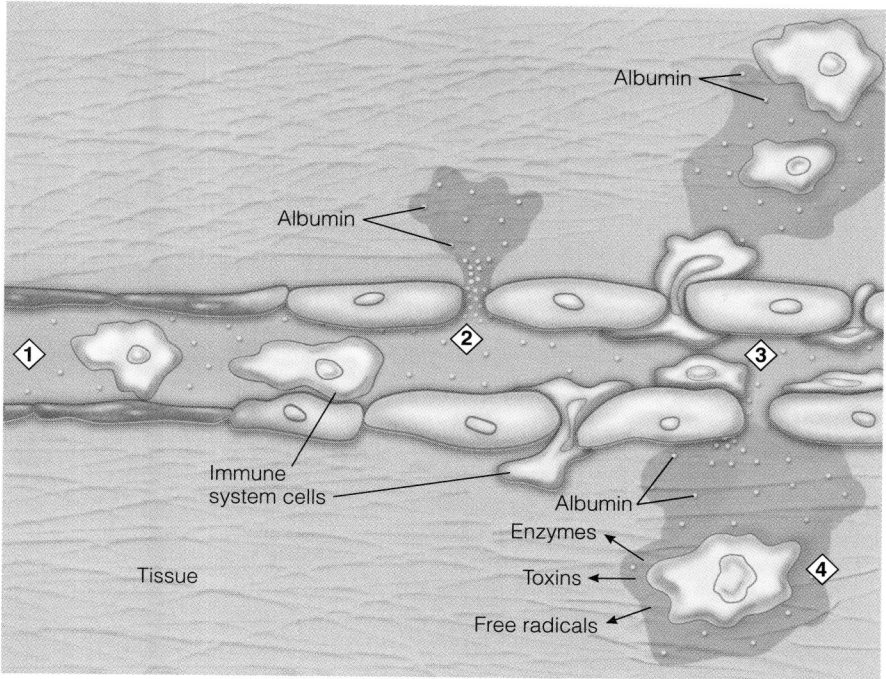

Figure 27–1

Inflammatory Responses

①Normal cells lining the blood vessels (red) lie close together, and immune system cells (blue) and albumin (yellow) do not cross into tissue.

②With inflammation, the cells lining the blood vessels swell (pink), gaps form between the cells, and immune system cells and albumin (and fluid along with it) enter tissues in the area and cause edema.

③Along with changes in blood vessels, inflammatory responses also result in the recruitment and adhesion of immune system cells. These responses cause blood flow to slow through the inflamed area.

④Immune system cells produce toxins (including free radicals and enzymes) that kill foreign invaders and destroy healthy tissue in the process.

are unaware that thousands of microbes mount attacks against them every day. Occasionally, though, an infection succeeds in making a person ill, and the immune system must then mount a vigorous counterattack. Serious infections greatly tax the body, and if the counterattack fails, death follows.

Inflammatory Responses As Chapter 15 described, tissues damaged by **trauma,** heat, chemicals, loss of blood flow, or infectious agents can disrupt organ function and render the body vulnerable to invading organisms. The body's response to tissue damage—the **inflammatory response**—serves to isolate and destroy foreign particles and repair tissue damage.

Blood Vessels and Immune System Factors At the site of injury, the walls of the blood vessels swell, the diameter of the blood vessels in the affected area shrinks, and blood flow around the injured area slows. Chemical signals from the immune system direct changes that cause blood-clotting, infection-fighting, and wound-healing factors to gather in the injured area. Figure 27–1 illustrates the changes in blood vessels that occur during inflammatory responses. Some **eicosanoids**—derivatives of omega-3 and omega-6 fatty acids—direct changes that alter the permeability of blood vessels and cause immune system cells to adhere to the area (procoagulant state). Some immune system cells generate free radicals and other toxins that destroy foreign particles; others remove dead cells. Together these responses protect the body from the spread of infection. In the final stages, scar tissue forms to replace damaged tissues.

Hormonal Responses

The hormonal changes characteristic of the immediate stress response shift the balance between insulin, which promotes the storage of carbohydrate and lipids and the synthesis of protein, and the **counterregulatory hormones,** which promote the breakdown of glycogen, the mobilization of fatty acids from lipids, and the synthesis of glucose from protein (see Table 27–1 on p. 686). Although insulin levels rise, the rise in the level of the counterregulatory hormones is greater still.

Eicosanoids derived from omega-6 fatty acids are **proinflammatory**—they direct changes that enhance inflammatory responses. Eicosanoids derived from omega-3 fatty acids are **anti-inflammatory**—they direct changes that modulate (slow) inflammatory responses.

trauma: bodily injury such as a gunshot wound, blow, or cut.

inflammatory response: the changes orchestrated by the immune system when tissues are injured by such forces as blows, wounds, foreign bodies (chemicals, microorganisms), loss of blood flow, heat, cold, electricity, or radiation.

eicosanoids (eye-COSS-uh-noyds): derivatives of 20-carbon fatty acids that regulate blood pressure, blood clotting, and other body functions. They include *prostaglandins* (PROS-tah-GLAND-ins), *thromboxanes* (throm-BOX-ains), and *leukotrienes* (LOO-ko-TRY-eens).

counterregulatory hormones: hormones such as glucagon, cortisol, and catecholamines that oppose insulin's actions and promote catabolism.

Energy- and Protein-Modified Diets for Acute Stress

Table 27–1 Metabolic Effects of Hormonal Changes during Acute Stress

Hormone	Alteration	Metabolic Effect
Catecholamines	Increase	Glucagon release increases.
		Insulin-to-glucagon ratio decreases.[a]
		Glycogen breakdown increases.
		Glucose production from amino acids increases.
		Mobilization of free fatty acids increases.
Cortisol	Increases	Mobilization of free fatty acids increases.
		Glucose production from amino acids increases.
Glucagon	Increases	Insulin-to-glucagon ratio decreases.[a]
		Glucose production from amino acids increases.
		Glycogen breakdown increases.
		Storage of glucose, amino acids, and fatty acids decreases.
Insulin	Increases	Breakdown of body fat inhibited; blood glucose levels rise despite the increase.
Antidiuretic hormone	Increases	Retention of water increases.
Aldosterone	Increases	Retention of sodium increases.
		Excretion of potassium increases.

NOTE: These changes are part of the immediate stress response. As adaptation occurs and recovery is in progress, hormone levels gradually return to normal.

[a]The net effect of a decrease in the ratio of insulin to glucagon is that catabolism predominates.

Negative Nitrogen Balance The hormonal changes that accompany acute stress result in negative nitrogen balance; that is, the body breaks down and uses more protein than it receives and synthesizes. The protein loss comes primarily from skeletal muscle, connective tissue, and the gut; protein synthesis in the liver and the tissues that produce immune system cells actually increases. Thus skeletal muscle, connective tissue, and gut tissue are sacrificed to supply the glucose and amino acids necessary to respond to stress.[2]

Other Effects As a result of the hormonal changes, the metabolic rate and blood glucose levels rise. High insulin levels suppress the mobilization of essential fatty acids from body stores, and clinical signs of essential fatty acid deficiencies may develop. Figure 27–2 on p. 687 illustrates the differences between the body's responses to simple fasting and to acute stress to show why the stressed body rapidly consumes energy and loses vital nutrients.

Other hormonal changes promote the retention of water and sodium and the excretion of potassium. Hypermetabolism generally peaks at about three to four days and subsides in about seven to ten days. With recovery, hormone levels gradually return to normal.

Consequences of Severe Stress

Unfortunately, the benefits conferred by inflammatory responses come with a cost to healthy tissue. Acute, severe stresses result in exaggerated or prolonged inflammatory responses that can significantly alter blood flow and lead to **shock.** Blood flow can be further impeded by the tiny blood clots that form under the direction of certain eicosanoids to help immune system cells "stick" to the damaged area. The altered blood flow and the release of free radicals and toxins to destroy foreign particles also damage healthy tissues and can lead to oxidative stress (see p. 388). Furthermore, the replacement of healthy tissue with scar tissue (fibrosis) can interfere with organ function and significantly impair health.

shock: a sudden drop in blood volume that disrupts the supply of oxygen to the tissues and organs and the return of blood to the heart. Shock is a critical event that requires immediate correction.

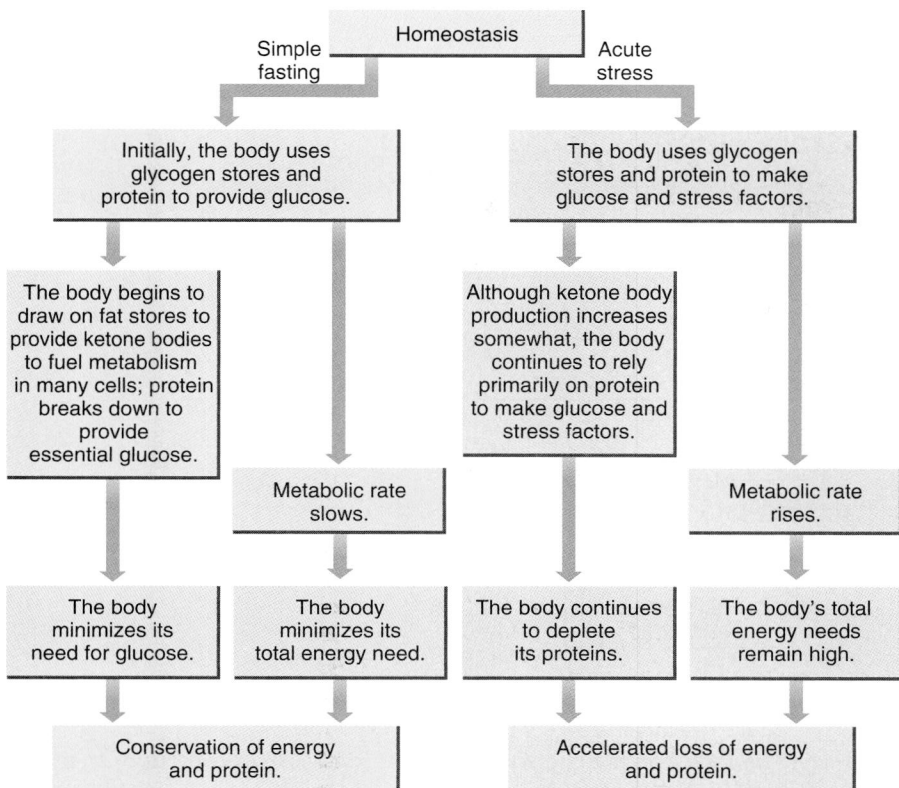

Figure 27–2
Metabolic Responses to Fasting and Stress

Homeostasis

Simple fasting / Acute stress

Simple fasting:

Initially, the body uses glycogen stores and protein to provide glucose.

→ The body begins to draw on fat stores to provide ketone bodies to fuel metabolism in many cells; protein breaks down to provide essential glucose.

→ Metabolic rate slows.

→ The body minimizes its need for glucose.

→ The body minimizes its total energy need.

→ Conservation of energy and protein.

Acute stress:

The body uses glycogen stores and protein to make glucose and stress factors.

→ Although ketone body production increases somewhat, the body continues to rely primarily on protein to make glucose and stress factors.

→ Metabolic rate rises.

→ The body continues to deplete its proteins.

→ The body's total energy needs remain high.

→ Accelerated loss of energy and protein.

Figure 27–3 on p. 688 shows how inflammatory responses can damage healthy tissues.

Systemic Inflammatory Response Syndrome The inflammatory and immune responses to stressors result in redness, swelling, heat, and pain at the injury site. Body temperature, heart rate, respiratory rate, and white blood cell count increase. The person experiences anorexia and malaise—a feeling of discomfort and uneasiness. The symptoms associated with the inflammatory response are collectively called the **systemic inflammatory response syndrome (SIRS).**

In addition to an elevated white blood cell count, laboratory tests reveal elevated blood glucose (hyperglycemia), blood urea nitrogen (from protein catabolism), and triglycerides; negative nitrogen balance; increased retention of fluid and sodium; and increased excretion of potassium. Blood concentrations of albumin, iron, and zinc fall.

Sepsis and Multiple Organ Failure If local defenses fail to disarm infectious agents, sepsis may develop. Sepsis triggers further and exaggerated inflammatory responses and may lead to multiple organ failure—the most common cause of life-threatening complications and death in critically ill people. In people who develop multiple organ failure (see Nutrition in Practice 27), the lungs often fail first, followed by the liver and kidneys. Although mortality is high in people with multiple organ failure, those who do recover often completely regain organ function.

With hypermetabolism, the body avidly consumes oxygen and produces carbon dioxide, and the lungs must work hard to keep the body's gases in balance. At the same time, the inflammatory responses to stress may increase the permeability of the lungs' **alveoli,** fluids may accumulate in the lungs' interstitial spaces, and pulmonary edema may develop. If the problem becomes severe enough, the lungs lose their elasticity and ability to exchange gases, and respiratory failure results. **Acute respiratory failure** can also occur if an embolus (blood clot)

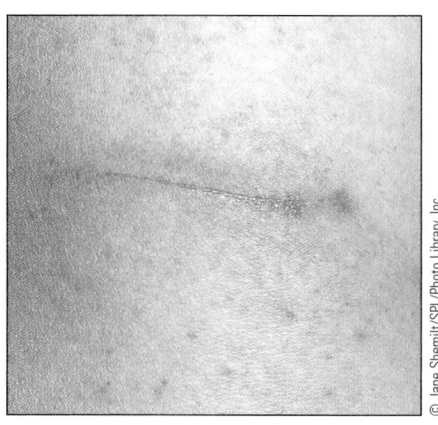

Inflammatory responses to tissue damage cause redness, swelling, heat, and pain.

© Jane Shemilt/SPL/Photo Library, Inc.

systemic inflammatory response syndrome (SIRS): the complex of symptoms (see the text) that occur as a result of immune and inflammatory factors in response to tissue damage. In severe cases, SIRS may progress to multiple organ failure.

alveoli (al-VEE-oh-lie): air sacs in the lungs. One sac is an *alveolus.*

acute respiratory failure: sudden failure of the lungs to exchange gases.

Energy- and Protein-Modified Diets for Acute Stress

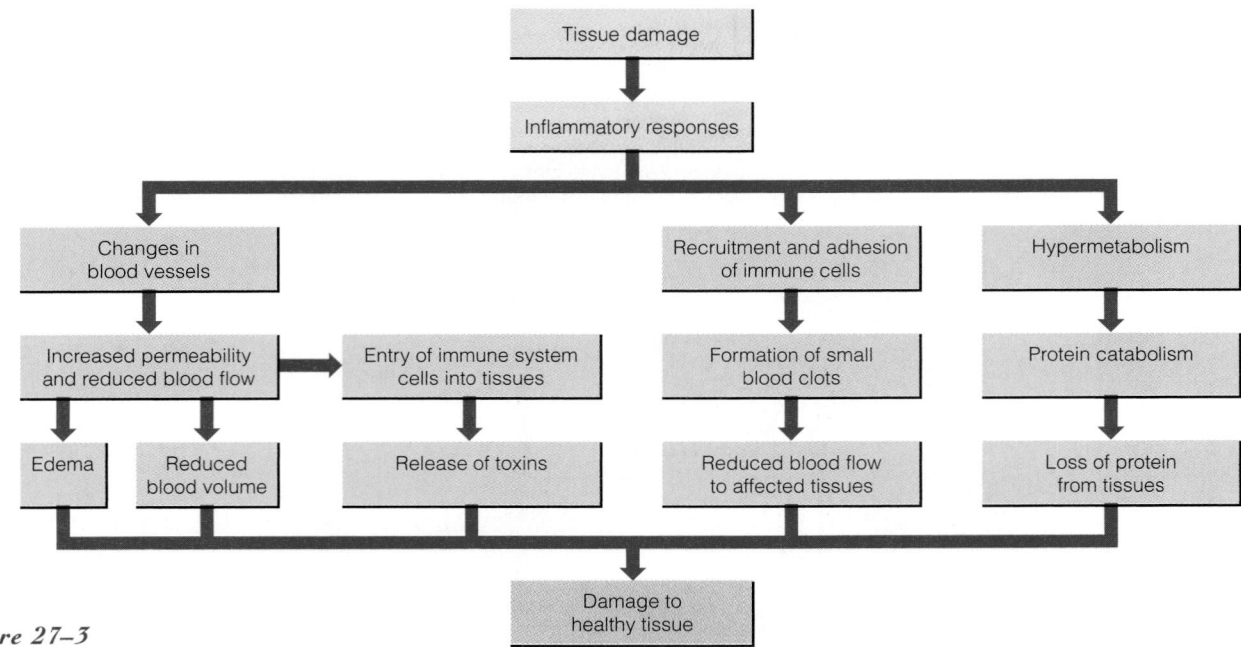

Figure 27–3
Inflammation and Tissue Damage

Reminder: *Hepatic encephalopathy* and *hepatic coma* result from severe liver disease and are believed to be related to the toxic effects of elevated blood ammonia levels on the central nervous system.

Reminder: The *glomerular filtration rate* is the rate at which the kidneys form filtrate. Chapter 26 provides more information.

acute liver failure: sudden failure of the liver.

acute renal failure: a sudden drop in the glomerular filtration rate (GFR) and urinary output.

lodges in the lungs or if toxic substances such as gastric contents (aspiration pneumonia) directly damage the lung tissue.

During stress, the liver redirects metabolism and produces many of the stress factors, including coagulation factors, necessary to recovery. With the accelerated breakdown of protein, the liver must step up its production of urea to prevent the buildup of ammonia. **Acute liver failure** associated with sepsis and shock rapidly destroys liver cells and may lead to hepatic encephalopathy, hepatic coma, and hemorrhage.

Acute renal failure is characterized by a sudden and precipitous drop in the glomerular filtration rate (GFR) and urine output. At the same time, negative nitrogen balance rapidly raises levels of nitrogen-containing waste products in the blood, and uremia quickly develops. Electrolyte levels rise dangerously high as the body's cells consume protein and release potassium, phosphorus, and magnesium in the process. Blood pressure also increases rapidly as the kidneys fail to excrete excess fluids and sodium. Rising blood pressure stresses the heart muscle, and potassium imbalances can alter the heart rate—either or both raise the likelihood of acute heart failure.

GI Tract Responses During stress, blood flow to the GI tract diminishes, and GI tract motility slows. Altered blood flow to the stomach may hinder the stomach's ability to protect itself from gastric acid. Consequently, ulcers may form in the stomach or small intestine. The cells of the intestinal tract gradually shrink and may lose some of their absorptive and immune functions as gut proteins are sacrificed to support the body's defenses. Nutrition in Practice 27 reviews the immune functions of the GI tract and discusses the ways that changes in the GI tract during severe stress raise the likelihood of infection and multiple organ failure.

Effects on Nutrition Status For a well-nourished person with a short-term stress, the duration of negative nitrogen balance (loss of protein) is acceptable because sufficient protein remains to support defense systems and maintain vital functions. Acute stresses that markedly elevate the metabolic rate over longer periods of time, however, rapidly deplete energy reserves and the body's protein tis-

Table 27–2 Possible Causes of Malnutrition Associated with Severe Stress

Reduced Nutrient Intake	Excessive Nutrient Losses	Raised Nutrient Needs
Anorexia	Blood loss	Fever
Emotional stress	Immobility	Hypermetabolism
Immobility	Malabsorption	Infection
Location of injury	Medications	Medications
Malaise	Urinary losses	Pressure sores
Medications	Vomiting	
Nausea/vomiting		

sues. Consequently, they may lead to **acute malnutrition** in a previously healthy person or worsen preexisting malnutrition. When a chronically malnourished person suffers an acute stress, or when acute malnutrition extends over long periods of time, the loss of lean body mass and the consequent depletion of vital proteins compromise the function of the immune system, heart, lungs, kidneys, and GI tract. Once organ systems begin to fail, recovery is severely compromised. Table 27–2 lists the consequences of stress that can lead to malnutrition.

Effects on Medications   Acute stress, and the malnutrition that may accompany it, can alter the body's use of medications and worsen health and nutrition status. Many medications are transported in the blood bound to blood proteins such as albumin, and low blood albumin is a symptom of both stress and protein-energy malnutrition (PEM). Without sufficient carriers, medications may be slow to reach their sites of action. Once medications do arrive at their target cells, the lack of carriers may delay the medications' transport to the liver and kidneys, where they are often detoxified and excreted. Thus medications may take a long time to work and then may remain active longer than intended. In addition, people with acute stresses often receive multiple medications, which complicates decisions about the correct dosage and raises the risk of diet-drug interactions. Furthermore, when medications are given in an oral form, the effects of both stress and malnutrition on GI tract function can alter the absorption of both medications and nutrients. The box on p. 696 describes diet-drug interactions of medications commonly used in the treatment of acute stresses.

In Summary   Acute stress sparks a series of hormonal and metabolic responses to reestablish the body's internal balance. These changes demand energy and drain the body of nutrients. As a result, the response to acute stress can lead a well-nourished person rapidly into PEM or lead a malnourished person perilously close to death, if not to death itself.

Medical Nutrition Therapy for Acute Stress

To support recovery from acute stress, the immediate concerns of the health care team are to restore blood flow and oxygen transport and to prevent or treat infection. Possible measures include providing intravenous (IV) solutions to correct fluid and electrolyte imbalances, giving transfusions, removing dead tissues **(debridement),** draining abscesses, and administering antibiotics. People who experience respiratory failure may require a **mechanical ventilator;** those who

Nutrition status may be affected not only by stress itself, but also by its consequences or treatment. These include:

- *Emotional stress.*
- *Immobility, which can lead to pressure sores, negative nitrogen and calcium balance, and the formation of kidney stones.*
- *The location of injuries, which can make it difficult or painful to eat.*
- *Diagnostic tests and medical procedures, which can interfere with nutrient intake.*
- *Medications.*

acute malnutrition: protein-energy malnutrition (PEM) that develops rapidly due to a sudden and dramatic demand for nutrients. The person suffering from acute malnutrition may be of normal weight or may be overweight, but serum protein levels are low. In contrast, **chronic malnutrition** develops as a consequence of insufficient intake of energy and protein over long periods of time and is characterized by underweight, depleted fat stores, and normal serum protein levels.

debridement (dee-BREED-ment): the removal of dead tissues resulting from burns and other wounds to speed healing and prevent infection.

mechanical ventilator: a machine that "breathes" for a person who can't. In normal respiration, air is drawn in when the lungs expand. With mechanical ventilation, air is forced into the lungs at regular intervals using pressure.

Energy- and Protein-Modified Diets for Acute Stress

develop renal failure often benefit from dialysis or continuous renal replacement therapy (see Nutrition in Practice 26). Supplying adequate energy and nutrients following stress helps to minimize nutrient losses, preserve organ function, promote wound healing, and maintain immune defenses—all of which support recovery. Once hypermetabolism subsides, nutrition support aims to restore nutrient deficits and promote positive energy and nitrogen balance.

Effect of Acute Stress on Nutrient Needs

Feeding the acutely stressed individual is a formidable challenge. Nourishment must be introduced cautiously, and adjustments must be made throughout the course of the stress. Both underfeeding and introducing nutrients too rapidly can result in serious complications that compromise recovery.

Underfeeding   Stress compromises nutrition status, and underfeeding hinders the functions of vital organs and impedes the production of the special factors necessary to respond to stress. Thus underfeeding raises the risk of infections, delays wound healing, and may compromise lung function.[3]

The Refeeding Syndrome   Reintroducing nutrients too rapidly can overtax organ systems and lead to many complications, including malabsorption, cardiac insufficiency, respiratory distress, congestive heart failure, convulsions, coma, and even death. Collectively, the complications that occur as a consequence of reintroducing nutrients too rapidly are called the **refeeding syndrome.** People with chronic malnutrition, anorexia nervosa, or chronic alcohol abuse and those who have received minimal nutrition support for a week or more, have received parenteral nutrition, or have experienced a significant weight loss are more likely than others to experience the refeeding syndrome.

Replenishing Fluids and Electrolytes

To restore circulation and prevent dehydration and electrolyte imbalances, the medical team must act quickly to stabilize the body's blood volume and electrolyte balance. This step is critical for, without adequate blood volume, oxygen, nutrients, and medications cannot be delivered to the cells, organ systems cannot function properly, and waste products cannot be eliminated from the body.

Fluids   Although dehydration poses many dangers, excess fluids stress the heart, which must pump a greater blood volume, and the kidneys, which must work hard to normalize fluid volume. Fluid restrictions may be necessary for people with acute respiratory, heart, or renal failure. The physician determines the person's fluid needs based on clinical measures such as blood pressure, heart rate, respiratory rate, urinary output, level of consciousness, and body temperature.

Electrolytes   Serum electrolytes are closely monitored, and adjustments are made as necessary. Along with fluids, sodium is provided carefully to avoid fluid imbalances. For people with respiratory, heart, liver, or kidney failure, sodium restrictions may be necessary to avoid further stress on these organ systems.

Catabolism causes a shift of potassium and phosphorus from the intracellular fluids, and blood levels rise. The problem is compounded in people with renal failure because the kidneys cannot excrete the excess, and so these electrolytes, especially potassium, can rapidly rise to dangerously high levels. During this time, potassium and phosphorus may need to be restricted temporarily. Once catabolism subsides, however, these electrolytes, along with glucose and amino acids, move back into the cells to begin rebuilding tissues. Without careful attention to replacement, blood levels of these electrolytes can plummet, resulting in life-threatening complications.

refeeding syndrome: the physiological and metabolic complications associated with reintroducing nutrients too rapidly in people with depleted nutrient stores due to chronic malnutrition or in those who have been underfed for several days. These complications can include malabsorption, cardiac insufficiency, respiratory distress, congestive heart failure, convulsions, coma, and possibly death.

Table 27-3 Energy Needs during Stress

To determine energy needs, multiply the basal energy expenditure (BEE) derived by the Harris-Benedict equation (see p. 438) by these factors, called *stress factors:*

Type of Stress	Stress Factor	Percentage of BEE
Most types	1.0 to 1.2	100 to 120
Head injuries	1.3 to 1.5	130 to 150

Energy needs for burn injuries depend on the percentage of the total body surface area (TBSA) affected.[a]

Percentage of TBSA of Burn	Stress Factor	Percentage of BEE
<20	1.2 to 1.4	120 to 140
20–25	1.6	160
26–30	1.7	170
31–35	1.8	180
36–40	1.9	190
41–45	2.0	200
>45	2.1	210

[a]Many formulas are available for estimating energy needs for burn injuries. A variety of formulas are described and referenced in D. J. Rodriguez, Nutrition in major burns: State of the art, *Support Line,* August 1995, pp. 1–8; American Dietetic Association and Dietitians of Canada, *Manual of Clinical Dietetics* (Chicago: American Dietetic Association, 2000), pp. 222–223.

SOURCES: F. B. Cerra and coauthors, Applied nutrition in ICU patients: A consensus statement of the American College of Chest Physicians, *Chest* 111 (1997): 769–778; M. M. McMahon, Nutrition support of hospitalized patients, presented at the Eighth Annual Advances and Controversies in Clinical Nutrition, Mayo Clinic Foundation, April 5–7, 1998; E. Weeks and M. Elia, Observations on patterns of 24-hour energy expenditure changes in body composition and gastric emptying in head injured patients receiving nasogastric tube feeding, *Journal of Parenteral and Enteral Nutrition* 20 (1996): 31–37.

Energy Needs

Stress raises energy needs, but supplying too much energy further elevates the metabolic rate, which, in turn, increases oxygen use and carbon dioxide production. Then the heart and lungs, already working hard as a consequence of stress, must work even harder to keep the body's gases in balance. If these vital organs are weakened by pre-existing malnutrition or if their function is already compromised by stress, the heart and/or lungs may fail. Providing too much energy can also lead to fatty liver (see Chapter 23), which enlarges the liver and can interfere with its function.

Estimating Energy Needs　Energy needs during severe stress depend on both the type and the degree of stress, as well as the individual's metabolic rate and nutrition status. Indirect calorimetry provides the best estimate of energy needs during acute stress, especially when the stress is particularly severe or complicated by organ failure.[4] As Chapter 17 described, however, clinicians often rely on estimates, such as the Harris-Benedict equation to estimate basal energy needs (see Table 17–1 on p. 438). Additional kcalories (stress factors) are added to the basal energy estimate to meet the demands of stress. Table 27–3 shows how different types of stress raise energy needs. For clients who are likely to develop the refeeding syndrome, fewer kcalories may be provided at first and then gradually increased.

Adjusting Energy Needs for Obesity　The effects of obesity on the heart, lungs, and immune function, as well as the insulin resistance typical of obesity, add to the problems of the obese person who suffers a severe stress. Although it is critical to provide nutrition support to those who are obese, clinical research to determine the best approach for feeding obese people with severe stress is lacking. However, some researchers report that providing less energy to obese people is beneficial and safe, as long as protein needs are met.[5]

Reminder: *Indirect calorimetry* is an indirect estimate of resting energy expenditure made by measuring the amount of carbon dioxide expired (carbon dioxide production) and the amount of oxygen inspired (oxygen consumption).

Many clinicians estimate energy needs in critically ill people based on body weight and provide 25 to 35 kcal/kg body weight for people who are not obese and 21 kcal/kg body weight for those who are obese.

Energy- and Protein-Modified Diets for Acute Stress

Estimate Energy and Protein Needs following Stress

Andrew is a 24-year-old male, who is 6 feet, 2 inches tall with a usual body weight of 180 pounds. He was severely injured in a car accident and is in the intensive care unit. Use the Harris-Benedict equation to estimate basal energy expenditure (BEE) as follows:

Weight in kilograms = 180 lb ÷ 2.2 kg/lb = 81.8 kg.

Height in centimeters = 74 in × 2.54 cm/in = 188 cm.

BEE = 66.5 + (13.8 × wt in kg) + (5 × ht in cm) − (6.8 × age in yr).

BEE = 66.5 + (13.8 × 81.8 kg) + (5 × 188 cm) − (6.8 × 24) = 66.5 + 1128.8 + 1278 − 163.2 = 2310 kcal

Additional kcalories are needed for critical illness. Since energy needs in critical illness increase by 100 to 120 percent (see Table 27-3), multiply by 1.0 to 1.2 (the mathematical equivalent):

2310 kcal × 1.0 = 2310 kcal.

2310 kcal × 1.2 = 2772 kcal.

Thus Andrew's energy needs are between 2310 and 2772 kcalories.

Protein needs for Andrew can be estimated at between 1.5 and 2.0 grams of protein per kilogram of body weight per day:

81.8 kg × 1.5 g/kg = 123 g.

81.8 kg × 2.0 g/kg = 164 g.

Andrew needs an estimated 123 to 164 grams of protein daily. Some clinicians subtract the kcalories provided by protein from the total energy needs and then provide the remaining kcalories as carbohydrate and fat. Others meet the energy needs with carbohydrate and fat and provide the protein in addition to these kcalories.

The estimates for energy and protein are just that: estimates only. Clinicians monitor weight, serum proteins, and sometimes nitrogen balance (see Appendix D) to determine whether the estimates should be adjusted.

Protein Needs

The greater the stress, the more body protein is broken down (up to the limit of the body's capabilities), the more nitrogen is excreted in the urine, and the greater the need for protein. Immediately following stress, negative nitrogen balance cannot be fully corrected even when adequate amounts of protein are supplied.

Estimating Protein Needs Providing too much protein places additional stress on the liver, as it increases its production of urea to handle the excess nitrogen. Then the kidneys must excrete it. To meet protein needs, stressed people with normal kidney and liver function generally need about 1.5 to 2.0 grams of protein per kilogram of body weight per day. People with severe burns may need more—their exact needs depend on the extent of the burn, but may be as high as 2.5 grams per kilogram of body weight per day. The "How to" box shows an example of how to estimate energy and protein needs following acute stress.

For people with acute renal failure, clinicians often continue to provide a high protein intake and use dialysis or continuous renal replacement therapy to remove excess nitrogen-containing waste products. In general, people with renal failure and a relatively mild stress who are not on dialysis receive about 0.6 to 1.0 gram of protein per kilogram of body weight per day. Those on dialysis receive from 1.0 to 1.5 grams of protein per kilogram of body weight per day, depending on the degree of stress. As Nutrition in Practice 26 describes, protein needs are higher (1.5 to 2.0 grams per kilogram of body weight per day) for people on continuous renal replacement therapy.

Amino Acids An amino acid receiving wide attention in relation to stress is glutamine. Glutamine is a nonessential amino acid in healthy people During stress, however, the body's demand for glutamine may outpace its ability to synthesize glutamine in adequate amounts. Consequently, glutamine may become a conditionally essential amino acid. Glutamine provides fuel for the intestinal cells and plays important roles in maintaining intestinal immune function and promoting wound healing.[6] Glutamine supplementation may reduce the incidence of infection in acutely stressed people and in those undergoing bone marrow transplants.[7]

Another nonessential amino acid, arginine, and nucleotides (nitrogen-containing components of RNA and DNA) may also be important during acute stress. Arginine plays an important role in immune defenses and wound healing.[8] Further research is necessary, however, to determine if supplementing arginine and nucleotides provides benefits for people experiencing severe stress.

Carbohydrates and Lipids

Nonprotein energy sources spare protein, and the amount of energy provided as carbohydrate or lipid has ramifications for stressed people. Carbohydrates provide a readily usable source of energy. During stress, however, the body can metabolize only a fixed amount (about 500 grams per day) of glucose. Excess glucose serves no benefit, but contributes to hyperglycemia, which can adversely affect fluid and electrolyte balances and may also raise the likelihood of infections in stressed people.[9] Lipids provide energy and essential fatty acids, but given in excess, they can also tax metabolic functions and hamper immune responses and lung function.

Overfeeding generates excess carbon dioxide, and glucose metabolism generates more carbon dioxide than fat metabolism does—an important consideration for people with acute respiratory failure. When total energy needs are not exceeded, however, the effect of glucose and fat metabolism on carbon dioxide production is less important.[10] Ideally, clinicians use indirect calorimetry to determine the appropriate mix of carbohydrate and fat to meet those needs.

People recovering from respiratory failure who are being weaned from mechanical ventilators have high energy and protein needs. The person's lungs were weak before mechanical ventilation began, and they become weaker with disuse. As mechanical ventilation decreases, the lungs must do more work—a stressful and energy-consuming process.

Estimating Carbohydrate and Lipid Needs Clinicians often supply nonprotein kcalories through a mixture of about 70 percent carbohydrates and 30 percent lipids.[11] For clients with burns, restricting fat to 15 to 20 percent of the nonprotein kcalories may help prevent respiratory infections and speed healing. Some clinicians provide about half the estimated carbohydrate kcalories and withhold lipids for the first few days following a severe stress.[12]

Fatty Acids Much research today focuses on the best source of fatty acids for clients experiencing acute, severe stress. Standard IV lipid emulsions and many enteral formulas are rich sources of omega-6 fatty acids. When given in excess of essential fatty acid requirements, however, omega-6 fatty acids lead to the production of eicosanoids that accelerate inflammatory responses.[13] Alternative lipid sources such as fish oils (a rich source of omega-3 fatty acids) may be a better choice. Omega 3-fatty acids lead to the production of factors that slow inflammatory responses.[14] Still other research suggests that olive oil–based lipids (a rich source of monounsaturated fatty acids) may also be beneficial in modulating inflammatory responses.[15]

The type of fat may also have an effect on the altered lipid metabolism that accompanies severe stress. Lipid emulsions containing both long-chain and medium-chain triglycerides (LCT and MCT) may help normalize lipid levels more quickly than lipid emulsions containing only LCT.[16] For people experiencing severe stress, MCT provide a readily available source of energy.

Micronutrients

Vitamin and mineral needs during stress are highly variable, and specific requirements are unknown. The needs for many B vitamins increase when energy and protein intakes increase, and these vitamins are important cofactors in many metabolic reactions. Other micronutrients that play specific roles in repairing tissues include vitamin A, vitamin C, iron, zinc, copper, and selenium. With oxidative stress, antioxidant nutrients become depleted, and supplementing these nutrients may prove to be beneficial. To date, little research has been done to determine if supplementing antioxidants can help reverse oxidative stress.[17]

During stress, blood iron levels may fall as iron moves from the circulation into tissues for storage. Unless a true iron deficiency is identified, iron is not given to compensate, however, because the shift robs invading organisms of the iron they need to grow and thus helps prevent the spread of infection.

In Summary Recovery from acute stress depends, in part, on the body's receiving the energy and nutrients it needs to mount a defense, repair damaged tissue, and replenish nutrients. Providing nutrients too rapidly, however, can stress organ systems and further compromise recovery.

Delivery of Nutrients following Stress

Selecting the appropriate amounts and types of nutrients to help people recover from stress is only part of medical nutrition therapy. Just as important is supplying nutrients in a form that best serves the body's ability to recover.

Oral Diets   Well-nourished clients who are experiencing mild stresses and are expected to eat within seven to ten days following stress receive simple IV solutions to maintain fluid and electrolyte balances and provide minimal kcalories. Once GI tract motility returns, they begin an oral diet that often progresses from clear liquids to regular foods as tolerated (see Chapter 18). Many people have great difficulty eating enough food to meet nutrient needs following an acute stress, however. Use the suggestions in the box on p. 443 in Chapter 17 to help improve the client's intake. Enteral formulas provided orally often help supplement energy and nutrient intake.

Tube Feedings and Parenteral Nutrition   People with preexisting malnutrition, those who are not expected to be able to eat within seven to ten days, and those who have especially high nutrient needs benefit from tube feedings or parenteral nutrition. To prevent abdominal distention, nausea, vomiting, and the possible aspiration of foods or formula into the lungs, oral or gastric feedings have to wait until gastric motility is restored. Because peristalsis returns more quickly to the small intestine than to the stomach, feeding formula directly into the small intestine through a tube is not only possible, but provides advantages over parenteral nutrition. Early feeding (initiated within 48 hours following stress) stimulates intestinal blood flow, function, and adaptation and may minimize hypermetabolism and help maintain the GI tract's immune and absorptive functions. Most significantly, however, early enteral feeding reduces septic complications.[18] Feeding below the stomach may also reduce the risk of aspiration.[19] Early intestinal feedings are not possible in all cases, however, particularly if blood flow to the intestine is disrupted either because of the person's medical condition (shock, injury) or as a consequence of medications.[20] Additionally, some clients may need both enteral feedings and parenteral nutrition until they are able to meet all nutrient needs enterally. Once oral feeding is possible, tube feedings or parenteral nutrition is gradually discontinued (see Chapters 20 and 21).

Case Study

JOURNALIST WITH A THIRD-DEGREE BURN

Mr. Sampson, a 48-year-old journalist, has been admitted to the emergency room. He suffered a severe burn covering over 40 percent of his body when he was trapped in a burning building. His wife told the nurse that Mr. Sampson's height is 6 feet and that he weighs about 175 pounds. The physician ordered lab work, including serum proteins; the results are not back yet.

■ Identify Mr. Sampson's immediate postinjury needs. What measures might be taken to meet those needs? What additional concerns might you have if Mr. Sampson was malnourished before experiencing the burn?

■ Considering Mr. Sampson's condition, what problems might the health care team encounter in getting information from him about his preburn nutrition status?

■ Calculate Mr. Sampson's energy and protein needs to support burn healing (use 2 to 3 grams of protein per kilogram of body weight). What other nutrients might be of concern (consider proportions of energy nutrients and also nutrients that might bolster immune responses or speed wound healing)?

■ What are the possible benefits of early enteral nutrition to Mr. Sampson? How might these benefits be particularly important following a severe burn injury?

Special Formulas Clinicians eager to improve a client's outcome often rely on enteral formulas designed to meet nutrient needs during stress. Many such formulas are of high nutrient density. Some contain extra vitamins A and C, zinc, and other nutrients designed to promote wound healing. Formulas designed to improve immune function often contain added glutamine, arginine, nucleotides, and omega-3 fatty acids. Although such formulas appear to be beneficial for specific situations, further research is necessary to determine if their impact on recovery justifies their use.

Special formulas also are available to meet the needs of clients with organ failure. Chapter 23 described formulas that may be used in the treatment of liver failure, Chapter 25 described those used in heart failure, and Chapter 26 described those used in kidney failure. Often people with acute respiratory failure are unable to meet their nutrient needs with an oral diet. For these people, intestinal feedings may be preferred to gastric feedings to reduce the risk of aspiration. Special nutrient-dense enteral pulmonary formulas that deliver nutrient needs in a smaller fluid volume and provide less carbohydrate and sodium and more fat are available, but other nutrient-dense formulas can also effectively meet nutrient needs.[21]

In Summary Following stress, enteral nutrition, whether orally or by tube, is provided whenever possible. When tube feedings are necessary, they can be provided earlier if the tube is placed in the intestine rather than the stomach, because motility returns to the intestine more quickly than to the stomach following stress. The accompanying case study of a client with burns tests your knowledge of nutrition and acute stress, and the Diet-Drug Interactions box identifies the nutrition-related side effects of drugs that may be used in the treatment of stress. The Nutrition Assessment Checklist summarizes information necessary to monitor the nutrition care of stressed clients.

Diet-Drug Interactions

Analgesics

See box on p. 486.

Antianxiety Agents

See box on p. 486.

Antidiarrheals

See box on pp. 486, 557.

Antiemetics/Antinauseants

See box on p. 486.

Anti-infective Agents

Numerous *anti-infective agents* with a variety of nutrition-related interactions may be used to treat infections related to acute stresses. Consult a pharmacist or drug guide for specific interactions.

Anti-inflammatory Agents

See box on p. 486.

Antisecretory Agents

See box on p. 464.

Sedatives

Sedation itself may lead to reduced food intake. In addition, *sedatives* may cause nausea. Of note is the sedative *propofol*, which is a lipid-based sedative that contains 1.1 kcalorie per milliliter and can provide a significant amount of energy from fat. When propofol is provided to acutely stressed clients, enteral and parenteral formulas must be adjusted to avoid overfeeding.

Nutrition Assessment Checklist

FOR PEOPLE EXPERIENCING ACUTE STRESS

Medical History

Check the medical record regularly to determine:

- ☐ Type of stress
- ☐ Severity of stress
- ☐ Stage of stress
- ☐ Route of feeding (oral/tube feeding/parenteral)
- ☐ If any organ system is compromised

Review the medical record for complications that may be related to underfeeding or overfeeding:

- ☐ Dehydration/fluid overload
- ☐ Hyperglycemia
- ☐ Electrolyte imbalances
- ☐ Acid-base imbalances
- ☐ Fatty liver

Medications

Record all medications and note:

- ☐ Signs that medication dosage may be inappropriate
- ☐ Signs of nutrient deficiencies
- ☐ kCalories provided from propofol, if prescribed

For people on medications who are able to eat an oral diet:

- ☐ Note any medications that may result in reduced food intake or nutrient imbalances
- ☐ Provide pain medications at times when they will be effective during meals, when appropriate
- ☐ Provide antinauseants or antiemetics at times when they will be effective during meals, when appropriate

Food/Nutrient Intake

If the client is not meeting nutrition goals:

- ☐ Calculate nutrient intake from table foods, enteral formulas, and/or parenteral solutions
- ☐ Determine the contribution of nutrients from the sedative propofol, if prescribed
- ☐ Investigate causes, including incorrect determination of nutrient needs or need to alter prescription based on client's unique responses

For clients eating an oral diet:

- ☐ Note the problems the client is having eating
- ☐ Consider interventions to correct problems with eating and take action

Anthropometrics

Measure baseline height and weight and daily weights. Note that body weight changes erratically in acutely ill clients due to the large amounts of fluids required for resuscitation. Once clients are in the adaptive stage of stress, if their weights are not meeting goals:

- ☐ Reestimate energy needs
- ☐ Check to see if the client is receiving the diet (or formula) that has been prescribed
- ☐ Consider the need to change the energy prescription to meet weight goals
- ☐ Consider kcalories contributed from propofol, if prescribed

Laboratory Tests

Laboratory tests that often change as a consequence of stress itself and require careful interpretation as indicators of nutrition status during acute stress include:

- ☐ Albumin
- ☐ Transferrin
- ☐ Prealbumin
- ☐ Total lymphocyte count (white blood cell counts are often elevated)

Check laboratory tests for signs of:

- ☐ Dehydration/fluid overload
- ☐ Hypertriglyceridemia
- ☐ Electrolyte imbalances
- ☐ Acid-base imbalances
- ☐ Vitamin and trace element deficiencies
- ☐ Essential fatty acid deficiencies
- ☐ Response to protein intake (nitrogen balance studies, prealbumin, and retinol-binding protein), when available
- ☐ Organ dysfunction or return to normal organ function

Physical Signs

Regularly assess vital signs including:

- ☐ Blood pressure
- ☐ Pulse
- ☐ Temperature
- ☐ Respiration

Look for physical signs of:

- ☐ PEM (muscle mass and strength; body fat)
- ☐ Essential fatty acid deficiencies
- ☐ Dehydration/fluid overload
- ☐ Nutrient deficiencies and excesses

1. The acute stresses described in this chapter are characterized by:
   a. chronic malnutrition.
   b. tissue damage and hypermetabolism.
   c. reduced protein synthesis in the liver.
   d. hormonal changes that protect skeletal muscle.

2. Cytokines are proteins that:
   a. repair damaged tissues.
   b. destroy microorganisms.
   c. direct stress responses.
   d. oppose insulin's action and result in the catabolism of protein in skeletal muscle, connective tissue, and the gut.

3. Which of the following metabolic changes accompany acute stress?
   a. Proteins are broken down in the liver.
   b. Protein synthesis in the liver decreases.
   c. The body conserves protein as it does in simple fasting.
   d. Proteins are broken down in skeletal muscle, connective tissue, and the gut.

4. If the immune system fails to control a local infection following an acute stress, the most serious complication that can occur from the list below is:
   a. sepsis.
   b. catabolism of protein.
   c. essential fatty acid deficiency.
   d. low serum albumin and transferrin.

5. Which of the following statements with respect to nutrition and severe stress is true?
   a. A person with preexisting malnutrition who suffers an acute stress does not require immediate attention to energy and nutrient needs.
   b. A previously well-nourished person can develop acute malnutrition if the stress is extreme or prolonged.
   c. A person with either acute or preexisting malnutrition has the energy reserves and protein needed to respond successfully to stress.
   d. A person with malnutrition who suffers an acute stress always needs tube feedings or TPN.

6. All of the following parameters help to assess a person's fluid needs during stress **except:**
   a. urinary output.
   b. nitrogen balance.
   c. blood pressure.
   d. body temperature.

7. The electrolytes that may rise in the blood following an acute stress and then plummet as stress resolves include:
   a. sodium and potassium.
   b. calcium and magnesium.
   c. potassium and phosphorus.
   d. phosphorus, iron, and zinc.

8. Which of the following statements correctly describes an appropriate diet for acute stress?
   a. The diet is always high in kcalories.
   b. The diet has little effect on blood glucose levels.
   c. The amounts of carbohydrates, lipids, and proteins that supply energy make little difference.
   d. Although the diet must supply adequate energy, energy needs are generally not exceptionally high immediately following stress, except for some people with extensive burns or head injuries.

9. The use of oral diets in the immediate poststress period is not possible because acute stress:
   a. slows gastric motility.
   b. decreases the appetite.
   c. alters plasma amino acid levels.
   d. results in increased blood flow to the GI tract.

10. The major reason early enteral nutrition must be introduced by tube following a severe stress is that:
    a. appetite is depressed.
    b. hydrolyzed diets are necessary.
    c. gastric feeding is not medically possible.
    d. oral diets cannot meet energy and nutrient needs.

Critical Thinking

1. Depending on the type of stress, a stressed person who weighs 150 pounds needs about _____ grams of protein per day.
   a. 100 to 130
   b. 120 to 150
   c. 150 to 180
   d. 180 to 200

2. If the person in question 1 is suffering a mild stress and develops renal failure but is not placed on dialysis, what are the person's approximate daily protein needs?
   a. 20 to 50 grams
   b. 40 to 70 grams
   c. 60 to 90 grams
   d. 150 to 180 grams

3. Which of the following nutrition measures is most important for the person with acute respiratory failure?
   a. high volumes of fluids and electrolytes to clear the lungs of fluids
   b. a diet prescription that is lower in carbohydrate and higher in fat than a typical diet following acute stress
   c. attention to ensuring an adequate, but not excessive energy intake
   d. a high-kcalorie, high-protein diet to support weaning from mechanical ventilation

Answers to these questions can be found in Appendix G.

1. Returning to Andrew from the "How to" box on p. 692, use the alternative method for calculating energy needs during stress shown in the margin on p. 691. Compare these results with those shown in the box. Discuss possible reasons for the differences you uncover. Do you think the differences are clinically significant, and if so, why?

2. Again returning to Andrew from the "How to" box on p. 692 and assuming that he can tolerate an intact enteral formula, find at least three formulas in Appendix F that the health care team might select for tube feeding. Determine the volume of each formula that would be needed to meet Andrew's energy and protein needs. Would this volume also meet the recommendations for vitamins and minerals?

3. Susan, a 45-year-old woman, was admitted to the hospital following a car accident in which she broke several bones, ruptured a portion of her small intestine, and suffered a severe burn. She has been in the hospital for several weeks and is now eating table foods. Aside from the nutrient demands imposed by the stresses she withstood, describe how the following factors can impair her nutrition status:

- Susan's injuries are painful.
- Susan's medications cause extreme drowsiness.
- Susan is depressed.
- Susan is often out of her room for X rays and other diagnostic tests when her menus and food trays arrive.
- Susan's food intake is often restricted for diagnostic tests she will be receiving.

How might these problems be resolved to improve Susan's ability to eat?

4. Reviewing the information about the metabolic changes directed by hormones in response to acute stress (pp. 685–686), suggest a reason why a previously well-nourished person who develops acute malnutrition as a consequence of stress might have normal or excessive body fat stores.

Nutrition on the Net

For further study of the topics of this chapter, access these websites.

Find updates and quick links to these and other nutrition-related sites at our website:
www.wadsworth.com/nutrition

Read complete articles and learn more about critical care by visiting the American Association of Critical Care Nurses and the Canadian Association of Critical Care Nurses:
www.aacn.org and **www.execulink.com/~caccn**

Learn more about resources for clients and families dealing with burn injuries by visiting the American Burn Association and the Alberta Burn Rehabilitation Society:
www.ameriburn.org and **www.burnrehab.com**

Notes

1. H. R. Chang and B. Bistrian, The role of cytokines in the catabolic consequences of infection and injury, *Journal of Parenteral and Enteral Nutrition* 22 (1998): 156–166.

2. L. Moldawer, Cytokines and the cachexia response to acute inflammation, *Support Line,* April 1996, pp. 1–6.

3. S. A. McClave and coauthors, Are patients fed appropriately according to their caloric requirements? *Journal of Parenteral and Enteral Nutrition* 22 (1998): 375–381.

4. S. A. McClave and D. A. Spain, Indirect calorimetry should be used, *Nutrition in Clinical Practice* 13 (1998): 143–145.

5. L. E. Trombley, T. Reinhard, and D. Klurfeld, Energy expenditure in the critically ill obese patient, *Support Line,* April 2001, pp. 18–23; P. S. Choban, J. C. Burge, and L. Flancbaum, Nutrition support of obese hospitalized patients, *Nutrition in Clinical Practice* 12 (1997): 149–154; J. C. Burge and coauthors, Efficacy of hypocaloric total parenteral nutrition in hospitalized obese patients: A prospective, double-blind, randomized trial, *Journal of Parenteral and Enteral Nutrition* 18 (1994): 203–207.

6. H. Saito, S. Furukawa, and T. Matsuda, Glutamine as an immunoenhancing nutrient, *Journal of Parenteral and Enteral Nutrition* (supplement) 23 (1999): 59–61.

7. A. P. Houdijk, R. J. Nijveldt, and P. A. van Leeuwen, Glutamine-enriched enteral feeding in trauma patients: Reduced infectious morbidity is not related to changes in endocrine and metabolic responses, *Journal of Parenteral and Enteral Nutrition* (supplement) 23 (1999): 52–58; A. P. Houdijk and coauthors, Randomised trial of glutamine-enriched enteral nutrition on infectious morbidity in patients with multiple trauma, *Lancet* 352 (1998): 772–776; P. R. Schloerb and M. Amare, TPN with glutamine in bone marrow transplantation and other clinical applications (a randomized, double-blind study), *Journal of Parenteral and Enteral Nutrition* 17 (1993): 407–413; T. R. Ziegler and coauthors, Clinical and metabolic efficacy of glutamine-supplemented parenteral nutrition after bone marrow transplantation, *Annals of Internal Medicine* 116 (1992): 821–828.

8. J. B. Ochoa and coauthors, Effects of L-arginine on the proliferation of T lymphocyte subpopulations, *Journal of Parenteral and Enteral Nutrition* 25 (2001): 23–29; E. M. Mathus-Vliegen, Nutritional status, nutrition, and pressure ulcers, *Nutrition in Clinical Practice* 16 (2001): 286–291.

9. G. Van den Berghe and coauthors, Intensive insulin therapy in critically ill patients, *New England Journal of Medicine* 345 (2001): 1359–1367; M. M. McMahon, Nutrition support of hospitalized

patients, presented at the Eighth Annual Advances and Controversies in Clinical Nutrition, Mayo Clinic Foundation, April 5–7, 1998.

10. A. M. Malone, Is a pulmonary enteral formula warranted for patients with pulmonary dysfunction? *Nutrition in Clinical Practice* 12 (1997): 168–171.

11. McMahon, 1998.

12. L. S. Schlesselman, Trends in nutrition, coverage of the American Society of Health-System Pharmacists 35th Midyear Clinical Meeting, available at **www.medscape.com**, site visited July 17, 2001.

13. D. F. Driscoll, Intravenous lipid emulsions: 2001, *Nutrition in Clinical Practice* 16 (2001): 215–228; Z. A. Gonzalez, Practical aspects of nutritional modulation of the immune response, *Nutrition in Clinical Practice* 15 (2000): 45–47; C. Lo and coauthors, Fish oil modulates macrophage P44/P42 mitogen-activated protein kinase activity by lipopolysaccharide, *Journal of Parenteral and Enteral Nutrition* 24 (2000): 159–163.

14. Lo and coauthors, 2000.

15. D. Granato and coauthors, Effects of parenteral lipid emulsions with different fatty acid composition on immune cell functions *in vitro, Journal of Parenteral and Enteral Nutrition* 24 (2000): 113–118.

16. S. Hailer, K. W. Jauch, and G. Wolfram, Influence of different fat emulsions with 10 or 20% MCT/LCT or LCT on lipoproteins in plasma of patients after abdominal surgery, *Annals of Nutrition and Metabolism* 42 (1998): 170–180.

17. K. M. Oldham and P. E. Bowen, Oxidative stress in critical care: Is antioxidant supplementation beneficial? *Journal of the American Dietetic Association* 98 (1998): 1001–1008.

18. S. J. Lewis and coauthors, Early enteral feeding versus "nil by mouth" after gastrointestinal surgery: Systematic review and meta-analysis of controlled trials, *British Medical Journal* 323 (2001): 773–776.

19. D. K. Heyland and coauthors, Effect of postpyloric feeding on gastroesophageal regurgitation and pulmonary microaspiration: Results of a randomized controlled trial, *Critical Care Medicine* 29 (2001): 1495–1501.

20. K. A. Kles, M. A. Wallig, and K. A. Tappenden, Luminal nutrients exacerbate intestinal hypoxia in the hypoperfused jejunum, *Journal of Parenteral and Enteral Nutrition* 25 (2001): 246-253; G. Minard, Early enteral feeding—Is it safe for every patient? *Nutrition in Clinical Practice* 13 (1998): 79–80.

21. Malone, 1997.

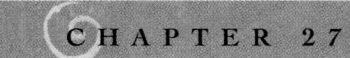

CHAPTER 27

Nutrition in Practice

▪ NUTRITION AND MULTIPLE ▪ ORGAN FAILURE ▪

Multiple organ failure is a critical, and often fatal, complication related to the inflammatory responses that occur as a consequence of uncontrolled infection or severe tissue injury. It generally develops days or weeks after the initial stress and may develop in organs not related to the site of injury. This Nutrition in Practice addresses the causes of multiple organ failure and discusses the possible roles of nutrition and the GI tract in protecting against its development. The glossary on p. 702 defines terms used in this discussion.

How can severe stress lead to multiple organ failure?

As Chapter 27 described, the inflammatory responses to tissue injury serve to protect the body from invading organisms and repair tissue damage, but the responses can also lead to further tissue damage. Most often, the body's inflammatory responses are **localized,** controlled, and self-limiting; the responses target the area of injury, gradually diminish, and eventually end once the danger is over. If the insult to the body is extensive, if further injuries occur, or if infection overwhelms immune defenses, the inflammatory responses may become **systemic,** prolonged, and exaggerated. In these cases, the harmful effects of the responses may overwhelm the beneficial effects, and virtually any organ can be affected.

The exact reasons why inflammatory responses run amok and lead to multiple organ failure remains unclear—most believe that many factors may be involved. Tissues especially vulnerable to the damaging effects of systemic inflammatory responses include the GI tract, lungs, liver, and kidneys.[1] Evidence suggests that changes that occur in the GI tract as a consequence of severe stress may play a pivotal role.[2]

Why are the GI tract, lungs, liver, and kidneys most often affected?

Again, the answer to this question is yet to be found. However, the GI tract and lungs are areas where infectious agents can enter the body, and perhaps that is why they are especially vulnerable during exaggerated inflammatory responses. As for the liver and kidneys, these organs play pivotal roles in maintaining homeostasis, and perhaps they are overwhelmed by the meta-bolic disruptions imposed by exaggerated inflammatory responses.

How can changes in the GI tract lead to multiple organ failure?

As Nutrition in Practice 15 described, the GI tract plays an important role in preventing microbes from entering the body. The closely packed intestinal villi provide a physical barrier to infectious agents, and intestinal immune system cells destroy invaders and stimulate immune responses. During severe stress, changes in the circulatory system divert blood flow away from the GI tract. The disruption in the delivery of oxygen and nutrients to the intestine, coupled with the breakdown of protein in the gut to support hypermetabolism, damages intestinal cells and impairs their functions.

The provision of fluid and electrolytes following stress (reperfusion) restores blood flow to the intestine. Reperfusion supplies needed oxygen and nutrients, but also stimulates the intestinal production of proinflammatory factors (cytokines) that further damage intestinal cells and stimulate and prolong inflammatory responses throughout the body.[3] Figure NP27–1 on p. 702 illustrates the effects and consequences of severe stress on the GI tract and the potential for the development of multiple organ failure. One way that changes in the GI tract may contribute to multiple organ failure, then, is by exaggerating and prolonging inflammatory responses, which damage tissue both by disrupting blood flow to the tissue and by producing toxins (see Figure 27–3 on p. 688).

Another way that changes in the intestine incurred by stress can lead to further tissue damage and multiple organ failure involves the bacterial flora in the colon. Recall that the colon normally houses a bacterial population that does not stimulate inflammatory responses. During stress, however, intestinal edema (from inflammation and consequent increased permeability of blood vessels) and reduced peristalsis set the stage for the overgrowth of bacteria from the colon to the small intestine and stomach. The growth of bacteria in the stomach and small intestine triggers further inflammatory responses and also damages the cells of those organs. An added danger of bacterial overgrowth

Nutrition in Practice

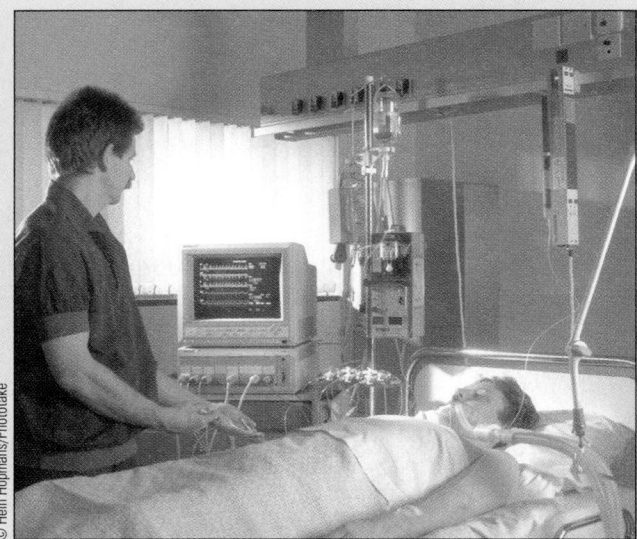

In multiple organ failure, the harmful effects of inflammatory responses overwhelm the beneficial effects.

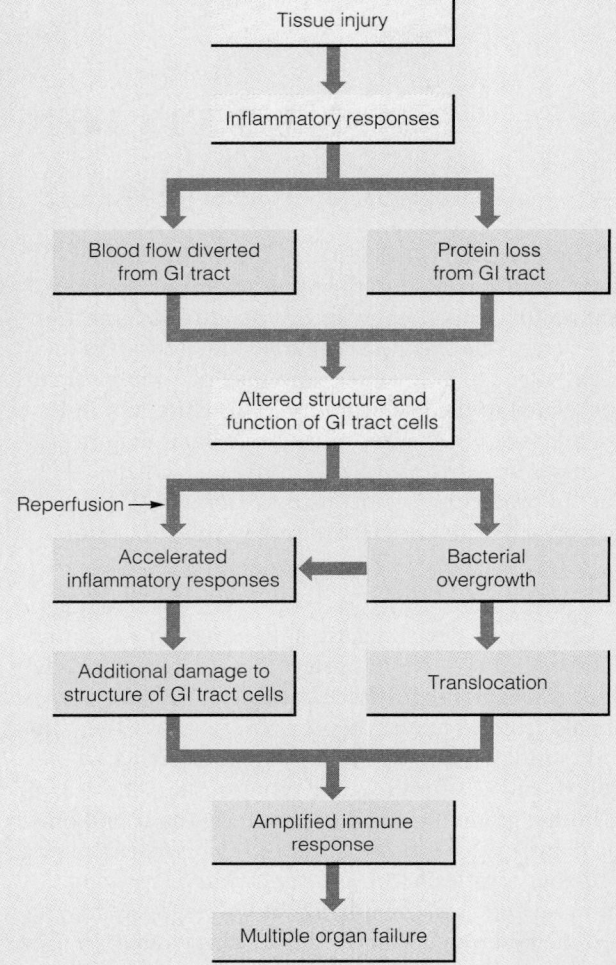

Figure NP27–1

Effects and Possible Consequences of Severe Stress on the GI Tract

of the stomach is that aspiration of gastric secretions into the lungs (a relatively common occurrence in critically ill people) raises the risk of respiratory infections and acute respiratory failure.

Early research into the potential connections between the GI tract and multiple organ failure centered on translocation—the passage of bacteria from the intestine into the body. According to this theory, conditions that compromise the intestinal tract's physical barrier, alter its immune responses, or change the composition of the normal bacterial flora potentially raise the likelihood that infectious agents or **endotoxins** will cross the intestinal cells (see Table NP27–1 on p. 703). Once infectious agents cross the intestinal barrier, they can lead to sepsis and the exaggerated systemic inflammatory responses that can result in tissue damage and organ failure.

Translocation of bacteria from the intestine into the circulation has been demonstrated in animals, where it

appears to occur readily. Animal studies suggest that bacterial translocation from the intestine can lead to peritonitis, pneumonia, and sepsis.[4] Translocation also occurs in humans, but its significance has been difficult

Glossary

endotoxins: products of infectious agents that damage tissues and trigger inflammatory responses.

localized: confined to the area of injury

multiple organ failure: failure of two or more organ systems related to prolonged or exaggerated inflammatory responses; also called *multiple organ dysfunction syndrome.*

systemic: throughout the body as opposed to the area of injury.

Nutrition in Practice

Table NP27–1 Conditions That Increase the Likelihood of Translocation
Altered structure and function of GI tract barrier
Prolonged fasting or lack of enteral nutrients
Injury to the GI tract
Altered blood flow to GI tract
Inflammatory responses
Malnutrition
Changes in bacterial flora
Lack of enteral nutrients
Decreased GI tract motility
Use of broad-spectrum antibiotics
Compromised function of immune factors
Malnutrition
Hypermetabolism

SOURCE: Adapted from M. T. DeMeo, The role of enteral nutrition in maintaining the structural and functional integrity of the GI tract, in *Enteral Nutrition Support,* Report of the First Ross Conference on Enteral Devices, Ross Laboratories, 1996, pp. 4–8.

to determine despite many attempts. A study of 279 people undergoing surgery found that bacterial overgrowth increased the incidence of both bacterial translocation and sepsis.[5] For the most part, however, studies in humans have provided inconsistent results and have failed to confirm a role for translocation as a cause for sepsis and multiple organ failure.

What are the connections between nutrition and GI tract immunity?

In clinical practice, measures designed to maintain the structure and function of the GI tract cells are widely used to preserve both their absorptive and immune functions. Without the stimulation of nutrients delivered enterally, the cells of the GI tract shrink and lose some of their functions. As Chapter 27 described, enteral nutrition provides benefits when compared to parenteral nutrition, including lower rates of infection and improved blood flow to the gut.[6] With improved blood flow, the risk of continuing tissue damage and bacterial overgrowth may also be reduced. The timing of enteral nutrition and its composition may be important in providing maximum benefits.

How do the timing and composition of enteral nutrition affect the integrity of the intestinal cells?

Negative nitrogen balance occurs following a stress regardless of nutrient availability, but as Chapter 27 discussed, delivery of enteral nutrients within a few days helps prevent acute protein malnutrition, preserves lean body mass, and may preserve immune

function. During stress, the amino acid glutamine, which provides fuel to both intestinal cells and immune system cells, may become conditionally essential. Glutamine added to parenteral nutrition solutions has been shown to help preserve intestinal structure. More recently, animal studies show that adding glutamine to parenteral solutions helps preserve some immune functions both within and outside the intestine.[7] Studies have also shown that glutamine-enriched parenteral nutrition can help modulate negative nitrogen balance in septic rats.[8] Glutamine-enriched enteral diets resulted in fewer incidences of pneumonia, bacteremia, and sepsis than a similar diet without glutamine in people with multiple trauma.[9] Enteral formulas supplemented with arginine, nucleotides, and omega-3 fatty acids have also been shown to have protective effects on immune function following severe stress.[10]

Evidence suggesting a pivotal role for the intestine in the development of sepsis and multiple organ failure following severe stress is intriguing, but more research is necessary to confirm potential connections. Perhaps the intestine plays a role in some cases and not in others. While waiting for more information, nutrition measures aimed at maintaining the integrity of the GI tract certainly have a low risk of causing any harm.

Notes

1. K. H. Cheever, Early enteral feeding of patients with multiple trauma, *Critical Care Nurse* 19 (1999): 40–51.

2. H. T. Hassoun and coauthors, Post-injury multiple organ failure: The role of the gut, *Shock* 15 (2001): 1–10.

3. F. A. Moore, The role of the gastrointestinal tract in postinjury multiple organ failure, *American Journal of Surgery* 178 (1999): 449–453.

4. T. J. Babineau and G. L. Blackburn, Time to consider early gut feeding, *Critical Care Medicine* 22 (1994): 191–193.

5. J. MacFie and coauthors, Gut origin of sepsis: A prospective study investigating associations between bacterial translocation, gastric microflora, and septic morbidity, *Gut* 45 (1999): 223–228.

6. K. Takagi and coauthors, Modulating effects of the feeding route on stress response and endotoxin translocation in severely stressed patients receiving thoracic esophagectomy, *Nutrition* 16 (2000): 355–360; T. R. Ziegler, C. Gatzen, and D. W. Wilmore, Strategies for attenuating protein-catabolic responses in the critically ill, *Annual Review of Medicine* 45 (1994): 459–480; Babineau and Blackburn, 1994; D. K. Heyland, D. J. Cook, and D. H. Guyatt, Does the formulation of enteral feeding influence infectious morbidity and mortality rates in the critically ill

patient? A critical review of the evidence, *Critical Care Medicine* 22 (1994): 1192–1202.

7. K. A. Kudsk and coauthors, Glutamine-enriched parenteral nutrition maintains intestinal interleukin-4 and muscosal immunoglobulin A levels, *Journal of Parenteral and Enteral Nutrition* 24 (2000): 270–275.

8. S. L. Yeh and coauthors, Effects of glutamine-supplemented total parenteral nutrition on cytokine production and T cell population in septic rats, *Journal of Parenteral and Enteral Nutrition* 25 (2001): 269–274.

9. A. P. Houdijk and coauthors, Randomised trial of glutamine-enriched enteral nutrition on infectious morbidity in patients with multiple trauma, *Lancet* 352 (1998): 772–776.

10. L. Gianotti and coauthors, A prospective, randomized clinical trial on perioperative feeding with an arginine-, omega-3 fatty acid-, and RNA-enriched enteral diet: Effect on host response and nutritional status, *Journal of Parenteral and Enteral Nutrition* 23 (1999): 314–320; M. Braga and coauthors, Immune and nutritional effects of early enteral nutrition after major abdominal operations, *European Journal of Surgery* 162 (1996): 105–112; M. Kemen and coauthors, Early postoperative enteral nutrition with a diet enriched with arginine-, omega-3 fatty acids-, and ribonucleic acid-supplemented diet *versus* control in cancer patients: An immunologic evaluation of impact, *Critical Care Medicine* 23 (1995): 652–659.

Jennie Oppenheimer/Studio Zocolo

CHAPTER 28 | ENERGY- AND PROTEIN-
MODIFIED DIETS FOR
WASTING SYNDROME

Many chronic diseases lead to malnutrition and wasting, especially in the end stages.

*C*hapter 27 described the use of energy-modified, high-protein diets in acute stresses, the kinds that place immediate demands on nutrient stores. This chapter focuses on high-kcalorie, high-protein diets for chronic stresses noted for their association with wasting, including chronic obstructive pulmonary disease (COPD), cancer, and human immunodeficiency virus (HIV). Previous chapters described many other chronic diseases associated with wasting including inflammatory bowel diseases (Chapter 19); disorders that require gastric or extensive intestinal resections (Chapter 22); chronic liver disease, chronic pancreatitis, and cystic fibrosis (Chapter 23); chronic heart failure (Chapter 25); and chronic renal diseases (Chapter 26).

People who develop wasting swiftly fall into a downward spiral. Once weight loss and wasting have been set in motion, the debilitation and general poor health that follow make it even more difficult for the person to eat. The body is unable to respond to the reduced nutrient supply as it does during fasting, and it rapidly depletes its nutrient stores.

kCalorie-Modified, High-Protein Diets in Wasting versus Stress

kCalorie-modified, high-protein diets serve to minimize weight loss and wasting in both chronic wasting disorders and acute stresses. In both cases, loss of appetite, altered metabolism, and accelerated nutrient losses compromise nutrition status, reducing the supply of nutrients at a time when they are greatly needed. In both cases, diet therapy aims to limit the loss of lean body mass, prevent nutrient deficiencies, and preserve organ function. Acute stresses, however, rapidly alter nutrient needs in short periods of time and then resolve or lead to exhaustion, whereas chronic conditions develop gradually and persist for long periods. Chronic conditions require long-term treatment (including medications) and continue to tax nutrient stores over the entire period. Once protein-energy malnutrition (PEM) develops in a person with a chronic disease, the likelihood of complications and death increases.[1] Chronic stresses, for example, render the body vulnerable to acute stresses, especially infections. Whereas an acute stress often strikes the body with little warning, the wasting associated with the chronic disorders frequently does not develop until the disease progresses to its end stages. Alert health care professionals institute nutrition therapy early to improve the quality of life and prevent early death associated with malnutrition.

As Chapter 27 explained, cytokines mediate the body's response to acute stress. Like acute stresses, chronic diseases associated with wasting involve tissue damage and trigger inflammatory responses, and cytokines play a role in the development of PEM.

Energy and Protein Energy and protein needs for a wasting disorder vary depending on the stage of the disorder, its treatment, related complications, and the person's current nutrition status. When the person's disease has progressed to the point of unintentional weight loss, the prescribed diet often provides energy at about 120 to 150 percent of the energy expenditure needs and 1.5 to 2.0 grams of protein per kilogram of body weight.

Reduced food intake is a major contributor to weight loss in wasting syndromes, and many symptoms and other problems can play a role. The "How to" box on pp. 708–709 provides suggestions for helping clients handle food-related problems based on symptoms and other complications and offers strategies for adding energy and protein to the diet.

Vitamins and Minerals Vitamin and mineral needs for wasting syndromes are highly variable, and little information is available concerning specific needs. Often vitamin and mineral supplements are prescribed to ensure an adequate intake.

In Summary Acute stresses dramatically alter nutrient needs over short periods of time, whereas chronic stresses gradually change nutrient needs but have effects over long periods of time. Therefore, malnutrition in chronic stresses frequently occurs in the later stages of the disorder and may be preventable with vigilant attention to nutrition.

Chronic Obstructive Pulmonary Diseases

Diet Order
High-kcalorie, high-protein diet. Encourage fluids.

Chronic obstructive pulmonary disease (COPD) describes several conditions characterized by persistent obstruction of air flow through the lungs. The two major types of COPD are **emphysema** and chronic **bronchitis.** Unlike acute respiratory failure associated with severe stress, lung failure resulting from COPD is gradual. COPD ranks as the fourth leading cause of death in the United States. Smoking is the primary risk factor for COPD, and most people with COPD are smokers. Other risk factors include exposure to environmental pollution (including exposure of nonsmokers to cigarette smoke) and, possibly, repeated respiratory tract infections. Experts believe that COPD may be largely preventable through avoidance of smoking and control of environmental pollution. Once COPD develops, it has no cure, but people with COPD who smoke can preserve lung function if they quit smoking before the disorder has progressed too far.

Consequences of COPD Regardless of the type of COPD, the lungs gradually lose their functional surface area and strength, making it difficult for them to deliver oxygen to the blood and remove carbon dioxide from it. As lung function becomes increasingly compromised, pulmonary infections, respiratory failure, and heart failure can follow.

COPD and Nutrition Status People with COPD frequently experience weight loss and PEM.[2] The extent of malnutrition is associated with the severity of the pulmonary disease. Body weight is a predictor of survival in people with COPD, and low body weight is associated with reduced muscle mass, respiratory function, and immune competence.[3] Weight loss, which may be rapid and dramatic, may occur for many reasons, including the following:

- Anorexia and poor food intake. These may be caused by depression and anxiety; chronic mouth breathing, which can alter tastes for foods; difficulty breathing while preparing and eating food; and gastric discomfort from swallowing air while eating.[4]
- High energy expenditures associated with labored breathing.
- Medications, including anti-inflammatory agents, diuretics, and antibiotics, which alter nutrient requirements and compromise nutrition status.
- Use of oxygen masks, because people cannot eat while using them.
- Repeated infections, which raise nutrient needs and deplete nutrient stores. Poor nutrition status, in turn, opens the way for further infection—a vicious cycle.

Health care professionals urge clients not to smoke to reduce the incidence of COPD or preserve lung function in those who already have the disorder.

chronic obstructive pulmonary disease (COPD): one of several disorders, including emphysema and bronchitis, that interfere with respiration.

emphysema (EM-fih-SEE-mah): a type of COPD in which the lungs lose their elasticity and the victim has difficulty breathing; often occurs along with bronchitis.

bronchitis (bron-KYE-tis): inflammation of the lungs' air passages.

Energy- and Protein-Modified Diets for Wasting Syndrome

How to

Help Clients Handle Food-Related Problems

*F*or people with chronic diseases, many problems can interfere with eating. It is important to find out what problems clients are having with food. The following list offers possible solutions to improve food intake based on answers to the question "Are you having any problems with eating or food?" For all clients, explain how eating can help them feel better. Not all of the suggestions will work for each client; encourage clients to experiment and find the ones that work best.

I just don't have an appetite.

- Eat small meals and snacks at regular times each day.
- Eat the most food at the time of day when you feel the best.
- Use nutrient-dense foods for meals and snacks. (Suggestions are provided later.)
- Eat nutrient-dense foods first.
- Indulge in favorite foods throughout the day.
- Avoid drinking large amounts of liquids with meals.
- Eat in a pleasant and relaxed environment.
- Listen to your favorite music or enjoy a program on TV while you eat.
- Eat with family and friends.
- Serve foods attractively.
- Take a walk before you eat.

I am too tired to fix meals and eat.

- Let friends and family members prepare food for you.
- Use foods that are easy to prepare and eat like sandwiches, frozen dinners, meals from take-out restaurants, instant breakfast drinks, liquid formulas, and supplements in candy bar and pudding form.
- Eat nutrient-dense foods first.

Foods just don't taste right.

- Brush your teeth or use a mouthwash before you eat.
- Rinse your mouth with water to which lemon juice has been added before eating or as needed.
- Add sauces and seasonings to meats.
- Eat meats cold or at room temperature.
- Use eggs, fish, poultry, and dairy products instead of meats.
- Try new foods and experiment with herbs and spices.
- Use plastic, rather than metal, eating utensils.
- Ask your doctor about zinc supplements. If you have a deficiency, your tastes may change.

I am nauseated a lot.

- Avoid foods and odors that make you feel nauseated.
- Drink cold or carbonated beverages at the first hint of nausea.
- If nausea is a problem at specific times of the day or after specific treatments, avoid eating immediately before, during, or after these times.
- Try drinking ginger or peppermint teas, which provide relief for some people.

I can't stand some of the foods I really used to like.

- Save your favorite foods for times when you are not feeling nauseated or sick to your stomach.
- Maintain a food-free "window" of an hour or so before and after you have treatments or take medications that cause nausea or vomiting.

I am having problems chewing and swallowing food.

- Experiment with food consistencies to find the ones you can handle best. Thin liquids, true solids, and sticky foods (like peanut butter) are often difficult to swallow.
- Add sauces and gravies to dry foods.
- Drink fluids with meals to ease chewing and swallowing.
- Try using a straw to drink liquids.
- Tilt your head forward and backward to see if you can swallow easier with the head positioned differently.

I have sores in my mouth and they hurt when I eat.

- Use cold or frozen foods; they are often soothing.
- Try soft, soothing foods like ice cream, milkshakes, bananas, applesauce, mashed potatoes, cottage cheese, and macaroni and cheese.
- Avoid foods that irritate mouth sores like citrus fruits and juices, tomatoes and tomato-based products, spicy foods,

Treatment of COPD People with COPD are encouraged to quit smoking, receive vaccinations to prevent influenza (flu) and pneumonia, and take antibiotics at the first sign of bacterial infection. Medications for COPD frequently include bronchodilators to limit respiratory muscle spasms and corticosteroids to reduce inflammation of lung tissue (see the Diet-Drug Interactions box on p. 724). People with severe COPD may need long-term oxygen therapy to help relieve symptoms.

Help Clients Handle Food-Related Problems—continued

foods that are very salty, foods with seeds (like poppy seeds and sesame seeds) that can be trapped in the sore, and coarse foods like raw vegetables and toast.

- Ask your doctor about using a local anesthetic solution like lidocaine before eating to reduce pain.
- Use a straw for drinking liquids, to bypass the sores.

My mouth is really dry.

- Rinse your mouth with warm salt water or mouthwash frequently, and drink liquids between meals.
- Ask your doctor or pharmacist about medications that can help with dry mouth.
- Use sour candy or gum to stimulate the flow of saliva.
- Make sure you brush your teeth and floss regularly to prevent cavities and oral infections.

I am having trouble with diarrhea.

- Drink plenty of fluids. Salty broths and soups, diluted fruit juices, and sports drinks are good choices. For severe diarrhea, try commercially prepared oral rehydration formulas.
- Eat foods with soluble fiber, and temporarily limit foods with insoluble fibers (see Table 19–5 on p. 479).
- Temporarily avoid foods that cause gas like dried beans and peas, broccoli, cucumbers, and onions.
- Take lactase enzyme replacements when you use dairy products because you may also

experience lactose intolerance while you are having diarrhea. You may be able to tolerate low-fat yogurt.

- Avoid high-fat foods and foods made with lots of sugar.
- Avoid caffeine.
- Eat smaller meals more often.
- Check with your doctor about using digestive enzyme replacements if you have diarrhea for a long time.

I am having trouble with constipation.

- Drink plenty of fluids. Try warm fluids, especially in the morning.
- Eat whole-grain breads and cereals, nuts, fresh fruits, prunes, prune juice, and raw vegetables. Avoid refined carbohydrates like white bread, white rice, and pasta.
- Exercise regularly.

I need to gain weight, but my blood lipids are elevated.

- To gain weight, you will need to eat more fat, but the type of fat you use is important. Use unhydrogenated margarine in tubs (some are available that contain no *trans*-fatty acids), but don't use the low-kcalorie types.
- Use more monounsaturated fats like olive, canola, and peanut oils for baking and frying and in salad dressings and dips. Snack on avocados, nuts, and peanut butter. Make guacamole dip and spread it on vegetables and low-fat crackers. Spread peanut butter on celery, cucumbers, fruit, and low-fat crackers.

- To increase protein in your diet without adding too much of the fat you need to avoid, eat larger-than-usual servings of chicken and fish, especially salmon. Add chicken, fish, or low-fat and fat-free cheeses to sauces, soups, casseroles, and vegetables. Use instant breakfast drinks made with low-fat milk, or use commercially available supplements that contain no more than 30 percent fat. Add low-fat milk or milk powder to meat loaves, casseroles, soups, puddings, and cereals. Use low-fat yogurt.
- Use plenty of dried fruits. Add nuts and dried fruits to desserts, cereals, and salads.

I need help figuring out how to eat more energy and protein. (Note: All of the suggestions directly above apply here as well, but for some people the need for energy and protein outweighs the need to restrict the type of fat.)

- Use whole milk and regular yogurt instead of the low-fat or fat-free varieties.
- Use plenty of butter, margarine, mayonnaise, cream cheese, oil, and salad dressings on breads, sandwiches, potatoes, vegetables, salads, pasta, and rice.
- Use yogurt, sour cream, or a sour cream dip with vegetables.
- Add whipping cream to desserts and hot chocolate, or use it to lighten coffee.
- Use cream instead of milk with cereal.

Medical Nutrition Therapy For people with COPD, attention to nutrition can improve immune responses, respiratory muscle function, and endurance. People newly diagnosed with COPD or those with mild cases who are also overweight benefit from safe weight-loss programs. Excessive body weight increases the work of the lungs in maintaining respiration. Unintentional weight loss is more likely to be a problem as the disorder progresses. Malnourished clients replete nutrient stores best

Energy- and Protein-Modified Diets for Wasting Syndrome

People with cancer take comfort from the support of others and from the knowledge that medical science is waging an unrelenting battle in their defense.

when refed gradually, with a goal of providing a high-kcalorie, high-protein diet without overfeeding. As Chapter 27 described, overfeeding and excessive carbohydrate intakes produce high levels of carbon dioxide, which the already stressed lungs must work hard to expel.

Meeting energy and protein needs from easy-to-eat foods provided in frequent, small meals often works best. Advise clients to minimize stress during meals, eat slowly, and chew foods thoroughly to avoid swallowing air while eating. When appetite is poor, clients may benefit from enteral formulas provided either orally or by tube. Pulmonary formulas that provide more kcalories from fat and fewer from carbohydrate than standard formulas are available, but there is little evidence to suggest that pulmonary formulas are superior to standard enteral formulas for managing COPD.[5]

Clients with COPD should be encouraged to drink liquids. As with eating, clients may find it difficult to drink enough fluids, and dehydration may impair their ability to cough and clear the lungs of pulmonary secretions.

In Summary Chronic diseases of the lungs progressively diminish lung function and can eventually progress to acute events that are fatal. COPD often leads to malnutrition, and malnutrition contributes to respiratory muscle dysfunction, impaired immunity, and increased risk of infections.

Cancer

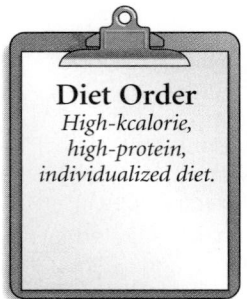

Diet Order
High-kcalorie, high-protein, individualized diet.

The thought of cancer often strikes fear in people, and, indeed, cancer ranks just below cardiovascular disease as a cause of death. As with cardiovascular disease, however, the prognosis for most people with cancer today is far brighter than in the past. Identification of risk factors, new techniques for early detection, and innovative therapies offer hope and encouragement.

Cancer is not a single disorder; instead, there are many different **cancers.** They have different characteristics, occur in different locations in the body, take different courses, and require different treatments. Whereas an isolated, nonspreading type of skin cancer may be removed in a physician's office with no observable effect on nutrition status, advanced cancers (especially those of the GI tract, pancreas, and liver) can seriously impair nutrition status.

How Cancers Develop

Cancers develop from mutations in the genes that ordinarily monitor replicating DNA for chemical errors. The affected cells seemingly have no brakes to halt cell division. As the abnormal mass of cells, called a **tumor,** grows, blood vessels form to supply the tumor with the nutrients it needs to support its growth.

Effects of Cancer on Healthy Tissue Eventually, the tumor invades more and more healthy tissue and interferes with its functions. The consequences of cancer depend on its location, severity, and treatment. Cancer of the lungs, for exam-

cancers: diseases that result from the unchecked growth of cells.

tumor: a new growth of tissue forming an abnormal mass with no function; also called a **neoplasm** (NEE-oh-plazm).

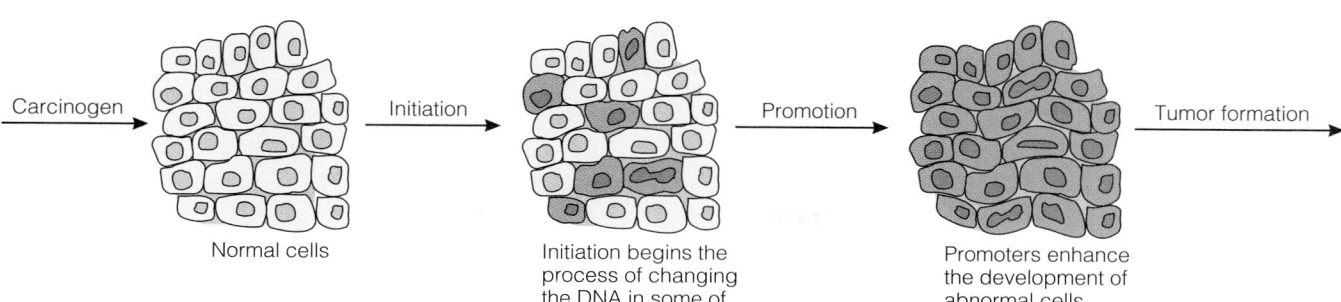

Carcinogen → Normal cells

Initiation → Initiation begins the process of changing the DNA in some of the cells.

Promotion → Promoters enhance the development of abnormal cells.

Tumor formation →

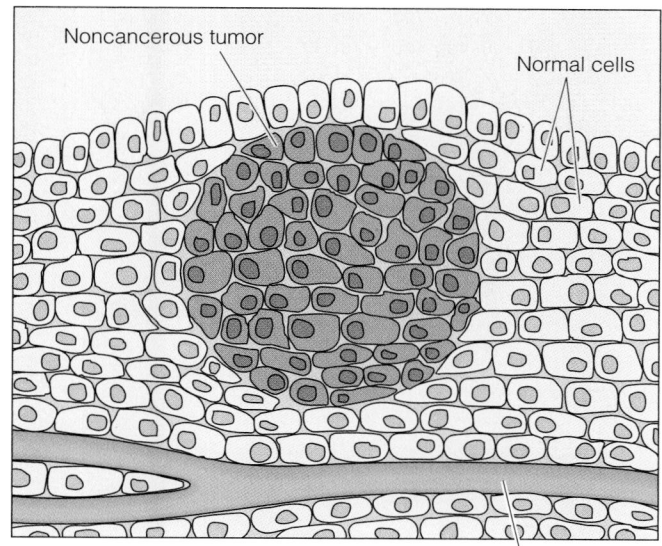

Noncancerous tumor

Normal cells

Blood vessel

A noncancerous (benign) tumor usually grows within a self-contained capsule. It does not invade nearby tissue, nor does it spread.

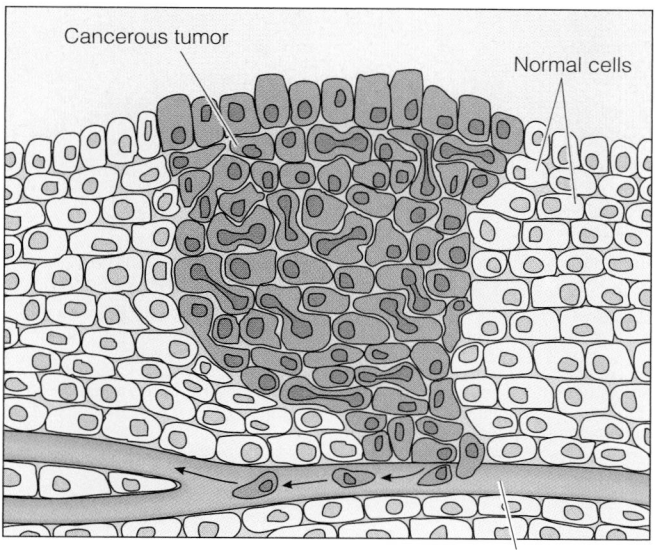

Cancerous tumor

Normal cells

Blood vessel

A cancerous (malignant) tumor usually grows out of control and may spread to other parts of the body through the blood or lymph systems.

ple, will have different consequences than cancer of the kidneys. The tumor may **metastasize** to other parts of the body and affect still other healthy tissues. In leukemia (cancer of the blood-forming cells of the bone marrow), the cancer cells do not form a tumor, but rather circulate with the blood through other tissues where they can accumulate. Figure 28–1 illustrates how a tumor develops and shows the difference between a **benign** and a **malignant** tumor.

Causes of Cancer Although the exact reasons for cancer development remain unclear in most cases, genetic, immune system, environmental, and dietary factors, or some combination of these, may all lead to the mutations that give rise to cancers. Table 28–1 on p. 712 shows examples of diet-related factors associated with an increased risk of certain cancers. The recommendations for reducing cancer risks are quite similar to the recommendations for heart health and are summarized in Table 28–2 on p. 713.

Wasting Associated with Cancer

Loss of appetite, weight loss, depletion of lean body mass and serum proteins, and debilitation typify the **cancer cachexia syndrome,** which occurs in as many as 80 percent of people with cancer before they die.[6] Weight loss is frequently evident at the time cancer is diagnosed, and severe malnutrition, often found in the later stages of cancer, may be the ultimate cause of death. Studies have shown

Figure 28–1

Tumor Formation
Cancer develops in stages: (1) exposure to a carcinogen; (2) entry of the carcinogen into normal cells; (3) initiation, in which initiators begin the process of altering the cells' genetic material; (4) promotion, in which promoters accelerate the process; and (5) formation of a tumor as the cells multiply out of control.

metastasize (meh-TAS-tah-size): to spread by the movement of cancer cells from one part of the body to another.

benign (bee-NINE): describes tumors that stop growing without intervention or can be removed surgically and most often pose no threat to health.
   benign = mild

malignant (ma-LIG-nant): describes tumors that multiply out of control, threaten health, and require treatment.
   malignus = of bad kind

Energy- and Protein-Modified Diets for Wasting Syndrome

Table 28–1 Diet-Related Factors Associated with Specific Cancers[a]

Cancer Type	High Incidence Associated with:	Protective Effect Associated with:
Oral cavity and pharynx	Alcohol, low intake of vitamin A; excessive vitamin A from supplements	Fruits and vegetables
Esophageal	Alcohol, pickled vegetables, obesity	Fruits and vegetables, vitamins A and C, riboflavin, selenium
Stomach	Smoked foods, salted fish, cured meats, pickled vegetables, alcohol, pernicious anemia, *possibly* grilled and barbecued meats	Fruits and vegetables, whole grains, vitamins A and C
Colorectal	Dietary fat (particularly animal fat), meat, obesity, central fat, and alcohol	Fruits and vegetables; *possibly* fiber; calcium from supplements or low-fat dairy foods, and folate
Pancreatic	Meat and dietary fat, obesity	Fruits, vegetables, and fiber
Liver	Alcohol, iron overload	
Lung		Fruits and vegetables
Breast	Alcohol, obesity, central fat, *possibly* dietary fat	Fruits and vegetables
Endometrial	Obesity, dietary fat (particularly animal fat)	*Possibly* fruits and vegetables
Cervical		Fruits and vegetables
Ovarian		Fruits and vegetables
Prostate	Dietary fat (particularly animal fat), *possibly* vitamin A supplements and low-calcium, high-fructose diets	Fruits and vegetables, *possibly* selenium and vitamin E supplements
Bladder	The herb Arisocholia fanchic, found in some herbal weight-loss products	Fruits and vegetables, adequate intake of fluids

[a]Factors unrelated to diet are not shown. Often these factors are related more strongly to cancer risk than diet factors. The risk of stomach cancer, for example, is more closely associated with *Heliobacter pylori* infections than with diet factors.

SOURCE: American Cancer Society, Cancer Resource Center, www3.cancer.org/cancerinfo, visited March 27, 2001.

Reminder: Chapter 27 described immune system responses to tissue damage, which include the secretion of cytokines. Some of the cytokines identified as mediators of cancer cachexia include tumor necrosis factor (cachetin), interleukin-1, interleukin-6, interferon-γ, and differentiation factor.

that once lean body mass is significantly depleted, regardless of the cause, death will follow.[7] Without adequate energy and nutrients, the body is poorly equipped to maintain immune defenses, support organ function, absorb nutrients, mend damaged tissues, and utilize medications.

Many factors appear to play a role in the wasting associated with cancer. Cytokines, produced as inflammatory responses mount a battle to halt tissue damage caused by tumor cells, appear to play an important role.[8] In one study, poor appetite and weight loss occurred more often in people with cancer who had heightened inflammatory responses.[9] The combination of poor appetite, accelerated and altered metabolism, and the diversion of nutrients to support tumor growth simultaneously reduces the supply of energy and nutrients and raises the demand for them. Figure 28–2 on p. 714 summarizes some of the factors that contribute to the cancer cachexia syndrome.

Anorexia and Reduced Food Intake Anorexia is a major contributor to wasting associated with cancer. Some factors that can contribute to anorexia or otherwise reduce food intake in the person with cancer include:

- *Chronic nausea and early satiety.* People with cancer frequently experience nausea and a premature feeling of fullness after eating small amounts of food.

- *Fatigue.* People with cancer often tire easily and lack energy to prepare and eat meals. Once wasting is evident, these tasks become even more difficult for the person to handle.

- *Pain.* People in pain may have little interest in eating, particularly if eating makes the pain worse.

- *Psychological stress.* The very diagnosis of cancer can cause so much distress that eating becomes unimportant. Stress can be compounded by the person's fear and anxiety about the medical, personal, and financial concerns

cancer cachexia (ka-KEKS-ee-ah) **syndrome:** loss of lean body mass, depletion of serum proteins, and debilitation that frequently accompany cancer.

Table 28–2   Recommendations for Reducing Cancer Risk[a]

Choose a diet rich in a variety of plant-based foods.

- Eat seven or more servings a day of a variety of whole grains, legumes, and starchy vegetables.
- Eat five or more servings a day of other vegetables and fruits all year round.
- Limit consumption of processed foods and refined sugar.

Maintain a healthy weight and be physically active.

- Avoid being underweight or overweight and limit weight gain during adulthood to less than 11 pounds (5 kilograms).
- If occupational activity is low or moderate, take an hour's brisk walk or participate in a similar exercise daily.
- Exercise vigorously for at least one hour each week.

Drink alcohol in moderation, if at all.

- Avoid alcohol consumption.
- If alcohol is consumed, limit it to less than two drinks a day for men and one for women.

Select foods low in fat and salt.

- Limit consumption of fatty foods, particularly those of animal origin. If red meat is eaten, limit intake to less than 3 ounces daily.
- Choose modest amounts of vegetable oils.
- Limit consumption of salted foods and use of cooking and table salt.
- Use herbs and spices to season foods.

Prepare and store foods safely.

- Use refrigeration and other appropriate methods to preserve perishable foods as purchased and at home.
- Do not eat charred food.
- Consume meat and fish grilled in direct flame and cured and smoked meats only occasionally.

Most importantly, do not smoke or use tobacco in any form.[b]

[a]American Institute for Cancer Research, Food, *Nutrition and the Prevention of Cancer: A Global Perspective,*1997.
[b]One additional recommendation is in order: vary food choices. This last suggestion is based on an important concept that applies specifically to the prevention of cancer initiation—dilution. Switching from food to food dilutes the negative qualities of a food.

Cruciferous vegetables, such as cauliflower, broccoli, and brussels sprouts, contain nutrients and phytochemicals that may inhibit cancer development.

© Polara Studios Inc.

created by the diagnosis. Once wasting is evident, people may become depressed by their inability to perform routine tasks and by their physical appearance, and depression can also lead to reduced food intake.

- *Obstructions.* A tumor may partially or completely obstruct any portion of the GI tract and interfere with chewing and swallowing; cause delayed gastric emptying, early satiety, nausea, or vomiting; or make oral diets impossible.

- *Cancer therapy.* Therapy for cancer including medications, **chemotherapy, radiation therapy,** surgery, and **bone marrow transplants** can dramatically affect food intake by causing nausea, vomiting, altered taste perceptions, diminished taste sensitivity (mouth blindness), inflammation of the mouth (stomatitis) and esophagus (esophagitis), mouth ulcers, mouth dryness, food aversions (strong dislikes for certain foods), and depression. (A later section provides more information about treatments for cancer.)

Metabolic Alterations   Altered metabolism causes people with cancer to use nutrients in inefficient ways that demand more energy and waste vital protein tissues. This explains why some people with cancer fail to regain lean body mass even when they are receiving adequate energy and nutrients. Some people with cancer are hypermetabolic and have high nutrient needs; others may become hypermetabolic

chemotherapy: the use of drugs to arrest or destroy cancer cells. Drugs used for chemotherapy are called **chemotherapeutic** or **antineoplastic agents.**

radiation therapy: the use of radiation to arrest or destroy cancer cells.

bone marrow transplants: the replacement of diseased bone marrow in a recipient with healthy bone marrow from a donor; sometimes used as a treatment for breast cancer, leukemia, lymphomas, and certain blood disorders.

Energy- and Protein-Modified Diets for Wasting Syndrome

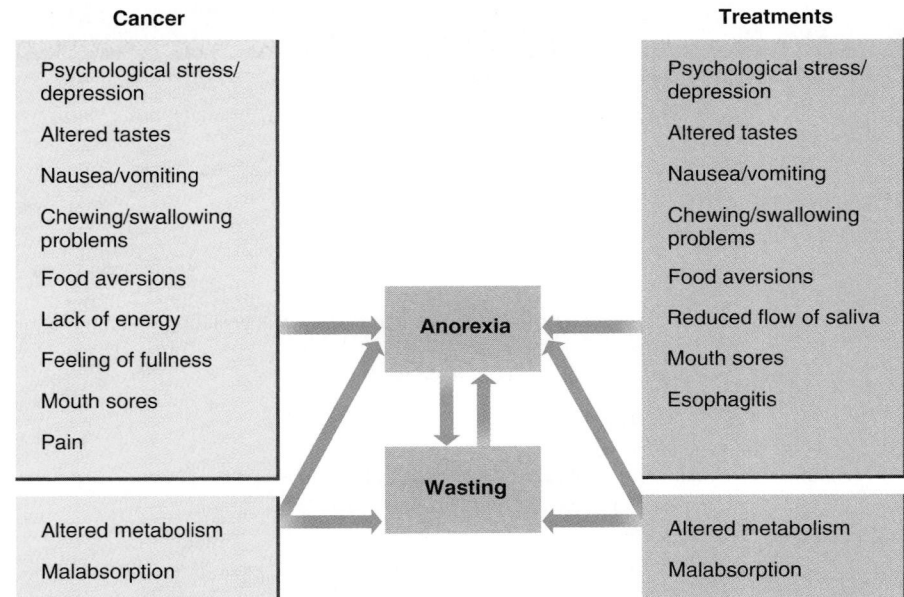

Figure 28–2

The Cancer Cachexia Syndrome: Contributing Factors

Anorexia and wasting contribute to each other. Cancer and its treatments make both problems worse.

Cancer	Treatments
Psychological stress/depression	Psychological stress/depression
Altered tastes	Altered tastes
Nausea/vomiting	Nausea/vomiting
Chewing/swallowing problems	Chewing/swallowing problems
Food aversions	Food aversions
Lack of energy	Reduced flow of saliva
Feeling of fullness	Mouth sores
Mouth sores	Esophagitis
Pain	

Anorexia

Wasting

Altered metabolism	Altered metabolism
Malabsorption	Malabsorption

Altered metabolism may raise nutrient needs in cancer due to:

- *Hypermetabolism.*
- *Inefficient use of energy and nutrients.*
- *Insulin resistance.*
- *Nutrient deficiencies created by chemotherapy.*
- *Secondary infections.*

Reminder: Chapter 15 described how the medication methotrexate resembles folate (see Figure 15–3 on p. 403) and works by depriving cells (tumor cells and healthy cells alike) of the folate they need to grow.

radiation enteritis: inflammation and scarring of the intestinal cells caused by exposure to radiation.

tissue rejection: destruction of healthy donor cells by the recipient's immune system, which recognizes the donor cells as foreign; also called **graft-versus-host disease (GVHD).**

Chapter Twenty-Eight

714

---

as a consequence of surgery or infection. Cancer and its treatments tax the immune system, raising the likelihood of infections, which, in turn, create additional demands for energy and nutrients. People who are hypermetabolic are also insulin resistant, a condition that interferes with the body's ability to obtain the energy it needs. Chemotherapy can interfere with normal metabolic pathways and create nutrient deficiencies as well.

Nutrient Losses   Excessive nutrient losses can develop as a consequence of cancer itself or the treatment for it. Causes of nutrient losses include vomiting, inadequate digestion, diarrhea, and malabsorption, and often several of these conditions occur at the same time. Some tumors, some antineoplastic agents, sometimes radiation therapy, and bone marrow transplants can lead to vomiting and diarrhea, electrolyte imbalances, and dehydration. Cancers of the pancreas, liver, or small intestine often lead to malabsorption. Radiation therapy to the small intestine can cause **radiation enteritis,** which can lead to malabsorption, chronic blood loss, fluid and electrolyte imbalances, and, sometimes, intestinal obstructions and fistulas. Intestinal function may return after radiation therapy ends, but for some the changes are permanent. Severe diarrhea and malabsorption, with fluid losses exceeding 10 liters a day, can occur in people who undergo bone marrow transplants and then reject the transplanted tissue.

Treatments for Cancer

The primary treatments for cancer—radiation therapy, chemotherapy, surgery, or any combination of the three—aim to annihilate cancer cells, relieve pain, alleviate symptoms, and prevent tumor growth. Table 28–3 summarizes the nutrition-related side effects of radiation therapy and chemotherapy. Table 28–4 on p. 716 shows the potential effects of various surgeries for cancers on nutrition status. Preparation for a bone marrow transplant includes high doses of chemotherapy and sometimes whole-body radiation to eradicate cancer cells. Thus the nutrition-related side effects of these treatments apply to bone marrow transplants as well. In addition, immunosuppressants, given to help prevent **tissue rejection** following a bone marrow transplant, have multiple effects on nutrition status (see the Diet-Drug Interactions box on p. 724).

Table 28–3   Possible Causes of Wasting Associated with Radiation and Chemotherapy

	Reduced Nutrient Intake	Accelerated Nutrient Losses	Altered Metabolism
Radiation	Anorexia Damage to teeth and jaws Dysphagia Esophagitis Mouth ulcers Nausea Reduced salivary secretions Taste alterations Thick salivary secretions Vomiting	Blood loss from intestine and bladder Diarrhea Fistulas Intestinal obstructions Malabsorption Radiation enteritis Vomiting	Fluid and electrolyte imbalances as a consequence of vomiting, diarrhea or malabsorption. Secondary effects of malnutrition, infection, or tissue damage (inflammation)
Chemotherapy	Abdominal pain Anorexia Mouth ulcers Nausea Taste alterations Vomiting	Diarrhea Intestinal ulcers Malabsorption Vomiting	Fluid and electrolyte imbalances Hyperglycemia Interference with vitamins or other metabolites Negative nitrogen and calcium balance Secondary effects of malnutrition, infection, or tissue damage (inflammation)

Medications to Combat Anorexia and Wasting   To help people with advanced cancers combat anorexia, medications that stimulate the appetite have gained wide use. One medication, megestrol acetate, stimulates the appetite and promotes weight gain. Dronabinol (a medication containing the principal psychoactive ingredient in marijuana) works as both an appetite stimulant and an antiemetic, although side effects including euphoria and confusion limit its use in some cases.

Even when diets supply seemingly adequate amounts of energy and nutrients, wasting may continue. People may be able to regain weight, but the weight gain is often in the form of fat rather than lean body mass. Thus diet alone may be ineffective in treating cachexia, and drug therapy may be necessary.[10] Under investigation are medications, notably anabolic steroids, growth hormone, insulin-like growth factor, and thalidomide, which may help to restore lean body mass.[11]

Other Medications   Depending on the organ systems affected by cancer and by the side effects of treatments, many medications may be used in treatment. Medications commonly used to treat symptoms of cancer include antiemetics, antidiarrheals, analgesics, and sedatives.

Alternative Therapies   Many people with cancer turn to complementary therapies (see Chapter 15) in their fight against the disease. Up to 83 percent of people with cancer use at least one complementary therapy; 63 percent use dietary supplements (vitamins and herbs).[12] Although most people use complementary therapies because they believe such therapies will improve the quality of their lives, boost their immune systems, prolong their lives, or relieve uncomfortable symptoms, many (38 percent) believe that complementary therapies will cure their cancer.

Clinical studies are vitally needed to find out what interactions, if any, dietary supplements may have with antineoplastic agents or whether the supplements may have other negative effects on the body. Taking high doses of folate, for example, might reduce the effectiveness of methotrexate. Even when it seems logical that a supplement might be helpful and carry few risks, the reverse might be true. Although logic suggested that beta-carotene might be protective against lung cancer

Megestrol acetate is believed to exert its effects on appetite by altering the synthesis and release of certain cytokines. Megestrol acetate is also used as an antineoplastic agent in the treatment of breast cancer.

The herbs and other dietary supplements often used by people with cancer may have unknown effects on cancer treatments.

Energy- and Protein-Modified Diets for Wasting Syndrome

Table 28–4	Possible Effects of Surgeries for Cancers on Nutrition Status
Head and Neck Resection	
Difficulty in chewing/swallowing	Inability to chew/swallow
Esophageal Resection	
Diarrhea	Reduced gastric motility
Fistula formation	Steatorrhea (fat malabsorption)
Reduced gastric acid secretion	Stenosis (constriction)
Gastric Resection	
Dumping syndrome	Lack of gastric acid
General malabsorption	Vitamin B_{12} malabsorption
Hypoglycemia	
Intestinal Resection	
Blind loop syndrome	Hyperoxaluria
Diarrhea	Malabsorption
Fluid and electrolyte imbalances	Steatorrhea
Pancreatic Resection	
Diabetes mellitus	Malabsorption

because a high intake of fruits and vegetables (good sources of beta-carotene) protects against lung cancer, the logic did not hold up under the scrutiny of clinical research. Not only did beta-carotene not protect against lung cancer, it increased the risk of developing the disease.[13] Other researchers have voiced concerns over the potential effects of antioxidant supplements on chemotherapy and radiation therapy, because these therapies are believed to stop cancer growth, in part, by generating free radicals.[14] Since antioxidant supplement use is widespread among people with cancer, an answer derived from careful research appears warranted.

Medical Nutrition Therapy

Although nutrition cannot change the ultimate outcome of cancer, attention to nutrition can help people maintain their strength and nutrition status and can bolster the immune system while they undergo stressful treatments and cope with their diseases. Compared to malnourished people with cancer, well-nourished people with cancer enjoy a better quality of life—that is, they feel better, function better, are stronger and more active, and eat more.[15] Malnourished people with cancer may also have a poorer response to cancer treatments and a shorter survival time than people without malnutrition.[16] At a minimum, meeting nutrient needs eliminates the additional stresses imposed by malnutrition.

Early Nutrition Intervention   For people with cancer, early nutrition intervention is a high priority. The initial nutrition assessment evaluates the individual's current nutrition status and establishes baseline parameters from which to monitor changes. Early intervention helps to prepare the person for the stresses ahead and helps detect and correct deficiencies before the task becomes monumental.

Oral Diets   For people with cancer, medical nutrition therapy must account for the organ systems involved, the type and severity of the cancer (see Table 28–5), and the specific symptoms that each person is experiencing. Dietitians regularly monitor weight changes and assess food intake information to make appropriate

Table 28–5 Dietary Considerations for Specific Cancers

Cancer Sites	Dietary Considerations
Brain/nervous system	Physical feeding disabilities (see Nutrition in Practice 18); chewing and swallowing problems (see Chapter 18).
Head/neck	Chewing and swallowing problems.
Mouth/esophagus	Chewing and swallowing problems; vomiting; if obstructed, tube feeding below the obstruction may be necessary.
Stomach	Nausea, vomiting, early satiety; if obstructed, tube feeding below the obstruction or TPN may be necessary; if resection is performed, a postgastrectomy diet (see Chapter 22) may be needed; bacterial overgrowth (Chapter 22) may occur.
Intestine	If obstructed, tube feeding or TPN may be necessary; resections or inflammation may cause multiple nutrition problems (see Chapter 22); fat- and lactose-restricted diet may be useful.
Liver	Protein-, sodium-, and fluid-restricted diet may be necessary (see Chapter 23).
Pancreas	Fat-restricted diet and enzyme replacements may be necessary (see Chapter 23); diabetic diet may be necessary if insulin production is affected (see Chapter 24).
Kidneys	Protein-, electrolyte-, and fluid-controlled diet may be necessary (see Chapter 26).

NOTE: The considerations listed here are specific to the type of cancer; they do not include other nutrition-related effects of treatment.

recommendations. The "How to" box on pp. 708–709 outlines strategies for overcoming symptoms that interfere with food intake. Clients who are unable to eat adequate amounts of table foods often benefit from nutrient-dense formula supplements. Limited research suggests that fish oils may help minimize weight loss in people with cancer.[17]

Not all people with cancer lose weight. Women diagnosed with breast cancer often gain, rather than lose weight, and this weight gain can be distressing. Weight gain may continue for some time after diagnosis. By encouraging physical activity, regularly assessing weight and energy intakes, and recommending strategies to correct problems early, health care professionals can help clients avoid unnecessary weight gain.[18]

Tube Feedings and TPN   In general, a tube feeding or total parenteral nutrition (TPN) is not routinely recommended for an adequately nourished or mildly malnourished person with cancer who is unable to eat. Most studies have failed to show that the use of tube feedings or parenteral nutrition reduces complications, lowers mortality rates, or shortens hospital stays for people with cancer. When anorexia persists, however, or when a person is severely malnourished and is about to undergo aggressive cancer therapy, nutrition support can maintain functional and nutrition status.[19] Each case is decided individually, and the use of special nutrition support is more likely when the person's chances of recovery or of significant response to treatment are high, or when the type of cancer is associated with a high risk of death from malnutrition.[20] People requiring head and neck resections, for example, may need long-term tube feedings and may need to continue tube feedings at home. Considering the many negative effects that cancers can have on GI tract absorptive and immune functions, tube feedings are strongly preferred to parenteral nutrition, whenever possible. People with severe radiation enteritis, however, may require home TPN.

Nutrition Support and Bone Marrow Transplants   The potent and intensive chemotherapy (and sometimes radiation) used to prepare the person for a bone marrow transplant severely compromises the GI tract and makes it difficult to

Energy- and Protein-Modified Diets for Wasting Syndrome

717

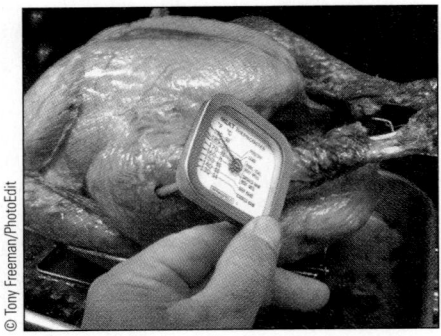

Cooking meats, poultry, and fish to the correct internal temperature is an important safeguard against food-borne illness.

provide enteral nutrition either orally or by tube in the immediate posttransplant period. (See Table 28–3 for the effects of radiation and chemotherapy on the GI tract.) Infections and bleeding commonly occur following a bone marrow transplant, and infections raise nutrient needs, while bleeding increases nutrient losses. Nutrition complications can be severe and debilitating, especially for people who reject the transplant and develop GI complications, such as severe diarrhea and vomiting that are unresponsive to therapy.

TPN often provides nutrients until oral intake is possible. Some researchers have found that adding glutamine to the TPN solution results in fewer infections and shorter hospital stays for people undergoing bone marrow transplants, but other researchers failed to confirm a benefit of glutamine-supplemented TPN.[21] In some cases, oral intake fails to meet nutrient needs, and TPN is required permanently.

Early oral feedings are cautiously introduced. The diet is individualized to minimize uncomfortable symptoms and side effects. In all cases, care must be taken to provide foods least likely to expose clients to food-borne infectious agents. In addition, clients who develop graft-versus-host diseases often start with limited amounts of lactose-free, low-residue, low-fat liquids to maximize absorption and minimize nausea, vomiting, and fat malabsorption. Gradually, lactose-free, low-residue, low-fat solid foods consisting of foods less likely to irritate the GI tract are introduced. Fiber, lactose, and fat are gradually added to the diet as individual tolerances allow. Clients are advised to follow safe food-handling practices to minimize the risks of food-borne illnesses (see pp. 270–280 in Chapter 11). Because the transplant recipient also receives immunosuppressants, which often incur negative nitrogen and calcium balances, the final goal is to provide a high-kcalorie, high-protein, high-calcium diet. In addition, physicians often prescribe calcium and vitamin D supplements. Individuals with persistent diarrhea are encouraged to eat high-potassium foods.

Ethical Issues   Every malnourished person with cancer or an HIV infection who cannot eat an adequate diet orally is a potential candidate for a tube feeding or TPN. This chapter describes uses of tube feedings or TPN as they are applied to people with cancer and HIV infections who have a chance of recovery (from cancer) or a reasonable life expectancy. When incurable cancer or HIV infection has reached its final stages, however, the person, caregivers, and the health care team need to make some important decisions about the use of a tube feeding or parenteral nutrition support. Nutrition in Practice 28 provides more information about medical ethics and describes some factors that go into the decision-making process.

In Summary   Cancers develop when genes that normally regulate cell division fail to function properly. Once cancers develop, the effects on nutrition status depend on the type of cancer, its severity, and the types of treatment. Wasting is a frequent complication of many types of advanced cancers because cancer and its treatment often result in inadequate nutrient intake, wasteful metabolism of nutrients, and excessive nutrient losses. Attention to nutrition can help prevent wasting and preserve quality of life.

---

human immunodeficiency virus (HIV): a virus that progressively hampers the function of the immune system and leaves its host defenseless against other infections and cancer and eventually causes AIDS. The most common HIV is HIV-1—the virus described in this chapter.

acquired immune deficiency syndrome (AIDS): the severe complications associated with the end stages of HIV infection. A person with an HIV infection is diagnosed with AIDS when AIDS-defining illnesses (certain infections or severe wasting) develop.

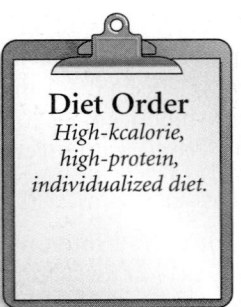

Diet Order
*High-kcalorie,
high-protein,
individualized diet.*

HIV Infection

For many years, the devastating effects of infection by the **human immunodeficiency virus (HIV),** the virus that causes **acquired immune deficiency syndrome (AIDS),** seemed unstoppable. Although the disease still has no cure, remarkable progress has been made in extending the life expectancy of people with HIV infections. In the United States, deaths from AIDS declined markedly from 1995 to 2000.[22]

How HIV Develops

HIV attacks the immune system and leaves its victims defenseless against **opportunistic illnesses**—illnesses that would cause few, if any, symptoms in a person with a healthy immune system. The infection progresses in stages. The virus gradually destroys cells with a specific protein called CD4+ on their surfaces. Among the cells most affected are the **CD4+ T-cells,** lymphocytes that are essential components of the immune system.

HIV Stages At first, the number of CD4+ T-cells declines gradually. As the infection continues, though, depletion of CD4+ T-cells progressively impairs immune function, and in the final stages, fatal complications develop.

Monitoring HIV Progression With improved treatments for HIV infection, the progression of the disorder has been slowed dramatically. On average, it takes about ten years for an HIV infection to progress to AIDS (the final stage of infection). Clinicians monitor the course of HIV infections by measuring concentrations of CD4+ T-cells and circulating virus (viral load) and by monitoring clinical symptoms.

The countless lives touched by AIDS serve as a potent reminder of the need to continue the search for a cure.

Consequences of HIV Infection

In the initial stages of HIV infection, the individual remains symptom-free. As the infection progresses, early symptoms may include fatigue, skin rashes, fevers, diarrhea, joint pain, night sweats, weight loss, oral lesions and infections, and other opportunistic illnesses that are not life-threatening. Coinfection with hepatitis C, which often leads to chronic hepatitis, is common in people with HIV infections. In the final stages, CD4+ T-cell counts fall markedly, and the person develops frequent and eventually fatal complications (**AIDS-defining illnesses),** such as severe weight loss; recurrent bacterial pneumonia; serious infections of the central nervous system, GI tract, and skin; cancers; and severe diarrhea. Improved treatments for HIV infections have significantly reduced the incidence of many opportunistic illnesses and severe malnutrition.[23]

The cluster of symptoms and disorders that sometimes occurs before AIDS develops is called **AIDS-related complex (ARC).**

Lipodystrophy Prior to improved treatments for HIV infections, severe wasting and debilitation were common manifestations of the disease. Although involuntary weight loss and wasting remain a problem for many people with AIDS, some people who are on aggressive therapies (described later) and are not experiencing complications may experience alterations in fat metabolism and distribution (**lipodystrophy).** People with HIV-related lipodystrophy tend to accumulate fat around the abdominal area (central fat) and lose fat from the face, arms, legs, and buttocks.[24] Thus they may appear to be quite thin except for a "pot belly." Fat not only accumulates under the skin in the abdominal area, but also is interspersed between the organs located there. Breast enlargement, thick necks, fat at the top of the back (**buffalo hump),** and multiple benign growths composed of fat (**lipomas)** have also been observed. Some people with HIV-related lipodystrophy also develop hyperlipidemia (especially elevated triglycerides) and insulin resistance consistent with the metabolic syndrome.[25] The reasons why lipodystrophy develops are unclear, and different factors may lead to lipodystrophy in different cases. Many believe lipodystrophy is related to antiviral medications, but other factors may also be important, including increasing age and body mass index, immune function, and duration of medication use.[26] Some people develop lipodystrophy even when they are not taking these medications.

opportunistic illnesses: illnesses that normally would not occur or that would cause only minor problems in a healthy population, but can cause great harm when the immune system is compromised.

CD4+ T-cells: white blood cells (lymphocytes) that have a specific protein receptor on their surfaces and are a necessary component of the immune system. Nutrition in Practice 15 describes T-cells.

AIDS-defining illnesses: very low CD4+ T-cell counts, wasting, and other complications that mark the final stages of an HIV infection.

lipodystrophy (LIP-oh-DISS-tro-fee): the redistribution of fat that can occur as a consequence of HIV infections as well as other disorders. The accumulation of abdominal fat associated with HIV infections is sometimes called *protease paunch.*

buffalo hump: the accumulation of fat at the top of the back.

lipomas (lih-POE-mahs): benign tumors composed of fat.

Weight Loss and Wasting Although weight loss is uncommon in the early stages of HIV infection, in the later stages, the person with AIDS may experience severe wasting. Loss of lean body mass may begin before weight loss becomes

Energy- and Protein-Modified Diets for Wasting Syndrome

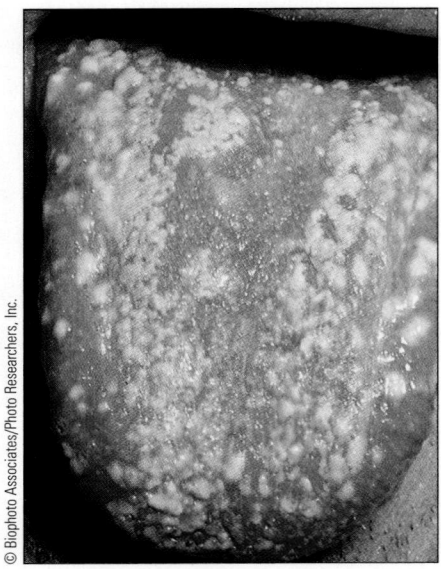

The oral infection thrush is easily identified by the characteristic milky white patches on the tongue.

The Centers for Disease Control identifies wasting as an AIDS-defining illness and defines it as the involuntary loss of greater than 10% of body weight accompanied by chronic diarrhea, lethargy, and/or fever lasting longer than 30 days. However, this definition does not account for wasting not associated with fever or diarrhea and loss of lean body mass as opposed to loss of weight.

Reminder: When the person with an HIV infection develops cancer, the effects of cancer and its treatments also affect nutrition status.

Reminder: *H. pylori* is the bacterium that lives in the stomach and is related to the development of gastritis, peptic ulcers, and gastric cancer (see Chapter 18).

thrush: a fungal infection of the mouth and esophagus, caused by *Candida albicans,* that coats the tongue with a milky film and leads to mouth ulcers, altered taste sensations, and pain on chewing and swallowing. The technical term for this infection is *candidiasis.*

herpes virus: a virus that can lead to mouth lesions and may also affect the lower GI tract, causing diarrhea.

Kaposi's (cap-OH-seez) **sarcoma:** a type of cancer rare in the general population but common in people with HIV infections.

evident.[27] People with HIV infections often lose weight rapidly during periods when they are experiencing complications that alter their appetites or interfere with the absorption or metabolism of nutrients; then they regain weight during periods when they are feeling better.[28] Others experience a gradual but steady weight loss, which may be related to inadequate energy intake and gastrointestinal complications. Much as in cancer, the causes of wasting associated with HIV infection are multifactorial: anorexia and inadequate food intake, altered metabolism, excessive nutrient losses, and nutrient-medication interactions.

Anorexia and Reduced Nutrient Intake   Anorexia and reduced food intake play a major role in the development of wasting associated with both HIV infections and cancer. For people with HIV infections, anorexia and inadequate nutrient intakes may occur as a consequence of:

- *Psychological stress and pain.* As in cancer, fear, depression, and anxiety over the HIV diagnosis, the prognosis, and the medical, personal, and financial problems that lie ahead, as well as the pain associated with complications of the disorder, can destroy the appetite.

- *Oral infections and mouth sores.* Infections and fever cause anorexia. In addition, HIV-related oral infections frequently plague people at all stages of HIV infection.[29] **Thrush,** which occurs in 90 percent of people with HIV infections, can alter taste sensitivity, reduce the flow of saliva, and cause pain on swallowing. Oral infections caused by **herpes virus** can cause painful mouth ulcers that interfere with chewing and swallowing. People with HIV infections who require treatments for cancer may develop mouth ulcers and taste alterations as a consequence.

- *Respiratory infections.* Pneumonia and tuberculosis, frequent complications associated with HIV infection, cause fever and pain that contribute to anorexia. The person who must use an oxygen mask to improve breathing may find it difficult to eat.

- *GI tract complications and altered organ function.* In addition to oral infections, *H. pylori* infections are common in people with HIV infection. The person may experience belching, heartburn, reflux esophagitis, gastritis, and ulcers, and these may cause nausea and interfere with eating. Intestinal complications, including infections, diarrhea, and constipation, and hepatitis can contribute to reduced food intake and food aversions.

- *Fatigue, lethargy, and dementia.* Fatigue is a common complication of HIV infections, even in the early stages. Fatigue may be a consequence of anemia, which is also a frequent complication of HIV infections, or it may develop for other reasons. In the later stages of HIV infections, lethargy and dementia frequently occur and may interfere with food intake. The individual may not care or even remember to eat.

- *Cancer.* As previously described, cancer often leads to anorexia. **Kaposi's sarcoma,** a cancer associated with HIV infection, can cause lesions and obstructions in the esophagus that make eating painful.

- *Medical treatments.* Medications used to treat HIV infection, associated infections, and cancer often cause anorexia, taste alterations, nausea, and vomiting and reduce food intake. Food aversions may also arise.

Metabolic Alterations and Nutrient Losses   The metabolic alterations that occur as a result of HIV infections and their consequences, including repeated infections and cancer, contribute to wasting in people with HIV infections. In addition, nutrient losses can occur as a consequence of diarrhea and malabsorption, which often develop late in the course of HIV infections. About 60 percent of people with HIV infections in the United States experience significant diarrhea and malabsorption. Both the structure and the function of the intestinal cells are altered as a consequence of GI tract infections. Fat malabsorption is common.[30] Food, water, and enteral formulas may serve as a source of infectious agents, and people with advanced HIV infections are highly susceptible to food-borne illness.

The prolonged use of antibiotics to treat infections and antisecretory agents and antacids to relieve nausea can lead to bacterial overgrowth in the upper small intestine, further contributing to malabsorption. Treatments common among people with HIV infections, especially anti-infective agents, chemotherapy, and radiation therapy, can accelerate nutrient losses due to vomiting, diarrhea, and malabsorption. Megadoses of vitamin C and other home remedies that some people with HIV infections use may also cause diarrhea.

In most cases, diarrhea is recurrent, and typical losses average less than 1 liter of diarrheal fluids daily. Diarrhea caused by parasites, however, may be severe and unresponsive to medications—the person may lose 10 or more liters of diarrheal fluids daily. Once malnutrition is under way, it, too, contributes to malabsorption.

The diarrhea and malabsorption associated with HIV infection are called **HIV enteropathy** (EN-ter-OP-a-thee).

Treatments for HIV Infection

Treatments for HIV infection focus on slowing the course of the infection, controlling symptoms, and alleviating pain. Drug therapy often includes a combination of medications that disrupt HIV at different stages of replication. The combined drug therapy called **highly active antiretroviral therapy (HAART)** has made a remarkable difference in the treatment of HIV infection. In 1996, for the first time since 1990, AIDS was no longer among the top ten causes of death in the United States, falling from 8th to 14th place. Antiviral agents can have many interactions with diet as the box on p. 724 describes.

Medications to Combat Anorexia and Wasting The medications megestrol acetate and dronabinol (described on p. 715) may be prescribed to stimulate the appetite and help people with HIV infections to gain weight. More recently, human growth hormone has been approved to treat wasting in HIV infection. Growth hormone helps restore lean body mass. Testosterone and its derivatives may also be beneficial and are currently being studied.[31] Testosterone levels often decrease as an HIV infection progresses, and people with low testosterone may lose lean body mass even when their weights remain stable. When testosterone levels are low, administering testosterone can increase lean body mass and improve strength and quality of life. The combination of megestrol acetate and testosterone can improve the appetite and increase weight and lean body mass.

Growth hormone and testosterone are called **anabolic agents.**

Other Medications Still other medications serve to control infections (anti-infective agents) and complications. Antibiotics, antinauseants, antisecretory agents, oral antidiabetic agents, antilipemics, and antidiarrheals are examples. Like people with cancer, people with HIV infections are likely to use complementary therapies, including dietary supplements.

Medical Nutrition Therapy

Nutrition assessment and counseling should begin as soon as a diagnosis of HIV infection has been made. The initial assessment provides baseline data from which to monitor progress throughout the course of the disease. For people with HIV infections on aggressive combination drug therapies, assessment should include an evaluation of body composition.[32] Clinics that treat clients with HIV infections may use bioelectrical impedance analysis to estimate how much muscle and lean tissue, fat, and water a person's body contains (see Appendix D). Although the technique does have drawbacks—it can't tell where fat is, for example—it provides a simple and convenient analysis. Measurements repeated at intervals allow clinicians to make adjustments to diet and drug therapies.

highly active antiretroviral therapy (HAART): a combination of antiviral agents that disrupt various stages of replication of HIV.

Energy- and Protein-Modified Diets for Wasting Syndrome

Nutrition provides an edge in maintaining quality of life and encouraging independence.

For specific information about safe filtered and bottled water, visit **www.thebody.com/ cdc/crypto1297.html.**

cryptosporidiosis (KRIP-toe-spo-rid-ee-OH-sis): a food-borne illness caused by the parasite *Cryptosporidium parvum.* Most people develop few or minor problems from this infection, but people with HIV infections, and especially those with AIDS, can develop long-lasting and serious problems.

Oral Diets Whether dietary changes and/or drug therapy can help reverse the metabolic changes associated with lipodystrophy remains to be determined. For people on aggressive combination drug therapies who have hyperlipidemia, diet advice often mimics that for all clients with elevated lipids: achieve or maintain a desirable weight, reduce fat to less than 30 percent of total kcalories, reduce saturated fat, limit trans-fatty acids, and replace saturated fats with monounsaturated fats and omega-3 fatty acids (see Chapter 25). People with elevated triglycerides may benefit from limiting simple sugars. People with glucose intolerance or diabetes benefit from eating a consistent intake of carbohydrate throughout the day and eating mostly complex carbohydrates. A combination of aerobic activity and resistance training may also play an important role in preventing central obesity and loss of lean body mass.

For clients who are losing weight, a high-kcalorie (up to 150 percent of the basal energy expenditure), high-protein (at least 1.5 grams or more per kilogram of body weight) oral diet can help to halt weight loss and restore weight. Enteral formulas and supplements in the form of easy-to-eat bars and pudding can help boost nutrient intake when intake from table foods is inadequate. Liquid formulas may be especially useful for the person who is too tired to eat and prepare meals. All people with HIV infections experience anorexia from time to time. Almost all experience bouts of diarrhea and constipation as well. It is important to find out what problems the person is having and make appropriate suggestions. The "How to" box on pp. 708–709 offers other suggestions for individualizing the diet to improve nutrition status.

Vitamins and Minerals Vitamin and mineral deficiencies are also common in people with HIV infections. Many deficiencies have been documented, even in people who are consuming recommended amounts of micronutrients. Still, little information is available concerning specific needs. Anemia, fatigue, and neuropathy, common clinical findings in people with HIV infections, are associated with vitamin B and iron deficiencies, and such deficiencies are also common in people with HIV infections. People with fat malabsorption may also have deficiencies of fat-soluble vitamins. Deficiencies of minerals, especially zinc and selenium, are also common. A vitamin and mineral supplement that contains at least 100 percent of recommended intakes is frequently prescribed to ensure an adequate intake. Some clinicians recommend two multivitamin and mineral tablets daily.

Some people with HIV infections have elevated blood levels of iron, and excess iron may be harmful to the liver and interfere with immune function. For women with HIV infections who are not menstruating or pregnant, the physician should determine if iron supplementation is appropriate. Some clinicians recommend a variety of dietary supplements, including antioxidant nutrients, and supplemental nonessential amino acids, including glutamine, cysteine, and carnitine, although research to support the use of these supplements is lacking.

Food Safety The depressed immune system in people with HIV infections raises the likelihood of infections from microorganisms in food and water. To prevent infections from food-borne microorganisms, clients are provided with instructions for the safe handling and preparation of foods (see pp. 270–280 in Chapter 11). Water can also be a source of food-borne illnesses, especially **cryptosporidiosis.** Water quality varies throughout the United States, and local health departments should be consulted to determine if the local tap water is safe for people with HIV infections to drink. If not, or to take additional safety measures, water used for cooking and making ice cubes should be boiled for one minute. Some types of filtered and bottled waters are also safe, but not all.

Tube Feedings and TPN People with HIV infections may need aggressive nutrition support whenever they are unable to consume oral diets that are sufficient to prevent nutrition complications and unintentional weight loss. Tube feedings are preferred whenever the GI tract is functional. Tube feedings can be

Case Study

TRAVEL AGENT WITH HIV INFECTION

Three years ago, Mr. Sands, a 34-year-old travel agent, sought medical help when he began feeling run-down and developed a painful white coating over his mouth and tongue. The presence of thrush and anemia alerted Mr. Sands's physician to the possibility of an HIV infection. When Mr. Sands tested positive for an HIV infection, he and his family and friends were devastated by the news, but those closest to him have remained supportive. Mr. Sands has gained weight and developed lipodystrophy and hyperlipidemia during the three years since he began combination drug therapy. Mr. Sands is 6 feet tall and currently weighs 190 pounds. He occasionally develops diarrhea and sometimes anorexia.

- What is Mr. Sands's percent ideal body weight (%IBW)? What is lipodystrophy, and what is its typical pattern in people with HIV infections? Describe an appropriate diet for Mr. Sands.

- What suggestions can you give Mr. Sands for the times he has trouble with diarrhea or anorexia? What factors can lead to diarrhea or contribute to anorexia in people with HIV infections?

- Describe how HIV infection can lead to wasting as the disease progresses to the later stages.

- How will Mr. Sands's diet change if wasting becomes a problem for him?

given at night to supplement oral diets during the day. Preventing bacterial contamination of the formula is particularly important. TPN is generally reserved for people with HIV infections who are unable to tolerate enteral nutrition, but need to maintain their nutrition status while undergoing a therapy that is expected to improve their condition.

People with GI tract obstructions, severe vomiting, or GI infections affecting the entire small intestine may benefit from TPN. As with enteral formulas, the concern for infection related to TPN is magnified in people with HIV infections. The accompanying case study discusses nutrition concerns of a person with HIV infection. The Diet-Drug Interactions box provides more information about the medications described in this chapter. The Nutrition Assessment Checklist applies to people with COPD, cancer, or HIV infections.

In Summary In some cases, people with HIV infections develop lipodystrophy, hyperlipidemia, and insulin resistance. For others, especially those with AIDS, weight loss and wasting are the primary concerns. Reduced food intake, metabolic alterations, and increased nutrient losses all play a role in wasting associated with HIV infections. Health care professionals serve people with HIV infections best by identifying nutrition problems early and offering solutions before nutrition status seriously deteriorates.

This chapter brings to a close your introduction to normal and clinical nutrition. Congratulations! You have received an abundance of information since you first began your study. The normal nutrition chapters of this text provided you with current recommendations to promote optimal health. You learned how the body transforms foods into nutrients and how those nutrients support the body's well-being. The clinical chapters addressed you as a future health care professional, concerned with the well-being of others during times of illness. We hope this text has served you well and that you will remember, when selecting food for yourself or when making recommendations for others, to honor the body.

Diet-Drug Interactions

Anabolic Agents

Testosterone, testosterone derivatives *(oxandrolone, nandrolone,* and *oxymetholone),* and *growth hormone* may be used to promote weight gain, specifically a gain of lean body mass. These medications can be taken without regard to food, and nutrition-related side effects are uncommon.

Antidiabetic Agents

For *antidiabetic agents,* see p. 616.

Antidiarrheals

For *antidiarrheals,* see pp. 486, 557.

Anti-infectives

Many medications may be used to treat the infections associated with both cancer and HIV infections (consult a drug guide for specific examples). The antiviral agents specifically used to treat HIV infections are described below.

Antilipemics

For *antilipemics,* see p. 647.

Antinauseants

For *antinauseants,* see p. 486.

Antineoplastics

About 80 antineoplastic agents are in use today. Many of these agents can cause nausea and vomiting. Most often nausea and vomiting begin a few hours after a treatment and resolve shortly thereafter. In some cases, the problems may last for a few days. The suggestions in the "How to" box on pp. 708–709 can help alleviate these problems. Antinauseants may also be prescribed. Mouth sores, taste alterations, fatigue, anemia, diarrhea, and constipation are also associated with many agents. *Megestrol acetate* is used as both an antineoplastic agent and an appetite stimulant. Megestrol acetate and some antineoplastic agents that are hormones (including testosterone) lead to weight gain. Fluid and sodium retention and edema can also occur, although these side effects are not common.

Antisecretory Agents

For *antisecretory agents,* see p. 464.

Antivirals

The nonnucleoside reverse transcriptase inhibitors include *delavirdine* and *nevirapine,* which must be taken one hour apart from antacids and without regard to food. Neither is associated with significant nutrition-related side effects. The nucleoside reverse transcriptase inhibitors include *lamivudine (3TC), zidovudine (AZT), stavudine, zalcitabine (ddC),* and *didanosine (ddl).* Lamivudine, zidovudine, and stavudine can be taken without regard to food; zidovudine can cause nausea and vomiting. Clients who use zalcitabine and didanosine should avoid aluminum- and magnesium-containing antacids. Didanosine can lead to nausea and vomiting and should be taken one hour before or two hours after meals. The protease inhibitors include *indinavir, saquinavir, ritonavir,* and *nelfinavir.* Indinavir should be taken one hour before or two hours after meals; however, people who experience nausea and abdominal pain can eat a light meal (less than 300 kcalories, 6 grams of protein, and 3 grams of fat). Indinavir can lead to kidney stones, so clients taking this medication are encouraged to drink liquids (6 liters a day). Saquinavir should be taken within two hours of a meal; people taking saquinavir should not take garlic supplements. Ritonavir and nelfinavir are better absorbed along with food and should be taken with meals. Ritonavir can cause nausea, vomiting, diarrhea, taste alterations, and abdominal pain. The most common side effect of nelfinavir is mild-to-moderate diarrhea, although nausea and abdominal pain may also occur.

Bronchodilators

Bronchodilators can dry the mouth and throat and lead to nausea and vomiting.

Corticosteroids

See *anti-inflammatory agents* on p. 486.

Immunosuppresants

For *immunosuppressants,* see pp. 584, 674.

Laxatives

For *laxatives,* see pp. 486, 584.

Nutrition Assessment Checklist

FOR PEOPLE WITH CANCER OR HIV INFECTIONS

Medical History

Check the medical record to determine:

- ☐ Stage of COPD
- ☐ Type and extent of cancer
- ☐ Stage of HIV infection

Review the medical record for complications that may alter medical nutrition therapy including:

- ☐ Malnutrition/wasting
- ☐ Altered organ function
- ☐ Anorexia
- ☐ Nausea/vomiting
- ☐ Mouth ulcers/thrush
- ☐ Taste alterations
- ☐ Dry mouth
- ☐ Diarrhea/malabsorption
- ☐ Constipation

Medications

For all clients:

- ☐ Make a note of all medications the client is taking and remain alert for possible diet-medication interactions.
- ☐ Give antinauseants or pain medications at times when they will be most effective during meals.
- ☐ Ask about use of alternative therapies, including herbal preparations and megadoses of vitamins.

For clients with cancer who require chemotherapy:

- ☐ Recommend strategies to prevent food aversions (see the box on pp. 708–709).
- ☐ Offer suggestions for handling complications associated with medications.

For clients on antivirals for HIV infections:

- ☐ Note that some antivirals can be taken without regard to foods, some are better absorbed when taken with foods, and still others must be taken on an empty stomach.
- ☐ Help clients work out an acceptable medication schedule that considers the client's lifestyle and the timing of the medication dose with food intake.
- ☐ Offer suggestions for handling complications associated with medications.

Food/Nutrient Intake

For clients with poor food intakes and weight loss:

- ☐ Determine the reason(s) for reduced food intake.

- ☐ Offer suggestions for improving intake based on specific problems the client is having.
- ☐ Provide interventions before weight loss progresses too far.

For clients with HIV infections who experience weight gain and hyperlipidemia and/or glucose intolerance:

- ☐ Assess diet for total energy; total fat and types and amounts of specific fats; and total carbohydrate, fiber, and simple sugars.
- ☐ Recommend a low-fat, low–saturated fat, energy-controlled diet for hyperlipidemia.
- ☐ Recommend a low-fat, low–saturated fat, energy-controlled diet for glucose intolerance or diabetes with a consistent carbohydrate intake from day to day.

Anthropometrics

Take baseline height and weight measurements, monitor weight regularly, and make dietary adjustments promptly, if necessary. Baseline and periodic body composition measurements are important for people with HIV infections who take combination antiviral medications.

Laboratory Tests

Note that albumin and serum proteins may be reduced for people with cancer or HIV infections, especially those who are experiencing wasting. Check laboratory tests for indications of:

- ☐ Changes in organ function
- ☐ Anemia
- ☐ Dehydration

For people with HIV infections, the progression of the HIV infection is evaluated by checking:

- ☐ CD4+ T-cell counts
- ☐ Viral loads

Physical Signs

Look for physical signs of:

- ☐ Wasting and PEM
- ☐ Fluid status (especially for those with fever, vomiting, or diarrhea)
- ☐ Mouth ulcers

1. With respect to nutrient needs, chronic stresses differ from acute stresses because chronic stresses:
   a. are less serious than acute stresses.
   b. tax nutrient stores over long periods of time.
   c. alter nutrient intake.
   d. alter the metabolism and excretion of nutrients.

2. Factors that frequently contribute to wasting in people with COPD include:
   a. anorexia, medications, and respiratory infections.
   b. dehydration, fat malabsorption, and lactose intolerance.
   c. essential fatty acid deficiencies and potassium imbalances.
   d. vitamin K deficiencies, dysphagia, and low-fiber, low-potassium diets.

3. Which of the following statements describes wasting associated with cancer?
   a. Wasting always accompanies cancer and HIV infection.
   b. Altered metabolism plays no role in the development of wasting.
   c. Anorexia and reduced food intake are major factors in the development of wasting.
   d. Unlike wasting associated with acute stresses, cytokines do not contribute to wasting associated with cancer.

4. Oral diets after bone marrow transplants may:
   a. limit calcium.
   b. test the person's tolerance for lactose.
   c. restrict high-protein foods.
   d. restrict foods likely to carry food-borne illnesses.

5. A tube feeding or TPN is most likely to benefit people with cancer or HIV infection if:
   a. they do not wish to prolong their lives.
   b. no further treatments are available to them.
   c. they have been told they have no other alternatives.
   d. malnutrition may have an undesirable effect on their ability to receive additional treatments.

6. Mouth sores in people with HIV infections are often due to:
   a. dehydration.
   b. oral infections.
   c. malabsorption.
   d. food-borne illnesses.

7. Which of the following statements is true with respect to the diarrhea that often occurs as a consequence of HIV infection?
   a. Megadoses of vitamin C can help resolve it.
   b. HIV and secondary infections can play roles in its development.
   c. Fibers from wheat bran and raw vegetables are useful in controlling it.
   d. Diarrhea can always be corrected with appropriate nutrition therapy and fluid replacement.

8. Which people are most prone to infections arising from foods, enteral formulas, and TPN?
   a. people with HIV infections or those who have undergone bone marrow transplants
   b. people who have undergone chemotherapy or surgery
   c. people with oral infections or cancer of the GI tract
   d. people who have undergone radiation therapy

9. The changes in body fat seen in some people with HIV infections include:
   a. increased central and peripheral fat.
   b. decreased central and peripheral fat.
   c. increased central and decreased peripheral fat.
   d. decreased central and increased peripheral fat.

Critical Thinking

1. The goal for energy and protein intake in a man with a wasting syndrome who weighs 140 pounds and has a basal energy expenditure of 1500 kcalories is about _____ kcalories per day and _____ grams or more of protein per day.
   a. 1500 and 95
   b. 2000 and 150
   c. 2300 and 100
   d. 3000 and 100

2. Which of the following nutrition-related measures represents the best practice for correcting wasting associated with COPD?
   a. Provide a diet with enough energy to promote a rapid weight gain before increasingly severe symptoms make it impossible for the person to gain weight.
   b. Provide a diet with enough energy to promote a gradual weight gain without overtaxing the lungs.
   c. Provide a sodium-restricted diet.
   d. Provide a low-fat diet.

3. Practical advice for a person with wasting who has trouble preparing and eating foods due to fatigue might include:
   a. prepare dinner for friends.
   b. prepare fresh vegetables every day.
   c. make homemade ice cream for snacks.
   d. keep premixed breakfast drinks or formula supplements in the refrigerator for snacks.

4. Which of the following statements is true?
   a. People with HIV infections may gain weight.
   b. People with cancer never have to worry about gaining weight.
   c. People with cancer never have to worry about losing weight.
   d. People with HIV infections never have to worry about losing weight.

Answers to these questions can be found in Appendix G.

Clinical Applications

1. Many disorders can lead to wasting. For some of these disorders, such as fat malabsorption (Chapter 22), diet is a cornerstone of treatment. For others, such as congestive heart failure, chronic obstructive pulmonary disease (COPD), cancer, and HIV infection, nutrition plays a supportive role. What determines whether nutrition plays a major or a supportive role in the treatment of a disorder? Review the effects of PEM on p. 390. Carefully consider how severe malnutrition can further debilitate people with wasting due to COPD, cancer, or HIV infection.

2. Consider problems associated with nutrition in a 36-year-old woman with a malignant brain tumor affecting her ability to move the right side of her body (including the tongue) and to speak coherently. She has an expected length of survival of six months and is taking a pain medication that makes her nauseated and sleepy. What would be a realistic goal of nutrition support? If she is right-handed, how can her impairment interfere with eating? What suggestions might you have for overcoming this problem? How might nutrition be affected by her problems with communication? Describe ways that the medications she is taking can affect her nutrition status. Would tube feedings or TPN be appropriate for this woman? Why or why not?

Nutrition on the Net

For further study of the topics of this chapter, access these websites.

Find updates and quick links to these and other nutrition-related sites at our website:
www.wadsworth.com/nutrition

Learn more about COPD by visiting the National Heart, Lung, and Blood Institute, the American Lung Association, and the Canadian Lung Association:
www.nhlbi.nih.gov/nhilbi/nhlbi.htm, www. lungusa.org, and **www.lung.ca**

Uncover more about cancer, including risk factors, prevention, screening, detection, treatments (including nutrition), complementary therapies, support networks, and links to online cancer resources, by visiting the American Cancer Society, the National Cancer Institute, CancerSource, the American Institute for Cancer Research, and the American Association for Cancer Research:
www.cancer.org, www.nci.nih/gov, www. cancersource.com, www.aicr.org, and **www.aacr.org**

Find virtually anything you would like to know about HIV infections and AIDS by visiting The Body: A Multimedia AIDS and HIV Information Resource:
www.thebody.com

Be sure to check out the forums on a variety of topics (including wasting, lipodystrophy, and controlling symptoms) to gain valuable insights into what it is like to live with an HIV infection.

View statistics about the incidence of HIV infections and AIDS, as well as basic facts about the disease and its prevention and detection, by visiting the Centers for Disease Control:
www.cdc.gov

Uncover links to online resources for HIV infection and AIDS by visiting the AIDS Education Global Information System:
www.aegis.com.

Learn about complementary therapies for COPD, cancer, and AIDS by visiting The Natural Pharmacist:
www.TNP.com

Review safe food-handling practices by visiting Healthfinder and searching for "food safety":
www.healthfinder.gov/searchoptions/ topicsaz.htm

Notes

1. G. Akner and T. Cederholm, Treatment of protein-energy malnutrition in chronic nonmalignant disorders, *American Journal of Clinical Nutrition* 74 (2001): 6–24.

2. I. Throsdottir, I. Gunnarsdottir, and B. Eriksen, Screening method evaluated by nutritional status measurements can be used to detect malnourishment in chronic obstructive pulmonary disease, *Journal of the American Dietetic Association* 101 (2001): 648–654.

3. K. Gray-Donald and coauthors, Nutritional status and mortality in chronic obstructive pulmonary disease, *American Journal of Respiratory and Critical Care Medicine* 153 (1996): 961–966.

4. K. M. Chapman and L. Winter, COPD: Using nutrition to prevent respiratory function decline, *Geriatrics* 51 (1996): 37–42.

5. A. M. Malone, Is a pulmonary enteral formula warranted for patients with pulmonary dysfunction? *Nutrition in Clinical Practice* 12 (1997): 168–171.

6. As cited in A. M. Herrington, J. D. Herrington, and C. A. Church, Pharmacologic options for the treatment of cachexia, *Nutrition in Clinical Practice* 12 (1997): 101–113.

7. D. P. Kotler and coauthors, Magnitude of body cell mass depletion and timing of death from wasting in AIDS, *American Journal of Clinical Nutrition* 50 (1989): 444–447.

8. M. Puccio and L. Nathanson, The cancer cachexia syndrome, *Seminars in Oncology* 24 (1997): 277–278.

9. P. O'Gorman, D. C. McMillan, and C. S. McArdle, Impact of weight loss, appetite, and the inflammatory response on quality of life in gastrointestinal cancer patients, *Nutrition and Cancer* 32 (1998): 76–80.

10. K. Mulligan and A. S. Bloch, Energy expenditure and protein metabolism in human immunodeficiency virus infection and cancer cachexia, *Seminars in Oncology* (supplement 6) 25 (1998): 82–91.

11. T. M. Nash, Use of anabolic agents in patients with cancer cachexia, *Support Line*, June 1999, pp. 14–18; Herrington, Herrington, and Church, 1997.

12. M. A. Richardson and coauthors, Complementary/alternative medicine use in a comprehensive cancer center and implications for oncology, *Journal of Clinical Oncology* 18 (2000): 2505–2514.

13. D. Albanes and coauthors, Alpha-tocopherol and beta-carotene supplements and lung cancer incidence in the Alpha-Tocopherol Beta-Carotene Cancer Prevention Study: Effects of base-line characteristics and study compliance, *Journal of the National Cancer Institute* 88 (1996): 1560–1570.

14. D. Labriola and R. Livingstone, Possible interactions between dietary antioxidants and chemotherapy, *Oncology* 13 (1999): 1003–1008.

15. O'Gorman, McMillan, and McArdle, 1998.

16. S. Mercadante, Parenteral versus enteral nutrition in cancer patients: Indications and practice, *Supportive Care in Cancer* 6 (1998): 85–93.

17. P. Bougnoux, n-3 polyunsaturated fatty acids and cancer, *Current Opinion in Clinical Nutrition and Metabolic Care* 2 (1999): 121–126; M. D. Barber and coauthors, Fish oil–enriched nutritional supplement attenuates progression of the acute-phase response in weight-losing patients with advanced cancer, *Journal of Nutrition* 129 (1999): 1120–1125; J. M. Daly and coauthors, Enteral nutrition with supplemental arginine, RNA, and omega-3 fatty acids in patients after operation: Immunologic, metabolic, and clinical outcome, *Surgery* 112 (1992): 56–67.

18. C. L. Rock, Factors associated with weight gain in women after diagnosis of breast cancer, *Journal of the American Dietetic Association* 99 (1999): 1212–1218, 1221.

19. Mercadante, 1998.

20. M. Marian, Cancer cachexia: Prevalence, mechanisms, and interventions, *Support Line*, April 1998, pp. 3–12.

21. P. R. Schloerb and B. S. Skikne, Oral and parenteral glutamine in bone marrow transplantation: A randomized, double-blind study, *Journal of Parenteral and Enteral Nutrition* 23 (1999): 117–122; P. R. Schloerb and M. Amare, Total parenteral nutrition with glutamine in bone marrow transplantation and other clinical applications (randomized, double-blind study), *Journal of Parenteral and Enteral Nutrition* 17 (1993): 407–413; T. R. Ziegler and coauthors, Clinical and metabolic efficacy of glutamine-supplemented parenteral nutrition after bone marrow transplantation, *Annals of Internal Medicine* 116 (1992): 821–828.

22. Centers for Disease Control and Prevention, Presented at the National HIV Prevention Conference, from **hiv.medscape.com**, site visited August 21, 2001.

23. A. Mocroft and coauthors, Changes in AIDS-defining illnesses in a London clinic, 1987–1998, *Journal of Acquired Immune Deficiency Syndrome* 21 (1999): 401–407; Centers for Disease Control and Prevention, Surveillance for AIDS-defining opportunistic illness, 1992–1997, *Morbidity and Mortality Weekly Report*, April 19, 1999.

24. K. Fisher, Wasting and lipodystrophy in patients infected with HIV: A practical approach in clinical practice, *The AIDS Reader* 11 (2001): 132–137.

25. A. L. Graber, Syndrome of lipodystrophy, hyperlipidemia, insulin resistance, and diabetes in treated patients with human immunodeficiency virus infection, *Endocrine Practice* 7 (2001): 430–437; L. A. Kosmiski and coauthors, Fat distribution and metabolic changes are strongly correlated and energy expenditure is increased in HIV lipodystrophy syndrome, *AIDS* 15 (2001): 1993–2000.

26. K. A. Lichtenstein and coauthors, Clinical assessment of HIV-associated lipodystrophy in an ambulatory population, *AIDS* 15 (2001): 1389–1398; The aetiology of antiretroviral-associated lipodystrophy remains elusive posing problems for its prevention and treatment, *Drug and Therapy Perspectives* 17 (2001): 11–15.

27. Fisher, 2001.

28. L. M. Kruse, Nutritional assessment and management for patients with HIV disease, *The AIDS Reader* 8 (1998): 121–130, available online from Medscape at **www.medscape.com**.

29. A. D. Lecours, Understanding common oral lesions associated with HIV, *Clinician Reviews* 11 (2001): 96–106.

30. M. A. Poles and coauthors, HIV-related diarrhea is multifactorial and fat malabsorption is commonly present, independently of HAART, *American Journal of Gastroenterology* 96 (2001): 1831–1837.

31. Centers for Disease Control and Prevention, The use of testosterone in AIDS wasting syndrome, *AIDS Clinical Care* 11 (1999): 25; J. C. Loss, The use of anabolic agents in HIV disease, *Support Line*, June 1999, pp. 23–28.

32. C. R. Steinhart, HIV-associated wasting in the era of HAART: A practice-based approach to diagnosis and treatment, *The AIDS Reader* 11 (2001): 557–560, 566–569.

Nutrition in Practice

■ ETHICAL ISSUES IN NUTRITION CARE ■

As Chapter 28 described, every person with cancer or an HIV infection who cannot eat an oral diet is a potential candidate for a tube feeding or total parenteral nutrition (TPN). Tube feedings and TPN can meet nutrient needs and support recovery at times when table foods cannot. Like other medical technologies, however, the availability of special nutrition support forces health care professionals and society to face **ethical** issues. Such treatments can prolong life by merely delaying death; the remaining life may be of low quality. (Terms relating to ethical issues appear in the accompanying glossary.)

Is it ever morally or legally appropriate to withhold or withdraw nutrition support?

In attempting to answer a question such as this one, ethics experts weigh such factors as:

- The client's right to make decisions concerning his or her own well-being (**autonomy**), even if withholding treatment will result in death.

- The potential benefits (**beneficence**) of the treatment versus the potential risks (**maleficence**) the treatment poses to the client's health or quality of life.

- The client's right to be fully informed of a treatment's benefits and risks in a fair and honest manner.

- The ability of caregivers or family to promote the client's well-being without selfish intent when the client cannot speak for herself or himself.

- The client's right to expect that health care professionals and caregivers will honor his or her stated wishes about health care.

When sifting through ethical dilemmas, answers rarely come easily. It may be difficult to determine whether clients truly comprehend the complexity and potential finality of their decisions, and it may be equally as difficult to determine whether a treatment will ultimately present more risks or benefits. In reality, the answers often lie tangled in personal values, charged emotions, and legal conflict.

Isn't there a moral obligation to give nutrition support whenever it will prolong life?

Not necessarily. Health care professionals readily recommend whatever form of nutrition is necessary to support clients who have any reasonable chance of recovering from a disease. Clearly, health care professionals cannot rightfully withhold nutrition support because of poor judgment or negligence. The decision of whether to feed a client becomes less clear, however, when clients have a **terminal illness,** when they are in a **persistent vegetative state,** or when they (or their caregivers) simply refuse specialized nutrition support because they feel the quality of their lives is poor.[1] Do we (as a society) allow them such choices? Are health care professionals morally and legally obligated to comply with, or to deny, such requests? Furthermore, who determines when clients are **competent** to speak for themselves? If a client is **comatose** or incompetent to make such a decision, who, if anyone, should be allowed to make life-and-death decisions for the client? These are but a few of the questions that have evolved along with advances in medical technology and nutrition support; a discipline known as "medical ethics" has developed out of the need to discuss and solve problems such as these.

Who decides what will be done in cases where there are ethical dilemmas?

Most often the client (when competent) and the client's family decide what to do in cases where there are ethical dilemmas. They often make their decisions based on extensive consultation with the physician and the health care team. Health care professionals must provide decision makers with the information they need to fully understand the disease, its treatments, the potential benefits of treatments, and the risks so that they can make a truly informed choice. When the client or client's family makes a decision that the health care team or the facility housing the client believes may make the team or facility liable for malpractice, the conflict of interests may give rise to court cases. Then the court is charged with defining the problems and solving them.

How have the courts resolved nutrition support issues?

One of the most widely publicized cases involving nutrition support issues was that of Nancy Cruzan. Cruzan was a young woman who suffered permanent and irreversible brain damage after a car crash in 1983. For eight years, she was in a persistent vegetative

Nutrition in Practice

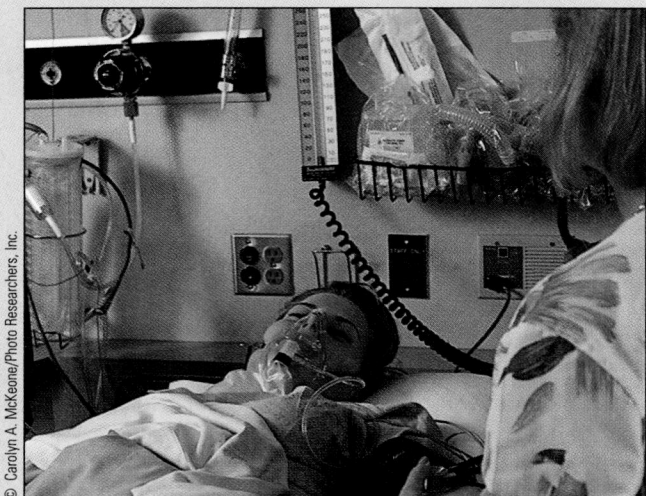

When is it morally and legally appropriate to withhold or withdraw special nutrition support?

state—awake but unaware. Her physicians and parents held no hope for her recovery, yet given food and water, she might have lived for another 30 years. Her parents requested permission to discontinue tube feeding, but the Missouri Supreme Court rejected their request. The court held that Cruzan never definitively stated her "right to die" wishes and that her parents had no **legal** right to make such a request for her. The court stated that preserving life, no matter what its quality, takes precedence over all other considerations.

Cruzan's parents appealed the decision, and in 1989, the U.S. Supreme Court agreed to hear their arguments. The Cruzans tried to convince the Supreme Court that their once independent and vivacious daughter would not want to live in a vegetative state. No one questioned that Cruzan's parents knew their daughter's wishes better than anyone and had the highest and most loving motives. The question for the Court was whether families (or anyone) can make life-and-death decisions on behalf of incompetent persons. The Supreme Court recognized that competent adults have the right to stop life-sustaining treatment, but upheld the Missouri Supreme Court's decision to require "clear and convincing" evidence that the incompetent person would refuse treatment.[2] It took still another round of court battles before additional evidence convinced the Missouri Supreme Court of Nancy Cruzan's wishes. Finally, her feeding tube was removed, and she died from dehydration two weeks later.

Additional ethical concerns are likely to arise as the population of elderly clients grows and new medical advances capable of sustaining life are developed. End-of-life decisions touch all of us, because they influence the extent to which our society views life-sustaining treatment as optional not only for our clients, but for ourselves and our families.

What is the prevailing opinion about a person's right to make decisions about life-sustaining treatments?

The emerging ethical, medical, and legal consensus seems to support the view that individual rights out-

Glossary

advance directives: the means by which competent adults record their preferences for future medical interventions. A living will and durable power of attorney are types of advance directives.

artificial feedings: parenteral and enteral nutrition; feeding by a route other than the normal ingestion of food.

autonomy: independence.
 auto = self
 nomos = law

beneficence (be-NEF-eh-sens): doing good.

comatose: in a state of deep unconsciousness from which the person cannot be aroused.

competent: having sufficient mental ability to understand a treatment, weigh its risks and benefits, and comprehend the consequences of refusing or accepting the treatment.

durable power of attorney: a legal document in which one competent adult authorizes another competent adult to make

decisions for her or him in the event of incapacitation. The phrase "durable power" means that the agent's authority survives the client's incompetence; "attorney" refers to an attorney-in-fact (not an attorney-at-law).

ethical: in accordance with moral principles or professional standards. Socrates described *ethics* as "how we ought to live."

legal: established by law.

living will: a document signed by a competent adult that specifically states whether the person wishes aggressive treatment in the event of terminal illness or irreversible coma from which he or she is not expected to recover.

maleficence (mah-LEF-eh-sens): doing harm.

persistent vegetative state: exhibiting motor reflexes but without the ability to regain cognitive behavior, communicate, or interact purposefully with the environment.

terminal illness: a progressive, irreversible disease that will lead to death in the near future.

Nutrition in Practice

weigh those of the state. Competent individuals have a legal right to refuse medical treatment—including nourishment and hydration—even when medical experts consider that treatment necessary to sustain life. Most people believe that it is acceptable to withhold nourishment and hydration when a competent person desires to forgo it, when an incompetent person has given **advance directives** about it, or when a person is in a permanently unconscious state.[3] In other words, even when treatment is lifesaving and its refusal may bring an earlier death, clients' rights remain paramount.

What are advance directives?

Advance directives allow clients to express ahead of time their preferences regarding medical treatments, including **artificial feedings,** should terminal illness, coma, or incompetency develop. One form of an advance directive is a **living will.** Another is the **durable power of attorney.** All states have statutes governing the use of advance directives; health care professionals should be aware of the regulations of the states where they work. Some states' laws specifically address the conditions under which nutrition support can be withheld or withdrawn.

What is a living will?

A living will allows a competent adult to express clear written directions regarding medical treatment in the event that the person is unable to make the necessary decisions at that time. The living will may specify that no extraordinary treatments should be administered, or alternatively, it may declare that every effort should be made to maintain life. People who prefer that nourishment and hydration be continued or discontinued need to write this specification into their living wills to ensure that their wishes are known.

What's the difference between a living will and a durable power of attorney?

A durable power of attorney allows a competent adult to designate another competent adult (usually a relative or close friend) as an agent to make health care decisions in the event of incapacitation. In essence, it says, "I give this person the right to make health care decisions on my behalf should I become unable to make them."

When people give another person a durable power of attorney, they should also have a living will. They should ensure that the person they appoint clearly understands their health care preferences, knows the contents of their living will (when available), and can be trusted to make decisions that reflect those preferences.

How well do advance directives work?

People often fail to complete advance directives, and when they do, compliance with them is not guaranteed. Advance directives are not given to the physician 62 percent of the time, and even when they are, they are frequently misinterpreted.[4] The Study to Understand Prognoses and Preferences for Outcomes and Risks of Treatments (SUPPORT) found that among people with terminal diseases who were hospitalized, only 10 percent had advance directives.[5] Physicians were aware of these directives only 25 percent of the time. Other researchers found that among elderly clients admitted to intensive care, only 5 percent had advance directives.[6] Forty percent of the clients in the study who died had received resuscitation despite advance directives that instructed health care professionals to forgo this procedure.

Can any measures be taken to promote compliance with advance directives?

People and health care professionals can both take measures to improve compliance with advance directives. People who wish to ensure that they receive medical care consistent with their individual wishes need to complete advance directives. Furthermore, they need to discuss these directives and hypothetical scenarios with their physicians, family members, and the person who holds the durable power of attorney before medical conditions arise that will necessitate others making decisions for them. They must ensure that their physician has a copy of any advanced directive and that the copy is part of their medical record. Clients should expect that physicians, other health care professionals, and health care facilities will comply with their preferences and not ignore them or merely tolerate them grudgingly. Clients should also be reminded that living wills are changeable at any time during the progression of their disease.

The person who plans ahead for future care in the case of a terminal illness or irreversible state of unconsciousness relieves others of the guilt and some of the anxiety of having to make decisions. Imagine the anguish a family member may go through in directing the health care team to stop nutrition support, knowing that doing so will hasten death. That decision is a little easier if the family member knows it is what the individual would want, or if a legal document takes the decision out of the family's hands altogether.

Health care professionals should encourage clients to complete advance directives and reassure clients that their decisions to refuse life-sustaining treatments, including nutrition support, will not mean that other care will be withheld. The client who makes such a choice still deserves meticulous physical care, pain management, and compassionate emotional support.[7] Any health care professional who is unwilling to abide by the client's stated preferences should arrange for continuing care by another equally qualified professional and then withdraw from that client's care.

Ethical questions have no easy answers, yet decisions must be made. Each case requires careful, individualized consideration. Most hospitals and extended care facilities have established ethics committees to deal with issues such as those presented here. Health care team members should ensure that their disciplines are represented on such committees and should become familiar with their profession's ethics policies and guidelines.

Nutrition on the Net

For further study of the topics of this Nutrition in Practice, access these websites.

Find updates and quick links to these and other nutrition-related sites at our website:
www.wadsworth-com/nutrition

Learn more about medical ethics by visiting the American Society of Law, Medicine, and Ethics; Choice in Dying; and the American Medical Association:
www.aslme.org, www.choices.org, and **www.ama-assn.org**

Find links to living wills at:
www.midspring.com/~scottr/will.html

Notes

1. R. DeChicco, A. Trew, and D. L. Seidner, What to do when the patient refuses a feeding tube, *Nutrition in Clinical Practice* 12 (1997): 228–230.

2. *Cruzan v. Director, Missouri Department of Health,* 497 U.S. 261, 110 S.Ct. 2841, 111 L.Ed.2d 224 (1990).

3. R. Burck, Feeding, withdrawing, and withholding: Ethical perspectives, *Nutrition in Clinical Practice* 11 (1996): 243–253.

4. D. R. Lustbader, Medical-legal issues in the ICU: When the court speaks, 30th International Educational and Scientific Symposium of the Society of Critical Care Medicine, February 10, 2001, available from Medscape at **www.medscape.com/ medscape/cno/2001/SCCM/Story.cfm?story_id=2077,** site visited March 28, 2001.

5. The SUPPORT Principal Investigators, A controlled trial to improve care for seriously ill hospitalized patients: The Study to Understand Prognoses and Preferences for Outcomes and Risks of Treatment (SUPPORT), *Journal of the American Medical Association* 274 (1995): 1591–1598.

6. M. D. Goodman, M. Tarnoff, and G. J. Slotman, Effect of advance directives on the management of elderly critically ill patients, *Critical Care Medicine* 26 (1998): 701–704.

7. R. L. Craig, Hospice care focus: Nutritional intervention, *Nutrition in Clinical Practice* (supplement) 15 (2001): 66–69; V. M. Herrmann, Ethics in nutrition support, presented at the Eighth Annual Advances and Controversies in Clinical Nutrition, Dallas, Texas, April 5–7, 1998.

Appendixes

A TABLE OF FOOD COMPOSITION

This edition of the table of food composition includes a wide variety of foods from all food groups. It is updated yearly to reflect nutrient changes for current foods, remove outdated foods, and add foods that are new to the marketplace.*

The nutrient database for this appendix is compiled from a variety of sources, including the USDA Standard Release database (Release 14), literature sources, and manufacturers' data. The USDA database provides data for a wider variety of foods and nutrients than other sources. Because laboratory analysis for each nutrient can be quite costly, manufacturers tend to provide data only for those nutrients mandated on food labels. Consequently, data for their foods are often incomplete; any missing information is designated in this table as a blank space. Keep in mind that a blank space means only that the information is unknown and should not be interpreted as a zero.

Whenever using nutrient data, remember that many factors influence the nutrient contents of foods, including the mineral content of the soil, the diet of the animal or the fertilizer of the plant, the season of harvest, the method of processing, the length and method of storage, the method of cooking, the method of analysis, and the moisture content of the sample analyzed. With so many factors involved, users must view nutrient data as a close approximation of the actual amount.

For updates, corrections, and a list of 3000 additional foods and codes found in the diet analysis software that accompanies this text, visit **www.wadsworth.com/nutrition** and click on *Diet Analysis*.

Fats   Total fats, as well as the breakdown to saturated, monounsaturated, and polyunsaturated fats, are listed in the table. The fatty acids seldom add up to the total due to rounding and to other fatty acid components that are not included in these basic categories, such as *trans*-fatty acids and glycerol. *Trans*-fatty acids can comprise a large share of the total fat in margarine and shortening (hydrogenated oils) and in any foods that include them as ingredients.

Vitamin A   The vitamin A data in this table are reported in micrograms retinol activity equivalents, if available, oth-

erwise the data are reported in micrograms retinol equivalents. (An asterisk is used to designate retinol equivalents.) In 2001, the DRI committee established retinol activity equivalents as the preferred unit of measure for vitamin A. This unit reflects recent research suggesting a new, lower rate of conversion of carotenoids to vitamin A in the body.

Vitamin E   Databases, including this one, currently report vitamin E in milligrams α-tocopherol equivalents, a measure of vitamin E activity. This measure derives from all eight naturally occurring forms of the vitamin, but recent evidence has determined that the body derives vitamin E activity only from the α-tocopherol form. The 2000 DRI values for vitamin E are based only on the α-tocopherol form. Future editions of this table will include new vitamin E values as they become available.

Bioavailability   Keep in mind that the availability of nutrients from foods depends not only on the quantity provided by a food, but also on the amount absorbed and used by the body—the bioavailability. The bioavailability of folate from fortified foods, for example, is greater than from naturally occurring sources. (Note that this appendix has been updated with data to reflect the folate fortification of grain products.) Similarly, the body can make niacin from the amino acid tryptophan, but niacin values in this table (and most databases) report preformed niacin only. Chapter 8 provides conversion factors and additional details.

Using the Table   The items in this table have been organized into several categories, which are listed at the head of each right-hand page. Page numbers have been provided, and each group has been color-coded to make it easier to find individual items.

In an effort to conserve space, the following abbreviations have been used in the food descriptions and nutrient breakdowns:

- diam = diameter
- ea = each
- enr = enriched
- f/ = from
- frzn = frozen
- g = grams
- liq = liquid
- pce = piece
- pkg = package
- w/ = with

*This food composition table has been prepared for Wadsworth Publishing Company and is copyrighted by ESHA Research in Salem, Oregon—the developer and publisher of the Food Processor and Genesis nutritional software programs. The nutritional data are supported by over 1300 references. Because the list of sources is so extensive, it is not provided here, but is available from the publisher.

- w/o = without
- t = trace
- 0 = zero (no nutrient value)
- blank space = information not available

• *Caffeine Sources* • Caffeine occurs in several plants, including the familiar coffee bean, the tea leaf, and the cocoa bean from which chocolate is made. Most human societies use caffeine regularly, most often in beverages, for its stimulant effect and flavor. Caffeine contents of beverages vary depending on the plants they are made from, the climates and soils where the plants are grown, the grind or cut size,

the method and duration of brewing, and the amounts served. The accompanying chart shows that in general, a cup of coffee contains the most caffeine; a cup of tea, less than half as much; and cocoa or chocolate, less still. As for cola beverages, they are made from kola nuts, which contain caffeine, but most of their caffeine is added, using the purified compound obtained from decaffeinated coffee beans.

The FDA lists caffeine as a multipurpose GRAS substance that may be added to foods and beverages. Drug manufacturers use caffeine in many kinds of drugs: stimulants, pain relievers, cold remedies, diuretics, and weight-loss aids.

Caffeine Content of Beverages, Foods, and Over-the-Counter Drugs

Beverages and Foods	Average (mg)	Range (mg)
Coffee (5-oz cup)		
Brewed, drip method	130	110–150
Brewed, percolator	94	64–124
Instant	74	40–108
Decaffeinated, brewed or instant	3	1–5
Tea (5-oz cup)		
Brewed, major U.S. brand	40	20–90
Brewed, imported brands	60	25–110
Instant	30	25–50
Iced (12-oz can)	70	67–76
Soft drinks (12-oz can)		
Dr. Pepper	40	
Colas and cherry cola		
Regular		30–46
Diet		2–58
Caffeine-free		0–trace
Jolt	72	
Mountain Dew, Mello Yello	52	
Fresca, Hires Root Beer, 7-Up, Sprite, Squirt, Sunkist Orange	0	
Cocoa beverage (5-oz cup)	4	2–20
Chocolate milk beverage (8 oz)	5	2–7
Milk chocolate candy (1 oz)	6	1–15
Dark chocolate, semisweet (1 oz)	20	5–35
Baker's chocolate (1 oz)	26	
Chocolate flavored syrup (1 oz)	4	

Drugs[a]	Average (mg)
Cold remedies (standard dose)	
Dristan	0
Coryban-D, Triaminicin	30
Diuretics (standard dose)	
Aqua-ban, Permathene H2Off	200
Pre-Mens Forte	100
Pain relievers (standard dose)	
Excedrin	130
Midol, Anacin	65
Aspirin, plain (any brand)	0
Stimulants	
Caffedrin, NoDoz, Vivarin	200
Weight-control aids (daily dose)	
Prolamine	280
Dexatrim, Dietac	200

Note: A pharmacologically active dose of caffeine is defined as 200 milligrams.
[a]Because products change, contact the manufacturer for an update on products you use regularly.

Table A-1

Food Composition   (Computer code number is for Wadsworth Diet Analysis program)   (For purposes of calculations, use "0" for t, <1, <.1, <.01, etc.)

A

Computer Code Number	Food Description	Measure	Wt (g)	H₂O (%)	Ener (kcal)	Prot (g)	Carb (g)	Dietary Fiber (g)	Fat (g)	Fat Breakdown (g)		
										Sat	Mono	Poly
	BEVERAGES											
	Alcoholic:											
	Beer:											
1	Regular (12 fl oz)	1½ c	356	92	146	1	13	1	0	0	0	0
2	Light (12 fl oz)	1½ c	354	95	99	1	5	0	0	0	0	0
1506	Non alcohol beer (12 fl oz)	1 ea	360	98	32	1	5	0	0	0	0	0
	Gin, rum, vodka, whiskey:											
3	80 proof	1½ fl oz	42	67	97	0	0	0	0	0	0	0
4	86 proof	1½ fl oz	42	64	105	0	<1	0	0	0	0	0
5	90 proof	1½ fl oz	42	62	110	0	0	0	0	0	0	0
	Liqueur:											
1359	Coffee liqueur, 53 proof	1½ fl oz	52	31	175	<1	24	0	<1	.1	t	.1
1360	Coffee & cream liqueur, 34 proof	1½ fl oz	47	46	154	1	10	0	7	4.5	2.1	.3
1361	Creme de menthe, 72 proof	1½ fl oz	50	28	186	0	21	0	<1	t	t	.1
	Wine, 4 fl oz:											
6	Dessert, sweet	½ c	118	72	181	<1	14	0	0	0	0	0
7	Red	½ c	118	88	85	<1	2	0	0	0	0	0
8	Rose'	½ c	118	89	84	<1	2	0	0	0	0	0
9	White medium	½ c	118	90	80	<1	1	0	0	0	0	0
1592	Nonalcoholic	1 c	232	98	14	1	3	0	0	0	0	0
1593	Nonalcoholic light	1 c	232	98	14	1	3	0	0	0	0	0
1409	Wine cooler, bottle (12 fl oz)	1½ c	340	90	170	<1	20	<1	<1	t	t	t
1595	Wine cooler, cup	1 c	227	90	113	<1	13	<1	<1	t	t	t
	Carbonated:											
10	Club soda (12 fl oz)	1½ c	355	100	0	0	0	0	0	0	0	0
11	Cola beverage (12 fl oz)	1½ c	372	89	153	0	39	0	0	0	0	0
12	Diet cola w/aspartame (12 fl oz)	1½ c	355	100	4	<1	<1	0	0	0	0	0
13	Diet soda pop w/saccharin (12 fl oz)	1½ c	355	100	0	0	<1	0	0	0	0	0
14	Ginger ale (12 fl oz)	1½ c	366	91	124	0	32	0	0	0	0	0
15	Grape soda (12 fl oz)	1½ c	372	89	160	0	42	0	0	0	0	0
16	Lemon-lime (12 fl oz)	1½ c	368	89	147	0	38	0	0	0	0	0
17	Orange (12 fl oz)	1½ c	372	88	179	0	46	0	0	0	0	0
18	Pepper-type soda (12 fl oz)	1½ c	368	89	151	0	38	0	<1	.3	0	0
19	Root beer (12 fl oz)	1½ c	370	89	152	0	39	0	0	0	0	0
20	Coffee, brewed	1 c	237	99	5	<1	1	0	<1	t	0	t
20592	Coffee, cappuccino w/lowfat milk	1½ c	244		110	8	11	0	3	2.5		
20639	Coffee, cappuccino w/whole milk	1½ c	244		140	7	11	0	7	4.5		
20668	Coffee, latte w/lowfat milk	1½ c	366		170	12	17	0	6	4		
21	Coffee, prepared from instant	1 c	238	99	5	<1	1	0	<1	t	0	t
	Fruit drinks, noncarbonated:											
22	Fruit punch drink, canned	1 c	248	88	117	0	29	<1	0	0	0	0
1358	Gatorade	1 c	241	93	60	0	15	0	0	0	0	0
23	Grape drink, canned	1 c	250	87	125	<1	32	<1	0	0	0	0
1304	Koolade sweetened with sugar	1 c	262	90	97	0	25	0	0	0	0	0
1356	Koolade sweetened with nutrasweet	1 c	240	95	43	0	11	0	0	0	0	0
26	Lemonade, frzn concentrate (6-oz can)	¾ c	219	52	396	1	103	1	<1	.1	t	.1
27	Lemonade, from concentrate	1 c	248	89	99	<1	26	<1	0	0	0	0
28	Limeade, frzn concentrate (6-oz can)	¾ c	218	50	408	<1	108	1	<1	t	t	.1
29	Limeade, from concentrate	1 c	247	89	101	0	27	<1	<1	t	t	t
24	Pineapple grapefruit, canned	1 c	250	88	118	<1	29	<1	<1	t	t	.1
25	Pineapple orange, canned	1 c	250	87	125	3	29	<1	0	0	0	0
20559	Powerade	1 c	247	92	72	0	19	0	0	0	0	0
20737	Snapple, fruit punch	1 c	252	88	110	0	29	0	0	0	0	0
20761	Snapple, tropical	1 c	252	89	110	0	27	0	0	0	0	0
	Fruit and vegetable juices: see Fruit and Vegetable sections											
	Ultra Slim Fast, ready to drink, can:											
	Chocolate Royale	1 ea	350	83	220	10	40	5	3	1	1.5	.5
	French Vanilla	1 ea	350	84	220	10	40	5	3	.5	1.5	.5
	Strawberries n' cream	1 ea	350	84	220	10	40	5	2	.5	1.5	.5
2427	Water, bottled: La Croix	1 c	236	100	0	0	0	0	0	0	0	0

Chol (mg)	Calc (mg)	Iron (mg)	Magn (mg)	Pota (mg)	Sodi (mg)	Zinc (mg)	VT-A (µg)	Thia (mg)	VT-E (mg)	Ribo (mg)	Niac (mg)	V-B6 (mg)	Fola (µg)	VT-C (mg)
0	18	.11	21	89	18	.07	0	.02	0	.09	1.61	.18	21	0
0	18	.14	18	64	11	.11	0	.03	0	.11	1.39	.12	14	0
0	25	.04	32	90	18	.04	0	.02	0	.09	1.63	.18	22	0
0	0	.02	0	1	<1	.02	0	<.01	0	<.01	<.01	0	0	0
0	0	.02	0	1	<1	.02	0	<.01	0	<.01	<.01	0	0	0
0	0	.02	0	1	<1	.02	0	<.01	0	<.01	<.01	0	0	0
0	1	.03	2	16	4	.02	0	<.01	0	.01	.07	0	0	0
7	8	.06	1	15	43	.07	20*	0	.12	.03	.04	.01	0	0
0	0	.03	0	0	2	.02	0	0	0	0	<.01	0	0	0
0	9	.28	11	109	11	.08	0	.02	0	.02	.25	0	0	0
0	9	.51	15	132	6	.11	0	.01	0	.03	.1	.04	2	0
0	9	.45	12	117	6	.07	0	<.01	0	.02	.09	.03	1	0
0	11	.38	12	94	6	.08	0	<.01	0	.01	.08	.02	0	0
0	21	.93	23	204	16	.19	0	0	0	.02	.23	.05	2	0
0	21	.93	23	204	16	.19	0	0	0	.02	.23	.05	2	0
0	19	.92	18	153	29	.2	<1	.02	.02	.02	.15	.04	4	6
0	13	.62	12	102	19	.13	<1	.01	.01	.02	.1	.03	3	4
0	18	.04	4	7	75	.35	0	0	0	0	0	0	0	0
0	11	.11	4	4	15	.04	0	0	0	0	0	0	0	0
0	14	.11	4	0	21	.28	0	.02	0	.08	0	0	0	0
0	14	.14	4	7	57	.18	0	0	0	0	0	0	0	0
0	11	.66	4	4	26	.18	0	0	0	0	0	0	0	0
0	11	.3	4	4	56	.26	0	0	0	0	0	0	0	0
0	7	.26	4	4	40	.18	0	0	0	0	.05	0	0	0
0	19	.22	4	7	45	.37	0	0	0	0	0	0	0	0
0	11	.15	0	4	37	.15	0	0	0	0	0	0	0	0
0	18	.18	4	4	48	.26	0	0	0	0	0	0	0	0
0	5	.12	12	128	5	.05	0	0	0	0	.53	0	0	0
15	250	0			110		98							2
30	250	0			105		73							2
25	400	0			170		122							4
0	7	.12	10	86	7	.07	0	0	0	<.01	.67	0	0	0
0	20	.52	5	62	55	.3	2	.05	0	.06	.05	0	2	73
0	0	.12	2	26	96	.05	0	.01	0	0	0	0	0	0
0	7	.25	10	87	2	.07	<1	.02	0	.02	.25	.05	2	40
0	42	.13	3	3	37	.08	0	0	0	<.01	<.01	0	0	31
0	17	.65	5	50	50	.26	1	.02	0	.05	.05	0	5	77
0	15	1.58	11	147	9	.17	11	.06	0	.21	.16	.05	22	39
0	7	.4	5	37	7	.1	2	.01	0	.05	.04	.01	5	10
0	11	.22	9	129	0	.09	0	.02	0	.02	.22	0	9	26
0	7	.07	2	32	5	.05	0	<.01	0	<.01	.05	0	2	7
0	17	.77	15	153	35	.15	5	.07	0	.04	.67	.1	27	115
0	12	.67	15	115	7	.15	66	.07	0	.05	.52	.12	27	56
0	0	0		32	28		0							0
0					10									0
0					10									
5	400	2.7	140	600	220	2.25	350*	.52	20	.59	7	.7	120	60
5	400	2.7	140	600	220	2.25	350*	.52	20	.59	7	.7	120	60
5	400	2.7	140	600	220	2.25	350*	.52	20	.59	7	.7	120	60
0					5									

*This value is expressed in retinol equivalents (RE). All other values are in retinol activity equivalents (RAE).

Table A–1

Food Composition   (Computer code number is for Wadsworth Diet Analysis program)   (For purposes of calculations, use "0" for t, <1, <.1, <.01, etc.)

A

Computer Code Number	Food Description	Measure	Wt (g)	H₂O (%)	Ener (kcal)	Prot (g)	Carb (g)	Dietary Fiber (g)	Fat (g)	Fat Breakdown (g)		
										Sat	Mono	Poly
	BEVERAGES—Continued											
1357	Water, bottled: Perrier (6½ fl oz)	1 ea	192	100	0	0	0	0	0	0	0	0
1594	Water, bottled: Tonic water	1½ c	366	91	124	0	32	0	0	0	0	0
	Tea:											
30	Brewed, regular	1 c	237	100	2	0	1	0	0	0	0	0
1662	Brewed, herbal	1 c	237	100	2	0	<1	0	<1	t	t	t
32	From instant, sweetened	1 c	259	91	88	<1	22	0	<1	t	t	t
31	From instant, unsweetened	1 c	237	100	2	0	<1	0	0	0	0	0
	DAIRY											
	Butter: see Fats and Oils, #158,159,160											
	Cheese, natural:											
33	Blue	1 oz	28	42	99	6	1	0	8	5.2	2.2	.2
34	Brick	1 oz	28	41	104	7	1	0	8	5.2	2.4	.2
35	Brie	1 oz	28	48	93	6	<1	0	8	4.9	2.2	.2
36	Camembert	1 oz	28	52	84	6	<1	0	7	4.3	2	.2
37	Cheddar:	1 oz	28	37	113	7	<1	0	9	5.9	2.6	.3
38	1" cube	1 ea	17	37	68	4	<1	0	6	3.6	1.6	.2
39	Shredded	1 c	113	37	455	28	1	0	37	23.8	10.6	1.1
1406	Low fat, low sodium	1 oz	28	65	48	7	1	0	2	1.2	.6	.1
	Cottage:											
2425	Fat Free	1 c	230	83	160	26	12	0	0	0	0	0
984	Low Sodium, low fat	1 c	225	83	162	28	6	0	2	1.4	.6	.1
40	Creamed, large curd	1 c	225	79	232	28	6	0	10	6.4	2.9	.3
41	Creamed, small curd	1 c	210	79	216	26	6	0	9	6	2.7	.3
42	With fruit	1 c	226	72	280	22	30	0	8	4.9	2.2	.2
43	Low fat 2%	1 c	226	79	203	31	8	0	4	2.8	1.2	.1
44	Low fat 1%	1 c	226	82	163	28	6	0	2	1.5	.7	.1
46	Cream	1 tbs	15	54	52	1	<1	0	5	3.3	1.5	.2
983	low fat	1 tbs	15	64	35	2	1	0	3	1.7	.7	.1
47	Edam	1 oz	28	42	100	7	<1	0	8	4.9	2.3	.2
48	Feta	1 oz	28	55	74	4	1	0	6	4.2	1.3	.2
49	Gouda	1 oz	28	41	100	7	1	0	8	4.9	2.2	.2
50	Gruyere	1 oz	28	33	116	8	<1	0	9	5.3	2.8	.5
51	Gorgonzola	1 oz	28	43	97	6	1	0	8	5		
1676	Limburger	1 oz	28	48	92	6	<1	0	8	4.7	2.4	.1
53	Monterey Jack	1 oz	28	41	104	7	<1	0	8	5.3	2.4	.3
54	Mozzarella, whole milk	1 oz	28	54	79	5	1	0	6	3.7	1.8	.2
55	Mozzarella, part-skim milk, low moisture	1 oz	28	49	78	8	1	0	5	3	1.4	.1
56	Muenster	1 oz	28	42	103	7	<1	0	8	5.3	2.4	.2
2422	Neufchatel	1 oz	28	62	73	3	1	0	7	4.1	1.9	.2
1399	Nonfat cheese (Kraft Singles)	1 oz	28	61	44	6	4	0	0	0	0	0
59	Parmesan, grated:	1 oz	28	18	128	12	1	0	8	5.3	2.4	.2
57	Cup, not pressed down	1 c	100	18	456	42	4	0	30	19.1	8.7	.7
58	Tablespoon	1 tbs	6	18	27	2	<1	0	2	1.1	.5	t
60	Provolone	1 oz	28	41	98	7	1	0	7	4.8	2.1	.2
61	Ricotta, whole milk	1 c	246	72	428	28	7	0	32	20.4	8.9	.9
62	Ricotta, part-skim milk	1 c	246	74	339	28	13	0	19	12.1	5.7	.6
63	Romano	1 oz	28	31	108	9	1	0	8	4.8	2.2	.2
64	Swiss	1 oz	28	37	105	8	1	0	8	5	2	.3
976	low fat	1 oz	28	60	50	8	1	0	1	.9	.4	t
	Pasteurized processed cheese products:											
65	American	1 oz	28	39	105	6	<1	0	9	5.5	2.5	.3
66	Swiss	1 oz	28	42	93	7	1	0	7	4.5	2	.2
67	American cheese food, jar	½ c	57	43	187	11	4	0	14	8.8	4.1	.4
68	American cheese spread	1 tbs	15	48	43	2	1	0	3	2	.9	.1
982	Velveeta cheese spread, low fat, low sodium, slice	1 pce	34	62	61	8	1	0	2	1.5	.7	.1
	Cream, sweet:											
69	Half & half (cream & milk)	1 c	242	81	315	7	10	0	28	17.3	8	1
70	Tablespoon	1 tbs	15	81	19	<1	1	0	2	1.1	.5	.1

Chol (mg)	Calc (mg)	Iron (mg)	Magn (mg)	Pota (mg)	Sodi (mg)	Zinc (mg)	VT-A (µg)	Thia (mg)	VT-E (mg)	Ribo (mg)	Niac (mg)	V-B6 (mg)	Fola (µg)	VT-C (mg)
0	27	0	0	0	2	0	0	0	0	0	0	0	0	0
0	4	.04	0	0	15	.37	0	0	0	0	0	0	0	0
0	0	.05	7	88	7	.05	0	0	0	.03	0	0	12	0
0	5	.19	2	21	2	.09	0	.02	0	.01	0	0	2	0
0	5	.05	5	49	8	.08	0	0	0	.05	.09	<.01	10	0
0	5	.05	5	47	7	.07	0	0	0	<.01	.09	<.01	0	0
21	148	.09	6	72	391	.74	64*	.01	.18	.11	.28	.05	10	0
26	189	.12	7	38	157	.73	85*	<.01	.14	.1	.03	.02	6	0
28	51	.14	6	43	176	.67	51*	.02	.18	.15	.11	.07	18	0
20	109	.09	6	52	236	.67	71*	.01	.18	.14	.18	.06	17	0
29	202	.19	8	27	174	.87	78*	.01	.1	.1	.02	.02	5	0
18	123	.12	5	17	106	.53	47*	<.01	.06	.06	.01	.01	3	0
119	815	.77	32	111	702	3.51	314*	.03	.41	.42	.09	.08	20	0
6	197	.2	8	31	6	.86	17*	.01	.05	.01	.02	.02	5	0
10	240	0		380	1000									0
9	137	.31	11	194	29	.85	25	.04	.25	.36	.29	.16	27	0
34	135	.31	11	189	911	.83	108*	.05	.27	.37	.28	.15	27	0
31	126	.29	10	176	851	.78	101*	.04	.25	.34	.26	.14	25	0
25	108	.25	9	151	915	.65	81*	.04	.2	.29	.23	.12	23	0
18	156	.36	14	217	918	.95	45*	.05	.14	.42	.32	.17	29	0
9	138	.32	11	194	918	.86	25*	.05	.25	.37	.29	.15	27	0
16	12	.18	1	18	44	.08	57*	<.01	.14	.03	.01	.01	2	0
8	17	.25	1	25	44	.11	33*	<.01	.07	.04	.02	.01	3	0
25	205	.12	8	53	270	1.05	71*	.01	.21	.11	.02	.02	4	0
25	138	.18	5	17	312	.81	36*	.04	.01	.24	.28	.12	9	0
32	196	.07	8	34	229	1.09	49*	.01	.1	.09	.02	.02	6	0
31	283	.05	10	23	94	1.09	84*	.02	.1	.08	.03	.02	3	0
30	170	.18			280		43*							0
25	139	.04	6	36	224	.59	88*	.02	.18	.14	.04	.02	16	0
25	209	.2	8	23	150	.84	71*	<.01	.09	.11	.03	.02	5	0
22	145	.05	5	19	104	.62	67*	<.01	.1	.07	.02	.02	2	0
15	205	.07	7	27	148	.88	53*	.01	.13	.1	.03	.02	3	0
27	201	.11	8	37	176	.79	88*	<.01	.13	.09	.03	.02	3	0
21	21	.08	2	32	112	.15	84*	<.01	.26	.05	.03	.01	3	0
7	221	0		88	398	.88	84*			.15				0
22	385	.27	14	30	521	.89	48*	.01	.22	.11	.09	.03	2	0
79	1376	.95	51	107	1862	3.19	173*	.04	.8	.39	.31	.1	8	0
5	83	.06	3	6	112	.19	10*	<.01	.05	.02	.02	.01	<1	0
19	212	.15	8	39	245	.9	74*	<.01	.1	.09	.04	.02	3	0
125	509	.93	27	258	207	2.85	330*	.03	.86	.48	.26	.11	29	0
76	669	1.08	37	308	308	3.3	278*	.05	.52	.45	.19	.05	32	0
29	298	.22	11	24	336	.72	39*	.01	.2	.1	.02	.02	2	0
26	269	.05	10	31	73	1.09	71*	.01	.14	.1	.03	.02	2	0
10	269	.05	10	31	73	1.09	18*	.01	.05	.1	.02	.02	2	0
26	172	.11	6	45	400	.84	81*	.01	.13	.1	.02	.02	2	0
24	216	.17	8	60	384	1.01	64*	<.01	.19	.08	.01	.01	2	0
36	327	.48	18	159	678	1.7	125*	.02	.4	.25	.08	.08	4	0
8	84	.05	4	36	202	.39	28*	.01	.11	.06	.02	.02	1	0
12	233	.15	8	61	2	1.13	22*	.01	.17	.13	.03	.03	3	0
89	254	.17	24	315	99	1.23	259*	.08	.27	.36	.19	.09	7	2
6	16	.01	1	19	6	.08	16*	<.01	.02	.02	.01	.01	<1	<1

*This value is expressed in retinol equivalents (RE). All other values are in retinol activity equivalents (RAE).

Table A–1

Food Composition   (Computer code number is for Wadsworth Diet Analysis program)   (For purposes of calculations, use "0" for t, <1, <.1, <.01, etc.)

A

Computer Code Number	Food Description	Measure	Wt (g)	H₂O (%)	Ener (kcal)	Prot (g)	Carb (g)	Dietary Fiber (g)	Fat (g)	Sat	Mono	Poly
	DAIRY—Continued											
71	Light, coffee or table:	1 c	240	74	468	6	9	0	46	28.8	13.4	1.7
72	Tablespoon	1 tbs	15	74	29	<1	1	0	3	1.8	.8	.1
73	Light whipping cream, liquid:	1 c	239	63	698	5	7	0	74	46.2	21.7	2.1
74	Tablespoon	1 tbs	15	63	44	<1	<1	0	5	2.9	1.4	.1
75	Heavy whipping cream, liquid:	1 c	238	58	821	5	7	0	88	54.8	25.4	3.3
76	Tablespoon	1 tbs	15	58	52	<1	<1	0	6	3.4	1.6	.2
77	Whipped cream, pressurized:	1 c	60	61	154	2	7	0	13	8.3	3.8	.5
78	Tablespoon	1 tbs	4	61	10	<1	<1	0	1	.6	.3	t
79	Cream, sour, cultured:	1 c	230	71	492	7	10	0	48	30	13.9	1.8
80	Tablespoon	1 tbs	14	71	30	<1	1	0	3	1.8	.8	.1
2423	Fat free	1 tbs	15	79	12	1	2	0	0	0	0	0
	Cream products-imitation and part dairy:											
81	Coffee whitener, frozen or liquid	1 tbs	15	77	20	<1	2	0	1	1.4	t	0
82	Coffee whitener, powdered	1 tsp	2	2	11	<1	1	0	1	.6	t	0
83	Dessert topping, frozen, nondairy:	1 c	75	50	239	1	17	0	19	16.3	1.2	.4
84	Tablespoon	1 tbs	5	50	16	<1	1	0	1	1.1	.1	t
85	Dessert topping, mix with whole milk:	1 c	80	67	151	3	13	0	10	8.5	.7	.2
86	Tablespoon	1 tbs	5	67	9	<1	1	0	1	.5	t	t
88	Dessert topping, pressurized	1 c	70	60	185	1	11	0	16	13.2	1.3	.2
87	Tablespoon	1 tbs	4	60	11	<1	1	0	1	.8	.1	t
91	Sour cream, imitation:	1 c	230	71	478	6	15	0	45	40.9	1.3	.1
92	Tablespoon	1 tbs	14	71	29	<1	1	0	3	2.5	.1	t
89	Sour dressing, part dairy:	1 c	235	75	418	8	11	0	39	31.2	4.6	1.1
90	Tablespoon	1 tbs	15	75	27	<1	1	0	2	2	.3	.1
	Milk, fluid:											
93	Whole milk	1 c	244	88	149	8	11	0	8	5.1	2.3	.3
94	2% lowfat milk	1 c	244	89	122	8	12	0	5	2.9	1.3	.2
95	2% milk solids added	1 c	245	89	125	9	12	0	5	2.9	1.4	.2
96	1% lowfat milk	1 c	244	90	102	8	12	0	3	1.6	.7	.1
97	1% milk solids added	1 c	245	90	105	9	12	0	2	1.5	.7	.1
98	Nonfat milk, vitamin A added	1 c	245	91	86	8	12	0	<1	.3	.1	t
99	Nonfat milk solids added	1 c	245	90	91	9	12	0	1	.4	.2	t
100	Buttermilk, skim	1 c	245	90	98	8	12	0	2	1.3	.6	.1
	Milk, canned:											
101	Sweetened condensed	1 c	306	27	982	24	166	0	27	16.8	7.4	1
103	Evaporated, nonfat	1 c	256	79	200	19	29	0	1	.3	.2	t
	Milk, dried:											
104	Buttermilk, sweet	1 c	120	3	464	41	59	0	7	4.3	2	.3
105	Instant, nonfat, vit A added-makes 1 qt	1 ea	91	4	326	32	47	0	1	.4	.2	t
106	Instant nonfat, vit A added, cup	1 c	68	4	243	24	35	0	<1	.3	.1	t
107	Goat milk	1 c	244	87	168	9	11	0	10	6.5	2.7	.4
108	Kefir	1 c	233	88	149	8	11	0	8			
	Milk beverages and powdered mixes:											
	Chocolate:											
109	Whole	1 c	250	82	208	8	26	2	8	5.3	2.5	.3
110	2% fat	1 c	250	84	180	8	26	1	5	3.1	1.5	.2
111	1% fat	1 c	250	84	158	8	26	1	2	1.5	.7	.1
	Chocolate-flavored beverages:											
112	Powder containing nonfat dry milk:	1 oz	28	1	101	3	22	<1	1	.7	.4	t
113	Prepared with water	1 c	275	86	138	4	30	3	2	.9	.5	t
114	Powder without nonfat dry milk:	1 oz	28	1	98	1	25	2	1	.5	.3	t
115	Prepared with whole milk	1 c	266	81	226	9	31	1	9	5.5	2.6	.3
116	Eggnog, commercial	1 c	254	74	343	10	34	0	19	11.3	5.7	.9
974	2% low-fat eggnog	1 c	254	85	191	12	17	0	8	3.7	2.7	.7
1027	Instant Breakfast, envelope, powder only:	1 ea	37	7	131	7	24	<1	1	.2	.1	.1
1028	Prepared with whole milk	1 c	281	77	280	15	36	<1	9	5.3		
1029	Prepared with 2% milk	1 c	281	78	252	15	36	<1	5	3.1		
1283	Prepared with 1% milk	1 c	281	79	233	15	36	<1	3	1.8		
1284	Prepared with nonfat milk	1 c	282	80	216	16	36	<1	1	.7		
117	Malted milk, chocolate, powder:	3 tsp	21	1	79	1	18	<1	1	.5	.2	.1
118	Prepared with whole milk	1 c	265	81	228	9	30	<1	9	5.5	2.6	.4
1661	Ovaltine with whole milk	1 c	265	81	225	9	29	<1	9	5.5	2.5	.4

Chol (mg)	Calc (mg)	Iron (mg)	Magn (mg)	Pota (mg)	Sodi (mg)	Zinc (mg)	VT-A (µg)	Thia (mg)	VT-E (mg)	Ribo (mg)	Niac (mg)	V-B6 (mg)	Fola (µg)	VT-C (mg)
158	230	.1	22	293	96	.65	437*	.08	.36	.35	.14	.08	5	2
10	14	.01	1	18	6	.04	27*	<.01	.02	.02	.01	<.01	<1	<1
265	165	.07	17	232	81	.6	705*	.06	1.43	.3	.1	.07	10	1
17	10	<.01	1	15	5	.04	44*	<.01	.09	.02	.01	<.01	1	<1
326	155	.07	17	179	90	.55	1001*	.05	1.5	.26	.09	.06	10	1
21	10	<.01	1	11	6	.03	63*	<.01	.09	.02	.01	<.01	1	<1
46	61	.03	7	88	78	.22	124*	.02	.36	.04	.04	.02	2	0
3	4	<.01		6	5	.01	8*	<.01	.02	<.01	<.01	<.01	<1	0
101	267	.14	25	331	122	.62	449*	.08	1.31	.34	.15	.04	25	2
6	16	.01	2	20	7	.04	27*	<.01	.08	.02	.01	<.01	2	<1
0	27	0		42	15									0
0	1	<.01	0	29	12	<.01	1	0	.24	0	0	0	0	0
0		.02		16	4	.01	<1	0	<.01	<.01	0	0	0	0
0	4	.09	1	13	19	.02	32	0	.14	0	0	0	0	0
0		.01		1	1	<.01	2	0	.01	0	0	0	0	0
8	72	.03	8	121	53	.22	39*	.02	.11	.09	.05	.02	3	1
<1	4	<.01		8	3	.01	2*	<.01	.01	.01	<.01	<.01	<1	<1
0	3	.01	1	13	43	.01	33*	0	.12	0	0	0	0	0
0		<.01		1	2	0	2*	0	.01	0	0	0	0	0
0	7	.9	14	370	235	2.71	0	0	.34	0	0	0	0	0
0		.05	1	22	14	.16	0	0	.02	0	0	0	0	0
12	266	.07	23	381	113	.87	5*	.09	.28	.38	.17	.04	28	2
1	17	<.01	1	24	7	.06	<1*	.01	.02	.02	.01	<.01	2	<1
34	290	.12	32	371	120	.93	76*	.09	.24	.39	.2	.1	12	2
19	298	.12	34	376	122	.95	139*	.09	.17	.4	.21	.1	12	2
20	314	.12	34	397	127	.98	140*	.1	.17	.42	.22	.11	12	2
10	300	.12	34	381	124	.95	144*	.09	.1	.41	.21	.1	12	2
10	314	.12	34	397	127	.98	145*	.1	.1	.42	.22	.11	12	2
5	301	.1	27	407	127	.98	149*	.09	.1	.34	.22	.1	12	2
5	316	.12	37	419	130	1	149*	.1	.1	.43	.22	.11	12	2
10	284	.12	27	370	257	1.03	20*	.08	.15	.38	.14	.08	12	2
104	869	.58	80	1135	389	2.88	248*	.27	.64	1.27	.64	.16	34	8
10	742	.74	69	850	294	2.3	300	.11	0	.79	.44	.14	23	3
83	1420	.36	132	1910	620	4.82	65*	.47	.48	1.89	1.05	.41	56	7
16	1120	.28	106	1551	500	4.01	646	.38	.02	1.59	.81	.31	45	5
12	837	.21	80	1159	373	3	483	.28	.01	1.19	.61	.23	34	4
27	327	.12	34	498	122	.73	137	.12	.22	.34	.68	.11	2	3
		.3	33	373	107									
30	280	.6	32	418	150	1.03	72*	.09	.22	.4	.31	.1	12	2
17	285	.6	32	423	150	1.03	143*	.09	.12	.41	.31	.1	12	2
7	288	.6	32	425	153	1.03	148*	.09	.07	.41	.32	.1	12	2
1	91	.33	23	199	141	.41	1*	.03	.04	.16	.16	.03	0	<1
3	129	.47	33	270	198	.6	<1	.04	.06	.21	.22	.04	0	1
0	10	.88	27	165	59	.43	<1	.01	.11	.04	.14	<.01	2	<1
32	301	.8	53	497	165	1.28	77*	.1	.21	.43	.32	.1	13	2
150	330	.51	48	419	137	1.17	203*	.09	.58	.48	.27	.13	3	4
194	270	.71	32	368	155	1.26	197*	.11	.58	.55	.21	.15	30	2
4	105	4.74	84	350	142	3.16	554*	.31	5.31	.07	5.27	.42	105	28
38	396	4.87	117	721	262	4.09	430*	.41	5.55	.47	5.47	.52	118	31
23	402	4.87	118	727	264	4.12	469*	.41	5.48	.48	5.48	.53	118	31
14	406	4.87	118	731	266	4.12	469*	.41	5.4	.48	5.48	.53	118	31
9	407	4.83	112	755	268	4.14	469*	.4	5.3	.42	5.47	.52	118	31
1	13	.48	15	130	53	.17	4*	.04	.08	.04	.42	.03	4	<1
34	305	.61	48	498	172	1.09	79*	.13	.26	.44	.62	.13	16	3
34	384	3.76	53	620	244	1.17	901*	.73	.32	1.26	10.9	1.02	32	34

*This value is expressed in retinol equivalents (RE). All other values are in retinol activity equivalents (RAE).

A

Table A-1

Food Composition

(Computer code number is for Wadsworth Diet Analysis program)   (For purposes of calculations, use "0" for t, <1, <.1, <.01, etc.)

Computer Code Number	Food Description	Measure	Wt (g)	H₂O (%)	Ener (kcal)	Prot (g)	Carb (g)	Dietary Fiber (g)	Fat (g)	Fat Breakdown (g) Sat	Mono	Poly
	DAIRY—Continued											
119	Malted mix powder, natural:	3 tsp	21	2	87	2	16	<1	2	.9	.4	.3
120	Prepared with whole milk	1 c	265	81	236	10	27	0	10	5.9	2.8	.6
121	Milk shakes, chocolate	1 c	166	71	211	6	34	1	6	3.8	1.8	.2
122	Milk shakes, vanilla	1 c	166	75	184	6	30	1	5	3.1	1.4	.2
	Milk desserts:											
134	Custard, baked	1 c	282	79	296	14	30	0	13	6.6	4.3	1
1548	Low-fat frozen dessert bars	1 ea	81	72	88	2	19	0	1	.2	.1	.4
	Ice cream, vanilla (about 10% fat):											
124	Hardened	1 c	132	61	265	5	31	0	14	9	4.2	.5
126	Soft serve	1 c	172	60	370	7	38	0	22	12.9	6	.8
	Ice cream, rich vanilla (16% fat):											
128	Hardened	1 c	148	57	357	5	33	0	24	14.8	6.9	.9
1724	Ben & Jerry's	½ c	108	60	250	4	22	0	16	11		
	Ice milk, vanilla (about 4% fat):											
130	Hardened	1 c	132	68	183	5	30	0	6	3.5	1.6	.2
131	Soft serve (about 3.3% fat)	1 c	176	70	222	9	38	0	5	2.9	1.3	.2
	Pudding, canned (5 oz can = .55 cup):											
135	Chocolate	1 ea	142	69	189	4	32	1	6	1	2.4	2
136	Tapioca	1 ea	142	74	169	3	27	<1	5	.9	2.2	1.9
137	Vanilla	1 ea	142	71	185	3	31	<1	5	.8	2.2	1.9
	Puddings, dry mix with whole milk:											
138	Chocolate, instant	1 c	294	74	326	9	55	3	9	5.4	2.7	.5
139	Chocolate, regular, cooked	1 c	284	74	315	9	51	3	10	5.9	2.8	.4
140	Rice, cooked	1 c	288	72	351	9	60	<1	8	5.1	2.3	.3
141	Tapioca, cooked	1 c	282	74	321	8	55	0	8	5.1	2.3	.3
142	Vanilla, instant	1 c	284	73	324	8	56	0	8	4.9	2.4	.4
143	Vanilla, regular, cooked	1 c	280	75	311	8	52	0	8	5.1	2.4	.4
133	Sherbet (2% fat)	1 c	198	66	273	2	60	0	4	2.3	1	.2
20440	Rice Milk	1 c	245	89	120	<1	25	0	2	.2	1.3	.3
20590	Rice/Soy Milk, blend	1 c	241	88	120	7	18	0	3	.5		
144	Soy Milk	1 c	245	93	81	7	4	3	5	.5	.8	2
2301	Soy Milk, fortified, fat free	1 c	240	88	110	6	22	1	0	0	0	0
	Yogurt, fat free:											
2851	Strawberry, container	1 ea	227	86	120	8	22	0	0	0	0	0
2424	Vanilla	1 c	245	76	223	12	43	0	<1	.3	.1	t
	Yogurt, frozen, low-fat											
1584	Cup 1 c	144	65	229	6	35	0	8	4.9	2.3	.3	
1512	Scoop	1 ea	79	74	78	4	15	0	<1	.1	t	t
	Yogurt, lowfat:											
1172	Fruit added with low-calorie sweetener	1 c	241	86	122	11	19	1	<1	.2	.1	t
145	Fruit added	1 c	245	74	250	11	47	0	3	1.7	.7	.1
146	Plain	1 c	245	85	154	13	17	0	4	2.4	1	.1
147	Vanilla or coffee flavor	1 c	245	79	208	12	34	0	3	2	.8	.1
148	Yogurt, made with nonfat milk	1 c	245	85	137	14	19	0	<1	.3	.1	t
149	Yogurt, made with whole milk	1 c	245	88	149	8	11	0	8	5.1	2.2	.2
	EGGS											
	Raw, large:											
150	Whole, without shell	1 ea	50	75	74	6	1	0	5	1.5	1.9	.7
151	White	1 ea	33	88	16	3	<1	0	0	0	0	0
152	Yolk 1 ea	17	49	61	3	<1	0	5	1.6	2	.7	
	Cooked:											
153	Fried in margarine	1 ea	46	69	91	6	1	0	7	1.9	2.7	1.3
154	Hard-cooked, shell removed	1 ea	50	75	77	6	1	0	5	1.6	2	.7
155	Hard-cooked, chopped	1 c	136	75	211	17	2	0	14	4.4	5.5	1.9
156	Poached, no added salt	1 ea	50	75	74	6	1	0	5	1.5	1.9	.7
157	Scrambled with milk & margarine	1 ea	61	73	101	7	1	0	7	2.2	2.9	1.3
1681	Egg substitute, liquid:	½ c	126	83	106	15	1	0	4	.8	1.1	2
1254	Egg Beaters, Fleischmann's	½ c	122		60	12	2	0	0	0	0	0
1262	Egg substitute, liquid, prepared	½ c	105	80	107	11	2	0	6	1.1	2.1	2.1

A

Chol (mg)	Calc (mg)	Iron (mg)	Magn (mg)	Pota (mg)	Sodi (mg)	Zinc (mg)	VT-A (µg)	Thia (mg)	VT-E (mg)	Ribo (mg)	Niac (mg)	V-B6 (mg)	Fola (µg)	VT-C (mg)
4	63	.15	19	159	104	.21	18*	.11	.08	.19	1.1	.09	10	1
37	355	.26	53	530	223	1.14	95*	.2	.32	.59	1.31	.19	21	3
22	188	.51	28	332	161	.68	38*	.1	.12	.41	.27	.08	7	1
18	203	.15	20	289	136	.6	53*	.07	.1	.3	.31	.09	5	1
245	316	.85	39	431	217	1.49	169*	.09	.68	.64	.24	.14	28	1
1	81	.04	9	107	44	.26	38	.03	.07	.11	.06	.03	3	1
58	169	.12	18	263	106	.91	154*	.05	0	.32	.15	.06	7	1
157	225	.36	21	304	105	.89	265*	.08	.64	.31	.16	.08	15	1
90	173	.07	16	235	83	.59	272*	.06	0	.24	.12	.06	7	1
75	100	.36			60		150*							0
18	183	.13	20	279	112	.58	62*	.08	0	.35	.12	.09	8	1
21	276	.11	25	389	123	.93	51*	.09	0	.35	.21	.08	11	2
4	128	.72	30	256	183	.6	16*	.04	.17	.22	.49	.04	4	3
1	119	.33	11	138	226	.38	0	.03	.13	.14	.44	.03	4	1
10	125	.18	11	160	192	.35	9*	.03	.17	.2	.36	.02	0	0
32	300	.85	53	488	835	1.23	62*	.1	.18	.41	.28	.11	12	3
34	315	1.02	43	463	293	1.28	74*	.09	.17	.49	.29	.1	11	2
32	297	1.09	37	372	314	1.09	58*	.22	.17	.4	1.28	.1	11	2
34	293	.17	34	372	341	.96	76*	.08	.23	.4	.21	.11	11	2
31	287	.2	34	364	812	.94	71*	.09	.17	.39	.21	.1	11	2
34	300	.14	36	381	448	.98	76*	.08	.17	.4	.21	.09	11	2
12	107	.28	16	190	91	.95	28*	.05	.16	.15	.12	.05	10	6
0	20	.2	10	69	86	.24	<1	.08	1.76	.01	1.91	.04	91	1
0	13	1.08	40	270	85	.9	0	.09		.1	.4	.08	26	
0	10	1.42	47	345	29	.56	4	.39	.02	.17	.36	.1	5	0
0	400	1.44		20	60		0	.07			.1		3	0
5	350	0		380	160		0							5
4	436	.21	42	559	168	2.13	4	.1	.01	.52	.27	.11	27	2
3	206	.43	20	304	125	.6	82*	.05	.07	.32	.41	.11	9	1
1	137	.07	13	175	53	.67	1	.03	<.01	.16	.08	.04	8	1
3	369	.61	41	550	139	1.83	6*	.1	.17	.45	.5	.11	32	26
10	372	.17	37	478	142	1.81	27*	.09	.07	.44	.23	.1	22	2
15	448	.2	42	573	172	2.18	39*	.11	.1	.52	.28	.12	27	2
12	419	.17	39	537	162	2.03	32*	.1	.07	.49	.26	.11	27	2
5	488	.22	47	625	189	2.38	5*	.12	0	.57	.3	.13	29	2
32	296	.12	29	380	113	1.45	73*	.07	.22	.35	.18	.08	17	1
213	24	.72	5	60	63	.55	95	.03	.52	.25	.04	.07	23	0
0	2	.01	4	47	54	<.01	0	<.01	0	.15	.03	<.01	1	0
218	23	.6	2	16	7	.53	99	.03	.54	.11	<.01	.07	25	0
211	25	.72	5	61	162	.55	114	.03	.75	.24	.03	.07	17	0
212	25	.59	5	63	62	.52	84	.03	.52	.26	.03	.06	22	0
577	68	1.62	14	171	169	1.43	228	.09	1.43	.7	.09	.16	60	0
212	24	.72	5	60	140	.55	95	.02	.52	.21	.03	.06	17	0
215	43	.73	7	84	171	.61	119*	.03	.8	.27	.05	.07	18	<1
1	67	2.65	11	416	223	1.64	272	.14	.62	.38	.14	<.01	19	0
0	40	2.16		170	250	1.2	120*		1.61	1.7		.16	64	0
1	82	1.85	11	337	201	1.25	233*	.09	.83	.29	.12	.01	11	<1

*This value is expressed in retinol equivalents (RE). All other values are in retinol activity equivalents (RAE).

Table A–1

Food Composition

(Computer code number is for Wadsworth Diet Analysis program)   (For purposes of calculations, use "0" for t, <1, <.1, <.01, etc.)

Computer Code Number	Food Description	Measure	Wt (g)	H₂O (%)	Ener (kcal)	Prot (g)	Carb (g)	Dietary Fiber (g)	Fat (g)	Fat Breakdown (g) Sat	Mono	Poly
	FATS AND OILS											
158	Butter: Stick	½ c	114	16	817	1	<1	0	92	57.5	26.7	3.4
159	Tablespoon:	1 tbs	14	16	100	<1	<1	0	11	7.1	3.3	.4
8025	Unsalted	1 tbs	14	18	100	<1	<1	0	11	7.1	3.3	.4
160	Pat (about 1 tsp)	1 ea	5	16	36	<1	<1	0	4	2.5	1.2	.2
1682	Whipped	1 tsp	3	16	21	<1	<1	0	2	1.5	.7	.1
	Fats, cooking:											
1363	Bacon fat	1 tbs	14		125	0	0	0	14	6.3	5.9	1.1
1362	Beef fat/tallow	1 c	205	0	1849	0	0	0	205	102	85.7	8.2
1364	Chicken fat	1 c	205		1845	0	0	0	205	61.1	91.6	42.8
161	Vegetable shortening:	1 c	205	0	1812	0	0	0	205	51.3	91.2	53.5
162	Tablespoon	1 tbs	13	0	115	0	0	0	13	3.2	5.8	3.4
163	Lard:	1 c	205	0	1849	0	0	0	205	81.1	87	28.3
164	Tablespoon	1 tbs	13	0	117	0	0	0	13	5.1	5.5	1.8
	Margarine:											
165	Imitation (about 40% fat), soft:	1 c	232	58	800	1	1	0	90	17.9	36.4	32
166	Tablespoon	1 tbs	14	58	48	<1	<1	0	5	1.1	2.2	1.9
167	Regular, hard (about 80% fat):	½ c	114	16	820	1	1	0	92	18	40.8	29
168	Tablespoon	1 tbs	14	16	101	<1	<1	0	11	2.2	5	3.6
169	Pat	1 ea	5	16	36	<1	<1	0	4	.8	1.8	1.3
170	Regular, soft (about 80% fat):	1 c	227	16	1625	2	1	0	183	31.3	64.7	78.5
171	Tablespoon	1 tbs	14	16	100	<1	<1	0	11	1.9	4	4.8
2056	Saffola, unsalted	1 tbs	14	20	100	0	0	0	11	2	3	4.5
2057	Saffola, reduced fat	1 tbs	14	37	60	0	0	0	8	1.3	2.7	4.4
172	Spread (about 60% fat), hard:	1 c	227	37	1225	1	0	0	138	32	59	41.1
173	Tablespoon	1 tbs	14	37	76	<1	0	0	9	2	3.6	2.5
174	Pat	1 ea	5	37	27	<1	0	0	3	.7	1.2	1
175	Spread (about 60% fat), soft:	1 c	227	37	1225	1	0	0	138	29.1	71.5	31.3
176	Tablespoon	1 tbs	14	37	76	<1	0	0	9	1.8	4.4	1.9
2160	Touch of Butter (47% fat)	1 tbs	14		60	0	0	0	7	1.5	3.1	1.5
	Oils:											
1585	Canola:	1 c	218	0	1927	0	0	0	218	15.5	128	64.5
1586	Tablespoon	1 tbs	14	0	124	0	0	0	14	1	8.2	4.1
177	Corn:	1 c	218	0	1927	0	0	0	218	27.7	52.8	128
178	Tablespoon	1 tbs	14	0	124	0	0	0	14	1.8	3.4	8.2
179	Olive:	1 c	216	0	1909	0	0	0	216	29.2	159	18.1
180	Tablespoon	1 tbs	14	0	124	0	0	0	14	1.9	10.3	1.2
1683	Olive, extra virgin	1 tbs	14		126	0	0	0	14	2	10.8	1.3
181	Peanut:	1 c	216	0	1909	0	0	0	216	36.5	99.8	69.1
182	Tablespoon	1 tbs	14	0	124	0	0	0	14	2.4	6.5	4.5
183	Safflower:	1 c	218	0	1927	0	0	0	218	13.5	31.3	163
184	Tablespoon	1 tbs	14	0	124	0	0	0	14	.9	2	10.4
185	Soybean:	1 c	218	0	1927	0	0	0	218	31.4	50.8	126
186	Tablespoon	1 tbs	14	0	124	0	0	0	14	2	3.3	8.1
187	Soybean/cottonseed:	1 c	218	0	1927	0	0	0	218	39.2	64.3	105
188	Tablespoon	1 tbs	14	0	124	0	0	0	14	2.5	4.1	6.7
189	Sunflower:	1 c	218	0	1927	0	0	0	218	22.5	42.5	143
190	Tablespoon	1 tbs	14	0	124	0	0	0	14	1.4	2.7	9.2
	Salad dressings/sandwich spreads:											
191	Blue cheese, regular	1 tbs	15	32	76	1	1	0	8	1.5	1.8	4.2
1040	Low calorie	1 tbs	15	80	15	1	<1	0	1	.4	.3	.4
1684	Caesar's	1 tbs	12		55	1	<1	<1	5	.9		
192	French, regular	1 tbs	16	38	69	<1	3	0	7	1.5	1.3	3.5
193	Low calorie	1 tbs	16	69	21	<1	3	0	1	.1	.2	.5
194	Italian, regular	1 tbs	15	38	70	<1	2	0	7	1	1.7	4.2
195	Low calorie	1 tbs	15	82	16	<1	1	<1	1	.2	.3	.9
	Kraft, Deliciously Right:											
2150	1000 Island	1 tbs	16	64	34	0	3	0	2	.2		
2153	Bacon & tomato	1 tbs	16		31	1	2	0	3	.5		
2154	Cucumber ranch	1 tbs	16	76	31	0	1	0	3	.5		
2151	French	1 tbs	16		25	0	3	0	1	.2		
2152	Ranch	1 tbs	16		52	0	3	0	5	.8		

Chol (mg)	Calc (mg)	Iron (mg)	Magn (mg)	Pota (mg)	Sodi (mg)	Zinc (mg)	VT-A (µg)	Thia (mg)	VT-E (mg)	Ribo (mg)	Niac (mg)	V-B6 (mg)	Fola (µg)	VT-C (mg)
250	27	.18	2	30	942	.06	860*	.01	1.8	.04	.05	<.01	3	0
31	3	.02		4	116	.01	106*	<.01	.22	<.01	.01	0	<1	0
31	3	.02		4	2	.01	106*	<.01	.22	<.01	.01	0	<1	0
11	1	.01		1	41	<.01	38*	0	.08	<.01	<.01	0	<1	0
7	1	<.01		1	25	<.01	23*	0	.05	<.01	<.01	0	<1	0
14		<.01			76	<.01	0	0	.31	0	0	0	0	0
223	0	0	0	0	0	0	0	0	5.54	0	0	0	0	0
174	0	0	0	0	0	0	0	0	5.54	0	0	0	0	0
0	0	0	0	0	0	0	0	0	17	0	0	0	0	0
0	0	0	0	0	0	0	0	0	1.08	0	0	0	0	0
195		0			<1	.23	0*	0	2.46	0	0	0	0	0
12		0			<1	.01	0*	0	.16	0	0	0	0	0
0	42	0	5	58	2227	0	1853*	.01	5.41	.05	.03	.01	2	<1
0	3	0		3	134	0	112*	<.01	.33	<.01	<.01	<.01	<1	<1
0	34	.07	3	48	1075	0	911*	.01	14.6	.04	.03	.01	1	<1
0	4	.01		6	132	0	112*	<.01	1.79	<.01	<.01	<.01	<1	<1
0	1	<.01		2	47	0	40*	<.01	.64	<.01	<.01	0	<1	<1
0	61	0	5	86	2449	0	1813*	.02	27.2	.07	.04	.02	2	<1
0	4	0		5	151	0	112*	<.01	1.68	<.01	<.01	<.01	<1	<1
	0	0			0		51*							0
	0	0			115		51*							0
0	48	0	5	68	2256	0	1813*	.02	11.4	.06	.04	.01	2	<1
0	3	0		4	139	0	112*	<.01	.7	<.01	<.01	<.01	<1	<1
0	1	0		1	50	0	40*	0	.25	<.01	<.01	0	<1	<1
0	48	0	5	68	2256	0	1813*	.02	20.5	.06	.04	.01	2	<1
0	3	0		4	139	0	112*	<.01	1.26	<.01	<.01	<.01	<1	<1
0	0	0		0	110	0	100*		1.27					0
0	0	0	0	0	0	0	0	0	45.7	0	0	0	0	0
0	0	0	0	0	0	0	0	0	2.93	0	0	0	0	0
0	0	0	0	0	0	0	0	0	46	0	0	0	0	0
0	0	0	0	0	0	0	0	0	2.96	0	0	0	0	0
0	0	.82	0	0	0	.13	0	0	26.8	0	0	0	0	0
0	0	.05	0	0	0	.01	0	0	1.74	0	0	0	0	0
									1.74					
0	0	.06	0	0	0	.02	0	0	27.9	0	0	0	0	0
0	0	<.01	0	0	0	<.01	0	0	1.81	0	0	0	0	0
0	0	0	0	0	0	0	0	0	93.9	0	0	0	0	0
0	0	0	0	0	0	0	0	0	6.03	0	0	0	0	0
0	0	.04	0	0	0	0	0	0	39.7	0	0	0	0	0
0	0	<.01	0	0	0	0	0	0	2.55	0	0	0	0	0
0	0	0	0	0	0	0	0	0	61.5	0	0	0	0	0
0	0	0	0	0	0	0	0	0	3.95	0	0	0	0	0
0	0	0	0	0	0	0	0	0	110	0	0	0	0	0
0	0	0	0	0	0	0	0	0	7.08	0	0	0	0	0
3	12	.03	0	6	164	.04	10*	<.01	1.4	.01	.01	.01	1	<1
<1	13	.07	1	1	180	.04	<1	<.01	.13	.01	.01	<.01	<1	<1
12	22	.2	3	20	210	.12	5*	<.01	.72	.02	.49	.01	2	1
0	2	.06	0	13	219	.01	10	<.01	1.35	<.01	0	<.01	1	0
0	2	.06	0	13	126	.03	10	0	.19	0	0	0	0	0
0	1	.03	0	2	118	.02	4*	<.01	1.55	<.01	0	<.01	1	0
1		.03	0	2	118	.02	0	0	.22	0	0	0	0	0
5	0	0		29	165		0							0
1	0	0		21	155		0		.75					0
0	0	0		10	248		0							0
0	0	0		7	130		50*		.42					0
0	0	0		5	165		0*		1.31					0

*This value is expressed in retinol equivalents (RE). All other values are in retinol activity equivalents (RAE).

Table A–1

Food Composition   (Computer code number is for Wadsworth Diet Analysis program)   (For purposes of calculations, use "0" for t, <1, <.1, <.01, etc.)

Computer Code Number	Food Description	Measure	Wt (g)	H₂O (%)	Ener (kcal)	Prot (g)	Carb (g)	Dietary Fiber (g)	Fat (g)	Fat Breakdown (g) Sat	Mono	Poly
	FATS AND OILS—Continued											
199	Mayo type, regular	1 tbs	15	40	58	<1	4	0	5	.7	1.3	2.7
1030	Low calorie	1 tbs	14	54	36	<1	3	0	3	.4	.6	1.4
	Mayonnaise:											
197	Imitation, low calorie	1 tbs	15	63	35	<1	2	0	3	.5	.7	1.6
196	Regular (soybean)	1 tbs	14	15	100	<1	<1	0	11	1.6	3.2	5.8
1488	Regular, low calorie, low sodium	1 tbs	14	63	32	<1	2	0	3	.5	.6	1.5
1493	Regular, low calorie	1 tbs	15	63	35	<1	2	0	3	.5	.7	1.6
198	Ranch, regular	1 tbs	15	39	80	0	<1	0	8	1.2		
2251	Low calorie	1 tbs	14	72	29	<1	1	<1	3	.4	.3	.8
1685	Russian	1 tbs	15	34	74	<1	2	0	8	1.1	1.8	4.4
1502	Salad dressing, low calorie, oil free	1 tbs	15	88	4	<1	1	<1	<1	0	0	0
	Salad dressing, no cholesterol											
1605	Miracle Whip	1 tbs	15	57	48	0	2	0	4	.6	1	2.5
203	Salad dressing, from recipe, cooked	1 tbs	16	69	25	1	2	0	2	.5	.6	.3
200	Tartar sauce, regular	1 tbs	14	34	74	<1	1	<1	8	1.5	2.6	4.1
1503	Low calorie	1 tbs	14	63	31	<1	2	<1	3	.4	.6	1.4
201	Thousand island, regular	1 tbs	16	46	60	<1	2	0	6	1	1.3	3.2
202	Low calorie	1 tbs	15	69	24	<1	2	<1	2	.2	.4	.9
204	Vinegar and oil	1 tbs	16	47	72	0	<1	0	8	1.5	2.4	3.9
	Wishbone:											
2180	Creamy Italian, lite	1 tbs	15	72	26	<1	2		2	.4	.9	.7
2166	Italian, lite	1 tbs	16	90	6	0	1		<1	0	.2	.1
8427	Ranch, lite	1 tbs	15	56	50	0	2	0	4	.7		
	FRUITS and FRUIT JUICES											
	Apples:											
	Fresh, raw, with peel:											
205	2¾" diam (about 3 per lb w/cores)	1 ea	138	84	81	<1	21	4	<1	.1	t	.1
206	3¼" diam (about 2 per lb w/cores)	1 ea	212	84	125	<1	32	6	1	.1	t	.2
207	Raw, peeled slices	1 c	110	84	63	<1	16	2	<1	.1	t	.1
208	Dried, sulfured	10 ea	64	32	156	1	42	6	<1	t	t	.1
209	Apple juice, bottled or canned	1 c	248	88	117	<1	29	<1	<1	t	t	.1
210	Applesauce, sweetened	1 c	255	80	194	<1	51	3	<1	.1	t	.1
211	Applesauce, unsweetened	1 c	244	88	105	<1	27	3	<1	t	t	t
	Apricots:											
212	Raw, w/o pits (about 12 per lb w/pits)	3 ea	105	86	50	1	12	3	<1	t	.2	.1
	Canned (fruit and liquid):											
213	Heavy syrup	1 c	240	78	199	1	51	4	<1	t	.1	t
214	Halves	3 ea	120	78	100	1	26	2	<1	t	t	t
215	Juice pack	1 c	244	87	117	2	30	4	<1	t	t	t
216	Halves	3 ea	108	87	52	1	13	2	<1	t	t	t
217	Dried, halves	10 ea	35	31	83	1	22	3	<1	t	.1	t
218	Dried, cooked, unsweetened, w/liquid	1 c	250	76	213	3	55	8	<1	t	.2	.1
219	Apricot nectar, canned	1 c	251	85	141	1	36	2	<1	t	.1	t
	Avocados, raw, edible part only:											
220	California	1 ea	173	73	306	4	12	8	30	4.5	19.4	3.5
221	Florida	1 ea	304	80	340	5	27	16	27	5.3	14.8	4.5
222	Mashed, fresh, average	1 c	230	74	370	5	17	11	35	5.6	22.1	4.5
	Bananas, raw, without peel:											
223	Whole, 8¾" long (175g w/peel)	1 ea	118	74	109	1	28	3	1	.2	t	.1
224	Slices	1 c	150	74	138	2	35	4	1	.3	.1	.1
1285	Bananas, dehydrated slices	½ c	50	3	173	2	44	4	1	.3	.1	.2
225	Blackberries, raw	1 c	144	86	75	1	18	8	1	t	.1	.3
	Blueberries:											
226	Fresh	1 c	145	85	81	1	20	4	1	t	.1	.2
227	Frozen, sweetened	10 oz	284	77	230	1	62	6	<1	t	.1	.2
228	Frozen, thawed	1 c	230	77	186	1	50	5	<1	t	t	.1
3239	Breadfruit	1 c	220	71	227	2	60	11	<1	.1	.1	.1
	Cherries:											
229	Sour, red pitted, canned water pack	1 c	244	90	88	2	22	3	<1	.1	.1	.1
230	Sweet, red pitted, raw	10 ea	68	81	49	1	11	2	1	.1	.2	.2
231	Cranberry juice cocktail, vitamin C added	1 c	253	85	144	0	36	<1	<1	t	t	.1
1411	Cranberry juice, low calorie	1 c	237	95	45	0	11	0	0	0	0	0
232	Cranberry-apple juice, vitamin C added	1 c	245	83	164	<1	42	<1	0	0	0	0

Chol (mg)	Calc (mg)	Iron (mg)	Magn (mg)	Pota (mg)	Sodi (mg)	Zinc (mg)	VT-A (µg)	Thia (mg)	VT-E (mg)	Ribo (mg)	Niac (mg)	V-B6 (mg)	Fola (µg)	VT-C (mg)
4	2	.03		1	107	.03	13*	<.01	.6	<.01	<.01	<.01	1	0
4	2	.03		1	99	.02	9	<.01	.6	<.01	0	<.01	1	0
4	0	0	0	1	75	.02	0	0	.96	0	0	0	0	0
8	3	.07	5		79	.02	12*	0	.32	0	<.01	.08	1	0
3	0	0	0	1	15	.01	1	0	.53	<.01	0	0	<1	0
4	0	0	0	1	75	.02	0	0	.96	0	0	0	0	0
5	0	0			105		0							0
5	5	.03		5	118		1*		.7					<1
3	3	.09		24	130	.06	31*	.01	1.53	.01	.09	<.01	1	1
0	1	.04	2	7	256	<.01	<1	0	0	0	<.01	<.01	<1	<1
0	0	<.01	0	0	102	0	2	0	.64	0	0	0	0	0
9	13	.08	1	19	117	.06	20*	.01	.3	.02	.04	<.01	1	<1
7	3	.13		11	99	.02	9*	<.01	2.24	<.01	0	<.01	1	<1
3	1	.06		4	82	.02	1	0	.84	<.01	.01	<.01	<1	<1
4	2	.1		18	112	.02	15*	<.01	.18	<.01	<.01	<.01	1	0
2	2	.09		17	150	.02	14	<.01	.18	<.01	0	<.01	1	0
0	0	0	0	1	<1	0	0	0	1.41	0	0	0	0	0
<1	0	0			148		0	0	.56	0	0			0
0	1	0			255		2*	0	.24	0	0			<1
2	0	0			120		0							0
0	10	.25	7	159	0	.05	3	.02	.44	.02	.11	.07	4	8
0	15	.38	11	244	0	.08	5	.04	.68	.03	.16	.1	6	12
0	4	.08	3	124	0	.04	2	.02	.09	.01	.1	.05	0	4
0	9	.9	10	288	56	.13	0	0	.35	.1	.59	.08	0	2
0	17	.92	7	295	7	.07	<1	.05	.02	.04	.25	.07	0	2
0	10	.89	8	156	8	.1	1	.03	.03	.07	.48	.07	3	4
0	7	.29	7	183	5	.07	4	.03	.02	.06	.46	.06	2	3
0	15	.57	8	311	1	.27	137	.03	.93	.04	.63	.06	9	10
0	22	.72	17	336	10	.26	148	.05	2.14	.05	.9	.13	5	7
0	11	.36	8	168	5	.13	74	.02	1.07	.03	.45	.06	2	4
0	29	.73	24	403	10	.27	206	.04	2.17	.05	.84	.13	5	12
0	13	.32	11	178	4	.12	91	.02	.96	.02	.37	.06	2	5
0	16	1.65	16	482	3	.26	127	<.01	.52	.05	1.05	.05	3	1
0	40	4.18	42	1222	7	.65	295	.01	1.25	.07	2.36	.28	0	4
0	18	.95	13	286	8	.23	166	.02	.2	.03	.65	.05	3	2
0	19	2.04	71	1096	21	.73	53	.19	2.32	.21	3.32	.48	114	14
0	33	1.61	103	1483	15	1.28	93	.33	2.37	.37	5.84	.85	161	24
0	25	2.35	90	1377	23	.97	70	.25	3.08	.28	4.42	.64	143	18
0	7	.37	34	467	1	.19	5	.05	.32	.12	.64	.68	22	11
0	9	.46	43	594	1	.24	6	.07	.4	.15	.81	.87	28	14
0	11	.57	54	746	1	.3	8	.09	0	.12	1.4	.22	7	3
0	46	.82	29	282	0	.39	11	.04	1.02	.06	.58	.08	49	30
0	9	.25	7	129	9	.16	7	.07	1.45	.07	.52	.05	9	19
0	17	1.11	6	170	3	.17	6	.06	2.02	.15	.72	.17	20	3
0	14	.9	5	138	2	.14	5	.05	1.63	.12	.58	.14	16	2
0	37	1.19	55	1078	4	.26	4	.24	2.46	.07	1.98	.22	31	64
0	27	3.34	15	239	17	.17	91	.04	.32	.1	.43	.11	19	5
0	10	.26	7	152	0	.04	7	.03	.09	.04	.27	.02	3	5
0	8	.38	5	45	5	.18	<1	.02	0	.02	.09	.05	0	90
0	21	.09	5	52	7	.05	<1	.02	0	.02	.08	.04	0	76
0	17	.15	5	66	5	.1	<1	.01	0	.05	.15	.05	0	78

*This value is expressed in retinol equivalents (RE). All other values are in retinol activity equivalents (RAE).

Table A–1

Food Composition　　(Computer code number is for Wadsworth Diet Analysis program)　　(For purposes of calculations, use "0" for t, <1, <.1, <.01, etc.)

A

Computer Code Number	Food Description	Measure	Wt (g)	H₂O (%)	Ener (kcal)	Prot (g)	Carb (g)	Dietary Fiber (g)	Fat (g)	Sat	Mono	Poly
	FRUITS and FRUIT JUICES—Continued											
233	Cranberry sauce, canned, strained	1 c	277	61	418	1	108	3	<1	t	.1	.2
234	Dates, whole, without pits	10 ea	83	22	228	2	61	6	<1	.2	.1	t
235	Dates, chopped	1 c	178	22	490	4	131	13	1	.3	.3	.1
236	Figs, dried	10 ea	190	28	485	6	124	23	2	.4	.5	1.1
	Fruit cocktail, canned, fruit and liq:											
237	Heavy syrup pack	1 c	248	80	181	1	47	2	<1	t	t	.1
238	Juice pack	1 c	237	87	109	1	28	2	<1	t	t	t
	Grapefruit:											
	Raw 3¾" diam (half w/rind = 241g)											
239	Pink/red, half fruit, edible part	1 ea	123	91	37	1	9	2	<1	t	t	t
240	White, half fruit, edible part	1 ea	118	90	39	1	10	1	<1	t	t	t
241	Canned sections with light syrup	1 c	254	84	152	1	39	1	<1	t	t	.1
	Grapefruit juice:											
242	Fresh, white, raw	1 c	247	90	96	1	23	<1	<1	t	t	.1
243	Canned, unsweetened	1 c	247	90	94	1	22	<1	<1	t	t	.1
244	Sweetened	1 c	250	87	115	1	28	<1	<1	t	t	.1
	Frozen concentrate, unsweetened:											
246	Diluted with 3 cans water	1 c	247	89	101	1	24	<1	<1	t	t	.1
	Grapes, raw European (adherent skin):											
247	Thompson seedless	10 ea	50	81	35	<1	9	<1	<1	.1	t	.1
248	Tokay/Emperor, seeded types	10 ea	50	81	35	<1	9	<1	<1	.1	t	.1
	Grape juice:											
249	Bottled or canned	1 c	253	84	154	1	38	<1	<1	.1	t	.1
	Frozen concentrate, sweetened:											
251	Diluted with 3 cans water, vit C added	1 c	250	87	128	<1	32	<1	<1	.1	t	.1
1410	Low calorie	1 c	253	84	154	1	38	<1	<1	.1	t	.1
3636	Jackfruit, fresh, sliced	1 c	165	73	155	2	40	3	<1	.1	.1	.1
252	Kiwi fruit, raw, peeled (88g with peel)	1 ea	76	83	46	1	11	3	<1	t	t	.2
253	Lemons, raw, without peel and seeds (about 4 per lb whole)	1 ea	58	89	17	1	5	2	<1	t	t	.1
	Lemon juice:											
254	Fresh:	1 c	244	91	61	1	21	1	0	0	0	0
255	Tablespoon	1 tbs	15	91	4	<1	1	<1	0	0	0	0
256	Canned or bottled, unsweetened:	1 c	244	92	51	1	16	1	1	.1	t	.2
257	Tablespoon	1 tbs	15	92	3	<1	1	<1	<1	t	t	t
258	Frozen, single strength, unsweetened:	1 c	244	92	54	1	16	1	1	.1	t	.2
2298	Tablespoon	1 tbs	15	92	3	<1	1	<1	<1	t	t	t
	Lime juice:											
260	Fresh:	1 c	246	90	66	1	22	1	<1	t	t	.1
261	Tablespoon	1 tbs	15	90	4	<1	1	<1	<1	t	t	t
262	Canned or bottled, unsweetened	1 c	246	93	52	1	16	1	1	.1	.1	.2
3758	Pomelos, raw	1 ea	609	89	231	5	59	6	<1			
263	Mangos, raw, edible part (300g w/skin & seeds)	1 ea	207	82	135	1	35	4	1	.1	.2	.1
	Melons, raw, without rind and contents:											
264	Cantaloupe, 5" diam (2⅓ lb whole with refuse), orange flesh	½ ea	276	90	97	2	23	2	1	.2	t	.3
265	Honeydew, 6½ diam (5¼ lb whole with refuse), slice = ⅒ melon	1 pce	160	90	56	1	15	1	<1	t	t	.1
266	Nectarines, raw, w/o pits, 2" diam	1 ea	136	86	67	1	16	2	1	.1	.2	.3
	Oranges, raw:											
267	Whole w/o peel and seeds, 2⅝" diam (180g with peel and seeds)	1 ea	131	87	62	1	15	3	<1	t	t	t
268	Sections, without membranes	1 c	180	87	85	2	21	4	<1	t	t	t
	Orange juice:											
269	Fresh, all varieties	1 c	248	88	112	2	26	<1	<1	.1	.1	.1
270	Canned, unsweetened	1 c	249	89	105	1	24	<1	<1	t	.1	.1
3480	Calcium fortified	1 c	247	89	110	1	27	0	0	0	0	0
271	Chilled	1 c	249	88	110	2	25	<1	1	.1	.1	.2
	Frozen concentrate:											
273	Diluted w/3 parts water by volume	1 c	249	88	112	2	27	<1	<1	t	t	t
1345	Orange juice, from dry crystals	1 c	248	88	114	0	29	0	0	0	0	0
274	Orange and grapefruit juice, canned	1 c	247	89	106	1	25	<1	<1	t	t	t

Chol (mg)	Calc (mg)	Iron (mg)	Magn (mg)	Pota (mg)	Sodi (mg)	Zinc (mg)	VT-A (µg)	Thia (mg)	VT-E (mg)	Ribo (mg)	Niac (mg)	V-B6 (mg)	Fola (µg)	VT-C (mg)
0	11	.61	8	72	80	.14	3	.04	.28	.06	.28	.04	3	6
0	27	.95	29	541	2	.24	2	.07	.08	.08	1.83	.16	11	0
0	57	2.05	62	1160	5	.52	4	.16	.18	.18	3.92	.34	23	0
0	274	4.24	112	1352	21	.97	12	.13	0	.17	1.32	.43	15	2
0	15	.72	12	218	15	.2	25	.04	.72	.05	.93	.12	7	5
0	19	.5	17	225	9	.21	37	.03	.47	.04	.95	.12	7	6
0	13	.15	10	159	0	.09	16	.04	.31	.02	.23	.05	15	47
0	14	.07	11	175	0	.08	1	.04	.29	.02	.32	.05	12	39
0	36	1.02	25	328	5	.2	0	.1	.63	.05	.62	.05	23	54
0	22	.49	30	400	2	.12	1	.1	.12	.05	.49	.11	25	94
0	17	.49	25	378	2	.22	1	.1	.12	.05	.57	.05	25	72
0	20	.9	25	405	5	.15	0	.1	.12	.06	.8	.05	25	67
0	20	.35	27	336	2	.12	1	.1	.12	.05	.54	.11	10	83
0	5	.13	3	92	1	.02	2	.05	.35	.03	.15	.05	2	5
0	5	.13	3	92	1	.02	2	.05	.35	.03	.15	.05	2	5
0	23	.61	25	334	8	.13	1	.07	0	.09	.66	.16	8	<1
0	10	.25	10	52	5	.1	1	.04	.12	.06	.31	.1	2	60
0	23	.61	25	334	8	.13	1	.07	0	.09	.66	.16	8	<1
0	56	.99	61	500	5	.69	25	.05	.25	.18	.66	.18	23	11
0	20	.31	23	252	4	.13	7	.01	.85	.04	.38	.07	29	74
0	15	.35	5	80	1	.03	1	.02	.14	.01	.06	.05	6	31
0	17	.07	15	303	2	.12	2	.07	.22	.02	.24	.12	32	112
0	1	<.01	1	19	<1	.01	<1	<.01	.01	<.01	.01	.01	2	7
0	27	.32	19	249	51	.15	2	.1	.22	.02	.48	.1	24	60
0	2	.02	1	15	3	.01	<1	.01	.01	<.01	.03	.01	1	4
0	19	.29	19	217	2	.12	1	.14	.22	.03	.33	.15	24	77
0	1	.02	1	13	<1	.01	<1	.01	.01	<.01	.02	.01	1	5
0	22	.07	15	268	2	.15	1	.05	.22	.02	.25	.11	20	72
0	1	<.01	1	16	<1	.01	<1	<.01	.01	<.01	.01	.01	1	4
0	29	.57	17	185	39	.15	2	.08	.17	.01	.4	.07	20	16
0	24	.67	36	1315	6	.49	0	.21	.55	.16	1.34	.22	158	371
0	21	.27	19	323	4	.08	403	.12	2.32	.12	1.21	.28	29	57
0	30	.58	30	853	25	.44	444	.1	.41	.06	1.58	.32	47	116
0	10	.11	11	434	16	.11	3	.12	.24	.03	.96	.09	10	40
0	7	.2	11	288	0	.12	50	.02	1.21	.06	1.35	.03	5	7
0	52	.13	13	237	0	.09	14	.11	.31	.05	.37	.08	39	70
0	72	.18	18	326	0	.13	19	.16	.43	.07	.51	.11	54	96
0	27	.5	27	496	2	.12	25	.22	.22	.07	.99	.1	74	124
0	20	1.1	27	436	5	.17	22	.15	.22	.07	.78	.22	45	86
0	300	0		430	15		0	0	0		0	0	40	78
0	25	.42	27	473	2	.1	10	.28	.47	.05	.7	.13	45	82
0	22	.25	25	473	2	.12	10	.2	.47	.04	.5	.11	110	97
0	62	.2	2	50	12	.1	275	<.01	0	.04	0	0	144	121
0	20	1.14	25	390	7	.17	15	.14	.17	.07	.83	.06	35	72

*This value is expressed in retinol equivalents (RE). All other values are in retinol activity equivalents (RAE).

Table A-1

Food Composition (Computer code number is for Wadsworth Diet Analysis program) (For purposes of calculations, use "0" for t, <1, <.1, <.01, etc.)

Computer Code Number	Food Description	Measure	Wt (g)	H_2O (%)	Ener (kcal)	Prot (g)	Carb (g)	Dietary Fiber (g)	Fat (g)	Fat Breakdown (g) Sat	Mono	Poly
	FRUITS and FRUIT JUICES—Continued											
	Papayas, raw:											
275	½" slices	1 c	140	89	55	1	14	3	<1	.1	.1	t
276	Whole, 3" diam by 5⅛" w/o seeds and skin (1 lb w/refuse)	1 ea	304	89	119	2	30	5	<1	.1	.1	.1
1031	Papaya nectar, canned	1 c	250	85	143	<1	36	1	<1	.1	.1	.1
	Peaches:											
277	Raw, whole, 2½" diam, peeled, pitted (about 4 per lb whole)	1 ea	98	88	42	1	11	2	<1	t	t	t
278	Raw, sliced	1 c	170	88	73	1	19	3	<1	t	.1	.1
	Canned, fruit and liquid:											
279	Heavy syrup pack:	1 c	262	79	194	1	52	3	<1	t	.1	.1
280	Half	1 ea	98	79	72	<1	19	1	<1	t	t	t
281	Juice pack:	1 c	248	87	109	2	29	3	<1	t	t	t
282	Half	1 ea	98	87	43	1	11	1	<1	t	t	t
283	Dried, uncooked	10 ea	130	32	311	5	80	11	1	.1	.4	.5
284	Dried, cooked, fruit and liquid	1 c	258	78	199	3	51	7	1	.1	.2	.3
	Frozen, slice, sweetened:											
285	10-oz package,vitamin C added	1 ea	284	75	267	2	68	5	<1	t	.1	.2
286	Cup, thawed measure, vitamin C added	1 c	250	75	235	2	60	4	<1	t	.1	.2
1032	Peach nectar, canned	1 c	249	86	134	1	35	1	<1	4.7	19	2.6
	Pears:											
	Fresh, with skin, cored:											
287	Bartlett, 2½" diam (about 2½ per lb)	1 ea	166	84	98	1	25	4	1	t	.1	.2
288	Bosc, 2⅛" diam (about 3 per lb)	1 ea	139	84	82	1	21	3	1	t	.1	.1
289	D'Anjou, 3" diam (about 2 per lb)	1 ea	209	84	123	1	32	5	1	t	.2	.2
	Canned, fruit and liquid:											
290	Heavy syrup pack:	1 c	266	80	197	1	51	4	<1	t	.1	.1
291	Half	1 ea	76	80	56	<1	15	1	<1	t	t	t
292	Juice pack:	1 c	248	86	124	1	32	4	<1	t	t	t
293	Half	1 ea	76	86	38	<1	10	1	<1	t	t	t
294	Dried halves	10 ea	175	27	459	3	122	13	1	.1	.2	.3
1033	Pear nectar, canned	1 c	250	84	150	<1	39	1	<1	t	t	t
	Pineapple:											
295	Fresh chunks, diced	1 c	155	86	76	1	19	2	1	t	.1	.2
	Canned, fruit and liquid:											
	Heavy syrup pack:											
296	Crushed, chunks, tidbits	½ c	127	79	99	<1	26	1	<1	t	t	.1
297	Slices	1 ea	49	79	38	<1	10	<1	<1	t	t	t
298	Juice pack, crushed, chunks, tidbits	1 c	250	84	150	1	39	2	<1	t	t	.1
299	Juice pack, slices	1 ea	47	84	28	<1	7	<1	<1	t	t	t
300	Pineapple juice, canned, unsweetened	1 c	250	86	140	1	34	<1	<1	t	t	.1
	Plantains, yellow fleshed, without peel:											
301	Raw slices (whole=179g w/o peel)	1 c	148	65	181	2	47	3	1	.2	t	.1
302	Cooked, boiled, sliced	1 c	154	67	179	1	48	4	<1	.1	t	.1
	Plums:											
303	Fresh, medium, 2⅛" diam	1 ea	66	85	36	1	9	1	<1	t	.3	.1
304	Fresh, small, 1½" diam	1 ea	28	85	15	<1	4	<1	<1	t	.1	t
	Canned, purple, with liquid:											
305	Heavy syrup pack:	1 c	258	76	230	1	60	3	<1	t	.2	.1
306	Plums	3 ea	138	76	123	<1	32	1	<1	t	.1	t
307	Juice pack:	1 c	252	84	146	1	38	3	<1	4.8	3.4	12
308	Plums	3 ea	138	84	80	1	21	1	<1	2.6	1.8	6.6
1698	Pomegranate, fresh	1 ea	154	81	105	1	26	1	<1	.1	.1	.1
	Prunes, dried, pitted:											
309	Uncooked (10 = 97g w/pits, 84g w/o pits)	10 ea	84	32	201	2	53	6	<1	t	.3	.1
310	Cooked, unsweetened, fruit & liq (250g w/pits)	1 c	248	70	265	3	70	16	1	t	.4	.1
311	Prune juice, bottled or canned	1 c	256	81	182	2	45	3	<1	7.4	5.2	17.3
	Raisins, seedless:											
312	Cup, not pressed down	1 c	145	15	435	5	115	6	1	.2	t	.2
313	One packet, ½ oz	½ oz	14	15	42	<1	11	1	<1	t	t	t
	Raspberries:											

A

Chol (mg)	Calc (mg)	Iron (mg)	Magn (mg)	Pota (mg)	Sodi (mg)	Zinc (mg)	VT-A (µg)	Thia (mg)	VT-E (mg)	Ribo (mg)	Niac (mg)	V-B6 (mg)	Fola (µg)	VT-C (mg)
0	34	.14	14	360	4	.1	20	.04	1.57	.04	.47	.03	53	86
0	73	.3	30	781	9	.21	43	.08	3.4	.1	1.03	.06	116	188
0	25	.85	7	77	12	.37	14	.01	.05	.01	.37	.02	5	7
0	5	.11	7	193	0	.14	26	.02	.69	.04	.97	.02	3	6
0	8	.19	12	335	0	.24	46	.03	1.19	.07	1.68	.03	5	11
0	8	.71	13	241	16	.24	43	.03	2.33	.06	1.61	.05	8	7
0	3	.26	5	90	6	.09	16	.01	.87	.02	.6	.02	3	3
0	15	.67	17	317	10	.27	47	.02	3.72	.04	1.44	.05	7	9
0	6	.26	7	125	4	.11	19	.01	1.47	.02	.57	.02	3	4
0	36	5.28	55	1294	9	.74	140	<.01	0	.28	5.69	.09	0	6
0	23	3.38	33	826	5	.46	26	.01	0	.05	3.92	.1	0	10
0	9	1.05	14	369	17	.14	40	.04	2.53	.1	1.85	.05	9	268
0	7	.92	12	325	15	.12	35	.03	2.23	.09	1.63	.04	7	236
0	12	.47	10	100	17	.2	32	.01	.02	.03	.72	.02	2	13
0	18	.41	10	208	0	.2	2	.03	.83	.07	.17	.03	12	7
0	15	.35	8	174	0	.17	1	.03	.69	.06	.14	.02	10	6
0	23	.52	12	261	0	.25	2	.04	1.05	.08	.21	.04	15	8
0	13	.58	11	173	13	.21	0	.03	1.33	.06	.64	.04	3	3
0	4	.17	3	49	4	.06	0	.01	.38	.02	.18	.01	1	1
0	22	.72	17	238	10	.22	1	.03	1.24	.03	.5	.03	2	4
0	7	.22	5	73	3	.07	<1	.01	.38	.01	.15	.01	1	1
0	59	3.68	58	933	10	.68	<1	.01	0	.25	2.4	.13	0	12
0	12	.65	7	32	10	.17	<1	<.01	.25	.03	.32	.03	2	3
0	11	.57	22	175	2	.12	2	.14	.15	.06	.65	.13	17	24
0	18	.48	20	132	1	.15	1	.11	.13	.03	.36	.09	6	9
0	7	.19	8	51	<1	.06	<1	.04	.05	.01	.14	.04	2	4
0	35	.7	35	305	2	.25	5	.24	.25	.05	.71	.18	12	24
0	7	.13	7	57	<1	.05	1	.04	.05	.01	.13	.03	2	4
0	42	.65	32	335	2	.27	1	.14	.05	.05	.64	.24	57	27
0	4	.89	55	739	6	.21	84	.08	.4	.08	1.02	.44	33	27
0	3	.89	49	716	8	.2	70	.07	.22	.08	1.16	.37	40	17
0	3	.07	5	114	0	.07	11	.03	.4	.06	.33	.05	1	6
0	1	.03	2	48	0	.03	4	.01	.17	.03	.14	.02	1	3
0	23	2.17	13	235	49	.18	33	.04	1.81	.1	.75	.07	8	1
0	12	1.16	7	126	26	.1	18	.02	.97	.05	.4	.04	4	1
0	25	.86	20	388	3	.28	127	.06	1.76	.15	1.19	.07	8	7
0	14	.47	11	213	1	.15	70	.03	.97	.08	.65	.04	4	4
0	5	.46	5	399	5	.18	0	.05	.85	.05	.46	.16	9	9
0	43	2.08	38	626	3	.44	84	.07	1.22	.14	1.65	.22	3	3
0	57	2.75	50	828	5	.59	38	.06	0	.25	1.79	.54	0	7
0	31	3.02	36	707	10	.54	<1	.04	.03	.18	2.01	.56	0	10
0	71	3.02	48	1088	17	.39	1	.23	1.02	.13	1.19	.36	4	5
0	7	.29	5	105	2	.04	<1	.02	.1	.01	.11	.03	<1	<1

*This value is expressed in retinol equivalents (RE). All other values are in retinol activity equivalents (RAE).

Table A–1

Food Composition

(Computer code number is for Wadsworth Diet Analysis program) (For purposes of calculations, use "0" for t, <1, <.1, <.01, etc.)

Computer Code Number	Food Description	Measure	Wt (g)	H₂O (%)	Ener (kcal)	Prot (g)	Carb (g)	Dietary Fiber (g)	Fat (g)	Fat Breakdown (g) Sat	Mono	Poly
	FRUITS and FRUIT JUICES—Continued											
314	Fresh	1 c	123	87	60	1	14	8	1	t	.1	.4
315	Frozen, sweetened	10 oz	284	73	293	2	74	12	<1	t	t	.3
316	Cup, thawed measure	1 c	250	73	258	2	65	11	<1	t	t	.2
317	Rhubarb, cooked, added sugar	1 c	240	68	278	2	75	5	<1	t	t	.1
	Strawberries:											
318	Fresh, whole, capped	1 c	144	92	43	1	10	3	1	t	.1	.3
	Frozen, sliced, sweetened:											
319	10-oz container	10 oz	284	73	273	2	74	5	<1	t	.1	.2
320	Cup, thawed measure	1 c	255	73	245	1	66	5	<1	t	t	.2
	Tangerines, without peel and seeds:											
321	Fresh (2⅜" whole) 116g w/refuse	1 ea	84	88	37	1	9	2	<1	t	t	t
322	Canned, light syrup, fruit and liquid	1 c	252	83	154	1	41	2	<1	t	t	t
323	Tangerine juice, canned, sweetened	1 c	249	87	125	1	30	<1	<1	t	t	.1
	Watermelon, raw, without rind and seeds:											
324	Piece, ¹⁄₁₆th wedge	1 pce	286	92	91	2	20	1	1	.1	.3	.4
325	Diced	1 c	152	92	49	1	11	1	1	.1	.2	.2
	BAKED GOODS: BREADS, CAKES, COOKIES, CRACKERS, PIES											
42100	Bagel, cinnamon raisin, 3½" diam.	1 ea	71	32	195	7	39	2	1	.2	.1	.5
326	Bagel, plain, enriched, 3½" diam.	1 ea	71	33	195	7	38	2	1	.2	.1	.5
1663	Bagel, oat bran	1 ea	110	33	281	12	59	4	1	.2	.3	.5
42617	Bagel, whole wheat	1 ea	110	28	291	12	62	10	2	.3	.2	.6
	Biscuits:											
327	From home recipe	1 ea	60	29	212	4	27	1	10	2.6	4.2	2.5
328	From mix	1 ea	57	29	191	4	28	1	7	1.6	2.4	2.4
329	From refrigerated dough	1 ea	74	27	276	4	34	1	13	8.7	3.4	.5
330	Bread crumbs, dry, grated (see 364, 365 for soft crumbs)	1 c	108	6	427	13	78	3	6	1.3	2.6	1.2
2087	Bread sticks, brown & serve	1 ea	57	34	150	7	28	1	1	.5	.5	.5
	Breads:											
331	Boston brown, canned, 3¼" slice	1 pce	45	47	88	2	19	2	1	.1	.1	.3
	Cracked wheat (¼ cracked-wheat & ¾ enriched wheat flour):											
333	Slice (18 per loaf)	1 pce	25	36	65	2	12	1	1	.2	.5	.2
334	Slice, toasted	1 pce	23	30	65	2	12	1	1	.2	.5	.2
	French/Vienna, enriched:											
337	Slice, 4¾ x 4½"	1 pce	25	34	68	2	13	1	1	.2	.3	.2
336	French, slice, 5 x 2½"	1 pce	25	34	68	2	13	1	1	.2	.3	.2
	French toast: see Mixed Dishes, and Fast Foods, #691											
2083	Honey wheatberry	1 pce	38	38	100	3	18	2	1	0	.5	1
	Italian, enriched:											
339	Slice, 4½ x 3¼ x ¾"	1 pce	30	36	81	3	15	1	1	.3	.2	.4
	Mixed grain, enriched:											
341	Slice (18 per loaf)	1 pce	26	38	65	3	12	2	1	.2	.4	.2
342	Slice, toasted	1 pce	24	32	65	3	12	2	1	.2	.4	.2
	Oatmeal, enriched:											
344	Slice (18 per loaf)	1 pce	27	37	73	2	13	1	1	.2	.4	.5
345	Slice, toasted	1 pce	25	31	73	2	13	1	1	.2	.4	.5
346	Pita pocket bread, enr, 6½" round	1 ea	60	32	165	5	33	1	1	.1	.1	.3
	Pumpernickel(⅔ rye & ⅓ enr wheat flr):											
348	Slice, 5 x 4 x ⅜"	1 pce	26	38	65	2	12	2	1	.1	.2	.3
349	Slice, toasted	1 pce	29	32	80	3	15	2	1	.1	.3	.4
	Raisin, enriched:											
351	Slice (18 per loaf)	1 pce	26	34	71	2	14	1	1	.3	.6	.2
352	Slice, toasted	1 pce	24	28	71	2	14	1	1	.3	.6	.2
353	Rye, light (⅓ rye & ⅔ enr wheat flr): 1-lb loaf	1 ea	454	37	1175	39	219	26	15	2.8	5.9	3.6
354	Slice, 4¾ x 3¾ x ⁷⁄₁₆"	1 pce	32	37	83	3	15	2	1	.2	.4	.3
355	Slice, toasted	1 pce	24	31	68	2	13	2	1	.2	.3	.2

Chol (mg)	Calc (mg)	Iron (mg)	Magn (mg)	Pota (mg)	Sodi (mg)	Zinc (mg)	VT-A (µg)	Thia (mg)	VT-E (mg)	Ribo (mg)	Niac (mg)	V-B6 (mg)	Fola (µg)	VT-C (mg)
0	27	.7	22	187	0	.57	8	.04	.55	.11	1.11	.07	32	31
0	43	1.85	37	324	3	.51	9	.05	1.28	.13	.65	.1	74	47
0	37	1.63	32	285	2	.45	7	.05	1.13	.11	.57	.08	65	41
0	348	.5	29	230	2	.19	8	.04	.48	.05	.48	.05	12	8
0	20	.55	14	239	1	.19	2	.03	.2	.09	.33	.08	26	82
0	31	1.68	20	278	9	.17	3	.04	.4	.14	1.14	.08	43	118
0	28	1.5	18	250	8	.15	3	.04	.36	.13	1.02	.08	38	106
0	12	.08	10	132	1	.2	39	.09	.2	.02	.13	.06	17	26
0	18	.93	20	197	15	.6	106	.13	.86	.11	1.12	.11	13	50
0	45	.5	20	443	2	.07	52	.15	.22	.05	.25	.08	12	55
0	23	.49	31	332	6	.2	53	.23	.43	.06	.57	.41	6	27
0	12	.26	17	176	3	.11	28	.12	.23	.03	.3	.22	3	15
0	13	2.7	20	105	229	.8	0*	.27	.11	.2	2.19	.04	64	<1
0	52	2.53	21	72	379	.62	0	.38	.03	.22	3.24	.04	62	0
0	13	3.39	34	127	558	.99	<1	.36	.15	.37	3.26	.05	89	<1
0	32	3.52	116	379	592	2.52	0	.34	.99	.28	5.75	.3	66	<1
2	141	1.74	11	73	348	.32	14*	.21	.78	.19	1.77	.02	37	<1
2	105	1.17	14	107	544	.35	15*	.2	.23	.2	1.72	.04	30	<1
5	89	1.64	9	87	584	.29	24*	.27	.44	.18	1.63	.03	6	0
0	245	6.61	50	239	931	1.32	<1	.83	.62	.47	7.4	.11	118	0
0	60	2.7			290		0*	.22		.1	1.6			0
<1	31	.94	28	143	284	.22	5*	.01	.25	.05	.5	.04	5	0
0	11	.7	13	44	135	.31	0	.09	.15	.06	.92	.08	15	0
0	11	.7	13	44	135	.31	0*	.07	.14	.05	.83	.07	7	0
0	19	.63	7	28	152	.22	0	.13	.07	.08	1.19	.01	24	0
0	19	.63	7	28	152	.22	0	.13	.07	.08	1.19	.01	24	0
0	20	.72			200		0	.12	.24	.07	.8			0
0	23	.88	8	33	175	.26	0	.14	.11	.09	1.31	.01	28	0
0	24	.9	14	53	127	.33	0	.11	.17	.09	1.13	.09	21	<1
0	24	.9	14	53	127	.33	0	.08	.16	.08	1.02	.08	16	<1
0	18	.73	10	38	162	.27	1*	.11	.16	.06	.85	.02	17	0
0	18	.73	10	38	163	.28	<1*	.09	.09	.06	.77	.02	13	<1
0	52	1.57	16	72	322	.5	0	.36	.02	.2	2.78	.02	57	0
0	18	.75	14	54	174	.38	0	.08	.11	.08	.8	.03	21	0
0	21	.91	17	66	214	.47	0	.08	.17	.09	.89	.04	20	0
0	17	.75	7	59	101	.19	0	.09	.13	.1	.9	.02	23	<1
0	17	.76	7	59	102	.19	<1	.07	.2	.09	.81	.02	18	<1
0	331	12.8	182	754	2996	5.18	5*	1.97	1.68	1.52	17.3	.34	390	2
0	23	.91	13	53	211	.36	<1*	.14	.12	.11	1.22	.02	27	<1
0	19	.74	10	44	174	.3	<1	.09	.15	.08	.9	.02	17	<1

*This value is expressed in retinol equivalents (RE). All other values are in retinol activity equivalents (RAE).

Table A–1

Food Composition   (Computer code number is for Wadsworth Diet Analysis program)   (For purposes of calculations, use "0" for t, <1, <.1, <.01, etc.)

Computer Code Number	Food Description	Measure	Wt (g)	H₂O (%)	Ener (kcal)	Prot (g)	Carb (g)	Dietary Fiber (g)	Fat (g)	Fat Breakdown (g) Sat	Mono	Poly
	BAKED GOODS: BREADS, CAKES, COOKIES, CRACKERS, PIES—Continued											
	Wheat (enr wheat & whole-wheat flour):											
357	Slice (18 per loaf)	1 pce	25	37	65	2	12	1	1	.2	.4	.2
358	Slice, toasted	1 pce	23	32	65	2	12	1	1	.2	.4	.2
	White, enriched:											
360	Slice	1 pce	42	35	120	3	21	1	2	.5	.5	1.2
361	Slice, toasted	1 pce	38	29	119	3	21	1	2	.5	.5	1.2
	Whole Wheat:											
367	Slice (16 per loaf)	1 pce	28	38	69	3	13	2	1	.3	.5	.3
368	Slice, toasted	1 pce	25	30	69	3	13	2	1	.3	.5	.3
	Bread stuffing, prepared from mix:											
369	Dry type	1 c	200	65	356	6	43	6	17	3.5	7.6	5.2
370	Moist type, with egg and margarine	1 c	232	65	390	9	51	5	17	3.4	7.4	4.9
	Cakes, prepared from mixes using enrich flour and veg shortening, w/frostings made from margarine:											
372	Angel Food, ¹⁄₁₂ of cake	1 pce	28	33	72	2	16	<1	<1	t	t	.1
373	Boston cream pie, ⅛ of cake	1 pce	92	45	232	2	39	1	8	2.2	4.2	.9
375	Coffee Cake, ⅙ of cake	1 pce	56	30	178	3	30	1	5	1	2.2	1.8
	Devil's food, chocolate frosting:											
377	Piece, ¹⁄₁₆ of cake	1 pce	64	23	235	3	35	2	10	3	5.6	1.2
378	Cupcake, 2½" diam	1 ea	42	23	154	2	23	1	7	2	3.7	.8
380	Gingerbread, ⅑ of cake	1 pce	67	33	207	3	34	1	7	1.8	3.8	.9
	Yellow, chocolate frosting, 2 layer:											
382	Piece, ¹⁄₁₆ of cake	1 pce	64	22	243	2	35	1	11	3	6.1	1.3
	Cakes from home recipes w/enr flour: Carrot cake, made with veg oil, cream cheese frosting:											
384	Piece, ¹⁄₁₆ of cake, 2¼ x 3¼" slice	1 pce	111	21	484	5	52	1	29	5.4	7.2	15.1
	Fruitcake, dark:											
386	Piece, ¹⁄₃₂ of cake, ⅔" arc	1 pce	43	25	139	1	26	2	4	.5	1.8	1.4
	Sheet, plain, made w/veg shortening,											
388	no frosting, ⅑ of cake	1 pce	86	24	313	4	48	<1	12	3.3	5.7	2.8
	Sheet, plain, made w/margarine,											
390	uncooked white frosting, ⅑ of cake	1 pce	64	22	239	2	38	<1	9	1.5	3.9	3.3
	Cakes, commerical:											
402	Cheesecake, ¹⁄₁₂ of cake	1 pce	80	46	257	4	20	<1	18	7.9	6.9	1.3
394	Pound cake, ¹⁄₁₇ of loaf, 2" slice	1 pce	28	25	109	2	14	<1	6	3.2	1.6	.3
	Snack cakes:											
395	Chocolate w/creme filling,Ding Dong	1 ea	50	20	188	2	30	<1	7	1.4	2.8	2.6
396	Sponge cake w/creme filling,Twinkie	1 ea	43	20	157	1	27	<1	5	1.1	1.7	1.4
1677	Sponge cake, ¹⁄₁₂ of 12" cake	1 pce	38	30	110	2	23	<1	1	.3	.4	.2
398	White, white frosting, 2 layer, ¹⁄₁₆	1 pce	71	20	266	2	45	1	10	4.3	3.8	1
	Yellow, chocolate frosting, 2 layer:											
400	Slice, ¹⁄₁₆ of cake	1 pce	64	22	243	2	35	1	11	3	6.1	1.3
1332	Bagel Chips	5 pce	70	3	298	6	52	4	7	1.3	2.1	3.4
2225	Bagel chips, onion garlic, toasted	1 oz	28		181	5	30	3	7	1.6	4.9	0
1035	Cheese puffs/Cheetos	1 c	20	1	111	2	11	<1	7	1.3	4.1	1
	Cookies made with enriched flour:											
	Brownies with nuts:											
403	Commercial w/frosting, 1½ x 1¾ x ⅞"	1 ea	61	14	247	3	39	1	10	2.6	5.5	1.4
1902	Fat free fudge, Entenmann's	1 pce	40	24	110	2	27	1	0	0	0	0
	Chocolate chip cookies:											
405	Commercial, 2¼" diam	4 ea	60	12	275	2	35	2	15	4.4	7.8	2.1
406	Home recipe, 2¼" diam	4 ea	64	6	312	4	37	2	18	5.2	6.6	5.4
407	From refrigerated dough, 2¼" diam	4 ea	64	13	284	3	39	1	13	4.3	6.7	1.4
408	Fig bars	4 ea	64	16	223	2	45	3	5	.7	1.9	1.8
2052	Fruit bar, no fat	1 ea	28		90	2	21	0	0	0	0	0
2162	Fudge, fat free, Snackwell	1 ea	16	14	53	1	12	<1	<1	.1	.1	t
409	Oatmeal raisin, 2⅝" diam	4 ea	60	6	261	4	41	2	10	1.9	4.1	3
410	Peanut butter, home recipe, 2⅝"diam	4 ea	80	6	380	7	47	2	19	3.5	8.7	5.8

PAGE KEY: A–2 = Beverages   A–4 = Dairy   A–8 = Eggs   A–10 = Fat/Oil   A–12 = Fruit   A–18 = Bakery   A–24 = Grain   *Table of Food Composition*   **A–21**
A–30 = Fish   A–32 = Meats   A–36 = Poultry   A–38 = Sausage   A–38 = Mixed/Fast   A–44 = Nuts/Seeds   A–46 = Sweets
A–50 = Vegetables/Legumes   A–60 = Vegetarian   A–62 = Misc   A–64 = Soups/Sauces   A–66 = Fast   A–80 = Convenience   A–86 = Baby foods

A

Chol (mg)	Calc (mg)	Iron (mg)	Magn (mg)	Pota (mg)	Sodi (mg)	Zinc (mg)	VT-A (µg)	Thia (mg)	VT-E (mg)	Ribo (mg)	Niac (mg)	V-B6 (mg)	Fola (µg)	VT-C (mg)
0	26	.83	11	50	133	.26	0	.1	.13	.07	1.03	.02	19	0
0	26	.83	11	50	132	.26	0	.08	.14	.06	.93	.02	15	0
1	24	1.25	8	61	151	.27	9*	.17	.36	.16	1.51	.02	38	<1
1	24	1.24	8	61	150	.27	8*	.13	.46	.14	1.35	.02	12	<1
0	20	.92	24	71	148	.54	0	.1	.24	.06	1.07	.05	14	0
0	20	.93	24	71	148	.54	0	.08	.29	.05	.97	.04	10	0
0	64	2.18	24	148	1086	.56	162	.27	2.8	.21	2.95	.08	202	0
0	148	3.8	35	304	1069	.74	160*	.39	2.78	.33	3.69	.12	39	4
0	39	.15	3	26	210	.02	0	.03	.03	.14	.25	.01	10	0
34	21	.35	6	36	132	.15	21*	.37	.97	.25	.18	.02	14	<1
27	76	.8	10	63	236	.25	22*	.09	.93	.1	.85	.03	27	<1
27	27	1.41	22	128	214	.44	16*	.02	1.08	.08	.37	.03	11	<1
18	18	.92	14	84	140	.29	10*	.01	.71	.06	.24	.02	7	<1
23	46	2.22	11	161	307	.27	11*	.13	.92	.12	1.05	.02	7	<1
35	24	1.33	19	114	216	.4	21	.08	1.45	.1	.8	.02	14	0
60	28	1.39	20	124	273	.54	426*	.15	4.68	.17	1.12	.08	13	1
2	14	.89	7	66	116	.12	2*	.02	.71	.04	.34	.02	8	<1
56	55	1.3	12	68	258	.3	41*	.14	1.22	.15	1.12	.03	6	<1
35	40	.68	4	34	220	.16	12*	.06	1.22	.04	.32	.02	17	0
44	41	.5	9	72	166	.41	117*	.02	1.26	.15	.16	.04	14	<1
62	10	.39	3	33	111	.13	44*	.04	.18	.06	.37	.01	11	0
8	36	1.68	20	61	213	.25	2*	.11	1.68	.15	1.21	.01	14	0
7	19	.55	3	37	157	.12	2*	.07	.87	.06	.53	.01	12	<1
39	27	1.03	4	38	93	.19	17*	.09	.1	.1	.73	.02	15	0
6	34	.57	4	41	166	.11	23*	.07	1.28	.09	.64	.01	4	<1
35	24	1.33	19	114	216	.4	21	.08	1.45	.1	.8	.02	14	0
0	9	1.39	39	167	419	.87	0	.13	1.71	.12	1.62	.15	46	<1
0	0	2.37			461		0	.37	<.01	.22	3.29			0
1	12	.47	4	33	210	.08	7*	.05	1.02	.07	.65	.03	24	<1
10	18	1.37	19	91	190	.44	4*	.16	1.27	.13	1.05	.02	13	0
0	0	1.08		90	140		0	.01						
0	9	1.45	21	56	196	.28	0*	.07	1.74	.12	.97	.1	23	0
20	25	1.57	35	143	231	.59	105*	.12	1.84	.11	.87	.05	21	<1
15	16	1.44	15	115	134	.32	11*	.12	1.48	.12	1.26	.02	36	0
0	41	1.86	17	132	224	.25	3*	.1	.8	.14	1.2	.05	17	<1
0	0	.36			95		0	.01						0
0	3	.29	5	26	71	.08	<1*	.02	<.01	.02	.26	<.01		0
20	60	1.59	25	143	323	.52	98*	.15	1.5	.1	.75	.04	18	<1
25	31	1.78	31	185	414	.66	125*	.18	3.04	.17	2.81	.07	44	<1

*This value is expressed in retinol equivalents (RE). All other values are in retinol activity equivalents (RAE).

Table A–1

Food Composition   (Computer code number is for Wadsworth Diet Analysis program)   (For purposes of calculations, use "0" for t, <1, <.1, <.01, etc.)

A

Computer Code Number	Food Description	Measure	Wt (g)	H₂O (%)	Ener (kcal)	Prot (g)	Carb (g)	Dietary Fiber (g)	Fat (g)	Fat Breakdown (g)		
										Sat	Mono	Poly
	BAKED GOODS: BREADS, CAKES, COOKIES, CRACKERS, PIES—Continued											
411	Sandwich-type, all	4 ea	40	2	189	2	28	1	8	1.5	3.4	2.9
412	Shortbread, commercial, small	4 ea	32	4	161	2	21	1	8	1.9	4.3	1
413	Shortbread, home recipe, large	2 ea	22	3	120	1	12	<1	7	4.5	2.1	.3
414	Sugar from refrigerated dough,2" diam	4 ea	48	5	232	2	31	<1	11	2.8	6.2	1.4
1874	Vanilla sandwich, Snackwell's	2 ea	26	4	109	1	21	1	2	.5	.8	.2
415	Vanilla wafers	10 ea	40	5	176	2	29	1	6	1.5	2.6	1.5
42672	Cornbread, 2.5 x 2.5 x 1.5" piece	1 pce	65	49	152	4	23	2	5	1.6	2.4	.5
416	Corn chips	1 c	26	1	140	2	15	1	9	1.2	2.5	4.3
	Crackers:											
417	Cheese-enriched	10 ea	10	3	50	1	6	<1	3	.9	1.2	.2
418	Cheese with peanut butter-enriched	4 ea	28	4	135	4	16	1	6	1.5	3.3	1.3
	Fat free-enriched:											
2161	Cracked pepper, Snackwell	1 ea	14	2	61	1	10	<1	2	.3	.6	.2
2159	Wheat, Snackwell	7 ea	15	1	60	2	12	1	<1	.1	.1	.1
2075	Whole wheat, herb seasoned	5 ea	14	5	50	2	11	2	0	0	0	0
2077	Whole wheat, onion	5 ea	14	4	50	2	11	2	0	0	0	0
419	Graham-enriched	2 ea	14	4	59	1	11	<1	1	.2	.6	.5
420	Melba toast, plain-enriched	1 pce	5	5	19	1	4	<1	<1	t	t	.1
1514	Rice cakes, unsalted-enriched	2 ea	18	6	70	1	15	1	<1	.1	.2	.2
421	Rye wafer, whole grain	2 ea	22	5	73	2	18	5	<1	t	t	.1
422	Saltine-enriched	4 ea	12	4	52	1	9	<1	1	.4	.8	.2
1971	Saltine, unsalted tops-enriched	2 ea	6		25	1	4	0	<1			
423	Snack-type, round like Ritz-enriched	3 ea	9	3	45	1	5	<1	2	.3	1	.9
424	Wheat, thin-enriched	4 ea	8	3	38	1	5	<1	2	.3	.9	.2
425	Whole-wheat wafers	2 ea	8	3	35	1	5	1	1	.3	.5	.5
426	Croissants, 4½ x 4 x 1¾"	1 ea	57	23	231	5	26	1	12	6.6	3.1	.6
1699	Croutons, seasoned	½ c	20	4	93	2	13	1	4	1	1.9	.5
	Danish pastry:											
428	Round piece, plain, 4¼" diam, 1" high	1 ea	88	21	349	5	47	<1	17	3.5	10.6	1.6
429	Ounce, plain	1 oz	28	21	111	2	15	<1	5	1.1	3.4	.5
430	Round piece with fruit	1 ea	94	29	335	5	45		16	3.3	10.1	1.6
	Desserts, 3 x 3" piece:											
1348	Apple crisp	1 pce	78	61	127	1	25	1	3	.6	1.2	.9
1353	Apple cobbler	1 pce	104	57	199	2	35	2	6	1.2		
1349	Cherry crisp	1 pce	138	75	157	2	27	2	5	.9		
1352	Cherry cobbler	1 pce	129	66	197	2	34	1	6	1.2		
1350	Peach crisp	1 pce	139	73	166	1	30	2	5	.8		
1351	Peach cobbler	1 pce	130	64	203	2	36	2	6	1.2		
	Doughnuts:											
431	Cake type, plain, 3¼" diam	1 ea	47	21	198	2	23	1	11	1.7	4.4	3.7
432	Yeast-leavened, glazed, 3¾" diam	1 ea	60	25	242	4	27	1	14	3.5	7.7	1.7
	English muffins:											
433	Plain, enriched	1 ea	57	42	134	4	26	2	1	.1	.2	.5
434	Toasted	1 ea	52	37	133	4	26	2	1	.1	.2	.5
1504	Whole wheat	1 ea	66	46	134	6	27	4	1	.2	.3	.6
1414	Granola bar, soft	1 ea	28	6	124	2	19	1	5	2	1.1	1.5
1415	Granola bar, hard	1 ea	25	4	118	3	16	1	5	.6	1.1	3
1985	Granola bar, fat free, all flavors	1 ea	42	10	140	2	35	3	0	0	0	0
	Muffins, 2½" diam, 1½" high:											
	From home recipe:											
435	Blueberry	1 ea	57	39	165	4	23	1	6	1.4	1.6	3.1
436	Bran, wheat	1 ea	57	35	164	4	24	4	7	1.5	1.8	3.6
437	Cornmeal	1 ea	57	32	183	4	25	2	7	1.6	1.8	3.5
	From commercial mix:											
438	Blueberry	1 ea	50	36	150	3	24	1	4	.7	1.8	1.5
439	Bran, wheat	1 ea	50	35	138	3	23	2	5	1.2	2.3	.7
440	Cornmeal	1 ea	50	30	161	4	25	1	5	1.4	2.6	.6
1864	Nabisco Newtons, fat free, all flavors	1 ea	23		69	1	16		0	0	0	0
	Pancakes, 4" diam:											
441	Buckwheat, from mix w/ egg and milk	1 ea	30	54	62	2	8	1	2	.6	.6	.8
442	Plain, from home recipe	1 ea	38	53	86	2	11	1	4	.8	.9	1.7
443	Plain, from mix; egg, milk, oil added	1 ea	38	53	74	2	14	<1	1	.2	.3	.3

Chol (mg)	Calc (mg)	Iron (mg)	Magn (mg)	Pota (mg)	Sodi (mg)	Zinc (mg)	VT-A (µg)	Thia (mg)	VT-E (mg)	Ribo (mg)	Niac (mg)	V-B6 (mg)	Fola (µg)	VT-C (mg)
0	10	1.55	18	70	242	.32	0*	.03	1.38	.07	.83	.01	17	0
6	11	.88	5	32	146	.17	4*	.11	1.03	.1	1.07	.03	19	0
20	4	.58	3	15	102	.09	67*	.08	.18	.06	.64	<.01	2	0
15	43	.88	4	78	225	.13	5*	.09	1.54	.06	1.16	.01	25	0
<1	17	.61	5	28	95	.16	<1*	.05		.07	.69	.01		0
20	19	.95	6	39	125	.14	3	.11	.51	.13	1.24	.03	20	0
22	70	.82	13	105	356	.38	32*	.12	.64	.16	.93	.05	6	<1
0	33	.34	20	37	164	.33	1	.01	.35	.04	.31	.06	5	0
1	15	.48	4	14	99	.11	3*	.06	.26	.04	.47	.05	8	0
1	22	.82	16	69	278	.3	10*	.11	1.05	.1	1.83	.42	25	<1
0	24	.51	4	16	117	.11	<1*	.04	0	.05	.75	.01	11	<1
<1	28	.58	7	43	169	.21	<1*	.04		.07	.73	.02		0
0	0				80		100*							2
0	0				80		100*							2
0	3	.52	4	19	85	.11	0	.03	.29	.04	.58	.01	8	0
0	5	.18	3	10	41	.1	0	.02	<.01	.01	.21	<.01	6	0
0	2	.27	24	52	5	.54	<1	.01	.02	.03	1.41	.03	4	0
0	9	1.31	27	109	175	.62	<1	.09	.31	.06	.35	.06	10	<1
0	14	.65	3	15	156	.09	0	.07	.19	.05	.63	<.01	15	0
0		.36	5	50					.1					
0	11	.32	2	12	76	.06	0	.04	.41	.03	.36	<.01	7	0
0	4	.35	5	15	64	.13	0	.04	.32	.03	.4	.01	1	0
0	4	.25	8	24	53	.17	0	.02	.09	.01	.36	.01	2	0
38	21	1.16	9	67	424	.43	106*	.22	.24	.14	1.25	.03	35	<1
1	19	.56	8	36	248	.19	2*	.1	.44	.08	.93	.02	18	0
27	37	1.8	14	96	326	.48	5*	.25	.79	.19	2.2	.05	55	3
9	12	.57	4	30	104	.15	2*	.08	.25	.06	.7	.02	17	1
19	22	1.4	14	110	333	.48	24*	.29	.85	.21	1.8	.06	31	2
0	22	.58	5	76	142	.12	24*	.07		.06	.6	.03	4	2
1	66	.87	6	86	345	.16	57*	.09	.92	.08	.77	.03	11	<1
0	29	2.15	12	164	73	.15	209*	.06	.95	.08	.6	.05	15	3
1	73	1.89	10	113	352	.19	175*	.09	1.02	.11	.88	.04	17	2
0	23	.95	13	198	69	.2	129*	.05	2.46	.05	1.05	.03	10	5
1	68	1	10	139	349	.23	116*	.09	2.15	.09	1.22	.02	13	3
17	21	.92	9	60	257	.26	8*	.1	1.8	.11	.87	.03	22	<1
4	26	1.22	13	65	205	.46	2*	.22	1.84	.13	1.71	.03	26	<1
0	99	1.43	12	75	264	.4	0	.25	.1	.16	2.21	.02	46	0
0	98	1.41	11	74	262	.39	0	.2	.09	.14	1.98	.02	38	<1
0	175	1.62	47	139	420	1.06	0	.2	.46	.09	2.25	.11	32	0
<1	29	.72	21	91	78	.42	0	.08	.34	.05	.14	.03	7	0
0	15	.74	24	84	73	.51	4*	.07	.33	.03	.39	.02	6	<1
0	0	3.6			5		100*							0
22	107	1.29	9	69	251	.31	16*	.15	1.03	.16	1.26	.02	7	1
20	106	2.39	44	181	335	1.57	136*	.19	1.31	.25	2.29	.18	30	4
26	147	1.49	13	82	333	.35	23*	.17	1.08	.18	1.36	.05	10	<1
23	12	.56	5	39	219	.19	11*	.07	.7	.16	1.12	.04	5	<1
34	16	1.27	28	73	234	.57	15*	.1	.75	.12	1.44	.09	8	0
31	37	.97	10	65	398	.32	22*	.12	.75	.14	1.05	.05	28	<1
					77									
20	77	.56	17	70	160	.35	20*	.05	.62	.08	.4	.04	5	<1
22	83	.68	6	50	167	.21	20*	.08	.36	.11	.59	.02	14	<1
5	48	.59	8	66	239	.15	3*	.08	.32	.08	.65	.03	14	<1

A

*This value is expressed in retinol equivalents (RE). All other values are in retinol activity equivalents (RAE).

Table A–1

Food Composition (Computer code number is for Wadsworth Diet Analysis program) (For purposes of calculations, use "0" for t, <1, <.1, <.01, etc.)

Computer Code Number	Food Description	Measure	Wt (g)	H_2O (%)	Ener (kcal)	Prot (g)	Carb (g)	Dietary Fiber (g)	Fat (g)	Fat Breakdown (g) Sat	Mono	Poly
	BAKED GOODS: BREADS, CAKES, COOKIES, CRACKERS, PIES—Continued											
1468	Pan dulce, sweet roll w/topping	1 ea	79	21	291	5	48	1	9	1.8	4	2.5
	Piecrust,with enriched flour, vegetable shortening, baked:											
444	Home recipe, 9″ shell	1 ea	180	10	949	11	85	3	62	15.5	27.3	16.4
	From mix:											
445	Piecrust for 2-crust pie	1 ea	320	10	1686	20	152	5	111	27.6	48.5	29.2
446	1 pie shell	1 ea	160	11	802	11	81	3	49	12.3	27.6	6.2
	Pies, 9″ diam; pie crust made with vegetable shortening, enriched flour:											
448	Apple, ⅙ of pie	1 pce	117	52	277	2	40	2	13	4.4	5.1	2.6
450	Banana cream, ⅙ of pie	1 pce	144	48	387	6	47	1	20	5.4	8.2	4.7
452	Blueberry, ⅙ of pie	1 pce	147	51	360	4	49	2	17	4.3	7.5	4.5
454	Cherry, ⅙ of pie	1 pce	180	46	486	5	69	3	22	5.4	9.6	5.8
456	Chocolate cream, ⅙ of pie	1 pce	199	63	358	8	47	2	16	5.9		
458	Custard, ⅙ of pie	1 pce	105	61	221	6	22	2	12	2.5	5	3.9
460	Lemon meringue, ⅙ of pie	1 pce	113	42	303	2	53	1	10	2	3	4.1
462	Peach, ⅙ of pie	1 pce	139	47	354	4	53	3	15			
464	Pecan, ⅙ of pie	1 pce	113	19	452	5	65	4	21	4	12.1	3.6
466	Pumpkin, ⅙ of pie	1 pce	109	58	229	4	30	3	10	1.9	4.4	3.4
467	Pies, fried, commercial: Apple	1 ea	85	40	266	2	33	1	14	6.5	5.8	1.2
468	Pies, fried, commercial: Cherry	1 ea	128	38	404	4	54	3	21	3.1	9.5	6.9
	Pretzels, made with enriched flour:											
469	Thin sticks, 2¼″ long	1 oz	28	3	107	3	22	1	1	.2	.4	.3
470	Dutch twists	10 pce	60	3	229	5	47	2	2	.4	.8	.7
471	Thin twists, 3¼ x 2¼ x ¼″	10 pce	60	3	229	5	47	2	2	.4	.8	.7
	Rolls & buns, enriched, commercial:											
472	Cloverleaf rolls, 2½″ diam, 2″ high	1 ea	28	32	84	2	14	1	2	.5	1	.3
473	Hot dog buns	1 ea	40	34	114	3	20	1	2	.5	.3	1
474	Hamburger buns	1 ea	43	34	123	4	22	1	2	.5	.4	1.1
475	Hard roll, white, 3¾″ diam, 2″ high	1 ea	57	31	167	6	30	1	2	.3	.6	1
476	Submarine rolls/hoagies, 11¼ x 3 x 2½″	1 ea	135	34	386	11	68	4	7	1.6	3.4	1.2
	Rolls & buns, enriched, home recipe:											
477	Dinner rolls 2½″ diam, 2″ high	1 ea	35	29	112	3	19	1	3	.7	1.1	.7
	Sports/fitness bar:											
2043	Forza energy bar	1 ea	70	18	231	10	45	4	1			
2042	Power bar	1 ea	65		230	10	45	3	2			
2041	Tiger sports bar	1 ea	65	14	260	11	33	2	9	1.9		
478	Toaster pastries, fortified (Poptarts)	1 ea	52	12	204	2	37	1	5	.8	2.1	2
2132	Toaster strudel pastry—cream cheese	1 ea	54	30	200	3	24	<1	10	3		
2134	Toaster strudel pastry—french toast	1 ea	54	30	200	3	24	<1	10	3		
	Tortilla chips:											
1271	Plain	10 pce	18	2	90	1	11	1	5	.9	2.8	.7
1036	Nacho flavor	1 c	26	2	129	2	16	1	7	1.3	3.9	.9
1037	Taco flavor	1 pce	18	2	86	1	11	1	4	.8	2.6	.6
	Tortillas:											
479	Corn, enriched, 6″ diam	1 ea	26	44	58	1	12	1	1	.1	.2	.3
480	Flour, 8″ diam	1 ea	49	27	159	4	27	2	3	.9	1.8	.5
1301	Flour, 10″ diam	1 ea	72	27	234	6	40	2	5	1.3	2.7	.8
481	Taco shells	1 ea	14	6	65	1	9	1	3	.5	1.3	1.2
	Waffles, 7″ diam:											
482	From home recipe	1 ea	75	42	218	6	25	1	11	2.1	2.6	5.1
483	From mix, egg/milk added	1 ea	75	42	218	5	26	1	10	1.7	2.7	5.2
1510	Whole grain, prepared from frozen	1 ea	39	43	105	4	13	1	4	1.2	1.7	1.1
	GRAIN PRODUCTS: CEREAL, FLOUR, GRAIN, PASTA and NOODLES, POPCORN											
38070	Amaranth	1 c	195	10	729	28	129	30	13	3.2	2.8	5.6
484	Barley, pearled, dry, uncooked	1 c	200	10	704	20	155	31	2	.5	.3	1.1
485	Barley, pearled, cooked	1 c	157	69	193	4	44	6	1	.1	.1	.3

Chol (mg)	Calc (mg)	Iron (mg)	Magn (mg)	Pota (mg)	Sodi (mg)	Zinc (mg)	VT-A (µg)	Thia (mg)	VT-E (mg)	Ribo (mg)	Niac (mg)	V-B6 (mg)	Fola (µg)	VT-C (mg)
26	13	1.84	9	57	75	.35	67*	.23	1.22	.21	2.02	.03	19	<1
0	18	5.2	25	121	976	.79	0	.7	9.94	.5	5.95	.04	121	0
0	32	9.25	45	214	1734	1.41	0	1.25	17.7	.89	10.6	.08	214	0
0	96	3.44	24	99	1166	.62	0	.48	8.83	.3	3.8	.09	112	0
0	13	.53	8	76	311	.19	35*	.03	.08	.03	.31	.04	26	4
73	108	1.5	23	238	346	.69	101*	.2	2.12	.3	1.52	.19	39	2
0	10	1.81	12	73	272	.29	6*	.22	3.09	.19	1.76	.05	34	1
0	18	3.33	16	139	344	.36	86*	.27	3.42	.22	2.3	.06	49	2
18	171	1.47	28	284	347	.82	33*	.17	1.85	.4	1.22	.06	28	1
35	84	.61	12	111	252	.55	70*	.04	1.96	.22	.31	.05	21	1
51	63	.69	17	101	165	.55	59*	.07	2.45	.24	.73	.03	15	4
4	12	2.22	16	280	228	.33	42*	.16	2.19	.13	2.55	.04	28	3
36	19	1.18	20	84	479	.64	53*	.1	2.09	.14	.28	.02	30	1
22	65	.86	16	168	307	.49	405*	.06	1.85	.17	.2	.06	22	1
13	13	.88	8	51	325	.17	33*	.1	.37	.08	.98	.03	4	1
0	28	1.56	13	83	479	.29	22*	.18	.55	.14	1.82	.04	23	2
0	10	1.21	10	41	480	.24	0	.13	.06	.17	1.47	.03	48	0
0	22	2.59	21	88	1029	.51	0	.28	.13	.37	3.15	.07	103	0
0	22	2.59	21	88	1029	.51	0	.28	.13	.37	3.15	.07	103	0
<1	33	.88	6	37	146	.22	0	.14	.25	.09	1.13	.01	27	<1
0	56	1.27	8	56	224	.25	0	.19	.62	.12	1.57	.02	38	<1
0	60	1.36	9	61	241	.27	0	.21	.67	.13	1.69	.02	41	<1
0	54	1.87	15	62	310	.54	0	.27	.19	.19	2.42	.02	54	0
0	188	4.28	27	190	756	.84	0	.65	.62	.42	5.31	.06	36	0
13	21	1.04	7	53	145	.24	28*	.14	.35	.14	1.21	.02	15	<1
0	300	6.3	160	220	65	5.25		1.5	27.1	1.7	20	2	400	60
0	300	5.4	140	150	110	5.25		1.5		1.7	20	2	400	60
0	557	5.01	186		139		279*	2.37		1.11	5.57	1.11		11
0	13	1.81	9	58	218	.34	100*	.15	1.19	.19	2.05	.2	34	<1
10	0	1.08			220		0							0
10	0	1.08			220		0							0
0	28	.27	16	35	95	.27	2	.01	.24	.03	.23	.05	2	0
1	38	.37	21	56	184	.31	5	.03	.35	.05	.37	.07	4	<1
1	28	.36	16	39	142	.23	8	.04	.24	.04	.36	.05	4	<1
0	45	.36	17	40	42	.24	0	.03	.04	.02	.39	.06	30	0
0	61	1.62	13	64	234	.35	0	.26	.45	.14	1.75	.02	60	0
0	90	2.38	19	94	344	.51	0	.38	.66	.21	2.57	.04	89	0
0	22	.35	15	25	51	.2	5	.03	.42	.01	.19	.05	1	0
52	191	1.73	14	119	383	.51	49*	.2	1.73	.26	1.55	.04	34	<1
38	93	1.22	15	134	458	.35	19*	.15	1.5	.19	1.23	.07	9	<1
37	102	.81	15	90	132	.44	30*	.08	.55	.13	.77	.04	7	<1
0	298	14.8	519	714	41	6.2	0	.16	2.01	.41	2.51	.43	96	8
0	58	5	158	560	18	4.26	2	.38	.26	.23	9.21	.52	46	0
0	17	2.09	34	146	5	1.29	1	.13	.08	.1	3.24	.18	25	0

*This value is expressed in retinol equivalents (RE). All other values are in retinol activity equivalents (RAE).

A

Table A–1

Food Composition   (Computer code number is for Wadsworth Diet Analysis program)   (For purposes of calculations, use "0" for t, <1, <.1, <.01, etc.)

Computer Code Number	Food Description	Measure	Wt (g)	H₂O (%)	Ener (kcal)	Prot (g)	Carb (g)	Dietary Fiber (g)	Fat (g)	Sat	Mono	Poly
	GRAIN PRODUCTS: CEREAL, FLOUR, GRAIN, PASTA and NOODLES, POPCORN—Continued											
2009	Breakfast bars, fat free, all flavors	1 ea	38	25	110	2	26	3	0	0	0	0
	Breakfast bar, Snackwell:											
2165	Apple-cinnamon	1 ea	37	16	119	1	29	1	<1	.1	t	.1
2164	Blueberry	1 ea	37	16	121	1	29	1	<1	t	t	.1
2163	Strawberry	1 ea	37	16	120	1	29	1	<1	t	t	.1
	Breakfast cereals, hot, cooked:w/o salt added											
	Corn grits (hominy) enriched:											
486	Regular/quick prep w/o salt, yellow:	1 c	242	85	145	3	31	<1	<1	.1	.1	.2
487	Instant, prepared from packet, white	1 ea	137	82	89	2	21	1	<1	t	t	.1
	Cream of wheat:											
488	Regular, quick, instant	1 c	239	87	129	4	27	1	<1	.1	.1	.3
489	Mix and eat, plain, packet	1 ea	142	82	102	3	21	<1	<1	t	t	.2
1664	Farina cereal, cooked w/o salt	1 c	233	88	117	3	25	3	<1	t	t	.1
490	Malt-O-Meal, cooked w/o salt	1 c	240	88	122	4	26	1	<1	.1	.1	t
494	Maypo	1 c	216	83	153	5	29	5	2	.4	.7	.8
	Oatmeal or rolled oats:											
	Regular, quick, instant,nonfortified											
491	cooked w/o salt	1 c	234	85	145	6	25	4	2	.4	.7	.9
	Instant, fortified:											
492	Plain, from packet	½ c	118	85	70	3	12	2	1	.2	.4	.4
493	Flavored, from packet	½ c	109	76	106	3	21	2	1	.2	.5	.5
	Breakfast cereals, ready to eat:											
495	All-Bran	1 c	62	3	164	8	47	20	2	.4	.4	1.1
1306	Alpha Bits	1 c	28	1	110	2	24	1	1	.1	.2	.2
1307	Apple Jacks	1 c	33	3	127	2	29	1	<1	.1	.1	.2
1308	Bran Buds	1 c	90	3	248	8	72	36	2	.4	.4	1.3
1305	Bran Chex	1 c	49	2	156	5	39	8	1	.2	.3	.7
1309	Honey BucWheat Crisp	1 c	38	5	147	4	31	3	1	.2	.2	.5
1310	C.W. Post, plain	1 c	97	2	421	9	73	7	13	1.7	6	4.7
1311	C.W. Post, with raisins	1 c	103	4	446	9	74	14	15	11	1.7	1.4
496	Cap'n Crunch	1 c	37	2	147	2	32	1	2	.5	.4	.3
1312	Cap'n Crunchberries	1 c	35	2	140	2	30	1	2	.5	.3	.3
1313	Cap'n Crunch, peanut butter	1 c	35	2	146	3	28	1	3	.7	1.1	.7
497	Cheerios	1 c	23	3	84	2	17	2	1	.3	.5	.2
1314	Cocoa Krispies	1 c	41	2	159	2	36	1	1	.8	.1	.2
1316	Cocoa Pebbles	1 c	32	3	127	1	28	1	1	1.2	.1	t
1315	Corn Bran	1 c	36	3	120	2	30	6	1	.3	.3	.4
1317	Corn Chex	1 c	28	2	105	2	24	<1	<1	.1	.1	.2
498	Corn Flakes, Kellogg's	1 c	28	3	102	2	24	1	<1	.1	t	.1
499	Corn Flakes, Post Toasties	1 c	28	3	101	2	24	1	<1	0	t	t
1340	Corn Pops	1 c	31	3	118	1	28	<1	<1	.1	.1	t
1318	Cracklin' Oat Bran	1 c	65	4	266	5	47	8	8	3.4	3.8	.9
1038	Crispy Wheat `N Raisins	1 c	43	7	150	3	35	3	1	.1	.1	.1
1319	Fortified Oat Flakes	1 c	48	3	180	8	36	1	1	.2	.3	.4
500	40% Bran Flakes, Kellogg's	1 c	39	4	128	4	31	6	1	.2	.2	.5
501	40% Bran Flakes, Post	1 c	47	4	150	4	38	8	1	.2	.2	.7
502	Froot Loops	1 c	32	2	125	2	28	1	1	.4	.2	.3
518	Frosted Flakes	1 c	41	3	158	2	37	1	<1	.1	3.4	.1
1320	Frosted Mini-Wheats	1 c	55	5	186	5	45	6	1	.2	.1	.6
1321	Frosted Rice Krispies	1 c	35	2	132	2	32	<1	<1	.1	.1	.1
1324	Fruit & Fibre w/dates	1 c	57	9	193	5	43	8	3	.4	.5	1.5
1322	Fruity Pebbles	1 c	32	3	128	1	28	<1	1	.3	.6	.4
503	Golden Grahams	1 c	39	3	150	2	33	1	1	.2	.4	.2
504	Granola, homemade	½ c	61	5	285	9	32	6	15	2.9	4.8	6.4
	Granola, low fat	½ c	47	3	181	5	38	3	3	0		
1670	Granola, low fat, commercial	½ c	45	5	165	4	36	3	2	.8	.5	.9
505	Grape Nuts	½ c	55	3	197	6	45	5	1	.2	.2	.6
1326	Grape Nuts Flakes	1 c	39	3	142	4	32	3	1	.2	.3	.6
1665	Heartland Natural with raisins	1 c	110	5	468	11	76	6	16	4	4.2	6.2
1327	Honey & Nut Corn Flakes	1 c	37	2	150	3	31	1	2	.3	.8	.6
506	Honey Nut Cheerios	1 c	33	2	126	3	27	2	1	.3	.5	.2

Chol (mg)	Calc (mg)	Iron (mg)	Magn (mg)	Pota (mg)	Sodi (mg)	Zinc (mg)	VT-A (µg)	Thia (mg)	VT-E (mg)	Ribo (mg)	Niac (mg)	V-B6 (mg)	Fola (µg)	VT-C (mg)
0	20	.72			25		20*							1
<1	17	5	6	68	103	3.88	260*	.39		.44	5.2	.52		<1
<1	14	4.83	5	43	107	3.85	260*	.39		.44	5.2	.52		<1
<1	14	4.82	6	47	102	3.83	260*	.39		.44	5.2	.52		2
0	0	1.55	10	53	0	.17	7	.24	.05	.14	1.96	.06	75	0
0	8	8.19	11	38	289	.21	0	.15	.03	.08	1.38	.05	47	0
0	50	10.3	12	45	139	.33	0	.24	.02	0	1.43	.03	108	0
0	20	8.09	7	38	241	.24	376	.43	.01	.28	4.97	.57	101	0
0	5	1.17	5	30	0	.16	0	.19	.02	.12	1.28	.02	54	0
0	5	9.6	5	31	2	.17	0	.48	.02	.24	5.76	.02	5	0
0	112	7.56	45	190	233	1.34	633	.65	1.51	.65	8.42	.86	9	26
0	19	1.59	56	131	2	1.15	2	.26	.23	.05	.3	.05	9	0
0	109	4.2	28	66	190	.58	302*	.35	.14	.19	3.65	.49	65	0
0	112	4.45	34	91	169	.66	306*	.35	.14	.25	3.92	.51	100	<1
0	219	9.3	266	706	126	7.75	466	.81	1.14	.87	10.4	1.05	186	31
0	8	2.66	16	54	178	1.48	371	.36	.02	.42	4.93	.5	99	0
0	4	4.95	10	35	148	4.13	248	.43	.05	.46	5.51	.56	116	16
0	60	13.5	250	809	599	19.4	676	1.17	1.42	1.26	15	1.53	270	45
0	29	14	69	216	345	6.48	5	.64	.56	.26	8.62	.88	173	26
0	54	10.9	43	142	361	.68	914	.9	8.99	1.03	12.1	1.88	11	36
<1	47	15.4	67	198	167	1.64	1284*	1.26	.68	1.46	17.1	1.75	342	0
<1	50	16.4	74	261	161	1.64	1363	1.34	.72	1.55	18.1	1.85	364	0
0	7	6.17	13	47	286	5.14	2	.51	.18	.58	6.86	.68	137	0
0	9	6.06	13	49	256	5.4	6*	.5	.25	.57	6.73	.67	135	<1
0	3	5.83	24	80	264	4.86	2	.49	.19	.55	6.48	.65	130	0
0	42	6.21	25	68	218	2.88	288	.29	.16	.33	3.83	.38	153	11
0	5	2.38	15	79	278	1.97	298	.49	.19	.57	6.6	.66	123	20
0	4	1.99	12	47	173	1.65	248	.41		.47	5.52	.55	110	0
0	27	10.1	19	75	338	5	3	.1	.19	.56	6.67	.67	134	0
0	93	8.4	8	30	270	.35	0	.35	.09	0	4.67	.47	186	6
0	1	8.68	3	25	298	.17	210	.36	.03	.39	4.68	.48	99	14
0	1	5.4	4	33	266	.13	225	.37	.67	.43	5	.5	100	0
0	2	1.86	2	23	123	1.55	233	.4	.03	.43	5.18	.53	109	15
0	29	2.41	90	301	231	1.95	299	.5	.43	.56	6.63	.66	181	20
0	54	3.52	33	180	223	.85	293	.29	.45	.33	3.91	.39	157	0
0	68	13.7	58	228	220	2.54	636	.62	.34	.72	8.45	.86	169	0
0	19	10.9	81	236	304	5.04	488	.51	7.22	.58	6.72	.66	138	20
0	26	12.7	101	290	344	2.35	353	.59		.67	7.83	.78	157	0
0	4	4.51	9	34	150	4	225	.42	.12	.45	5.34	.54	96	15
0	1	5.95	4	27	264	.2	298	.49	.05	.57	6.6	.66	123	20
0	20	15.4	56	183	2	1.6	0	.38	.49	.44	5.39	.49	110	0
0	2	2.42	8	27	256	.42	303	.49	.03	.56	6.72	.66	140	20
0	30	10.1	81	335	270	3.02	725	.75	1.32	.85	10.1	1	201	0
0	2	2.13	6	35	187	1.78	267	.44		.5	5.93	.59	118	0
0	19	5.85	12	69	357	4.88	293	.49	.29	.55	6.5	.65	130	19
0	49	2.56	109	328	15	2.48	1	.45	7.86	.17	1.25	.19	52	1
0		2.71	36	143	90	5.64	226*	.56	7.57	.64	7.52	.75	151	
0	20	1.58	37	127	101	2.84	169	.27	4.53	.31	3.74	.36	90	1
0	19	15.4	55	169	336	1.14	214	.36		.4	4.74	.47	95	0
0	15	10.9	40	133	188	1.61	303	.5	.1	.57	6.72	.67	135	0
0	66	4.02	141	415	226	2.83	3	.32	.77	.14	1.54	.2	44	1
0	4	3.03	3	40	249	.26	152	.26	.09	.3	3.37	.33	74	10
0	22	4.95	32	94	285	4.13	248	.41	.34	.47	5.5	.55	220	16

*This value is expressed in retinol equivalents (RE). All other values are in retinol activity equivalents (RAE).

Table A–1

Food Composition

(Computer code number is for Wadsworth Diet Analysis program)   (For purposes of calculations, use "0" for t, <1, <.1, <.01, etc.)

Computer Code Number	Food Description	Measure	Wt (g)	H₂O (%)	Ener (kcal)	Prot (g)	Carb (g)	Dietary Fiber (g)	Fat (g)	Fat Breakdown (g)		
										Sat	Mono	Poly
	GRAIN PRODUCTS: CEREAL, FLOUR,											
	GRAIN, PASTA and NOODLES, POPCORN—Continued											
1328	HoneyBran	1 c	35	2	119	3	29	4	1	.3	.1	.3
1329	HoneyComb	1 c	22	1	87	1	20	1	<1	.1	.1	.2
1330	King Vitaman	1 c	21	2	81	2	18	1	1	.2	.3	.2
1039	Kix	1 c	19	2	72	1	16	1	<1	.1	.1	t
1331	Life	1 c	44	4	167	4	35	3	2	.3	.6	.8
507	Lucky Charms	1 c	32	2	124	2	27	1	1	.2	.4	.2
1323	Mueslix Five Grain	1 c	82	8	289	6	63	6	5	.7	2	1.8
508	Nature Valley Granola	1 c	113	4	510	12	74	7	20	2.6	13.3	3.8
1666	Nutri Grain Almond Raisin	1 c	40	6	147	3	31	3	2	.1	1	1.2
1336	100% Bran	1 c	66	3	178	8	48	19	3	.6	.6	1.9
509	100% Natural cereal, plain	1 c	104	2	462	11	71	8	17	7.4	7.4	2.2
1337	100% Natural with apples & cinnamon	1 c	104	2	477	11	70	7	20	15.5	1.8	1.3
1338	100% Natural with raisins & dates	1 c	110	3	496	12	72	7	20	13.6	3.7	1.7
510	Product 19	1 c	30	3	110	3	25	1	<1	t	.1	.2
1339	Quisp	1 c	30	3	121	1	25	1	2	.5	.4	.2
511	Raisin Bran, Kellogg's	1 c	61	8	186	6	47	8	1	.1	.2	.8
512	Raisin Bran, Post	1 c	59	9	187	5	46	8	1	.2	.2	.7
1667	Raisin Squares	1 c	71	9	241	6	55	7	2	.2	.1	.6
1041	Rice Chex	1 c	33	2	125	2	29	<1	<1	t	t	.1
513	Rice Krispies, Kellogg's	1 c	28	2	111	2	25	<1	<1	t	t	t
514	Rice, puffed	1 c	14	4	54	1	12	<1	<1	t	t	t
515	Shredded Wheat	1 c	43	5	154	5	35	4	1	.1	.1	.4
516	Special K	1 c	31	3	115	6	22	1	<1	t	0	.2
517	Super Golden Crisp	1 c	33	1	123	2	30	<1	<1	.1	.1	.1
519	Honey Smacks	1 c	36	3	137	2	31	1	1	.4	.1	.3
1341	Tasteeos	1 c	24	2	94	3	19	3	1	.2	.2	.2
1342	Team	1 c	42	4	164	3	36	1	1	.1	.2	.3
520	Total, wheat, with added calcium	1 c	40	3	140	4	32	4	1	.2	.2	.1
521	Trix	1 c	28	2	114	1	24	1	2	.4	.8	.3
1344	Wheat Chex	1 c	46	2	159	5	37	5	1	.2	.2	.4
1043	Wheat cereal, puffed, fortified	1 c	12	4	44	2	9	1	<1	t	t	.1
522	Wheaties	1 c	29	3	106	3	23	2	1	.2	.2	.1
523	Buckwheat flour, dark	1 c	120	11	402	15	85	12	4	.8	1.1	1.1
525	Buckwheat, whole grain, dry	1 c	170	10	583	22	122	17	6	1.3	1.8	1.8
526	Bulgar, dry, uncooked	1 c	140	9	479	17	106	26	2	.3	.2	.8
527	Bulgar, cooked	1 c	182	78	151	6	34	8	<1	.1	.1	.2
	Cornmeal:											
528	Whole-ground, unbolted, dry	1 c	122	10	442	10	94	9	4	.6	1.2	2
530	Degermed, enriched, dry	1 c	138	12	505	12	107	10	2	.3	.6	1
38041	Degermed, enriched, baked	1 c	138	12	505	12	107	10	2	.3	.6	1
38076	Couscous, cooked	1 c	157	73	176	6	36	2	<1	t	t	.1
38329	Cracked wheat	1 c	120	10	407	16	87	14	2	.4	.3	.9
	Macaroni, cooked:											
532	Enriched	1 c	140	66	197	7	40	2	1	.1	.1	.4
533	Whole wheat	1 c	140	67	174	7	37	4	1	.1	.1	.3
534	Vegetable, enriched	1 c	134	68	172	6	36	6	<1	t	t	.1
535	Millet, cooked	1 c	240	71	286	8	57	3	2	.4	.4	1.2
7508	Natto	1 c	175	55	371	31	25	9	19	2.8	4.2	10.9
	Noodles (see also Pasta and Spaghetti):											
1507	Cellophane noodles, cooked	1 c	190	79	160	<1	39	<1	<1	t	t	t
1995	Cellophane noodles, dry	1 c	140	13	491	<1	121	1	<1	t	t	t
537	Chow Mein, dry	1 c	45	1	237	4	26	2	14	2	3.5	7.8
536	Egg noodles, cooked, enriched	1 c	160	69	213	8	40	2	2	.5	.7	.7
538	Spinach noodles, dry	3½ oz	100	8	372	13	75	11	2	.2	.2	.6
1343	Oat bran, dry	¼ c	24	7	59	4	16	4	2	.3	.6	.7
	Pasta, cooked:											
1418	Fresh	2 oz	57	69	75	3	14	1	1	.1	.1	.2
1417	Linguini/Rotini	1 c	140	66	197	7	40	2	1	.1	.1	.4
	Popcorn:											
539	Air popped, plain	1 c	8	4	31	1	6	1	<1	t	.1	.2
1042	Microwaved, low fat, low sodium	1 c	6	3	25	1	4	1	1	.1	.2	.2

Chol (mg)	Calc (mg)	Iron (mg)	Magn (mg)	Pota (mg)	Sodi (mg)	Zinc (mg)	VT-A (µg)	Thia (mg)	VT-E (mg)	Ribo (mg)	Niac (mg)	V-B6 (mg)	Fola (µg)	VT-C (mg)
0	16	5.57	46	151	202	.9	463	.45	.81	.52	6.16	.63	23	19
0	4	2.05	8	26	163	1.14	171	.28		.32	3.79	.38	76	0
0	3	5.92	18	58	176	2.65	212	.26	1.42	.3	3.53	.35	71	8
0	28	5.13	6	26	167	2.38	238	.24	.05	.27	3.17	.32	127	9
0	134	12.3	43	109	240	5.5	1	.55	.22	.62	7.33	.73	147	0
0	35	4.8	21	58	217	4	240	.4	.14	.45	5.33	.53	213	16
0	67	8.94	82	369	107	7.46	747	.75	8.94	.84	9.84	.99	197	1
0	85	3.53	107	375	183	2.27	0	.35	7.97	.12	1.25	.16	17	0
0	122	1	9	143	142	2.72	0	.28	4	.32	3.64	.36	80	0
0	46	8.12	312	652	457	5.74	0	1.58	1.53	1.78	20.9	2.11	47	63
1	100	3.11	109	457	28	2.5	1*	.36	1.19	.17	1.84	.19	26	<1
0	157	2.89	72	514	52	2	3	.33	.73	.57	1.87	.11	17	1
0	160	3.12	124	538	47	2.11	3	.31	.77	.65	2.09	.16	45	0
0	3	18	12	40	216	15	225	1.5	22.2	1.71	20	2.01	390	60
0	6	5.09	15	40	216	4.25	2	.42	.16	.48	5.66	.56	113	0
0	35	5	89	437	354	4.15	250	.43	.55	.49	5.55	.55	122	0
0	27	10.8	88	357	360	2.25	225	.38		.42	5	.5	100	0
0	24	21.7	62	335	4	1.99	0	.5	.38	.57	6.67	.64	142	0
0	110	9.58	10	38	310	0	0	.4	0	.02	5.32	.53	213	6
0	5	.7	12	27	206	.46	371	.52	.03	.59	6.92	.69	138	15
0	1	.41	4	16	1	.15	0	.06	.01	.01	.87	0	1	0
0	16	1.81	57	155	4	1.42	0	.11	.23	.12	2.26	.11	21	0
0	5	8.71	18	55	250	3.75	225	.53	.08	.59	7.01	.71	93	15
0	7	2.08	20	48	51	1.75	437	.43	.12	.49	5.81	.59	116	0
0	4	2.4	21	56	68	.47	300	.5	.18	.58	6.66	.68	133	20
0	11	6.86	26	71	183	.69	318	.31	.17	.36	4.22	.43	85	13
0	6	12	12	71	260	.58	556*	.55	.1	.63	7.39	.76	7	22
0	344	24	43	129	265	20	500	2	31.3	2.27	26.8	2.67	533	80
0	30	4.2	3	16	184	3.5	210	.35	.56	.4	4.67	.47	93	14
0	92	13.8	51	178	412	1.13	0	.34	.68	.06	4.6	.46	368	6
0	3	.56	16	44	1	.37	<1	.05	.08	.03	1.43	.02	4	0
0	53	7.83	31	101	215	.68	218	.36	.36	.41	4.83	.48	97	14
0	49	4.87	301	692	13	3.74	0	.5	1.24	.23	7.38	.7	65	0
0	31	3.74	393	782	2	4.08	0	.17	1.75	.72	11.9	.36	51	0
0	49	3.44	230	574	24	2.7	0	.32	.22	.16	7.16	.48	38	0
0	18	1.75	58	124	9	1.04	0	.1	.05	.05	1.82	.15	33	0
0	7	4.21	155	350	43	2.22	29	.47	.82	.24	4.43	.37	30	0
0	7	5.7	55	224	4	.99	28	.99	.45	.56	6.95	.35	258	0
0	7	5.7	55	224	4	.99	28	.74	.45	.48	6.6	.27	52	0
0	13	.6	13	91	8	.41	0	.1	.02	.04	1.54	.08	24	0
0	41	4.67	166	486	6	3.53	0	.54	.5	.26	7.64	.41	53	0
0	10	1.96	25	43	1	.74	0	.29	.04	.14	2.34	.05	98	0
0	21	1.48	42	62	4	1.13	0	.15	.14	.06	.99	.11	7	0
0	15	.66	25	41	8	.59	3	.15	.05	.08	1.44	.03	87	0
0	7	1.51	106	149	5	2.18	0	.25	.14	.2	3.19	.26	46	0
0	380	15.1	201	1275	12	5.3	0	.28	.02	.33	0	.23	14	23
0	13	.85	3	3	8	.2	0	.04	.06	0	.06	.01	1	0
0	35	3.04	4	14	14	.57	0	.21	.18	0	.28	.07	3	0
0	9	2.13	23	54	198	.63	2	.26	.07	.19	2.68	.05	40	0
53	19	2.54	30	45	11	.99	10	.3	.08	.13	2.38	.06	102	0
0	58	2.13	174	376	36	2.76	23	.37	.04	.2	4.55	.32	48	0
0	14	1.3	56	136	1	.75	0	.28	.41	.05	.22	.04	12	0
19	3	.65	10	14	3	.32	3	.12	.09	.09	.56	.02	36	0
0	10	1.96	25	43	1	.74	0	.29	.08	.14	2.34	.05	98	0
0	1	.21	10	24	<1	.27	1	.02	.01	.02	.16	.02	2	0
0	1	.14	9	14	29	.23	<1	.02	.06	.01	.12	.01	1	0

*This value is expressed in retinol equivalents (RE). All other values are in retinol activity equivalents (RAE).

Table A–1

Food Composition   (Computer code number is for Wadsworth Diet Analysis program)   (For purposes of calculations, use "0" for t, <1, <.1, <.01, etc.)

Computer Code Number	Food Description	Measure	Wt (g)	H₂O (%)	Ener (kcal)	Prot (g)	Carb (g)	Dietary Fiber (g)	Fat (g)	Fat Breakdown (g) Sat	Mono	Poly
	GRAIN PRODUCTS: CEREAL, FLOUR,											
	GRAIN, PASTA and NOODLES, POPCORN—Continued											
540	Popped in vegetable oil/salted	1 c	11	3	55	1	6	1	3	.5	.9	1.5
541	Sugar-syrup coated	1 c	35	3	151	1	28	2	4	1.3	1	1.6
38079	Quinoa, dry	1 c	170	9	636	22	117	10	10	1	2.6	4
	Rice:											
542	Brown rice, cooked	1 c	195	73	216	5	45	4	2	.4	.6	.6
8858	Mexican rice, cooked	1 c	250		349	17	66	2	4		2	
2216	Spanish rice, cooked	1 c	246	85	130	3	28	2	1	0		
	White, enriched, all types:											
543	Regular/long grain, dry	1 c	185	12	675	13	148	2	1	.3	.4	.3
544	Regular/long grain, cooked	1 c	158	68	205	4	44	1	<1	.1	.1	.1
545	Instant, prepared without salt	1 c	165	76	162	3	35	1	<1	.1	.1	.1
	Parboiled/converted rice:											
546	Raw, dry	1 c	185	10	686	13	151	3	1	.3	.3	.3
547	Cooked	1 c	175	72	200	4	43	1	<1	.1	.1	.1
1486	Sticky Rice (Glutinous), cooked	1 c	174	77	169	4	37	2	<1	.1	.1	.1
548	Wild rice, cooked	1 c	164	74	166	7	35	3	1	.1	.1	.3
1700	Rice and pasta (Rice-a-Roni), cooked	1 c	202	72	246	5	43	5	6	1.1	2.3	1.9
549	Rye flour, medium	1 c	102	10	361	10	79	15	2	.2	.2	.8
1044	Soy flour, low-fat	1 c	88	3	325	45	30	9	6	.9	1.3	3.3
	Spaghetti pasta:											
550	Without salt, enriched	1 c	140	66	197	7	40	2	1	.1	.1	.4
551	With salt, enriched	1 c	140	66	197	7	40	2	1	.1	.1	.4
552	Whole-wheat spaghetti, cooked	1 c	140	67	174	7	37	6	1	.1	.1	.3
1302	Tapioca-pearl, dry	1 c	152	11	544	<1	135	1	<1	t	t	t
553	Wheat bran, crude	1 c	58	10	125	9	37	25	2	.4	.4	1.3
554	Wheat germ, raw	1 c	115	11	414	27	60	15	11	1.9	1.6	6.9
555	Wheat germ, toasted	1 c	113	6	432	33	56	15	12	2.1	1.7	7.5
1669	Wheat germ, with brown sugar & honey	1 c	113	3	420	30	66	11	9	1.5	1.2	5.5
556	Rolled wheat, cooked	1 c	240	84	149	5	33	4	1	.1	.1	.5
557	Whole-grain wheat, cooked	1 c	150	86	84	4	20	3	<1	.1	.1	.2
	Wheat flour (unbleached):											
	All-purpose white flour, enriched:											
558	Sifted	1 c	115	12	419	12	88	3	1	.2	.1	.5
559	Unsifted	1 c	125	12	455	13	95	3	1	.2	.1	.5
560	Cake or pastry, enriched, sifted	1 c	96	13	348	8	75	2	1	.1	.1	.4
561	Self-rising, enriched, unsifted	1 c	125	11	443	12	93	3	1	.2	.1	.5
562	Whole wheat, from hard wheats	1 c	120	10	407	16	87	15	2	.4	.3	.9
	MEATS: FISH and SHELLFISH											
1045	Bass, baked or broiled	4 oz	113	69	165	27	0	0	5	1.1	2.1	1.5
1046	Bluefish, baked or broiled	4 oz	113	63	180	29	0	0	6	1.3	2.6	1.5
1686	Catfish, breaded/flour fried	4 oz	113	49	329	21	14	<1	20	4.5	9.1	5.2
	Clams:											
563	Raw meat only	1 ea	145	82	107	18	4	0	1	.1	.1	.4
564	Canned, drained	1 c	160	64	237	41	8	0	3	.3	.3	.9
1290	Steamed, meat only	10 ea	95	64	141	24	5	0	2	.2	.2	.5
	Cod:											
565	Baked	4 oz	113	76	119	26	0	0	1	.2	.1	.3
566	Batter fried	4 oz	113	67	197	20	8	<1	9	1.8	3.6	3
567	Poached, no added fat	4 oz	113	77	116	25	0	0	1	.1	.1	.3
	Crab, meat only:											
1048	Blue crab, cooked	1 c	118	77	120	24	0	0	2	.3	.3	.8
1049	Dungeness crab, cooked	1 c	118	73	130	26	1	0	1	.2	.3	.5
568	Blue crab, canned	1 c	135	76	134	28	0	0	2	.3	.3	.6
1587	Crab, imitation, from surimi	4 oz	113	74	115	14	11	0	1	.3	.2	.8
569	Fish sticks, breaded pollock	2 ea	56	46	152	9	13	0	7	1.8	2.8	1.8
572	Flounder/sole, baked	4 oz	113	73	132	27	0	0	2	.4	.3	.7
1599	Grouper, baked or broiled	4 oz	113	73	133	28	0	0	1	.3	.3	.5
573	Haddock, breaded, fried	4 oz	113	60	247	23	10	<1	12	2.6	5.3	3.7
1050	Haddock, smoked	4 oz	113	71	131	28	0	0	1	.2	.2	.4

Chol (mg)	Calc (mg)	Iron (mg)	Magn (mg)	Pota (mg)	Sodi (mg)	Zinc (mg)	VT-A (µg)	Thia (mg)	VT-E (mg)	Ribo (mg)	Niac (mg)	V-B6 (mg)	Fola (µg)	VT-C (mg)
0	1	.31	12	25	97	.29	1	.01	.01	.01	.17	.02	2	<1
2	15	.61	12	38	72	.2	3*	.02	.42	.02	.77	.01	1	0
0	102	15.7	357	1258	36	5.61	0	.34	8.28	.67	4.98	.38	83	0
0	19	.82	84	84	10	1.23	0	.19	.43	.05	2.98	.28	8	0
0					1162		120*							
0	0	0			1340		0							0
0	52	7.97	46	213	9	2.02	0	1.07	.24	.09	7.76	.3	427	0
0	16	1.9	19	55	2	.77	0	.26	.08	.02	2.33	.15	92	0
0	13	1.04	8	7	5	.4	0	.12	.08	.08	1.45	.02	68	0
0	111	6.59	57	222	9	1.78	0	1.1	.24	.13	6.72	.65	427	0
0	33	1.98	21	65	5	.54	0	.44	.09	.03	2.45	.03	87	0
0	3	.24	9	17	9	.71	0	.03	.05	.02	.5	.04	2	0
0	5	.98	52	166	5	2.2	0	.08	.38	.14	2.11	.22	43	0
2	16	1.9	24	85	1147	.57	0	.25	.27	.16	3.6	.2	89	<1
0	24	2.16	76	347	3	2.03	0	.29	1.36	.12	1.76	.27	19	0
0	165	5.27	202	2261	16	1.04	2	.33	.17	.25	1.9	.46	361	0
0	10	1.96	25	43	1	.74	0	.29	.08	.14	2.34	.05	98	0
0	10	1.96	25	43	140	.74	0	.29	.38	.14	2.34	.05	98	0
0	21	1.48	42	62	4	1.13	0	.15	.07	.06	.99	.11	7	0
0	30	2.4	2	17	2	.18	0	.01	0	0	0	.01	6	0
0	42	6.13	354	686	1	4.22	0	.3	1.35	.33	7.87	.76	46	0
0	45	7.2	275	1025	14	14.1	0	2.16	20.7	.57	7.83	1.5	323	0
0	51	10.3	362	1070	5	18.8	0	1.89	20.5	.93	6.32	1.11	398	7
0	56	9.1	307	1089	12	15.7	6	1.51	24.9	.78	5.34	.56	376	0
0	17	1.49	53	170	0	1.15	0	.17	.48	.12	2.14	.17	26	0
0	9	.88	35	99	1	.73	0*	.12	.3	.03	1.5	.08	12	0
0	17	5.34	25	123	2	.8	0	.9	.07	.57	6.79	.05	177	0
0	19	5.8	27	134	2	.87	0	.98	.07	.62	7.38	.05	193	0
0	13	7.03	15	101	2	.59	0	.86	.06	.41	6.52	.03	148	0
0	423	5.84	24	155	1587	.77	0	.84	.07	.52	7.29	.06	193	0
0	41	4.66	166	486	6	3.52	0	.54	1.48	.26	7.64	.41	53	0
98	116	2.16	43	515	102	.94	40	.1	.84	.1	1.72	.16	19	2
86	10	.7	47	539	87	1.18	156	.08	.71	.11	8.19	.52	2	0
91	62	1.88	36	391	240	1.17	33*	.46	2.87	.21	3.78	.22	16	1
49	67	20.3	13	455	81	1.99	131	.12	1.45	.31	2.56	.09	23	19
107	147	44.7	29	1004	179	4.37	274	.24	1.6	.68	5.37	.18	46	35
64	87	26.6	17	597	106	2.59	162	.14	1.86	.4	3.19	.1	28	21
62	16	.55	47	276	88	.65	16	.1	.34	.09	2.84	.32	9	1
56	33	.81	28	437	104	.57	10*	.08	1.49	.11	2.58	.37	10	2
52	10	.37	30	484	90	.56	9	.02	.32	.05	2.45	.45	7	3
118	123	1.07	39	382	329	4.98	2	.12	1.18	.06	3.89	.21	60	4
90	70	.51	68	481	446	6.45	37	.07	1.33	.24	4.28	.2	50	4
120	136	1.13	53	505	450	5.43	3*	.11	1.35	.11	1.85	.2	58	4
23	15	.44	49	102	950	.37	23	.04	.11	.03	.2	.03	2	0
63	11	.41	14	146	326	.37	17	.07	.77	.1	1.19	.03	25	0
77	20	.38	65	389	119	.71	12	.09	2.14	.13	2.46	.27	10	0
53	24	1.29	42	537	60	.58	56	.09	.71	.01	.43	.4	11	0
87	70	2.02	49	376	194	.63	28*	.11	1.93	.12	4.94	.31	15	<1
87	55	1.58	61	469	862	.56	25	.05	.45	.05	5.73	.45	17	0

*This value is expressed in retinol equivalents (RE). All other values are in retinol activity equivalents (RAE).

Table A–1

Food Composition

(Computer code number is for Wadsworth Diet Analysis program)   (For purposes of calculations, use "0" for t, <1, <.1, <.01, etc.)

Computer Code Number	Food Description	Measure	Wt (g)	H₂O (%)	Ener (kcal)	Prot (g)	Carb (g)	Dietary Fiber (g)	Fat (g)	Fat Breakdown (g) Sat	Mono	Poly
	MEATS: FISH and SHELLFISH—Continued											
	Halibut:											
17291	Baked	4 oz	113	72	158	30	0	0	3	.5	1.1	1.1
1051	Smoked	4 oz	113	64	203	34	0	0	4	.6	1.2	1.5
1054	Raw	4 oz	113	78	124	23	0	0	3	.4	.8	.8
575	Herring, pickled	4 oz	113	55	296	16	11	0	20	2.7	13.5	1.9
1052	Lobster meat, cooked w/moist heat	1 c	145	76	142	30	2	0	1	.2	.2	.1
1687	Ocean perch, baked/broiled	4 oz	113	73	137	27	0	0	2	.4	.9	.6
576	Ocean perch, breaded/fried	4 oz	113	58	255	23	10	<1	13	2.7	5.8	3.9
1056	Octopus, raw	4 oz	113	80	93	17	2	0	1	.3	.2	.3
	Oysters:											
577	Raw, Eastern	1 c	248	85	169	17	10	0	6	1.9	.8	2.4
578	Raw, Pacific	1 c	248	82	201	23	12	0	6	1.3	.9	2.2
	Cooked:											
579	Eastern, breaded, fried, medium	5 ea	73	65	144	6	8	<1	9	2.3	3.4	2.4
580	Western, simmered	5 ea	125	64	204	24	12	0	6	1.3	.9	2.2
581	Pollock, baked, broiled, or poached	4 oz	113	74	128	27	0	0	1	.3	.2	.6
	Salmon:											
582	Canned pink, solids and liquid	4 oz	113	69	157	22	0	0	7	1.7	2	2.3
583	Broiled or baked	4 oz	113	62	244	31	0	0	12	2.2	6	2.7
584	Smoked	4 oz	113	72	132	21	0	0	5	1	2.3	1.1
585	Atlantic sardines, canned, drained, 2 = 24g	4 oz	113	60	235	28	0	0	13	1.7	4.4	5.8
586	Scallops, breaded, cooked from frozen	6 ea	93	58	200	17	9	<1	10	2.5	4.2	2.7
1588	Scallops, imitation, from surimi	4 oz	113	74	112	14	12	0	<1	.1	.1	.2
1688	Scallops, steamed/boiled	½ c	60	77	65	10	1	0	2	.3	.7	.6
	Shrimp:											
587	Cooked, boiled, 2 large=11g	16 ea	88	77	87	18	0	0	1	.3	.2	.4
588	Canned, drained	½ c	64	73	77	15	1	0	1	.2	.2	.5
589	Fried, 2 large=15g, breaded	12 ea	90	53	218	19	10	<1	11	1.9	3.4	4.6
1057	Raw, large, about 7g each	14 ea	98	76	104	20	1	0	2	.3	.2	.7
1589	Shrimp, imitation, from surimi	4 oz	113	75	114	14	10	0	2	.3	.2	.8
1053	Snapper, baked or broiled	4 oz	113	70	145	30	0	0	2	.4	.4	.7
1060	Squid, fried in flour	4 oz	113	65	198	20	9	0	8	2.1	3.1	2.4
1590	Surimi	4 oz	113	76	112	17	8	0	1	.2	.2	.5
1058	Swordfish, raw	4 oz	113	76	137	22	0	0	5	1.2	1.7	1
1059	Swordfish, baked or broiled	4 oz	113	69	175	29	0	0	6	1.6	2.2	1.3
590	Trout, baked or broiled	4 oz	113	70	170	26	0	0	7	1.8	2	2.1
	Tuna, light, canned, drained solids:											
591	Oil pack	1 c	145	60	287	42	0	0	12	2.2	4.3	4.2
592	Water pack	1 c	154	75	179	39	0	0	1	.4	.2	.5
1061	Bluefin tuna, fresh	4 oz	113	68	163	26	0	0	6	1.4	1.8	1.6
	MEATS: BEEF, LAMB, PORK and others											
	BEEF, cooked, trimmed to ½" outer fat:											
	Braised, simmered, pot roasted:											
	Relatively fat, choice chuck blade:											
593	Lean and fat, piece 2½ x 2½ x ¾"	4 oz	113	47	393	30	0	0	29	11.5	12.5	1.1
594	Lean only	4 oz	113	55	297	35	0	0	16	6.3	7	.5
	Relatively lean, like choice round:											
595	Lean and fat, pce 4 ⅛ x 2½ x ¾"	4 oz	113	52	311	32	0	0	19	7.2	8.3	.7
596	Lean only	4 oz	113	57	249	36	0	0	11	3.6	4.7	.4
	Ground beef, broiled, patty 3 x ⅝":											
597	Extra lean, about 16% fat	4 oz	113	54	299	32	0	0	18	7	7.8	.7
598	Lean, 21% fat	4 oz	113	53	316	32	0	0	20	7.8	8.7	.7
	Roasts, oven cooked, no added liquid:											
	Relatively fat, prime rib:											
601	Lean and fat, piece 4⅛ x 2¼ x ½"	4 oz	113	46	425	25	0	0	35	14.2	15.2	1.2
602	Lean only	4 oz	113	58	271	31	0	0	16	7	7.9	.7
	Relatively lean, choice round:											
603	Lean and fat, piece 2½ x 2½ x ¾"	4 oz	113	59	272	30	0	0	16	6.2	6.8	.6
604	Lean only	4 oz	113	65	198	33	0	0	6	2.3	2.7	.2
1701	Steak, rib, broiled, lean	4 oz	113	58	250	32	0	0	13	5.1	5.3	.4

Chol (mg)	Calc (mg)	Iron (mg)	Magn (mg)	Pota (mg)	Sodi (mg)	Zinc (mg)	VT-A (µg)	Thia (mg)	VT-E (mg)	Ribo (mg)	Niac (mg)	V-B6 (mg)	Fola (µg)	VT-C (mg)
46	68	1.21	121	651	78	.6	61	.08	1.23	.1	8.05	.45	16	0
59	87	1.56	154	833	2260	.78	86	.11	1.11	.14	10.8	.64	22	0
36	53	.95	94	509	61	.47	53	.07	.96	.08	6.61	.39	14	0
15	87	1.38	9	78	983	.6	292	.04	1.13	.16	3.73	.19	2	0
104	88	.57	51	510	551	4.23	38	.01	1.45	.1	1.55	.11	16	0
61	155	1.33	44	396	108	.69	16	.15	1.84	.15	2.75	.3	11	1
71	151	1.88	39	335	201	.75	23*	.18	2.87	.2	2.98	.24	13	1
54	60	5.99	34	396	260	1.9	51	.03	1.36	.04	2.37	.41	18	6
131	112	16.5	117	387	523	225	74	.25	2.11	.24	3.42	.15	25	9
124	20	12.7	55	417	263	41.2	201	.17	2.11	.58	4.98	.12	25	20
59	45	5.07	42	178	304	63.6	66	.11	1.66	.15	1.2	.05	23	3
125	20	11.5	55	378	265	41.6	183	.16	2.21	.55	4.52	.11	19	16
108	7	.32	82	437	131	.68	26	.08	.23	.09	1.86	.08	5	0
62	241	.95	38	368	626	1.04	19	.03	1.53	.21	7.39	.34	17	0
98	8	.62	35	424	75	.58	71	.24	1.42	.19	7.54	.25	6	0
26	12	.96	20	198	886	.35	29*	.03	1.53	.11	5.33	.31	2	0
160	432	3.3	44	449	571	1.48	76	.09	.34	.26	5.93	.19	14	0
57	39	.76	55	310	432	.99	20	.04	1.77	.1	1.4	.13	34	2
25	9	.35	49	116	898	.37	23	.01	.12	.02	.35	.03	2	0
19	15	.15	33	171	112	.56	22*	.01	.8	.04	.61	.08	7	1
172	34	2.72	30	160	197	1.37	58	.03	.45	.03	2.28	.11	4	2
111	38	1.75	26	134	108	.81	11	.02	.59	.02	1.76	.07	1	1
159	60	1.13	36	203	310	1.24	50	.12	1.35	.12	2.76	.09	16	1
149	51	2.36	36	181	145	1.09	53	.03	.8	.03	2.5	.1	3	2
41	21	.68	49	101	797	.37	23	.03	.12	.04	.19	.03	2	0
53	45	.27	42	590	64	.5	40	.06	.71	<.01	.39	.52	7	2
294	44	1.14	43	315	346	1.97	12	.06	2.09	.52	2.94	.07	16	5
34	10	.29	49	127	162	.37	23	.02	.28	.02	.25	.03	2	0
44	5	.91	30	325	102	1.3	41	.04	.56	.11	10.9	.37	2	1
56	7	1.18	38	417	130	1.66	46	.05	.71	.13	13.3	.43	2	1
78	97	.43	35	506	63	.58	17	.17	.57	.11	6.52	.39	21	2
26	19	2.02	45	300	513	1.31	33	.05	1.74	.17	18	.16	7	0
46	17	2.36	42	365	521	1.19	26	.05	.82	.11	20.5	.54	6	0
43	9	1.15	56	285	44	.68	740	.27	1.13	.28	9.78	.51	2	0
112	11	3.45	21	275	67	7.57	0	.08	.26	.27	3.54	.32	10	0
120	15	4.16	26	297	80	11.6	0	.09	.16	.32	3.02	.33	7	0
108	7	3.53	25	319	56	5.55	0	.08	.21	.27	4.21	.37	11	0
108	6	3.91	28	348	58	6.19	0	.08	.2	.29	4.61	.41	12	0
112	10	3.13	28	417	93	7.27	0	.08	.2	.36	6.61	.36	12	0
114	14	2.77	27	394	101	7.01	0	.07	.23	.27	6.75	.34	12	0
96	12	2.61	21	334	71	5.92	0	.08	.27	.19	3.8	.26	8	0
91	11	2.95	28	425	84	7.84	0*	.09	.14	.24	4.64	.34	9	0
81	7	2.07	27	406	67	4.87	0	.09	.23	.18	3.92	.4	7	0
78	6	2.2	30	446	70	5.36	0	.1	.12	.19	4.24	.43	8	0
90	15	2.9	30	445	78	7.9	0	.11	.16	.25	5.42	.45	9	0

*This value is expressed in retinol equivalents (RE). All other values are in retinol activity equivalents (RAE).

Table A–1

Food Composition   (Computer code number is for Wadsworth Diet Analysis program)   (For purposes of calculations, use "0" for t, <1, <.1, <.01, etc.)

Computer Code Number	Food Description	Measure	Wt (g)	H₂O (%)	Ener (kcal)	Prot (g)	Carb (g)	Dietary Fiber (g)	Fat (g)	Fat Breakdown (g) Sat	Mono	Poly
	MEATS: BEEF, LAMB, PORK and others—Continued											
	Steak, broiled, relatively lean,											
606	choice sirloin, lean only	4 oz	113	62	228	34	0	0	9	3.5	3.8	.3
	Steak, broiled, relatively fat,											
	choice T-bone:											
1063	Lean and fat	4 oz	113	52	349	26	0	0	26	10.3	11.6	.9
1064	Lean only	4 oz	113	61	232	30	0	0	11	4.1	5.1	.3
	Variety meats:											
1086	Brains, panfried	4 oz	113	71	221	14	0	0	18	4.2	4.5	2.6
599	Heart, simmered	4 oz	113	64	198	32	<1	0	6	1.9	1.4	1.5
600	Liver, fried	4 oz	113	56	245	30	9	0	9	3	1.8	1.9
1062	Tongue, cooked	4 oz	113	56	320	25	<1	0	23	10.1	10.7	.9
607	Beef, canned, corned	4 oz	113	58	283	31	0	0	17	7	6.7	.7
608	Beef, dried, cured	1 oz	28	56	46	8	<1	0	1	.4	.5	.1
	LAMB, domestic, cooked:											
	Chop, arm, braised (5.6 oz raw w/bone):											
609	Lean and fat	1 ea	70	44	242	21	0	0	17	6.9	7.1	1.2
610	Lean only	1 ea	55	49	153	19	0	0	8	2.8	3.4	.5
	Chop, loin, broiled (4.2oz raw w/bone):											
611	Lean and fat	1 ea	64	52	202	16	0	0	15	6.3	6.2	1.1
612	Lean only	1 ea	46	61	99	14	0	0	4	1.6	2	.3
1067	Cutlet, avg of lean cuts, cooked	4 oz	113	54	330	28	0	0	23	9.9	9.8	1.7
	Leg, roasted, 3 oz = 4 ⅛ x 2¼ x ½":											
613	Lean and fat	4 oz	113	57	292	29	0	0	19	7.8	7.9	1.3
614	Lean only	4 oz	113	64	216	32	0	0	9	3.1	3.8	.6
615	Rib, roasted, lean and fat	4 oz	113	48	406	24	0	0	34	14.4	14.1	2.4
616	Rib, roasted, lean only	4 oz	113	60	262	30	0	0	15	5.4	6.6	1
1065	Shoulder, roasted, lean and fat	4 oz	113	56	312	25	0	0	23	9.5	9.2	1.8
1066	Shoulder, roasted, lean only	4 oz	113	63	231	28	0	0	12	4.6	4.9	1.1
	Variety meats:											
1069	Brains, pan-fried	4 oz	113	76	164	14	0	0	11	2.9	2.1	1.2
1068	Heart, braised	4 oz	113	64	209	28	2	0	9	3.5	2.5	.9
1070	Sweetbreads, cooked	4 oz	113	60	264	26	0	0	17	7.7	6.2	.8
1071	Tongue, cooked	4 oz	113	58	311	24	0	0	23	8.8	11.3	1.4
	PORK, cured, cooked (see also Sausages and Lunch Meats)											
617	Bacon, medium slices	3 pce	19	13	109	6	<1	0	9	3.3	4.5	1.1
1087	Breakfast strips, cooked	2 pce	23	27	106	7	<1	0	8	2.9	3.8	1.3
618	Canadian-style bacon	2 pce	47	62	87	11	1	0	4	1.3	1.9	.4
	Ham, roasted:											
619	Lean and fat, 2 pces 4⅛ x 2¼ x ¼"	4 oz	113	65	201	26	0	0	10	3.5	5	1.6
620	Lean only	4 oz	113	68	164	24	2	0	6	2	3	.6
621	Ham, canned, roasted, 8% fat	4 oz	113	69	154	24	1	0	6	1.8	2.8	.5
	PORK, fresh, cooked:											
	Chops, loin (cut 3 per lb with bone):											
1291	Braised, lean and fat	1 ea	89	58	213	24	0	0	12	4.5	5.4	1
1292	Lean only	1 ea	80	61	163	23	0	0	7	2.7	3.3	.6
622	Broiled, lean and fat	1 ea	82	58	197	23	0	0	11	3.9	4.8	.8
623	Broiled, lean only	1 ea	74	61	149	22	0	0	6	2.2	2.7	.4
624	Panfried, lean and fat	1 ea	78	53	216	23	0	0	13	4.7	5.5	1.5
625	Panfried, lean only	1 ea	63	59	152	16	0	0	9	3.2	3.9	1.2
626	Leg, roasted, lean and fat	4 oz	113	55	308	30	0	0	20	7.3	8.9	1.9
627	Leg, roasted, lean only	4 oz	113	61	233	35	0	0	9	3.2	4.3	.9
628	Rib, roasted, lean and fat	4 oz	113	56	288	31	0	0	17	6.7	7.9	1.4
629	Rib, roasted, lean only	4 oz	113	59	252	32	0	0	13	4.9	5.9	1
630	Shoulder, braised, lean and fat	4 oz	113	48	372	32	0	0	26	9.6	11.7	2.6
631	Shoulder, braised, lean only	4 oz	113	54	280	36	0	0	14	4.7	6.5	1.3
1088	Spareribs, cooked, yield from 1 lb raw with bone	4 oz	113	40	449	33	0	0	34	12.6	15.2	3.1
1095	Rabbit, roasted (1 cup meat=140g)	4 oz	113	61	223	33	0	0	9	2.7	2.4	1.8
	VEAL, cooked:											
632	Cutlet, braised or broiled, 4⅛ x 2¼ x ½"	4 oz	113	52	321	34	0	0	19	7.6	7.6	1.3

Chol (mg)	Calc (mg)	Iron (mg)	Magn (mg)	Pota (mg)	Sodi (mg)	Zinc (mg)	VT-A (µg)	Thia (mg)	VT-E (mg)	Ribo (mg)	Niac (mg)	V-B6 (mg)	Fola (µg)	VT-C (mg)
101	12	3.8	36	455	75	7.37	0	.15	.16	.33	4.84	.51	11	0
76	9	3.06	26	363	72	5.03	0	.1	.15	.24	4.46	.37	8	0
67	7	3.58	32	427	80	6	0	.12	.16	.28	5.23	.44	9	0
2254	10	2.51	17	400	179	1.53	0	.15	2.37	.29	4.27	.44	7	4
218	7	8.49	28	263	71	3.54	0	.16	.81	1.74	4.6	.24	2	2
545	12	7.1	26	411	120	6.16	12123	.24	.72	4.68	16.3	1.62	249	26
121	8	3.83	19	203	68	5.42	0	.03	.4	.4	2.43	.18	6	1
97	14	2.35	16	154	1136	4.03	0	.02	.17	.17	2.75	.15	10	0
12	2	1.26	9	124	972	1.47	0	.02	.04	.06	1.53	.1	3	0
84	17	1.67	18	214	50	4.26	0	.05	.1	.17	4.66	.08	13	0
67	14	1.49	16	186	42	4.02	0	.04	.1	.15	3.48	.07	12	0
64	13	1.16	15	209	49	2.23	0	.06	.08	.16	4.54	.08	11	0
44	9	.92	13	173	39	1.9	0	.05	.07	.13	3.15	.07	11	0
110	12	2.26	25	340	77	4.67	0	.12	.15	.32	7.48	.16	19	0
105	12	2.24	27	354	75	4.97	0	.11	.17	.3	7.45	.17	23	0
101	9	2.4	29	382	77	5.58	0	.12	.2	.33	7.16	.19	26	0
110	25	1.81	23	306	82	3.94	0	.1	.11	.24	7.63	.12	17	0
99	24	2	26	356	91	5.05	0	.1	.17	.26	6.96	.17	25	0
104	23	2.23	26	284	75	5.91	0	.1	.16	.27	6.95	.15	24	0
98	21	2.41	28	299	77	6.83	0	.1	.2	.29	6.51	.17	28	0
2308	14	1.9	16	232	151	1.54	0	.12	1.73	.27	2.79	.12	6	14
281	16	6.24	27	212	71	4.16	0	.19	.79	1.34	4.93	.34	2	8
452	14	2.4	21	329	59	3.03	0	.02	.78	.24	2.89	.06	15	23
214	11	2.97	18	179	76	3.38	0	.09	.36	.47	4.17	.19	3	8
16	2	.31	5	92	303	.62	0	.13	.1	.05	1.39	.05	1	0
24	3	.45	6	107	483	.85	0	.17	.07	.08	1.75	.08	1	0
27	5	.38	10	183	727	.8	0	.39	.12	.09	3.25	.21	2	0
67	9	1.51	25	462	1695	2.79	0	.82	.29	.37	6.95	.35	3	0
60	9	1.67	16	324	1359	3.25	0	.85	.29	.23	4.55	.45	3	0
34	7	1.04	24	393	1282	2.52	0	1.17	.29	.28	5.53	.51	6	0
71	19	.95	17	333	43	2.12	2	.56	.23	.23	3.93	.33	3	1
63	14	.9	16	310	40	1.98	2	.53	.21	.21	3.67	.31	3	<1
67	27	.66	20	294	48	1.85	2	.87	.27	.24	4.3	.35	5	<1
61	23	.63	20	278	44	1.76	1	.85	.31	.23	4.1	.35	4	<1
72	21	.71	23	332	62	1.8	2	.89	.2	.24	4.37	.37	5	1
52	14	.67	16	230	49	2.44	1	.46	.16	.23	2.8	.26	3	<1
106	16	1.14	25	398	68	3.34	3	.72	.29	.35	5.17	.45	11	<1
108	8	1.29	33	442	73	3.4	3	.91	.46	.4	5.56	.38	3	<1
82	32	1.06	24	476	52	2.33	2	.82	.41	.34	6.91	.37	3	<1
80	29	1.11	25	494	53	2.41	2	.86	.55	.36	7.25	.38	3	<1
123	20	1.82	21	417	99	4.72	3	.61	.29	.35	5.89	.4	5	<1
129	9	2.2	25	458	115	5.62	2	.68	.29	.41	6.71	.46	6	<1
137	53	2.09	27	362	105	5.2	3	.46	.29	.43	6.19	.4	5	0
93	21	2.57	24	433	53	2.57	0	.1	.96	.24	9.53	.53	12	0
133	32	1.23	27	316	90	4.1	0	.04	.45	.34	10.2	.29	16	0

*This value is expressed in retinol equivalents (RE). All other values are in retinol activity equivalents (RAE).

Table A–1

Food Composition   (Computer code number is for Wadsworth Diet Analysis program)   (For purposes of calculations, use "0" for t, <1, <.1, <.01, etc.)

Computer Code Number	Food Description	Measure	Wt (g)	H₂O (%)	Ener (kcal)	Prot (g)	Carb (g)	Dietary Fiber (g)	Fat (g)	Fat Breakdown (g) Sat	Mono	Poly
	MEATS: BEEF, LAMB, PORK and others—Continued											
633	Rib roasted, lean, 2 pieces 4⅛ x 2¼ x ¼"	4 oz	113	60	258	27	0	0	16	6.1	6.1	1.1
634	Liver, panfried	4 oz	113	67	186	24	3	0	8	2.9	1.7	1.2
1096	Venison (deer meat), roasted	4 oz	113	65	179	34	0	0	4	1.4	1	.7
	MEATS: POULTRY and POULTRY PRODUCTS											
	CHICKEN, cooked:											
	Fried, batter dipped:											
635	Breast	1 ea	280	52	728	70	25	1	37	9.9	15.3	8.6
636	Drumstick	1 ea	72	53	193	16	6	<1	11	3	4.6	2.7
637	Thigh	1 ea	86	51	238	19	8	<1	14	3.8	5.8	3.3
638	Wing	1 ea	49	46	159	10	5	<1	11	2.9	4.4	2.5
	Fried, flour coated:											
639	Breast	1 ea	196	57	435	62	3	<1	17	4.8	6.9	3.8
1212	Breast, without skin	1 ea	172	60	322	57	1	0	8	2.2	3	1.8
640	Drumstick	1 ea	49	57	120	13	1	<1	7	1.8	2.7	1.6
641	Thigh	1 ea	62	54	162	17	2	<1	9	2.5	3.6	2.1
1099	Thigh, without skin	1 ea	52	59	113	15	1	0	5	1.4	2	1.3
642	Wing	1 ea	32	49	103	8	1	<1	7	1.9	2.8	1.6
	Roasted:											
643	All types of meat	1 c	140	64	266	40	0	0	10	2.9	3.7	2.4
644	Dark meat	1 c	140	63	287	38	0	0	14	3.7	5	3.2
645	Light meat	1 c	140	65	242	43	0	0	6	1.8	2.2	1.4
646	Breast, without skin	1 ea	172	65	284	53	0	0	6	1.7	2.1	1.3
647	Drumstick, without skin	1 ea	44	67	76	12	0	0	2	.7	.8	.6
1703	Leg, without skin	1 ea	95	65	181	26	0	0	8	2.2	2.9	1.9
648	Thigh	1 ea	62	59	153	15	0	0	10	2.7	3.8	2.1
1100	Thigh, without skin	1 ea	52	63	109	13	0	0	6	1.6	2.2	1.3
649	Stewed, all types	1 c	140	67	248	38	0	0	9	2.6	3.3	2.2
656	Canned, boneless chicken	4 oz	113	69	186	25	0	0	9	2.5	3.6	2
1102	Gizzards, simmered	1 c	145	67	222	39	2	0	5	1.5	1.3	1.5
1101	Hearts, simmered	1 c	145	65	268	38	<1	0	11	3.3	2.9	3.3
2300	Liver, simmered: Ounce	3 oz	85	68	133	21	1	0	5	1.6	1.1	.8
1098	Liver, simmered: Piece = 20g	6 ea	120	68	188	29	1	0	7	2.2	1.6	1.1
	DUCK, roasted:											
1293	Meat with skin, about 2.7 cups	½ ea	382	52	1287	72	0	0	108	36.9	49.3	13.9
651	Meat only, about 1.5 cups	½ ea	221	64	444	52	0	0	25	9.2	8.2	3.2
	GOOSE, domesticated, roasted:											
1294	Meat only, about 4.2 cups	½ ea	591	57	1406	171	0	0	75	26.9	25.6	9.1
1295	Meat with skin, about 5.5 cups	½ ea	774	52	2360	195	0	0	170	53.2	79.3	19.5
	TURKEY:											
	Roasted, meat only:											
652	Dark meat	4 oz	113	63	211	32	0	0	8	2.7	1.8	2.4
653	Light meat	4 oz	113	66	177	34	0	0	4	1.2	.6	1
654	All types, chopped or diced	1 c	140	65	238	41	0	0	7	2.3	1.4	2
1103	Ground, cooked	4 oz	113	59	266	31	0	0	15	3.8	5.5	3.6
1106	Gizzard, cooked	2 ea	134	65	218	39	1	0	5	1.5	1	1.5
1107	Heart, cooked	4 ea	64	64	113	17	1	0	4	1.1	.8	1.1
1108	Liver, cooked	1 ea	75	66	127	18	3	0	4	1.4	1.1	.8
	POULTRY FOOD PRODUCTS (see also items in Sausage & Lunchmeats section):											
1567	Chicken patty, breaded, cooked	1 ea	75	49	213	12	11	<1	13	4.1	6.4	1.6
659	Turkey and gravy, frozen package	3 oz	85	85	57	5	4	0	2	.7	.8	.4
	Turkey breast, Louis Rich:											
1104	Barbecued	2 oz	56	72	57	11	2	0	<1	.2	.2	.1
1943	Hickory smoked	1 pce	80	73	80	16	2	0	1	0		
1947	Honey roasted	1 pce	80	73	80	16	3	0	1	.5		
1945	Oven roasted	1 pce	80		70	16	0	0	1	0		
661	Turkey patty, breaded, fried	2 oz	57	50	161	8	9	<1	10	2.7	4.3	2.7
662	Turkey, frozen, roasted, seasoned	4 oz	113	68	175	24	3	0	7	2.1	1.4	1.9
1704	Turkey roll, light meat	1 pce	28	72	41	5	<1	0	2	.6	.7	.5

A

PAGE KEY: A–2 = Beverages   A–4 = Dairy   A–8 = Eggs   A–10 = Fat/Oil   A–12 = Fruit   A–18 = Bakery   A–24 = Grain
A–30 = Fish   A–32 = Meats   A–36 = Poultry   A–38 = Sausage   A–38 = Mixed/Fast   A–44 = Nuts/Seeds   A–46 = Sweets
A–50 = Vegetables/Legumes   A–60 = Vegetarian   A–62 = Misc   A–64 = Soups/Sauces   A–66 = Fast   A–80 = Convenience   A–86 = Baby foods

Chol (mg)	Calc (mg)	Iron (mg)	Magn (mg)	Pota (mg)	Sodi (mg)	Zinc (mg)	VT-A (µg)	Thia (mg)	VT-E (mg)	Ribo (mg)	Niac (mg)	V-B6 (mg)	Fola (µg)	VT-C (mg)
124	12	1.1	25	333	104	4.62	0	.06	.4	.3	7.89	.28	15	0
634	8	2.96	21	232	60	10.8	9095	.15	.38	2.19	9.58	.55	858	35
127	8	5.05	27	379	61	3.11	0	.2	.28	.68	7.58	.42	5	0
238	56	3.5	67	563	770	2.66	56	.32	2.97	.41	29.5	1.2	42	0
62	12	.97	14	134	194	1.68	19	.08	.88	.15	3.67	.19	13	0
80	15	1.25	18	165	248	1.75	25	.1	1.05	.19	4.91	.22	16	0
39	10	.63	8	68	157	.68	17	.05	.52	.07	2.58	.15	9	0
174	31	2.33	59	508	149	2.16	29	.16	1.12	.26	26.9	1.14	12	0
157	27	1.96	53	475	136	1.86	12	.14	.72	.21	25.4	1.1	7	0
44	6	.66	11	112	44	1.42	12	.04	.41	.11	2.96	.17	5	0
60	9	.92	15	147	55	1.56	18	.06	.52	.15	4.31	.2	7	0
53	7	.76	13	135	49	1.45	11	.05	.3	.13	3.7	.2	5	0
26	5	.4	6	57	25	.56	12	.02	.18	.04	2.14	.13	2	0
125	21	1.69	35	340	120	2.94	22	.1	.36	.25	12.8	.66	8	0
130	21	1.86	32	336	130	3.92	31	.1	.36	.32	9.17	.5	11	0
119	21	1.48	38	346	108	1.72	13	.09	.36	.16	17.4	.84	6	0
146	26	1.79	50	440	127	1.72	10	.12	.45	.2	23.6	1.03	7	0
41	5	.57	11	108	42	1.4	8	.03	.11	.1	2.67	.17	4	0
89	11	1.24	23	230	86	2.72	18	.07	.25	.22	6	.35	8	0
58	7	.83	14	138	52	1.46	30	.04	.16	.13	3.95	.19	4	0
49	6	.68	12	124	46	1.34	10	.04	.13	.12	3.39	.18	4	0
116	20	1.64	29	252	98	2.79	21	.07	.36	.23	8.56	.36	8	0
70	16	1.79	14	156	568	1.59	38	.02	.24	.15	7.15	.4	5	2
281	14	6.02	29	260	97	6.35	81	.04	1.73	.35	5.76	.17	77	2
351	28	13.1	29	191	70	10.6	13	.1	2.32	1.07	4.06	.46	116	3
536	12	7.2	18	119	43	3.69	4176	.13	1.22	1.48	3.78	.49	655	13
757	17	10.2	25	168	61	5.21	5895	.18	1.73	2.1	5.34	.7	924	19
321	42	10.3	61	779	225	7.11	241	.66	2.67	1.03	18.4	.69	23	0
197	26	5.97	44	557	144	5.75	51	.57	1.55	1.04	11.3	.55	22	0
567	83	17	148	2293	449	18.7	71	.54	9.16	2.3	24.1	2.78	71	0
704	101	21.9	170	2546	542	20.3	163	.6	13.5	2.5	32.3	2.86	15	0
96	36	2.63	27	328	89	5.04	0	.07	.72	.28	4.12	.41	10	0
78	21	1.53	32	345	72	2.31	0	.07	.1	.15	7.73	.61	7	0
106	35	2.49	36	417	98	4.34	0	.09	.46	.25	7.62	.64	10	0
115	28	2.18	27	305	121	3.23	0	.06	.38	.19	5.45	.44	8	0
311	20	7.29	25	283	72	5.57	74	.04	.21	.44	4.12	.16	70	2
145	8	4.41	14	117	35	3.37	5	.04	.1	.56	2.08	.2	51	1
470	8	5.85	11	146	48	2.32	2805	.04	2.18	1.07	4.46	.39	500	1
45	12	.94	15	185	399	.78	11	.07	1.46	.1	5.04	.23	8	<1
15	12	.79	7	52	471	.59	6	.02	.3	.11	1.53	.08	3	0
25	14	.61	16	173	592	.58	0	.02		.06	5.28	.22	2	0
35	0	.72			1060		0							0
35	0	.72			940		0							0
35	0				910		0							0
35	8	1.25	9	157	456	.82	6	.06	1.36	.11	1.31	.11	16	0
60	6	1.84	25	337	768	2.87	0	.05	.43	.18	7.09	.3	6	0
12	11	.36	4	70	137	.44	0	.02	.04	.06	1.96	.09	1	0

*This value is expressed in retinol equivalents (RE). All other values are in retinol activity equivalents (RAE).

Table A–1

Food Composition

(Computer code number is for Wadsworth Diet Analysis program)   (For purposes of calculations, use "0" for t, <1, <.1, <.01, etc.)

Computer Code Number	Food Description	Measure	Wt (g)	H₂O (%)	Ener (kcal)	Prot (g)	Carb (g)	Dietary Fiber (g)	Fat (g)	Fat Breakdown (g)		
										Sat	Mono	Poly
	MEATS: SAUSAGES and LUNCHMEATS (see also Poultry Food Products)											
1072	Beerwurst/beer salami, beef	1 oz	28	53	92	3	<1	0	8	3.6	3.9	.3
1074	Beerwurst/beer salami, pork	1 oz	28	61	67	4	1	0	5	1.8	2.5	.7
1075	Berliner sausage	1 oz	28	61	64	4	1	0	5	1.7	2.2	.4
	Bologna:											
1297	Beef	1 pce	23	55	72	3	<1	0	7	2.8	3.2	.3
2115	Beef, light, Oscar Mayer	1 pce	28	65	55	3	2	0	4	1.6	2	.1
663	Beef & pork	1 pce	28	54	88	3	1	0	8	3	3.7	.7
2155	Healthy Favorites	1 pce	23		22	3	1	0	<1	0		
1298	Pork	1 pce	23	61	57	4	<1	0	5	1.6	2.2	.5
2114	Regular, light, Oscar Mayer	1 pce	28	65	56	3	2	0	4	1.6	2	.4
664	Turkey	1 pce	28	65	56	4	<1	0	4	1.4	1.3	1.2
1970	Turkey, Louis Rich	1 pce	56	67	115	6	1	0	10	2.9	3.6	2.6
665	Braunschweiger sausage	2 pce	57	48	205	8	2	0	18	6.2	8.5	2.1
1073	Bratwurst-link	1 ea	70	51	226	10	2	0	19	6.9	9.3	2
666	Brown & serve sausage links, cooked	2 ea	26	45	102	4	1	0	10	3.4	4.4	1
1089	Cheesefurter/cheese smokie	2 ea	86	52	281	12	1	0	25	9	11.7	2.6
2157	Chicken breast, Healthy Favorites	4 pce	52		40	9	1	0	0	0	0	0
1556	Chorizo, pork & beef	1 ea	60	32	273	14	1	0	23	8.6	11	2.1
1090	Corned beef loaf, jellied	1 pce	28	69	43	6	0	0	2	.7	.7	.1
	Frankfurters:											
1077	Beef, large link, 8/package	1 ea	57	55	180	7	1	0	16	6.9	7.8	.8
1078	Beef and pork, large link, 8/package	1 ea	45	54	144	5	1	0	13	4.8	6.1	1.2
667	Beef and pork, small link, 10/pkg	1 ea	45	54	144	5	1	0	13	4.8	6.1	1.2
668	Turkey frankfurter, 10/package	1 ea	45	63	102	6	1	0	8	2.6	2.5	2.2
1968	Turkey/chicken frank 8/pkg	1 ea	43	67	81	5	2	0	6	1.6	2.4	1.4
	Ham:											
669	Ham lunchmeat, canned, 3 x 2 x ½"	1 pce	21	52	70	3	<1	0	6	2.3	3	.7
670	Chopped ham, packaged	2 pce	42	64	96	7	0	0	7	2.4	3.4	.9
2156	Honey ham, Healthy Favorites	4 pce	52	73	55	9	2	0	1	.4	.8	.1
2113	Oscar Mayer lower sodium ham	1 pce	21	73	23	3	1	0	1	.3	.4	.1
673	Turkey ham lunchmeat	2 pce	57	71	73	11	<1	0	3	1	.7	.9
1091	Kielbasa sausage	1 pce	26	54	81	3	1	0	7	2.6	3.4	.8
1092	Knockwurst sausage, link	1 ea	68	55	209	8	1	0	19	6.9	8.7	2
1093	Mortadella lunchmeat	2 pce	30	52	93	5	1	0	8	2.8	3.4	.9
1097	Olive loaf lunchmeat	2 pce	57	58	134	7	5	0	9	3.3	4.5	1.1
1952	Turkey breast, fat free	1 pce	28	76	23	4	1	0	<1	.1	.1	t
1080	Turkey pastrami	2 pce	57	71	80	10	1	0	4	1	1.2	.9
1969	Turkey salami	1 pce	28	72	41	4	<1	0	3	.8	.9	.7
1081	Pepperoni sausage	2 pce	11	27	55	2	<1	0	5	1.8	2.3	.5
1094	Pickle & pimento loaf	2 pce	57	57	149	7	3	0	12	4.5	5.5	1.5
1082	Polish sausage	1 oz	28	53	91	4	<1	0	8	2.9	3.8	.9
674	Pork sausage, cooked, link, small	2 ea	26	45	96	5	<1	0	8	2.8	4.1	.8
1079	Pork sausage, cooked, patty	4 oz	113	45	417	22	1	0	35	12.1	17.7	3.3
675	Salami, pork and beef	2 pce	57	60	143	8	1	0	11	4.6	5.2	1.1
677	Salami, pork and beef, dry	3 pce	30	35	125	7	1	0	10	3.7	5.1	1
676	Salami, turkey	2 pce	57	66	112	9	<1	0	8	2.3	2.6	2
	Sandwich spreads:											
1300	Ham salad spread	2 tbsp	30	63	65	3	3	0	5	1.5	2.2	.8
678	Pork and beef	2 tbsp	30	60	70	2	4	<1	5	1.8	2.3	.8
1296	Chicken/turkey	2 tbsp	26	66	52	3	2	0	4	.9	.8	1.6
1084	Smoked link sausage, beef and pork	1 ea	68	52	228	9	1	0	21	7.2	9.6	2.2
1083	Smoked link sausage, pork	1 ea	68	39	265	15	1	0	22	7.7	10	2.6
1085	Summer sausage	2 pce	46	51	154	7	<1	0	14	5.5	6	.6
1076	Turkey breakfast sausage	1 pce	28	60	64	6	0	0	5	1.6	1.8	1.2
679	Vienna sausage, canned	2 ea	32	60	89	3	1	0	8	3	4	.5
	MIXED DISHES and FAST FOODS											
	MIXED DISHES:											
1445	Almond Chicken	1 c	242	77	280	22	16	3	14	1.9	6.1	5.6
1981	Baked beans, fat free, honey	½ c	120	73	110	7	24	7	0	0	0	0
1454	Bean cake	1 ea	32	23	130	2	16	1	7	1	2.9	2.6
680	Beef stew w/vegetables, homemade	1 c	245	82	218	16	15	2	10	4.9	4.5	.5
1109	Beef stew w/vegetables, canned	1 c	245	82	194	14	17	2	8	2.4	3.1	.3

Chol (mg)	Calc (mg)	Iron (mg)	Magn (mg)	Pota (mg)	Sodi (mg)	Zinc (mg)	VT-A (µg)	Thia (mg)	VT-E (mg)	Ribo (mg)	Niac (mg)	V-B6 (mg)	Fola (µg)	VT-C (mg)
17	3	.42	3	49	288	.68	0	.02	.05	.03	.95	.05	1	0
16	2	.21	4	71	347	.48	0	.15	.06	.05	.91	.1	1	0
13	3	.32	4	79	363	.69	0	.11	.06	.06	.87	.06	1	0
13	3	.38	3	36	226	.5	0	.01	.04	.02	.55	.03	1	0
13	4	.34	4	44	314	.53	0							0
15	3	.42	3	50	285	.54	0	.05	.06	.04	.72	.05	1	0
7		.18			255									
14	3	.18	3	65	272	.47	0	.12	.06	.04	.9	.06	1	0
15	14	.39	6	46	312	.45	0							0
28	23	.43	4	56	246	.49	0	.01	.15	.05	.99	.06	2	0
44	68	.9	10	103	484	1.14	0	.03		.1	2.15	.1		0
89	5	5.34	6	113	652	1.6	2405	.14	.2	.87	4.77	.19	25	0
44	34	.72	11	197	778	1.47	0	.17	.19	.16	2.31	.09	3	0
16	2	.62	4	70	248	.3	0*	.21	.06	.09	.96	.06	1	0
58	50	.93	11	177	931	1.94	33*	.21	.27	.14	2.49	.11	3	0
25		.72			620									
53	5	.95	11	239	741	2.05	0	.38	.13	.18	3.08	.32	1	0
13	3	.57	3	28	267	1.15	0	0	.05	.03	.49	.03	2	0
35	11	.81	2	95	585	1.24	0	.03	.11	.06	1.38	.07	2	0
22	5	.52	4	75	504	.83	0	.09	.11	.05	1.19	.06	2	0
22	5	.52	4	75	504	.83	0	.09	.11	.05	1.19	.06	2	0
48	48	.83	6	81	642	1.4	0	.02	.28	.08	1.86	.1	4	0
40	56	.94	10	69	488	.8	0							0
13	1	.15	2	45	271	.31	0*	.08	.05	.04	.66	.04	1	<1
21	3	.35	7	134	576	.81	0	.26	.11	.09	1.63	.15	<1	0
24	6	.7	18	144	635	1.02	0							0
9	1	.3	5	197	174	.42	0							0
32	6	1.57	9	185	568	1.68	0	.03	.36	.14	2.01	.14	3	0
17	11	.38	4	70	280	.52	0	.06	.06	.06	.75	.05	1	0
39	7	.62	7	135	687	1.13	0	.23	.39	.09	1.86	.12	1	0
17	5	.42	3	49	374	.63	0	.04	.07	.05	.8	.04	1	0
22	62	.31	11	169	846	.79	11*	.17	.14	.15	1.05	.13	1	0
9	3	.31	8	57	334	.24	0							0
31	5	.95	8	148	596	1.23	0	.03	.12	.14	2.01	.15	3	0
21	11	.35	6	60	281	.65	0							0
9	1	.15	2	38	224	.27	0	.03	.02	.03	.55	.03	<1	0
21	54	.58	10	194	792	.8	2	.17	.14	.14	1.17	.11	3	0
20	3	.4	4	66	245	.54	0	.14	.06	.04	.96	.05	1	<1
22	8	.33	4	94	336	.65	0*	.19	.07	.07	1.18	.09	1	<1
94	36	1.42	19	408	1462	2.84	0*	.84	.29	.29	5.11	.37	2	2
37	7	1.52	9	113	607	1.22	0	.14	.12	.21	2.03	.12	1	0
24	2	.45	5	113	558	.97	0	.18	.08	.09	1.46	.15	1	0
47	11	.92	9	139	572	1.03	0	.04	.32	.1	2.01	.14	2	0
11	2	.18	3	45	274	.33	0	.13	.52	.04	.63	.04	<1	0
11	4	.24	2	33	304	.31	3*	.05	.52	.04	.52	.04	1	0
8	3	.16	3	48	98	.27	11*	.01	.57	.02	.43	.03	1	<1
48	7	.99	8	129	643	1.43	0	.18	.15	.12	2.19	.12	1	0
46	20	.79	13	228	1020	1.92	0	.48	.17	.17	3.08	.24	3	1
34	6	1.17	6	125	571	1.18	0	.07	.1	.15	1.98	.12	1	0
23	5	.51	6	75	188	.96	0	.03	.14	.08	1.4	.08	1	0
17	3	.28	2	32	305	.51	0	.03	.07	.03	.52	.04	1	0
40	69	1.97	60	549	526	1.62	37*	.08	3.8	.2	9.48	.44	26	7
0	40	2.7			135				225					12
0	3	.67	6	58	1	.16	0	.07	1.24	.05	.55	.02	9	0
64	29	2.94	40	613	292	5.29	568*	.15	.49	.17	4.66	.28	37	17
34	29	2.21	39	426	1006	4.24	262*	.07	.34	.12	2.45	.2	31	7

*This value is expressed in retinol equivalents (RE). All other values are in retinol activity equivalents (RAE).

Table A–1

Food Composition   (Computer code number is for Wadsworth Diet Analysis program)   (For purposes of calculations, use "0" for t, <1, <.1, <.01, etc.)

A

Computer Code Number	Food Description	Measure	Wt (g)	H₂O (%)	Ener (kcal)	Prot (g)	Carb (g)	Dietary Fiber (g)	Fat (g)	Sat	Mono	Poly
	MIXED DISHES and FAST FOODS MIXED DISHES:—Continued											
1116	Beef, macaroni, tomato sauce casserole	1 c	226	76	255	16	26	2	10			
2295	Beef fajita	1 ea	223	65	399	23	36	3	18	5.5	7.6	3.5
1265	Beef flauta	1 ea	113	51	354	14	13	2	28	4.8	11.8	9.4
681	Beef pot pie, homemade	1 pce	210	55	517	21	39	3	30	8.4	14.7	7.3
1898	Broccoli, batter fried	1 c	85	74	122	3	9	2	9	1.3	2.1	4.9
1462	Buffalo wings/spicy chicken wings	2 pce	32	53	98	8	<1	<1	7	1.8	2.6	1.8
1675	Carrot raisin salad	½ c	88	58	203	1	21	2	14	2.1		
2248	Cheeseburger deluxe	1 ea	219	52	563	28	38		33	15	12.6	2
682	Chicken a la king, homemade	1 c	245	68	468	27	12	1	34	12.7	14.3	6.2
683	Chicken & noodles, homemade	1 c	240	71	367	22	26	2	18	5.9	7.1	3.5
684	Chicken chow mein, canned	1 c	250	89	95	6	18	2	1	0	.1	.8
685	Chicken chow mein, homemade	1 c	250	78	255	31	10	1	10	2.4	4.3	3.1
1266	Chicken fajita	1 ea	223	65	363	20	44	5	12	2.2	5.5	3.1
1264	Chicken flauta	1 ea	113	55	330	13	12	2	26	4.2	10.7	9.2
686	Chicken pot pie, homemade (⅓)	1 pce	232	57	545	23	42	3	31	10.9	14.5	5.8
1672	Chili con carne	½ c	127	77	128	12	11	2	4	1.7	1.7	.3
1112	Chicken salad with celery	½ c	78	53	268	11	1	<1	25	3.1		
1382	Chicken teriyaki, breast	1 ea	128	67	178	27	7	<1	4	.9	1.1	.9
687	Chili with beans, canned	1 c	256	76	287	15	30	11	14	6	6	.9
1479	Chinese Pastry	1 oz	28	46	67	1	13	<1	2	.2	.5	.8
688	Chop suey with beef & pork	1 c	220	63	421	22	31	4	24	4.7	8.3	9.2
690	Coleslaw	1 c	132	74	195	2	17	2	15	2.1	3.2	8.5
689	Corn pudding	1 c	250	76	273	11	32	4	13	6.3	4.3	1.7
1110	Corned beef hash, canned	1 c	220	67	398	19	23	1	25	11.9	10.9	.9
1255	Deviled egg (½ egg + filling)	1 ea	31	70	63	4	<1	0	5	1.2	1.7	1.5
	Egg Foo Yung Patty:											
1467	Meatless	1 ea	86	77	113	6	3	1	8	2	3.4	2.1
1458	With beef	1 ea	86	76	119	8	3	<1	8	2	2.9	2.2
1465	With chicken	1 ea	86	76	121	8	4	<1	8	1.9	2.8	2.3
1602	Egg roll, meatless	1 ea	64	70	101	3	10	1	6	1.2	2.9	1.3
1550	Egg roll, with meat	1 ea	64	66	113	5	9	1	6	1.4	3	1.3
1113	Egg salad	1 c	183	57	584	17	3	0	56	10.5	17.4	23.9
56102	Falafel	1 ea	17	35	57	2	5	1	3	.4	1.7	.7
691	French toast w/wheat bread, homemade	1 pce	65	54	151	5	16	<1	7	2	3	1.7
1355	Green Pepper, stuffed	1 ea	172	75	229	12	18	1	12	5.3		
1487	Hot & Sour Soup (Chinese)	1 c	244	87	162	15	5	1	8	2.7	3.4	1.2
2242	Hamburger deluxe	1 ea	110	49	279	13	27		13	4.1	5.3	2.6
1997	Hummous/hummus	¼ c	62	65	106	3	12	3	5	.8	2.2	2
16335	Kung Pao Chicken	1 c	162	54	431	29	11	2	31	5.2	13.9	9.7
	Lasagna:											
1346	With meat, homemade	1 pce	245	67	392	23	40	3	16	8	5.2	.8
1111	Without meat, homemade	1 pce	218	69	306	16	40	3	10	5.6	2.5	.6
1117	Frozen entree	1 ea	340	75	389	24	41	4	14	6.7	5.5	.8
1606	Lo mein, meatless	1 c	200	82	135	6	27	4	1	.1	.1	.3
1607	Lo mein, with meat	1 c	200	72	283	20	21	3	14	2.6	3.9	6
692	Macaroni & cheese, canned	1 c	240	80	228	9	26	1	10	4.2	3.1	1.4
693	Macaroni & cheese, homemade	1 c	200	70	302	15	30	1	14	8.5		
1115	Macaroni salad, no cheese	1 c	177	60	460	5	28	2	37	4		
1120	Meat loaf, beef	1 pce	87	63	182	16	4	<1	11	4.3		
1119	Meat loaf, beef and pork (⅓)	1 pce	87	60	210	14	5	<1	15	5.5		
1303	Moussaka (lamb & eggplant)	1 c	250	82	238	16	13	4	13	4.5		
1899	Mushrooms, batter fried	5 ea	70	63	155	2	11	1	12	1.5	3.6	6
715	Potato salad with mayonnaise and eggs	½ c	125	76	179	3	14	2	10	1.8	3.1	4.7
1674	Pizza, combination, 1⁄12 of 12" round	1 pce	79	48	184	13	21		5	1.5	2.5	.9
1673	Pizza, pepperoni, 1⁄12 of 12" round	1 pce	71	47	181	10	20		7	2.2	3.1	1.2
694	Quiche Lorraine ⅛ of 8" quiche	1 pce	176	53	526	15	25	1	41	18.8	14.3	5.2
1449	Ramen noodles-cooked	1 c	227	86	153	3	20	1	6	1.6	1.2	3.3
1671	Ravioli, meat	½ c	125	69	197	11	18	1	9	3	3.7	1
1597	Fried rice (meatless)	1 c	166	68	271	5	34	1	12	1.8	3.2	6.7
2142	Roast beef hash	½ c	117	66	230	9	11	1	16	7	5.8	3.2
	Spaghetti (enriched) in tomato sauce With cheese:											

Chol (mg)	Calc (mg)	Iron (mg)	Magn (mg)	Pota (mg)	Sodi (mg)	Zinc (mg)	VT-A (µg)	Thia (mg)	VT-E (mg)	Ribo (mg)	Niac (mg)	V-B6 (mg)	Fola (µg)	VT-C (mg)
39	26	2.69	40	522	882	3.13	49	.22	1.55	.21	4.31	.29	59	14
45	84	3.76	38	479	316	3.52	21	.39	1.74	.3	5.4	.38	23	27
37	51	1.87	28	313	68	3.45	21*	.06	4.65	.13	1.88	.23	10	19
44	29	3.78	6	334	596	3.17	519*	.29	3.78	.29	4.83	.24	29	6
15	66	.98	20	242	64	.38	98*	.08	2.85	.13	.73	.11	36	53
26	5	.4	6	59	25	.57	17*	.01	.27	.04	2.06	.13	1	<1
10	26	.74	14	317	118	.18	2905*	.08	2.4	.05	.64	.22	9	5
88	206	4.66	44	445	1108	4.6	129*	.39	1.18	.46	7.38	.28	81	8
186	127	2.45	20	404	760	1.8	272*	.1	.98	.42	5.39	.23	11	12
96	26	2.16	26	149	600	1.53	10*	.05		.17	4.32	.19	10	0
7	45	1.25	14	418	725	1.3	28*	.05	.05	.1	1	.09	12	12
77	57	2.5	28	473	718	2.12	50*	.07	.75	.22	4.25	.41	19	10
39	101	3.32	48	533	343	1.65	65*	.43	1.71	.33	6.12	.38	42	37
35	50	.95	27	268	71	1.13	26*	.05	4.36	.09	3.1	.22	8	18
72	70	3.02	25	343	594	2	735*	.32	3.25	.32	4.87	.46	29	5
67	34	2.6	23	347	505	1.79	42	.06	.81	.57	1.24	.16	23	1
48	16	.62	11	138	201	.79	31*	.03	6.27	.07	3.28	.34	8	1
82	27	1.71	35	309	1683	1.96	16*	.08	.35	.19	8.75	.47	12	3
43	120	8.78	115	934	1336	5.12	43	.12	1.89	.27	.92	.34	59	4
0	6	.18	7	25	3	.16	<1	.02	.26	<.01	.27	.04	1	0
43	39	4.19	54	519	950	3.48	103*	.36	1.8	.36	5.73	.39	44	20
7	45	.96	12	236	356	.26	66*	.05	5.28	.04	.11	.14	51	11
250	100	1.4	37	403	138	1.25	90*	1.03	.52	.32	2.47	.29	62	7
73	29	4.4	36	440	1188	3.3	0*	.02	.48	.2	4.62	.43	20	0
122	15	.35	3	37	50	.3	50	.02	.6	.15	.02	.05	13	0
185	31	1.04	12	117	317	.7	86*	.04	1.22	.26	.43	.09	29	5
166	25	1.01	11	139	131	1.01	86*	.05	1.06	.23	.65	.13	22	3
167	27	.82	11	136	132	.76	87*	.05	1.1	.23	.89	.12	22	3
30	14	.81	9	97	274	.25	16*	.08	.85	.11	.8	.05	13	3
37	15	.83	10	124	273	.46	16*	.16	.8	.12	1.28	.09	10	2
581	74	1.81	13	181	464	1.45	262	.09	7.66	.66	.09	.46	61	0
0	9	.58	14	99	50	.25	<1	.02	.19	.03	.18	.02	16	<1
76	64	1.09	11	86	311	.44	81*	.13	.31	.21	1.06	.05	15	<1
37	16	1.73	20	245	233	2.38	23	.14	.78	.11	2.75	.31	42	59
34	29	1.9	29	384	1010	1.51	2*	.27	.15	.25	5	.2	13	1
26	63	2.63	22	227	504	2.06	4	.23	.82	.2	3.69	.12	52	2
0	31	.97	18	108	151	.68	1	.06	.62	.03	.25	.25	37	5
64	49	1.96	63	428	907	1.5	58*	.15	3.9	.15	13.2	.59	43	8
58	270	3.07	50	460	391	3.33	158*	.24	1.16	.34	4.2	.25	20	14
32	265	2.35	44	373	364	1.82	157*	.23	1.09	.28	2.51	.17	17	14
55	265	3.82	64	759	839	3.7	360*	.29	3.45	.39	5.07	.32	28	12
0	46	2.03	33	386	564	.92	65	.23	.35	.24	2.82	.19	48	12
42	29	2.07	42	332	142	1.83	8*	.41	2.06	.28	5.02	.36	53	11
24	199	.96	31	139	730	1.2	73*	.12	.14	.24	.96	.02	8	<1
41	257	.82	41	204	493	1.36	153*	.28	.7	.35	1.57	.11	74	1
27	30	1.55	20	155	360	.54	37*	.18	10.3	.1	1.45	.33	70	3
84	29	1.6	14	187	145	3.22	23*	.05	.32	.22	2.62	.15	14	1
82	29	1.39	13	181	289	2.56	23*	.11	.33	.2	2.46	.12	13	1
96	75	1.74	40	565	460	2.57	109*	.16	.98	.31	4.14	.23	46	6
2	15	1.22	7	154	112	.42	6*	.11	2.34	.26	2.25	.04	8	1
85	24	.81	19	318	661	.39	41*	.1	2.33	.07	1.11	.18	9	12
20	101	1.53	18	179	382	1.11	101*	.21		.17	1.96	.09	32	2
14	65	.94	9	153	267	.52	55*	.13		.23	3.05	.06	37	2
221	231	1.88	24	239	221	1.5	279*	.26	2.02	.49	2.01	.1	19	1
<1	13	.39	10	49	802	.18	2	.02	2.34	.01	.25	.01	3	<1
85	35	2.15	20	264	90	1.7	87*	.16	1.32	.22	2.99	.14	14	4
43	28	1.94	23	128	261	.92	20*	.21	2.51	.11	2.24	.15	22	4
40	10	.9	22	362	695	2.99	0*	.09		.12	2.33	.3	12	0

*This value is expressed in retinol equivalents (RE). All other values are in retinol activity equivalents (RAE).

Food Composition

(Computer code number is for Wadsworth Diet Analysis program)　　(For purposes of calculations, use "0" for t, <1, <.1, <.01, etc.)

Computer Code Number	Food Description	Measure	Wt (g)	H₂O (%)	Ener (kcal)	Prot (g)	Carb (g)	Dietary Fiber (g)	Fat (g)	Fat Breakdown (g)		
										Sat	Mono	Poly
	MIXED DISHES and FAST FOODS MIXED DISHES:—Continued											
695	Canned	1 c	250	80	190	5	38	2	1	0	.4	.5
696	Home recipe	1 c	250	77	260	9	37	2	9	2		
	With meatballs:											
697	Canned	1 c	250	78	258	12	28	6	10	2.1	3.9	3.9
698	Home recipe	1 c	248	71	370	19	29	3	18	5		
716	Spinach souffle	1 c	136	74	219	11	3	3	18	7.1	6.8	3.1
2995	Spring roll, vegetable	1 ea	63	50	157	4	20	1	7	.9		
56313	Sushi, fish and vegetable	1 c	166	65	232	9	47	2	1	.2	.2	.2
56314	Sushi, vegetable seaweed	1 c	166	71	194	4	43	1	<1	.1	.1	.1
1553	Sweet & sour pork	1 c	226	77	231	15	25	2	8	2.1	3.1	2.3
1263	Sweet & sour chicken breast	1 ea	131	79	118	8	15	1	3	.5	.9	1.4
56916	Tabouli	1 c	160	77	199	3	16	4	15	2	10.8	1.4
1994	Thai lemongrass vegetables, svg	1 ea	187	75	238	8	19	4	16	2.8		
2426	Thai peanut chicken, svg	1 ea	309	81	271	19	26	3	10	2		
1515	Three bean salad	1 c	150	81	140	4	15	5	8	1.1	1.7	4.4
717	Tuna salad	1 c	205	63	383	33	19	0	19	3.2	5.9	8.4
1121	Tuna noodle casserole, homemade	1 c	202	75	238	17	25	1	7			
1270	Waldorf salad	1 c	137	58	411	4	12	3	41	4.2		
56111	Wonton, meat filled	1 ea	19	45	55	3	5	<1	3	.8	1.2	.3
	FAST FOODS and SANDWICHES (see end of this appendix for additional Fast Foods)											
699	Burrito, beef & bean	1 ea	116	52	255	11	33	3	9	4.2	3.5	.6
700	Burrito, bean	1 ea	109	53	225	7	36	4	7	3.5	2.4	.6
2106	Burrito, chicken con queso	1 ea	299	73	350	14	60	6	6	2.5		
701	Cheeseburger with bun, regular	1 ea	154	55	359	18	28		20	9.2	7.2	1.5
702	Cheeseburger with bun, 4-oz patty	1 ea	166	51	417	21	35		21	8.7	7.8	2.7
703	Chicken patty sandwich	1 ea	182	47	515	24	39	1	29	8.5	10.4	8.4
704	Corndog	1 ea	175	47	460	17	56		19	5.2	9.1	3.5
1922	Corndog, chicken	1 ea	113	52	271	13	26		13			
705	Enchilada	1 ea	163	63	319	10	28		19	10.6	6.3	.8
706	English muffin with egg, cheese, bacon	1 ea	146	57	308	18	28	2	13	5	5	1.7
	Fish sandwich:											
707	Regular, with cheese	1 ea	183	45	523	21	48	<1	29	8.1	8.9	9.4
708	Large, no cheese	1 ea	158	47	431	17	41	<1	23	5.2	7.7	8.2
709	Hamburger with bun, regular	1 ea	107	45	275	12	35	2	10	3.6	3.4	1
710	Hamburger with bun, 4-oz patty	1 ea	215	51	576	32	39		32	12	14.1	2.8
711	Hotdog/frankfurter with bun	1 ea	98	54	242	10	18		14	5.1	6.8	1.7
	Lunchables:											
2129	Bologna & American cheese	1 ea	128		450	18	19	0	34	15		
2130	Ham & cheese	1 ea	128		320	22	19	0	17	8		
2117	Honey ham & Amer. w/choc pudding	1 ea	176		390	18	34	<1	20	9		
2118	Honey turkey & cheddar w/Jello	1 ea	163		320	17	27	<1	16	9		
2131	Pepperoni & American cheese	1 ea	128		480	20	19	0	36	17		
2125	Salami & American cheese	1 ea	128		430	18	18	0	32	15		
2127	Turkey & cheddar cheese	1 ea	128		360	20	20	1	22	11		
712	Pizza, cheese, ⅛ of 15″ round	1 pce	63	48	140	8	20	1	3	1.5	1	.5
	SANDWICHES:											
	Avocado, chesse, tomato & lettuce:											
1276	On white bread, firm	1 ea	210	62	429	14	35	5	27	7.7		
1278	On part whole wheat	1 ea	201	63	402	14	30	6	27	7.8		
1277	On whole wheat	1 ea	214	63	424	15	33	8	28	8.2		
	Bacon, lettuce & tomato sandwich:											
1137	On white bread, soft	1 ea	124	52	318	10	29	2	17	4.1		
1139	On part whole wheat	1 ea	124	53	314	11	26	3	19	4.6		
1138	On whole wheat	1 ea	137	52	339	12	29	5	20	4.9		
	Cheese, grilled:											
1140	On white bread, soft	1 ea	119	37	399	17	30	1	23	11.9		
1142	On part whole wheat	1 ea	119	37	402	18	26	2	25	13.1		
1141	On whole wheat	1 ea	132	38	432	20	30	4	27	13.8		
1596	Chicken fillet	1 ea	182	47	515	24	39	1	29	8.5	10.4	8.4

Chol (mg)	Calc (mg)	Iron (mg)	Magn (mg)	Pota (mg)	Sodi (mg)	Zinc (mg)	VT-A (µg)	Thia (mg)	VT-E (mg)	Ribo (mg)	Niac (mg)	V-B6 (mg)	Fola (µg)	VT-C (mg)
7	40	2.75	21	303	955	1.12	120*	.35	2.13	.27	4.5	.13	6	10
7	80	2.25	26	408	955	1.3	215*	.25	2.75	.17	2.25	.2	8	12
22	52	3.25	20	245	1220	2.39	100*	.15	1.5	.17	2.25	.12	5	5
66	99	3.38	44	469	1108	3.5	90*	.26	2.32	.31	4.47	.28	67	14
184	230	1.35	38	201	763	1.29	675*	.09	1.22	.3	.48	.12	80	3
3	58	1.77	12	75	261	.38	70*	.18	1.29	.14	1.87	.03	31	1
11	25	2.33	27	218	93	.84	136*	.28	.62	.07	2.96	.16	15	4
0	21	1.65	21	106	5	.75	33	.21	.13	.04	1.99	.15	11	3
38	28	1.44	34	386	838	1.47	31*	.55	1.09	.21	3.63	.41	10	20
23	15	.84	21	185	506	.67	22*	.06	.67	.08	3.09	.18	6	12
0	29	1.25	36	245	799	.48	34	.07	2.16	.05	1.14	.11	31	28
0	144	2.85	48	457	724	1.08	523	.18	2.27	.12	1.45	.29	47	90
36	43	2.69	49	374	882	1.24	183*	.22	1.59	.12	8.24	.48	72	68
0	35	1.48	27	246	520	.58	19*	.08	1.74	.1	.44	.04	56	4
27	35	2.05	39	365	824	1.15	55*	.06	1.95	.14	13.7	.17	16	5
41	34	2.3	31	182	686	1.21	8*	.18	1.07	.15	7.8	.2	55	1
21	45	.97	37	258	234	.7	38*	.09	8.6	.05	.53	.36	33	5
20	4	.4	4	51	10	.32	6*	.09	.18	.06	.63	.04	3	<1
24	53	2.46	42	329	670	1.93	16	.27	.7	.42	2.71	.19	58	1
2	57	2.27	44	328	495	.76	8	.32	.87	.3	2.04	.15	44	1
35	40	1.8			590		300*							6
52	182	2.65	26	229	976	2.62	71*	.32	1.34	.23	6.38	.15	65	2
60	171	3.42	30	335	1050	3.49	65*	.35		.28	8.05	.18	61	2
60	60	4.68	35	353	957	1.87	31	.33	.55	.24	6.81	.2	100	9
79	102	6.18	17	263	973	1.31	18	.28	.7	.7	4.17	.09	103	0
64					668									
44	324	1.32	50	240	784	2.51	186*	.08	1.47	.42	1.91	.39	65	1
250	161	2.6	25	212	777	1.66	166*	.53	.9	.48	3.55	.16	73	2
68	185	3.5	37	353	939	1.17	97*	.46	1.83	.42	4.23	.11	91	3
55	84	2.61	33	340	615	.99	30	.33	.87	.22	3.4	.11	85	3
30	127	2.74	23	254	539	2.27	5	.29	.01	.24	3.95	.12	52	2
103	92	5.55	45	527	742	5.81	2	.34	1.61	.41	6.73	.37	84	1
44	23	2.31	13	143	670	1.98	0	.23	.27	.27	3.65	.05	48	<1
85	300	2.7			1620		60*							0
60	300	1.8			1770		80*							
55	250	2.7			1540		40*							
50	20	6			1360		80*							
95	250	2.7			1840		60*							
80	250	2.7			1740		60*							
70	300	1.8			1650		60*							
9	117	.58	16	110	336	.81	74*	.18		.16	2.48	.04	35	1
29	283	6.1	48	548	507	1.45	175*	.37	3.15	.43	3.76	.3	102	12
29	272	5.96	56	576	454	1.59	175*	.3	3.19	.37	3.47	.31	85	12
30	270	6.36	83	636	499	2.21	181*	.3	3.51	.36	3.67	.37	77	13
20	68	2.21	22	240	632	.98	49*	.41	2.32	.26	3.7	.16	66	6
22	61	2.22	32	288	625	1.21	53*	.36	2.55	.21	3.68	.18	53	6
23	51	2.54	61	346	690	1.87	56*	.37	2.91	.2	3.97	.25	45	7
53	407	2	27	162	1154	2.03	167*	.29	1.01	.4	2.37	.08	60	<1
57	430	2	38	208	1196	2.37	182*	.24	1.15	.36	2.25	.09	46	<1
60	439	2.33	68	264	1293	3.13	192*	.24	1.44	.35	2.46	.16	36	<1
60	60	4.68	35	353	957	1.87	31	.33	.55	.24	6.81	.2	100	9

*This value is expressed in retinol equivalents (RE). All other values are in retinol activity equivalents (RAE).

Table A–1

Food Composition   (Computer code number is for Wadsworth Diet Analysis program)   (For purposes of calculations, use "0" for t, <1, <.1, <.01, etc.)

Computer Code Number	Food Description	Measure	Wt (g)	H₂O (%)	Ener (kcal)	Prot (g)	Carb (g)	Dietary Fiber (g)	Fat (g)	Fat Breakdown (g) Sat	Mono	Poly
	MIXED DISHES and FAST FOODS MIXED DISHES:—Continued											
	Chicken salad:											
1143	On white bread, soft	1 ea	110	41	366	10	31	2	22	2.7		
1145	On part whole wheat	1 ea	110	41	369	11	27	3	24	3.1		
1144	On whole wheat	1 ea	123	41	398	13	31	5	26	3.4		
1146	Corned beef & swiss on rye	1 ea	156	47	427	28	22	6	26	9.5		
	Egg salad:											
1147	On white bread, soft	1 ea	117	43	379	9	31	1	24	3.8		
1149	On part whole wheat	1 ea	116	44	378	10	27	2	26	4.3		
1148	On whole wheat	1 ea	130	44	410	11	31	5	28	4.6		
	Ham:											
1279	On rye bread	1 ea	150	57	312	24	20	6	16	3.3		
1151	On white bread, soft	1 ea	157	52	364	24	30	2	16	3.3		
1153	On part whole wheat	1 ea	156	54	354	25	26	2	17	3.6		
1152	On whole wheat	1 ea	169	53	378	27	29	4	18	3.8		
	Ham & cheese:											
1280	On white bread, soft	1 ea	157	48	423	24	31	2	22	8.1		
1282	On part whole wheat	1 ea	156	49	417	25	26	2	24	8.7		
1281	On whole wheat	1 ea	170	48	445	27	30	4	25	9.2		
1150	Ham & swiss on rye	1 ea	150	53	359	24	21	6	21	7.1		
	Ham salad:											
1154	On white bread, soft	1 ea	131	47	361	10	37	1	19	4.2		
1156	On part whole wheat	1 ea	131	48	358	11	33	2	20	4.7		
1155	On whole wheat	1 ea	144	48	383	12	37	4	22	5		
1157	Patty melt: Ground beef & cheese on rye	1 ea	182	46	561	37	22	6	37	12.7		
	Peanut butter & jelly:											
1158	On white bread, soft	1 ea	101	27	348	11	47	3	14	2.7		
1160	On part whole wheat	1 ea	101	27	352	11	45	4	15	3.1		
1159	On whole wheat	1 ea	114	27	383	13	50	6	17	3.4		
1161	Reuben, grilled: Corned beef, swiss cheese, sauerkraut on rye	1 ea	239	60	555	22	27	5	40	17.3		
	Roast beef:											
713	On a bun	1 ea	139	49	346	21	33		14	3.6	6.8	1.7
1162	On white bread, soft	1 ea	157	46	405	29	34	1	16	2.9		
1164	On part whole wheat	1 ea	156	47	397	30	30	2	17	3.2		
1163	On whole wheat	1 ea	169	47	422	32	34	4	18	3.4		
	Tuna salad:											
1165	On white bread, soft	1 ea	122	46	326	13	35	1	14	1.9		
1167	On part whole wheat	1 ea	122	47	322	14	32	2	15	2.2		
1166	On whole wheat	1 ea	135	47	347	16	36	4	17	2.4		
	Turkey:											
1168	On white bread, soft	1 ea	156	54	346	24	29	1	14	1.9		
1170	On part whole wheat	1 ea	155	55	336	25	25	2	15	2.1		
1169	On whole wheat	1 ea	169	54	360	27	29	4	16	2.3		
	Turkey ham:											
1272	On rye bread	1 ea	150	60	280	21	20	6	13	2.5		
1273	On white bread, soft	1 ea	156	55	331	21	30	2	14	2.5		
1275	On part whole wheat	1 ea	156	52	363	23	28	4	19	3.8		
1274	On whole wheat	1 ea	169	56	343	24	29	4	15	3		
714	Taco	1 ea	171	58	369	21	27		21	11.4	6.6	1
	Tostada:											
1114	With refried beans	1 ea	144	66	223	10	26	7	10	5.4	3	.7
1118	With beans & beef	1 ea	225	70	333	16	30	4	17	11.5	3.5	.6
1354	With beans & chicken	1 ea	156	70	242	19	16	3	11	4.5		
	NUTS, SEEDS and PRODUCTS											
	Almonds:											
1365	Dry roasted, salted	1 c	138	3	824	30	27	16	73	5.6	46.4	17.4
718	Slivered, packed, unsalted	1 c	108	5	624	23	21	13	55	4.2	34.7	13.2
720	Whole, dried, unsalted	1 oz	28	5	162	6	6	3	14	1.1	9	3.4
721	Almond butter:	1 tbs	16	1	101	2	3	1	9	.9	6.1	2
4572	Salted	1 tbs	16	1	101	2	3	1	9	.9	6.1	2
722	Brazil nuts, dry (about 7)	1 c	140	3	918	20	18	8	93	22.6	32.2	33.8

Chol (mg)	Calc (mg)	Iron (mg)	Magn (mg)	Pota (mg)	Sodi (mg)	Zinc (mg)	VT-A (µg)	Thia (mg)	VT-E (mg)	Ribo (mg)	Niac (mg)	V-B6 (mg)	Fola (µg)	VT-C (mg)
30	76	2.21	20	146	483	.79	21*	.3	5.47	.24	4.09	.27	63	1
33	70	2.25	32	194	468	1.04	23*	.25	6.07	.2	4.15	.3	49	1
34	59	2.6	63	252	528	1.76	24*	.25	6.65	.18	4.47	.38	39	1
83	267	3.03	28	232	1469	3.59	58*	.2	2.59	.33	2.76	.18	32	<1
154	87	2.36	18	122	500	.76	50*	.31	4.24	.38	2.45	.21	74	0
167	80	2.38	29	165	479	.99	54*	.25	4.65	.34	2.3	.24	60	0
177	71	2.75	60	222	543	1.71	57*	.26	5.17	.33	2.54	.31	51	0
56	49	2.84	29	375	1155	2.66	7*	.82	2.52	.36	5.89	.4	27	<1
54	76	3.1	32	384	1237	2.61	6*	.9	2.52	.44	6.82	.39	60	<1
56	68	3.12	43	438	1251	2.92	7*	.88	2.71	.4	6.9	.42	44	<1
59	58	3.46	72	500	1336	3.66	7*	.9	3.05	.38	7.28	.49	34	<1
64	246	2.82	33	328	1366	2.71	74*	.7	2.57	.46	5.36	.31	61	<1
67	248	2.82	44	379	1388	3.03	78*	.67	2.77	.42	5.35	.34	45	<1
70	246	3.17	74	441	1487	3.79	82*	.68	3.13	.41	5.7	.41	36	<1
63	258	2.61	32	353	1326	2.93	57*	.61	2.64	.39	4.37	.31	27	<1
29	72	2.25	21	167	935	1.06	5*	.55	3.33	.28	3.69	.18	59	0
30	65	2.26	32	213	950	1.31	6*	.52	3.66	.23	3.65	.21	44	0
32	54	2.59	62	269	1029	2.02	6*	.53	4.07	.21	3.92	.28	34	0
113	222	4.17	39	391	715	7.04	95*	.25	3.51	.46	6.14	.37	37	<1
1	76	2.29	52	240	428	1.05	<1	.29	2.55	.23	5.44	.15	79	2
0	70	2.36	67	297	412	1.32	<1	.24	2.87	.18	5.68	.17	67	2
0	61	2.72	99	361	472	2.05	<1	.25	3.28	.16	6.12	.24	60	2
106	418	3.54	46	381	2166	3.89	175*	.18	2.59	.33	2.64	.33	56	14
51	54	4.23	31	316	792	3.39	21	.37	.19	.31	5.87	.26	57	2
43	76	4.14	30	436	1605	3.73	8*	.35	3.38	.36	6.79	.4	67	0
45	67	4.21	41	493	1642	4.11	8*	.29	3.62	.32	6.86	.44	51	0
47	57	4.6	70	557	1743	4.9	8*	.29	4	.3	7.24	.51	42	0
13	76	2.41	24	168	589	.68	15*	.3	2.77	.24	5.89	.13	63	1
13	69	2.45	36	215	578	.9	17*	.25	3.06	.19	6.07	.16	48	1
14	59	2.79	67	273	642	1.61	17*	.25	3.46	.17	6.47	.22	38	1
43	72	2.19	31	307	1589	1.33	8*	.31	3.27	.29	9.29	.42	60	0
45	63	2.15	42	356	1625	1.56	8*	.25	3.51	.24	9.52	.45	45	0
47	53	2.46	71	417	1735	2.25	8*	.25	3.91	.22	10.1	.53	35	0
55	51	4.04	27	343	1178	2.94	7*	.22	2.86	.33	4.32	.3	29	<1
52	77	4.21	29	351	1252	2.85	6*	.32	2.82	.41	5.29	.29	62	<1
50	49	2.53	65	408	1553	2.52	8*	.58	2.28	.27	7.98	.48	32	0
58	59	4.72	69	466	1361	3.95	7*	.26	3.4	.35	5.62	.39	37	<1
56	221	2.41	70	474	802	3.93	147*	.15	1.88	.44	3.21	.24	68	2
30	210	1.89	59	403	543	1.9	85*	.1	1.15	.33	1.32	.16	43	1
74	189	2.45	67	491	871	3.17	173*	.09	1.8	.49	2.86	.25	85	4
55	146	1.57	41	263	386	1.93	63*	.08	.66	.15	4.26	.31	25	5
0	367	6.22	395	1029	468	4.89	<1	.1	36.3	1.19	5.31	.17	45	0
0	268	4.64	297	786	1	3.63	1	.26	28.3	.88	4.24	.14	31	0
0	69	1.2	77	204	<1	.94	<1	.07	7.33	.23	1.1	.04	8	0
0	43	.59	48	121	2	.49	0	.02	3.25	.1	.46	.01	10	<1
0	43	.59	48	121	72	.49	0	.02	3.24	.1	.46	.01	10	<1
0	246	4.76	315	840	3	6.43	0	1.4	10.6	.17	2.27	.35	6	1

*This value is expressed in retinol equivalents (RE). All other values are in retinol activity equivalents (RAE).

Table A–1

Food Composition   (Computer code number is for Wadsworth Diet Analysis program)   (For purposes of calculations, use "0" for t, <1, <.1, <.01, etc.)

A

Computer Code Number	Food Description	Measure	Wt (g)	H₂O (%)	Ener (kcal)	Prot (g)	Carb (g)	Dietary Fiber (g)	Fat (g)	Fat Breakdown (g)		
										Sat	Mono	Poly
	NUTS, SEEDS and PRODUCTS—Continued											
	Cashew nuts, dry roasted:											
724	Salted	1 oz	28	2	161	4	9	1	13	2.6	7.6	2.2
4621	Unsalted	1 oz	28	2	161	4	9	1	13	2.6	7.6	2.2
725	Oil roasted:	1 c	130	4	749	21	37	5	63	12.4	36.9	10.6
726	Ounce	1 oz	28	4	161	5	8	1	13	2.7	7.9	2.3
4622	Unsalted:	1 c	130	4	749	21	37	5	63	12.4	36.9	10.6
4622	Ounce	1 oz	28	4	161	5	8	1	13	2.7	7.9	2.3
727	Cashew butter, unsalted	1 tbs	16	3	94	3	4	<1	8	1.6	4.7	1.3
4662	Cashew butter, salted	1 tbs	16	3	94	3	4	<1	8	1.6	4.7	1.3
728	Chestnuts, European, roasted (1 cup = approx 17 kernels)	1 c	143	40	350	5	76	7	3	.6	1.1	1.2
	Coconut, raw:											
729	Piece 2 x 2 x ½"	1 pce	45	47	159	1	7	4	15	13.4	.6	.2
730	Shredded/grated, unpacked	½ c	40	47	142	1	6	4	13	11.9	.6	.1
	Coconut, dried, shredded/grated:											
731	Unsweetened	1 c	78	3	515	5	19	13	50	44.6	2.1	.6
732	Sweetened	1 c	93	13	466	3	44	4	33	29.3	1.4	.4
4559	Coconut milk, canned	1 c	226	73	445	5	6	3	48	42.7	2	.5
734	Filberts/hazelnuts, chopped	1 oz	28	5	176	4	5	3	17	1.2	12.8	2.2
735	Macadamias, oil roasted, salted:	1 c	134	2	962	10	17	12	103	15.4	80.9	1.8
736	Ounce	1 oz	28	2	201	2	4	3	21	3.2	16.9	.4
1368	Macadamias, oil roasted, unsalted	1 c	134	2	962	10	17	12	103	15.4	80.9	1.8
	Mixed Nuts:											
737	Dry roasted, salted	1 c	137	2	814	24	35	12	70	9.4	43	14.7
738	Oil roasted, salted	1 c	142	2	876	24	30	13	80	12.4	45	18.9
1369	Oil roasted, unsalted	1 c	142	2	876	24	30	14	80	12.4	45	18.9
	Peanuts:											
740	Oil roasted, salted	1 oz	28	2	163	7	5	3	14	1.9	6.8	4.4
742	Dried, salted	1 oz	28	2	164	7	6	2	14	1.9	6.9	4.4
743	Peanut butter:	½ c	128	1	759	32	25	8	65	13.2	31.1	17.6
1371	Tablespoon	2 tbs	32	1	190	8	6	2	16	3.3	7.8	4.4
745	Pecan halves, dried, unsalted	1 oz	28	4	193	3	4	3	20	1.7	11.4	6
1372	Pecan halves, dry roasted, salted	¼ c	28	1	199	3	4	3	21	1.8	12.3	5.8
746	Pine nuts/pinons, dried	1 oz	28	6	176	3	5	3	17	2.6	6.4	7.2
747	Pistachios, dried, shelled	1 oz	28	4	156	6	8	3	12	1.5	6.5	3.8
1373	Pistachios,dry roasted,salted,shelled	1 c	128	2	727	27	34	13	59	7.1	31	17.8
748	Pumpkin kernels, dried, unsalted	1 oz	28	7	151	7	5	1	13	2.4	4	5.8
1374	Pumpkin kernels, roasted, salted	1 c	227	7	1184	75	30	9	96	18.1	29.7	43.6
749	Sesame seeds, hulled, dried	¼ c	38	5	223	10	4	4	21	2.9	7.9	9.1
8878	Soy nuts, BBQ	5 pce	28		119	12	9	4	4	1		
8877	Soy nuts, salted	5 pce	28		119	12	9	5	4	1		
	Sunflower seed kernels:											
750	Dry	¼ c	36	5	205	8	7	4	18	1.9	3.4	11.8
751	Oil roasted	¼ c	34	3	209	7	5	2	19	2	3.7	12.9
752	Tahini (sesame butter)	1 tbs	15	3	91	3	3	1	8	1.2	3.2	3.7
1334	Trail Mix w/chocolate chips	1 c	146	7	707	21	66	8	47	8.9	19.8	16.5
754	Black walnuts, chopped	1 oz	28	4	170	7	3	1	16	1	3.6	10.5
756	English walnuts, chopped	1 oz	28	4	183	4	4	2	18	1.7	2.5	13.2
	SWEETENERS and SWEETS (see also Dairy (milk desserts) and Baked Goods)											
757	Apple butter	2 tbs	36	56	62	<1	15	1	0	0	0	0
1124	Butterscotch topping	2 tbs	41	32	103	1	27	<1	<1	3.5	.6	0
1125	Caramel topping	2 tbs	41	32	103	1	27	<1	<1	3.5	.6	0
	Cake frosting, creamy vanilla:											
1127	Canned	2 tbs	39	13	163	<1	27	<1	7	1.9	3.4	.9
1123	From mix	2 tbs	39	12	165	<1	28	<1	6	1.3	2.6	2.2
	Cake frosting, lite:											
2061	Milk chocolate	1 tbs	16	18	58	<1	11	<1	1		.9	
2062	Vanilla	1 tbs	16	15	60	0	12	<1	1	0	1	
	Candy:											

Chol (mg)	Calc (mg)	Iron (mg)	Magn (mg)	Pota (mg)	Sodi (mg)	Zinc (mg)	VT-A (µg)	Thia (mg)	VT-E (mg)	Ribo (mg)	Niac (mg)	V-B6 (mg)	Fola (µg)	VT-C (mg)
0	13	1.68	73	158	179	1.57	0	.06	.16	.06	.39	.07	19	0
0	13	1.68	73	158	4	1.57	0	.06	.16	.06	.39	.07	19	0
0	53	5.33	332	689	814	6.18	0	.55	2.03	.23	2.34	.32	88	0
0	11	1.15	71	148	175	1.33	0	.12	.44	.05	.5	.07	19	0
0	53	5.33	332	689	22	6.18	0	.55	2.03	.23	2.34	.32	88	0
0	11	1.15	71	148	5	1.33	0	.12	.44	.05	.5	.07	19	0
0	7	.8	41	87	2	.83	0	.05	.25	.03	.26	.04	11	0
0	7	.8	41	87	98	.83	0	.05	.25	.03	.26	.04	11	0
0	41	1.3	47	847	3	.81	1	.35	1.72	.25	1.92	.71	100	37
0	6	1.09	14	160	9	.49	0	.03	.33	.01	.24	.02	12	1
0	6	.97	13	142	8	.44	0	.03	.29	.01	.22	.02	10	1
0	20	2.59	70	424	29	1.57	0	.05	1.05	.08	.47	.23	7	1
0	14	1.79	46	313	244	1.69	0	.03	1.26	.02	.44	.25	7	1
0	41	7.46	104	497	29	1.27	0	.05	1.47	0	1.44	.06	32	2
0	32	1.32	46	190	0	.69	1	.18	4.25	.03	.5	.16	32	2
0	60	2.41	157	441	348	1.47	1*	.28	.55	.15	2.71	.26	21	0
0	13	.5	33	92	73	.31	<1*	.06	.11	.03	.57	.05	4	0
0	60	2.41	157	441	9	1.47	1*	.28	.55	.15	2.71	.26	21	0
0	96	5.07	308	818	917	5.21	1	.27	8.22	.27	6.44	.41	68	1
0	153	4.56	334	825	926	7.21	1	.71	8.52	.31	7.19	.34	118	1
0	153	4.56	334	825	16	7.21	1	.71	8.52	.31	7.19	.34	118	1
0	25	.51	52	191	121	1.86	0	.07	2.07	.03	4	.07	35	0
0	15	.63	49	184	228	.93	0	.12	2.07	.03	3.79	.07	41	0
0	49	2.36	204	856	598	3.74	0	.11	12.8	.13	17.2	.58	95	0
0	12	.59	51	214	149	.93	0	.03	3.2	.03	4.29	.14	24	0
0	20	.71	34	115	0	1.27	1	.18	1.13	.04	.33	.06	6	<1
0	20	.78	37	119	107	1.42	2	.13	1.05	.03	.33	.05	4	<1
0	2	.86	65	176	20	1.2	<1	.35	.98	.06	1.22	.03	16	1
0	30	1.16	34	287	<1	.62	8	.24	1.28	.04	.36	.48	14	1
0	141	5.38	154	1333	518	2.94	34	1.08	5.45	.2	1.82	2.18	64	3
0	12	4.19	150	226	5	2.09	5	.06	.28	.09	.49	.06	16	1
0	98	33.9	1212	1829	1305	16.9	43	.48	2.27	.72	3.95	.2	129	4
0	50	2.96	132	155	15	3.9	1	.27	.86	.03	1.78	.05	36	0
0	59	1.07			415		0							0
0	59	1.07			148		0							0
0	42	2.44	127	248	1	1.82	1	.82	18.1	.09	1.62	.28	82	<1
0	19	2.28	43	164	1	1.77	1	.11	17.1	.09	1.4	.27	80	<1
0	21	.95	53	69	<1	1.57	1	.24	.34	.02	.85	.02	15	0
6	159	4.95	235	946	177	4.58	7*	.6	15.6	.33	6.43	.38	95	2
0	16	.86	57	147	<1	.96	4	.06	.73	.03	.19	.15	18	1
0	27	.81	44	123	1	.86	1	.09	.82	.04	.56	.15	27	<1
0	5	.11	2	33	1	.02	4*	.01	<.01	.01	.04	.02	<1	<1
<1	22	.08	3	34	143	.08	11*	<.01	0	.04	.02	.01	1	<1
<1	22	.08	3	34	143	.08	11*	<.01	0	.04	.02	.01	1	<1
0	1	.04		14	35	0	88*	0	1.84	<.01	<.01	0	0	0
0	4	.09	1	9	87	.04	42*	.01	.79	.01	.13	<.01	0	0
0							<1*							
0							0*	0					0	

*This value is expressed in retinol equivalents (RE). All other values are in retinol activity equivalents (RAE).

Table A-1

Food Composition

(Computer code number is for Wadsworth Diet Analysis program)   (For purposes of calculations, use "0" for t, <1, <.1, <.01, etc.)

Computer Code Number	Food Description	Measure	Wt (g)	H₂O (%)	Ener (kcal)	Prot (g)	Carb (g)	Dietary Fiber (g)	Fat (g)	Fat Breakdown (g)		
										Sat	Mono	Poly
	SWEETENERS and SWEETS (see also Dair (milk desserts) and Baked Goods)—Continued											
1128	Almond Joy candy bar	1 oz	28	10	131	1	16	1	7	4.8	1.8	.4
2069	Butterscotch morsels	¼ c	43	8	243	0	27	0	12	10.6		
758	Caramel, plain or chocolate	1 pce	10	8	38	<1	8	<1	1	.7	.1	t
1961	Chewing gum, sugarless	1 pce	3		6	0	2		0	0	0	0
	Chocolate (see also #784, 785, 971):											
	Milk chocolate:											
759	Plain	1 oz	28	1	144	2	17	1	9	5.2	2.8	.3
760	With almonds	1 oz	28	1	147	3	15	2	10	4.7	3.8	.6
761	With peanuts	1 oz	28	1	155	5	11	2	11	3.4	5.1	2.5
762	With rice cereal	1 oz	28	2	139	2	18	1	7	4.4	2.4	.2
763	Semisweet chocolate chips	1 c	168	1	805	7	106	10	50	29.8	16.7	1.6
764	Sweet Dark chocolate (candy bar)	1 ea	41	1	226	2	25	2	13	8.3	4.6	.4
765	Fondant candy, uncoated (mints, candy corn, other)	1 pce	16	7	57	0	15	0	0	0	0	0
1697	Fruit Roll-Up (small)	1 ea	14	11	49	<1	12	<1	<1	.1	.2	.1
766	Fudge, chocolate	1 pce	17	10	65	<1	13	<1	1	.9	.4	.1
767	Gumdrops	1 c	182	1	703	0	180	0	0	0	0	0
768	Hard candy-all flavors	1 pce	6	1	24	0	6	0	<1			
769	Jellybeans	10 pce	11	6	40	0	10	0	<1	t	t	t
1134	M&M's plain chocolate candy	10 pce	7	2	34	<1	5	<1	1	.9	.5	t
1135	M&M's peanut chocolate candy	10 pce	20	2	103	2	12	1	5	2.1	2.2	.8
1130	MARS almond bar	1 ea	50	4	234	4	31	1	11	3.6	5.3	2
1129	MILKY WAY candy bar	1 ea	60	6	254	3	43	1	10	4.7	3.6	.4
1708	Milk chocolate-coated peanuts	1 c	149	2	773	19	74	7	50	21.8	19.3	6.4
1709	Peanut brittle, recipe	1 c	147	2	666	11	102	3	28	7.4	12.5	6.9
1132	Reese's peanut butter cup	2 ea	50	2	271	5	27	2	16	5.5	6.5	2.8
1133	Skor English toffee candy bar	1 ea	39	3	217	2	22	1	13	8.5	4.3	.5
1131	Snickers candy bar (2.2oz)	1 ea	62	5	297	5	37	2	15	5.6	6.5	3
23082	Chewing gum	1 pce	3	3	10	0	3	0	<1	t	t	t
1482	Fruit juice bar (2.5 fl oz)	1 ea	77	78	63	1	16	0	<1	t	0	t
771	Gelatin dessert/Jello, prepared	½ c	135	85	80	2	19	0	0	0	0	0
1702	SugarFree	½ c	117	98	8	1	1	0	0	0	0	0
772	Honey:	1 c	339	17	1030	1	279	1	0	0	0	0
773	Tablespoon	1 tbs	21	17	64	<1	17	<1	0	0	0	0
774	Jams or preserves:	1 tbs	20	29	54	<1	14	<1	<1	0	t	t
775	Packet	1 ea	14	30	39	<1	10	<1	<1	t	t	0
776	Jellies:	1 tbs	19	29	54	<1	13	<1	<1	.2	1.3	.2
777	Packet	1 ea	14	29	40	<1	10	<1	<1	.2	1	.1
1136	Marmalade	1 tbs	20	33	49	<1	13	<1	0	0	0	0
770	Marshmallows	1 ea	7	16	22	<1	6	<1	<1	t	t	t
1126	Marshmallow creme topping	2 tbs	38	20	122	<1	30	<1	<1	t	t	t
778	Popsicle/ice pops	1 ea	128	80	92	0	24	0	0	0	0	0
23171	Rice crispie bar	1 ea	28	13	107	1	20	<1	3	.6	1.3	.8
	Sugars:											
779	Brown sugar	1 c	220	2	827	0	214	0	0	0	0	0
780	White sugar, granulated:	1 c	200	0	774	0	200	0	0	0	0	0
781	Tablespoon	1 tbs	12	0	46	0	12	0	0	0	0	0
782	Packet	1 ea	6	0	23	0	6	0	0	0	0	0
783	White sugar, powdered, sifted	1 c	100		389	0	99	0	<1	t	t	t
	Sweeteners:											
1711	Equal, packet	1 ea	1	12	4	<1	1	0	<1	0	t	t
1712	Sweet 'N Low, packet	1 ea	1		4	0	1	0	0	0	0	0
	Syrups:											
	Chocolate:											
785	Hot fudge type	2 tbs	43	22	151	2	27	1	4	1.7	1.7	.1
784	Thin type	2 tbs	38	29	93	1	25	1	<1	.3	.2	t
25003	Molasses	2 tbs	41	26	109	0	28	0	0	0	0	0
1710	Light cane	2 tbs	41	24	103	0	27	0	0	0	0	0
787	Pancake table syrup (corn and maple)	2 tbs	40	24	115	0	30	0	0	0	0	0

Chol (mg)	Calc (mg)	Iron (mg)	Magn (mg)	Pota (mg)	Sodi (mg)	Zinc (mg)	VT-A (µg)	Thia (mg)	VT-E (mg)	Ribo (mg)	Niac (mg)	V-B6 (mg)	Fola (µg)	VT-C (mg)
1	17	.39	18	69	41	.22	1*	.01	.63	.04	.13	.02		<1
0	0	0		79	45		0	.03		.03	.02			0
1	14	.01	2	21	24	.04	1*	<.01	.05	.02	.02	<.01	<1	<1
				0	0									
6	53	.39	17	108	23	.39	8	.02	.35	.08	.09	.01	2	<1
5	63	.46	25	124	21	.37	4*	.02	.53	.12	.21	.01	3	<1
3	32	.52	34	150	11	.68	5*	.08	1.3	.05	2.12	.04	23	0
5	48	.21	14	96	41	.31	3*	.02	.3	.08	.13	.02	3	<1
0	54	5.26	193	613	18	2.72	2	.09	2	.15	.72	.06	5	0
<1	11	.98	45	123	3	.59	2*	.01	.18	.03	.16	.01	1	0
0		.01		3	6	.01	0*	0	0	<.01	0	0	0	0
0	4	.14	3	41	9	.03	1	.01	.04	<.01	.01	.04	1	1
2	7	.08	4	17	10	.07	8*	<.01	.02	.01	.02	<.01	<1	<1
0	5	.73	2	9	80	0	0	0	0	<.01	<.01	0	0	0
0		.02			2	<.01	0	0	0	0	0	0	0	0
0		.12	4	3		.01	0	0	0	0	0	0	0	0
1	7	.08	3	19	4	.07	4*	<.01	.06	.01	.02	<.01	<1	<1
2	20	.23	15	69	10	.46	5*	.02	.49	.03	.75	.02	7	<1
8	84	.55	36	163	85	.55	25*	.02	2.33	.16	.47	.03	9	<1
8	78	.46	20	145	144	.43	19*	.02	.39	.13	.21	.03	6	1
13	155	1.95	140	748	61	2.89	0	.17	3.8	.26	6.33	.31	12	0
19	44	2.03	73	306	664	1.43	69*	.28	2.41	.08	5.14	.15	103	0
2	39	.6	44	176	159	.91	9*	.12	2.04	.08	2.31	.07	27	<1
20	51	.19	13	93	108	.3	27*	.01	.53	.13	.03	.01		<1
8	58	.47	45	201	165	1.46	24*	.06	.95	.09	2.6	.05	25	<1
0	0	0	0		<1	0	0	0	0	0	0	0	0	0
0	4	.15	3	41	3	.04	1	.01	0	.01	.12	.02	5	7
0	3	.04	1	1	57	.04	0	0	0	<.01	<.01	<.01	0	0
0	2	.01	1	0	56	.03	0	0	0	<.01	<.01	<.01	0	0
0	20	1.42	7	176	14	.75	0	0	0	.13	.41	.08	7	2
0	1	.09		11	1	.05	0	0	0	.01	.02	<.01	<1	<1
0	4	.2	1	18	2	.01	<1*	<.01	.02	.01	.04	<.01	2	<1
0	3	.07	1	11	4	.01	<1	0	0	<.01	<.01	<.01	5	1
0	2	.04	1	12	5	.01	<1	0	0	<.01	.01	<.01	<1	<1
0	1	.03	1	9	4	.01	<1	0	0	<.01	<.01	<.01	<1	<1
0	8	.03		7	11	.01	<1	<.01	0	<.01	.01	<.01	7	1
0		.02			3	<.01	<1	0	0	0	<.01	0	<1	0
0	1	.08	1	2	19	.01	<1	0	0	0	.03	<.01	<1	0
0	0	0	1	5	15	.03	0	0	0	0	0	0	0	0
0	2	.51	4	12	123	.15	85*	.1	.41	.11	1.31	.13	27	4
0	187	4.2	64	761	86	.4	0	.02	0	.01	.18	.06	2	0
0	2	.12	0	4	2	.06	0	0	0	.04	0	0	0	0
0		.01	0		<1	<.01	0	0	0	<.01	0	0	0	0
0		<.01	0		<1	<.01	0	0	0	<.01	0	0	0	0
0	1	.06	0	2	1	.03	0	0	0	0	0	0	0	0
0		<.01			<1	0	0	0	0	0	0	0	0	0
0	0	0			0	0	0							0
1	35	.56	22	156	149	.29	2*	.02	1.26	.09	.13	.03	2	<1
0	5	5.15	25	183	58	.28	494*	<.01	.01	.31	12.8	.01	2	<1
0	84	1.94	99	600	15	.12	0	.02	0	0	.38	.27	0	0
0	68	1.76	100	376	6	.12	0*	.03	0	.02	.08	.27	0	0
0		.04	1	1	33	.02	0	<.01	0	<.01	.01	0	0	0

*This value is expressed in retinol equivalents (RE). All other values are in retinol activity equivalents (RAE).

Table A-1

Food Composition

(Computer code number is for Wadsworth Diet Analysis program)   (For purposes of calculations, use "0" for t, <1, <.1, <.01, etc.)

Computer Code Number	Food Description	Measure	Wt (g)	H₂O (%)	Ener (kcal)	Prot (g)	Carb (g)	Dietary Fiber (g)	Fat (g)	Fat Breakdown (g) Sat	Mono	Poly
	VEGETABLES AND LEGUMES											
788	Alfalfa sprouts	1 c	33	91	10	1	1	1	<1	t	t	.1
1815	Amaranth leaves, raw, chopped	1 c	28	92	6	1	1	<1	<1	t	t	t
1816	Amaranth leaves, raw, each	1 ea	14	92	3	<1	1	<1	<1	t	t	t
1817	Amaranth leaves, cooked	1 c	132	91	28	3	5	2	<1	.1	.1	.1
1987	Arugula, raw, chopped	½ c	10	92	2	<1	<1	<1	<1	t	t	t
789	Artichokes, cooked globe (300g with refuse)	1 ea	120	84	60	4	13	6	<1	t	t	.1
1177	Artichoke hearts, cooked from frozen	1 c	168	86	76	5	15	8	1	.2	t	.4
1176	Artichoke hearts, marinated	1 c	130	80	116	5	14	5	7	0		
2021	Artichoke hearts, in water	½ c	100	91	37	2	6	0	0	0	0	0
	Asparagus, green, cooked: From fresh:											
790	Cuts and tips	½ c	90	92	22	2	4	1	<1	.1	t	.1
791	Spears, ½" diam at base	4 ea	60	92	14	2	3	1	<1	t	t	.1
	From frozen:											
792	Cuts and tips	½ c	90	91	25	3	4	1	<1	.1	t	.2
793	Spears, ½" diam at base	4 ea	60	91	17	2	3	1	<1	.1	t	.1
794	Canned, spears, ½" diam at base	4 pce	72	94	14	2	2	1	<1	.1	t	.2
795	Bamboo shoots, canned, drained slices	1 c	131	94	25	2	4	2	1	.1	t	.2
1795	Bamboo shoots, raw slices	1 c	151	91	41	4	8	3	<1	.1	t	.2
1798	Bamboo shoots, cooked slices	1 c	120	96	14	2	2	1	<1	.1	t	.1
	Beans (see also alphabetical listing this section):											
1990	Adzuki beans, cooked	½ c	115	66	147	9	28	8	<1	t	t	t
796	Black beans, cooked	½ c	86	66	114	8	20	7	<1	.1	t	.2
	Canned beans (white/navy):											
803	With pork and tomato sauce	½ c	127	73	124	7	25	6	1	.5	.6	.2
804	With sweet sauce	½ c	130	71	144	7	27	7	2	.7	.8	.2
805	With frankfurters	½ c	130	69	185	9	20	9	9	3.1	3.7	1.1
	Lima beans:											
797	Thick seeded (Fordhooks), cooked from frozen	½ c	85	73	85	5	16	5	<1	.1	t	.1
798	Thin seeded (Baby), cooked from frozen	½ c	90	72	94	6	17	5	<1	.1	t	.1
799	Cooked from dry, drained	½ c	94	70	108	7	20	7	<1	.1	t	.2
1998	Red Mexican, cooked f/dry	½ c	112	70	127	8	24	9	<1	.1	.1	.2
	Snap bean/green string beans cuts and french style:											
800	Cooked from fresh	½ c	63	89	22	1	5	2	<1	t	t	.1
801	Cooked from frozen	½ c	68	91	19	1	4	2	<1	t	t	.1
802	Canned, drained	½ c	68	93	14	1	3	1	<1	t	t	t
1713	Snap bean, yellow, cooked f/fresh	½ c	63	89	22	1	5	2	<1	t	t	.1
	Bean sprouts (mung):											
806	Raw	½ c	52	90	16	2	3	1	<1	t	t	t
807	Cooked, stir fried	½ c	62	84	31	3	7	1	<1	t	t	t
808	Cooked, boiled, drained	½ c	62	93	13	1	3	<1	<1	t	t	t
1788	Canned, drained	½ c	63	96	8	1	1	<1	<1	t	t	t
	Beets, cooked from fresh:											
809	Sliced or diced	½ c	85	87	37	1	8	2	<1	t	t	.1
810	Whole beets, 2" diam	2 ea	100	87	44	2	10	2	<1	t	t	.1
	Beets, canned:											
811	Sliced or diced	½ c	79	91	24	1	6	1	<1	t	t	t
812	Pickled slices	½ c	114	82	74	1	19	3	<1	t	t	t
813	Beet greens, cooked, drained	½ c	72	89	19	2	4	2	<1	t	t	t
	Broccoli, raw:											
817	Chopped	½ c	44	91	12	1	2	1	<1	t	t	.1
818	Spears	1 ea	31	91	9	1	2	1	<1	t	t	.1
	Broccoli, cooked from fresh:											
819	Spears	1 ea	180	91	50	5	9	5	1	.1	t	.3
820	Chopped	½ c	78	91	22	2	4	2	<1	t	t	.1
	Broccoli, cooked from frozen:											
821	Spear, small piece	½ c	92	91	26	3	5	3	<1	t	t	.1

Chol (mg)	Calc (mg)	Iron (mg)	Magn (mg)	Pota (mg)	Sodi (mg)	Zinc (mg)	VT-A (µg)	Thia (mg)	VT-E (mg)	Ribo (mg)	Niac (mg)	V-B6 (mg)	Fola (µg)	VT-C (mg)
0	11	.32	9	26	2	.3	3	.02	.01	.04	.16	.01	12	3
0	60	.65	15	171	6	.25	41	.01	.22	.04	.18	.05	24	12
0	30	.32	8	85	3	.13	20	<.01	.11	.02	.09	.03	12	6
0	276	2.98	73	846	28	1.16	183	.03	.66	.18	.74	.23	75	54
0	16	.15	5	37	3	.05	12	<.01	.04	.01	.03	.01	10	1
0	54	1.55	72	425	114	.59	11	.08	.23	.08	1.2	.13	61	12
0	35	.94	52	444	89	.6	13	.1	.32	.26	1.54	.15	200	8
0	0	0		488		0								46
0	0	1.35		0	250	6								4
0	18	.66	9	144	10	.38	24	.11	.34	.11	.97	.11	131	10
0	12	.44	6	96	7	.25	16	.07	.23	.08	.65	.07	88	6
0	21	.58	12	196	4	.5	37	.06	1.13	.09	.93	.02	122	22
0	14	.38	8	131	2	.34	25	.04	.75	.06	.62	.01	81	15
0	11	1.32	7	124	207	.29	19	.04	.31	.07	.69	.08	69	13
0	10	.42	5	105	9	.85	1	.03	.5	.03	.18	.18	4	1
0	20	.75	5	805	6	1.66	2	.23	1.51	.11	.91	.36	11	6
0	14	.29	4	640	5	.56	0	.02	.8	.06	.36	.12	2	0
0	32	2.3	60	612	9	2.04	1	.13	.11	.07	.82	.11	139	0
0	23	1.81	60	305	1	.96	<1	.21	.07	.05	.43	.06	128	0
9	71	4.17	44	381	559	7.44	8	.07	.69	.06	.63	.09	29	4
9	79	2.16	44	346	437	1.95	7	.06	.7	.08	.46	.11	48	4
8	62	2.25	36	306	559	2.43	10	.07	.61	.07	1.17	.06	39	3
0	19	1.16	29	347	45	.37	8	.06	.25	.05	.91	.1	18	11
0	25	1.76	50	370	26	.49	8	.06	.58	.05	.69	.1	14	5
0	16	2.25	40	478	2	.89	0	.15	.17	.05	.4	.15	78	0
0	42	1.87	48	371	6	.87	<1	.13	.08	.07	.38	.11	94	2
0	29	.81	16	188	2	.23	21	.05	.09	.06	.39	.03	21	6
0	33	.6	16	86	6	.33	14	.02	.09	.06	.26	.04	16	3
0	18	.61	9	74	178	.2	12	.01	.09	.04	.14	.02	22	3
0	29	.81	16	188	2	.23	3	.05	.18	.06	.39	.03	21	6
0	7	.47	11	77	3	.21	1	.04	<.01	.06	.39	.05	32	7
0	8	1.18	20	136	6	.56	1	.09	.01	.11	.74	.08	43	10
0	7	.4	9	63	6	.29	<1	.03	.01	.06	.51	.03	18	7
0	9	.27	6	17	88	.18	1	.02	.01	.04	.14	.02	6	<1
0	14	.67	20	259	65	.3	2	.02	.25	.03	.28	.06	68	3
0	16	.79	23	305	77	.35	2	.03	.3	.04	.33	.07	80	4
0	12	1.44	13	117	153	.17	<1	.01	.24	.03	.12	.04	24	3
0	12	.47	17	169	301	.3	1	.01	.15	.05	.29	.06	31	3
0	82	1.37	49	654	174	.36	184	.08	.22	.21	.36	.09	10	18
0	21	.39	11	143	12	.18	34	.03	.73	.05	.28	.07	31	41
0	15	.27	8	101	8	.12	24	.02	.51	.04	.2	.05	22	29
0	83	1.51	43	526	47	.68	125	.1	3.04	.2	1.03	.26	90	134
0	36	.65	19	228	20	.3	54	.04	1.32	.09	.45	.11	39	58
0	47	.56	18	166	22	.28	87	.05	.95	.07	.42	.12	28	37

*This value is expressed in retinol equivalents (RE). All other values are in retinol activity equivalents (RAE).

Table A–1

Food Composition   (Computer code number is for Wadsworth Diet Analysis program)   (For purposes of calculations, use "0" for t, <1, <.1, <.01, etc.)

Computer Code Number	Food Description	Measure	Wt (g)	H₂O (%)	Ener (kcal)	Prot (g)	Carb (g)	Dietary Fiber (g)	Fat (g)	Fat Breakdown (g) Sat	Mono	Poly
	VEGETABLES AND LEGUMES—Continued											
822	Chopped	½ c	92	91	26	3	5	3	<1	t	t	.1
1603	Broccoflower-steamed	½ c	78	90	25	2	5	2	<1	t	t	.1
823	Brussels sprouts, cooked from fresh	½ c	78	87	30	2	7	2	<1	.1	t	.2
824	Brussels sprouts, cooked from frozen	½ c	78	87	33	3	6	3	<1	.1	t	.2
	Cabbage, common varieties:											
825	Raw, shredded or chopped	1 c	70	92	17	1	4	2	<1	t	t	.1
826	Cooked, drained	1 c	150	94	33	2	7	3	1	.1	t	.3
	Cabbage, Chinese:											
1178	Bok Choy, raw, shredded	1 c	70	95	9	1	2	1	<1	t	t	.1
827	Bok Choy, cooked, drained	1 c	170	96	20	3	3	3	<1	t	t	.1
1937	Kim chee style	1 c	150	92	31	2	6	2	<1	t	t	.1
828	Pe Tsai, raw, chopped	1 c	76	94	12	1	2	2	<1	t	t	.1
1796	Pe Tsai, cooked	1 c	119	95	17	2	3	3	<1	t	t	.1
	Cabbage, red, coarsely chopped:											
829	Raw	1 c	89	92	24	1	5	2	<1	t	t	.1
830	Cooked, drained	1 c	150	94	31	2	7	3	<1	t	t	.1
831	Cabbage, savoy, coarsely chopped, raw	1 c	70	91	19	1	4	2	<1	t	t	t
1785	Cabbage, savoy, cooked	1 c	145	92	35	3	8	4	<1	t	t	.1
1896	Capers	1 ea	5			<1	<1		<1			
	Carrots, raw:											
832	Whole, 7½ x 1⅛"	1 ea	72	88	31	1	7	2	<1	t	t	.1
833	Grated	½ c	55	88	24	1	6	2	<1	t	t	t
	Carrots, cooked, sliced, drained:											
834	From raw	½ c	78	87	35	1	8	3	<1	t	t	.1
835	From frozen	½ c	73	90	26	1	6	3	<1	t	t	t
836	Carrots, canned, sliced, drained	½ c	73	93	18	<1	4	1	<1	t	t	.1
837	Carrot juice,canned	1 c	236	89	94	2	22	2	<1	.1	t	.2
5625	Cassava, cooked	1 c	137	59	221	2	53	2	<1	.1	.1	.1
	Cauliflower, flowerets:											
838	Raw	½ c	50	92	12	1	3	1	<1	t	t	t
839	Cooked from fresh, drained	½ c	62	93	14	1	3	2	<1	t	t	.1
840	Cooked, from frozen, drained	½ c	90	94	17	1	3	2	<1	t	t	.1
	Celery, pascal type, raw:											
841	Large outer stalk,8 x 1 ½"(root end)	1 ea	40	95	6	<1	1	1	<1	t	t	t
842	Diced	1 c	120	95	19	1	4	2	<1	t	t	.1
1789	Celeriac/celery root, cooked	1 c	155	92	42	1	9	2	<1	.1	.1	.2
1179	Chard, swiss, raw, chopped	1 c	36	93	7	1	1	1	<1	t	t	t
1180	Chard, swiss, cooked	1 c	175	93	35	3	7	4	<1	t	t	t
1855	Chayote fruit, raw	1 ea	203	94	39	2	9	3	<1	.1	t	.1
1856	Chayote fruit, cooked	1 c	160	93	38	1	8	4	1	.1	.1	.3
	Chickpeas (see Garbanzo Beans #854)											
	Collards, cooked, drained:											
843	From raw	½ c	95	92	25	2	5	3	<1	t	t	.2
844	From frozen	½ c	85	88	31	3	6	2	<1	.1	t	.2
	Corn, yellow, cooked, drained:											
845	From raw, on cob, 5" long	1 ea	77	73	72	2	17	2	1	.1	.2	.3
846	From frozen, on cob, 3½" long	1 ea	63	73	59	2	14	2	<1	.1	.1	.2
847	Kernels, cooked from frozen	½ c	82	77	66	2	16	2	<1	.1	.1	.2
	Corn, canned:											
848	Cream style	½ c	128	79	92	2	23	2	1	.1	.2	.3
849	Whole kernel, vacuum pack	½ c	105	77	83	3	20	2	1	.1	.2	.2
	Cowpeas (see Black-eyed peas #814-816)											
850	Cucumber slices with peel	7 pce	28	96	4	<1	1	<1	<1	t	t	t
1948	Cucumber, kim chee style	1 c	150	91	31	2	7	2	<1	t	t	.1
	Dandelion Greens:											
851	Raw	1 c	55	86	25	1	5	2	<1	.1	t	.2
852	Chopped, cooked, drained	1 c	105	90	35	2	7	3	1	.2	t	.3
853	Eggplant, cooked	1 c	99	92	28	1	7	2	<1	t	t	.1
1714	Endive, fresh, chopped	1 c	50	94	8	1	2	2	<1	t	t	t
856	Escarole/curly endive-chopped	1 c	50	94	8	1	2	2	<1	t	t	t
854	Garbanzo beans (Chickpeas), cooked	1 c	164	60	269	14	45	12	4	.4	1	1.9

Chol (mg)	Calc (mg)	Iron (mg)	Magn (mg)	Pota (mg)	Sodi (mg)	Zinc (mg)	VT-A (µg)	Thia (mg)	VT-E (mg)	Ribo (mg)	Niac (mg)	V-B6 (mg)	Fola (µg)	VT-C (mg)
0	47	.56	18	166	22	.28	87	.05	1.52	.07	.42	.12	51	37
0	25	.55	16	251	18	.39	3	.06	.23	.07	.59	.14	38	49
0	28	.94	16	247	16	.26	28	.08	.66	.06	.47	.14	47	48
0	19	.58	19	254	18	.28	23	.08	.45	.09	.42	.22	79	36
0	33	.41	10	172	13	.13	5	.03	.07	.03	.21	.07	30	22
0	46	.25	12	146	12	.13	10	.09	.15	.08	.42	.17	30	30
0	73	.56	13	176	45	.13	105	.03	.08	.05	.35	.14	46	31
0	158	1.77	19	631	58	.29	218	.05	.2	.11	.73	.28	70	44
0	145	1.28	27	375	995	.36	213	.07	.24	.1	.75	.34	88	80
0	58	.24	10	181	7	.17	46	.03	.09	.04	.3	.18	60	20
0	38	.36	12	268	11	.21	58	.05	.14	.05	.59	.21	63	19
0	45	.44	13	183	10	.19	2	.04	.09	.03	.27	.19	19	51
0	55	.52	16	210	12	.22	2	.05	.18	.03	.3	.21	19	52
0	24	.28	20	161	20	.19	35	.05	.07	.02	.21	.13	56	22
0	43	.55	35	267	35	.33	64	.07	.15	.03	.03	.22	67	25
0	2	.05			105		1							0
0	19	.36	11	233	25	.14	1012	.07	.33	.04	.67	.11	10	7
0	15	.27	8	178	19	.11	773	.05	.25	.03	.51	.08	8	5
0	24	.48	10	177	51	.23	957	.03	.33	.04	.39	.19	11	2
0	20	.34	7	115	43	.17	646	.02	.31	.03	.32	.09	8	2
0	18	.47	6	131	177	.19	503	.01	.31	.02	.4	.08	7	2
0	57	1.09	33	689	68	.42	1292	.22	.02	.13	.91	.51	9	20
0	21	.35	28	337	18	.45	1	.1	.26	.06	1.06	.11	24	18
0	11	.22	7	152	15	.14	<1	.03	.02	.03	.26	.11	28	23
0	10	.2	6	88	9	.11	1	.03	.02	.03	.25	.11	27	27
0	15	.37	8	125	16	.12	1	.03	.04	.05	.28	.08	37	28
0	16	.16	4	115	35	.05	3	.02	.14	.02	.13	.03	11	3
0	48	.48	13	344	104	.16	8	.05	.43	.05	.39	.1	34	8
0	40	.67	19	268	95	.31	0	.04	.31	.06	.66	.16	5	6
0	18	.65	29	136	77	.13	59	.01	.68	.03	.14	.04	5	11
0	102	3.96	151	961	313	.58	275	.06	3.31	.15	.63	.15	16	31
0	34	.69	24	254	4	1.5	6	.05	.24	.06	.95	.15	189	16
0	21	.35	19	277	2	.5	4	.04	.19	.06	.67	.19	29	13
0	113	.44	16	247	9	.4	149	.04	.84	.1	.55	.12	88	17
0	179	.95	25	213	42	.23	254	.04	.42	.1	.54	.1	65	22
0	2	.47	22	193	3	.48	8	.13	.07	.05	1.17	.17	24	4
0	2	.38	18	158	3	.4	7	.11	.06	.04	.96	.14	19	3
0	3	.29	16	121	4	.33	9	.07	.07	.06	1.07	.11	25	3
0	4	.49	22	172	365	.68	6	.03	.11	.07	1.23	.08	58	6
0	5	.44	24	195	286	.48	13	.04	.09	.08	1.23	.06	51	9
0	4	.07	3	40	1	.06	3	.01	.02	.01	.06	.01	4	1
0	13	7.23	12	176	1531	.76	25	.04	.24	.04	.69	.16	34	5
0	103	1.71	20	218	42	.23	385	.1	1.38	.14	.44	.14	15	19
0	147	1.89	25	244	46	.29	614	.14	2.63	.18	.54	.17	14	19
0	6	.35	13	246	3	.15	3	.07	.03	.02	.59	.08	14	1
0	26	.41	7	157	11	.39	51	.04	.22	.04	.2	.01	71	3
0	26	.41	7	157	11	.39	51	.04	.22	.04	.2	.01	71	3
0	80	4.74	79	477	11	2.51	2	.19	.57	.1	.86	.23	282	2

*This value is expressed in retinol equivalents (RE). All other values are in retinol activity equivalents (RAE).

Table A–1

Food Composition (Computer code number is for Wadsworth Diet Analysis program)   (For purposes of calculations, use "0" for t, <1, <.1, <.01, etc.)

Computer Code Number	Food Description	Measure	Wt (g)	H₂O (%)	Ener (kcal)	Prot (g)	Carb (g)	Dietary Fiber (g)	Fat (g)	Fat Breakdown (g) Sat	Mono	Poly
	VEGETABLES AND LEGUMES—Continued											
1939	Grape leaf, raw:	1 ea	3	73	3	<1	1	<1	<1	t	t	t
7914	Cup	1 c	14	73	13	1	2	2	<1	t	t	.1
855	Great northern beans, cooked	1 c	177	69	209	15	37	12	1	.2	t	.3
857	Jerusalem artichoke, raw slices	1 c	150	78	114	3	26	2	<1	0	t	t
1794	Jicama	1 c	120	90	46	1	11	6	<1	t	t	.1
	Kale, cooked, drained:											
858	From raw	1 c	130	91	36	2	7	3	1	.1	t	.3
859	From frozen	1 c	130	90	39	4	7	3	1	.1	t	.3
860	Kidney beans, canned	1 c	256	77	218	13	40	16	1	.1	.1	.5
1181	Kohlrabi, raw slices	1 c	135	91	36	2	8	5	<1	t	t	.1
861	Kohlrabi, cooked	1 c	165	90	48	3	11	2	<1	t	t	.1
1183	Leeks, raw, chopped	1 c	89	83	54	1	13	2	<1	t	t	.1
1182	Leeks, cooked, chopped	1 c	104	91	32	1	8	1	<1	t	t	.1
862	Lentils, cooked from dry	1 c	198	70	230	18	40	16	1	.1	.1	.3
1288	Lentils, sprouted, stir fried	1 c	124	69	125	11	26	5	1	.1	.1	.2
1289	Lentils, sprouted, raw	1 c	77	67	82	7	17	3	<1	t	.1	.2
	Lettuce:											
	Butterhead/Boston types:											
863	Head, 5" diameter	¼ ea	41	96	5	1	1	<1	<1	t	t	t
864	Leaves, inner or outer	4 ea	30	96	4	<1	1	<1	<1	t	t	t
	Iceberg/crisphead:											
867	Chopped or shredded	1 c	55	96	7	1	1	1	<1	t	t	.1
865	Head, 6" diameter	1 ea	539	96	65	5	11	8	1	.1	t	.5
866	Wedge, ¼ head	1 ea	135	96	16	1	3	2	<1	t	t	.1
868	Looseleaf, chopped	½ c	28	94	5	<1	1	1	<1	t	t	t
869	Romaine, chopped	½ c	28	95	4	<1	1	<1	<1	t	t	t
870	Romaine, inner leaf	3 pce	30	95	4	<1	1	1	<1	t	t	t
1930	Luffa, cooked (Chinese okra)	1 c	178	90	57	3	13	4	<1	.1	t	.1
6777	Manioc, raw	1 c	206	60	330	3	78	4	1	.2	.2	.1
	Mushrooms:											
871	Raw, sliced	½ c	35	92	9	1	1	<1	<1	t	t	t
872	Cooked from fresh, pieces	½ c	78	91	21	2	4	2	<1	t	t	.1
1962	Stir fried, shitake slices	½ c	73	83	40	1	10	2	<1	t	t	t
873	Canned, drained	½ c	78	91	19	1	4	2	<1	t	t	.1
1951	Mushroom caps, pickled	8 ea	47	92	11	1	2	<1	<1	t	t	.1
	Mustard greens:											
874	Cooked from raw	½ c	70	94	10	2	1	1	<1	t	.1	t
875	Cooked from frozen	½ c	75	94	14	2	2	2	<1	t	.1	t
876	Navy beans, cooked from dry	1 c	182	63	258	16	48	12	1	.3	.1	.4
	Okra, cooked:											
877	From fresh pods	8 ea	85	90	27	2	6	2	<1	t	t	t
878	From frozen slices	1 c	184	91	51	4	11	5	1	.1	.1	.1
1236	Batter Fried from fresh	1 c	92	67	175	2	14	2	12	1.7	3.1	7.1
1930	Chinese, (Luffa), cooked	1 c	178	90	57	3	13	4	<1	.1	t	.1
	Onions:											
879	Raw, chopped	½ c	80	90	30	1	7	1	<1	t	t	t
880	Raw, sliced	½ c	58	90	22	1	5	1	<1	t	t	t
881	Cooked, drained, chopped	½ c	105	88	46	1	11	1	<1	t	t	.1
882	Dehydrated flakes	¼ c	14	4	49	1	12	1	<1	t	t	t
1934	Onions, pearl, cooked	½ c	93	88	41	1	9	1	<1	t	t	.1
	Spring/green onions, bulb and top,											
883	chopped	½ c	50	90	16	1	4	1	<1	t	t	t
884	Onion rings, breaded, heated f/frozen	2 ea	20	28	81	1	8	<1	5	1.7	2.2	1
1917	Palm hearts, cooked slices	1 c	146	69	150	4	39	2	<1	.1	t	.1
885	Parsley, raw, chopped	½ c	30	88	11	1	2	1	<1	t	.1	t
888	Parsnips, sliced, cooked	½ c	78	78	63	1	15	3	<1	t	.1	t
	Peas:											
	Black-eyed, cooked:											
814	From dry, drained	½ c	86	70	100	7	18	6	<1	.1	t	.2
815	From fresh, drained	½ c	82	75	79	3	17	4	<1	.1	t	.1
816	From frozen, drained	½ c	85	66	112	7	20	5	1	.1	.1	.2
889	Edible pod peas, cooked	½ c	80	89	34	3	6	2	<1	t	t	.1
890	Green, canned, drained:	½ c	85	82	59	4	11	3	<1	.1	t	.1

Chol (mg)	Calc (mg)	Iron (mg)	Magn (mg)	Pota (mg)	Sodi (mg)	Zinc (mg)	VT-A (µg)	Thia (mg)	VT-E (mg)	Ribo (mg)	Niac (mg)	V-B6 (mg)	Fola (µg)	VT-C (mg)
0	11	.08	3	8	<1	.02	40	<.01	.06	.01	.07	.01	2	<1
0	51	.37	13	38	1	.09	189	.01	.28	.05	.33	.06	12	2
0	120	3.77	88	692	4	1.56	<1	.28	.53	.1	1.21	.21	181	2
0	21	5.1	25	644	6	.18	1	.3	.28	.09	1.95	.12	19	6
0	14	.72	14	180	5	.19	1	.02	.55	.03	.24	.05	14	24
0	94	1.17	23	296	30	.31	481	.07	1.11	.09	.65	.18	17	53
0	179	1.22	23	417	19	.23	413	.06	.23	.15	.87	.11	18	33
0	61	3.23	72	658	873	1.41	0	.27	.13	.22	1.17	.06	131	3
0	32	.54	26	473	27	.04	3	.07	.65	.03	.54	.2	22	84
0	41	.66	31	561	35	.51	3	.07	2.76	.03	.64	.25	20	89
0	52	1.87	25	160	18	.11	4	.05	.82	.03	.36	.21	57	11
0	31	1.14	15	90	10	.06	3	.03	.63	.02	.21	.12	25	4
0	38	6.59	71	731	4	2.51	1	.33	.22	.14	2.1	.35	358	3
0	17	3.84	43	352	12	1.98	2	.27	.11	.11	1.49	.2	83	16
0	19	2.47	28	248	8	1.16	2	.18	.07	.1	.87	.15	77	13
0	13	.12	5	105	2	.07	20	.02	.18	.02	.12	.02	30	3
0	10	.09	4	77	1	.05	15	.02	.13	.02	.09	.01	22	2
0	10	.27	5	87	5	.12	9	.02	.15	.02	.1	.02	31	2
0	102	2.7	48	852	48	1.19	89	.25	1.51	.16	1.01	.22	302	21
0	26	.67	12	213	12	.3	22	.06	.38	.04	.25	.05	76	5
0	19	.39	3	74	3	.08	27	.01	.12	.02	.11	.01	14	5
0	10	.31	2	81	2	.07	36	.03	.12	.03	.14	.01	38	7
0	11	.33	2	87	2	.07	39	.03	.13	.03	.15	.01	41	7
0	112	.8	101	573	9	.98	52	.23	1.23	.1	1.55	.33	81	29
0	33	.56	43	558	29	.7	3	.18	.39	.1	1.76	.18	56	42
0	2	.36	3	130	1	.26	0	.03	.04	.15	1.41	.04	4	1
0	5	1.36	9	278	2	.68	0	.06	.09	.23	3.48	.07	14	3
0	2	.32	10	85	3	.97	0	.03	.09	.12	1.1	.12	15	<1
0	9	.62	12	101	332	.56	0	.07	.09	.02	1.24	.05	9	0
0	2	.51	5	140	2	.28	0	.03	.05	.16	1.42	.03	6	1
0	52	.49	10	141	11	.08	106	.03	1.41	.04	.3	.07	51	18
0	76	.84	10	104	19	.15	168	.03	1.31	.04	.19	.08	52	10
0	127	4.51	107	670	2	1.93	<1	.37	.73	.11	.97	.3	255	2
0	54	.38	48	274	4	.47	25	.11	.59	.05	.74	.16	39	14
0	177	1.23	94	431	6	1.14	47	.18	1.27	.23	1.44	.09	269	22
2	61	1.26	36	190	122	.5	39*	.18	3.04	.14	1.44	.12	38	10
0	112	.8	101	573	9	.98	52	.23	1.23	.1	1.55	.33	81	29
0	16	.18	8	126	2	.15	0	.03	.1	.02	.12	.09	15	5
0	12	.13	6	91	2	.11	0	.02	.07	.01	.09	.07	11	4
0	23	.25	12	174	3	.22	0	.04	.14	.02	.17	.13	16	5
0	36	.22	13	227	3	.26	0	.07	.19	.01	.14	.22	23	10
0	20	.22	10	154	3	.19	0	.04	.12	.02	.15	.12	14	5
0	36	.74	10	138	8	.19	10	.03	.06	.04	.26	.03	32	9
0	6	.34	4	26	75	.08	2	.06	.14	.03	.72	.01	13	<1
0	26	2.47	15	2636	20	5.45	5	.07	.73	.25	1.25	1.06	30	10
0	41	1.86	15	166	17	.32	78	.03	.54	.03	.39	.03	46	40
0	29	.45	23	286	8	.2	0	.06	.78	.04	.56	.07	45	10
0	21	2.16	46	239	3	1.11	1	.17	.24	.05	.43	.09	179	<1
0	105	.92	43	343	3	.84	32	.08	.18	.12	1.15	.05	104	2
0	20	1.8	42	319	4	1.21	3	.22	.33	.05	.62	.08	120	2
0	34	1.58	21	192	3	.3	5	.1	.31	.06	.43	.11	23	38
0	17	.81	14	147	214	.6	33	.1	.32	.07	.62	.05	37	8

*This value is expressed in retinol equivalents (RE). All other values are in retinol activity equivalents (RAE).

Table A–1

Food Composition   (Computer code number is for Wadsworth Diet Analysis program)   (For purposes of calculations, use "0" for t, <1, <.1, <.01, etc.)

Computer Code Number	Food Description	Measure	Wt (g)	H₂O (%)	Ener (kcal)	Prot (g)	Carb (g)	Dietary Fiber (g)	Fat (g)	Fat Breakdown (g) Sat	Mono	Poly
	VEGETABLES AND LEGUMES—Continued											
5267	Unsalted	½ c	124	86	66	4	12	4	<1	.1	t	.2
891	Green, cooked from frozen	½ c	80	80	62	4	11	4	<1	t	t	.1
1786	Snow peas, raw	½ c	49	89	21	1	4	1	<1	t	t	t
1787	Snow peas, raw	10 ea	34	89	14	1	3	1	<1	t	t	t
892	Split, green, cooked from dry	½ c	98	69	116	8	21	8	<1	.1	.1	.2
1187	Peas & carrots, cooked from frozen	½ c	80	86	38	2	8	2	<1	.1	t	.2
1186	Peas & carrots, canned w/liquid	½ c	128	88	49	3	11	3	<1	.1	t	.2
	Peppers, hot:											
893	Hot green chili, canned	½ c	68	92	14	1	3	1	<1	t	t	t
894	Hot green chili, raw	1 ea	45	88	18	1	4	1	<1	t	t	t
1715	Hot red chili, raw, diced	1 tbs	9	88	4	<1	1	<1	<1	t	t	t
1988	Jalapeno, raw	1 ea	45	90	11	<1	2		<1			
895	Jalapeno, chopped, canned	½ c	68	89	18	1	3	2	1	.1	t	.3
1918	Jalapeno wheels, in brine (Ortega)	2 tbs	29		10	0	2		0	0	0	0
	Peppers, sweet, green:											
896	Whole pod (90g with refuse), raw	1 ea	119	92	32	1	8	2	<1	t	t	.1
897	Cooked, chopped (1 pod cooked = 73g)	½ c	68	92	19	1	5	1	<1	t	t	.1
	Peppers, sweet, red:											
1286	Raw, chopped	½ c	75	92	20	1	5	1	<1	t	t	.1
1807	Raw, each	1 ea	74	92	20	1	5	1	<1	t	t	.1
1287	Cooked, chopped	½ c	68	92	19	1	5	1	<1	t	t	.1
	Peppers, sweet, yellow:											
1872	Raw, large	1 ea	186	92	50	2	12	2	<1	.1	t	.2
1873	Strips	10 pce	52	92	14	1	3	<1	<1	t	t	.1
898	Pinto beans, cooked from dry	½ c	85	64	116	7	22	7	<1	.1	.1	.2
1191	Poi - two finger	½ c	120	72	134	<1	33	<1	<1	t	t	.1
	Potatoes:											
	Baked in oven, 4¾"x2⅓" diam											
899	With skin	1 ea	202	71	220	5	51	5	<1	.1	t	.1
900	Flesh only	1 ea	156	75	145	3	34	2	<1	t	t	.1
901	Skin only	1 ea	58	47	115	2	27	5	<1	t	t	t
	Baked in microwave, 4¾"x 2⅓"dm:											
902	With skin	1 ea	202	72	212	5	49	5	<1	.1	t	.1
903	Flesh only	1 ea	156	74	156	3	36	2	<1	t	t	.1
904	Skin only	1 ea	58	63	77	3	17	3	<1	t	t	t
	Boiled, about 2½ diam:											
905	Peeled after boiling	1 ea	136	77	118	3	27	2	<1	t	t	.1
906	Peeled before boiling	1 ea	135	77	116	2	27	2	<1	t	t	.1
	French fried, strips 2-3½" long:											
907	Oven heated	10 ea	50	35	167	2	20	2	9	3	5.7	.7
908	Fried in vegetable oil	10 ea	50	38	158	2	20	2	8	1.9	4.7	.7
1188	Fried in veg and animal oil	10 ea	50	38	158	2	20	2	8	1.9	4.7	.7
909	Hashed browns from frozen	1 c	156	56	340	5	44	3	18	7	8	2.1
	Mashed:											
910	Home recipe with whole milk	½ c	105	78	81	2	18	2	1	.3	.2	.1
911	Home recipe with milk and marg	½ c	105	76	111	2	17	2	4	1.1	1.9	1.3
912	Prepared from flakes; water, milk, margarine, salt added	½ c	110	76	124	2	16	3	6	1.6	2.5	1.7
	Potato products, prepared:											
	Au gratin:											
913	From dry mix	½ c	123	79	114	3	16	1	5	3.2	1.4	.2
914	From home recipe, using butter	½ c	122	74	161	6	14	2	9	4.3	3.2	1.3
	Scalloped:											
915	From dry mix	½ c	122	79	113	3	16	1	5	3.2	1.5	.2
916	From home recipe, using butter	½ c	123	81	106	4	13	2	5	1.7	1.7	.9
	Potato Salad (see Mixed Dishes #715)											
1192	Potato Puffs, cooked from frozen	½ c	64	53	142	2	19	2	7	3.3	2.8	.5
918	Pumpkin, cooked from fresh, mashed	½ c	123	94	25	1	6	1	<1	t	t	t
919	Pumpkin, canned	½ c	123	90	42	1	10	4	<1	.2	t	t
1891	Radicchio, raw, shredded	½ c	20	93	5	<1	1	<1	<1	t	t	t
1894	Radicchio, raw, leaf	10 ea	80	93	18	1	4	1	<1	t	t	.1
920	Red radishes	10 ea	45	95	9	<1	2	1	<1	t	t	t

Chol (mg)	Calc (mg)	Iron (mg)	Magn (mg)	Pota (mg)	Sodi (mg)	Zinc (mg)	VT-A (µg)	Thia (mg)	VT-E (mg)	Ribo (mg)	Niac (mg)	V-B6 (mg)	Fola (µg)	VT-C (mg)
0	22	1.26	21	124	11	.87	24	.14	.47	.09	1.04	.08	36	12
0	19	1.26	23	134	70	.75	27	.23	.14	.08	1.18	.09	47	8
0	21	1.02	12	98	2	.13	3	.07	.19	.04	.29	.08	21	29
0	15	.71	8	68	1	.09	2	.05	.13	.03	.2	.05	14	20
0	14	1.26	35	355	2	.98	<1	.19	.38	.05	.87	.05	64	<1
0	18	.75	13	126	54	.36	310	.18	.26	.05	.92	.07	21	6
0	29	.96	18	128	333	.74	369	.09	.24	.07	.74	.11	23	8
0	5	.34	10	127	798	.12	21	.01	.47	.03	.54	.1	7	46
0	8	.54	11	153	3	.13	17	.04	.31	.04	.43	.12	10	109
0	2	.11	2	31	1	.03	48	.01	.06	.01	.09	.02	2	22
			2		2		30*		.37					53
0	16	1.28	10	131	1136	.23	58	.03	.47	.03	.27	.13	10	7
0				55	390		10*		.2					21
0	11	.55	12	211	2	.14	37	.08	.82	.04	.61	.29	26	106
0	6	.31	7	113	1	.08	20	.04	.47	.02	.32	.16	11	51
0	7	.34	7	133	1	.09	214	.05	.52	.02	.38	.19	16	143
0	7	.34	7	131	1	.09	211	.05	.51	.02	.38	.18	16	141
0	6	.31	7	113	1	.08	128	.04	.47	.02	.32	.16	11	116
0	20	.86	22	394	4	.32	22	.05	1.28	.05	1.66	.31	48	341
0	6	.24	6	110	1	.09	6	.01	.36	.01	.46	.09	13	95
0	41	2.22	47	398	2	.92	<1	.16	.8	.08	.34	.13	146	2
0	19	1.06	29	220	14	.26	1	.16	.22	.05	1.32	.33	25	5
0	20	2.75	54	844	16	.65	0	.22	.1	.07	3.32	.7	22	26
0	8	.55	39	610	8	.45	0	.16	.06	.03	2.18	.47	14	20
0	20	4.08	25	332	12	.28	0	.07	.02	.06	1.78	.36	13	8
0	22	2.5	54	903	16	.73	0	.24	.1	.06	3.46	.69	24	30
0	8	.64	39	641	11	.51	0	.2	.06	.04	2.54	.5	19	24
0	27	3.45	21	377	9	.3	0	.04	.02	.04	1.29	.28	10	9
0	7	.42	30	515	5	.41	0	.14	.07	.03	1.96	.41	14	18
0	11	.42	27	443	7	.36	0	.13	.07	.03	1.77	.36	12	10
0	6	.83	11	270	307	.2	0	.04	.25	.02	1.33	.11	11	3
0	9	.38	17	366	108	.19	0*	.09	.25	.01	1.63	.12	14	5
6	9	.38	17	366	108	.19	0*	.09	.25	.01	1.63	.12	14	5
0	23	2.36	26	680	53	.5	0	.17	.3	.03	3.78	.2	11	10
2	27	.28	19	314	318	.3	6*	.09	.05	.04	1.17	.24	8	7
2	27	.27	19	303	310	.28	21*	.09	.31	.04	1.13	.23	8	6
4	54	.24	20	256	365	.2	23*	.12	.77	.05	.74	.01	8	11
18	102	.39	18	269	540	.29	38*	.02	1.48	.1	1.15	.05	9	4
18	145	.78	24	483	528	.84	46*	.08	.64	.14	1.21	.21	13	12
13	44	.46	17	248	416	.3	26*	.02	.18	.07	1.26	.05	12	4
7	70	.7	23	465	412	.49	23*	.08	.4	.11	1.3	.22	13	13
0	19	1	12	243	477	.19	1	.12	.03	.05	1.38	.15	11	4
0	18	.7	11	283	1	.28	66	.04	1.3	.1	.51	.05	11	6
0	32	1.71	28	253	6	.21	1356	.03	1.3	.07	.45	.07	15	5
0	4	.11	3	60	4	.12	<1	<.01	.45	.01	.05	.01	12	2
0	15	.46	10	242	18	.5	1	.01	1.81	.02	.2	.05	48	6
0	9	.13	4	104	11	.13	<1	<.01	0	.02	.13	.03	12	10

*This value is expressed in retinol equivalents (RE). All other values are in retinol activity equivalents (RAE).

Table A–1

Food Composition

(Computer code number is for Wadsworth Diet Analysis program)   (For purposes of calculations, use "0" for t, <1, <.1, <.01, etc.)

Computer Code Number	Food Description	Measure	Wt (g)	H₂O (%)	Ener (kcal)	Prot (g)	Carb (g)	Dietary Fiber (g)	Fat (g)	Fat Breakdown (g)		
										Sat	Mono	Poly
	VEGETABLES AND LEGUMES—Continued											
1793	Daikon radishes (Chinese) raw	½ c	44	95	8	<1	2	1	<1	t	t	t
921	Refried beans, canned	½ c	126	76	118	7	20	7	2	.6	.7	.2
1375	Rutabaga, cooked cubes	½ c	85	89	33	1	7	2	<1	t	t	.1
922	Sauerkraut, canned with liquid	½ c	118	93	22	1	5	3	<1	t	t	.1
923	Seaweed, kelp, raw	½ c	40	82	17	1	4	1	<1	.1	t	t
924	Seaweed, spirulina, dried	½ c	8	5	23	5	2	<1	1	.2	.1	.2
1866	Shallots, raw, chopped	1 tbs	10	80	7	<1	2	<1	<1	t	t	t
1557	Snow Peas, stir fried	½ c	83	89	35	2	6	2	<1	t	t	.1
925	Soybeans, cooked from dry	½ c	86	63	149	14	9	5	8	1.1	1.7	4.4
1996	Soybeans, dry roasted	½ c	86	1	387	34	28	7	19	2.7	4.1	10.5
	Soybean products:											
926	Miso	½ c	138	41	284	16	39	7	8	1.2	1.8	4.7
	Soy milk (see #144 and #2301 under Dairy)											
	Tofu (soybean curd)											
7540	Extra firm, silken	½ c	126	88	69	9	3	<1	2	.4	.4	1.3
7542	Firm, silken	½ c	126	87	78	9	3	<1	3	.5	.7	1.9
927	Regular	½ c	124	87	76	8	2	<1	5	.7	1	2.6
7541	Soft, silken	½ c	124	89	68	6	4	<1	3	.4	.6	1.9
	Spinach:											
928	Raw, chopped	½ c	15	92	3	<1	1	<1	<1	t	t	t
929	Cooked, from fresh, drained	½ c	90	91	21	3	3	2	<1	t	t	.1
930	Cooked from frozen (leaf)	½ c	95	90	27	3	5	3	<1	t	t	.1
931	Canned, drained solids:	½ c	107	92	25	3	4	3	1	.1	t	.2
5149	Unsalted	½ c	107	92	25	3	4	3	1	.1	t	.2
	Spinach souffle (see Mixed Dishes)											
	Squash, summer varieties, cooked w/skin:											
932	Varieties averaged	½ c	90	94	18	1	4	1	<1	.1	t	.1
933	Crookneck	½ c	90	94	18	1	4	1	<1	.1	t	.1
934	Zucchini	½ c	90	95	14	1	4	1	<1	t	t	t
	Squash, winter varieties, cooked:											
	Average of all varieties, baked:											
935	Mashed	1 c	245	89	96	2	21	7	2	.3	.1	.6
936	Cubes	1 c	205	89	80	2	18	6	1	.3	.1	.5
937	Acorn, baked, mashed	½ c	123	83	69	1	18	5	<1	t	t	.1
1218	Acorn, boiled, mashed	½ c	122	90	41	1	11	3	<1	t	t	t
	Butternut squash:											
938	Baked cubes	½ c	103	88	41	1	11	3	<1	t	t	t
1219	Baked, mashed	½ c	103	88	41	1	11	3	<1	t	t	t
1193	Cooked from frozen	½ c	120	88	47	1	12	3	<1	t	t	t
1194	Hubbard, baked, mashed	½ c	120	85	60	3	13	3	1	.2	.1	.3
1195	Hubbard, boiled, mashed	½ c	118	91	35	2	8	3	<1	.1	t	.2
1196	Spaghetti, baked or boiled	½ c	77	92	21	<1	5	1	<1	t	t	.1
1189	Succotash, cooked from frozen	½ c	85	74	79	4	17	3	1	.1	.1	.4
	Sweet potatoes:											
939	Baked in skin, peeled, 5 x 2" diam	1 ea	114	73	117	2	28	3	<1	t	t	.1
940	Boiled without skin, 5 x 2" diam	1 ea	151	73	159	2	37	3	<1	.1	t	.2
941	Candied, 2½ x 2"	1 pce	105	67	144	1	29	3	3	1.4	.7	.2
	Canned:											
942	Solid pack	½ c	128	74	129	3	30	2	<1	.1	t	.1
943	Vacuum pack, mashed	½ c	127	76	116	2	27	2	<1	.1	t	.1
944	Vacuum pack, 3¾ x 1"	2 pce	80	76	73	1	17	1	<1	t	t	.1
1940	Taro shoots, cooked slices	1 c	140	95	20	1	4	1	<1	t	t	t
1941	Taro, tahitian, cooked slices	1 c	137	86	60	6	9	1	1	.2	.1	.4
	Tomatillos:											
1877	Raw, each	1 ea	34	92	11	<1	2	1	<1	t	.1	.1
1875	Raw, chopped	1 c	132	92	42	1	8	3	1	.2	.2	.5
	Tomatoes:											
945	Raw, whole, 2⅗" diam	1 ea	123	94	26	1	6	1	<1	.1	.1	.2
946	Raw, chopped	1 c	180	94	38	2	8	2	1	.1	.1	.2
947	Cooked from raw	1 c	240	92	65	3	14	2	1	.1	.2	.4
948	Canned, solids and liquid:	1 c	240	94	46	2	10	2	<1	t	t	.1
5741	Unsalted	1 c	240	94	46	2	10	2	<1	t	t	.1
1879	Tomatoes, sundried:	1 c	54	15	139	8	30	7	2	.2	.3	.6

Chol (mg)	Calc (mg)	Iron (mg)	Magn (mg)	Pota (mg)	Sodi (mg)	Zinc (mg)	VT-A (µg)	Thia (mg)	VT-E (mg)	Ribo (mg)	Niac (mg)	V-B6 (mg)	Fola (µg)	VT-C (mg)
0	12	.18	7	100	9	.07	0	.01	0	.01	.09	.02	12	10
10	44	2.09	42	336	377	1.47	0	.03	0	.02	.4	.18	14	8
0	41	.45	20	277	17	.3	24	.07	.13	.03	.61	.09	13	16
0	35	1.73	15	201	780	.22	1	.02	.12	.03	.17	.15	28	17
0	67	1.14	48	36	93	.49	2	.02	.35	.06	.19	<.01	72	1
0	10	2.28	16	109	84	.16	2	.19	.4	.29	1.03	.03	8	1
0	4	.12	2	33	1	.04	6	.01	.01	<.01	.02	.03	3	1
0	36	1.73	20	166	3	.22	5	.11	.32	.06	.47	.13	28	42
0	88	4.42	74	443	1	.99	<1	.13	1.68	.24	.34	.2	46	1
0	120	3.4	196	1173	2	4.1	1	.37	3.96	.65	.91	.19	176	4
0	91	3.78	58	226	5032	4.58	6	.13	.01	.34	1.19	.3	45	0
0	39	1.5	34	194	79	.76	0	.1	.18	.04	.3	.01		0
0	40	1.3	34	244	45	.77	0	.13	.24	.05	.31	.01		0
0	138	1.38	33	149	10	.79	<1	.06	.01	.05	.66	.06	55	<1
0	38	1.02	36	223	6	.64	0	.12	.25	.05	.37	.01		0
0	15	.41	12	84	12	.08	50	.01	.28	.03	.11	.03	29	4
0	122	3.21	78	419	63	.68	369	.09	.86	.21	.44	.22	131	9
0	139	1.44	66	283	82	.66	370	.06	.91	.16	.4	.14	103	12
0	136	2.46	81	370	29	.49	470	.02	1.39	.15	.41	.11	105	15
0	136	2.46	81	370	29	.49	470	.02	1.39	.15	.41	.11	105	15
0	24	.32	22	173	1	.35	13	.04	.11	.04	.46	.06	18	5
0	24	.32	22	173	1	.35	13	.04	.11	.04	.46	.08	18	5
0	12	.31	20	228	3	.16	11	.04	.11	.04	.38	.07	15	4
0	34	.81	20	1070	2	.64	436	.21	.29	.06	1.72	.18	69	23
0	29	.68	16	896	2	.53	365	.17	.25	.05	1.44	.15	57	20
0	54	1.14	53	538	5	.21	26	.2	.15	.02	1.08	.24	23	13
0	32	.68	32	321	4	.13	16	.12	.15	.01	.65	.14	13	8
0	42	.62	30	293	4	.13	361	.07	.17	.02	1	.13	20	16
0	42	.62	30	293	4	.13	361	.07	.17	.02	1	.13	20	16
0	23	.7	11	160	2	.14	200	.06	.16	.05	.56	.08	19	4
0	20	.56	26	430	10	.18	362	.09	.14	.06	.67	.21	19	11
0	12	.33	15	253	6	.12	237	.05	.14	.03	.39	.12	12	8
0	16	.26	8	90	14	.15	4	.03	.09	.02	.62	.08	6	3
0	13	.76	20	225	38	.38	10	.06	.31	.06	1.11	.08	28	5
0	32	.51	23	397	11	.33	1243	.08	.32	.14	.69	.27	26	28
0	32	.85	15	278	20	.41	1287	.08	.42	.21	.97	.37	17	26
8	27	1.19	12	198	73	.16	220	.02	3.99	.04	.41	.04	12	7
0	38	1.7	31	269	96	.27	968	.03	.35	.11	1.22	.3	14	7
0	28	1.13	28	396	67	.23	507	.05	.32	.07	.94	.24	22	33
0	18	.71	18	250	42	.14	319	.03	.2	.05	.59	.15	14	21
0	20	.57	11	482	3	.76	3	.05	1.4	.07	1.13	.16	4	26
0	204	2.14	70	854	74	.14	121	.06	3.7	.27	.66	.16	10	52
0	2	.21	7	91	<1	.07	2	.01	.13	.01	.63	.02	2	4
0	9	.82	26	354	1	.29	7	.06	.5	.05	2.44	.07	9	15
0	6	.55	13	273	11	.11	38	.07	.47	.06	.77	.1	18	23
0	9	.81	20	400	16	.16	56	.11	.68	.09	1.13	.14	27	34
0	14	1.34	34	670	26	.26	89	.17	.91	.14	1.8	.23	31	55
0	72	1.32	29	530	355	.38	72	.11	.77	.07	1.76	.22	19	34
0	72	1.32	29	545	24	.38	72	.11	.91	.07	1.76	.22	19	34
0	59	4.91	105	1850	1131	1.07	23	.28	<.01	.26	4.89	.18	37	21

*This value is expressed in retinol equivalents (RE). All other values are in retinol activity equivalents (RAE).

Table A–1

Food Composition   (Computer code number is for Wadsworth Diet Analysis program)   (For purposes of calculations, use "0" for t, <1, <.1, <.01, etc.)

A

Computer Code Number	Food Description	Measure	Wt (g)	H₂O (%)	Ener (kcal)	Prot (g)	Carb (g)	Dietary Fiber (g)	Fat (g)	Fat Breakdown (g)		
										Sat	Mono	Poly
	VEGETABLES AND LEGUMES—Continued											
1881	Pieces	10 pce	20	15	52	3	11	2	1	.1	.1	.2
1885	Oil pack, drained	10 pce	30	54	64	2	7	2	4	.6	2.6	.6
2020	Tomato, raw	1 ea	62	94	13	1	3	1	<1	t	t	.1
949	Tomato juice, canned:	1 c	243	94	41	2	10	1	<1	t	t	.1
5397	Unsalted	1 c	243	94	41	2	10	2	<1	t	t	.1
	Tomato products, canned:											
950	Paste-no added salt	1 c	262	74	215	10	51	11	1	.2	.2	.6
951	Puree-no added salt	1 c	250	87	100	4	24	5	<1	.1	.1	.2
952	Sauce-with salt	1 c	245	89	73	3	18	3	<1	.1	.1	.2
953	Turnips, cubes, cooked from fresh	1 c	156	94	33	1	8	3	<1	t	t	.1
	Turnip greens, cooked:											
954	From fresh, leaves and stems	1 c	144	93	29	2	6	5	<1	.1	t	.1
955	From frozen, chopped	1 c	164	90	49	5	8	6	1	.2	t	.3
956	Vegetable juice cocktail, canned	1 c	242	94	46	2	11	2	<1	t	t	.1
	Vegetables, Mixed:											
957	Canned, drained	½ c	81	87	38	2	7	2	<1	t	t	.1
958	Frozen, cooked, drained	½ c	91	83	54	3	12	4	<1	t	t	.1
1818	Water chestnuts, Chinese, raw	½ c	62	73	60	1	15	2	<1	t	t	t
	Water chestnuts, canned:											
959	Slices	½ c	70	86	35	1	9	2	<1	t	t	t
960	Whole	4 ea	28	86	14	<1	3	1	<1	t	0	t
1190	Watercress, fresh, chopped	½ c	17	95	2	<1	<1	<1	<1	t	t	t
	VEGETARIAN FOODS:											
7509	Bacon strips, meatless	3 ea	15	49	46	2	1	<1	4	.7	1.1	2.3
1511	Baked beans, canned	½ c	127	73	118	6	26	6	1	.1	t	.2
7526	Bakon Crumbles	¼ c	7	8	31	2	2	1	2	0		
7548	Chicken, breaded, fried, meatless	1 pce	57	70	97	6	3	3	7	1	1.6	3.9
7547	Chicken slices, meatless	2 ea	60	59	132	10	4	3	8	1.3	2	4.3
7557	Chili w/meat substitute	½ c	107	65	141	19	15	5	2	.3	.6	.9
7549	Fish stick, meatless	2 ea	57	45	165	13	5	3	10	1.6	2.5	5.3
7550	Frankfurter, meatless	1 ea	51	58	102	10	4	2	5	.8	1.2	2.6
7504	GardenBurger, patty	1 ea	71	58	130	8	18	5	3	1	1.5	.5
7505	GardenSausage, patty	1 ea	71	59	130	7	18	4	3	2	.7	.3
7551	Luncheon slice, meatless	1 sl	67	46	188	17	6	3	11	1.7	2.6	5.6
7560	Meatloaf, meatless	1 ea	71	58	142	15	6	3	6	1	1.5	3.3
1171	Nuteena	1 ea	55	58	162	6	6	2	13	5.1	5.8	1.7
7556	Pot pie, meatless	1 ea	227	60	510	14	41	5	32	8.6	12.4	9.6
7554	Soyburger, patty	1 ea	71	58	142	15	6	3	6	1	1.5	3.3
7562	Soyburger w/cheese, patty	1 ea	135	51	308	20	30	4	12	3.6	3.6	3.6
8832	Soyburger, veggie, patty	1 ea	85	76	70	11	7	2	0	0	0	0
7517	Soy protein isolate	1 oz	28.35	5	96	23	2	2	1	.1	.2	.5
7564	Tempeh	1 c	166	60	320	31	16	9	18	3.7	5	6.3
7670	Vegan burger, patty	1 ea	78	71	83	13	7	4	<1	.1	.3	.1
8842	Veggie slices, soy	1 pce	15	68	17	4	1	0	0	0	0	0
8830	Veggie ground soy	⅓ c	55	70	60	12	3	3	0	0	0	0
	Vegetarian foods, Green Giant:											
7677	Breakfast links	3 ea	68	65	114	12	5	4	5		4.3	
7676	Breakfast patties	2 ea	57	65	95	10	5	3	4		3.6	
	Burger, harvest, patty:											
7673	Italian	1 ea	90	67	140	17	8	5	4	1.5	.5	.5
7674	Original	1 ea	90	65	138	18	7	6	4	1	2.1	.3
7675	Southwestern	1 ea	90	68	140	16	9	5	4	1.5	0	.5
	Vegetarian Foods, Loma Linda											
7727	Chik nuggets, frozen	5 pce	85	47	245	12	13	5	16	2.5	4	8.8
7753	Chik-fried, frozen	1 pce	57	51	178	11	1	1	15	1.9	3.7	8.7
7744	Franks, big, canned	1 ea	51	58	118	12	2	1	7	.8	1.5	3.7
7747	Linketts, canned	1 ea	35	60	72	7	1	1	4	.7	1.2	2.5
1173	Redi-burger, patty	1 ea	85	59	172	16	5		10	1.5	2.4	5.8
7755	Swiss stake w/gravy, canned	1 pce	92	71	120	9	8	4	6	.8	1.5	3.3
1174	Vege-Burger, patty	1 ea	55	71	66	10	2	2	2	.4	.6	.5
	Vegetarian foods, Morningstar Farms:											
7672	Better-n-burgers, svg	1 ea	78	71	83	13	7	4	<1	.1	.3	.1

Chol (mg)	Calc (mg)	Iron (mg)	Magn (mg)	Pota (mg)	Sodi (mg)	Zinc (mg)	VT-A (µg)	Thia (mg)	VT-E (mg)	Ribo (mg)	Niac (mg)	V-B6 (mg)	Fola (µg)	VT-C (mg)
0	22	1.82	39	685	419	.4	9	.11	<.01	.1	1.81	.07	14	8
0	14	.8	24	470	80	.23	19	.06	.16	.11	1.09	.1	7	30
0	3	.28	7	138	6	.06	19	.04	.24	.03	.39	.05	9	12
0	22	1.41	27	535	877	.34	68	.11	2.21	.07	1.64	.27	49	44
0	22	1.41	27	535	24	.34	68	.11	2.21	.07	1.64	.27	49	44
0	92	5.08	134	2454	231	2.1	320	.41	11.3	.5	8.44	1	58	111
0	42	3.1	60	1065	85	.55	160	.18	6.3	.13	4.29	.38	27	26
0	34	1.89	47	909	1482	.61	120	.16	3.43	.14	2.82	.38	22	32
0	34	.34	12	211	78	.31	0	.04	.05	.04	.47	.1	14	18
0	197	1.15	32	292	42	.2	396	.06	2.48	.1	.59	.26	170	39
0	249	3.18	43	367	25	.67	654	.09	4.79	.12	.77	.11	64	36
0	27	1.02	27	467	653	.48	142	.1	.77	.07	1.76	.34	51	67
0	22	.85	13	236	121	.33	472	.04	.49	.04	.47	.06	19	4
0	23	.75	20	154	32	.45	195	.06	.33	.11	.77	.07	17	3
0	7	.04	14	362	9	.31	0	.09	.74	.12	.62	.2	10	2
0	3	.61	3	83	6	.27	0	.01	.35	.02	.25	.11	4	1
0	1	.24	1	33	2	.11	0	<.01	.14	.01	.1	.04	2	<1
0	20	.03	4	56	7	.02	40	.01	.17	.02	.03	.02	2	7
0	3	.36	3	25	220	.06	1	.66	1.04	.07	1.13	.07	6	0
0	63	.37	41	376	504	1.78	11	.19	.67	.08	.54	.17	30	4
0	16	.28			163		0							0
0	13	.97	7	171	228	.37	0	.4	1.11	.27	2.68	.28	32	0
0	21	.78	10	198	474	.42	0	.66	1.61	.24	3.18	.42	46	0
0	54	4.38	36	366	355	1.27	37	.12	1.28	.07	1.22	.15	82	6
0	54	1.14	13	342	279	.8	0	.63	2.25	.51	6.84	.85	58	0
0	17	.92	9	76	219	.61	0	.56	.98	.61	8.16	.5	40	0
11	84	0	30	193	290	.89	10*	.11	.2	.15	1.08	.08	10	0
10	80	.5	28	143	300	.84	24*	.1	.2	.15	1.01	.07	10	<1
0	27	1.54	15	188	576	1.07	0	.64	2.01	.37	7.37	.74	67	0
0	21	1.49	13	128	391	1.28	0	.64	1.23	.43	7.1	.85	55	0
0	9	.27	33	166	119	.46	0	.1		.35	1.04	.45	49	0
19	68	2.96	32	378	486	1.08	785*	.82	4.23	.44	5.17	.41	58	10
0	21	1.49	13	128	391	1.28	0	.64	1.23	.43	7.1	.85	55	0
9	158	2.94	27	242	922	1.91	36*	.8	1.58	.59	8.32	.84	64	1
0	60	1.8		300	520		0							0
0	50	4.11	11	23	285	1.14	0	.05	0	.03	.41	.03	50	0
0	184	4.48	134	684	15	1.89	0	.13	.03	.59	4.38	.36	40	0
0	80	2.66	15	398	351	.69	0	.23	.01	.51	3.77	.18	225	0
0				37	133									
0	40	9		220	250	3.75	0	.22		.17	.4	.2		0
0							0*	0		.18	.09	.27		<1
0							0*	0		.15	.07	2.28		<1
0	80	2.7			370	6.75	0	.3		.14	4	.3		0
0	102	3.85	70	432	411	8.07	0	.31	1.56	.2	6.3	.39	22	0
0	80	2.7			370	6.75	0	.3		.14	4	.3		0
2	40	1.4		153	709	.43	0	.67		.3	2.89	.45		0
4	2	.63		76	503	.2	0	.98		.46	2.1	.35		0
0	10	.99		61	224	1.2	0	.28		.68	5.78	.67		
1	4	.39		29	160	.46	0	.13		.22	.64	.29		0
1	12	1.06	16	121	455	1.11	0	.14		.3	1.9	.51	21	0
2	24	.31		225	433	.41	0	1.25		.65	5.41	1		0
0	8	.5	12	30	114	.58	0	.2		.25	.78	.31	15	0
0	80	2.66	15	398	351	.69	0	.23	.01	.51	3.77	.18	225	0

*This value is expressed in retinol equivalents (RE). All other values are in retinol activity equivalents (RAE).

Table A–1

Food Composition (Computer code number is for Wadsworth Diet Analysis program) (For purposes of calculations, use "0" for t, <1, <.1, <.01, etc.)

Computer Code Number	Food Description	Measure	Wt (g)	H₂O (%)	Ener (kcal)	Prot (g)	Carb (g)	Dietary Fiber (g)	Fat (g)	Fat Breakdown (g) Sat	Mono	Poly
	VEGETARIAN FOODS—Continued											
7766	Better-n-eggs	¼ c	57	88	26	5	<1	0	<1	.1	.1	.1
57436	Breakfast links	2 pce	45	67	64	9	2	1	2	.4	.5	1
7752	Breakfast strips	2 pce	16	43	56	2	2	1	4	.7	.9	2.6
7725	Burger crumbles, svg	1 ea	55	60	116	11	3	3	6	1.6	2.3	2.5
7726	Burger, spicy black bean	1 ea	78	60	115	12	15	5	1	.2	.2	.4
7665	Chik pattie	1 ea	71	54	153	9	14	3	6	.9	1.6	3.6
7724	Frank, deli	1 ea	57	52	141	13	5	3	8	1.1	2.5	4.2
7722	Garden vege pattie	1 ea	67	60	119	11	10	4	4	.5	1.1	2.2
7746	Grillers	1 ea	64	56	139	15	5	2	6	1.1	1.5	3.1
7664	Prime pattie	1 ea	64	64	94	16	4	3	2	.2	.4	.6
	Vegetarian foods, Worthington:											
7634	Beef style, meatless, frzn	3 pce	55	58	113	9	4	3	7	1.2	2.7	2.6
7732	Burger, meatless, patty	¼ c	55	71	60	9	2	1	2	.3	.5	1.1
1846	Chik slices, canned	2 pce	60	78	62	6	1	1	4	.6	.9	2.3
1833	Chili, canned	½ c	106	73	136	9	10	4	7	1.1	1.7	4.1
1835	Choplets, slices, canned	2 pce	92	72	93	17	3	2	2	.9	.3	.3
7608	Corned beef style, meatless, frzn	4 pce	57	55	138	10	5	2	9	1.9	4.1	3.1
1831	Country stew, canned	1 c	240	81	208	13	20	5	9	1.6	2.3	4.8
7632	Egg Roll, meatless, frzn	1 ea	85	53	181	6	20	2	8	1.7	4.5	2.3
1838	Numete, slices, canned	1 pce	55	58	132	6	5	3	10	2.4	4.4	2.7
1839	Prime stakes, slices, canned	1 pce	92	71	136	9	4	4	9	1.4	2.9	4.9
1840	Protose, slices, canned	1 pce	55	53	131	13	5	3	7	1	3	2.4
7606	Roast, dinner, meatless, frzn	1 ea	85	63	180	12	5	3	12	2.2	5	5.2
1842	Saucette links, canned	1 pce	38	62	86	6	1		6	1.1	1.6	3.8
1844	Savory slices, canned	1 pce	28	66	48	3	2	1	3	1.2	1.3	.6
7735	Stakelets, frzn	1 pce	71	58	145	12	6	2	8	1.4	2.7	3.9
1847	Turkee slices, canned	1 pce	33	64	68	5	1	1	5	.8	1.9	2.1
	MISCELLANEOUS											
	Baking Powders for home use:											
	Sodium aluminum sulfate:											
962	With monocalcium phosphate monohydrate	1 tsp	5	2	6	<1	2	0	0	0	0	0
963	With monocalcium phosphate monohydrate, calcium sulfate	1 tsp	5	5	3	0	1	<1	0	0	0	0
964	Straight Phosphate	1 tsp	5	4	3	<1	1	<1	0	0	0	0
965	Low sodium	1 tsp	5	6	5	<1	2	<1	<1	t	0	t
1204	Baking soda	1 tsp	5		0	0	0	0	0	0	0	0
966	Basil, dried	1 tbsp	5	6	13	1	3	2	<1	t	t	.1
2068	Cajun seasoning	1 tsp	3	5	6	<1	1	<1	<1			
961	Carob flour	1 c	103	4	229	5	91	41	1	.1	.2	.2
967	Catsup:	¼ c	61	67	63	1	17	1	<1	t	t	.1
968	Tablespoon	1 tbsp	15	67	16	<1	4	<1	<1	t	t	t
1200	Cayenne/red pepper	1 tbsp	5	8	16	1	3	1	1	.2	.1	.4
969	Celery seed	1 tsp	2	6	8	<1	1	<1	<1	t	.3	.1
1203	Chili powder:	1 tbsp	8	8	25	1	4	3	1	.2	.3	.6
970	Teaspoon	1 tsp	3	8	9	<1	2	1	<1	.1	.1	.2
	Chocolate:											
971	Baking, unsweetened, square	1 oz	28	1	146	3	8	4	15	9.1	5.2	.5
	For other chocolate items, see Sweeteners & Sweets											
972	Cilantro/Coriander, fresh	1 tbsp	1	92		<1	<1	<1	<1	0	t	0
2287	Cinnamon	1 tsp	2	10	5	<1	2	1	<1	t	t	t
1197	Cornstarch	1 tbsp	8	8	30	<1	7	<1	<1	.7	t	t
2239	Curry powder	1 tsp	2	10	6	<1	1	1	<1	t	.1	.1
1202	Dill weed, dried	1 tbsp	3	7	8	1	2	<1	<1	t	.1	t
975	Garlic cloves	1 ea	3	59	4	<1	1	<1	<1	t	0	t
2238	Garlic powder	1 tsp	3	6	10	<1	2	<1	<1	t	t	t
977	Gelatin, dry, unsweetened: Envelope	1 ea	7	13	23	6	0	0	<1	t	.3	.5
978	Ginger root, slices, raw	2 pce	5	82	3	<1	1	<1	<1	t	t	t
1198	Horseradish, prepared	1 tbsp	15	85	7	<1	2	<1	<1	t	t	.1
1997	Hummous/hummus	1 c	246	65	421	12	50	12	21	3.1	8.7	7.8
1909	Mustard, country dijon	1 tsp	5		5	<1	<1	0	0	0	0	0

PAGE KEY: A–2 = Beverages A–4 = Dairy A–8 = Eggs A–10 = Fat/Oil A–12 = Fruit A–18 = Bakery A–24 = Grain
A–30 = Fish A–32 = Meats A–36 = Poultry A–38 = Sausage A–38 = Mixed/Fast A–44 = Nuts/Seeds A–46 = Sweets
A–50 = Vegetables/Legumes A–60 = Vegetarian A–62 = Misc A–64 = Soups/Sauces A–66 = Fast A–80 = Convenience A–86 = Baby foods

Table of Food Composition

Chol (mg)	Calc (mg)	Iron (mg)	Magn (mg)	Pota (mg)	Sodi (mg)	Zinc (mg)	VT-A (µg)	Thia (mg)	VT-E (mg)	Ribo (mg)	Niac (mg)	V-B6 (mg)	Fola (µg)	VT-C (mg)
1	24	.83		60	98	.6	32	.05		.36	0	.11		0
1	9	1.77	16	46	355	.35	0	5.43		.15	2.5	.38	12	0
<1	3	.33		16	228	.06	0	.67		.05	.75	.08		0
0	40	3.2	1	89	238	.82	0	4.96	.35	.18	1.49	.27		0
1	56	1.84	44	269	499	.93	7	8.06	.36	.14	0	.21		0
2	14	1.31		202	581	.35	0	.7		.15	2.26	.18		0
1	22	.77	5	63	545	.48	0	.18	1.59	.03	0	.01		0
1	48	1.21	29	180	382	.58	38	6.47	.98	.1	0	0	29	0
2	22	2.5		122	269	.67	0	11.8		.2	4.86	.48		0
1	46	2.14		142	247	.74	0*	.51		.25	.92	.41		2
0	4	2.63		44	624	.22	0	.89		.34	6.46	.56		0
0	4	1.73		25	269	.38	0	.13		.1	1.96	.24		0
1	9	.73		111	257	.26	0	.06		.05	.37	.08		0
0	20	1.49		195	523	.57	0	.02		.03	1.04	.31		0
0	6	.37		40	500	.65	0	.05		.05	0	.05		0
1	6	1.17		58	524	.26	0	10.6		.07	1.36	.3		0
2	51	5.09		270	826	1.03	108	1.85		.29	4.22	.86		0
1	15	.57		96	384	.31	0	1.22		.19	0	.03		0
0	10	1.12		155	272	.56	0	.08		.06	.54	.2		0
2	12	.38		82	445	.38	0	.12		.13	1.98	.38		0
<1	1	1.84		50	283	.7	0	.18		.13	1.34	.24		0
2	36	2.87		38	566	.64	0	2.13		.25	6.02	.6		0
1	9	1.15		25	205	.26	0	.59		.08	.09	.13		0
<1		.47		14	179	.08	0	.08		.06	.48	.1		0
2	49	.99		95	484	.5	0	1.51		.12	3.1	.26		0
1	3	.47		16	203	.11	0	1.13		.05	.39	.09		0
0	97	0		7	547	0	0*	0	0	0	0	0	0	0
0	294	.55	1	1	530	<.01	0	0	0	0	0	0	0	0
0	368	.56	2		395	<.01	0	0	0	0	0	0	0	0
0	217	.41	1	505	4	.04	0	0	<.01	0	0	0	0	0
0	0	0	0	0	1368	0	0	0	0	0	0	0	0	0
0	106	2.1	21	172	2	.29	23	.01	.08	.02	.35	.06	14	3
				29	474									
0	358	3.03	56	852	36	.95	1	.05	.65	.47	1.95	.38	30	<1
0	12	.43	13	293	723	.14	31	.05	.89	.04	.83	.11	9	9
0	3	.1	3	72	178	.03	8	.01	.22	.01	.2	.03	2	2
0	7	.39	8	101	1	.12	104	.02	.24	.05	.43	.1	5	4
0	35	.9	9	28	3	.14	<1	.01	.02	.01	.06	.01	<1	<1
0	22	1.14	14	153	81	.22	140	.03	.08	.06	.63	.15	8	5
0	8	.43	5	57	30	.08	52	.01	.03	.02	.24	.06	3	2
0	21	1.77	87	233	4	1.12	1	.02	.34	.05	.31	.03	2	0
0	1	.02		5	<1	<.01	3	<.01	.02	<.01	.01	<.01	1	<1
0	25	.76	1	10	1	.04	<1	<.01	0	<.01	.03	<.01	1	1
0		.04			1	<.01	0	0	0	0	0	0	0	0
0	10	.59	5	31	1	.08	1	<.01	.01	.01	.07	.01	3	<1
0	53	1.46	13	99	6	.1	9	.01		.01	.08	.04		1
0	5	.05	1	12	1	.03	0	.01	0	<.01	.02	.04	<1	1
0	2	.08	2	33	1	.08	0	.01	0	<.01	.02	.08	<1	1
0	4	.08	2	1	14	.01	0	<.01	0	.02	.01	0	2	0
0	1	.02	2	21	1	.02	0	<.01	.01	<.01	.03	.01	1	<1
0	8	.06	4	37	47	.12	<1	<.01	<.01	<.01	.06	.01	9	4
0	123	3.86	71	428	600	2.71	2	.23	2.46	.13	1.01	.98	145	19
0				10	120									

*This value is expressed in retinol equivalents (RE). All other values are in retinol activity equivalents (RAE).

Table A-1

Food Composition   (Computer code number is for Wadsworth Diet Analysis program)   (For purposes of calculations, use "0" for t, <1, <.1, <.01, etc.)

A

Computer Code Number	Food Description	Measure	Wt (g)	H₂O (%)	Ener (kcal)	Prot (g)	Carb (g)	Dietary Fiber (g)	Fat (g)	Fat Breakdown (g) Sat	Mono	Poly
	MISCELLANEOUS—Continued											
2019	Mustard, gai choy chinese	1 tbsp	16	94	3	<1	1		<1			
979	Mustard, prepared (1 packet = 1 tsp)	1 tsp	5	80	4	<1	<1	<1	<1	t	.2	t
	Miso (see #926 under Vegetables and Legumes, Soybean products)											
980	Olives, green	5 ea	20	78	23	<1	<1	<1	3	.3	1.9	.2
981	Olives, ripe, pitted	5 ea	22	80	25	<1	1	1	2	.3	1.7	.2
26008	Onion powder	1 tsp	2	5	7	<1	2	<1	<1	t	t	t
2237	Oregano, ground	1 tsp	2	7	6	<1	1	1	<1	.1	t	.1
2236	Paprika	1 tsp	2	10	6	<1	1	<1	<1	t	t	.2
887	Parsley, freeze dried	¼ c	1	2	3	<1	<1	<1	<1	t	t	t
	Parsley, fresh (see #885 and #886)											
985	Pepper, black	1 tsp	2	11	5	<1	1	1	<1	t	t	t
	Pickles:											
986	Dill, medium, 3¾ x 1¼" diam	1 ea	65	92	12	<1	3	1	<1	t	t	t
987	Fresh pack, slices, 1½" diam x ¼"	2 pce	15	79	11	<1	3	<1	<1	0	0	t
988	Sweet, medium	1 ea	35	65	41	<1	11	<1	<1	t	t	t
989	Pickle relish, sweet	1 tbsp	15	63	21	<1	5	<1	<1	t	t	t
	Popcorn (see Grain Products #539-541)											
917	Potato chips:	10 pce	20	2	107	1	11	1	7	2.2	2	2.4
44076	Unsalted	1 oz	28	2	150	2	15	1	10	3.1	2.8	3.4
1201	Sage, ground	1 tsp	1	8	3	<1	1	<1	<1	.1	t	t
1347	Salsa, from recipe	1 tbsp	15	95	3	<1	1	<1	<1	t	t	t
2218	Salsa, pico de gallo, medium	1 tbsp	15	92	2	0	1	<1	0	0	0	0
990	Salt	1 tsp	6		0	0	0	0	0	0	0	0
	Salt Substitutes:											
1205	Morton, salt substitute	1 tsp	6		0	<1		0	0	0	0	
1207	Morton, light salt	1 tsp	6		0	<1		0	0	0	0	
2067	Seasoned salt, no MSG	1 tsp	5	5	0	0	0	0	0	0	0	0
991	Vinegar, cider	½ c	120	94	17	0	7	0	0	0	0	0
2172	Balsamic	1 tbsp	15	64	21	0	5	0	0	0	0	0
2176	Malt	1 tbsp	15	90	5	0	1	0	0	0	0	0
2182	Tarragon	1 tbsp	15	95	3	0	<1	0	0	0	0	0
2181	White wine	1 tbsp	15	89	5	0	1	0	0	0	0	0
	Yeast:											
992	Baker's, dry, active, package	1 ea	7	8	21	3	3	1	<1	t	.2	t
993	Brewer's, dry	1 tbsp	8	5	23	3	3	3	<1	t	t	0
	SOUPS, SAUCES, and GRAVIES											
	SOUPS, canned, condensed:											
	Unprepared, condensed:											
1210	Cream of celery	1 c	251	85	181	3	18	2	11	2.8	2.6	5
1215	Cream of chicken	1 c	251	82	233	7	18	<1	15	4.2	6.5	3
1216	Cream of mushroom	1 c	251	81	259	4	19	1	19	5.1	3.6	8.9
1220	Onion	1 c	246	86	113	8	16	2	3	.5	1.5	1.3
	Prepared w/equal volume of whole milk:											
994	Clam chowder, New England	1 c	248	85	164	9	17	1	7	2.9	2.3	1.1
1209	Cream of celery	1 c	248	86	164	6	14	1	10	3.9	2.5	2.6
995	Cream of chicken	1 c	248	85	191	7	15	<1	11	4.6	4.5	1.6
996	Cream of mushroom	1 c	248	85	203	6	15	<1	14	5.1	3	4.6
1214	Cream of potato	1 c	248	87	149	6	17	<1	6	3.8	1.7	.6
1213	Oyster stew	1 c	245	89	135	6	10	0	8	5	2.1	.3
997	Tomato	1 c	248	85	161	6	22	3	6	2.9	1.6	1.1
	Prepared with equal volume of water:											
998	Bean with bacon	1 c	253	84	172	8	23	9	6	1.5	2.2	1.8
999	Beef broth/bouillon/consomme'	1 c	240	98	17	3	<1	0	1	.3	.2	t
1000	Beef noodle	1 c	244	92	83	5	9	1	3	1.1	1.2	.5
1001	Chicken noodle	1 c	241	92	75	4	9	1	2	.7	1.1	.6
1002	Chicken rice	1 c	241	94	60	4	7	1	2	.5	.9	.4
1208	Chili beef	1 c	250	85	170	7	21	9	7	3.3	2.8	.3
1003	Clam chowder, Manhattan	1 c	244	92	78	2	12	1	2	.4	.4	1.3
1004	Cream of chicken	1 c	244	91	117	3	9	<1	7	2.1	3.3	1.5
1005	Cream of mushroom	1 c	244	90	129	2	9	<1	9	2.4	1.7	4.2
1006	Minestrone	1 c	241	91	82	4	11	1	3	.6	.7	1.1

Chol (mg)	Calc (mg)	Iron (mg)	Magn (mg)	Pota (mg)	Sodi (mg)	Zinc (mg)	VT-A (µg)	Thia (mg)	VT-E (mg)	Ribo (mg)	Niac (mg)	V-B6 (mg)	Fola (µg)	VT-C (mg)
0	4	.1	2	6	63	.03	0*	0	.09	0	0	<.01	0	0
0	12	.32	4	11	480	.01	6*	0	.6	0	0	<.01	<1	0
0	19	.73	1	2	192	.05	4	<.01	.66	0	.01	<.01	0	<1
0	7	.05	2	19	1	.05	0	.01	<.01	<.01	.01	.03	3	<1
0	31	.88	5	33	<1	.09	7	.01	.03	.01	.12	.02	5	1
0	4	.47	4	47	1	.08	61	.01	.01	.03	.31	.04	2	1
0	2	.54	4	63	4	.06	32	.01	.06	.02	.1	.01	15	1
0	9	.58	4	25	1	.03	<1	<.01	.02	<.01	.02	.01	<1	<1
0	6	.34	7	75	833	.09	11	.01	.1	.02	.04	.01	1	1
0	5	.27	1	30	101	0	2*	0	.02	<.01	0	<.01	0	1
0	1	.21	1	11	329	.03	2	<.01	.06	.01	.06	<.01	<1	<1
0	3	.12	1	30	107	.01	1*	0	.02	<.01	0	0	0	1
0	5	.33	13	255	119	.22	0	.03	.98	.04	.76	.13	9	6
0	7	.46	19	357	2	.3	0	.05	1.37	.05	1.07	.18	13	9
0	16	.28	4	11	<1	.05	3	.01	.02	<.01	.06	.01	3	<1
0	1	.05	1	23	1	.01	3	.01	.04	<.01	.06	.01	2	2
0													0	
0	1	.02			2325	.01	0	0	0	0	0	0	0	0
	33			3018	<1									
	2		4	1560	1170									
				15	1583									
0	7	.72	26	120	1	0	0	0	0	0	0	0	0	0
	2	.07		10	3		<1*	.07		.07	.07			<1
	2	.07		13	4		<1*	.07		.07	.07			1
		.07		2	1		<1*	.07		.07	.07			<1
	1	.07		12	1		<1*	.07		.07	.07			<1
0	4	1.16	7	140	3	.45	<1	.16	.01	.38	2.78	.11	164	<1
0	17	1.38	18	151	10	.63	0*	1.25		.34	3.03	.4	313	0
28	80	1.26	13	246	1900	.3	60*	.06	.38	.1	.66	.02	5	<1
20	68	1.2	5	176	1972	1.26	113*	.06	.33	.12	1.64	.03	3	<1
3	65	1.05	10	168	1736	1.18	0	.06	2.61	.17	1.62	.02	8	2
0	54	1.35	5	138	2115	1.23	0	.07	.57	.05	1.21	.1	29	2
22	186	1.49	22	300	992	.79	40*	.07	.15	.24	1.03	.13	10	3
32	186	.69	22	310	1009	.2	67*	.07	.97	.25	.44	.06	7	1
27	181	.67	17	273	1046	.67	94*	.07	.25	.26	.92	.07	7	1
20	179	.59	20	270	918	.64	37*	.08	1.34	.28	.91	.06	10	2
22	166	.55	17	322	1061	.67	67*	.08	.1	.24	.64	.09	10	1
32	167	1.05	20	235	1041	10.3	44*	.07	.49	.23	.34	.06	10	4
17	159	1.81	22	449	744	.3	109*	.13	2.6	.25	1.52	.16	20	68
3	81	2.05	45	402	951	1.04	44	.09	.08	.03	.57	.04	33	2
0	14	.41	5	130	782	0	0	<.01	0	.05	1.87	.02	5	0
5	15	1.1	5	100	952	1.54	63*	.07	0	.06	1.07	.04	19	<1
7	17	.77	5	55	1106	.39	72*	.05	.07	.06	1.39	.03	22	<1
7	17	.75	0	101	815	.26	65*	.02	.05	.02	1.13	.02	0	<1
12	42	2.13	30	525	1035	1.4	150*	.06	.17	.07	1.07	.16	17	4
2	27	1.63	12	188	578	.98	98*	.03	.73	.04	.82	.1	10	4
10	34	.61	2	88	986	.63	56*	.03	.19	.06	.82	.02	2	<1
2	46	.51	5	100	881	.59	0	.05	1.24	.09	.72	.01	5	1
2	34	.92	7	313	911	.75	117	.05	.07	.04	.94	.1	36	1

*This value is expressed in retinol equivalents (RE). All other values are in retinol activity equivalents (RAE).

Table A–1

Food Composition   (Computer code number is for Wadsworth Diet Analysis program)   (For purposes of calculations, use "0" for t, <1, <.1, <.01, etc.)

Computer Code Number	Food Description	Measure	Wt (g)	H₂O (%)	Ener (kcal)	Prot (g)	Carb (g)	Dietary Fiber (g)	Fat (g)	Fat Breakdown (g) Sat	Mono	Poly
	SOUPS, SAUCES, and GRAVIES—Continued											
1211	Onion	1 c	241	93	58	4	8	1	2	.3	.7	.7
1007	Split pea & ham	1 c	253	82	190	10	28	2	4	1.8	1.8	.6
1008	Tomato	1 c	244	90	85	2	17	<1	2	.4	.4	1
1009	Vegetable beef	1 c	244	92	78	6	10	<1	2	.9	.8	.1
1010	Vegetarian vegetable	1 c	241	92	72	2	12	<1	2	.3	.8	.7
	Ready to serve:											
1707	Chunky chicken soup	1 c	251	84	178	13	17	2	7	2	3	1.4
	SOUPS, dehydrated:											
	Prepared with water:											
1299	Beef broth/bouillon	1 c	244	97	19	1	2	0	1	.3	.3	t
1376	Chicken broth	1 c	244	97	22	1	1	0	1	.3	.4	.4
1013	Chicken noodle	1 c	252	94	58	2	9	<1	1	.3	.5	.4
1122	Cream of chicken	1 c	261	91	107	2	13	<1	5	3.4	1.2	.4
1014	Onion	1 c	246	96	27	1	5	1	1	.1	.3	.1
1217	Split pea	1 c	255	87	125	7	21	3	1	.4	.7	.3
1015	Tomato vegetable	1 c	253	94	56	2	10	<1	1	.4	.3	.1
	Unprepared, dry products:											
1011	Beef bouillon, packet	1 ea	6	3	14	1	1	0	1	.3	.2	t
1012	Onion soup, packet	1 ea	39	4	115	5	21	4	2	.5	1.4	.3
	SAUCES											
	From dry mixes, prepared with milk:											
1016	Cheese sauce	1 c	279	77	307	17	23	1	17	9.3	5.3	1.6
1017	Hollandaise	1 c	259	84	240	5	14	<1	20	11.6	5.9	.9
1018	White sauce	1 c	264	81	240	10	21	<1	13	6.4	4.7	1.7
	From home recipe:											
1206	Lowfat cheese sauce	¼ c	61	74	81	6	4	<1	5	1.8	1.8	.8
1019	White sauce, medium	¼ c	72	77	102	2	6	<1	8	2.3		
	Ready to serve:											
2202	Alfredo sauce, reduced fat	¼ c	69	64	144	5	9	0	9	6.2		
1020	Barbeque sauce	1 tbsp	16	81	12	<1	2	<1	<1	t	.1	.1
1706	Chili sauce, tomato base	1 tbsp	17	68	18	<1	4	<1	<1	t	t	t
2126	Creole sauce	¼ c	62	89	25	1	4	1	1	.1	.2	.3
2124	Hoisin sauce	1 tbsp	17	47	35	<1	7	0	1	0		
2199	Pesto sauce	2 tbsp	16	34	83	2	1	<1	8	1.5		
1021	Soy sauce	1 tbsp	16	71	8	1	1	<1	<1	t	t	t
2123	Szechuan sauce	1 tbsp	16	71	21	<1	3	<1	1	.1	.3	.4
1380	Teriyaki sauce	1 tbsp	18	68	15	1	3	<1	0	0	0	0
	Spaghetti sauce, canned:											
1377	Plain	1 c	249	75	271	5	40	8	12	1.7	6.1	3.3
1378	With meat	1 c	250	85	178	7	19	4	8	1.8	3.3	1.8
1379	With mushrooms	½ c	123	84	108	2	13	1	3	.4	1.5	.8
	GRAVIES											
	Canned:											
1022	Beef	1 c	233	87	123	9	11	1	5	2.7	2.2	.2
1023	Chicken	1 c	238	85	188	5	13	1	14	3.4	6.1	3.6
1024	Mushroom	1 c	238	89	119	3	13	1	6	1	2.8	2.4
1025	From dry mix, brown	1 c	258	92	75	2	13	<1	2	.8	.7	.1
1026	From dry mix, chicken	1 c	260	91	83	3	14	<1	2	.5	.9	.4
	FAST FOOD RESTAURANTS											
	ARBY'S											
1402	Bac'n cheddar deluxe	1 ea	231	59	512	21	39	<1	31	8.7	12.7	10.1
	Roast beef sandwiches:											
1403	Regular	1 ea	155	54	326	21	35	2	14	6.9		
1404	Junior	1 ea	89	50	200	11	23	1	8	3.4	3.3	1.6
1405	Super	1 ea	254	61	467	23	50	3	22	8.3		
1407	Beef 'n cheddar	1 ea	194	53	451	22	42	2	22	8.8	8.8	5
1408	Chicken breast sandwich	1 ea	204	49	539	23	46	2	29	4.9	12	12.5
1412	Ham'n cheese sandwich	1 ea	169	57	338	23	35	1	13	4.5	5.1	3.3
1726	Italian sub sandwich	1 ea	297	57	743	28	47	3	50	14.3	23.5	12.7
1413	Turkey sandwich, deluxe	1 ea	218	68	292	26	37	3	6	.6	2.5	2.6
1680	Turkey sub sandwich	1 ea	277	62	570	23	46	2	33	8.1	11.7	13.7
	Milkshakes:											

Chol (mg)	Calc (mg)	Iron (mg)	Magn (mg)	Pota (mg)	Sodi (mg)	Zinc (mg)	VT-A (µg)	Thia (mg)	VT-E (mg)	Ribo (mg)	Niac (mg)	V-B6 (mg)	Fola (µg)	VT-C (mg)
0	26	.67	2	67	1053	.6	0	.03	.29	.02	.6	.05	14	1
8	23	2.28	48	400	1006	1.32	45*	.15	.15	.08	1.47	.07	3	2
0	12	1.76	7	264	695	.24	34	.09	2.49	.05	1.42	.11	15	66
5	17	1.12	5	173	791	1.54	95	.04	.32	.05	1.03	.08	10	2
0	22	1.08	7	210	822	.46	301*	.05	.79	.05	.92	.05	10	1
30	25	1.73	8	176	889	1	131*	.08	.18	.17	4.42	.05	5	1
0	10	.02	7	37	1361	.07	1	<.01	.02	.02	.36	0	0	0
0	15	.07	5	24	1483	0	12*	.01	.02	.03	.19	0	2	0
10	5	.5	8	33	577	.2	5*	.2	.1	.08	1.09	.02	18	0
3	76	.26	5	214	1184	1.57	123*	.1	.16	.2	2.61	.05	5	1
0	12	.15	5	64	849	.05	<1	.03	.1	.06	.48	0	2	<1
3	20	.94	43	224	1147	.56	5*	.21	.13	.14	1.26	.05	41	0
0	8	.63	20	104	1146	.18	10	.06	.81	.05	.79	.05	10	6
1	4	.06	3	27	1018	0	<1*	<.01	.01	.01	.27	.01	2	0
2	55	.58	25	260	3493	.23	<1	.11	.42	.24	1.99	.04	6	1
53	569	.28	47	552	1565	.97	117*	.15	.33	.56	.32	.14	13	2
52	124	.9	8	124	1564	.7	220*	.04	.26	.18	.06	.5	22	<1
34	425	.26	264	444	797	.55	92*	.08	1.58	.45	.53	.07	16	3
9	167	.24	10	101	307	.74	56*	.03	.51	.14	.16	.03	4	<1
8	75	.21	9	100	82	.26	62*	.05	.98	.12	.28	.03	4	1
31	103	0	8	93	618		154*	0		.1	0		0	0
0	3	.14	3	28	130	.03	14*	<.01	.18	<.01	.14	.01	1	1
0	3	.14	2	63	227	.05	24*	.01	.05	.01	.27	.02	1	3
0	35	.31	9	187	339	.1	12	.03	.61	.02	.53	.07	9	0
0	0	0			250		0*							0
4	64	.09		26	137		26*	0		.02	0		2	0
0	3	.32	5	29	914	.06	0	.01	0	.02	.54	.03	3	0
0	2	.12	2	13	218	.02	10*	<.01	.07	<.01	.1	.01	1	<1
0	4	.31	11	40	690	.02	0	<.01	0	.01	.23	.02	4	0
0	70	1.62	60	956	1235	.52	306*	.14	4.98	.15	3.76	.88	54	28
15	53	2.1	43	742	982	1.27	176*	.13	2.98	.12	3.47	.31	25	19
0	15	1	15	332	494	.34	120	.08	1.35	.08	.93	.16	12	9
7	14	1.63	5	189	1304	2.33	0	.07	.14	.08	1.54	.02	5	0
5	48	1.12	5	259	1373	1.9	264*	.04	.38	.1	1.05	.02	5	0
0	17	1.57	5	252	1356	1.67	0	.08	.19	.15	1.6	.05	29	0
3	67	.23	10	57	1075	.31	0*	.04	.05	.08	.81	0	0	0
3	39	.26	10	62	1133	.32	0*	.05	.05	.15	.78	.03	3	3
38	110	4.32		491	1094	3	40*	.34		.46	9.6			11
44	59	3.55	16	422	879	3.75	0	.28		.48	11	.2	14	0
28	41	1.86	8	201	483	1.5		.18		.25	6.6	.1	7	
47	62	3.73	25	533	1098	3.73	30*	.39		.58	12.4	.3	21	1
49	98	3.53		321	1146	2.94		.42		.63	9.8			1
88	78	1.77	30	330	1137	.15		.22		.54	8.99	.38	18	4
89	149	2.68	31	380	1441	.89	40*	.82		.37	7.75	.31	26	1
114	238	2.57		565	2322		97*	.91		.49	8.19			2
45	90	2.02	33	394	1157	1.69	45*	.09		.46	17.2	.58	22	1
90	181	.33		500	1964		20*	13.2		.54	18.8			2

*This value is expressed in retinol equivalents (RE). All other values are in retinol activity equivalents (RAE).

Table A–1

Food Composition   (Computer code number is for Wadsworth Diet Analysis program)   (For purposes of calculations, use "0" for t, <1, <.1, <.01, etc.)

Computer Code Number	Food Description	Measure	Wt (g)	H₂O (%)	Ener (kcal)	Prot (g)	Carb (g)	Dietary Fiber (g)	Fat (g)	Fat Breakdown (g) Sat	Mono	Poly
	FAST FOOD RESTAURANTS—Continued											
1419	Chocolate	1 ea	340	71	411	9	72	0	14	6.8	5.5	1.3
1420	Jamocha	1 ea	326	72	386	8	67	0	12	5.7	5.3	1.3
1421	Vanilla	1 ea	312	72	369	8	65	0	12	5.5	4.4	1.9
1728	Salad, roast chicken	1 ea	400	90	152	19	14	6	2	0		
1729	Sports drink, Upper Ten	1 ea	358	88	169	0	42		0	0	0	0
	Source: Arby's											
	BURGER KING											
1423	Croissant sandwich, egg,sausage&cheese	1 ea	176	46	600	22	25	1	46	16		
	Whopper sandwiches:											
1425	Whopper	1 ea	270	58	640	27	45	3	39	11		
1426	Whopper with cheese	1 ea	294	57	730	33	46	3	46	16		
	Sandwiches:											
1629	BK broiler chicken sandwich	1 ea	248	59	550	30	41	2	29	6		
1432	Cheeseburger	1 ea	138	48	380	23	28	1	19	9		
1434	Chicken sandwich	1 ea	229	45	710	26	54	2	43	9		
1427	Double beef	1 ea	351	57	870	46	45	3	56	19		
1428	Double beef & cheese	1 ea	375	56	960	52	46	3	63	24		
1433	Double cheeseburger with bacon	1 ea	218	48	640	44	28	1	39	18		
1431	Hamburger	1 ea	126	48	330	20	28	1	15	6		
1437	Ocean catch fish fillet	1 ea	255	51	700	26	56	3	41	6		
1435	Chicken tenders	1 ea	88	50	230	16	14	2	12	3		
1439	French fries (salted)	1 svg	116	40	370	5	43	3	20	5		
1630	French toast sticks	1 svg	141	33	500	4	60	1	27	7		
1440	Onion rings	1 svg	124	51	310	4	41	6	14	2	8	4
1441	Milk shakes, chocolate	1 ea	284	75	320	9	54	3	7	4		
1442	Milk shakes, vanilla	1 ea	284	75	300	9	53	1	6	4		
1443	Fried apple pie	1 ea	113	47	300	3	39	2	15	3		
	Source: Burger King Corporation											
	CHICK-FIL-A											
	Sandwiches:											
69153	Chargrilled chicken	1 ea	150	54	280	27	36	1	3	1		
69152	Chicken	1 ea	167	61	290	24	29	1	9	2		
69155	Chicken salad	1 ea	167	55	320	25	42	1	5	2		
69154	Chicken salad club	1 ea	232	62	390	33	38	2	12	5		
	Salads:											
52139	Carrot and raisin	1 ea	76	53	150	5	28	2	2	0		
52136	Chicken plate	1 ea	468	85	290	21	40	6	5	0		
52134	Chicken garden, charbroiled	1 ea	397	89	170	26	10	5	3	1		
52135	Chick-n-strips	1 ea	451	86	290	32	21	5	9	2		
52138	Cole slaw	1 ea	79	70	130	6	11	1	6	1		
52137	Tossed salad	1 ea	130	85	70	5	13	1	0	0	0	0
15263	Chicken nuggets, svg	1 ea	110	51	290	28	12	0	14	3		
15262	Chicken-n- strips, svg	1 ea	119	59	230	29	10	0	8	2		
50885	Hearty breast of chicken soup, svg	1 ea	215	86	110	16	10	1	1	0		
7973	Waffle potato fries, svg	1 ea	85	28	290	1	49	0	10	4		
46489	Cheesecake, svg	1 ea	88	52	270	13	7	0	21	9		
49134	Fudge nut brownie, svg	1 ea	74	8	350	10	41	0	16	3		
20601	Icedream, svg	1 ea	127	74	140	11	16	0	4	1		
48214	Lemon pie, svg	1 ea	99	56	280	1	19	0	22	6		
	Source: Chick-Fil-A											
	DAIRY QUEEN											
	Ice cream cones:											
1446	Small vanilla	1 ea	142	63	230	6	38	0	7	4.5		
1447	Regular vanilla	1 ea	213	64	355	9	57	0	10	6.4		
1448	Large vanilla	1 ea	253	65	410	10	65	0	12	8		
1450	Chocolate dipped	1 ea	220	58	490	8	59	1	24	12.5		
1453	Chocolate sundae	1 ea	234	62	400	8	71	0	10	6		

Chol (mg)	Calc (mg)	Iron (mg)	Magn (mg)	Pota (mg)	Sodi (mg)	Zinc (mg)	VT-A (µg)	Thia (mg)	VT-E (mg)	Ribo (mg)	Niac (mg)	V-B6 (mg)	Fola (µg)	VT-C (mg)
38	428	.62	48	410	317	1.5	60*	.12		.68	.8	.14	14	2
37	411	.59	36	525	320	1.48	60*	.12		.68	.8	.14	14	2
35	393	.85	36	686	283	1.49	60*	.12		.68	4	.14	37	2
38	76	1.71		877	667		970*	.31		.54	5.6			<1
0				0	40									
260	150	3.6			1140		80*							0
90	80	4.5			870		100*	.33		.41	7	.35		9
115	250	4.5			1350		150*	.34		.48	7	.33		9
80	60	5.4			480		60*							6
65	100	2.7			770		60*							0
60	100	3.6			1400		0							0
170	80	7.2			940		100*	.34		.56	10			9
195	250	7.2			1420		150*	.35		.63	10			9
145	200	4.5			1240		80*	.31		.42	6			0
55	40	1.8			530		20*	.28		.31	4.89			0
90	60	2.7			980		20*							1
35	0	.72			530		0							0
0	0	1.08			240		0							4
0	60	2.7			490		0							0
0	100	1.44			810		0							0
20	200	1.8			230		60*	.13		.55	.13			0
20	300	0			230		60*	.11		.57	.13			4
0	0	1.44			230		0							6
40	0	1.8			640		80*							1
50	0	1.8			870		80*							0
10	0	1.44			810		80*							0
70	80	1.8			980		140*							8
6	40	3.6			650		200*							6
35	60	3.24			570		160*							13
25	60	2.52			650		240*							13
20		2.52			430		200*							13
15	20	3.6			430		140*							6
0	0	3.6			0		100*							8
60	0	1.44			770		80*							0
20	0	1.08			380		60*							8
45	0	2.52			760		140*							2
5	0	0			960		0							0
10	20	.72			510		60*	.03		.19	.19		18	0
30	0	.72			650		80*	.12		.21	.97		29	0
40	80	1.08			240		100*							0
5	150	1.08			550		150*							5
20	200	1.08		250	115		100*	.05		.28				1
32	269	1.94		390	172		161*	.09		.38	.16	.13		3
40	350	1.8		451	200		200*	.11		.4	.2			2
30	250	1.8		409	190		150*	.08		.36	.15	.13		2
30	250	1.44		383	210		150*	.08		.34	.39	.18		0

*This value is expressed in retinol equivalents (RE). All other values are in retinol activity equivalents (RAE).

Table A–1

Food Composition

(Computer code number is for Wadsworth Diet Analysis program)   (For purposes of calculations, use "0" for t, <1, <.1, <.01, etc.)

Computer Code Number	Food Description	Measure	Wt (g)	H₂O (%)	Ener (kcal)	Prot (g)	Carb (g)	Dietary Fiber (g)	Fat (g)	Fat Breakdown (g) Sat	Mono	Poly
	FAST FOOD RESTAURANTS—Continued											
1455	Banana split	1 ea	369	67	510	8	96	3	12	8		
1456	Peanut buster parfait	1 ea	305	51	730	16	99	2	31	17		
1457	Hot fudge brownie delight	1 ea	305	52	710	11	102	1	29	14	12	2
1459	Buster bar	1 ea	149	45	450	10	41	2	28	12		
1645	Breeze, strawberry, regular	1 ea	383	70	460	13	99	1	1	1	0	0
1460	Dilly bar	1 ea	85	55	210	3	21	0	13	7	3	3
1461	DQ ice cream sandwich	1 ea	61	46	150	3	24	1	5	2		
1463	Milk shakes, regular	1 ea	397	71	520	12	88	<1	14	8	2	2
1464	Milk shakes, large	1 ea	461	71	600	13	101	<1	16	10	2	2
1466	Milk shakes, malted	1 ea	418	68	610	13	106	<1	14	8	2	2
1470	Misty slush, small	1 ea	454	88	220	0	56	0	0	0	0	0
2250	Starkiss	1 ea	85	75	80	0	21	0	0	0	0	0
	Yogurt:											
1641	Yogurt cone, regular	1 ea	198	65	260	9	56	0	1	.5		
1643	Yogurt sundae, strawberry	1 ea	234	69	280	8	61	1	<1	0		
	Sandwiches:											
1481	Cheeseburger, double	1 ea	219	55	540	35	30	2	31	16		
1480	Cheeseburger, single	1 ea	152	55	340	20	29	2	17	8		
1474	Chicken	1 ea	191	56	430	24	37	2	20	4		
1647	Chicken fillet, grilled	1 ea	184	64	310	24	30	3	10	2.5		
1475	Fish fillet sandwich	1 ea	170	57	370	16	39	2	16	3.5		
1476	Fish fillet with cheese	1 ea	184	56	420	19	40	2	21	6	7	8
1477	Hamburger, single	1 ea	138	56	290	17	29	2	12	5	6	1
1478	Hamburger, double	1 ea	212	62	440	30	29	2	22	10		
	Hotdog:											
1483	Regular	1 ea	99	57	240	9	19	1	14	5		
1484	With cheese	1 ea	113	55	290	12	20	1	18	8	8	2
1485	With chili	1 ea	128	61	280	12	21	2	16	6		
1489	French fries, small	1 ea	112	41	350	4	42	3	18	3.5		
1490	French fries, large	1 ea	128	40	390	5	52	6	18	4	8	6
1491	Onion rings	1 ea	113	46	320	5	39	3	16	4		
	Source: International Dairy Queen											
	HARDEES'S											
	Sandwiches:											
56414	Cheeseburger	1 ea	120	47	300	15	34		13	6.5	4.6	1.9
1734	Frisco burger hamburger	1 ea	242	46	760	36	43		50	18		
56412	Hamburger	1 ea	107	49	260	11	33		9	3.6	3.6	1.8
56415	Quarter pound cheeseburger	1 ea	184	51	490	27	37		25	11.5	11.5	1.9
56422	Chicken sandwich	1 ea	187	56	400	19	48		14	3	4	6
6146	French fries, svg	1 ea	96	48	240	4	33		10	2.5	4.2	3.3
	JACK IN THE BOX											
	Breakfast items:											
1492	Breakfast jack sandwich	1 ea	126	54	280	17	28	1	12	5	4.7	2.3
1494	Sausage crescent	1 ea	181	41	660	20	37	0	48	15		
1495	Supreme crescent	1 ea	164	43	530	21	37	0	34	10	17	7
1496	Pancake platter	1 ea	231	45	610	15	87	0	22	9	7.6	3.5
1497	Scrambled egg platter	1 ea	213	52	560	18	50	0	32	9	16.6	4.4
	Sandwiches:											
1654	Bacon cheeseburger	1 ea	274	51	760	39	39	2	50	17	21.2	11.8
1499	Cheeseburger	1 ea	116	48	300	14	31	2	13	6	5	2
1739	Chicken caesar pita sandwich	1 ea	237	59	520	27	44	4	26	6		
1655	Chicken sandwich	1 ea	164	53	400	15	38	3	21	3		
1656	Chicken sandwich, sourdough ranch	1 ea	225	57	490	29	45	1	21	6		
1505	Chicken supreme	1 ea	305	49	830	33	66	3	49	7		
1583	Double cheeseburger	1 ea	158	48	440	24	31	2	24	11	10.3	2.7
1651	Grilled sourdough burger	1 ea	233	49	690	34	37	2	45	15	20.8	9.2
1498	Hamburger	1 ea	104	50	250	12	30	2	9	3.5	3.9	1.6
1500	Jumbo jack burger	1 ea	271	62	550	27	43	2	30	10	12.4	7.6
1501	Jumbo jack burger with cheese	1 ea	296	60	640	31	44	2	38	15	14.4	8.6
1740	Monterey roast beef sandwich	1 ea	238	57	540	30	40	3	30	9		

Chol (mg)	Calc (mg)	Iron (mg)	Magn (mg)	Pota (mg)	Sodi (mg)	Zinc (mg)	VT-A (µg)	Thia (mg)	VT-E (mg)	Ribo (mg)	Niac (mg)	V-B6 (mg)	Fola (µg)	VT-C (mg)
30	250	1.8		860	180		200*	.15		.25	.4	.2		15
35	300	1.8		660	400		150*	.15		.51	3	.22		1
35	300	5.4		510	340		80*	.15		.68	.3	.18		1
15	150	1.08		400	280		80*	.09		.17	3	.08		0
10	450	2.7		530	270		0	.13		.73				9
10	100	.36		170	75		60*	.03		.14		.06		0
5	60	.72		105	115		40*	.03		.25	.4	.05		0
45	400	1.44		570	230		80*	.12		.59	.8	.19		<1
50	450	1.44		660	260		200*	.15		.68	.8			<1
45	400	1.44		570	230		80*	.12		.59	.8	.19		<1
0	0	0			20		0							0
0	0	0			10		0							0
5	250	1.8		265	160		0	.08		.35				2
5	300	1.44		323	160		0	.08		.45				6
115	250	4.5		426	1130		150*	.29		.49	6.78			4
55	150	3.6		263	850		100*	.29		.33	3.89			4
55	40	1.8		350	760		0	.37		.34	11			0
50	200	2.7		330	1040		0	.3		1.02	12			0
45	40	1.8		280	630		0	.3		.22	3			0
60	100	1.8		290	850		80*	.3		.25	5			0
45	60	2.7		252	630		40*	.29		.25	3.88			4
90	60	4.5		444	680		60*	.32		.45	7.49			6
25	60	1.8		170	730		20*	.22		.14	2			4
40	150	1.8		180	950		60*	.22		.17	2			4
35	60	1.8		262	870		80*	.23		.14	3			4
0	20	.72		678	630		0	.14		.05	3.15			4
0	0	1.44		780	200		0*	.15		.07	3			9
0	20	1.44		120	180		0	.12		.07	.53			0
25	178	3		210	690									
70					1280									
20	111	3		200	460									
35	248	5		350	980									
55	123	3		290	1100									
0	12	1		350	100		0							
190	150	3.6		120	750		80*	.49		.43	3.12			10
240	100	1.8		160	860		80*	.7		.59	5.3			0
225	100	1.8		165	1060		80*	.7		.58	4.5			4
100	100	1.8		310	890		80*	.03		.85	7			6
380	150	4.5		450	1060		150*			.66	5			9
135	250	4.5		530	1570		150*	.27		.54	9.96	.44		9
40	150	3.6		180	840		40*	.24		.24	3.16			0
55	250	2.7		490	1050		80*							2
40	100	2.7		200	770		40*							5
65	150	1.8		340	1060									0
65	200	3.6		250	2140		100*	.49		.4	13.7			9
80	250	4.5		290	1100		80*	.16		.35	6.23			1
105	200	4.5		480	1180		150*	.68		.5	8.36	.34		9
30	100	3.6		155	610		0	.16		.28	2.14			0
75	150	4.5		490	880		100*	.43		.34	2.1			9
105	250	4.5		530	1340		150*	.44		.54	1.96			9
75	300	3.6		500	1270		80*							5

*This value is expressed in retinol equivalents (RE). All other values are in retinol activity equivalents (RAE).

Table A–1

Food Composition   (Computer code number is for Wadsworth Diet Analysis program)   (For purposes of calculations, use "0" for t, <1, <.1, <.01, etc.)

Computer Code Number	Food Description	Measure	Wt (g)	H₂O (%)	Ener (kcal)	Prot (g)	Carb (g)	Dietary Fiber (g)	Fat (g)	Fat Breakdown (g) Sat	Mono	Poly
	FAST FOOD RESTAURANTS—Continued											
1508	Tacos, regular	1 ea	90	66	170	7	12	2	10	3.5		
1509	Tacos, super	1 ea	138	63	270	12	19	4	17	6		
	Teriyaki bowl:											
1679	Beef	1 ea	440	62	640	28	124	7	3	1		
1668	Chicken	1 ea	502	68	670	26	128	3	4	1		
1516	French fries	1 ea	113	40	350	4	46	3	16	4		
1517	Hash browns	1 ea	57	52	170	1	14	1	12	2	9.6	.4
1518	Onion rings	1 ea	120	30	450	7	50	3	25	5	18.9	1.1
	Milkshakes:											
1519	Chocolate	1 ea	332	62	630	11	85	1	27	16		
1520	Strawberry	1 ea	382	67	640	10	85	0	28	15		
1521	Vanilla	1 ea	332	64	610	12	73	0	31	18		
1522	Apple turnover	1 ea	107	40	340	4	41	2	18	4	12.3	1.7
	Source: Jack in the Box Restaurant, Inc											
	KENTUCKY FRIED CHICKEN											
	Rotisserie gold:											
1472	Dark qtr, no skin	1 ea	117	66	217	27	0	0	12	3.5		
1473	Dark qtr, w/skin	1 ea	146	62	333	30	1		24	6.6		
1513	White qtr with wing, w/skin	1 ea	176	65	335	40	1		19	5.4		
1525	White qtr with wing, no skin	1 ea	117	63	199	37	0	0	6	1.7		
	Original Recipe:											
1253	Center breast	1 ea	103	52	260	25	9	<1	14	3.8	7.8	2
1251	Side breast	1 ea	153	53	400	29	16	1	24	6	14.4	3.6
1250	Drumstick	1 ea	61	56	140	13	4	0	9	2	5.3	1.7
1252	Thigh	1 ea	91	54	250	16	6	1	18	4.5	10.2	3.3
1249	Wing	1 ea	47	48	140	9	5	0	10	2.5	5.8	1.7
	Hot & spicy:											
1451	Center breast	1 ea	180	48	505	38	23	1	29	8		
1452	Side breast	1 ea	120	43	400	22	16		28	6		
1430	Thigh	1 ea	107	45	355	19	13	1	26	7		
1471	Wing	1 ea	55	37	210	10	9	1	15	4		
	Extra Crispy Recipe:											
1261	Center breast	1 ea	168	48	470	39	17	1	28	8	16.7	3.3
1259	Side breast	1 ea	116	40	400	21	19	<1	27	5.5	12.9	2.3
1258	Drumstick	1 ea	67	48	195	15	7	1	12	3	7.4	1.6
1260	Thigh	1 ea	118	45	380	21	14	1	27	7	15.8	4.2
1257	Wing	1 ea	55	35	220	10	10	1	15	4	8.9	2.1
1390	Baked beans	½ c	167	71	203	6	35	6	3	1.1	1.4	.7
1526	Breadstick	1 ea	33	30	110	3	17	0	3	0		
1388	Buttermilk biscuit	1 ea	56	37	180	4	20	1	10	2.5	5.4	2.1
1391	Chicken little sandwich	1 ea	47	35	169	6	14	<1	10	2	4.7	3.4
1269	Coleslaw	1 svg	142	70	232	2	26	3	13	2	3.5	7
1527	Cornbread	1 ea	56	26	228	3	25	1	13	2		
1268	Corn-on-the-cob	1 ea	162	73	150	5	35	2	1	0	.6	.9
1429	Chicken, hot wings	1 svg	135	41	471	27	18	2	33	8		
1386	Kentucky fries	1 svg	77	42	228	3	26	3	12	3.2		
1381	Kentucky nuggets	6 ea	95	48	284	16	15	<1	18	4		
1534	Macaroni & cheese	1 svg	153	74	180	7	21	2	8	3		
1387	Mashed potatoes & gravy	1 svg	136	82	120	1	17	2	6	1	3.6	1.4
1530	Pasta salad	1 svg	108	78	135	2	14	1	8	1		
1389	Potato salad	½ c	160	73	230	4	23	3	14	1.7	4.5	7.8
1383	Potato wedges	1 svg	135	65	278	5	28	5	13	4		
1535	Red beans & rice	1 svg	112	76	114	4	18	3	3	1		
1529	Vegetable medley salad	1 ea	114	77	126	1	21	3	4	1		
	Source: Kentucky Fried Chicken Corp											
	LONG JOHN SILVER'S											
1528	Chicken plank dinner, 3 piece	1 ea	399	56	890	32	101		44	9.5	24.8	9.4
1531	Clam chowder	1 ea	198	86	140	11	10	1	6	1.8	2.5	1.7
1532	Clam dinner	1 ea	361	46	990	24	114		52	10.9	31.3	9.9

Chol (mg)	Calc (mg)	Iron (mg)	Magn (mg)	Pota (mg)	Sodi (mg)	Zinc (mg)	VT-A (µg)	Thia (mg)	VT-E (mg)	Ribo (mg)	Niac (mg)	V-B6 (mg)	Fola (µg)	VT-C (mg)
15	100	1.08	40	235	390	1.38	60*	.08		.2	1.15	.15		<1
30	200	1.44	49	365	630	1.8	100*	.13		.09	1.53	.2		2
25	150	4.5		430	930		1000*							6
15	100	4.5		620	1730		1300*							24
0	10	.72		590	710		0	.19		.03	3.94			6
0	10	.18		100	250		0	.05			1			0
0	40	2.7		150	780		40*	.34		.2	3.03			18
85	350	.36		720	330		150*							0
85	350	0		620	300		150*							0
95	400	0		730	320		150*							0
0	10	1.8		85	510		20*	.19		.12	1.75			10
128	10	.18			772		15*							1
163	10	.18			980		15*							1
157	10	.18			1104		15*							1
97	10	.18			667		15*							1
92	30	.72			609		15*	.09		.17	11.5			
135	40	1.08			1116		20*							1
75	20	.72			422		20*							1
95	20	.72			747		20*							1
55	20	.36			414		20*							1
162	60	1.08			1170		20*							1
80	40	1.08			850		15*							6
126	20	.72			630		20*							1
55	20	.72			350		20*							1
160	20	1.08			874		20*							1
75	20	.72			710		15*	.09		.1	8.5			
77	20	.72			375		20*							1
118	20	1.08			625		20*							1
55	20	.36			415		20*							1
5	86	1.93			814		21							1
0	30	.18			15		0*							0
0	20	1.08			560		20*							1
18	23	1.7			331		5*	.16		.12	2.2			
8	30	.36			285		90*							34
42	60	.72			194		10*							
0	20	.36			20		5							4
150	40	1.44			1230		20*							1
4	11	.98			535		0*							0
66	2	.1			865		15*	.02		.02	1	.05		<1
10	150	.36			860		200*							1
1	20	.36			440		20*							1
1	20	1.08			663		110*							7
15	20	2.7			540		90*							1
5	20	1.79			744		5							1
4	10	.72			315									
0	20	.36			240		375*							5
55	200	4.5		1170	2000	3	40*	.52		.51	16			9
20	200	1.8		380	590	.6	150*	.09		.25	2			
75	200	4.5		910	1830	3	40*	.75		.42	12			12

*This value is expressed in retinol equivalents (RE). All other values are in retinol activity equivalents (RAE).

Food Composition (Computer code number is for Wadsworth Diet Analysis program) (For purposes of calculations, use "0" for t, <1, <.1, <.01, etc.)

Computer Code Number	Food Description	Measure	Wt (g)	H₂O (%)	Ener (kcal)	Prot (g)	Carb (g)	Dietary Fiber (g)	Fat (g)	Fat Breakdown (g) Sat	Mono	Poly
	FAST FOOD RESTAURANTS—Continued											
	Fish, batter fried:											
1523	Fish & fryes (fries), 3 piece	1 ea	384	54	980	31	92		50	11.3	28.4	9.7
1524	Fish & fryes, 2 piece	1 ea	261	54	610	27	52		37	7.9	23.5	5.3
2240	Fish and lemon crumb dinner, 3 piece	1 ea	493	71	610	39	86		13	2.2	3.9	5.3
2241	Fish and lemon crumb dinner, 2 piece	1 ea	334	77	330	24	46		5	.9	1.6	1.2
1533	Fish & chicken dinner	1 ea	431	55	950	36	102		49	10.6	28.8	9.5
1537	Shrimp dinner, batter fried	1 ea	331	54	840	18	88		47	9.7	27.2	9.1
	Salads:											
1541	Cole slaw	1 ea	98	70	140	1	20	1	6	1	1.5	3.5
1539	Ocean chef salad	1 ea	234	89	110	12	13	2	1	.4	.4	.2
1540	Seafood salad	1 ea	278	79	380	15	12	2	31	5.1	8.2	17.5
1542	Fryes (fries) serving	1 ea	85	43	250	3	28	1	15	2.5	7.4	5.1
1543	Hush puppies	1 ea	24	40	70	2	10	<1	2	.4	1.3	.2
	Source: Long John Silver's, Lexington KY											
	McDONALD'S											
	Sandwiches:											
1221	Big mac	1 ea	216	51	560	26	45	3	31	10		
1226	Cheeseburger	1 ea	122	46	323	15	35	2	13	6		
1224	Filet-o-fish	1 ea	156	45	450	16	42	2	25	4.5		
1225	Hamburger	1 ea	108	49	262	13	34	2	9	3.5		
1444	McChicken	1 ea	189	52	491	17	42	2	29	5.4	8.5	10.2
1591	McLean deluxe	1 ea	214	64	345	23	37	2	12	4.4	3.6	1.2
1438	McLean deluxe with cheese	1 ea	228	63	398	26	38	2	16	6.8	4.6	1.3
1222	Quarter-pounder	1 ea	171	52	418	23	37	2	21	7.9		
1223	Quarter-pounder with cheese	1 ea	199	50	527	28	38	2	30	12.9		
1227	French fries, small serving	1 ea	68	40	210	3	26	2	10	1.5		
1228	Chicken McNuggets	4 pce	71	51	190	12	10	0	11	2.5		
	Sauces (packet):											
1229	Hot mustard	1 ea	28	60	60	1	7	1	3	0		
1230	Barbecue	1 ea	28	60	45	0	10	0	0	0	0	0
1231	Sweet & sour	1 ea	28	57	50	0	11	0	0	0	0	0
	Low-fat (frozen yogurt) milk shakes:											
1232	Chocolate	1 ea	295	72	360	11	60	1	9	6		
1233	Strawberry	1 ea	294	72	360	11	60	0	9	6		
1234	Vanilla	1 ea	293	72	360	11	59	0	9	6		
	Low-fat (frozen yogurt) sundaes:											
1237	Hot caramel	1 ea	182	56	360	7	61	0	10	6		
1235	Hot fudge	1 ea	179	59	340	8	52	1	12	9		
1267	Strawberry	1 ea	178	63	290	7	50	1	7	5		
1238	Vanilla	1 ea	90	64	150	4	23	0	4	3		
1241	Cookies, McDonaldland	1 ea	42	3	180	3	32	1	5	1		
1242	Cookies, chocolaty chip	1 ea	35	1	170	2	22	1	10	6		
1240	Muffin, apple bran, fat-free	1 ea	114	37	300	6	61	3	3	.5		
1239	Pie, apple	1 ea	77	34	260	3	34	1	13	3.5		
	Breakfast items:											
1243	English muffin with spread	1 ea	63	33	189	5	30	2	6	2.4	1.5	1.3
1244	Egg McMuffin	1 ea	137	57	292	17	27	1	12	4.5		
1245	Hotcakes with marg & syrup	1 ea	228	41	610	9	104	2	18	3.5		
1246	Scrambled eggs	1 ea	102	73	160	13	1	0	11	3.5		
1247	Pork sausage	1 ea	43	45	170	6	0	0	16	5		
1248	Hashbrown potatoes	1 ea	53	55	130	1	14	1	8	1.5		
1392	Sausage McMuffin	1 ea	112	42	360	13	26	1	23	8		
1393	Sausage McMuffin with egg	1 ea	163	52	443	19	27	1	28	10.1		
1394	Biscuit with biscuit spread	1 ea	84	32	290	5	34	1	15	3		
1395	Biscuit with sausage	1 ea	127	37	470	11	35	1	31	9		
1396	Biscuit with sausage & egg	1 ea	178	48	550	18	35	1	37	10		
1397	Biscuit with bacon, egg, cheese	1 ea	157	46	470	18	36	1	28	8		
	Salads:											
1398	Chef salad	1 ea	313	86	206	19	9	3	11	4.2	3	1.2
1400	Garden salad	1 ea	149	88	100	7	4	2	6	3		

Chol (mg)	Calc (mg)	Iron (mg)	Magn (mg)	Pota (mg)	Sodi (mg)	Zinc (mg)	VT-A (µg)	Thia (mg)	VT-E (mg)	Ribo (mg)	Niac (mg)	V-B6 (mg)	Fola (µg)	VT-C (mg)
70	200	4.5		1120	1530	3	40*	.45		.42	8			15
60	40	1.8		900	1480	1.2		.37		.34	8			9
125	200	5.4		990	1420	2.25	700*	.75		.59	24			6
75	80	1.8		440	640	.9	1000*	.3		.25	14			18
75	200	4.5		1280	2090	3	40*	.6		.59	14			9
100	200	3.6		840	1630	3	40*	.45		.42	9			9
15	60	.72		190	260	.6	40*	.06		.07	2			
40	100	3.6		95	730	.3	500*	.12		.14	3			21
55	150	4.5		130	980	.9	200*	.15		.25	3			21
0	200	.72		370	500	.3	0	.09			1.6			6
	40	.72		65	25	.3		.06		.03	.8			
85	250	4.5	45	455	1070	4.8	60*	.49	1.01	.44	6.07	.25	49	4
40	202	2.72	27	281	827	2.62	60*	.33	.46	.31	3.81	.15	24	2
50	150	1.8	34	286	870	.76	40*	.34	1.64	.24	2.78	.07	32	0
30	151	2.73	24	260	585	2.25	22*	.33	.23	.26	3.81	.14	21	2
52	128	2.5	32	319	797	1.06	29*	.91	6.16	.24	7.74	.38	37	1
59	131	4.29	40	537	811	4.9	74*	.42	.63	.34	7.16	.28	44	8
73	139	4.29	43	559	1046	5.26	115*	.42	.85	.39	7.16	.29	47	8
70	149	4.47	33	405	815	4.66	20*	.39	.36	.32	6.78	.24	27	2
94	299	4.48			1283		99*	.39	.81	.43	6.78	.27	33	2
0	9	.36	26	469	135	.32	0	.05	.83	0	1.94	.24	26	9
40	9	.72	17	204	340	.67	0	.08	.94	.11	5.01	.2		0
5	7	.72		27	240		4*	.01		.01	.14			0
0	3	0		45	250		0	.01		.01	.15			4
0	2	.14		7	140		60*	0		.01	.07			0
40	350	.72		543	250		73*	.12		.51	.4	.1		1
40	350	.72		542	180		73*	.12		.51	.4	.11		6
40	350	.36		533	250		73*	.12		.51	.31			1
35	250				180		100*							1
30	250	.72			170		100*							1
30	200	.36			95		100*							1
20	100	.36			75		60*							1
0	20	1.8	8	46	190	.29	0	.18	.74	.12	1.5	.02		0
20	20	1.08	15	89	120	.25	40*	.09	.58	.1	.92			0
0	100	1.44	20	117	380	.5	0	.22	0	.22	2.01	.04	8	1
0	20	1.08	7	63	200	.21		.18	1.38	.11	1.42	.03	8	24
13	103	1.59	13	69	386	.42	33*	.25	.13	.31	2.61	.04	57	1
237	201	2.72	24	199	796	1.56	101*	.49	.85	.44	3.32	.15	33	1
25	150	1.08	28	293	680	.54	80*	.25	1.23	.26	1.91	.09	<1	<1
425	40	1.08	10	126	170	1.06	135	.07	.92	.51	.06	.12	44	0
35	7	.36	7	102	290	.78	0	.18	.26	.06	1.7	.09		0
0	7	.36	11	213	330	.15	0	.08	.58	.02	.9	.08	8	2
45	200	1.8	22	191	740	1.51	40*	.56	.66	.27	3.76	.13	16	0
257	252	2.72	26	251	895	2.06	101*	.59	1.11	.49	3.79	.19	30	0
0	60	1.8	10	116	780	.33	2*	.32	.89	.26	2.46	.03	5	0
35	80	2.7	16	221	1080	1.08	2*	.51	1.14	.31	4.19	.13	5	0
245	100	2.7	21	283	1160	1.69	60*	.53	1.6	.57	4.14	.19	28	0
235	150	2.7	21	253	1250	1.7	100*	.4	1.54	.59	3.43	.13	31	0
179	157	1.81	40	605	727	2.16	1179*	.33	1.45	.37	4.32	.36	100	22

*This value is expressed in retinol equivalents (RE). All other values are in retinol activity equivalents (RAE).

Table A–1

Food Composition   (Computer code number is for Wadsworth Diet Analysis program)   (For purposes of calculations, use "0" for t, <1, <.1, <.01, etc.)

Computer Code Number	Food Description	Measure	Wt (g)	H₂O (%)	Ener (kcal)	Prot (g)	Carb (g)	Dietary Fiber (g)	Fat (g)	Fat Breakdown (g) Sat	Mono	Poly
	FAST FOOD RESTAURANTS—Continued											
1401	Chunky chicken salad	1 ea	296	87	164	23	8	3	5	1.3	1.6	1
	Source: McDonald's Corporation.											
	PIZZA HUT											
	Pan pizza:											
1657	Cheese	2 pce	216	50	569	24	55	4	27	11.8		
1658	Pepperoni	2 pce	208	48	549	22	55	4	27	9.8		
1659	Supreme	2 pce	273	54	657	27	59	6	35	12.3		
1660	Super supreme	2 pce	286	55	680	28	60	6	36	12		
	Thin 'n crispy pizza:											
1649	Cheese	2 pce	174	49	409	20	45	4	18	10.2		
1623	Pepperoni	2 pce	168	50	394	19	44	4	19	8.3		
1622	Supreme	2 pce	232	57	496	24	46	4	26	11.9		
1620	Super supreme	2 pce	247	58	532	25	44	4	28	11.4		
	Hand tossed pizza:											
1619	Cheese	2 pce	216	51	489	24	57	4	20	10.2		
1618	Pepperoni	2 pce	208	52	502	23	50	4	23	10.8		
1648	Supreme	2 pce	273	56	568	32	60	6	24	10		
1617	Super supreme	2 pce	286	56	599	29	64	7	27	11.1		
	Personal pan pizza:											
1610	Pepperoni	1 ea	255	49	615	26	69	5	28	10.9		
1609	Supreme	1 ea	327	57	721	33	70	6	34	12	14.7	5.6
	Source: Pizza Hut.											
	SUBWAY											
	Deli style sandwich:											
69104	Bologna	1 ea	171	64	292	10	38	2	12	4		
69102	Ham	1 ea	171	69	234	11	37	2	4	1		
69103	Roast beef	1 ea	180	69	245	13	38	2	4	1		
69105	Seafood and crab:	1 ea	178	66	298	12	37	2	11	2		
69106	With light mayo	1 ea	178	68	256	12	37	2	7	2		
69108	Tuna:	1 ea	178	63	354	11	37	2	18	3		
69107	With light mayo	1 ea	178	67	279	11	38	2	9	2		
69101	Turkey	1 ea	180	69	235	12	38	2	4	1		
	Sandwiches, 6 inch:											
	B.L.T.:											
69135	On white bread	1 ea	191	67	311	14	38	3	10	3		
69136	On wheat bread	1 ea	198	65	327	14	44	3	10	3		
	Chicken taco sub:											
69131	On white bread	1 ea	286	70	421	24	43	3	16	5		
69132	On wheat bread	1 ea	293	69	436	25	49	4	16	5		
	Club :											
69117	On white bread	1 ea	246	73	297	21	40	3	5	1		
69118	On wheat bread	1 ea	253	71	312	21	46	3	5	1		
	Cold cut trio:											
69113	On white bread	1 ea	246	71	362	19	39	3	13	4		
69114	On wheat bread	1 ea	253	68	378	20	46	3	13	4		
	Ham:											
69115	On white bread	1 ea	232	73	287	18	39	3	5	1		
69116	On wheat bread	1 ea	232	71	293	18	44	3	5	1		
	Italian B.M.T.											
69139	On white bread	1 ea	246	66	445	21	39	3	21	8		
69140	On wheat bread	1 ea	253	64	460	21	45	3	22	7		
	Meatball:											
69129	On white bread	1 ea	260	70	404	18	44	3	16	6		
69130	On wheat bread	1 ea	267	67	419	19	51	3	16	6		
	Melt with turkey, ham, bacon, cheese:											
69127	On white bread	1 ea	251	70	366	22	40	3	12	5		
69128	On wheat bread	1 ea	258	68	382	23	46	3	12	5		

Chol (mg)	Calc (mg)	Iron (mg)	Magn (mg)	Pota (mg)	Sodi (mg)	Zinc (mg)	VT-A (µg)	Thia (mg)	VT-E (mg)	Ribo (mg)	Niac (mg)	V-B6 (mg)	Fola (µg)	VT-C (mg)
75	150	1.08			120		300*							15
76	54	1.62	44	673	318	1.52	1973*	.51	1.28	.2	8.46	.52	83	30
20	393	3.53			1158		295*							5
29	196	3.53			1196		196*							5
41	308	3.69			1375		205*							12
50	300	3.6			1560		200*							12
20	409	2.95			1207		307*							5
31	207	2.99			1265		207*							5
40	297	3.57			1407		198*							18
47	285	3.42			1596		190*							17
20	408	2.93			1324		306*							5
36	359	3.23			1416		269*							43
60	232	4.6	87	589	1769	5.48	192*	.82		.66	8.45			14
44	333	3.99			1618		222*							13
30	298	4.46			1418		248*							6
66	276	5.19	74	603	1757	4.69	240*	.73		.82	9.91	.4		14
20	39	3			744		113*							14
14	24	3			773		113*							14
13	23	3			638		113*							14
17	24	3			544		113*							14
16	24	3			556		118*							14
18	26	3			557		116*							14
16	26	3			583		126*							14
12	26	3			944		113*							14
16	27	3			945		120*							15
16	33	3			957		120*							15
52	118	4			1264		209*		3.08					18
52	124	4			1275		209*		2.83					18
26	29	4			1341		120*							15
26	35	4			1352		120*							15
64	49	4			1401		130*							16
64	55	4			1412		130*							16
28	28	3			1308		120*							15
27	34	2.91			1280		117*							15
56	44	4			1652		151*							15
56	50	4			1664		151*							15
33	32	4			1035		142*							16
33	39	4			1046		142*							16
42	93	4			1735		155*							15
42	100	3			1746		156*							15

*This value is expressed in retinol equivalents (RE). All other values are in retinol activity equivalents (RAE).

Table A–1

Food Composition

(Computer code number is for Wadsworth Diet Analysis program)   (For purposes of calculations, use "0" for t, <1, <.1, <.01, etc.)

Computer Code Number	Food Description	Measure	Wt (g)	H₂O (%)	Ener (kcal)	Prot (g)	Carb (g)	Dietary Fiber (g)	Fat (g)	Fat Breakdown (g)		
										Sat	Mono	Poly
	FAST FOOD RESTAURANTS—Continued											
	Pizza sub:											
69133	On white bread	1 ea	250	66	448	19	41	3	22	9		
69134	On wheat bread	1 ea	257	65	464	19	48	3	22	9		
	Roast beef:											
69121	On white bread	1 ea	232	72	288	19	39	3	5	1		
69122	On wheat bread	1 ea	239	70	303	20	45	3	5	1		
	Roasted chicken breast:											
69125	On white bread	1 ea	246	70	332	26	41	3	6	1		
69126	On wheat bread	1 ea	253	68	348	27	47	3	6	1		
	Seafood and crab:											
69145	On white bread:	1 ea	246	69	415	19	38	3	19	3		
69147	With light mayo	1 ea	246	72	332	19	39	3	10	2		
69146	On wheat bread:	1 ea	253	67	430	20	44	3	19	3		
69148	With light mayo	1 ea	253	70	347	20	45	3	10	2		
	Spicy italian:											
69123	On white bread	1 ea	232	64	467	20	38	3	24	9		
69124	On wheat bread	1 ea	239	62	482	21	44	3	25	9		
	Steak and cheese:											
69119	On white bread	1 ea	257	68	383	29	41	3	10	6		
69120	On wheat bread	1 ea	264	67	398	30	47	3	10	6		
	Tuna:											
69141	On white bread:	1 ea	246	62	527	18	38	3	32	5		
69143	with light mayo	1 ea	246	70	376	18	39	3	15	2		
69142	On wheat bread:	1 ea	253	62	542	19	44	3	32	5		
69144	With light mayo	1 ea	253	68	391	19	46	3	15	2		
	Turkey:											
69111	On white bread	1 ea	232	73	273	17	40	3	4	1		
69112	On wheat bread	1 ea	239	71	289	18	46	3	4	1		
	Turkey breast and ham:											
69137	On white bread	1 ea	232	73	280	18	39	3	5	1		
69138	On wheat bread	1 ea	239	71	295	18	46	3	5	1		
	Veggie delite:											
69109	On white bread	1 ea	175	71	222	9	38	3	3	0		
69110	On wheat bread	1 ea	182	69	237	9	44	3	3	0		
	Salads:											
52128	B.L.T.	1 ea	276	91	140	7	10	2	8	3		
52124	B.M.T., classic italian	1 ea	331	86	274	14	11	1	20	7		
52127	Chicken taco	1 ea	370	87	250	18	15	2	14	5		
52115	Club	1 ea	331	91	126	14	12	1	3	1		
52120	Cold cut trio	1 ea	330	89	191	13	11	1	11	3		
52123	Ham	1 ea	316	91	116	12	11	1	3	1		
52129	Meatball	1 ea	345	88	233	12	16	2	14	5		
52131	Melt	1 ea	336	88	195	16	12	1	10	4		
52121	Pizza	1 ea	335	86	277	12	13	2	20	8		
52126	Roast beef	1 ea	316	92	117	12	11	1	3	1		
52119	Roasted chicken breast	1 ea	331	89	162	20	13	1	4	1		
52117	Seafood and crab:	1 5	331	88	244	13	10	2	17	3		
52116	With light mayo	1 5	331	90	161	13	11	2	8	1		
52130	Steak and cheese	1 ea	342	87	212	22	13	1	8	5		
52122	Tuna:	1 ea	331	84	356	12	10	1	30	5		
52118	With light mayo	1 ea	331	89	205	12	11	1	13	2		
52114	Turkey breast	1 ea	316	92	102	11	12	1	2	1		
52125	With ham	1 ea	316	92	109	11	11	1	3	1		
52113	Veggie delite	1 ea	260	94	51	2	10	1	1	0		
	Cookies:											
47662	Brazil nut and chocolate chip	1 ea	48	12	229	3	27	1	12	3.5		
47655	Chocolate chip:	1 ea	48	14	209	2	29	1	10	3.5		
47658	With M&M's	1 ea	48	14	209	2	29	1	10	3		
47659	Chocolate chunk	1 ea	48	14	209	2	29	1	10	3.5		
47656	Oatmeal raisin	1 ea	48	15	199	3	29	1	8	2		
47657	Peanut butter	1 ea	48	13	219	3	26	1	12	2.5		
47660	Sugar	1 ea	48	11	229	2	28	0	12	3		

Table of Food Composition

PAGE KEY: A–2 = Beverages   A–4 = Dairy   A–8 = Eggs   A–10 = Fat/Oil   A–12 = Fruit   A–18 = Bakery   A–24 = Grain
A–30 = Fish   A–32 = Meats   A–36 = Poultry   A–38 = Sausage   A–38 = Mixed/Fast   A–44 = Nuts/Seeds   A–46 = Sweets
A–50 = Vegetables/Legumes   A–60 = Vegetarian   A–62 = Misc   A–64 = Soups/Sauces   A–66 = Fast   A–80 = Convenience   A–86 = Baby foods

Chol (mg)	Calc (mg)	Iron (mg)	Magn (mg)	Pota (mg)	Sodi (mg)	Zinc (mg)	VT-A (µg)	Thia (mg)	VT-E (mg)	Ribo (mg)	Niac (mg)	V-B6 (mg)	Fola (µg)	VT-C (mg)
50	103	4			1609		238*							16
50	110	3			1621		238*							16
20	25	4			928		120*							15
20	32	3			939		120*							15
48	35	3			967		123*							15
48	42	3			978		123*							15
34	28	3			849		121*							15
32	28	3			873		131*							15
34	34	3			860		121*							15
32	34	3			884		131*							15
57	40	4			1592		169*							15
57	47	4			1604		169*							15
70	88	5			1106		175*							18
70	95	5			1117		176*							18
36	32	3			875		125*							15
32	32	3			928		146*							15
36	38	3			886		126*							15
32	38	3			940		146*							15
19	30	4			1391		120*							15
19	37	3			1403		120*							15
24	29	3			1350		120*							15
24	36	3			1361		120*							15
0	25	3			582		120*							15
0	32	3			593		120*							15
16	24	1			672		273*							32
56	41	2			1379		303*							32
52	115	3			990		361*							35
26	26	2			1067		273*							32
64	46	2			1127		282*							33
28	25	2			1034		273*							32
33	30	2			761		295*							33
42	90	2			1461		308*							32
50	100	2			1336		390*							33
20	23	2			654		273*							32
48	32	2			693		276*							32
34	25	2			575		273*							32
32	25	2			599		284*							32
70	86	3			832		328*							35
36	29	2			601		278*							32
32	29	2			654		298*							32
19	28	2			1117		273*							32
24	27	2			1076		273*							32
0	23	1			308		136*							32
10	32	1.99			115		0							0
10	16	1.99			139		0							0
15	16	1			139		0							0
10	16	1			139		0							0
15	32	1			159		0							0
0	16	1			179		0							0
20	0	.72			179		0							0

*This value is expressed in retinol equivalents (RE). All other values are in retinol activity equivalents (RAE).

Table A–1

Food Composition (Computer code number is for Wadsworth Diet Analysis program) (For purposes of calculations, use "0" for t, <1, <.1, <.01, etc.)

Computer Code Number	Food Description	Measure	Wt (g)	H₂O (%)	Ener (kcal)	Prot (g)	Carb (g)	Dietary Fiber (g)	Fat (g)	Sat	Mono	Poly
	FAST FOOD RESTAURANTS—Continued											
47661	White chip macadamia	1 ea	48	12	229	2	28	1	12	2.5		
	Source: Subway International											
	TACO BELL											
	Breakfast burrito:											
1601	Bacon breakfast burrito	1 ea	99	48	291	11	23		17	4		
1627	Country breakfast burrito	1 ea	113	55	220	8	26	2	14	5		
1626	Fiesta breakfast burrito	1 ea	92	44	280	9	25	2	16	6		
1625	Grande breakfast burrito	1 ea	177	56	420	13	43	3	22	7		
1604	Sausage breakfast burrito	1 ea	106	49	303	11	23		19	6		
	Burritos:											
1544	Bean with red sauce	1 ea	198	58	380	13	55	13	12	4		
1545	Beef with red sauce	1 ea	198	57	432	22	42	4	19	8	6.7	.7
1546	Beef & bean with red sauce	1 ea	198	57	412	17	50	5	16	6	6.1	2.1
1569	Big beef supreme	1 ea	298	64	520	24	54	11	23	10		
1552	Chicken burrito	1 ea	171	58	345	17	41		13	5		
1547	Supreme with red sauce	1 ea	255	64	440	17	51	10	19	8		
1571	7 layer burrito	1 ea	283	61	530	16	66	13	23	7		
1538	Chilito	1 ea	156	49	391	17	41		18	9		
1549	Chilito, steak	1 ea	257	62	496	26	47		23	10		
	Tacos:											
1551	Taco	1 ea	78	58	180	9	12	3	10	4		
1554	Soft taco	1 ea	90	63	220	9	12	3	10	4		
1536	Soft taco supreme	1 ea	142	64	260	12	23	3	14	7		
1568	Soft taco, chicken	1 ea	121	63	200	14	21	2	7	2.5		
1572	Soft taco, steak	1 ea	128	63	230	15	20	2	10	2.5		
1555	Tostada with red sauce	1 ea	177	67	300	10	31	12	15	5		
1558	Mexican pizza	1 ea	220	53	570	21	42	8	35	10		
1559	Taco salad with salsa	1 ea	539	71	850	30	65	16	52	15		
1560	Nachos, regular	1 ea	99	40	320	5	34	3	18	4		
1561	Nachos, bellgrande	1 ea	312	51	770	21	84	17	39	11		
1562	Pintos & cheese with red sauce	1 ea	120	68	190	9	18	10	9	4		
1563	Taco sauce, packet	1 ea	11	94	3	<1	<1	<1	<1			
1564	Salsa	1 ea	10	28	27	1	6		<1	0	0	0
1565	Cinnamon twists	1 ea	28	6	140	1	19	0	6	0		
1628	Caramel roll	1 ea	85	19	353	6	46		16	4		
	Source: Taco Bell Corporation											
	WENDY'S											
	Hamburgers:											
1566	Single on white bun, no toppings	1 ea	133	44	358	24	31	1	15	6.1		
1570	Cheeseburger, bacon	1 ea	166	55	384	20	34	2	19	7.3	6.7	2.6
1730	Chicken sandwich, grilled	1 ea	189	64	303	24	36	2	7	1.6	.8	1.9
	Baked potatoes:											
1573	Plain	1 ea	284	71	310	7	71	7	0	0	0	0
1574	With bacon & cheese	1 ea	380	69	525	16	78	7	17	4	5.9	7.2
1575	With broccoli & cheese	1 ea	411	74	478	8	80	7	14	2.7	4.3	6.8
1576	With cheese	1 ea	383	68	571	14	78	7	23	8.4	7.1	7.2
1577	With chili & cheese	1 ea	439	69	625	20	83	9	24	8.9	6.3	7.1
1578	With sour cream & chives	1 ea	439	71	515	10	102	10	8	5.2	2	.4
1579	Chili	1 ea	227	79	206	15	21	5	7	2.4		
1582	Chocolate chip cookies	1 ea	57	6	270	3	36	1	13	6		
1580	French fries	1 ea	130	41	390	5	50	5	19	3	11.9	2.4
1581	Frosty dairy dessert	1 ea	227	67	333	8	56	0	8	5.4	2.1	.3
	Source: Wendy's International											
	CONVENIENCE FOODS and MEALS											
	BUDGET GOURMET											
1695	Chicken cacciatore	1 ea	312	80	300	20	27		13			

PAGE KEY: A–2 = Beverages   A–4 = Dairy   A–8 = Eggs   A–10 = Fat/Oil   A–12 = Fruit   A–18 = Bakery   A–24 = Grain
A–30 = Fish   A–32 = Meats   A–36 = Poultry   A–38 = Sausage   A–38 = Mixed/Fast   A–44 = Nuts/Seeds   A–46 = Sweets
A–50 = Vegetables/Legumes   A–60 = Vegetarian   A–62 = Misc   A–64 = Soups/Sauces   A–66 = Fast   A–80 = Convenience   A–86 = Baby foods

Table of Food Composition   **A–81**

Chol (mg)	Calc (mg)	Iron (mg)	Magn (mg)	Pota (mg)	Sodi (mg)	Zinc (mg)	VT-A (µg)	Thia (mg)	VT-E (mg)	Ribo (mg)	Niac (mg)	V-B6 (mg)	Fola (µg)	VT-C (mg)
10	16	1			139		0							0
181	80	1.8			652		310*							
195	80	1.08			690		250*							0
25	80	.72			580		150*							0
205	100	1.8			1050		500*							0
183	80	1.8			661		320*							
10	150	2.7		495	1100		450*	.04		2.02	1.98	.31		0
57	160	3.96		380	1303		530*	.4		2.14	3.44	.32		1
32	170	3.78	50	442	1221	2.67	450*	.49		.41	3.09	.59	38	1
55	150	2.7			1520		600*							5
57	140	2.52			854		440*							1
35	150	9	50	422	1230		500*	.4		2.1	2.89	.35		5
25	200	3.6			1280		300*							6
47	300	3.06			980		950*							
78	200	2.7			1313		970*							2
25	80	1.08		159	330		100*	.05		.14	1.2	.12		0
25	80	1.08		192	330		100*	.38		.22	2.68	.98		0
35	100	1.8			590		150*							4
35	80	.72			540		60*							1
25	80	1.44			1020		40*							0
15	150	1.8		455	650		500*	.06		.19	.71	.29		1
45	250	3.6	79	403	1040	5.3	400*	.32		.33	2.92	1.1	59	5
60	300	6.3		966	1780	1.54	1600*	.47		.7	4.42	.51	9	24
5	100	.72		149	570	1.57	60*	.16		.15	.64	.18	9	0
35	200	3.6		733	1310		150*	.11		.37	2.36			4
15	150	1.8	103	360	650	2.03	250*	.05		.14	.4	.2	64	0
0	0	.08		11	92		37*	0			.02			<1
0	50	.6		376	709		168*	.02		.14	0			10
0	0	.36		22	190		40*	.08		.03	.57	.03		0
15	60	1.44			312		330*							4
65	110	4.13		296	580		0	.43		.38	6.7			3
57	167	3.49	38	307	874	5.94	76*	.3		.31	6.43		28	9
65	89	2.92		428	746		43*							10
0	20	3.6	75	1187	25	.74	0	.31	.14	.12	4.3	.8	31	36
26	116	4.36	87	1385	821	2.75	103*	.24		.19	5.04	.94	36	39
5	112	6.24	93	1266	471	.97	105*	.34		.29	4.5	.97	74	79
33	312	4.06	85	1293	641	.67	172*	.25		.28	3.6	.88	36	39
41	301	5.07	122	1435	776	4.15	201*	.33		.29	4.5	.99	55	39
20	89	5.5	99	1723	108	1.28	49*	.32		.19	4.25	1.11	45	54
29	83	2.86		488	797		82*	.11		.15	2.66			4
30	10	1.8	13	89	120	.41	0*	.05		.06	.36	.03	5	0
0	20	1.08	55	845	120	.62	0*	.18		.04	3.6	.33	40	6
37	311	1.11	46	585	197	.97	163*	.11		.47	.32	.13	17	0
60	150	1.8			810		40*	.23		.51	5			21

*This value is expressed in retinol equivalents (RE). All other values are in retinol activity equivalents (RAE).

Table A–1

Food Composition   (Computer code number is for Wadsworth Diet Analysis program)   (For purposes of calculations, use "0" for t, <1, <.1, <.01, etc.)

Computer Code Number	Food Description	Measure	Wt (g)	H₂O (%)	Ener (kcal)	Prot (g)	Carb (g)	Dietary Fiber (g)	Fat (g)	Fat Breakdown (g) Sat	Mono	Poly
	CONVENIENCE FOODS and MEALS—Continued											
1692	Linguini & shrimp	1 ea	284		364	14	55	3	10	3.1		
1691	Scallops & shrimp	1 ea	326	79	320	16	43		9			
2245	Seafood newburg	1 ea	284	74	350	17	43		12			
1693	Sirloin tips with country gravy	1 ea	284	80	310	16	21		18			
1694	Sweet & sour chicken with rice	1 ea	284		354	14	64	2	4	1.2		
1689	Teriyaki chicken	1 ea	340	77	360	20	44		12			
1690	Veal parmigiana	1 ea	340	75	440	26	39		20			
1696	Yankee pot roast	1 ea	312	77	380	27	22		21			
	Source: The All American Gourmet Co.											
	HAAGEN DAZS											
1755	Ice cream bar, vanilla almond	1 ea	87	42	304	5	21	1	22	11.5	8.2	2.5
	Sorbet:											
1758	Lemon	½ c	113	72	120	0	31	<1	0	0	0	0
1760	Orange	½ c	113	68	140	0	36		0	0	0	0
1759	Raspberry	½ c	105	71	120	0	30	2	0	0	0	0
	Yogurt, frozen:											
1753	Chocolate	½ c	98		171	8	26		4		2	
1754	Strawberry	½ c	98		171	6	27		4		2	
	Yogurt extra, frozen:											
1752	Brownie nut	½ c	101		220	8	29		9		5	
1751	Raspberry rendezvous	½ c	101		132	4	26		2		1	
	Source: Pillsbury											
	HEALTHY CHOICE											
	Entrees:											
2112	Fish, lemon pepper	1 ea	303	78	320	14	50	5	7	2		
1624	Lasagna	1 ea	383	75	420	26	59	6	9	3		
2111	Meatloaf, traditional	1 ea	340	78	316	15	52	6	5	2.5	1.9	.6
2104	Zucchini lasagna	1 ea	396	83	290	13	49	5	4	2.6		
2110	Dinner, pasta shells marinara	1 ea	340	74	380	25	55	5	6	3.5		
	Low-Fat ice cream:											
973	Brownie	½ c	71	60	120	3	22	1	2	1	.3	.7
259	Butter pecan	½ c	71	60	120	3	22	1	2	1	.3	.7
650	Chocolate chip	½ c	71	62	120	3	21	1	2	1	1	0
1608	Cookie & cream	½ c	71	62	120	3	21	1	2	1	1	0
45	Rocky road	½ c	71	52	140	3	28	1	2	1	1	0
1621	Vanilla	½ c	71	66	100	3	18	1	2	1	1	0
391	Vanilla fudge	½ c	71	62	120	3	21	1	2	1.5		
	Source: ConAgra Frozen Foods, Omaha, NE											
	HEALTH VALLEY											
	Soups, fat-free:											
2001	Beef broth, no salt added	1 c	240	98	18	5	0	0	0	0	0	0
2073	Beef broth, w/salt	1 c	240	98	30	5	2	0	0	0	0	0
2016	Black bean & vegetable	1 c	240	85	110	11	24	12	0	0	0	0
2017	Chicken broth	1 c	240	97	30	6	0	0	0	0	0	0
2018	14 garden vegetable	1 c	240	90	80	6	17	4	0	0	0	0
2015	Lentil & carrot	1 c	240	85	90	10	25	14	0	0	0	0
2014	Split pea & carrot	1 c	240	89	110	8	17	4	0	0	0	0
2013	Tomato vegetable	1 c	240	90	80	6	17	5	0	0	0	0
	Source: Health Valley											
	LA CHOY											
2100	Egg rolls, mini, chicken	1 svg	106		108	3	13	1	5	1.3		
2099	Egg rolls, mini, shrimp	1 svg	106		98	3	14	1	3	.8		
	Source: Beatrice/Hunt-Wesson											

Chol (mg)	Calc (mg)	Iron (mg)	Magn (mg)	Pota (mg)	Sodi (mg)	Zinc (mg)	VT-A (µg)	Thia (mg)	VT-E (mg)	Ribo (mg)	Niac (mg)	V-B6 (mg)	Fola (µg)	VT-C (mg)
31	50	2.26		565			75*							5
70	150	.72		690			150*			.26	3			12
70	100	.72		660			40*	.23		.26	2			
40	60	.36		570			150*	.15		.17	4	.28		2
24	71	3.18		519			354*							25
55	80	1.4		610			300*	.15		.34	6			12
165	30	4.5		1160			1000*	.45		.6	6			6
70	150	1.8		690			600*	.15		.43	7			6
74	123	.59		180	66		82*			.15				0
0	0	0		30	5		0							4
0													0	
0	0	0		56	0		0							2
40				147			20*				.17			
50				147			20*			.03	.17			
55				152			20*				.14			
20				61			0*	0			.1			
30	20	1.08			480		100*							30
35	150	3.6		500	580		100*	.3		.26	2			6
37	48	2.24			459		149*							55
10	207	1.86			321		258*							0
25	400	1.8			390		100*							0
5	100	0		268	55		40*							0
2	100	0		211	60		40*							0
5	100	0		240	50		40*							0
5	100	0		254	90		40*	.03		.15				0
5	100	0		168	60		40*	.03		.15				0
5	100	0		254	50		60*	.05		.22				0
2	100	0		296	50		40*							0
0				196	74						.98			
0	0	0		196	160		0				.98			5
0	40	3.6		676	280		2000*	.34		.11	1.35	.22	135	9
0	20	1.8		147	170		0			.03	2.45			1
0	40	1.8		406	250		2000*	.26		.08	2.25	.18	27	15
0	60	5.4		439	220		2000*	.1		.16	5.63	.45	27	2
0	40	5.4		439	230		2000*	.1		.16	5.63	.45		9
0	40	5.4		609	240		2000*	.1		.08	2.25	.13	<1	9
8	10	.56			335		10*							0
5	10	.56			377		52*							0

*This value is expressed in retinol equivalents (RE). All other values are in retinol activity equivalents (RAE).

Table A–1

Food Composition (Computer code number is for Wadsworth Diet Analysis program) (For purposes of calculations, use "0" for t, <1, <.1, <.01, etc.)

A

Computer Code Number	Food Description	Measure	Wt (g)	H₂O (%)	Ener (kcal)	Prot (g)	Carb (g)	Dietary Fiber (g)	Fat (g)	Fat Breakdown (g) Sat	Mono	Poly
	CONVENIENCE FOODS and MEALS—Continued											
	LEAN CUISINE											
	Dinners:											
1639	Baked cheese ravioli	1 ea	241	76	260	12	38	4	7	3.5	1.5	.5
1632	Chicken chow mein	1 ea	255	78	240	14	37	3	3	1	1.5	.5
1633	Lasagna	1 ea	291	78	293	19	37	4	8	4.4	1.9	1
1634	Macaroni & cheese	1 ea	255	77	261	13	38	2	6	3.6	1.3	.4
1631	Spaghetti w/meatballs	1 ea	269	75	299	18	39	5	8	2.1	2.7	1.3
	Pizza:											
1636	French bread sausage pizza	1 ea	170	75	210	8	24	1	9	3.5		
	Source: Stouffer's Foods Corp, Solon OH											
	TASTE ADVENTURE SOUPS											
1905	Black bean	1 c	242	8	807	51	148	36	4	.9	.3	1.5
1904	Curry lentil	1 c	241	4	795	66	135	71	3	.4	.5	1.2
1906	Lentil chili	1 c	242		411	24	75	14	1			
1903	Split pea	1 c	244	4	807	58	143	60	3	.4	.6	1.2
	Source: Taste Adventure Soups											
	WEIGHT WATCHERS											
	Cheese, fat-free slices:											
1978	Cheddar, sharp	2 pce	21	65	30	5	2	0	0	0	0	0
1980	Swiss	2 pce	21	65	30	5	2	0	0	0	0	0
1977	White	2 pce	21	65	30	5	2	0	0	0	0	0
1979	Yellow	2 pce	21	65	30	5	2	0	0	0	0	0
	Dinners:											
2029	Chicken chow mein	1 ea	255	81	200	12	34	3	2	.5		
1646	Oven fried fish	1 ea	218	78	230	15	25	2	8	2.1	4.2	1.7
1972	Margarine, reduced fat	1 tbsp	14	49	59	0	0	0	7	1.5		
	Pizza:											
1653	Cheese	1 ea	163	48	390	23	49	6	12	4	3	1
1650	Deluxe combination pizza	1 ea	186	56	380	23	47	6	11	3.5	5	2
1652	Pepperoni pizza	1 ea	158	48	390	23	46	4	12	4	5	2
	Desserts:											
1644	Chocolate brownie	1 ea	91	75	95	3	17	2	2	.5	1	.5
2024	Chocolate eclair	1 ea	60	48	143	2	24	1	4	.8	1.1	1.9
2247	Chocolate mousse	1 ea	78	44	190	6	33	3	4	1.5		
1642	Strawberry cheesecake	1 ea	111	62	180	7	28	2	5	2	1	2
2027	Triple chocolate cheesecake	1 ea	89	52	199	7	32	1	5	2.5		
	Source: Weight Watchers											
	SWEET SUCCESS:											
	Drinks, prepared:											
1776	Chocolate chip	1 c	265	81	180	15	30	6	3	1.6		
1777	Chocolate fudge	1 c	265	81	180	15	30	6	2			
1774	Chocolate mocha	1 c	265	81	180	15	30	6	1	1		
1778	Milk chocolate	1 c	265	81	180	15	30	6	2	1		
1775	Vanilla	1 c	265	81	180	15	33	6	1	.6		
	Drinks, ready to drink:											
2147	Chocolate mint	1 c	265	82	167	10	32	5	3	0		
2148	Strawberry	1 c	265	82	167	10	32	5	3	0		
	Shakes:											
1771	Chocolate almond	1 c	250	82	158	9	30	5	2	0	2.1	.2
1773	Chocolate fudge	1 c	250	82	158	9	30	5	2	0	2.1	.2
1768	Chocolate mocha	1 c	250	82	158	9	30	5	2	0	.6	1.8
1769	Chocolate raspberry truffle	1 c	250	82	158	9	30	5	2	0	2.2	.2
1770	Vanilla creme	1 c	250	82	158	9	30	5	2	0	2.1	.3
	Snack bars:											
1767	Chocolate brownie	1 ea	33	9	120	2	23	3	4	2	.5	.6

Chol (mg)	Calc (mg)	Iron (mg)	Magn (mg)	Pota (mg)	Sodi (mg)	Zinc (mg)	VT-A (µg)	Thia (mg)	VT-E (mg)	Ribo (mg)	Niac (mg)	V-B6 (mg)	Fola (µg)	VT-C (mg)
35	150	1.44	42	450	590	1.5	100*	.06		.25	1.2	.2	48	5
35	40	.72	30	300	590	1.1	20*	.15		.17	5			0
29	244	1.41	44	596	576	2.9	98*	.15		.25	3	.32		6
18	180	.65		423	567		0	.12		.25	1.2			0
5	94	2.37		539	465		0							6
10	100	2.7	39	165	630	2.2	30*	.45		.51	.05	.07		1
0	296	12.2	405	3521	1978	8.67	39	2.12	.12	.47	4.8	.84	1042	1
0	140	22	256	2169	2182	8.51	48	1.12	.71	.59	6.33	1.25	1005	16
				1476	1016									
0	140	10.7	272	2324	1728	7.14	19	1.71	2.75	.51	6.84	.42	646	5
0	99	0		64	306		56*							0
0	99	0		74	276		56*							0
0	99	0		64	306		56*							0
0	99	0		64	306		56*							0
25	40	.72		360	430		300*							36
25	20	1.44		370	450		40*	.09		.14	1.6			0
0	0	0		5	128		49*							0
35	700	1.8		290	590		80*	.3		.51	3	.06		6
40	500	3.6		370	550		150*	.3		.51	3	.2		5
45	450	1.8		320	650		80*	.23		.51	3			5
2	40	.54		115	80		0	.03		.01	.1	.01		0
31					189		0							
5	60	1.8		320	150		0							0
15	80	.36		115	230		40*	.06		.07	1.6			2
10	80	1.08		169	199		0							0
6	500	6.3	140	600	288	5.25	350*	.52	7.05	.59	7	.7	140	21
6	500	6.3	140	750	336	5.25	350*	.52	7.05	.59	7	.7	140	21
6	500	6.3	140	800	336	5.25	350*	.52	7.05	.59	7	.7	140	21
6	500	6.3	140	750	336	5.25	350*	.52	7.05	.59	7	.7	140	21
6	500	6.3	140	830	312	5.25	250*	.52	7.05	.59	7	.7	140	21
5	419	5.3	117	469	201	4.51	294*	.45	5.86	.5	5.83	.58	117	17
5	419	5.3	117	310	175	4.51	294*	.45	5.86	.5	5.83	.58	117	17
5	396	5	110	443	190	4.25	277*	.42	7.5	.47	5.5	.55	110	16
5	396	5	110	443	175	4.25	277*	.42	7.5	.47	5.5	.55	110	16
5	396	5	110	403	175	4.25	277*	.42	7.5	.47	5.5	.55	110	16
5	383	5	110	443	175	4.25	277*	.42	7.5	.47	5.5	.55	110	16
5	396	5	110	293	175	4.25	277*	.42	7.5	.47	5.5	.55	110	16
3	150	2.71	60	140	45	.59	150*	.22	3.01	.25	3	.3	60	9
3	150	2.71	60	110	40	.59	150*	.22	3.01	.25	3	.3	60	9

*This value is expressed in retinol equivalents (RE). All other values are in retinol activity equivalents (RAE).

Table A–1

Food Composition (Computer code number is for Wadsworth Diet Analysis program)   (For purposes of calculations, use "0" for t, <1, <.1, <.01, etc.)

Computer Code Number	Food Description	Measure	Wt (g)	H₂O (%)	Ener (kcal)	Prot (g)	Carb (g)	Dietary Fiber (g)	Fat (g)	Fat Breakdown (g) Sat	Mono	Poly
	CONVENIENCE FOODS and MEALS—Continued											
1766	Chocolate chip	1 ea	33	9	120	2	23	3	4	2	.4	.5
1921	Oatmeal raisin	1 ea	33	9	120	2	23	3	4	2		
1765	Peanut butter	1 ea	33	9	120	2	23	3	4	2	.6	.6
	Source: Foodway National Inc, Boise, ID											
	BABY FOODS											
1720	Apple juice	½ c	125	88	59	0	15	<1	<1	t	t	t
1721	Applesauce, strained	1 tbsp	16	89	7	<1	2	<1	<1	t	t	t
1716	Carrots, strained	1 tbsp	14	92	4	<1	1	<1	<1	t	t	t
1718	Cereal, mixed, milk added	1 tbsp	15	75	17	1	2	<1	1	.3		
1719	Cereal, rice, milk added	1 tbsp	15	75	17	1	3	<1	1	.3		
1723	Chicken and noodles, strained	1 tbsp	16	86	11	<1	1	<1	<1	.1	.1	.1
1722	Peas, strained	1 tbsp	15	87	6	1	1	<1	<1	t	t	t
1717	Teething biscuits	1 ea	11	6	43	1	8	<1	<1	.2	.2	.1

Chol (mg)	Calc (mg)	Iron (mg)	Magn (mg)	Pota (mg)	Sodi (mg)	Zinc (mg)	VT-A (µg)	Thia (mg)	VT-E (mg)	Ribo (mg)	Niac (mg)	V-B6 (mg)	Fola (µg)	VT-C (mg)
3	150	2.71	60		30	.59	150*	.22	3.01	.25	3	.3	60	9
3	150	2.71	60	125	35	.59	150*	.22	3.01	.25	3	.3	60	9

Chol (mg)	Calc (mg)	Iron (mg)	Magn (mg)	Pota (mg)	Sodi (mg)	Zinc (mg)	VT-A (µg)	Thia (mg)	VT-E (mg)	Ribo (mg)	Niac (mg)	V-B6 (mg)	Fola (µg)	VT-C (mg)
0	5	.71	4	114	4	.04	1	.01	.75	.02	.1	.04	0	72
0	1	.03		11	<1	<.01	<1*	<.01	.1	<.01	.01	<.01	<1	6
0	3	.05	1	27	5	.02	160*	<.01	.07	.01	.06	.01	2	1
2	33	1.56	4	30	7	.11	4*	.06		.09	.87	.01	2	<1
2	36	1.83	7	28	7	.1	4*	.07		.07	.78	.02	1	<1
3	4	.1	2	22	4	.09	17	.01	.03	.01	.12	.01	2	<1
0	3	.14	2	17	1	.05	8*	.01	.08	.01	.15	.01	4	1
0	29	.39	4	35	40	.1	1*	.03	.05	.06	.48	.01	5	1

*This value is expressed in retinol equivalents (RE). All other values are in retinol activity equivalents (RAE).

UNITED STATES: EXCHANGE LISTS

◆ Appendix C presents Canada's Choice System for Meal Planning.

Chapter 24 introduced the exchange system. ◆ This appendix provides additional details. The exchange system groups together foods that have about the same amount of carbohydrate, protein, fat, and kcalories. Then any food on a list can be "exchanged" for any other on that same list (see Tables B-1 through B-9).

TABLE B-1 U.S. Exchange System: Starch List

1 starch exchange = 15 g carbohydrate, 3 g protein, 0–1 g fat, and 80 kcal
Note: In general, a starch serving is ½ c cereal, grain, pasta, or starchy vegetable; 1 oz of bread; ¾ to 1 oz snack food.

Serving Size	Food
Bread	
½ (1 oz)	Bagels
2 slices (1½ oz)	Bread, reduced-kcalorie
1 slice (1 oz)	Bread, white (including French and Italian), whole-wheat, pumpernickel, rye
2 (⅔ oz)	Bread sticks, crisp, 40 x ½"
½	English muffins
½ (1 oz)	Hot dog or hamburger buns
½	Pita, 6" across
1 (1 oz)	Plain rolls, small
1 slice (1 oz)	Raisin bread, unfrosted
1	Tortillas, corn, 6" across
1	Tortillas, flour, 7–8" across
1	Waffles, 4½" square, reduced-fat
Cereals and Grains	
½ c	Bran cereals
½ c	Bulgur, cooked
½ c	Cereals, cooked
¾ c	Cereals, unsweetened, ready-to-eat
3 tbs	Cornmeal (dry)
⅓ c	Couscous
3 tbs	Flour (dry)
¼ c	Granola, low-fat
¼ c	Grape nuts
½ c	Grits, cooked
½ c	Kasha
¼ c	Millet
¼ c	Muesli
½ c	Oats
½ c	Pasta, cooked
1½ c	Puffed cereals
½ c	Rice milk
⅓ c	Rice, white or brown, cooked
½ c	Shredded wheat
½ c	Sugar-frosted cereal
3 tbs	Wheat germ
Starchy Vegetables	
⅓ c	Baked beans
½ c	Corn
1 (5 oz)	Corn on cob, medium
1 c	Mixed vegetables with corn, peas, or pasta

Serving Size	Food
½ c	Peas, green
½ c	Plantains
1 small (3 oz)	Potatoes, baked or boiled
½ c	Potatoes, mashed
1 c	Squash, winter (acorn, butternut)
½ c	Yams, sweet potatoes, plain
Crackers and Snacks	
8	Animal crackers
3	Graham crackers, 2½" square
¾ oz	Matzoh
4 slices	Melba toast
24	Oyster crackers
3 c	Popcorn (popped, no fat added or low-fat microwave)
¾ oz	Pretzels
2	Rice cakes, 4" across
6	Saltine-type crackers
15–2" (¾ oz)	Snack chips, fat-free (tortilla, potato)
2–5 (¾ oz)	Whole-wheat crackers, no fat added
Dried Beans, Peas, and Lentils	
½ c	Beans and peas, cooked (garbanzo, lentils, pinto, kidney, white, split, black-eyed)
⅔ c	Lima beans
3 tbs	Miso 🖉
Starchy Foods Prepared with Fat	
Count as 1 starch + 1 fat exchange.	
1	Biscuit, 2½" across
½ c	Chow mein noodles
1 (2 oz)	Cornbread, 2" cube
6	Crackers, round butter type
1 c	Croutons
16–25 (3 oz)	French-fried potatoes
¼ c	Granola
1 (1½ oz)	Muffin, small
2	Pancake, 4" across
3 c	Popcorn, microwave
3	Sandwich crackers, cheese or peanut butter filling
⅓ c	Stuffing, bread (prepared)
2	Taco shell, 6" across
1	Waffle, 4½" square
4–6 (1 oz)	Whole-wheat crackers, fat added

🖉 = 400 mg or more of sodium per serving.

TABLE B-2 U.S. Exchange System: Fruit List

1 fruit exchange = 15 g carbohydrate and 60 kcal
Note: In general, a fruit serving is 1 small to medium fresh fruit; ½ c canned or fresh fruit or fruit juice; ¼ c dried fruit.

Serving Size	Food	Serving Size	Food
1 (4 oz)	Apples, unpeeled, small	½ (8 oz) or 1 c cubes	Papayas
½ c	Applesauce, unsweetened	1 (6 oz)	Peaches, medium, fresh
4 rings	Apples, dried	½ c	Peaches, canned
4 whole (5½ oz)	Apricots, fresh	½ (4 oz)	Pears, large, fresh
8 halves	Apricots, dried	½ c	Pears, canned
½ c	Apricots, canned	¾ c	Pineapple, fresh
1 (4 oz)	Bananas, small	½ c	Pineapple, canned
¾ c	Blackberries	2 (5 oz)	Plums, small
¾ c	Blueberries	½ c	Plums, canned
⅓ melon (11 oz) or 1 c cubes	Cantaloupe, small	3	Prunes, dried
12 (3 oz)	Cherries, sweet, fresh	2 tbs	Raisins
½ c	Cherries, sweet, canned	1 c	Raspberries
3	Dates	1¼ c whole berries	Strawberries
1½ large or 2 medium (3½ oz)	Figs, fresh	2 (8 oz)	Tangerines, small
1½	Figs, dried	1 slice (13½ oz) or 1¼ c cubes	Watermelon
½ c	Fruit cocktail	**Fruit Juice**	
½ (11 oz)	Grapefruit, large	½ c	Apple juice/cider
¾ c	Grapefruit sections, canned	⅓ c	Cranberry juice cocktail
17 (3 oz)	Grapes, small	1 c	Cranberry juice cocktail, reduced-kcalorie
1 slice (10 oz) or 1 c cubes	Honeydew melon	⅓ c	Fruit juice blends, 100% juice
1 (3½ oz)	Kiwi	⅓ c	Grape juice
¾ c	Mandarin oranges, canned	½ c	Grapefruit juice
½ (5½ oz) or ½ c	Mangoes, small	½ c	Orange juice
1 (5 oz)	Nectarines, small	½ c	Pineapple juice
1 (6½ oz)	Oranges, small	⅓ c	Prune juice

TABLE B-3 U.S. Exchange System: Milk List

Serving Size	Food	Serving Size	Food
Nonfat and Low-Fat Milk		**Reduced-Fat Milk**	
1 nonfat/low-fat milk exchange = 12 g carbohydrate, 8 g protein, 0–3 g fat, 90 kcal		1 reduced-fat milk exchange = 12 g carbohydrate, 8 g protein, 5 g fat, 120 kcal	
1 c	Nonfat milk	1 c	2% milk
1 c	½% milk	¾ c	Plain low-fat yogurt
1 c	1% milk	1 c	Sweet acidophilus milk
1 c	Nonfat or low-fat buttermilk	**Whole Milk**	
½ c	Evaporated nonfat milk	1 whole milk exchange = 12 g carbohydrate, 8 g protein, 8 g fat, 150 kcal	
⅓ c dry	Dry nonfat milk	1 c	Whole milk
¾ c	Plain nonfat yogurt	½ c	Evaporated whole milk
1 c	Nonfat or low-fat fruit-flavored yogurt sweetened with aspartame or with a non-nutritive sweetener	1 c	Goat's milk
		1 c	Kefir

TABLE B-4	U.S. Exchange System: Other Carbohydrates List

1 other carbohydrate exchange = 15 g carbohydrate, or 1 starch, or 1 fruit, or 1 milk exchange

Food	Serving Size	Exchanges per Serving
Angel food cake, unfrosted	1⁄12 cake	2 carbohydrates
Brownies, small, unfrosted	2″ square	1 carbohydrate, 1 fat
Cake, unfrosted	2″ square	1 carbohydrate, 1 fat
Cake, frosted	2″ square	2 carbohydrates, 1 fat
Cookie, fat-free	2 small	1 carbohydrate
Cookies or sandwich cookies	2 small	1 carbohydrate, 1 fat
Cupcakes, frosted	1 small	2 carbohydrates, 1 fat
Cranberry sauce, jellied	1⁄4 c	2 carbohydrates
Doughnuts, plain cake	1 medium, (1½ oz)	1½ carbohydrates, 2 fats
Doughnuts, glazed	3¾″ across (2 oz)	2 carbohydrates, 2 fats
Fruit juice bars, frozen, 100% juice	1 bar (3 oz)	1 carbohydrate
Fruit snacks, chewy (pureed fruit concentrate)	1 roll (¾ oz)	1 carbohydrate
Fruit spreads, 100% fruit	1 tbs	1 carbohydrate
Gelatin, regular	½ c	1 carbohydrate
Gingersnaps	3	1 carbohydrate
Granola bars	1 bar	1 carbohydrate, 1 fat
Granola bars, fat-free	1 bar	2 carbohydrates
Hummus	⅓ c	1 carbohydrate, 1 fat
Ice cream	½ c	1 carbohydrate, 2 fats
Ice cream, light	½ c	1 carbohydrate, 1 fat
Ice cream, fat-free, no sugar added	½ c	1 carbohydrate
Jam or jelly, regular	1 tbs	1 carbohydrate
Milk, chocolate, whole	1 c	2 carbohydrates, 1 fat
Pie, fruit, 2 crusts	⅙ pie	3 carbohydrates, 2 fats
Pie, pumpkin or custard	⅛ pie	1 carbohydrate, 2 fats
Potato chips	12–18 (1 oz)	1 carbohydrate, 2 fats
Pudding, regular (made with low-fat milk)	½ c	2 carbohydrates
Pudding, sugar-free (made with low-fat milk)	½ c	1 carbohydrate
Salad dressing, fat-free 🖍	1⁄4 c	1 carbohydrate
Sherbet, sorbet	½ c	2 carbohydrates
Spaghetti or pasta sauce, canned 🖍	½ c	1 carbohydrate, 1 fat
Sweet roll or danish	1 (2½ oz)	2½ carbohydrates, 2 fats
Syrup, light	2 tbs	1 carbohydrate
Syrup, regular	1 tbs	1 carbohydrate
Syrup, regular	1⁄4 c	4 carbohydrates
Tortilla chips	6–12 (1 oz)	1 carbohydrate, 2 fats
Yogurt, frozen, low-fat, fat-free	⅓ c	1 carbohydrate, 0–1 fat
Yogurt, frozen, fat-free, no sugar added	½ c	1 carbohydrate
Yogurt, low-fat with fruit	1 c	3 carbohydrates, 0–1 fat
Vanilla wafers	5	1 carbohydrate, 1 fat

🖍 = 400 mg or more sodium per exchange.

The Exchange Lists are the basis of a meal planning system designed by a committee of the American Diabetes Association and The American Dietetic Association. While designed primarily for people with diabetes and others who must follow special diets, the Exchange Lists are based on principles of good nutrition that apply to everyone. Copyright © 1995 the American Diabetes Association. From Exchange Lists for Meal Planning. Reprinted with permission from The American Diabetes Association.

TABLE B-5 **U.S. Exchange System: Vegetable List**

1 vegetable exchange = 5 g carbohydrate, 2 g protein, 25 kcal

Note: In general, a vegetable serving is ½ c cooked vegetables or vegetable juice; 1 c raw vegetables. Starchy vegetables such as corn, peas, and potatoes are on the starch list.

Artichokes	Mushrooms
Artichoke hearts	Okra
Asparagus	Onions
Beans (green, wax, Italian)	Pea pods
Bean sprouts	Peppers (all varieties)
Beets	Radishes
Broccoli	Salad greens (endive, escarole, lettuce, romaine, spinach)
Brussels sprouts	
Cabbage	Sauerkraut 🖊
Carrots	Spinach
Cauliflower	Summer squash (crookneck)
Celery	Tomatoes
Cucumbers	Tomatoes, canned
Eggplant	Tomato sauce 🖊
Green onions or scallions	Tomato/vegetable juice 🖊
Greens (collard, kale, mustard, turnip)	Turnips
Kohlrabi	Water chestnuts
Leeks	Watercress
Mixed vegetables (without corn, peas, or pasta)	Zucchini

🖊 = 400 mg or more sodium per exchange.

The Exchange Lists are the basis of a meal planning system designed by a committee of the American Diabetes Association and The American Dietetic Association. While designed primarily for people with diabetes and others who must follow special diets, the Exchange Lists are based on principles of good nutrition that apply to everyone. Copyright © 1995 the American Diabetes Association. From Exchange Lists for Meal Planning. Reprinted with permission from The American Diabetes Association.

B

TABLE B-6 **U.S. Exchange System: Meat and Meat Substitutes List**

Note: In general, a meat serving is 1 oz meat, poultry, or cheese; ½ c dried beans (weigh meat and poultry and measure beans after cooking).

Serving Size	Food
Very Lean Meat and Substitutes	
1 very lean meat exchange = 7 g protein, 0–1 g fat, 35 kcal	
1 oz	Poultry: Chicken or turkey (white meat, no skin), Cornish hen (no skin)
1 oz	Fish: Fresh or frozen cod, flounder, haddock, halibut, trout; tuna, fresh or canned in water
1 oz	Shellfish: Clams, crab, lobster, scallops, shrimp, imitation shellfish
1 oz	Game: Duck or pheasant (no skin), venison, buffalo, ostrich
	Cheese with ≤1g fat/oz:
¼ c	Nonfat or low-fat cottage cheese
1 oz	Fat-free cheese
	Other:
1 oz	Processed sandwich meats with ≤1 g fat/oz (such as deli thin, shaved meats, chipped beef, 🖊 turkey ham)
2	Egg whites
¼ c	Egg substitutes, plain
1 oz	Hot dogs with ≤1 g fat/oz 🖊
1 oz	Kidney (high in cholesterol)
1 oz	Sausage with ≤1 g fat/oz
Count as one very lean meat and one starch exchange:	
½ c	Dried beans, peas, lentils (cooked)
Lean Meat and Substitutes	
1 lean meat exchange = 7 g protein, 3 g fat, 55 kcal	
1 oz	Beef: USDA Select or Choice grades of lean beef trimmed of fat (round, sirloin, and flank steak); tenderloin; roast (rib, chuck, rump); steak (T-bone, porterhouse, cubed), ground round
1 oz	Pork: Lean pork (fresh ham); canned, cured, or boiled ham; Canadian bacon; 🖊 tender-loin, center loin chop
1 oz	Lamb: Roast, chop, leg
1 oz	Veal: Lean chop, roast
1 oz	Poultry: Chicken, turkey (dark meat, no skin), chicken white meat (with skin), domestic duck or goose (well-drained of fat, no skin)
	Fish:
1 oz	Herring (uncreamed or smoked)
6 medium	Oysters
1 oz	Salmon (fresh or canned), catfish
2 medium	Sardines (canned)
1 oz	Tuna (canned in oil, drained)
1 oz	Game: Goose (no skin), rabbit
	Cheese:

Serving Size	Food
¼ c	4.5%-fat cottage cheese
2 tbs	Grated Parmesan
1 oz	Cheeses with ≤3 g fat/oz
	Other:
1½ oz	Hot dogs with ≤3 g fat/oz
1 oz	Processed sandwich meat with ≤3 g fat/oz (turkey pastrami or kielbasa)
1 oz	Liver, heart (high in cholesterol)
Medium-Fat Meat and Substitutes	
1 medium-fat meat exchange = 7 g protein, 5 g fat, and 75 kcal	
1 oz	Beef: Most beef products (ground beef, meatloaf, corned beef, short ribs, Prime grades of meat trimmed of fat, such as prime rib)
1 oz	Pork: Top loin, chop, Boston butt, cutlet
1 oz	Lamb: Rib roast, ground
1 oz	Veal: Cutlet (ground or cubed, unbreaded)
1 oz	Poultry: Chicken dark meat (with skin), ground turkey or ground chicken, fried chicken (with skin)
1 oz	Fish: Any fried fish product
	Cheese with ≤5 g fat/oz:
1 oz	Feta
1 oz	Mozzarella
¼ c (2 oz)	Ricotta
	Other:
1	Egg (high in cholesterol, limit to 3/week)
1 oz	Sausage with ≤5 g fat/oz
1 c	Soy milk
¼ c	Tempeh
4 oz or ½ c	Tofu
High-Fat Meat and Substitutes	
1 high-fat meat exchange = 7 g protein, 8 g fat, 100 kcal	
1 oz	Pork: Spareribs, ground pork, pork sausage
1 oz	Cheese: All regular cheeses (American 🖊, cheddar, Monterey Jack, swiss)
	Other:
1 oz	Processed sandwich meats with ≤8 g fat/oz (bologna, pimento loaf, salami)
1 oz	Sausage (bratwurst, Italian, knockwurst, Polish, smoked)
1 (10/lb)	Hot dog (turkey or chicken) 🖊
3 slices (20 slices/lb)	Bacon
Count as one high-fat meat plus one fat exchange:	
1 (10/lb)	Hot dog (beef, pork, or combination) 🖊
2 tbs	Peanut butter (contains unsaturated fat)

🖊 = 400 mg or more of sodium per serving.

TABLE B-7 U.S. Exchange System: Fat List

fat exchange = 5 g fat, 45 kcal

Note: In general, a fat serving is 1 tsp regular butter, margarine, or vegetable oil; 1 tbs regular salad dressing. Many fat-free and reduced-fat foods are on the Free Foods List.

Serving Size	Food
Monounsaturated Fats	
⅛ medium (1 oz)	Avocados
1 tsp	Oil (canola, olive, peanut)
8 large	Olives, ripe (black)
10 large	Olives, green, stuffed 🖊
6 nuts	Almonds, cashews
6 nuts	Mixed nuts (50% peanuts)
10 nuts	Peanuts
4 halves	Pecans
2 tsp	Peanut butter, smooth or crunchy
1 tbs	Sesame seeds
2 tsp	Tahini paste
Polyunsaturated Fats	
1 tsp	Margarine, stick, tub, or squeeze
1 tbs	Margarine, lower-fat (30% to 50% vegetable oil)
1 tsp	Mayonnaise, regular
1 tbs	Mayonnaise, reduced-fat
4 halves	Nuts, walnuts, English
1 tsp	Oil (corn, safflower, soybean)
1 tbs	Salad dressing, regular
2 tbs	Salad dressing, reduced-fat
2 tsp	Mayonnaise type salad dressing, regular 🖊
1 tbs	Mayonnaise type salad dressing, reduced-fat
1 tbs	Seeds: pumpkin, sunflower
Saturated Fats*	
1 slice (20 slices/lb)	Bacon, cooked
1 tsp	Bacon, grease
1 tsp	Butter, stick
2 tsp	Butter, whipped
1 tbs	Butter, reduced-fat
2 tbs (½ oz)	Chitterlings, boiled
2 tbs	Coconut, sweetened, shredded
2 tbs	Cream, half and half
1 tbs (½ oz)	Cream cheese, regular
2 tbs (1 oz)	Cream cheese, reduced-fat
	Fatback or salt pork†
1 tsp	Shortening or lard
2 tbs	Sour cream, regular
3 tbs	Sour cream, reduced-fat

🖊 = 400 mg or more sodium per exchange

*Saturated fats can raise blood cholesterol levels.

† Use a piece 1″ × 1″ × ¼″ if you plan to eat the fatback cooked with vegetables. Use a piece 2″ × 1″ × ½″ when eating only the vegetables with the fatback removed.

B

TABLE B-8 U.S. Exchange System: Free Foods List

Note: A serving of free food contains less than 20 kcalories; those with serving sizes should be limited to 3 servings a day whereas those without serving sizes can be eaten freely.

Serving Size	Food	Serving Size	Food
Fat-Free or Reduced-Fat Foods			Carbonated or mineral water
1 tbs	Cream cheese, fat-free	1 tbs	Cocoa powder, unsweetened
1 tbs	Creamers, nondairy, liquid		Coffee
2 tsp	Creamers, nondairy, powdered		Club soda
1 tbs	Mayonnaise, fat-free		Diet soft drinks, sugar-free
1 tsp	Mayonnaise, reduced-fat		Drink mixes, sugar-free
4 tbs	Margarine, fat-free		Tea
1 tsp	Margarine, reduced-fat		Tonic water, sugar-free
1 tbs	Mayonnaise type salad dressing, nonfat		
1 tsp	Mayonnaise type salad dressing, reduced-fat	**Condiments**	
	Nonstick cooking spray	1 tbs	Catsup
1 tbs	Salad dressing, fat-free		Horseradish
2 tbs	Salad dressing, fat-free, Italian		Lemon juice
¼ c	Salsa		Lime juice
1 tbs	Sour cream, fat-free, reduced-fat		Mustard
2 tbs	Whipped topping, regular or light	1½ large	Pickles, dill 🖊
Sugar-Free or Low-Sugar Foods			Soy sauce, regular or light 🖊
1 piece	Candy, hard, sugar-free	1 tbs	Taco sauce
	Gelatin dessert, sugar-free		Vinegar
	Gelatin, unflavored	**Seasonings**	
	Gum, sugar-free		Flavoring extracts
2 tsp	Jam or jelly, low-sugar or light		Garlic
	Sugar substitutes		Herbs, fresh or dried
2 tbs	Syrup, sugar-free		Pimento
Drinks			Spices
	Bouillon, broth, consommé 🖊		Hot pepper sauces
	Bouillon or broth, low-sodium		Wine, used in cooking
			Worcestershire sauce

🖊 = 400 mg or more of sodium per serving.

TABLE B-9	U.S. Exchange System: Combination Foods List

Food	Serving Size	Exchanges per Serving
Entrees		
Tuna noodle casserole, lasagna, spaghetti with meatballs, chili with beans, macaroni and cheese	1 c (8 oz)	2 carbohydrates, 2 medium-fat meats
Chow mein (without noodles or rice)	2 c (16 oz)	1 carbohydrate, 2 lean meats
Pizza, cheese, thin crust	¼ of 10" (5 oz)	2 carbohydrates, 2 medium-fat meats, 1 fat
Pizza, meat topping, thin crust	¼ of 10" (5 oz)	2 carbohydrates, 2 medium-fat meats, 2 fats
Pot pie	1 (7 oz)	2 carbohydrates, 1 medium-fat meat, 4 fats
Frozen Entrees		
Salisbury steak with gravy, mashed potato	1 (11 oz)	2 carbohydrates, 3 medium-fat meats, 3–4 fats
Turkey with gravy, mashed potato, dressing	1 (11 oz)	2 carbohydrates, 2 medium-fat meats, 2 fats
Entree with less than 300 kcalories	1 (8 oz)	2 carbohydrates, 3 lean meats
Soups		
Bean	1 c	1 carbohydrate, 1 very lean meat
Cream (made with water)	1 c (8 oz)	1 carbohydrate, 1 fat
Split pea (made with water)	½ c (4 oz)	1 carbohydrate
Tomato (make with water)	1 c (8 oz)	1 carbohydrate
Vegetable beef, chicken noodle, or other broth-type	1 c (8 oz)	1 carbohydrate
Fast Foods		
Burritos with beef	2	4 carbohydrates, 2 medium-fat meats, 2 fats
Chicken nuggets	6	1 carbohydrate, 2 medium-fat meats, 1 fat
Chicken breast and wing, breaded and fried	1	1 carbohydrate, 4 medium-fat meats, 2 fats
Fish sandwich/tartar sauce	1	3 carbohydrates, 1 medium-fat meat, 3 fats
French fries, thin	20–25	2 carbohydrates, 2 fats
Hamburger, regular	1	2 carbohydrates, 2 medium-fat meats
Hamburger, large	1	2 carbohydrates, 3 medium-fat meats, 1 fat
Hot dog with bun	1	1 carbohydrate, 1 high-fat meat, 1 fat
Individual pan pizza	1	5 carbohydrates, 3 medium-fat meats, 3 fats
Soft serve cone	1 medium	2 carbohydrates, 1 fat
Submarine sandwich	1 (60)	3 carbohydrates, 1 vegetable, 2 medium-fat meats, 1 fat
Taco, hard shell	1 (6 oz)	2 carbohydrates, 2 medium-fat meats, 2 fats
Taco, soft shell	1 (3 oz)	1 carbohydrate, 1 medium-fat meat, 1 fat

= 400 mg or more sodium per exchange.

NUTRITION RESOURCES

CONTENTS

People interested in nutrition often want to know where they can find reliable nutrition information. Wherever you live, there are several sources you can turn to:

- The Department of Health may have a nutrition expert.
- The local extension agent is often an expert.
- The food editor of your local paper may be well informed.
- The dietitian at the local hospital had to fulfill a set of qualifications before he or she became an RD. (See Nutrition in Practice 1).
- There may be knowledgeable professors of nutrition or biochemistry at a nearby college or university.

In addition, you may be interested in building a nutrition library of your own. Books you can buy, journals you can subscribe to, and addresses you can contact for general information are given below.

Books

For students seeking to establish a personal library of nutrition references, the authors of this text recommend the following books:

- *Present Knowledge in Nutrition,* 8th ed. (Washington, D.C.: International Life Sciences Institute—Nutrition Foundation, 2001).

This 760-page paperback has a chapter on each of 65 topics, including energy, obesity, each of the nutrients, several diseases, malnutrition, growth and its assessment, immunity, alcohol, fiber, exercise, drugs, and toxins. Watch for an update; new editions come out every few years.

- M. E. Shils and coeditors, *Modern Nutrition in Health and Disease,* 9th ed. (Baltimore: Williams & Wilkins, 1999).

This reference book contains encyclopedic articles on the nutrients, foods, diet, metabolism, malnutrition, age-related needs, and nutrition in disease.

- Committee on Dietary Reference Intakes, *Dietary Reference Intakes for Cal-* *cium, Phosphorus, Magnesium, Vitamin D, and Fluoride* (Washington, D.C.: National Academy Press, 1997).
- Committee on Dietary Reference Intakes, *Dietary Reference Intakes for Thiamin, Riboflavin, Niacin, Vitamin B_6, Folate, Vitamin B_{12}, Pantothenic Acid, Biotin, and Choline* (Washington, D.C.: National Academy Press, 1998).
- Committee on Dietary Reference Intakes, *Dietary Reference Intakes for Vitamin C, Vitamin E, Selenium, and Carotenoids* (Washington, D.C.: National Academy Press, 2000).
- Committee on Dietary Reference Intakes, *Dietary Reference Intakes for Vitamin A, Vitamin K, Arsenic, Boron, Chromium, Copper, Iodine, Iron, Manganese, Molybdenum, Nickel, Silicon, Vanadium, and Zinc* (Washington, D.C.: National Academy Press, 2001).

These reports review the function of each nutrient, dietary sources, and deficiency and toxicity symptoms as well as provide recommendations for intakes. Watch for additional reports on the Dietary Reference Intakes for the remaining nutrients and other food components.

- Committee on Diet and Health, *Diet and Health Implications for Reducing Chronic Disease Risk* (Washington, D.C.: National Academy Press, 1989).

This 749-page book presents the integral relationship between diet and chronic disease prevention. Its nutrient chapters provide evidence on how diet influences disease development, and its disease chapters review the dietary patterns implicated in each chronic disease.

- S. S. Gropper, *The Biochemistry of Human Nutrition: A Desk Reference,* 2nd ed. (Belmont, Calif.: Wadsworth/ Thomson Learning, 2000).

This 263-page paperback presents the biochemical concepts necessary for an understanding of nutrition. It is a handy reference book for those who need a refresher in the basics of biochemistry or for those who are learning biochemistry for the first time.

We also recommend two of our own books that explore current topics in nutrition, health, and the life span:

- S. R. Rolfes, L. K. DeBruyne, and E. N. Whitney, *Life Span Nutrition: Conception through Life,* 2nd ed. (Belmont, Calif.: West/Wadsworth, 1998).
- E. N. Whitney, C. B. Cataldo, L. K. DeBruyne, and S. R. Rolfes, *Nutrition for Health and Health Care,* 2nd ed. (Belmont, Calif.: Wadsworth/ Thomson Learning, 2001).

Journals

Nutrition Today is an excellent magazine for the interested layperson. It makes a point of raising controversial issues and providing a forum for conflicting opinions. Six issues per year are published. Order from Williams & Wilkins, 12107 Insurance Way, Hagerstown, MD 21740.

The *Journal of the American Dietetic Association,* the official publication of the ADA, contains articles of interest to dietitians and nutritionists, news of legislative action on food and nutrition, and a very useful section of abstracts of articles from many other journals of nutrition and related areas. There are 12 issues per year, available from the American Dietetic Association (see "Addresses," later).

Nutrition Reviews, a publication of the International Life Sciences Institute, does much of the work for the library researcher, compiling recent evidence on current topics and presenting extensive bibliographies. Twelve issues per year are available from Nutrition Reviews, P.O. Box 1897, Lawrence, KS 66044-8897.

Nutrition and the M.D. is a monthly newsletter that provides up-to-date, easy-to-read, practical information on nutrition for health care providers. It is available from Lippincott-Williams & Wilkens, 16522 Hunters Green Parkway, Hagerstown, MD 21740.

Other journals that deserve mention here are *Food Technology, Journal of Nutrition, American Journal of Clinical*

Nutrition, Nutrition Research, and *Journal of Nutrition Education. FDA Consumer,* a government publication with many articles of interest to the consumer, is available from the Food and Drug Administration (see "Addresses," below). Many other journals of value are referred to throughout this book.

Addresses

Many of the organizations listed below will provide publication lists free on request. Government and international agencies and professional nutrition organizations are listed first, followed by organizations in the following areas: aging, alcohol and drug abuse, consumer organizations, fitness, food safety, health and disease, infancy and childhood, pregnancy and lactation, trade and industry organizations, weight control and eating disorders, and world hunger.

U.S. Government

- Federal Trade Commission (FTC)
  Public Reference Branch
  600 Pennsylvania Avenue NW
  Washington, DC 20580
  (202) 326-2222
  www.ftc.gov

- Food and Drug Administration (FDA)
  Office of Consumer Affairs, HFE 1
  Room 16-85
  5600 Fishers Lane
  Rockville, MD 20857
  (301) 443-1726
  www.fda.gov

- FDA Consumer Information Line
  (888) INFO-FDA
  (888) 463-6332

- FDA Center for Food Safety &
  Applied Nutrition, HFS 150
  200 C Street SW
  Washington, DC 20204
  (202) 205-4561; fax (202) 205-4564
  www.cfsan.fda.gov

- Food and Nutrition Information
  Center
  National Agricultural Library,
  Room 304
  10301 Baltimore Avenue
  Beltsville, MD 20705-2351
  fax (301) 504-6409
  www.nal.usda.gov/fnic

- National Institutes of Health (NIH)
  9000 Rockville Pike
  Bethesda, MD 20892
  (301) 496-2433
  www.nih.gov

- Superintendent of Documents
  U.S. Government Printing Office
  Washington, DC 20402
  (202) 512-1530
  www.access.gpo.gov/su_docs

- U.S. Department of Agriculture
  (USDA)
  14th Street and Independence
  Avenue SW
  Washington, DC 20250
  (202) 720-2791
  www.fns.usda.gov/fncs

- USDA Center for Nutrition Policy and
  Promotion (Dietary Guidelines and
  Food Guide Pyramid)
  1120 20th Street NW, Suite 200
  North Lobby
  Washington, DC 20036
  (800) 687-2258 or (202) 418-2312
  www.cnpp.usda.gov

- USDA Food Safety and Inspection Service
  Food Safety Education Office
  1400 Independence Avenue SW,
  Room 2932-S
  Washington, DC 20250
  (202) 690-0351
  www.usda.gov/fsis

- U.S. Department of Education
  (DOE)
  Accreditation and State Liaison
  Accrediting Agency Evaluation
  1990 K Street NW, #7105
  Washington, DC 20006-8509
  (202) 219-7011
  www.ed.gov/offices/OPE/
  accreditation

- U.S. Department of Health and
  Human Services
  200 Independence Avenue SW
  Washington, DC 20201
  (202) 619-0257
  www.os.dhhs.gov

- U.S. Environmental Protection
  Agency (EPA)
  1200 Pennsylvania Avenue NW
  Washington, DC 20460
  (202) 260-2090
  www.epa.gov

- Assistant Secretary of Health
  Office of Public Health and
  Science
  Department of Health and Human Sciences
  200 Independence Avenue SW,
  Room 725-H
  Washington, DC 20201
  (202) 690-7694

Canadian Government

Federal

- Bureau of Nutritional Sciences
  Food Directorate Health Protection
  Branch, 3-West
  Sir Frederick Banting Research Centre,
  AL0904A Tunney's Pasture, Ottawa,
  Ontario K1A 0K9
  (613) 957-2991 fax (613) 941-5366
  www.hc-sc.gc.ca

- Canadian Food Inspection Agency
  Agriculture and Agri-Food Canada
  59 Camelot Drive
  Nepean, Ontario K1A 0Y9
  (613) 225-CFIA or (613) 225-2342
  fax (613) 228-6653
  www.agr.ca

- Office of Nutrition Policy and
  Promotion, Health Products and Food
  Branch AL 90761, 7th Floor—Jeanne
  Mance Bldg., Tunney's Pasture, Ottawa,
  Ontario K1A 1B4
  www.hc-sc.gc.ca/hppb/nutrition

Provincial and Territorial

- Population Health Strategies Branch
  Alberta Health 23rd Floor,
  TELUS Plaza,
  North Tower
  10025 Jasper Ave
  Edmonton, AB T5J 2N3
  www.health.gov.ab.ca

- Preventive Services Branch
  Ministry of Health
  1520 Blanshard Street
  Victoria, BC V8W 3C8
  www.gov.bc.ca/healthservices

- Manitoba Health
  Health Programs Branch
  300 Carlton Street
  Winnipeg, MB R3B 3M9
  www.gov.mb.ca/health

- Public Health Mgmt. Services
  Health and Community Services
  P.O. Box 5100 520 King Street
  Fredericton, NB E3B 5G8
  www.gnb.ca/0051/index-e.asp

  Department of Health
  Newfoundland and Labrador
  P.O. Box 8700
  Confederation Building, West Bank
  St. John's, NF A1B 4J6
  www.gov.nf.ca/health

- Health Promotion Unit
  Population Health Division
  Dept. of Health and Social Services
  Gov't of the Northwest Territories

Centre Square Tower, 6th Floor
P.O. Box 1320
Yellowknife, NT X1A 2L9
www.gov.nt.ca/agendas/health

- Nova Scotia Dept. of Health
  P.O. Box 488
  Halifax, NS B3J 2R8
  www.gov.ns.ca/health

- Ontario Ministry of Health and
  Long-Term Care
  8th FLR
  5700 Yonge Street
  North York, ON M2M 4K5
  www.gov/on.ca/MOH

- Health Promotion
  Government of Nunavut
  Dept. of Health and Social Services
  P.O. Box 800
  Iqaluit, NT X0A 0H0
  www.gov.nu.ca/hss.htm

  PEI, Dept. of Health and Social
  Services
  11 Kent Street, 2nd floor
  Jones Building, Box 2000
  Charlottetown, PEI C1A 7N8
  www.gov/pe.ca/hss

- Ministère de la Santé et des Services
  Sociaux,
  Service de la promotion des saines
  habitudes de vie et dépistage
  1075, Chemin Ste-Foy, 3e étage
  Québec (Québec) G1S 2M1
  www.msss.gouv.qc.ca

- Health Promotion Unit
  Population Health Branch
  Saskatchewan Health
  3475 Albert Street
  Regina, SK S4S 6X6
  www.health.gov.sk.ca

- Government of Yukon
  Dept. of Health and Social Services
  #5 Hospital Road
  Whitehorse, YT Y1A 3H7
  www.hss.gov.yk.ca

International Agencies

- Food and Agriculture Organization of
  the United Nations (FAO)
  Liaison Office for North America
  2175 K Street, Suite 300
  Washington, DC 20437
  (202) 653-2400; fax (202) 653-5760
  www.fao.org

- International Food Information Council Foundation
  1100 Connecticut Avenue NW,
  Suite 430

Washington, DC 20036
(202) 296-6540
ificinfo.health.org

- UNICEF
  3 United Nations Plaza
  New York, NY 10017
  (212) 326-7000
  www.unicef.org

- World Health Organization (WHO)
  Regional Office
  525 23rd Street NW
  Washington, DC 20037
  (202) 974-3000
  www.who.org

Professional Nutrition Organizations

- American Society of Nutritional
  Sciences
  9650 Rockville Pike
  Bethesda, MD 20814
  (301) 530-7050; fax (301) 571-1892
  www.nutrition.org

- American Dietetic Association (ADA)
  216 West Jackson Boulevard, Suite 800
  Chicago, IL 60606-6995
  (800) 877-1600; (312) 899-0040
  www.eatright.org

- ADA, Consumer Nutrition Hotline
  (800) 366-1655

- American Society for Clinical Nutrition
  9650 Rockville Pike
  Bethesda, MD 20814-3998
  (301) 530-7110; fax (301) 571-1863
  www.faseb.org/ascn

- Dietitians of Canada
  480 University Avenue, Suite 604
  Toronto, Ontario, Canada M5G 1V2
  (416) 596-0857; fax (416) 596-0603
  www.dietitians.ca

- International Life Sciences Institute
  1126 Sixteenth Street NW
  Washington, DC 20036
  (202) 659-0074
  www.ilsi.org

- National Academy of Sciences/
  National Research Council
  (NAS/NRC)
  2101 Constitution Avenue, NW
  Washington, DC 20418
  (202) 334-2000
  www.nas.edu

- National Institute of Nutrition
  265 Carling Avenue, Suite 302
  Ottawa, Ontario K1S 2E1
  (613) 235-3355; fax (613) 235-7032
  www.nin.ca

- Society for Nutrition Education
  1001 Connecticut Avenue NW,
  Suite 528
  Washington, DC 20036
  (202) 452-8534; fax (202) 452-8536
  www.sne.org

Aging

- Administration on Aging
  330 Independence Avenue SW
  Washington, DC 20201
  (202) 619-0724
  www.aoa.dhhs.gov

- American Association of Retired Persons (AARP)
  601 E Street NW
  Washington, DC 20049
  (800) 424-3410
  www.aarp.org

- National Aging Information Center
  330 Independence Avenue SW
  Washington, DC 20201
  (202) 619-7501
  www.aoa.dhhs.gov/naic

- National Institute on Aging
  Public Information Office
  31 Center Drive, MSC 2292
  Bethesda, MD 20892
  (800) 222-2225
  www.nih.gov/nia

Alcohol and Drug Abuse

- Al-Anon Family Groups, Inc.
  1600 Corporate Landing Parkway
  Virginia Beach, VA 23454-5617
  (888) 4AL-ANON or (888) 425-2666
  (757) 563-1600; fax (757) 563-1655
  www.al-anon.alateen.org

- Alcohol & Drug Abuse Information Line
  Adcare Hospital
  (800) 252-6465

- Alcoholics Anonymous (AA)
  Grand Central Station
  P.O. Box 459
  New York, NY 10163
  (212) 870-3400
  www.alcoholics-anonymous.org

- Narcotics Anonymous (NA)
  P.O. Box 9999
  Van Nuys, CA 91409
  (818) 773-9999; fax (818) 700-0700
  www.wsoinc.com

- National Clearinghouse for Alcohol
  and Drug Information (NCADI)
  P.O. Box 2345
  Rockville, MD 20847-2345
  (800) 729-6686
  www.health.org

- National Council on Alcoholism and Drug Dependence (NCADD)
12 West 21st Street
New York, NY 10010
(800) NCA-CALL or (800) 622-2255
(212) 206-6770; fax (212) 645-1690
www.ncadd.org

- U.S. Center for Substance Abuse Prevention
1010 Wayne Avenue, Suite 850
Silver Spring, MD 20910
(301) 459-1591 ext. 244
fax (301) 495-2919
www.covesoft.com/csap.html

Consumer Organizations

- Center for Science in the Public Interest (CSPI)
1875 Connecticut Avenue NW, Suite 300
Washington, DC 20009-5728
(202) 332-9110; fax (202) 265-4954
www.cspinet.org

- Choice in Dying, Inc.
1035 30th Street NW
Washington, DC 20007
(800) 989-WILL or (800) 989-9455
(202) 338-9790; fax (202) 338-0242
www.choices.org

- Consumer Information Center
Pueblo, CO 81009
(888) 8 PUEBLO or (888) 878-3256
www.pueblo.gsa.gov

- Consumers Union of US Inc.
101 Truman Avenue
Yonkers, NY 10703-1057
(914) 378-2000
www.consumersunion.org

- National Council Against Health Fraud, Inc. (NCAHF)
P.O. Box 141
Fort Lee, NJ 07024
(212) 723-2955
www.ncahf.org

Fitness

- American College of Sports Medicine
401 West Michigan Street
Indianapolis, IN 46206-1440
(317) 637-9200
www.acsm.org

- American Council on Exercise (ACE)
5820 Oberlin Drive, Suite 102
San Diego, CA 92121
(800) 825-3636
www.acefitness.org

- Shape Up America!
6707 Democracy Boulevard,
Suite 306
Bethesda, MD 20817
(301) 493-5368
www.shapeup.org

Food Safety

- Alliance for Food & Fiber Food Safety Hotline
(800) 266-0200

- FDA Center for Food Safety and Applied Nutrition Outreach and Information
200 C Street SW
Washington, DC 20204
www.cfsan.fda.gov

- National Lead Information Center
(800) LEAD-FYI or (800) 532-3394
(800) 424-LEAD or (800) 424-5323

- National Pesticide Telecommunications Network (NPTN)
Oregon State University
333 Weniger Hall
Corvallis, OR 97331-6502
(800) 858-7378
www.ace.orst.edu/info/nptn

- USDA Meat and Poultry Hotline
(800) 535-4555

- U.S. EPA Safe Drinking Water Hotline
(800) 426-4791

Health and Disease

- Alzheimer's Disease Education and Referral Center
P. O. Box 8250
Silver Spring, MD 20907-8250
(800) 438-4380
www.alzheimers.org

- Alzheimer's Disease Information and Referral Service
919 North Michigan Avenue, Suite 1100
Chicago, IL 60611
(800) 272-3900
www.alz.org

- American Academy of Allergy, Asthma, and Immunology
611 East Wells Street
Milwaukee, WI 53202
(414) 272-6071; fax (414) 272-6070
www.aaaai.org

- American Cancer Society
National Home Office
1599 Clifton Road NE
Atlanta, GA 30329-4251
(800) ACS-2345 or (800) 227-2345
www.cancer.org

- American Council on Science and Health
1995 Broadway, 2nd Floor
New York, NY 10023-5860
(212) 362-7044; fax (212) 362-4919
www.acsh.org

- American Dental Association
211 East Chicago Avenue
Chicago, IL 60611
(312) 440-2500; fax (312) 440-2800
www.ada.org

- American Diabetes Association
1701 North Beauregard Street
Alexandria, VA 22311
(800) 232-3472 or (703) 549-1500
www.diabetes.org

- American Heart Association
Box BHG, National Center
7272 Greenville Avenue
Dallas, TX 75231
(800) 242-8721
www.americanheart.org

- American Institute for Cancer Research
1759 R Street NW
Washington, DC 20009
(800) 843-8114 or (202) 328-7744
fax (202) 328-7226
www.aicr.org

- American Medical Association
515 North State Street
Chicago, IL 60610
(312) 464-5000
www.ama-assn.org

- American Public Health Association (APHA)
800 I Street NW
Washington, DC 20001-3710
(282) 777-2742
www.apha.org

- American Red Cross
National Headquarters
8111 Gatehouse Road
Falls Church, VA 22042
(703) 206-8143
www.redcross.org

- Arthritis Foundation
(800) 283-7800
www.arthritis.org

- Canadian Diabetes Association
15 Toronto Street, Suite 800
Toronto, ON M5C 2E3
(800) BANTING or (800) 226-8464
(416) 363-3373
www.diabetes.ca

- Canadian Public Health Association
400-1565 Carling Avenue

Ottawa, Ontario K1Z 8R1
(613) 725-3769; fax (613) 725-9826
www.cpha.ca

- Centers for Disease Control
  and Prevention (CDC)
  1600 Clifton Road NE
  Atlanta, GA 30333
  (404) 639-3311
  www.cdc.gov

- The Food Allergy Network
  10400 Eaton Place, Suite 107
  Fairfax, VA 22030-2208
  (800) 929-4040 or (703) 691-2713
  www.foodallergy.org

- Internet Health Resources
  www.ihr.com

- Mayo Clinic Health Oasis
  www.mayohealth.org

- National AIDS Hotline (CDC)
  (800) 342-AIDS (English)
  (800) 344-SIDA (Spanish)
  (800) 2437-TTY (Deaf)
  (900) 820-2437

- National Cancer Institute
  Office of Cancer Communications
  Building 31, Room 10A31
  31 Center Drive MSC 2580
  Bethesda, MD 20892
  (800) 4-CANCER or
  (800) 422-6237
  www.nci.nih.gov

- National Diabetes Information Clearinghouse
  31 Center Drive MSC 2560
  Bethesda, MD 20892-2560
  (301) 654-3327
  www.niddk.nih.gov

- National Digestive Disease Information
  Clearinghouse (NDDIC)
  31 Center Drive MSC 2560
  Bethesda, MD 20892-2560
  (301) 654-3810
  www.niddk.nih.gov

- National Health Information Center
  (NHIC)
  P.O. Box 1133
  Washington, DC 20013
  (800) 336-4797
  nhic-nt.health.org

- National Heart, Lung, and Blood Institute Information Center
  P.O. Box 30105
  Bethesda, MD 20824-0105
  (301) 592-8573
  www.nhlbi.nih.gov

- National Institute of Allergy and Infectious Diseases

Office of Communications
Building 31, Room 7A50
31 Center Drive, MSC 2520
Bethesda, MD 20892-2520
(301) 496-5717
www.niaid.nih.gov

- National Institute of Dental Research
  (NIDR)
  National Institute of Health
  Bethesda, MD 20892-2190
  www.nidr.nih.gov

- National Osteoporosis Foundation
  1232 22 Street NW
  Washington, DC 20037
  (202) 223-2226
  www.nof.org

- Office of Aids Research
  www.nih.gov/od/oar

- Office of Disease Prevention
  and Health Promotion
  odphp.osophs.dhhs.gov

- Osteoporosis and Related Bone Diseases
  (800) 624-BONE or (800) 624-2663
  www.osteo.org

Infancy and Childhood

- American Academy of Pediatrics
  141 Northwest Point Boulevard
  Elk Grove Village, IL 60007-1098
  (847) 434-4000; fax (847) 434-8000
  www.aap.org

- Birth Defect Research for Children
  930 Woodcock Road, Suite 225
  Orlando, FL 32803
  (407) 895-0802; fax (407) 895-0824
  www.birthdefects.org

- Canadian Paediatric Society
  100-2204 Walkley Road
  Ottawa, ON K1G 4G8
  (613) 526-9397; fax (613) 526-3332
  www.cps.ca

- National Center for Education in
  Maternal & Child Health
  2000 15th Street North, Suite 701
  Arlington, VA 22201-2617
  (703) 524-7802
  www.ncemch.org

Pregnancy and Lactation

- American College of Obstetricians and
  Gynecologists Resource Center
  409 12th Street SW
  Washington, DC 20090
  (202) 638-5577
  www.acog.org

- La Leche International, Inc.
  1400 N. Meacham Road

Schaumburg, IL 60173
(847) 519-7730
www.lalecheleague.org

- March of Dimes Birth Defects
  Foundation
  1275 Mamaroneck Avenue
  White Plains, NY 10605
  (888) MoDimes or (888) 663-4637
  www.modimes.org

Trade and Industry Organizations

- Beech-Nut Nutrition Corporation
  100 South 4th Street
  St. Louis, MO 63102
  (800) 523-6633
  www.beechnut.com

- Borden Foods Nutrition Department
  180 East Broad Street
  Columbus, OH 43215
  (800) 426-7336

- Campbell Soup Company
  Consumer Response Center
  Campbell Place, Box 26B
  Camden, NJ 08103-1701
  (800) 257-8443
  www.campbellssoup.com

- General Mills, Inc.
  Number One General Mills Boulevard
  Minneapolis, MN 55440
  (800) 328-6787
  www.generalmills.com

- Kellogg Company
  P.O. Box 3599
  Battle Creek, MI 49016-3599
  (800) 962-1413
  www.kelloggs.com

- Kraft Foods
  Consumer Response and Information
  Center
  One Kraft Court
  Glenview, IL 60025
  (800) 323-0768
  www.kraftfoods.com

- Mead Johnson Nutritionals
  2400 West Lloyd Expressway
  Evansville, IN 47721
  (800) 247-7893
  www.meadjohnson.com

- Nabisco Consumer Affairs
  100 DeForest Avenue
  East Hanover, NJ 07936
  (800) NABISCO or (800) 932-7800
  www.nabisco.com

- National Dairy Council
  10255 West Higgins Road, Suite 900

Rosemond, IL 60018-5616
(847) 803-2000
www.dairyinfo.com

- NutraSweet/KELCO
P.O. Box 2986
Chicago, IL 60654-0986
www.equal.com

- Pillsbury Company
2866 Pillsbury Center
Minneapolis, MN 55402
(800) 767-4466
www.pillsbury.com

- Procter and Gamble Company
One Procter and Gamble Plaza
Cincinnati, OH 45202
(513) 983-1100
www.pg.com/info

- Ross Laboratories, Abbot Laboratory
625 Cleveland Avenue
Columbus, OH 43215
(800) 227-5767
www.abbot.com

- Sunkist Growers
Consumer Affairs
Fresh Fruit Division
14130 Riverside Drive
Sherman Oaks, CA 91423
www.sunkist.com

- United Fresh Fruit and Vegetable Association
727 North Washington Street
Alexandria, VA 22314
(703) 836-3410

- USA Rice Federation
4301 North Fairfax Drive,
Suite 305
Arlington, VA 22203
Phone: (703) 351-8161
www.usarice.com

Weight Control and Eating Disorders

- American Anorexia & Bulimia
Association, Inc.
165 West 46th Street #1108

New York, NY 10036
(212) 575-6200
www.aabaine.org

- American Obesity Association
1250 24 Street NW, Suite 300
Washington, DC 20037
(800) 98-OBESE or (800) 986-2373
www.obesity.org

- Anorexia Nervosa and Related Eating
Disorders (ANRED)
P.O. Box 5102
Eugene, OR 97405
(541) 344-1144
www.anred.com

- National Association of Anorexia Nervosa and Associated Disorders, Inc.
(ANAD)
P.O. Box 7
Highland Park, IL 60035
(847) 831-3438; fax (847) 433-4632
www.anad.org

- National Eating Disorder Information
Centre
Toronto General Hospital
200 Elizabeth Street, CW 1-211
Toronto, Ontario M5G 2C4
(416) 340-4156; fax (416) 340-4736
www.nedic.ca

- National Institute of Diabetes and
Digestive Diseases Weight-Control
Information Network (WIN)
31 Center Drive MSC 2560
Bethesda, MD 20892
(800) WIN-8098
www.niddk.nih.gov/health/
nutrit/win.htm

- Overeaters Anonymous (OA)
World Service Office
6075 Zenith Court NE
Rio Rancho, NM 87124
(505) 891-2664; fax (505) 891-4320
www.overeatersanonymous.org

- TOPS (Take Off Pounds Sensibly)
4575 South Fifth Street
P.O. Box 07360

Milwaukee, WI 53207-0360
(800) 932-8677 or (414) 482-4620
www.tops.org

- Weight Watchers International, Inc.
Consumer Affairs Department/IN
175 Crossways Park West
Woodbury, NY 11797
(800) 651-6000; fax (516) 390-1632
www.weightwatchers.com

World Hunger

- Bread for the World
50 F Street NW, Suite 500
Washington, DC 20001
(800) 82-BREAD or (800) 822-7323
(202) 639-9400; fax (202) 639-9401
www.bread.org

- Center on Hunger, Poverty and Nutrition Policy
Tufts University School of Nutrition
11 Curtis Avenue
Medford, MA 02155
(617) 627-3956

- Freedom from Hunger
P.O. Box 2000
1644 DaVinci Court
Davis, CA 95616
(530) 758-6241
www.freefromhunger.org

- Oxfam America
26 West Street
Boston, MA 02111-1206
(800) 77-OXFAM or (800) 776-9326
www.oxfamamerica.org

- Worldwatch Institute
1776 Massachusetts Avenue NW
Washington, DC 20036
(202) 452-1999
www.worldwatch.org

C

NUTRITION ASSESSMENT: SUPPLEMENTAL INFORMATION

Chapter 16 described data from nutrition assessments that help evaluate clients' nutrition status and nutrient needs. This appendix provides additional information that may be useful for complete assessments.

Diet-Drug Interactions

Chapter 15 described diet-drug interactions and later chapters provided a series of diet-drug interaction boxes. Table D-1 shows where you can find examples of diet-drug interactions for different classes of medications.

TABLE D-1	Locating Examples of Diet-Drug Interactions
Medication	**Page Number(s)**
Anabolic agents	000
Analgesics	486
Antacids	464
Antianxiety agents	486
Antibiotics	464
Anticoagulants	647
Antidepressants	486
Antidiabetic agents	616
Antidiarrheals	486, 557
Antiemetics, antinauseants	486
Anti-GERD, see *antisecretory agents*	
Antigout agents	674
Antihypertensives	647
Anti-infectives	584
Antilipemics	647
Antineoplastic agents	724
Antisecretory agents	464
Antiviral agents	724
Appetite stimulants	000
Bronchodilators	724
Cardiac glycosides	647
Diuretics	647
Gallstone solubilizers	584
Immunosuppressants	466, 584, 674
Laxatives	486, 584
Phosphate binders	674
Potassium binders	674
Sedatives	696
Other	
Exchange Resins	674
Infliximab	486
Interferon	584
Mesalamine	486
Pancreatic enzyme replacements	557
Penicillamine	584
Sulfasalazine	486

Growth Charts and Body Measurements

Growth charts, shown in Figures D-1 (A and B) through D-6 (A and B), allow health care professionals to evaluate the growth and development of children from birth to 20 years of age. The growth charts for plotting body mass index-for-age are shown in Figure NP13-1 on p. 000. Percentile charts divide the measures of a population into 100 equal divisions. Thus, half of the population falls above the 50th percentile, and half falls below. The use of percentile measures allows for comparisons among children of the same age and gender. For example, a six-month-old female infant whose weight is at the 75 percentile weighs more than 75 percent of the female infants her age.

• **Fatfold Measures** • Fatfold measures provide a good estimate of total body fat and a fair assessment of the fat's location. Approximately half the fat in the

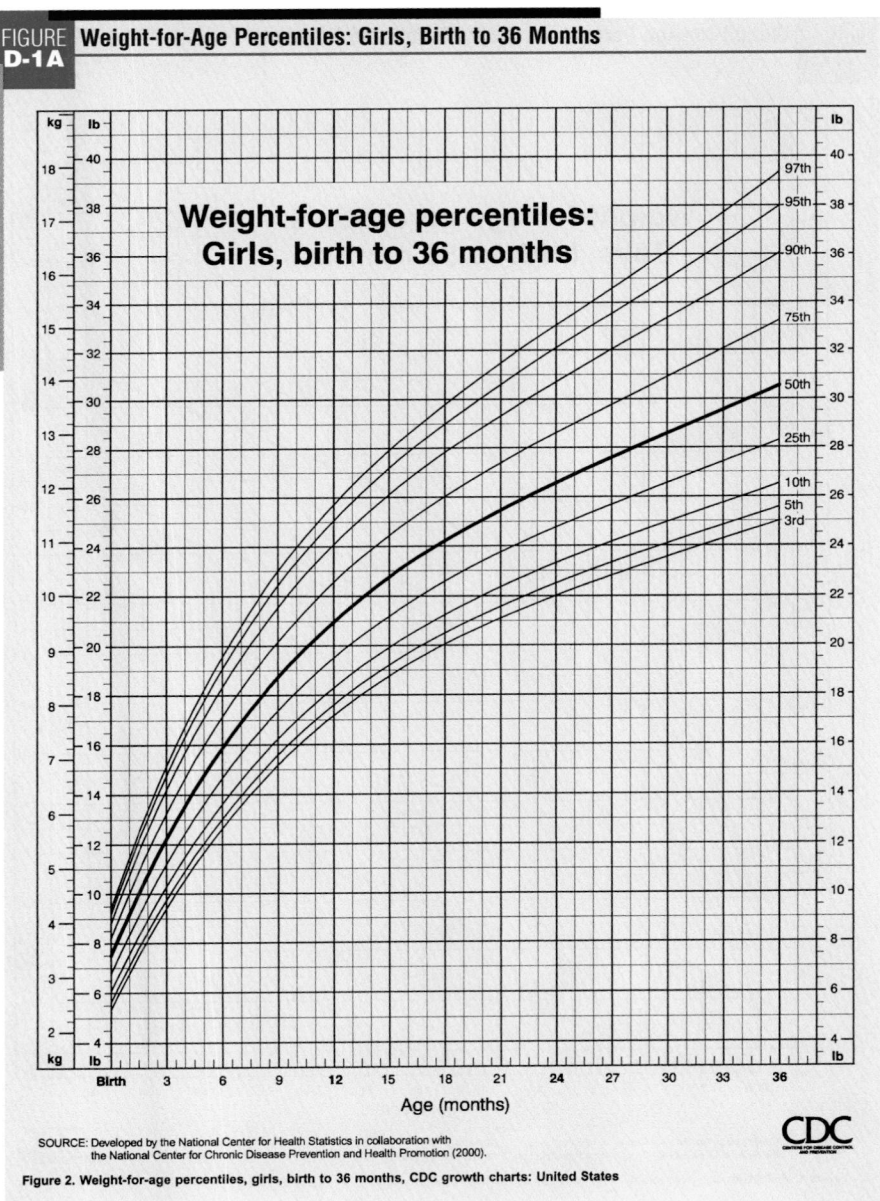

FIGURE D-1A **Weight-for-Age Percentiles: Girls, Birth to 36 Months**

Weight-for-age percentiles:
Girls, birth to 36 months

97th
95th
90th
75th
50th
25th
10th
5th
3rd

Age (months)

SOURCE: Developed by the National Center for Health Statistics in collaboration with the National Center for Chronic Disease Prevention and Health Promotion (2000).

CDC

Figure 2. Weight-for-age percentiles, girls, birth to 36 months, CDC growth charts: United States

Common sites for fatfold measures: ◆
- Triceps
- Subscapular (below shoulder blade)
- Suprailae (above hip bone)
- Abdomen
- Upper thigh

body lies directly beneath the skin, and the thickness of this subcutaneous fat reflects total body fat. In some parts of the body, such as the back and the back of the arm over the triceps muscle, this fat is loosely attached;◆ a person can pull it up between the thumb and forefinger to obtain a measure of fatfold thickness. To measure fatfold, a skilled assessor follows a standard procedure using reliable calipers (illustrated in Figure D-7 on p. D-8) and then compares the measurement with standards. Triceps fatfold measures greater than 15 millimeters in men or 25 millimeters in women suggest excessive body fat.

Fatfold measurements correlate directly with the risk of heart disease. They assess central obesity and its associated risks better than do weight measures alone. If a person gains body fat, the fatfold increases proportionately; if the person loses fat, it decreases. Measurements taken from central-body sites (around the abdomen) better reflect changes in fatness than those taken from upper sites

(continues on p. D-8)

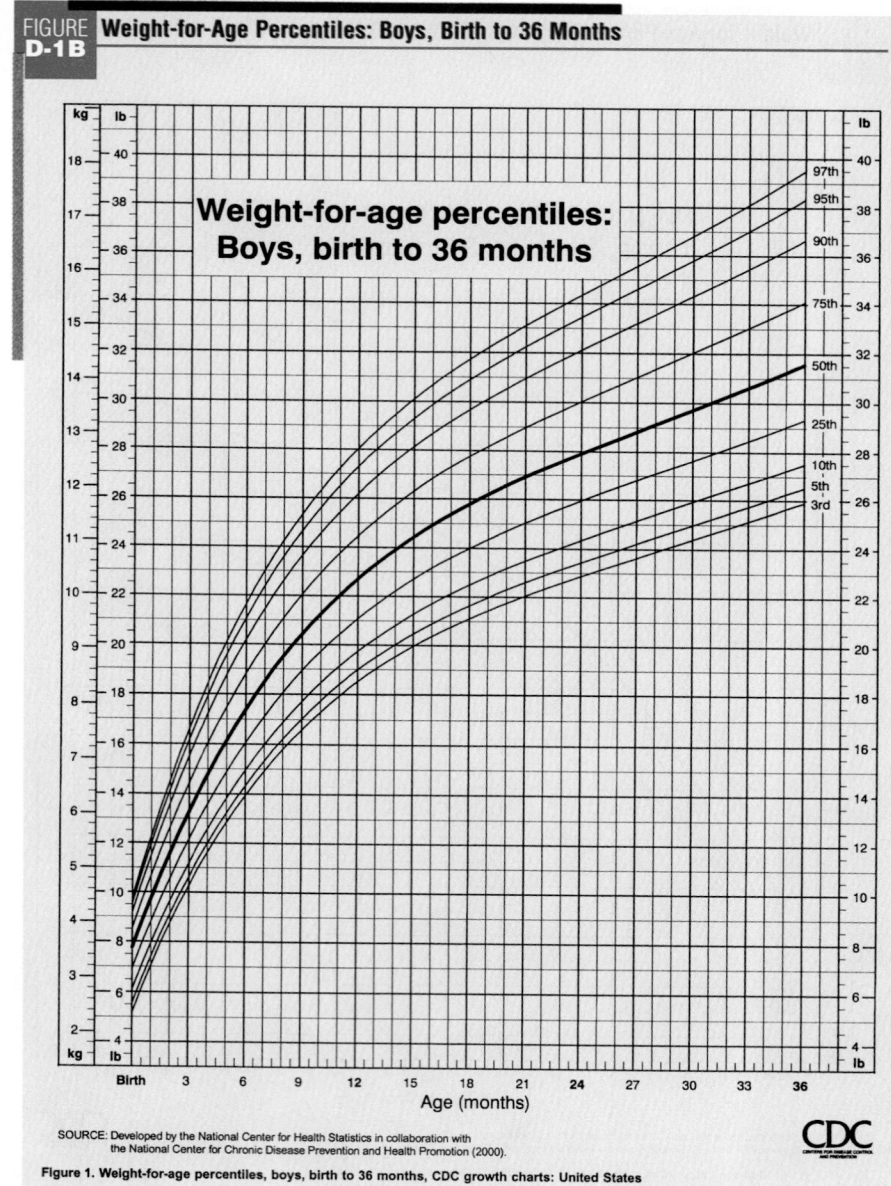

FIGURE D-1B

Weight-for-Age Percentiles: Boys, Birth to 36 Months

SOURCE: Developed by the National Center for Health Statistics in collaboration with the National Center for Chronic Disease Prevention and Health Promotion (2000).

Figure 1. Weight-for-age percentiles, boys, birth to 36 months, CDC growth charts: United States

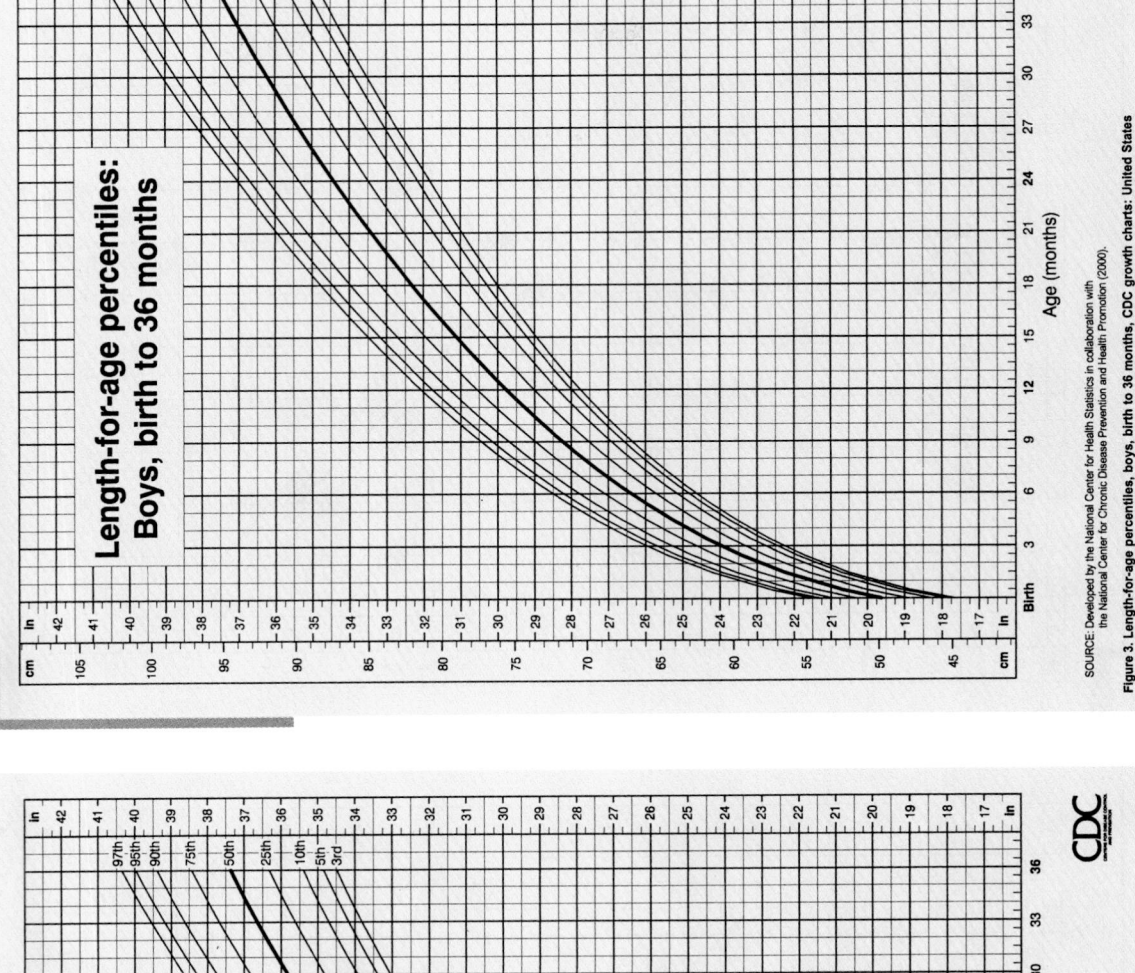

FIGURE **D-2B** Length-for-Age Percentiles: Boys, Birth to 36 Months

Length-for-age percentiles: Boys, birth to 36 months

SOURCE: Developed by the National Center for Health Statistics in collaboration with the National Center for Chronic Disease Prevention and Health Promotion (2000).

Figure 3. Length-for-age percentiles, boys, birth to 36 months, CDC growth charts: United States

FIGURE **D-2A** Length-for-Age Percentiles: Girls, Birth to 36 Months

Length-for-age percentiles: Girls, birth to 36 months

SOURCE: Developed by the National Center for Health Statistics in collaboration with the National Center for Chronic Disease Prevention and Health Promotion (2000).

Figure 4. Length-for-age percentiles, girls, birth to 36 months, CDC growth charts: United States

D

D

FIGURE D-3A Weight-for-Length Percentiles: Girls, Birth to 36 Months

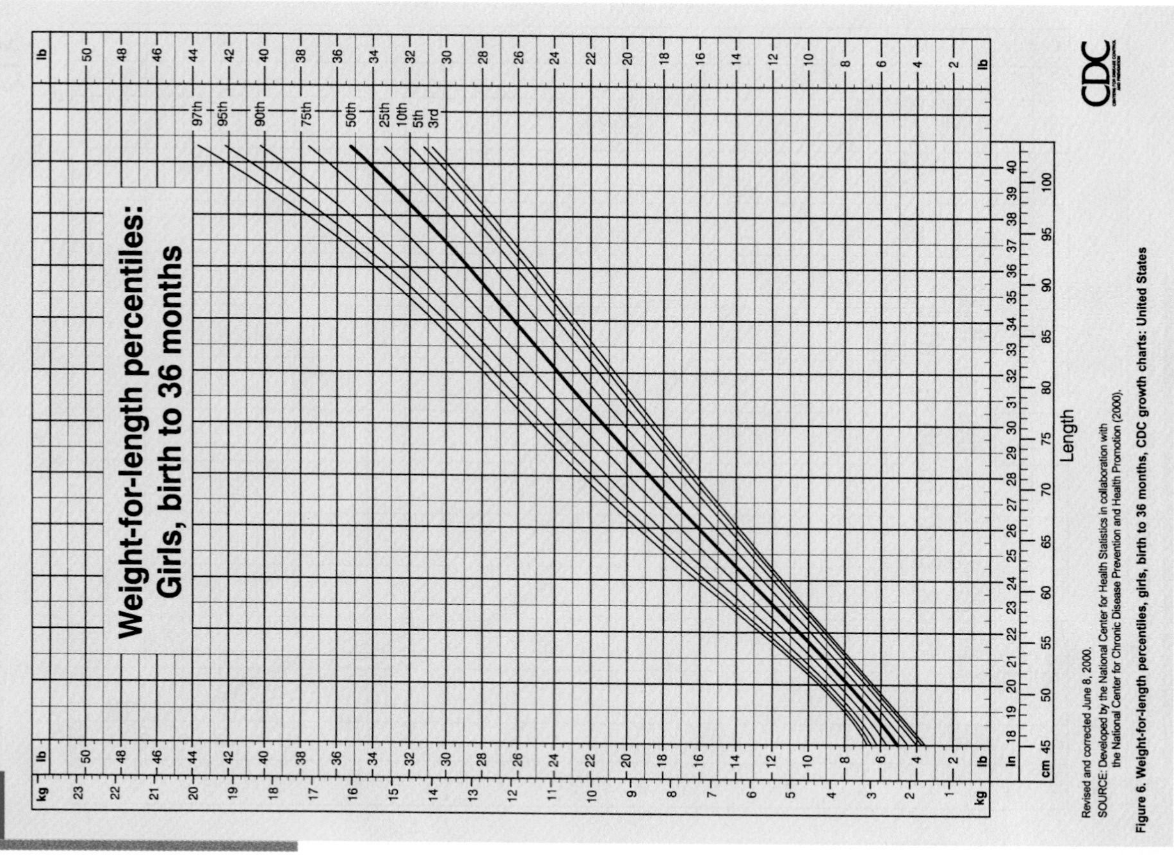
Weight-for-length percentiles:
Girls, birth to 36 months
97th
95th
90th
75th
50th
25th
10th
5th
3rd
kg
lb
in
cm
Length
CDC
Revised and corrected June 8, 2000.
SOURCE: Developed by the National Center for Health Statistics in collaboration with
the National Center for Chronic Disease Prevention and Health Promotion (2000).
Figure 6. Weight-for-length percentiles, girls, birth to 36 months, CDC growth charts: United States

FIGURE D-3B Weight-for-Length Percentiles: Boys, Birth to 36 Months

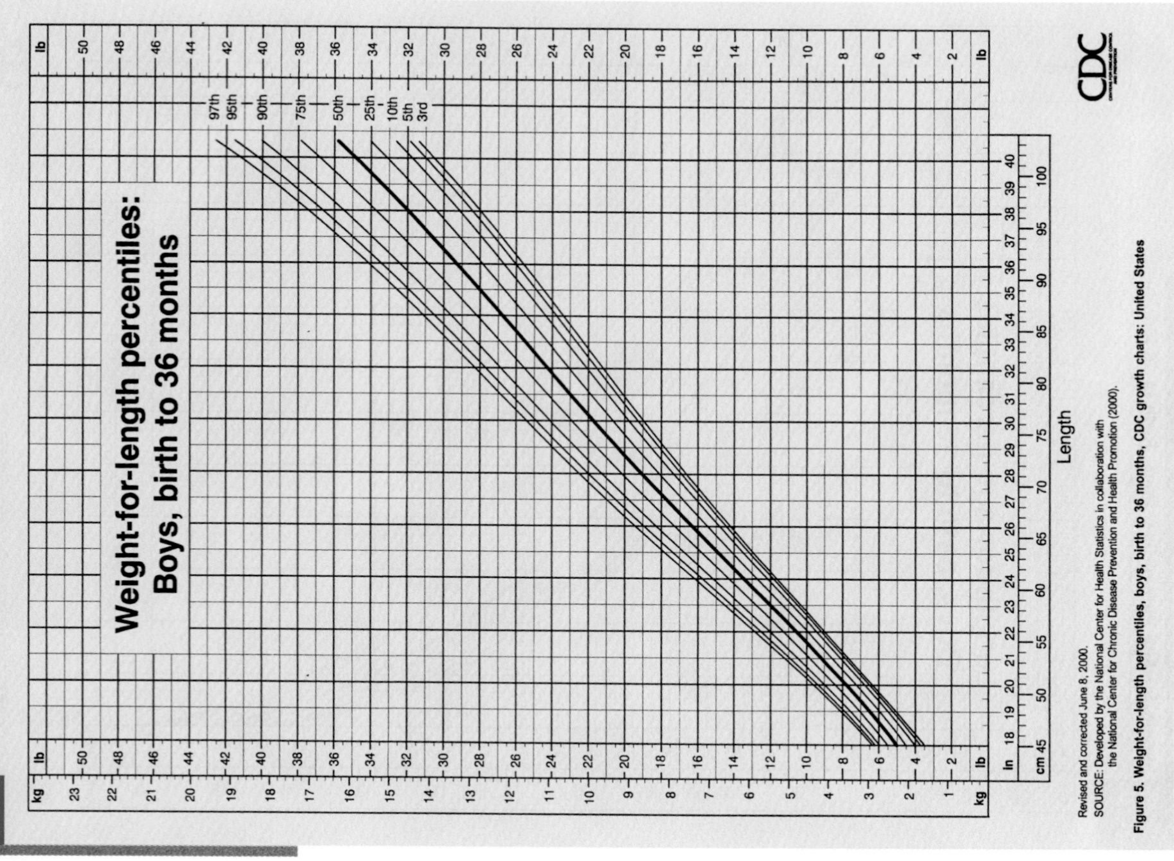
Weight-for-length percentiles:
Boys, birth to 36 months
97th
95th
90th
75th
50th
25th
10th
5th
3rd
kg
lb
in
cm
Length
CDC
Revised and corrected June 8, 2000.
SOURCE: Developed by the National Center for Health Statistics in collaboration with
the National Center for Chronic Disease Prevention and Health Promotion (2000).
Figure 5. Weight-for-length percentiles, boys, birth to 36 months, CDC growth charts: United States

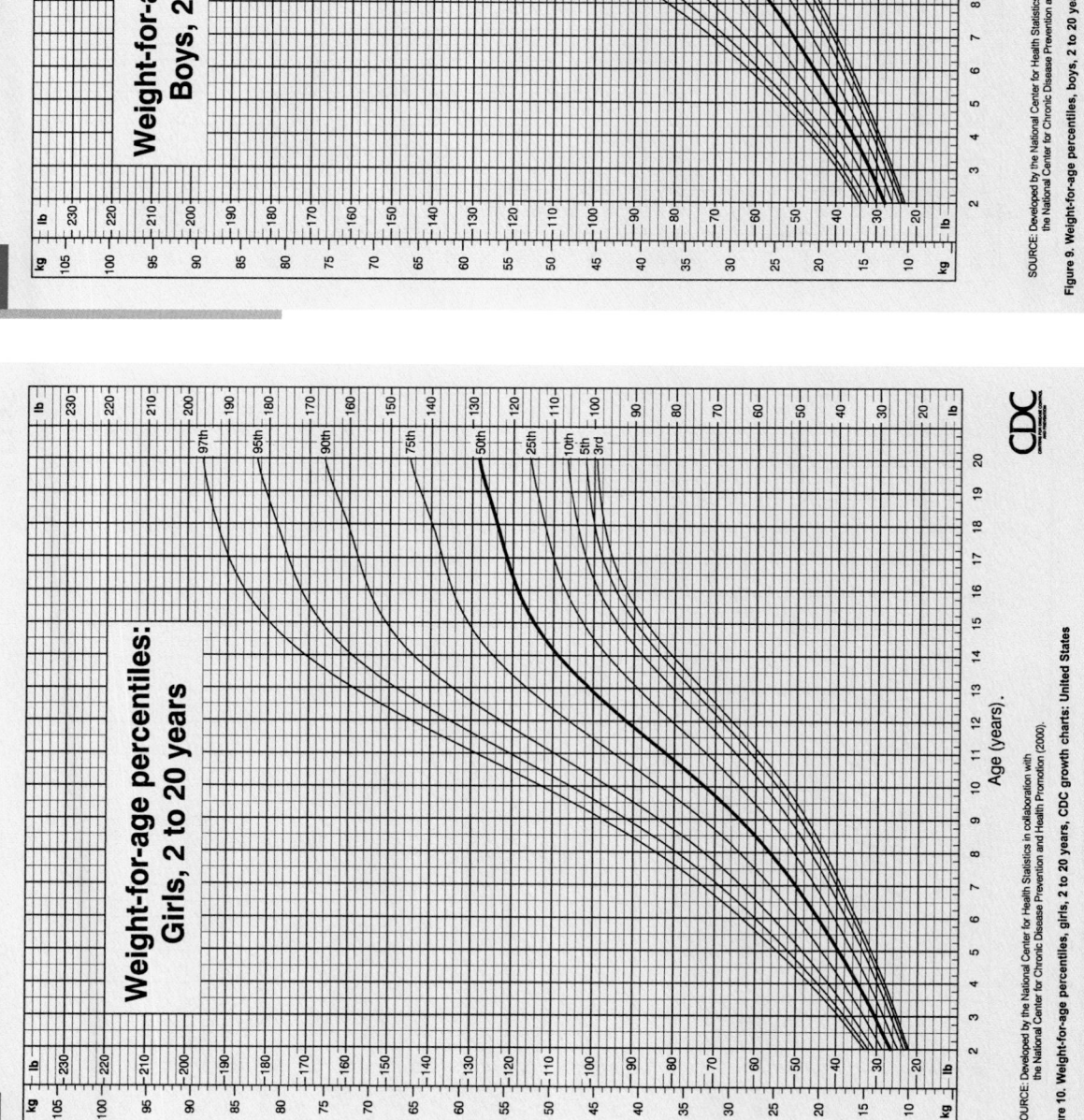

FIGURE
D-4B Weight-for-Age Percentiles: Boys, 2 to 20 Years

Weight-for-age percentiles:
Boys, 2 to 20 years

SOURCE: Developed by the National Center for Health Statistics in collaboration with
the National Center for Chronic Disease Prevention and Health Promotion (2000).

Figure 9. Weight-for-age percentiles, boys, 2 to 20 years, CDC growth charts: United States

FIGURE
D-4A Weight-for-Age Percentiles: Girls, 2 to 20 Years

Weight-for-age percentiles:
Girls, 2 to 20 years

SOURCE: Developed by the National Center for Health Statistics in collaboration with
the National Center for Chronic Disease Prevention and Health Promotion (2000).

Figure 10. Weight-for-age percentiles, girls, 2 to 20 years, CDC growth charts: United States

D

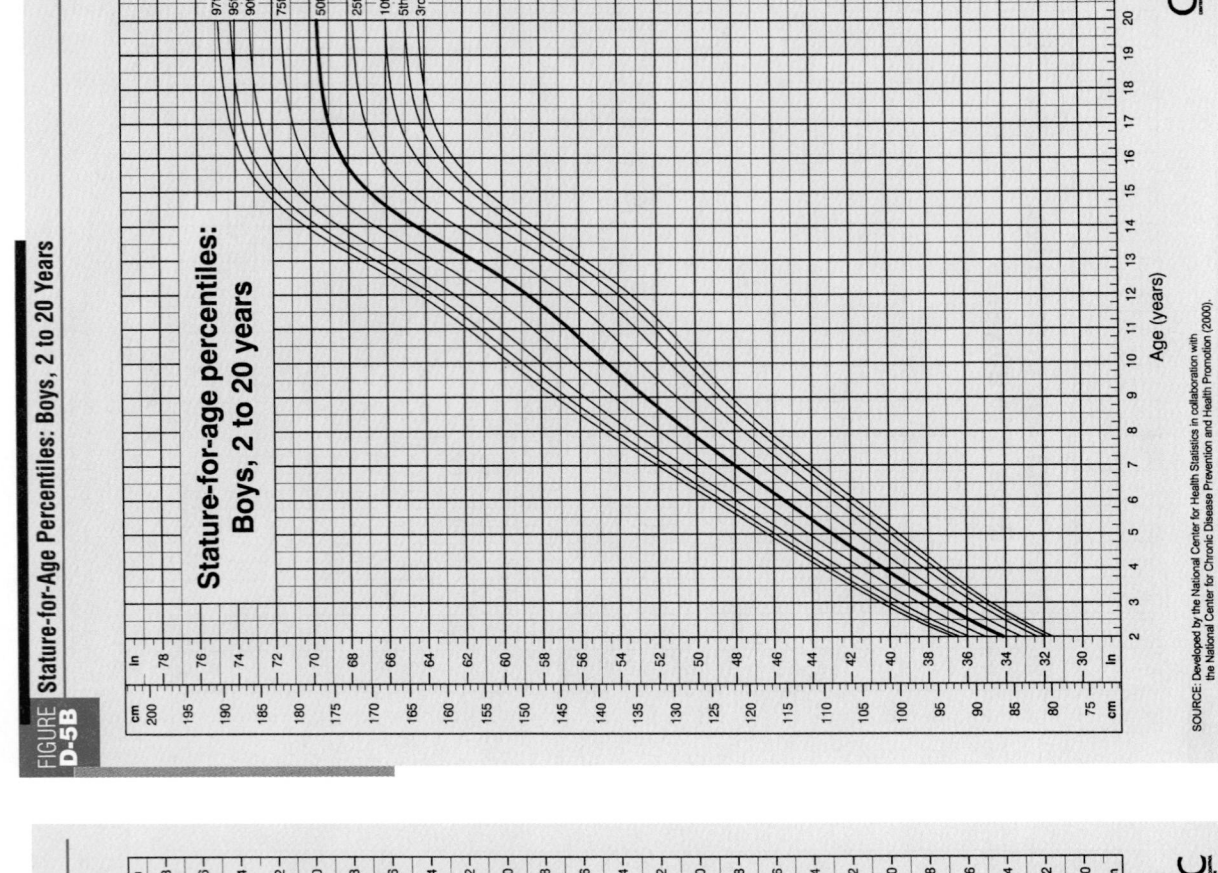

Stature-for-age percentiles:
Boys, 2 to 20 years

SOURCE: Developed by the National Center for Health Statistics in collaboration with
the National Center for Chronic Disease Prevention and Health Promotion (2000).

Figure 11. Stature-for-age percentiles, boys, 2 to 20 years, CDC growth charts: United States

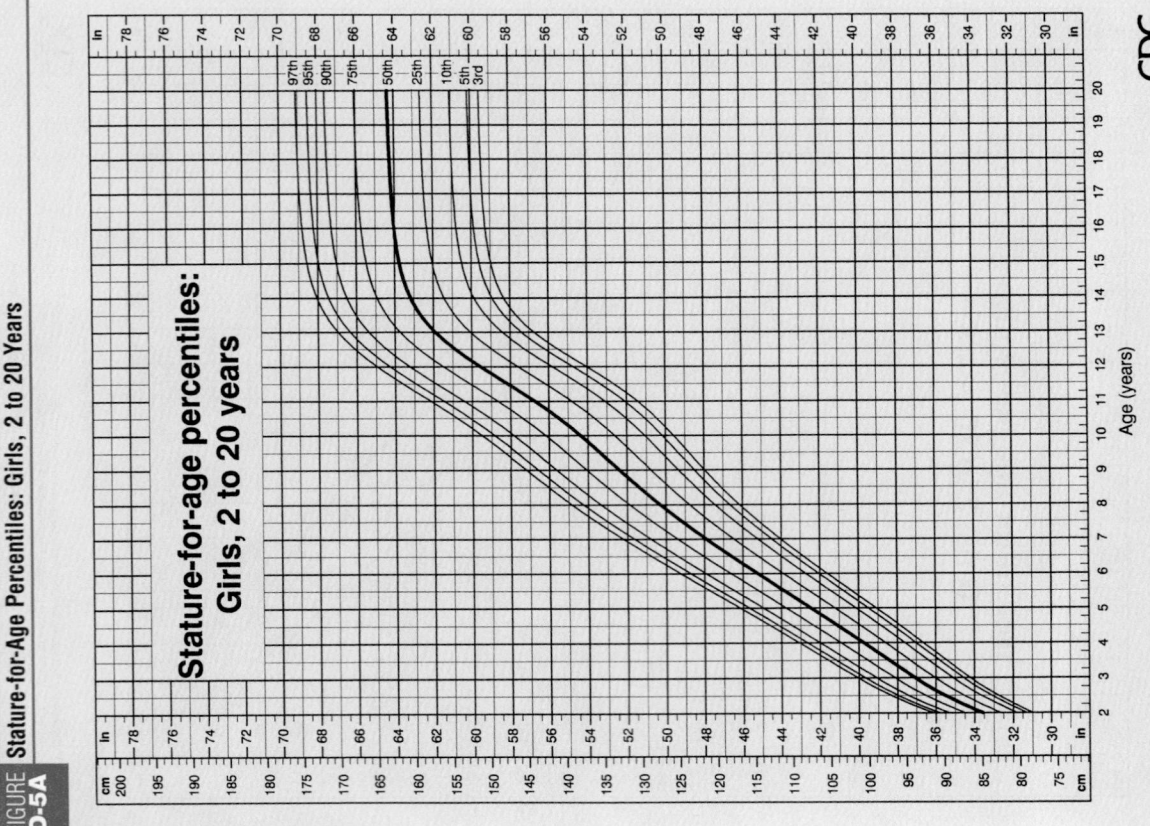

Stature-for-age percentiles:
Girls, 2 to 20 years

SOURCE: Developed by the National Center for Health Statistics in collaboration with
the National Center for Chronic Disease Prevention and Health Promotion (2000).

Figure 12. Stature-for-age percentiles, girls, 2 to 20 years, CDC growth charts: United States

FIGURE D-6B · Weight-for-Stature Percentiles: Boys, 2 to 20 Years

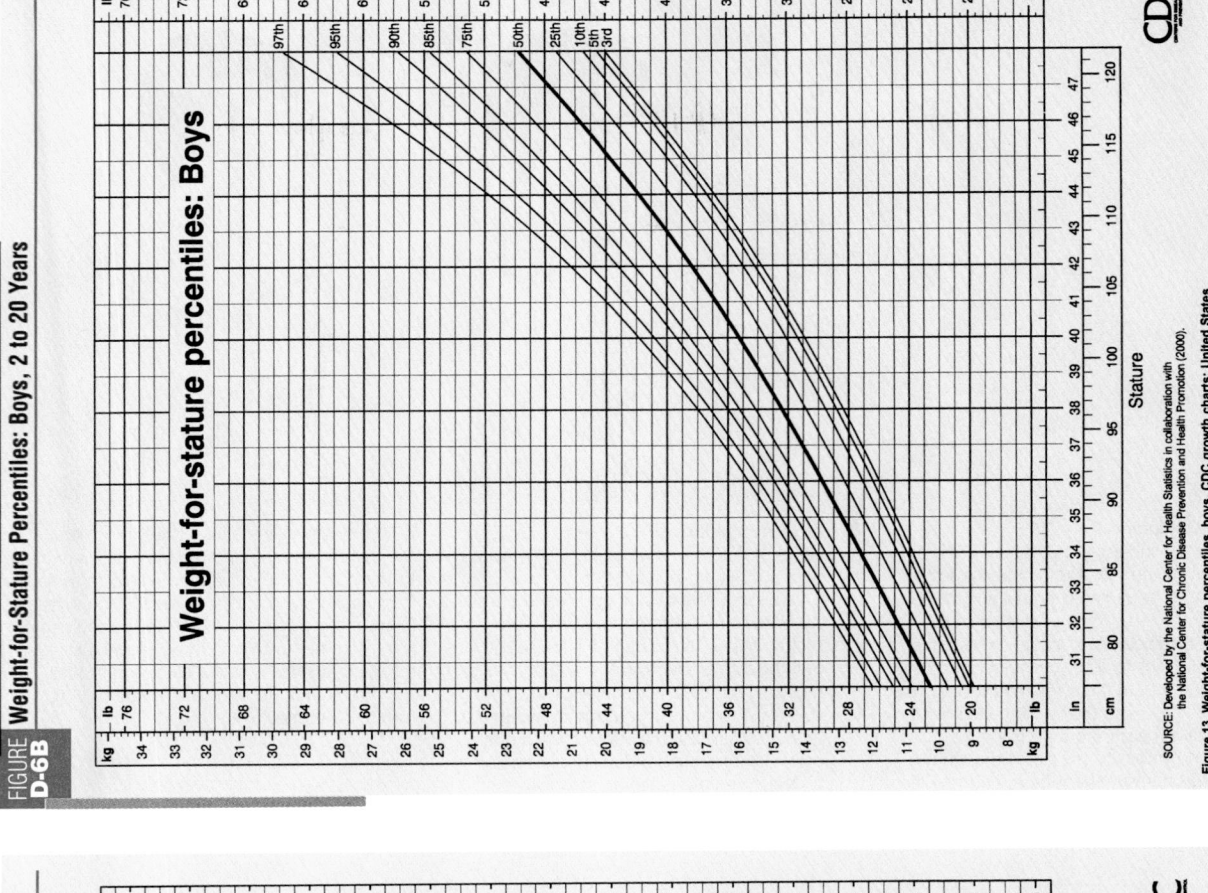

Weight-for-stature percentiles: Boys

SOURCE: Developed by the National Center for Health Statistics in collaboration with the National Center for Chronic Disease Prevention and Health Promotion (2000).

Figure 13. Weight-for-stature percentiles, boys, CDC growth charts: United States

FIGURE D-6A · Weight-for-Stature Percentiles: Girls, 2 to 20 Years

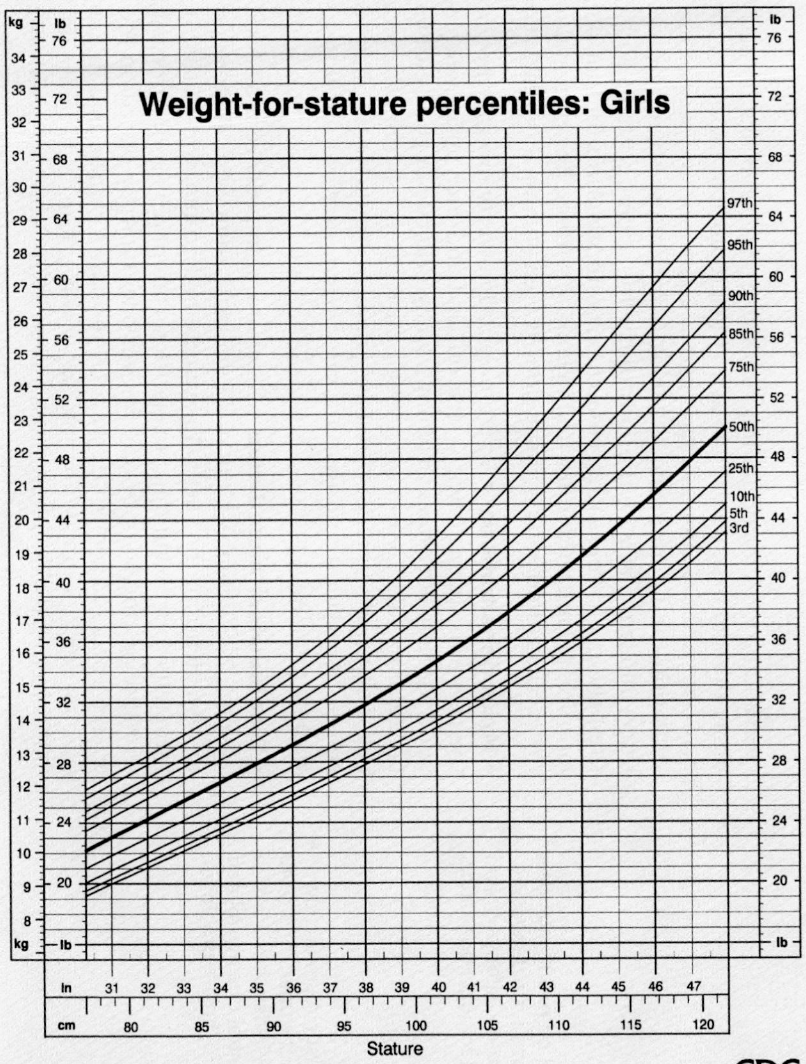

Weight-for-stature percentiles: Girls

SOURCE: Developed by the National Center for Health Statistics in collaboration with the National Center for Chronic Disease Prevention and Health Promotion (2000).

Figure 14. Weight-for-stature percentiles, girls, CDC growth charts: United States

D

FIGURE D-7 — How to Measure the Triceps Fatfold

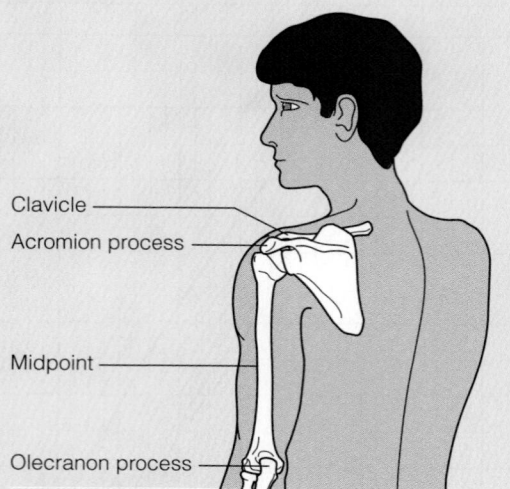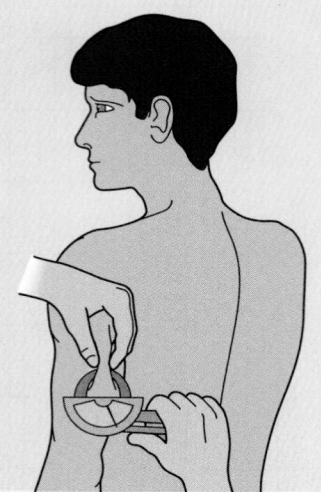

Clavicle
Acromion process
Midpoint
Olecranon process

A. Find the midpoint of the arm:
1. Ask the subject to bend his or her arm at the elbow and lay the hand across the stomach. (If he or she is right-handed, measure the left arm, and vice versa.)
2. Feel the shoulder to locate the acromion process. It helps to slide your fingers along the clavicle to find the acromion process. The olecranon process is the tip of the elbow.
3. Place a measuring tape from the acromion process to the tip of the elbow. Divide this measurement by 2, and mark the midpoint of the arm with a pen.

B. Measure the fatfold:
1. Ask the subject to let his or her arm hang loosely to the side.
2. Grasp a fold of skin and subcutaneous fat between the thumb and forefinger slightly above the midpoint mark. Gently pull the skin away from the underlying muscle. (This step takes a lot of practice. If you want to be sure you don't have muscle as well as fat, ask the subject to contract and relax the muscle. You should be able to feel if you are pinching muscle.)

3. Place the calipers over the fatfold at the midpoint mark, and read the measurement to the nearest 1.0 millimeter in two to three seconds. (If using plastic calipers, align pressure lines, and read the measurement to the nearest 1.0 millimeter in two to three seconds.)
4. Repeat steps 2 and 3 twice more. Add the three readings, and then divide by 3 to find the average.

Figure 16-3 on p. 000 includes ◆ photos of triceps fatfold measurement, hydro-densitometry, and bioelectrical impedance.

Reminder: A waist circumference of ◆ > 35 inches in women and > 40 inches in men is associated with a high risk of central obesity-related health problems.

(arm and back). A major limitation of the fatfold test is that fat may be thicker under the skin in one area than in another. A pinch at the side of the waistline may not yield the same measurement as a pinch on the back of the arm. This limitation can be overcome by taking fatfold measurements at several (often three) different places on the body (including upper-, central-, and lower-body sites) and comparing each measurement with standards for that site. Multiple measures are not always practical in clinical settings, however, and most often, the triceps fatfold measurement alone is used because it is easily accessible.◆

• Hydrodensitometry • To estimate body density using hydrodensitometry, the person is weighed twice—first on land and then again when submerged under water. Underwater weighing usually generates a good estimate of body fat and is useful in research, although the technique has drawbacks: it requires bulky, expensive, and nonportable equipment. Furthermore, submerging some people (especially those who are very young, very old, ill, or fearful) under water is not always practical.

• Bioelectrical Impedance • Bioelectrical impedance provides a means to estimate body composition using portable equipment that incurs minimal inconvenience to the client. To evaluate body composition using the bioelectrical impedance technique, an electrical current of very low intensity is briefly sent through the body by way of electrodes typically placed on the wrist and ankles. The measurement of electrical resistance is then used in a mathematical equation to estimate the total body water, lean body mass, and body fat. Since electrolyte-containing fluids, which readily conduct electrical current, are found primarily in lean tissues, the leaner the person, the less resistance to the current. To provide reliable results, bioelectrical impedance requires standardized procedures, calibrated instruments, and mathematical equations that are population- and diagnosis-specific.

Laboratory Tests of Nutrition Status

As Chapter 17 pointed out, blood and urine tests provide valuable information about nutrition status. Table D-2 shows laboratory tests that help assess vitamin and mineral status.

TABLE D-2	Biochemical Tests Useful for Assessing Vitamin and Mineral Status
Nutrient	**Assessment Tests**
Vitamins	
Vitamin A	Serum retinol, retinol-binding protein
Thiamin[a]	Erythrocyte (red blood cell) transketolase activity, erythrocyte thiamin pyrophosphate
Riboflavin[a]	Erythrocyte glutathione reductase activity
Vitamin B_6[a]	Urinary xanthurenic acid excretion after tryptophan load test, erythrocyte transaminase activity, plasma pyridoxal 5'-phosphate (PLP)
Niacin	Plasma or urinary metabolites NMN (N-methyl nicotinamide) or 2-pyridone, or preferably both expressed as a ratio
Folate[b]	Serum folate, erythrocyte folate (reflects liver stores)
Vitamin B_{12}[b]	Serum vitamin B_{12}, serum and urinary methylmalonic acid, Schilling test
Biotin	Urinary biotin, urinary 3-hydroxyisovaleric acid
Vitamin C	Plasma vitamin C[c], leukocyte vitamin C
Vitamin D	Serum vitamin D
Vitamin E	Serum α-tocopherol, erythrocyte hemolysis
Vitamin K	Serum vitamin K, plasma prothrombin; blood-clotting time (prothrombin time) is not an adequate indicator
Minerals	
Phosphorus	Serum phosphate
Sodium	Serum sodium
Chloride	Serum chloride
Potassium	Serum potassium
Magnesium	Serum magnesium, urinary magnesium
Iron	Hemoglobin, hematocrit, serum ferritin, total iron-binding capacity (TIBC), erythrocyte protoporphyrin, serum iron, transferrin saturation
Iodine	Serum thyroxine or thyroid-stimulating hormone (TSH), urinary iodine
Zinc	Plasma zinc, hair zinc
Copper	Erythrocyte superoxide dismutase, serum copper, serum ceruloplasmin
Selenium	Erythrocyte selenium, glutathione peroxidase activity

[a]Urinary measurements for these vitamins are common, but may be of limited use. Urinary measurements reflect recent dietary intakes and may not provide reliable information concerning the severity of a deficiency.
[b]Folate assessments should always be conducted in conjunction with vitamin B_{12} assessments (and vice versa) to help distinguish the cause of common deficiency symptoms.
[c]Vitamin C shifts between the plasma and the white blood cells known as leukocytes; thus a plasma determination may not accurately reflect the body's pool. A measurement of leukocyte vitamin C can provide information about the body's stores of vitamin C. A combination of both tests may be more reliable than either one alone.
SOURCE: Adapted from H. E. Sauberlich, *Laboratory Tests for the Assessment of Nutritional Status* (Boca Raton, Fla.: CRC Press, 1999).

D

D

• **Nutrition-Related Anemia** • Anemia, a symptom of a wide variety of nutrition- and nonnutrition-related disorders, is characterized by a reduced number of red blood cells. Iron, folate, and vitamin B_{12} deficiencies caused by inadequate intake, poor absorption, or abnormal metabolism of these nutrients are the most common nutritional anemias. Some nonnutrition-related causes of anemia include massive blood loss, infections, hereditary blood disorders, such as sickle-cell anemia, and chronic liver or kidney disease. Table D-3 lists laboratory tests that help to define anemia and distinguish among its major nutrition-related causes. Table D-4 on p. D-11 provides standards for tests used to detect iron status (hemoglobin, hematocrit, serum ferritin, total iron binding capacity, serum iron transferrin saturation, erythrocyte protoporphyrin). Table D-5 on p. D-11 shows standards for assessing folate and vitamin B_{12} status.

• **Nitrogen Balance Studies** • Nitrogen balance studies provide information about protein status and helps assessors determine a person's protein requirements. Nitrogen balance studies require that food intake be recorded and urine be collected during the same time period, often for 24 hours. Much effort is wasted if the proper techniques are not conscientiously followed or communications between health care professionals and clients are not clear. Nurses or assistants caring for clients must collect and save all urine samples and record food intake data or instruct clients on how to perform these tasks. Each urine sample is added to a collection container and refrigerated until the collection is complete. If even one urine sample is spilled or discarded, the test is invalid. Furthermore, any protein provided intravenously or by tube must be included as part of the protein intake. Nitrogen intake is calculated from the protein intake, and urine urea nitrogen (UUN) is measured from the urine sample. Nitrogen balance equals the nitrogen intake minus the nitrogen output.

TABLE D-3   Laboratory Tests Useful in Evaluating Nutrition-Related Anemias

Test or Test Result	What It Reflects
General Tests for Anemia	
Hemoglobin (Hg)	Total amount of hemoglobin in the red blood cells (RBC)
Hematocrit (Hct)	Percentage of RBC in the total blood volume
Red blood cell (RBC) count	Number of RBC
Mean corpuscular volume (MCV)	RBC size; helps to determine if anemia is microcytic (iron deficiency) or macrocytic (folate or vitamin B_{12} deficiency)
Mean corpuscular hemoglobin hypochromic concentration (MCHC)	Hemoglobin concentration within the average RBC; helps to determine if anemia is (iron deficiency) or normochromic (folate or vitamin B_{12} deficiency)
Bone marrow aspiration	The manufacture of blood cells in different developmental states
Iron Deficiency	
Serum ferritin	Early deficiency state (iron stores diminish)
Transferrin saturation	Progressing deficiency state (transport iron decreases)
Erythrocyte protoporphyrin	Later deficiency state (hemoglobin production falters)
Folate Deficiency	
Serum folate	Progressing deficiency state
RBC folate	Later deficiency state
Vitamin B_{12} Deficiency	
Serum vitamin B_{12}	Progressing deficiency state
Schilling test	Absorption of vitamin B_{12}

In most cases, protein contains about 16 percent nitrogen. Thus, assessors multiply the protein intake by 16 percent (0.16) or divide by 6.25—the mathematical equivalent—to calculate nitrogen intake. To determine nitrogen output, assessors must consider that in addition to nitrogen lost as UUN, the person also loses nitrogen through the skin, feces, and other sources of urinary nitrogen. To account for these losses, the assessor adds 4 grams of nitrogen to the UUN. Thus, nitrogen balance is calculated as follows:

$$\text{Nitrogen balance} = \frac{\text{24-hour protein intake (g)}}{6.25} - [\text{24-hour UUN (g)} + 4\text{ g}].$$

For example, if a woman's 24-hour protein intake is 90 grams and her 24-hour UUN is 17 grams, nitrogen balance can be calculated as follows:

$$\text{Nitrogen balance} = \frac{90\text{ g protein}}{6.25} - (17 + 4)$$

$$\text{Nitrogen balance} = 14.4\text{ g} - 21\text{ g} = -6.6\text{ g}.$$

The woman in this example is in negative nitrogen balance; that is, she is losing about 6.5 grams of nitrogen more than she is consuming.

TABLE D-4　Criteria for Assessing Iron Status

Test	Age (yr)	Gender	Deficiency Value
Hemoglobin (g/dL)	0.5–10	M–F	<11
	11–15	M	<12
		F	<11.5
	>15	M	<13
		F	<12
	Pregnancy		<11
Hematocrit (%)	0.5–4	M–F	<32
	5–10	M–F	<33
	11–15	M	<35
		F	<34
	>15	M	<40
		F	<36
Serum ferritin (μg/L)	0.5–15	M–F	<10
	>15	M–F	<12
Total iron-binding capacity (μg/dL)	>15	M–F	>400
Serum iron (μg/dL)	>15	M–F	<60
Transferrin saturation (%)	0.5–4	M–F	<12
	5–10	M–F	<14
	>10	M–F	<16
Erythrocyte protoporphyrin (μg/dL RBC)	0.5–4	M–F	>80
	>4	M–F	>70

TABLE D-5　Criteria for Assessing Folate and Vitamin B$_{12}$

	Deficient	Borderline	Acceptable
Serum folate (ng/ml)[a]	<3.0	3.0–5.9	>6.0
Erythrocyte folate (ng/ml)[a]	<140	140–159	>160
Serum vitamin B$_{12}$ (pg/ml)	<150	150–200	≥201
Serum methyl-malonic acid (nmol/L)	<376	—	—

[a]To convert folate values (ng/ml) to international standard units (nmol/L), multiply by 2.266.
Note: A nanogram (ng) is one-billionth of a gram; a picogram (pg) is one-trillionth of a gram.

AIDS TO CALCULATIONS

E

Many mathematical problems have been worked out as examples at appropriate places in the text. This appendix aims to help with the use of the metric system and with problems not fully explained elsewhere.

Conversion Factors

Conversion factors are useful mathematical tools in everyday calculations, including those encountered in the study of nutrition. A conversion factor is a fraction in which the numerator (top) and the denominator (bottom) express the same quantity in different units. One example used many times throughout this text is that 2.2 pounds (lb) and 1 kilogram (kg) are equivalent; they express the same weight. The conversion factor used to change pounds to kilograms or vice versa is:

$$\frac{2.2 \text{ lb}}{1 \text{ kg}} \text{ or } \frac{1 \text{ kg}}{2.2 \text{ lb}}.$$

Because both factors equal 1, measurements can be multiplied by the factor without changing the value of the measurement. Thus, the units can be changed.

To perform a conversion, use the factor with the unit you are seeking in the numerator (top) of the fraction. The following example illustrates the usefulness of conversion factors in the study of nutrition.

1. If you know that a 4-ounce hamburger contains 7 grams of saturated fat and you would like to determine how many grams (g) of saturated fat are contained in a 3-ounce (oz) hamburger, the conversion factor is:

$$\frac{7 \text{ g saturated fat}}{4 \text{ oz hamburger}}.$$

2. Multiply 3 ounces of hamburger by the conversion factor:

$$3 \text{ oz hamburger} \times \frac{7 \text{ g saturated fat}}{4 \text{ oz hamburger}} =$$

$$\frac{3 \times 7}{4} = \frac{21}{4} = 5 \text{ g saturated fat (rounded off to the nearest whole number).}$$

Percentages

A percentage is a comparison between a number of items (perhaps your intake of energy) and a standard number (perhaps the number of kcalories recommended for your age and sex—the energy RDA). The standard number is the number you divide by. The answer you get after the division must be multiplied by 100 to be stated as a percentage (*percent* means "per 100").

For example, if you wish to determine what percentage of the RDA your energy intake is, first find your energy RDA (see Appendix C). We'll use 2200 kcalories to demonstrate.

1. Total your energy intake for a day—for example, 1500 kcalories.
2. Divide your kcalorie intake by the RDA kcalories:

   1500 kcal (your intake) ÷ 2200 kcal (RDA) = 0.68.

3. Multiply your answer by 100 to state it as a percentage:

   $$0.68 \times 100 = 68 = 68\%.$$

In some problems in nutrition, the percentage may be more than 100. For example, suppose your daily intake of vitamin A is 3200 RE and your RDA (male) is 1000 RE. The following calculations show your vitamin A intake as a percentage of the RDA:

$$3200 \div 1000 = 3.2.$$

$$3.2 \times 100 = 320\% \text{ of RDA.}$$

Sometimes the comparison is between a part of a whole (for example, your kcalories from protein) and the total amount (your total kcalories). For example, suppose you wish to know what percentage of your total kcalories for the day come from protein, fat, and carbohydrate.

1. Find the total grams of protein, fat, and carbohydrate you consumed—for example, 60 grams protein, 80 grams fat, and 310 grams carbohydrate.
2. Multiply the number of grams by the number of kcalories from 1 gram of each energy nutrient (conversion factors):

$$60 \text{ g protein} \times \frac{4 \text{ kcal}}{1 \text{ g protein}} = 240 \text{ kcal.}$$

$$80 \text{ g fat} \times \frac{9 \text{ kcal}}{1 \text{ g fat}} = 720 \text{ kcal.}$$

$$310 \text{ g carbohydrate} \times \frac{4 \text{ kcal}}{1 \text{ g carbohydrate}} = 1240 \text{ kcal.}$$

$$240 + 720 + 1240 = 2200 \text{ kcal.}$$

CONTENTS

3. Find the percentage of total kcalories from each energy nutrient:

- Protein: 240 ÷ 2200 = 0.109 × 100 = 10.9 = 11% of kcal.
- Fat: 720 ÷ 2200 = 0.327 × 100 = 32.7 = 33% of kcal.
- Carbohydrate: 1240 ÷ 2200 = 0.563 × 100 = 56.3 = 56% of kcal.
- 11% + 33% + 56% = 100% of kcal (total).

The percentages total 100 percent, but sometimes they total 99 or 101 because of rounding off. This is a reasonable error.

Ratios

A ratio is a comparison of two or three values in which one of the values is reduced to 1. A ratio compares identical units and so is expressed without units. For example, to find the potassium-to-sodium ratio of your diet, you first need to know how many milligrams of potassium and sodium you consumed, say, 3000 milligrams potassium and 2500 milligrams sodium.

1. Divide the potassium milligrams by the sodium milligrams:

   3000 mg potassium ÷ 2500 mg sodium = 1.2.

2. The potassium-to-sodium ratio is usually expressed as correct to one decimal point: 1.2.

The potassium-to-sodium ratio of your diet is 1.2:1 (read as "one point two to one" or simply "one point two"). A ratio greater than 1 means that the first value (in this case, milligrams of potassium) is greater than the second (sodium). When the second value is larger, the ratio is less than 1.

Weights and Measures

Length
1 inch (in) = 2.54 centimeters.
1 foot (ft) = 30.48 centimeters.
1 meter (m) = 39.37 inches.

To find degrees Fahrenheit (t_F) when you know degrees Celsius (t_C)*, multiply by 9/5 and then add 32:

$$9/5\ t_C + 32 = t_F.$$

To find degrees Celsius (t_C) when you know degrees Fahrenheit (t_F), multiply by 5/9 after subtracting 32:

$$5/9\ (t_F - 32) = t_C.$$

Volume
1 liter (L) = 1.06 quarts (qt) or 0.85 imperial quart.
1 liter = 1000 milliliters (ml).
1 milliliter = 0.03 fluid ounces.
1 gallon = 3.79 liters.
1 quart = 0.95 liter or 32 fluid ounces.
1 cup (c) = 8 fluid ounces.
1 tablespoon (tbs) = 15 milliliters.
3 teaspoons (tsp) = 1 tablespoon.
1 teaspoon = about 5 grams or 5 milliliters.
16 tablespoons = 1 cup.
4 cups = 1 quart.

Weight
1 ounce (oz) = approximately 28 grams (g).
16 ounces = 1 pound (lb).
1 pound = 454 grams.
1 kilogram (kg) = 1000 grams or 2.2 pounds.
1 gram = 1000 milligrams (mg).
1 milligram = 1000 micrograms (μg).

Energy units
1 kcalorie (kcal) = 4.2 kilojoules (kJ).
1 millijoule (mJ) = 240 kcalories.
1 kilojoule = 0.24 kcalories.
1 g carbohydrate = 4 kcal = 17 kJ.
1 g fat = 9 kcal = 37 kJ.
1 g protein = 4 kcal = 17 kJ.
1 g alcohol = 7 kcal = 29 kJ.

*Also known as *centigrade*.

ENTERAL FORMULAS

The staggering number of enteral formulas available allows health care professionals to meet a variety of their clients' medical needs, but also complicates the process of selecting an appropriate formula. The first step in narrowing the choice of formulas is to determine the client's ability to digest and absorb nutrients. Table F-1 on pp. F-1 through F-4 lists examples of intact formulas for clients with the ability to digest and absorb nutrients, and Table F-2 on pp. F-5 through F-6 provides examples of hydrolyzed formulas for clients with limited ability to digest and absorb nutrients. Products promoted to the general public and intended primarily as oral supplements, including Carnation Instant Breakfast® (Nestlé), Boost® (Mead Johnson), and Ensure

(Ross) are not included as examples. Each formula is listed only once, although the formula may have more than one use. A high-protein formula, for example, may also be a fiber-containing formula. Tables F-3 through F-5 on p. F-7 list modular formulas. Although this appendix provides many examples, the list of formulas is not complete. The information reflects the manufacturer's literature and does not suggest endorsement by the authors. Manufacturers frequently add new formulas and stop production or rename formulas. Be aware that formula composition changes periodically. Consult manufacturer's literature and web sites for updates. The following products are listed in this appendix:

- B. Braun[a]
    Hepatic-Aid® II
    Immun-Aid®

- Mead Johnson Nutritionals[b]
    Casec®
    Choice DM®
    Comply®
    Criticare HN®
    Deliver® 2.0
    Isocal®
    Isocal® HN
    Kindercal® TF
    Kindercal® TF with Fiber
    Lipisorb®
    Magnacal® Renal
    MCT Oil®
    Microlipid®
    Moducal®
    Protain XL®
    Respalor®
    TraumaCal®
    Ultracal®

- Nestlé Clinical Nutrition[c]
    Crucial®
    Glytrol®
    Nutren ®1.0
    Nutren® 1.0 with Fiber
    Nutren® 1.5
    Nutren® 2.0
    Nutren® Junior
    Nutren® Junior with Fiber
    NutriHep®
    NutriRenal®
    NutriVent®
    Peptamen®
    Peptamen® 1.5
    Peptamen® Junior
    Peptamen® VHP
    ProBalance®
    Reabilan®
    Reabilan® HN
    Renalcal® Diet
    Replete®
    Replete® with Fiber

- Novartis Nutrition Corporation[d]
    Compleat® Modified
    Compleat® Pediatric
    Diabetisource®
    Fibersource®
    Fibersource® HN
    Impact®
    Impact® Glutamine
    Impact® 1.5
    Impact® with Fiber
    Isosource® Standard
    Isosource® HN
    Isosource® VHN
    Isosource® 1.5
    Novasource® Pulmonary
    Novasource® Renal
    Novasource® 2.0
    Resource® Diabetic
    Sandosource® Peptide
    Tolerex®
    Vivonex® Pediatric
    Vivonex® Plus
    Vivonex® T.E.N.

- Ross Laboratories[e]
    Advera®
    Alitraq®
    Glucerna®
    Jevity®
    Jevity® Plus
    Nepro®
    Optimental®
    Osmolite®
    Osmolite® HN
    Osmolite® HN Plus
    Oxepa®
    PediaSure®
    PediaSure® with Fiber
    Perative®
    Polycose®
    ProMod®
    Promote®
    Promote® with Fiber
    Pulmocare®
    Suplena®
    TwoCal® HN
    Vital® HN

---

[a]*Partners in Caring* (1998), B. Braun Medical Inc., Irvine, CA 92614.
[b]Mead Johnson Nutritionals, www.meadjohnson.com, visited April 6, 2001.
[c]Nestlé Clinical Nutrition, www.nestleclinicalnutrition.com, visited April 6, 2001.
[d]*Pocket Guide* (2001), Novartis Nutrition, Minneapolis, MN 55440.
[e]Ross Products, www.ross.com, visited April 6, 2001.

TABLE **F-1** Intact Protein Formulas

Product[a]	Volume to Meet 100% RDI (ml)	Energy (kcal/ml)	Protein or Amino Acids (g/L)	Carbohydrate (g/L)	Fat (g/L)	Osmolality[b] (mOsm/kg)	Notes
Lactose-Free, Standard Formulas							
Compleat® Modified	1500	1.07	43	140	37	300	4.3 g fiber/L
Isocal®	1890	1.06	34	135	44	270	20% fat from MCT
Isosource Standard®	1165	1.20	43	170	39	490	50% fat from MCT
Nutren® 1.0	1500	1.00	40	127	38	315	25% fat from MCT
Osmolite®	2000	1.06	37	151	35	300	20% fat from MCT
Lactose-Free, Fiber-Containing Formulas							
Fibersource® Standard	1165	1.20	43	170	39	490	10 g fiber/L
Impact® with Fiber	1500	1.00	56	140	28	375	10 g fiber/L, enriched with arginine, nucleotides, and omega-3 fatty acids
Jevity®	1321	1.06	44	155	35	300	14 g fiber/L
Nutren® 1.0 with Fiber	1500	1.00	40	127	38	320	14 g fiber/L
ProBalance®	1000	1.20	54	156	41	350	10 g fiber/L
Promote® with Fiber	1000	1.00	63	138	28	370	14 g fiber/L
Replete® with Fiber	1000	1.00	63	113	34	310	14 g fiber/L
Ultracal®	1120	1.06	45	142	39	360	14.4 g fiber/L

[a]Formulas come in ready-to-use (liquid) form unless specified under "Notes."
[b]Osmolality may vary, depending on the flavorings added to a product.
MCT = Medium-chain triglycerides.

TABLE F-1 Intact Protein Formulas (continued)

Product[a]	Volume to Meet 100% RDI (ml)	Energy (kcal/ml)	Protein or Amino Acids (g/L)	Carbohydrate (g/L)	Fat (g/L)	Osmolality[b] (mOsm/kg)	Notes
Lactose-Free, High-kCalorie Formulas							
Comply®	830	1.50	60	180	61	460	
Deliver® 2	1000	2.00	75	200	102	640	
Isosource® 1.5	933	1.50	68	170	65	650	8 g fiber/L
Novasource® 2.0	948	2.00	90	220	88	790	
Nutren® 1.5	1000	1.50	60	169	68	430	50% fat from MCT
Nutren® 2.0	750	2.00	80	196	106	745	75% fat from MCT
Osmolite® HN Plus	1000	1.20	56	158	39	360	19% fat from MCT
TwoCal HN®	947	2.00	84	219	91	730	
Lactose-Free, High-Protein Formulas							
Fibersource® HN	1165	1.20	53	160	39	490	7 g fiber/L
Isocal® HN	1180	1.06	45	124	45	270	20% fat from MCT
Isocal® HN Plus	1000	1.20	54	156	40	390	Part of fat from MCT
Isosource® HN	1165	1.20	53	160	39	401	
Isosource® VHN	1250	1.00	62	130	29	300	10g fiber/L
Jevity® Plus	1000	1.20	56	173	39	450	12 g fiber/L
Lipisorb®	1180	1.35	57	161	57	630	85% fat from MCT
Osmolite® HN	1321	1.06	44	144	35	300	Low-residue, 20% fat from MCT
Promote®	1000	1.00	63	130	26	340	Fat primarily from polyunsaturated and monounsaturated sources, 20% fat from MCT
Ultracal® HN Plus	1000	1.20	54	156	40	370	30% fat from MCT, 11 g fiber/L
Special-Use Formulas: Pediatric (1 to 10 years)							
Compleat® Pediatric	Varies[c]	1.00	38	130	39	380	Blenderized formula, 4 g fiber/L
Kindercal® TF	Varies[c]	1.06	30	135	44	345	20% fat from MCT
Kindercal® TF with Fiber	Varies[c]	1.06	30	135	44	345	6 g fiber/L
Nutren Junior®	Varies[c]	1.00	30	128	42	350	25% fat from MCT
Nutren Junior® with Fiber	Varies[c]	1.00	30	128	42	350	25% fat from MCT, 6 g fiber/L
PediaSure®	Varies[c]	1.00	30	110	50	335	
PediaSure® Enteral Formula with Fiber	Varies[c]	1.00	30	114	45	345	7 g fiber/L

[c]Depends on age of child.

TABLE F-1 Intact Protein Formulas (continued)

Product[a]	Volume to Meet 100% RDI (ml)	Energy (kcal/ml)	Protein or Amino Acids (g/L)	Carbohydrate (g/L)	Fat (g/L)	Osmolality[b] (mOsm/kg)	Notes
Special-Use Formulas: Glucose Intolerance							
Choice dm® TF	946	1.06	45	119	51	300	14 g fiber/L
Diabetisource®	1500	1.00	50	90	49	360	4 g fiber/L
Glucerna®	1422	1.00	42	96	54	355	14 g fiber/L
Glytrol®	1400	1.00	45	100	48	380	15 g fiber/L; 20% fat from MCT
Resource® Diabetic	1180	1.06	63	100	47	320	13 g fiber/L
Special-Use Formulas: Immune System Support							
Immun-Aid®	2000	1.00	80	120	22	460	Powder form; enriched with arginine, glutamine, branched-chain amino acids, nucleic acids, omega-3 fatty acids, vitamins A, C, E, and B₆ and trace elements
Impact®	1500	1.00	56	130	28	375	Enriched with arginine, nucleic acids, and omega-3 fatty acids
Impact® 1.5	1250	1.50	84	140	69	550	Same as above
Impact® Glutamine	1000	1.30	78	150	43	630	Same as above and enriched with glutamine: 10 g fiber/L
Special-Use Formulas: Renal Insufficiency[d]							
Magnacal® Renal	—	2.00	75	200	101	570	High in monounsaturated fat, 20% fat from MCT; intended for use once hemodialysis has been instituted
Nepro®	—	2.00	70	222	96	665	High-calcium, low-phosphorus; intended for use once dialysis has been instituted
Novasource® Renal	—	2.00	74	200	100	700	Low in electrolytes; intended for use once dialysis has been instituted
NutriRenal®	—	2.00	70	205	104	N/A	50% fat from MCT; enriched with vitamins C and B₆, folate, zinc, and selenium; intended for use once dialysis has been instituted
Renalcal® Diet	—	2.00	35	290	82	NA	Contains histidine, 70% fat from MCT; intended for use before dialysis is instituted
Suplena®	—	2.00	30	255	96	600	Low in electrolytes; intended for use before dialysis is instituted

[d] Renal formulas are intentionally low in essential minerals and thus do not meet the RDI in commonly administered volumes.

F

Intact Protein Formulas (continued)

Product[a]	Volume to Meet 100% RDI (ml)	Energy (kcal/ml)	Protein or Amino Acids (g/L)	Carbohydrate (g/L)	Fat (g/L)	Osmolality[b] (mOsm/kg)	Notes
Special-Use Formulas: Respiratory Insufficiency							
Novasource® Pulmonary	933	1.50	75	150	68	650	8 g fiber/L
NutriVent®	1000	1.50	68	100	94	330	55% kcal from fat, 40% fat from MCT
Oxepa	947	1.50	63	106	94	493	55% kcal from fat, enriched with antioxidant nutrients
Pulmocare®	1136	1.50	63	106	93	475	55% kcal from fat, 20% fat from MCT, enriched with antioxidant nutrients
Respalor®	1000	1.50	75	146	68	400	41% kcal from fat; 30% fat from MCT; enriched with vitamins C and E, B-complex vitamins, zinc and trace elements
Special-Use Formulas: Wound Healing							
Protain XL®	—	1.00	57	145	30	340	9 g fiber/L, 20% fat from MCT, enriched with vitamins A and C and zinc
Replete®	1000	1.00	63	113	34	300	Enriched with vitamins A and C and zinc; 25% fat from MCT
TraumaCal®	2000	1.50	82	142	68	560	Enriched with vitamins C, B-complex, and E and copper and zinc

TABLE **F-2** Hydrolyzed Protein Formulas

Product	Volume to Meet 100% RDI (ml)	Energy (kcal/ml)	Protein or Amino Acids (g/L)	Carbohydrate (g/L)	Fat (g/L)	Osmolality (mOsm/kg)	Notes
Special-Use Hydrolyzed Formulas: Hepatic Insufficiency							
Hepatic Aid® II	N/A	1.20	44	169	36	560	Powder form; free amino acids, high in branched-chain amino acids, low in aromatic amino acids, no added vitamins or electrolytes
NutriHep®	1000	1.50	40	290	21	790	Free amino acids, 50% branched-chain amino acids, 50% aromatic amino acids, contains vitamins and minimal electrolytes
Special-Use Hydrolyzed Formulas: HIV Infection or AIDS							
Advera®	1184	1.28	60	216	23	680	78% hydrolyzed and 22% intact protein, low-fat, fiber added, enriched with vitamins E, C, B_6, B_{12}, and folate
Special-Use Hydrolyzed Formulas: Immune System Support							
Alitraq®	1500	1.00	53	165	16	575	Powder form; 47% free amino acids, 42% small peptides, enriched with glutamine and arginine
Crucial®	1000	1.50	94	135	68	490	Enriched with arginine, glutamine, antioxidant nutrients, and zinc
Perative®	1155	1.30	67	177	37	385	Enriched with arginine and b-carotene
Vivonex® Plus	1800	1.00	45	190	7	650	Powder form; 100% free amino acids, enriched with glutamine, arginine, and branched-chain amino acids

F

TABLE F-2 Hydrolyzed Protein Formulas (continued)

Product	Volume to Meet 100% RDI (ml)	Energy (kcal/ml)	Protein or Amino Acids (g/L)	Carbohydrate (g/L)	Fat (g/L)	Osmolality[a] (mOsm/kg)	Notes
Special-Use Hydrolyzed Formulas: Malabsorption							
Criticare HN®	1890	1.06	38	220	5	650	50% free amino acids, 50% small peptides
Optimental®	1422	1.5	51	139	28	540	Contains MCT and arginine; enriched with vitamins C and E and beta-carotene
Peptamen®	1500	1.00	40	127	39	270	70% fat from MCT; contains glutamine
Peptamen® 1.5	1000	1.5	60	191	59	450	70% fat from MCT; contains glutamine
Peptamen VHP®	1500	1.00	63	105	39	300	70% fat from MCT; contains glutamine
Reabilan®	2000	1.00	32	132	41	350	50% fat from MCT
Reabilan® HN	1500	1.33	58	158	54	490	50% fat from MCT
SandoSource® Peptide	1750	1.00	50	160	17	490	60% free amino acids and small peptides
Tolerex® acids	1800	1.00	21	230	1.5	550	Powder form; 100% free amino acids
Vital® HN	1500	1.00	42	185	11	500	Powder form; 87% hydrolyzed proteins, 13% essential amino acids
Vivonex® T.E.N.	2000	1.00	38	210	3	630	Powder form; 100% free amino acids, enriched with branched-chain amino acids
Special-Use Hydrolyzed Formulas: Pediatric (1 to 10 years)							
Peptamin Junior®	varies[b]	1.0	30	138	39	260	60% fat from MCT; contains glutamine
Vivonex® Pediatric	varies[b]	0.8	24	130	24	360	Powder form; 100% free amino acids

[a]Osmolality may vary depending on the flavorings added to a product.
[b]Depends on age of child.

TABLE F-3 Protein Modules

Product	Form	Major Protein Source	Energy (kcal/g)	Protein (g/100 g)
Casec®	powder	Calcium caseinate	3.8	90
ProMod®	powder	Whey protein	4.2	75

TABLE F-4 Carbohydrate Modules

Product	Form	Major Carbohydrate Source	Energy (kcal/ml or g)
Moducal®	powder	Hydrolyzed cornstarch	3.8 kcal/g
Polycose Liquid®	liquid	Hydrolyzed cornstarch	2.0 kcal/ml
Polycose Powder®	powder	Hydrolyzed cornstarch	3.8 kcal/g

TABLE F-5 Fat Modules

Product	Form	Major Fat Source	Energy (kcal/ml)	Fat (g/100 ml)
MCT Oil®	liquid	Medium-chain triglycerides	7.7	86
Microlipid®	liquid	Safflower oil	4.5	51

F

ANSWERS TO SELF CHECK QUESTIONS

Chapter 1

1. d	5. c	8. c
2. d	6. b	9. c
3. c	7. b	10. b
4. d		

Critical Thinking. 1. A. 20 kcal protein, 120 kcal carbohydrate, 99 kcal fat, = 239 kcal total, B. 8% protein, 50% carbohydrate, 41% fat, = 99% total

Chapter 2

1. b	5. a	8. a
2. b	6. a	9. d
3. c	7. b	10. c
4. a		

Critical Thinking 1. c, 2. b

Chapter 3

1. b	5. d	8. b
2. d	6. a	9. a
3. a	7. a	10. c
4. a		

Critical Thinking 1. b, 2. d

Chapter 4

1. c	5. a	8. b
2. b	6. d	9. c
3. d	7. c	10. d
4. d		

Critical Thinking 1. b, 2. d

Chapter 5

1. b	5. c	8. c
2. a	6. a	9. d
3. d	7. d	10. d
4. a		

Chapter 6

1. d	5. c	8. d
2. d	6. b	9. d
3. b	7. c	10. b
4. b		

Chapter 7

1. b	5. a	8. d
2. c	6. c	9. c
3. c	7. d	10. b
4. b		

Critical Thinking 1. b

Chapter 8

1. d	5. d	8. c
2. d	6. d	9. c
3. c	7. b	10. a
4. a		

Critical Thinking 1. c

Chapter 9

1. b	5. c	8. d
2. a	6. d	9. d
3. a	7. a	10. b
4. d		

Critical Thinking 1. b, 2. b

Chapter 10

1. b	5. d	8. b
2. a	6. d	9. a
3. b	7. c	10. d
4. c		

Critical Thinking 1. d, 2. c, 3. c

Chapter 11

1. a	5. d	8. a
2. d	6. b	9. a
3. d	7. d	10. d
4. d		

Critical Thinking 1. a

Chapter 12

1. a	5. d	8. d
2. d	6. b	9. d
3. c	7. c	10. c
4. b		

Critical Thinking 1. d, 2. d

Chapter 13

1. c	5. b	8. c
2. b	6. c	9. b
3. b	7. d	10. c
4. a		

Critical Thinking 1. a

Chapter 14

1. d	5. d	8. b
2. c	6. d	9. a
3. d	7. b	10. b
4. d		

Critical Thinking 1. c

Chapter 15

1. b	5. d	9. c
2. c	6. d	10. a
3. a	7. a	11. b
4. b	8. c	12. b

Critical Thinking 1. b, 2. c, 3. d

Chapter 16

1. d	5. b	8. a
2. c	6. d	9. c
3. b	7. c	10. d
4. a		

Critical Thinking 1. d, 2. c, 3. b, 4. a

Chapter 17

1. b	5. d	8. c
2. a	6. a	9. b
3. c	7. d	10. c
4. b		

Critical Thinking 1. b, 2. a, 3. d

Chapter 18

1. b	5. c	8. a
2. a	6. d	9. a
3. a	7. b	10. b
4. d		

Critical Thinking 1. a, 2. d, 3. c

Chapter 19

1. b	5. d	9. c
2. a	6. c	10. a
3. c	7. a	11. d
4. d	8. b	

Critical Thinking 1. a, 2. d

Chapter 20

1. a	4. b	7. c
2. d	5. c	8. c
3. b	6. a	9. d

Critical Thinking 1. d, 2. a, 3. b, 4. c

Chapter 21

1. d	5. c	8. a
2. a	6. a	9. d
3. b	7. c	10. c
4. b		

Critical Thinking 1. a, 2. c, 3. b

Chapter 22

1. d	4. d	7. c
2. a	5. b	8. b
3. c	6. c	

Critical Thinking 1. d, 2. b, 3. a, 4. c

Chapter 23

1. b	5. a	9. c
2. d	6. c	10. d
3. c	7. b	11. d
4. a	8. b	

Critical Thinking 1. c, 2. a, 3. d

Chapter 24

1. c	5. b	8. d
2. a	6. c	9. b
3. d	7. b	10. a
4. a		

Critical Thinking 1. b, 2. c, 3. a, 4. d

Chapter 25

1. a	5. d	8. b
2. c	6. c	9. b
3. b	7. c	10. a
4. c		

Critical Thinking 1. a, 2. d, 3. b, 4. a

Chapter 26

1. a	5. b	8. b
2. d	6. c	9. b
3. a	7. c	10. d
4. d		

Critical Thinking 1. d, 2. c, 3. b, 4. a

Chapter 27

1. b	5. b	8. d
2. c	6. b	9. a
3. d	7. c	10. c
4. a		

Critical Thinking 1. a, 2. b, 3. c

Chapter 28

1. b	4. d	7. b
2. a	5. d	8. a
3. c	6. b	9. c

Critical Thinking 1. c, 2. b, 3. d, 4. a

Glossary

1,25-dihydroxy vitamin D: the active form of vitamin D.

Acceptable Daily Intake (ADI): the amount of a sweetener that individuals can safely consume each day over the course of a lifetime without adverse effects. It includes a 100-fold safety factor.

acesulfame (AY-sul-fame) **potassium:** a 0-kcalorie artificial sweetener that tastes 200 times as sweet as sucrose; also known as acesulfame-K because K is the chemical symbol for potassium. Acceptable Daily Intake (ADI) = 15 milligrams/kilogram body weight.

acetyl CoA (ASS-uh-teel or uh-SEET-ul co-AY): a compound made up of acetic acid (formed from the breakdown of pyruvate) with a molecule of CoA attached to it. **CoA** (co-AY) is a nickname for a small molecule (coenzyme A) that participates in metabolism.

achalasia (ack-ah-LAY-zee-ah): failure of the lower esophageal sphincter to relax and allow foods to pass from the esophagus to the stomach; formally called **achalasia of the cardia** or **cardiospasm.**

acid-base balance: the balance maintained between acid and base concentrations in the blood and body fluids.

acidosis: too much acid in the blood and body fluids.

acids: compounds that release hydrogen ions in a solution.

acquired immune deficiency syndrome (AIDS): the severe complications associated with the end stages of HIV infection. A person with an HIV infection is diagnosed with AIDS when AIDS-defining illnesses (certain infections or severe wasting) develop.

acute: developing rapidly and associated with symptoms that are severe, but limited in duration.

acute liver failure: sudden failure of the liver.

acute malnutrition: protein-energy malnutrition (PEM) that develops rapidly due to a sudden and dramatic demand for nutrients. The person suffering from acute malnutrition may be of normal weight or may be overweight, but serum protein levels are low. In contrast, **chronic malnutrition** develops as a consequence of insufficient intake of energy and protein over long periods of time and is characterized by underweight, depleted fat stores, and normal serum protein levels.

acute renal failure: a sudden drop in the glomerular filtration rate (GFR) and urinary output.

acute respiratory failure: sudden failure of the lungs to exchange gases.

additives: substances that are added to foods, but normally are not consumed by themselves as foods.

Adequate Intake (AI): a value that is used as a guide for nutrient intake when scientific evidence is insufficient for determination of an RDA.

adipose tissue: the body's fat, which consists of masses of fat-storing cells called adipose cells.

adolescence: the period of growth from the beginning of puberty until full maturity. Timing of adolescence varies from person to person.

advance directives: the means by which competent adults record their preferences for future medical interventions. A living will and durable power of attorney are types of advance directives.

adynamic or **paralytic ileus:** ileus caused by the failure of the intestinal muscles to function.

aerobic (air-ROE-bic): requiring oxygen. Aerobic activity strengthens the heart and lungs by requiring them to work harder than normal to deliver oxygen to the tissues.

AIDS-defining illnesses: very low CD4+ T-cell counts, wasting, and other complications that mark the final stages of an HIV infection.

alcohol abuse: a pattern of drinking that includes failure to fulfill work, school, or home responsibilities; drinking in situations that are physically dangerous (as in driving while intoxicated); recurring alcohol-related legal problems (as in aggravated assault charges); or continued drinking despite ongoing health or social problems that are caused by or worsened by alcohol use.

alcohol addiction: a pattern of drinking that includes a strong craving for alcohol, a loss of control and an inability to stop drinking once begun, withdrawal symptoms (nausea, sweating, shakiness, and anxiety) after heavy drinking, and the need for increasing amounts of alcohol in order to feel "high"; also called *alcoholism.*

alcohol-related birth defects (ARBD): a condition caused by prenatal alcohol exposure. ARBD is diagnosed when there is a history of substantial, regular maternal alcohol intake or heavy episodic drinking and birth defects known to be associated with alcohol exposure.

alcohol-related neurodevelopmental disorder (ARND): a condition caused by prenatal alcohol exposure. ARND is diagnosed when there is a confirmed history of substantial, regular maternal alcohol intake or heavy episodic drinking and behavioral, cognitive, or central nervous system abnormalities known to be associated with alcohol exposure.

aldosterone (al-DOS-ter-own or AL-dough-STEER-own): a hormone secreted from the adrenal glands that signals the kidneys to retain sodium and fluid and also regulates chloride and potassium concentrations.

alitame (AL-ih-tame): a low-kcalorie artificial sweetener that is composed of two amino acids (alanine and aspartic acid) and tastes 2000 times as sweet as sucrose; FDA approval pending.

alkalosis: too much base in the blood and body fluids.

allergen: any substance that triggers an inappropriate immune response to a substance not normally harmful to the body.

alpha-lactalbumin (lackt-AL-byoo-min): the chief protein in human breast milk, as casein (CAY-seen) is the chief protein in cow's milk.

alveoli (al-VEE-oh-lie): air sacs in the lungs. One sac is an *alveolus.*

Alzheimer's disease: a progressive, degenerative disease that attacks the brain and impairs thinking, behavior, and memory.

amenorrhea (ay-MEN-oh-REE-ah): the absence of or cessation of menstruation. **Primary amenorrhea** is menarche delayed beyond 16 years of age. **Secondary amenorrhea** is the absence of three to six consecutive menstrual cycles.

amino (a-MEEN-oh) **acids:** building blocks of protein. Each has an amino group and an acid group attached to a central carbon, which also carries a distinctive side chain.

amniotic (am-nee-OTT-ic) **sac:** the "bag of waters" in the uterus, in which the fetus floats.

amylase (AM-uh-lace): an enzyme that splits amylose (a form of starch). Amylase is a carbohydrase. The ending *-ase* indicates an enzyme; the root tells what it digests. Other examples: protease, lipase.

anabolic steroids: drugs related to the male sex hormone, testosterone, that stimulate the development of lean body mass.

anabolism (an-ABB-o-lism): reactions in which small molecules are put together to build larger ones. Anabolic reactions consume energy.

anaerobic (AN-air-ROE-bic): not requiring oxygen. Anaerobic activity may require strength but does not work the heart and lungs very hard for a sustained period.

anaphylactic (an-AFF-ill-LAC-tic) **shock:** a life-threatening whole-body allergic reaction to an offending substance.

aneurysm (AN-you-riz-um): a ballooning out of a portion of a blood vessel (usually an artery) due to weakness of the vessel's wall.

angina (an-JYE-nah or AN-ji-nah): a painful feeling of tightness or pressure in and around the heart, often radiating to the back, neck, and arms; caused by a lack of oxygen to an area of heart muscle.

anorexia nervosa: an eating disorder characterized by a refusal to maintain a minimally normal body weight, self-starvation to the extreme, and a disturbed perception of body weight and shape; seen (usually) in adolescent girls and young women.

anthropometric (an-throw-poe-MEH-trick) **measurements:** measurements of the physical characteristics of the body, such as height and weight.

antibodies: large proteins of the blood and body fluids, produced in response to invasion of the body by unfamiliar molecules (mostly proteins) called *antigens.* Antibodies inactivate the invaders and so protect the body.

antidiuretic hormone (ADH): the hormone secreted by the pituitary gland that increases the reabsorption of water by the kidneys.

antigen: any substance that triggers an immune system response, including bacteria, viruses, fungi, parasites, worn-out cells, and malignant cells.

antimicrobial agents: substances used as food additives that prevent the growth of illness-causing microorganisms in foods.

antioxidant (anti-OX-ih-dant): a compound that protects other compounds from oxygen by itself reacting with oxygen. *Oxidation* is a potentially damaging effect of normal cell chemistry involving oxygen.

anus (AY-nus): the terminal sphincter muscle of the GI tract.

Apgar scores: a system of scoring an infant's physical condition right after birth. Heart rate, respiration, muscle tone, response to stimuli, and color are ranked 0, 1, or 2. A low total score indicates that medical attention is required to facilitate survival.

appendix: a narrow blind sac extending from the beginning of the colon. The appendix stores lymphocytes.

appetite: the psychological desire to eat; a learned motivation that is experienced as a pleasant sensation that accompanies the sight, smell, or thought of appealing foods.

artery: a vessel that carries blood away from the heart.

arthritis: inflammation of a joint, usually accompanied by pain, swelling, and structural changes.

artificial colors: certified food colors, added to enhance appearance (*certified* means approved by the FDA).

artificial feedings: parenteral and enteral nutrition; feeding by a route other than the normal ingestion of food.

artificial flavors, flavor enhancers: chemicals that mimic natural flavors and those that enhance flavor.

artificial sweeteners: noncarbohydrate, nonkcaloric synthetic sweetening agents; sometimes called **nonnutritive sweeteners.**

ascites (ah-SIGH-teez): a type of edema characterized by the accumulation of fluid in the abdominal cavity.

ascorbic acid: one of the two active forms of vitamin C.

aspartame (ah-SPAR-tame or ASS-par-tame): a low-kcalorie artificial sweetener that is composed of two amino acids (phenylalanine and aspartic acid) and tastes 200 times as sweet as sucrose. ADI = 50 milligrams/kilogram body weight.

aspiration: the drawing of food, saliva, or gastric secretions into the lungs.

aspiration pneumonia: an infection of the lungs caused by inhaling fluids regurgitated from the stomach.

asthma: a respiratory disease in which lung tissue becomes irritated and inflamed, interfering with breathing.

atherosclerosis (ATH-er-oh-skler-OH-sis): the buildup of plaque along the inner walls of the arteries, which narrows the lumen of the artery and restricts blood flow to the tissue it supplies.

atrophic gastritis (a-TRO-fik gas-TRI-tis): a condition characterized by chronic inflammation of the stomach accompanied by a diminished size and functioning of the mucosa and glands.

atrophy (AT-tro-fee): a decrease in size (for example, of a muscle) because of disuse.

autoimmune disorders: disorders that result from immune system defenses attacking the body's own cells.

autonomy: independence.

Barrett's esophagus: changes in the cells of the esophagus associated with chronic reflux that raise the risk of cancer of the esophagus.

basal metabolism: the energy needed to maintain life when a person is at complete rest after a 12-hour fast. Basal metabolism is normally the largest part of a person's daily energy expenditure.

bases: compounds that accept hydrogen ions in a solution.

B-cells: lymphocytes that produce antibodies.

behavior modification: the changing of behavior by the manipulation of *antecedents* (cues or environmental factors that trigger behavior), the behavior itself, and *consequences* (the penalties or rewards attached to behavior).

belching: the expulsion of gas from the stomach through the mouth.

beneficence (be-NEF-eh-sens): doing good.

benign (bee-NINE): describes tumors that stop growing without intervention or can be removed surgically and most often pose no threat to health.

beriberi: the thiamin-deficiency disease; characterized by loss of sensation in the hands and feet, muscular weakness, advancing paralysis, and abnormal heart action.

beta-carotene: a vitamin A precursor made by plants and stored in human fat tissue; an orange pigment.

BHA, BHT: preservatives commonly used to slow the development of off-flavors, odors, and color changes caused by oxidation.

bicarbonate: an alkaline secretion of the pancreas; part of the pancreatic juice. (Bicarbonate also occurs widely in all cell fluids.)

bifidus (BIFF-id-us, by-FEED-us) **factors:** factors in colostrum and breast milk that favor the growth of the "friendly" bacterium *Lactobacillus* (lack-toe-ba-SILL-us) *bifidus* in the infant's intestinal tract. These bacteria prevent other, less desirable intestinal inhabitants from flourishing.

bile: an emulsifier that prepares fats and oils for digestion; made by the liver, stored in the gallbladder, and released into the small intestine when needed.

biliary colic: pain associated with gallstones that have entered the common bile duct.

biliary tract: the gallbladder and bile ducts.

bilirubin: a pigment in the bile whose blood concentration may rise as a result of medical conditions that affect the liver.

binge eating disorder: an eating disorder whose criteria are similar to those of bulimia nervosa, excluding purging or other compensatory behaviors.

bioaccumulation: the accumulation of a contaminant in the tissues of living things at higher and higher concentrations along the food chain.

bioavailability: the rate and extent to which a nutrient is absorbed and used.

biotechnology: the use of biological systems or organisms to create or modify products; also called *biogenetic engineering.*

bland diets: diets that aim to minimize gastric acid secretion and limit gastric irritants.

blind loop syndrome: the problems of fat and vitamin B_{12} malabsorption that result from the overgrowth of bacteria in a bypassed segment of the intestine.

body composition: the proportions of muscle, bone, fat, and other tissue that make up a person's total body weight.

body mass index (BMI): an index of a person's weight in relation to height, determined by dividing the weight (in kilograms) by the square of the height (in meters).

bolus (BOH-lus): the portion of food swallowed at one time.

bolus feeding: delivery of about 300 to 400 ml of formula over 15 minutes or less.

bone marrow transplants: the replacement of diseased bone marrow in a recipient with healthy bone marrow from a donor; sometimes used as a treatment for breast cancer, leukemia, lymphomas, and certain blood disorders.

botulism: an often-fatal food poisoning caused by botulinum toxin, a toxin produced by the *Clostridium botulinum* bacterium that grows without oxygen in nonacidic canned foods.

branched-chain amino acids: the amino acids leucine, isoleucine, and valine, which are present in large amounts in skeletal muscle tissue; falsely promoted as fuel for exercising muscles.

bronchitis (bron-KYE-tis): inflammation of the lungs' air passages.

brown sugar: refined white sugar crystals to which manufacturers have added molasses syrup with natural flavor and color; 91 to 96 percent pure sucrose.

buffalo hump: the accumulation of fat at the top of the back.

buffers: compounds that can reversibly combine with hydrogen ions to help keep a solution's acidity or alkalinity constant.

bulimia (byoo-LEEM-ee-uh) **nervosa:** recurring episodes of binge eating combined with a morbid fear of becoming fat, usually followed by self-induced vomiting or purging.

bulk-forming agents: laxatives composed of fibers that work like dietary fibers. Attracts water in the intestine to form a bulky stool, which stimulates peristalsis. Metamucil and Fiberall are examples.

caffeine: a natural stimulant found in many common foods and beverages, including coffee, tea, and chocolate; may enhance endurance by stimulating fatty acid release but also causes fluid losses. High doses cause headaches, trembling, rapid heart rate, and other undesirable side effects.

calcium rigor: hardness or stiffness of the muscles caused by high blood calcium.

calcium tetany: intermittent spasms of the extremities due to nervous and muscular excitability caused by low blood calcium.

cancer cachexia (ka-KEKS-ee-ah) **syndrome:** loss of lean body mass, depletion of serum proteins, and debilitation that frequently accompany cancer.

cancers: diseases that result from the unchecked growth of cells.

capillaries: small vessels that branch from an artery. Capillaries connect arteries to veins. Oxygen, nutrients, and waste materials are exchanged across capillary walls.

capitation: prepayment of a set fee per client in exchange for medical services.

carbohydrates: energy nutrients composed of monosaccharides.

carbohydrate-to-insulin ratio: the grams of carbohydrate covered by one unit of insulin. The lower the ratio, the more insulin is needed to cover carbohydrate intake.

cardiac cachexia (ka-KEKS-ee-ah): chronic PEM that develops as a consequence of heart failure. Research suggests that cytokines play a role in the development of PEM in the late stages of CHF.

cardiac output: the volume of blood discharged by the heart each minute.

cardiomegaly (CAR-dee-oh-MEG-ah-lee): enlargement of the heart.

cardiorespiratory conditioning: improvements in heart and lung function and increased blood volume, brought about by aerobic training.

cardiorespiratory endurance: the ability to perform large-muscle dynamic exercise of moderate-to-high intensity for prolonged periods.

cardiovascular disease (CVD): a general term for all diseases of the heart and blood vessels. Atherosclerosis is the main cause of CVD. When the arteries that carry blood to the heart muscle become occluded, the heart suffers damage known as **coronary heart disease (CHD).**

cariogenic (KARE-ee-oh-JEN-ik): conducive to dental decay.

carnitine (CAR-ne-teen): a non-protein amino acid made in the body from lysine that helps transport fatty acids across the mitochondrial membrane. Carnitine supposedly "burns" fat and spares glycogen during endurance events, but in reality it does neither.

catabolism (ca-TAB-o-lism): reactions in which large molecules are broken down to smaller ones. Catabolic reactions usually release energy.

cataracts: thickenings of the eye lenses that impair vision and can lead to blindness.

cathartics (ca-THART-ics): strong laxatives.

CD4+ T-cells: white blood cells (lymphocytes) that have a specific protein receptor on their surfaces and are a necessary component of the immune system. Nutrition in Practice 15 describes T-cells.

celiac (SEE-lee-ack) **disease:** a sensitivity to a part of the protein gluten that causes flattening of the intestinal villi and malabsorption; also called *gluten-sensitive enteropathy* or *celiac sprue.*

cell-mediated immunity: immunity conferred by T-cells traveling to the invasion site to fight specific antigens.

central obesity: excess fat on the abdomen and around the trunk of the body.

central total parenteral nutrition (central TPN): the provision of an IV solution that meets nutrient needs delivered into a central vein.

central veins: the large-diameter veins located close to the heart.

cerebral cortex: the outer surface of the cerebrum, which is the largest part of the brain.

cesarean section: surgical childbirth, in which the infant is delivered through an incision in the woman's abdomen.

CHD risk equivalents: disorders that raise the risk of heart attacks, strokes, and other complications associated with cardiovascular diseases to the same degree as existing CHD. These disorders include symptomatic carotid artery disease, peripheral arterial disease, abdominal aortic aneurysm, and diabetes mellitus.

chemotherapy: the use of drugs to arrest or destroy cancer cells. Drugs used for chemotherapy are called **chemotherapeutic** or **antineoplastic agents.**

cholangitis (KOH-lan-JYE-tis): inflammation of the bile ducts.

cholecystectomy (KOH-lee-sis-TEK-toe-mee): surgical removal of the gallbladder.

cholecystitis (KOH-lee-sis-TIE-tis): inflammation of the gallbladder.

choledocholithiasis (koh-LED-oh-koh-lih-THIGH-ah-sis): the presence of gallstones in the common bile duct.

cholelithiasis (KOH-lee-lih-THIGH-ah-sis): the formation or presence of stones in the gallbladder or common bile duct.

choline: a nonessential nutrient that can be made in the body from an amino acid.

chromium (CROW-mee-um) **picolinate:** a trace mineral supplement; falsely promoted as building muscle, enhancing energy, and burning fat. **Picolinate** (pick-oh-LYN-ate) is a derivative of the

amino acid tryptophan that seems to enhance chromium absorption.

chronic: developing gradually and associated with symptoms that progressively worsen with time.

chronic or **congestive heart failure (CHF):** a syndrome in which the heart gradually weakens and can no longer adequately pump blood through the circulatory system.

chronic diseases: degenerative diseases characterized by deterioration of the body organs; also called chronic, **noncommunicable diseases (NCD).** Examples include heart disease, cancer, and diabetes.

chronic obstructive pulmonary disease (COPD): one of several disorders, including emphysema and bronchitis, that interfere with respiration.

chronic renal failure: the gradual and irreversible deterioration of kidney function.

chronological age: a person's age in years from his or her date of birth.

chylomicrons (kye-lo-MY-crons): the lipoproteins that transport lipids from the intestinal cells into the body. The cells of the body remove the lipids they need from the chylomicrons, leaving chylomicron remnants to be picked up by the liver cells.

chyme (KIME): the semiliquid mass of partly digested food expelled by the stomach into the duodenum (the top portion of the small intestine).

cirrhosis (sih-ROW-sis): an advanced form of liver disease in which scar tissue replaces liver cells that have permanently lost their function.

clear-liquid diets: diets consisting of foods that are mainly liquid and transparent at body temperature.

clinical pathways, critical pathways, or **care maps:** charts or tables that outline a plan of care for a specific diagnosis, treatment, or procedure, with a goal of providing the best possible outcome at the lowest cost. The plan, developed by the health care team after a careful study of each facility's unique client population, is regularly reassessed and improved.

clinically severe obesity: a BMI of 40 or greater or a BMI of 35 or greater and one or more serious conditions such as hypertension. A less preferred term used to describe the same condition is **morbid obesity.**

closed feeding systems: enteral formula delivery systems in which the formula comes prepackaged in a container that is ready to be attached to a feeding tube for administration.

club drug: any of a wide variety of drugs used by young adults at all-night dance parties such as "raves."

coenzyme (co-EN-zime): a small molecule that works with an enzyme to promote the enzyme's activity. Many coenzymes have B vitamins as part of their structure.

cofactor: a mineral element that, like a coenzyme, works with an enzyme to facilitate a chemical reaction.

cognitive therapy: psychological therapy aimed at changing undesirable behaviors by changing underlying thought processes contributing to these behaviors. In anorexia nervosa, a goal is to replace false beliefs about body weight, eating, and self-worth with health-promoting beliefs.

collagen: the characteristic protein of connective tissue.

collaterals: small branches of a blood vessel that develop when blood flow through the liver is obstructed; also called **shunts.**

colon or **large intestine:** the last portion of the intestine, which absorbs water. Its main segments are the ascending colon, the transverse colon, the descending colon, and the sigmoid colon.

colonic irrigation: the popular, but potentially harmful practice of "washing" the large intestine with a powerful enema machine.

colostomy (co-LOSS-toe-me): surgery that creates an opening from any portion of the colon through the abdominal wall and out through the skin.

colostrum (co-LAHS-trum): a milklike secretion from the breast that is rich in protective factors. Colostrum is present during the first day or so after delivery, before milk appears.

comatose: in a state of deep unconsciousness from which the person cannot be aroused.

competent: having sufficient mental ability to understand a treatment, weigh its risks and benefits, and comprehend the consequences of refusing or accepting the treatment.

complementary proteins: two or more proteins whose amino acid assortments complement each other in such a way that the essential amino acids missing from one are supplied by the other.

complete formulas: liquid diets designed to supply all needed nutrients when consumed in sufficient volume.

complete protein: a protein containing all the amino acids essential in human nutrition in amounts adequate for human use.

complex carbohydrates: long chains of sugars arranged as starch or fiber; also called polysaccharides.

concentrated fruit juice sweetener: a concentrated sugar syrup made from dehydrated, deflavored fruit juice, commonly grape juice; used to sweeten products that can then claim to be "all fruit."

conditionally essential amino acid: an amino acid that is normally nonessential but must be supplied by the diet in special circumstances when the need for it becomes greater than the body's ability to produce it.

conditioning: the physical effect of training; improved flexibility, strength, and endurance.

confectioners' sugar: finely powdered sucrose; 99.9 percent pure.

consistency-modified diets: diets that include foods modified in firmness, such as liquid, pureed, chopped, tender-cooked, or soft whole foods (such as bananas).

constipation: the condition of having infrequent or difficult bowel movements.

contaminant: a substance that makes a food impure and unsuitable for ingestion.

continuous ambulatory peritoneal dialysis (CAPD): the most common type of peritoneal dialysis that people use at home.

continuous feedings: slow delivery of formula in constant amounts over an 8- to 24-hour period.

continuous renal replacement therapy (CRRT): a slow and continuous type of dialysis used in the treatment of acute renal failure.

cool-down: Five to ten minutes of light activity, such as walking or stretching, following a vigorous workout to return the body's core gradually to near-normal temperature.

corn sweeteners: corn syrup and sugars derived from corn.

corn syrup: a syrup containing mostly glucose,

produced by the action of enzymes on cornstarch. See also *high-fructose corn syrup (HFCS)*.

cornea (KOR-nee-uh): the hard, transparent membrane covering the outside of the eye.

coronary heart disease (CHD): heart damage that results from an inadequate supply of blood to the heart.

counterregulatory hormones: hormones that oppose insulin actions. The counterregulatory hormones include glucagon, epinephrine, norepinephrine, cortisol, and growth hormone.

creatine (KREE-ah-tin): a nitrogen-containing compound that combines with phosphate to form the high-energy compound creatine phosphate (or phosphocreatine) in muscles. Claims that creatine enhances energy use and muscle strength need further confirmation.

cretinism (CREE-tin-ism): an iodine-deficiency disease characterized by mental and physical retardation.

critical period: a finite period during development in which certain events occur that will have irreversible effects on later developmental stages; usually a period of rapid cell division.

Crohn's disease: inflammation and ulceration along the length of the GI tract, often with granulomas.

cross-contamination: the contamination of a food through exposure to utensils, hands, or other surfaces that were previously in contact with a contaminated food.

cryptosporidiosis (KRIP-toe-spo-rid-ee-OH-sis): a food-borne illness caused by the parasite *Cryptosporidium parvum*. Most people develop few or minor problems from this infection, but people with HIV infections, and especially those with AIDS, can develop long-lasting and serious problems.

cyclamate (SIGH-kla-mate): a 0-kcalorie artificial sweetener that tastes 30 times as sweet as sucrose; FDA approval pending in the United States; available in Canada in grocery stores but only as a tabletop sweetener, not as an additive.

cyclic parenteral nutrition: the continuous administration of a parenteral solution for 8 to 12 hours with time periods when no nutrients are infused.

cystic fibrosis (SIS-tic fie-BRO-sis): a hereditary disorder characterized by the production of thick mucus that affects many organs including the pancreas, lungs, liver, heart, gallbladder, and small intestine.

cystinuria (SIS-tin-NEW-ree-ah): an inherited metabolic disorder that is characterized by the excessive urinary excretion of cystine, lysine, arginine, and ornithine and commonly leads to kidney stone formation.

cytokines (SIGH-toe-kynes): proteins that help regulate immune system responses. Cytokines trigger hypermetabolism and cause anorexia, fever, and discomfort.

Daily Food Guide: the USDA's food group plan for ensuring dietary adequacy that assigns foods to five major food groups.

Daily Values: reference values developed by the FDA specifically for use on food labels.

deamination: removal of the amino (NH2) group from a compound such as an amino acid.

debridement (dee-BREED-ment): the removal of dead tissues resulting from burns and other wounds to speed healing and prevent infection.

defecate (DEF-uh-cate): to move the bowels and eliminate waste.

deficient: in regard to nutrient intake, the amount below which almost all healthy people can be expected, over time, to experience deficiency symptoms.

dehydration: the loss of water from the body that occurs when water output exceeds water input. The symptoms progress rapidly from thirst, to weakness, to exhaustion and delirium and end in death if not corrected.

delusions (dee-LOO-shuns): inappropriate beliefs not consistent with the individual's own knowledge and experience.

dementia (dee-MEN-she-ah): irreversible loss of mental function.

denaturation (dee-nay-cher-AY-shun): the change in a protein's shape brought about by heat, acid, or other agents. Past a certain point, denaturation is irreversible.

dental caries (KARE-eez): the gradual decay and disintegration of a tooth.

dental plaque (PLACK): a gummy mass of bacteria that grows on teeth and can lead to dental caries and gum disease.

dextrose: an older name for glucose and the name used for glucose in intravenous solutions.

dextrose monohydrate: a form of glucose that contains a molecule of water and is stable in IV solutions. IV dextrose solutions provide 3.4 kcal/g.

DHEA (dehydroepliandrosterone) and **androstenedione:** hormones made in the adrenal glands that serve as precursors to the male hormone testosterone; falsely promoted as burning fat, building muscle, and slowing aging. Side effects include acne, aggressiveness, and liver enlargement.

diabetes (DYE-uh-BEET-eez) **mellitus** (MELL-ih-tus or mell-EYE-tus): a group of metabolic disorders of glucose regulation and utilization.

dialysate (dye-AL-ih-SATE): a solution used during dialysis to draw wastes and fluids from the blood.

dialysis (dye-AL-ih-sis): removal of waste from the blood through a semipermeable membrane using the principles of simple diffusion and osmosis. The two main types are hemodialysis and **peritoneal dialysis** (see Nutrition in Practice 26).

dialyzer (dye-ah-LYES-er): the machine used for hemodialysis; also called an **artificial kidney.**

diaphragm (DYE-a-fram): the dome-shaped muscle forming a partition between the chest cavity and the abdominal cavity; the muscle responsible for breathing.

diarrhea: the frequent passage of watery bowel movements.

diet history: a comprehensive record of eating-related behaviors and the foods a person eats.

diet manual: a book that describes the foods allowed and restricted on a diet, outlines the rationale and indications for use of each diet, and provides sample menus.

diet order: a statement of the client's diet prescription that the physician writes in the health record.

diet pills: pills that depress the appetite temporarily; often, physician-prescribed amphetamines (speed). It is generally agreed that these medications are of little value for weight loss and that their use can cause a dangerous dependency.

dietary adequacy: the characteristic of a diet that provides all the essential nutrients, fiber, and energy necessary to maintain health and body weight.

dietary balance: providing foods of a number of types in balance with one another such that foods rich in one nutrient do not crowd out foods that are rich in another nutrient.

dietary folate equivalents (DFE): the amount of folate available to the body from naturally occurring sources, fortified foods, and supplements, accounting for differences in bioavailability from each source.

Dietary Reference Intakes (DRI): a set of values for the dietary nutrient intakes of healthy people in the United States and Canada. These values are used for planning and assessing diets.

dietetic technicians registered (D.T.R.'s): professionals who have earned an associate degree or higher; have completed a dietetic technician program approved by the American Dietetic Association (ADA); have passed a national registration exam; and assist in planning, implementing, and evaluating nutritional care.

dietetics: the practical application of nutrition, including the assessment of nutrition status, recommendation of appropriate diets, nutrition education, and the planning and serving of meals.

differentiation: the development of specific functions different from those of the original.

diffusion: movement of solutes from an area of high concentration to one of low concentration.

digestion: the process by which complex food particles are broken down to smaller absorbable particles.

dioxins: toxic organic compounds containing chlorine that arise in industry as (among other things) by-products of the bleaching process.

dipeptide: two amino acids bonded together.

disaccharides: pairs of sugar units bonded together.

disufiram-like reaction: nausea, vomiting, headache, cramps, flushing of the skin, and a rapid heartbeat that can occur when some medications are taken along with alcohol. The medication disulfiram produces these effects when combined with alcohol to discourage alcohol abusers from using alcohol.

diuretic abuse: use of diuretics to promote water excretion by dieters who believe their weight excesses are due to water accumulation. Diuretics promote water loss, not fat loss, and their use can cause dehydration and mineral imbalances.

diuretics (dye-yoo-RET-ics): medications that promote the excretion of water through the kidneys. Not all diuretics increase the urinary loss of potassium. Some, called potassium-sparing diuretics, are less likely to result in a potassium deficiency (see Chapter 25).

diverticula (dye-ver-TIC-you-lah): sacs or pouches that develop in the weakened areas of the intestinal wall (like bulges in an inner tube of a bike tire where the wall is weak).

diverticulitis (DYE-ver-tic-you-LYE-tis): infected or inflamed diverticula.

diverticulosis (DYE-ver-tic-you-LOH-sis): the condition of having diverticula.

DNA: deoxyribonucleic acid; the double-stranded chemical that carries hereditary information. Genes are composed of DNA.

drug history: an account of all prescription and over-the-counter medications, dietary supplements, and illegal drugs a client uses as well as the amounts the client takes and the length of time the client has taken the drugs.

dry weight: weight after excess fluids have been removed from the body.

dumping syndrome: the symptoms that result from the rapid entry of undigested food into the jejunum: sweating, weakness, and diarrhea shortly after eating and hypoglycemia later.

duodenum (doo-oh-DEEN-um or doo-ODD-ah-num): the top portion of the small intestine (about "12 fingers' breadth" long, in ancient terminology).

durable power of attorney: a legal document in which one competent adult authorizes another competent adult to make decisions for her or him in the event of incapacitation. The phrase "durable power" means that the agent's authority survives the client's incompetence; "attorney" refers to an attorney-in-fact (not an attorney-at-law).

duration: length of time (for example, the time spent in each activity session).

dysentery (DIS-en-terry): an infection of the gastrointestinal tract caused by an amoeba or bacterium that gives rise to severe diarrhea.

dysphagia (dis-FAY-gee-ah): difficulty in swallowing.

dysuria (dis-YOU-ree-ah): painful or difficult urination.

eating disorder: a disturbance in eating behavior that jeopardizes a person's physical and psychological health.

eclampsia: a severe stage of preeclampsia in which convulsions occur.

ecstasy: a street or slang term used for the drug methylenedioxymethamphetamine (MDMA). Ecstasy is chemically similar to amphetamine (speed) and mescaline (a hallucinogen).

edema (eh-DEEM-uh): the swelling of body tissue caused by leakage of fluid from the blood vessels and accumulation of the fluid in the interstitial spaces.

eicosanoids (eye-COSS-uh-noyds): derivatives of 20-carbon fatty acids that regulate blood pressure, blood clotting, and other body functions. They include prostaglandins (PROS-tah-GLAND-ins), thromboxanes (throm-BOX-ains), and leukotrienes (LOO-ko-TRY-eens).

electrolyte: a salt that dissolves in water and dissociates into charged particles called ions.

electrolyte solutions: solutions that can conduct electricity.

embolism (EM-boh-lizm): the obstruction of a blood vessel by an embolus, causing sudden tissue death.

embolus (EM-boh-lus): a traveling blood clot.

emetic (em-ETT-ic): an agent that causes vomiting.

emotions: mental states such as love or hate that arise from subjective experiences rather than conscious efforts.

emphysema (EM-fih-SEE-mah): a type of COPD in which the lungs lose their elasticity and the victim has difficulty breathing; often occurs along with bronchitis.

emulsifiers: substances that mix with both fat and water and that permanently disperse the fat in the water, forming an emulsion.

enamel: the hard, white, dense substance that forms a covering for the crown of a tooth.

endotoxins: products of infectious agents that damage tissues and trigger inflammatory responses.

end-stage renal disease (ESRD): the severe stage of chronic renal failure in which dialysis or a kidney transplant is necessary to sustain life. In ESRD, the GRF falls to less than about 25 ml/min, and the BUN may rise as high as 150 to 250 mg/dL.

enemas: solutions inserted into the rectum and colon to stimulate a bowel movement and empty the lower large intestine.

energy metabolism: all the reactions by which the body obtains and spends the energy from food or body stores.

energy-yielding nutrients: the fuel nutrients, those that yield energy the body can use. The three energy-yielding nutrients are carbohydrate, protein, and fat.

engorgement: overfilling of the breasts with milk.

enrichment: the addition to a food of nutrients to meet a specified standard. In the case of refined bread or cereal, five nutrients have been added: thiamin, riboflavin, niacin, and folate in amounts approximately equivalent to, or higher than, those originally present and iron in amounts to alleviate the prevalence of iron-deficiency anemia.

enteral (EN-ter-all) nutrition: the provision of nutrients using the GI tract. Enteral nutrition includes both oral diets and tube feedings.

enteral formulary: the formulas available for use in a health care facility.

enteral formulas: liquid diets designed to be delivered through the GI tract, either orally or by tube.

enterostomy (EN-ter-OSS-toe-mee): an opening into the stomach or jejunum through which a feeding tube can be passed.

environmental tobacco smoke (ETS): the combination of exhaled smoke (mainstream smoke) and smoke from lighted cigarettes, pipes, or cigars (sidestream smoke) that enters the air and may be inhaled by other people.

enzymes: protein catalysts. A catalyst is a compound that facilitates chemical reactions without itself being changed in the process.

EPA, DHA: omega-3 fatty acids made from linolenic acid. The full name for EPA is eicosapentaenoic (EYE-cosa-PENTA-ee-NO-ick) acid. The full name for DHA is docosahexaenoic (DOE-cosa-HEXA-ee-NO-ick) acid

epigastric: the region of the body just above the stomach.

epiglottis (epp-ee-GLOT-tiss): a cartilage structure in the throat that prevents fluid or food from entering the trachea when a person swallows.

epinephrine: one of the stress hormones secreted whenever emergency action is needed; prescribed therapeutically to relax the bronchioles during allergy or asthma attacks.

epithelial (ep-i-THEE-lee-ul) cells: cells on the surface of the skin and mucous membranes.

epithelial tissue: tissue composing the layers of the body that serve as selective barriers between the body's interior and the environment (examples are the cornea, the skin, the respiratory lining, and the lining of the digestive tract).

ergogenic (ER-go-JEN-ick) aids: substances or techniques used in an attempt to enhance physical performance.

erythrocyte (er-REETH-ro-cite) hemolysis (he-MOLL-uh-sis): rupture of the red blood cells, caused by vitamin E deficiency.

erythropoietin (eh-RITH-row-POY-eh-tin): a hormone, secreted by the kidneys in response to oxygen depletion or anemia, that stimulates the bone marrow to produce red blood cells.

esophageal stricture: narrowing of the inner diameter of the esophagus from inflammation and scarring.

esophageal ulcers: lesions or sores in the lining of the esophagus.

esophagus (e-SOFF-uh-gus): the food pipe; the conduit from the mouth to the stomach.

essential fatty acids: fatty acids that the body requires but cannot make in amounts sufficient to meet its physiological needs.

essential nutrients: nutrients a person must obtain from food because the body cannot make them for itself in sufficient quantities to meet physiological needs.

Estimated Average Requirement (EAR): the amount of a nutrient that will maintain a specific biochemical or physiological function in half the people of a given age and gender group.

ethanol: the type of alcohol found in wine, beer, and other alcoholic drinks.

ethical: in accordance with moral principles or professional standards. Socrates described ethics as "how we ought to live."

ethnic diets: foodways and cuisines typical of national origins, races, cultural heritages, or geographic locations.

euphoria (you-FOR-ee-ah): a sense of happiness and physical well-being.

extracellular fluid: fluid residing outside the cells; includes the fluid between the cells (interstitial fluid), plasma, and the water of structures such as the skin and bones. Extracellular fluid accounts for about one-third of the body's water.

fad diets: diets based on exaggerated or false theories of weight loss. Such diets are usually inadequate in energy and nutrients. Most fad diets, including the currently popular Atkins Diet and Zone Diet, advocate essentially the same high-protein, low-carbohydrate diet. Such diets may offer short-term weight-loss success to some who try them, but they fail to produce long-lasting results for most people. Furthermore, high-protein, low-carbohydrate diets are often high in fat and low in fiber, vitamins, and some minerals. Long-term use of such diets may produce adverse side effects such as nausea, fatigue, constipation, and low blood pressure. Some fad diets are more dangerous to health than obesity itself.

fasting hypoglycemia: low blood glucose that develops gradually, does not occur in response to food intake, and primarily affects the brain and central nervous system.

fat replacers: ingredients that replace some or all of the functions of fat and may or may not provide energy. In this text, the term fat replacer is used interchangeably with fat substitute, which technically applies only to an ingredient that replaces all of the functions of fat and provides no energy.

fatfold measure: a clinical estimate of total body fatness in which the thickness of a fold of skin on the back of the arm (over the triceps muscle), below the shoulder blade (subscapular), or in other places is measured with a caliper. (The older, less preferred, term is skinfold test.)

fats: lipids that are solid at room temperature (70°F or 25°C).

fatty acids: organic compounds composed of a chain of carbon atoms with hydrogens attached and an acid group at one end.

fatty liver: an accumulation of triglycerides in the liver resulting from many disorders, including exposure to excessive alcohol, excessive weight gain, and diabetes mellitus; also called hepatic steatosis, steatohepatitis, and fatty infiltration of the liver.

fatty streaks: accumulations of cholesterol and other lipids along the walls of the arteries.

fecal impaction: a dry, compacted mass of fecal matter in the colon or rectum.

female athlete triad: a potentially fatal triad of medical problems: disordered eating, amenorrhea, and osteoporosis.

fetal alcohol syndrome (FAS): the cluster of symptoms seen in an infant or child whose mother consumed excessive alcohol during her pregnancy. FAS includes, but is not limited to, brain damage, growth retardation, mental retardation, and facial abnormalities.

fetor hepaticus (FEE-tor he-PAT-eh-cuss): a pungent odor of the breath that may develop in people with impending hepatic coma.

fetus (FEET-us): the developing infant from eight weeks after conception until its birth.

fibers: a general term denoting in plant foods the polysaccharides cellulose, hemicellulose, pectins, gums, and mucilages, as well as the nonpolysaccharide lignins, that are not attacked by human digestive enzymes.

fibrinogen (fie-BRIN-oh-jen): a protein produced by the liver that is essential to blood clotting.

fibrocystic breast disease: a harmless condition in which the breasts develop lumps, sometimes associated with caffeine consumption. In some, it responds to abstinence from caffeine; in others, it can be treated with vitamin E.

fibrosis (figh-BROH-sis): abnormal growth of scar tissue.

fibrous plaques: mounds of lipid material, mixed with smooth muscle cells and calcium, which develop in the artery walls in atherosclerosis.

filtrate: in the kidneys, the fluid that passes from the blood through the capillary walls of the glomeruli, eventually forming urine.

fistulas (FIS-chew-lahs): abnormal openings formed between two organs or between an internal organ and the skin.

fitness: the characteristics that enable the body to perform physical activity; more broadly, the ability to meet routine physical demands with enough reserve energy to rise to a physical challenge; or the body's ability to withstand stress of all kinds.

flapping tremor: uncontrolled movement of a muscle group causing the outstretched arm and hand to flap like a wing; occurs in disorders that cause encephalopathy; also called **asterixis** (AS-ter-ICK-sis).

flatus (FLAY-tuss): gas in the intestinal tract or the expelling of gas from the intestinal tract, especially through the anus.

flexibility: the capacity of the joints to move through a full range of motion; the ability to bend and recover without injury.

fluid and electrolyte balance: maintenance of the necessary amounts and types of fluid and minerals in each compartment of the body fluids.

fluorapatite (floor-APP-uh-tite): the stabilized form of bone and tooth crystal, in which fluoride has replaced the hydroxy portion of hydroxyapatite.

fluorosis (floor-OH-sis): mottling of the tooth enamel from ingestion of too much fluoride during tooth development.

follicle (FOLL-i-cul): a group of cells in the skin from which a hair grows.

food allergies: adverse reactions to foods that involve an immune response; also called *food-hypersensitivity reactions.*

Food and Drug Administration (FDA): an agency of the Public Health Service within the Department of Health and Human Services. The FDA is responsible for ensuring the safety and whole-

someness of all foods processed and sold in interstate commerce except meat, poultry, and eggs (which are under the jurisdiction of the USDA); inspecting food plants and imported foods; and setting standards for food composition.

food aversions: strong desires to avoid particular foods.

food bank: a facility that provides food to the hungry.

food consumption surveys: surveys that measure the amounts and kinds of food people consume (using diet histories), estimate the nutrient intakes, and compare them with a standard such as the DRI.

food cravings: deep longings for particular foods.

food frequency questionnaire: a tool for gathering food intake data that asks clients about the types and amounts of foods they routinely eat.

food group plans: diet-planning tools that sort foods of similar origin and nutrient content into groups and then specify that people should eat certain numbers of servings from each group.

food insecurity: limited or uncertain access to foods of sufficient quality or quantity to sustain a healthy and active life.

food intolerance: an adverse response to a food or food additive that does not involve the immune system.

food poverty: hunger occurring when encough food exists in an area but some of the people cannot obtain it because they lack money, are being deprived for political reasons, live in a country at war, or suffer from other problems such as lack of transportation.

food records: logs that list all the foods eaten over a period of time and that may also include records of behaviors and symptoms, physical activity, and medications; also called **eating** or **food diaries.**

food recovery: collecting food for distribution to low-income people who are hungry.

food rescue: collecting prepared foods from commercial kitchens.

food-borne illness: an illness transmitted to human beings through food or water; caused by a poisonous substance *(food intoxication)* or an infectious agent *(food-borne infection)*; also called *food poisoning.*

fortification: the addition to a food of nutrients that were either not originally present or present in insignificant amounts. Fortification can be used to correct or prevent a widespread nutrient deficiency, to balance the total nutrient profile of a food, or to restore nutrients lost in processing.

fossil fuels: coal, oil, and natural gas, which all come from the fossilized remains of plant life of earlier times. These are nonrenewable fuels that pollute. (Renewable, or alternative, fuels, such as solar and wind energy, pollute less or not at all.)

free radicals: highly reactive chemical forms that can cause destructive changes in nearby compounds, sometimes setting up a chain reaction.

frequency: the number of occurrences per unit of time (for example, the number of activity sessions per week).

fructose: a monosaccharide; sometimes known as fruit sugar. It is abundant in fruits, honey, and saps.

full-liquid diets: diets consisting of both clear and opaque liquid foods and near-liquid foods.

fulminant (FULL-mih-nant) **liver failure:** liver failure that rapidly progresses to a life-threatening stage.

functional foods: foods that contain physiologically active compounds that may provide health benefits beyond their nutrient contributions.

galactose: a monosaccharide; part of the disaccharide lactose.

galactosemia (ga-LAK-toe-SEE-me-ah): an inborn error of metabolism in which enzymes that normally metabolize galactose to compounds the body can handle are missing and an alternative metabolite accumulates in the tissues, causing damage.

gallbladder: the organ that stores and concentrates bile. When it receives the signal that fat is present in the duodenum, the gallbladder contracts and squirts bile through the bile duct into the duodenum.

gallstone: a hard mass formed when crystals of cholesterol or bile pigments precipitate together.

gastric banding: a surgical means of producing weight loss by restricting stomach size with a constricting band or pouch; used in people whose severe obesity brings extreme health risks.

gastric bypass: surgery that reroutes food from the stomach to the lower part of the small intestine; creates a chronic, lifelong state of malabsorption by preventing normal digestion and absorption of nutrients.

gastric decompression: the removal of pressure and gas from the stomach.

gastric glands: exocrine glands in the stomach wall that secrete gastric juice into the stomach.

gastric juice: the digestive secretion of the gastric glands containing a mixture of water, hydrochloric acid, and enzymes. The principal enzymes are pepsin (acts on proteins) and lipase (acts on emulsified fats).

gastric partitioning: surgery for severe obesity that limits the functional size of the stomach.

gastric residual: the volume of formula that remains in the stomach from a previous feeding.

gastritis: inflammation of the stomach lining.

gastroesophageal reflux disease (GERD): gastroesophageal reflux is the backflow of gastric juices into the esophagus. GERD is diagnosed when reflux occurs frequently or produces significant heartburn or indigestion.

gastrointestinal motility: spontaneous motion in the digestive tract accomplished by involuntary muscular contractions.

gastrointestinal (GI) tract: the digestive tract. The principal organs are the stomach and intestines.

gastroplasty: surgery that partitions the stomach by stapling off a "pouch" or otherwise modifying the stomach, thereby reducing total food intake.

gastrostomy (gas-TROSS-toe-mee): an opening in the stomach made surgically or under local anesthesia through which a feeding tube can be passed. The technique for creating a gastrostomy under local anesthesia is called **percutaneous endoscopic gastrostomy,** or **PEG** for short. When the feeding tube is guided from such an opening into the jejunum, the procedure is called **percutaneous endoscopic jejunostomy (PEJ),** a misnomer because the enterostomy is in the stomach rather than the jejunum.

gatekeepers: with respect to nutrition, key persons who control other people's access to foods and thereby exert a profound impact on their nutrition. Examples are the spouse who buys and cooks the food, the parent who feeds the children, and the caregiver in a day-care center.

gene expression: the production of a protein by the gene that codes for it.

gene therapy: treatment that replaces a defective gene with a normal gene.

generally recognized as safe (GRAS) list: a list,

established by the FDA, of food additives long in use and believed safe.

genes: the portion of a chromosome that carries the blueprint for making a protein.

genetic (gen-ET-ick) **code:** The hereditary information programmed into the DNA of a living organism.

genome (GEE-nome): the complete set of chromosomes, which contain instructions for making an organism and controlling its functions.

genomics (gee-NOM-icks): pertaining to the genes.

genotype (GEEN-oh-type): the complete genetic makeup of an individual.

gestational diabetes: the detection of abnormal glucose tolerance during pregnancy.

glands: single cells or groups of cells that secrete materials for special uses in the body. Glands may be *exocrine glands,* secreting their materials "out" (into the digestive tract or onto the surface of the skin), or *endocrine glands,* secreting their materials "in" (into the blood).

gliadin (GLY-ah-din): the fraction of gluten that causes the toxic effects in celiac disease. Corresponding protein fractions in barley, rye, and possibly oats also have these effects.

glomerular filtration rate (GFR): the rate at which the kidneys form filtrate, usually measured by determining the amount of creatinine excreted in 24 hours. The normal GFR is about 130 ml/min for males and 120 ml/min for females.

glomerulus (glow-MARE-you-lus): a cup-shaped membrane enclosing a tuft of capillaries within a nephron. (The plural is *glomeruli.*)

glucagon (GLOO-ka-gon): a hormone that is secreted by special cells in the pancreas in response to low blood glucose concentration and elicits release of glucose from storage.

glucose: a monosaccharide, the sugar common to all disaccharides and polysaccharides; also called *blood sugar* or *dextrose.*

glucose polymers: compounds that supply glucose, not as single molecules, but linked in chains somewhat like starch. The objective is to attract less water from the body into the digestive tract.

gluten (GLUE-ten): a protein found in wheat.

glycated proteins: proteins that have glucose attached to them.

glycemic effect: a measure of the extent to which a food raises the blood glucose concentration and elicits an insulin response, as compared with pure glucose.

glycerol (GLISS-er-ol): an organic compound, three carbons long, that can form the backbone of triglycerides and phospholipids.

glycogen (GLY-co-gen): a polysaccharide composed of glucose, made and stored by liver and muscle tissues of human beings and animals as a storage form of glucose. Glycogen is not a significant food source of carbohydrate and is not counted as one of the complex carbohydrates in foods.

glycolysis (gligh-COLL-uh-sis): the metabolic breakdown of glucose to pyruvate.

goiter (GOY-ter): an enlargement of the thyroid gland due to an iodine deficiency, malfunction of the gland, or overconsumption of a thyroid antagonist. Goiter caused by iodine deficiency is *simple goiter.*

gout (GOWT): an inherited metabolic disorder that results in excessive uric acid in the blood and urine and the deposition of uric acid in and around the joints, which causes acute arthritis and joint inflammation.

granulated sugar: common table sugar, crystalline sucrose; 99.9 percent pure.

hazard: a state of danger; used to refer to any circumstance in which harm is possible under normal conditions of use.

Hazard Analysis Critical Control Points (HACCP): a systematic plan for identifying and correcting potential microbial hazards in the manufacturing, distribution, and commercial use of food products; commonly referred to as "HAASS-ip."

health: a range of states with physical, mental, emotional, spiritual, and social components. At a minimum, *health* means freedom from physical disease, mental disturbances, emotional distress, spiritual discontent, social maladjustment, and other negative states. At a maximum, *health* means "wellness."

health claims: statements that characterize the relationship between a nutrient or other substance in food and a disease or health-related condition.

health history: an account of the client's current and past health status and risk factors for disease. Traditionally, the health history has been called the *medical history.* The term *health history* now seems more appropriate, however, because the contents describe the client's health status, and the goal of medical care is health promotion and disease prevention.

health maintenance organizations (HMOs): managed care organizations that limit the subscriber's choice of health care professionals to those affiliated with the organization and control access to services by directing care through a primary care physician.

heartburn: a burning sensation felt behind the sternum that is caused by the presence of gastric juices in the esophagus; also called **pyrosis** (pie-ROE-sis).

heat stroke: an acute and life-threatening reaction to heat buildup in the body.

heavy metal: any of a number of mineral ions such as mercury and lead, so called because they are of relatively high atomic weight. Many heavy metals are poisonous.

Heimlich (HIME-lick) **maneuver (abdominal thrust maneuver):** a technique for dislodging an object from the trachea of a choking person; named for the physician who developed it.

Helicobacter pylori: a bacterium that may lead to gastritis and peptic ulcers and may raise the risk of cancer of the stomach.

hematuria (HE-mah-TOO-ree-ah): blood in the urine.

hemochromatosis (heem-oh-crome-a-TOE-sis): iron overload characterized by deposits of iron-containing pigment in many tissues, with tissue damage. Hemochromatosis is a hereditary defect in iron metabolism.

hemodialysis: removal of fluids and wastes from the blood by passing it through a dialyzer.

hemofiltration: removal of fluids and wastes from the blood by using ultrafiltration and fluid replacement.

hemoglobin: the oxygen-carrying protein of the red blood cells.

hemorrhagic (hem-oh-RAJ-ik) **disease:** the vitamin K–deficiency disease in which blood fails to clot.

hemorrhoids (HEM-oh-royds): painful swelling of the veins surrounding the rectum.

hemosiderosis (heem-oh-sid-er-OH-sis): a condition characterized by the deposition of the iron-storage protein hemosiderin in the liver and other tissues.

hepatic coma: a state of unconsciousness that results from severe liver disease.

hepatic encephalopathy (en-SEF-ah-LOP-ah-thee): mental changes associated with liver disease that may include irritability, short-term memory loss, and an inability to concentrate.

hepatitis (hep-ah-TIE-tis): inflammation of the liver.

hepatorenal syndrome: the combined symptoms of liver and renal failure that occur as a consequence of severe liver damage.

herbal laxatives: laxatives containing senna, aloe, rhubarb root, castor oil, or buckthorn, which are commonly sold as "dieter's tea." Such products cause nausea, vomiting, diarrhea, fainting, and, in some users, possibly death.

herbal products: substances extracted from plants to mimic medications that suppress appetite. Such substances may have dangerous side effects. St. John's wort, for example, is often prepared in combination with the herbal stimulant ephedrine, extracted from the Chinese plant ma huang. Ephedrine has been implicated in several cases of heart attacks and seizures (see Chapter 15).

herpes virus: a virus that can lead to mouth lesions and may also affect the lower GI tract, causing diarrhea.

hiatal hernias: protrusions of a portion of the stomach through the esophageal hiatus of the diaphragm. There are several types of hiatal hernias, but the type most commonly associated with reflux is a **sliding hiatal hernia.**

hiccups (HICK-ups): repeated cough-like sounds and jerks that are produced when an involuntary spasm of the diaphragm muscle sucks air down the windpipe; also spelled *hiccoughs.*

high-density lipoproteins (HDL): the type of lipoproteins that transport cholesterol back to the liver from peripheral cells; composed primarily of protein.

high-fructose corn syrup (HFCS): the predominant sweetener used in processed foods today. HFCS is mostly fructose; glucose makes up the balance.

highly active antiretroviral therapy (HAART): a combination of antiviral agents that disrupt various stages of replication of HIV.

homeostasis: the maintenance of constant internal conditions (such as chemistry, temperature, and blood pressure) by the body's control system.

honey: sugar (mostly sucrose) formed from nectar gathered by bees. An enzyme splits the sucrose into glucose and fructose. Composition and flavor vary, but honey always contains a mixture of sucrose, fructose, and glucose.

hormones: chemical messengers. Hormones are secreted by a variety of glands in the body in response to altered conditions. Each travels to one or more target tissues or organs and elicits specific responses to restore normal conditions.

house diet: a menu preselected by the dietary department.

Human Genome Project (HGP): an extensive research project, funded through the National Institutes of Health and the Department of Energy, that aims to identify and locate all the genes in the human body.

human immunodeficiency virus (HIV): a virus

that progressively hampers the function of the immune system and leaves its host defenseless against other infections and cancer and eventually causes AIDS. The most common HIV is HIV-1—the virus described in this chapter.

humoral immunity: immunity conferred by B-cells, which produce antibodies that travel through the blood to the invasion site.

hunger: the physiological need to eat, experienced as a drive to obtain food; an unpleasant sensation that demands relief.

hydrochloric acid (HCl): an acid composed of hydrogen and chloride atoms; normally produced by the gastric glands.

hydrogenation (high-dro-gen-AY-shun): a chemical process by which hydrogens are added to monounsaturated or polyunsaturated fats to reduce the number of double bonds, making the fats more saturated (solid) and more resistant to oxidation (protecting against rancidity). Hydrogenation produces *trans*-fatty acids.

hydrolyzed formulas: liquid diets that contain broken-down molecules of protein, such as amino acids and short peptide chains; also called **monomeric formulas.**

hyperactivity: inattentive and impulsive behavior that is more frequent and severe than is typical of others a similar age; professionally called **attention-deficit/hyperactivity disorder (ADHD).**

hyperammonemia (HIGH-per-AM-moe-KNEE-me-ah): elevated blood ammonia.

hypercalcemia (HIGH-per-cal-SEE-me-ah): excessive calcium in the blood.

hypercalciuria (HIGH-per-kal-see-YOU-ree-ah): excessive urinary excretion of calcium.

hyperkalemia (HIGH-per-kay-LEE-me-ah): excessive potassium in the blood.

hyperoxaluria (HIGH-per-OX-all-YOU-ree-ah): excessive urinary excretion of oxalate.

hyperperistalsis (HY-per-pear-ih-STALL-sis): rapid movement through the intestine.

hypertension: elevated blood pressure.

hypertonic formula: a formula with an osmolality greater than that of blood serum.

hypertrophy (high-PURR-tro-fee): an increase in size (for example, of a muscle) in response to use.

hypoallergenic formulas: clinically tested infant formulas that do not provoke reactions in 90% of infants or children with confirmed cow's milk allergy. Like all infant formulas, hypoallergenic formulas must demonstrate nutritional suitability to support infant growth and development. Extensively hydrolyzed and free amino acid–based formulas are examples.

hypoglycemia (HY-poe-gly-SEE-me-ah): low blood sugar. The type of hypoglycemia that occurs following gastric surgery is called *reactive* or *postgastrectomy* hypoglycemia.

hyponatremia (HIGH-poe-nay-TREE-mee-ah): low levels of sodium in the blood.

hypothalamus (high-poh-THALL-uh-mus): a part of the brain that helps regulate many body balances, including fluid balance.

hypothermia: a below-normal body temperature.

idiopathic (ID-ee-oh-PATH-ick): arising from an unknown cause.

ileocecal (ill-ee-oh-SEEK-ul) **valve:** the sphincter muscle separating the small and large intestines.

ileostomy (ILL-ee-OSS-toe-me): surgery that creates an opening from the ileum through the abdominal wall and out through the skin.

ileum (ILL-ee-um): the last segment of the small intestine.

ileus (ILL-ee-us): an intestinal obstruction.

immune system: the body's system of defense against harmful substances.

impaired glucose tolerance: inability to maintain normal blood glucose levels without excessive insulin production. Some people with impaired glucose tolerance have fasting glucose levels somewhat higher than normal but not high enough to diagnose diabetes. Others have normal blood glucose levels most of the time, but when given a large amount of glucose, their blood glucose rises too high.

inborn error of metabolism: an inherited flaw evident as a metabolic disorder or disease present from birth.

incidental additives: substances that can get into food not through intentional introduction but as a result of contact with the food during growing, processing, packaging, storing, or some other stage before the food is consumed; also called *accidental* or *indirect additives.*

incomplete protein: a protein lacking or low in one or more of the essential amino acids.

indemnity insurance: traditional fee-for-service insurance.

indirect calorimetry (kal-oh-RIM-eh-tree): an indirect estimate of resting energy needs made by measuring the ratio of carbon dioxide expired to the amount of oxygen inspired and using the results in a mathematical equation.

inflammation: the immune system's response to any type of tissue injury. The symptoms of inflammation include redness, swelling, and pain.

inflammatory bowel diseases (IBD): diseases characterized by inflammation of the bowel.

inflammatory response: the changes orchestrated by the immune system when tissues are injured by such forces as blows, wounds, foreign bodies (chemicals, microorganisms), loss of blood flow, heat, cold, electricity, or radiation.

inorganic: not containing carbon or pertaining to living things.

insoluble fibers: the tough, fibrous structures of fruits, vegetables, and grains; indigestible food components that do not dissolve in water.

insulin: a hormone secreted by the pancreas in response to high blood glucose. It promotes cellular glucose uptake for use or storage.

insulin resistance: the condition in which the cells fail to respond to insulin as they do in healthy people.

insulin resistance of adipose tissue: failure of the enzyme that releases free fatty acids from fat cells to adequately reduce its activity in the presence of insulin.

insulin resistance of skeletal muscle: failure of the muscle cells to take up glucose from the blood in response to insulin as they normally would.

intensity: the degree of exertion while exercising (for example, the amount of weight lifted or the speed of running).

intermittent claudication: severe calf pain caused by inadequate blood supply; it occurs when walking and subsides during rest.

intermittent feedings: delivery of about 250 to 400 ml of formula over 30 minutes or more.

intestinal flora: the bacterial inhabitants of the GI tract.

intestinal juice: the secretion of the intestinal glands; contains enzymes for the digestion of carbohydrate and protein and a minor enzyme for fat digestion.

intra-abdominal fat: fat stored within the abdominal cavity in association with the internal abdominal organs, as opposed to the fat stored directly under the abdominal skin (subcutaneous fat).

intravenous: through a vein.

intrinsic: inside the system. Anemia that reflects a vitamin B_{12} deficiency caused by lack of intrinsic factor is known as **pernicious anemia.**

invert sugar: a mixture of glucose and fructose formed by splitting sucrose in a chemical process; sold only in liquid form, sweeter than sucrose. Invert sugar is used as an additive to help preserve food freshness and prevent shrinkage.

iron deficiency: having depleted iron stores.

iron-deficiency anemia: a blood iron deficiency characterized by small, pale red blood cells; also called **microcytic hypochromic anemia.**

iron overload: toxicity from excess iron.

irradiation: the application of ionizing radiation to foods to reduce insect infestation or microbial contamination or to slow the ripening or sprouting process; also called *cold pasteurization.*

irritable bowel syndrome: an intestinal disorder of unknown cause characterized by abdominal discomfort, cramping, diarrhea or constipation, or alternating diarrhea and constipation; also called *spastic colon.*

isotonic formula: a formula with an osmolality similar to that of blood serum (300 mOsm/kg).

IV catheter: a thin tube inserted into a peripheral or central vein. Additional tubing connects the IV solution to the catheter.

jaundice (JON-dis): a characteristic yellowing of the skin, whites of the eyes, mucous membranes, and body fluids resulting from the accumulation of bilirubin in the blood.

jejunostomy (JEE-ju-NOSS-toe-mee): an opening in the jejunum made surgically or under local anesthesia through which a feeding tube can be passed. The technique for creating a jejunostomy under local anesthesia is called a **direct endoscopic jejunostomy (DEJ).** Note: Some clinicians also refer to this procedure as a PEJ, which is a more accurate use of the term than the more common use described above.

jejunum (je-JOON-um): the first two-fifths of the small intestine beyond the duodenum.

Kaposi's (cap-OH-seez) **sarcoma:** a type of cancer rare in the general population but common in people with HIV infections.

kcalorie (energy) control: management of food energy intake.

kcalorie count: a determination of a client's food intake from a direct observation of how much the client eats.

kcalories: units by which energy is measured. One kcalorie (kcal) is the amount of heat necessary to raise the temperature of 1 kilogram (kg) of water 1°C. Most people speak of these units simply as calories, but on paper the word *calorie* is prefaced by a *k* for kilocalorie. We use kcalories and kcal throughout this book.

keratin (KERR-uh-tin): a water-insoluble protein; the normal protein of hair and nails. Keratin-producing cells may replace mucus-producing cells in vitamin A deficiency.

ketones (KEY-tones): acidic, fat-related compounds formed from the incomplete breakdown of fat when carbohydrate is not available; technically known as *ketone bodies.*

kidney stones: crystals of salts or other components that form a concentrated mass in the kidney; also called **renal calculi.**

kwashiorkor (kwash-ee-OR-core or kwash-ee-or-CORE): a disease related to protein-energy malnutrition (PEM).

lactation: production and secretion of breast milk for the purpose of nourishing an infant.

lactic acid: a product of the incomplete breakdown of glucose during anaerobic metabolism. When oxygen becomes available, lactic acid can be completely broken down for energy or converted back to glucose.

lactoferrin (lack-toe-FERR-in): a factor in breast milk that binds iron and keeps it from supporting the growth of the infant's intestinal bacteria.

lacto-ovo vegetarians: people who include milk or milk products and eggs, but omit meat, fish, shellfish, and poultry from their diets.

lactose: a disaccharide composed of glucose and galactose; commonly known as milk sugar.

lacto-vegetarians: people who include milk or milk products, but exclude meat, poultry, fish, shellfish, and eggs from their diets.

latent: the period in the course of a disease when the conditions are present but the symptoms have not begun to appear.

lecithins: one type of phospholipid.

legal: established by law.

leptin: a protein produced by fat cells under the direction of the obesity gene that increases satiety and energy expenditure.

letdown reflex: the reflex that forces milk to the front of the breast when the infant begins to nurse.

levulose: an older name for fructose.

life expectancy: the average number of years lived by people in a given society.

life span: the maximum number of years of life attainable by a member of a species.

light beer: beer that contains the same amount of alcohol as regular beer, but with fewer kcalories.

linoleic acid, linolenic acid: polyunsaturated fatty acids, essential for human beings.

lipids: a family of compounds that includes triglycerides (fats and oils), phospholipids, and sterols.

lipodystrophy (LIP-oh-DISS-tro-fee): the redistribution of fat that can occur as a consequence of HIV infections as well as other disorders. The accumulation of abdominal fat associated with HIV infections is sometimes called *protease paunch.*

lipomas (lih-POE-mahs): benign tumors composed of fat.

lipoprotein lipase (LPL): an enzyme mounted on the surface of fat cells (and other cells). It hydrolyzes triglycerides in the blood into fatty acids and glycerol for absorption into the cells. There they are metabolized or reassembled for storage.

lipoproteins: clusters of lipids associated with proteins that serve as transport vehicles for lipids in the lymph and blood.

living will: a document signed by a competent adult that specifically states whether the person wishes aggressive treatment in the event of terminal illness or irreversible coma from which he or she is not expected to recover.

localized: confined to the area of injury

longevity: long duration of life.

low birthweight (LBW): a birthweight less than 5½ lb (2500 g); indicates probable poor health in the newborn and poor nutrition status of the mother during pregnancy. Normal birthweight for a full-term baby is 6½ to 8½ lb (about 3000 to 4000 g). Low-birthweight infants are of two different types. Some are **premature;** they are born early and are of a weight **appropriate for gestational age (AGA).** Others have suffered growth failure in the uterus; they may or may not be born early, but they are **small for gestational age (SGA).**

low-carbohydrate diets: diets designed to bring about metabolic responses similar to those of fasting. Without sufficient carbohydrate, the body cannot use its fat in the normal way, and ketosis results. Many physiological hazards accompany low-carbohydrate diets: high blood cholesterol, mineral imbalances, hypoglycemia, and more.

low-density lipoproteins (LDL): the type of lipoproteins derived from VLDL as cells remove triglycerides from them. LDL carry cholesterol and triglycerides from the liver to the cells of the body and are composed primarily of cholesterol.

lower esophageal sphincter (SFINK-ter): the sphincter muscle at the junction between the esophagus and the stomach (also called *cardiac sphincter*).

lymph (LIMF): the body fluid found in lymphatic vessels. Lymph consists of all the constituents of blood except red blood cells.

lymph tissues: tissues that contain lymphocytes.

lymphatic system: a loosely organized system of vessels and ducts that conveys the products of digestion toward the heart.

lymphocytes: cells made in lymph tissues that travel throughout the lymphatic and circulatory systems.

mad cow disease: an often-fatal illness of cattle affecting the nerves and brain; also called *bovine spongiform encephalopathy (BSE).*

maleficence (mah-LEF-eh-sens): doing harm.

malignant (ma-LIG-nant): describes tumors that multiply out of control, threaten health, and require treatment.

malnutrition: any condition caused by deficient or excess energy or nutrient intake or by an imbalance of nutrients.

maltose: a disaccharide composed of two glucose units; sometimes known as malt sugar.

managed care: a health care delivery system that aims to provide cost-effective health care by coordinating services and limiting access to services.

maple sugar: a sugar (mostly sucrose) purified from concentrated sap of the sugar maple tree. Maple sugar is expensive compared with other sweeteners.

marasmus (ma-RAZZ-mus): a disease related to PEM. Marasmus results from severe deprivation, or impaired absorption, of protein, energy, vitamins, and minerals.

margin of safety: in reference to food additives, a zone between the concentration normally used and that at which a hazard exists. For common table salt, for example, the margin of safety is ⅕ (five times the concentration normally used would be hazardous).

mastitis: infection of a breast.

mechanical ileus: ileus caused by a physical obstruction.

mechanical soft diets: diets that exclude only those foods that a person cannot chew or swallow; also called **dental soft diets** or **edentulous diets.**

mechanical ventilator: a machine that "breathes" for a person who can't. In normal respiration, air is drawn in when the lungs expand. With mechanical ventilation, air is forced into the lungs at regular intervals using pressure.

medical nutrition therapy: the provision of a client's nutrient, dietary, and nutrition education needs based on a comprehensive nutrition assessment.

medications: chemicals (drugs) that alter one or more body functions that are marketed only with approval of the FDA and only after research shows that they are safe and effective.

medium-chain triglycerides (MCT): triglycerides containing fatty acids with 8 to 12 carbon atoms; they require minimal lipase and no bile for absorption.

metabolic stress: the state in which a body's internal balance (homeostasis) is upset by a threat to a person's physical well-being (**stressor**). The terms **acute stress** and **severe stress** are used in this chapter to refer to pathological stresses that rapidly and markedly raise the body's metabolic rate and significantly upset its normal internal balance.

metabolic syndrome: the combination of insulin resistance, hyperinsulinemia, obesity, hypertension, elevated LDL and triglycerides, and reduced HDL that is frequently associated with type 2 diabetes and cardiovascular disease; also called *syndrome X* and *insulin-resistance syndrome.*

metabolism: the sum total of all the chemical reactions that go on in living cells.

metastasize (meh-TAS-tah-size): to spread by the movement of cancer cells from one part of the body to another.

microvascular: pertaining to the capillaries. Retinopathy, nephropathy, and neuropathy are microvascular complications.

microvilli (MY-cro-VILL-ee or MY-cro-VILL-eye): tiny, hairlike projections on each cell of every villus that can trap nutrient particles and transport them into the cells. The singular form is **microvillus.**

moderate exercise: activity that can be sustained comfortably for 60 minutes or so.

moderation: providing enough, but not too much of a substance.

modified diets: diets adjusted to meet medical needs. Such diets may be adjusted in consistency, level of energy and nutrients, amount of fluid, or number of meals, or by the inclusion or elimination of certain foods.

modular formulas: formulas that provide a single nutrient and are designed to be added to other formulas to alter nutrient composition or combined to create a highly individualized formula.

molasses: a thick brown syrup, left over from sugarcane juice during sugar refining. Blackstrap molasses contains iron, which comes from the machinery used to process it. This iron is not as well absorbed as the iron in meats and other foods.

molybdenum (mo-LIB-duh-num): a trace element.

monosaccharides: single sugar units.

monounsaturated fatty acid: a fatty acid that has one point of unsaturation; for example, the oleic acid found in olive oil.

mood disorders: mental illness characterized by episodes of severe depression or excessive excitement (mania) or both.

mouth ulcers: lesions or sores in the lining of the mouth. Certain medications, radiation therapy, and some disorders, such as oral herpes infections, can cause mouth ulcers.

MSG symptom complex: the acute, temporary, and self-limiting reactions experienced by sensitive people upon ingesting a large dose of MSG. The name *MSG symptom complex,* given by the FDA, replaces the former *Chinese restaurant syndrome.*

mucous membrane: membrane composed of mucus-secreting cells that lines the surfaces of body tissues. (Reminder: *Mucus* is the smooth, slippery substance secreted by these cells.)

mucus (MYOO-cuss): a mucopolysaccharide (a relative of carbohydrate) secreted by cells of the stomach wall that protects the cells from exposure to digestive juices (and other destructive agents). The cellular lining of the stomach wall with its coat of mucus is known as the mucous membrane. (The noun is *mucus;* the adjective is *mucous.*)

multiple daily injections: delivery of a mixture of insulins by injection three or more times daily.

multiple organ failure: failure of two or more organ systems related to prolonged or exaggerated inflammatory responses; also called *multiple organ dysfunction syndrome.*

muscle endurance: the ability of a muscle to contract repeatedly within a given time without becoming exhausted.

muscle strength: the ability of muscles to work against resistance.

muscular dystrophy (DIS-tro-fee): a hereditary disease in which the muscles gradually weaken; its most debilitating effects arise in the lungs. This disease should not be confused with nutritional muscular dystrophy, a vitamin E–deficiency disease of animals characterized by gradual paralysis of the muscles.

mutation: an alteration in a gene such that an altered protein is produced.

mutual supplementation: the strategy of combining two incomplete protein sources so that the amino acids in one food make up for those lacking in the other food. Such protein combinations are sometimes called *complementary proteins.*

myocardial (my-oh-CAR-dee-al) **infarction** (in-FARK-shun) or **MI:** sudden tissue death caused by blockages of vessels that feed the heart muscle; also called **heart attack, cardiac arrest,** or **acute heart failure.**

myoglobin: the oxygen-carrying protein of the muscle cells.

nasoduodenal (ND): from the nose to the duodenum.

nasoenteric: from the nose to the stomach or intestine. Nasoenteric feedings include nasogastric, nasoduodenal, and nasojejunal feedings. Most clinicians use nasoenteric to refer to nasoduodenal and nasojejunal feedings only.

nasogastric (NG): from the nose to the stomach.

nasojejunal (NJ): from the nose to the jejunum.

natural killer cells: lymphocytes that confer nonspecific immunity. Natural killer cells destroy viruses and tumor cells.

nephrons (NEF-rons): the working units of the kidneys. Each nephron consists of a glomerulus and a tubule.

nephrotic (neh-FRAUT-ic) **syndrome:** the cluster of clinical findings that occur when glomerular function falters, including proteinuria, low serum albumin, edema, and elevated blood lipids.

neural tube defects (NTD): malformations of the brain, spinal cord, or both during embryonic development.

Glossary
G–10

neurons: nerve cells; the structural and functional units of the nervous system. Neurons initiate and conduct nerve transmissions.

niacin equivalents (NE): the amount of niacin present in food, including the niacin that can theoretically be made from its precursor tryptophan present in the food.

night blindness: the slow recovery of vision after exposure to flashes of bright light at night; an early symptom of vitamin A deficiency.

nitric oxide (NO): a substance produced by the vascular endothelium that causes blood vessels to dilate and inhibits clot formation.

nitrites: salts added to food to prevent botulism. An example is sodium nitrite.

nitrogen balance: the amount of nitrogen consumed (N in) as compared with the amount of nitrogen excreted (N out) in a given period of time. The laboratory scientist can estimate the protein in a sample of food, body tissue, or excreta by measuring the nitrogen in it.

nitrosamines (nigh-TROHS-uh-meens): derivatives of nitrites that may form when nitrites combine with amines.

nonpathogenic (non-path-oh-GEN-ick): not causing disease.

nonperishable food collection: collecting processed foods from wholesalers and markets.

nonspecific immunity: immunity directed at foreign substances in general, rather than specific antigens.

nursing bottle tooth decay: extensive tooth decay due to prolonged tooth contact with formula, milk, fruit juice, or other carbohydrate-rich liquid offered to an infant in a bottle.

nursing diagnoses: identification of the client's actual or potential health problems that require intervention by the nurse.

nutrient claims: statements that characterize the quantity of a nutrient in a food.

nutrient density: a measure of the nutrients a food provides relative to the energy it provides. The more nutrients and the fewer kcalories, the higher the nutrient density.

nutrients: substances obtained from food and used in the body to provide energy and structural materials and to serve as regulating agents to promote growth, maintenance, and repair. Nutrients may also reduce the risks of some diseases.

nutrition: the science of foods and the nutrients and other substances they contain, and of their ingestion, digestion, absorption, transport, metabolism, interaction, storage, and excretion. A broader definition includes the study of the environment and of human behavior as it relates to these processes.

nutrition assessment: a comprehensive evaluation of a person's nutrition status, completed by a registered dietitian. Chapter 16 provides more information.

nutrition care plan: a strategy for meeting nutrient and nutrition education needs identified through nutrition assessment.

nutrition screening: a tool for quickly identifying clients at risk for malnutrition so that they can receive complete nutrition assessments.

nutrition status surveys: surveys that evaluate people's nutrition status using nutrition assessment methods.

nutritionists: persons who specialize in the study of nutrition. Some nutritionists are registered dietitians, but others are self-described experts

whose training may be minimal or nonexistent. Some states make the term meaningful by allowing it to apply only to people who have master's (M.S.) or doctoral (Ph.D.) degrees from institutions accredited to offer such degrees in nutrition or related fields.

nutritive sweeteners: sweeteners that yield energy, including both the sugars and the sugar alcohols.

obesity: a chronic disease characterized by excessively high body fat in relation to lean body tissue.

oils: lipids that are liquid at room temperature (70°F or 25°C).

olestra: a synthetic fat made from sucrose and fatty acids that provides 0 kcalories per gram; formerly known as **sucrose polyester.**

omega-3 fatty acids: polyunsaturated fatty acids in which the endmost double bond is three carbons back from the end of the carbon chain; relatively newly recognized as important in nutrition. Linolenic acid is an example.

omega-6 fatty acids: a polyunsaturated fatty acid with its endmost double bond six carbons back from the end of its carbon chain; long recognized as important in nutrition. Linoleic acid is an example.

open feeding systems: enteral formula delivery systems that require formula to be transferred from its original packaging to a feeding container before it can be administered through a feeding tube.

opportunistic illnesses: illnesses that normally would not occur or that would cause only minor problems in a healthy population, but can cause great harm when the immune system is compromised.

oral antidiabetic agents: medications taken by mouth to lower blood glucose levels in people with type 2 diabetes.

organic: carbon containing. The four organic nutrients are carbohydrate, fat, protein, and vitamins.

organically grown crops: crops grown and processed according to USDA regulations defining the use of fertilizers, herbicides, insecticides, fungicides, preservatives, and other chemical ingredients.

orogastric: from the mouth to the stomach. This method is often used to feed infants because they breathe through their noses, and a nasogastric tube can hinder the infant's breathing.

osmolality (OZ-moh-LAL-eh-tee): a measure of the concentration of particles in a solution, expressed as the number of milliosmoles (mOsm) per kilogram.

osmosis: movement of water from an area of low solute concentration to one of high solute concentration.

osteoarthritis: a painful, chronic disease of the joints caused when the cushioning cartilage in a joint breaks down; joint structure is usually altered, with loss of function; also called *degenerative arthritis.*

osteomalacia (os-tee-oh-mal-AY-shuh): a bone disease characterized by softening of the bones; symptoms include bending of the spine and bowing of the legs. The disease occurs most often in adult women.

osteoporosis (oss-tee-oh-pore-OH-sis): literally, porous bones; reduced density of the bones. Also known as *adult bone loss,* it is a condition in which the bones become porous and fragile. The causes of osteoporosis are multiple.

outcome measures: indicators that describe an observable change and used to evaluate the effects of interventions.

overnutrition: overconsumption of food energy or nutrients sufficient to cause disease or increased susceptibility to disease; a form of malnutrition.

overt: out in the open, full-blown.

overweight: body weight above some standard of acceptable weight that is usually defined in relation to height (such as BMI).

oxidation (OKS-ee-day-shun): the process of a substance combining with oxygen.

oxidative stress: a condition in which the production of oxidants and free radicals exceeds the body's ability to neutralize them.

pancreas: a gland that secretes enzymes and digestive juices into the duodenum. (This is its exocrine function; it also has the endocrine function of secreting insulin and other hormones into the blood.)

pancreatic (pank-ree-AT-ic) **juice:** the exocrine secretion of the pancreas, containing enzymes for the digestion of carbohydrate, fat, and protein. Juice flows from the pancreas into the small intestine through the pancreatic duct. The pancreas also has an endocrine function, the secretion of insulin and other hormones.

pancreatic enzyme replacements: extracts of pork or beef pancreatic enzymes that are taken as supplements to aid digestion.

pancreatitis (PAN-cree-ah-TIE-tis): inflammation of the pancreas.

paranoia (PAR-ah-NOY-ah): mental illness characterized by delusions of persecution.

parenteral (par-EN-ter-all) **nutrition:** the provision of nutrients bypassing the intestine.

pasteurization: the treatment of milk with heat sufficient to kill certain pathogens (disease-causing microbes) commonly transmitted through milk; not a sterilization process. Pasteurized milk retains bacteria that cause milk spoilage. Raw milk, even if labeled "certified," transmits many food-borne diseases to people each year and should be avoided.

pathogenic: causing disease

pathogens: disease-causing microorganisms.

pathological stresses: disruptions to the body's internal balance that lie beyond its normal and healthy functioning.

pellagra (pell-AY-gra): the niacin-deficiency disease. Symptoms include the "4 Ds": diarrhea, dermatitis, dementia, and, ultimately, death.

pepsin: a protein-digesting enzyme (gastric protease) in the stomach. It circulates as a precursor, pepsinogen, and is converted to pepsin by the action of stomach acid.

peptic ulcer: a lesion or erosion of the cells of the mucosa of the lower esophagus, stomach, or small intestine. Ulcers may also develop in the mouth, upper esophagus, and large intestine, or on the skin.

perforation (per-foe-RAY-shun): a hole or tear.

peripheral fat: located directly beneath the skin.

peripheral parenteral nutrition (PPN): the provision of an IV solution that meets nutrient needs delivered into the peripheral veins.

peripheral veins: the small-diameter veins that carry blood from the arms and legs.

peripherally inserted central catheter (PICC): a catheter inserted into a peripheral vein and advanced into a central vein.

perishable food salvage: collecting perishable produce from wholesalers and markets.

peristalsis (peri-STALL-sis): successive waves of involuntary muscular contractions passing along the walls of the GI tract that push the contents along.

peritoneal dialysis: removal of wastes and fluids from the body by using the peritoneal membrane as a semipermeable membrane.

peritonitis (pear-ih-toe-NIGH-tus): infection and inflammation of the membrane lining the abdominal cavity caused by leakage of infectious organisms through a perforation in an abdominal organ.

persistent vegetative state: exhibiting motor reflexes but without the ability to regain cognitive behavior, communicate, or interact purposefully with the environment.

persistent: of a stubborn or enduring nature; with respect to food contaminants, the quality of remaining unaltered and unexcreted in plant foods or in the bodies of animals and human beings.

personal history: an account of socioeconomic and psychosocial factors that affect a person's nutrient needs or ability or willingness to follow nutrition advice.

pesticides: chemicals used to control insects, diseases, weeds, fungi, and other pests on crops and around animals. Used broadly, the term includes herbicides (to kill weeds), insecticides (to kill insects), and fungicides (to kill fungi).

pH: the concentration of hydrogen ions. The lower the pH, the stronger the acid. Thus pH 2 is a strong acid; pH 6 is a weak acid; pH 7 is neutral; and a pH above 7 is alkaline.

phagocytes: large white blood cells that confer nonspecific immunity. Phagocytes engulf and destroy foreign substances. Phagocytes that travel in the blood are called monocytes; when monocytes embed themselves in tissues, they grow larger and are called macrophages. Other types of phagocytes include neutrophils, polymorphonuclear leukocytes, and basophils.

phospholipids: one of the three main classes of lipids. These compounds are similar to triglycerides, but have choline (or another compound) and a phosphorus-containing acid in place of one of the fatty acids.

physiological age: a person's age as estimated from her or his body's health and probable life expectancy.

physiological stresses: disruptions to the body's internal balance caused by processes necessary to sustain life.

phytates: nonnutrient components of grains, legumes, and seeds. Phytates can bind minerals such as iron, zinc, calcium, and magnesium in insoluble complexes in the intestine, and the body excretes them unused.

phytobezoar (FIGH-toe-BEE-zor): a mass of plant matter (fibers, leaves, and skins) that forms a ball and may block the outlet from the stomach to the intestine. Trichobezoars contain hair and nails. People with some psychiatric disorders may chew and swallow their hair and nails.

phytochemicals: nonnutrient compounds in plant-derived foods that have biological activity in the body.

pica (PIE-ka): a craving for nonfood substances; also known as geophagia (jee-oh-FAY-jee-uh) when referring to clay-eating behavior.

pituitary (pit-TOO-ih-tary) **gland:** in the brain, the "king gland" that regulates the operation of many other glands.

PKU, phenylketonuria (FEN-el-KEY-toe-NEW-ree-ah): an inborn error of metabolism in which phenylalanine, an essential amino acid, cannot be converted to tyrosine. Alternative metabolites of phenylalanine (phenylketones) accumulate in the tissues, causing damage, and overflow into the urine.

placenta (pla-SEN-tuh): an organ that develops inside the uterus early in pregnancy, in which maternal and fetal blood circulate in close proximity and exchange materials. The fetus receives nutrients and oxygen across the placenta; the mother's blood picks up carbon dioxide and other waste materials to be excreted via her lungs and kidneys.

plaques (PLACKS): mounds of lipid material (mostly cholesterol) covered with fibrous connective tissue and embedded in artery walls that may harden with time.

plasminogen (plaz-MIN-oh-jen) **activator inhibitor-1 (PAL-1):** a substance important in blood clotting.

platelets: tiny, disc-shaped bodies in the blood that are important in clot formation.

polycystic (POL-ee-SIS-tik) **ovary syndrome:** a disorder of women characterized by ovaries enlarged with fluid-filled sacs and elevated levels of male hormones (androgens).

polypeptide: ten or more amino acids bonded together. An intermediate strand of between four and ten amino acids is an oligopeptide.

polyunsaturated fatty acids (PUFA): fatty acids with two or more points of unsaturation. For example, linoleic acid has two such points, and linolenic acid has three. Thus polyunsaturated fat is composed of triglycerides containing a high percentage of PUFA.

portal hypertension: elevated blood pressure in the portal vein caused by obstructed blood flow through the liver.

postgastrectomy (post-gas-TREK-toe-me): following surgery that removes all (total gastrectomy) or part (subtotal or partial gastrectomy) of the stomach.

potassium iodide: a medication approved by the FDA as safe and effective for the prevention of thyroid cancer caused by radioactive iodine known to be released during radiation emergencies.

prebiotics: nondigestible food ingredients that favor the growth and activity of harmless bacteria.

precursors: compounds that can be converted into other compounds; with regard to vitamins, compounds that can be converted into active vitamins; also known as provitamins.

preeclampsia: a condition characterized by hypertension, fluid retention, and protein in the urine.

preferred provider organizations (PPOs): managed care organizations that encourage subscribers to select health care providers from a group that has contracted with the organization to provide services at lower costs.

preformed vitamin A: vitamin A in its active form.

pregame meal: a meal eaten three to four hours before athletic competition.

preservatives: antimicrobial agents, antioxidants, chelating agents, radiation, and other additives that retard spoilage or preserve desired qualities, such as softness in baked goods.

pressure sores: the breakdown of skin and underlying tissues due to constant pressure and lack of oxygen to the affected area; also called **decubitus** (dee-CUE-bih-tus) **ulcers** or **bedsores.**

primary prevention: efforts aimed at preventing CHD.

probiotics: foods and dietary supplements that contain live bacteria.

progressive overload principle: the training principle that a body system, in order to improve, must be worked at frequencies, durations, or intensities that gradually increase physical demands.

proof: a description of the amount of alcohol in a product. Liquor that is 80 proof is 40% alcohol. Liquor that is 100 proof is 50% alcohol.

protein isolates: proteins that have been separated from a food. Examples include casein from milk and albumin from egg.

protein-energy malnutrition (PEM): a deficiency of protein and food energy; the world's most widespread malnutrition problem, including both marasmus and kwashiorkor.

proteins: compounds composed of carbon, hydrogen, oxygen, and nitrogen atoms arranged into strands of amino acids. Some amino acids also contain sulfur atoms.

protein-sparing effect: the effect of carbohydrate in providing energy that allows protein to be used for other purposes.

proteinuria (pro-teen-YOUR-ee-ah): the loss of protein in the urine. The loss of albumin in the urine is *albuminuria*.

puberty: the period in life in which a person becomes physically capable of reproduction.

purine (PU-reen): an end product of nucleotide metabolism that eventually breaks down to form uric acid.

pyloric (pie-LORE-ic) **sphincter:** the sphincter muscle separating the stomach from the small intestine (also called *pylorus or pyloric valve*).

pyloroplasty (pie-LOOR-oh-PLAS-tee): surgery that enlarges the pyloric sphincter.

pyruvate (PIE-roo-vate): pyruvic acid, a 3-carbon compound derived from glucose, glycerol, and certain amino acids in metabolism.

radiation enteritis: inflammation and scarring of the intestinal cells caused by exposure to radiation.

radiation therapy: the use of radiation to arrest or destroy cancer cells.

rancid: the term used to describe fats when they have deteriorated, usually by oxidation. Rancid fats often have an "off" odor.

raw sugar: the first crop of crystals harvested during sugar processing. Raw sugar cannot be sold in the United States because it contains too much filth (dirt, insect fragments, and the like). Sugar sold domestically as raw sugar has actually gone through about half of the refining steps.

reactive hypoglycemia: low blood glucose that develops two to five hours following a meal. Also called **postprandial hypoglycemia.**

Recommended Dietary Allowances (RDA): a set of values reflecting the average daily amounts of nutrients considered adequate to meet the known nutrient needs of practically all healthy people; a goal for dietary intake by individuals.

rectum: the muscular terminal part of the GI tract extending from the sigmoid colon to the anus. The rectum stores waste prior to elimination.

refeeding syndrome: the physiological and metabolic complications associated with reintroducing nutrients too rapidly in people with depleted nutrient stores due to chronic malnutrition or in those who have been underfed for several days. These complications can include malabsorption, cardiac insufficiency, respiratory distress, congestive heart failure, convulsions, coma, and possibly death.

refined grain: a product from which the bran, germ, and husk have been removed, leaving only the endosperm.

reflux esophagitis (eh-sof-ah-JYE-tis): inflammation of the esophagus caused by the reflux of gastric juices.

registered dietitians (R.D.'s): dietitians who have graduated from a university or college after completing a program of dietetics that has been accredited by the American Dietetic Association (or Dietitians of Canada). The dietitian must serve in an approved internship or coordinated program to practice the necessary skills, pass the association registration examination, and maintain competency through continuing education. Many states require licensing for practicing dietitians. Licensed dietitians (L.D.'s) have met all *state* requirements to offer nutrition advice.

renal: pertaining to the kidneys.

renal colic: the severe pain that accompanies the movement of a kidney stone through the ureter to the bladder.

renal insufficiency: the stage of renal failure in which renal function is reduced but not to a life-threatening degree.

renin (REN-in): an enzyme, secreted by the kidneys in response to a reduced blood flow, that triggers the release of the hormone aldosterone.

requirement: the lowest continuing intake of a nutrient that will maintain a specified criterion of adequacy.

residue, intestinal: the total amount of material in the colon; includes dietary fiber and undigested food, intestinal secretions, bacterial cell bodies, and cells shed from the intestinal mucosa.

residues: whatever remains. In the case of pesticides, those amounts that remain on or in foods when people buy and use them.

retina (RET-in-uh): the layer of light-sensitive nerve cells lining the back of the inside of the eye; consists of rods and cones.

retinol activity equivalents (RAE): a measure of vitamin A activity; the amount of retinol that the body will derive from a food containing preformed retinol or its precursor beta-carotene.

retinol-binding protein (RBP): the specific protein responsible for transporting retinol. Measurement of the blood concentration of RBP is a sensitive test of vitamin A status.

rheumatoid arthritis: a disease of the immune system involving painful inflammation of the joints and related structures.

rickets: the vitamin D–deficiency disease in children.

rooting reflex: a reflex that causes an infant to turn toward whichever cheek is touched, in search of a nipple.

routine progressive diets: diets that advance from liquids to solids as a client's tolerances permit.

saccharin (SAK-ah-ren): a 0-kcalorie artificial sweetener that tastes 500 times as sweet as sucrose; approved in the United States, but available in Canada only in pharmacies and only as a sweetener, not as an additive. ADI = 5 milligrams/kilogram body weight.

saliva: the secretion of the salivary glands. The principal enzyme is salivary amylase.

salivary glands: exocrine glands that secrete saliva into the mouth.

salts: compounds composed of charged particles (ions). An example of a salt is potassium chloride (K^+Cl^-).

sarcopenia (SAR-koh-PEE-nee-ah): loss of skeletal muscle mass, strength, and quality.

saturated fatty acid: a fatty acid carrying the maximum possible number of hydrogen atoms (having no points of unsaturation). A saturated fat is a triglyceride that contains three saturated fatty acids.

schizophrenia (SKITZ-oh-FREN-ee-ah): mental illness characterized by an altered concept of reality and, in some cases, delusions and hallucinations.

scurvy: the vitamin C–deficiency disease.

secondary prevention: efforts aimed at preventing complications associated with CHD or CHD risk equivalents.

sedentary: physically inactive (literally, "sitting down a lot").

segmentation: a periodic squeezing or partitioning of the intestine by its circular muscles that both mixes and slowly pushes the contents along.

semipermeable membrane: a membrane with pores that allows some particles to pass through the membrane but not others.

semivegetarians: people who include some, but not all, groups of animal-derived foods in their diets; they usually exclude meat and may occasionally include poultry, fish, and shellfish; also called *partial vegetarians.*

senile dementia: the loss of brain function beyond the normal loss of physical adeptness and memory that occurs with aging.

senility (see-NIL-ih-tee): mental or physical weakness associated with old age.

sepsis: the spread of an infection from a local area into the blood, which alters blood flow to vital organs and can lead to multiple organ failure and death.

set-point theory: the theory that proposes that the body tends to maintain a certain weight by means of its own internal controls.

shock: a sudden drop in blood volume that disrupts the supply of oxygen to the tissues and organs and the return of blood to the heart. Shock is a critical event that requires immediate correction.

short-bowel or **short-gut syndrome:** severe malabsorption that may occur when the absorptive surface of the small bowel is reduced, resulting in diarrhea, weight loss, bone disease, hypocalcemia, hypomagnesemia, and anemia.

silent stones: gallstones that cause no symptoms.

simple carbohydrates: the monosaccharides (glucose, fructose, and galactose) and the disaccharides (sucrose, lactose, and maltose); also called **sugars.**

small intestine: a 10-foot length of small-diameter (1-inch) intestine that is the major site of digestion of food and absorption of nutrients.

soaps: chemical compounds formed between a basic mineral (such as calcium) and unabsorbed fatty acids.

soluble fibers: indigestible food components that readily dissolve in water and often impart gummy or gel-like characteristics to foods. An example is pectin from fruit, which is used to thicken jellies.

specific immunity: immunity directed at specific organisms. The B-cells and T-cells confer this type of immunity.

sphincter (SFINK-ter): a circular muscle surrounding, and able to close, a body opening.

sports anemia: a transient condition of low hemoglobin in the blood, associated with the early stages of sports training or other strenuous activity.

standard or **regular diets:** diets that include all foods and meet the nutrient needs of healthy people.

standard formulas: liquid diets that contain complete molecules of protein; also called **intact** or **polymeric formulas.**

starch: a plant polysaccharide composed of glucose and digestible by human beings.

steatorrhea (STEE-ah-toe-REE-ah): fatty diarrhea characterized by loose, foamy, foul-smelling stools. Soaps that form between unabsorbed fatty acids and minerals give steatorrhea its foamy appearance.

sterile: free of microorganisms, such as bacteria.

steroids (STARE-oids): medications used to reduce tissue inflammation, to suppress the immune response, or to replace certain steroid hormones in people who cannot synthesize them.

sterol esters: compounds belonging to the sterol family of lipids, derived from plants, that have been shown experimentally to reduce blood cholesterol when consumed in place of other fats in a low-fat diet.

sterols: one of the three main classes of lipids. Sterols include cholesterol, vitamin D, and the sex hormones (such as testosterone).

stoma (STOH-ma): a surgically formed opening. After a colostomy or ileostomy, a stoma is formed by bringing the cutoff end of the intestine through the abdominal wall, rerouting the excretion of wastes.

stroke volume: the amount of oxygenated blood ejected from the heart toward body tissues at each beat.

stroke: an event in which the blood flow to a part of the brain is cut off; also called a **cerebral vascular accident (CVA).**

structured lipids: triglycerides that have been artificially constructed to include both LCT and MCT fatty acids attached to the glycerol backbone.

structure-function claims: statements that describe how a product may affect a structure or function of the body; for example, "calcium builds strong bones." Structure-function claims do not require FDA authorization.

struvite (STREW-vite): crystals of magnesium ammonium phosphate formed by the action of bacterial enzymes.

sucralose (SUE-kra-lose): a 0-kcalorie artificial sweetener that tastes 600 times as sweet as sucrose. ADI = 5 milligrams/kilogram body weight.

sucrose: a disaccharide composed of glucose and fructose; commonly known as table sugar, beet sugar, or cane sugar.

sugar alcohols: sugarlike compounds; like sugars, they are sweet to taste but yield 2 to 3 kcal per gram, slightly less than sucrose. Examples are maltitol, mannitol, sorbitol, isomalt, lactitol, and xylitol.

sushi: a Japanese dish consisting of vinegar-flavored rice, seafood, and colorful vegetables, typically wrapped in seaweed. Some sushi is wrapped in raw fish; other sushi contains only cooked ingredients.

sustainable: able to continue indefinitely; using resources at such a rate that the earth can keep on replacing them and producing pollutants at a rate with which the environment and human cleanup efforts can keep pace so that no net accumulation of pollution occurs.

systemic: throughout the body as opposed to the area of injury.

systemic inflammatory response syndrome (SIRS): the complex of symptoms (see the text) that occur as a result of immune and inflammatory factors in response to tissue damage. In severe cases, SIRS may progress to multiple organ failure.

tannins: compounds in tea (especially black tea) and coffee that bind iron.

T-cells: lymphocytes that react to specific antigens by traveling directly to the invasion site. Some T-cells (cytotoxic T-cells) kill invaders; others (helper/inducer T-cells) activate immune responses; still others (suppressor T-cells) turn off immune responses.

teratogenic (ter-AT-oh-jen-ik): causing abnormal fetal development and birth defects.

terminal illness: a progressive, irreversible disease that will lead to death in the near future.

thermic effect of food: an estimation of the energy required to process food (digest, absorb, transport, metabolize, and store ingested nutrients).

thrombosis (throm-BOH-sis): the formation or presence of a blood clot in the vascular system. A *coronary thrombosis* occurs in a coronary artery, and a *cerebral thrombosis* occurs in an artery that feeds the brain.

thrombus (THROM-bus): a blood clot that blocks a blood vessel or a cavity of the heart.

thrush: a fungal infection of the mouth and esophagus, caused by *Candida albicans,* that coats the tongue with a milky film and leads to mouth ulcers, altered taste sensations, and pain on chewing and swallowing. The technical term for this infection is *candidiasis.*

tissue rejection: destruction of healthy donor cells by the recipient's immune system, which recognizes the donor cells as foreign; also called **graft-versus-host disease (GVHD).**

tocopherol (tuh-KOFF-er-ol): a general term for several chemically related compounds, one of which has vitamin E activity.

Tolerable Upper Intake Level (UL): the maximum amount of a nutrient that appears safe for most healthy people and beyond which there is an increased risk of adverse health effects.

tolerance level: the maximum amount of a residue permitted in a food when a pesticide is used according to label directions.

tonsillectomy (tawn-sill-ECK-tah-me): surgical removal of the tonsils.

total nutrient admixtures: intravenous solutions that contain all nutrients, including lipid emulsions; also called **three-in-one (3-in-1) admixtures** or **all-in-one admixtures.**

toxicity: the ability of a substance to harm living organisms. All substances are toxic if the concentration is high enough.

toxins: poisons. Toxins produced by bacteria come in two varieties: enterotoxins, which act in the GI tract, and neurotoxins, which act on the nervous system.

trachea (TRAKE-ee-uh): the windpipe; the passageway from the mouth and nose to the lungs.

training: practicing an activity regularly, which leads to conditioning. (Training is what you do; conditioning is what you get.)

trans-fatty acids: fatty acids with an unusual configuration around the double bond.

transferrin (trans-FERR-in): the body's iron-carrying protein.

transient ischemic attack (TIA): a temporary reduction in blood flow to the brain, which causes temporary symptoms that vary depending on the part of the brain that is affected. Common symptoms include light-headedness, visual disturbances, paralysis, staggering, numbness, or dysphagia.

translocation: the movement of bacteria into the body through the intestine.

transnasal: through the nose. A **transnasal feeding tube** is one that is inserted through the nose.

trauma: bodily injury such as a gunshot wound, blow, or cut.

triglycerides (try-GLISS-er-ides): one of the three main classes of lipids; the chief form of fat in foods and the major storage form of fat in the body; composed of glycerol with three fatty acids attached.

tripeptide: three amino acids bonded together.

tube feedings: liquid formulas delivered through a tube placed in the stomach or intestine.

tubule: a tubelike structure that surrounds the glomerulus and descends through the nephron. A pressure gradient between the glomerular capillaries and the tubule returns needed materials to the blood and moves wastes into the tubule to be excreted in the urine.

tumor: a new growth of tissue forming an abnormal mass with no function; also called a **neoplasm** (NEE-oh-plazm).

turbinado (ter-bih-NOD-oh) **sugar:** raw (brown) sugar from which the filth has been washed; legal to sell in the United States.

24-hour recall: a record of foods eaten by a person in the previous 24 hours.

type 1 diabetes: the less common type of diabetes in which the person produces no insulin at all.

type 2 diabetes: the more common type of diabetes that develops gradually and is associated with obesity and insulin resistance.

ulcerative colitis (ko-LYE-tis): inflammation and ulceration of the colon.

ultrafiltration: removal of fluids and small- to medium-size molecules from the blood by using pressure to transfer the blood across a semipermeable membrane.

ultrahigh temperature treatment (UHT): a process of sterilizing food by exposing it for a short time to temperatures above those normally used in processing.

umbilical (um-BIL-ih-cul) **cord:** the ropelike structure through which the fetus's veins and arteries reach the placenta; the route of nourishment and oxygen into the fetus and the route of waste disposal from the fetus.

undernutrition: underconsumption of food energy or nutrients severe enough to cause disease or increased susceptibility to disease; a form of malnutrition.

unsaturated fatty acid: a fatty acid with one or more points of unsaturation where hydrogens are missing (includes monounsaturated and polyunsaturated fatty acids).

uremia (you-REE-me-ah): abnormal accumulation of nitrogen-containing substances, especially urea, in the blood; also called *azotemia* (AZE-oh-TEE-me-ah).

uremic frost: the appearance of urea crystals on the skin.

uremic syndrome: the cluster of clinical findings associated with the buildup of nitrogen-containing waste products in the blood; may include fatigue, diminished mental alertness, agitation, muscle twitches, cramps, anorexia, nausea, vomiting, inflammation of the membranes of the mouth, unpleasant taste in the mouth, itchy skin, skin hemorrhages, gastritis, GI bleeding, and diarrhea.

usual intake method: a record of the foods eaten by a person in a typical day.

uterus (YOO-ter-us): the womb, the muscular organ within which the infant develops before birth.

vagotomy (vay-GOT-oh-mee): surgery that severs the nerves that stimulate gastric acid secretion.

varices (VAIR-ih-seez): blood vessels that have become twisted and distended.

variety (dietary): eating a wide selection of foods within and among the major food groups (the opposite of monotony).

vegans: people who exclude all animal-derived foods (including meat, poultry, fish, shellfish, eggs, cheese, and milk) from their diets; also called *strict vegetarians* or *total vegetarians*.

vein: a vessel that carries blood back to the heart.

very-low-density lipoproteins (VLDL): the type of lipoproteins made primarily by liver cells to transport lipids to various tissues in the body; composed primarily of triglycerides.

villi (VILL-ee or VILL-eye): fingerlike projections from the folds of the small intestine. The singular form is **villus.**

visceral fat: fat located between abdominal organs.

vitamin A: a fat-soluble vitamin. Its three chemical forms are *retinol* (the alcohol form), *retinal* (the aldehyde form), and *retinoic acid* (the acid form).

vitamins: essential, noncaloric, organic nutrients needed in tiny amounts in the diet.

VO$_2$ max: the maximum rate of oxygen consumption by an individual (measured at sea level).

voluntary activities: the component of a person's daily energy expenditure that involves conscious and deliberate muscular work—walking, lifting, climbing, and other physical activities. Voluntary activities normally require less energy in a day than basal metabolism does.

vomiting: expulsion of the contents of the stomach up through the esophagus to the mouth.

waist circumference: a measurement used to assess a person's abdominal fat.

warm-up: five to ten minutes of light activity, such as easy jogging or cycling, prior to a workout to prepare the body for more vigorous activity.

water balance: the balance between water intake and water excretion that keeps the body's water content constant.

water intoxication: the rare condition in which body water contents are too high. The symptoms may include confusion, convulsion, coma, and even death in extreme cases.

weight training: the use of free weights or weight machines to provide resistance for developing muscle strength and endurance; also called *resistance training*. A person's own body weight may also be used to provide resistance as when a person does push-ups, pull-ups, or abdominal crunches.

wellness: maximum well-being; the top range of health states; the goal of the person who strives toward realizing his or her full potential physically, mentally, emotionally, spiritually, and socially.

Wernicke's (VER-nih-keys) **encephalopathy:** a brain dysfunction and disturbances of motor coordination caused by severe thiamin deficiency; also called **Wernicke's-Korsakoff syndrome.**

whey protein: a by-product of cheese production; falsely promoted as increasing muscle mass.

white sugar: pure sucrose, produced by dissolving, concentrating, and recrystallizing raw sugar.

Zollinger-Ellison syndrome: marked hypersecretion of gastric acid and consequent peptic ulcers caused by a tumor of the pancreas.

Index

Daily Values for Food Labels

The Daily Values are standard values developed by the Food and Drug Administration (FDA) for use on food labels. Daily Values for protein, vitamins, and minerals reflect average allowances based on the RDA. Daily Values for nutrients and food components, such as fat and fiber, that do not have an established RDA but do have important relationships with health are based on recommended calculation factors as noted.

Nutrient	Amount
Protein[a]	50 g
Thiamin	1.5 mg
Riboflavin	1.7 mg
Niacin	20 mg NE
Biotin	300 µg
Pantothenic acid	10 mg
Vitamin B_6	2 mg
Folate	400 µg
Vitamin B_{12}	6 µg
Vitamin C	60 mg
Vitamin A	5000 IU
Vitamin D	400 IU
Vitamin E	30 IU
Vitamin K	80 µg
Calcium	1000 mg
Iron	18 mg
Zinc	15 mg
Iodine	150 µg
Copper	2 mg
Chromium	120 µg
Selenium	70 µg
Molybdenum	75 µg
Manganese	2 mg
Chloride	3400 mg
Magnesium	400 mg
Phosphorus	1000 mg

[a]The Daily Values for protein vary for different groups of people: pregnant women, 60 g; nursing mothers, 65 g; infants under 1 year, 14 g; children 1 to 4 years, 16 g.

Food Component	Amount	Calculation Factors
Fat	65 g	30% of kcalories
Saturated fat	20 g	10% of kcalories
Cholesterol	300 mg	Same regardless of kcalories
Carbohydrate (total)	300 g	60% of kcalories
Fiber	25 g	11.5 g per 1000 kcalories
Protein	50 g	10% of kcalories
Sodium	2400 mg	Same regardless of kcalories
Potassium	3500 mg	Same regardless of kcalories

Note: Daily Values were established for adults and children over 4 years old. The values for energy-yielding nutrients are based on 2000 kcalories a day.

Glossary of Nutrient Measures

kcal: kcalories; a unit by which energy is measured (Chapter 1 provides more details).

g: grams; a unit of weight equivalent to about 0.03 ounces.

mg: milligrams; one-thousandth of a gram.

µg: micrograms; one-millionth of a gram.

IU: international units; an old measure of vitamin activity determined by biological methods (as opposed to new measures that are determined by direct chemical analyses). Many fortified foods and supplements use IU on their labels.
- For vitamin A, 1 IU = 0.3 µg retinol, 3.6 µg β-carotene, or 7.2 µg other vitamin A carotenoids.
- For vitamin D, 1 IU = 0.025 µg cholecalciferol.
- For vitamin E, 1 IU = 0.67 natural α-tocopherol (other conversion factors are used for different forms of vitamin E).

mg NE: milligrams niacin equivalents; a measure of niacin activity (Chapter 10 provides more details).
- 1 NE = 1 mg niacin.
  = 60 mg tryptophan (an amino acid).

µg DFE: micrograms dietary folate equivalents; a measure of folate activity (Chapter 10 provides more details).
- 1 µg DFE = 1 µg food folate.
  = 0.6 µg fortified food or supplement folate.
  = 0.5 µg supplement folate taken on an empty stomach.

µg RAE: micrograms retinol activity equivalents; a measure of vitamin A activity (Chapter 11 provides more details).
- 1 µg RE = 1 µg retinol.
  = 12 µg β-carotene.
  = 24 µg other vitamin A carotenoids.

Photo and Art Credits